ICL Contribution to Landslide Disaster Risk Reduction

Series Editor

Kyoji Sassa, The International Consortium on Landslides, ICL, Kyoto, Japan

The ICL Contribution to Landslide Disaster Risk Reduction book-series publishes integrated research on all aspects of landslides. The volumes present summaries on the progress of landslide sciences, disaster mitigation and risk preparation. The contributions include landslide dynamics, mechanisms and processes; volcanic, urban, marine and reservoir landslides; related tsunamis and seiches; hazard assessment and mapping; modeling, monitoring, GIS techniques; remedial or preventive measures; early warning and evacuation and a global landslide database.

More information about this series at http://www.springer.com/series/16332

Kyoji Sassa • Matjaž Mikoš • Shinji Sassa •
Peter T. Bobrowsky • Kaoru Takara •
Khang Dang
Editors

Understanding and Reducing Landslide Disaster Risk

Volume 1 Sendai Landslide Partnerships and Kyoto Landslide Commitment

Springer

Editors
Kyoji Sassa
International Consortium
on Landslides
Kyoto, Japan

Shinji Sassa
Port and Airport Research Institute
National Institute of Maritime, Port
and Aviation Technology
Yokosuka, Japan

Kaoru Takara
Graduate School of Advanced Integrated
Studies in Human Survivability (Shishu-kan)
Kyoto University
Kyoto, Japan

Matjaž Mikoš
Faculty of Civil and Geodetic Engineering
University of Ljubljana
Ljubljana, Slovenia

Peter T. Bobrowsky
Geological Survey of Canada
Sidney, BC, Canada

Khang Dang
International Consortium on Landslides
Kyoto, Japan

University of Science
Vietnam National University
Hanoi, Vietnam

Associate Editor
Doan Huy Loi
International Consortium on Landslides
Kyoto, Japan

ISSN 2662-1894 ISSN 2662-1908 (electronic)
ICL Contribution to Landslide Disaster Risk Reduction
ISBN 978-3-030-60195-9 ISBN 978-3-030-60196-6 (eBook)
https://doi.org/10.1007/978-3-030-60196-6

Cover illustration: Landslides in Hiroshima, Japan after the heavy rainfall in July 2018 (International Consortium on Landslides. All Rights Reserved)

This Springer imprint is published by the registered company Springer Nature Switzerland AG
The registered company address is: Gewerbestrasse 11, 6330 Cham, Switzerland

ICL and Springer created a new book series "ICL Contribution to Landslide Disaster Risk Reduction" in 2019 which is registered as ISSN 2662-1894 (print version) and ISSN 2662-1908 (electronic version). The first books in this series are six volume of books "Understanding and Reducing Landslide Disaster Risk" containing the recent progress of landslide science and technologies from 2017 to 2020.

Editor-in-Chief: Kyoji Sassa
Assistant Editor-in-Chief: Željko Arbanas

Organizational Structure of the Fifth World Landslide Forum

Organizers

International Consortium on Landslides (ICL)
Global Promotion Committee of International Programme on Landslides (IPL-GPC), including: United Nations Educational, Scientific and Cultural Organization (UNESCO), World Meteorological Organization (WMO), Food and Agriculture Organization (FAO), United Nations Office for Disaster Risk Reduction (UNDRR), United Nations University (UNU), International Science Council (ISC), World Federation of Engineering Organizations (WFEO), International Union of Geological Sciences (IUGS), International Union of Geodesy and Geophysics (IUGG)
Kyoto University (KU), Japan Landslide Society (JLS), Japanese Geotechnical Society (JGS), Japan Society for Natural Disaster Science (JSNDS) and Japan Association for Slope Disaster Management (JASDiM)

Co-sponsors

Cabinet Office (Disaster Management Bureau) of Japan; Ministry of Foreign Affairs of Japan (MOFA); Ministry of Education, Culture, Sports, Science and Technology-Japan (MEXT); Ministry of Land Infrastructure, Transport and Tourism (MLIT); Ministry of Agriculture, Forestry and Fisheries (MAFF); Science Council of Japan (SCJ); Japan International Cooperation Agency (JICA); Japan Society of Civil Engineers (JSCE); Japanese Society of Irrigation, Drainage and Rural Engineering (JSIDRE); Japan Society of Erosion Control Engineering; Japan Society of Engineering Geology.

Supporting Organizations with Finance

Tokyo Geographical Society
International Union of Geological Sciences (IUGS)
Association for Disaster Prevention Research, Kyoto, Japan

Organizing Committee

Honorary Chairpersons
Audrey Azoulay, Director-General of UNESCO*
Mami Mizutori, Special Representative of the United Nations Secretary-General for Disaster Risk Reduction*

Petteri Taalas, Secretary-General of WMO*
Qu Dongyu, Director-General of FAO*
David Malone, Under-Sectretary General of the Unitred Nations and Rector of UNU
Daya Reddy, President of ISC
Gong Ke, President of WFEO
Qiuming Cheng, President of IUGS
Kathryn Whaler, President of IUGG
Qunli Han, Executive Director of Integrated Research on Disaster Risk (IRDR)
Walter Ammann, President and CEO of Global Risk Forum GRF Davos, Switzerland
Juichi Yamagiwa, President of Kyoto University, Japan
Angelo Borrelli, Head of the National Civil Protection Department, Italian Presidency of the Council of Ministers, Italy
Darko But, Director General of the Administration for Civil Protection and Disaster Relief of the Republic of Slovenia, Slovenia
Akifumi Nakao, Director, International Cooperation Division, Disaster Management Bureau, Cabinet Office, Japan
Kazuyuki Imai, Director General of Sabo Department, Ministry of Land Infrastructure, Transport and Tourism, Japan*
Chungsik Yoo, President of the International Geosynthetics Society
Rafig Azzam, President of the International Association for Engineering Geology and the Environment
(*to be confirmed)

Chairpersons
Kyoji Sassa, Professor Emeritus, Kyoto University; Secretary General of ICL
Peter T. Bobrowsky, Geological Survey of Canada; President of ICL
Kaoru Takara, Kyoto University, Japan; Executive Director of ICL

Members
Željko Arbanas (University of Rijeka, Croatia)
Snježana Mihalić Arbanas (University of Zagreb, Croatia)
Nicola Casagli (University of Firenze, Italy)
Fausto Guzzetti (Department of Civil Protection, Italy)
Matjaž Mikoš (University of Ljubljana, Slovenia)
Paola Reichenbach (Research Institute for Geo-Hydrological Protection, National Research Council, Italy)
Shinji Sassa (Port and Airport Research Institute, Japan)
Alexander Strom (Geodynamics Research Center LLC, Russia)
Binod Tiwari (California State University, Fullerton, USA)
Veronica Tofani (University of Firenze, Italy)
Vít Vilímek (Charles University in Prague, Czech Republic)
Fawu Wang (Tongji University, China)

Chairpersons of Local Organizing Committee
Kaoru Takara (Kyoto University)
Daisuke Higaki (Japan Landslide Society)
Ikuo Towhata (Japanese Geotechnical Society)

Secretary Generals
Ryosuke Uzuoka (Disaster Prevention Research Institute, Kyoto University)
Kazuo Konagai (International Consortium on Landslides)
Khang Dang (International Consortium on Landslides)

International Scientific Committee

Beena Ajmera, North Dakota State University, USA
Snježana Mihalić Arbanas, University of Zagreb, Croatia
Željko Arbanas, Faculty of Civil Engineering, University of Rijeka, Croatia
Amin Askarinejad, Technische Universiteit Delft, Delft, The Netherlands
Peter T. Bobrowsky, Geological Survey of Canada, Sidney, Canada
Michele Calvello, University of Salerno, Italy
Giovanna Capparelli, Universita degli Studi della Calabria, Rende, Italy
Nicola Casagli, University of Florence, Italy
Yifei Cui, Tsinghua University, Beijing, China
Sabatino Cuomo, University of Salerno, Fisciano, Italy
Khang Dang, International Consortium on Landslides, Kyoto, Japan
Elias Garcia-Urquia, National Autonomous University of Honduras, Tegucigalpa, Honduras
Stefano Luigi Gariano, Research Institute for Geo-Hydrological Protection, CNR, Perugia, Italy
Daniele Giordan, Research Institute for Geo-Hydrological Protection, CNR, Italy
Fausto Guzzetti, Department of Civil Protection, Italy
Baator Has, Asia Air Survey, Tokyo, Japan
Hans-Balder Havenith, Universite de Liege, Liege, Belgium
D. P. Kanungo, Central Building Research Institute (CBRI), Roorkee, Uttarakhand, India
Oded Katz, Geological Survey of Israel, Jerusalem, Israel
Kazuo Konagai, International Consortium on Landslides, Kyoto, Japan
Doan Huy Loi, International Consortium on Landslides, Kyoto, Japan
Ping Lu, Tongji University, Shanghai, China
Olga Mavrouli, University of Twente, Enschede, The Netherlands
Matjaž Mikoš, Faculty of Civil and Geodetic Engineering, University of Ljubljana, Slovenia
Alessandro C. Mondini, Research Institute for Geo-Hydrological Protection, CNR, Italy
Veronica Pazzi, Department of Earth Science, University of Florence, Florence, Italy
Dario Peduto, Department of Civil Engineering, University of Salerno, Fisciano, Italy
Paola Reichenbach, Research Institute for Geo-Hydrological Protection, CNR, Italy
Paola Salvati, Research Institute for Geo-Hydrological Protection, CNR, Italy
Katsuo Sasahara, Kochi University, Japan
Kyoji Sassa, International Consortium on Landslides, Kyoto, Japan
Shinji Sassa, Port and Airport Research Institute, Japan
Andrea Segalini, University of Parma, Italy
Hendy Setiawan, Universitas Gadjah Mada, Yogyakarta, Indonesia
Alexander Strom, Geodynamics Research Center LLC, Moscow, Russia
Kaoru Takara, Kyoto University, Japan
Faraz Tehrani, Deltares, Delft, The Netherlands
Binod Tiwari, California State University, Fullerton, California, USA
Veronica Tofani, University of Florence, Italy
Ryosuke Uzuoka, Kyoto University, Kyoto, Japan
Vít Vilímek, Faculty of Science, Charles University, Prague, Czech Republic
Fawu Wang, College of Civil Engineering, Tongji University, Shanghai, China
Gonghui Wang, Kyoto University, Kyoto, Japan
Mike Winter, Winter Associates Limited, Kirknewton, UK
Hiromitsu Yamagishi, Hokkaido Research Center of Geology (HRCG), Sapporo, Japan

Local Organizing Committee

Shinro Abe, Okuyama Boring Co., Ltd.
Kiminori Araiba, Fire and Disaster Management College
Shiho Asano, Forestry and Forest Products Research Institute
Has Baator, Asia Air Survey Co., Ltd.

Hiromu Daimaru, Forestry and Forest Products Research Institute
Khang Dang, International Consortium on Landslides
Mitusya Enokida, Japan Conservation Engineers & Co., Ltd.
Kumiko Fujita, International Consortium on Landslides
Kazunori Hayashi, Okuyama Boring Co., Ltd.
Daisuke Higaki, The Japan Landslide Society
Kiyoharu Hirota, Kokusai Kogyo Co., Ltd.
Kazuo Konagai, International Consortium on Landslides
Taketoshi Marui, MARUI & Co., Ltd.
Satoshi Nishikawa, Nagoya University
Keisuke Oozone, OYO Corporation
Katsuo Sasahara, Kochi University
Kyoji Sassa, International Consortium on Landslides
Shinji Sassa, Port and Airport Research Institute
Go Sato, Teikyo Heisei University
Nobuyuki Shibasaki, Nippon Koei Co., Ltd.
Nobuo Sugiura, Japan Association for Slope Disaster Management
Kaoru Takara, Kyoto University
Keisuke Takimoto, GODAI KAIHATSU Corporation
Yoko Tomita, Public Works Research Institute
Ikuo Towhata, The Japanese Geotechnical Society
Kenichi Tsukahara, Kyushu University
Ryosuke Tsunaki, Sabo & Landslide Technical Center
Taro Uchida, Saitama University
Mie Ueda, International Consortium on Landslides
Ryosuke Uzuoka, Kyoto University
Fawu Wang, Tongji University
Hiroshi Yagi, Yamagata University
Hiromitsu Yamagishi, Shin Engineering Consultants Co., Ltd.
Maki Yano, OSASI Technos Inc.

Foreword by Mami Mizutori

More landslides can be expected as climate change exacerbates rainfall intensity. The long-term trend of the last 40 years has seen the number of major recorded extreme weather events almost double, notably floods, storms, landslides, and wildfires.

Landslides are a serious geological hazard. Among the host of natural triggers are intense rainfall, flooding, earthquakes or volcanic eruption, and coastal erosion caused by storms that are all too often tied to the El Niño phenomenon. Human triggers including deforestation, irrigation or pipe leakage, and mine tailings, or stream and ocean current alteration can also spark landslides. Landslides can also generate tsunamis, as Indonesia experienced in 2018.

Globally, landslides cause significant economic loss and many deaths and injuries each year. Therefore, it is important to understand the science of landslides: why they occur, what factors trigger them, the geology associated with them, and where they are likely to happen.

Landslides with high death tolls are often a result of failures in risk governance, poverty reduction, environmental protection, land use and the implementation of building codes. Understanding the interrelationships between earth surface processes, ecological systems, and human activity is the key to reducing landslide risk.

The Sendai Framework for Disaster Risk Reduction, the global plan to reduce disaster losses adopted in 2015, emphasizes the importance of tackling these risk drivers through improved governance and a better understanding of disaster risk.

One important vehicle for doing that is the Sendai Landslide Partnerships 2015–2025 for global promotion of understanding and reduction of landslide risk facilitated by the International Consortium on Landslides (ICL) and signed by the leaders of 22 global stakeholders, including the UN Office for Disaster Risk Reduction (UNDRR), during the Third UN World Conference on Disaster Risk Reduction in Sendai, Japan.

The Sendai Landslide Partnerships—featured on the Sendai Framework Voluntary Commitments online platform—helps to provide practical solutions and tools, education, and capacity building, to reduce landslide risks.

The work done by the Sendai Partnerships can be of value to many stakeholders including civil protection, planning, development and transportation authorities, utility managers, agricultural and forest agencies, and the scientific community.

UNDRR fully supports the work of the Sendai Landslide Partnerships and ICL and looks forward to an action-oriented outcome from the 5th World Landslide Forum to be held in November 2020 in Kyoto, Japan. Successful efforts to reduce disaster losses are a major contribution to achieving the overall 2030 Agenda for Sustainable Development.

Mami Mizutori
United Nations Special Representative of the
Secretary-General for Disaster Risk Reduction

Foreword by the Assistant Director-General for the Natural Sciences Sector of UNESCO for the Book of the 5th *World Landslide Forum*

As the world slowly recovers from the COVID-19 global pandemic, and looking back at the way this crisis developed, it becomes evident that as a global community we were not prepared for an event of this scale. Although not commonly perceived as such, biological hazards such as epidemics are included in the Sendai Framework for Disaster Risk Reduction 2015–2030. In that sense, the preparedness approach for a pandemic is very similar to that of a geophysical natural hazard such as landslides.

Although natural hazards are naturally occurring phenomena, the likelihood of their occurrence and of associated disasters is rising. Climate change, urban pressure, under-development and poverty and lack of preparedness are increasingly transforming these natural hazards into life-threatening disasters with severe economic impacts. Therefore, Disaster Risk Reduction (DRR) is gaining momentum on the agenda of the UN system of Organizations including UNESCO. While the Sendai Framework for Disaster Risk Reduction 2015–2030 is the roadmap for DRR, other global agendas including the Sustainable Development Goals, the Paris Climate Agreement and the New Urban Agenda have targets which cannot be attained without DRR.

In shaping its contribution to those global agendas, UNESCO is fully committed in supporting its Member States in risk management, between its different mandates and disciplines and with relevant partners. The International Consortium on Landslides (ICL) is UNESCO's key partner in the field of landslide science. The Organization's support to the Consortium is unwavering. Since ICL was established in 2002, the two organizations have a long history of cooperation and partnership and UNESCO has been associated with almost all of ICL activities. I am very glad that ICL and UNESCO are mutually benefitting from their collaboration.

The 5th World Landslide Forum (WLF5) is expected to represent a milestone in the history of landslide science particularly for scientists and practitioners. One of the major outcomes of WLF5 will be the Kyoto 2020 Commitment for global promotion of understanding and reducing landslide disaster risk (KLC2020). This commitment is expected to strengthen and expand the activities of the Sendai Landslide Partnership 2015–2025. With UNESCO already engaged as a partner, the adoption of this international commitment will raise global awareness on landslide risk and mobilize wider partnerships that draw together stakeholders from all levels of society, across different regions, sectors and disciplines.

It is my great pleasure to congratulate the organizers for holding this event and assure you that UNESCO is fully committed in contributing to its success. As part of that contribution, our Organization is proud to host a session on landslides and hazard assessment at UNESCO-designated sites such as natural World Heritage sites, biosphere reserves and UNESCO Global Geoparks. This session aims to assess landslide impacts on our shared cultural and natural heritage, providing the best opportunity to generate public awareness and capacity development for landslide disaster reduction.

I am confident that WLF5 will contribute to further advance the knowledge of both scientists and practitioners regarding landslide disaster risk reduction. This book paves the way for the science, knowledge and know-how which will feature in the deliberations of the Forum. UNESCO commends all of the contributors to this publication. I look forward to an enhanced collaboration between UNESCO and ICL in future activities and undertakings.

Shamila Nair-Bedouelle
Assistant Director-General for Natural Sciences
UNESCO

Preface

Understanding and Reducing Landslide Disaster Risk

Book Series: ICL Contribution to Landslide Disaster Risk

The International Consortium on Landslides (ICL) was established in pursuance of the 2002 Kyoto Declaration "Establishment of an International Consortium on Landslides," with its Statutes adopted in January 2002. The Statutes define the General Assembly of ICL as follows: in order to report and disseminate the activities and achievements of the Consortium, a General Assembly shall be convened every 3 years by inviting Members of the International Consortium on Landslides, individual members within those organizations, and all levels of cooperating organizations and individual researchers, engineers, and administrators. The General Assembly developed gradually prior to, during and after its first meeting in 2005. In the light of the 2006 Tokyo Action Plan, the Assembly was further facilitated at, and following the First World Landslide Forum held in November 2008. On the occasion of each of its triennial forums, ICL publishes the latest progress of landslide science and technology for the benefit of the whole landslide community including scientists, engineers, and practitioners in an understandable form. Full color photos of landslides and full color maps are readily appreciated by those from different disciplines. We have published full color books on landslides at each forum. In 2019, ICL created a new book series "ICL Contribution to Landslide Disaster Risk Reduction" ISSN 2662-1894 (print version) and ISSN 2662-1908 (electronic version). Six volumes of full color books *Understanding and Reducing Landslide Disaster Risk* will be published in 2020 as the first group of books of this series.

The Letter of Intent 2005 and the First General Assembly 2005

The United Nations World Conference on Disaster Reduction (WCDR) was held in Kobe, Japan, 18–22 January 2005. At this Conference, ICL organized session 3.8 "New international Initiatives for Research and Risk Mitigation of Floods (IFI) and Landslides (IPL)" on 19 January 2005 and adopted a "Letter of Intent" aimed at providing a platform for a holistic approach in research and learning on 'Integrated Earth System Risk Analysis and Sustainable Disaster Management'. This Letter was agreed upon and signed, during the first semester of 2005, by heads of seven global stakeholders including the United Nations Educational, Scientific and Cultural Organization (UNESCO), the World Meteorological Organization (WMO), the Food and Agriculture Organization of the United Nations (FAO), the United Nations International Strategy for Disaster Risk Reduction (UNISDR-currently UNDRR), the United Nations University (UNU), the International Council for Science (ICSU-Currently ISC), and the World Federation of Engineering Organizations (WFEO).

The first General Assembly of ICL was held at the Keck Center of the National Academy of Sciences in Washington D.C., USA, on 12–14 October 2005. It was organized after the aforementioned 2005 World Conference on Disaster Reduction (WCDR). ICL published the first full color book reporting on Consortium activities for the initial 3 years, 2002–2005 titled "Landslides-Risk analysis and sustainable disaster management". In the preface of this book, the Letter of Intent for Integrated Earth System Risk Analysis and Sustainable Disaster Management was introduced. Results of the initial projects of the International Programme on Landslides (IPL) including IPL C101-1 Landslide investigation in Machu Picchu World Heritage, Cusco, Peru and previous agreements and MoU between UNESCO, ICL and the Disaster Prevention Research Institute of Kyoto University including UNESCO/KU/ICL UNITWIN Cooperation programme were published as well in this book.

The 2006 Tokyo Action Plan and the First World Landslide Forum 2008

Based on the Letter of Intent, the 2006 Tokyo Round-Table Discussion—"Strengthening Research and Learning on Earth System Risk Analysis and Sustainable Disaster Management within UN-ISDR as Regards Landslides"—towards a dynamic global network of the International Programme on Landslides (IPL) was held at the United Nations University, Tokyo, on 18–20 January 2006. The 2006 Tokyo Action Plan—Strengthening research and learning on landslides and related earth system disasters for global risk preparedness—was adopted. The ICL exchanged Memoranda of Understanding (MoUs) concerning strengthening cooperation in research and learning on earth system risk analysis and sustainable disaster management within the framework of the United Nations International Strategy for Disaster Reduction regarding the implementation of the 2006 Tokyo action plan on landslides with UNESCO, WMO, FAO, UNISDR (UNDRR), UNU, ICSU (ISC) and WFEO, respectively in 2006. A set of these MoUs established the International Programme on Landslides (IPL) as a programme of the ICL, the Global Promotion Committee of IPL to manage the IPL, and the triennial World Landslide Forum (WLF), as well as the concept of the World Centres of Excellence on Landslide Risk Reduction (WCoE).

The First World Landslide Forum (WLF1) was held at the Headquarters of the United Nations University, Tokyo, Japan, on 18–21 November 2008. 430 persons from 49 countries/regions/UN entities were in attendance. Both Hans van Ginkel, Under Secretary-General of the United Nations/Rector of UNU who served as chairperson of the Independent Panel of Experts to endorse WCoEs, and Salvano Briceno, Director of UNISDR who served as chairperson of the Global Promotion Committee of IPL, participated in this Forum. The success of WLF1 paved the way to the successful second and third World Landslide Forum held in Italy and China respectively.

The Second World Landslide Forum 2011 and the Third World Landslide Forum 2014

The Second World Landslide Forum (WLF2)—Putting Science into Practice—was held at the Headquarters of the Food and Agriculture Organization of the United Nations (FAO) on 3–9 October 2011. It was jointly organized by the IPL Global Promotion Committee (ICL, UNESCO, WMO, FAO, UNDRR, UNU, ISC, WFEO) and two ICL members from Italy: the Italian Institute for Environmental Protection and Research (ISPRA) and the Earth Science Department of the University of Florence with support from the Government of Italy and many Italian landslide-related organizations. It attracted 864 participants from 63 countries.

The Third World Landslide Forum (WLF3) was held at the China National Convention Center, Beijing, China, on 2–6 June 2014. A high-level panel discussion on an initiative to create a safer geoenvironment towards the UN Third World Conference on Disaster Risk Reduction (WCDRR) in 2015 and forward was moderated by Hans van Ginkel, Chair of Independent Panel of Experts for World Centers of Excellence (WCoE). In a special address to this high-level panel discussion, Irina Bokova, Director-General of UNESCO, underlined that countries should be united to work against natural disasters and expressed commitment that UNESCO would like to further deepen cooperation with ICL. Ms. Bokova awarded certificates to 15 World Centres of Excellence.

The Sendai Landslide Partnerships 2015 and the Fourth World Landslide Forum 2017

The UN Third World Conference on Disaster Risk Reduction (WCDRR) was held in Sendai, Japan, on 14–18 March 2015. ICL organized the Working Session "Underlying Risk Factors" together with UNESCO, the Japanese Ministry of Land, Infrastructure, Transport and Tourism (MLIT) and other competent organizations. The session adopted ISDR-ICL Sendai Partnerships 2015–2025 (later changed to Sendai Landslide Partnerships) for global promotion of understanding and reducing landslide disaster risk as a Voluntary Commitment to the World Conference on Disaster Risk Reduction, Sendai, Japan, 2015 (later changed to Sendai Framework for Disaster Risk Reduction). After the session on 16 March 2015, the Partnerships was signed by Margareta Wahlström, Special Representative of the UN Secretary-General for Disaster Risk Reduction, Chief of UNISDR (UNDDR), and other representatives from 15 intergovernmental, international, and national organizations. Following the Sendai Landslide Partnerships, the Fourth World Landslide Forum was held in Ljubljana, Slovenia from 29 May to 2 June in 2017. On that occasion, five volumes of full color books were published to disseminate the advances of landslide science and technology. The high-level panel discussion on 30 May and the follow-up round table discussion on 31 May adopted the 2017 Ljubljana Declaration on Landslide Risk Reduction. The Declaration approved the outline of the concept of "Kyoto 2020 Commitment for global promotion of understanding and reducing landslide disaster risk" to be adopted at the Fifth World Landslide Forum in Japan, 2020.

The Fifth World Landslide Forum 2020 and the Kyoto Landslide Commitment 2020

The Fifth World Landslide Forum was planned to be organized on 2–6 November 2020 at the National Kyoto International Conference Center (KICC) and the preparations for this event were successfully ongoing until the COVID-19 pandemic occurred over the world in early 2020. The ICL decided to postpone the actual Forum to 2–6 November 2021 at KICC in Kyoto, Japan. Nevertheless, the publication of six volumes of full color books *Understanding and Reducing Landslide Disaster Risk* including reports on the advances in landslide science and technology from 2017 to 2020 is on schedule. We expect that this book will be useful to the global landslide community.

The Kyoto Landslide Commitment 2020 will be established during the 2020 ICL-IPL Online Conference on 2–6 November 2020 on schedule. Joint signatories of Kyoto Landslide Commitment 2020 are expected to attend a dedicated session of the aforementioned Online Conference, scheduled on 5 November 2020 which will also include and feature the Declaration of the launching of KLC2020. Landslides: Journal of the International Consortium on Landslides is the common platform for KLC2020. All partners may contribute and publish news and reports of their activities such as research, investigation, disaster reduction administration in the category of News/Kyoto Commitment. Online access or/and hard copy of the Journal will be sent to KLC2020 partners to apprise them of the updated information from other partners. As of 21 May 2020, 63 United Nations, International and national organizations have already signed the KLC2020.

Call for Partners of KLC2020

Those who are willing to join KLC2020 and share their achievements related to understanding and reducing landslide disaster risk in their intrinsic missions with other partners are invited to inform the ICL Secretariat, the host of KLC2020 secretariat (secretariat@iclhq.org). The ICL secretariat will send the invitation to the aforementioned meeting of the joint signatories and the declaration of the launching of the KLC2020 on 5 November 2020.

Eligible Organizations to be Partners of the KLC2020

1. ICL member organizations (full members, associate members and supporters)
2. ICL supporting organization from UN, international or national organizations and programmes
3. Government ministries and offices in countries having more than 2 ICL on-going members
4. International associations /societies that contribute to the organization of WLF5 in 2021 and WLF6 in 2023
5. Other organizations having some aspects of activities related to understanding and reducing landslide disaster risk as their intrinsic missions.

Kyoji Sassa
Chair of WLF5/
Secretary-General of ICL
Kyoto, Japan

Peter T. Bobrowsky
President of ICL
Sidney, Canada

Kaoru Takara
Executive Director of ICL
Kyoto, Japan

Appendix: World Landslide Forum Books

WLF	Place/participants	Title	Editors	Publisher/pages
WLF0 (1st General Assembly) 2005	Washington D.C., USA 59 from 17 countries/UNs	Landslides-Risk Analysis and Sustainable Disaster Management	Kyoji Sassa, Hiroshi Fukuoka, Fawu Wang, Goghui Wang	Springer/377 pages ISBN: 978-3-540-2864-6
WLF1 2008	Tokyo, Japan 430 from 49 countries/regions/UNs	Landslides-Disaster Risk Reduction	Kyoji Sassa, Paolo Canuti	Springer/649 pages ISBN: 978-3-540-69966-8
WLF2 2011	Rome, Italy 864 from 63 countries	Landslide Science and Practice Vol. 1 Landslide inventory and Sustainability and Hazard Zoning	Claudia Margottini, Paolo Canuti, Kyoji Sassa	Springer/607 pages ISBN: 978-3-642-31324-0
		Vol. 2 Early Warning, Instrumentation and Monitoring		Springer/685 pages ISBN: 978-3-642-31444-5
		Vol. 3 Spatial Analysis and Modelling		Springer/440 pages ISBN: 978-3-642-31309-7
		Vol. 4 Global Environmental Change		Springer/431 pages ISBN: 978-3-642-31336-3
		Vol. 5 Complex Environment		Springer/354 pages ISBN: 978-3-642-31426-1
		Vol. 6 Risk Assessment, Management and Mitigation		Springer/789 pages ISBN: 978-3-642-31318-9
		Vol. 7 Social and Economic Impact and Policies		Springer/333 pages ISBN: 978-3-642-31312-7
WLF3 2014	Beijing, China 531 from 45 countries/regions/UNs	Landslide Science for a Safer Geoenvironment Vol. 1 The International Programme on Landslides (IPL)	Kyoji Sassa, Paolo Canuti, Yueping Yin	Springer/493 pages ISBN: 978-3-319-04998-4
		Vol. 2 Methods of Landslide Studies		Springer/851 pages ISBN: 978-3-319-05049-2
		Vol. 3 Targeted Landslides		Springer/717 pages ISBN: 978-3-319-04995-3
WLF4 2017	Ljubljana, Slovenia 588 from 59 countries/regions/UNs	Advancing Culture of Living with Landslides Vol. 1 ISDR-ICL Sendai Partnerships 2015-2025	Kyoji Sassa, Matjaž Mikoš, Yueping Yin	Springer/585 pages ISBN: 978-319-53500-5

(continued)

WLF	Place/participants	Title	Editors	Publisher/pages
		Vol. 2 Advances in Landslide Science	Matjaž Mikoš, Binod Tiwari, Yueping Yin, Kyoji Sassa	Springer/1197 pages ISBN: 978-319-53497-8
		Vol. 3 Advances in Landslide Technology	Matjaž Mikoš, Željko Arbanas, Yueping Yin, Kyoji Sassa	Springer/621 pages ISBN: 978-3-319-53486-2
		Vol. 4 Diversity of Landslide Forms	Matjaž Mikoš, Nicola Casagli, Yueping Yin, Kyoji Sassa	Springer/707 pages ISBN: 978-3-319-53484-8
		Vol. 5 Landslides in Different Environments	Matjaž Mikoš, Vít Vilímek, Yueping Yin, Kyoji Sassa	Springer/557 pages ISBN: 978-3-319-53482-4
WLF5	2020 (publication) 2021 (Forum)	Understanding and Reducing Landslide Disaster Risk Vol. 1 Sendai Landslide Partnerships and Kyoto Landslide Commitment	Kyoji Sassa, Matjaž Mikoš, Shinji Sassa, Peter T. Bobrowsky, Kaoru Takara, Khang Dang	Springer In Process
		Vol. 2 From mapping to hazard and risk zonation	Fausto Guzzetti, Snježana Mihalić Arbanas, Paola Reichenbach, Kyoji Sassa, Peter T. Bobrowsky, Kaoru Takara	
		Vol. 3 Monitoring and early Warning	Nicola Casagli, Veronica Tofani, Kyoji Sassa, Peter T. Bobrowsky, Kaoru Takara	
		Vol. 4 Testing, modelling and risk assessment	Binod Tiwari, Kyoji Sassa, Peter T. Bobrowsky, Kaoru Takara	
		Vol. 5 Catastrophic landslides and Frontier of Landslide Science	Vit Vilimek, Fawu Wang, Alexander Strom, Kyoji Sassa, Peter T. Bobrowsky, Kaoru Takara	
		Vol. 6 Specific topics in landslide science and applications	Željko Arbanas, Peter T. Bobrowsky, Kazuo Konagai, Kyoji Sassa, Kaoru Takara	

Contents

Part I

Forum Lectures and Special Lectures

On the Prediction of Landslides and Their Consequences

Fausto Guzzetti

Abstract

The general assumptions and the most popular methods used to assess landslide hazard and for risk evaluation have not changed significantly in recent decades. Some of these assumptions have conceptual weakness, and the methods have revealed limitations. In this work, I deal with populations of landslides i.e. numerous landslides caused in an area by a single trigger (e.g. a rainstorm, an earthquake, a rapid snowmelt event), or by multiple events in a short or long period. Following an introduction on what we need to predict to assess landslide hazard and risk, I introduce the strategies and the main methods currently used to detect and map landslides, to predict populations of landslides in space and time, and to anticipate the numerosity and size characteristics of the expected landslides. For landslide detection and mapping, I consider traditional methods based on the visual interpretation of aerial photographs, and modern approaches that exploit the visual, semi-automatic or automatic analysis of remotely sensed images. For landslide spatial prediction, I discuss the results of a global review of statistical, classification-based methods for landslide susceptibility assessment. For the temporal prediction, leveraging on a global analysis of geographical landslide forecasting and early warning systems, I discuss short term forecast capabilities and their limitations. Next, I discuss long term landslide projections considering the impact of climate variations on landslide projections. For landslide numerosity and size characteristics, I discuss existing statistics of landslide area and volume obtained from large populations of event-triggered landslides. This is followed by an analysis of the landslide consequences, with emphasis on a spatial-temporal model of societal landslide risk in Italy. I end offering recommendations on what I think we should do to make significant progress in our collective ability to predict the hazard posed by populations of landslides, and to mitigate their risk.

Keywords

Prediction • Maps • Susceptibility • Hazard • Model • Consequences • Risk

Introduction

Landslides are present on all continents and play an important role in the evolution of landscapes (Densmore et al. 1997; Hovius et al. 1997, 2000; Lavé and Burbank 2004; Malamud et al. 2004a; Guzzetti et al. 2009; Chen et al. 2014; Bucci et al. 2016). In many areas, landslides pose a serious threat to people, private and public properties, society, and the environment (Brabb 1989, 1991; Nadim et al. 2006; Kirschbaum et al. 2009; Petley 2012; Guthrie 2013; Dowling and Santi 2014; Pereira et al. 2015, 2017; Badoux et al. 2016; Grahn and Jaldell 2017; Froude and Petley 2018; Herrera et al. 2018; Salvati et al. 2018). It is therefore not surprising that landslide hazard assessment and risk evaluation, at the boundary between science, technology and application (Fig. 1), are increasingly popular among scientists, practitioners, decision makers, and even concerned citizens.

Despite numerous attempts and unquestionable progress, the general assumptions and the most popular methods and techniques used to detect and map landslides, to assess landslide susceptibility and hazard, and for risk evaluation and mitigation, have not changed significantly in recent

F. Guzzetti (✉)
Consiglio Nazionale delle Ricerche, Istituto di Ricerca per la Protezione Idrogeologica, Via Della Madonna Alta 126, 06128 Perugia, Italy
e-mail: fausto.guzzetti@irpi.cnr.it; fausto.guzzetti@protezionecivile.it

Civil Protection Department, Office of the Prime Minister, Via Vitorchiano 2, 00189 Rome, Italy

K. Sassa et al. (eds.), *Understanding and Reducing Landslide Disaster Risk*, ICL Contribution to Landslide Disaster Risk Reduction, https://doi.org/10.1007/978-3-030-60196-6_1

decades. Today, some of these assumptions reveal theoretical weakness, and the adopted methods and techniques show practical and operational limits. The complete spectrum of landslide hazard and risk studies is vast, spanning from the hazard caused by a single landslide or a small portion of a natural or engineered slope, to the risk posed by multiple landslides of different types to the population of a region or nation (Guzzetti 2000; Badoux et al. 2016; Pereira et al. 2017; Salvati et al. 2018; Rossi et al. 2019), to the assessment of the potential occurrence of landslides at the continental (Van Den Eeckhaut et al. 2012a; Broeckx et al. 2018; Wilde et al. 2018) or the global (Nadim et al. 2006) scales.

This paper describes and expands the topics covered in the talk entitled *On the prediction of landslides and their consequences* that I was invited to deliver at the 5th World Landslide Forum (WLF5) in Kyoto, Japan. Based on my competency and personal experience, in the talk and in this work, I do not cover the entire spectrum of landslide hazards and risk. Instead, I concentrate on the hazards and risk posed by populations of landslides, i.e. the occurrence of many landslides that sculpt (literally!) landscapes, and pose serious threats to people, property, and the environment. This is a complex, multi-faceted topic with scientific, technological, and practical aspects (Fig. 1), which I address based on the assumption that our ability to predict landslides and their consequences measures our ability to understand the underlying physical (e.g. geological, geomorphological, hydrological, mechanical, climatic, tectonic) processes that control or condition landslides, as well as their spatial and temporal occurrence. With an important corollary worth mentioning: prediction must be scientifically based (Guzzetti 2015).

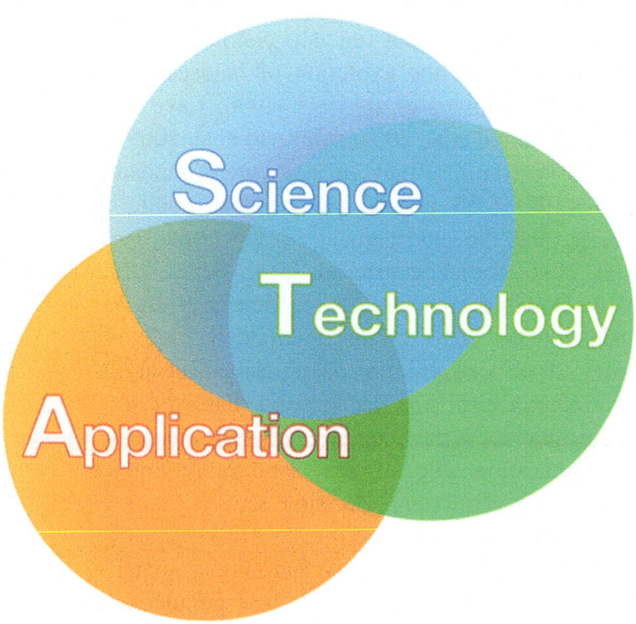

Fig. 1 Conceptual representation of the three realms of science, technology and application, which underpin modern landslide hazard and risk assessment

The article builds upon previous work I have published in the last three decades on (i) landslide detection and mapping (Guzzetti et al. 2012) and the often neglected quantification of the uncertainty associated with landslide mapping (Carrara et al. 1992; Ardizzone et al. 2002) and the quality of landslide maps (Santangelo et al. 2015); (ii) the types, comparison, and use of landslide inventory maps (Guzzetti et al. 2000; Galli et al. 2008; Guzzetti et al. 2012); (iii) the modelling of landslide susceptibility adopting statistical classification methods (Carrara et al. 1991; Guzzetti et al. 1999; Reichenbach et al. 2018), or physically based approaches (Guzzetti et al. 2002a; Alvioli et al. 2014, 2016, 2018; Raia et al. 2014); (iv) the definition and application of a conceptual framework for landslide hazard assessment based on the pioneering work of Varnes and the IAEG Commission on Landslides and other Mass-Movements (1984) (Guzzetti 2005; Guzzetti et al. 2015, 2006a); (v) the objective definition of rainfall thresholds for possible landslide occurrence (Guzzetti et al. 2007, 2008) and the quantification of their related epistemic and aleatory uncertainties (Brunetti et al. 2010; Peruccacci et al. 2012, 2017; Gariano et al. 2015a; Melillo et al. 2015, 2016), and the use of the thresholds for the design and operation of geographical landslide early warning systems (Guzzetti et al. 2020); (vi) the determination and analysis of landslide size statistics obtained from event inventory maps (Guzzetti et al. 2002b, 2009; Malamud et al. 2004b; Brunetti et al. 2009; Stark and Guzzetti 2009; Taylor et al. 2018); (vii) the vulnerability to landslides of various types of elements at risk (Galli and Guzzetti 2007), including the population (Salvati et al. 2018); and on (viii) the definition and quantitative measurement of landslide risk (Cardinali et al. 2002; Reichenbach et al. 2005), with emphasis on the risk to the population (Guzzetti 2000; Guzzetti et al. 2005a; Salvati et al. 2010, 2018), and the perception of landslide risk (Salvati et al. 2014).

The experience gained working, mostly in Italy, with the CNR IRPI Geomorphology Research Group (http://geomorphology.irpi.cnr.it) in Perugia has conditioned my own understanding of landslide processes, and has influenced my ideas and opinions on the methods and techniques best suited to detect and map landslides, to ascertain landslide susceptibility and hazard, and to evaluate landslide vulnerability and risk. I acknowledge that this has somewhat biased the article. However, I maintain that the topics are of interest to a broad audience and the recommendations are general.

The paper is organized as follows. After a brief presentation of the meaning of *landslide prediction*, I discuss key findings and inherent limitations for the detection and mapping of landslides using remotely sensed imagery (Guzzetti et al. 2012). This is followed by the discussion of some of the results of a review of statistically-based landslide susceptibility methods and models (Reichenbach et al. 2018). Based on two reviews of landslide-climate studies (Gariano and Guzzetti, 2016) and of geographical landslide early warning systems (Guzzetti et al. 2020), I then discuss the temporal prediction of landslides. Next, I present what I consider key results on the prediction of the size and number of landslides. This is followed by a description of landslide vulnerability concepts, and the presentation of a recent spatial-temporal model for the prediction of societal landslide risk to the population of Italy (Rossi et al. 2019). I conclude offering recommendations on what I maintain should be done to advance our collective ability to predict the hazards posed by populations of landslides, and to mitigate the associated risk.

Landslide Prediction

For landslides, like for other natural phenomena that can have harmful consequences or unwanted impacts—e.g. on the environment, population, society, the economy—it is important to clarify what we want (or need) to predict, and what we can (and cannot) predict. In the case of landslides, it is important to clarify the meaning of the term *landslide prediction*. I do so with three examples. First, if I say, *tomorrow there will be a landslide*, I am sure to be right. This is a trivial prediction, because I have not said *where* the landslide is expected, and every day landslides occur somewhere in the world. Second, if I am (slightly) more specific and I say, *tomorrow there will be a landslide in Italy* or in any other region, nation, or landscape dominated by mass-wasting processes, knowing that, for instance, it will be raining hard, I am (almost) certain of being correct. Every year there are thousands of rainfall-induced landslides in Italy, and in landscapes forced by rainfall or other weather-related processes. These two predictions are almost certainly correct, but they are not useful for hazard assessment or for the design or implementation of risk mitigation strategies. Third, if I say, *due to the expected rainfall, in the next 12 h, there will be 50 landslides with a volume larger than 100,000* m^3, *in a given area*, I will anticipate *where*, *when*, *how many*, and *how large* we should expect the landslides. The third is a useful prediction, which we presently struggle to do.

Predicting *where*, *when*, *how many*, and *how large* landslides are expected in an area and during a period of time is at the base of landslide hazard assessment (Varnes and the IAEG Commission on Landslides and other Mass-Movements 1984; Guzzetti 2005; Guzzetti et al. 2005a). I maintan that there is nothing in the literature, and in our collective understanding of populations of lanslides, that prevent us to predict *where*, *when*, *how many*, and *how large* landslides are expected, or to predict the consequences of population of landslides. In the next three sections, I discuss the strategies, the methods, and the techniques adopted to determine where landslides occur and where they can be expected, how to predict when or how frequently landslides can occur, and how large and numerous they can be following a triggering event.

Where Landslides Occur and Where They can be Expected

Let me start with *where* landslides occur, and *where* they can be expected. The first topic pertains to the realm of landslide detection and mapping, and the production of landslide inventory maps (Guzzetti et al. 2012), and the second is in the domain of landslide susceptibility modelling and terrain zonation (Reichenbach et al. 2018).

Landslide Detection and Mapping

In 2012, Guzzetti et al. (2012) estimated that less than 1% of the landmasses was covered by landslide maps. Despite new encouraging national efforts (e.g. in Vietnam), and attempts to prepare landslide inventory maps for entire continents (e.g. Africa, Broeckx et al. 2018), the situation has not changed much, and information on past and present landslides remains limited and unsystematic. Even in Italy, a country for which information on landslides is abundant compared to other areas (Trigila et al. 2010; Van Den Eeckhaut et al. 2012b; ISPRA 2018), no effort is underway for the systematic mapping of landslides following major landslide triggering events such as intense or prolonged rainfall periods, rapid snowmelt events, or earthquakes. The lack of accurate and updated landslide information limits the ability to model landslide hazard, evaluate landslide risk, assess the performance of prediction models, and design, implement, and verify adequate landslide mitigation strategies.

Using a combination of landslide mapping methods and techniques, the Italian National and regional geological surveys have collectively mapped more than 620,000 landslides in Italy (Trigila et al. 2010; ISPRA 2018) (Fig. 2). This is a result of the multi-decadal IFFI project (an Italian acronym for *Inventario dei Fenomeni Franosi in Italia*, Italian Landslide Inventory), which, despite some inherent limitation (Van Den Eeckhaut et al. 2012a; Marchesini et al. 2014), is the largest, most advanced, and best organized collection of geographical information on landslides for a nation. Not counting the large alluvial plains and other flat areas where landslides do not occur, this is an average of about 3.2 landslides per km^2.

At CNR IRPI, in a period of about 30 years, my colleagues mapped more than 124,000 landslides in various parts of Italy covering a total area of 30,285 km^2, 11% of the hills, mountains, and high coastal areas potentially subject to landslides in the country (Fig. 3). This is equivalent to an average density of 4.1 landslides per km^2. The difference between the average landslide density measured by IFFI (3.2 landslides per km^2) (Fig. 2) and the average density measured by our mappings (4.1 landslides per km^2) (Fig. 3) is significant, and highlights the incompleteness of the national inventory, particularly in terms of event and multi-temporal landslide information.

For years, populations of landslides were mapped through the visual interpretation of aerial photographs, with some—typically limited—field checks. Today, satellite images

Fig. 2 Landslide inventory for Italy showing more than 620,000 landslides in 302,100 km^2. Map prepared in the framework of the IFFI Italian National project (Trigila et al. 2010). Curtesy of A. Trigila (ISPRA 2018)

replace aerial photographs, and digital stereoscopes the traditional analogue instruments (Guzzetti et al. 2012; Murillo-García et al. 2014; Fiorucci et al. 2018) (Fig. 4).

Experience gained in various parts of the world (Guzzetti et al. 2012; Jackson et al. 2012) concludes that preparing landslide inventory maps—and particularly multi-temporal inventories—for large and very large areas using traditional methods i.e. through the visual interpretation of stereoscopic aerial photographs aided by field surveys, is time consuming and requires the availability of experts capable of recognizing landslides on the imagery. The process is error prone, and the results depend on the skills and experience of the investigators (Roth 1983; Carrara et al. 1992; Galli et al. 2008; Guzzetti et al. 2012). For these reasons, there is mounting interest for the development of methods and techniques for the semi-automatic or automatic detection and mapping of landslides using remote sensing data, chiefly satellite imagery and very-high resolution LiDAR terrain elevation data (Ardizzone et al. 2007; Guzzetti et al. 2012; Van Den Eeckhaut et al. 2012b). Limiting the discussion to methods that exploit optical (multi-spectral) satellite images, most of them (i) work better in vegetated areas and where event landslides alter significantly the vegetation and soil coverage (e.g. stripping the soil and vegetation, and exposing the bedrock); and (ii) tend to map too many landslides, particularly when using single, post-event imagery, because the detection algorithms interpret features as landslides that have the same radiometric signal of landslides, but in fact are

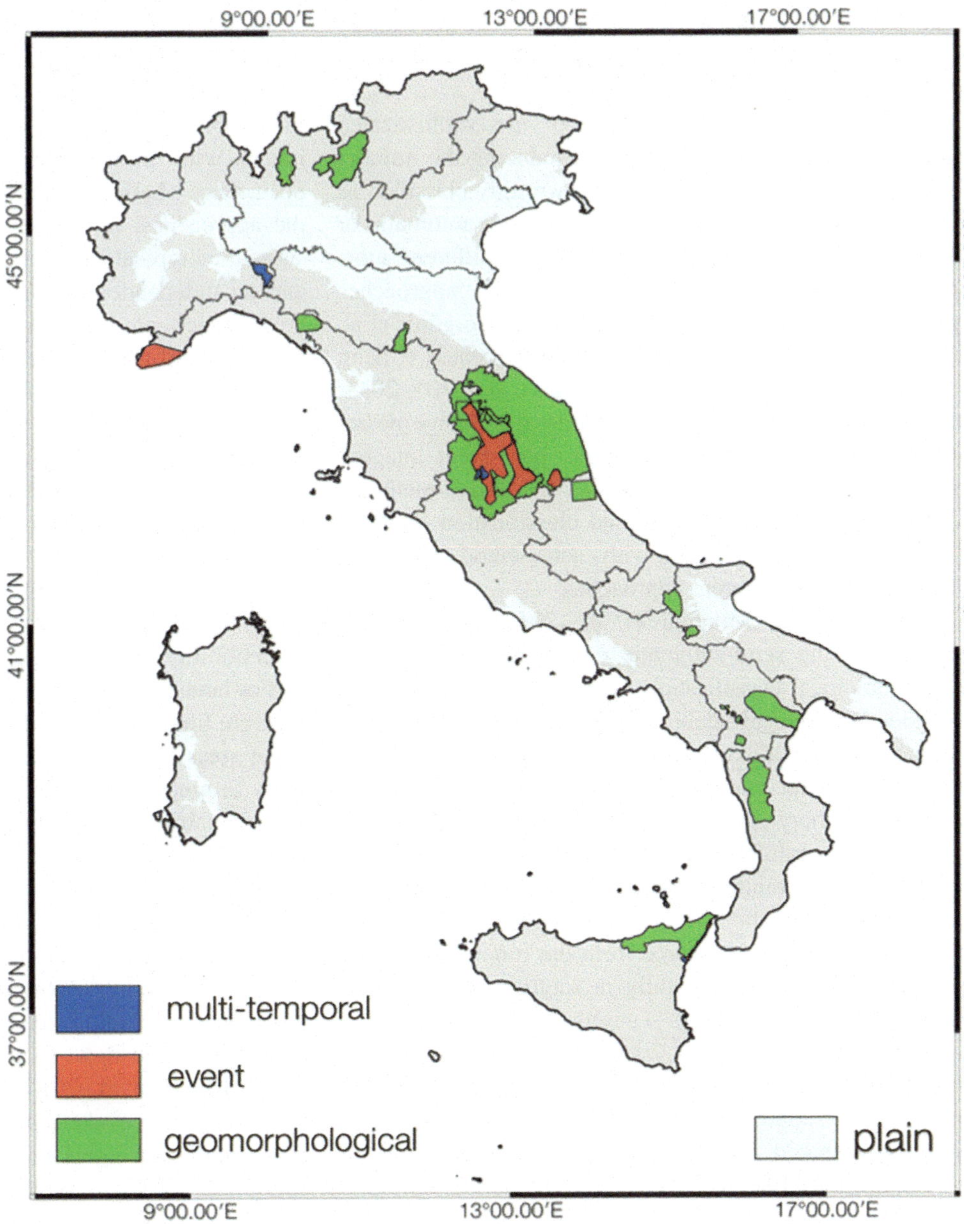

Fig. 3 Area coverage of geomorphological (green), event (red), and multi-temporal (blue) landslide inventory maps in Italy prepared by the Geomorphology Research Group at CNR IRPI (http://geomorphology.irpi.cnr.it). The 40 inventories collectively show more than 124,600 landslides in 30,285 km^2. The 29 geomorphological inventories show 100,197 landslides in 28,987 km^2; the 6 event inventories show 12,782 landslides in 4,5311 km^2; and the 5 multi-temporal inventories show 11,648 landslides in 527 km^2

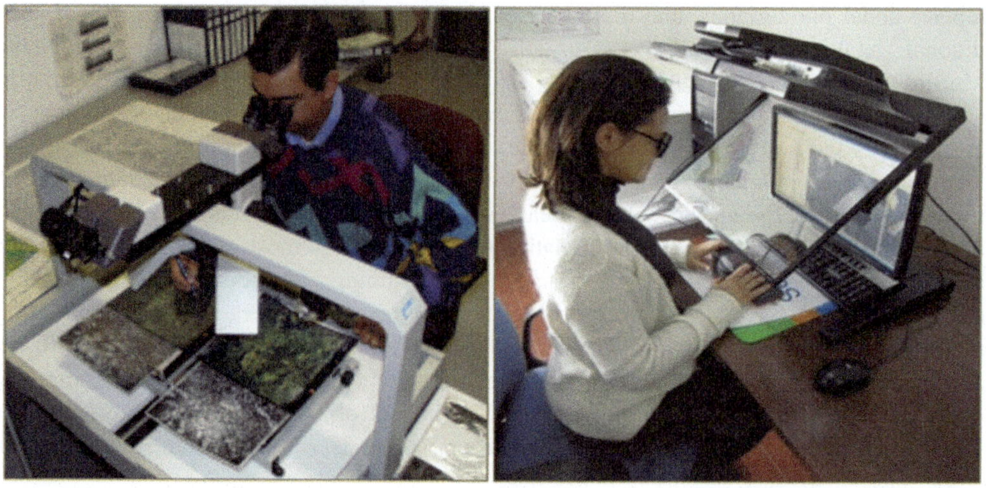

Fig. 4 Left, dual, zoom stereoscope designed by CNR IRPI for the visual interpretation of traditional aerial photographs. Right, modern digital stereoscope used for the visual interpretation of digital stereoscopic aerial photographs and satellite images

not landslides for instance, wet dirt roads, areas affected by surface erosion other than landslides, abandoned quarries, or inundated rice fields.

To address the issue, investigators use sophisticated image processing techniques aimed at reducing the number of false positives that is, image pixels wrongly classified as landslides (Guzzetti et al. 2012). Several semi-automatic or automatic approaches have been tested in different morphometric, geological, and climatic settings. The approaches exploit (Mondini et al. 2017): (i) pixel based (Borghuis et al. 2007; Mondini and Chang 2014) or object oriented (Cheng and Han 2016; Lu et al. 2011; Martha et al. 2010, 2011, 2012; Stumpf and Kerle 2011) methods; (ii) change detection (Yang and Chen 2010) or single, post-event imagery (Borghuis et al. 2007; Mondini et al. 2013, 2014) methods; and (iii) supervised or unsupervised classification methods (Mondini et al. 2011a, 2011b), with each method combining at least one of the different approaches.

As an exemple, Mondini et al. (2013) proposed an approach for the semi-automatic detection, mapping, and geomorphological classification of rainfall-induced shallow landslides from optical satellite images. The approach used a Bayesian framework to combine a classification of a post-event multispectral satellite image with information on the landslide morphometric signature. The approach is in two steps. First, rainfall-induced landslides are detected and mapped from the multispectral image, and the failed areas are separated from the stable terrain. Next, landslides are classified, separating the source from the runout areas. The approach was tested successfully in southern Taiwan, in a catchment where landslides caused by the devastating Typhoon Morakot in August 2009 were numerous. Results showed that the event landslide inventory prepared using the semi-automatic approach matched well a pre-existing event inventory prepared through the visual interpretation of post-event ortho-photographs.

Mondini and Chang (2014) and Mondini et al. (2017) proposed alternative (and complementary) automatic and semi-automatic landslide detection and mapping approaches to combine the spectral signal of a multispectral satellite image with information on the likelihood of landslide occurrence obtained from a landslide susceptibility map. In the approaches, image analysis detects areas with spectral signatures typical of *bare soil*, considered event landslides, and an independent landslide susceptibility map filters out areas which are not landslide prone. Figure 5 shows an application of the approach in a 256 km^2 area in the Kaoping watershed, in southern Taiwan, where landslides caused by Typhoon Morakot were numerous. For this experiment, Mondini et al. (2017) used a FORMOSAT-2 image acquired in 2009, after Typhoon Morakot. A relevant innovation of the approach consisted of the fact that the final landslide event inventory is probabilistic that is, the inventory shows the probability that a single pixel represents (or does not represent) an event landslide. This may prove useful for susceptibility mapping and hazard modelling.

For landslide detection and mapping, optical sensors have a severe limitation: they do not see below the clouds that in some areas can persist for long periods, limiting the possibility to prepare landslide maps immediately after an event. This explains the mounting interest for the use of radar, and chiefly synthetic aperture radar (SAR) sensors, which have the ability to penetrate the clouds, to detect and map event landslides. Mondini et al. (2019) have shown 32 examples of attempted detections and mapping of landslides using ESA Sentinel-1, C-band SAR images. Landslides were rapid mass movements including debris flow, debris slide, earth flow, earth slide, mud flow, rock avalanche, rock slide, and slide types (Hungr et al. 2014), as snow avalanches, and were caused by earthquakes, rainfall, snowmelt, human actions, and unknown triggers in all continents, except Antarctica, from May 2015 to September 2018. Figure 6 shows an

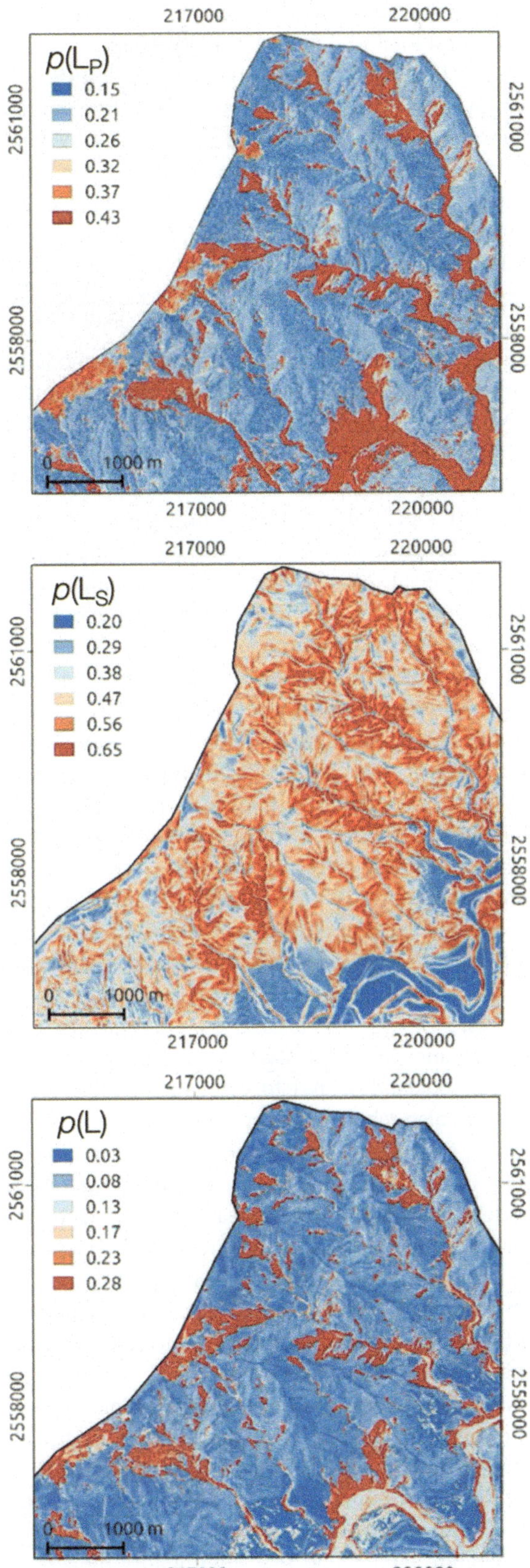

Fig. 5 Kaoping watershed, southern Taiwan. Upper map shows the probability of the presence of an event landslide (bare soil) in a FORMOSAT-2 image taken after Typhoon Morakot in August 2009. Central map shows landslide susceptibility for the same area. Lower map shows final event landslide inventory obtained combining the preliminary event inventory with the landslide susceptibility map. Modified after Mondini et al. (2017)

example for the 7.2 × 10^6 m^3 Villa Santa Lucia landslide of 16/12/2017 in the Los Lagos Region of southern Chile, which killed 22 persons (Duhart et al. 2018).

For landslide detection, Mondini et al. (2019) used the backscattering log-ratio of two SAR images taken after and before the individual slope failures. Results showed that for 27 of the 32 cases (84%) the landslides—ranging in size from 0.01 to 21 km^2—were easy to identify, visually. One case (3%) was difficult to identify visually, and for the remaining four cases (13%) the landslides were impossible to recognize visually, for different reasons. Overall, results are encouraging and promise to exploit satellite SAR images for the systematic detection of landslides. This may help the systematic validation of geographical landslide early warning systems in areas where landslide information is limited, or missing (Guzzetti et al. 2020).

I conclude this section on landslide detection and mapping reminding that in their review of methods and techniques for the production of landslide inventory maps, Guzzetti et al. (2012) identified a "*lack of standards and accepted, properly defined best practices, or operational protocols*", for the preparation, validation, and update of landslide maps. They also stressed "*lack of standards limits the credibility and usefulness of landslide maps, with adverse effects on the derivative products and analyses*". Eight years later the situation has not changed much, unfortunately.

Susceptibility Modelling and Zonation

Landslide scientists, practitioners, stakeholders, and decision makers are not only interested in knowing where landslides have occurred in the past, or where they are today, but they are also interested in knowing where landslides can occur in the future. For this purpose, we estimate landslide *susceptibility* (Brabb 1984) or, the spatial, time-independent probability of landslides occurring in an area depending on local terrain conditions (Guzzetti et al. 1999; Guzzetti 2005).

Inspection of the literature reveals that a number of different approaches have been proposed to evaluate landslide susceptibility, including: (i) geomorphological mapping, (ii) analysis of landslide inventories, (iii) heuristic terrain and susceptibility zoning, (iv) physically-based numerical modelling, and (v) statistically-based classification methods

Fig. 6 Villa Santa Lucia landslide of 16/12/2017, in Los Lagos Region, southern Chile. Upper panel shows backscattering log-ratio of two ESA Sentinel-1, C-band images taken on 24 and 12 December 2017 i.e. after and before the landslide. Whitish area crossing the image is the landslide. Lower panel shows colour-composite from a satellite image taken in January 2019, several weeks after the landslide (image source: 2019 Planet Labs Inc.) Modified after Mondini et al. (2019)

(Aleotti and Chowdhury 1999; Guzzetti et al. 1999; Guzzetti 2005; Reichenbach et al. 2018). Limiting the discussion to the statistically-based classification methods, the most popular approach to evaluate landslide susceptibility in a broad range of scales and physiographical settings, Reichenbach and her colleagues (2018) analyzed a large number (565) of peer-reviewed articles published in the 34-year period between 1983 and 2016, a sufficiently long coverage to evaluate trends and changes in the modelling methods. The systematic analysis of such a large literature collection revealed that more than 600 areas were studied in 63 nations, with area sizes in the range from a few to hundreds of thousands of km^2 and most of the areas extending to about 100 km^2, for a total of 4.6 × 10^6 km^2, 0.03% of the Earth's land area. A key result of the work was that, despite the large number of studies, and the (apparently) large number of study areas, most of the world is not covered by landslide susceptibility models and zonings, and that for only two continents i.e. Africa (Broeckx et al. 2018) and Europe (Wilde et al. 2018), synoptic-scale landslide susceptibility maps are available. The finding highlights the need for new susceptibility evaluations, particularly where such evaluations are not available.

Statistically-based landslide susceptibility modelling is performed typically using *classification* techniques (Carrara 1983; Michie et al. 1994; Chung and Fabbri 1999; Huabin et al. 2005; Chacón et al. 2006; Fell et al. 2008a, b; Galli et al. 2008; van Westen et al. 2008; Pardeshi et al. 2013; Reichenbach et al. 2018). In their review of statistically-based landslide susceptibility modelling efforts, Reichenbach et al. (2018) found 163 different model classification types, with the more *traditional* methods (e.g. logistic regression, discriminant analysis) less and less popular over time in favour of more modern *machine learning* techniques (Michie et al. 1994) (Fig. 7a). For statistically-based landslide susceptibility modelling and zonation, Rossi and Reichenbach (2016) proposed the LAND-slide Susceptibility Evaluation (LAND-SE) open source software for the R free software environment for statistical computing and graphics (R Core Team 2015). I recommend using this software for landslide susceptibility modelling, as use of a single software tool will facilitate model comparison. If necessary, interested investigators can expand the capabilities of LAND-SE by adding new modelling tools.

The thematic variables used to explain the known distribution of landslides, and to estimate landslide susceptibility, have also changed over time, with an increasing preference for morphometric variables (Fig. 7b). It is important to note that these two clear temporal trends do not obey scientific causes. The increasing preference for *machine learning* methods is not justified by the lack of complexity and the limited number of variables used in most of the published susceptibility models, and the growing preference for morphometric variables depends on the increasing availability of digital terrain models, and not on a specific improved understanding of the landslide phenomena.

Adopting the Susceptibility Quality Level (SQL) criteria and related index scheme proposed by Guzzetti et al. (2006b), Reichenbach et al. (2018) measured and ranked the quality of a large number of susceptibility models available in the literature, and found that high quality models (violet (7) and red (6) in Fig. 7c) remain an exception in the literature, even in recent years, and that the number of low

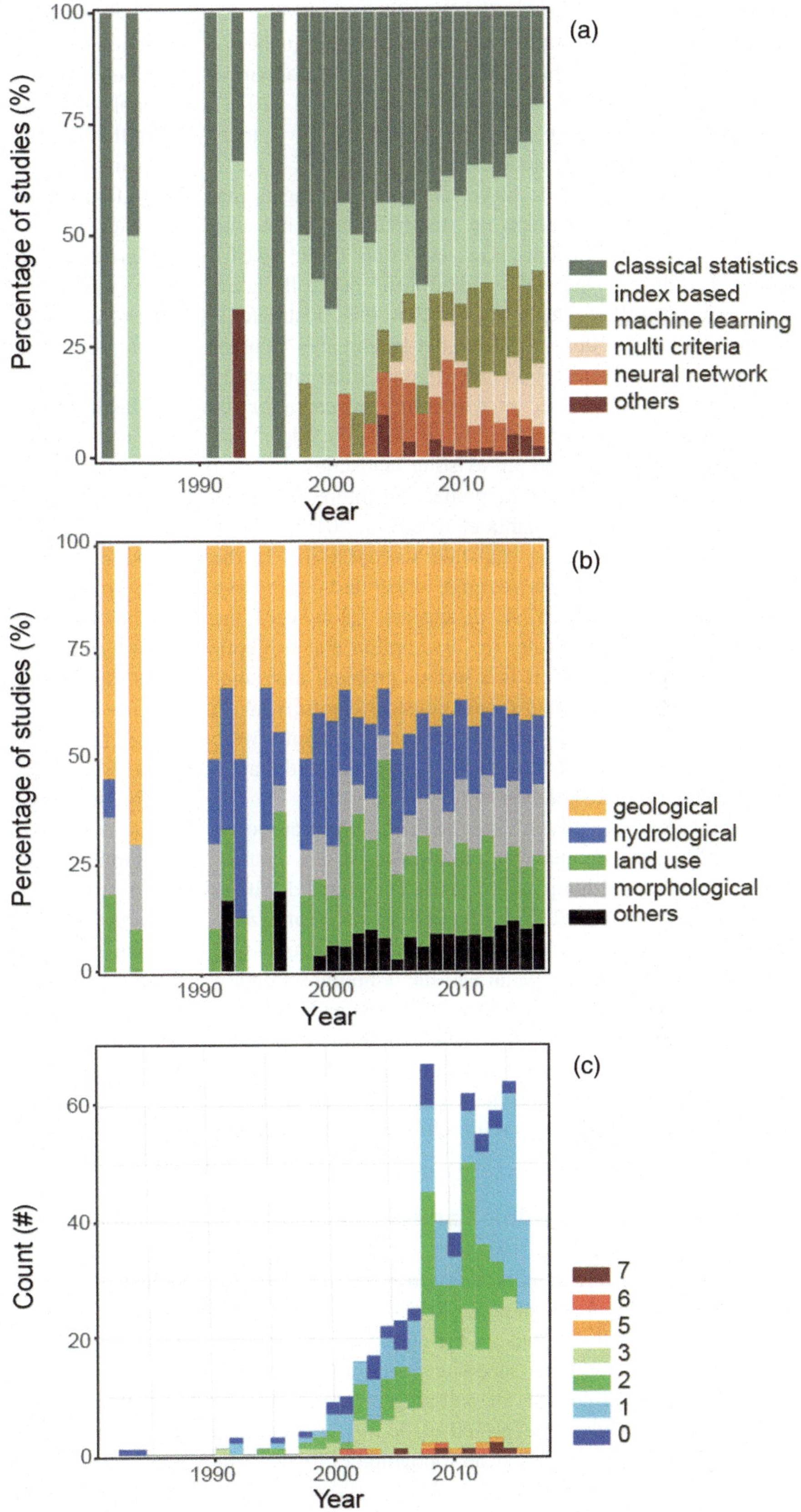

Fig. 7 **a** Temporal distribution of the percentage of susceptibility model types, per year. **b** Temporal distribution of thematic variables used for susceptibility modelling, per year. **c** Temporal distribution of the Susceptibility Quality Level (SQL) index Guzzetti et al. (2006b), in seven classes, for 565 articles in a literature database. Zero is the lowest and 7 is the highest SQL. Charts modified after Reichenbach et al. (2018)

quality models (dark blue (0) and light blue (1) in Fig. 7c) are increasing in recent times. I maintain that the quality of the published susceptibility models is a serious problem in the landslide literature. Following Reichenbach et al. (2018), I recommend not to prepare, publish, and use low-quality (SQL = 0) susceptibility model assessments.

A postulate of landslide susceptibility modelling is that susceptibility is time-invariant i.e. susceptibility is not controlled or influenced by time (Guzzetti et al. 1999, 2005b). Recent findings in the well-studied area of Collazzone, Umbria, central Italy, have shown that this postulate is incorrect, at least in this study area but presumably in other similar areas in Italy and elsewhere. In the Collazzone study area, when a landslide occurs it becomes an *attractor* for future landslides that tend to cluster inside, at the edges, and in the neighborhood of the pre-existing landslides. In other words, a landslide has a *heritage* effect on future landslides in the same general area (Samia et al. 2017a, 2017b, 2018), and this locally conditions landslide susceptibility. In the Collazzone study area this *heritage* effect lasts for about 10-15 years, after which it disappears. However, the empirical observation violates the assumption that susceptibility is time invariant. This is a serious problem that landslide susceptibility modelers should consider carefully. With this respect, encouraging—albeit preliminary—results were obtained by Samia et al. (2019) who proposed a dynamic path dependent landslide susceptibility modelling in the same study area. Recently, Lombardo et al. (2020) proposed a novel Bayesian modelling framework for the spatial, temporal, and spatial-temporal occurrence of landslides in an area. The authors tested the approach in the Collazzone study area, and showed that their spatial-temporal model was capable of predicting the spatial and the temporal evolution of landslide intensity (Lombardo et al. 2018) and susceptibility in the study area. This novel modelling framework opens to the possibility of modelling landslide hazard directly; a significant advancement over previous landslide hazard modelling frameworks (Guzzetti et al. 2005a, 2006a).

Predicting Non-susceptible Landslide Areas

In the literature, a few authors have proposed *non-susceptibility* terrain zonation (Godt et al. 2012a; Marchesini et al. 2014). The approach is based on the empirical observation that topography conditions landslide susceptibility (Godt et al. 2012b), and it consists of outlining areas where susceptibility to landslides is expected to be *negligible* or nil (Marchesini et al. 2014), based on topographic information obtained from a Digital Elevation Model (DEM). These efforts are statistically-based, but are conceptually different from the works that determine landslide susceptibility using the classification approaches discussed earlier.

Godt et al. (2012a) were first to propose a model for the zonation of non-susceptible landslide areas for the conterminous United States. Their model exploited two terrain variables, relative relief (R) and terrain slope (S) calculated from the SRTM digital elevation data (Farr et al. 2007), and information on the topographic (R,S) settings at the location of 16,000 landslides shown in landslide maps for five geographical areas in the conterminous United States. The topographic information at the point location of the landslides was used to determine a topographic threshold below which susceptibility to landslides was expected to be *negligible*, and above which *some* landslide susceptibility was expected (Godt et al. 2012b).

Marchesini et al. (2014) modified the approach proposed by Godt et al. (2012a) to prepare a map of non-susceptible landslide areas in Italy (Fig. 8). For this purpose, they used landslide information from 13 geomorphological, event, and multi-temporal landslide inventories covering 26,992 km^2, showing 93,538 landslides, for a total landslide area of 2726 km^2, 10.1% of the total studied area.

The modifications introduced by Marchesini et al. (2014) consisted in (i) considering the entire area of the landslides, instead of the single landslide point locations, and in (ii) adopting a quantile non-linear regression model to determine the topographic threshold used to separate areas where susceptibility to landslides was considered negligible, from areas where some susceptibility to landslides was expected. The first modification increased substantially the number of the empirical data points used to construct the regression model, and it allowed us to consider all the different terrain conditions present in a landslide. The second modification allowed us to decide the expected error in the threshold-based non-susceptibility model. Marchesini et al. (2014) further applied their non-susceptibility threshold model for Italy, calibrated in an area of 0.27×10^6 km^2, to the land masses surrounding the Mediterranean Sea, covering an area of 5.77×10^6 km^2. Results outlined an area of about 3.65×10^6 km^2 (63% of the studied area) where susceptibility to landslides was considered negligible. In the remaining 37% of the land masses surrounding the Mediterranean Sea some susceptibility to landslides was expected. Using independent landslide information for two areas in Spain, the authors validated their synoptic, non-susceptibility landslide model. For this purpose, they computed the fraction of landslide area that overlaid the modelled non-susceptible areas, and found a matching of 6.1%, only slightly larger than the values obtained for Italy, and of the expected proportion of landslide cells in non-susceptible areas (5.0%). The authors concluded that the model was capable to identify areas where landslide susceptibility is expected to be negligible in the Mediterranean region.

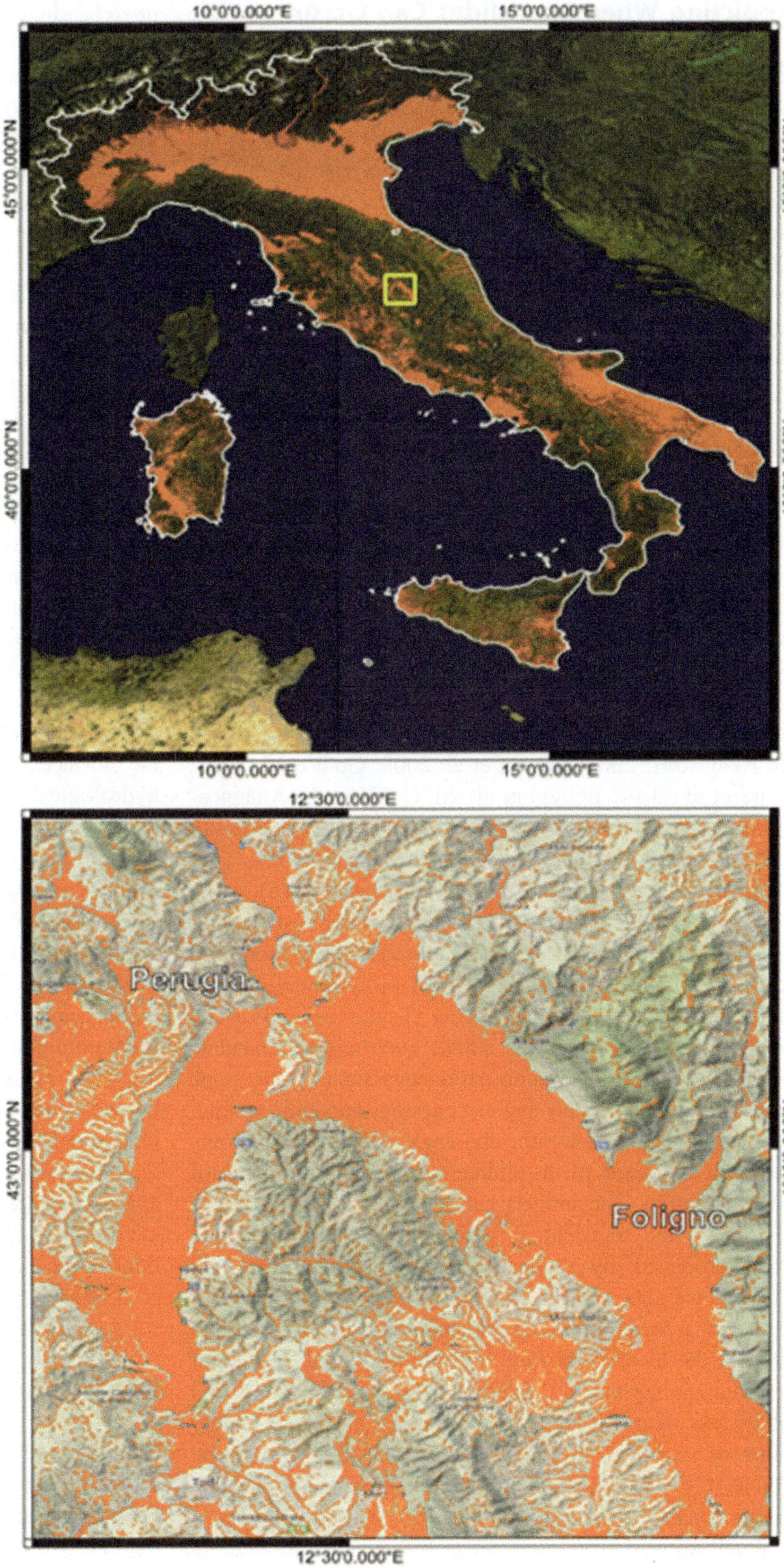

Fig. 8 Upper map shows non-susceptible landslide areas in Italy (orange). Yellow box shows location of lower map portraying an enlargement for an area in Umbria, Central Italy Modified after Marchesini et al. (2014)

Predicting When Landslides Can Occur

The temporal prediction of landslides—that is, predicting *when* or *how frequently* landslides occur—can be for short (hours to days) or long (decades to millennia) periods. Limiting the discussion to populations of rainfall-induced landslides—and hence, excluding from the discussion landslides triggered by rapid snow melting (Cardinali et al. 2000; Bíl and Müller 2008; Martelloni et al. 2013) or earthquake shaking (Fan et al. 2019)—for short term forecasts investigators use *process-based* models and empirical *rainfall thresholds* (Guzzetti 1998).

Process-Based Models

For the assessment of landslide susceptibility, process-based models, also known as deterministic or physically-based models—although the *physics* in the models is elementary—rely upon the understanding of the physical (mechanical, hydrological) laws that control the slope stability/instability conditions (Montgomery and Dietrich 1994; Terlien 1998; Borga et al. 1998, 2002a, b; Baum et al. 2002, 2008, 2010; Crosta and Frattini 2003; Savage et al. 2003, 2004; Brien and Reid 2007, 2008; Simoni et al. 2008; Godt et al. 2008; Vieira et al. 2010; Bellugi et al. 2011, 2015a, b; Anagnostopoulos and Burlando 2012; Baum and Godt 2013; Mergili et al. 2014a, b; Raia et al. 2014; Anagnostopoulos et al. 2015; Milledge et al. 2015; Alvioli and Baum 2016).

When applied to the prediction of—mostly shallow—rainfall-induced landslides, process-based models attempt to extend spatially the simplified slope stability models widely adopted in geotechnical engineering (Taylor 1948; Wu and Sidle 1995; Wyllie and Mah 2004). Such models calculate the stability of a slope using parameters such as the slope material internal friction angle, cohesion, and unit weight, the groundwater unit weight, the angle of the terrain slope, the depth and angle of the sliding surface, and the depth of the water table. The models can also consider—jointly or separately—the transient or persistent effects of external forces produced, by vegetation, seismic acceleration, and external weights. Most commonly, model computation results in a *factor of safety*, an index expressing the ratio between the local stabilizing and driving forces. Values of the index greater than 1.0 indicate stability of the slope, and values less than 1.0 identify unstable conditions. A safety factor of exactly 1.0 indicates the meta-stable condition produced by the equivalence of the stabilizing and the driving forces.

When applied over large areas for regional modelling, local stability conditions are generally evaluated by means of a static stability model, such as the well-known *infinite slope model*, where the local equilibrium along a potential slip surface is considered. For simplicity, the slip surface is assumed planar, at a fixed depth, and parallel to the topographic surface, and some assumed value of the pore fluid pressure is selected. Some models consider more complex geometries of the slip surface, including 3-D concave upward geometries (Brien and Reid 2007, 2008; Reid et al. 2015), or they can test a large number of potential slip surfaces with different geometries searching for the most unstable surface (Mergili et al. 2014a, b).

Other approaches couple the infinite slope stability model with more or less complex rainfall infiltration models (Ward et al. 1981, 1982; Okimura and Kawatani 1987; Benda and Zhang 1990; Dunne 1991; Hammond et al. 1992; van Asch et al. 1999; Montgomery and Dietrich 1994; Dietrich et al. 1995; Terlien et al. 1995; Wu and Sidle 1995; Iverson 2000; Dymond et al. 1999; Borga et al. 2002a; Crosta and Frattini 2003; Crosta and Dal Negro 2003; D'Odorico and Fagherazzi 2003; Lan et al. 2005; Bellugi et al. 2011, 2015a, b; Anagnostopoulos and Burlando 2012; Baum and Godt 2013; Raia et al. 2014; Anagnostopoulos et al. 2015; Milledge et al. 2015; Alvioli and Baum 2016).

The most advanced distributed models for the stability of slopes and the forecast of shallow landslides take as input the surface and sub-surface information on lithological, hydrological and geo-mechanical conditions, including the depth of the shear surface and of the water table at the beginning of the simulation, and a measured or inferred spatial-temporal rainfall pattern. These models run in incremental, regular or irregular time steps, and estimate jointly the location and the time of the expected rainfall-induced landslides. In this respect, the results of such models are superior to a simple susceptibility assessment. Most commonly, spatially-distributed models for the stability of slopes are based on a raster (*grid-based*) representation of the landscape, and exploit GIS-raster technology to implement the models, which generally relay heavily on a digital representation of the terrain (a DEM). Alternative approaches are based on topographic units and stream tube elements, which are hydrological, vector-based representations of the terrain (Montgomery and Dietrich 1994; Dietrich et al. 2001).

Rainfall Thresholds

Rainfall thresholds are a second popular tool for the short-term temporal prediction of rainfall-induced landslides. For rainfall-induced landslides, a threshold may define the rainfall, soil moisture, or hydrological conditions that when reached or exceeded are likely to trigger landslides (Crozier 1997; Reichenbach et al. 1998; Guzzetti et al. 2007; Segoni et al. 2018a). Analysis of the literature reveals that rainfall thresholds are defined physically, to obtain process-based or conceptual thresholds, or empirically, to obtain historical or statistical thresholds (Corominas 2000; Aleotti 2004; Wieczorek and Glade 2005; Guzzetti et al. 2007, 2008; Segoni et al. 2018b), and that most of the empirical thresholds are defined drawing—typically visually, without any mathematical or statistical criterion or formal procedure—lower-bound lines to clouds of empirical rainfall conditions known to have caused landslides plotted in a Cartesian plane, in linear, semi-logarithmic, or logarithmic coordinates (Guzzetti et al. 2007). Where information on the rainfall and climate conditions that did not result in landslides is available for the same area and period for which the information is available for events known to have caused landslides, the empirical thresholds are often defined as *optimal* separators of the rainfall conditions that have and have not resulted in landslides (Onodera et al. 1974; Lumb 1975; Jibson 1989; Corominas and Moya 1999; Zêzere et al. 2005; Pedrozzi 2004; Giannecchini 2005; Giannecchini et al. 2016). The number or abundance of the rainfall-induced landslides can also be considered to construct the empirical thresholds (Guzzetti et al. 2007, 2008), albeit figures on the number of the triggered landslides are known to be affected by uncertainties and may be severely underestimated, biasing the thresholds.

Different types of empirical rainfall thresholds for the possible initiation of landslides have been proposed in the literature, including (i) intensity-duration (ID) thresholds, (ii) thresholds based on the total event rainfall (E), (iii) rainfall event-duration (ED) thresholds, and (iv) rainfall event-intensity (EI) thresholds (Endo 1970; Onodera et al. 1974; Caine 1980; Govi and Sorzana 1980; Innes 1983; Keefer et al. 1987; Cannon 1988; Crosta 1998; Wilson et al. 1992; Corominas 2000; Crosta and Frattini 2001; Aleotti 2004; Cannon and Gartner 2005; Guzzetti et al. 2007, 2008; Hong and Adler 2008; Brunetti et al. 2010, 2018; Tiranti and Rabuffetti 2010; Martelloni et al. 2011; Rosi et al. 2012, 2015, 2016; Vennari et al. 2014; Vessia et al. 2014; Segoni et al. 2014a, b, 2015a, b; 2018a; Gariano et al. 2015; Melillo et al. 2015, 2018; Rossi et al. 2017; Palladino et al. 2018).

Guzzetti et al. (2007) listed 25 rainfall and climate variables used in the literature for the definition of empirical rainfall thresholds for the initiation of landslides, including climate variables such as the mean annual precipitation (MAP), the average number of rainy-days in a year (RDs) and the rainy day normal (RDN = MAP/RDs), all of which have been used to define normalized thresholds (Guidicini and Iwasa 1977; Govi and Sorzana 1980; Cannon 1988; Wilson et al. 1992; Wilson and Jayko 1997; Terlien 1998; Aleotti 2004; Guzzetti et al. 2007, 2008). The large number of meteorological and climate variables used to establish empirical rainfall thresholds, on the one side measures the (apparent) variability of the rainfall conditions that can result in landslides; but on the other side reveals our limited collective ability to forecast landslides using rainfall thresholds. Use of a large number of variables also limits the ability to compare thresholds established for the same and for distant areas. Albeit the literature on rainfall thresholds dates back to at least the 1970's (Endo 1970; Onodera et al. 1974; Lumb 1975), and has grown significantly in the last three decades (Guzzetti et al. 2007, 2008; Segoni et al. 2018b), inspection of the literature reveals a number of potential problems in the way empirical rainfall thresholds are established and used for landslide forecasting and early warning (Guzzetti et al. 2020). Here, I discuss what I consider to be some of the most serious problems with the definition of rainfall thresholds for possible landslide occurrence.

Inspection of the literature reveals that ID thresholds are the most common type of thresholds (Guzzetti et al. 2007, 2008; Segoni et al. 2018). However, ID thresholds have at least two inherent problems. First, rainfall intensity (I), typically measured in mm per hour or mm per day, depends on the duration (D) of the rainfall period. Thus, along the two axes of the Cartesian space used to establish an ID threshold, the same dimension (i.e. time, T) is shown; and this is unwise when searching for correlations—and possible relations (Dondi and Moser 2015; Guzzetti 2015)—between two (or more) variables. Second, intensity (I) has the dimensions of a velocity (i.e. length/time, $L\,T^{-1}$), which measures the rate of change with respect to a frame of reference as a function of time. For rainfall intensity, the rate of change of rainfall as a function of time. When used to establish ID thresholds, the rainfall duration (D) spans multiple orders of magnitude, from less than an hour to several months (Guzzetti et al. 2007, 2008) and this results in mixing the *instantaneous* and the *average* velocity, that is, the instantaneous (mm of rain in a few minutes to an hour) and the *average* (mean rainfall rate in a long or very long period) rainfall rate of change. This is also unwise, because it mixes in the same analysis measures that capture very different processes that drive instantaneous and average rainfall rates. Use of ID thresholds is also unnecessary for two other practical reasons. First, the information given by ID thresholds is easily and completely given by ED thresholds (Peruccacci et al. 2014). Second, for any practical or operational purpose related to rainfall monitoring for landslide forecasting and early warning, measuring the total, cumulated amount of rainfall in a period or event (E) is a simpler and more direct alternative to calculating rainfall rates in the same period or event. I recommend that landslide investigators abandon intensity-duration (ID) thresholds, in favour of event-duration (ED) thresholds.

Inspection of the literature further reveals that the number of empirical data points used to define a threshold are often (very) limited, and that the landslide and rainfall data are often of dubious or unknown quality. In their pioneering global studies, Caine (1980) and Innes (1983) used information on only 73 and 35 events, respectively. In the early days of the definition of empirical rainfall thresholds use of small datasets was justified; but it should not be accepted in modern studies, which should also consider the quality and representativeness of the empirical data used to define the thresholds (Guzzetti et al. 2007, 2008); an issue rarely considered in the literature. In an attempt to address the issue, Brunetti et al. (2010) and Peruccacci et al. (2017) defined a *protocol* or a set of pre-defined, unambiguous rules, to collect landslide and rainfall data used to establish ED rainfall thresholds for Italy. I encourage investigators to use the same or a similar protocol to collect landslide and rainfall data to define reliable thresholds.

A related question concerns the definition of a *rainfall event* that is, the rainfall period for which to calculate the duration (D) and the total cumulative precipitation (E) that caused (or did not cause) landslides. Inspection of the literature reveals that investigators typically do not explain clearly how they define a rainfall event in their studies, and are often not consistent and systematic in the calculation of the rainfall (D, E) conditions that represent a rainfall event. In an attempt to address the issue, Melillo et al. (2015, 2018) proposed an algorithm, which they coded into open source software, for the automatic selection and matching of landslide and rainfall data from historical records, for the definition of rainfall thresholds, and for the quantification of their associated uncertainties, using a set of standard, pre-defined, open criteria. Using this algorithm and software, Peruccacci et al. (2017) determined ED thresholds for different environmental conditions in Italy, including morphometry, lithology, soil type, land cover, climate, and mean annual precipitation (Fig. 9).

Use of a standard, objective, and automatic procedure allowed to determine reproducible thresholds, and to confront the thresholds, visually and quantitatively. Visual inspection of the upper plot in Fig. 9 reveals the considerable spread of the thresholds; a measure of the variability of the thresholds in the considered environments. I encourage landslide investigators to use the same or a similar algorithm and software for the selection and matching of landslide and rainfall records, for the definition of rainfall thresholds, and for the quantification of the associated uncertainties.

Like other empirical models, rainfall thresholds are affected by uncertainty that is important to quantify. The lower plot of Fig. 9 portrays four thresholds for four different MAP regions in Italy, with their associated uncertainty shown by the coloured areas (Peruccacci et al. 2017). Visual inspection of the plot reveals that the uncertainties are significant, and that only three of the four thresholds can be separated, when the uncertainties are considered. I stress that the joint visual comparison of the two plots in Fig. 9 reveals the inherent difficulty in separating the natural variability (aleatory) from the epistemic (modelling) uncertainty

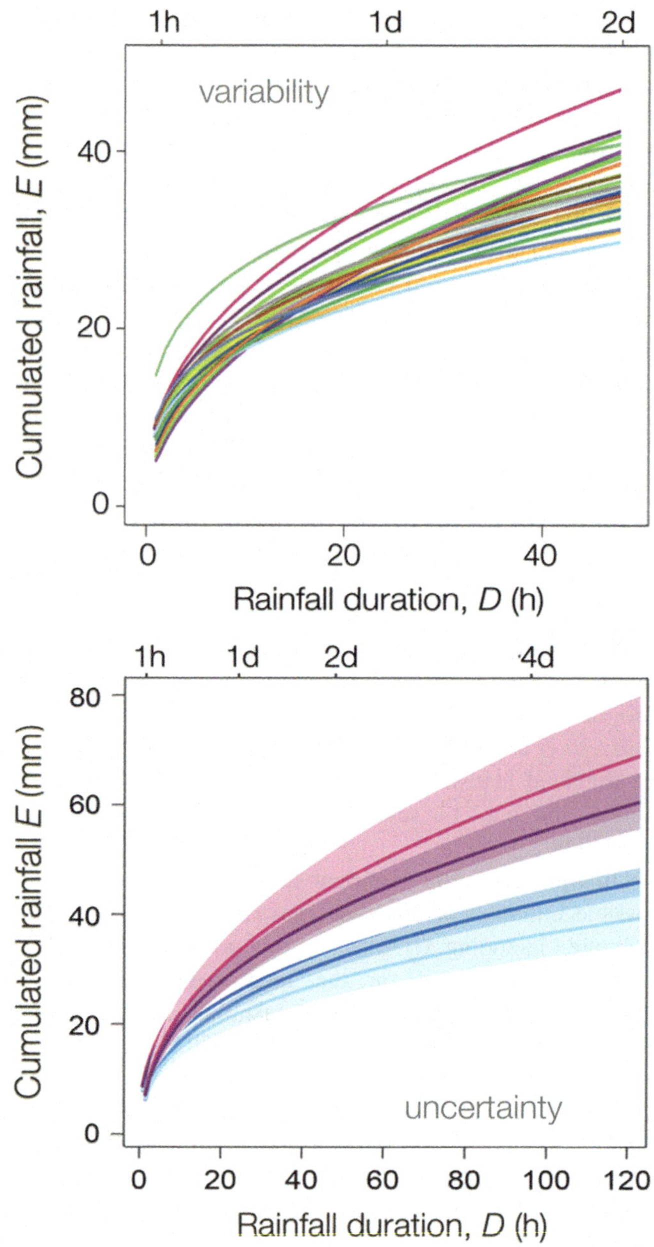

Fig. 9 Upper plot shows cumulated rainfall—rainfall duration (ED) thresholds for 26 different environmental conditions in Italy, including morphometry, lithology, soil type, land cover, climate, and mean annual precipitation. The spread of the thresholds measures the variability of the thresholds. Lower plot portrays four thresholds for different mean annual precipitation (MAP) regions, with their associated uncertainty shown by the shadowed areas. Modified after Peruccacci et al. (2017)

associated to the thresholds. This is an issue that requires further investigation, particularly if the thresholds are used for operational landslide forecasting and early warning (Guzzetti et al. 2020).

Geographical Landslide Early Warning Systems

Operational forecasting of population of rainfall or weather-induced landslides over large and very large areas is performed using geographical landslide early warning systems (LEWSs), which are devices, systems, or sets of capacities used to generate and disseminate timely and meaningful information to enable individuals, communities, and organizations threatened by landslide hazards to act promptly and appropriately to avoid or to reduce the impact of the landslide threats (UNISDR 2006; Di Biagio and Kjekstad 2007; Medina-Cetina and Nadim 2008; Huggel et al. 2010; Alfieri et al. 2012; Stähli et al. 2015; United Nations 2016; Calvello 2017; Greco and Pagano 2017; Piciullo et al. 2018; Segoni et al. 2018a; Guzzetti et al. 2020).

The idea of a geographical LEWS is not new. In 1975, the U.S. Geological Survey scientist Russell H. Campbell, working in Southern California to study debris flow events occurred in the late 1960's, wrote:

> A warning system […] could be constructed [using] three major elements, each of which is partly or wholly operative at the present time: (1) a system of rain gauges, recording the rainfall on an hourly basis; (2) a weather-mapping system capable of recognizing centers of high-intensity rainfall in the storm area and, at frequent intervals, plotting the location of these centers with respect to location of gauges with adequate registry for accurate transfer to slope maps or topographic maps; and (3) an administrative and communications network to collate the data, recognize when critical factors have been exceeded in a particular area, and inform the residents there. Such a system is probably well within the capability of existing technology.

Campbell (1975) description of a regional LEWS and his lucid vision were extraordinary, considering that it took the global landslide community three decades to implement the vision in just a few regions globally (Piciullo et al. 2018; Guzzetti et al. 2020).

Together with six co-authors at CNR IRPI, and with the help of 22 colleagues world-wide who provided articles, reports, and information on specific LEWSs, we examined 26 geographical systems globally, including 19 regional systems, 6 national systems, and one global system covering from 60° north to 60° south (Guzzetti et al. 2020). In our analysis, we considered LEWSs at different operational levels; including design, experimental, operational, and dismissed systems. We found that the examined LEWSs operate (or operated) (i) in several topographic settings; (ii) in all Köppen-Geiger climate types (Peel et al. 2007), with the majority of the systems in the temperate climate (18), followed by the tropical (6) and cold (3) climates; (iii) in all geological domains (Chorlton 2007), and primarily in sedimentary rocks (51%), followed by volcanic (20%) and metamorphic (19%) rocks; and (iv) in all seismicity zones (Giardini et al. 2003), with more LEWSs in the medium (36%) and high seismicity (28%) zones, followed by the low (20%) and the very high (16%) zones. We concluded that LEWSs can operate virtually everywhere in the world, with no morphological, climatic, geological, seismic, or tectonic constrains. This is good news for operational landslide forecasting and early warning. However, a visual comparison of the location and coverage of the active LEWSs globally with the distribution and abundance of non-seismically-induced fatal landslides from 2004 to 2017 (Froude and Petley 2018) revealed that many geographical regions where fatal landslides are numerous, and hence where landslide risk to the population is high or very high, are not covered by LEWSs. I consider this a serious problem that the global landslide community should address.

We further found a lack of standards for the design, implementation, management, and validation of LEWSs, and for the delivery of the landslide forecasts to various audiences, including authorities and the public. More specifically, we found that most LEWSs have undergone some form of validation or verification, but that no accepted standard exists to check the performances and the forecasting skills of the LEWSs. This limits greatly the possibility to compare the performance and the forecasting skill of different LEWSs. A known problem with the quantification of the LEWS forecasting skill is the need for information on landslide occurrence, or the lack of occurrence, in the area and period of the landslide forecast. Since no instrumental device or system exists for the automatic detection and mapping of landslides over large or very large areas, information on landslide occurrence is typically scant, biased towards inhabited areas, and often incomplete. The lack of accurate and reasonably complete information on where and when landslides occur limits greatly the ability to verify the landslide forecasts quantitatively (Gariano et al. 2015; Piciullo et al. 2017). The issue can be addressed—or at least mitigated—by improving landslide detection/mapping techniques that exploit remote sensing data and technologies (Guzzetti et al. 2012, 2020; Mondini et al. 2019).

We further found that the 26 considered LEWSs used (at least) 10 different approaches to prepare and disseminate their landslide advisories to different audiences, including internal dissemination to the landslide forecasting teams, to notify local, regional or national authorities, and to inform the general public, using advisory schemes based on two to five levels. We also noticed a confusion with the language used in the advisories; which can be confusing and makes it difficult to compare the perceived performance of the systems by different audiences.

Overall, the review of the literature on past and existing geographical LEWSs, and our experience in designing and operating national and regional LEWSs in Italy, allowed to conclude that (Guzzetti et al. 2020):

> Operational forecasting of weather-induced landslides is possible [...] and it can contribute to mitigate landslide risk, reducing fatalities and economic loses. However, LEWSs remain complex, and operational landslide forecasting a difficult and uncertain task." [and] that "standard and shared performance evaluations approaches will allow the discovery of problems in existing LEWSs and possible fixes, and will increase the credibility and authority of the LEWSs; a key aspect for landslide risk reduction.

Long-Term Landslide Projections

For long-term planning and decision-making, as well as for geomorphological, geological, environmental, and landscape evolution studies, it is important to investigate the long-term (decades to millennia) behaviour of single landslides, and of landslide populations. This proves difficult, for at least three main reasons. First, there is generally a lack of reliable historical and pre-historical records of landslides and their consequences (Guzzetti et al. 1994, 2005b; Ibsen and Brunsden 1996; Glade 2001; Guzzetti and Tonelli 2004; Bach Kirschbaum et al. 2009; Salvati et al. 2010, 2018; Van Den Eeckhaut and Hervás 2012; Hurst et al. 2013; Gariano and Guzzetti 2016; Froude and Petley 2018). This is largely due to the lack of instrumental methods for the systematic detection of landslides (Guzzetti et al. 2012). Second, the historical landslide records are typically *sparse*, with only a few days with landslides and many days without landslides (Rossi et al. 2010a; Witt et al. 2010; Hurst et al. 2013; Rossi et al. 2019). Third, like other records of Nat Hazards (Boschi et al. 1997; Glade et al. 2001; Simkin et al. 2001; Doocy et al. 2003a, b; Guzzetti et al. 2005; Schulte and Mooney 2005; Witham 2005; Geyer and Marti 2008; EEA 2010; NGDC/WDC 2011; Stucchi et al. 2013), historical landslide records are typically *non stationary* (Guzzetti 2000; Guzzetti et al. 2005b), with the lack of stationarity due chiefly to incompleteness of the sources—and hence of the record—and to environmental, geological, and climate variations, at different spatial and temporal scales. To complicate the matter, the two mentioned sources of non-stationarity are difficult to separate. All this complicates the analysis and the interpretation of historical records of landslides and their consequences (Guzzetti et al. 2005b; Rossi et al. 2010b, 2019; Witt et al. 2010; Gariano et al. 2015; Froude and Petley 2018), and the possibility to project the findings in the future. One cause of non-stationarity of the landslide records is climate change (Crozier 1997, 2010; Dikau and Schrott 1999; Sidle and Ochiai 2006; McInnes et al. 2007; Coe and Godt 2012; Gariano and Guzzetti 2016; Mal et al. 2018). In a recent review of the literature, Gariano and Guzzetti (2016) found that landslide–climate studies were only slightly more than 100 in the 27-year period from 1992 to 2018; significantly less numerous than studies in other fields of landslide research (Wu et al. 2015). Inspection of the geographical distribution of the landslide–climate studies (Fig. 10) revealed a bias in the distribution of the studies, with most of the works in the mountains of Europe and Canada, only a few works in South America and Asia, almost no studies in Africa, and no studies in India, Japan, and Australia. The scarcity of landslide–climate studies has many causes,

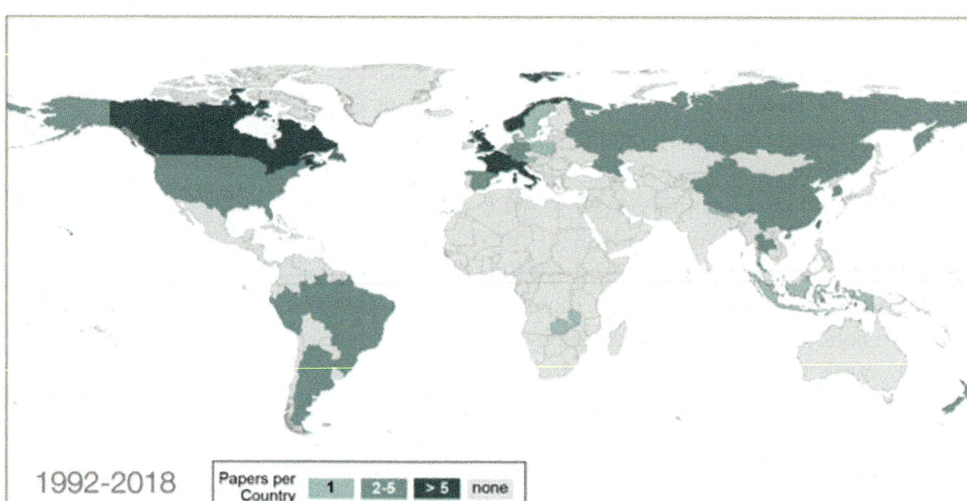

Fig. 10 Map shows number of climate-landslide studies in different nations from 1992 to 2018. Updated from Gariano and Guzzetti (2016)

including the complexity of the problem, the different scales of climate and landslides, and the fact that different landslide types respond differently to climate changes.

In the literature, landslide–climate studies can be loosely grouped into three main classes, namely: (i) modelling studies, (ii) historical studies, and (iii) analyses of landslide paleo-evidences. The various approaches exploit—more or less strictly—a single logical framework which is based on two coupled modelling chains: a climate modelling chain, and a slope stability modelling chain, which have inherent advantages and limitations (Gariano and Guzzetti 2016). The *modelling* studies are the most common, and their popularity is increasing. They encompass three main sub-categories that is, modelling of single slopes or landslides using slope stability methods, statistical analyses of landslide catalogues, and regional landslide studies using a variety of modelling techniques (Buma and Dehn 1998; Dehn 1999; Collison et al. 2000; Schmidt and Glade 2003; Dixon and Brook 2007; Tacher and Bonnard 2007; Jakob and Lambert 2009; Jomelli et al. 2009; Chang and Chiang 2011; Coe 2012; Melchiorre and Frattini 2012; Comegna et al. 2013; Rianna et al. 2014; Gassner et al. 2015; Villani et al. 2015; Ciabatta et al. 2016; Turkington et al. 2016; Gariano et al. 2017; Alvioli et al. 2018). Modelling studies are particularly well suited to obtain landslide–climate projections spanning periods from a few to several decades.

The *historical studies* attempt to find statistical relationships linking historical landslide and climate records, chiefly rainfall and temperature records (Evans and Clague 1994; Rebetez et al. 1997; Flageollet et al. 1999; Jomelli et al. 2004; Chiarle et al. 2007; Polemio and Petrucci 2010; Huggel et al. 2012; Stoffel et al. 2014; Chiarle et al. 2015; Gariano et al. 2015; Polemio and Lonigro 2015; Paranunzio et al. 2016). The approach works best with past records, but it can be used for qualitative and quantitative landslide projections (Gariano et al. 2015). Lastly, a few investigators have used landslide *paleo-evidence* to search for possible relationships between long-term climate and environmental changes and the temporal distribution of landslides (Innes 1983, 1985; Bovis and Jones 1992; González Díez et al. 1996; Lateltin et al. 1997; Margielewski 1998; Trauth et al. 2000; Schmidt and Dikau 2004; Soldati et al. 2004; Stoffel and Beniston 2006; Matthews et al. 2009; Borgatti and Soldati 2010; Yin et al. 2014; Sewell et al. 2015). Paleo-evidence studies focus typically on long and very long periods in the distant past, but can be used to infer scenarios of future landslide activity and its variations.

In their review of the landslide–climate studies Gariano and Guzzetti (2016) collected information on the expected and/or projected variations in the abundance and/or activity of four main landslide types—including rock fall/rock avalanche, debris flow, shallow landslide, deep-seated landslide—driven by the projected climate changes landslide–climate studies, which they used to produce a global, synoptic map showing the geographical distribution of the expected changes in the abundance and/or activity of the considered landslide types (Fig. 11). Determining if and where landslide risk—and particularly the risk to the population—is expected to increase in response to the projected climate and environmental changes, remains a difficult task. Following Gariano and Guzzetti (2016), we infer that where global warming is expected to increase the frequency and the intensity of the rainfall events, the number of people exposed to landslide risk is also expected to increase. This is because intense rainfall is the main trigger of shallow, rapid-moving landslides (e.g. soil slips, debris flows, rock falls, minor rock slides), which are the primary cause of landslide fatalities (Guzzetti et al. 2005b; Froude and Petley 2018). Conversely, where the seasonal rainfall amounts are projected to decrease, deep-seated landslides are expected to decrease their activity and to occur less frequently, reducing landslide risk.

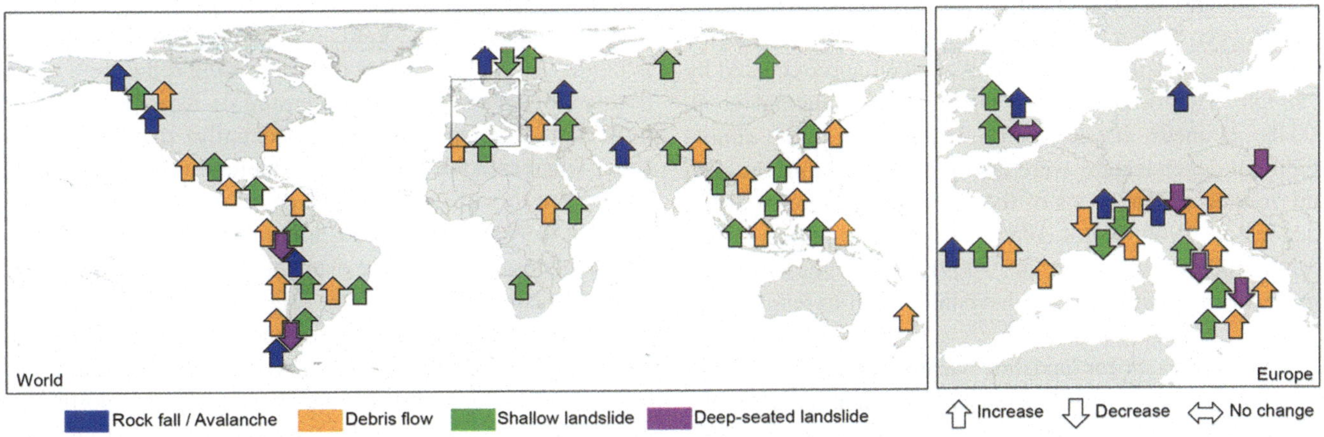

Fig. 11 Maps show areas of expected variations in the abundance/activity of four landslide types (rock fall/rock avalanche, debris flow, shallow landslide, deep-seated landslide) driven by projected climate change. Modified after Gariano and Guzzetti (2016)

Size and Number of Landslides

Looking at a good quality landslide inventory map, and particularly at an event or multi-temporal inventory, one notices that landslides of the slide type (excluding rock falls and debris flows) are not all of the same size (length, width, area, volume), and do not exhibit a *typical size* as one could expect based on geomorphological or geological considerations. Instead, a typical event or multi-temporal inventory has only a few very large landslides, many medium size landslides, very many small landslides, and a somewhat smaller number of very small landslides. This empirical evidence results in typical, non-cumulative frequency or probability distributions of landslide area or volume that increase rapidly with landslide size up to a maximum value —known as the *rollover* (Malamud et al. 2004b)—and then decreases rapidly, first approaching and then following a power law (Hovius et al. 1997; Guzzetti et al. 2002b; Malamud et al. 2004b). Early investigators (Hovius et al. 1997; Brardinoni and Church 2004) argued that the *rollover* resulted from the incompleteness of the landslide inventory for small and very small landslides. Guzzetti et al. (2002b) and Malamud et al. (2004b) confirmed that for geomorphological inventories (Guzzetti et al. 2012) the rollover was due to the incompleteness of the landslide maps for small and very small failures, but showed that for good quality event inventories the rollover was a real, physical characteristic of the landslide size distribution. Two mathematical distributions, the *double Pareto* (Stark and Hovius 2001) and the *inverse Gamma* (Malamud et al. 2004b) distributions were shown to approximate well the empirical frequency and probability distributions of landslide area and volume of landslides of the slide type, and Katz and Aharanov (2006), Stark and Guzzetti (2009), and Klar et al. (2011) proposed different, albeit similar, conceptual and mechanical models to explain the empirical distributions of landslide areas observed in nature.

The volume of rock falls, rock slides, and rock avalanches has also been shown to exhibit empirical distributions, which do not show a typical *rollover* and are well approximated by a simple, Pareto, power law distribution (Malamud et al. 2004b; Brunetti et al. 2009). In the literature, simple empirical dependencies were also proposed to link the area and volume of landslides of the slide type (Simonett 1967; Rice et al. 1969; Innes 1983; Guthrie and Evans 2004; Korup 2005; Guzzetti et al. 2009), and to describe the empirical distribution of the landslide width-to-length ratio (Taylor et al. 2018).

What is relevant for landslide hazard and risk assessment is that the various dependencies that approximate the empirical distributions of the area and volume of landslides of the slide type (Stark and Hovius 2001; Malamud et al. 2004b), of rock fall volume (Malamud et al. 2004b; Brunetti et al. 2009), of the landslide area-volume ratio (Simonett 1967; Rice et al. 1969; Innes 1983; Guthrie and Evans 2004; Korup 2005; Guzzetti et al. 2009), and of the landslide width-to-length ratio (Taylor et al. 2018), can be used to predict the expected *size* (area, volume, area-volume ratio, width-to-length ratio) of the expected landslides. The empirical distributions of the area and volume of the landslides can also be used to predict the numerosity of event triggered landslides in an area (Malamud et al. 2004b). This information is crucial to assess landslide hazard (Guzzetti 2005; Guzzetti et al. 2005a, 2006b), and is instrumental to design landslide impact scenarios.

Landslide Consequences

Anticipating the landslide consequences means predicting the vulnerability to landslides of various elements at risk (Alexander 1999; Guzzetti 2005; Galli and Guzzetti 2007), and assessing landslide risk (Cruden and Fell 1997; Glade et al. 2005; Guzzetti 2005; Porter and Morgenstern 2013). Of particular importance is the assessment of landslide risk to the population (Guzzetti et al. 2005b; Salvati et al. 2010, 2018; Rossi et al. 2019).

Vulnerability

Little is known about the *vulnerability to landslides* of various types of elements at risk (Galli and Guzzetti 2007). The lack of reliable information about the vulnerability to landslides limits our ability to determine and mitigate landslide risk (Glade et al. 2005; Guzzetti 2005). Varnes and the IAEG Commission on Landslides and other Mass-Movements (1984) defined landslide vulnerability as the degree of loss to a given element or a set of elements at risk resulting from the occurrence of a landslide of a given magnitude in an area. VanDine et al. (2004) defined vulnerability to landslides as a measure of the robustness or the fragility of an element, or a measure of its exposure to, or protection from an expected (potentially) damaging landslide, whereas Alexander (2005) defined it as the ability of an element to withstand a landslide of a given type and size. Einstein (1998) defined landslide vulnerability as the probability of loss to a specific element given a landslide or, in mathematical language, $V_L = P[D_L \geq 0 | L]$ where D_L is the damage to an element given the occurrence of a landslide, L. Based on this equation, vulnerability is the probability of loss to a specific element, or the proportion of damage to an element, given the occurrence of a landslide (Galli and Guzzetti 2007).

Vulnerability to landslides can be ranked using monetary (quantitative) or heuristic (qualitative) scales (Galli and Guzzetti 2007). Using monetary values, vulnerability is expressed in terms of the element value that can be (i) monetary as in, the price or value of the asset or the cost of replacing it; (ii) intrinsic as in, the extent to which an asset is important; and (iii) utilitarian or the usefulness of an asset, or the monetary value of its usage over a period (Alexander 2000, 2005). Vulnerability of structures and infrastructures is described typically as (i) aesthetic or minor, where the functionality of buildings, roads, and railways is not compromised and the damage can be repaired rapidly and at low cost; (ii) functional or medium, where the functionality of structures or infrastructures is compromised, and the damage takes time and large resources to be repaired; and (iii) structural or total, where buildings and transportation routes are severely or completely damaged, and they require extensive work to be fixed or replaced (Reichenbach et al. 2005; Galli and Guzzetti 2007).

When using a heuristic approach, vulnerability to people can be described by the number of expected casualties (none, few, numerous, very numerous), or by the type of expected damage to the population: (i) no damage, where damage to the population is not expected; (ii) direct damage, where fatalities (dead and missing persons) or casualties (fatalities and injured people) are expected; (iii) indirect damage, where only socio-economic damage is expected; and (iv) temporary damage, where temporary loss of private homes is foreseen resulting in evacuees and homeless people (Reichenbach et al. 2005; Galli and Guzzetti 2007).

Risk to the Population

Landslide risk analysis seeks to determine the probability that a single landslide or a landslide event will cause damage. For this purpose, it investigates the relationships between the frequency of the damaging landslides or landslide events, and the magnitude of the consequences (Fell and Hartford 1997; Guzzetti et al. 2005b; Salvati et al. 2010). To assess landslide risk to the population, two approaches are used. The first approach evaluates *individual risk* or the risk imposed by a landslide or a landslide event to any single individual. Individual risk is expressed typically using mortality rates, which are given by the number of deaths and missing persons per 100,000 of any given population over a period (Salvati et al. 2010). The second approach evaluates *societal risk* or the risk imposed by a landslide or a landslide event on society. Societal risk is established constructing and analysing *frequency–consequence* plots, where the number of fatalities or casualties caused by each landslide or landslide event is plotted versus the frequency of the landslide or landslide event, typically on a log-log scale (Fell and Hartford 1997; Guzzetti et al. 2005b; Salvati et al. 2010). Both approaches provide quantitative estimates of landslide risk and can be used to compare landslide risk to the risk posed by other natural and human-induced hazards (Guzzetti et al. 2005b; Salvati et al. 2010, 2018).

Italy is one of the few countries in the world for which a long and accurate catalogue of landslides with human consequences is available (Van Den Eeckhaut and Hervás 2012). This catalogue was updated repeatedly, and was used to define landslide risk to the population in terms of both individual and societal risk levels (Guzzetti 2000; Guzzetti et al. 2005b; Salvati et al. 2010, 2013, 2016). The catalogue was also used to study the effects of gender and age on landslide fatalities (Salvati et al. 2018), and to confront the perception of landslide risk to the actual landslide risk to the population of Italy (Salvati et al. 2014). Recently, Rossi et al. (2019) used the catalogue of landslides with human consequences in Italy from 1861 to 2015 to prepare a predictive model of societal landslide risk in Italy. To construct their model, Rossi et al. (2019) computed three risk variables inside a circular kernel of radius r = 1o km moved across a regular grid of 10 km × 10 km. The three variables showed (i) the maximum number of landslide fatalities caused by a single landslide inside the circular kernel, F; (ii) the total number of landslides with fatalities in the circular kernel, E; and (iii) the proportion of landslides with few, many, and very many fatalities, measured by the scaling exponent s of a Zipf distribution used to model the frequency–magnitude distribution of landslide fatalities, with the magnitude measured by the number of fatalities. Then, the three variables $\{F,E,s\}$ were assigned to the red, green and blue bands of a composite image to obtain a single, synoptic view of landslide risk to the population of Italy (Fig. 12).

To test their societal landslide risk model, Rossi et al. (2019) prepared different risk scenarios for fatal landslides of increasing magnitudes (for a different number of landslide fatalities), and for different return periods. Then, they validated their scenarios confronting the anticipated return period of the fatal events against information on 130 fatal landslides between 1000 and 1860, and 11 fatal landslides between January 2016 and August 2018, not used to construct the societal model and the scenarios. The anticipated return periods were in good agreement with the occurrence of fatal landslides in the two validation periods. This proved the reliability of the spatial-temporal model for societal landslide risk for Italy (Rossi et al. 2019).

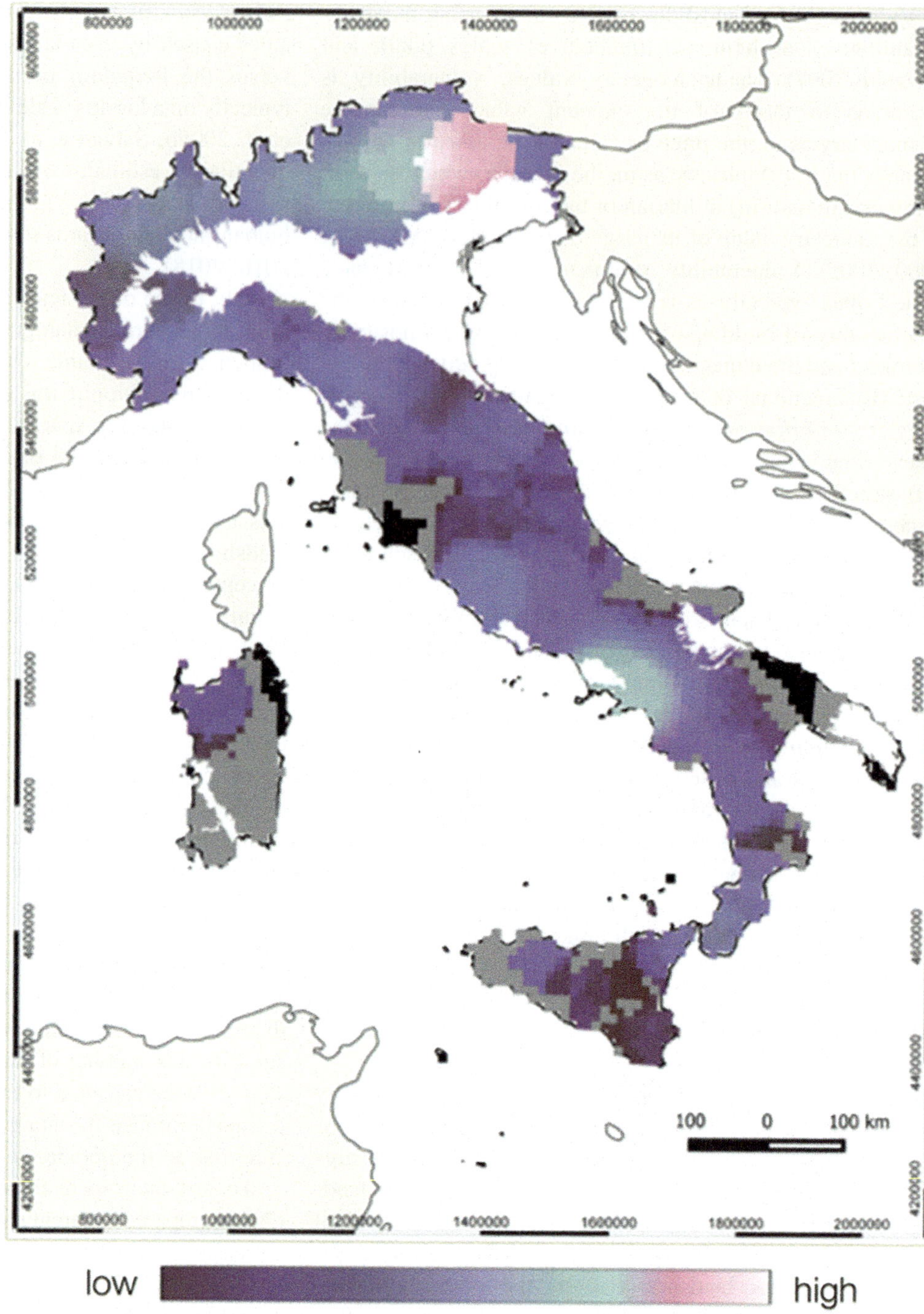

Fig. 12 Map shows a societal landslide risk model for Italy Modified after Rossi et al. (2019)

Final Remarks

I based the article on the idea of predicting landslides, because (i) I am convinced that it is possible, (ii) it measures our ability to understand the landslide phenomena, and (iii) it can help mitigate landslide risk, reducing damage and fatalities. However, I acknowledge that landslide prediction, in all the forms discussed in this work, remains a difficult, complex and uncertain task. Figure 13 portrays a space-time chart that summarizes the geographical and temporal domains of landslide prediction. In the chart, the grey area shows the domain of the problems—and of the potential conceptual and practical solutions for landslide prediction—discussed in the article. On the right side of Fig. 13, rectangles shown in shades of red exemplify sub-domains for (i) event landslides, the production of event inventory maps, and the prediction of the size statistics and numerosity of event landslides (Malamud et al. 2014b; Guzzetti et al. 2012); for (ii) the assessment of landslide risk to the population (Salvati et al. 2018; Rossi et al. 2019), and for (iii) the long-term geomorphological evolution of landslide-dominated landscapes, the production and use of geomorphological and historical landslide inventories (Guzzetti et al. 2012), and the assessment of landslide susceptibility (Reichenbach et al. 2018) and hazard (Guzzetti et al. 2005a). Similarly, on the left side of Fig. 13, rectangles in the shades of blue show sub-domains for (iv) operational landslide forecasting and early warning (Guzzetti et al. 2020); and for (v) long-term landslide-climate projections (Gariano and Guzzetti 2016). Lastly, the 3D chart in the lower-left corner of Fig. 13 exemplifies a sub-domain for the prediction of the stability/instability conditions of single landslides or slopes through geotechnical, geomechanical, and hydrological modelling; a broad and complex field not covered in this work.

Inspection of the literature, and personal experience, suggest that the sub-domains shown in Fig. 13 are typically studied by different communities as if they were distinctive and somewhat distant topics. This is not because there are physical (phenomenological) scale breaks between the sub-domains, but because everyone looks at the problem in their own way and at their own preferred scale. This is a mistake. We must close the gaps between the scales looking at the problem from an integrated perspective.

To study natural hazards and their consequences, and specifically landslide hazards and risk, we need a *new science*. A science that is not the algebraic sum of the available knowledge, but is able to consider the complex interactions between all the physical, environmental, economic, social, and human elements that characterize landslides and their consequences (Guzzetti 2018). In medicine, a highly developed field of science with self-evident societal relevance—and a field not distant from the field of natural hazards and risk (Guzzetti 2015)—the new paradigm of *convergence science* is emerging, proposed by the Nobel Laureate Phil Sharp and the former MIT President Susan Hockfield (Sharp and Hockfield 2017). In a report, Sharp et al. (2016) explain:

> Convergence comes as a result of the sharing of methods and ideas by chemists, physicists, computer scientists, engineers, mathematicians, and life scientists across multiple fields and industries. It is the integration of insights and approaches from historically distinct scientific and technological disciplines.

To advance the current capacities to predict landslides and to anticipate their consequences, and to improve our ability to defend ourselves from landslides, the landslide scientific and technical community should adopt the converge science paradigm. There are clear (albeit not fully understood) relationships between landslide populations, the landscapes that host them, and the hazards and risk they pose. Landslides somehow self-organize themselves to go from stable to unstable conditions, and this is because natural landscapes respond to energy inputs (topography, precipitation, seismic shaking, volcanic activity, climate and environmental changes), and they do so in a non-stationary context, at multiple spatial and temporal scales, adapting to meteorological, geomorphological, climatic, seismic, land and environmental changes.

To advance our landslide prediction capacities, the landslide community needs to (i) combine current regional, geographical models with local, mechanics and hydraulics models; (ii) develop new models capable of operating at different spatial and temporal scales; (iii) construct new, advanced models capable of predicting jointly multiple landslide characteristics (geometry, dynamics, time, number, size); (iv) improve the landslide propagation models, and apply them over large and very large areas; (v) advance the —currently rather poor—understanding of intermediate, complex geo-materials at the fuzzy interface between soil

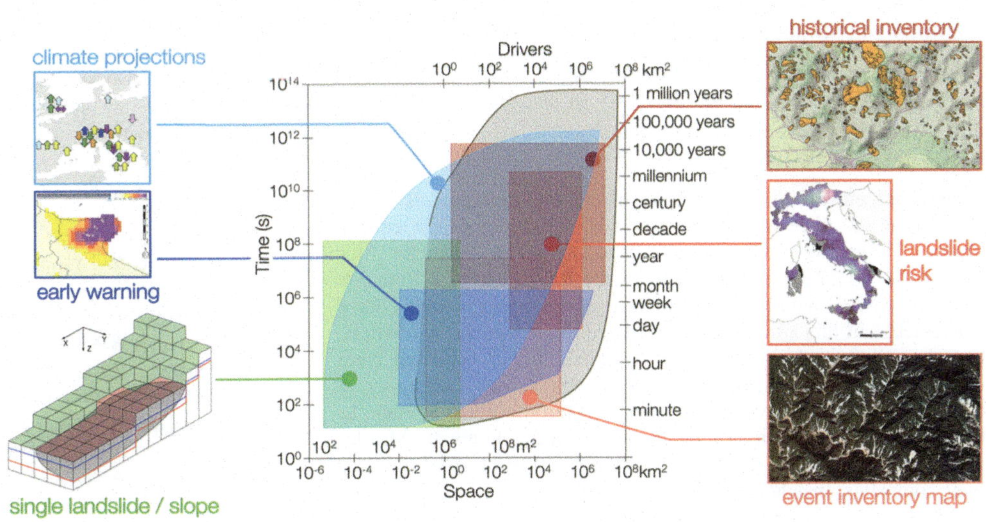

Fig. 13 Space-time chart shows domain of landslide predictions. Grey area shows domain of the landslide predictions discussed in the article, and colored rectangles show various sub-domains of landslide predictions

and rock, including flysch deposits; and the common needs to (vi) improve and expand the monitoring capacities. Technologies for enhanced modelling and improved, distributed monitoring exist, and if it is easy to predict that in the near future they will become more effective, reliable and intelligent, less expensive and easier to manage. But we need to know what we want to model and measure, and why. A key, underrated and largely open issue is the management of the ever-increasing amount of modelling and monitoring data that are, and will become available. A related issue is the proper, timely and effective communication of the modelling results and the monitoring data to different audiences, including scientists, practitioners, decision and policy makers, and the public. In this respect, openness and transparency are becoming crucial aspects for science and its application to society and decision making.

To achieve these challenging goals, the landslide community needs to establish common standards for collecting, storing, managing and distributing data, for the preparation of maps and the construction of models, and for their comparison and validation. Lack of standards hinders comparison and reduces the credibility of the products and the results (Guzzetti et al. 2012, 2020; Reichenbach et al. 2018). Further, the landslide community needs more, and more accurate data, including: (i) event and multi-temporal landslide inventories; (ii) ground deformation maps and space-time series of ground deformations; (iii) monitoring data and models on the landslide triggers, including weather, seismic, and anthropogenic triggers; (iv) data and spatial-temporal models on climate and environmental changes that may control or influence landslide occurrence and abundance; (v) data and models on the vulnerability to landslides of various elements at risk, including the population; and (vi) data and models for risk assessment. Essential are also common, shared approaches to design and implement landslide mitigation and adaptation strategies, and to measure the strategy short and long-term efficacy; and methods and strategies to assess and measure the perception of landslide risk. Lastly, the landslide community must have the capacity to ask, and the modesty to accept, new, difficult and challenging scientific questions.

Acknowledgements The paper distils results of research work I have done in the last three decades mainly with colleagues of the Geomorphology Research Group of the Research Institute for Geo-Hydrological Protection (IRPI), of the Italian National Research Council (CNR). Without their enduring efforts, the research I have conducted, and this article, would not have been possible. The article, and the talk on which the article is based, further benefited from comments, suggestions, and discussions with several colleagues at CNR IRPI, and particularly Paolo Allasia, Mauro Cardinali, Stefano L. Gariano, Daniele Giordan, Piernicola Lollino, Lorenzo Marchi, Alessandro C. Mondini, and Paola Reichenbach. I thank all of them for their comments, insight and wisdom.

References

Aleotti P (2004) A warning system for rainfall-induced shallow failures. Eng Geol 73:247–265

Aleotti P, Chowdhury R (1999) Landslide hazard assessment: summary review and new perspectives. B Eng Geol Environ 58(1):21–44

Alexander ED (1999) Vulnerability. In: Alexander ED, Fairbridge RW (eds) Encyclopedia of environmental science. Kluwer Academic Publishers, pp 663–664

Alexander ED (2000) Confronting catastrophe: new perspectives on natural disasters. Terra Publishing

Alexander ED (2005) Vulnerability to landslides. In: Glade T, Anderson MG, Crozier MJ (eds) Landslide risk assessment. John Wiley, pp 175–198

Alfieri L, Salamon P, Pappenberger F, Wetterhall F, Thielen J (2012) Operational early warning systems for water-related hazards in Europe. Environ Sci Policy 21:35–49

Alvioli M, Baum RL (2016) Parallelization of the TRIGRS model for rainfall-induced landslides using the message passing interface. Environ Modell Soft 81:122–135

Alvioli M, Guzzetti F, Rossi M (2014) Scaling properties of rainfall induced landslides predicted by a physically based model. Geomorphology 213:38–47

Alvioli M, Marchesini I, Reichenbach P, Rossi M, Ardizzone F, Fiorucci F, Guzzetti F (2016) Automatic delineation of geomorphological slope units with r.slopeunits v1.0 and their optimization for landslide susceptibility modeling. Geosci Model Dev 9:3975–3991

Alvioli M, Melillo M, Guzzetti F, Rossi M, Palazzi E, von Hardenberg J, Brunetti MT, Peruccacci S (2018) Implications of climate change on landslide hazard in Central Italy. Sci Total Environ 630:1528–1543

Anagnostopoulos GG, Burlando P (2012) An Object-oriented computational framework for the simulation of variably saturated flow in soils, using a reduced complexity model. Environ Modell Soft 38:191–202

Anagnostopoulos GG, Fatichi S, Burlando P (2015) An advanced process-based distributed model for the investigation of rainfall-induced landslides: the effect of process representation and boundary conditions. Water Resour Res 51:7501–7523

Ardizzone F, Cardinali M, Carrara A, Guzzetti F, Reichenbach P (2002) Impact of mapping errors on the reliability of landslide hazard maps. Nat Hazard Earth Sys 2:3–14

Ardizzone F, Cardinali M, Galli M, Guzzetti F, Reichenbach P (2007) Identification and mapping of recent rainfall-induced landslides using elevation data collected by airborne LiDAR. Nat Hazard Earth Sys 7(6):637–650

Bach Kirschbaum D, Adler RF, Hong Y, Hill S, Lerner-Lam A (2009) A global landslide catalog for hazard applications: method, results, and limitations. Nat Hazards 52:561–575

Badoux A, Andres N, Techel F, Hegg C (2016) Natural hazard fatalities in Switzerland from 1946 to 2015. Nat Hazard Earth Sys 16 (12):2747–2768

Baum RL, Godt JW (2013) Correction to "Estimating the timing and location of shallow rainfall-induced landslides using a model for transient, unsaturated infiltration". J Geophys Res-Earth 118 (3):1999–1999

Baum RL, Savage WZ, Godt JW (2002) TRIGRS—a fortran program for transient rainfall infiltration and grid-based regional slope-stability analysis. Open-File Report 02-0424

Baum RL, Savage WZ, Godt JW (2008) TRIGRS - a Fortran Program for Transient Rainfall Infiltration and Grid-based Regional Slope-stability Analysis. Version 2.0. US Geological Survey Open-File Report 2008–1159

Baum RL, Godt JW, Savage WZ (2010) Estimating the timing and location of shallow rainfall-induced landslides using a model for transient, unsaturated infiltration. J Geophys Res-Earth 115(F3): F03013

Bellugi D, Dietrich WE, Stock JD, McKean JA, Kazian B, Hargrove P (2011) Spatially Explicit Shallow Landslide Susceptibility Mapping Over Large Areas. It J Eng Geol Environ 399–407

Bellugi D, Milledge DG, Dietrich WE, McKean JA, Perron JT, Sudderth EB, Kazian B (2015a) A spectral clustering search algorithm for predicting shallow landslide size and location. J Geophys Res-Earth 120:300–324

Bellugi D, Milledge DG, Dietrich WE, Perron JT, McKean J (2015b) Predicting shallow landslide size and location across a natural landscape: Application of a spectral clustering search algorithm. J Geophys Res-Earth 120:2015JF003520

Benda L, Zhang W (1990) Accounting for the stochastic occurrences of landslides when predicting sediment yields. Proceedings Fiji Symposium, IAHS-AISH Publication 192:115–127

Bíl M, Müller I (2008) The origin of shallow landslides in Moravia (Czech Republic) in the spring of 2006. Geomorphology 99 (1):246–253

Borga M, Dalla Fontana G, De Ros D (1998) Shallow landslide hazard assessment using a physically based model and digital elevation data. Environ Geol 35:81–88

Borga M, Dalla Fontana G, Cazorzi F (2002a) Analysis of topographic and climatic control on rainfall-triggered shallow landsliding using a quasi-dynamic wetness index. J Hydrol 268:56–71

Borga M, Dalla Fontana G, De Ros D, Marchi L (2002b) Assessment of shallow landsliding by using a physically based model of hillslope stability. Hydrol Process 17(2):505–508

Borgatti L, Soldati M (2010) Landslides as a geomorphological proxy for climate change: a record from the Dolomites (northern Italy). Geomorphology 120:56–64

Borghuis AM, Chang K, Lee HY (2007) Comparison between automated and manual mapping of typhoon-triggered landslides from SPOT-5 imagery. Int J Remote Sens 28:1843–1856

Boschi E, Guidoboni E, Ferrari G, Valensise G, Gasperini P (1997) Catalogo dei forti terremoti in Italia dal 461 a.C. al 1990. ING-SGA, Bologna

Bovis MJ, Jones P (1992) Holocene history of earthflow mass movements in southcentral British Columbia: the influence of hydroclimatic changes. Can J Earth Sci 29:1746–1754

Brabb EE (1984) Innovative approaches to landslide hazard mapping. Proceedings 4th International Symposium of Landslides. Toronto 1:307–324

Brabb EE (1989) Landslides: Extent and economic significance in the United States. Proceedings 28th International Geological Congress, Symposium on Landslides, pp 25–50

Brabb EE (1991) The world landslide problem. Episodes 14(1):52–61

Brardinoni F, Church M (2004) Representing the landslide magnitude-frequency relation: Capilano River basin, British Columbia. Earth Surf Proc Land 29:115–124

Brien DL, Reid ME (2007) Modeling 3-D slope stability of coastal Bluffs using 3-D Ground-Water flow. US Geological Survey, Scientific Investigations Report 2007–509, Washington DC

Brien DL, Reid ME (2008) Assessing deep-seated landslide susceptibility using 3-D groundwater and slope-stability analyses, southwestern Seattle. Geological Society America, Washington DC, Washington. Landslides and Engineering Geology of the Seattle Area

Broeckx J, Vanmaercke M, Duchateau R, Poesen J (2018) A data-based landslide susceptibility map of Africa. Earth-Sci Rev 185:102–121

Brunetti MT, Guzzetti F, Rossi M (2009) Probability distributions of landslide volumes. Nonlinear Proc Geoph 16:179–188

Brunetti MT, Peruccacci S, Rossi M, Melillo M, Rossi M, Guzzetti F (2010) Rainfall thresholds for the possible occurrence of landslides in Italy. Nat Hazard Earth Sys 10:447–458

Brunetti MT, Melillo M, Peruccacci S, Ciabatta L, Brooca L (2018) How far are we from the use of satellite rainfall products in landslide forecasting? Remote Sens Environ 210:65–75

Bucci F, Santangelo M, Cardinali M, Fiorucci F, Guzzetti F (2016) Landslide distribution and size in response to quaternary fault activity: the Peloritani range, NE Sicily, Italy. Earth Surf Proc Land 41:711–720

Buma J, Dehn M (1998) A method for predicting the impact of climate change on slope stability. Environ Geol 35(2–3):190–196

Caine N (1980) The rainfall intensity-duration control of shallow landslides and debris flows. Geografiska Annaler A 62:23–27

Calvello M (2017) Early warning strategies to cope with landslide risk. Rivista Italiana Geotecnica 2:63–69

Campbell RH (1975) Soil slips, debris flows, and rainstorms in the santa monica mountains and vicinity. US Geological Survey Professional Paper, Southern California, p 851

Cannon SH (1988) Regional rainfall-threshold conditions for abundant debris-flow activity. In: Ellen SD, Wieczorek GF (eds) Landslides, floods, and marine effects of the storm of January 3–5, 1982, in the San Francisco Bay Region, California. US Geological Survey Professional Paper 1434, pp 35–42

Cannon SH, Gartner JE (2005) Wildfire-related debris flow from a hazards perspective. In: Jakob M, Hungr O (eds) Debris flow hazards and related phenomena. Springer, Berlin, pp 363–385

Cardinali M, Ardizzone F, Galli M, Guzzetti F, Reichenbach P (2000) Landslides triggered by rapid snow melting: the December 1996-January 1997 event in Central Italy. Proceedings EGS Plinius Conf, Maratea, pp 439–448

Cardinali M, Reichenbach P, Guzzetti F, Ardizzone F, Antonini G, Galli M, Cacciano M, Castellani M, Salvati P (2002) A geomorphological approach to the estimation of landslide hazards and risks in Umbria, Central Italy. Nat Hazard Earth Sys 2:57–72

Carrara A (1983) Multivariare models for landslide hazard evaluation. Math Geol 15(3):403–426

Carrara A, Cardinali M, Detti R, Guzzetti F, Pasqui V, Reichenbach P (1991) GIS techniques and statistical models in evaluating landslide hazard. Earth Surf Proc Land 16(5):427–445

Carrara A, Cardinali M, Guzzetti F (1992) Uncertainty in assessing landslide hazard and risk. ITC Journal 2:172–183

Chacón J, Irigaray C, Fernandez T, El Hamdouni R (2006) Eng Geol maps: landslides and geographical information systems. B Eng Geol Environ 65(4):341–411

Chang S-H, Chiang K-T (2011) The potential impact of climate change on typhoon-triggered landslides in Taiwan, 2010-2099. Geomorphology 133:143–151

Chen A, Darbon J, Morel J-M (2014) Landscape evolution models: a review of their fundamental equations. Geomorphology 219:68–86

Cheng G, Han J (2016) A survey on object detection in optical remote sensing images. ISPRS J Photogramm 117:11–28

Chiarle M, Iannotti S, Mortara G, Deline P (2007) Recent debris flow occurrences associated with glaciers in the Alps. Global Planet Change 56:123–136

Chiarle M, Coviello V, Arattano M, Silvestri P, Nigrelli G (2015) High elevation rock falls and their climatic control: a case study in the Conca di Cervinia (NW Italian Alps). In: Lollino G, Manconi A, Clague J, Shan W, Chiarle M (eds) Eng Geol for Society and Territory, 1, Climate Change Engineering Geology, pp 423–443

Chorlton LB (2007) Generalized geology of the world: bedrock domains and major faults in GIS format: a small-scale world geology map with an extended geological attribute database. Geological Survey Canada, OF, p 5529

Chung C-JF, Fabbri AG (1999) Probabilistic prediction models for landslide hazard mapping. ISPRS J Photogramm 65:1389–1399
Ciabatta L, Camici S, Brocca L, Ponziani F, Stelluti F, Berni N, Moramarco T (2016) Assessing the impact of climate-change scenarios on landslide occurrence in Umbria Region, Italy. J Hydrol 541(A):285–295
Coe JA (2012) Regional moisture balance control of landslide motion: Implications for landslide forecasting in a changing climate. Geology 40(4):323–326
Coe JA, Godt JW (2012) Review of approaches for assessing the impact of climate change on landslide hazards. In: Eberhardt E, Froese C, Turner AK, Leroueil S (eds) Landslides and engineered slopes, protecting society through improved understanding: proceedings 11th international and 2nd North American symposium on landslides and engineered slopes, banff, 1. Taylor & Francis, London, pp 371–377
Collison A, Wade S, Griffiths J, Dehn M (2000) Modelling the impact of predicted climate change on landslide frequency and magnitude in SE England. Eng Geol 5:205–218
Comegna L, Picarelli L, Bucchignani E, Mercogliano P (2013) Potential effects of incoming climate changes on the behaviour of slow active landslides in clay. Landslides 10(4):373–391
Corominas J (2000) Landslides and climate. In: Bromhead E, Dixon N, Ibsen ML (eds) Proceedings 8th international symposium on landslides. Balkema, Cardiff, 4:1–33
Corominas J, Moya J (1999) Reconstructing recent land- slide activity in relation to rainfall in the Llobregat River basin, Eastern Pyrenees, Spain. Geomorphology 30:79–93
Crosta GB (1998) Regionalization of rainfall thresholds: an aid to landslide hazard evaluation. Environ Geol 35(2–3):131–145
Crosta GB, Dal Negro P (2003) Observations and modelling of soil slip-debris flow initiation processes in pyroclastic deposits: the Sarno 1988 event. Nat Hazard Earth Sys 3(1–2):53–69
Crosta GB, Frattini P (2001) Rainfall thresholds for triggering soil slips and debris flow. In: Mugnai A, Guzzetti F, Roth G (eds) Mediterranean storms. Proc 2nd EGS Plinius Conference on Mediterranean Storms. Siena, pp 463–487
Crosta GB, Frattini P (2003) Distributed modelling of shallow landslides triggered by intense rainfall. Nat Hazard Earth Sys 3:81–93
Crozier MJ (1997) The climate landslide couple: a Southern Hemisphere perspective. Paleoclimate Res 19(12):329–350
Crozier MJ (2010) Deciphering the effect of climate change on landslide activity: a review. Geomorphology 124:260–267
Cruden DM, Fell R (1997) Landslide Risk Assessment. CRC Press
D'Odorico P, Fagherazzi S (2003) A probabilistic model of rainfall-triggered shallow landslides in hollows: a long-term analysis. Water Resour Res 39:9
Dehn M (1999) Application of an analog downscaling technique to the assessment of future landslide activity - a case study in the Italian Alps. Clim Res 13:103–113
Densmore AL, Anderson RS, McAdoo BG, Ellis MA (1997) Hillslope evolution by bedrock landslides. Science 275(5298):369–372
Di Biagio E, Kjekstad O (2007) In: Early warning, instrumentation and monitoring landslides. 2nd Regional Training Course, RECLAIM II. Phuket
Dietrich EW, Reiss R, Hsu M-L, Montgomery DR (1995) A process-based model for colluvial soil depth and shallow landsliding using digital elevation data. Hydrol Process 9:383–400
Dietrich WE, Bellugi D, Real de Asua R (2001) Validation of the Shallow Landslide Model, SHALSTAB, for forest management. In: Wigmosta MS, Burges SJ (eds) Water science and application. American Geophysical Union, Washington DC, pp 195–227
Dikau R, Schrott L (1999) The temporal stability and activity of landslides in Europe with respect to climatic change (TESLEC): main objectives and results. Geomorphology 30:1–12
Dixon N, Brook E (2007) Impact of predicted climate change on landslide reactivation: case study of Mam Tor, UK. Landslides 4:137–147
Dondi F, Moser F (2015) University and the risk society. Toxicol Environ Chem 86(9):997–1012
Doocy S, Daniels A, Murray S, Kirsch TD (2013a) The human impact of floods: a historical review of events 1980–2009 and systematic literature review. 1st edn, PLOS Currents Disasters
Doocy S, Daniels A, Packer C, Dick A, Kirsch TD (2013b) The human impact of earthquakes: a historical review of events 1980–2009 and systematic literature review. 1st edn, PLoS Currents Disasters
Dowling CA, Santi PM (2014) Debris flows and their toll on human life: a global analysis of debris-flow fatalities from 1950 to 2011. Nat Hazards 71(1):203–227
Duhart P, Sepúlveda V, Garrido N, Mella M, Quiroz D, Hermosilla G, Roa H (2018) The Santa LucÍa landslide disaster, Chaitén-Chile (43.4°SL): origin and effects. Proceedings XV Congreso Geológico Chileno, Concepción, Chile
Dunne T (1991) Stochastic aspects of the relations between climate, hydrology and landform evolution. Tran Japan Geomorph Union 12:1–24
Dymond JR, Jessen MR, Lovell LR (1999) Computer simulation of shallow landsliding in New Zealand hill country. Int J Appl Earth Obs 1(2):122–131
EEA (2010) Mapping the impacts of Nat Hazards and technological accidents in Europe: an overview of the last decade. European Environmental Agency Technical Report 13. Office for Official Publications of the European Union, Luxembourg
Einstein HH (1998) Special lecture: landslide risk assessment procedure. In: Bonnard C (ed) Landslides. Balkema, Rotterdam, pp 1075–1090
Endo T (1970) Probable distribution of the amount of rainfall causing landslides, Annual Report 1968. Hokkaido Branch, Forestry Experimental Station, Sapporo, pp 122–136
Evans SG, Clague JJ (1994) Recent climatic changes and catastrophic geomorphic processes in mountain environments. Geomorphology 10:107–128
Fan X, Scaringi G, Korup O et al (2019) Earthquake-Induced Chains of Geologic Hazards: Patterns, Mechanisms, and Impacts. Rev Geoph 83
Farr TG, Rosen PA, Caro E, Crippen R, Duren R, Hensley S, Kobrick M, Paller M, Rodriguez E, Roth L, Seal D, Shaffer S, Shimada J, Umland J, Werner M, Oskin M, Burbank D, Alsdorf D (2007) The shuttle radar topography mission, Rev Geophy 45: RG2004
Fell R, Hartford D (1997) Landslide risk management. In: Cruden D, Fell R (eds), Landslide risk assessment. Balkema, Rotterdam, pp 51–109
Fell R, Corominas J, Bonnard C, Cascini L, Leroi E, Savage WZ (2008a) Guidelines for landslide susceptibility, hazard and risk zoning for land use planning. Eng Geol 102:85–98
Fell R, Corominas J, Bonnard C, Cascini L, Leroi E, Savage WZ (2008b) Guidelines for landslide susceptibility, hazard and risk zoning for land-use planning. Commentary. Eng Geol 102:99–111
Fiorucci F, Giordan D, Santangelo M, Dutto F, Rossi M, Guzzetti F (2018) Criteria for the optimal selection of remote sensing optical images to map event landslides. Nat Hazard Earth Sys 18:405–417
Flageollet J-C, Maquaire O, Martin B, Weber D (1999) Landslides and climatic conditions in the Barcelonnette and Vars basins (Southern French Alps, France). Geomorphology 30:65–78

Froude MJ, Petley DN (2018) Global fatal landslide occurrence from 2004 to 2016. Nat Hazard Earth Sys 18:2161–2181

Galli M, Guzzetti F (2007) Vulnerability to landslides in Umbria, central Italy. Environ Manage 40:649–664

Galli M, Ardizzone F, Cardinali M, Guzzetti F, Reichenbach P (2008) Comparing landslide inventory maps. Geomorphology 94:268–289

Gariano SL, Guzzetti F (2016) Landslides in a changing climate. Earth-Sci Rev 162:227–252

Gariano SL, Brunetti MT, Iovine G, Melillo M, Peruccacci S, Terranova O, Vennari C, Guzzetti F (2015a) Calibration and validation of rainfall thresholds for shallow landslide forecasting in Sicily, southern Italy. Geomorphology 228:653–665

Gariano SL, Petrucci O, Guzzetti F (2015b) Changes in the occurrence of rainfall-induced landslides in Calabria, southern Italy, in the 20th century. Nat Hazard Earth Sys 15:2313–2330

Gariano SL, Petrucci O, Rianna G, Santini M, Guzzetti F (2017) Impacts of past and future land changes on landslides in southern Italy. Reg Environ Change 1–13

Gassner C, Promper C, Begueria S, Glade T (2015) Climate change impact for spatial landslide susceptibility. In: Lollino G, Manconi A, Clague J, Shan W, Chiarle M (eds) Eng Geol for Society and Territory, vol 1, Climate Change and Engineering Geology, Springer, pp 429–433

Geyer A, Martí J (2008) The new worldwide collapse caldera database (CCDB): a tool for studying and understanding caldera processes. J Volcanol Geoth Res 175:334–354

Giannecchini R (2005) Rainfall triggering soil slips in the southern Apuane Alps (Tuscany, Italy). Adv Geosc 2:21–24

Giannecchini R, Galanti Y, D'Amato Avanzi G, Barsanti M (2016) Probabilistic rainfall thresholds for triggering debris flows in a human-modified landscape. Geomorphology 257:94–107

Giardini D, Grünthal G, Shedlock KM, Zhang P (2003) The GSHAP global seismic hazard map. In: Lee W, Kanamori H, Jennings P, Kisslinger C (eds) International handbook earthquake engineering seismology, International Geophysics Series 81 B. Academic Press, Amsterdam, pp 1233–1239

Glade T (2001) Landslide hazard assessment and historical landslide data—an inseparable couple. In: Glade T, Albini P, Francés F (eds) The use of historical data in natural hazard assessments. Kluwer AP, Dordrecht, pp 153–168

Glade T, Albini P, Francés F (eds) (2001) The use of historical data in natural hazard assessments. Kluwer AP, Dordrecht

Glade T, Anderson MG, Crozier MJ (eds) (2005) Landslide hazard and risk. John Wiley Sons, ISBN-13:978-0471486633

Godt JW, Coe JA, Baum RL, Highland LM, Keaton JR, Roth Jr RJ (2012) Prototype landslide hazard map of the conterminous United States. In: Eberhardt E, Froese C, Turner K, Leroueil S (eds) Landslides and engineered slopes: protecting society through improved understanding. Taylor & Francis Group, London

Godt JW, Coe JA, Baum RL, Highland LM, Keato JR, Roth RJ (2012) Prototype landslide hazard map of the conterminous United States. In: Eberhardt E, Froese C, Turner K, Leroueil S (eds), Landslides and engineered slopes: protecting society through improved understanding. Taylor & Francis Group, London

González Díez A, Salas L, Díaz de Terán JR, Cendrero A (1996) Late Quaternary climate changes and mass movement frequency and magnitude in the Cantabrian regionm, Spain. Geomorphology 15 (3–4):291–309

Govi M, Sorzana PF (1980) Landslide susceptibility as function of critical rainfall amount in Piedmont basin (North-Western Italy). Studia Geomorphologica Carpatho-Balcanica 14:43–60

Grahn T, Jaldell H (2017) Assessment of data availability for the development of landslide fatality curves. Landslides 14(3):1113–1126

Greco R, Pagano L (2017) Basic features of the predictive tools of early warning systems for water-related Nat Hazards: examples for shallow landslides. Nat Hazard Earth Sys 17:2213–2227

Guidicini G, Iwasa OY (1977) Tentative correlation between rainfall and landslides in a humid tropical environment. Bull Int Assoc Eng Geol 16:13–20

Guthrie R (2013) Socio-economic significance—Canadian technical guidelines and best practices related to landslides: a national initiative for loss reduction. Geological Survey of Canada, Open File 7311:19

Guthrie RH, Evans SG (2004) Magnitude and frequency of landslides triggered by a storm event, Loughborough Inlet, British Columbia. Nat Hazard Earth Sys 4:475–483

Guzzetti F (1998) Preface. Hydrological triggers of diffused landsliding. Environ Geol 35(2–3):79–80

Guzzetti F (2000) Landslide fatalities and the evaluation of landslide risk in Italy. Eng Geol 58:89–107

Guzzetti F (2005) Landslide hazard and risk assessment, Ph.D Thesis, Mathematisch-Naturwissenschaftlichen Fakultat der Rheinischen Friedrich-Wilhelms-Universitat, University of Bonn, Bonn

Guzzetti F (2015) Forecasting Nat Hazards, performance of scientists, ethics, and the need for transparency. Toxicol Environ Chem 98:1043–1059

Guzzetti F (2018) Rischi naturali: l'urgenza di una scienza nuova. In: Caporale C, Maffei L, Marchis V, de Martin JC (eds) Europa: Le sfide della scienza. Istituto Enciclopedia Italiana, Roma, pp 127–133

Guzzetti F, Tonelli G (2004) Information system on hydrological and geomorphological catastrophes in Italy (SICI): a tool for managing landslide and flood hazards. Nat Hazard Earth Sys 4:213–223

Guzzetti F, Cardinali M, Reichenbach P (1994) The AVI project: a bibliographical and archive inventory of landslides and floods in Italy. Env Manag 18:623–633

Guzzetti F, Carrara A, Cardinali M, Reichenbach P (1999) Landslide hazard evaluation: a review of current techniques and their application in a multi-scale study, Central Italy. Geomorphology 31:181–216

Guzzetti F, Cardinali M, Reichenbach P, Carrara A (2000) Comparing landslide maps: a case study in the upper Tiber River Basin, Central Italy. Env Manag 25(3):247–363

Guzzetti F, Crosta GB, Detti R, Agliardi F (2002a) STONE: a computer program for the three-dimensional simulation of rock-falls. Comput Geosci 28:1079–1093

Guzzetti F, Malamud BD, Turcotte DL, Reichenbach P (2002b) Power-law correlations of landslide areas in central Italy. Earth Planet Sc Lett 195:169–183

Guzzetti F, Reichenbach P, Cardinali M, Galli M, Ardizzone F (2005a) Probabilistic landslide hazard assessment at the basin scale. Geomorphology 72:272–299

Guzzetti F, Stark CP, Salvati P (2005b) Evaluation of flood and landslide risk to the population of Italy. Env Manag 36:15–36

Guzzetti F, Galli M, Reichenbach P, Ardizzone F, Cardinali M (2006a) Landslide hazard assessment in the Collazzone area, Umbria, central Italy. Nat Hazard Earth Sys 6:115–131

Guzzetti F, Reichenbach P, Ardizzone F, Cardinali M, Galli M (2006b) Estimating the quality of landslide susceptibility models. Geomorphology 81:166–184

Guzzetti F, Peruccacci S, Rossi M, Stark CP (2007) Rainfall thresholds for the initiation of landslides in central and southern Europe. Meteorol Atmos Phys 98:239–267

Guzzetti F, Peruccacci S, Rossi M, Stark CP (2008) The rainfall intensity-duration control of shallow landslides and debris flows: an update. Landslides 5:3–17

Guzzetti F, Ardizzone F, Cardinali M, Rossi M, Valigi D (2009) Landslide volumes and landslide mobilization rates in Umbria, central Italy. Earth Planet Sc Lett 279:222–229

Guzzetti F, Mondini AC, Cardinali M, Fiorucci F, Santangelo M, Chang K-T (2012) Landslide inventory maps: new tools for an old problem. Earth-Sci Rev 112:42–66

Guzzetti F, Gariano SL, Peruccacci S, Brunetti MT, Marchesini I, Rossi M, Melillo M (2020) Geographical landslide early warning systems. Earth-Sci Rev 2020:102973

Hammond CJ, Prellwitz RW, Miller SM (1992) Landslide hazard assessment using Monte Carlo Simulation. In: Bell DH (ed) Proc 6th international symposium landslides. Balkema, Christchurch, pp 959–964

Herrera G, Mateos RM, García-Davalillo JC et al (2018) Landslide databases in the Geological Surveys of Europe. Landslides 15:359–379

Hong Y, Adler RF (2008) Predicting global landslide spatiotemporal distribution: integrating landslide susceptibility zoning techniques and real-time satellite rainfall estimates. Int J Sediment Res 23 (3):249–257

Hovius N, Stark CP, Allen PA (1997) Sediment flux from a mountain belt derived by landslide mapping. Geology 25:231–234

Hovius N, Stark CP, Hao-Tsu C, Jinn-Chuan L (2000) Supply and removal of sediment in a landslide-dominated mountain belt: Central Range. Taiwan. J Geol 108:73–89

Huabin W, Gangjun L, Gonghui W (2005) GIS-based landslide hazard assessment: an overview. Prog Phys Geog 29(4):548–567

Huggel C, Khabarov N, Obersteiner M, Ramírez JM (2010) Implementation and integrated numerical modeling of a landslide early warning system: a pilot study in Colombia. Nat Hazards 52:501–518

Huggel S, Clague JJ, Korup O (2012) Is climate change responsible for changing landslide activity in high mountains? Earth Surf Proc Land 37:77–91

Hungr O, Leroueil S, Picarelli L (2014) The Varnes classification of landslide types, an update. Landslides 11:167–194

Hurst MD, Ellis MA, Royse KR, Lee KA, Freeborough K (2013) Controls on the magnitude-frequency scaling of an inventory of secular landslides. Earth Surf Dynam 1:67–78

Ibsen ML, Brunsden D (1996) The nature, use and problems of historical archives for the temporal occurrence of landslides, with specific reference to the south coast of Britain, Ventnor, Isle of Wight. Geomorphology 15:241–258

Innes JL (1983) Debris flows. Prog Phys Geog 7:469–501

Innes JL (1985) Lichenometric dating of debris-flow deposits on alpine colluvial fans in southwest Norway. Earth Surf Proc Land 10:519–524

ISPRA - Istituto Superiore per la Protezione e la Ricerca Ambientale (2018) Landslides and floods in Italy: hazard and risk indicators. Summary report 2018, ISPRA, Rome

Iverson RM (2000) Landslide triggering by rain infiltration. Water Resour Res 36(7):1897–1910

Jackson LE Jr, Bobrowsky PT, Bichler A (2012) Identification, maps and mapping—Canadian technical guide lines and best practices related to landslides: a national initiative for loss reduction. Geological Survey of Canada, Open File 7059:33

Jakob M, Lambert S (2009) Climate change effects on landslides along the southwest coast of British Columbia. Geomorphology 107:275–284

Jibson RW (1989) Debris flow in southern Porto Rico. Geological Society America, Special paper 236:29–55

Jomelli V, Pech VP, Chochillon C, Brunstein D (2004) Geomorphic variations of debris flows and recent climatic change in the French Alps. Clim Change 64:77–102

Jomelli V, Brunstein D, Déqué M, Vrac M, Grancher D (2009) Impacts of future climatic change (2070-2099) on the potential occurrence of debris flows: a case study in the Massif des Ecrins (French Alps). Clim Change 97:171–191

Keefer DK, Wilson RC, Mark RK, Brabb EE, Brown WM-III, Ellen SD, Harp EL, Wieczorek GF, Alger CS, Zatkin RS (1987) Real-time landslide warning during heavy rainfall. Science 238:921–925

Kirschbaum D, Adler R, Hong Y, Lerner-Lam A (2009) Evaluation of a preliminary satellite-based landslide hazard algorithm using global landslide inventories. Nat Hazard Earth Sys 9(3):673–686

Korup O (2005) Distribution of landslides in southwest New Zealand. Landslides 2:43–51

Lan HX, Lee CF, Zhou CH, Martin CD (2005) Dynamic characteristics analysis of shallow landslides in response to rainfall event using GIS. Environ Geol 47:254–267

Lateltin O, Beer C, Raetzo H, Caron C (1997) Landslides in flysch terranes of Switzerland: causal factors and climate change. Eclogae Geol Helv 90:401–406

Lavé J, Burbank D (2004) Denudation processes and rates in the transverse ranges, southern California: erosional response of a transitional landscape to external and anthropogenic forcing. J Geophys Res 109:F01006

Lombardo L, Opitz T, Huser R (2018) Point process-based modeling of multiple debris flow landslides using INLA: an application to the 2009 Messina disaster. Stoc Env Res Risk A 32(7):2179–2198

Lombardo L, Opitz T, Ardizzone F, Guzzetti F, Huser R (2020) Space-time landslide predictive modelling, Earth-Sci Rev 209:103318

Lu P, Stumpf A, Kerle N, Casagli N (2011) Object-oriented change detection for landslide rapid mapping. Geosci Remote Sens Lett IEEE 8:701–705

Lumb P (1975) Slope failure in Hong Kong. Q J Eng Geol 8:31–65

Mal S, Singh RB, Huggel C, Grover A (2018) Introducing linkages between climate change, extreme events, and disaster risk reduction. Springer, Cham, pp 1–14

Malamud BD, Turcotte DL, Guzzetti F, Reichenbach P (2004a) Landslides, earthquakes and erosion. Earth Planet Sc Lett 229:45–59

Malamud BD, Turcotte DL, Guzzetti F, Reichenbach P (2004b) Landslide inventories and their statistical properties. Earth Surf Proc Land 29:687–711

Marchesini I, Ardizzone F, Alvioli M, Rossi M, Guzzetti F (2014) Non-susceptible landslide areas in Italy and in the Mediterranean region. Nat Hazard Earth Sys 14:2215–2231

Margielewski W (1998) Landslide phases in the polish outer carpathians and their relation to the climatic changes in the late glacial and the holocene. Quatern Stud Poland 15:37–53

Martelloni G, Segoni S, Fanti R, Catani F (2011) Rainfall thresholds for the forecasting of landslide occurrence at regional scale. Landslides 9(4):485–495

Martelloni G, Segoni S, Lagomarsino D, Fanti R, Catani F (2013) Snow accumulation/melting model (SAMM) for integrated use in regional scale landslide early warning systems. Hydrol Earth Syst Sc 17:1229–1240

Martha TR, Kerle N, Jetten V, van Westen CJ, Kumar KV (2010) Characterising spectral, spatial and morphometric properties of landslides for semi-automatic detection using object-oriented methods. Geomorphology 116:24–36

Martha TR, Kerle N, Van Westen CJ, Jetten V, Kumar KV (2011) Segment optimization and data-driven thresholding for knowledge-based landslide detection by object-based image analysis. Geosci Remote Sens IEEE Trans 49:4928–4943

Martha TR, Kerle N, Van Westen CJ, Jetten V, Kumar KV (2012) Object-oriented analysis of multi-temporal panchromatic images for creation of historical landslide inventories. ISPRS J Photogramm 67:105–119

Matthews JA, Dahl SO, Dresser PQ, Berrisford MS, Lie O, Nesje A, Owen G (2009) Radiocarbon chronology of holocene colluvial (debris-flow) events at Sletthamm, Jotunheimen, southern Norway: a window on the changing frequency of extreme climatic events and their landscape impact. Holocene 19(8):1107–1129

McInnes R, Jakeways J, Fairbank H, Mathie E (eds) (2007) Landslides and Climate Change: challenges and solutions. Proceedings of the international conference on landslides and climate change. Taylor & Francis, Ventnor

Medina-Cetina Z, Nadim F (2008) Stochastic design of an early warning system. Georisk 2:223–236

Melchiorre C, Frattini P (2012) Modelling probability of rainfall-induced shallow landslides in a changing climate, Otta, Central Norway. Clim Change 113:413–436

Melillo M, Brunetti MT, Peruccacci S, Gariano SL, Guzzetti F (2015) An algorithm for the objective reconstruction of rainfall events responsible for landslides. Landslides 12:311–320

Melillo M, Brunetti MT, Peruccacci S, Gariano SL, Guzzetti F (2016) Rainfall thresholds for the possible landslide occurrence in Sicily (Southern Italy) based on the automatic reconstruction of rainfall events. Landslides 13(1):165–172

Melillo M, Brunetti MT, Peruccacci S, Gariano SL, Roccati A, Guzzetti F (2018) A tool for the automatic calculation of rainfall thresholds for landslide occurrence. Environ Modell Softw 105:230–243

Mergili M, Marchesini I, Alvioli M, Metz M, Schneider-Muntau B, Rossi M, Guzzetti F (2014a) A strategy for GIS-based 3-D slope stability modelling over large areas. Geosci Model Dev 7:2969–2982

Mergili M, Marchesini I, Rossi M, Guzzetti F, Fellin W (2014b) Spatially distributed three-dimensional slope stability modelling in a raster GIS. Geomorphology 206:178–195

Michie D, Spiegelhalter DJ, Taylor CC (eds) (1994) Machine Learning, Neural and Statistical Classification

Milledge DG, Bellugi D, McKean JA, Densmore AL, Dietrich WE (2015) A multidimensional stability model for predicting shallow landslide size and shape across landscapes. J Geophys Res-Earth 119:2481–2504

Mondini AC, Chang K-T (2014) Combining spectral and geoenvironmental information for probabilistic event landslide mapping. Geomorphology 213:183–189

Mondini AC, Chang K-T, Yin H-Y (2011a) Combining multiple change detection indices for mapping landslides triggered by typhoons. Geomorphology 134:440–451

Mondini AC, Guzzetti F, Reichenbach P, Rossi M, Cardinali M, Ardizzone F (2011b) Semi-automatic recognition and mapping of rainfall induced shallow landslides using optical satellite images. Remote Sens Environ 115:1743–1757

Mondini AC, Marchesini I, Rossi M, Chang K-T, Pasquariello G, Guzzetti F (2013) Bayesian framework for mapping and classifying shallow landslides exploiting remote sensing and topographic data. Geomorphology 201:135–147

Mondini AC, Chang K-T, Chiang S-H, Schlögel R, Notarnicola C, Saito H (2017) Automatic mapping of event landslides at basin scale in Taiwan using a Montecarlo approach and synthetic land cover fingerprints. Int J Appl Earth Obs 63:112–121

Mondini AC, Santangelo M, Rocchetti M, Rossetto E, Manconi A, Monserrat O (2019) Sentinel-1 SAR amplitude imagery for rapid landslide detection. Remote Sensing 11:760

Montgomery DR, Dietrich WE (1994) A physically based model for the topographic control on shallow landsliding. Water Resour Res 30:1153–1171

Murillo-García FG, Alcántara-Ayala I, Ardizzone F, Cardinali M, Fiourucci F, Guzzetti F (2014) Satellite stereoscopic pair images of very high resolution: a step forward for the development of landslide inventories. Landslides 12:277–291

Nadim F, Kjekstad O, Peduzzi P, Herold C, Jaedicke C (2006) Global landslide and avalanche hotspots. Landslides 3:159–173

NGDC/WDC (2011) Historical tsunami database, Boulder, USA. National Geophysical Data Center, World Data Center

Okimura T, Kawatani T (1987) Mapping of the potential surface-failure sites on granite slopes. In: Gardiner E (ed) International geomorphology 1986. Part I, John Wiley, Chichester, pp 121–138

Onodera T, Yoshinaka R, Kazama H (1974) Slope failures caused by heavy rainfall in Japan. Proceedings 2nd international congress international association engineering geololgy. San Paulo, 11:1–10

Palladino MR, Viero A, Turconi L et al (2018) Rainfall thresholds for the activation of shallow landslides in the Italian Alps: the role of environmental conditioning factors. Geomorphology 303:53–67

Paranunzio R, Laio F, Chiarle M, Nigrelli G, Guzzetti F (2016) Climate anomalies associated with the occurrence of rockfalls at high-elevation in the Italian Alps. Nat Hazard Earth Sys 16:2085–2106

Pardeshi SD, Autade SE, Pardeshi SS (2013) Landslide hazard assessment: recent trends and techniques. SpringerPlus 2(1):523

Pedrozzi G (2004) Triggering of landslides in Canton Ticino (Switzerland) and prediction by the rainfall intensity and duration method. Bull Eng Geol Environ 63(4):281–291

Peel MC, Finlayson BL, McMahon TA (2007) Updated world map of the Köppen-Geiger climate classification. Hydrol Earth Syst Sc 11:1633–1644

Pereira S, Zêzere JL, Quaresma ID, Santos PP, Santos M (2015) Mortality patterns of hydro-geomorphologic disasters. Risk Anal 1–23

Pereira S, Zêzere JL, Quaresma I (2017) Landslide societal risk in Portugal in the period 1865–2015. World Landslide Forum, pp 491–499

Peruccacci S, Brunetti MT, Luciani S, Vennari C, Guzzetti F (2012) Lithological and seasonal control on rainfall thresholds for the possible initiation of landslides in central Italy. Geomorphology 139–140:79–90

Peruccacci S, Brunetti MT, Luciani S, Calzolari MC, Bartolini D, Guzzetti F (2014) Topographic and pedological rainfall thresholds for the prediction of shallow landslides in Central Italy. World Landslide Forum 3, Beijing. Springer, pp 299–304

Peruccacci S, Brunetti MT, Gariano SL, Melillo M, Rossi M, Guzzetti F (2017) Rainfall thresholds for possible landslide occurrence in Italy. Geomorphology 290:39–57

Petley D (2012) Global patterns of loss of life from landslides. Geology 40(10):927–930

Piciullo L, Gariano SL, Melillo M, Brunetti MT, Peruccacci S, Guzzetti F, Calvello M (2017) Definition and performance of a threshold-based regional early warning model for rainfall-induced landslides. Landslides 14:995–1008

Piciullo L, Calvello M, Cepeda JM (2018) Territorial early warning systems for rainfall-induced landslides. Earth-Sci Rev 179:228–247

Polemio M, Lonigro T (2015) Trends in climate, short-duration rainfall, and damaging hydrogeological events (Apulia, Southern Italy). Nat Hazards 75:515–540

Polemio M, Petrucci O (2010) Occurrence of landslide events and the role of climate in the twentieth century in Calabria, southern Italy. Q J Eng Geol Hydroge 43:403–415

Porter M, Morgenstern N (2013) Landslide risk evaluation—canadian technical guidelines and best practices related to landslides: a national initiative for loss reduction. Geol Surv Can, Open File 7312:21

Raia S, Alvioli M, Rossi M, Baum R, Godt J, Guzzetti F (2014) Improving predictive power of physically based rainfall-induced shallow landslide models: a probabilistic approach. Geosci Model Dev 7(2):495–514

Rebetez M, Lugon R, Baeriswyl P-A (1997) Climatic change and debris flows in high mountain regions: the case study of the Ritigraben torrent (Swiss Alps). Clim Change 36:371–389

Reichenbach P, Cardinali M, De Vita P, Guzzetti F (1998) Regional hydrological thresholds for landslides and floods in the Tiber River Basin (central Italy). Environ Geol 35:146–159

Reichenbach P, Galli M, Cardinali M, Guzzetti F, Ardizzone F (2005) Geomorphological Mapping to assess landslide risk: concepts, methods and applications in the umbria region of central Italy. In: Glade T, Anderson MG, Crozier MJ (eds) Landslide Hazard and Risk. John Wiley, pp 429–468

Reichenbach P, Rossi M, Malamud BD, Mihir M, Guzzetti F (2018) A review of statistically-based landslide susceptibility. Earth-Sci Rev 180:60–91

Reid M, Christian S, Brien D, Henderson S (2015) Scoops 3d software to analyze three-dimensional slope stability throughout a digital landscape. Chap. A1. US Geological Survey Techniques and Methods, Book 14

Rianna G, Zollo AL, Tommasi P, Paciucci M, Comegna L, Mercogliano P (2014) Evaluation of the effects of climate changes on landslide activity of Orvieto clayey slope. Proced Earth Plan Sci 9:54–63

Rice RM, Corbett ES, Bailey RG (1969) Soil slips related to vegetation, topography, and soil in Southern California. Water Resour Res 5 (3):647–659

Rosi A, Segoni S, Catani F, Casagli N (2012) Statistical and environmental analyses for the definition of a regional rainfall threshold system for landslide triggering in Tuscany (Italy). J Geogr Sci 22(4):617–629

Rosi A, Lagomarsino D, Rossi G, Segoni S, Battistini A, Casagli N (2015) Updating EWS rainfall thresholds for the triggering of landslides. Nat Hazards 78:297–308

Rosi A, Peternel T, Jemec-Auflič M, Komac M, Segoni S, Casagli N (2016) Rainfall thresholds for rainfall-induced landslides in Slovenia. Landslides 13:1571–1577

Rossi M, Reichenbach P (2016) LAND-SE: a software for statistically based landslide susceptibility zonation, version 1.0. Geosci Model Dev 9:3533–3543

Rossi M, Guzzetti F, Reichenbach P, Mondini AC, Peruccacci S (2010a) Optimal landslide susceptibility zonation based on multiple forecasts. Geomorphology 114:129–142

Rossi M, Witt A, Guzzetti F, Malamud BD, Peruccacci S (2010b) Analysis of historical landslide time series in the Emilia-Romagna region, northern Italy. Earth Surf Proc Land 35:1123–1137

Rossi M, Kirschbaum D, Valigi D, Mondini AC, Guzzetti F (2017) Comparison of satellite rainfall estimates and rain gauge measurements in italy, and impact on landslide modeling. Climate 5:90

Rossi M, Guzzetti F, Salvati P, Donnini M, Napolitano E, Bianchi C (2019) A predictive model of societal landslide risk in Italy. Earth-Sci Rev 196:102849

Roth RA (1983) Factors affecting landslide susceptibility in San Mateo County. California. Assoc Eng Geols Bull 20(4):353–372

Salvati P, Bianchi C, Rossi M, Guzzetti F (2010) Societal landslide and flood risk in Italy. Nat Hazard Earth Sys 10:465–483

Salvati P, Marchesini I, Balducci V, Bianchi C, Guzzetti F (2013) A new digital catalogue of harmful landslides and floods in Italy. In: Margottini C, Canuti P, Sassa K (eds) Landslide science and practice. Springer, Berlin, pp 409–414

Salvati P, Bianchi C, Fiorucci F, Giostrella P, Marchesini I, Guzzetti F (2014) Perception of flood and landslide risk in Italy: a preliminary analysis. Nat Hazard Earth Sys 14:2589–2603

Salvati P, Rossi M, Bianchi C, Guzzetti F (2016) Landslide risk to the population of Italy and its geographical and temporal variations. In: Chavez M, Ghil M, Urrutia- Fucugauchi J (eds) Extreme events: observations, modeling, and economics. John Wiley, Geophysical Monograph Series, pp 177–194

Salvati P, Petrucci O, Rossi M, Bianchi C, Pasqua AA, Guzzetti F (2018) Gender, age and circumstances analysis of flood and landslide fatalities in Italy. Sci Total Environ 610–611:867–879

Samia J, Temme AJAM, Bregt AK, Wallinga J, Guzzetti F, Ardizzone F, Rossi M (2017a) Characterization and quantification of path dependency in landslide susceptibility. Geomorphology 292:16–24

Samia J, Temme AJAM, Bregt AK, Wallinga J, Stuiver J, Guzzetti F, Ardizzone F, Rossi M (2017b) Do landslides follow landslides? Insights in path dependency from a multi-temporal landslide inventory. Landslides 14:547–558

Samia J, Temme AJAM, Bregt AK, Wallinga J, Guzzetti F, Ardizzone F, Rossi M (2018) Implementing landslide path dependency in landslide susceptibility modelling. Landslides 15:2129–2144

Samia J, Temme AJAM, Bregt AK, Wallinga J, Guzzetti F, Ardizzone F (2019) Dynamic path dependent landslide susceptibility modelling. Nat Hazard Earth Sys 2019

Santangelo M, Marchesini I, Bucci F, Cardinali M, Fiorucci F, Guzzetti F (2015) An approach to reduce mapping errors in the production of landslide inventory maps. Nat Hazard Earth Sys 15:2111–2126

Savage W, Godt J, Baum R (2003) A model for spatially and temporally distributed shallow landslide initiation by rainfall infiltration. Proceedings 3rd International Conference on Debris Flow Hazards, Davos, September 10–13, 2003, pp 179–187

Savage W, Godt J, Baum R (2004) Modeling time-dependent aerial slope stability. In: Lacerda W, Erlich M, Fontoura S, Sayao A (eds) Proceedings 9th International Symposium on Landslides. Balkema, London, pp 23–36

Schmidt J, Dikau R (2004) Modeling historical climate variability and slope stability. Geomorphology 60:433–447

Schmidt J, Glade T (2003) Linking global circulation model outputs to regional geomorphic models: a case study of landslide activity in New Zealand. Clim Res 25:135–150

Schulte SM, Mooney WD (2005) An updated global earthquake catalogue for stable continental regions: reassessing the correlation with ancient rifts. Geophys J Int 161:707–721

Segoni S, Rosi A, Rossi G, Catani F, Casagni N (2014a) Analysing the relationship between rainfalls and landslides to define a mosaic of triggering thresholds for regional scale warning systems. Nat Hazard Earth Sys 14:2637–2648

Segoni S, Rossi G, Rosi A, Catani F (2014b) Landslides triggered by rainfall: a semiautomated procedure to define consistent intensity-duration thresholds. Comput Geosci 63:123–131

Segoni S, Battistini A, Rossi G, Rosi A, Lagomarsino D, Catani F, Moretti S, Casagli N (2015a) Technical note: an operational landslide early warning system at regional scale based on space-time-variable rainfall thresholds. Nat Hazard Earth Sys 15:853–861

Segoni S, Lagomarsino D, Fanti R, Moretti S, Casagli N (2015b) Integration of rainfall thresholds and susceptibility maps in the

Emilia Romagna (Italy) regional-scale landslide warning system. Landslides 12:773–785
Segoni S, Piciullo L, Gariano SL (2018a) Preface: landslide early warning systems: monitoring systems, rainfall thresholds, warning models, performance evaluation and risk perception. Nat Hazard Earth Sys 18:3179–3186
Segoni S, Piciullo L, Gariano SL (2018b) A review of the recent literature on rainfall thresholds for landslide occurrence. Landslides 15:1483–1501
Sewell RJ, Parry S, Millis SW, Wang N, Rieser U, DeWit R (2015) Dating of debris flow fan complexes from Lantau Island, Hong Kong, China: The potential relationship between landslide activity and climate change. Geomorphology 248:205–227
Sharp P, Hockfield S (2017) Convergence: The future of health. Science 355:589.1–589
Sharp P, Jacks T, Hockfield S (2016) Convergence: the future of health. ConvergenceRevolution.net, Cambridge, Massachusetts
Sidle RC, Ochiai H (2006) Landslides: processes, prediction, and land use. Water Resource Monograph Series 18, AGU, Washington
Simkin T, Siebert L, Blong R (2001) Volcano fatalities-lessons from the historical record. Science 291:255
Simonett DS (1967) Landslide distribution and earthquakes in the Bewani and Torricelli Mountains, New Guinea. In: Jennings JN, Mabbutt JA (eds) Landform studies from Australia and New Guinea. Cambridge University Press, Cambridge, pp 64–84
Simoni S, Zanotti F, Bertoldi G, Rigon R (2008) Modeling the probability of occurrence of shallow landslides and channelized debris flows using GEOtop-FS. Hydrol Process 22:532–545
Soldati M, Corsini A, Pasuto A (2004) Landslides and climate change in the Italian Dolomites since the Late glacial. CATENA 55:141–161
Stähli M, Sättele M, Huggel C, McArdell BW, Lehmann P, Van Herwijnen A, Berne A, Schleiss M, Ferrari A, Kos A, Or D, Springman SM (2015) Monitoring and prediction in early warning systems for rapid mass movements. Nat Hazard Earth Sys 15:905–917
Stark CP, Guzzetti F (2009) Landslide rupture and the probability distribution of mobilized debris volumes. J Geophys Res 114: F00A02
Stark CP, Hovius N (2001) The characterization of landslide size distributions. Geophys Res Lett 28:1091–1094
Stoffel M, Beniston M (2006) On the incidence of debris flows from the early Little Ice Age to a future greenhouse climate: A case study from the Swiss Alps. Geophys Res Lett 33:L16404
Stoffel M, Tiranti D, Huggel C (2014) Climate change impacts on mass movements—case studies from the European Alps. Sci Total Environ 493:1255–1266
Stucchi M, Rovida A, Gomez Capera AA et al (2013) The SHARE European Earthquake Catalogue (SHEEC) 1000–1899. J Seismol 17:523–544
Stumpf A, Kerle N (2011) Object-oriented mapping of landslides using Random Forests. Remote Sens Environ 115:2564–2577
Tacher L, Bonnard C (2007) Hydromechanical modelling of a large landslide considering climate change conditions. In: McInnes R, Jakeways J, Fairbank H, Mathie E (eds) Landslides and climate change: challenges and solutions, proceedings international conference on landslides and climate change, Ventnor, Taylor & Francis, pp 131–141
Taylor D (1948) Fundamentals of soil mechanics. Wiley, New York
Taylor FE, Malamud BD, Witt A, Guzzetti F (2018) Landslide shape, ellipticity and length-to-width ratios. Earth Surf Proc Land 43:3164–3189
Terlien MTJ (1998) The determination of statistical and deterministic hydrological landslide-triggering thresholds. Environ Geol 35:124–130
Terlien MTJ, van Westen CJ, van Asch ThWJ (1995) Deterministic modelling in GIS-based landslide hazard assessment. In: Carrara A, Guzzetti F (eds) Geographical information systems in assessing nat hazards. Kluwer AP, Dordrecht, pp 57–77
Tiranti D, Rabuffetti D (2010) Estimation of rainfall thresholds triggering shallow implementation. Landslides 7:471–481
Trauth MH, Alonso RA, Haselton KR, Hermanns RL, Strecker MR (2000) Climate change and mass movements in the NW Argentine Andes. Earth Planet Sc Lett 179(2):243–256
Trigila A, Iadanza C, Spizzichino D (2010) Quality assessment of the Italian landslide inventory using GIS processing. Landslides 7:455–470
Turkington T, Remaître A, Ettema J, Hussin H, van Westen C (2016) Assessing debris flow activity in a changing climate. Climate Change
UNISDR (2006) Platform for the promotion of early warning—developing early warning systems: a checklist. In: Proc. of EWC III, Third International Conference on Early Warning, From Concept to Action, UN Secretariat of the International Strategy for Disaster Reduction, Bonn
United Nations (2016) Report of the open-ended intergovernmental expert working group on indicators and terminology relating to disaster risk reduction
van Asch Th WJ, Buma J, van Beek LPH (1999) A view on some hydrological triggering systems in landslides. Geomorphology 30 (1–2):25–32
Van Den Eeckhaut M, Hervás J (2012) State of the art of national landslide databases in Europe and their potential for assessing landslide susceptibility, hazard and risk. Geomorphology 139–140:545–558
Van Den Eeckhaut M, Hervás J, Jaedicke C, Malet J-P, Montanarella L, Nadim F (2012a) Statistical modelling of Europe-wide landslide susceptibility using limited landslide inventory data. Landslides 9:357–369
Van Den Eeckhaut M, Kerle N, Poesen JAW, Hervás J (2012b) Object-oriented identification of forested landslides with derivatives of single pulse LiDAR data. Geomorphology 173–174:30–42
van Westen CJ, Castellanos Abella EA (2008) Spatial data for landslide susceptibility, hazard, and vulnerability assessment: an overview. Eng Geol 102:112–131
VanDine DF, Moore G, Wise M, Vanbuskirk C, Gerath R (2004) Technical terms and methods. In: Wise M, Moore G, VanDine D (eds) Landslide risk case studies in forest development planning and operations. BC, Ministry of Forests, Forest Science Program, Land management handbook 56, pp 13–26
Varnes DJ, and the IAEG Commission on Landslides and other Mass-Movements (1984) Landslide hazard zonation: a review of principles and practice. UNESCO Press, Paris
Vennari C, Gariano SL, Antronico L, Brunetti MT, Iovine GGR, Peruccacci S, Terranova OG, Guzzetti F (2014) Rainfall thresholds for shallow landslide occurrence in Calabria, southern Italy. Nat Hazard Earth Sys 14:317–330
Vessia G, Parise M, Brunetti MT, Peruccacci S, Rossi M, Vennari C, Guzzetti F (2014) Automated reconstruction of rainfall events responsible for shallow landslides. Nat Hazard Earth Sys 14:2399–2408
Vieira BC, Fernandes NF, Filho OA (2010) Shallow landslide prediction in the Serra do Mar, São Paulo. Nat Hazard Earth Sys 10:1829–1837
Villani V, Rianna G, Mercogliano P, Zollo AL (2015) Statistical approaches versus weather generator to downscale rcm outputs to slope scale for stability assessment: a comparison of performances. Electr J Geotech Eng 20(4):1495–1515

Ward TJ, Li RM, Simons DB (1981) Use of a mathematical model for estimating potential landslide sites in steep forested drainage basins. IAHS Publication 132:21–41

Ward TJ, Li RM, Simons DB (1982) Mapping landslides in forested watersheds. J Geotech Eng-ASCE 8:319–324

Wilde M, Günther A, Reichenbach P, Malet J-P, Hervás J (2018) Pan-European landslide susceptibility mapping: ELSUS Version 2. J Maps 14:97–104

Wilson RC, Jayko AS (1997) Preliminary maps showing rainfall thresholds for debris-flow activity, San Francisco Bay Region, California. US Geological Survey Open-File Report 97–745 F

Wilson RC, Torikai JD, Ellen SD (1992) Development of rainfall thresholds for debris flows in the Honolulu District, Oahu. US Geological Survey Open-File Report 92–521

Witham CS (2005) Volcanic disasters and incidents: a new database. J Volcanol Geoth Res 148:191–233

Witt A, Malamud BD, Rossi M, Guzzetti F, Peruccacci S (2010) Temporal correlations and clustering of landslides. Earth Surf Proc Land 35:1138–1156

Wu W, Sidle RC (1995) A distributed slope stability model for steep forested basins. Water Resour Res 31:2097–2110

Wu X, Chen X, Zhan FB, Hong S (2015) Global research trends in landslides during 1991-2014: a bibliometric analysis. Landslides 12 (6):1215–1226

Wyllie DC, Mah CW (2004) Rock Slope Engineering: civil and mining. Spon, London

Yang X, Chen L (2010) Using multi-temporal remote sensor imagery to detect earthquake-triggered landslides. Int J Appl Earth Obs 12:488–496

Yin Z, Qin X, Yin Y, Zhao W, Wei G (2014) Landslide developmental characteristics and response to climate change since the last glacial in the upper reaches of the Yellow river. NE Tibetan plateau. Acta Geologica Sinica 88(2):635–646

Zêzere JL, Trigo RM, Trigo IF (2005) Shallow and deep landslides induced by rainfall in the Lisbon region (Portugal): assessment of relationships with the North Atlantic Oscillation. Nat Hazard Earth Sys 5:331–344

Design Recommendations for Single and Dual Debris Flow Barriers with and Without Basal Clearance

2020 Kyoto Japan

Charles Wang Wai Ng, Clarence Edward Choi, Haiming Liu, Sunil Poudyal, and Julian Shun Hang Kwan

Abstract

Debris flows pose threats to sustainable development in many countries worldwide, including China, Japan, Switzerland and USA. To mitigate these flows, rigid and flexible barriers are commonly installed along the predicted flow paths. To arrest large volumes of debris flow, several barriers may be installed in series to create a cascading effect to progressively decelerate and retain the debris. Barriers may even be designed with a basal clearance to allow small discharges to pass underneath the barrier to reduce the peak impact force. Despite the importance of barriers as life-saving assets, their design remains essentially empirical because of the highly heterogeneous and scale-dependent nature of debris flow. These features of debris flow have hindered an understanding of their fundamental impact mechanisms, thereby hampering the development of scientific design guidelines to enable robust and cost-effective barriers. This forum paper presents a collection of physical experiments modelling the impact mechanisms of the two extreme cases of water and dry granular flows, and two-phase debris flows against single and dual rigid barriers, and a single flexible barrier. Furthermore, the effects of a basal clearance on the impact dynamics of dry granular flow against a single rigid barrier are examined. Experiments were conducted at two different scales, including 5 m-long and 28 m-long flumes. Based on the observed impact mechanisms and measured data, a newly developed analytical framework for designing multiple rigid barriers was evaluated. Recommendations and procedures are provided for the design of single and multiple rigid barriers with and without a basal clearance.

Keywords

Debris flow • Impact • Rigid barrier • Flexible barrier • Multiple barriers • Physical modelling • Basal clearance

Introduction

Debris flows, which are mixtures of soil and water, surge down mountainsides under the influence of gravity at high velocities. Such flows often result in fatalities (Froude and Petley 2018) and damage to infrastructure (Jakob et al. 2012). To arrest these flows, a single large reinforced concrete rigid barrier (Fig. 1) is conventionally constructed at the end of a catchment. Such an approach may enable a debris flow to increase in velocity and volume via entrainment (Berger et al. 2011) before impacting the barrier. Consequently, a larger barrier with a higher design capacity is required. With the challenges of land scarcity in densely populated urban areas, such as Hong Kong, and the importance of preserving the natural environment, bulky reinforced concrete barriers are decreasingly viable in terms of sustainability. Given these challenges, smaller rigid barriers in series along a channel provide an ideal solution to create a cascading effect to dissipate flow energy and reduce the flow volume. However, there are no design guidelines on how to space barriers in series to optimise their design loads.

C. W. W. Ng · H. Liu · S. Poudyal
Department of Civil and Environmental Engineering, The Hong Kong University of Science and Technology, Hong Kong SAR, China
e-mail: cecwwng@ust.hk

H. Liu
e-mail: hliubc@connect.ust.hk

S. Poudyal
e-mail: spoudyal@connect.ust.hk

C. E. Choi (✉)
Department of Civil Engineering, The University of Hong Kong, Hong Kong SAR, China
e-mail: cechoi@hku.hk

J. S. H. Kwan
Civil Engineering and Development Department, Geotechnical Engineering Office, Hong Kong SAR, China
e-mail: juliankwan@cedd.gov.hk

K. Sassa et al. (eds.), *Understanding and Reducing Landslide Disaster Risk*, ICL Contribution to Landslide Disaster Risk Reduction,
https://doi.org/10.1007/978-3-030-60196-6_2

Another important design feature of rigid barriers is the basal clearance (Fig. 2), which is an opening at the base of a barrier (Piton and Recking 2015). These clearances are often included to prevent smaller discharges from being trapped, which over time depletes the design retention capacity intended for larger and more dangerous debris flow events. The basal clearance also regulates the volume of retained debris and allows discharge underneath the barrier. Sze and Lam (2017) summarised international guidelines and reported that basal clearances of up to 1.5 m are commonly used in barrier designs. However, it is evident from their report that the sizing of a basal clearance is essentially empirical. Without a scientific basis for design, basal clearances that are too large render a barrier ineffective at reducing the peak impact load, while clearances that are too small give rise to uncontrolled overflow and increase the impact and drag loads induced on a barrier. Evidently, research is required to shed light on an optimum basal clearance.

Aside from rigid barriers, flexible barriers have become an emerging structural countermeasure in debris flow hazard mitigation. Over the past decade, flexible barriers for rock fall have been impacted by debris flow and proved to be effective at arresting debris flows. However, the impact dynamics of a rock fall and debris flow are fundamentally different. Therefore, the design of flexible barriers for debris flows remains largely empirical. Compared with reinforced concrete rigid barriers, flexible barriers (Fig. 3) blend in well with their natural surroundings and are easy to install. More importantly, the deformation of flexible barriers is ideal for attenuating impact forces (DeNatale et al. 1999; Wendeler et al. 2006; Brighenti et al. 2013; Ng et al. 2016). To use flexible barriers more extensively in mountainous regions, research is required to elucidate the fundamental impact mechanisms so that design guidelines can be developed.

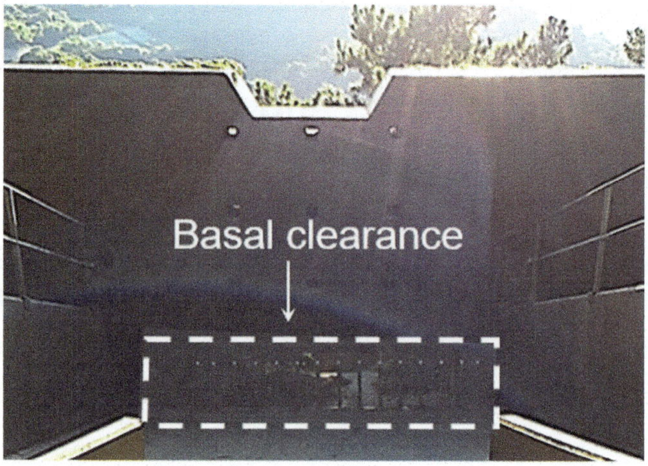

Fig. 2 Basal clearance under a rigid barrier (Tung Chung, Hong Kong)

In existing international design guidelines (Kwan 2012; Volkwein 2014), the impact force exerted by a debris flow is calculated as follows:

$$F = \alpha \rho v^2 h_0 w \tag{1}$$

where α is a dynamic impact coefficient; ρ is the flow density; v is the velocity of the flow; h_0 is the flow depth and w is the channel width. To ensure robust design loads, international guidelines (Lo 2000; ASI 2008) often prescribe high α values to account for the idiosyncrasies involved in natural materials and settings in the field. For instance, Kwan (2012) recommends an α of 2.5 for the design of rigid reinforced concrete barriers and an α of 2.0 for flexible barriers. These recommended values account for hard inclusions in the flow. Figure 4 shows a typical impact load

Fig. 1 Rigid barrier (Cheung Tung Road, Hong Kong) (DB 2020)

Fig. 3 Flexible barrier (Hiroshima, Japan)

model for the design of rigid barriers against debris flow impact in Hong Kong. In their guidelines, they generally recommend that each surge impacting a barrier should be designed using the maximum velocity and flow depth obtained from a debris mobility analysis (Kwan and Koo 2015). However, Koo et al. (2017) reported that these assumptions were over-conservative, and recommended a velocity attenuation model. Nevertheless, there is room to improve the estimation of design impact loads and optimise the design of barriers.

Further compounding the challenges in designing barriers is dealing with the scale-dependency and heterogeneous nature of debris flow. Iverson (2015) reported that small-scale two-phase debris flow models cannot replicate the timescales for pore pressure dissipation and the ratio between viscous and inertial stresses observed in field events. Therefore, unique physical modelling facilities are necessary to replicate the appropriate debris flow dynamics. Furthermore, the characteristics of debris flows are catchment specific (Rickenmann 1999), some flows may be more frictional, while some may be more viscous. Choi et al. (2015) reported that frictional and viscous flows exhibit entirely different impact mechanisms. Frictional flows exhibit a pileup mechanism (Koo et al. 2017), while viscous flows exhibit a vertical jet mechanism (Ng et al. 2019). One can imagine that complex two-phase debris flows must exhibit some characteristics of the two idealised flows. Without a clear understanding of these fundamental impact mechanisms, it remains unclear whether Eq. 1 is over-conservative, adequate or unsafe.

In this forum paper, physical experiments that model the complex impact mechanisms of dry granular, water, and two-phase debris flows on single and dual rigid barriers and a single flexible barrier are presented. In addition, the effects of a basal clearance on a single rigid barrier are examined. Experiments were conducted at two different scales, including 5 m-long and 28 m-long flumes. Based on the observed fundamental impact mechanisms, an analytical framework for designing multiple rigid barriers was developed and then evaluated using experimental data. Finally, recommendations are provided for the design impact load for single rigid barriers with and without a basal clearance, a dual rigid barrier system, and a single flexible barrier.

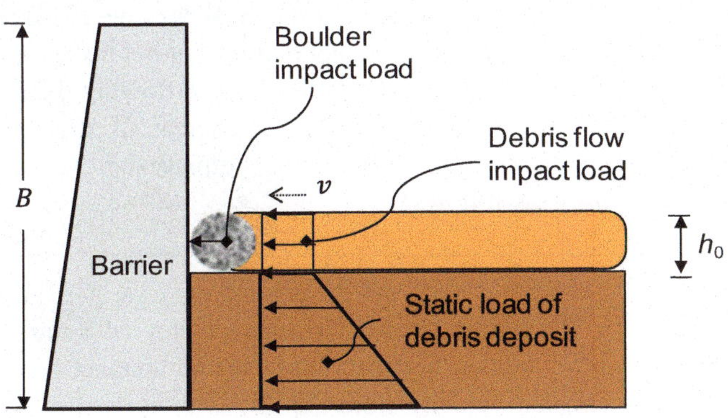

Fig. 4 Impact model proposed by Kwan (2012)

Analytical Framework for Dual Rigid Barriers

Conventionally, the design of multiple barriers in series is based on the estimated retention of the total debris volume (NILIM 2007; Faug et al. 2012). More recently, the impact dynamics between debris flows and barriers in series have been reported as important considerations in design (Kwan et al. 2015; Ng et al. 2018). Figure 5 shows a newly proposed analytical framework for multiple rigid barriers in series (Kwan et al. 2015). This framework includes a set of velocity-attenuation and impact equations for rigid barriers that captures the dissipation of kinetic energy as granular material is deposited in layers up to the crest of a barrier (Koo et al. 2017). The overflow then follows the trajectory of an inviscid jet from the barrier crest before landing on the channel bed. After landing on the channel bed, energy is dissipated before the granular material flows towards the next barrier in the channel (Ng et al. 2018; 2019). Details of the analytical framework for designing multiple rigid barriers are discussed below.

Velocity Attenuation Impact Model

When a granular flow impacts a rigid barrier, the material is arrested at the base of the barrier. Granular material then progressively deposits to the crest of the barrier (Koo et al. 2017; Ng et al. 2019). Figure 6 shows a granular flow with velocity v and depth h_0 on a channel inclined at θ. As the granular flow climbs on the wedge of deposited granular material, shearing occurs along their interface. The attenuated velocity from shear between the incoming flow and dead zone v_d and its corresponding velocity attenuation factor R_d can be calculated as follows (Koo et al. 2017; Ng et al. 2019):

$$v_d = v(1 - R_d) \quad (2)$$

and,

$$R_d = 1 - \sqrt{1 - \frac{2g(h_d + L_T \tan\phi)}{v^2}} \quad (3)$$

where g is the acceleration due to Earth's gravity, ϕ is the friction angle, L_T is the length of the free surface of the arrested granular material and h_d is the height of the deposited granular material. Equations 2 and 3 are used

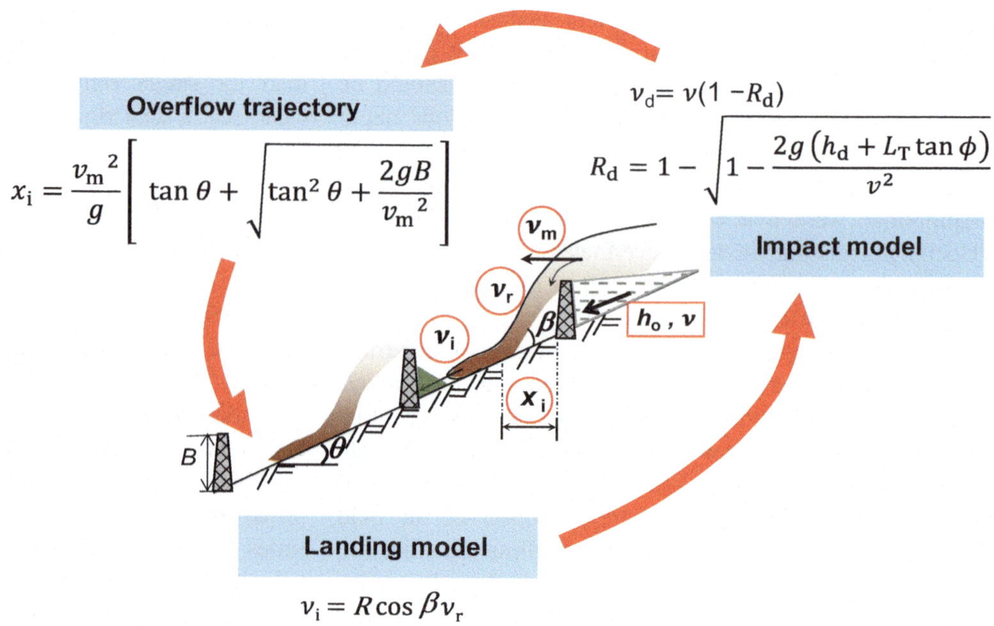

Fig. 5 Analytical multiple barrier framework (Kwan et al. 2015; Ng et al. 2018; Ng et al. 2019)

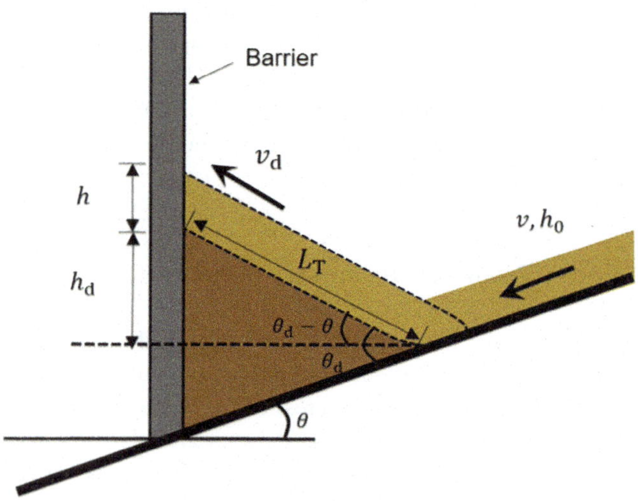

Fig. 6 A schematic diagram showing velocity attenuation mechanism (Koo et al. 2017)

repeatedly until the granular material has reached the crest of the barrier. The velocity attenuation model will be verified and discussed later.

Overflow Dynamics

Once the granular material reaches the crest of the barrier, overflow occurs. The overflowing material in the proposed framework is assumed to launch in the horizontal direction (Kwan et al. 2015). The overflow trajectory is then estimated by assuming a point mass as follows:

$$x_i = \frac{v_m^2}{g}\left[\tan\theta + \sqrt{\tan^2\theta + \frac{2gB}{v_m^2}}\right] \tag{4}$$

where v_m is the horizontal overflow velocity and B is the barrier height. The horizontal distance required to land on the channel x_i can then be used to determine the minimum spacing required between the barriers. Ideally, the overflow should be allowed to impact the channel bed before it is allowed to impact or overtop the next barrier in the channel to maximise the energy dissipated in the flow. In other words, the design spacing between barriers should be greater than x_i.

Landing

Energy is attenuated when overflow lands on the channel. Therefore, the flow velocity towards the next barrier v_i depends on the slope-parallel component of the landing velocity v_r and the angle of impact on the channel bed β. To account for changes in velocity after the overflowing debris impacts the channel bed, a landing coefficient C_r is introduced. The relationship between the velocity before and after landing is as follows:

$$v_i = C_r v_r \tag{5}$$

where $C_r = R\cos\beta$; R is the reduction factor of landing velocity due to friction between the flow and channel bed.

The attenuated velocity from Eq. (5) is then used as an input in Eqs. (2) and (3) to predict the impact dynamics on the second barrier (Ng et al. 2019). Equations (2)–(5) can be repeatedly used to design the third and subsequent barriers installed downstream in the channel. Details about the verification of the analytical framework by physical experiment results are illustrated later.

Physical Modelling of Flow-Barrier Interaction

To investigate the impact dynamics of debris flows against rigid and flexible barriers, and to evaluate the analytical framework for dual barriers, a series of physical model tests were carried out. Dry granular and water flows were modelled using a 5 m-long flume model, and two-phase debris flows were modelled using a 28 m-long flume model. The collection of flume experiment model flows impacting single rigid barrier with and without a basal clearance, a single flexible barrier, and dual rigid barriers is presented. Details of the test setups, instrumentation, test plans, and modelling procedures are discussed below.

Five Metre-Long Flume Modelling

A 5 m-long flume with a rectangular cross-section that has a width of 0.2 m and a depth of 0.5 m was used to study the impact dynamics of dry granular and water flows impacting single and dual rigid barriers. A storage container with a volume of 0.06 m^3 is located at the upstream end of the flume to hold the debris material, which is retained behind a remote-controlled gate. The debris material in the 5 m-long flume was modelled as water or Toyoura sand with an average particle diameter of 0.3 mm. Figure 7 shows a typical side view of the 5 m-long flume and the instrumentation layout. Flow kinematics were captured using high-speed cameras (Mikrotron motionblitz EoSens mini2) mounted at the side of the flume. Images with a resolution of 1300 × 1600 pixels were captured at a sampling rate of 640 frames per second. Images were then analysed using Particle Image Velocimetry (PIV) (White et al. 2003) to obtain velocity fields. Laser displacement sensors (Wenglor YT44 MGV80) were used to measure the flow depth along the centreline of the flume. Instrumented model rigid barriers were assembled by sandwiching load cells (KYOWA LUX200N) between an acrylic force plate and an aluminium reaction frame mounted inside the flume.

For the experiments modelling dual rigid barriers without basal clearance (Ng et al. 2018), the height of the first barrier was varied from 100 mm to 260 mm. The height of the second barrier was 500 mm to ensure that overflow did not occur. The first barrier was located at an inclined distance of 800 mm from the gate. The inclined distance between barriers was 1000 mm. For the experiments of the single rigid barrier with a basal clearance (Choi et al. 2020), clearances of 0 mm, 15 mm and 30 mm were modelled. The basal clearance heights (H_c) are normalised by the maximum flow depth (h_o). The single barrier was installed at 1100 mm from the gate. Details of the model rigid barrier with a basal clearance are shown in Fig. 8. The experiments were designed to investigate the effects of normalised basal clearances H_c/h_o ranging from 0.3 to 1.0 on the impact dynamics. Three different Froude numbers, governed by the flume inclination, are modelled. Specifically, Froude numbers of 3.2, 4.5 and 5.3, which are obtained at flume inclinations of 15°, 25°, and 35°, respectively. A summary of the test plan using the 5 m-long flume for the dual rigid barriers and rigid barrier with a basal clearance is given in Tables 1 and 2, respectively.

In each 5 m-long flume test, the appropriate barrier configuration is installed in the flume, then the debris material is placed in the storage container. The flume is then inclined to the appropriate inclination angle. Finally, the gate is released, and the debris is allowed to discharge downslope and impact the various barrier configurations installed in the flume.

Twenty-Eight-Metre-Long Flume Modelling

A 28 m-long flume (Fig. 9) is used to conduct physical experiments to gain new insight on the impact mechanisms of two-phase debris flows against dual rigid barriers and a single flexible barrier. The flume has a uniform rectangular cross-section with a width of 2 m and a depth of 1 m. The side walls are transparent on one side of the flume to enable the impact kinematics to be captured during experiments.

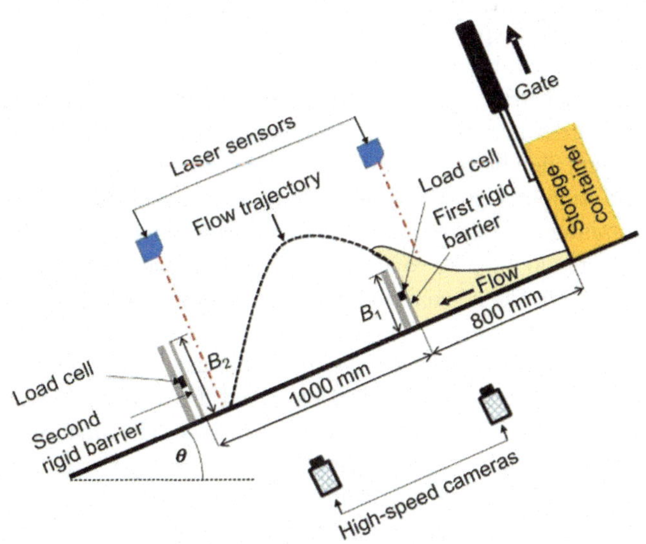

Fig. 7 A typical five-metre flume test setup with instrumentation of dual rigid barriers without basal clearance (not drawn to scale)

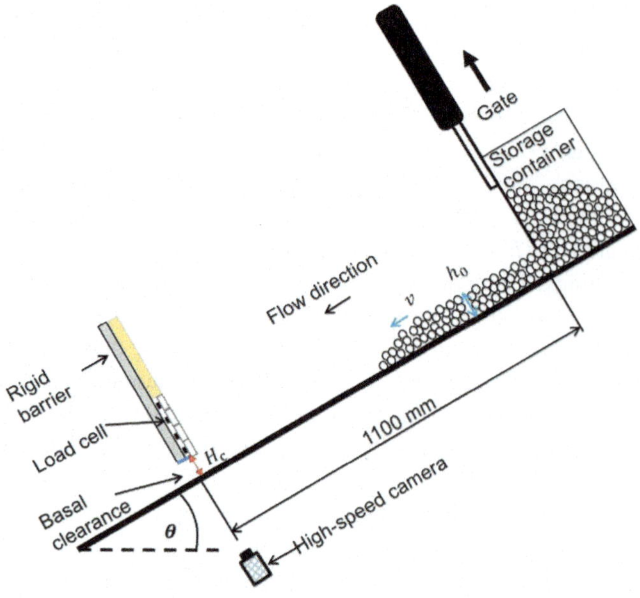

Fig. 8 A typical five-metre flume test setup with instrumentation of a single rigid barrier with basal clearance (not drawn to scale)

A storage container that can hold up to 10 m^3 is inclined at 30° at the top of the flume. The main flume is 15 m in length and has an inclination of 20°. The bottom part of the flume is horizontal and 8 m in length. A double gate system is used to retain debris material inside the storage container. The doors are secured and released using a mechanical arm, which is controlled by an electric motor. Laser and ultrasonic displacement sensors (Keyence IL600/IL1000 and Banner TUB30X) are mounted above the channel to measure the flow depth. Furthermore, high-speed cameras (Mikrotron EoSens 4CXP) are used in the 28 m-long flume tests to capture the impact kinematics and an unmanned aerial vehicle (DJI Phantom 3) is used to capture an aerial view of the experiment.

Dual rigid barriers (Fig. 10) are modelled to evaluate the multiple barrier framework for two-phase debris flows. The first barrier is a stiffened aluminium plate with a thickness of 20 mm, height of 500 mm and width of 2000 mm. Two steel supports at the flume sidewalls are used to fix the aluminium plate. A load cell is sandwiched between the support and the plate to measure the impact load. The second rigid barrier is a steel plate that is 1500 mm in height and 2000 mm in width. The steel plate is installed on a reinforced concrete structure. At each corner, a load cell is sandwiched between the steel plate and the reinforced concrete structure to measure the impact load. The first rigid barrier is installed at an inclined distance of 11 m from the gates and the second rigid barrier is installed at a curvilinear distance of 6 m from the first barrier.

A separate study was conducted to investigate the impact dynamics of debris flows on a flexible barrier. The flexible barrier has a ring net panel that is 2000 mm wide and 800 mm in height. The barrier is installed at an inclined distance of 13 m from the gate (Fig. 11). Each ring is 100 mm in diameter and made using steel wires that are 2 mm in diameter. The ring net panel is supported by two horizontal cables, which are anchored to the sidewalls of the flume. Tension load cells (TML TCLK50KNB) are used to measure the impact load of each cable. A mesh with 25 mm square openings made of 1 mm diameter stainless steel wire was overlaid onto the ring net to retain the debris material during and after impact. Dual spring elements (Ng et al. 2016; Fig. 12) are installed between ends of cable and anchors at flume side walls using eyebolts to replicate the loading response of an energy dissipating device used in prototype barriers. The dual spring element exhibits a bilinear load-displacement response. Each of the dual spring element comprises two compression springs—one stiff (k_1) and the other soft (k_2)—in series inside a cylinder. The springs are separated inside the cylinder by a coaxial separator. The flexible spring is preloaded to a specifiable load (P_{pre}) by adding a spacer between the spring and separator inside the cylinder. Before the applied load reaches the inflection point (P_{pre}), only the stiff spring resists the load and slope $K_1 = k_1$. After reaching the inflection point, the load is shared by both springs in series and the equivalent stiffness reduces to model the elongation of energy dissipating elements $K_2 = k_1k_2/(k_1 + k_2)$. The peak deformation of dual spring elements is preserved by a pneumatic locking system.

In each 28 m-long flume experiment, the two-phase debris flow material is representative of the typical debris flow material in East Asia. This mixture generally comprises 35% gravel (20 mm), 62.5% sand (0.6 mm), and 2.5% clay (< 2 μm) (Ng et al. 2019). Figure 13 shows a textural classification of debris flow mixtures consisting of different percentages of gravel, sand and fines (silt and clay). A comparison was conducted between field mapping data from 50 debris flow events that occurred in June 2008 in Hong Kong (Sze and Lam 2017), relevant experiments (Bugnion et al. 2012; Iverson et al. 2010) and natural debris flows (Takahashi 1991; Remaitre et al. 2003; Choi 2010; Tecca et al. 2007). The majority of debris flows shown are predominantly sand-gravel mixtures, with limited samples of clay-rich debris. For Hong Kong cases, tests for particle size distribution were carried out using soil samples collected at debris deposition zone. Particles of size larger than 20 mm were not sampled. Mixtures with fine contents greater than 20% are classified as muddy flows (Bonnet-Staub 1999). A solid fraction of 0.6—typical in field debris flows (Iverson 2015)—is adopted for the testing material. The initial density of the mix is approximately 2,000 kg/m^3. This density lies within the range observed for natural debris flows, which typically ranges from 1,700 kg/m^3 to 2,400 kg/m^3 (Iverson

Table 1 Test programme for dual rigid barriers using the 5 m-long flume

Test ID	Material	Upstream barrier height (mm)	Downstream barrier height (mm)	Flume inclination (°)
WCI0	Water	Nil	Nil	0
WCI05				5
WCI10				10
WCI15				15
SCI26	Sand			26
SCI35				35
SCI45				45
WH10I0	Water	100	500	0
WH10I05				5
WH10I10				10
WH10I15				15
WH18I0		180		0
WH18I05				5
WH18I10				10
WH18I15				15
WH26I0		260		0
WH26I0				0
WH26I05				5
WH26I10				10
WH26I15				15
SH10I26	Sand	100		26
SH10I35				35
SH10I45				45
SH18I26		180		26
SH18I35				35
SH18I45				45
SH26I26		260		26
SH26I35				35
SH26I45				45

Table 2 Test programme for rigid barrier with basal clearance

Test ID	Basal clearance (mm)	Flume inclination (°)
C15-I15	15	15
C15-I25	15	25
C15-I35	15	35
C30-I15	30	15
C30-I25	30	25
C30-I35	30	35

and George 2014). It is worthwhile to note that hard and large inclusions were not explicitly modelled.

Before each 28 m-long flume experiment, the barriers were installed in the flume. The gates were then closed and the debris flow mixture was prepared in the storage container. After the debris mixture was ready, the gates were opened and debris was allowed to discharge downstream.

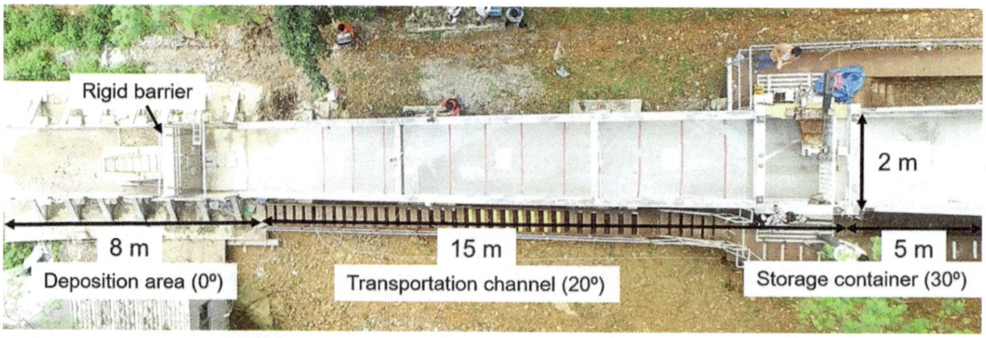

Fig. 9 Plan view of the 28 m-flume model

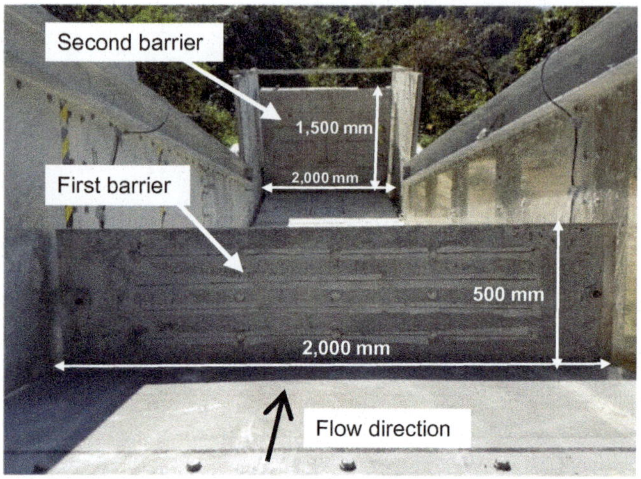

Fig. 10 Model dual rigid barriers

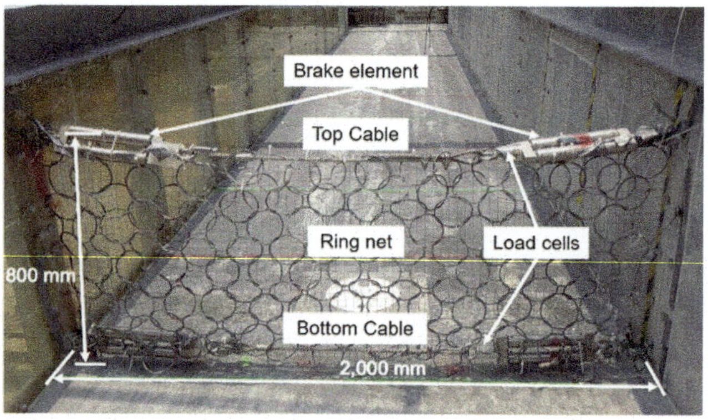

Fig. 11 Upstream view of model flexible barrier

Observed Impact Mechanisms

To develop impact models for design, the fundamental impact mechanisms must first be elucidated. This section examines the impact mechanisms of dry granular, water, and two-phase debris flows on rigid and flexible barriers using the 5 m-long flume and 28 m-long flume.

Single Rigid Barrier

Figure 14 shows a comparison of the impact kinematics of dry granular and water flows impacting against a single rigid barrier installed orthogonally to the channel bed. Tests were carried out in the 5 m-long flume. The flume is inclined at 26° and 5° for modelling dry granular and water flows, respectively. Figure 14a shows the kinematics of a dry granular flow impacting against a single rigid barrier. The kinematics, shown on the left, were captured using a high-speed camera and the corresponding velocity vectors from PIV analysis are shown on the right. At time $t = 0$ s, a tapered flow front reaches the barrier. Static deposits, called dead zone herein, accumulate at the base of the barrier and subsequent granular flow rides on top of the deposits. The granular flow then piles up along the face of the barrier ($t = 0.50$ s). The observed impact mechanism for dry granular flow on a rigid barrier is consistent with the layering mechanism proposed in the velocity attenuation model in Fig. 6. The granular material deposits in layers towards the crest of the barrier. As shearing occurs between the incoming flow and the deposited material forming the dead zone, the velocity of the incoming flow is attenuated. The attenuated flow eventually deposits and contributes to the dead zone. Velocity reduction is evident by examining the magnitude of the vectors deduced in the PIV analysis. The velocity reduces from 1.0 m/s to 0.5 m/s, which is a 50% decrease. At $t = 1.00$ s, the incoming flow is noticeably thinner due to a limited supply of granular material from the storage container. At $t = 1.50$ s, the barrier is filled to its crest. By using attenuated velocities to estimate the impact load using Eq. (1), lower impact loads can be obtained. The proposed velocity attenuation is implemented in design guidelines for rigid barriers in Hong Kong (Kwan and Koo 2015).

Figure 14b shows the kinematics of water flow impacting against a single rigid barrier (shown on the left), and the corresponding PIV analysis (shown on the right). The water flow front reaches the barrier at $t = 0$ s. The flow exhibits a

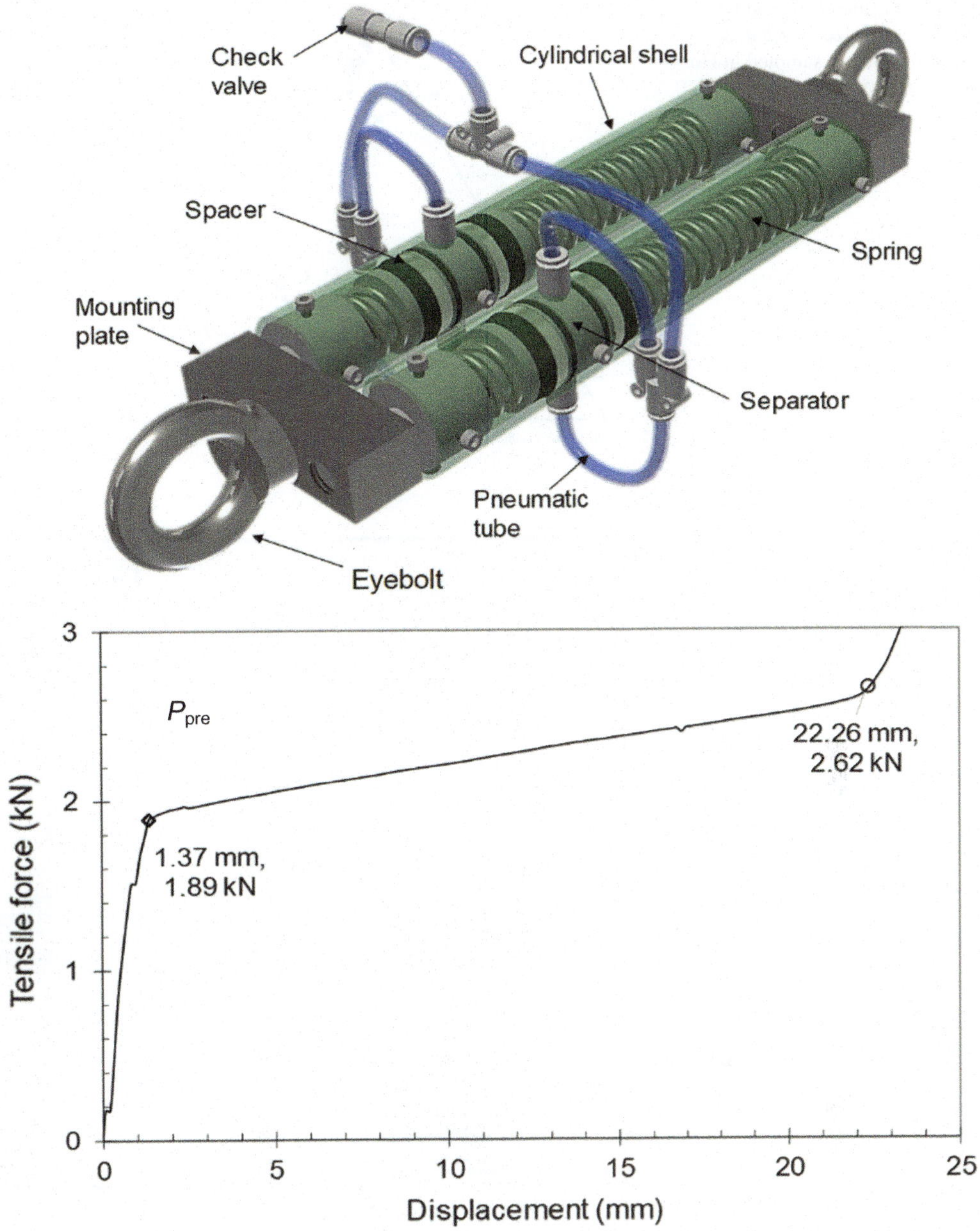

Fig. 12 Dual spring element model with force-displacement curve

vertical jet-like run-up (Choi et al. 2015) along the face of the rigid barrier upon impact at t = 0.15 s. The run-up height of the water exceeds the barrier height. The velocity vectors show a 90° redirection after impacting the barrier. At t = 0.5 s, the run-up rolls back towards the channel. Overspill is only observed at t = 1 s near the end of the impact process. Significant turbulence is observed as the water rolls-back and mixes with the incoming flow, further dissipating flow kinetic energy. Armanini et al. (2019) reported a similar impact mechanism in their experiments, which modelled water flows impacting against a rigid barrier. They showed that the peak impact pressure on the barrier is inversely related to the radius of curvature formed at the base of the barrier as the water runs-up. By comparing the impact mechanisms of dry granular and water flow on a single rigid barrier, distinct differences in the impact mechanisms between dry granular and water flows highlight the importance of flow material on the impact dynamics. Ng et al. (2016) compared the impact dynamics of dry granular flow with viscous flow on a rigid barrier using centrifuge model tests. They reported that for the same initial conditions, the impact load resulting from dry granular flows was about 2.5 times smaller than that of viscous flows. They explained that the impact dynamics of dry granular and water flows are governed by frictional and viscous stresses, respectively. It was reported that energy dissipated by shearing of frictional contacts in dry granular flow is more significant compared to viscous shearing in water flows. Furthermore, the bulk

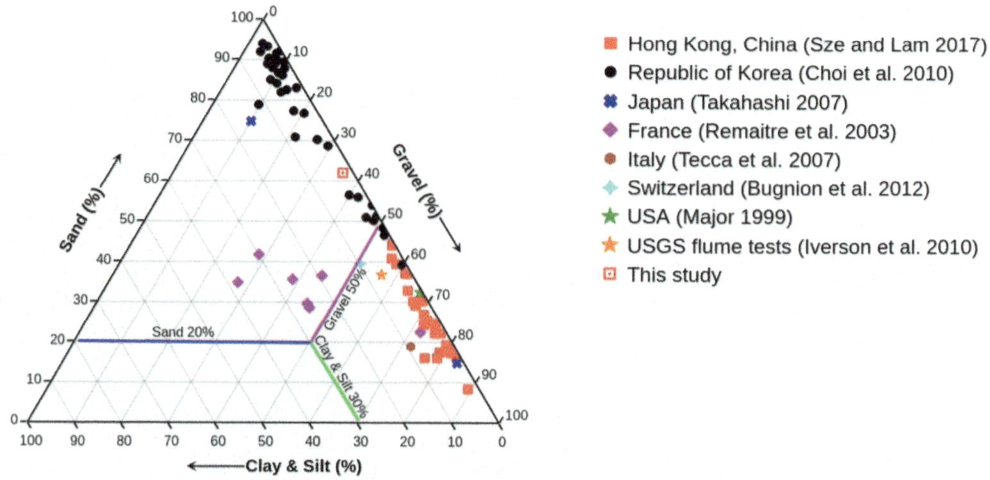

Fig. 13 Debris flow composition (particles of size larger than 20 mm were not sampled in Hong Kong cases)

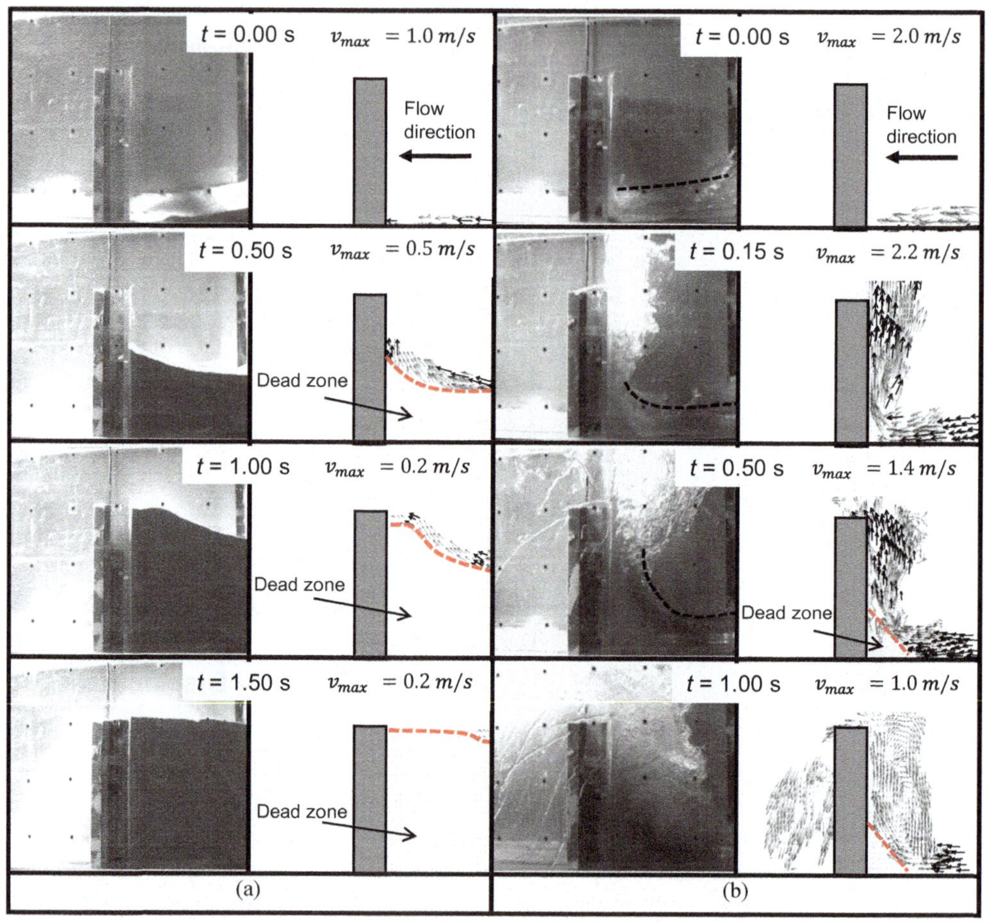

Fig. 14 Comparison of impact mechanisms in the 5 m-long flume for a single rigid barrier for: **a** dry granular flow; **b** water

compressibility of dry granular flow is much higher compared to water flows, resulting in more deformation and energy dissipation during impact. Therefore, granular flows induce lower impact forces. Song et al. (2017) reported a series of experiments modelling the impact dynamics of two-phase debris flows on a rigid barrier. As the volumetric solid content of the flow is increased, the impact mechanism transitioned from pileup to run-up. They also reported that the impact load increased with the volumetric solid content in the flow by up to 0.6, which is the typical volumetric solid content of typical debris flows observed in the field (Iverson 1997). Details of the measured impact forces in this study and their corresponding impact mechanisms are discussed later.

Single Flexible Barrier

Figure 15 shows a plan view of the impact kinematics of debris flow impacting against a single flexible barrier installed orthogonally to the channel bed. The test was carried out in the 28 m-long flume, which is inclined at an angle of 20°. Upon impact, the flow jumps along the face of the barrier at $t = 0.5$ s. Subsequent flow impacts the arrested material while some fines and fluid pass through the pervious flexible barrier. The dual spring elements were activated as the top cable deforms. The run-up follows the curvature of the deformed barrier and rolls back towards the upstream direction at $t = 1$ s. The original and deformed profiles of the top cable is shown using solid red and dashed white lines, respectively. As more debris deposits near the base of the barrier, the volume of material discharging through the barrier diminishes. The dual spring elements installed on the top and bottom cables were eventually fully mobilized. The bottom cables are no longer visible in the field of view due to the deformation of the flexible barrier along the flow direction. Overspill is observed at $t = 1.5$ s. Simultaneously, the rolling back motion of the debris flow front impacts the incoming flow and the dead zone increases in size. At $t = 2$ s the roll back diminishes and overspill continues. At $t = 5.0$ s, debris is retained by the flexible barrier with a horizontal free surface up to the fully deformed height of the barrier. The horizontal free surface of the deposit indicates a fluidized debris material.

The observed kinematics of debris flow impacting against the model flexible barrier exhibits characteristics of both run-up and pileup mechanism reported by Choi et al. (2015). Initially, the observed impact process in this study is reminiscent of the run-up mechanism. Near the end of impact, the observed impact mechanism resembles the pileup mechanism. More importantly, deformation and the perviousness of the barrier play important roles in the observed impact dynamics. The importance of barrier stiffness was demonstrated by Ng et al. (2020). They investigated the role of barrier stiffness on the peak impact force induced by dry granular flow impacting against a deformable barrier. They reported that barrier deformation during initial impact allows relative movement between the flow and a deformable

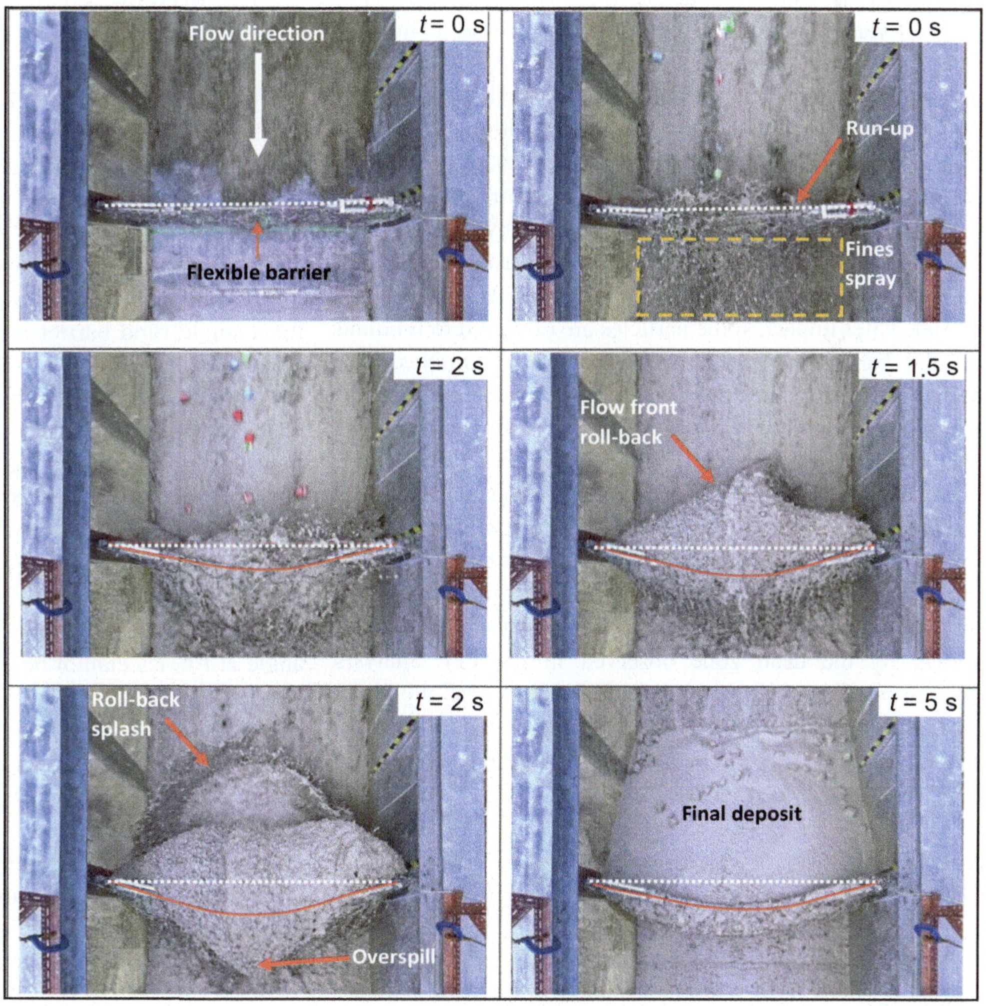

Fig. 15 Impact kinematics between two-phase debris flow and a flexible barrier in the 28 m-long flume

barrier, thereby extending the impact duration and attenuating the peak impact load. They showed that by reducing the bending stiffness of a typical 1-m thick concrete cantilevered barrier by five orders of magnitude (i.e. equivalent to a steel flexible barrier), the peak impact load was reduced by up to 40%. The importance of barrier perviousness was investigated by Liu (2019a). It was reported that compared to an impervious rigid barrier (Fig. 14), a pervious steel flexible barrier allows discharge through the barrier, thereby reducing the impact force induced on the barrier. The force induced on the flexible barrier is directly related to the rate of change in momentum. Thus, the effects of deformation and perviousness are both instrumental to the load attenuation mechanism of a flexible barrier (Volkwein 2014).

Rigid Barrier with Basal Clearance

Figure 16 shows the observed kinematics of dry granular flow, with 10 mm glass spheres, impacting against a model rigid barrier with a basal clearance. The kinematics, shown on the left, are captured using a high-speed camera, and the velocity fields, shown on the right, are analysed using PIV. The test was carried out using a 5 m-long flume inclined at 35°. The rigid barrier has a basal clearance with a ratio between basal clearance height and flow depth H_c/h_0 of 0.7. The thin tapered flow front initially passes through the basal clearance since the flow depth is less than the height of the basal clearance. Discharge underneath the barrier reduces the momentum transferred to the barrier, thereby reducing the impact force on the barrier. As the saltating front increases in depth (Fig. 16a), part of the flow impacts the base of the barrier. After impact, some particles are observed to rebound off the barrier and collide with incoming particles. The collisions lead to the rapid deceleration and redirection of subsequent flow upwards along the face of the barrier. This change in impact direction is shown by the PIV vectors. The measured maximum flow velocity from PIV analysis at this moment (t = 0.15 s; Fig. 16a) was 2.8 m/s, which reduced by 10% compared with initial impact velocity. As more granular flow impacts the barrier, granular material accumulates behind the rigid barrier above the basal clearance as shown by the dead zone observed in the PIV analysis (Fig. 16b). Concurrently, the vertical stress near the basal clearance increases with retained height of material and a reduction in discharge velocity underneath the barrier by up to 75% is observed (Fig. 16c).

Dual Rigid Barriers

Figure 17a shows the kinematics of dry granular flow impacting dual rigid barriers installed orthogonally to the channel bed. The flume is inclined at an angle of 35° to the horizontal. The flow front reaches the first barrier at t = 0 s and runs-up. Incoming flow material piles up in layers behind the barrier up to the height of the barrier. At t = 0.5 s the first barrier is filled, and cascading overflow is observed. The first barrier gets filled to the crest in a similar manner as that shown in Fig. 14 for a single rigid barrier and similar to the proposed velocity attenuation model (Fig. 6). At t = 1 s, the overflow lands on the flume base and flows downstream to impact the second barrier. The overflow impacts the flume base following a trajectory described by Eq. 4 at an angle of 30° measured relative to the flume base. The measured landing angle $\beta = 30^\circ$ corresponds to a landing factor $C_r = 0.87$ (Eq. 5 with R = 1). The energy dissipated during landing attenuates the flow velocity between the first and second barriers, thus reducing the impact load induced on the second barrier. As the impact process on the second barrier continues, the wedge of deposited dry granular material accumulates behind the second barrier and propagates upstream. The upstream movement of granular material eventually intercepts the overflow from the first barrier at t = 1.5 s. From t = 1.5 s to t = 3 s, overflow continues to impact the downstream deposition wedge, thereby enlarging the deposition volume. The end of impact process is marked at t = 25 s, when the dry granular material has come to rest.

Figure 17b shows the kinematics of water flow impacting against dual rigid barriers installed orthogonally to the channel bed. The flume is inclined at an angle of 10° to the horizontal. The flow front reaches the first barrier at t = 0 s and runs-up along the barrier face at t = 0.5 s. The initial impact kinematics are similar to those described in Fig. 14b for a single rigid barrier resisting water flow. However, the impact process in Fig. 17b is different compared to Fig. 14b for a single barrier once the barrier is filled to the crest and water overflows the first barrier. The difference between the impact kinematics of dry granular flow and water flow also emerges between t = 0.5 s–t = 1.0 s. Compared to dry granular overflow, water overflow lands on the flume at a steeper angle. As a result, water overflow lands closer to the first barrier compared to dry granular flow. Significant turbulence is observed for water overflow, thus the landing angle at t = 1 s cannot be deduced. At t = 1.5 s, the impact process at the first barrier ends as water upstream of the barrier comes to rest. The water that landed in between the two barriers flows downstream and impacts the second barrier. The impact process ends at t = 6 s with water flow arrested by the dual barriers.

Overflow kinematics and impact dynamics of dual rigid barriers for both dry granular and water flows observed in the above experiments are used to verify the proposed analytical framework and discussed in detail later.

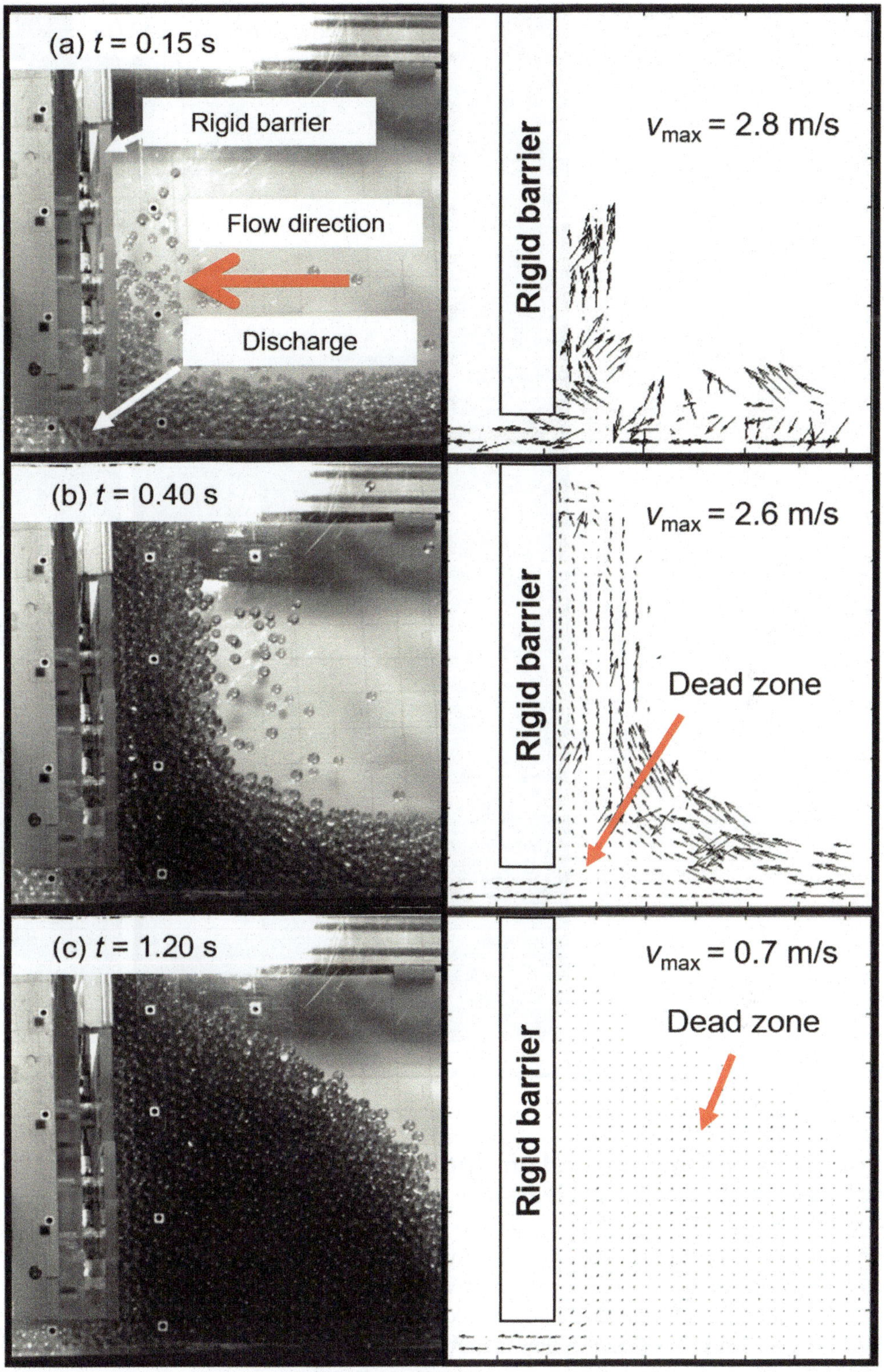

Fig. 16 Flume modelling of the rigid barrier with basal clearance of relative opening $H_c/h_0 = 0.7$ (with 35^0 flume inclination): Observed dry granular flow kinematics (left) and PIV analysis (right)at **a** $t = 0.15$ s; **b** $t = 0.40$ s; **c** $t = 1.20$ s (Choi et al. 2020)

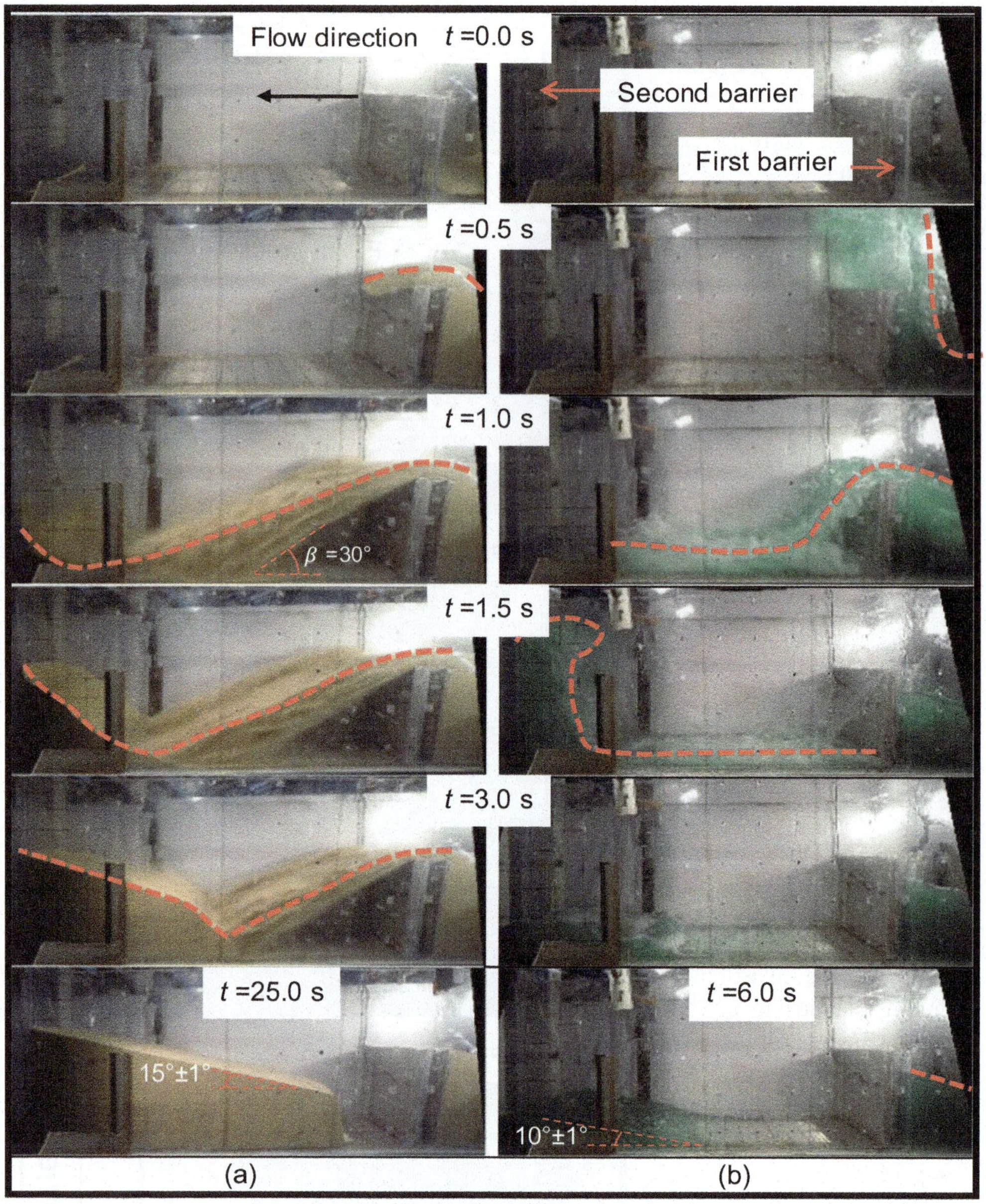

Fig. 17 Comparison of overflow and landing mechanisms in 5 m-long flume for dual-rigid barriers: **a** dry granular flow; **b** water

Estimating the Impact Load on Single Barriers with and Without Basal Clearance

The peak load induced on a barrier is the sum of the dynamic and static components of the flow. The relative contributions of the dynamic and static loads to the peak load is related to the Froude number (*Fr*) of the flow before impact (Faug 2015). Furthermore, the Froude number in turn is strongly influenced by the geomorphological settings. For example, in Hong Kong, debris flows travel at high velocities on steep terrain over short distances. These conditions may lead to higher Froude numbers ($Fr > 3$). In contrast, the Froude conditions in the Alps or Rockies tend to be lower ($Fr < 3$) because flows travel on gentler terrain.

Figure 18 shows the relationship between the Froude number and the peak load F_{peak} normalised by the theoretical static load $F_{static} = 0.5k\rho g h_0^2 w$(Armanini and Scotton 1993; Armanini 2009) of the flow before impact, which can be calculated as follows:

$$\frac{F_{peak}}{0.5k\rho g h_0^2 w} = 1 + \frac{2\alpha}{k} Fr^2 \qquad (6)$$

Existing design guidelines in Hong Kong adopt a static impact coefficient $k = 1$ (no internal shear strength) and $\alpha = 2.5$ for rigid barriers (Kwan 2012), and $k = 1$ and $\alpha = 2$ for flexible barriers (Kwan and Cheung 2012). It is worthwhile to note that the α recommended by Kwan (2012) and Kwan and Cheung (2012) accounts for hard and large

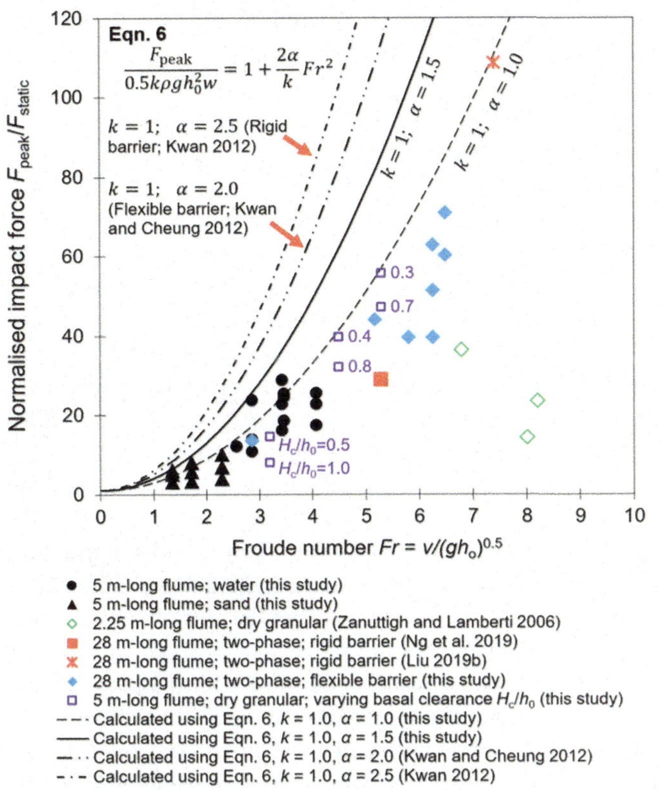

Fig. 18 Design Froude number for single rigid and flexible barrier against debris flow impact

inclusions. These design parameters are used in Eq. 6 and compared to the measured data. In addition, two theoretical bounding lines using Eq. 6 are shown. The first theoretical bounding line adopts $k = 1$ and $\alpha = 1$. The second theoretical bounding line using $k = 1$ and $\alpha = 1.5$. Details of the bounding lines are discussed below.

Compared to the existing design guidelines in Hong Kong for rigid barrier ($k = 1$ and $\alpha = 2.5$), the theoretical bounding line ($k = 1$ and $\alpha = 1.5$) conservatively estimates all the measured data in this study for dry granular, water, and two-phase debris flows, and for dry granular flow from Zanuttigh and Lamberti (2006). The data falls within a range of Froude numbers from 0 to 9. Thus, the theoretical bounding line ($k = 1$ and $\alpha = 1.5$) can provide load optimisation for rigid barrier design. A summary of the recommended design parameters is given in Table 3.

Compared to the existing design guidelines in Hong Kong for flexible barrier ($k = 1$ and $\alpha = 2$), the theoretical bounding line ($k = 1$ and $\alpha = 1$) conservatively estimates the impact force for all flexible barrier data points for two-phase debris flows and provide opportunities for load optimisation for flexible barrier design.

The dry granular flows in this study exhibited lower normalised impact forces compared to the water flows in the 5-m flume tests. This may be attributed to a higher degree of bulk compressibility and internal energy dissipation mainly due to friction in dry granular flows leading to higher energy dissipation during the impact process (Choi et al. 2015). Both of these features, compressibility and enduring frictional contacts, led to a pileup mechanism. Similarly, the dry granular flows reported in the literature (Zanuttigh and Lamberti 2006) for higher Froude numbers ($Fr > 4$) also tend to exhibit lower impact force compared to two-phase debris flows (Ng et al. 2019) in which the flows were almost incompressible.

A comparison of the normalised peak impact loads resulting from dry granular flow impacting rigid barriers with different basal clearances is shown. The basal clearance H_c is normalised by the maximum flow depth h_o. In the experiments, glass spheres with a uniform particle size of 10 mm were used. The impact load decreases with an enlarging basal clearance height under a same Froude condition. The theoretical bounding line ($k = 1$ and $\alpha = 1$) provides a conservative estimate of the impact load for basal clearances $H_c/h_o \leq 1.0$ and can be adopted for designing rigid barriers with a basal clearance (see Table 3).

Evaluation of Analytical Framework for Dual Rigid Barriers

Verification of the analytical framework for dual rigid barriers (Eqs. 2, 4, and 5) from physical experimental results from the 5 m-long and 28 m-long flumes is discussed below.

Run-up velocity (v_d)

Figure 19 shows a comparison of the measured and calculated normalised run-up heights (h_d/h_O) and normalised

Table 3 Design recommendations for estimating the impact loads for different barrier configurations

Design recommendations#	Dynamic impact coefficient (α)
Single rigid barrier without basal clearance	1.5
Single rigid barrier with basal clearance ($0.3 \leq H_c/h_O \leq 1.0$)	1.0
Single flexible barrier	1.0
The second barrier in a dual rigid barrier system	1.0

A static impact coefficient $k = 1.0$ is recommended to deduce static load

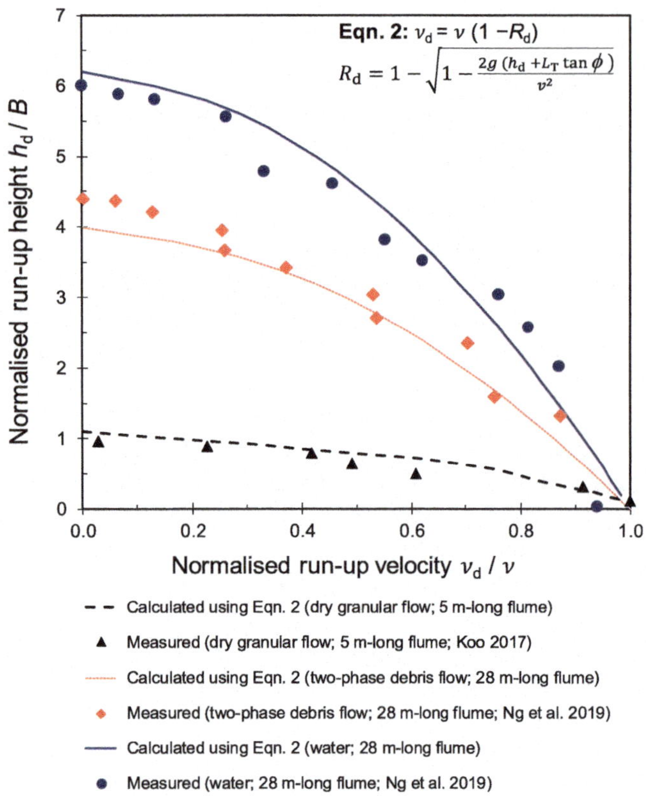

Fig. 19 Comparison of measured and calculated run-up velocity with barrier height (modified from Ng et al. 2018)

run-up velocities (v_d/v) for dry granular (Koo 2017), water (Ng et al. 2019) and two-phase debris flows (Ng et al. 2019) impacting a rigid barrier. Dry granular flow impacting a rigid barrier was modelled using the 5 m-long flume. The run-up velocity v_d is normalised by the flow velocity v before impacting the rigid barrier, whereas the run-up height h_d is normalised by the maximum flow depth h_0. Water and two-phase debris flows impacting a rigid barrier were modelled using the 28 m-long flume. The calculated run-up velocity was determined using the velocity attenuation model (Fig. 6). The initial flow velocity measured in the physical experiments is used to calculate the attenuation of velocity during impact by using Eq. 2. Therefore, the measured and calculated velocities at the base of the barrier are the same. The calculated velocities for the dry granular tests agree with the measured velocities. Similarly, the results of the two-phase debris flows from the 28 m-long flume tests for rigid barrier show reasonable agreement with the calculated values. Dry granular flows exhibit lower run-up heights compared to water flows. The run-up of two-phase debris flow lies closer to the water flows even with 60% solid fraction. This run-up height may be attributed to fluidization of the two-phase material due to sustained excess pore fluid pressure during impact (Ng et al. 2019). A summary of the parameters used to deduce the run-up velocity is given in Table 4.

Overflow Distance (x_i)

The launch angle of the overflow from the first barrier determines the trajectory and overflow distance. Figure 20 compares the measured and calculated overflow distances for dry granular flow, water and two-phase debris flows. The calculated values are obtained by using Eq. 4. The effects of barrier height and flume inclination on overflow distance are examined. The overflow distance x_i and barrier height B are normalised by the flow depth h_o before the flow impacts the rigid barrier. In the tests conducted with dry granular, the normalised barrier height B/h_0 varies as 1.1, 2.0 and 2.9 for a flume inclination of 26° and as 1.0, 1.9 and 2.7 for a flume inclination of 32°. These heights correspond to typical barrier designs observed in the field. The inclination of the flume was adjusted to vary the flow inertia before impact. In contrast with the tests for dry granular flow, the flume inclination was fixed to be 20° for the two-phase debris flow and water flow tests conducted in the 28 m-long flume. Normalised barrier heights B/h_0 of 5.5 and 5.0 are used for the two-phase debris and water flows, respectively. A comparison of the measured results of dry granular impacting different barrier heights shows that the overflow distance decreases with increasing barrier height. More energy is dissipated from shearing among grains and the conversion of kinetic energy to potential energy as the barrier height increases. High flume inclinations lead to more inertial flows before impact, thereby causing longer overflow distances. A comparison between measured and calculated overflow distances for dry granular flow shows that calculated distances using Eq. 4 can provide reasonable estimates and are on the conservative side.

The measured overflow distances downstream from the first barrier were 4.5 m and 3.3 m for the water and two-phase debris flows, respectively. The measured overflow distance for both flows is somewhat lower than that calculated, but on the conservative side. Evidently, Eq. 4 can provide a reasonable estimate of the overflow distance for all three types of flows investigated in this study. The minimum barrier spacing required between the successive barriers should ensure that the flow lands between two barriers.

Landing Factor (C_r)

The landing factor C_r (Eq. 5) accounts for the momentum loss from the impact between the flow and the flume bed. Physically, momentum is assumed to be completely

Table 4 Parameters for calculating run-up velocity (v_d)

Parameter	Value	Flow type	Method of determination: This study	Method of determination: Remark
Initial impact velocity, v (m/s)	2.0	Water	Measured	Velocity hydrograph from debris mobility analysis
	6.0	Debris flow		
	1.0	Dry granular		
Initial flow depth, h_0 (m)	0.060	Water	Measured	Flow depth hydrograph from debris mobility analysis
	0.065	Debris flow		
	0.090	Dry granular		
Friction angle, (°)	–	Water	–	–
	0	Debris flow	Deduced from deposition angle	Measured deposition angle is zero
	30	Dry granular	Measured	Material dependent
Angle of deposition θ_d (°)	0	Water	Measured	Assume same as
	0	Debris flow		
	30	Dry granular		
Height of the deposited granular material, h_d (m)	Varies	Water	Measured in stages	Initially assume $h_d = h_0$; Calculate h_d for successive stages by adding $h_o/\sin(\theta_d - \theta)$; θ is the channel slope Number of stages = B/h_0; Fig. 6
		Debris flow		
		Dry granular		
Length of the free surface of the arrested granular material, L_T (m)	Varies	Water	Measured in stages	Using Fig. 6 geometry calculate as $L_T = h_d/\sin(\theta_d - \theta)$
		Debris flow		
		Dry granular		

Acceleration due to gravity (g = 9.81 m/s^2)

destroyed when flow lands perpendicularly to the flume with $\beta = 90°$ or $C_r = 0$. In contrast to the perpendicular impact, no energy is dissipated when flow lands tangentially to the flume bed with $\beta = 0°$ or $C_r = 1$. Figure 21 compares the estimated landing factors C_r using Eq. 5 and those back-calculated from the flume experiments of dry granular flows in a 5 m-long flume (Koo 2017) and two-phase debris flows in a 28 m-long flume (Ng et al. 2019). The back-calculated C_r is for dry granular flows landing on the acrylic bed of the 5 m-long flume and a two-phase debris flow landing on the steel bed of the 28 m-long flume. The landing angle from the water flow is not included in the figure because the water flow was turbulent when it landed, thereby making it impossible to determine the landing velocity.

It is evident that the landing factor C_r increases for dry granular flow as the landing angle decreases, thereby implying that less energy is dissipated upon landing. A best fit line through the measured data is shown to reveal the contribution of the energy dissipated via basal friction upon landing. The projected best fit gives an R coefficient of 0.9 at = 0°. This result indicates that energy dissipation from the tangential shear between the flow and the bed is only 10% of the total energy dissipated upon landing. Nonetheless, a value of $R = 1$ (no energy loss from the tangential bed shear) provides an upper bound for the dry granular and two-phase debris flows. The results imply that Eq. 5 with $R = 1$ can be used to estimate the landing factor C_r.

The impact dynamics in a dual rigid barrier system can be calculated at different stages of impact process. The frontal

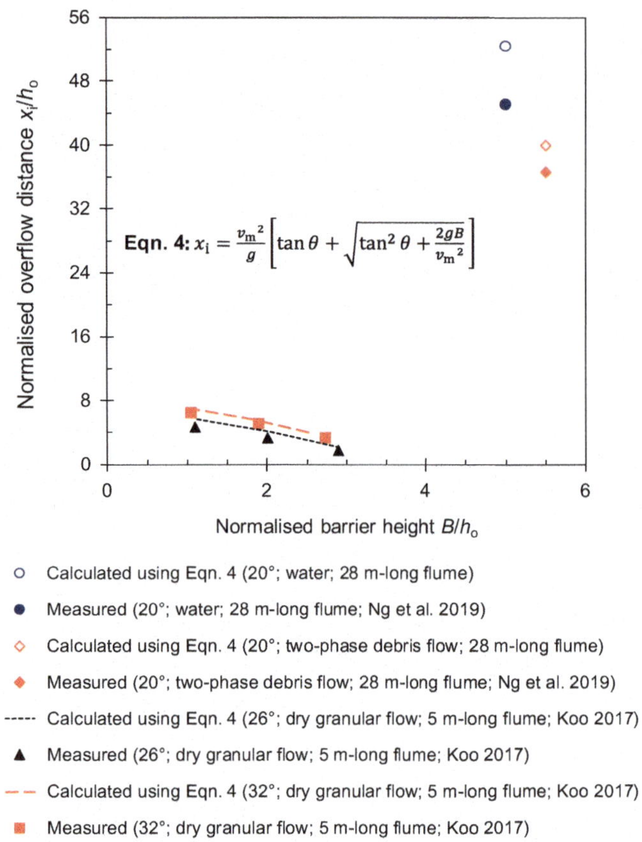

Fig. 20 Effects of flume inclination and barrier height on overflow distance (modified from Ng et al. 2018)

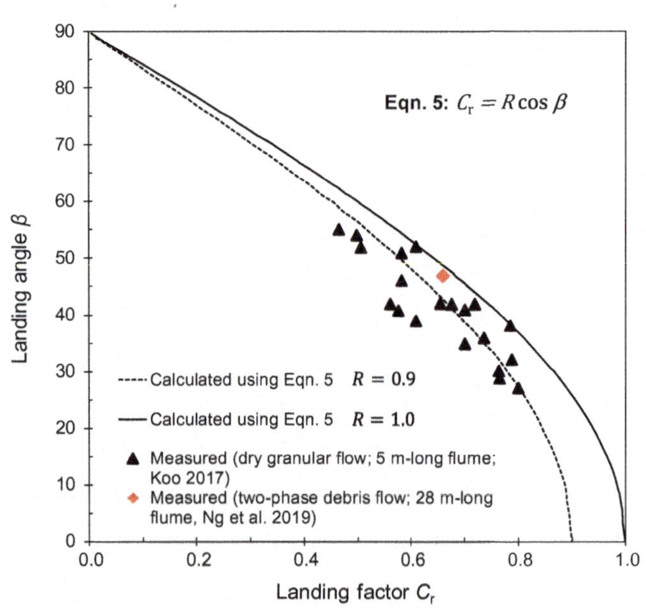

Fig. 21 Comparison of landing factor (modified from Ng et al. 2018)

velocity is an indicator of flow attenuation and acceleration at different stages of the impact process. Three different frontal velocities for the two-phase debris flow test are compared in Fig. 22 to evaluate the entire multiple barrier analytical framework in Eqs. (2)–(5). The calculated and measured velocities before impact, during overflow and after landing are compared.

The frontal velocity of the two-phase debris flow before impacting the first barrier was 6 m/s. This pre-impact velocity was adopted as the initial input for Eqs. (2)–(4). The resulting calculated overflow velocity at the crest of the barrier is 5.7 m/s, which is close to the measured velocity, which was 5.4 m/s. Furthermore, the measured velocity after landing was 3 m/s, and the measured landing angle was 48° ± 2° ($C_r = 0.66$). Notably, water overflow landing in the channel was turbulent, thereby making it impossible to measure the landing velocity correctly. Hence, the test result for water flow is not included.

Velocity reduction upon landing depends on the landing angle, the flow composition and the flume bed. Kwan et al. (2015) reviewed data from the field and laboratory tests. The data includes dry granular and two-phase debris flows impacting hard and/or soft beds. The reported velocity reduction factors C_r range from 0.3 to 0.75. The calculated post-landing velocity, by using a C_r of 0.7 and the measured velocity are compared. The calculated landing velocity is

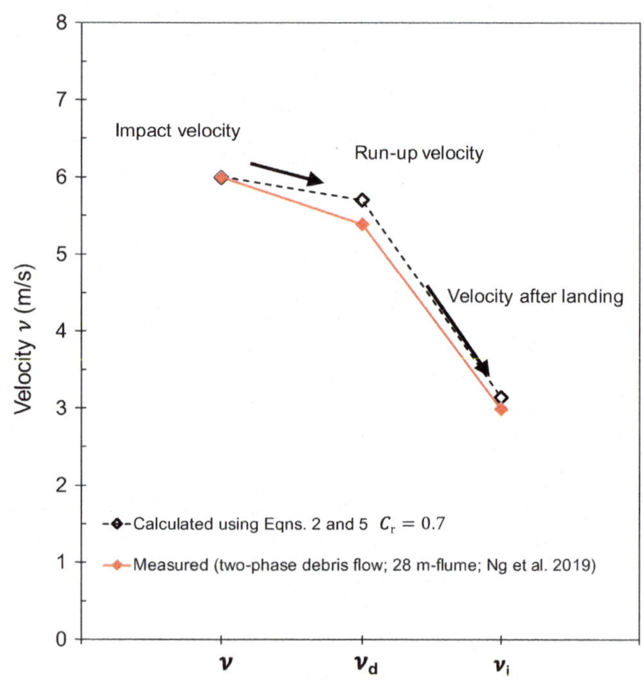

Fig. 22 Comparison of measured and calculated frontal velocity for dual rigid barrier system (modified from Ng et al. 2019)

3.2 m/s, which agrees well with the measured post-landing velocity (3 m/s) for a two-phase debris flow impacting a steel flume bed in the 28 m-long flume. Therefore, $C_r = 0.7$ is recommended to estimate the post-landing velocity.

Impact Force on Second Rigid Barrier

Design guidelines for multiple rigid barriers in series are not yet available. Current approaches for designing multiple barriers adopt a similar principle as that for a single barrier (Kwan 2012). The single barrier approach does not consider the effects of the upstream barriers on attenuating the impact forces on the downstream barriers. Figure 23 shows the relationship between the normalized peak impact force and flow Froude number for the second rigid barrier of a dual barrier system that is investigated in this study. Similarly to Fig. 18, the measured peak impact forces (F_{peak}) for dry granular flows and water are normalized by the theoretical static force (F_{static}) $0.5k\rho g h_0^2 w$. To obtain conservative impact forces on the second barrier, the flow depth before impacting the second barrier is assumed to remain constant and equal to the flow depth at the first barrier. Both water and dry granular flow data from the 5 m-long flume experiments are compared. Generally, the Froude numbers of the dry granular flows have lower Froude numbers compared to those of the water flows because more energy dissipation occurs via frictional shearing among grains (Choi et al. 2015). The normalized impact forces exerted by water flows are higher than those exerted by dry granular flows because water overflow lands closer to the first barrier compared to dry granular overflow, resulting in sufficient length for flow acceleration (Fig. 17). The measured impact forces are compared with the theoretical normalized peak impact force (Eq. 6). In comparison with the upper bound for a single rigid barrier (Fig. 18) where $k = 1$ and $\alpha = 1.5$, an upper bound with $k = 1$ and $\alpha = 1$ (refer to Table 3) provides a conservative estimate of the impact force exerted on the second barrier as shown in Fig. 23. The reduction of normalized peak impact forces for the second barrier is mainly attributed to the energy dissipation during impact on the first barrier and landing between the dual barriers.

In summary, the newly proposed analytical framework for designing a multiple rigid barrier system by considering the velocity attenuation during impact, overflow and landing has been verified by the experimental data shown in this study. The attenuated flow velocity at the crest of the first barrier (Fig. 19) is used to estimate the overflow velocity and distance (Fig. 20). Landing reduction factors (Fig. 21) are then implemented to obtain the flow velocity after landing (Fig. 22). This velocity is then adopted to estimate the impact force on the second barrier (Fig. 23). The analysis can be carried out repeatedly to predict the impact loads on the third and any subsequent barriers installed downstream in the channel.

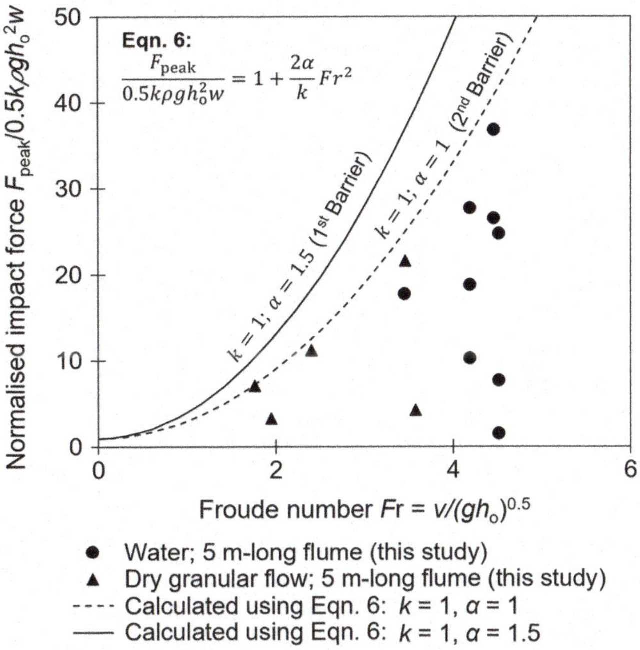

Fig. 23 Design impact force for first and second rigid barrier in dual barrier system

Summary and Conclusions

This forum paper presents a collection of physical experiments modelling the impact mechanisms against single and dual rigid barriers and a single flexible barrier. The flow types examined include dry granular, water and two-phase flows. Additionally, the effects of a basal clearance on the impact dynamics of dry granular flow on a single rigid barrier were examined. Experiments were conducted at two different scales, including 5 m-long and 28 m-long flumes. Based on the observed fundamental impact mechanisms, impact load measurements were compared with a newly developed analytical framework for designing multiple rigid barriers.

The design load for a single rigid barrier without a basal clearance should be estimated using an $\alpha = 1.5$. The static load should be estimated using $k = 1$. The design load for a single rigid barrier with a basal clearance ($0.3 \leq H_c/h_0 \leq 1.0$), a single flexible barrier, and the second rigid barrier in a dual rigid barrier system should be estimated using an $\alpha = 1.0$. It should be noted that large and hard inclusions were not explicitly considered in this study. Therefore, some α values proposed in other design guidelines are higher than those recommended in this study.

The newly proposed multiple barrier framework was verified using experimental data. The attenuated velocity v_d along the barrier height can be predicted reasonably well using the velocity attenuation model. The horizontal overflow distance x_i can be conservatively estimated by the proposed equation. The minimum barrier spacing should be kept larger than the calculated overflow distance to ensure a robust multiple barrier design. Based on the experimental data, a landing factor of $C_r = 0.7$ gives a conservative estimate of flow velocity before the second barrier. This estimated velocity serves as the initial input velocity for the design of the next barrier along the flow path.

Looking Ahead

Given that debris flow is scale-dependent, the development and construction of the largest testing facilities possible will be necessary to advance the current state of scientific and engineering understanding on utilising barriers to mitigate debris flows. The world's largest man-made flume is currently under construction. This facility is jointly developed between The Hong Kong University of Science and Technology and the Institute of Mountain Hazards and Environment of the Chinese Academy of Sciences. The model is 172 m in length, 6 m in width and has an inclination of up to 30° (Fig. 24). The storage container at the most upstream end of the slope can store a debris volume of 500 m^3. This facility will not only be used to evaluate the proposed multiple barrier framework in this study, but serve the local and international scientific and engineering community at large for decades to come.

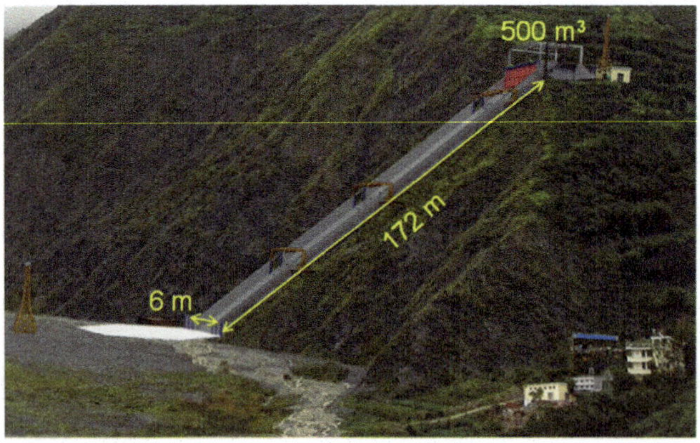

Fig. 24 Schematic of 172 m-long flume (Kunming, China)

Acknowledgements The authors are grateful for financial support from the theme-based research grant T22-603/15-N and area of excellence project grant AoE/E-603/18, as well as the general research fund grants 16212618, 16209717, and 16210219 provided by the Research Grants Council of the Government of Hong Kong Special Administrative Region, China. The authors are immensely grateful for the support from the National Natural Science Foundation of China (51,709,052). This paper is published with the permission of the Head of the Geotechnical Engineering Office and the Director of Civil Engineering and Development, Hong Kong SAR Government.

References

Armanini A (2009) Discussion on: Experimental analysis of the impact of dry avalanches on structures and implication for debris flow (Zanuttigh, Lamberti). J Hydraul Res 47(3):381–383

Armanini A, Scotton P (1993) On the dynamic impact of a debris flow on structures. In Proceedings of XXV IAHR Congress, Tokyo (Tech. Sess. B, III), 203–210

Armanini A, Rossi G, Larcher M (2019) Dynamic impact of a water and sediments surge against a rigid wall. J Hydraul Res 58(2):314–325

ASI (2008) ONR 24801. Austrian Standard Institute, Austria, Protection Works for Torrent Control - Actions on Structures (Draft), p 25

Berger C, McArdell BW, Schlunegger F (2011) Direct measurement of channel erosion by debris flows, Illgraben, Switzerland. J Geophys Res Earth Surf 116:F01002

Bonnet-Staub I (1999) Definition d'une typologie des deposits de laves torrentielles et identification de critres granulomtriques et geotechniques concernant les zones sources. J Bull Eng Geol Environ 57 (4):359–367

Brighenti R, Segalini A, Ferrero AM (2013) Debris flow hazard mitigation: a simplified analytical model for the design of flexible barriers. Comput Geotech 54:1–15

Bugnion L, McArdell BW, Bartelt P, Wendeler C (2012) Measurements of hillslope debris flow impact pressure on obstacles. Landslides 9(2):179–187

Development Bureau (2020) Geotechnical engineering office built a debris-resisting barrier in Tung Chung to reduce the risk of landslides on natural hillsides [online]. Available from https://www.devb.gov.hk/filemanager/tc/content_1044/20180204_08.html [cited 13 April. 2020]

Choi YJ (2010) A study on downstream process of debris flow mobilized from landslides, Master Thesis, Kangwon National University, Korea

Choi CE, Au-Yeung SCH, Ng CWW, Song D (2015) Flume investigation of landslide granular debris and water runup mechanisms. Géotech Lett 5(1):28–32

Choi CE, Ng CWW, Liu H, Wang Y (2020) Interaction between dry granular flow and rigid barrier with basal clearance: analytical and physical modelling. Can Geotech J 57(2):236–245

DeNatale JS, Iverson RM, Major JJ (1999) Experimental testing of flexible barriers for containment of debris flows. US Department of the Interior, US Geological Survey

Faug T, Caccamo P, Chanut B (2012) A scaling law for impact force of a granular avalanche flowing past a wall. Geophys Res Lett 39(23): L23401

Faug T (2015) Macroscopic force experienced by extended objects in granular flows over a very broad Froude-number range: macroscopic granular force on extended object. Eur Phys J E 38:13–16

Froude MJ, Petley DN (2018) Global fatal landslide occurrence from 2004 to 2016. Nat Hazards Earth Syst Sci 18:2161–2181

Iverson RM (1997) The physics of debris flows. Rev Geophys 35 (3):245–296

Iverson RM (2015) Scaling and design of landslide and debris-flow experiments. Geomorphology 244:9–20

Iverson RM, George DL (2014) A depth-averaged debris-flow model that includes the effects of evolving dilatancy. I. Physical basis. Proceedings of the royal society a: mathematical, Physical and Engineering Sciences 470(2170):20130819

Iverson RM, Logan M, LaHusen RG, Berti M (2010) The perfect debris flow? aggregated results from 28 large-scale experiments. J. Geophys. Res. Earth Surf. 115:F03005

Jakob M, Stein D, Ulmi M (2012) Vulnerability of buildings to debris flow impact. Nat Hazards 60:241–261

Koo RCH (2017) Mechanisms of interaction between dry sand flow and multiple rigid barriers: flume and finite-element modelling. Ph. D Thesis, The Hong Kong University of Science and Technology, Hong Kong, China

Koo RCH, Kwan JSH, Ng CWW, Lam C, Choi CE, Song D, Pun WK (2017) Velocity attenuation of debris flows and a new momentum-based load model for rigid barriers. Landslides 14(2):617–629

Kwan JSH (2012) Supplementary technical guidance on design of rigid debris-resisting barriers, Technical Note No. TN 2/2012. Hong Kong, SAR China, Geotechnical Engineering Office, Civil Engineering and Development Department, The HKSAR Government

Kwan JSH, Cheung RWM (2012) Suggestion on design approaches for flexible debris resisting barriers. Discussion note DN1/2012. Hong Kong, SAR China, Geotechnical Engineering Office, Civil Engineering and Development Department, The HKSAR Government

Kwan JSH, Koo RCH (2015) Enhanced Technical Guidelines for Design of Debris-resisting Barriers. GEO Report No. 333. Hong Kong, SAR China, Geotechnical Engineering Office, Civil Engineering and Development Department, The HKSAR Government

Kwan JSH, Koo RCH, Ng CWW (2015) Landslide mobility analysis for design of multiple debris-resisting barriers. Can Geotech J 52 (9):1345–1359

Liu HD (2019a) Effects on mesh size on the impact mechanism of debris flow impacting net barriers. Ph.D Thesis, The Hong Kong University of Science and Technology, Hong Kong, China

Liu HM (2019b) Impact mechanisms of debris flow against multiple rigid barriers with basal clearance. PhD Thesis, The Hong Kong University of Science and Technology, Hong Kong, China

Lo DOK (2000) Review of natural terrain landslide debris-resisting barrier design. Geotechnical Engineering Office, Hong Kong, SAR, China, p 91 (GEO report no. 104)

Ng CWW, Song D, Choi CE, Koo RCH, Kwan JSH (2016) A novel flexible barrier for landslide impact in centrifuge. Géotech Lett 6 (3):221–225

Ng CWW, Choi CE, Koo RCH, Goodwin GR, Song D, Kwan JSH (2018) Dry granular flow interaction with dual-barrier systems. Géotechnique 68(5):386–399

Ng CWW, Choi CE, Majeed U, Poudyal S, De Silva WARK (2019) Fundamental framework to design multiple rigid barriers for resisting debris flows. In: Proceedings of the 16th asian regional conference on soil mechanics and geotechnical engineering. 14th to 18th October 2019. Taipei, Taiwan, China

Ng CWW, Wang C, Choi CE, De Silva WARK, Poudyal S (2020) Effects of barrier deformability on load reduction and energy dissipation of granular flow impact. Comput Geotech 121:103445

NILIM (2007) Manual of technical standard for establishing Sabo master plan for debris flow and driftwood. Technical Note of NILIM No. 364. Natural Institute for Land and Infrastructure Management, Ministry of Land, Infrastructure and Transport, Japan. [In Japanese.]

Piton G, Recking A (2015) Design of sediment traps with open check dams. I: hydraulic and deposition processes. J Hydraul Eng 142 (2):04015045

Remaitre A, Malet JP, Maquaire O, Ancey C (2003) Study of a debris-flow event by coupling a geomorphological and a rheological investigation, example of the Faucon stream (Alpes-de-Haute-Provence, France), Debris-Flow Hazards Mitigation, Mechanics, Prediction, and Assessment, eds. Rickenmann and Chen. Millpress, Rotterdam, pp 375–385

Rickenmann D (1999) Empirical relationships for debris flows. Nat Hazards 19:47–77

Song D, Ng CWW, Choi CE, Zhou GGD, Kwan JSH, Koo RCH (2017) Influence of debris flow solid fraction on rigid barrier impact. Can Geotech J 54(10):1421–1434

Sze EHY, Lam HWK (2017) Some suggested detailing of flexible net barriers traversing a stream course for drainage purposes, GEO Technical Note TN 3/2017. Geotechnical Engineering Office, Hong Kong, SAR, China

Takahashi T (1991) Debris flows. IAHR Monograph, A. A. Balkema: Rotterdam

Tecca P, Genevois R, Deganutti A, Armento MC (2007) Numerical modelling of two debris-flows in the Dolomites (Northeastern Italian Alps), Debris-flow Hazards Mitigation, Mechanics, Prediction, and Assessment, Chen CL, Major JJ (eds). Millpress, Rotterdam, pp 179–188

Volkwein A (2014) Flexible debris flow barriers. Design and application. WSL Berichte, 18. Birmensdorf, Swiss Federal Institute for Forest, Snow and Landscape Research WSL

Wendeler C, McArdell BW, Rickenmann D, Volkwein A, Roth A, Denk M (2006) Field testing and numerical modeling of flexible debris flow barriers. In Proceedings of the 6th international conference on physical modelling in geotechnics, Hong Kong, pp 1573–1578

White DJ, Take WA, Bolton MD (2003) Soil deformation measurement using particle image velocimetry (PIV) and photogrammetry. Géotechnique 53(7):619–631

Zanuttigh B, Lamberti A (2006) Experimental analysis of the impact of dry avalanches on structures and implication for debris flows. J Hydraul Res 44(4):522–534

The Rockfall Failure Hazard Assessment: Summary and New Advances

Michel Jaboyedoff, Mariam Ben Hammouda, Marc-Henri Derron, Antoine Guérin, Didier Hantz, and François Noel

Abstract

The estimation of rockfall hazards is usually based only on hazards related to rockfall propagation. The rockfall failure hazard is not currently well defined, and only a few studies have truly addressed this topic. The basics of slope stability assessment are reviewed. Here, we propose a summary of the standard methods used to assess susceptibility to rock mass failure, mainly based on techniques from the mining industry or tunneling. Most of them are qualitative. Many susceptibility scales have been described. Due to computer power and the high-resolution topography in real 3D, topography analysis and standard kinematic tests have been adapted and improved to obtain rockfall susceptibility. Hazard assessments based on the power law are one of the best and only ways to obtain a real assessment of rockfall hazard failure; however, they present some drawbacks that must be solved. The most promising avenues of research for rockfall failure hazards are linked to rock mass strength degradation, which is currently observed using high-resolution 3D monitoring of cyclic deformations with hysteresis. These are the resulting movements caused by groundwater circulations, thermal cycles, earthquakes, rainfall, etc. In conclusion, the rockfall hazard will be improved by better understanding these processes in addition to the chemical weathering effect.

Keywords

Rockfall • Failure • Hazard • Degrading factor • Susceptibility

M. Jaboyedoff (✉)
Institute of Earth Sciences, University of Lausanne, Institute of Earth Sciences, Risk-Group, GEOPOLIS, 1015 Lausanne, Switzerland
e-mail: Michel.Jaboyedoff@unil.ch

M. Ben Hammouda · M.-H. Derron · A. Guérin · F. Noel
University of Lausanne, Institute of Earth Sciences, Risk-Group, GEOPOLIS, 1015 Lausanne, Switzerland
e-mail: Mariam.BenHammouda@unil.ch

M.-H. Derron
e-mail: Marc-Henri.Derron@unil.ch

A. Guérin
e-mail: guerina74200@gmail.com

F. Noel
e-mail: Francois.Noel@unil.ch

D. Hantz
Université Grenoble Alpes—ISTerre, 38400 Saint Martin D'Hère, France
e-mail: didier.hantz@univ-grenoble-alpes.fr

Introduction

Here, we use the term "rockfall failure hazard" to describe the failure hazard in relation to the volume of rocks that will lead to rockfall, which includes any type of mechanism such as slides, wedges, and toppling (Hungr et al. 2014). To date, rockfall failure hazards have not been studied in great detail. Most rockfall hazard studies are mainly based on rockfall trajectory modeling, arbitrarily fixing the failure frequency. Nevertheless, there are methods that are designed to provide an assessment the slope rock mass strength such as slope mass rating (SMR) (Romana 1988). Some recent studies, using new technologies such as light detection and ranging (LiDAR), structure from motion (SfM) photogrammetry, thermal imaging, passive seismic monitoring and InSAR, allow us to investigate the hazard (frequency-volume relationship) (Hantz 2011; Williams et al. 2018) loss of strengths of rock instability (Levy et al. 2010) and potential fatigue processes (Rouyet et al. 2017). Furthermore, the impact of rainfall, freezing and thaw cycles (D'Amato et al. 2016) or thermal effects (Collins and Stock 2016) on rock fatigue or rockfall triggering are now increasingly being studied. It is also clear that even for limited volumes, there can be some precursory movements leading to failure (Royán et al. 2013; Kromer et al. 2017).

K. Sassa et al. (eds.), *Understanding and Reducing Landslide Disaster Risk*, ICL Contribution to Landslide Disaster Risk Reduction,
https://doi.org/10.1007/978-3-030-60196-6_3

Here, we present the basic factors controlling the rock slope stability and examine the potential external factors that can lead to failure, especially the loss of strength. We start our review from the slope angle thresholds above which the rockfall is prone to observed strength degradation. The time dependence of this degradation is the ultimate objective of this research.

The Challenges of Rockfall Hazards

The main challenge is to improve the risk assessment of rockfall, which implies a better understanding of how rockfall hazards must be assessed.

The Rockfall Risk and Hazard

A rock instability (RI) is a potentially unstable rock compartment which is generally made up of several individual blocks of different volumes. The rockfall risk linked to several RIs on an object at risk W occupying the domain X can be written in a conceptual way (modified after Leroueil and Locat 1998; Jaboyedoff et al. 2001; Volkwein et al. 2011 Farvacque et al. 2019):

$$R(\mathrm{X}) = \sum_{RI} W(X) \times Exp(X) \int H(RI, \mathrm{X}, E(X)) \times V(E(X), W(X)) dE \quad (1)$$

where the unit of risk is a cost per year or the number of causalities per year considering several RIs and the object at risk:

- W(X) the object-at-risk's value or number of people located in the domain X, which is located in the potential impacted area;
- V(E(X), W(X)) is the vulnerability of the object (W(X)) at risk or the lethality of the person to the specific block intensity, i.e. energy E, at location X;
- Exp(X) is the exposure to a hazard, which means the presence of the object or a person in the domain X with value ranging from 0 to 1;
- H(RI,X,E(X)) dE is the incremental frequency for a range of intensity (energy) dE of the blocks coming from instability RI in domain X.

The rockfall hazard in domain X can be decomposed into two terms (Fig. 1) (Leroueil and Locat 1998; Jaboyedoff et al. 2001; Volkwein et al. 2011):

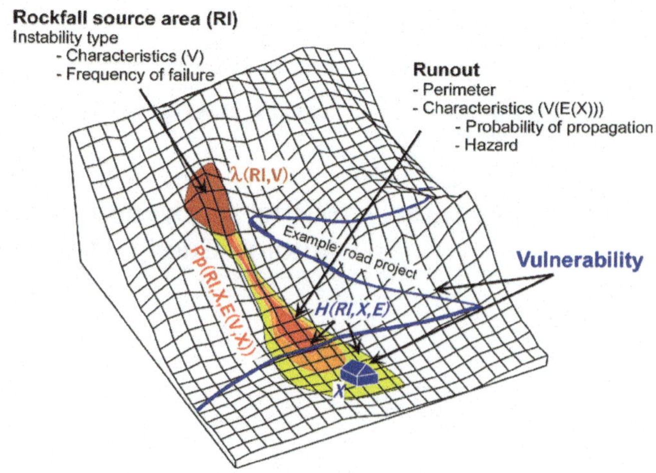

Fig. 1 The different parameters of rockfall hazards

$$H(RI, \mathrm{X}, E(X)) = \int \lambda(RI, V) \times Pp(RI, \mathrm{X}, E(V, X)|V) dV \quad (2)$$

where λ(RI) is the frequency of the failure (= temporal frequency) of the blocks for a given range of volumes dV coming off from the RI. Pp(RI, X, E(V, X)|V) is the probability of propagation of one RI for a given range of volumes dV providing the probability for a given energy density E by all the blocks reaching X. Such formalism can be more refined, but these formulas provide a good framework to tackle risks and hazards. The present paper is related only to the assessment of λ(RI, V).

Hazard Versus Susceptibility

The hazard, as indicated above, is deduced from a quantitative approach, providing the temporal frequency of hazard of a given intensity. The relationship intensity-frequency is an example of such quantification. However, formally, it is the intensity at a given location. For rockfalls, it is kinetic energy. For the rockfall source, the hazard is a temporal frequency depending on volume.

The susceptibility is based on an empirical scale of the likelihood of a danger at a given location based on the rating of the predisposing factors and summarizing them in a relative hazard scale (Ferrari et al. 2016; Fell et al. 2008). It is often used for qualifying the rockfall failure hazard of the sources based on various choices of internal parameters (IPs) and external factors (EFs). For instance, the SMR provides a preliminary assessment of slope stability (Romana, 1988).

Site-Specific or Regional Assessment

The hazard assessment process is very different if it is dedicated to a full slope or a specific site (localized hazard). Regionally, it is based on average knowledge and statistics, while for specific instability, the main mechanisms of instability (Fig. 2) can be deduced, and more detailed investigations can be performed based on instability geometry, mechanical parameters tests, calculations, movements monitoring, etc. Here, we consider both types.

Elements that Control the Rockfall Failure Hazard

The evolution with time of a slope system's stability can be described in terms of internal parameters (IPs) and external factors (EFs) (Fig. 3). The IPs can be considered as functions that evolve under the effect of the EFs. To ideally characterize a potential instability, we need to identify (Volkwein et al. 2011): (1) the pre-failure processes and (2) the areas sensitive to rapid strength degradation leading to slope failure (Jaboyedoff et al. 2005; Leroueil and Locat, 1998). IPs can be summarized as proposed by Volkwein et al. (2011):

(a) Morphology: slope types (e.g. slope angle, height of slope, profile), exposure, type of relief (depending on the controlling erosive processes), etc.

(b) Geology: rock types and weathering, variability of the geological structure, bedding, type of deposit, folded zone, etc.

(c) Fracturing: joint sets, trace lengths, spacing, fracturing intensity, etc.

(d) Mechanical properties of rocks and soil: cohesion, friction angle, etc.

(e) Activity: movements or rockfall, etc.

(f) Hydrogeology: permeability, joint permeability, etc.

It must be emphasized that the joint sets or discontinuities are assumed to be the main features controlling the stability (Hoek and Bray, 1981), which is supported by the fact that more intensely fractured rock produces more rockfall events (Coe and Harp, 2007). As stated in Volkwein et al. (2011), the EF actions on IPs are mainly:

- gravitational effects;
- water circulation: hydrology or hydrogeology, climate, precipitation in the form of rainfall or snow, infiltration rates, groundwater;
- weathering;
- erosion;
- seismicity;
- active tectonics;
- freezing and thawing, permafrost, which are increasingly invoked to explain rockfall activities (Frayssines and Hantz, 2006; Matsuoka and Sakai, 1999; Matsuoka, 2008; Gruner, 2008);
- heat and cooling cycles by sun exposure (Collins and Stock, 2016),
- rainfall impact (D'Amato et al. 2016);
- nearby instabilities;
- human activities (anthropogenic factors);
- etc.

These non-exhaustive lists of IPs and EFs include the key parameters used by the existing methods used to assess the hazard or susceptibility to rockfall failure.

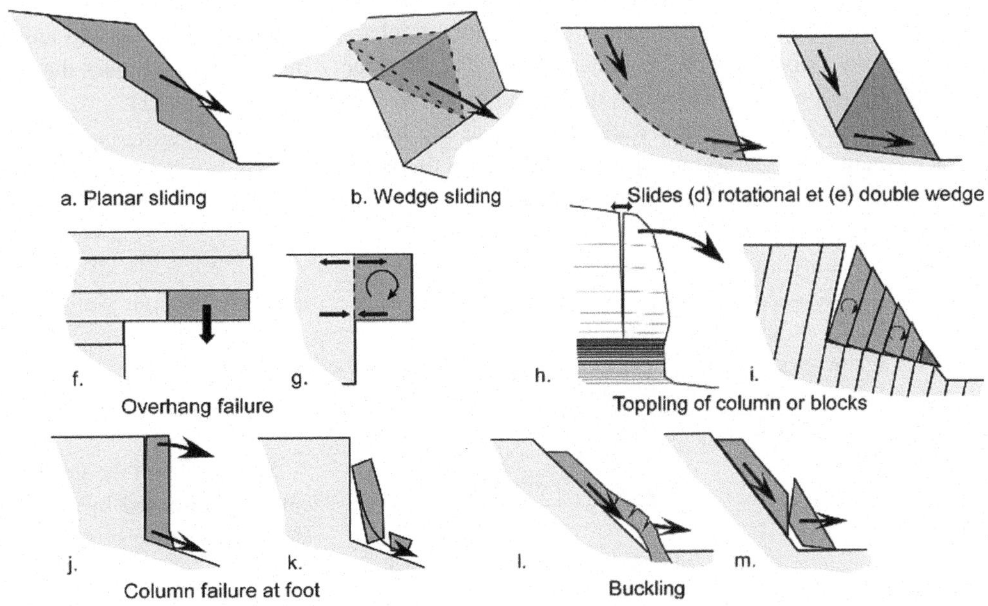

Fig. 2 Main rockfall mechanisms (adapted from: Hantz et al. 2003a)

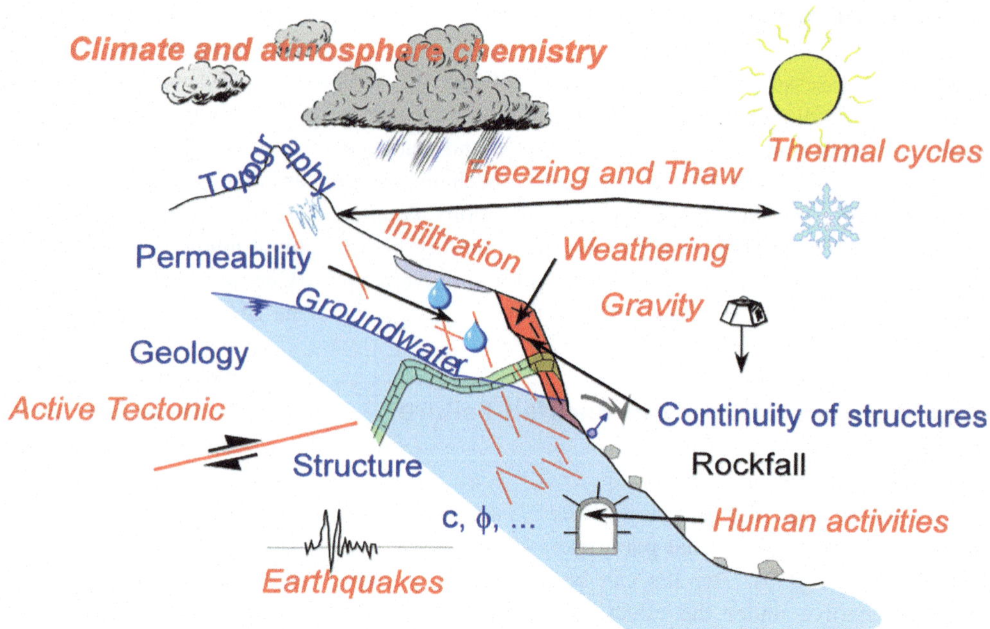

Fig. 3 External factors and intrinsic parameters for rockfall stability and triggers (after Volkwein et al. 2011)

Geotechnical Basics

To illustrate the parameters and factors controlling rock instabilities, the simple model of a sliding block permits us to understand the issues linked to the stability of rock slopes (Fig. 4). The factor of safety of a sliding block submitted to pore water pressure and seismic acceleration is given by (Wyllie, 2018):

$$FS = \frac{cA + (W(\cos\alpha - k_h \sin\alpha) - U - V\sin\alpha)\tan\phi}{(W(\sin\alpha + k_h\cos\alpha) + V\cos\alpha)} \quad (3)$$

where c is the cohesion, ϕ is the friction angle, k_h is the ratio of the horizontal seismic acceleration to the earth acceleration g, α is the slope angle of the planar failure, β is the slope angle of the slope, W is the weight of the sliding mass, U is the water force caused by the water pressure at the failure surface A, and V is the horizontal force applied by the water at the back of the instability within the back crack.

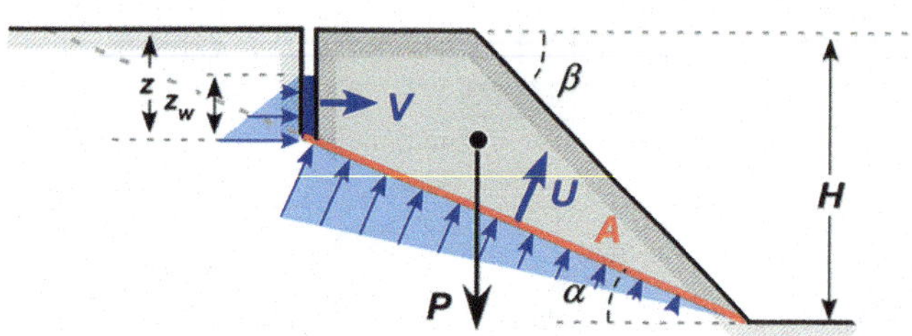

Fig. 4 Illustration of an idealized rockslide including back crack and pore water pressure (after Hoek and Bray, 1981)

This model clearly shows that if ϕ or c decreases, the stability is diminished. The increase in water pressure and seismic acceleration are factors that destabilize the slope. However, all EFs can influence the stability, and we can see that modifying the geometric properties of the rock mass ground may lead to important changes in the safety factors. Hoek and Bray (1981) proposed creating a graph that illustrates the relationship of a slope height with its slope angle based on the previous Eq. 3, assuming a dry slope, and that the back crack with a depth z is located to minimize the stability:

$$\frac{z}{H} = 1 - \sqrt{\frac{\tan\alpha}{\tan\beta}} \quad (4)$$

In addition, the dip angle of the failure surface follows a plane that maximises the cohesion assuming a fixed ϕ (Hoek and Bray, 1981):

$$\alpha = \frac{1}{2}(\beta + \phi) \quad (5)$$

inserting the unit weight of the rock γ, it leads to:

$$H = \frac{2c}{\gamma\left(FS - \frac{\tan\phi}{\tan\alpha}\right)} \times \frac{\left(1 - \left(\frac{z_c}{H}\right)\right)}{\sin^2\alpha\left(\left(1 - \left(\frac{z_c}{H}\right)^2\right)\frac{1}{\tan\alpha} - \frac{1}{\tan\beta}\right)} \quad (6)$$

Starting from the geometry from Fig. 5b, the slope height can be expressed by minimizing the stability regarding the slope angle α to obtain the Culman model (Taylor 1948):

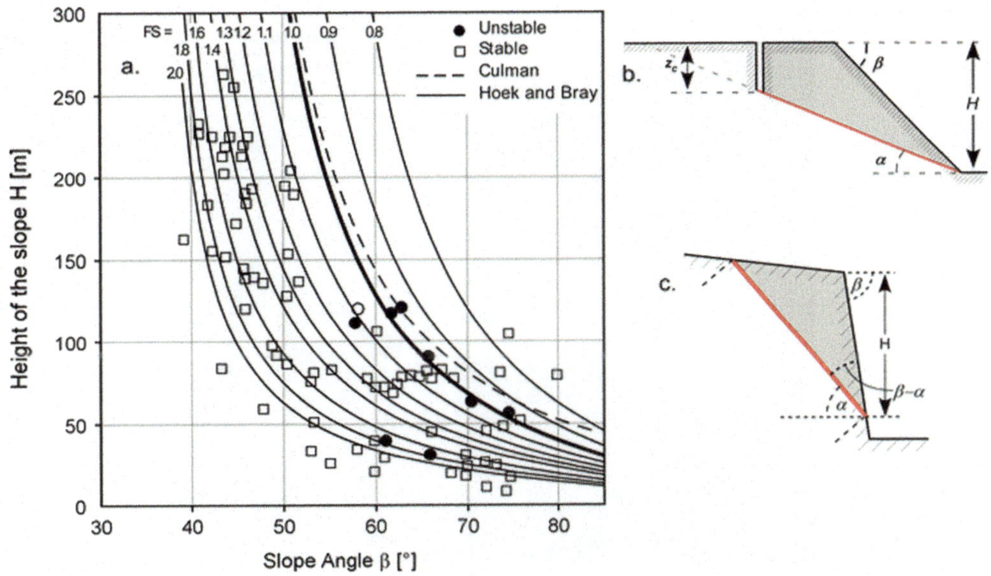

Fig. 5 **a** Stability relationship between the slope angle β of a slope and its height H for the slopes of the Rio Tinto mine (after Hoek and Bray, 1981) with $\phi = 35°$, c = 137.3 kPa, and rock mass density $\gamma = 2950$ kg/m^3. **b** Hoek and Bray model, **c** Culman model (modified from Jaboyedoff et al. 2021)

$$eqalignH = \frac{2c}{\gamma}\frac{\cos\phi\sin\beta}{\sin\left(\beta-\frac{\alpha+\beta}{2}\right)\sin\left(\frac{\alpha+\beta}{2}-\phi\right)} = \frac{4c}{\gamma}\frac{\cos\phi\sin\beta}{(1-\cos(\beta-\phi))} \quad (7)$$

These relationships permit to back-analyze cases in a simple way. This process demonstrates the strong dependence of stability on slope height for steep slopes. Figure 5a also shows that unstable slopes can have a very high factor of safety, demonstrating that c and ϕ are either evolving under EFs or have a high variability and that no groundwater is considered in this simple model. There are other types of back-analyses that provide apparent cohesion of discontinuities, knowing the dip angle of the failure plane, the friction angle and the height of the slope (Locat et al. 2000).

Furthermore, it is clear from Eq. 3 that groundwater can have a strong external effect, reducing the factor of safety by more than 30%. The earthquake effect can be assessed for sliding by a simple model based on the critical acceleration (Wilson and Keefer, 1985), and for specific cases, methods used to assess the failure hazard are often based on the Newmark (1965) method.

Field Surveys

The basic tool used to assess rockfall source hazard is based on field surveys. Experienced professionals can easily identify rocky compartments looking at a rock wall. Often, field observations are the basis for many of the methods used to qualify hazard or susceptibility.

In many cases, detailed mapping is required to assess the stability of the source areas, such as for the "gorge du Tarn" in France, where a sub-horizontal carbonate platform is incised by a deep gorge (Pauly and Payany, 2002). The rock mass is fractured, and high rock columns are created by regional joint sets, which are not simple to characterize as unstable (Fig. 6). The destabilization seems to be often created by "thermally induced wedging" (Bakun-Mazor et al. 2013). The stability of such columns and the probability of failure are very difficult to assess without an expert approach, which demonstrates that field work and conceptual models are of primary importance in such cases.

The identification of special geological and mechanical settings requires a detailed field survey to identify the location prone to rockfall initiation, such as the influence of folding on rockfall susceptibility. Coe and Harp (2007) demonstrated using both field observations and the rock mass quality (Q) index (Barton et al. 1974; Harp and Noble, 1993), in which the formation of folds decreases the friction angle of flexural slip along discontinuities by smoothing the roughness. In addition, the hinges have been shown to be more susceptible to rockfall than limbs because they are more fractured.

Field surveys are also important for characterizing the rockfall volume delineated by discontinuities (Agliardi and Crosta, 2003).

Fig. 6 Columnar instability from the Tarn Valley (France). **a** Column cut by discontinuities and rounded by dissolution. **b** Block wedging induced by thermal cycles. **c** Scheme of the progressive destabilization by vegetation growth in the back crack and thermal wedging

3D Techniques

Major advances in recent years for rockfall source characterization include the high-resolution DEM (digital elevation model) and 3D cloud points, which are produced by LiDAR (Light Detection and Ranging) (Tonon and Kottenstette, 2006; Abellán et al. 2014) and SfM (structure from motion) (Tonon and Kottenstette, 2006; Kromer et al. 2019). The LiDAR or laser scanner can be static or installed on vehicles, boats, drones or planes, and there are also handheld versions (Jaboyedoff and Derron, 2020). SfM is based on photogrammetric techniques using several images with large overlaps. It can be based on any type of cameras or pre-existing pictures (Guerin et al. 2020a). It allows structural characterization (Slob et al. 2002; Sturzenegger and Stead, 2009; Jaboyedoff et al. 2007; Gigli and Casagli, 2011; Abellán et al. 2014) and monitoring (Royan et al. 2014; Kromer et al. 2017; 2019). It can now be considered a basic tool for any study of rock outcrops (virtual outcrop), especially because SfM methods are very cheap. In addition, currently, the accessibility to very high-resolution devices ($\sim$ 50 μm) that can capture microtopography may have a strong impact on the rock joint roughness characterization (Mah et al. 2013).

Geometrical Methods

As mentioned, the first aspect of rock slope stability is linked to the presence of a steep slope, which is the primary factor influencing rockfall initiation. Another important aspect of rock slope hazards is to quantify volumes and to characterize their geometry, which is important for frequency estimation and/or instability mechanism understanding.

Using Slope

Strahler (1954) showed that the slope angle of uniform lithology follows a Gaussian distribution based on the digital elevation model (DEM). Following this idea, Rouiller et al. (1997) proposed decomposing the histogram of slope angles in several Gaussian distributions (Fig. 7) and attributing them a geomorphological meaning (Loye et al. 2009). Identifying the population of the steep slope, i.e. cliffs, it allows us to select the limit angle for which this population dominates, which is chosen as the limit to detect the zone prone to rockfall initiation. This appears to also work for DEMs with low resolution (25 m), as demonstrated in the Saasthal in the Swiss Alps, where most scars of past and present rock instabilities are located within slopes with angles higher than the defined threshold of 50° (Fig. 8). It is also extremely efficient for high-resolution DEMs (1 m) to create geomorphic maps (Loye et al. 2009). It has been used by several authors (Crosta et al. 2015; Corona et al. 2013; Lopez-Saez et al. 2016; Losasso et al. 2017; Farvacque et al. 2019). The extracted distribution was used to create a susceptibility scale for rock slope failure (Michoud et al. 2012). Fernandez-Hernandez et al. (2012) used this method to define a limiting slope of 50°, below which their susceptibility index I_s was null. They also added other criteria (see below).

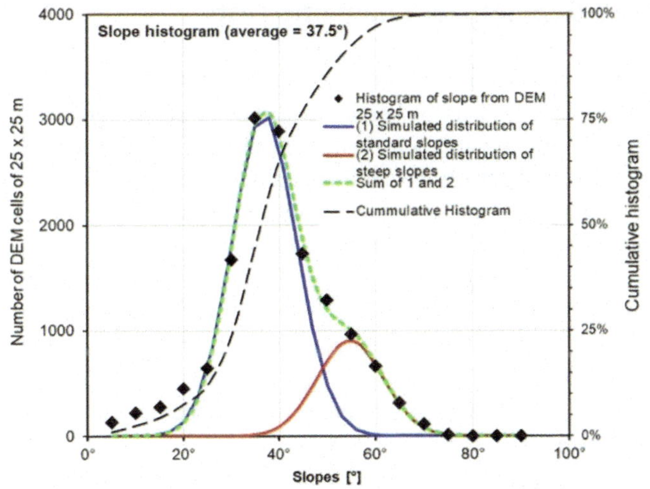

Fig. 7 Decomposition of the slope angle histogram extracted from a 25 m DEM of areas where scars of rock instabilities and present rock instabilities were mapped. Two main Gaussian slope angle distributions are identified; the rock slopes are assumed to be dominant above 50°

Empirical Approaches

There are very simple ways to create susceptibility maps based on geographic information systems (GIS), which permit the development of heuristic techniques (Van Westen 2006). This process consists of giving weights to spatialize IPs or EFs and aggregating them by summing or multiplying them to obtain a final weight. This process can be used for any type of IP or EF. Such susceptibility often depends on the available data and the objective of the study. GIS and related software permit the use of any type of map. For instance, in Switzerland, the 1:25,000 vectorized topographic maps include polygons of the cliff areas (Jaboyedoff and Labiouse, 2003; Loye et al. 2009).

Baillifard et al. (2003) developed a susceptibility (S_{ij}) that simply sums the number of criteria (k) present in raster maps for each pixel ij to detect the rock instability above the roads, assuming that $r_{ijk} = 0$ is null if the criterion is not reached and $r_{ijk} = 1$ if otherwise:

Fig. 8 Location of present instability (•) scars of past rockslides (□), which are mostly initiated in areas with slope angles higher than 50°

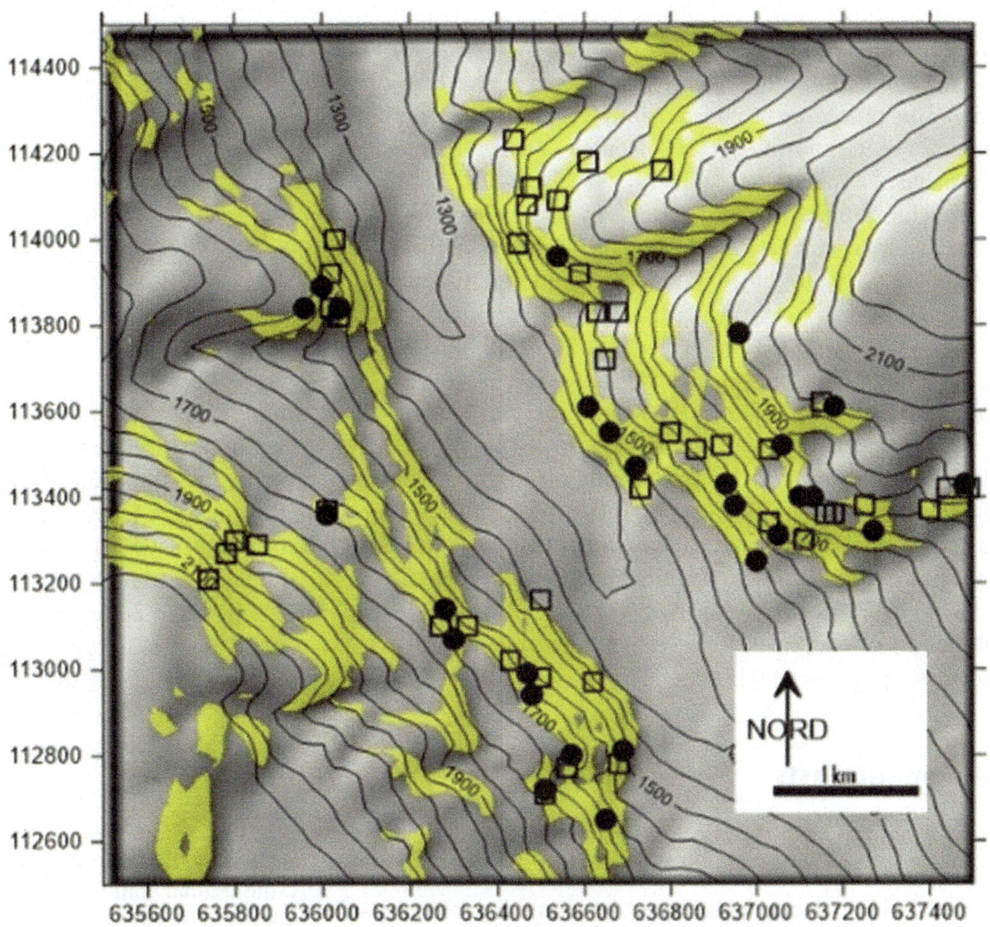

$$S_{ij} = \sum_{k=1}^{5} r_{ijk} \tag{8}$$

Five different maps were used: (1) cliff extracted from digitized topographic maps, (2) steep slope > 45° extracted from a raster at 25 m, (3) a scree slope with a buffer of 100 m, (4) a buffer of 150 m around important faults, and (5) the slope located above a road. The results showed that the past instabilities along roads in Valais (Switzerland) possessed a susceptibility $\geq$ 3. Saroglou (2019) applied the same strategy at the nationwide level of Greece but with a score for each criterion going from 3 to 1 for the highest susceptibility. A lower number represents a higher susceptibility. This rockfall susceptibility index (RSI) for pixel ij is given by:

$$RSI_{ij} = L_{ij} + R_{ij} + E_{ij} + F_{ij} \tag{9}$$

where L_{ij} is the class of the lithology, R_{ij} is the annual rainfall, E_{ij} is the earthquake impact and F_{ij} is the presence of a fault. Only five rockfalls among 43 were ranked within pixels with a high RSI. The identical approach was applied on El Hierro volcanic island (Canary Archipelago) assuming a susceptibility index calculated summing five parameters within the slope steepest than 50°: slope angle, topographic profile curvature, lithology, vegetation cover and dike density. This procedure provided mainly very high susceptibility indexes for known rockfall source areas (Fernandez-Hernández et al. 2012).

Such examples demonstrate that by using simple rules, it is possible to obtain a first screening of the instability. The major issue is to obtain a high-resolution DEM that is optimal.

Susceptibility to Earthquakes

Keefer (1993) proposed a very simple method to assess the failure of rock slopes induced by earthquakes. It is based on a dichotomous decision tree, allowing us to assess the hazard of failure (Fig. 9). Some additional parameters can be considered, such as the presence of vegetation that can reduce the effect of earthquakes for M > 6.5 and groundwater that increases the susceptibility.

Kinematic Tests

Traditionally, rock mechanics uses kinematic tests using stereonets (Hoek and Bray, 1981), looking at the average orientation of slopes and testing the feasibility of a set of discontinuities to slide on planes, wedges or topples, adding some criteria linked to the basal friction angle.

Pure Geometrical Kinematic Tests

Kinematic tests are usually used to verify if a mechanism is possible. It is usually based on a unique slope orientation representing the geometry within stereonets (Hoek and Bray, 1981). Since DEMs became available, kinematic tests have been adapted in computer codes (Wagner et al. 1988; Gokceoglu et al. 2000; Jaboyedoff et al. 2004a).

Often, kinematic tests include friction angles and empirical limitations in addition to geometrical criteria, but this can be an issue because these tests do not integrate the effects of water and joint strength degradation. As a consequence, as a first approximation, the mechanical properties are not used.

As a basic hypothesis, the most important parameter is the density of dangerous structures that intersect the topography. For that purpose, it is necessary to know the mean spacing L for the discontinuities and their mean trace length T, their orientations and the slope face orientation for each DEM cell or pixel (Fig. 10). Then, the average number of structures per cell can be calculated.

For potential planar failures, the direction of sliding must point out of the slope, i.e. the slope angle of discontinuity α_i must be lower than the apparent slope angle of cell β_c in the sliding direction ω_i, $\alpha_i < \beta_c$, and a geometrical restriction of the direction of sliding can be added (Hoek and Bray, 1981):

$$\omega_i - 20^\circ < \omega_c < \omega_i + 20^\circ \tag{10}$$

where ω_c is the dip direction or aspect of the cell. The number of potential planar failures N_p produced by a joint set is given by (Jaboyedoff et al. 2004a; Matasci et al. 2015):

$$N_p = \frac{A_c \sin\delta}{LT} \tag{11}$$

where A_c is the surface area of the topographic face of a cell given by the horizontal surface of the cell divided by cos β, which is the slope angle cosines, L the mean spacing and T the mean trace length of the discontinuities. δ is the angle between the normal vectors of the local slope and the discontinuity set. If N_p is lower than 1, it means that the instability size is larger than the cell size, i.e. the potential rockfall sources frequency is lower but the volume is larger. For N_t, the number of topplings, the normal must point downward in the slope $(90 - \alpha_i \leq \beta_c)$ with an additional condition that can be applied:

$$\omega_i + 180^\circ - 20^\circ < \omega_c < \omega_i + 180^\circ + 20 \tag{12}$$

The directions of wedge failure have the same restrictions as the planar failure replacing the dip direction and the discontinuity slope angle by the wedge direction of sliding (Fig. 11). If γ is the angle between the two joint sets with mean spacings L_i and L_j, and δ is the angle between the

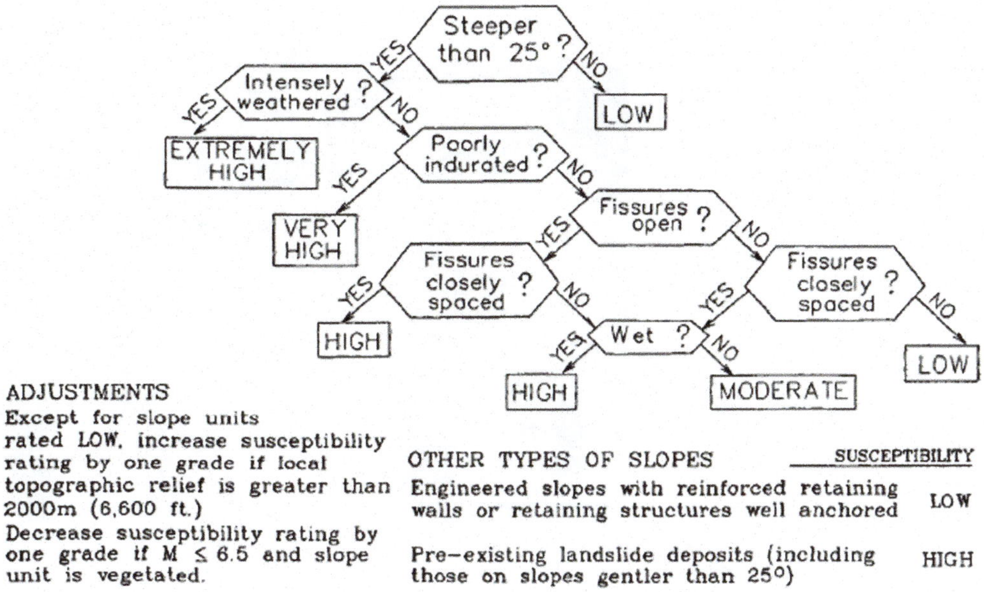

Fig. 9 Decision tree for susceptibility of rock slope failure (From Keefer, 1993)

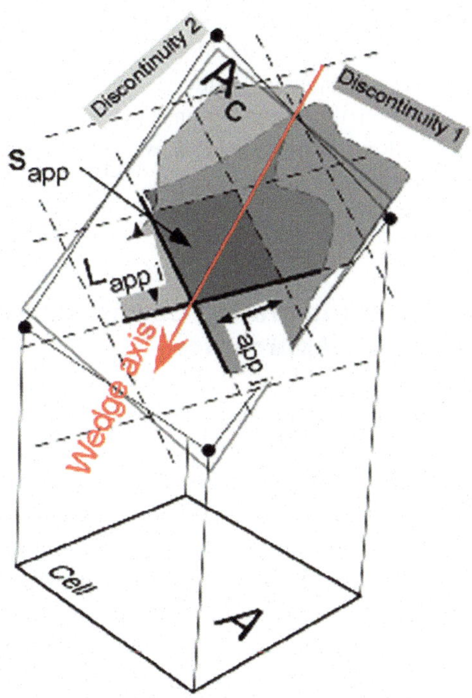

Fig. 10 Illustration of the procedure of counting the number of wedges within a cell of horizontal surface A and topographic surface area A_c of a cell. $L_{app\ i}$ and $L_{app\ j}$ are the apparent spacing of the discontinuity on the topographic surface

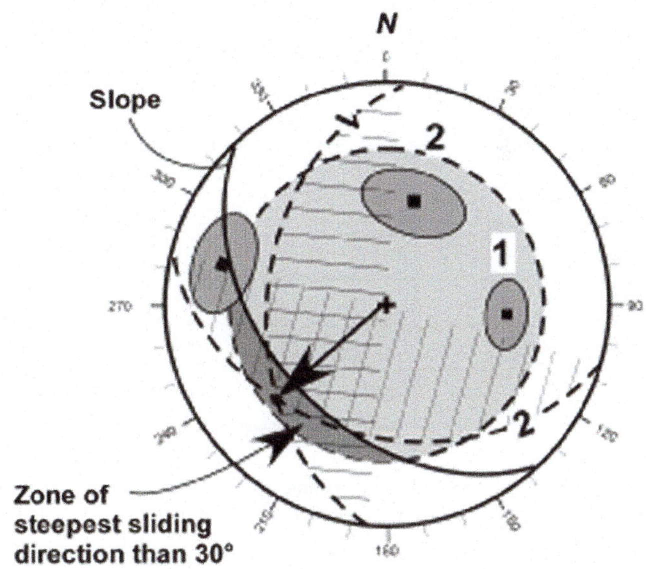

Fig. 11 Synthetic example of kinematic test for wedges, where the averages of discontinuities 1 and 2 are represented. They have a sliding dip larger than $\phi = 30°$. The gray ellipses represent the zones of variability of the directions of the poles of the discontinuities

wedge axis and the normal to the topography, then the number of wedges is given by (Matasci et al. 2015):

$$N_w = \frac{A_c \cos\delta \sin\gamma}{L_i \times L_j} \quad (13)$$

Using these three numbers (N_p, N_t and N_w), it is possible to obtain a susceptibility index S_c to rockfall failure in each cell of the DEM (Matasci et al. 2015a):

$$S_c = \sum N_p + \sum N_t + \sum N_w \quad (14)$$

Matasci et al. (2015a) were able to compare the actual rockfall activity with the estimated susceptibility (Figs. 12 and 13). This was done within small catchments in Ticino (Switzerland), where the rockfall was trapped and the total

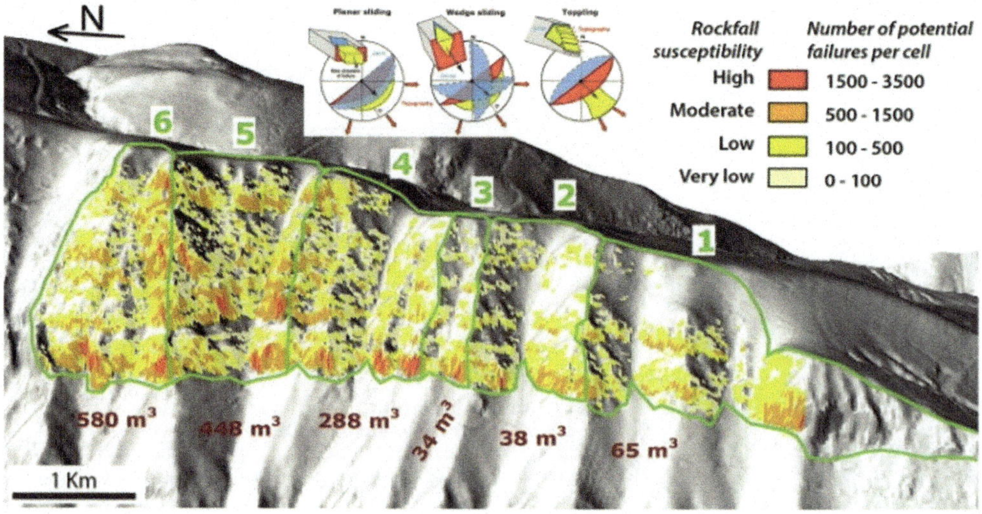

Fig. 12 Rockfall susceptibility map of the western side Monte Generoso (Ticino; Switzerland) draped on a 3D DEM, indicating the number of potential failures S_c and the volumes of the blocks trapped in the protective barriers for each catchment (from Matasci et al. 2015a; DEM, 2 m cell size, Swisstopo)

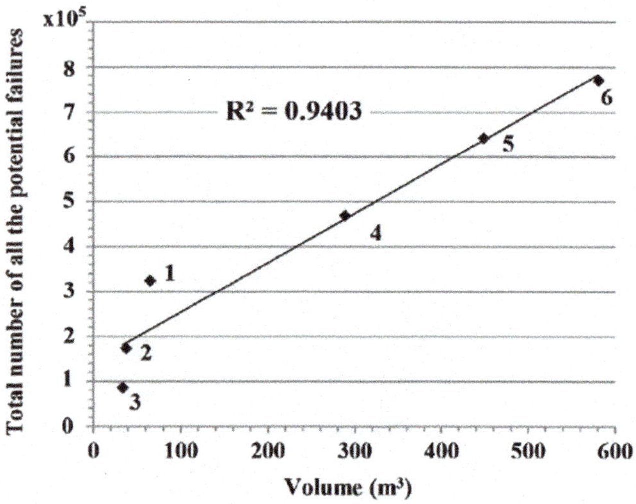

Fig. 13 Relationships of the volume of blocks trapped in the protective barriers for each catchment and the total number of potential failures (from Matasci et al. 2015a)

volume was measured for a period. The results showed the total number of potential structures within six different catchments:

$$S_{tot} = \sum S_c \tag{15}$$

And was well-correlated with the observed rockfall activity within the catchment. This technique was more efficient than the estimations based on the catchment surface areas or the GSI (Hoek, 1994).

Kinematic Tests Integrating Stress Tensor

Günther (2003) developed the kinematic test software SLOPEMAP, considering the outcropping of geological structures. It builds a digital structural model (DSM) based on field data and uses an interpolator that can include bedding, folds and joint sets (Günther et al. 2004). Assuming a vertical maximum stress in the slope within the cell of the raster, the stress tensor is deduced, and the stability is computed to detect the potential unstable zones. A module integrating shallow groundwater flow using similar techniques as shallow landslide techniques is also implemented (Günther et al. 2004).

Kinematic Tests Integrating Geomechanics and Probabilistic Approaches

As proposed by Hoek and Bray (1981) and Wyllie (2018), kinematic tests can include geo-mechanical properties. The basal friction angle ϕ for slides must be less steep than the dip angle of the sliding direction of planar or wedges (Fig. 11). The friction angle can also be introduced for toppling.

Several authors developed a probabilistic approach by assuming probability distribution functions for the parameters used to calculate the probability of failure, which corresponds to the probability P_{ff}, to obtain a safety factor below 1 for planar failure, wedges or topple (Scavia et al. 1990; Carter and Lajtai, 1992).

Using simulations of variable wedges, Park and West (2001) proposed decomposing the probability of failure P_f in two terms:

$$P_f = P_{fk} \times P_{ff} = \frac{N_k}{N_T} \times \frac{N_f}{N_k} = \frac{N_f}{N_T} \tag{16}$$

where N_k is the number of possible kinematic tests that are positive, N_T is the total number of tests, P_{fk} is the probability that the movement is mechanically possible and N_f is the total number of cases with a factor of safety below 1 among N_k, which corresponds to the probability P_{ff}.

Gokceoglu et al. (2000) introduced not only the friction angle in the kinematic test but also the probabilistic approach for the discontinuity orientations. They identified 3 joint sets in the Altindag (Turkey) and estimated their spatial variability. By simulations for each pixel of the DEM, they produced a probability of failure, P_{fk}, by simulating millions of kinematic tests based on the probabilistic approach of stability (Fig. 14).

Grenon and Hadjigeorgiou (2008) proposed a full Monte Carlo simulation of wedge stability based on an idealized slope. A discrete fracture network (DFN) model was created using the Veneziano model (Dershowitz and Einstein, 1988), and P_{fk} is estimated by evaluating the intersection of the fracture network with the slope (Fig. 15). The average trace lengths, the average spacings and the average orientations and the variances of 5 sets of discontinuities were measured in the field. They assumed that the spacings followed exponential negative distributions and that the orientations were simulated using a Fisher distribution. The results showed that the combination of the different couples of discontinuities did not have the same probability of failure but that it is necessary to use a large number of simulations to explore all possible intersections with topography.

Fig. 14 Map of the probability of wedge failure in Altindag (Turkey) with ϕ = 35° and 3 sets of discontinuities including their variabilities. (Modified after Gokceoglu et al. 2000)

Kinematic Test in Real 3D

Kinematic tests, including mechanics, are now used for cloud points, which means that real 3D topography is considered, including overhangs. Gigli et al. (2014) developed a tool to use point clouds and created a kinematic index (KI), which provides a probability of failure similar to Gokceoglu et al. (2000) but in 3D.

Matasci et al. (2018) modified the susceptibility developed for raster, assuming a density of discontinuities per unit area of the point cloud A = 1 m^2 and δ the angle of the normal vectors of the discontinuity and the topography.. The susceptibility for the planar joint i becomes:

$$S_{pi} = \frac{A \sin\delta}{L_i T_i} \tan \alpha_i \tag{17}$$

where α_i is the slope angle of the discontinuity, which comes from the inverse of the factor of safety for a planar surface FS = tanϕ/tan α. The number of wedges is given by (see eq. 13):

$$N_{wij} = \frac{A \cos \delta \sin \gamma}{L_i \times L_j} \tag{18}$$

However, the probability P that the two sets can intersect to form a wedge corresponds to the product of the ratios of the trace length T and the apparent spacing:

$$Lapp_i = \frac{L_i}{\sin \gamma} \tag{19}$$

$$\mathrm{P}_{ij} = \frac{T_i}{Lapp_j} \times \frac{T_j}{Lapp_i} \tag{20}$$

The wedge stability also depends on the factor K (Hoek and Bray, 1981), which considers the effect of the angle between joint sets:

$$K = \frac{\sin \theta_i + \sin \theta_j}{\sin(\theta_i + \theta_j)} \tag{21}$$

where θ is the slope of the joints in the plane perpendicular to the direction of sliding taken from the horizontal line. For wedge failure, the wedge susceptibility is inversely proportional to the safety factor, excluding the dependence on the friction angle:

$$S_{wij} = N_{wij} P_{ij} \frac{\tan \alpha_{ij}}{K} \tag{22}$$

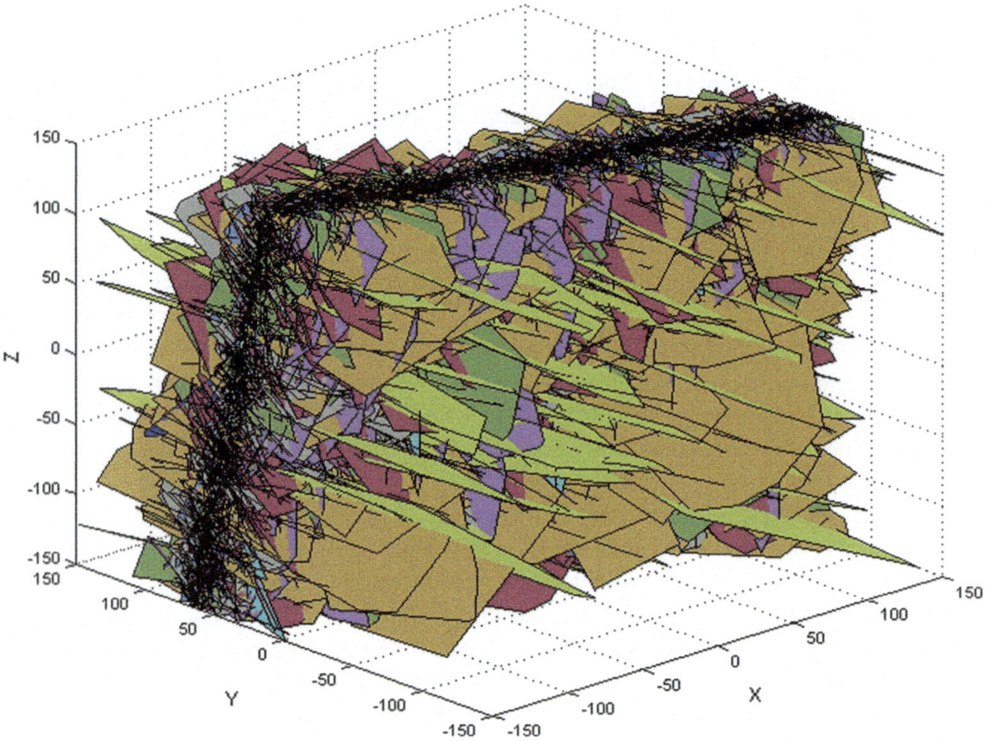

Fig. 15 Example of fractures in the DFN intercepting a plane-oriented (scale in meters). The simulations show that most wedges are small, i.e. ninety-nine percent of the 76,079 wedges are smaller than 155 m^3 in volume (modified from Grenon et al. 2014)

where α_{ij} is the dip angle of the intersection of the joints. As the kinematic test for toppling is very similar to the planar failure, if we consider the pole of the joint sets instead of the sliding direction, we obtain:

$$S_{tij} = \frac{\text{A} \sin\delta}{L_i T_i} \tan(90^\circ - \alpha) \tag{23}$$

When the slope is an overhang, then (23) is replaced by:

$$S_{tij} = \frac{\text{A}}{L_i T_i} \frac{1}{\tan\delta} \tag{24}$$

The lateral direction limitations of sliding are mostly identical to the standard one (see above). The final susceptibility per unit area at a given point of the 3D surface is given by.

$$S_{tot} = \sum S_{pi} + \sum S_{tij} + \sum S_{wij} \tag{25}$$

The susceptibility scale must be calibrated. It provides relevant results, demonstrating the potential of this method. The rockface of Les Drus (Mont-Blanc massif) shows good agreement with the observations as well as in the Yosemite Valley (Figs. 16 and 17). The number of instabilities increases with susceptibility.

Power Law and Inventories

Inventories of rockfalls are fundamental to obtaining access to frequencies of the events related to the magnitude, i.e. volumes in the case of rockfall sources. In addition, it is the basic tool used to identify the dependence of rockfall events, on EFs and IPs. There is an inventory of scars, fallen blocks or potential instabilities.

Volume Power Laws

Wieczorek et al. (1995) showed that the rockfall volume sources in Yosemite Valley followed a power law, which was confirmed by a recent study (Guerin et al. 2020a), and Hungr et al. (1999) showed it along a road. The inventories of volume of events are based on field analyses, and currently, by an analysis of the differences between point clouds acquired either by laser scanner or SfM (Fig. 18). Dussauge et al. (2003) gave the frequency of failure of volume larger than V as:

$$\text{N}(v > V) = \frac{N_0}{\Delta t} \left(\frac{\text{V}}{V_0}\right)^{-b} = aV^{-b} \tag{26}$$

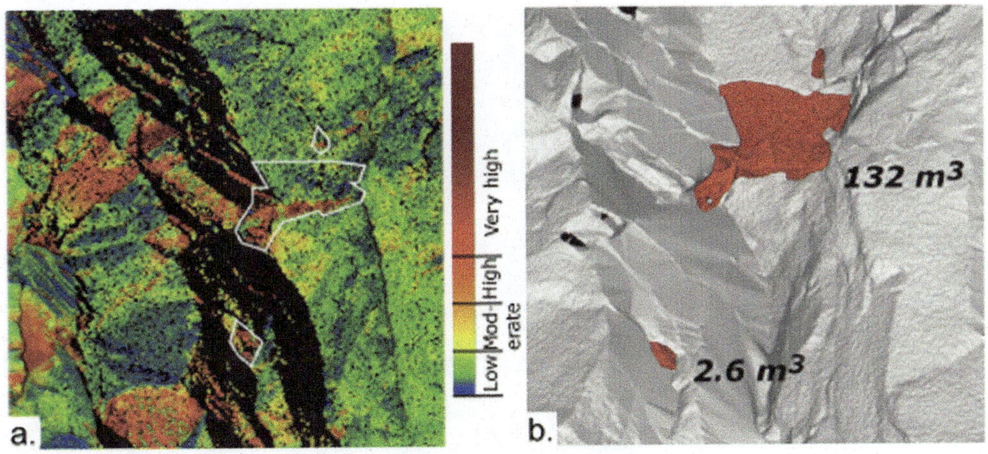

Fig. 16 Rockfall susceptibility **a** compared to the activity **b** between November 2011 and September 2014 in the Drus rock face (modified from Matasci et la. 2018)

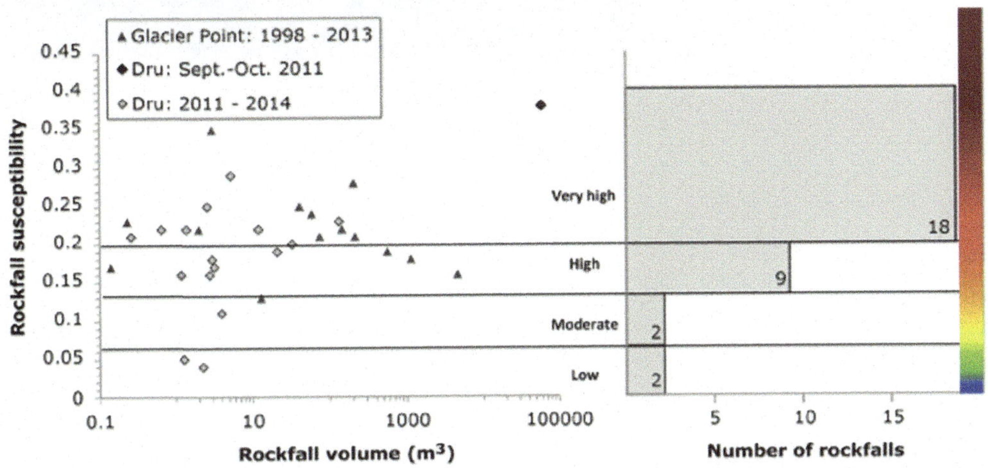

Fig. 17 Plot of the total rockfall susceptibility (S_{tot}) for past known events from the face of the Drus (Mont-Blanc, France) and in the Glacier Point cliff (Yosemite Valley) (modified from Matasci et al. 2018)

where N_0 is the number of failures larger than V_0 that occurred during period Δt and b is the exponent deduced from the inventory. In addition, the return period τ is given by:

$$\tau(v > V) = \frac{1}{N(v > V)} \quad (27)$$

This method allows us to assess rockfall hazards (Hungr et al. 1999; Dussauge-Peisser et al. 2002; Hantz, 2011). In addition, assuming a random process, it allows us to calculate the probability that n instabilities of volume larger than V occur during a period of time T, with $\lambda = 1/\tau$, is given by (Hantz et al. 2003a):

$$P(n, \Delta t) = \frac{(\lambda T)^n}{n!} e^{-\lambda T} \quad (28)$$

It must be reminded that the probability that at least one event occurs during T, which is given by (Hantz et al. 2003a):

$$P(n > 0) = 1 - e^{-\lambda T} \quad (29)$$

In addition, the probability and the frequency are equivalent, i.e. $P(n > 0, T) \approx \lambda T$, only if $\lambda T << 1$. This type of approach provides the quantification of cliff erosion retreat rate by integrating (modified after Hantz et al. 2003b):

$$V_{tot} = \int v dN = \int v(-abv^{-b-1}) dv = \frac{-ab}{(1-b)} v^{1-b} \Big|_{V_0}^{V_m} \quad (30)$$

It provides the volume per unit of time knowing the minimum (V_0) and the maximum volume (V_m), and the rate of retreat is simply given by volume dividing V_{tot} by the surface area of the cliff. The interest of such a method is also that if V_0 and V_m are known and the cliff retreat velocity is known by other methods (Domènech et al. 2018), it can provide an estimation of the rockfall failure hazard. It must be emphasized that other types of distribution laws are used, such as the generalized Pareto distribution (De Biagi et al. 2017).

Fig. 18 **a** Example of rockfall source detection in the Mont Saint-Eynard cliff, comparing August 2009 and November 2012 terrestrial laser scanner data. **b** Power law deduced from 344 scars > 0.05 m^3 (modified after Guerin et al. 2014)

The Drawback of Power Laws

The new techniques allow to monitor cliffs periodically at a high resolution, from every hour to several months. Observed rockfall scars are often the result of several events (Van Veen, et al. 2017; Williams et al. 2018, 2019). The b value increases as the interval between acquisitions diminishes (Fig. 19), But the retreat rate obtained in both cases is the same. In addition, if the small volumes are not fully recorded, the power law has a rollover for a small value. In addition, if the volume cannot be larger than a certain size, the distribution becomes steeper for large volumes.

Other Types of Inventories

Regional inventories can provide some basic information, such as the type of rock, the structures that control the mechanisms (Copons and Vilaplana, 2008), and the weather parameters (Corò et al. 2015; D'Amato et al. 2016).

As stated above, the inventories of rockfall events can be performed using a series of point clouds. However, in some areas, only one acquisition of the 3D topography exists. In some cases, when the structures are easy to recognize, such as joint sets that delineate block shapes well, it may be possible to extrapolate the magnitude frequency of the

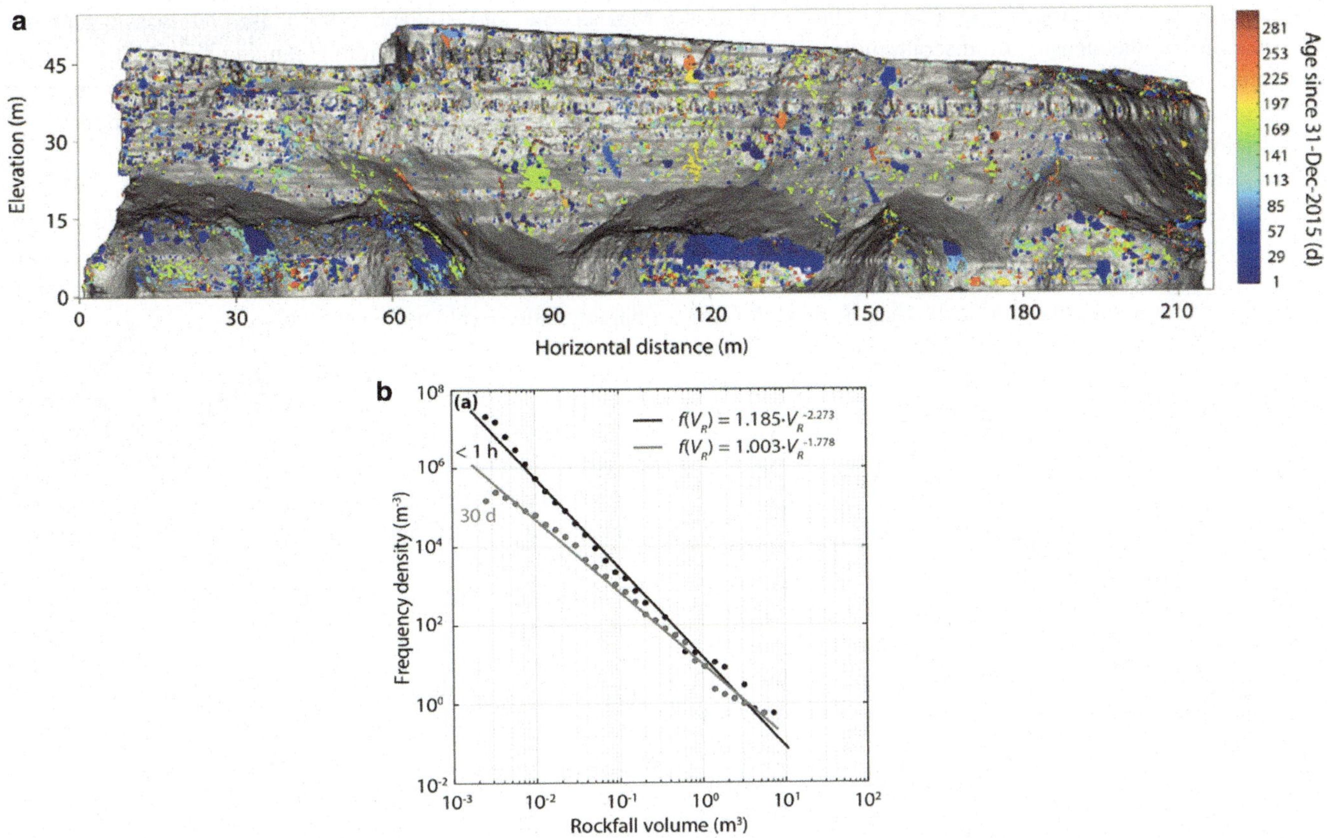

Fig. 19 **a** Rockfall scars identified nearly every hour for 10 months. The colors represent the time since 31 December. **b** Volume–frequency distributions calculated from the above inventory indicating the differences existing if the time-step interval varies from 1 h to 30 days (modified after Williams et al. 2018)

sources based on the basal sliding surfaces and the heights of the scars (Santana et al. 2012). This permitted a Monte Carlo simulation to deduce a power law for the volumes of the source areas. This study was prolonged by the analysis of the potential instability following the same scheme but this time looking at the discontinuities that delineate blocks that may fail by deducing the block sizes by both measurements and simulations using the distributions of the spacing of the discontinuities (Mavrouli et al. 2015).

The frequency of rockfalls can also be assessed by the rockfall boulders. The advantage is that there are usually a lot of deposited blocks, allowing to fit a power law for the block volume distribution (Ruiz-Carulla et al. 2015; Hantz et al. 2016). The disadvantage is that the length of the deposition period is usually not known, except for some recent blocks. De Biagi et al. (2017) used the temporal information given by these recent blocks to transform the block volume distribution into a block volume—temporal frequency relation. Hantz et al. (2016) combined the retreat rate of the cliff with the block volume distribution to obtain the block volume—temporal frequency relation. The analysis of the impacts on trees using a dendrogeomorphic approach can provide a return period for rockfall (Stoffel et al. 2011), which can be used as a proxy for rockfall failure. Bull and Brandon (1998) showed that lichenometry (the size of the lichens is growing with time) may provide an interesting tool for dating events. In New Zealand, these authors demonstrated that several clusters of rockfalls were triggered by several earthquakes.

Hazard Rating Based on Geomechanics Factors

In rock mechanics, different approaches have been developed to assess rock mass quality (Barton et al. 1974; Hoek and Brown, 1997; Bieniawski, 1973, 1993). Many methods developed for rockfall source hazards are based on such multiparameter rating systems derived from tunneling and mining engineering. such as rock mass rating (RMR) (Bieniawski, 1973, 1993), rock mass quality Q (Barton et al. 1974) or the geological strength index (GSI) (Hoek, 1994). The main parameters used by these different approaches are:

- Rock quality designation (RQD) (Deere, 1963), which is a measure of the density of discontinuities per unit of volume;
- The sets of discontinuity spacings;
- The joint roughness;
- The weathering state of the discontinuity surface;
- The strength of the intact rock;
- The groundwater.

Each parameter possesses a scale that is used for the rating. The RMR and GSI are equivalent (Hoek, 1994), but the GSI is simpler in its application. Brideau et al. (2007) applied such an approach to slope stability analysis. A similar method was developed by Selby (1980, 1982) for geomorphological applications. RMR was adapted within the SMR (Romana, 1988, 1993), which is more adapted for rock slopes by introducing parameters related to the joint-slope orientation relationship, but this can now be performed directly using kinematic tests with a DEM (see above). Effect such as blasting is also added. Some attempts have been made to use point clouds to calculate part of the SMR (Riquelme et al. 2016). These approaches are aimed to assess the instantaneous stability of future excavations but not the time evolution of the slopes. So, they can be used for rockfall susceptibility assessment, but they are not enough for true rockfall hazard assessment including temporal frequency.

Using similar considerations, Pierson et al. (1990) developed the rockfall hazard rating system (RHRS) but added the slope height, erosion, block size, and climatic conditions. However, its full version includes a risk aspect. It has been modified integrating SMR (Budetta, 2004). Several versions of the RHRS exist (Ferrari et al. 2016). To provide an example of RHRS for the part dedicated to rockfall hazard score to failure (PS), Santi et al. (2009) proposed a formula for cut slope in sedimentary rocks (Budetta and Nappi, 2013):

$$\mathrm{PS} = 277.2 + 1.67\,\mathrm{SH} + 1.74\,\mathrm{RF} + 1.78\,\mathrm{LF} + 1.42\,\mathrm{AS} + 1.63\mathrm{IN} + 1.35\,\mathrm{AP} \quad (31)$$

where SH is the slope height, RF is the rockfall frequency, LF is the launching features (block that may fall), AS is the slope aspect, IN is the degree of interbedding, and AP is the aperture of discontinuities. The two most important variables are RF and LF that indicates the presence of a favorable rock mass to fail.

Hudson (1992) proposed the rock engineering system (RES), which considers several parameters and creates a matrix of interaction between them to qualify the causes and the effects of each one to the other. The sums of this rating provide the weight of each parameter. To qualify a slope based on the early work of Cancelli and Crosta (1993), Mazzoccola and Hudson (1996) used 20 parameters that must be evaluated by a rating ranging from 0 to 3, using parameters such as lithology, potential instability or intact rock mass strength. This method has been applied, for instance, along coastal areas (Budetta et al. 2008).

Harp and Noble (1993) observed rockfall linked to earthquakes. They evaluated susceptibility based on the Q-value from Barton et al. (1974) and proposed a hazard rating using a modified Q-value based on six discontinuity characteristics evaluated in the field:

$$Q = \left[\frac{115 - 3.3J_v}{J_n}\right]\left[\frac{J_r}{J_a}\right]\left[\frac{J_w}{AF}\right] \quad (32)$$

where J_v is the total number of discontinuities, J_n is a number quantifying the number of joint sets (20–0.5), J_r is a number that characterizes the roughness of the discontinuities (0.5–4), J_a is the join state of alteration (4–0.75), J_w is the water reduction factor (0–1.0) and AF characterizes the joint apertures (1-15). Coe and Harp (2007) used this method to show the increase in susceptibility in fold hinges in folded limestones. In an area suffering earthquakes, Parise (2002) compared the Q-value with the method of Keefer (1993), which provided similar results. Harp and Noble (1993) proposed in their study that the number of rockfall with Q<Q' within a given surface area is given by:

$$N(Q < Q') = N_o(1 - e^{-aQ'}) \quad (33)$$

where N_o is the total number of rockfall events, for a given earthquakes sequence in California and a corresponds to the inverse of the average susceptibility Q. Such an approach is conceptually interesting because it links an observation to a statistical frequency distribution.

Back Analysis

Often, back analysis can be performed to obtain and permit information about the ranges of values such as friction angle, apparent cohesions or rock bridges. The most difficult type is to reconstruct the former topography and then choose the right mechanism and surfaces of failure, e.g. a scar of a former wedge of approximately 13,000 m^3 in Switzerland (Fig. 20). A simple back calculation indicates that if no cohesion exists and no rock bridges are considered, the friction angle must be as high as 40.5°. Some rock bridges can then be suspected (Fig. 20).

A simple back analysis of a wedge failure that occurred between 29-30 October 2011 in Les Drus (Mont-Blanc Massif, France) necessitates at least 3% of rock bridges using the concept of cohesion and friction angle equivalent (Jennings, 1972). This result was confirmed by the

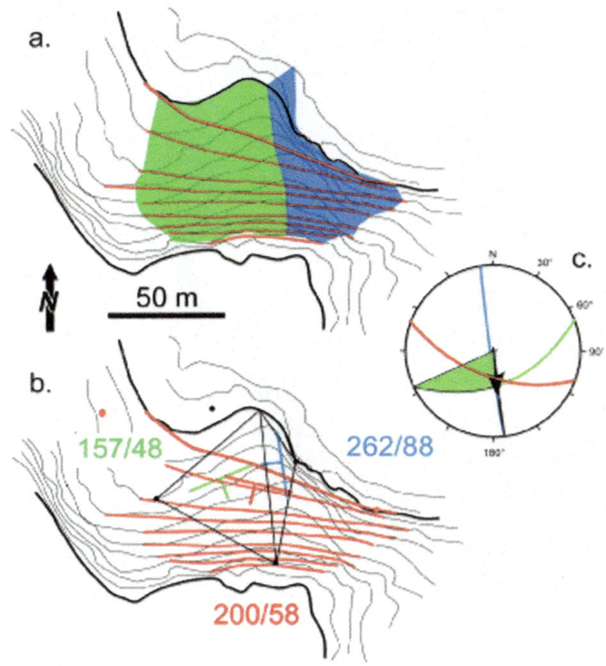

Fig. 20 Back analysis of a wedge in Saasthal (Switzerland), showing a manual reconstruction of the topography and the two joints involved. **a**. Identification of the surfaces in red (the paleo-topography) and in blue and green (the two planes). **b**. Point used to calculate the plane ordinations. **c**. Stereonet of wedge in lower hemisphere (Modified after Rouiller et al. 1997)

calculation of fresh rock failure in the scar (Matasci et al. 2015b) (Fig. 21).

Using Rockfall Modeling to Assess Failure Susceptibility

To assess the activity of rockfall, one method consists of establishing a link between rockfall sources and scree slopes. Menendez Duarte and Marquinez (2002) used the watershed area (GIS function) and the surface areas in pixels of the active scree slopes that are linked to the sources to deduce a susceptibility value. Based on the trajectories of rockfall using the software HY-STONE, Frattini et al. (2008) linked the screes to their potential sources, which allows us to identify the active cliffs if they are linked to active scree slopes. This method is of interest because with powerful computers and efficient numerical models, it permits the fast assessment of rockfall failure susceptibility. This method can also be used as a tool to test whether the rock sources reach any object, which permits the identification of zones that must be monitored or mitigated (Fig. 22).

Fragmentation at Source

The probability of propagation can be affected by rock fragmentation, which changes the rockfall hazard. Therefore, one of the aspects of the source characterization is to estimate the potential fragmentation by breakage. Ruiz-Carulla et al. (2017) proposed a rockfall fractal fragmentation model (RFFM) starting from the in situ block size distribution (IBSD), which provides the rockfall block size distribution (RBSD) (Fig. 23). It is based on three parameters: the probability of failure of the initial volume in smaller blocks, the survival rate of the blocks at each stage of breakage, and a factor that describes the fractal relationship between block sizes. By calibrating the three parameters, the authors were able to reproduce the observations made in Pyrenees (Fig. 23).

Remarks

The previous sections were dedicated to the assessment of the hazard failure or susceptibility to failure of rockfall sources by using geomorphic, geotechnical or observational

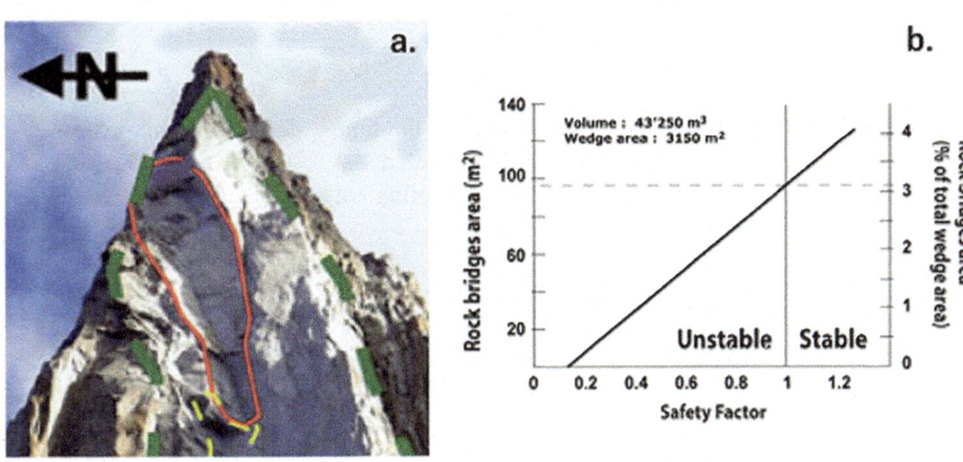

Fig. 21 A. wedge failure of 29–30 October 2011, with a volume of 43,000 m^3. B. Safety factor and percentage of rock bridges (modified after Matasci et al. 2015b)

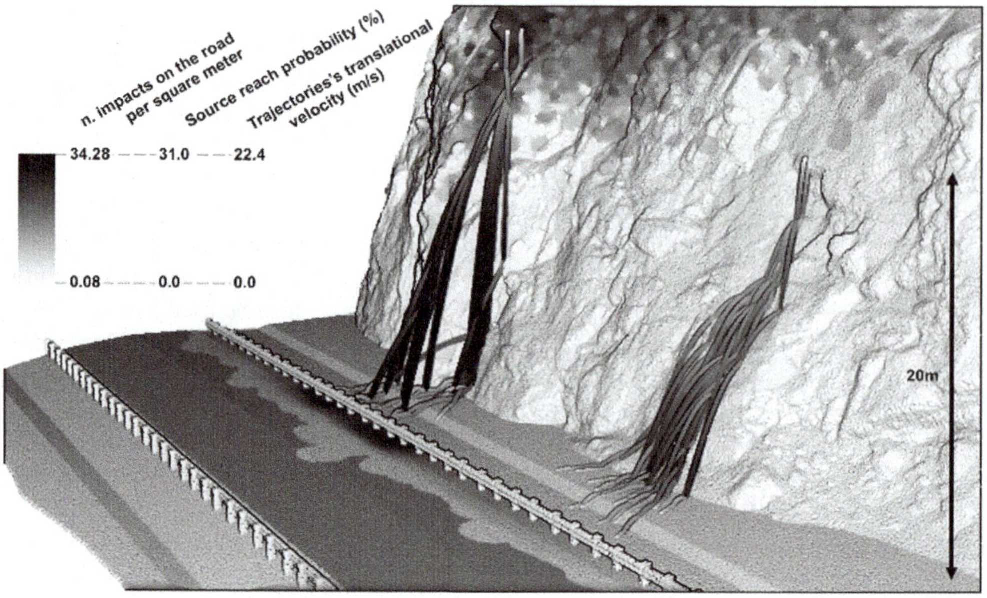

Fig. 22 Rockfall simulation based on point cloud (modified after Noël et al. 2016). The same grayscale is applied to three parameters: (1) the source area where dark gray indicates the probability of reaching the road; (2) the number of impacts per square meter reaching the road (grayscale on the road); and (3) the block velocities indicated by the trajectories

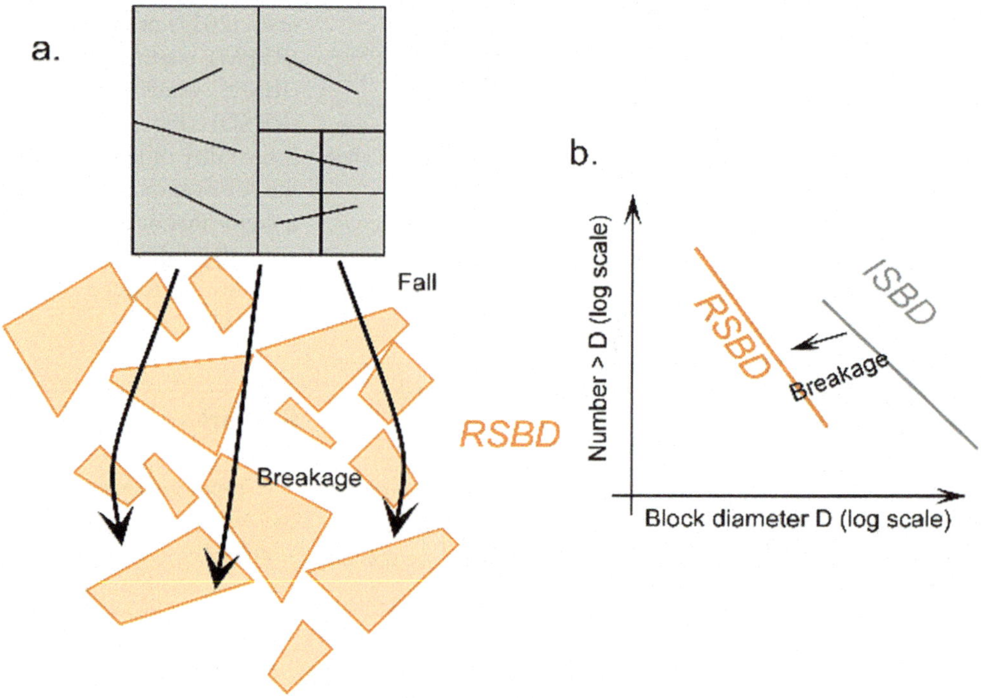

Fig. 23 Principle of rockfall breakage (modified after Ruiz-Carulla et al. 2017)

parameters. However, none of them were linked to the time evolution of the stability. In recent years, improvements in observational techniques have permitted monitoring of destabilization processes. It is not yet operational but is starting to provide new results and is opening new avenues of research.

Processes Mainly Driven by Groundwater and Precipitation

Weathering

Weathering is known to be one of the main factors decreasing strength. The degree of weathering is used to qualify the mechanical properties; nevertheless, it remains a qualitative and approximative qualification, since rock is diverse, but it must reflect the degree of degradation by chemical or physical weathering from fresh rock (Gonzalez de Vallejo and Ferrer 2011). It usually follows a scale from I to VI, i.e. from fresh rock to soils. It is also an input of the GSI classification (Hoek, 1994) but remains a fuzzy concept (Fig. 24).

Implicit weathering effects or fatigue are often used to model strength degradation by changing cohesion and/or friction angle (Eberhardt et al. 2004). Recent works show that strength weakening can occur for altered granite (Chen et al. 2019) by submitting them to wetting at ambient temperature and drying (105°) cycles, showing the that it increases with number of cycles.

Chemical weathering is recognized as an important weakening process for sedimentary rocks by the dissolution of minerals, but it is also possible to strengthen the rock by precipitation-inducing cementations (Chigira and Oyama, 2000). Matsukura and Hirose (2000) experimentally showed that the weathering of rocks depends on lithology and the environment controlling the chemistry of water. The situation can be diverse. Within samples taken in cataclastic material from fault zones, in increasingly weathered material approaching the la Clapière landslide (France), Lebourg et al. (2011) observed that the friction angle increased slightly and the cohesion decreased by leaching fine-grained material. In contrast, Jaboyedoff et al. (2004b) proposed that friction and cohesion may decrease in granite within the sliding surface by crushing and chemical weathering (feldspar dissolution and clay precipitation), inducing soil-like behavior. In some special environments, such as along coastal cliffs, the effect of seawater can induce salt crystallization, inducing a decrease in strength (Duperret et al. 2005).

Rainfall Effect

Rainfall has always been observed to affect rockfalls, but few papers have reported correlations. Nevertheless, most of the studies indicate a strong dependence of rockfall activity on rainfall intensities (Fig. 25). D'Amato et al. (2016) demonstrated for limestone cliff near Grenoble (France) (Fig. 18) that the frequency of rockfall occurrence was 26 times higher than that in "normal conditions" if the precipitation intensity was higher than 5 mm/h. This value is

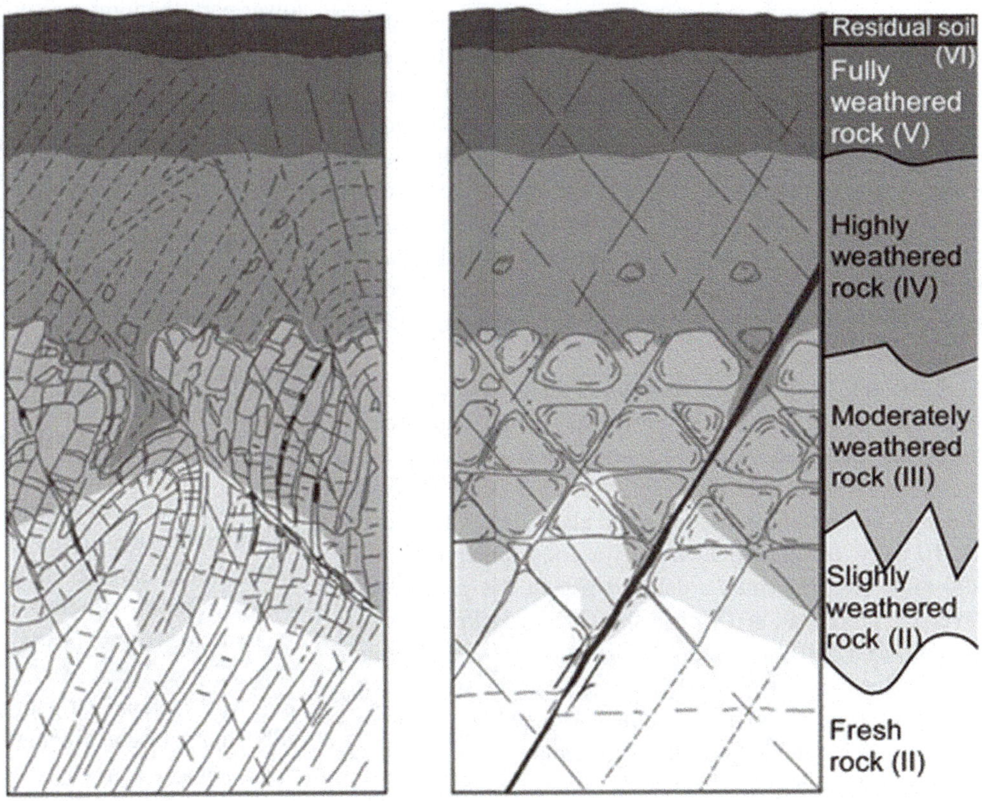

Fig. 24 Weathering profile and associated geotechnical weathering grades (Modified from Wyllie, 2018)

Fig. 25 Example of rockfalls that occurred during heavy rainfall in Korbous, Tunisia. **a**. Rockfall source that fell down on September 23rd, 2018, during rainfall reaching 300 mm/2 h. **b**. Rockfall of blocks on October 17–18, 2018 occurring during a 190 mm precipitation event

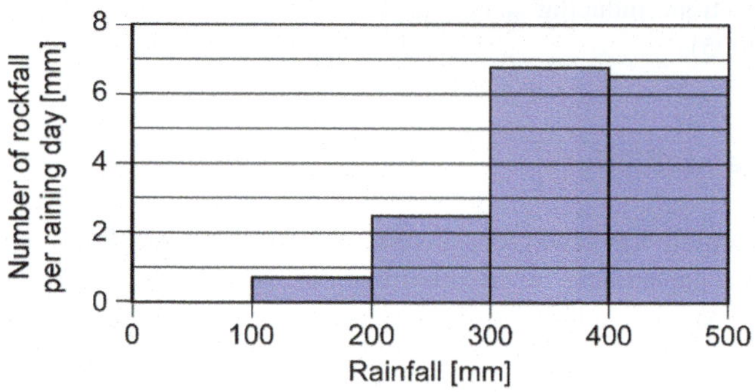

Fig. 26 Statistics of the average number of rockfalls per day with respect to the two antecedent days before the rockfall event (after Chau et al. 2003)

nearly four times greater than the effect of freezing and thawing in this environment. Based on an inventory of 245 rockfall events over 13 years in Hong Kong, Chau et al. (2003) showed that the threshold to trigger rockfall ranges from 150 to 200 mm in precipitation (Fig. 26) based on two days of antecedent rainfall statistics (for one day, the rockfall event is not significant in terms of the number of events). However, significant rockfall can develop when 100 mm rainfall occurs in one day. These authors conclude that the best indicator for rockfall hazard triggering is the two-day antecedent precipitation because of its good correlations with event frequencies.

Two case studies of rockfall disasters among other impacting roads showed in northern Taiwan that rockfall triggering occurred when 94.5 mm/h and nearly 200 mm precipitation were reached (Wei et al. 2014). In Hokkaido (Japan), rock toppling was triggered elven days after the end of 10 days of precipitation for a total precipitation of 290 mm. This indicates a delayed rise in the groundwater level by precipitation (Yamagishi, 2000). Helmstetter and Garambois (2010) showed that in rock slope instabilities such as La Clapière (France) rockfalls can be triggered by rainfall of more than 1 mm that also occurs after a delay of a few days. This demonstrated the interplay of direct effects with infiltration and groundwater.

Rock Breathing

Groundwater is known to have a great effect on stability. One of the potential sources of fatigue of a rock mass can be caused by cyclic variation in the groundwater table (Jaboyedoff et al. 2009). Recent observations based on GB-InSAR permit the observation of rock breathing within the rock face located close to the active rock instability of Mannen in Norway (Figs. 27 and 28) (Rouyet et al. 2017). The results showed inflation (movement outward from the rock cliff) during the snowmelt period, i.e. June to mid-July, and a deflation later during the summer (movement toward the rock cliff). The cause is most likely the water infiltration in the fracture network during the melting phase, inflating the rock mass by hydrostatic pressure. In addition, in the lower part, the joints are potentially obstructed by ice, which can increase the pore pressures. The infiltration is obvious at the top of the cliff by the presence of small lakes and syphons infiltrating down into the rock mass. Once the melt ends, the permeability is great enough to allow rather fast drainage, and the rock mass retracts. This process may follow a hysteresis, meaning that progressively, the rock mass expands irreversibly, progressively dislocating the rock mass.

Process Mainly Driven by Thermal Effects

Sunshine Effect and Associated Thermal Effects

It is well known that the sun can heat rock faces to rather high temperatures, but the real affect has been recognized and modeled only recently. In May 2000, Rochers de Valabres (France) suffered a 2000 m^3 rockslide, with no apparent triggering factors or source of destabilization, by “thermally induced wedging” (Bakun-Mazor et al. 2013) (Fig. 6). Later, investigations showed that the area suffered daily temperature fluctuations that induced daily cyclic movements of nearly 1 mm. The modeling based on

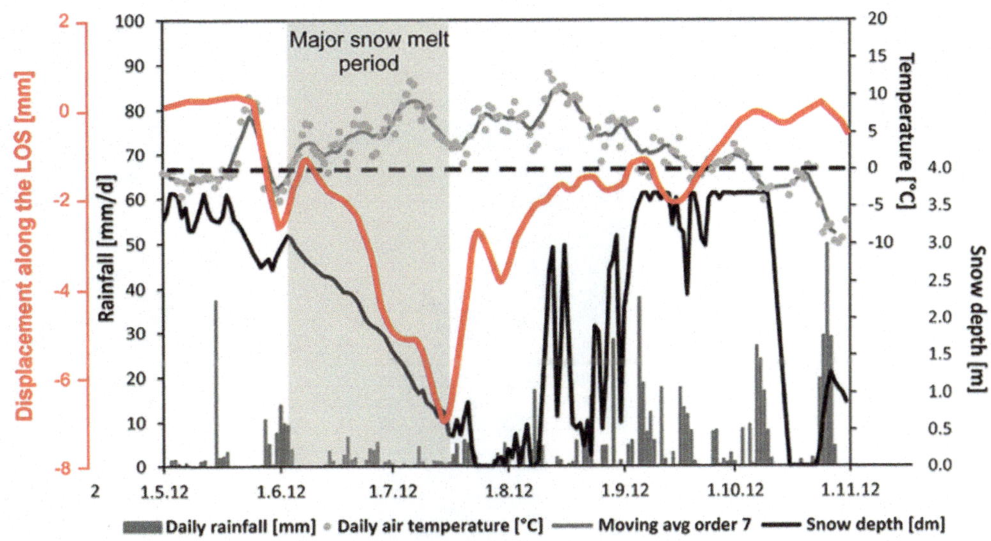

Fig. 27 Relationship of the movement along the line of site (LOS: the negative distance corresponds to movement toward the GB-InSAR) of point P1 (Modified after Rouyet al. 2009)

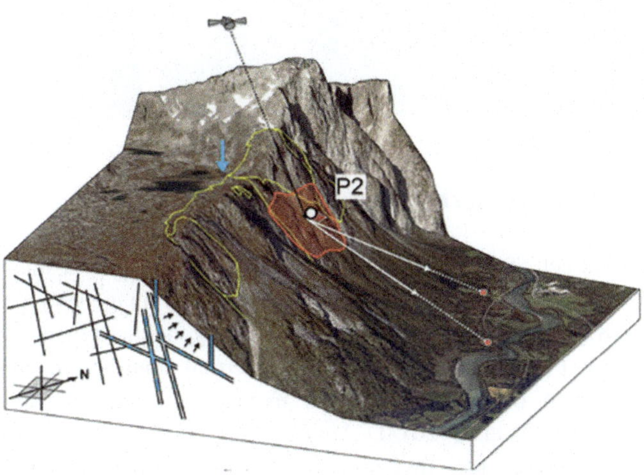

Fig. 28 Scheme explaining the inflation of the rock mass in the red zone with the approximate position of point P2 (modified after Rouyet al. 2009)

elastoplastic rheology showed a possible potential increase in the downward displacement in a few days that led to failure (Gunzburger et al. 2005).

Seasonal downward movement fluctuations have been shown above the Randa rockslide (Switzerland), which are correlated with ambient temperature (Jaboyedoff et al. 2004c; Gischig et al. 2011a). Observations and models show hysteresis of the displacements if the model includes slip weakening (Gischig et al. 2011a; 2011b). This model showed in that case study that stress variation significantly affected the slope up to 100 m depth.

A study in Yosemite Valley (USA) has shown that an exfoliation sheet was bent by up to 1 cm in summer in the center over a length of 19 m with increasing daily temperature by a delay of approximately 2 h (Collins and Stock, 2016). It was also demonstrated that seasonal variation existed, which showed a long-term increase in the bending deformation (Fig. 29). The difference in temperature of the Yosemite exfoliation sheet (Fig. 30) from 8 AM to 11:40 AM indicated a difference up to 24 °C (Guérin et al. 2020b).

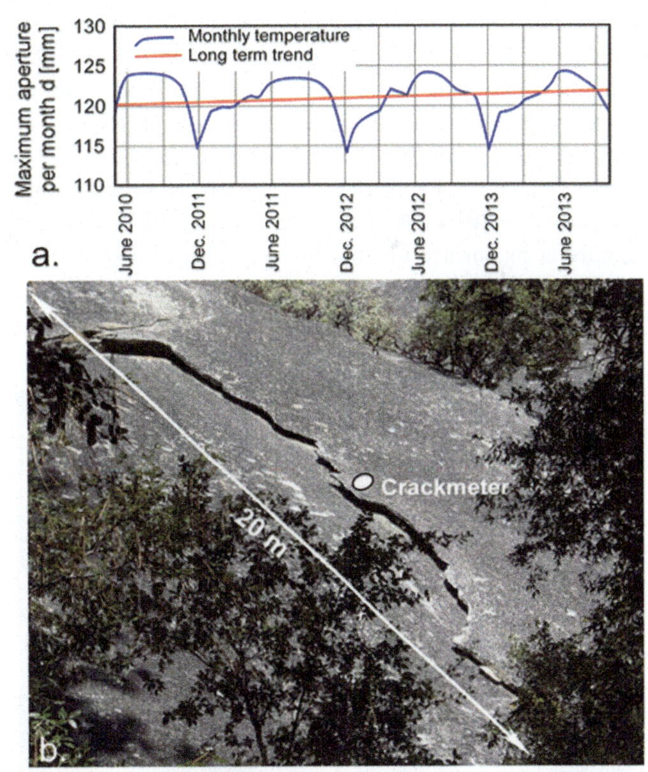

Fig. 29 **a** Monthly maximum aperture of the rock sheet in **b** showing the seasonal cycles and drift indicating the extension of the sheet (after Collins and Stock, 2016)

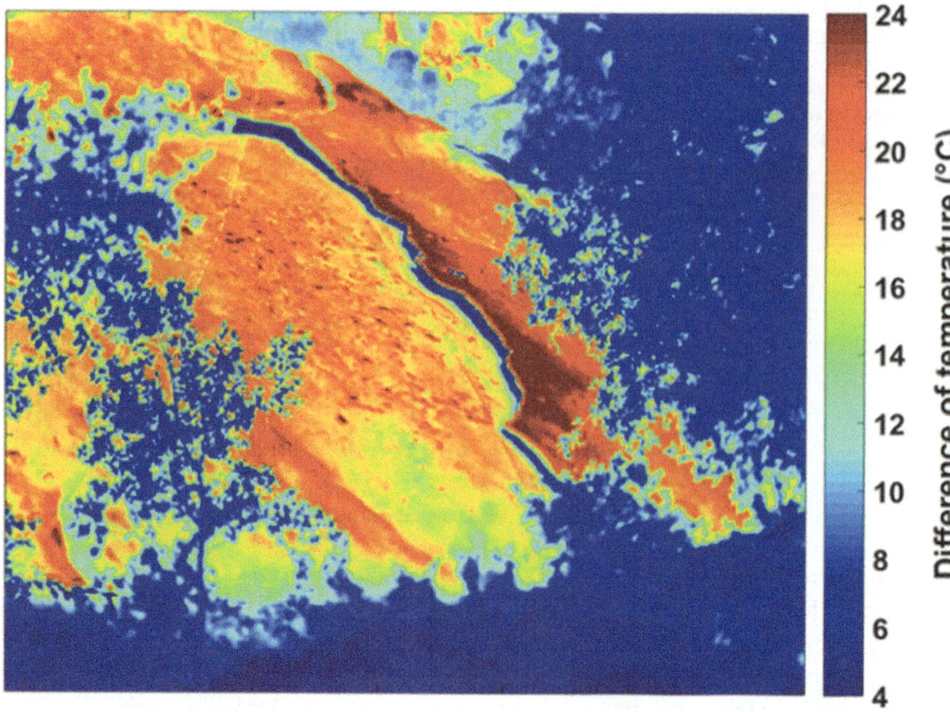

Fig. 30 Differential temperature image of the rock sheet of Fig. 29 between 8 AM and 12:00 AM on October 14, 2015 (modified after Guerin et al. 2020b)

Currently, the application of thermal imaging is emerging as a tool to monitor rock outcrops and to understand rock fatigue. For instance, Pappalardo et al. (2016) developed a cooling rate index (CRI) that corresponds to the ability to exchange heat with air. The CRI is correlated with the degree of fracturing or is anticorrelated to the RQD. Furthermore, it appears that thermal imaging can be a tool to identify weakness (Pappalrado et al. 2016) and rock bridges (Guerin et al. 2019). A rock flake can show a hotter temperature at its surface in front of rock bridges (Fig. 31).

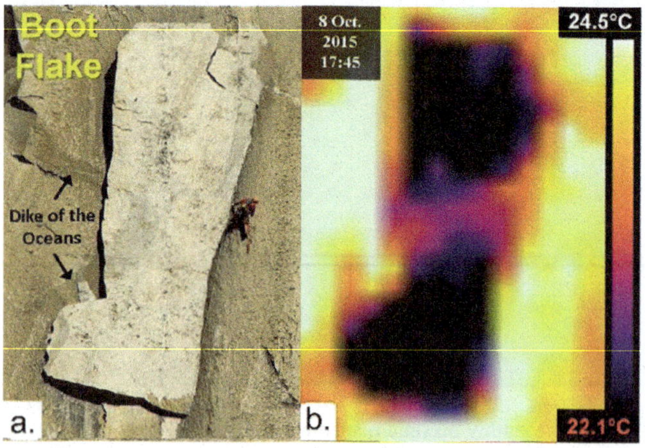

Fig. 31 Thermal image of Boot Flake on 8 October indicating a hotter zone in the middle lower part ha that corresponds to the rock bridges (modified after Guerin et al. 2019)

Freezing and Thawing

The effect of freezing has long been proposed as a destabilizing factor for rock slopes by ice-blocked drainage (Wyllie, 2018). The freezing and thawing cycles generate expansion during phase changes, cycles of thermal contraction and expansion of ice, and a final expansion before melting (D'Amato et al. 2016). D'Amato et al. (2016) showed that rockfall frequency increased by seven times compared to normal conditions in the limestone cliffs above Grenoble (Fig. 18) and observed that most of these rockfall initiations occurred during the melting phase of the ice because of its expansion, in addition to the effect of water pressure caused by melted ice water (Fig. 32).

Fig. 32 Ice in the Les Drus rock face visible after a rockfall. (modified after Matasci et al. 2015b)

In high alpine environments, as in the Mont-Blanc Massif (France), it is well established that the active layer (zone that can suffer temperature >0°) of the permafrost is deeper during hot periods, leading to an increase in rockfall activity triggered by ice melting, as in 2003 and 2015 (Ravanel et al. 2017). Out of these exceptional periods, global warming continues to degrade permafrost and trigger rockfalls (Ravanel et al. 2010).

Surface crack monitoring at nearly 3000 m a.s.l. in the Alps demonstrated that the freezing and thawing cycles, within the active layer or within the surface freezing and thaw zone without permafrost, lead to crack opening by frost wedging, i.e. freezing in cracks of snow melt water, rainfall, etc. The active layer thickness or frost-thawing layer plays an important role in rockfall volume initiation (Ravanel et al. 2017; Matsuoka and Sakai, 1999). The erosion rate measured in a rock cliff at nearly 3000 m a.s.l. in a non-permafrost zone rock wall facing SW reached 6.5 mm/year (Kenner et al. 2011).

The ice appears before failure; it becomes a cement before any failure, it maintains stability while rock bridges decrease in size.

Rock Stability Degradation

The unstable rock masses are linked to their ground either by rock bridges or maintained by friction. As it approaches failure, these links are broken or weakened. Seismometers installed on the unstable rock mass allow monitoring of the response to seismic noise, which provides information about its stiffness. Lévy et al. (2010) showed that the 1st mode of resonance of a limestone rock column of 21,000 m^3 decreases its resonance frequency with time. This frequency was correlated and mostly reversible with air temperature. The freezing and thaw cycles led to a decrease in the resonance frequency, which led to irreversible damage. Note that during the freezing period, the frequency can increase because of the stronger link of the rock mass with its substratum. Rainfall can decrease the frequency because the weight of the column increases or because the joints change their mechanical properties with water infiltration (no pressure is expected because there is no possibility to rise the pressure because of fractures).

Short-Term Rockfall Forecast

The previous sections were dedicated to the rockfall failure hazard in the long term, but here we look at the short-term hazard assessment, in other words, the forecast of the failure time. The traditional monitoring technics allow to measure the displacements of some specific points. Currently, it is possible to monitor failure in 3D. It is possible to identify a forecast over a few days. In 2006, monitoring using the LiDAR technique allowed us to forecast the collapse of a 170,000 m^3 rockfall after moving at 1.25 m/day in the Eiger area (Switzerland) (Oppikofer et al. 2008). GB-InSAR can also be used to monitor rockfall initiation (Intrieri et al. 2016), but the resolution is lower. Blocks in rock faces in Catalonia have failed, clearly showing an exponential acceleration before rock failure (Royán et al. 2014). Similar behavior has been observed in the final failure stage of a block failure within the Séchilienne rockslide, where the area was scanned every 30 min by a terrestrial laser scanner (Fig. 33). The identification of the toppling movement was easily performed by computing the difference between cloud points (Kromer et al. 2017).

Such a failure phase for small volumes compared to large rock instability is still under investigation because it seems it does not follow the inverse velocity model (Fukuzono 1990; Voight, 1989). Rose and Hungr (2007) pointed out that the limited volumes compared to deep seated rockslides do not follow the inverse velocity model.

Prior to failure, the seismicity shows an increase in the rate of seismic events and signal energy (Amitrano et al. 2005). The time of detachment shows a clear increase in seismic signal at relatively low frequency (Le Roy et al. 2019).

These studies are also fundamental because they can be analyzed to understand which patterns are leading to failure and which patterns are not.

New Techniques

Various attempts are in progress to assess rockfall failure susceptibility and hazards with new techniques. Some attempts have been made using fuzzy logic based on altitude difference, number of discontinuities, number of wedges, number of potential slides and known source area (Aksoy and Ercnoglu 2006). Different machine learning methods have been applied to identify rockfall source areas, which provided rather good results (Fanos et al. 2018). They were based on topographic indexes and other GIS data types. Losasso and Sdao (2018) used a neural network to assess rockfall source susceptibility based on lithology, DEM, slope angles, land use, etc. and the former scars to train the neurones.

The drawbacks of these methods are that they will favor the most important factors, such as slope angle, and are black boxes. However, it will be very interesting to use a large database with a large amount of information to try to extract the main trend of rockfall causes and to know the rockfall frequency in different contexts.

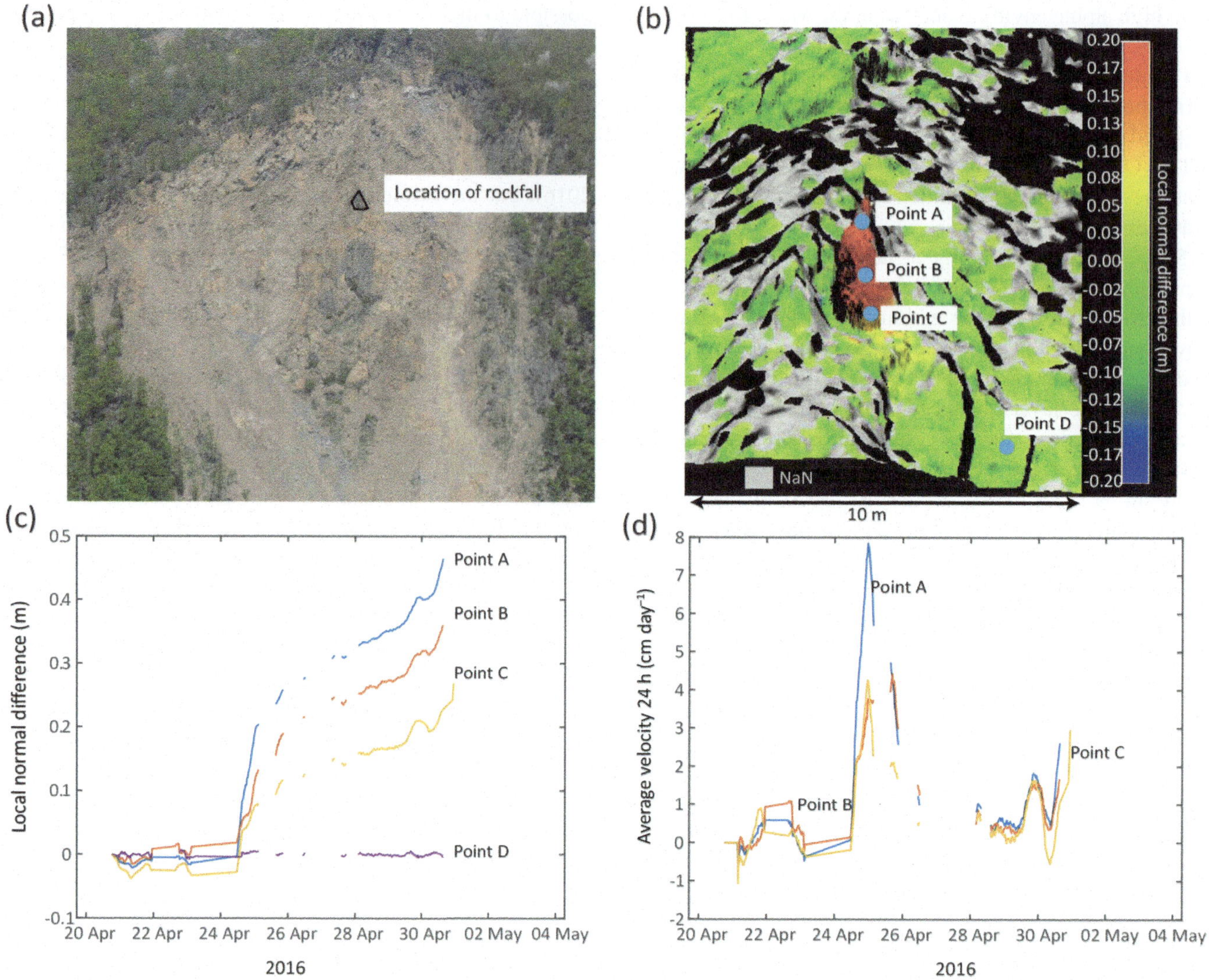

Fig. 33 Example of toppling a block of 80 m^3 failure showing the pre-failure movements. This was analyzed based on cloud points acquired every 4 h (from Kromer et al. 2017). **a**. Block location in the Séchillienne rockslide (France). **b**. 3D projection showing the deformation along the local normal (color scale). **c**. Time series of the displacement of points A, B and C located on the moving block; Point D is stable. **d**. Point A, B and C velocities averaged over 24 h

Discussion and Conclusion

In most cases, the rockfall hazard assessment is based on rockfall propagation Pp, and a few methods are dedicated to the failure frequency λ. When λ is quantified, it is usually by empirical methods, semi-empirical methods or oversimplified geomechanical models leading to a susceptibility scale. In the first part, this paper presented the main concepts and tools used to assess the hazard of rock failure, but we did not integrate the time dependence of the stability, except for the water table changes and earthquake accelerations.

There is no real-time dependence models of the stability in terms of its link to the degradation of strength. Furthermore, there is a need to consider more complex mechanisms of failure, mainly linked to rock bridge progressive failure (Lévy et al. 2010; Guerin et al. 2019). However, this cannot be done without improving our knowledge on the external factors that degrade the strength of rock masses or discontinuities (Chen et al. 2019). The approach will be very different if it is a site-specific or a regional study.

It is possible to extract more information from real 3D surface imaging than what is currently performed. It must permit us to characterize the degree of fracturing more precisely, which is beyond the scope of this paper. It can go up to microtopography with a potable laser scanner to characterize the roughness of joints in the field and probably detect its evolution under weathering effects.

Computers increase their capacity, and models based on simulations of discrete fracture networks (DFNs) (Grenon and Hadjigeorgiou, 2008), including high-resolution 3D topography coupled with time-dependent stability evolution, will certainly be the next step to assess failure hazards or susceptibility based on the principle of Park and West (2001). The effects of groundwater and earthquakes must be better assessed. However, in any case, this will be linked to the traditional field work, with systems such as GIS (Hoek, 1994) or others presented here or those that are being newly developed (Riquelme et al. 2016).

The inventory and its statistics must be developed by improving the probabilistic models (De Biagi et al. 2017). There are many drawbacks because power laws present some limitations, such as volume limits, that must be addressed. Another issue is to link the fracturing of the sources with the real volume involved in one event and the number of blocks that will be generated (Ruiz-Carulla et al. 2016).

The biggest challenges are linked to the quantification of the influence of external factors that lead to strength degradation. We know now that fatigue can be caused by cyclic slope movements, including hysteresis linked to groundwater water table level changes, thermal effects such isolation and freezing and thawing and earthquakes. In addition, until now, rainfall effects have been seen only as a triggering factor and not as a degrading factor, which must also be assessed in the future. This can be summarized by the degradation of joints and rock mass strengths and the progressive failure of rock bridges (Ruiz-Carulla et al. 2016).

One tool that can be used to better understand this process is to perform an increasing back analysis of events, carefully reconstructing past topography and rock bridges, including the effect of external factors such as insolation, water circulations, freezing and thawing, and weathering evolution. This additional information will permit us to conceptualize Bayesian probabilistic models and have enough inputs to use machine learning methods when the data are enough.

Acknowledgements The authors acknowledge the Swiss National Science Foundation (SNSF, grants #: 200021_127132, 200020_146426, and 200020_159221) for supporting this research. We are thankful to the American Journal Experts for their support, which improved the English language.

References

Abellán A, Oppikofer T, Jaboyedoff M, Rosser NJ, Lim M, Lato M (2014) State of science: terrestrial laser scanner on rock slopes instabilities. Earth Surf Process Land 39(1):80–97. https://doi.org/10.1002/esp.3493

Agliardi F, Crosta GB (2003) High resolution three-dimensional numerical modelling of rockfalls. Int J Rock Mech Min Sci 40 (4):455–471

Aksoy H, Ercanoglu M (2006) Determination of the rockfall source in an urban settlement area by using a rule-based fuzzy evaluation. Nat Hazards Earth Syst Sci 6(6):941–954

Amitrano D, Grasso JR, Senfaute G (2005) Seismic precursory patterns before a cliff collapse and critical point phenomena. Geophys Res Lett 32(8)

Baillifard F, Jaboyedoff M, Sartori M (2003) Rockfall hazard mapping along a mountainous road in Switzerland using a GIS-based parameter rating approach. Nat Hazards Earth Syst Sci 3:435–442. https://doi.org/10.5194/nhess-3-435-2003

Bakun-Mazor D, Hatzor YH, Glaser SD, Carlos Santamarina J (2013) Thermally vs. seismically induced block displacements in Masada rock slopes. Int J Rock Mech and Min Sci 61:196–211

Barton N, Lien R, Lunde J (1974) Engineering classification of rock masses for the design of tunnel support. Rock Mech 6(4):189–236

Bieniawski ZT (1973) Engineering classification of jointed rock masses. Trans S Afr Inst Civ Eng 15:335–344

Bieniawski ZT (1993) Classification of rock masses for engineering: the RMR system and future trends. Compr Rock Eng 3:553–573

Brideau M-A, Stead D, Roots C, Orwin J (2007) Geomorphology and engineering geology of a landslide in ultramafic rocks, Dawson City, Yukon. Eng Geol 89:171–194

Budetta P (2004) Assessment of rockfall risk along roads. Nat Hazards Earth Syst Sci 4:71–81

Budetta P, Nappi M (2013) Comparison between qualitative rockfall risk rating systems for a road affected by high traffic intensity. Nat Hazards Earth Syst Sci 13(6):1643–1653

Budetta P, Santo A, Vivenzio F (2008) Landslide hazard mapping along the coastline of the Cilento region (Italy) by means of a GIS-based parameter rating approach. Geomorphology 94(3):340–352

Bull WB, Brandon MT (1998) Lichen dating of earthquake-generated regional rockfall events, Southern Alps, New Zealand. Geol Soc Am Bull 110:60–84

Cancelli A, Crosta G (1993) Hazard and risk assessment in rockfall prone areas. In: Skip BO (ed) Risk reliability in ground engineering. Thomas Telford, London, pp 177–190

Carter BJ, Lajtai EZ (1992) Rock slope stability and distributed joint systems. Can Geotech J 29(1):53–60

Chau KT, Wong RHC, Liu J, Lee CF (2003) Rockfall hazard analysis for hong kong based on rockfall inventory. Rock Mech Rock Eng 36(5):383–408

Chen X, He P, Qin Z (2019) Strength weakening and energy mechanism of rocks subjected to Wet-Dry Cycles. Geotech Geol Eng 37(5):3915–3923

Chigira M, Oyama T (2000) Mechanism and effect of chemical weathering of sedimentary rocks. Eng Geol 55(1):3–14

Coe JA, Harp EL (2007) Influence of tectonic folding on rockfall susceptibility, American Fork Canyon, Utah, USA. Nat Hazards Earth Syst Sci 7(1):1–14

Collins BD, Stock GM (2016) Rockfall triggering by cyclic thermal stressing of exfoliation fractures. Nat Geosci 9(5):395–400

Copons R, Vilaplana JM (2008) Rockfall susceptibility zoning at a large scale: from geomorphological inventory to preliminary land use planning. Eng Geol 102:142–151

Corò D, Galgaro A, Fontana A, Carton A (2015) A regional rockfall database: the Eastern Alps test site. Environ Earth Sci 74(2):1731–1742

Corona C, Trappmann D, Stoffel M (2013) Parameterization of rockfall source areas and magnitudes with ecological recorders: when disturbances in trees serve the calibration and validation of simulation runs. Geomorphology 202:33–42

Crosta GB, Agliardi F, Frattini P, Lari S (2015) Key issues in rock fall modeling, hazard and risk assessment for rockfall protection.

Engineering Geology for Society and Territory—vol 2. Springer International Publishing, Cham, pp 43–58
D'Amato J, Hantz D, Guerin A, Jaboyedoff M, Baillet L, Mariscal A (2016) Influence of meteorological factors on rockfall occurrence in a middle mountain limestone cliff. Nat Hazards Earth Syst Sci 16 (3):719–735
De Biagi V, Napoli ML, Barbero M, Peila D (2017) Estimation of the return period of rockfall blocks according to their size. Nat Hazards Earth Syst Sci 17(1):103–113
Deere DU (1963) Technical description of rock cores for engineering purposes, Felsmechanik und In Qenieurgeologie (Rock Mechanics and Engineering Geology), 1:1, 16–22. 64
Dershowitz WS, Einstein HH (1988) Characterizing rock joint geometry with joint system models. Rock Mech Rock Eng 21:21–51
Domènech G, Corominas J, Mavrouli O, Merchel S, Abellán A, Pavetich S, Rugel G (2018) Calculation of the rockwall recession rate of a limestone cliff, affected by rockfalls, using cosmogenic chlorine-36. Case study of the Montsec Range (Eastern Pyrenees, Spain). Geomorphology 306:325–335
Duperret A, Taibi S, Mortimore RN, Daigneault M (2005) Effect of groundwater and sea weathering cycles on the strength of chalk rock from unstable coastal cliffs of NW France. Eng Geol 78(3):321–343
Dussauge C, Grasso JR, Helmstetter A (2003) Statistical analysis of rockfall volume distributions: implications for rockfall dynamics. J Geophys Res: Solid Earth, 108(B6)
Dussauge-Peisser C, Helmstetter A, Grasso J-R, Hantz D, Desvarreux P, Jeannin M, Giraud A (2002) Probabilistic approach to rock fall hazard assessment: potential of historical data analysis. Nat Hazards Earth Syst Sci 2:15–26. https://doi.org/10.5194/nhess-2-15-2002
Eberhardt E, Stead D, Coggan JS (2004) Numerical analysis of initiation and progressive failure in natural rock slopes—the 1991 Randa rockslide. Int J Rock Mech Min Sci 41(1):69–87
Fanos AM, Pradhan B, Mansor S, Yusoff ZM, Abdullah AFb (2018) A hybrid model using machine learning methods and GIS for potential rockfall source identification from airborne laser scanning data. Landslides 15(9):1833–1850
Farvacque M, Lopez-Saez J, Corona C, Toe D, Bourrier F, Eckert N (2019) How is rockfall risk impacted by land-use and land-cover changes? Insights from the French Alps. Global Planet Change 174:138–152
Fell R, Corominas J, Bonnard C, Cascini L, Leroi E, Savage WZ (2008) Guidelines for landslide susceptibility, hazard and risk zoning for land use planning. Eng Geol 102(3):85–98
Fernandez-Hernández M, Paredes C, Castedo R, Llorente M, de la Vega-Panizo R (2012) Rockfall detachment susceptibility map in El Hierro Island, Canary Islands. Spain Nat Hazards 64(2):1247–1271
Ferrari F, Giacomini A, Thoeni K (2016) Qualitative Rockfall Hazard Assessment: A Comprehensive Review of Current Practices. Rock Mech Rock Eng 49(7):2865–2922
Frattini P, Crosta G, Carrara A, Agliardi F (2008) Assessment of rockfall susceptibility by integrating statistical and physically-based approaches. Geomorphology 94(3):419–437
Frayssines M, Hantz D (2006) Failure mechanisms and triggering factors in calcareous cliffs of the Subalpine Ranges (French Alps). Eng Geol 86(4):256–270
Fukuzono T (1990) Recent studies on time prediction of slope failure. Landslide News 4:9–12
Gigli G, Casagli N (2011) Semi-automatic extraction of rock mass structural data from high resolution LIDAR point clouds. Int J Rock Mech Min Sci 48(2):187–198
Gigli G, Morelli S, Fornera S, Casagli N (2014) Terrestrial laser scanner and geomechanical surveys for the rapid evaluation of rock fall susceptibility scenarios. Landslides 11(1):1–14
Gischig VS, Moore JR, Evans KF, Amann F, Loew S (2011a) Thermomechanical forcing of deep rock slope deformation: 1. Conceptual study of a simplified slope. J Geophys Res: Earth Surf 116(F4)
Gischig VS, Moore JR, Evans KF, Amann F, Loew S (2011b) Thermomechanical forcing of deep rock slope deformation: 2. The Randa rock slope instability. J Geophys Res: Earth Surf 116(F4)
Gokceoglu C, Sonmez H, Ercanoglu M (2000) Discontinuity controlled probabilistic slope failure risk maps of the Altindag (settlement) region in Turkey. Eng Geol 55:277–296
González de Vallejo L, Ferrer M (2011) Geological Engineering. CRC Press/ Balkema, p 678
Grenon M, Hadjigeorgiou J (2008) A design methodology for rock slopes susceptible to wedge failure using fracture system modelling. Eng Geol 96(1–2):78–93
Grenon M, Matasci B, Jaboyedoff M, Stock GM (2014) Discrete fracture network modelling using Coltop3D for rockfall potential assessment at Glacier Point, Yosemite Valley. Proceedings of t DFNE 2014, the 1st international conference on discrete fracture network engineering, Vancouver
Gruner U (2008) Climatic and meteorological influences on rockfall and rockslides ("Bergsturz"). In: Protection of populated territories from floods, debris flow, mass movements and avalanches, 26–30 May 2008, 147–158
Günther A (2003) SLOPEMAP: programs for automated mapping of geometrical and kinematical properties of hard rock hill slopes. Comput Geosci 29(7):865–875
Günther A, Carstensen A, Pohl W (2004) Automated sliding susceptibility mapping of rock slopes. Nat Hazards Earth Syst Sci 4:95–102. https://doi.org/10.5194/nhess-4-95-2004
Guérin A, Hantz D, Rossetti J-P, Jaboyedoff M (2014) Brief communication "Estimating rockfall frequency in a mountain limestone cliff using terrestrial laser scanner". Nat Hazards Earth Syst Sci Discuss 2:123–135. https://doi.org/10.5194/nhessd-2-123-2014
Guérin A, Jaboyedoff M, Collins BD, Derron M-H, Stock GM, Matasci B, Boesiger M, Lefeuvre C, Podladchikov YY (2019) Detection of rock bridges by infrared thermal imaging and modeling. Sci Rep 9(1):13138
Guerin A, Stock GM, Radue MJ, Jaboyedoff M, Collins BD, Matasci B, Avdievitch N, Derron M-H (2020a) Quantifying 40 years of rockfall activity in Yosemite Valley with historical Structure-from-Motion photogrammetry and terrestrial laser scanning. Geomorphology 356:107069
Guerin A, Jaboyedoff M, Collins BD, Stock GM, Derron M-H, Abellán A, Matasci B (2020b) Remote thermal detection of exfoliation sheet deformation. Landslides. https://doi.org/10.1007/s10346-020-01524-1
Gunzburger Y, Merrien-Soukatchoff V, Guglielmi Y (2005) Influence of daily surface temperature fluctuations on rock slope stability: case study of the Rochers de Valabres slope (France). Int J Rock Mech Min Sci 42(3):331–349
Hantz D (2011) Quantitative assessment of diffuse rock fall hazard along a cliff foot. Nat Hazards Earth Syst Sci 11(5):1303–1309
Hantz D, Vengeon JM, Dussauge-Peisser C (2003a) An historical, geomechanical and probabilistic approach to rock-fall hazard assessment. Nat Hazards Earth Syst Sci 3:693–701. https://doi.org/10.5194/nhess-3-693-2003
Hantz D, Dussauge-Peisser C, Jeannin M, Vengeon JM (2003b) Rock fall hazard assessment: from qualitative to quantitative failure probability. In: Int conf on Fast Slope Movements, Naples, 263–267
Hantz D, Ventroux Q, Rossetti J-P, Berger F (2016) A new approach of diffuse rockfall hazard. Landslides and Engineered Slopes. CRC Press, Experience, Theory and Practice, Napoli, Italy, pp 1531–1535

Harp EL, Noble MA (1993) An engineering rock classification to evaluate seismic rock-fall susceptibility and its application to the Wasatch Front. Bull Ass Eng Geol 30(3):293–319

Helmstetter A, Garambois S (2010) Seismic monitoring of Séchilienne rockslide (French Alps): analysis of seismic signals and their correlation with rainfalls. J Geophys Res 115(F3)

Hoek E (1994) Strength of rock and rock masses. ISRM News J 2(2):4–16

Hoek E, Bray JW (1981) Rock slope engineering. The Institution of Mining and Metallurgy, London, p 358

Hoek E, Brown ET (1997) Practical estimates or rock mass strength. Int J Rock Mech Min Sci Geomech Abstr 34(8):1165–1186

Hudson JA (1992) Rock engineering systems: theory and practice. Ellis Horwood, London, p 185

Hungr O, Evans SG, Hazzard J (1999) Magnitude and frequency of rock falls along the main transportation corridors of southwestern British Columbia. Can Geotech J 36:224–238

Hungr O, Leroueil S, Picarelli L (2014) The varnes classification of landslide types, an update. Landslides 11(2):167–194

Intrieri E, Gigli G, Lombardi L, Raspini F, Salvatici T, Bertolini G (2016) Integration of ground-based interferometry and terrestrial laser scanning for rockslide and rockfall monitoring. Rend Online Soc Geol Ital 41:243–246

Jaboyedoff M, Labiouse V (2003) Preliminary assessment of rockfall hazard based on GIS data. 10th international congress on rock mechanics ISRM 2003—technology roadmap for rock mechanics. Johannesburgh, South Africa, pp 575–578

Jaboyedoff M, Derron MH (2020) Landslide analysis using laser scanner. In: Tarolli P (ed) Remote Sensing of Geomorphology. Elsevier

Jaboyedoff M, Baillifard F, Hantz D, Heidenreich B, Mazzoccola D (2001) Terminologie. In: Carere K,′ Ratto S, Zanolini FE (eds) Prevention des mouvements′ de versants et des instabilites de falaises, pp 48–57

Jaboyedoff M, Derron MH, Locat J, Michoud C (2021) Traité Mouvements Gravitaires. PPUR

Jaboyedoff M, Metzger R, Oppikofer T, Couture R, Derron MH., Locat J. Turmel D (2007) New insight techniques to analyze rock-slope relief using DEM and 3D-imaging cloud points: COLTOP-3D software. In Eberhardt E, Stead D, Morrison T (eds) Rock mechanics: Meeting Society's Challenges and demands, vol 1, Taylor & Francis. pp 61–68

Jaboyedoff M, Baillifard F, Philippossian F, Rouiller J-D (2004a) Assessing fracture occurrence using "weighted fracturing density": a step towards estimating rock instability hazard. Nat Hazards Earth Syst Sci 4:83–93. https://doi.org/10.5194/nhess-4-83-2004

Jaboyedoff M, Baillifard F, Bardou E, Girod F (2004b) Weathering, cycles of saturation-unsaturation, and strain effects as principal processes for rock mass destabilization. Q J Eng Geol Hydrogeol 37:95–103

Jaboyedoff M, Ornstein P, Rouiller JD (2004c) Design of a geodetic database and associated tools for monitoring rock-slope movements: the example of the top of Randa rockfall scar. Nat Hazards Earth Syst Sci 4(2):187–196

Jaboyedoff M, Baillifard F, Derron MH, Couture R, Locat J, Locat P (2005) Switzerland modular and evolving rock slope hazard assessment methods. In: Senneset K, Flaate KA, Larsen J (eds) Landslide and avalanches. ICFL

Jaboyedoff M, Couture R, Locat P (2009) Structural analysis of Turtle Mountain (Alberta) using digital elevation model: Toward a progressive failure. Geomorphology 103(1):5–16

Jennings JE (1972) An approach to the stability of rock slopes based on the theory of limiting equilibrium with a material exhibiting anisotropic shear strength, stability of rock slopes, proceedings of the 13th us symposium on rock mechanics (ed. ej cording), urbana, Illinois, New York

Keefer DK (1993) The susceptibility of rock slopes to earthquake-induced failure. Bull Assoc Eng Geol 30(3):353–361

Kenner R, Philips M, Danioth C, Denier C, Thee P, Zgraggen A (2011) Investigation of rock and ice loss in a recently deglaciated mountain rock wall using terrestrial laser scanning: gemsstock, Swiss Alps. Cold Reg Sci Technol 67:157–164

Kromer R, Walton G, Gray B, Lato M, Group R (2019) Development and optimization of an automated fixed-location time lapse photogrammetric rock slope monitoring system. Remote Sens 11(16):18

Kromer RA, Abellán A, Hutchinson DJ, Lato M, Chanut M-A, Dubois L, Jaboyedoff M (2017) Automated terrestrial laser scanning with near-real-time change detection – monitoring of the Séchilienne landslide. Earth Surf Dyn 5(2):293–310

Le Roy G, Helmstetter A, Amitrano D, Guyoton F, Le Roux-Mallouf R (2019) Seismic analysis of the detachment and impact phases of a rockfall and application for estimating rockfall volume and free-fall height. J Geophys Res: Earth Surf 124(11):2602–2622

Lebourg T, Mickael H, Hervé J, Samyr EBB, Thomas B, Swann Z, Emmanuel T, Maurin V (2011) Temporal evolution of weathered cataclastic material in gravitational faults of the La Clapiere deep-seated landslide by mechanical approach. Landslides 8 (2):241–252

Leroueil S, Locat J (1998) Slope movements—geotechnical characterization, risk assessment and mitigation. In: Maric B, Lisac L, Szavits-Nossan A (eds) Geotechnical Hazards. Balkema, Rotterdam, pp 95–106

Lévy C, Baillet L, Jongmans D, Mourot P, Hantz D (2010) Dynamic response of the Chamousset rock column (Western Alps, France). J Geophys Res 115(F4)

Locat J, Leroueil S, Picarelli L (2000) Some considerations on the role of geological history on slope stability and estimation of minimum apparent cohesion of a rock mass. In: Bromhead N. Dixon, Ibsen ML (eds) Landslides in research, theory and practice—proceedings of the 8th international symposium on landslides, Cardiff, 26–30 June 2000: 935–942

Lopez-Saez J, Corona C, Eckert N, Stoffel M, Bourrier F, Berger F (2016) Impacts of land-use and land-cover changes on rockfall propagation: Insights from the Grenoble conurbation. Sci Total Environ 547:345–355

Losasso L, Jaboyedoff M, Sdao F (2017) Potential rock fall source areas identification and rock fall propagation in the province of Potenza territory using an empirically distributed approach. Landslides, 1–10

Losasso L, Sdao F (2018) The artificial neural network for the rockfall susceptibility assessment. A case study in Basilicata (Southern Italy). Geomat Nat Hazards Risk 9(1):737–759

Loye A, Jaboyedoff M, Pedrazzini A (2009) Identification of potential rockfall source areas at a regional scale using a DEM-based geomorphometric analysis. Nat Hazards Earth Syst Sci 9(5):1643–1653

Mah J, Samson C, McKinnon SD, Thibodeau D (2013) 3D laser imaging for surface roughness analysis. Int J Rock Mech Min Sci 58:111–117

Matasci B, Jaboyedoff M, Loye A, Pedrazzini A, Derron MH, Pedrozzi G (2015a) Impacts of fracturing patterns on the rockfall susceptibility and erosion rate of stratified limestone. Geomorphology 241:83–97

Matasci B, Jaboyedoff M, Ravanel L, Deline P (2015b) Stability assessment, potential collapses and future evolution of the west face of the drus (3,754 m a.s.l., Mont Blanc Massif). 791–795

Matasci B, Stock GM, Jaboyedoff M, Carrea D, Collins BD, Guérin A, Matasci G, Ravanel L (2018) Assessing rockfall susceptibility in

steep and overhanging slopes using three-dimensional analysis of failure mechanisms. Landslides 15(5):859–878
Matsukura Y, Hirose T (2000) Five year measurements of rock tablet weathering on a forested hillslope in a humid temperate region. Eng Geol 55(1–2):69–76
Matsuoka N (2008) Frost weathering and rockwall erosion in the southeastern Swiss Alps: long-term (1994–2006) observations. Geomorphology 99(1–4):353–368
Matsuoka N, Sakai H (1999) Rockfall activity from an alpine cliff during thawing periods. Geomorphology 28:309–328
Mavrouli O, Corominas J, Jaboyedoff M (2015) Size distribution for potentially unstable rock masses and in situ rock blocks using LIDAR-generated digital elevation models. Rock Mech Rock Eng 48(4):1589–1604
Mazzoccola DF, Hudson JA (1996) A comprehensive method of rock mass characterisation for indicating natural slope instability. Q J Eng Geol Hydroge 29:37–56
Menendez Duarte R, Marquinez J (2002) The influence of environmental and lithologic factors on rockfall at a regional scale: an evaluation using GIS. Geomorphology 43:117–136
Michoud C, Derron MH, Horton P, Jaboyedoff M, Baillifard FJ, Loye A, Nicolet P, Pedrazzini A, Queyrel A (2012) Rockfall hazard and risk assessments along roads at a regional scale: example in Swiss Alps. Nat Hazards Earth Syst Sci 12(3):615–629
Newmark NM (1965) Effects of earthquakes on dams and embankments. Geotechnique 15(2):139–160
Noël F, Cloutier C, Turmel D, Locat J (2016) Using point clouds as topography input for 3D rockfall modeling. Landslides and Engineered Slopes. CRC Press, Experience, Theory and Practice, Napoli, Italy, pp 1531–1535
Oppikofer T, Jaboyedoff M, Keusen H-R (2008) Collapse at the eastern Eiger flank in the Swiss Alps. Nat Geosci 1(8):531–535
Pappalardo G, Mineo S, Zampelli SP, Cubito A, Calcaterra D (2016) InfraRed thermography proposed for the estimation of the Cooling Rate Index in the remote survey of rock masses. Int J Rock Mech Min Sci 83:182–196
Parise M (2002) Landslide hazard zonation of slopes susceptible to rock falls and topples. Nat Hazards Earth Syst Sci 2(1/2):37–49
Park H, West TR (2001) Development of a probabilistic approach for rock wedge failure. Eng Geol 59(3–4):233–251
Pauly J-C, Payany M (2002) Méthodologies mises en oeuvre pour la cartographie de l'aléa lié aux instabilités rocheuses sur basin géographique. Le cas des gorges du Tarn et de la Jonte (Lozère). Bull Lab Ponts et Chaussées 236:37–57
Pierson LA, Davis SA, Van Vickle R (1990) Rockfall hazard rating system implementation manual. U.S. Department of Transportation, Federal Highway Administration Report, FHWA-OREG-90-01
Ravanel L, Allignol F, Deline P, Gruber S, Ravello M (2010) Rock falls in the mont blanc massif in 2007 and 2008. Landslides 7 (4):493–501
Ravanel L, Magnin F, Deline P (2017) Impacts of the 2003 and 2015 summer heatwaves on permafrost-affected rock-walls in the Mont Blanc massif. Sci Total Environ 609:132–143
Riquelme A, Tomás R, Abellán A (2016) Characterization of Rock Slopes through Slope Mass Rating using 3D point clouds. Int J Rock Mech Min Sci. https://doi.org/10.1016/j.ijrmms.2015.12.008
Romana M (1988) Practice of SMR classification for slope appraisal, In: 5th International Symposium on Landslides, Balkema, Rotterdamm, Lausanne, Switzerland
Romana M (1993) A geomechanical classification for slopes: slope mass rating. Comprehensive Rock Engineering, Pergamon, Oxford
Rose ND, Hungr O (2007) Forecasting potential rock slope failure in open pit mines using the inverse-velocity method. Int J Rock Mech Min Sci 44(2):308–320
Rouiller J-D, Jaboyedoff M, Marro C, Phlippossian F, Mamin M (1998) Pentes instables dans le Pennique valaisan. Rapport final PNR31. VDF, Zürich, p 239
Rouyet L, Kristensen L, Derron M-H, Michoud C, Blikra LH, Jaboyedoff M, Lauknes TR (2017) Evidence of rock slope breathing using ground-based InSAR. Geomorphology 289:152–169
Royán MJ, Abellán A, Jaboyedoff M, Vilaplana JM, Calvet J (2014) Spatio-temporal analysis of rockfall pre-failure deformation using terrestrial LiDAR. Landslides 11(4):697–709
Ruiz-Carulla R, Corominas J, Mavrouli O (2016) A fractal fragmentation model for rockfalls. Landslides 14(3):875–889
Santana D, Corominas J, Mavrouli O, Garcia-Sellés D (2012) Magnitude–frequency relation for rockfall scars using a Terrestrial Laser Scanner. Eng Geol 145–146:50–64
Santi PM, Russell CP, Higgins JD, Spriet JI (2009) Modification and statistical analysis of the colorado rockfall hazard rating system. Eng Geol 104(1–2):55–65
Saroglou C (2019) GIS-Based Rockfall Susceptibility Zoning in Greece. Geosciences 9(4):163
Scavia C, Barla G, Bernaudo V (1990) Probabilistic stability analysis of block toppling failure in rock slopes. Int J Rock Mech Min Sci 27 (6):465–478
Selby MJ (1980) A rock mass strength classification for geomorphic purposes: with tests from Antartica and New Zealand, Z. Geomorphologie 24:31–51
Selby MJ (1982) Controls on the stability and inclinations of hillslopes formed on hard rock, Earth Surf. Proc Land 7:449–467
Slob S, Hack HRGK, Turner AK (2002) An approach to automate discontinuity measurements of rock faces using laser scanning techniques. In: Proceedings of ISRM international symposium on rock engineering for mountainous regions—EUROCK 2002, Funchal, pp 87–94
Stoffel M, Bollschweiler M, Vázquez-Selem L, Franco-Ramos O, Palacios D (2011) Dendrogeomorphic dating of rockfalls on low-latitude, high-elevation slopes: Rodadero, Iztaccíhuatl volcano, Mexico. Earth Surf Process Land 36(9):1209–1217
Strahler AN (1954) Quantitative geomorphology of erosional landscapes, In: Compt Rend 19th Intern Geol Cong 13:341–354
Sturzenegger M, Stead D (2009) Quantifying discontinuity orientation and persistence on high mountain rock slopes and large landslides using terrestrial remote sensing techniques. Nat Hazards Earth Syst Sci 9:267–287. https://doi.org/10.5194/nhess-9-267-2009
Taylor DW (1948) Fundamentals of soil mechanics, John Wiley & Sons, p 700
Tonon F, Kottenstette JT (2006) Laser and photogrammetric methods for rock face characterization. American Rock Mechanics Association, Alexandria
van Veen M, Hutchinson DJ, Bonneau DA, Sala Z, Ondercin M, Lato M (2018) Combining temporal 3-D remote sensing data with spatial rockfall simulations for improved understanding of hazardous slopes within rail corridors. Nat Hazards Earth Syst Sci 18 (8):2295–2308
van Westen CJ, van Asch TWJ, Soeters R (2006) Landslide hazard and risk zonation—why is it still so difficult? Bull Eng Geol Environ 65 (2):167–184
Voight B (1989) A relation to describe rate-dependent material failure. Science 243(4888):200–203
Volkwein A, Schellenberg K, Labiouse V, Agliardi F, Berger F, Bourrier F, Dorren LKA, Gerber W, Jaboyedoff M (2011) Rockfall characterisation and structural protection – a review. Nat Hazards Earth Syst Sci 11(9):2617–2651
Wagner A, Leite E, Olivier R (1988) Rock and debris-slides risk mapping in Nepal—a user-friendly PC system for risk mapping, In: Bonnard C (ed) 5th International symposium on landslides, vol 2, pp 1251–1258, A. A. Balkema, Rotterdam, Lausanne, Switzerland

Wei L-W, Chen H, Lee C-F, Huang W-K, Lin M-L, Chi C-C, Lin H-H (2014) The mechanism of rockfall disaster: A case study from Badouzih, Keelung, in northern Taiwan. Eng Geol 183:116–126

Wieczorek GF, Nishenko SP, Varnes DJ (1995) Analysis of rock falls in the Yosemite Valley, California, Proc. U.S. Symp Rock Mech, 35th, 85–89

Williams JG, Rosser NJ, Hardy RJ, Brain MJ, Afana AA (2018) Optimising 4-D surface change detection: an approach for capturing rockfall magnitude–frequency. Earth Surf Dyn 6(1):101–119

Williams JG, Rosser NJ, Hardy RJ, Brain MJ (2019) The Importance of Monitoring Interval for Rockfall Magnitude-Frequency Estimation. J Geophys Res: Earth Surf 124(12):2841–2853

Wilson RC, Keefer DK (1985) Predicting areal limits of earthquake induced landsliding, evaluating earthquake hazards in the Los Angeles Region. In U.S. Geological Survey Professional Paper, 317–345

Wyllie DC (2018) Rock slope engineering: civil applications, 5th Edn. CRC Press, p 568

Yamagishi H (2000) Recent landslides in western Hokkaido. Japan Pure Appl Geophys 157(6–8):1115–1134

Progress and Lessons Learned from Responses to Landslide Disasters

2020 Kyoto Japan

Brian D. Collins, Mark E. Reid, Jeffrey A. Coe, Jason W. Kean, Rex L. Baum, Randall W. Jibson, Jonathan W. Godt, Stephen L. Slaughter, and Greg M. Stock

Abstract

Landslides have the incredible power to transform landscapes and also, tragically, to cause disastrous societal impacts. Whereas the mechanics and effects of many landslide disasters have been analyzed in detail, the means by which landslide experts respond to these events has garnered much less attention. Herein, we evaluate nine landslide response case histories conducted by the U.S. Geological Survey over the past two decades and summarize the event history, the response conducted, and the lessons learned from each event. We group the responses into three categories—providing event context from past events, addressing ongoing hazards, and acquiring data for the future—and present the nine case studies accordingly. We also summarize the progress in landslide response that has been made over the past two decades, including insights and advancements on the preparation for such events, the use of new technologies, and the importance of clear communication between all parties during disasters. We believe that exchanging and sharing experiences such as these will promote more clear and successful approaches for responses to landslide disasters in the future.

Keywords

Landslide • Disaster • Response • Catastrophic

Introduction

Landslides are some of the world's most devastating hazards, capable of destroying everything in their path. Although some landslides can be mitigated with engineering measures or be avoided altogether to minimize economic impacts and loss of human life, many landslides occur with either such massive size or rapid mobility that consequences cannot be avoided. In these cases, undesired societal effects, whether they be related to loss of life, damaged or destroyed infrastructure, or environmental degradation, may result. These we term "landslide disasters" to describe landslides that cause devastating and sometimes tragic outcomes. Examples of landslides disasters are prevalent throughout the world and include many so-called catastrophic landslides (e.g. Schuster 1996a; Evans and DeGraff 2002). Catastrophic landslides causing disasters may be rapid and fatal events such as the 130,000 m^3 Las Colinas landslide in El Salvador that killed approximately 585 people (Jibson et al. 2004; Fig. 1), or they may be slower and non-fatal events such as the 21 million m^3 1983 Thistle, Utah USA landslide (Duncan et al. 1986; Fig. 2) that dammed a river and resulted in no direct loss of lives or injuries, but flooded an entire town and was nevertheless the costliest single landslide in North American history (Kaliser 1983, Schuster 1996b). Landslides need not be large however for them to be disastrous; debris flows of hundreds to a few thousand cubic

B. D. Collins (✉)
U.S. Geological Survey, Landslide Hazards Program, Geology, Minerals, Energy, and Geophysics Science Center, P.O. Box 158 Moffett Field, CA 94035, USA
e-mail: bcollins@usgs.gov

M. E. Reid
U.S. Geological Survey, Landslide Hazards Program, Volcano Science Center, Moffett Field, CA 94035, USA

J. A. Coe · J. W. Kean · R. L. Baum
U.S. Geological Survey, Natural Hazards Mission Area, Earthquake Hazards Program, Golden, CO 80401, USA

R. W. Jibson
U.S. Geological Survey, Earthquake Hazards Program, Geologic Hazards Science Center, Golden, CO 80401, USA

J. W. Godt · S. L. Slaughter
U.S. Geological Survey, Natural Hazards Mission Area, Landslide Hazards Program, Golden, CO 80401, USA

G. M. Stock
National Park Service, Yosemite National Park, El Portal, CA 95318, USA

K. Sassa et al. (eds.), *Understanding and Reducing Landslide Disaster Risk*,
ICL Contribution to Landslide Disaster Risk Reduction,
https://doi.org/10.1007/978-3-030-60196-6_4

Fig. 1 The 13 January 2001 Las Colinas landslide in Santa Tecla, El Salvador, was triggered by a M7.6 earthquake; the slide travelled more than 700 m, destroyed parts of a densely populated neighborhood, and killed approximately 585 people. Photo by Edwin L. Harp, USGS

Fig. 2 The 1983 Thistle landslide (Utah, USA) mobilized over several days and eventually dammed the Spanish Fork River, cutting off several transportation routes and leading to flooding of a town. Photo by Robert L. Schuster, USGS

meters in volume routinely cause significant impact to people and the built environment (Dowling and Santi 2014).

Society generally places the responsibility of responding to landslide disasters on emergency workers (i.e. first responders), and, where and when available, landslide experts. A lack of experts trained in landslide response is a fact of life in many areas of the world, and in these cases, outside expertise might need to be quickly mobilized. Whereas first responders are tasked with saving and protecting lives immediately affected by an event, the landslide expert, be it a geologist, hydrologist, engineer, or other specialist, may be asked to provide context for an event or to assess ongoing hazards. These responses are typically conducted to ensure that steps are taken to protect lives (including first responders) and property that remain in harm's way from ongoing landslide-related hazards.

Although the landslide expert might have experience evaluating potential and ongoing landslide hazards, developing a successful response strategy requires careful thought and scrutiny of possible options and resultant outcomes. Thus, being aware of what is and is not possible, and of what might happen, are defining characteristics for persons responding to landslide disasters. Further, working with first responders and government officials under stressful circumstances requires careful navigation and an ability to translate scientific understanding to actionable information. Thus, the landslide expert must also be able to communicate with a wide range of stakeholders including other technical experts, government officials, first responders, community members, landowners, and the media. A wide range of skills are therefore required for landslide response, and these are typically learned either through exposure to such events or by learning from others' experiences. It is to this purpose that we write.

Types of Landslide Response

In what follows, we present a summary of types of landslides and responses that landslide experts might encounter during disasters. This is achieved using nine case studies from U.S. Geological Survey (USGS) responses conducted over the past two decades (Fig. 3; Table 1) to illustrate the various roles in which an expert may be placed when responding to an event. Recognizing that this summary primarily represents our own experiences and context, we emphasize some general findings by organizing the case studies into three themes, loosely grouped by the temporal perspective on which a response was focused: past, present, or future. Responses focused on the past (Providing Event Context) require providing information or context on the event that just occurred. This is the subject of the first two case studies. Responses focused on the present, and commonly the most critical part of a response (Addressing Ongoing Hazards), may use real-time monitoring or assess ongoing potential for continued or renewed movement of a landslide. The next five case studies provide examples of these types of responses. Finally, a future-focused response, and the subject of the final two case studies (Acquiring Data), generally involves collecting crucial and perishable data with an eye toward better understanding landslide impacts and improving models for future hazard and risk

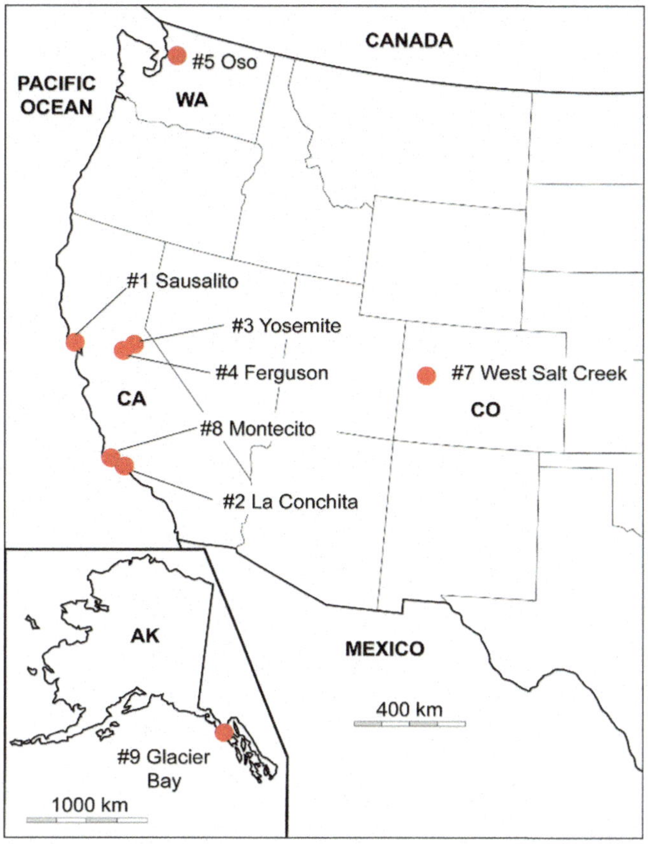

Fig. 3 Western United States showing locations of the case studies presented herein. Case study #6 (Gorkha, Nepal) not shown. AK = Alaska, CA = California, CO = Colorado, WA = Washington

assessments. Typically, landslide responses involve some aspect of each of these three perspectives; our classification is based only on which perspective dominated. Following each case study, we present some of the lessons learned from each event. We conclude by summarizing the progress that has been made in responding to landslide disasters in the nearly 20 years covered by the case studies.

The role of a landslide expert on the scene immediately following a disaster is commonly poorly defined given the chaos of the event. The expert may interact with first responders, government officials, community members and the media at various points in a response, and it may take some time to identify where best to dedicate resources to be most effective. Notably, the landslide expert may fill multiple rolls over the course of a response, and these rolls can overlap or morph during a response. An ability to recognize the need for change, and flexibility in responding to each disaster's needs is therefore imperative. The case studies and summary we present are not meant to be all inclusive, as we recognize that landslide experts from around the globe likely have their own insights and experiences on this subject. However, we intend this summary of our experiences to both help guide those new to the field, and to promote open dialogue with others working in the field of landslide response.

Factors Affecting Landslide Response

A wide range of landslide types, geologic conditions, triggering factors, as well as other circumstances can lead to a landslide disaster. The idiosyncrasies of each event dictate the type of response; thus, we present an overview of various factors that can influence initial response efforts.

Mechanism and Type

Landslides are triggered by a variety of mechanisms and have a wide array of types (Varnes 1978). Landslides can be triggered by strong earthquakes (e.g. the 1994 Northridge, California, USA earthquake, Harp and Jibson 1996; the 2004 Niigata, Japan earthquake, Sassa 2005; Kieffer et al. 2006), volcanic eruptions (e.g. the 1985 Nevado del Ruiz, Columbia lahars, Pierson et al. 1990) and collapses (e.g. the 2018 Anak Krakatau, Indonesia landslide, Walter et al. 2019), large storms (e.g. the 1999 Fukushima, Japan landslides, Wang et al. 2002; the 2011 Rio de Janeiro, Brazil landslides, Netto et al. 2013) and anthropogenic causes (e.g. the mining-induced Bingham Canyon, Utah, USA landslide, Pankow et al. 2014), among others. In many ways, all landslide responses are guided by the characteristics of the triggering event. For example, earthquakes generally trigger landslides over a broad region and thus prompt a different response than for a single anthropogenically triggered event. The mechanism and type of landslide(s) also affects the type of response through both the expertise required and the tools to be applied. Whereas debris flows may require process knowledge of hydrologic triggers to identify likely pathways for additional mobilization (e.g. Iverson et al. 2015—see Oso case study herein), a background in rock mechanics might be required to identify where ongoing failure susceptibility is highest following a rock fall event (e.g. Stock et al. 2018—see Yosemite case study herein). The choice of remote or in-situ monitoring tools during a rapid response also depends on the landslide type, and the type of instrumentation must be selected appropriately for each situation. Comprehensive monitoring techniques are outside the scope of this paper but are covered in many other publications (e.g. Keefer et al. 1987; Harp and Jibson 1995; Angeli et al. 2000; Reid et al. 2008, 2012; Collins et al. 2011; Scaioni 2015).

Table 1 Selected landslide disasters and responses in the United States (2002–2019) with international case studies discussed herein

Date[a] (YYYY-MM-DD)	Location[b]	Landslide Type[c]	Trigger[d]	Primary Response Type[e]	Volume (m^3) or Number[f]	Fatalities (Injuries)[g]	References
2002–07-02	Chuuk State, Micronesia	DF	P	B	[269]	43(n/a)	Harp et al. (2004)
2004–09-16	Blue Ridge Mts, North Carolina, USA	DF	P	C	2,800	5(2)	Wooten et al. (2008)
2005–01-10	La Conchita, California, USA (#2)	DF	P	A	200,000	10(10)	Jibson (2005)
2006–04-25	Ferguson, California, USA (#4)	RS	P	B	800,000	0(0)	Harp et al. (2008)
2012–2016	Glacier Bay, Alaska, USA (#9)	RA	PT	C	[24]	0(0)	Coe et al. (2018)
2013–04-10	Bingham Canyon, Utah, USA	RA	E	C	55×10^6	0(0)	Pankow et al. (2014)
2013–09-11	Front Range, Colorado, USA	DF	P	C	[1,138]	3(n/a)	Coe et al. (2014)
2013–12-12	Rockville, Utah, USA	RF	FT	A	1,050	2(0)	Lund et al. (2014)
2014–05-25	West Salt Creek, Utah, USA (#7)	RA	RS	B	54.5×10^6	3(0)	Coe et al. (2016)
2014–03-22	Oso, Washington, USA (#5)	DA	P	B	9.0×10^6	43(11)	Iverson et al. (2015)
2015–04-25	Gorkha, Nepal (#6)	RF, FS, DA	EQ	B	[25,000 +]	>440 (n/a)	Collins and Jibson (2015)
2015–08-18	Sitka, Alaska, USA	DF	P	B	[>40]	3(0)	Laprade (2017); Patton et al. (2019)
2017–05-20	Mud Creek, California, USA	RS	P	C	$\sim 3 \times 10^6$	0(0)	Warrick et al. (2019)
2017–09-20	Puerto Rico (Hurricane Maria)	DF	P	C	[71,431]	n/a	Bessette-Kirton et al. (2019); Hughes et al. (2019)
2017–09-27	Yosemite, California, USA (#3)	RF	TS[h]	B	10,300	1(2)	Stock et al. (2018)
2018–01-09	Montecito, California, USA (#8)	DF	P	C	680,000	23(167)	Kean et al. (2019)
2019–02-14	Sausalito, California, USA (#1)	DF	P	A	1,200	0(1)	Collins and Corbett (2019)

[a]For landslide sequences, only the first date of the event is listed

[b]# indicates a case study presented herein

[c]Landslide type: DF = debris flow; RS = rock slide; RA = rock avalanche; DA = debris avalanche; RF = rock fall

[d]Trigger: P = precipitation; PT = permafrost thaw; E = excavation; FT = freeze–thaw; RS = rain-on-snow; EQ = earthquake; TS = thermal stress

[e]Primary response type. A = Providing event context; B = Addressing ongoing hazards; C = Acquiring data for future needs. Some events involved multiple types

[f]Volume is given when known. For events with multiple landslides, the number of landslides is provided in brackets [XXX]

[g]For some events, the number of fatalities and injuries is not available (n/a) due to the widespread devastation of the event and/or the difficulties in assessing effects from secondary hazards resulting from landslides (for example, in some events, landslides blocked roads and prevented access to hospitals and food; these conditions may have resulted in landslide-related fatalities that are not always included in totals)

[h]The triggering mechanism for this event is presumptive and plausible, but may have also been affected by other subtle environmental factors

Number and Geographic Extent

Responding to a disaster consisting of multiple landslides (e.g. the 1982 San Francisco Bay, California, USA storm with over 18,000 debris flows, Ellen and Wieczorek 1988; the 2008 Wenchuan, China earthquake with nearly 200,000 landslides, Xu et al. 2014) is vastly different than responding to a single landslide. Initially, a key challenge is determining where best to commit limited resources within a geographically extensive disaster region while trying to prioritize lives and assets. The primary science role in these instances is thus to help decision makers focus on the landslides that either had or will have the greatest impact (e.g. from the potential of outbursts from landslide dams, Costa and Schuster 1988; Korup 2002; Satofuka et al. 2010). Further complicating these issues is the diversity of terrains and political subdivisions over which large disasters might extend. Geologic and geomorphic settings, as well as landslide types, are likely to differ spatially, and multiple local governments might be involved, which further challenges communication and logistics. On the other hand, the expectations of focusing all possible efforts on a single landslide can sometimes exceed those for widely distributed landslides. Installation of monitoring instrumentation might be expected on a single landslide that presents a continuing hazard but would be impractical for an event having a large number of landslides. Neither situation is necessarily more or less challenging, and the time spent responding to a disaster with one or with many landslides can be commensurate. Landslide experts, acting in concert with other responding government agencies, must therefore consider what can and cannot be achieved depending on the scale of the disaster and the resources available.

Size

Although not entirely indicative of the magnitude of a disaster, the relative size of a landslide (i.e. volume, areal extent, and/or runout) may dictate the overall response to an event. For example, the scale of a disaster is commonly modulated by landslide runout, with highly mobile debris flows being particularly consequential. As a result, response resources (personnel, mapping efforts, etc.) may be extended across variably affected terrain. The 1999 Caraballeda Fan debris flow (Vargas, Venezuela) that killed thousands stretched for several kilometers and covered a 1+ km^2 area with deposits as deep as 7 m, which required a widely distributed response effort (Wieczorek et al. 2001). Larger landslides may be more complex and involve multiple types of movement; this can require segmentation and delineation of various parts of the deposit that themselves may be of different styles and have differing potential future mobility and resulting hazard. Whereas main scarps may be prone to continual sloughing and retrogression from debris or rock falls, the toe of the same landslide may be susceptible to reactivation and generate debris flows. Larger landslides can also dissect transportation routes and thereby challenge both first responders and landslide experts in planning their respective approaches to the situation. The 2014 Oso landslide (Washington, USA, Iverson et al. 2015), blocked the main highway through the region and required initial landslide response from two different sides of the slide. All landslide responses should therefore consider how best to designate and assign resources early in the response to a disaster. Although greater landslide volume generally leads to greater loss of life and damage in a landslide disaster (e.g. Dowling and Santi 2014), landslides that cause such consequences need not always be large. Small volume rock falls of only a few to tens of cubic meters (Jibson and Baum 1999; Stock et al. 2013) have also led to highly destructive and fatal consequences requiring landslide response.

Potential for Continued Activity

The potential for additional slope instability, either locally or regionally, in the aftermath of a disastrous landslide event should be paramount in formulating an effective and safe landslide response. Large earthquakes are sometimes followed by aftershocks capable of triggering additional landslides (e.g. the 2015 Gorkha, Nepal earthquakes—see case study herein; Kargel et al. 2016; Tiwari et al. 2017), and the possibility of delayed landsliding, controlled by the time dependence of groundwater seepage, is well known (e.g. the 1991 Racha earthquake, Republic of Georgia; Jibson et al. 1994). Thus, identifying and anticipating possible hazardous conditions following a landslide disaster will affect response activities. For example, evacuating potentially affected populations might be warranted if the likelihood for additional landslide movement is high. The possibility for additional landslides or ongoing landslide movement requires landslide experts to evaluate the potential hazards to both themselves and first responders. Given that such responses commonly evolve rapidly and chaotically (e.g. with response personnel focused on the immediate needs of the situation), having an expert provide an overview of the broader potential dangers to personnel is both helpful and necessary to enhance overall scene safety.

Location

The scale of destructiveness of a landslide also depends on its location relative to population and infrastructure. Landslides occurring in urban areas generally cause greater damage and fatalities (e.g. the 2001 Las Colinas landslide in Santa Tecla, El Salvador, Evans and Bent 2004; Jibson et al. 2004), although landslides in rural areas are also capable of widespread destruction (e.g. the 2014 Oso landslide, Washington, USA, Iverson et al. 2015). Whereas urban landslide events generally allow rapid access for response, many landslide disasters occur in areas that are remote or are difficult to access. Responding to remote landslides may reduce the potential resources that can be dedicated to an event and can increase the response time. During the 2015 Gorkha Nepal, earthquake, landslides in many affected regions could only be reached by multi-day hikes or by helicopter, making for challenging decisions regarding how best to utilize time and available resources (Collins and Jibson 2015—see Nepal case study herein). In some cases, landslides may be virtually inaccessible on the ground despite their potential for severe consequences. For example, landslides with tsunamigenic potential in steep and glaciated parts of Alaska have largely been investigated using remote (air and satellite) imagery (Coe et al. 2018, 2019—see Glacier Bay case study herein). Experts must therefore consider how best to plan for a response to a remote location. With the advent of new technologies such as high-resolution multitemporal imagery, archived and rapidly collected InSAR data, and highly portable drones, the challenges to mounting an effective response to a remote area undergoing a landslide disaster are becoming more manageable.

Cultural Setting

Some disasters may require a host nation to request international assistance in managing a landslide crisis. This can arise when countries either lack landslide expertise or may be too overwhelmed by the immediate in-country humanitarian needs to mobilize their own resources. Given that even developed nations may have small numbers of landslide experts able to respond, these efforts can become collaborative involving personnel from multiple countries. During the response to the 2015 Gorkha, Nepal earthquakes, landslide expertise from 12 countries was utilized to rapidly construct a landslide inventory (Kargel et al. 2016) that immediately proved useful for in-country landslide response (e.g. Collins and Jibson 2015). Working across international borders can pose logistical barriers (i.e. visas, language barriers, unknown transportation conditions and/or restrictions, unfamiliarity with local government structure and authorities, etc.). Although these issues are surmountable, responding experts must be attentive to the cultural setting and norms of the country. Many details of a response are best handled through in-country collaborators and/or guides who can provide information on local customs, appropriate behavior, etc. Most importantly, international responses must recognize and be sensitive to the host country's specific needs. Usually, this can be accomplished by ensuring that responses remain focused on providing support to the in-country partners who remain the responsible parties for hazard communication. This is particularly important for post-disaster research-type responses (in either international or domestic settings), that seek to collect perishable data shortly after landslides have occurred for the purpose of future hazard preparedness. All responses should ensure that ethical guidelines for data collection are followed (e.g. Gaillard and Peek 2019) to provide the best response to the people most affected by a disaster.

Providing Event Context—Looking to the Past

Although landslides are common throughout the world, for any one community the occurrence of a landslide disaster could be a once in a lifetime event. First responders, community members, and local governments might not have experience or even knowledge of landslide hazards. Landslide experts responding to these disasters are thus in the unique and crucial position to disseminate the scientific understanding needed for emergency and planning personnel to grasp the magnitude and impact of the event, and manage expectations following a disaster. The cases studies presented in this section showcase how providing answers and context on landslides that have occurred in the past can be utilized to inform communities about events that have just happened.

Case Study 1. Providing Answers and Information: The 2019 Sausalito, California (USA) Debris Flow

Having the appropriate language to describe a landslide disaster can be paramount for a response to move forward in a calm and orderly manner. The 2019 Sausalito debris flow in California, USA (Fig. 3) destroyed several homes but there were no fatalities. This event offers a snapshot of how rapid response to a relatively modest landslide can benefit local municipalities by providing answers to immediate questions that may arise.

Event Description

In the early hours of 14 February 2019, high-intensity precipitation fell throughout the northern San Francisco Bay region of California following a month of wet weather that had saturated hillslopes throughout the region. The storm triggered isolated shallow landslides leading to road closures and minor damage, but one landslide destroyed two homes and damaged four others in the city of Sausalito (Fig. 4). The approximately 1200 m^3 landslide source, composed of

Fig. 4 The 2019 Sausalito (California, USA) debris flow destroyed two homes in a densely populated area north of the city of San Francisco. Views of **a** main scarp and source area, and **b** runout zone. Images adapted from Collins and Corbett (2019)

colluvial soil, initiated from a vegetated hillside above the city, flowed over a road, collided with homes, and travelled a total distance of 220 m. The debris flow carried a sleeping resident 100 m downhill and trapped the resident in the ruins of the home. Local fire and police responded within minutes and were able to extract the resident from the debris with only minor injuries. The lack of lives lost was probably due to two factors: (1) landslide debris likely undermined the bottom of the residence's structural system and allowed at least partial rafting of the structure on top of the debris flow, and (2) the bedroom of the resident was on the upper level of the structure. Although the city had experienced a similar (but more tragic) debris flow 36 years earlier in which one person was killed, the current first responders, community members, and government officials were facing an unfamiliar situation with respect to their immediate experiences.

Response

Among the major priorities that first responders needed to contend with were the unknown number of potential victims in the debris flow, natural gas leaks initiated from ruptured pipes, downed and live electrical power lines, and several blocked roads. Locally based USGS personnel arrived approximately 8 h after the event to document the landslide's characteristics as part of an ongoing project on landslide hazards in the region (Collins et al. 2012). However, upon arrival it was evident that local geologic and landslide expertise would be helpful to first responders. After contacting the incident commander (the local Director of Public Works), USGS offered assistance and began communicating with on-site first responders, primarily police and fire department staff. Initially, it was most helpful to provide immediate situational awareness, including evaluating concerns about secondary landslides that might mobilize from the source area and that could jeopardize search-and-rescue personnel. A subsequent site reconnaissance to look for items such as tension cracks and seepage zones allowed landslide experts to provide relevant information to city staff and emergency responders concerning the nature of the landslide and the potential for additional movement. Here, secondary landslides from the source area were a concern and spotters were set up to notify emergency responders in the case of reactivation. By being able to identify the type of movement (i.e. the event was a debris flow—a type of landslide), the cause (intense precipitation on saturated hillslope soils led to the landslide), and the commonality of such events (debris flows like this have occurred in other parts of the region in past years, and this event is one of several other landslides to occur from this storm), landslide experts were able to inform city staff and bring some order to a difficult initial response period.

In the days following the debris flow, the USGS continued to answer questions and provide information and suggestions to help manage traffic along the affected roadways and to engineer mitigation of the source area prior to the arrival of additional storm precipitation. During this phase, experts were able to take time to collect perishable data (terrestrial lidar models; Collins and Corbett 2019) for future analysis. Given that few cities know how to begin a landslide disaster clean-up (including such items as removal of hazardous waste resulting from oil and gasoline leaking from entangled vehicles), experts also provided names and contact information for personnel from other municipalities that had undergone similar landslide disasters, thus providing a path forward.

Lessons Learned

This event provided insight into the benefits of rapid response. Having a nearby presence to a disaster site clearly provides numerous advantages in that local landslide expertise can be quickly conveyed to first responders. Further, the presence of a landslide expert rapidly on scene can reassure city officials that (1) they are not facing an unprecedented event, and (2) they are surrounded by knowledgeable and competent experts who can provide technical assistance useful to keeping their personnel safe during the response. From a holistic viewpoint, being able to recognize people's critical needs during the situation and evolving one's role and messaging to those needs is fundamental to providing the best possible response.

Case Study 2. Providing Historical Context: The 2005 La Conchita, California (USA) Landslide

Responding to landslide disasters commonly requires looking to the past to place current events into context. Community members and government officials may desire information about the geologic, hydrologic, environmental, and sometimes anthropogenic history of an event to understand why a community has experienced a disaster and how to best move forward. This type of contextual information is especially relevant for landslide reactivations, particularly when a damaging or deadly landslide strikes a community that had experienced previous landslides that were far less damaging. Here, the landslide expert's role is to look at the record of landslides in the area and to place the current disaster within a framework that can assist community members in understanding the broader context of the event—outside the immediate tragedy. The 2005 La Conchita landslide in California, USA (Fig. 3) is an example of such a disaster, where reactivation of a landslide that had limited movement in 1995 resulted in a tragically different outcome 10 years later (Jibson 2005, 2006a). Here we show how a timely response to a disaster by landslide experts assisted with community decision making.

Event Description

The 2005 La Conchita landslide occurred on 10 January 2005 from a 180-m-high coastal bluff along the southern California coastline, approximately 25 km southeast of the city of Santa Barbara. The bluff had experienced prehistoric landsliding (Fig. 5) and consists of poorly indurated marine sediments, primarily shale, mudstone, siltstone and sandstone; an active fault trace extends across the slope face. In 1995, a 1.3-million-m^3 failure occurred from the coastal bluff due to rising groundwater associated with above-average seasonal rainfall. In 2005, 200,000 m^3 of the 1995 debris deposit remobilized, again following above-average precipitation. The 1995 landslide acted primarily as a complex slump-earth flow and debris travelled only about 30 m from the base of the slope; although it moved fairly quickly, people were able to evacuate and avoid injury. In 2005, the remobilized debris travelled more than 100 m from the slope toe as a rapid debris flow. The stark differences in mobility resulted in disparate consequences: the 1995 landslide destroyed and damaged 9 homes and caused no deaths, whereas the 2005 debris flow destroyed 36 homes and caused 10 deaths.

Response

Initially, the response to the 2005 landslide was conducted by emergency responders and the local (county) geologist, who arrived at the site just hours after the event. Upon request by the state geological survey (California Geological Survey, CGS), the USGS subsequently visited the site four days later to provide their interpretations of the event in an effort to disseminate scientific information about the landslide. Immediately after the site visit, representatives of the USGS and CGS along with the county geologist met with county government leaders to provide an overall assessment of the situation and respond to questions. Given that a previous damaging landslide had occurred 10 years earlier, the context and history of the 2005 event were extremely relevant. Within a couple of weeks, a report was published detailing the observations made of the landslide and a frank assessment of continuing landslide hazards (Jibson 2005). The report provided concise descriptions of the two recent (1995 and 2005) landslides along with information pertaining to their location within a steep, geologically active setting (Fig. 5). This information, along with supporting correspondence and photographs (Jibson 2006b), was subsequently made public and therefore equally available to various parties involved in legal action regarding causation of the landslide. The messages contained in the publications were also used by local and state officials to initiate discussion among community members on the relative hazard faced by residents of the La Conchita area. Although these discussions resulted in the community remaining where they were (due primarily to private property rights and a local acceptance of living with this hazard), the information

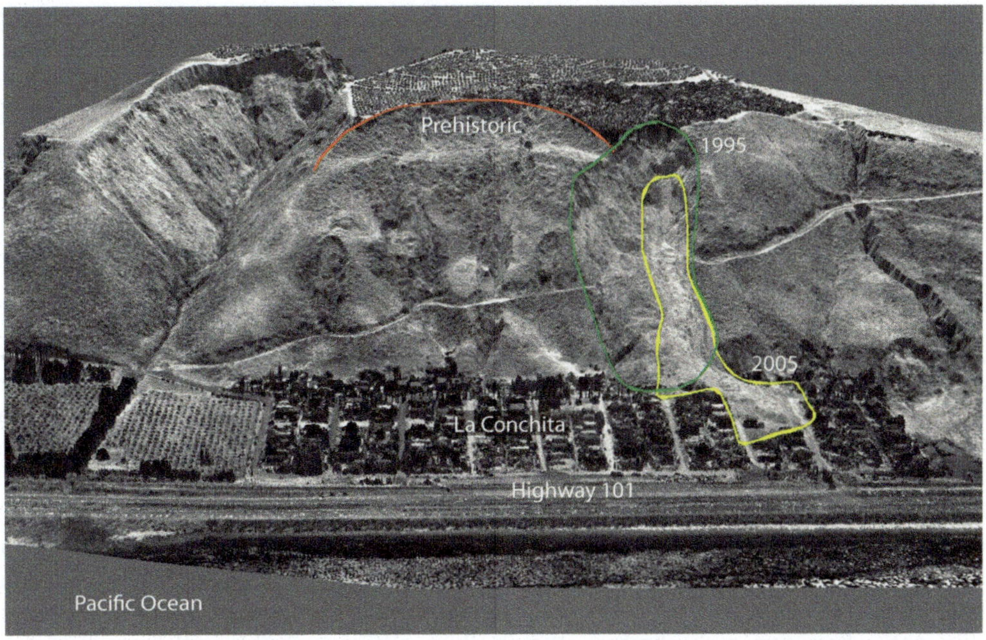

Fig. 5 Oblique lidar image of reactivations of the La Conchita (California, USA) landslide. Figure adapted from Jibson (2005). This information assisted affected parties with community decision making following the 2005 event

provided by this landslide response allowed the affected parties to make decisions using the best available information—obtained rapidly following the event.

Lessons Learned

The most beneficial result of this response was the ability to transmit to all parties—homeowners, emergency responders, and local government officials—a rapid assessment of the post-landside situation in the broader context of past and ongoing landslide hazards. Even if all parties were not pleased with the findings, they had credible, objective information with which to make decisions about future actions. Landslide disasters that involve private property owners, local government entities, and extensive media coverage, can become legally and politically sensitive. Having a non-local, federal-level landslide expert be part of the assessment team provided an additional degree of objectivity and insulation from the legal evaluation process. Given the political and legal complexities of the event, it was difficult for local experts to offer assessments without appearing to take sides or to have a conflict of interest. Several lawsuits were filed as a result of this disaster, and the USGS was involved in responding to subpoenas and U.S. Freedom of Information Act (FOIA) requests for several years following the event (Jibson 2006b). Although no personal liability was involved in these requests, data and expert testimony were sought by multiple parties, both officially and off the record. Experts who conduct post-disaster responses must be thoughtful and circumspect in what they speak, write, and offer to the media because of the legal and political issues that can arise following disasters.

Addressing Ongoing Hazards—Focusing on the Present

With many landslides, the potential for continuing activity may not be entirely over for some time afterward. Landslide source areas can remain active, and landslides can remobilize either entirely or partially from their deposits. The time frame for these additional concerns may be the hours to sometimes weeks that first responders are working in hazard zones but can also extend for years after an event. Landslide experts may thus be asked to evaluate the existence and/or probability of ongoing hazards and anticipate the consequences of these hazards. The following five case studies exemplify the different activities that might be conducted during an active landslide response.

Case Study 3. Analyzing Ongoing Landslide Potential: The 2017 El Capitan, Yosemite, California (USA) Rock Falls

Determining the potential for additional landslide hazards can be exceedingly challenging in an environment where few data or little landslide history is available. On the other hand, for locations having long histories of landslide activity and proactive research programs focused on assessing and reducing geohazards, experts can be well-positioned to provide rapid hazard assessments to officials charged with managing the crisis. The 2017 rock falls in Yosemite Valley

(Yosemite National Park, California, USA; Fig. 3) are an example of this type of response, where two days of rock falls impacted both visitors and infrastructure and where experts who were already conducting rock fall assessments in the park were called to the scene to evaluate the new hazards.

Event Description

Rock falls are common in Yosemite National Park where glacial erosion has created steep granitic cliffs extending 1 km above the valley floor and millions of visitors view the iconic landscape each year. Rock falls that affect people and infrastructure in the park are less common but do intermittently cause road closures and sometimes fatalities (Stock et al. 2013). On the afternoon of 27 September 2017, a 290 m^3 rock fall from one of Yosemite's most famous cliffs (900-m-tall "El Capitan") impacted the base of the wall (Fig. 6a), where two rock climbers were walking (Stock et al. 2018). One climber was killed and the other seriously injured. Six additional rock falls occurred from El Capitan during the next three hours as first responders raced to the scene to assist. The following afternoon, another rock fall occurred from the same area, this one an order of magnitude larger ($\sim$9800 m^3) than the previous day's seven events combined. A rock fragment from this rock fall struck a vehicle traveling on the road below and severely injured the driver. This prompted temporary closure of the main road out of Yosemite Valley.

Response

National Park Service (NPS) and USGS landslide experts specializing in rock falls quickly responded to the scene in the minutes and days following these events. Of primary concern was the potential for more rock falls from adjacent areas that might affect visitors and infrastructure. Photos of the rock-fall source area, taken from the ground and from a helicopter within a few hours after each event, were subsequently turned into three-dimensional (3-D) models using structure-from-motion (SfM) methods applied overnight by colleagues in Switzerland. Using existing baseline terrestrial lidar point-cloud data, these SfM models were then used to extract rock-fall volumes (Fig. 6b) and to estimate plausible volumes of potential subsequent rock falls from the newly active source area (Stock et al. 2018). Additional analysis of joint and fracture patterns in the imagery and 3-D models indicated low potential for an imminent rock fall that could reach the now-closed road below the cliff. As a result of these conclusions, and approximately 24 h following the last and largest rock fall on 28 September, the road was reopened, thereby lessening the significant burden on traffic-control resources within the heavily visited park. The rapid analyses enabled timely public messaging about these rock falls, including information specifically targeted at those visitors (i.e. rock climbers) that might venture into areas within the potential reach of additional rock falls in this location.

Lessons Learned

Rapid response for this event was aided by several factors. Foremost was having experts well-versed in the geomorphology of the area stationed regionally and near the event. However, collaborators capable of conducting analyses away from the scene (in this case in a different country and time zone) were also critical to the success of this response. On-scene landslide experts are commonly tasked with multiple roles, such as communicating with first responders and incident managers, that reduces the time available to conduct rigorous hazard analyses. Strong working relationships with outside collaborators can be critical to providing the best available information during a landslide response. Finally, the existence of baseline data prior to an event was also vitally important to this response. Terrestrial lidar data had been collected for most of Yosemite Valley's cliffs as part of other research projects (e.g. Collins and Stock 2012; Stock et al. 2014; Matasci et al. 2018) as had high-resolution imagery (https://www.xrez.com/yose_proj/Yose_result.html); this allowed rapid analysis and contextualization of the post-event SfM data.

Case Study 4. Partnering to Develop Mitigation Strategies: The 2006 Ferguson, California (USA) Landslide

When landslides impact multiple critical assets (e.g. roadways, utilities, etc.), the complexities of response can multiply as different government agencies and community interests develop their own mitigation strategies. Partnering to distribute responsibilities can be crucial when rapidly developing action plans, and these partnerships require clear communication between all involved partners. The 2006 Ferguson landslide in northern California, USA (Fig. 3) is an example where landslide experts from several agencies combined their experience to manage the threats posed by an active landslide. This slide had large potential impacts, including the possible formation of a landslide dam within a heavily travelled canyon.

Event Description

Following two years of above-average precipitation, a slate-phyllite rock-slide complex (the Ferguson rock slide) reactivated in April 2006 with isolated rocks tumbling from

Fig. 6 The 27–28 September 2017 Yosemite Valley (California, USA) rock falls totaled ~10,300 m^3, killing one, injuring two, and temporarily closing the main exit from Yosemite Valley. **a** Image of the 28 September event. Note cars in the bottom left of image for scale. **b** Rock-fall volumes from all events (including historical and subsequent events between 2010 and 2017); figure adapted from Stock et al. (2018). Photo in (a) by Philip Bay

the talus located at the toe of the landslide onto the roadway below. By May 2006, talus from movement of the slide blocked the entire highway for about 190 m and extended 10 m into the adjacent Merced River (Fig. 7a). Importantly, the highway serves as the only all-weather road into and out of Yosemite National Park, one of the most popular parks in the United States, because winter storms can block other routes in the surrounding mountainous terrain. Lack of highway access severely impacted towns upstream and downstream of the slide, and reduced the influx of tourists and their local spending. Furthermore, catastrophic failure of the 800,000 m^3 rock slide had the potential to dam the adjacent Merced River, thereby flooding homes upstream of the slide as well as downstream if a landslide dam subsequently failed (Harp et al. 2008). As slide movement continued in 2006, a multitude of agencies began developing mitigation strategies for the rock slide and its potential effects. Although the highway was temporarily rerouted after 92 days to the river bank opposite the slide, heightened concerns remained about whether or not the landslide might entirely block the canyon and dam the Merced River.

Response

Because the Ferguson rock slide affected the highway, the river, adjacent U.S. federal lands, a National Park, and nearby communities, each of which reflected different interests, multiple agencies and landslide experts examined the hazard and assessed likelihoods and uncertainties about future behavior. In response, some of these agencies undertook monitoring efforts designed to improve public safety of workers and travelers in the corridor, as well as to better understand the landslide's behavior, both to aid geotechnical design of a new roadway and to evaluate future motion. Because information was needed rapidly, agencies selected monitoring techniques with which they had experience in implementation and data interpretation. To monitor slide deformation, the maintainer of the highway (California Department of Transportation) used repeat surveying of monuments located on the slide as well as repeat scans from ground-based radar. The USGS, as the landslide science agency to U.S. federal organizations, used helicopter-deployed GPS spider units equipped with radio telemetry based on technology used to monitor active volcanoes (Reid et al. 2012; Fig. 7a). Finally, the U.S. Forest Service, as the landowner of the landslide, installed water-level sensors upstream and downstream of the slide, with data transmitted via satellite, to detect changes in river flow that could result from a landslide dam (DeGraff et al. 2015). Some of the near-real-time data obtained from these monitoring efforts were continuously shared between agencies via a dedicated website, and a color-coded web-based alert-level notification

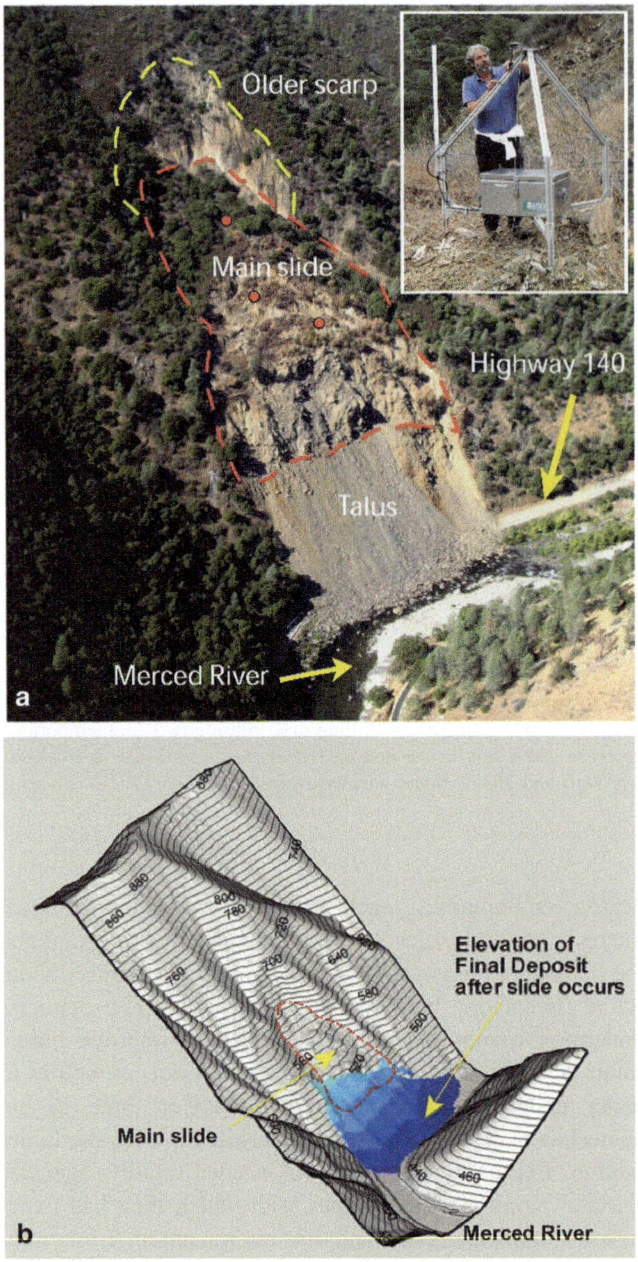

Fig. 7 **a** Overview of the Ferguson rock slide (California, USA) showing main slide with its toe located mid-slope. Talus from the slide buried highway 140 leading to Yosemite National Park and encroached into the Merced River. Red dots show locations of three GPS spider units (Reid et al. 2012). Inset shows one of the spider units with radio telemetry. **b** Simulated rock-slide deposit blocking river from potential rapid failure of entire slide mass (after Denlinger 2007). Slope length of main slide is approximately 200 m

system for slide activity was made publicly available (DeGraff et al. 2015). To address plausible runout scenarios from a potential catastrophic failure and its effects on the highway and river, the USGS performed a series of numerical simulations (Fig. 7b; Denlinger 2007). Although the Ferguson rock slide did not fail catastrophically and dam the river, parts of it did accelerate in response to modest rainstorms, and overall the slide moved more than 7 m downslope over several wet seasons (De Graff et al. 2015).

Lessons Learned

During the Ferguson rock-slide response, strong interagency communications and data sharing, combined with a clear alert notification system for the public, strengthened confidence in the assessment of both the current and projected activity of the rock slide. Clearly defined lines of responsibility also enabled smoother responses to changing conditions on the ground such as issuing alerts and controlling traffic. Different agencies contributed complimentary landslide monitoring techniques that reflected their expertise. Redundant monitoring systems for capturing landslide displacement enabled confident assessments of current slide activity. Relying on one monitoring technique alone would have led to some temporarily spurious interpretations, either due to sensor noise or misinterpretation of incoming data. In addition, relatively high-frequency displacement measurements from the GPS spiders (relayed every 15 to 30 min) were helpful for capturing the behavior of ongoing motion in response to precipitation during the years following the initial movement.

Case Study 5. Protecting First Responders Through Monitoring: The 2014 Oso, Washington (USA) Debris Avalanche

Landslide disasters sometimes require first responders to conduct search, rescue, and recovery operations in locations where they are exposed to continuing hazard from additional landslide movement. In these circumstances, landslide experts may be called in to evaluate the ongoing hazard and provide situational awareness to search operations. In some cases, near-real-time observations and monitoring may be suitable for identifying new instabilities in an effort to provide sufficient time for evacuation of people potentially in harm's way. The 2014 Oso, Washington, USA landslide (Fig. 3) is an example of an event where search operations were in the path of a potential landslide reactivation for several weeks. Landslide experts were tasked with deploying monitoring systems and alert protocols to ensure the safety of professional and volunteer searchers should the landslide reactivate.

Event Description

The 22 March 2014 Oso landslide occurred on a sunny morning following a month of anomalously high precipitation in western Washington State (Henn et al. 2015). The landslide initiated from a 180-m-tall terrace of glacial deposits (Badger and D'Ignazio 2018), entrained water from a 100-m-wide section of river adjacent to the slide, and travelled more than 1 km across the flat river flood plain, eventually reflecting off the opposite valley slope and burying a section of the main east–west roadway through the region (Collins and Reid 2020). The landslide deposit temporarily dammed the river for 25 h and impounded a lake that extended 3 km upstream (Magirl et al. 2015). The community of Steelhead Haven, built in the 1960s on the valley flats across the river from the landslide source (Keaton et al. 2014) and consisting of approximately 50 homes, was swept away, killing 43 people. With a volume of ~9 million m^3, the landslide moved an anomalously long distance given the style of sliding and height of the slope from which it originated (Iverson et al. 2015). Although the distal end of the landslide consisted of a highly mobile debris flow, most of the deposit covering the flood plain consisted of intact debris with hummocks up to 20 m high that were rafted on liquefied alluvial sediment beneath the overrunning landslide (Collins and Reid 2020). Initial search-and-rescue operations were followed by five weeks of recovery efforts, placing hundreds of responders in deep mud and debris downslope of a potentially unstable hillslope (Fig. 8a).

Fig. 8 **a** Downslope view from the top of the main scarp of the 2014 Oso (Washington, USA) landslide debris zone. Hundreds of rescue and recovery personnel worked at the distal deposits of the landslide for five weeks, which required continuous monitoring during daylight working hours. Yellow dot shows location of observation lookout. **b** View from the observation lookout at the distal end of the landslide where a surveyor took total-station measurements of control points along the main scarp, and a geologist was positioned to observe the slide and initiate evacuation orders if needed

Response

Geologists from the local (county) government were on-site the day of the disaster, and landslide experts from the state (Washington State Geological Survey) assisted with the response the following day. USGS hydrologists arrived one day later to help with reconnaissance and to deploy temporary stream gaging to monitor flooding from the potential failure of the landslide dam. The emergency response was initially coordinated by local officials and focused on rescuing survivors. As the mission shifted from rescue of survivors to recovery of victims, and the need to support hundreds of volunteer and professional searchers became apparent, responsibility for the response shifted to a US FEMA (Federal Emergency Management Agency) Incident Management Team (IMT).

Because search operations within the landslide deposit were expected to continue for an extended period, search personnel were exposed to potential impact from additional landslide activity. Landslide experts, organized in a team from different agencies, determined that the greatest threats were retrogressive failure of materials from the main scarp area, possible remobilization of landslide deposits remaining on the slope, and potential additional landsliding from adjacent hillslopes. To monitor the current state of landslide activity and to potentially identify precursory signs of large landslide activity, the team deployed a diverse monitoring network of instruments and human observers. Instrumentation included three spider units deployed via helicopter sling load (see previous Ferguson landslide case history and Fig. 7a; Reid et al. 2012) which contained high-resolution GPS receivers and geophones to monitor both 3-D surface displacements and vibrations potentially indicative of subsurface instability. Displacement monitoring also utilized three extensometers located upslope from the main scarp to detect potential retrogressive motion of the landslide. Periodic observations included weekly terrestrial lidar surveys of the main scarp area and hourly total-station measurements of 16 survey reflectors mounted at the edge of the main scarp. Crucial to maintaining overall situational awareness was stationing a geologist to directly observe the landslide area when field operations occurred (Fig. 8b). The geologist was in radio communication with the rest of the landslide monitoring team and could alert the IMT and search personnel if evacuation was needed. The overall monitoring and observational network was designed to include redundant observations thereby reducing the potential impacts of instrumentation failures and/or visual impairment from weather. Monitoring data were collected continuously and reviewed by members of the landslide team both on-site and remotely to evaluate any trends in landslide activity that might pose a threat to search operations. Monitoring activities were maintained for the duration of the search operations, which were conducted in daylight hours every day for more than five weeks.

During the first month of the response, at least six geologists from local, state, and federal agencies were stationed in the area, with one reporting directly to the IMT. This liaison geologist was a critical link between the expert team and the IMT; the liaison provided answers to questions about landslide conditions, translated monitoring observations, and communicated operational posture and planned activities. Many other staff from the agencies involved provided remote support from their home offices by analyzing and interpreting monitoring, topographic, and geologic data; providing modeling results to assist with search operations (Iverson et al. 2015); maintaining IT and other systems; attending to administrative requirements; and fielding inquiries from decision makers, the media, and the public. No additional catastrophic failure occurred during the monitoring period.

Lessons Learned

The Oso landslide response differed from other case studies primarily in that experts needed to provide continuous situational awareness to an ongoing search and recovery operation for a lengthy period of time. Key to the success of this mission was establishing a working trust among the representatives of the different governmental agencies that contributed expertise. These communication pathways extended to both local and state officials and throughout the organization of the IMT via the liaison geologist. Given the length of this response, both IMT personnel and landslide experts were rotated in two-week shifts. The imperfect alignment of rotation schedules resulted in communication challenges and some erosion of trust in the scientific information being provided to the IMT. The role of the geological liaison was key in maintaining continuity with the IMT. Success of the monitoring program, although never fully tested by a subsequent catastrophic event, was aided by the availability of GPS spider monitoring instrumentation that had already been prepped for deployment to another field site. This fortunate happenstance allowed for this equipment to be deployed in a matter of days. Rapid deployment for future responses requires plans and resources to maintain a ready-to-go cache of critical equipment and associated IT infrastructure for data capture and display. Gaps in scientific understanding mean that assessment of immediate landslide threats are fraught with uncertainty. The utility of precursory deformation and seismic measurements to provide adequate warning of landslide movement during the Oso response was uncertain, and additional effort to adapt research monitoring techniques to operational settings is needed.

Case Study 6. Assessing Community Safety: Landslides from the 2015 Gorkha, Nepal Earthquake

Responding to regional landslide events presents significant challenges due to the wide spatial extent that a disaster may encompass. These challenges are further amplified if the landslides affect people and infrastructure in remote areas. Understanding the scope of the disaster takes significant effort, and deciding where resources should best be focused requires careful and thoughtful consideration. The 2015 Gorkha earthquakes in Nepal caused widespread landslides that blocked roads and rivers and destroyed entire villages, some in very remote areas. An international consortium of remote-sensing and landslide experts rapidly formed a response team to map the landslide distribution and provide in-country expertise to affected communities.

Event Description

The 2015 Gorkha earthquake sequence occurred in the central Himalayan region of Nepal and consisted of a M7.8 earthquake on April 25, followed by a series of $M > 6$ aftershocks, including a M7.3 event on May 12. Given the steep topography affected by seismic shaking, landslides, mostly consisting of rock falls, rock slides, and debris avalanches, were widespread in the varied lithologic terrain of the affected region. More than 25,000 landslides occurred over a 30,000 km^2 region (Roback et al. 2018; Collins and Jibson 2015) of mostly rural terrain interspersed with small villages. Some villages were entirely destroyed as a result of the largest landslides (e.g. in Langtang, more than 350 perished from a debris avalanche; Kargel et al. 2016). In other locations, landsliding killed fewer but posed ongoing hazards due to potential reactivation from aftershocks and monsoon precipitation. In still other locations, landslides destroyed roads and blocked river valleys, and threatened villages located both upstream and downstream due to potential flooding and landslide-dam breaching (Fig. 9). In total, landslides were responsible for hundreds of deaths (Froude and Petley 2018) and continued to pose threats to the many villages spread throughout the seismically active area during the immediate months afterwards and during the following monsoon season.

Fig. 9 Identification of landslide dams from remotely sensed imagery and media reports guided on-site responses to remote, mountainous areas of ongoing hazard. Here, a landslide buried the evacuated village of Baisari (Myagdi District) and temporarily blocked the Kali Gandaki River. Figure adapted from Collins and Jibson (2015)

Response

Given the wide spatial extent of landsliding in this event, the initial in-country response was aided by out-of-country remote sensing efforts to assist with identifying areas that had damaging landslides. Several international teams (e.g. Kargel et al. 2016; Williams et al. 2018) divided the affected region into subregions to distribute the work-load of labor-intensive satellite-imagery-based landslide mapping. Results from this mapping, including the locations of blocked roads, dammed rivers, and landslide-affected villages, were quickly made public via various web portals to aid response efforts. As part of these efforts, a USGS team traveled to Nepal to provide assessments of ongoing landslide hazards (Collins and Jibson 2015). In cooperation with in-country response organizations (e.g. ICIMOD, https://www.icimod.org/), landslide experts traveled to twelve priority areas that had previously been identified by the remote-sensing teams as having the greatest potential for significant post-earthquake and post-landslide hazards. Travel to affected areas was conducted by helicopter, thereby surmounting the numerous logistical and time-demanding challenges of overland access to the remote villages affected by landslides. In total, 17 sites were assessed over 5 days, and 3200 km of flights lines were traversed which resulted in the identification of 69 valley-blocking landslides (Fig. 9). The results of on-site assessments (i.e. whether a landslide showed indications of ongoing hazard to a village) were communicated directly to village officials and citizens via a translator; the locations of potentially hazardous landslide dams were communicated to in-country government agencies stationed in the city of Kathmandu.

Lessons Learned

The collaboration between external remote-sensing specialists and in-country landslide experts was critical for this response. Navigating and determining how and with whom to establish collaborations with in a foreign country during a disaster requires proactive outreach and clear communication. Open and altruistic data sharing of rapidly disseminated imagery (e.g. disseminated through portals such as Google Crisis, https://www.google.org/crisismap/) and landslide locality maps, such as occurred for the Gorkha event, should be applauded and emulated in future responses. Although the response time of landslide identification using remotely sensed data needs improvement (Williams et al. 2018), these data still proved useful in the 2015 Gorkha event assessments. For an event that affected tens of thousands of square kilometers, remote-sensing-based allocation of response resources combined with efficient transportation methods (a helicopter) was crucial in providing available support where it was most needed. The use of helicopters in steep and varied topography should be conducted with the utmost caution however and preferably with multiple pilots familiar with the terrain; during the first month following the earthquakes, two helicopters assisting with humanitarian efforts crashed causing additional fatalities as a result of this event.

Case Study 7. Establishing Long-Term Monitoring Partnerships: The 2014 West Salt Creek, Colorado (USA) Rock Avalanche and 2016 Outburst Flood

Landslide agencies are rarely able to make long-term monitoring commitments for every landslide to which they respond. Given the frequency of landslide disasters each year (e.g. Froude and Petley 2018), allocated resources from any one agency can be quickly expended. Although local authorities may not be trained in landslide response, they are often excellent partners for long-term monitoring due to their inherent and dedicated engagement with affected communities. The 2014 West Salt Creek rock avalanche (Colorado, USA; Fig. 3) provides an example of a fatal and catastrophic landslide that continued to pose hazards over the years following initial failure. Partnerships with agencies from the local to federal level provided a mechanism for long-term monitoring of a large (tens of millions of m^3), post-rock-avalanche slump block that remained near the head of the slope, as well as a lake it had impounded.

Event Description

The 25 May 2014 West Salt Creek rock avalanche (White et al. 2015; Coe et al. 2016) occurred on the north flank of Grand Mesa in western Colorado (Fig. 10). The landslide, originating from a preexisting shale and mudstone rock slide, was triggered by a rain-on-snow event during otherwise unremarkable and below-average daily and seasonal meteorological conditions. Eight movement phases defined the complex motion of the landslide, beginning with a debris flow that originated from the existing rock-slide slope, followed about 12 h later by the main rock avalanche, and then subsequently composed of a medley of movement phases related to rearrangement of the newly formed rock-avalanche deposit (Coe et al. 2016). The resulting amalgamation of deposits, as well as eyewitness accounts, provided indications of precursor motion (noises, blocked irrigation ditches) as early as 12 h before the main rock-avalanche event. Three local workers investigating an irrigation ditch blocked by the initial debris flow were killed by the ensuing rock avalanche. Subsequent overtopping and incision of a slump-block dam near the main scarp two years later (27 May 2016) resulted in an outburst flood and debris flows totaling about 150,000 m^3 onto the downstream rock-avalanche deposit, gas well pads, and roads and ranches located as far as 13 km from the lake (Baum et al. 2016; Coe et al. 2017).

Response

The initial response by the local volunteer fire department occurred within hours of the event after nearby residents saw the deposit and realized the three workers were missing. Within the next few days, representatives from multiple local, state, and federal agencies gathered at a local emergency response command center to assess any ongoing landslide and flood hazards at the site and evaluate the risk to downslope residents. The assembled group included law enforcement personnel, weather forecasters, landslide geologists, dam-safety officials, and flood hydrologists, as well as the administrator of the nearby town and land-management officials. This diverse technical group conducted field assessments, decided which roads should be closed, held daily technical briefings and planning sessions, developed monitoring and emergency-response thresholds and plans, provided media briefings, and held multiple town meetings to inform residents downstream from the landslide and growing lake near the main scarp. These meetings continued for two weeks following the event, during which time it was determined that local county and federal government personnel had the equipment, capabilities, and interest to conduct longer term monitoring at the site. Over the next several months, local personnel installed monitoring equipment, including real-time cameras, GPS receivers, and a water-level gage in the lake. Some of this equipment was still being used five years after the installation. Technical personnel from multiple agencies continue to meet every winter to review and evaluate the monitoring data and the existing response plan prior to the beginning of the spring (March-June) snowmelt season when the landslide hazard is generally higher.

Lessons Learned

The response at West Salt Creek was successful for several reasons. First, and most importantly, a group of people with a wide range of expertise advised and empowered local authorities to begin immediate monitoring and to continue it into the foreseeable future. Second, the advisory group did not disband after the initial response in the spring of 2014

Fig. 10 Overview of the west salt creek (Colorado, USA) landslide—total length is approximately 4.6 km. Note dirt roads in bottom right corner of image for scale. Yellow arrow in upper left corner indicates view of inset image. Inset shows the main scarp, lake, and slump block on 7 June 2015, about 1 year before the outburst flood. Real-time camera in foreground. Maximum relief from the surface of the lake to the top of the main scarp is about 100 m. Main image adapted from Coe et al. (2016)

but instead continued to provide support and assistance to the local community for multiple years both after the 2014 rock avalanche, and then again, before and after the subsequent 2016 outburst flood. Lastly, the initial presence of the response group, and their availability to the public, kept the local community involved and updated on the hazard and risk associated with the site. Additionally, follow-up publications (White et al. 2015; Coe et al. 2016) provided useful information on the landslide triggering mechanism, guidelines for interpreting comparable older deposits elsewhere, and constraints for hazard mapping and modeling in similar geomorphic settings in the future.

Acquiring Data—Improving the Future

Many landslide disasters provide opportunities for advancing our knowledge of landslide processes. Whether it be for calibration of numerical models or for overall hazard recognition, the data sets that disasters provide can have tangible societal benefits through their ability to inform what might be expected during future landslide events. Importantly, responses focused on data collection must be sensitive to the specific disaster situation and not interfere with other potentially more crucial rescue or recovery operations. Responses that occur in a data-collection mode may be initiated rapidly after an event to obtain perishable targeted data or can take place some time afterward via more comprehensive investigations. The case studies presented in this section illustrate this range of approaches and highlight how forward-looking responses can provide new insights for enhancing progress in landslide hazard reduction.

Case Study 8. Identifying Landslide Mobility Flow Paths: The 2018 Montecito, California (USA) Debris Flows

Unlike many catastrophic landslides that occur with little warning, debris flows from recent burn areas can be anticipated to a certain extent. Decreases in soil infiltration capacity and consumption of protective vegetation by fire make hillslopes more vulnerable to intense precipitation and subsequent surficial erosion that can then transform into dangerous debris flows. In the United States, emergency-response teams are routinely assembled after wildfires to assess assets at risk (e.g. roads and homes) from post-fire flooding and debris flows. The 2018 Montecito debris-flow event (California, USA; Fig. 3) is an example of a disaster that occurred while a coordinated effort to plan for potential debris-flow impacts was already underway. A wildfire had just burned a 1140 km^2 region of steep, rugged terrain located adjacent to several heavily populated communities, and state and federal teams had completed a preliminary hazard assessment for most of the burn area. Resultant hazard maps (https://landslides.usgs.gov/hazards/postfire_debrisflow/detail.php?objectid=178) identified the likelihood and potential volume of debris flows and the rainfall thresholds for early warning of debris-flow initiation, but they did not identify specifically where potential debris would travel. Full recognition of the potential impacts was not possible due in part to the lack of a fully calibrated operational tool to rapidly forecast potential debris-flow inundation paths immediately after a fire. Consequently, a major focus of the Montecito response was to collect data to develop and test models of debris-flow inundation, which could improve future post-fire hazard assessment capabilities.

Fig. 11 Data collection following the 9 January 2018 Montecito (California, USA) debris flows which were triggered by intense rainfall on recently burned steep slopes. **a** Example of peak flow depth measurements, which were commonly several meters above the flood plain. **b** Mobile electronic devices allowed rapid and organized data collection of deposit thickness and boulder sizes

Event Description

The 9 January 2018 Montecito debris flows were triggered within minutes of an intense burst of rainfall (13 mm in 5 min) on steep slopes and channels burned weeks earlier by the Thomas Fire (Oakley et al. 2018; WERT 2018; Kean et al. 2019), at that time the largest wildfire in California history. This was the first storm of the winter rainy season, and the dry conditions that promoted the Thomas fire suggest that antecedent soil moisture was very low. The source material of the debris flows came from widespread surface soil erosion on the hillslopes and deep scour of the channel network, including mobilization of large (up to 6-m-diameter) boulders. Most of the societal impacts resulted from debris flows originating from drainage basins above the coastal town of Montecito, located just east of the city of Santa Barbara. The debris flows traveled 3 km over coalescing urbanized alluvial fans, killing 23 people and damaging or destroying more than 400 homes. Peak flow depths ranged from 5 m near the fan apex to less than a 1 m near the ocean (Fig. 11a). The event resulted in a nearly two-week closure of a major four-lane highway (U.S. Hwy 101) and cost an estimated $US 633,000,000 in debris removal and damage (Lukashov et al. 2019).

Response

Debris-flow hazard maps and subsequent risk assessment by state and federal emergency-response teams helped local authorities to develop emergency response and evacuation plans in the case that the region experienced a large storm event. Forecasts of debris-flow-generating rainfall (Staley et al. 2017) by the U.S. National Weather Service (NWS) beginning five days before the event provided crucial lead time for authorities to put these plans into action. One hour and fifteen minutes before the event, NWS detected an intense band of rainfall approaching Montecito and issued a warning that flows were imminent (NWS 2018). First responders were staged ahead of time and were able to react quickly as the debris flows inundated the Montecito area. The day after the event, members of the California Geological Survey (CGS, who were already on site mapping debris-flow potential; WERT 2018) and USGS (who helped conduct the debris-flow hazard assessment) teamed up to map the characteristics of the six runout paths that caused the most damage. The mapping took place alongside recovery operations in the first 12 days after the event. The team used GIS applications installed on portable electronic devices to synthesize field observations and measurements by multiple mappers into a single geodatabase (Fig. 11b). Data collection focused on documenting the maximum flow depth, deposit thickness, deposit grain size including boulder dimensions, and descriptions of flow avulsions. These observations were complimented by post-event imagery and lidar that delineated the spatial extent of inundation and provided a base layer for mapping. In addition, the team measured soil infiltration properties to constrain models of debris-flow initiation, flow superelevation around channel bends to estimate flow velocity, and peak flow depths on damaged buildings to develop debris-flow fragility curves (Kean et al. 2019).

Lessons Learned

The hazard maps, forecasts, and warnings of debris-flow potential by collaborative efforts between state and federal burned area emergency-response teams, the USGS, and NWS were critical to staging emergency responders so that they could begin rescues immediately after the debris flows. Prior research and collaboration to set rainfall thresholds for this event allowed NWS to quickly issue storm warnings, and post-fire debris flow models run after the fire illustrated the elevated landslide susceptibility. It is likely the loss of life would have been greater without the advance planning of researchers and the fast action of first responders. Nevertheless, the compliance with evacuation orders was low, suggesting the public did not fully recognize the threat of debris flows. Public recognition of the threat could possibly have been improved if detailed maps of potential debris-flow inundation zones were available prior to the event. Although there were insufficient tools and time to create these maps prior to the storm, the forecasts gave landslide experts time to prepare and mobilize in order to collect the perishable data following the event that are needed to develop these types of products. Close coordination between state (CGS) and federal (USGS) experts before and after the event also helped facilitate a rapid response that ensured collection of valuable ephemeral data. Synthesis of the field data was greatly aided by advances in mobile computing, which helped standardize and integrate the data collected by multiple mappers. These efforts provide the data needed for the development and testing of models that can be used for rapidly predicting runout characteristics and thus better forecast targeted impacts.

Case Study 9. Documenting Changing Patterns of Landslide Behavior: The 2012–2016 Glacier Bay, Alaska (USA) Rock Avalanches

There are times when the potential impact of future landslides in a region warrants advanced investigation. In these cases, landslides that themselves may cause no direct societal effects can provide insight into potentially more damaging scenarios. One such situation is the possibility of very large permafrost-degradation-initiated landslides from climate-change effects in arctic and alpine regions. The occurrence of several massive rock avalanches in the Glacier Bay region of Alaska, USA (Fig. 3) prompted landslide experts to investigate not only their causes but also their prevalence over time. The potential for localized tsunami-generating landslides poses hazards to both local communities and ship traffic along these coastlines (e.g. Coe et al. 2019). As such, this type of response provides the data needed to proactively develop possible scenarios and plans for future similar events in the region.

Event Description

Several large rock avalanches (each between 5 and 22 km^2 in total area and tens of millions of cubic meters in volume) occurred within a four-year period in Alaska's Glacier Bay National Park and Preserve (Bessette-Kirton and Coe 2016; Coe et al. 2018; Fig. 12a). These landslides, which initiated in a variety of lithologies and entrained significant volumes of ice and snow, travelled long distances—up to 10.5 km for the 2016 Lamplugh rock avalanche (Bessette-Kirton et al. 2018). Although none caused fatalities or damage to infrastructure (with none entering the ocean or Glacier Bay proper), the landslides signaled a seemingly anomalous cluster of events both in space (i.e. size and location) and time. Their timing also suggests that temperature might have played a role—the two largest (2012 Lituya Mountain and 2016 Lamplugh Glacier; Coe et al. 2018) both occurred

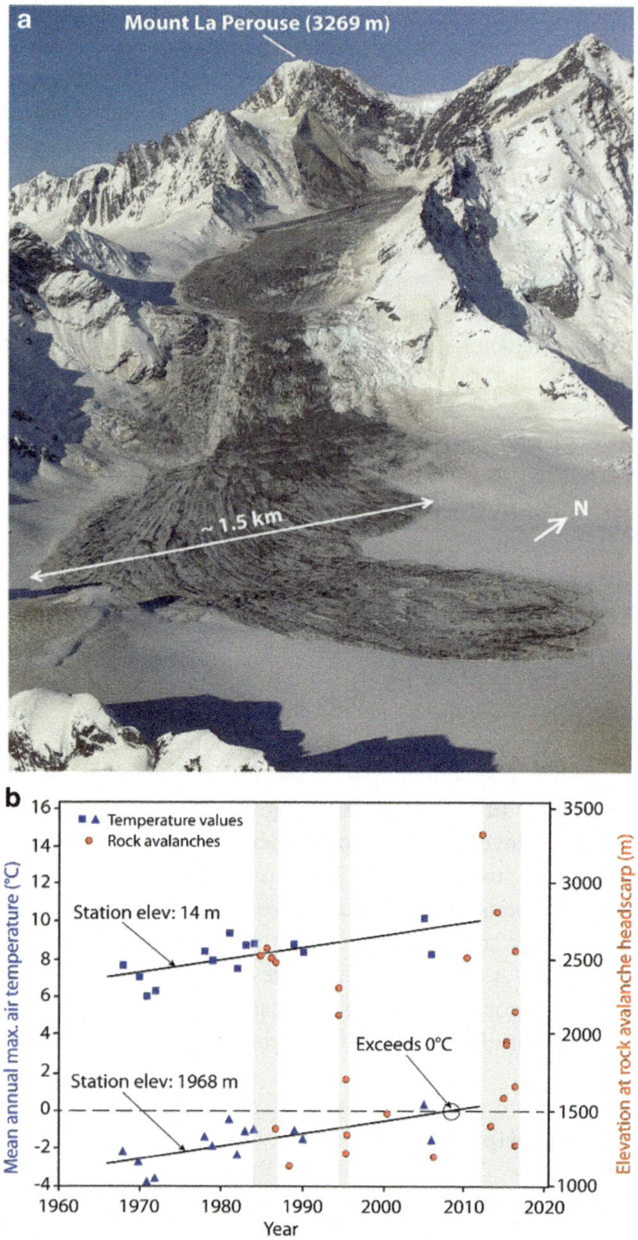

Fig. 12 **a** The 2014 Mount La Perouse rock avalanche (Alaska, USA) is one of a cluster of landslides that occurred in Glacier Bay National Park and Preserve between 2012 and 2016. Photo taken by Marten Geertsema and figure from Coe et al. (2018). **b** The response to clustered rock avalanche events in Alaska (shaded time ranges) included comparison with temperature trends in the region. Here, increased rock avalanche occurrence during 2012–2016 (ten rock avalanches; red circles in the wide gray vertical band on the right) coincides with high-elevation (1,968 m) temperatures exceeding the 0°C threshold and with increasing elevation, suggesting permafrost-degradation effects may have played a role in their occurrence. All three rock avalanche clusters (gray vertical bands) coincide with above average winter temperatures compared to long-term averages (figure adapted from Coe et al. 2018)

during warm months (June). Given their location within a National Park that is accessed nearly exclusively by cruise ships and smaller boats, similar events in the future that occur closer to the surrounding coastlines could have important ramifications for park visitors and nearby communities that might be affected by landslide-generated tsunamis (Coe et al. 2019).

Response

Analysis of satellite imagery provided the most effective response for these events due both to their remote location and their wide spatial extent. Seismic detection was initially used to identify some events (e.g. Ekström and Stark 2013; Petley 2014, 2016; Dufresne et al. 2019), with subsequent verification by airborne and satellite imagery providing additional details (Geertsema 2012; Coe et al. 2018), in particular, aerial extent, runout distances, and in some cases volumes (Bessette-Kirton et al. 2018). Using 30-m-resolution satellite imagery available over a 33-year period, landslide experts identified 24 rock avalanches within a 5000 km^2 area (Bessette-Kirton and Coe 2016); several events clustered in time, the most striking being the 2012–2016 cluster in which the largest rock avalanches occurred (Coe et al. 2018). By eliminating an earthquake-trigger for these landslides, connections with other triggers were investigated; the most compelling related to increased temperatures in the region. For example, the 2012–2016 cluster of 10 rock avalanches occurred 2 years after mean annual maximum temperatures recently exceeded 0 °C (Fig. 12B) and partially during a multi-year period of record-breaking high air temperatures. This correlation suggests climate-change-related permafrost degradation as a possible trigger (Coe et al. 2018) with the potential for additional hazards in the future (Coe et al. 2019).

Lessons Learned

Proactive recognition of future landslide disasters is possible when landslide experts conduct targeted comprehensive assessments. These assessments might not necessarily need to focus on damaging events; rather, patterns of non-damaging events can provide information of significant importance in planning for possible landslide disasters. This case study was conducted entirely using remotely sensed imagery which demonstrates the power and utility of satellite technology. Importantly, the decision to use lower resolution products (i.e. 30-m Landsat) allowed for a longer 33-year data set because higher resolution products were only available for more recent events. In addition, resolution consistency eliminated detection and mapping biases, and thereby made the results more robust for detecting changes and trends in landslide frequency and magnitude during the period of record.

Progress in Responding to Landslide Disasters

The way landslide experts respond to landslide disasters has evolved considerably over the past 20 years. Although the general approaches remain the same (communication of information, assessment of threats, utilization of monitoring instrumentation, etc.), experiences from previous events and advancements in technology now provide landslide experts with new methods and tools for increasing their effectiveness when a landslide disaster strikes. Here we summarize some of the progress that has been made with our ability to respond to landslide disasters.

Preparation

As first responders are well-aware, preparation is instrumental in providing the best possible response during a disaster. Similarly, landslide experts charged with responding to landslides must be well prepared. This, of course, is difficult given that landslides can have very different styles and occur in widely varying terrain, thereby requiring substantially different approaches from location to location (see Factors Affecting Landslide Response section). However, the advent of cloud-based imagery platforms and off-the-shelf data-acquisition software and sensor hardware provides considerable resources that can be applied to nearly every response scenario.

Archives of pre-event imagery are available worldwide and can be retrieved within seconds using open software platforms (e.g. the U.S. National Map—https://viewer.nationalmap.gov/advanced-viewer/ or Google Earth). These can provide site familiarity and potentially recent landslide history as well as assist with travel and access logistics. Similarly, open platforms such as Google Crisis can provide rapidly disseminated pre- and post-event satellite imagery useful for situational awareness during the early stages of a landslide response. The 2015 Gorkha earthquake landslide response (case study 6) benefited from these preparatory resources. In many locations, substantial pre-event data exist in other forms, including high-resolution lidar data. These data can be key for obtaining event volumes and for identifying mobility vectors. Archives of pre-event airborne lidar data were used in the initial response to the 2014 Oso landslide (case study 5; see also Haugerud 2014), and archives of airborne and ground-based lidar of nearly all of the cliffs in Yosemite Valley were similarly essential in the 2017 El Capitan rock-fall response (case study 3).

Software and hardware used for rapid data collection during a landslide response still require considerable experience for efficient data collection, and in some situations traditional analog methods (e.g. pen and notebook) can be more dependable, particularly when internet or cellular data are not available. Regardless of the methods used, responses benefit from data collection and instrumentation equipment that is ready to deploy, with personnel trained in their use and subsequent data analysis, whether on-site or remotely. For the 2014 Oso landslide response (case study 5), GPS receivers and seismometers were deployed rapidly for monitoring, but this was fortuitous because these units happened to be available along with landslide experts who had experience with their installation and data analysis. First-responder agencies provided helicopters for the installations, and trained personnel made their time available to both analyze data and to train others what to look for during the real-time monitoring periods when recovery efforts were ongoing.

Advances in Technology

Advancements in technology were an integral and, in some cases, necessary component of many of the investigated case studies. These included both monitoring equipment (e.g. the GPS-seismometer spiders used for the 2006 Ferguson rock slide and 2014 Oso landslide responses, case studies 4 and 5) and field equipment (the GIS-enabled data collectors used in the 2018 Montecito, California debris flow response, case study 8). In addition, new analytical methods and advanced computing power developed over the past 20 years allowed the assessment of mobility patterns and potential ongoing hazards in ways never before achieved so rapidly following an event. Numerical debris-flow runout modeling from the 2014 Oso landslide (case study 5) helped direct first responders to the most likely places in the debris field where victims were located (Iverson et al. 2015). Additionally, structure-from-motion and lidar point-cloud analyses were conducted overnight during the 2017 Yosemite rock-fall response (case study 3) to provide incident management personal information on recent and expected rock-fall volumes so that decisions could be made regarding ongoing hazards and road reopening.

High-resolution satellite imagery was essential in allowing several of the responses to even be performed. Although automated landslide detection from satellite imagery could benefit from further advances and optimal utilization of available resources (Williams et al. 2018), the opportunities afforded by these data are indispensable. During the response to the 2015 Gorkha earthquake landslides (case study 6), most of the landslides visited by experts during helicopter sorties would not have been located without the availability of rapidly collected, high-resolution, satellite imagery that had been analyzed prior to the field investigation. Further, the recognition of many of the Glacier Bay rock avalanches (case study 9) would not have been possible without satellite imagery. Notably, several of these rock avalanches were initially identified by still another new approach, seismic detection (e.g. Ekström and Stark 2013) that has also assisted with some landslide responses. Continued advances in technology, including important monitoring techniques such as satellite and ground-based radar (e.g. Casagli et al. 2010; Meier et al. 2016; Burrows et al. 2019), are likely to be of increasing importance in future landslide responses.

Communications and Collaborations

Purposeful and open communication was a key factor in all the presented case studies. This included interagency collaboration as well as communicating with the public. Open, fact-supported, and timely communication is necessary to manage expectations, build trust and confidence, and keep partners and officials up to date on latest developments. Responses to landslide emergencies and disasters generally involve multiple agencies, typically from different levels of government as well as private contractors and volunteers. Cultivating good working relationships with emergency management agencies and others having relevant expertise before disasters strike can improve the efficiency of operations during the response and recovery phases of disasters. Although important in all the case studies, previously established working relationships were particularly valuable in Yosemite (case study 3), Ferguson (case study 4), and Montecito (case study 8) to facilitate access and share workloads. In some cases, new working relationships may need to be forged, particularly for international responses in which collaboration is critical to accomplishing the mission. This was relevant during the Gorkha, Nepal earthquake response (case study 6) but can also be important in domestic responses in which local agencies may have little experience with landslide disasters and no prior relationship with landslide experts (e.g. during the Sausalito and West Salt Creek landslide responses; case studies 1 and 7).

Communicating with the public during landslide emergencies and in the aftermath of landslide disasters is important and challenging. Keeping messages to the public fact-based, simple, timely, and focused on what the public needs to know will help avoid misunderstandings, panic, and public mistrust. In briefing public officials as well as the general public, landslide experts should refrain from speculation and not be afraid to make clear distinctions between what is and is not known. This was important in the response to the La Conchita landslide (case study 2). Sharing the content of public web pages, blogs, press releases, and other general communications with officials in charge before going public with the information is a basic courtesy that will help maintain a positive working relationship and help avoid confusing or contradictory information. This is particularly important in this age of rapid communication, in which any news, especially news that might cause controversy, embarrass public officials, incite panic, or promote excitement, can spread quickly and cause significant damage to the operation and the reputations of the agencies or individuals involved if misinterpreted. For the same reasons, landslide experts must be circumspect when interacting with news reporters and members of the public. Landslide experts are typically not trained in public or media relations, such that learning to redirect questions to focus on the basic facts is a valuable skill to develop. Having a single or limited points of contact from a landslide response team who are skilled in talking to the media can help manage the flow of information during landslide responses.

In many situations, the communication focus is on the cause of a landslide disaster. Sometimes, such as in the case of earthquake triggering, the cause is self-evident and uncontroversial. In other situations, such as the 2005 La Conchita (case study 2) and 2014 Oso (case study 5) disasters, the cause can become controversial and even adversarial. Both political and legal issues can arise. Landslide experts must refrain from becoming ensnared with or advocates for competing factions in such cases. Sometimes determining causation will require detailed and time-consuming analysis. In other situations, the cause might be clear to the experts but unpopular with some faction. The expert needs to remain above the fray but be willing to clearly state objectively determined facts, even if those facts are considered controversial by some. Clearing up misconceptions and establishing a clear, concise narrative of the situation is a critical and sometimes very difficult role an expert must play. Responders must therefore be both assertive and capable of relaying unwelcome information; they should not be afraid to admit what they do not know or cannot determine with certainty. Tactful and respectful communication will make the responders' messages more readily accepted and improve working relationships.

General Experiences

Familiarity with the incident command and/or government-community structure facilitates efficient communication and can assist with obtaining the necessary support and cooperation to make the most complete assessment possible given the circumstances during and following a landslide disaster. Despite the similarity of an overall hierarchy in each of the presented case studies, roles and relationships among the various entities differed from place to place and situation to situation. For example, the organizational response structure differed considerably between international and domestic responses and, for domestic responses, between local emergencies (e.g. at Sausalito—case study 1) and those elevated to a level at which the national government's emergency management became involved (e.g. at Oso—case study 5). Early in the response effort, it is important to learn about the roles of each agency to understand who is in charge, their expectations, what resources are available, how to obtain necessary access to the site and resources, and security and reporting protocols.

Developing an accurate three-dimensional overview of the site and understanding the spatial and temporal relations between the landslide and its surroundings are necessary for making sound recommendations. Remote sensing can be very helpful in achieving this understanding, but it is not always an adequate substitute for on-site investigation. The common current practice of delivering optical imagery in orthorectified raster format, rather than as stereo pairs, can sometimes hamper geomorphological interpretation. Despite their value, remote sensing and other advanced technological tools can also easily become a distraction, causing the landslide experts to miss opportunities to view the landslide features with their own eyes or to lose focus on the real job of making an accurate assessment of the landslide(s). Discerning and accurately assessing the landslide hazard requires directly observing the site from multiple vantage points close enough to observe geological details such as incipient cracks and scarps, seeps, and other terrain and geologic features that contain clues to the landslide mechanism or the landslide's future behavior. Furthermore, an on-site landslide expert provides both technical expertise and confidence to the local government or incident command, particularly in an evolving situation when repeated or follow-up assessments are needed. An on-site presence also allows the landslide expert to better relate to and adopt the best tone when addressing any human consequences of a disaster.

Landslide response can be exhausting, stressful, intimidating, and, in the cases where aircraft support is needed, sometimes dangerous, due to uncertainty, changing conditions, long hours, and intense expectations of the incident commanders, government officials, and public. In high-fatality situations, and particularly when bodies are still being recovered, separating scientific judgment from human emotion can be profoundly difficult. Consequently, responders must be in sound physical and mental health and be adaptable to changes. Scientific organizations responding to disasters should consider the long-term mental health consequences of experts responding to challenging disasters and the development of "incident stress".

Responses can evolve, and landslide experts' roles can change as conditions change. For example, during the Oso response (case study 5), the first landslide experts on site were tasked with defining what had happened and identifying the potential for danger to rescue workers. Later, as instrumentation was installed, the role changed to monitoring and interpreting incoming data continually for signs of new movements to ensure the safety of ongoing recovery operations in the deposit area. Similarly, the Montecito response (case study 8) began as a multiagency collaborative assessment of ongoing hazard from post-fire debris flows. After heavy rain resulted in catastrophic debris flows, the response broadened to include the collection of data and observations pertaining to ephemeral deposits and conditions.

Despite improvements in the technology and tools available for monitoring and assessing landslides and in managing emergencies, the core job of landslide response is little changed. The job involves answering basic questions about the characteristics of the event (location, nature, extent, timing) and evaluating the impacts of landslides that have either just occurred or may soon possibly occur. This requires using sound methods combined with knowledge of the fundamentals of landslide science and then clearly communicating facts and anticipated landslide behavior to decision makers and the public.

Conclusions

Responding as an expert to landslide disasters is challenging and demanding both because of the societal impacts one must face and because of the lack of guidelines for how best to approach what may initially be an undefined or chaotic situation. The very nature of landslides, whose size, style, and spatial extent differs considerably, suggests that all responses will be different; nevertheless, our review of nine case studies, taken from nearly 20 years of experience, suggests that some commonalities exist with respect to both best practices (preparation, methods, etc.) and also with approaches to communication and collaborations.

The number and extent of landslide disasters are likely to grow worldwide as a result of civilization expansion into landslide-prone areas. As the best source of knowledge on landslide processes and hazards, landslide experts will continue to be called in to assist when help is needed. It is therefore in our best interest to be prepared for such events, whether it be from formulating response strategies, preparing instrumentation for rapid deployment, or cultivating relationships with others in the emergency response community. Learning from others' experiences, such as those presented here, should aid with this preparation. We hope that the lessons and summary of progress on responses to landslide disasters detailed in this paper will be useful to those who conduct similar responses. As such, we encourage others to record and disseminate their own experiences, so that landslide experts responding to these types of events are better prepared for what they encounter and can make the best and strongest contributions possible.

Acknowledgements The responses to the landslide case studies described herein were each a collaborative effort consisting of partners from all levels of government as well as academia, including many other colleagues at the USGS and other agencies. Whereas these are too lengthy to list here, we provide are sincerest gratitude for the partnerships that resulted from these endeavours. In many cases, the responses detailed in the case studies are as much a part of their work as it is ours. We thank Corina Cerovski-Darriau (USGS), Alex Grant (USGS), and an anonymous reviewer for providing helpful feedback on earlier versions of this paper and thank the World Landslide Forum organizers for the invitation and opportunity to present this work. Any use of trade, firm, or product names is for descriptive purposes only and does not imply endorsement by the U.S. Government.

References

Angeli M-G, Pasuto A, Silvano S (2000) A critical review of landslide monitoring experiences. Eng Geol 55(30):133–147. https://doi.org/10.1016/S0013-7952(99)00122-2

Badger TC, D'Ignazio M (2018) First-time landslides in Vashon advance glaciolacustrine deposits, Puget Lowland, U.S.A. Eng Geol 243:294–307. https://doi.org/10.1016/j.enggeo.2018.07.011

Baum RL, Coe JA, Geiger JW et al (2016) Outburst flood creates opportunity to look inside the West Salt Creek rock avalanche deposit, Mesa County, Colorado. Geol Soc Am Abstr Programs 48 (7). https://doi.org/10.1130/abs/2016AM-283901

Bessette-Kirton EK, Coe JA (2016) Inventory of rock avalanches in western Glacier Bay National Park and Preserve, Alaska, 1984–2016: a baseline data set for evaluating the impact of climate change on avalanche magnitude, mobility, and frequency. U.S. Geological Survey Data Release. https://doi.org/10.5066/F7C827F8

Bessette-Kirton EK, Coe JA, Zhou W (2018) Using stereo satellite imagery to account for ablation, entrainment, and compaction in volume calculations for rock avalanches on glaciers: application to the 2016 Lamplugh rock avalanche in Glacier Bay National Park, Alaska. J Geophys Res Earth Surf 123(4):622–641. https://doi.org/10.1002/2017JF004512

Bessette-Kirton EK, Cerovski-Darriau C, Schulz WH et al (2019) Landslides triggered by Hurricane Maria: assessment of an extreme event in Puerto Rico. GSA Today 29(6):4–10. https://doi.org/10.1130/GSATG383A.1

Burrows K, Walters RJ, Milledge D et al (2019) A new method for large-scale landslide classification from satellite radar. Remote Sens 11(3):237. https://doi.org/10.3390/rs11030237

Casagli N, Catani F, Del Ventisette C et al (2010) Monitoring, prediction, and early warning using ground-based radar interferometry. Landslides 7:291–301. https://doi.org/10.1007/s10346-010-0215-y

Coe JA, Kean JW, Godt JW et al (2014) New insights into debris-flow hazards from an extraordinary event in the Colorado front range. GSA Today 24(1):4–10. https://doi.org/10.1130/GSATG214A.1

Coe JA, Baum RL, Allstadt KE et al (2016) Rock-avalanche dynamics revealed by large-scale field mapping and seismic signals at a highly mobile avalanche in the West Salt Creek valley, western Colorado. Geosphere 12(2):607–631. https://doi.org/10.1130/GES01265.1

Coe JA, Bessette-Kirton EK, Baum RL et al (2017) The West Salt Creek rock avalanche, sag pond, and outburst floods: morphological and sedimentological signatures from a series of extraordinary events. Geol Soc Am Abstr Programs 49(6). https://doi.org/10.1130/abs/2017AM-306710

Coe JA, Bessette-Kirton EK, Geertsema M (2018) Increasing rock-avalanche size and mobility in Glacier Bay National Park and Preserve, Alaska detected from 1984 to 2016 Landsat imagery. Landslides 15(3):393–407. https://doi.org/10.1007/s10346-017-0879-7

Coe JA, Schmitt, RG, Bessette-Kirton EK (2019) An initial assessment of areas where landslides could enter the west arm of Glacier Bay, Alaska and implications for tsunami hazards. Alaska Park Sci 18(1): 26–37. https://www.nps.gov/articles/aps-18-1-4.htm

Collins BD, Corbett SC (2019) Terrestrial lidar data of the February 14, 2019, Sausalito Boulevard landslide, Sausalito, California. U.S. Geol Surv Data Ser 1112. 12. https://doi.org/10.3133/ds1112

Collins BD, Jibson RW (2015) Assessment of existing and potential landslide hazards resulting from the April 25, 2015 Gorkha, Nepal earthquake sequence (ver. 1.1, August 2015). U.S. Geological Survey Open-File Report 2015–1142. p 50. https://doi.org/10.3133/ofr20151142

Collins BD, Reid ME (2020) Enhanced landslide mobility by basal liquefaction: the 2014 State Route 530 (Oso), Washington, landslide. GSA Bull 132(3–4):451–476. https://doi.org/10.1130/B35146.1

Collins BD, Stock GM (2012) Lidar-based rock-fall hazard characterization of cliffs. In: Hryciw RD, Athanasopoulos-Zekkos A, Yesiller N (eds) ASCE Geotechnical Special Publication no. 225, Geotechnical Engineering State of the Art and Practice. pp 3021–3030. https://doi.org/10.1061/9780784412121.309

Collins BD, Baum RL, Mrozek T et al (2011) Evaluation of landslide monitoring in the Polish Carpathians. U.S. Geological Survey Open-File Report 2011–1001. p 30. https://doi.org/10.3133/ofr20111001

Collins BD, Stock J, Foster K et al (2012) Monitoring subsurface hydrologic response for determining thresholds of precipitation-induced shallow landslides in the San Francisco Bay area. In: Eberhardt E, Froese C, Turner K, Leroueil S (eds) Landslides and engineered slopes: protecting society through improved understanding, Proceedings of the 11th International Symposia on Landslides. Banff, Alberta, Canada, pp 1249–1255

Costa JE, Schuster RL (1988) The formation and failure of natural dams. Geol Soc Am Bull 100(7):1054–1068. https://doi.org/10.1130/0016-7606(1988)100%3c1054:TFAFON%3e2.3.CO;2

DeGraff JV, Gallegos AJ, Reid ME et al (2015) Using monitoring and modeling to define the hazard posed by the reactivated Ferguson rock slide, Merced Canyon, California. Nat Hazards 76(2):769–789. https://doi.org/10.1007/s11069-014-1518-4

Denlinger RP (2007) Simulations of potential runout and deposition of the Ferguson rockslide, Merced River Canyon, California. U.S. Geological Survey Open-File Report 2007–1275. p 25. https://pubs.usgs.gov/of/2007/1275

Dowling CA, Santi PM (2014) Debris flows and their toll on human life: a global analysis of debris-flow fatalities from 1950 to 2011. Nat Hazards 71(1):203–227. https://doi.org/10.1007/s11069-013-0907-4

Dufresne A, Wolken GJ, Hibert C et al (2019) The 2016 Lamplugh rock avalanche, Alaska: deposit structures and emplacement dynamics. Landslides 16(12):2301–2319. https://doi.org/10.1007/s10346-019-01225-4

Duncan JM, Fleming RW, Patton FD (1986) Report of the Thistle slide committee to State of Utah, Department of Natural Resources, Division of Water Rights. U.S. Geological Survey Open-File Report 86–505. p 100. https://pubs.usgs.gov/of/1986/0505/report.pdf

Ellen SD, Wieczorek GF (eds) (1988) Landslides, floods, and marine effects of the storm of January 3–5, 1982, in the San Francisco Bay Region, California. U.S. Geological Survey Professional Paper 1434, p 310 + 14 plates. https://pubs.usgs.gov/pp/1988/1434/

Ekström G, Stark CP (2013) Simple scaling of catastrophic landslide dynamics. Science 339(6126):1416–1419. https://doi.org/10.1126/science.1232887

Evans SG, Bent AL (2004) The Las Colinas landslide, Santa Tecla: a highly destructive flowslide triggered by the January 13, 2001, El Salvador earthquake. In: Rose WI, Bommer JJ, López DL et al (eds) Natural hazards in El Salvador. Geological Society of America Special Vol 375, pp 25–37 https://doi.org/10.1130/0-8137-2375-2.39

Evans SG, DeGraff JV (2002) Catastrophic landslides: effects, occurrence, and mechanisms. Geological Society of America, Boulder, Colorado. (ISBN 0-8137-4115-7). p 411

Froude MJ, Petley DN (2018) Global fatal landslide occurrence from 2004 to 2016. Nat Hazards Earth Sys Sci 18:2161–2181. https://doi.org/10.5194/nhess-18-2161-2018

Gaillard JC, Peek L (2019) Disaster-zone research needs a code of conduct. Nature. Comment 20 November 2019. https://www.nature.com/articles/d41586-019-03534-z

Geertsema M (2012) Initial observations of the 11 June 2012 rock/ice avalanche, Lituya Mountain, Alaska. The first meeting of cold region landslides network, Harbin, China, p 5. https://doi.org/10.13140/2.1.2473.5682

Harp EL, Jibson RW (1995) Seismic instrumentation of landslides: Building a better model of dynamic landslide behavior. Bull Seismol Soc Am 85: 93–99. https://pubs.geoscienceworld.org/bssa/article-lookup/85/1/93

Harp EL, Jibson RW (1996) Landslides triggered by the 1994 Northridge, California, earthquake. Bull Seismol Soc Am 86(1B): S319–S332

Harp EL, Reid ME, Michael JA (2004) Hazard analysis of landslides triggered by Typhoon Chata'an on July 2, 2002, in Chuuk State, Federated states of Micronesia. U.S. Geological Survey Open-File Report 2004–1348. p 24 + 2 plates. https://pubs.usgs.gov/of/2004/1348

Harp EL, Reid ME, Godt JW et al (2008) Ferguson rock slide buries California state highway near Yosemite National Park. Landslides 5 (3):331–337. https://doi.org/10.1007/s10346-008-0120-9

Haugerud RA (2014) Preliminary interpretation of pre-2014 landslide deposits in the vicinity of Oso, Washington. U.S. Geological Survey Open-File Report 2014–1065. p 4 https://doi.org/10.3133/ofr20141065

Henn B, Cao Q, Lettenmaier DP et al (2015) Hydroclimatic conditions preceding the March 2014 Oso Landslide. J Hydrometeorology 16:1243–1249. https://doi.org/10.1175/JHM-D-15-0008.1

Hughes KS, Bayouth García D, Martínez Milian GO et al (2019) Map of slope-failure locations in Puerto Rico after Hurricane María. US Geological Survey Data Release. https://doi.org/10.5066/P9BVMD74

Iverson RM, George DL, Allstadt K et al (2015) Landslide mobility and hazards—implications of the 2014 Oso disaster. Earth Planet Sci Lett 412:197–208. https://doi.org/10.1016/j.epsl.2014.12.020

Jibson RW (2005) Landslide hazards at La Conchita, California. U.S. geological survey open-file report 2005–1067. p 12. https://pubs.usgs.gov/of/2005/1067/508of05-1067.html

Jibson RW (2006) The La Conchita landslide. Landslides 3(1):73–78. https://doi.org/10.1007/s10346-005-0011-2

Jibson RW (2006b) Photographs, correspondence, and presentations related to the La Conchita, California, landslide. U.S. Geological Survey Open-File Report 2006–1278. https://pubs.usgs.gov/of/2006/1278/

Jibson RW, Baum RL (1999) Assessment of landslide hazards in Kaluanui and Maakua gulches, Oahu, Hawaii, following the 9 May 1999 Sacred Falls landslide. U.S. Geological Survey Open-File Report 99–364. p 9. https://doi.org/10.3133/ofr99364

Jibson RW, Prentice CS, Borissoff BA et al (1994) Some observations of landslides triggered by the 29 April 1991 Racha earthquake, Republic of Georgia. Bull Seismol Soc Am 84(4):963–973

Jibson RW, Crone AJ, Harp EL et al (2004) Landslides triggered by the 13 January and 13 February 2001 earthquakes in El Salvador. Natural Hazards in El Salvador. In: Rose WI, Bommer JJ, López DL et al (eds) Geological Society of America, Boulder, Colorado. https://doi.org/10.1130/SPE375

Kaliser BN (1983) Geologic hazards of 1983, Thistle landslide: Utah's first presidential disaster declaration. Survey Notes, Utah Geological and Mineral Survey. 17(2): 3–8. https://ugspub.nr.utah.gov/publications/Survey_Notes/SNT17-2.pdf

Kargel JS, Leonard GJ, Shugar DH et al (2016) Geomorphic and geologic controls of geohazards induced by Nepal's 2015 Gorkha earthquake. Science. 351(6269): 140, p 12. + 22p. supplement. https://doi.org/10.1126/science.aac8353

Kean JW, Staley DM, Lancaster JT et al (2019) Inundation, flow dynamics, and damage in the 9 January 2018 Montecito debris-flow event, California, USA: opportunities and challenges for post-wildfire risk assessment. Geosphere 15(4):1140–1163. https://doi.org/10.1130/GES02048.1

Keaton JR, Wartman J, Anderson SA et al (2014) The 22 March 2014 Oso landslide, Snohomish County, Washington: Geotechnical Extreme Events Reconnaissance Association Report GEER-036, p 186. https://www.geerassociation.org/index.php/component/geer_reports/?view=geerreports&id=30

Keefer DK, Wilson RC, Mark RK et al (1987) Real-time landslide warning during heavy rainfall. Science 238(4829):921–925. https://doi.org/10.1126/science.238.4829.921

Kieffer DS, Jibson RW, Rathje EM et al (2006) Landslides triggered by the 2004 Niigata Ken Chuetsu, Japan, earthquake. Earthquake Spectra 22:S47–S73. https://doi.org/10.1193/1.2173021

Korup O (2002) Recent research on landslide dams—a literature review with special attention to New Zealand. Progress Phys Geogr: Earth Environ 26(2):206–235. https://doi.org/10.1191/0309133302pp333ra

Laprade WT (2017) Sitka debris flow, August 2015. Landslides: putting experience, knowledge and emerging technologies into practice—Proceedings of the 3rd North American Symposium on Landslides. In: DeGraff JV, Shakoor A (eds) Association of Environmental and Engineering Geologists Special Publication 27. (ISBN: 978-0-9897253-7-8). pp 607–614

Lukashov SG, Lancaster JT, Oakley NS et al (2019) Post-fire debris flows of 9 January 2018, Thomas fire, southern California: initiation areas, precipitation and impacts. Proceedings of the 7th International Conference on Debris-flow Hazards Mitigation. In: Kean JW, Coe JA, Santi PM, Guillen BK (eds). Association of Environmental and Engineering Geologists Special Publication 28. pp 774–781. https://doi.org/10.25676/11124/173051

Lund WR, Knudsen TR, Bowman SD (2014) Investigation of the December 12, 2013, fatal rock fall at 368 West Main Street, Rockville, Utah. Utah Geological Survey, Report of Investigation

273. p 24. https://ugspub.nr.utah.gov/publications/reports_of_investigations/ri-273.pdf

Magirl CS, Keith MK, Anderson SW et al (2015) Preliminary assessment of aggradation potential in the North Fork Stillaguamish River downstream of the State Route 530 landslide near Oso, Washington. U.S. Geological Survey Scientific Investigations Report 2015–5173. p 20. https://doi.org/10.3133/sir20155173

Matasci B, Stock GM, Jaboyedoff M et al (2018) Assessing rockfall susceptibility in steep and overhanging slopes using three-dimensional analysis of failure mechanisms. Landslides 15 (5):859–878. https://doi.org/10.1007/s10346-017-0911-y

Meier L, Jacquemart M, Blattmann B et al (2016) Radar-based warning and alarm systems for alpine mass movements. In: Kienholz H (ed) Proceedings of INTERPRAEVENT 2016, 30 May-2 June 2016. Lucerne, Switzerland. pp 960–968. https://www.interpraevent.at/palm-cms/upload_files/Publikationen/Proceedings/IP_2016.pdf

Netto ALC, Sato AM, Avelar AS et al (2013) January 2011: the extreme landslide disaster in Brazil. In: Margottini C, Canuti P, Sassa K (eds) Landslide Science and Practice, vol. 6. Springer, Berlin, Heidelberg. pp 377–384. https://doi.org/10.1007/978-3-642-31319-6_51

NWS (National Weather Service) (2018) Flash flood and debris flow event, Montecito, California, January 9, 2018. National Weather Service, Los Angeles, Oxnard, California. https://noaa.maps.arcgis.com/apps/MapJournal/index.html?appid=541c23aa483b48978d1bc9904a6fb14d

Oakley NS, Cannon F, Munroe R et al (2018) Brief communication: meteorological and climatological conditions associated with the 9 January 2018 post-fire debris flows in Montecito and Carpinteria, California, USA. Nat Hazards Earth Sys Sci 18:3037–3043. https://doi.org/10.5194/nhess-18-3037-2018

Pankow KL, Moore JR, Hale JM et al (2014) Massive landslide at Utah copper mine generates wealth of geophysical data. GSA Today 24 (1):4–9. https://doi.org/10.1130/GSATG191A.1

Patton AI, Roering JJ, Orland E (2019) Hydrologic monitoring and morphometric analysis to characterize debris flow initiation at multiple temporal scales in Southeast Alaska. AGU 2019 Fall Meeting. Abstract NH43B-05

Petley D (2014) Mount La Perouse: Sunday's rock avalanche in Alaska has been found. The Landslide Blog. American Geophysical Union, February 22, 2014: electronic document available at https://blogs.agu.org/landslideblog/2014/02/22/mount-la-perouse-rock-avalanche/

Petley D (2016) Lamplugh Glacier rock avalanche: a massive new landslide in Alaska on Tuesday. The Landslide Blog. American Geophysical Union, July 3, 2016: electronic document available at https://blogs.agu.org/landslideblog/2016/07/03/lamplugh-glacier-rock-avalanche-1/

Pierson TC, Janda RJ, Thouret JC, Borrero CA (1990) Perturbation and melting of snow and ice by the 13 November 1985 eruption of Nevado del Ruiz, Colombia, and consequent mobilization, flow and deposition of lahars. J Volcanol Geoth Res 41(1–4):17–66. https://doi.org/10.1016/0377-0273(90)90082-Q

Reid ME, Baum RL, LaHusen RG et al (2008) Capturing landslide dynamics and hydrologic triggers using near-real-time monitoring. In: Chen Z, Zhang J, Li Z, Wu F, Ho K (eds) Landslides and engineered slopes from the past to the future, Proceedings of the Tenth International Symposium on Landslides and Engineered Slopes, Xi'an, China, vol 1. Taylor and Francis, London. pp 179–191

Reid ME, LaHusen RG, Baum RL et al (2012) Real-time monitoring of landslides. U.S. Geological Survey Fact Sheet 2012–3008. p 4. https://doi.org/10.3133/fs20123008

Roback K, Clark MK, West AW et al (2018) The size, distribution, and mobility of landslides caused by the 2015 Mw7.8 Gorkha, Nepal earthquake. Geomorphology 301:121–138. https://doi.org/10.1016/j.geomorph.2017.01.030

Sassa K (2005) Landslide disasters triggered by the 2004 mid-Niigata prefecture earthquake in Japan. Landslides 2(2):135–142. https://doi.org/10.1007/s10346-005-0054-4

Satofuka Y, Mori T, Mizuyama T et al (2010) Prediction of floods caused by landslide dam collapse. J Disaster Res 5(3): 288–295. https://www.sff.or.jp/content/uploads/H22JDR03.pdf

Scaioni M (2015) Modern technologies for landslide monitoring and prediction. Springer, Berlin, Heidelberg. p 249. https://doi.org/10.1007/978-3-662-45931-7

Schuster RL (1996a) The 25 most catastrophic landslides of the 20th century. In: Chacon J, Irigaray C, Fernandez T (eds) Landslides—Proceedings of the 8th International Conference and Field Trip on Landslides, 27–28 September 1996, Granada, Spain. Balkema, Rotterdam. pp 53–62

Schuster RL (1996b) Socioeconomic significance of landslides. Landslides-investigation and mitigation. In: Turner AK, Schuster RL (eds) Washington D.C., National Academy Press, Transportation Research Board Special Report 247. pp 12–35

Staley DM, Negri JA, Kean JW et al (2017) Prediction of spatially explicit rainfall intensity–duration thresholds for post-fire debris-flow generation in the western United States. Geomorphology 278:149–162. https://doi.org/10.1016/j.geomorph.2016.10.019

Stock GM, Collins BD, Santaniello DJ et al (2013) Historical rock falls in Yosemite National Park, California (1857–2011). U.S. Geological Survey Data Series 746. p 17 + data files. https://pubs.usgs.gov/ds/746/

Stock GM, Luco N, Collins, BD et al (2014) Quantitative rock-fall hazard and risk assessment for Yosemite Valley, Yosemite National Park, California. U.S. Geological Survey Scientific Investigations Report 2014–5129. p 52. https://doi.org/10.3133/sir20145129

Stock GM, Guerin A, Avdievitch N et al (2018) Rapid 3-D analysis of rockfalls in Yosemite Valley. GSA Today, Groundwork 28(8):28–29. https://doi.org/10.1130/GSATG374GW.1

Tiwari B, Ajmera B, Dhital S (2017) Characteristics of moderate- to large-scale landslides triggered by the Mw 7.8 2015 Gorkha earthquake and its aftershocks. Landslides 14(4):1297–1318. https://doi.org/10.1007/s10346-016-0789-0

Varnes DJ (1978) Slope movement types and processes. In: Schuster RL, Krizek RJ (eds) Landslides, analysis and control, Special Report 176. Transportation Research Board, National Academy of sciences, Washington, DC. pp 11–33

Walter TR, Haghighi MH, Schneider FM et al (2019) Complex hazard cascade culminating in the Anak Krakatau sector collapse. Nat Commun 10:4339. https://doi.org/10.1038/s41467-019-12284-5

Wang FW, Sassa K, Wang G (2002) Mechanism of a long-runout landslide triggered by the August 1998 heavy rainfall in Fukushima prefecture, Japan. Eng Geol 63(1–2):169–185. https://doi.org/10.1016/S0013-7952(01)00080-1

Warrick JA, Ritchie AC, Schmidt KM et al (2019) Characterizing the catastrophic 2017 Mud Creek Landslide, California, using repeat structure-from-motion (SfM) photogrammetry. Landslides 16 (6):1201–1219. https://doi.org/10.1007/s10346-019-01160-4

WERT (Watershed Emergency Response Team) (2018) Thomas fire—Watershed emergency Response Team final report, CA-VNC-103156: Sacramento, California, CAL FIRE. https://cdfdata.fire.ca.gov/incidents/incidents_details_info?incident_id=1922

White JL, Morgan ML, Berry KA (2015) The West Salt Creek landslide: a catastrophic rockslide and rock/debris avalanche in

Mesa County, Colorado. Colorado Geological Survey Bulletin 55. p 45 + 2 plates. https://coloradogeologicalsurvey.org/publications/west-salt-creek-landslide-catastrophic-rockslide-avalanche-mesa-colorado/

Wieczorek GF, Larsen MC, Eaton LS et al (2001) Debris-flow and flooding hazards associated with the December 1999 storm in coastal Venezuela and strategies for mitigation. U.S. Geological Survey Open-File Report 01–0144. https://pubs.usgs.gov/of/2001/ofr-01-0144/

Williams JG, Rosser NJ, Kincey ME et al (2018) Satellite-based emergency mapping using optical imagery: experience and reflections from the 2015 Nepal earthquakes. Nat Hazards Earth Sys Sci 18:185–205. https://doi.org/10.5194/nhess-18-185-2018

Wooten RM, Gillon KA, Witt AC et al (2008) Geologic, geomorphic, and meteorological aspects of debris flows triggered by Hurricanes Frances and Ivan during September 2004 in the Southern Appalachian mountains of Macon County, North Carolina (southeastern USA). Landslides 5(1):31–44. https://doi.org/10.1007/s10346-007-0109-9

Xu C, Xu X, Yao X et al (2014) Three (nearly) complete inventories of landslides triggered by the May 12, 2008 Wenchuan Mw 7.9 earthquake of China and their spatial distribution statistical analysis. Landslides. 11(3): 441–461. https://doi.org/10.1007/s10346-013-0404-6

Behind-the-Scenes in Mitigation of Landslides and Other Geohazards in Low Income Countries—in Memory of Hiroshi Fukuoka

2020 Kyoto Japan

Claudio Margottini

Abstract

Landslide science and geohazard mitigation in low income countries requires strong planning, but mainly a fundamental day by day management of what is behind-the-scenes. This is relevant since unexpected problems can appear anytime, and in the interim, most of the mitigation measures require a strong interconnection with local communities. This is a problem since it requires flexibility and capacity of adaptation to unpredictable matters. On the other hand, it is a great opportunity since it shifts our way of working from the traditional approach we study in University, to something that is more social, historical and anthropologically oriented. In this view, to work in low income countries does not require solving only the problem, but also helps the local community to understand how to do it. This means transferring scientific knowledge into the investigation phase (IP), as well as deeply collaborating in the mitigation phase (MP). More specifically, recovering and empowering local knowledge very often based on coherence between environment/heritage conservation, local materials, local climate, recovering and empowering traditional restoration techniques based on long term experience, as well as adapting to local socio-economy. Only with this approach is it possible to work in many countries of the world, they may be poor in science, but very rich in history and dignity. That's why humbleness and competence should guide any interventions. This article is dedicated to Hiroshi Fukuoka. He had clear concepts in mind in relation to this topic. Fukuoka operated to not only to make his high level of knowledge available, but he also offered cooperation and friendship, joining deep knowledge with immense social acceptance of his studies.

Keywords

Landslide • Geohazards • Mitigation • Behind-the-scenes • Low income countries

Introduction

This article describes the collateral activities and the human relationships to be organised and established when working in developing countries, in order to mitigate the consequences of landslides and other geohazards. An explanation about what is behind-the-scene is given before science and technology play their role. I would like to clarify that this is more of a socio-anthropological discussion than a scientific one. This perspective reflects most of the fieldwork of Hiroshi Fukuoka.

The effects of landslides are excessively severe in developing countries. Land use planning is weak in these areas, and a proper management of hazard is usually missing. Very often, the main activity is limited to emergency management during crises.

This is also evident in the case of the conservation of cultural heritage affected by landslides. Heritage sites represent the expression of cultural identity of a geographical region and, for this reason they need special attention, specifically in the prevention against impact phase.

The article describes most of the case studies where the author was involved. It covers various geohazards, but mainly landslides. The requesting agency was primarily UNESCO, who are interested in mitigating natural hazards in World Heritage Sites.

Figure 1 reports the distribution of case studies. They span from Mongolia to Easter Island, in an ideal East–West line covering the world.

C. Margottini (✉)
Embassy of Italy in Egypt, Cairo, Egypt
e-mail: claudio.margottini@gmail.com

C. Margottini
UNESCO Chair for Prevention and Sustainable Management of Geo-Hydrological Hazards, Florence University, Florence, Italy

K. Sassa et al. (eds.), *Understanding and Reducing Landslide Disaster Risk*,
ICL Contribution to Landslide Disaster Risk Reduction,
https://doi.org/10.1007/978-3-030-60196-6_5

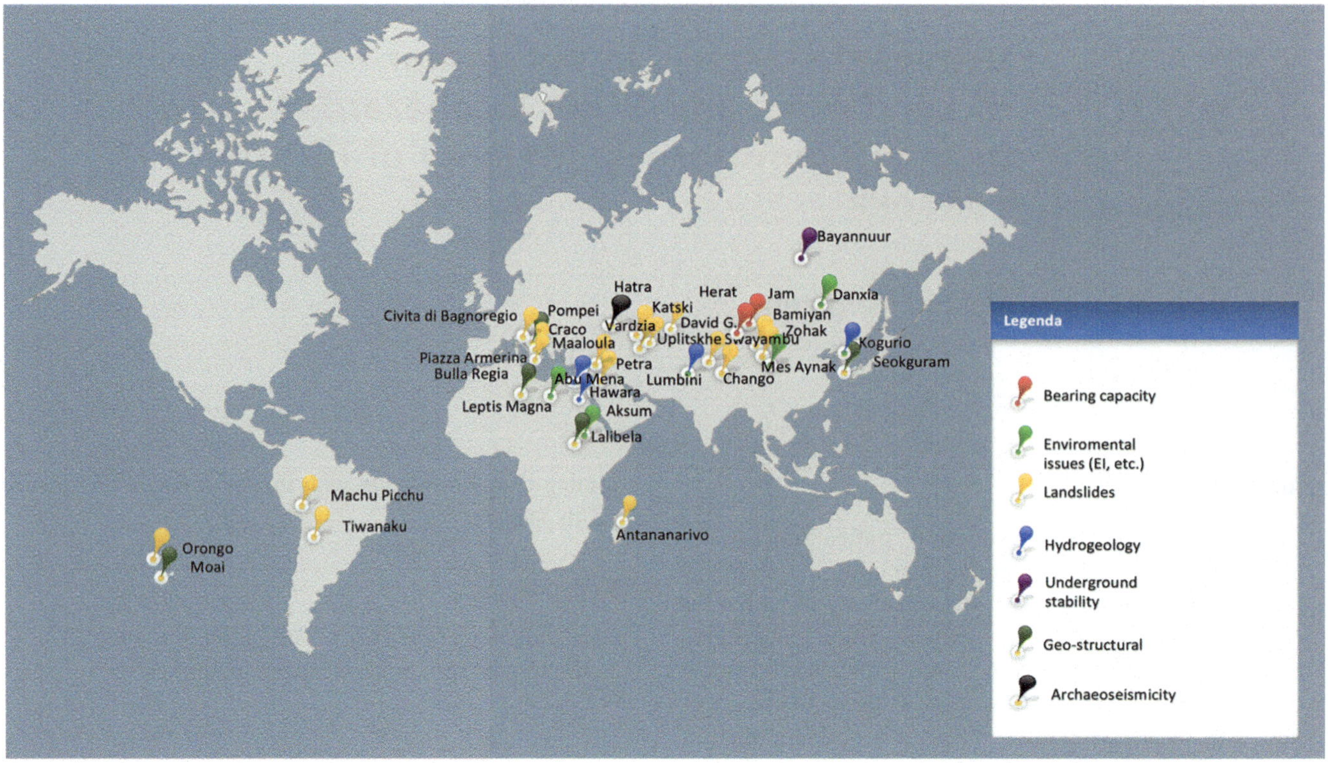

Fig. 1 Case studies investigated in the last 30 years

The largest part of these case studies is in low income countries areas where the prevention of natural disaster is unfortunately, not a priority.

As an example, to work and operate in countries such as Afghanistan is not easy. The mitigation of natural threats does not follow the same procedure and approach typical in other countries. In these areas, mitigation of landslides and other geohazards must be coupled with a possible local economic impact, as well as a political and social acceptance of the proposed solutions. This is the most difficult part of the work; to develop projects that solve issues, but also positively impact local communities, in terms of economy. To achieve this, there is a need to combine deep knowledge in understanding the threats with solutions that do not heavily impact the environment, as well as ensuring long term efficacy. This should be mainly based on traditional knowledge and local expertise. In other words, sustainable.

To do so, any mitigation project has to provide effective solutions—very often innovative—to be combined with original architecture and environment. In a pathway that harmonises advanced solutions with historical heritage. A mitigation project (sometimes pioneering), with a design that minimises the impact of new intervention. This can possibly be achieved by recovering traditional materials and/or solutions and/or the re-use of original construction techniques. In low income countries, the usage of original construction techniques ensures maintenance by local workers, hence enriching the local economy and finally, enhancing sustainability. This is the added value in terms of social impact.

Hiroshi Fukuoka had very clear concepts in mind in relation to this.

In Action

Bearing in mind high-income countries, there is a huge difference when operating in developing ones. In landslide disaster reduction the major concern is related to missing reference data. For example, missing advanced technology in monitoring and in mitigation, missing proper maintenance, missing people skilled for the job required or, at least, motivated to do so.

In Afghanistan, for example, we trained local people for simple monitoring of an anchor strain gauge (just reading the values in the data logger), but unfortunately, changing employees in local offices or the limited understanding of reason for such measurements did not allow any kind of measurements to take place.

It's not always likely to be as negative as described earlier. One of the most positive examples was in North Korea where the measuring of piezometric levels and the structural deformation of cracks in heritage tombs of Koguryo Empires was under the responsibility of the tomb keepers. They strictly followed the instruction and the timing of measurements, without any delay or gaps.

A better situation is where local institutions cooperate in the projects, even if they have a limited skill in the topic. A positive example is in Georgia, in the Ilia University in Tbilisi, where local experts were mainly improving their competence in seismic hazard and instrumental seismicity. Due to the need to install a ground-based radar for monitoring an unstable archaeological slope, they easily covered this part of the work. Even though they invested in monitoring unstable rock slopes with various geophysical techniques, they developed a science at the state of art. Now they are a reference point for central and local administrations.

The few examples show the difficulties, but also the success, to operate in countries with a limited experience in landslide science, sometimes also just out of a period of war.

After three decades of activities in various parts of the world, it became normal to identify a working approach that, with minor differences, was applied in most of the faced case studies. Following is the description of such an approach with some anthropological notes, hereto considered more interesting for understanding the way of working in some of the poorest countries of the world. Technical and scientific details on the projects can be recovered in the many publications and books from ICL.

Fig. 2 The library and the ancient books of the S. Caterina Monastery in Sinai (Egypt)

Fig. 3 Original image (left) and preliminary reconstruction (right) of 3D model of Eastern Bamiyan Buddha, from ancient photos (Margottini et. al. in press)

Desk Study

Collection of existing data is the first point in any scientific work. It includes investigation on the origin of the structure (or heritage) potentially affected by a geohazard, typology, construction techniques, restoration history, evolution in time, etc.

When operating in countries with limited available data, any possible information can be precious. Data often comes from local histories, then collected in the few existing libraries as in the case of the Monastery of S. Caterina, in Sinai (Egypt) where a large collection of ancient books, in many languages, are available today, even if they are at risk of terrorist attacks (Fig. 2).

Similarly, ancient photos can now provide a source of valuable information, as in the case of the Buddhas destroyed in Bamiyan (Afghanistan), where modern geomatic techniques allow the reconstruction of a 3D model of a destroyed statue, from digitisation of historical images (Fig. 3).

Diagnosis of Present Conditions

Investigations in landslide science and engineering geology cover a large variety of disciplines. When the studies impact Cultural Heritage, other professions need to be involved. However, it is not the target of this article to describe scientific methodologies and techniques. Some examples can be found in 2017, as well as many other articles.

Returning to a crucial point from behind-the-scenes, it is important to note in field surveys how one can reach the site. This is not a simple statement since in some countries the distance is still based on time and is not perceived in terms of distance. To go from Herat (Afghanistan) to the Heritage site of Jam, the necessary time is 12 h, only 221 km apart.

Roads do not exist in some areas (Fig. 4) and refuelling is done by hands, from a barrel (Fig. 5).

In Tiwanaku, Bolivia, the local heads of communities requested our support for the identification of the best location for a new earth dam to serve the area. But to walk at 4,000 m asl for a field survey is complicated. Thus, local people have developed a very long tradition, consisting of barely tasting coca leaves, by keeping it in their mouth for a while. It works as vasodilatation, as well as for fixing the oxygen in the blood. The setting up of a museum on coca was taking place in Tiwanaku.

Fig. 4 The road to the world heritage site of Jam (Afghanistan)

Fig. 5 Refuelling

We started early in the morning and every one/two km there was a special ceremony, including praying together, drinking sprite/cola and getting coca leaves in a mix of religious and pagan tradition (Fig. 6). It was a very touching moment as well as a very useful one, since we walked many kms without any issues arising.

A field survey was not always implemented, as in previous cases. Wild animals cause major fears, including snakes in Machu Picchu (Peru) or monkeys looking for food in Swayambhu, Kathmandu (Nepal) (Fig. 7). Nepal is where one of the monuments was damaged by a landslide triggered by the Gorka Earthquake of January 2016.

There is even the risk of landmines, unfortunately still very abundant in Afghanistan (Fig. 8).

Security conditions are problematic in countries just coming out of war. I cannot forget my nickname in the first mission in Afghanistan, in 2002. I was Whisky 1–05. That is the code for the survey that was used by security every

Fig. 6 At 4,000 m asl, in Tiwanaku (Bolivia) supporting the local community with finding the best location for an earth dam. Fieldwork was "supported" by the use of coca leaves, as it has been done in the Andes for millennia

Fig. 7 Animals, to take care of during fieldwork. A snake in Machu Picchu (Peru) and a monkey in Swayambhu, Kathmandu (Nepal)

Fig. 8 Mark for landmine in Afghanistan. Red means that a landmine was detected and white that the site was cleared

evening, via radio. They used to call each one of us and we had to respond promptly, also via radio, otherwise it meant that we were not in safe conditions and they had to intervene with military forces.

Whisky was the code for UNESCO, 1 stands for the head of the office, 05 for myself, the fifth on his list.

Security was very often a major issue, as in Mes Aynak (Afghanistan) one of the largest copper mines of the region. That was also the most important training field for Al Qaeda. To reach the site, affected by future excavation destroying many archaeological sites, required a proper convoy from Kabul. It was possible to survey only because of the continued presence of the army, ensuring our safety (Fig. 9).

Despite animals and landmines, fieldwork also requires time, due to the difficult access of some areas where a geological survey was necessary. To cross a river, the only possibility sometimes is an artisanal bridge, composed of three wires, and moved by people (Fig. 10).

Complexity in the operational phase also affects geophysical prospects and other kinds of investigations. In Afghanistan, for instance, it is not allowed to use small charges of explosive for passive seismic refraction. To implement such measurements, it is necessary to understand the geological profile of the Jam valley where the leaning minaret of Jam, from XII c. is located. It was necessary to hire a small truck and to construct a very heavy barrel, plenty of concrete and iron as a vibration tool. In this way the vibration was effective for deep investigations (Fig. 11). Unfortunately, no boreholes were possible to deploy on site. Since the material is composed of mountain alluvial deposits with large boulders, no drilling machine equipped with wire-line or similar hammer is available in Afghanistan.

A different situation occurred in Easter Island (Rapa Nui, Chile), where geophysical investigations were implemented to understand the level of weathering of Moai statues and the presence of discontinuities (Fig. 12). An important project was planned to give visibility to the local populations, to demonstrate their huge capacity of adaptation to climate change. For instance, they survived during the Little Ice Age (14th c. through the mid-19th c.). This view is in contrast to many legends that consider the island suffering for the destruction of ecosystems to construct the Moai statues.

Fig. 9 Mes Aynak (Afghanistan). The Afghan army secured the fieldwork

Fig. 10 Crossing the Hari Rud river (Afghanistan, at the border with Turkmenistan)

The survey was performed by means of a laser scanner, georadar and seismic tomography, without any impact of the Moai. Nonetheless, the local indigenous population, who have the responsibility of the heritage, according to Chilean law, deliberate to investigate only statues having no religious value (Margottini et al. 2013), that partially reduced the outcome of the project.

A more complicated issue is the organisation and execution of geotechnical boreholes. In some countries, drilling machines do not exist. If they do, they are very basic, with no possibility to have undisturbed samples. This generates serious concern, especially when the target is a heritage site,

Fig. 11 Triggering of mechanical vibrations for the generation of seismic waves

Fig. 12 Seismic tomography in the Moai statues in Easter Island, Rapa Nui, Chile

as it is crucial to have a precise understanding of its underground physical and mechanical properties.

At the Koguryo heritage site (North Korea), specifically in the Susanri tomb, a network of piezometers was required to investigate the water table fluctuation and the relationship with the periodical flooding of the earth mound tomb. The boreholes for the pipes were deployed with a very simple machine, as continuous coring (approx) requires the load of a person on top to have the necessary torque (Fig. 13). Also, falling head permeability tests were executed in the wells.

A better drilling machine was used for the investigation of the water table in Lumbini (Nepal) (Fig. 14). Falling head permeability tests were executed during the drill. In this case, since the material was mainly a partially cohesive silt and clay, with minor sand, it was possible to core continuously and more efficiently. At least a pocket penetrometer and a vane-test could be used to test the collected samples (Fig. 15).

Finally, some selected samples were transferred to a soil mechanic laboratory in Kathmandu, to get physical properties of materials.

Apart from very simple permeability tests, only in one site, in Herat (Afghanistan), it was possible to realise SPT tests in the boreholes. The equipment was made available from an Afghan drilling company.

The project aimed at the emergency consolidation, conservation and restoration of the leaning 5th Minaret in Herat (Afghanistan).

Boreholes and SPTs were applied all around the site, together with a single station measurement of seismic noise. Inclined boreholes were drilled to understand the depth of the foundation (Fig. 16).

Finally, the geotechnical laboratory of the drilling company was able to obtain physical properties, as well as permeability tests of reconstituted samples.

Fig. 13 Rotary core drilling machine, including top man, in Koguryo heritage site (North Korea)

Fig. 14 Rotary core drilling machine in Lumbini (Nepal)

Monitoring

One of the most important steps in investigating an unstable slope or a landslide, is to predict its evolution.

Fig. 15 Boxes with collected cores

Fig. 16 Inclined boreholes to identify the depth of the foundation of the leaning 5th Minaret in Herat (Afghanistan)

Geomorphological studies may help with this, but monitoring is certainly the most powerful tool.

Monitoring can vary from basic to very advanced, depending on the chosen technology and local capacity to manage it. There are of course hundreds of methods and techniques available and in use around the world for monitoring slope instability. An important distinction should be made regarding real time continuous monitoring vs stepped periodic continuous monitoring. Due to the need of long-term data, except when using satellite remote sensing, monitoring is deeply dependent on local expertise and knowledge. The selection of the most appropriate monitoring system, in a given project, is strongly dependent on the local capacity to manage it.

Understanding the evolution of Bamiyan cliff (Afghanistan) and long-term efficacy of executed mitigation works, was accompanied with various monitoring networks. They consist of simple and low cost manuals on many relevant joints and discontinuities were distributed over the 2 km of the cliff, as well as an automatic alarm system able to monitor the working area and record the possible movements of the major cracks during the working period. Comprising 15 potentiometric crack gauges (0–50 mm, 4–20 mA); a monitor via strain gauge for passive anchors, with a series of 5 spot weldable strain gauges, in two of the installed anchors.

All the monitoring networks provided useful data, with the only exception of the two monitored stranded anchors. A local technician was trained to read the measurement on a data logger (Fig. 17). Unfortunately, he did not execute the measurement. After some time he went to the US and remained there, he did not transfer the procedure to anybody else. And the data logger disappeared.

A very worthy project was run in North Korea. The structural monitoring of one of the most beautiful tombs, Annak III, belonging to the Kogurio empire, was required. Due to the lack of electricity and local capability, a very dense network of joint measurement with a manual crack gauge was implemented. Pins for measurement were installed with a special temporary dental cement—very strong but highly fragile—so easy to be removed without damaging the site. Responsibility for the measurement was

Fig. 17 Training course for local technician, to read a data logger for strain gauge monitor

assigned to the tomb keeper (Fig. 18). He was proud and happy of this role and, in fact, did a perfect job.

A similar positive experience was in the water table monitoring in the area of Susanri tomb, also belonging to the Koguryo UNESCO heritage site (North Korea). The responsibility was given to the tomb keeper and the records were precise and accurate. The most I got from this experience was learning that the sense of responsibility of North Korean workers is extremely high.

A large project of rock slope monitoring was implemented in Petra (Jordan) (Fig. 19). Considering the morphological setting and slope instability processes, the following monitoring techniques were designed, implemented and installed for the monitoring of the Siq slopes in Petra (Margottini et al. 2016):

- Satellite SqueeSAR™ analysis with permanent scatters techniques, to evaluate potential regional deformation pattern of the site and possible Siq border effects;
- Automated geotechnical crack-gauge network, with wireless connection, to monitor main cracks and isolated potentially unstable blocks, with a low environmental impact technology;
- High resolution total station network measuring a prisms network, robotic individual reflectorless points network and robotic reflectorless grid network in the Siq slopes, for monitoring slope/blocks deformation;
- Digital photogrammetry; and
- Manual crack gauge network on 25 main discontinuities.

The result was quite good for the remote sensing monitoring (Alberti et al. 2017) and for geotechnical crack gauge monitoring. The robotic reflectorless monitoring was planned to be implemented with the practical work of a local University. The training was realized by international experts. The activity is at the state of art of scientific knowledge and clearly requires appropriate know-how in the implementation and in the elaboration of data, before rock mechanic interpretation is attempted.

Fig. 18 Manual monitoring of the structural deformation of the tomb Annak III, Koguryo UNESCO heritage site (North Korea)

Fig. 19 H.E. the princess of Jordan and the director general of UNESCO, visiting the monitoring system of Petra in 2012

Unfortunately, despite the initial availability and the training from the total station manufacturer and international experts, the local capability to manage such a complex monitoring system was not appropriate. After finishing, this specific type of monitoring was abandoned.

A very positive result was achieved in Georgia. Thanks to the Director of the National Agency for Cultural Heritage Preservation of Georgia, Dr. Nika Antidze had a very smart approach developed in the conservation of many Georgian Heritages affected by rock instability.

Georgia is a beautiful country, at the border of Europe, Asia and Arab Countries. Georgians officially adopted Christianity in the early fourth century and, during its history, suffered from invasions and attacks from many different groups. Due to this location and due to the geomorphology of Caucasus, most Castles, Monasteries and Churches are located at the top of rocky cliffs, to enhance their military security.

Today, such sites are affected by rock slope instabilities. One of the most endangered sites, in the UNESCO tentative list, is the heritage site of Vardzia. The rock-cut city of Vardzia is a cave monastery site in south western Georgia, excavated from the slopes of the Erusheti mountain on the left bank of the Mtkvari river. The main period of construction was the second half of the twelfth century. The caves stretch along the cliff for some 800 m and up to 50 m into the rocky wall (Fig. 20). The monastery consists of more than 600 hidden rooms over 13 floors. The main site was carved from the cliff layer of volcanic and pyroclastic rocks, at an elevation of 1300 m above sea level. The cave city included a church, a royal hall, and a complex irrigation system.

The site is affected by many examples of slope instabilities such as rock fall, topple and rock slide, with volumes ranging from few to 1000 s of cubic meters (Margottini et al. 2015a). Many phenomena occurred in recent years, such as in May 1968 when a huge protective wall was constructed just above the main church, or the very recent rock slide of spring 2011, 2014, 2015, 2016 and 2017. All mass movements, occurred after the anthropization of the site, destroyed many remarkable remains of Georgian history.

A monitoring system was believed to be essential for understanding the evolution of the entire cliff and to establish an alert system. The most advanced and effective technology, covering the whole site, was supposed to be a ground-based radar monitoring system. Clearly, the equipment requires a relevant expertise and knowledge in the field of signal analysis and interpretation (Fig. 21).

The approach we followed in many projects was to increase local capacity building. In Georgia the Ilia University of Tbilisi, was a fundamental local partner in our project. With relevant expertise in seismology, they implemented their knowledge in the field of radar monitoring and now they are able to manage a very complex radar system.

In the meantime, Ilia University also integrated the ground based radar with a weather station, a strong motion accelerometer, a microclimate network in some of the caves, and photo and video monitoring. Now the site is one of the best monitored in the Caucasus region, one visited by the Georgian Prime Minister (Fig. 22).

Monitoring is, very often, the most powerful tool as an alarm system. When working at a very endangered site, with labourers and technicians, there is a strong need to operate safely and to understand the real time evolution of unstable blocks. In Bamiyan (Afghanistan), to operate safety, all the cracks in the working site were monitored with (Fig. 23):

- 15 potentiometric crack gauges, 0–50 mm, 4–20 mA, fitted with couplings and connecting cable (tot. length 350 m);
- 1 acquisition unit Datalogger Geolog with 24 channels.

Fig. 20 The rock-cut city of vardzia (Georgia)

- 1 dedicated laptop;
- data management software and alarm system (light and sound).

Power was supplied by a generator.

In one or two cases, at the very beginning, during the calibration of the monitoring system, the light was switching on. Workers immediately escaped from the site for fear of an immediate collapse.

Conservation Policies and Management

Mitigation projects in low income countries is a complex endeavour. It requires the capacity to deeply understand geological processes and to predict their evolution. This is in order to implement mitigation that has, possibly, low environmental impact and that can be managed by local people. Hopefully, making reference to indigenous traditional knowledge (Margottini et al. 2017; Margottini and Spizzichino, in press).

In this view, to work in low income countries is not only required to solve the problem, but also helps the local community in understanding how to do so. This means transferring scientific knowledge in the investigation phase (IP), as well as working together in the mitigation phase (MP). An MP is then ruled by effective solutions as much as possible coming from traditional knowledge or by measures suitable to be managed by local communities.

On the other hand, IP requires high technology and expertise. Generally, it is advisable to import in the country only what is locally missing, then work together. An MP, when possible and depending on the problem, should enhance traditional knowledge, in order to create new jobs and opportunities, in combination with the local scientific community and mainly, with local society.

This means recovering and empowering local knowledge. Very often based on coherence between conservation, local materials, local climate, recovering and empowering traditional restoration techniques based on long-term experience and adaptation to local socio-economy.

To work with this perspective requires a huge effort, a large amount of relational activities, not in terms of science application, but of something more complicated that is behind-the-scenes.

A clear example were the investigations on the geotechnical reason for the leaning of the Jam minaret in Afghanistan (Bruno et al. 2017). Prof. Andrea Bruno initially discovered the site in 1970. During the first visit to the site, in 2004, we felt the need to reach an agreement with the local community, clearly interested in working with us to get a temporary salary (Fig. 24).

To hire local people means enhancing security, since the entire community is involved and all of them are interested

Fig. 21 Output from ground based radar in vardzia (Georgia) in august 2018

Fig. 22 The georgian prime minister H.E. Irakli Garibashvili visited vardzia site, in october 2014

not only in the immediate salary, but also in a positive approach with national and international experts that can hardly visit a remote site as the one in Jam. In general, they understood that any future fieldwork would include them, but they also believed themselves to be in some way responsible for the preservation of the monument, as it is part of their historical identity. In these occasions, particular attention must be given to the hierarchic role of the various people and to relate accordingly. A local facilitator, in this case from UNESCO, was essential.

In the consolidation of the Bamiyan cliff (Afghanistan) more than 30 local workers were hired. In this case, the support of the local UNOPS office was fundamental. The official salary was 3 US dollars per day, per person, in 2003

Fig. 23 Field office in Bamiyan (Afghanistan): sketch of the alarm system

Fig. 24 A meeting in Jam, Afghanistan, discussing with local authorities about necessary activities and their involvement

Fig. 25 A young soldier in Bamiyan (Afghanistan)

and 5 US dollars in 2005. We were in fact approached by a local army group that wanted the job. We peacefully explained to them that according to UNESCO rules, it is not allowed to hire armed groups. We finally negotiated to obtain some support from them as guards during the fieldwork. We agreed on three people per day.

At the end, we recognised some of the previous people in uniform among those who later joined the UNOPS group. Instead of the three military guards, one young boy, maybe less than 14 years old, joined. (Fig. 25) 24/7. We accepted the situation, and everything went smoothly.

A strong involvement of local community also occurred in Tiwanaku Heritage site (Bolivia) a Pre-Columbian archaeological site in western Bolivia near Lake Titicaca. The earth pyramid of Akapana, was initially covered by a large block of stone, later on used as quarry in colonial period. The internal part is composed of earth, currently affected by various kinds of soil erosion and landslides (Margottini 2013a).

The involvement of local community was essential, since they obtain their income from the maintenance of the archaeological site. They belong to 5 different tribes and the agreement they have with the Ministry of Culture, gives to them the right to work on the site, with a rotation of 5 years.

On the other side, local experts wish for a mitigation project, possibly recovering the original shape of the pyramid. Due to possible limits and constraints and, after several meetings with UNESCO consultants and local managers, the final adopted measures were the use of small gabions (filled with local stone sandstone—Arenisca). In order to reduce and mitigate the impact of the measures (sustainable mitigations) and to recreate the original shape of the Akapana Pyramid, it was decided that any adopted solutions must be located on the ancient terraced wall path (rocky staircase), see Fig. 26.

Earth terraces were also created to maintain the original tradition. The material was compacted by hand and by foot, with the help of an old system based on compaction with a flat wood panel (Fig. 27).

Ensuring surfaces and terraces were impermeable was implemented by using an old Inca technique, which is still often used. This consists of using a combination of straw, mud and cactus-juice (penka) (Fig. 28).

Local involvement was used in Shahr-e-Zohak (Margottini et al. 2015b). In this specific case, it was demonstrated that the use of local shrubs should reduce soil erosion (Fig. 29). Since this bush is locally used as biomass for heating in winter, a special nursery was agreed upon. It was to be made with the support of the local population. In this way the plant production was jointly used to contrast soil erosion, and offered to the local community to constrain fire. The men kindly agreed to not tear off the bushes growing on the archaeological site, where the soil erosion mitigation fields are located. This project has not been implemented yet.

The site of Petra was investigated in the period 2012–2015. The purpose of the project was to limit the risk of rock fall in the narrow 1.8 km canyon (Siq) where, in high season, around 6,000 tourists are daily walking the site (Margottini and Spizzichino 2017).

Among the many designs for a proper anchoring of unstable blocks (Spizzichino et al. 2016), a preliminary cleaning of the upper part of the Siq was believed to be essential. For this purpose, a special group of local Bedouins were trained (Fig. 30). Thus, this kind of maintenance should have a regular periodicity. If the local authorities are

Fig. 26 Gabions with exposed surfaces composed by the same typology of local stones, originally covering most of the pyramid (Margottini and Spizzichino, in printing)

Fig. 27 Rill terracing and compaction of soil in Tiwanaku (Margottini and Spizzichino, in press)

Fig. 30 Petra (Jordan) scaling unstable blocks by local climbers in 2016 (*source* Facebook UNESCO Amman)

Fig. 28 From cactus to cactus-juice, to penka (Margottini and Spizzichino, in press)

Fig. 29 The bushes to be used against soil erosion

not aware about the importance of rock degradation, i.e. during the rainy season, the risk of new collapses may appear again. In fact, in a place like Petra, more than big block for which some precursory event can be estimated, the major risk for people is due to small falls and isolated stones falling from 20–30-40 m high. A relevant fall occurred in May 2015 and only by chance, it did not affect a group of tourists. A single small block, unfortunately, killed an Italian tourist in 2019.

Images of the successful experience in Vardzia (Georgia) were previously provided. This positive experience was also documented in the Mitigation Phase (MP) of the project.

After the first mission in the country, it became clear that no company with expertise in the sector of rock consolidation (installation of anchors, bolts, etc.) existed. Thus, climbers are present, and very often clean the cliff from falling boulders. In collaboration with the National Agency for Cultural Heritage Preservation of Georgia, it was decided to create a local company, in order to operate in the consolidation of unstable slopes affecting cultural heritages.

An excellent specialist, retired from TREVI, after having spent most of his life in such countries, managing practical works, accepted to work with us and transfer most of his knowledge to the newly created company. The required knowledge spanned from the very basic elements (e.g. how to pump cement from the ground to the injection point on a scaffolding) (Fig. 31), the execution of a wooden scaffolding, to the drilling and placing of bolts with low impact (Figs. 32 and 33).

Another positive experience was at Bamiyan (Afghanistan), where the local community was involved in the mitigation of falls and the restoration of the remaining heritage of the destroyed Buddha statue in the Eastern Niche. The local community, after a special training with ICOMOS technicians, was able to support the most complex part of the work (Praxenthaler 2014). Figure 34 shows a close up of such restoration works.

Now, these people can easily find other opportunities for the restoration of heritage in a country very rich in monuments, but poor in skilled expertise.

A special project was developed in North Korea, for the dewatering of some relevant tombs belonging to the UNESCO World Heritage Site of Koguryo. The tombs, covered by earth mounds, were affected by water infiltration during the rainy season. It is not clear what the reason was

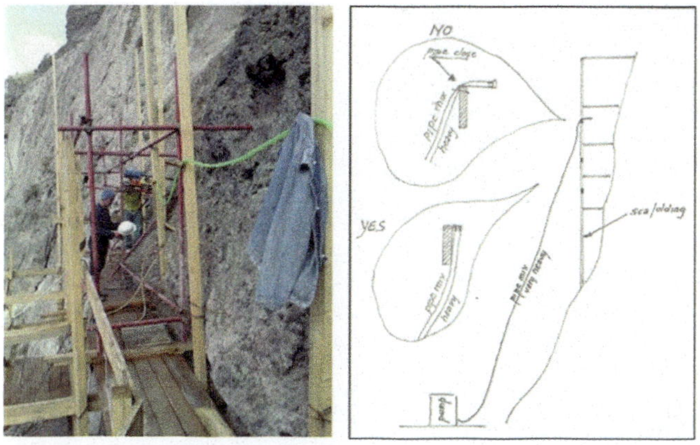

Fig. 31 Scheme for pumping cement from the ground to the passive rock bolt

Fig. 32 The first wooden scaffolding in vardzia in 2015. Metal scaffolding was not available

Fig. 33 Scaling of unstable blocks with local climbers (left) and cover-up the head of passive rock bolt (right)

(water table rising, leaking from surface, etc.). Finally, after years of investigation, it was demonstrated that earth mounds covering the tombs were completely reconstructed in 1960. Unfortunately, such reconstruction did not follow the original purpose of the earth mound, that is, impermeability and protection of the tomb. In fact, the earth mound

Fig. 34 Consolidation of rock fall and restoration of remaining heritages, in the back wall of the eastern Buddha niche in Bamiyan (Afghanistan)

was originally designed for a proper function and is today considered as an integral part of the archaeological tomb and not as a landscape add-on. As a consequence, after our studies, it was decided to place a filter 10 cm below the modern earth mound. It was placed to collect the water and, with a drainage trench, dispose of the water out of the area of influence on the tomb.

Thus, two major issues dominating the MP were the availability of special material in North Korea to be used as a filter, and the capacity to install such special materials.

After a first investigation it was clear that no filter material was available in North Korea. It was then requested by many companies, clearly asking to deliver the material in PyongYang (North Korea). Only one company declared to be able to deliver the requested material in Nampo, the harbour of PyongYang. UNESCO made a contract with Maccaferri to deliver the necessary material in North Korea (geocomposite MacDrain©, soil mattress© and drainage pipes). Considering the relevance of the project, Maccaferri also decided to send the Director of Asia branch and a technical director to support the installation of materials. This was the second successful step, since the only available resource in North Korea was a workforce, using very simple tools (Figs. 35 and 36).

The result was excellent, the flooding disappeared but, due to the design allowing a minimal infiltration, the relative humidity inside the tomb remained around 80–90%, ensuring the proper conservation of mural paintings.

A similar hydrogeological problem was faced in the Shoroon Bumbagar, Bayannur (Mongolia). The excavation and the initial activities, particularly the cutting of the earth mound, the emptying of the tomb, the installation of wooden supports, and the construction of a shelter above the tomb, altered the internal conditions and the surrounding environmental conditions, which had enabled the tomb and its murals to be preserved for some 1400 years. The archaeological excavation of the Shoroon Bumbagar tomb was

Fig. 35 Excavating the earth mound to set the geocomposite MacDrain©, soil mattress © and drainage pipes, all from Maccaferri

Fig. 36 Available tools

carried out in July and August 2011 by a Mongolian-Kazakh team of archaeologists from the Institute of History of the Mongolian Academy of Sciences and Eurasian University, Astana, Kazakhstan.

It is likely that the earth mound was put in place to prevent water infiltration into the tomb, so as to enable the tomb to remain intact for as long as possible. The earth mound was built using a combination of layers, with different degree of permeability, that is, pressed earth/clay in a very sophisticated manner, and not by simple debris accretion. Because the excavation removed part of the earth mound to access the tomb, the environmental conditions that allowed the tomb to survive intact until discovery were altered and water started to infiltrate, generating salt formation on the paintings (Fig. 37).

Moreover, the construction of a protective shelter, without any system to discharge rainy water and snowmelt, increased the infiltration inside the tomb, generating salt formation and damaging, very seriously, one of the best heritage sites of Mongolia.

After having identified the causes for water infiltration (Fig. 38), local conservators are now working on waterproofing of the tomb and the restoration of damaged paintings.

Fig. 37 Salt crystallization on the painting in Shoroon Bumbagar tomb

The two previous case studies clearly demonstrated the relevance of an interdisciplinary approach, even in archaeological excavation.

Sometimes, threats arrive from people, on a deliberate basis. There was a case in Syria, where the rocky cliff above the Monastery of Sant Tecla was under restoration (Margottini 2013b). With the arrival of DAESH the works were stopped and even the Monastery was damaged (Fig. 39).

Similar to the previous example, was the increased leaning of the 5th minaret in Heart (Afghanistan). The evolution in time of the tilt of the tower may help in understand the reason for such a process. According to the limited available information and photos it is possible to say that the out-of-plumb was only 0.90 m in 1974 (Bruno 1981). The topographic investigation implemented by Santana and Stevens (2002) show that the out-of-plumb is 2.43 m in 2002 and that a large horizontal crack is present at the height of 3 m from the base. Despite a different position

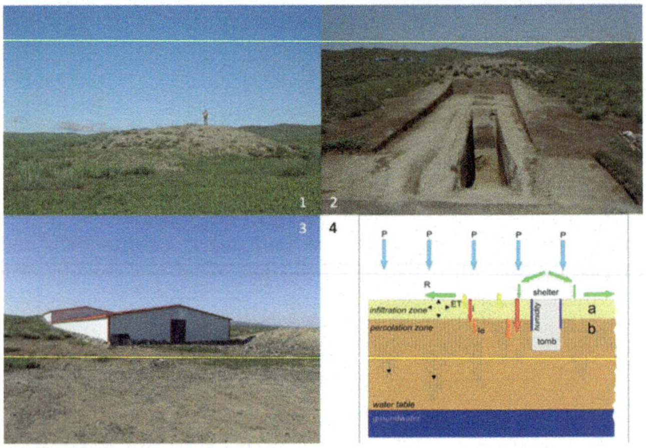

Fig. 38 The original landscape with the Shoroon Bumbagar tomb (1), the excavation of the entrance (2), the shelter (3), the hydrogeological model for water infiltration

Fig. 39 The Christian church of Maaloula (Syria), heavily damaged by DAESH

to get the photo it is quite evident that the Minaret suffered from an import out-of-plumb between the picture 1975 (Bruno 1981) and the picture of 1993 (Fig. 40).

According to some eye-witnesses, during 1985–1990, the complex of the musalla of Gauhar Shad, where the minaret is located, was exactly at the conflict line between Russians and Afghan Mujahidin. As a consequence, no maintenance plan was implemented on the site, with special reference to Hari Rud river, that crosses the whole area. The result was a general and continuous flooding of the site, with water flowing almost below the minaret (Fig. 41).

The bearing capacity of local material is, under present conditions, about 5 times of the ultimate conditions; the allowable one is, in common practice, generally 3 time the ultimate one. Simulating the worst-case conditions, in terms of water saturation and geotechnical properties, the acceleration of leaning can clearly be explained in terms of water saturation of the soil, as occurred in the period 1985–1990.

Also in that case, humans were responsible for the damage to a very important monument in the history of Afghanistan.

Fig. 41 The area flooded permanently during the war period between 1985 and 1990. The blue lines show the most depressed areas, one of which flowing very close to the minaret. The actual river Hari Rud is on the center-right side (blue). In that period the minaret increased the out-of-plumb from 0.90 m to 2.43 m

Discussion

The experience developed in decades of activities in low and medium income countries, sometimes very poor, pointed out some general findings and personal feelings. Such thoughts do not refer to scientific or technical work but to whatever is behind-the-scenes.

- Very often we have to find a solution (operational or mitigation) that is not common. It may be unusual or unexpected, but still solves the problem;
- Our work is useful when solving the problem but also when local communities can take advantage of it;
- Mitigation projects should be implemented and managed by local communities, when possible. This enhances long term maintenance;
- There is the need to enhance traditional knowledge and sustainable practices;
- There is a need to use local materials and workers from the local community (as much as possible);

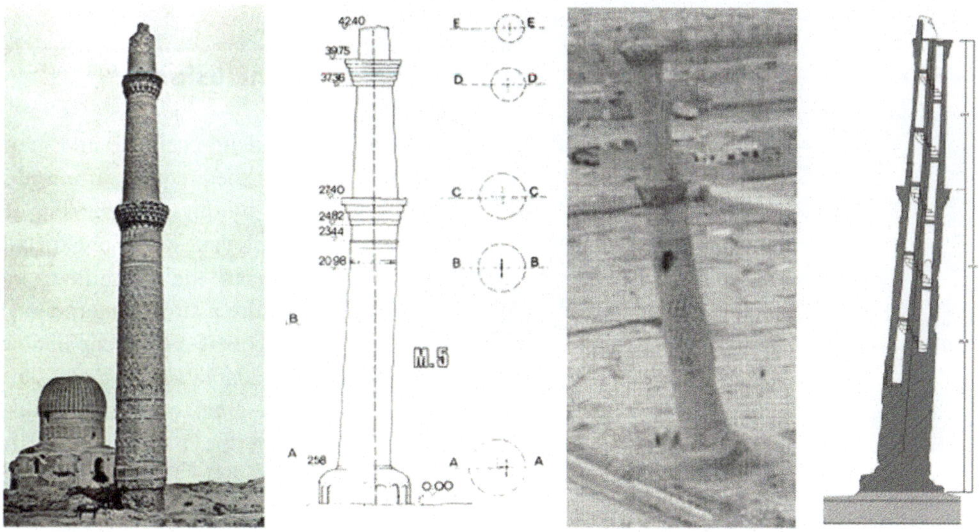

Fig. 40 The 5th Minaret in Herat (Afghanistan) in 1915, the survey of 1974 (Bruno 1981) the photo of 1993, the survey of 2002 (Santana and Stevens 2002). The position of photographer/view is clearly different in each image

- There is the need to trust to science, then to make awareness an important point of our work. There is very little awareness among the institutions and persons responsible for the management and conservation of Cultural Heritage sites in relation to the geohazard and disaster risks to which they are exposed. Accordingly, they had very little preparation to deal with this situation. We have to push this forward.

In practical terms, the respect of local communities who have a very long history and expertise in their oral traditions, is essential. We have to learn how to work with them, with humbleness and competence. It is correct to say that, very often, in such countries, human relationships are more relevant than professional relationships.

We have many examples of such an approach. In Antananarivo (Madagascar), the surveys on landslides triggered by monsoon was conducted together by the National Office for the Management of Risks and Catastrophes—BNGRC, RC Heritage from France and the Région Ile-de-France (Paris), with the office in Antananarivo, in a very fruitful cooperation (Fig. 42).

Fig. 42 Field work on landslides triggered by monsoon in Antananarivo (Madagascar)

Fig. 43 Italy versus Peru, celebrating the successful mission in Machu Picchu, in 2004

Another example of close cooperation is the tournament we used to organise every year during our mission in Machu Picchu, Italy versus Peru, which undoubtedly, ended with a group dinner (Fig. 43).

The cooperation in Tiwanaku (Bolivia) was much appreciated and the heads of the local community (Malcom) stayed with us during the fieldwork (Fig. 44).

Last but not least, to closely work with the local population it also means to support them in increasing the awareness of landslides. This was done, as an example, in Civita di Bagnoregio (Italy), currently in the tentative list of UNESCO. In past years, to enhance the importance of the site, many testimonials were invited to visit the village, including his Royal Highness Charles, Prince of Wales (Fig. 45).

It's not always simple as it was in Machu Picchu or Tiwanaku. Sometimes you have to face complex situations, especially when you discover that your amenities in the hotel include helmets and bulletproof vests (Fig. 46).

Some other times, you have to deal with real poverty, not the type you see on television or read about in a magazine. You see babies who have nothing, yet are happy. That is when you realize how lucky your children are (Fig. 47).

Conclusion

Landslide science and geohazard mitigation in low income countries require strong planning, but mainly a fundamental day by day management of what is behind-the-scenes. This is relevant since unexpected problems can appear anytime and, in the meantime, most of the mitigation measures require a strong interconnection with the local communities.

This is a problem since it requires flexibility and capacity for adaptation to unpredicted matters. On the other hand, it is a great opportunity since it shifts the way of working from the traditional approach we study in University to something that is more social, historical and anthropologically oriented.

Fig. 44 Field work in Tiwanaku area (Bolivia) with the great support of Malcom, the heads of local community

Fig. 45 Guiding his royal highness Charles, prince of wales, at Civita di Bagnoregio, famous as "the dying town" due to the numerous and continuous landslides affecting the site

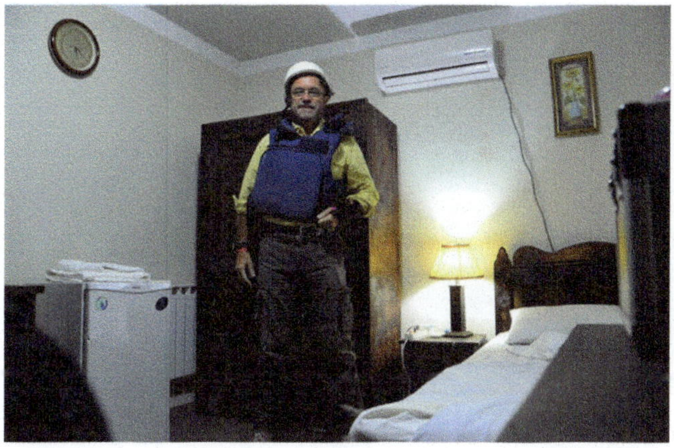

Fig. 46 A hotel in Kabul, 2015, with a helmet and a bulletproof vest provided, just in case

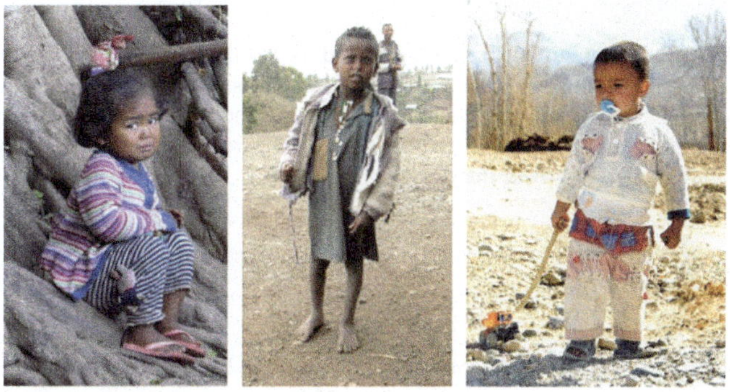

Fig. 47 Babies in Antananarivo (Madagascar), Lalibela (Ethiopia) and Bamiyan (Afghanistan)

Finally, when the target of our work is cultural heritage, another sector has to be involved, that is the conservation science sector.

In conclusion, according to personal experience, it is possible to say that working in such countries requires expertise at the border between earth sciences, social sciences and conservation sciences. In a broader sense, this is probably the basic statement for a new discipline, that we might call "Cultural Engineering Geology". In traditional terms, it is engineering geology and geotechnical engineering for the conservation of cultural heritage (Fig. 48). A new discipline, surely necessary in many developing countries, but also of basic importance in many advanced countries that have very rich heritage to manage and protect for future generations.

I know that Hiroshi Fukuoka had this approach in his life and his daily work. I am sure, he is now supporting us in pushing for a science that is based on the respect and cooperation with the less advantaged population of the world, affected by landslide disasters.

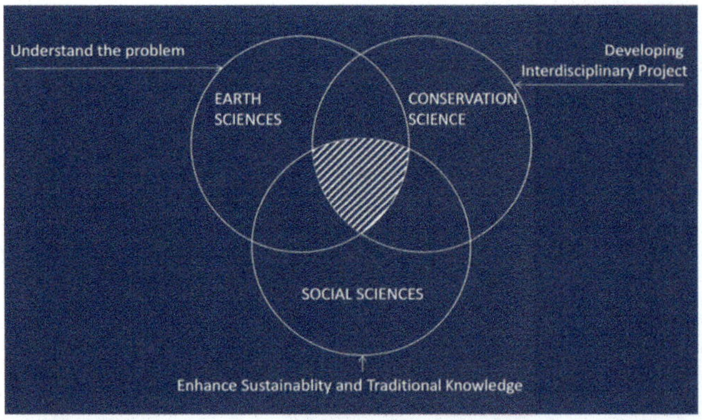

Fig. 48 A general scheme for engineering geology and geotechnical engineering for the conservation of cultural heritage. A new discipline?

In other words, to work with humbleness and competence.

Acknowledgements The author would like to thank the many people and organisations who made possible such lifelong experience. It is not possible to thank all of them, but they are undoubtedly part of my enlarged family.A special thanks to the reviewer of the article, who significantly improved the clarity of the paper.

References

Alberti S, Ferretti A, Leoni G, Margottini C, Spizzichino D (2017) Surface deformation data in the archaeological site of Petra from medium-resolution satellite radar images and SqueeSAR™ algorithm. J Cult Heritage 25:10–20

Bruno A (1981) Background and justification for the project and its immediate objectives. In: UNESCO/UNDP Restoration of monuments in Herat. Restricted UNDP/AFG/75/022 technical report. S. Mauro (TO), Italy

Bruno A, Margottini C, Orlando D, Spizzichino D (2017) Safeguarding the leaning minaret of Jam (Afghanistan) in a conflict scenario: state of art and further needs. In: Viggiani C, Flora A, Lancellotta R (eds) Geotechnical problems in historical towers. Taylor and Francis

Margottini C (2013a) Surface erosion and mass movement constraints in the conservation of Akapana pyramid mound (Tiwanaku, Bolivia). In: Margottini C, Canuti P, Sassa K (eds) Landslide science in practice: risk assessment and mitigation. Springer, Berlin, Heidelberg, New York, ISBN: 978-3-642-31318-9

Margottini C (2013b). Large rock slide and falls in the cradle of Christianity: Maaloula (Damascus, Syria). In: Margottini C, Canuti P, Sassa K (eds) Landslide science in practice: risk assessment and mitigation. Springer Inc., Heidelberg, Berlin, New York, ISBN: 978-3-642-31318-9

Margottini C, Spizzichino D, Orlando L, Marsella M, Renzi B, Sonnessa A, Pandolfi O, Soddu P (2013) Geotechnical and geophysical characterization of Moai statues—rapa nui easter Island (Chile). In: geotechnical engineering for the preservation of monuments and historic sites—Proceedings of the 2nd international symposium on geotechnical engineering for the preservation of monuments and historic sites. pp 539–547, ISBN: 9781138000551, Napoli; Italy, 30–31 Maggio 2013

Margottini C, Antidze N, Corominas J, Crosta GB, Frattini P, Gigli G, Giordan D, Iwasaky I, Lollino G, Manconi A, Marinos P, Scavia C, Sonnessa A, Spizzichino D, Vacheisvili N (2015a) Landslide hazard assessment, monitoring and conservation of Vardzia Byzantine monastery complex. Georgia Landslides Springer 12(1):193–204

Margottini C, Fidolini F, Iadanza C, Trigila A, Ubelman Y (2015b) The conservation of Shari-e-Zohak archaeological site (Central Afghanistan): geomorphological processes and ecosystem-based mitigation strategy. Geomorphology 239(2015):73–90

Margottini C, Gigli G, Ruther H, Spizzichino D (2016) Advances in geotechnical investigations and monitoring in rupestrian settlements inscribed in the UNESCO's world heritage list. The fourth Italian workshop on landslides, procedia earth and planetary science 16 (2016):35–51. (www.sciencedirect.com)

Margottini C, Bobrowsky P, Gigli G, Ruther H, Spizzichino D, Vlcko J (2017) Rupestrian world heritage sites: instability investigation and sustainable mitigation. In: Proceedings of the 4th world landslide forum, Ljubljana 2017, Springer

Margottini C, Spizzichino D (2017) Historical "infrastructures" to access cultural heritages: engineering geology for the sustainable conservation of Petra Siq. Innov Infrastruct Solut 2:25

Margottini C, Bruno A, Massari G, Ruther H, Casagli N, Tofani V, Tincolini F (in press) The renaissance of Bamiyan (Afghanistan). Some proposals for the revitalisation of the Bamiyan valley. Springer

Margottini C, Spizzichino D (in press) Traditional knowledge and local expertise in landslide risk mitigation of world heritages sites. In: ICL contribution to landslide disaster risk reduction. Springer

Praxenthaler B (2014) Safeguarding the clay plaster remains of the eastern Buddha statues and the rear side of the niche. In: Margottini C (ed) After the destruction of giant Buddha statues in Bamiyan (Afghanistan) in 2001, Springer, pp 265–283

Santana M, Stevens T (2002) Topographic survey of heart complex. UNESCO internal report

Spizzichino D, Margottini C, Chiessi V, Boldini D (2016) Assessment of the stability conditions of a large-volume sandstone block in the northern sector of the Siq of Petra. Landslides and engineered slopes. Experience, theory and practice—aversa et al. (eds) © 2016 associazione geotecnica Italiana, Rome, Italy, ISBN 978-1-138-02988-0

The Impact of Climate Change on Landslide Hazard and Risk

2020 Kyoto Japan

Luciano Picarelli, Suzanne Lacasse, and Ken K. S. Ho

Abstract

The gradual increase in temperature recorded all over the world over the past fifty years and the more frequent occurrence of extreme weather events demonstrate that climate change no longer is just an idea born in the mind of some crazy scientists, but is a reality and a challenge that mankind needs to manage within the next few years. Climate change can have severe consequences for mankind, including the increase of geo-hydrological hazards. In some areas of the world landslide risk will grow significantly, even potentially adopting new forms, which the landslide profession and the exposed population community need to be prepared. Unfortunately, in spite of the commitment and the efforts of many mindful scientists, who are well aware of the potential risk increase, the response of stakeholders and politicians is still too slow. This paper describes the effects of climate change on landslide susceptibility, hazard and risk, summarizes the state of slope safety preparedness around the world and proposes steps for enhanced landslide risk management.

Keywords

Climate change • Extreme weather events • Snowmelt • Landslide hazard • Risk mitigation

L. Picarelli (✉)
Chair, Joint Technical Committee Natural Slopes and Landslides, 80131 Napoli, Italy
e-mail: picarelliluciano@gmail.com

S. Lacasse
Norwegian Geotechnical Institute, Oslo, Norway
e-mail: suzanne.lacasse@ngi.no

K. K. S. Ho
Geotechnical Engineering Office, Civil Engineering and Development Department, The Government of the Hong Kong SAR, Hong Kong, China
e-mail: kenho@cedd.gov.hk

Introduction

The gradual increase of average temperature worldwide, the subsequent melting of glaciers and increasing desertification of temperate areas, and the more frequent occurrence of extreme weather events are confirming the impending climate change and its increasing impact. The multi-faceted consequences of this on human life are clearly emerging. The need for involvement and competence of both science and politics in the mitigation of the climate change impacts and in the enhancement of societal preparedness and resilience is then becoming increasingly critical.

One of the impacts of climate change is the increase in landslide hazard, which largely is a function of soil–atmosphere interaction mechanisms. Considering the difficulties inherent in the analysis of such a complex problem, cooperation among different scientific fields, such as physics of the atmosphere, hydrology, geology and geotechnical engineering, is necessary.

In 2015 the Joint Technical Committee on Natural Slopes and Landslides (JTC1) of the Federation of the Geo-engineering Societies, the ISSMGE, the ISRM, the IAEG and the IGS (FedIGS), organized an international Forum, where 17 reports focused on the state-of-the-art on slope safety preparedness for the impact of climate change in 21 countries in the world (Ho et al. 2017a). That experience suggests that further similar scientific initiatives should be undertaken in the near future to stimulate and encourage perhaps shy and sometimes reluctant governments to pursue common goals and actions to reduce landslide risk.

Causes and Impact of Climate Change

The most evident effect of climate change is the increase in temperature, which over the last century, especially since the seventies, was recorded in every corner of the Earth. The

K. Sassa et al. (eds.), *Understanding and Reducing Landslide Disaster Risk*, ICL Contribution to Landslide Disaster Risk Reduction,
https://doi.org/10.1007/978-3-030-60196-6_6

global average temperature rise in the century has been about 1°, with obviously some variations from region to region.

Based on recent studies and reports issued by the UN Intergovernmental Panel on Climate Change IPCC), the main causes of global warming, which clearly demonstrate the important role played by human activity, are:

- Carbon dioxide emissions from power plants burning fossil fuel and gasoline for transportation, which have increased by over 40% since pre-industrial times;
- Methane emissions from animals, agriculture such as rice paddies, and from Arctic seabeds;
- Deforestation, especially of tropical forests;
- Increase in usage of chemical fertilizers.

The first tangible consequences of global warming are permafrost degradation in cold areas, gradual melting of the polar caps and retreat of glaciers everywhere in the world; this, in turn, leads to sea level increase and coastal erosion.

The rise in temperature also results in an increase of water vapor in the atmosphere, which could increase precipitations. However, the relationship between temperature increase and rainfall regime is a complex issue, due to the influence of a series of local factors. As a matter of fact, a clear change in rainfall distribution is not always evident; in addition, the precipitation regime varies significantly from region to region. As an example, while a significant increase of the yearly rainfall is evident in northern regions such as in parts of Canada (Cloutier et al. 2017) and of northern U.S.A. (Walsh et al. 2014), available data suggest that in other parts of the world the yearly rainfall is decreasing. This is, for example, the situation across the Mediterranean basin, where the number of wet days is diminishing as well. However, the combined effect of these two trends is an increase of the intensity of rainstorms (Picarelli et al. 2017). Evidences about the increasing intensity of rainstorms have been reported for extensive areas in China and Hong Kong (Ho et al. 2017b), for Taiwan (Lin et al. 2017) and for Japan (Wang and Towhata 2017).

These considerations are continuously confirmed by the extreme weather events which are being recorded in many parts of the world. Wong and Mok (2009), for example, remarked that, in the last century in Hong Kong the return period of hourly rainfall of 100 mm or more has decreased from 37 to 18 years. Moreover, the peak hourly rainfall, which in 1886 was 88 mm, has been updated five times from 101 mm in 1926, 108 mm in 1966, 110 mm in 1992, 115 mm in 2006 up to 146 mm in 2008 (Ho et al. 2017b). The increase is exponential and the time interval between two consecutive peaks is shortening. Also, in Taiwan the number of typhoons, especially those that are classified as "intense", has been increasing (Lin et al. 2017).

Examples of recent extreme weather events are the 2008 Hong Kong rainstorm (384 mm rainfall in 4 h, Ho et al. 2020), the 2009 typhoon Morakot in Taiwan (400 victims, Tsou et al. 2011), the catastrophic precipitations in 2011 in the Nova Friburgo region, Brazil (about 1000 casualties and 500 missing, Coelho Netto et al. 2013), the 2013 Wipha event in Japan (40 casualties, Yang et al. 2015), the 2017 Mocoa event in Colombia (333 casualties, Garcia-Delgado et al. 2019).

Effects of Climate Change on Landslide Susceptibility

Increased landslide frequency is commonly listed as an impact of climate change, especially for many types of rapidly-moving landslides, such as debris avalanches, debris flows, flowslides and mudflows (Evans and Clague 1994; Geertsema et al. 2006; Jaedicke et al. 2008). However, the relationship between increased precipitation due to climate change and rapid flow-like landslide is not fully understood due to the paucity of clear and unquestionable data. In contrast, some clues are provided by events occurring in zones subjected to physical processes that may reasonably be correlated with climate change. As this last involves a significant increase in temperature, direct consequences are melting of snow and ice, permafrost degradation, coastal storm waves and tides and wildfires. Since, all these phenomena can have a direct influence on landslide occurrence, some clues about the relationship between climate change and landslide susceptibility are provided by events occurring in zones subjected to such physical processes.

This is the case in frozen areas, where the melting of perennial ice and of permafrost modifies the landscape, some physical and mechanical soil properties, such as moisture content, and factors, such as stress field, which govern the stability of slopes. As an example, the retreat of glaciers, which buttress soils and rocks, can adversely affect the slope stability conditions through removal of toe support, slope relaxation and opening of joints and fissures (Evans and Clague 1994; Haeberli et al. 1997). The retreat of glaciers also exposes unvegetated glaciogenic sediments, which become susceptible to landslides and, in particular, to flow-type movements (Chiarle et al. 2007).

The degradation of permafrost in rocks and soils is another cause of landslide. As an example, Ravanel and Deline (2011) documented 42 rock falls in the French Alps between 1949 and 2009, 70% of which occurred between 1989 and 2009 with a peak (11, i.e. 26% of the total) during the warm 2003 year. Permafrost represents an impervious barrier to water infiltration (Kenney and Lau 1984). As the temperature increases, the barrier may vanish or water infiltration may occur earlier than usual leading to soil saturation and decrease in shear strength.

Thaw flows also are becoming a dominant driver of landscape changes in the Arctic area (Kokelj et al. 2015). In such areas retrogressive rates, ranging from 6 to 15 m/y, have been recorded (Wang et al. 2016). In a zone of the Qinghai Tibetan Plateau the average increase rates of thaw slump areas in the periods 1997–2009, 2009–2015 and 2015–2017 was about 62, 60 and 157 m^2/y respectively (Mu et al. 2020) with annual headwall retreat rates of approximately 1.3, 1.6 and 2 m/y during the same three periods. Similar data on the effects of deglaciation in the Northern Alps were reported by Borgatti and Soldati (2010).

Failure of glacially dammed lakes or formation and subsequent failure of glacial lakes due to warming are other events that may occur due to temperature rise, leading in turn to landslides. Deglaciated areas can then become the seat of a variety of landslides (Geertsema et al. 2006).

The relationship between landslides and other phenomena due to climate change is more uncertain due to lack of reliable data and the concurrence of different factors. However, useful clues may be found locally. For instance, geological evidence suggests that frequent landslide activity may have taken place in Hong Kong during the early Holocene as a result of the rapid rise of the sea level (Sewell et al. 2006). Similar phenomena may take place in coastal areas presently.

Prediction of Climate Trends

Going beyond controversial and questionable evidences, the potential effects of climate change on precipitation regimes and landslide hazard can be provided by prediction of future precipitation trends.

Climate trend predictions are generally carried out by two-stage approaches today. An Atmosphere Ocean General Circulation Model (AOGCM) incorporating the key global physical and chemical processes induced by anthropogenic actions is used first to define a general framework of climate trends. In the second stage, the outputs of the AOGCM are transferred into higher scale solutions, which can account for local factors including orography.

The local climate projections are based on various techniques, such as dynamic downscaling and statistical downscaling. Dynamic downscaling is carried out through Regional Climate Models (RCMs) of finer spatial resolution than the AOGCM, which is used to define the boundary condition. The computer resource intensive RCM models often use only one AOGCM as input thus leading to generally fairly uncertain projections. Generating a large ensemble of projections through a single AOGCM and a single RCM is however possible, even though it can be expensive and time consuming.

Statistical downscaling is an alternate and less expensive method based on the use of quantitative relationships between large-scale atmospheric variables and regional/local-scale variables. The downscaling assumes, however, that the current statistical relationships are valid for future climate conditions. To reduce errors, the statistical models are usually calibrated against past weather factors.

With these two stage approaches climate predictions can depict near future scenarios based on different hypotheses about greenhouse gas emissions (RCP scenarios) through data that are provided periodically in IPCC Reports. Naturally, the magnitude of the predicted climate change depends strongly on the adopted assumptions.

Table 1 presents an example of prediction of climate change impact on hydro-meteorological factors in Norway under IPCC scenario RCP8.5, which is the most pessimistic one (L'Heureux et al. 2018; Hanssen-Bauer et al. 2017).

Based on such analyses, by year 2100 temperatures are expected to rise by an average of 4.5 °C and annual precipitation might increase by 18% with more frequent intense rainfalls than before 2010.

Potential Effects of Climate Change on Landslide Hazard

As shown, in some cases the relationship between temperature increase and landslide frequency is quite clear and some effects are right away evident in cold countries due to permafrost thawing and snowmelt. In contrast, the impact of climate change on precipitation regimes and rainfall-induced landslides is less clear. The role of climate projections is just to predict the potential effects of variations in the precipitation regime on landslide susceptibility. As an example, referring to the future scenarios in Norway (Table 1) and in particular, comparing slope stability on a regional basis in 2000 with that in 2100, for an increase in rainfall by 20%, Jaedicke et al. (2008) computed an increase in slope failure probability by as much as 25% in the mountainous areas. In addition, extensive areas would have increased slope failure probability between 15 and 25%.

However, current projections tend to focus on the changes in the long-term mean conditions, rather than on the frequency and magnitude of extreme events, which can generate extreme local precipitations. This has major practical significance, as experience shows that changes in mean annual precipitation correlate poorly with the temporal and spatial scale of shallow landslides, which are generally triggered by short to medium-duration intense rainstorms. Moreover, the hydrologic response in a slope depends

Table 1 Climate impact on temperature, precipitation, flooding and sea level in Norway by 2100 under IPCC Scenario RCP8.5 (Hanssen-Bauer et al. 2017)

Hydro-meteo-rological factor	Anticipated change by year 2001
Temperature	Increase by about 4.5° (3.3 to 6.4 °C)
Annual precipitation	Increase by about 18% (7 to 23%): highest in spring and lowest in summer
Rainfall	More intense and more frequent rainfall; summer draught in some regions (especially southeastern Norway)
Flooding	Increased and more frequent floods induced by rainfall; decreased and less frequent floods caused by snowmelt
Mean sea level	Increase by 15–55 cm, depending on location along the Norwegian coast

strongly on the nature and morphology of the outcrop, which may consist of steep variously fractured rock or of coarse grained deposit, of gently sloping fine grained soils or other intermediate materials. Similar rainfall events may therefore lead to different consequences.

Based on such considerations, in the past years, some researchers have focused on the potential effects of the modification of the main components of the hydrological balance in different lithological contexts. As an example, Naden and Watts (2001) used observed weather forcing data, perturbed according to future climate scenarios, in order to analyze the potential variations of the average monthly soil moisture in five areas in the UK thus highlighting the potential response of different soils.

Following a similar approach, Rianna et al. (2016) examined the potential effects of expected climate changes on soil-atmosphere interaction in the climatic environment of a hilly area located some tens of kilometres northeast of Naples and on the potential consequences on soil moisture. The analysis focused on the 2071–2100 time interval and used two different gas emission scenarios, i.e. the RCP8.5 "pessimistic scenario" characterised by an increase, in 2100, of the total radiative forcing of 8.5 W/m^2 with respect to the pre-industrial phase, and a RCP4.5 "stabilization scenario", characterised by an increase of 4.5 W/m^2. The main atmospheric processes on a global scale were simulated by the GCM CMCC_CM model (Scoccimarro et al. 2011), which works on an average horizontal resolution of 80 km. The output was then downscaled at a horizontal resolution of about 8 km through the statistical RCM model COSMO_CLM (Rockel et al. 2008). The results were then integrated into the hydrological HELP model (Schroeder et al. 1994) in order to assess the effects of rainfall and evapo-transpiration on the water content profile in different soil deposits. The analysis comprised a simulation of 1D vertical water infiltration into a virtual 50 m deep horizontal layer with a bare or vegetated ground surface and a lowermost free draining boundary. Five different soils, from coarse sand to clay, were considered, each characterized by typical values of initial porosity, water retention curve and permeability function.

The analysis showed that in the selected area the precipitations should increase slightly in winter (+7% for RCP8.5) and decrease in the warm season. The cumulative precipitation decrease should be remarkable in summer (20% and 65% for RCP4.5 and RCP8.5 respectively). Moreover, due to general warming, the potential evapotranspiration would grow with a rate strongly related to the selected concentration scenario.

Figure 1 compares the degree of saturation for the two different IPCC scenarios in the period 2071–2100 with the values during the period 1981–2010, called the "control period. In particular, the figure shows the estimated daily average and peak values in the uppermost 4 m of vegetated soil surfaces. To assess the influence of the lower boundary condition, a further configuration was considered consisting of a 4 m coarse sand column (denoted "coarse sand (2)", Fig. 1) overlying a deposit of infinite thickness with hydraulic conductivity three orders of magnitude lower.

Based on the results of the analyses, a remarkable reduction in the moisture content can be expected to occur for less pervious soils. In contrast, in coarse sand the estimated growth of cumulated precipitations during the wet season (from November to March), only partly balanced by higher evaporation, should lead to average values comparable or slightly higher than those estimated for the control (current) period. The effects of the modified precipitation regime are even more evident if considering the peak values. In fact, in the wet season higher moisture contents than the present ones might be reached in fine to coarse sands, especially for the case "coarse sand (2)".

Even though the analysis was carried out for simple situations and neglecting the role of morphology and of other influent factors, such as the initial state of stress, in slope stability problems, these data confirm that the effects of climate change on rainfall-induced landslide hazard may well differ from site to site, and strongly depend on geology and geomorphological context. As an example, referring again to the Mediterranean basin and reminding that in such an area climate change could lead to a decrease in the accumulated yearly precipitations and to more frequent and more intense single extreme events, landslide hazard could

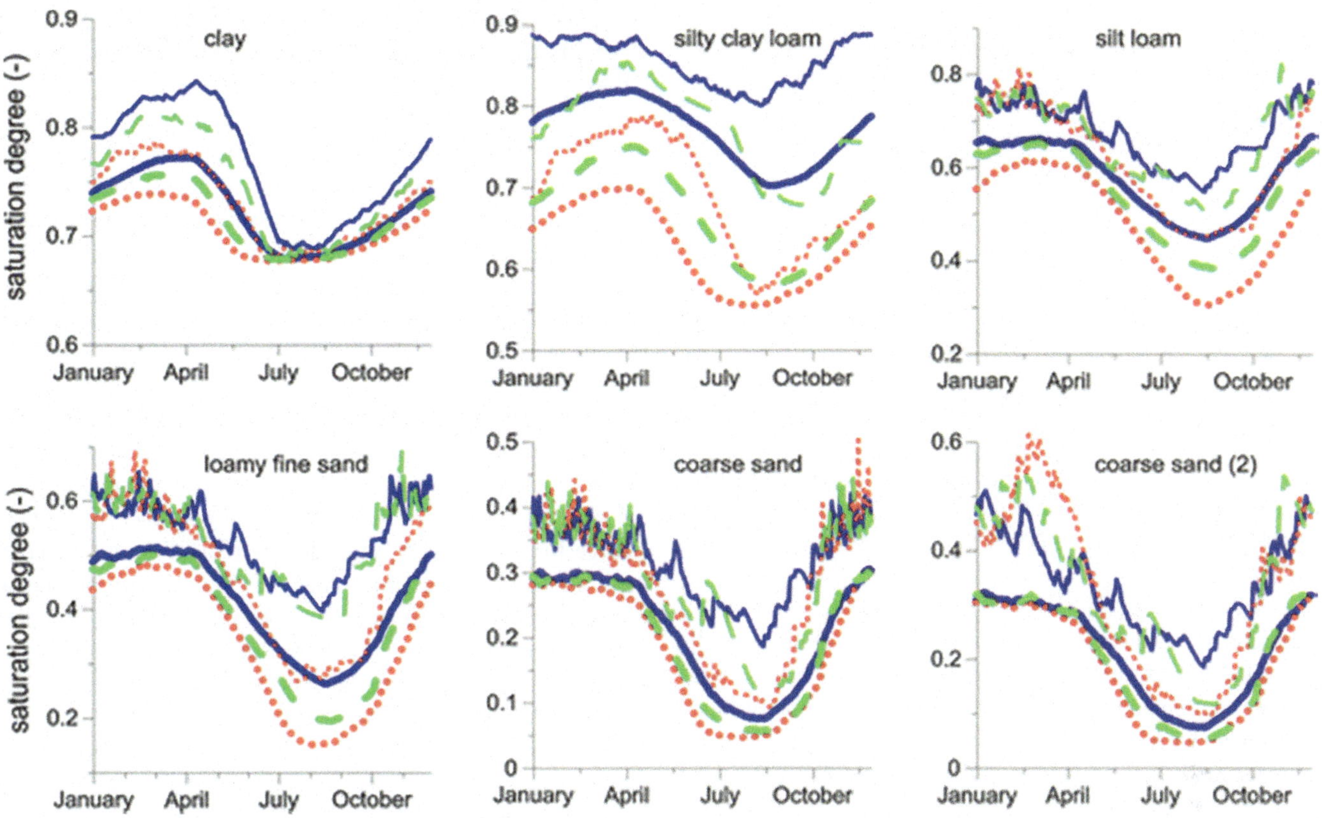

Fig. 1 Effects of climate change on the degree of saturation of a 4 m deep virtual soil column (Rianna et al. 2016). Legend: blue, control period 1981–2010; green, 2071–2100 (RCP4.5); red, 2071–2100 (RCP8.5). For each color, thicker lines are for average saturation degree, thinner lines for peak daily saturation degree

significantly increase in zones occupied by shallow deposits of quite high permeability (Damiano and Mercogliano 2013) and decrease in areas covered by soil of lower permeability, such as fine-grained soils.

Based again on projections of future precipitations for the same vast area considered above (see Rianna et al. 2016), in which the Cervinara debris avalanche in 1999 killed five people (Fiorillo et al. 2001), Rianna et al. (2017) calculated the cases, in the period 2071–2100, where the combined effect of the accumulated precipitations in the last 29 days (P29d) and the rainfall on the 30th day (P1d) would lead to slope failure. The analysis used the topography, hydraulic and mechanical soil properties and pore pressure regime data available for the Cervinara slope (Picarelli et al. 2007; Comegna et al. 2016).

The results of the analysis are reported in Fig. 2. The blue symbols represent the results of slope stability analyses based on the recorded precipitations in the time interval 1981–2010, which lead to slope failure. These calculations can reproduce the actual event (December, 1999, encircled blue dot), which was characterized by a very high accumulated daily precipitation P1d (the highest one according to official records covering the last century (Comegna et al. 2016). The other blue symbols indicate further potential slope failures. In this regard, it is worth to mention that, just on the same days, as many landslides were triggered in the same area, not on the same slope but in similar geomorphological contexts. These events were characterized by higher antecedent precipitations (P29d > 380 mm) and by lower triggering one-day rainfall than in the case of December 1999. Adopting as input the simulated weather forcing in the control period (1981–2010 again, red symbols), a couple of potential triggering rainfall combinations were found to lead the slope at hand to failure. Finally, a significant result of the projections to the 2071–2100 time interval is the dramatic increase of critical events (9 versus 4), which are characterized by either very high antecedent precipitations or very high triggering one-day rainfall, in line with the events recorded during the control period.

As shown in Fig. 1, in fine-grained (less permeable) soils outcropping in the same vast area considered above, the projected combination of decrease in accumulated precipitations and increase of single extreme events could produce an opposite effect.

Comegna et al. (2013) examined the potential effects of this scenario in a zone characterized by the presence of

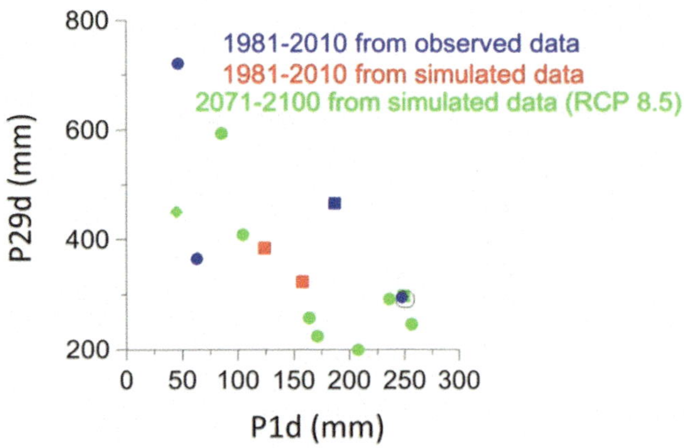

Fig. 2 Precipitations potentially triggering a landslide in the Cervinara area (Rianna et al. 2017). Legend: blue, critical recorded precipitations in the control period; red, critical precipitations simulated during the same period; green, critical precipitations simulated in the time window 2071–2010

numerous slow-moving active earthflows. Exploiting available data on historic precipitations, soil properties, pore pressure regime and long-term displacements of one of those earthflows (Di Maio et al. 2013), they calculated the accumulated yearly precipitations and the potential future pore pressure fluctuations in the same earthflow, assuming the present annual rainfall distribution; then they estimated the landslide velocity based on a simple relationship between displacement rate and mobilized shear stress proposed by Vulliet (1986) for viscous fine-grained soils.

Figure 3 reports the results of the analysis. In this case, due to the expected decrease in cumulative precipitations and increase in temperature, the future average groundwater level should display a small progressive reduction in the range of 6–8 mm per decade depending on the assumed annual precipitation distribution scenario (Fig. 3a). Combining pore pressures and landslide velocity through the empirical relationship discussed above, the future displacement rate should consequently decrease. The decrease should range between about 1.5 mm/decade and 3 mm/decade, even though phases of moderate acceleration during winter and spring will take place every year. However, in spite of such changes in velocity, the cumulative displacement will follow a substantial linear trend. The total displacement over about fifty years should attain a value of between 77 and 86 cm.

Such results demonstrate that the expected climatic changes in the investigated area should not play a significant role for fine-grained sloping soils due to the moderate decrease in the annual precipitation and the limited effect of temperature increase on evaporation and groundwater level. In any case, the global impact of climate change should be favourable, at least for the climatic and geomorphological context considered herein.

A State-of-the-Art on Slope Safety Preparedness for the Impact of Climate Change: The 2015 International Forum

The 2015 JTC1 International Forum was aimed at taking stock of the approaches in different countries to assess the impact of climate change on landslide hazard and establishing adequate measures for risk management, mitigation strategies and risk adaptation (Ho et al. 2017a). Key contributions to these topics have also been provided by recent European collaborative research projects, such as SafeLand and CHANGES.

The Forum demonstrated that several among the 21 contributing countries, especially those that have been recently struck by extreme events (Italy, Taiwan, Hong Kong, Brazil), are well aware of the need and urgency in face of climate change for coordinated landslide risk reduction and management strategy. In particular, many of them have undertaken downscaling studies to assess the potential impact of climate change. These studies predict increasing landslide hazards and associated risk. As shown, the expected effects, however, vary from one region to the other, depending on the nature of the potential triggers (deglaciation, floods, rainfall, permafrost thawing) and on lithological and geomorphological contexts. Cold countries (e.g. Canada, Norway, Russia) are well aware that the global warming will result in increased landslide hazards and risk to the population. Moreover, the countries with widely varying climates (e.g. Canada, China, U.S.) present a variety of vulnerabilities. In these countries, the impact of climate change may be reflected by an increased number and magnitude of landslides in some parts, whilst other parts might be little affected.

However, the countries are at different stages of preparedness for climate change impact. Hong Kong has come furthest in terms of holistic and systematic assessment of impact of extreme weather events on slope safety. All countries need to develop an improved understanding of the interrelationship among climate change effects, hydro-mechanical properties, failure modes, and debris mobility. In fact, while some of them have already developed for some time a systematic organization for the assessment of the impacts of extreme weather events and are further developing procedures to enhance emergency preparedness (e.g. Hong Kong), other places need to take a first step for setting up adequate infrastructure to address the challenges posed by geo-hydrological hazards and landslide risk.

However, all surveyed countries acknowledge the difficulty in preparing realistic scenarios for improved prediction and improved preparedness because of the considerable uncertainties involved. As a matter of fact, the development of early warning systems as a mitigation measure, is considered progressive and effective for rik mitigation,

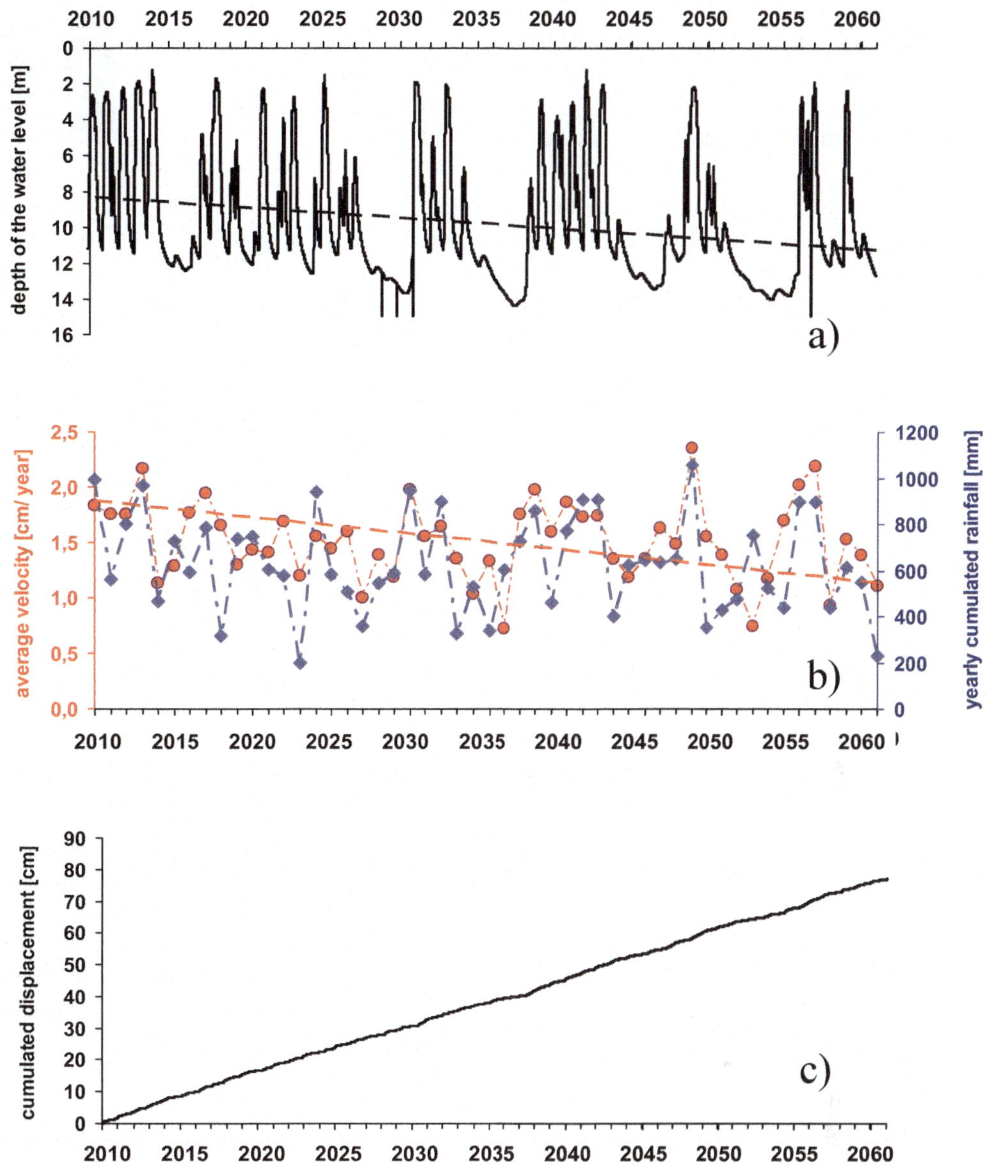

Fig. 3 Effects of the climate scenario in the time interval 2010–2060 on the hydrological response of a slow earthflow, assuming the annual rainfall distribution as in 2004 (Comegna et al. 2013): **a** piezometric level at a depth of 15 m; **b** annual cumulative rainfall and displacement rate; **c** cumulative displacement

especially by high population density regions or cities with frequent exposure to landslide hazards.

Certainly, international cooperation for increased competence and enhanced preparedness would help and this should be the direction to take in the future globalised world.

Needs for Improved Climate Change Related Landslide Risk Management

The four issues of most concern to the parties involved in the Forum with regard to climate change related landslide risk and emergency management were:

- Widespread landslides, including possibly massive movements, with increased propensity, scale and mobility of the moving masses;
- Large number of landslides affecting wide populations, and breakdown of infrastructure;
- Lack of understanding of extreme landslide scenarios, partly due to the scarcity of earlier occurrences.
- Generally inadequate preparedness for the emergency management of extreme landslide scenarios, which aggravates the consequence.

The identified needs for climate change related landslide risk management include:

- Quantifying the uncertainties, which can be fraught by differences in practice because of lack in knowledge about the actual uncertainties, including 'unknown unknowns';
- Insights from scenario-based assessment and stress testing;
- Institutional issues and addressing the policy gap;
- Application of improved technology and methods;
- National and international cooperation;
- Coordination with flood risk management experts;
- Enhanced public communication and community resilience;
- Formulating strategies for a way forward;
- How to account for multi-hazards or cascading hazards?

Figure 4 illustrates a sequence of events with cascading hazards that, ideally, need to be included in a scenario model and need to be mitigated against. Intense rainfall can trigger a complex chain of mass movement processes similar to that in the figure, with strong interactions and interdependencies of mass movement processes (landslide, rock fall, debris flow, landslide dam, flood and aggraded riverbed). The natural geo-phenomena form the outer periphery of the chain. The arrows illustrate the processes and transition that lead to increased hazards. Under heavy rainfall, the colluvium materials on steep hill slopes can reactivate, move downwards and become channel sediment deposits. The materials in the channels can gradually run out as channelized debris flows under the same or subsequent rainstorms. The materials from a hillslope slide, a rock fall or a debris flow can block a river and form a landslide dam. Flooding due to overtopping occurs if the "natural" dam breaches or fails due to piping. The debris transported by the flood elevates the riverbed. A flood and erosion at the toe of a hillslope can also prompt additional slope failures as a result of soil saturation or erosion by the flood. The evolution will not stop before the entire system reaches a new balance.

How can one develop a system to keep account of cascading events and the cascading and interrelationship among the hazards and the risks posed by each of the events? Stress testing presents an innovative approach for dealing with such complex situations with low probability of occurrence and high consequences. It involves testing a system beyond normal operational capacity, often to a breaking point. A stress testing framework includes:

(1) identifying future critical rainstorm scenarios under the changing climate; (2) evaluating the response of the slope safety system to the critical rainstorm scenarios; (3) assessing the risks posed by the multi-hazard processes; (4) evaluating the bottlenecks of the slope safety system; and (5) proposing strategies for improving system performance.

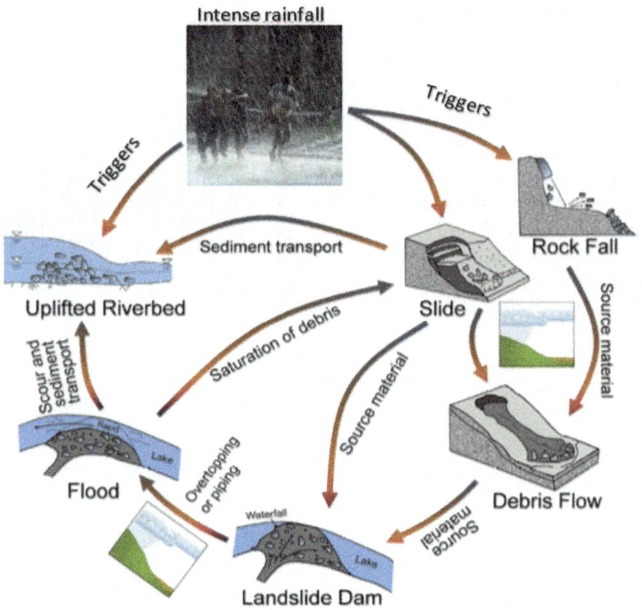

Fig. 4 Example of cascading hazards due to intense rainfall (after Zhang et al. 2017)

In the aftermath of the Tōhoku earthquake and Fukushima Dai-ichi accident, stress tests were imposed on all nuclear power plants in Europe by the West European Nuclear Regulation Association.

Stress testing of the slope safety system was carried out by the Geotechnical Engineering Office of the Hong Kong SAR Government following a record-breaking rainstorm in 2008, which triggered an unprecedented number of debris flows and hillslope failures on natural hillsides (Ho et al. 2017b).

Recommendations for Improved Climate Change Related Landslide Risk Management

Responsible landslide risk management refers to the coordinated activities to assess, direct and control the risks posed to society by flows and landslides. Risk management integrates the recognition of risk and the development of appropriate strategies for the treatment (reduction) of the risk. Risk management needs to be a systematic application of management policies, procedures and practices, including context definition, identification, analysis, evaluation, communication, consultation, implementation and monitoring of mitigation measures.

In view of climate change impacts and extreme weather events, a "modern" landslide risk reduction strategy should encompass the following key aspects of investigation, adaptation, response, resilience and recovery (Ho et al. 2017b):

- Avoidance, with land-use planning, warning or alert systems, and public education;
- Prevention, e.g. enforcing investigation, design, construction, supervision and maintenance;
- Mitigation, with implementation of engineering measures to reduce the impact of landslides;
- Preparedness, human resource management, emergency preparedness and response systems and community training;
- Response, including humanitarian relief, emergency solutions, settlement of evacuees etc.;
- Recovery to bring the affected area back to normal and do mitigation works.

When working with climate adaptation, one needs to work in a multi-disciplinary network. Technical solutions alone are not enough. The best approach for climate change adaptation is to learn about it in a network. More cooperation among municipalities and sectors is the most significant development need for the implementation of climate adaptation and resilience solutions. The goal should be to become a climate-resistant and future-proof region, well-prepared and adapted for the climate crisis. Learning about climate adaptation is also about changing attitude, ambition and values.

A novel approach to landslide risk mitigation is the use of nature-based solutions (NBSs). NBS is an umbrella term for 'ecosystem services', 'green–blue infrastructure' and 'natural capital'. The premise for NBSs is that nature itself is a source of ideas and solutions for mitigating risk from climate-driven natural hazards. Nature's designs are elegant, effective and frugal and adapted to their environment. Water is usually the main culprit in triggering landslides and debris flows. Planting suitable tree species and adapting vegetation to control surface erosion and strengthen the surficial soil layers exposed to excessive amounts of runoff water during extreme precipitation events is one of the few established NBSs for landslide risk mitigation. Several large ongoing research projects in EU's H2020 framework programme, e.g. PHUSICOS (www.phusicos.eu), focus on developing innovative NBSs for landslide control, including comprehensive frameworks and site trials for monitoring and verifying the performance of NBSs, i.e. a direct application of Peck's observational method (Peck 1969). The monitoring would enable the identification of the intensity of extreme events that the NBSs could withstand.

A cultural shift is needed for improved landslide risk management. The profession needs to move from consideration of hazard to consequence, from hazard evaluation to preparedness and risk reduction, from science-driven to multi-disciplinary solutions, from working for communities to working with communities, and to risk-informed decision making.

Conclusions

The paper describes the potential impact of climate change on landslide hazard, landslide susceptibility and landslide risk. Examples of climate impact on landslide hazards and the state of preparedness for climate change-related landslide risk around the world are also summarized. Finally, the paper proposes steps for enhanced landslide risk management.

Climate is defined in terms of variations in atmospheric conditions (temperature, precipitation, wind velocity and direction, cloudiness, atmospheric pressure, air humidity and chemistry of the air and contained dust). Landslide and rock slide hazards may increase due to changes in atmospheric conditions, for example with increased intensity and frequency of rainfall.

In a context of (1) more extreme weather and more intense rainfall, (2) human population increasing at an exponential rate, and thereby increased exposure and vulnerability, (3) human population moving towards coastal areas (www.Livesciences.com) and (4) increase in news reporting on natural disasters, especially since the invention of the Internet, the risk associated with climate-driven natural hazards and the public perception of and aversion to these risks will increase.

The effects of climate change will take different forms depending on local climatic zone and lithological and geomorphological context. The landslide hazard will surely increase in cold areas due to permafrost degradation and ice and snow melting, and this is, by the way, already evident. In more temperate zones, the landslide hazard will depend on the effects of climate change on the precipitation regime. Extreme weather events, which should become even more frequent than today, should generate floods, and consequent effects on the stability of slopes, but also rapid landslides as debris avalanches and debris flows. In contrast, in some areas, especially those covered by fine-grained soils, the landslide hazard could reduce.

However, the mentioned survey (Ho et al. 2017a) revealed that most of these expect increasing landslide hazards and increasing risk to population. As discussed above, the concerns varied greatly, depending on weather patterns, soil types, past landslide history and the nature of the triggers (earthquake, floods, rainfall, etc.). The countries/regions are, however, at different stages of preparedness for climate change impact and slope safety. Hong Kong has come furthest in terms of holistic and systematic assessment of impact of extreme weather events on slope safety through stress testing.

A cultural shift is needed for improved landslide risk management. The profession needs to move from hazard evaluation to preparedness and risk reduction, adoption of nature-based solutions in concert with the communities, and risk-informed decision making.

As a closing perspective, science and engineering helps to predict hazards. Knowing the hazards and the risk facilitates risk-informed decisions, and the consequences of landslide hazards can be reduced with innovative approach and technology. Human factors can turn unexpected events into catastrophes, and tomorrow's disasters are exponentially building on the disasters of today. Focus on disaster risk reduction is required because the social costs tend to persist over a person's lifetime, whereas material costs are a one-off. The time to innovate for enhanced landslide risk reduction is now.

Acknowledgments This paper is published with the permission of the Head of the Geotechnical Office and the Director of Civil Engineering and Development, Hong Kong SAR Government.

References

Borgatti L, Soldati M (2010) Landslides as a geomorphological proxy for climate change: a record from the Dolomites (northern Alps). Geomorphology 120:56–64

Chiarle M, Iannotti S, Mortara G, Deline P (2007) Recent debris flow occurrences associated with glaciers in the Alps. Glob Planet Change 56(1):123–136

Cloutier C, Locat J, Geertsema M, Jakob M, Schnorbus M (2017) Potential impacts of climate change on landslides occurrence in Canada. In Ho K, Lacasse S, Picarelli L (eds) Preparedness for climate change impact on slope safety. Taylor and Francis Group, London, pp 71–104. ISBN 978 1 138-03230-9. p 571

Coelho Netto AL, Sato AM, Avelar AS, Vianna LG, Araújo LS, Ferreira DL, Silva RP (2013) January 2011: the extreme landslide disaster in Brazil. Landslide science and practice. Springer, Berlin-Hedelberg, pp 377–384

Comegna L, Picarelli L, Bucchignani E, Mercogliano P (2013) Potential effects of incoming climate changes on the behaviour of slow active landslides in clay. Landslides 10:373–391

Comegna L, Damiano E, Olivares L, Picarelli L (2016) Considerations on the failure of the Cervinara slope. In: Aversa S, Cascini L, Picarelli L, Scavia C (eds) Proceedings of the 12th International Symposium on Landslides. Naples, 12–19 June, 2016, vol 2, pp 663–670. CRC Press, Taylor and Francis Group

Damiano E, Mercogliano P (2013) Potential effects of climate change on slope stability in unsaturated pyroclastic soils. Proceedings of the 2nd world landslide forum. Landslide science and practice: global environmental change, vol 4. Rome, 2011, pp 15–25

Di Maio C, Vassallo R, Vallario M (2013) Plastic and viscous shear displacements of a deep and very slow landslide in stiff clay formation. Eng Geol 162:53–66

Evans SG, Clague JJ (1994) Recent climatic change and catastrophic geomorphic processes in mountain environments. Geomorphology 10:107–128

Fiorillo F, Guadagno FM, Aquino S, De Blasio A (2001) The December 1999 cervinara landslides: further debris flows in the pyroclastic deposits of Campania (Southern Italy). Bull Eng Geol Env 60(3):171–184

Garcia-Delgado H, Machuca S, Medina E (2019) Dynamic and geomorphic characterizations of the Mocoa debris flow (March 31, 2017, Putamayo department, Southern Colombia). Landslides 16(3):597–609

Geertsema M, Clague JJ, Schwab JW, Evans SG (2006) An overview of recent large catastrophic landslides in northern British Columbia, Canada. Eng Geol 83:120–143

Haerbeli W, Wegmann M, Vonder Mühll D (1997) Slope stability problems related to glacier shrinkage and permafrost degradation in the Alps. Eclogae Geol Helv 90:407–414

Hanssen-Bauer I, Førland EJ, Haddeland I, Hisdal H, Lawrence D, Mayer S, Nesje A, Nilsen JEO, Sandven S, Sandø AB, Sorteberg A, Ådlandsvik B (2017) Climate in Norway 2100—a knowledge base for climate adaptation. NCCS report no. 1/2017. Norway

Ho KKS, Lacasse S, Picarelli L (eds) (2017a). Preparedness for climate change impact on slope safety. pp 1–42. Taylor and Francis Group, London. ISBN 978 1 138-03230-9. p 571

Ho KKS, Sun HW, Wong ACW, Yam CF (2017b) Enhancing slope safety preparedness for extreme rainfall and potential climate change impacts in Hong Kong. In: Ho KKS, Lacasse S, Picarelli L (eds) Preparedness for climate change impact on slope safety. pp 105–150. Taylor and Francis Group, London. ISBN 978 1 138-03230-9. p 571

Ho KKS, Sun HW, Lam EJ, Lo FLC (2020) Engineering preparedness against impact of climate change on slope safety in Hong Kong. Submitted to 5th world landslide forum, Kyoto

Jaedicke C, Solheim A, Blikra LH, Stalsberg K, Sorteberg A, Aaheim A, Kronholm K, Vikhamar-Schuler D, Isaksen K, Sletten K, Kristensen K (2008) Spatial and temporal variations of norwegian geohazards in a changing climate, the geoextreme project. Nat Hazards Earth Sys Sci 8(4):893–904

Kenney TC, Lau KC (1984) Temporal changes of groundwater pressure in a natural slope of non fissured clay. Can Geotech J 21 (1):138–146

Kokelj SV, Tunnicliff D, Lacelle D, Lanz TC, Fraser RF (2015) Retrogressive thaw slumps: from slope process to the landscape sensitivity of northwestern Canada. Proceedings of the 68th Canadian geotechnical conference, GeoQuébec 2015, Québec City

L'Heureux JS, Høydal O A, Paniagua Lopez AP, Lacasse S (2018) Impact of climate change and human activity on quick clay landslide occurrence in Norway. Proceedings of the 2nd JTC1 workshop triggering and propagation of rapid flow-like landslides, Hong Kong, www.hkges.org/JTC1_2nd/paper.html

Lin M-L, Lin S-C, Lin Y C (2017) Review of landslide occurrence and climate change in Taiwan. In: Ho KKS, Lacasse S, Picarelli L (eds) Preparedness for climate change impact on slope safety. pp 409–436. Taylor and Francis Group, London. ISBN 978 1 138-03230-9. p 571

Mu C, Shang J, Zhang T, Fan C, Wang S, Peng X, Zhong W, Zhang F, Mu M, Jia L (2020) Acceleration of thaw slump during 1997–2017 in the Qilian Mountains of the northern Qinghai-Thibetan plateau. Landslides, accepted for publication

Naden PS, Watts CD (2001) Estimating climate-induced change in soil moisture at the landscape scale: an application to five areas of ecological interest in the UK. Clim Change 49(4):411–440

Peck RB (1969) Advantages and limitations of the observational method in applied soil mechanics. Géotechnique 19(2):171–187

Picarelli L, Evangelista A, Rolandi G, Paone A, Nicotera M V, Olivares L, Scotto di Santolo A, Lampitiello S, Rolandi M (2007) Mechanical properties of pyroclastic soils in Campania region. Proceedings of the 2nd International workshop on characterisation and engineering properties of natural soils. Singapore, vol 4, pp 2331–2384. ISBN 978 0 415 40267 5

Picarelli L, Comegna L, Gariano SL, Guzzetti F, Mercogliano P, Rianna G, Santini M, Tommasi P (2017) Potential climate changes in Italy and consequences for land stability. In: Ho KKS, Lacasse S, Picarelli L (eds) Preparedness for climate change impact on slope safety. pp 151–197. Taylor and Francis Group, London. ISBN 978 1 138-03230-9. p 571

Ravanel L, Deline P (2011) Climate influence on rockfalls in high-alpine steep rockwalls: the north side of the Aiguilles de Chamonix (Mont Blanc massif) since the end of the 'Little Ice Age.' Holocene 21:357–365

Rianna G, Comegna L, Mercogliano P, Picarelli L (2016) Potential effects of climate changes on soil-atmosphere interaction and landslide hazard. Nat Hazards 84:1487–1499

Rianna G, Comegna L, Gariano S L, Guzzetti F, Mercogliano P, Picarelli L, Tommasi P (2017) Potential effects of climate changes on landslide activity in different geomorphological contexts. Proceedings of the 4th World Landslide Forum, Ljubljana, vol 4, pp 243–250. Springer, Berlin-Hedelberg

Rockel B, Will A, Hense A (2008) The regional climate model COSMO–CLM (CCLM). Meteorol Z 17(4):347–348

Schroeder PR, Dozier TS, Zappi PA, McEnroe BM, Sjostrom JW, Peyton RL (1994) The hydrologic evaluation of landfill performance (HELP) model: engineering documentation for version 3. EPA/600/R-94/168b. U.S environmental protection agency office of research and development, Washington, DC

Scoccimarro E, Gualdi S, Bellucci A, Sanna A, Fogli P, Manzini E, Vichi M, Oddo P, Navarra A (2011) Effects of tropical cyclones on ocean heat transport in a high resolution coupled general circulation model. J Clim 24:4368–4384

Sewell RJ, Barrows TT, Campbell SDG, Fifield FK (2006) Exposure dating (^{10}Be, ^{26}Al) of natural terrain landslides in Hong Kong, China. Geol Soc Am Spec Pap 2006(415):131–146

Tsou CY, Feng ZY, Chigira M (2011) Catastrophic landslides induced by Typhoon Morakot, Shiaolin Taiwan. Geomorphology 127:166–178

Vulliet L (1986) Modélisation des pentes naturelles en movement. PhD Thesis, École Polytechnique Fédérale de Lausanne

Walsh J, Wuebbles D, Hayhoe K, Kossin K, Kunkel K, Stephens G, Thorne P, Vose R, Wehner M, Willis J, Anderson D, Doney S, Feely R, Hennon P, Kharin V, Knutson T, Landerer F, Lenton T, Kennedy J, Somerville R (2014) Our changing climate. Climate change impacts in the United States. In: Melillo JM, Richmond TC, Yohe GW (eds) The third national climate assessment. U.S. global change research program, pp 19–67

Wang B, Paudel B, Hao QL (2016) Behaviour of retrogressive thaw slumps in northern Canada—three-year monitoring results from 18 sites. Landslides 13(1):1–8

Wang G, Towhata I (2017). Global warming and landslide disaster: evidence from Japan. In: Ho KKS, Lacasse S, Picarelli L (eds) Preparedness for climate change impact on slope safety. pp 199–213. Taylor and Francis Group, London. ISBN 978 1 138-03230-9. p 571

Wong MC, Mok KY (2009) Trends in Hong Kong climate parameters relevant to engineering design. Proceedings of HKIE civil division conference 2009. January 2009, HKO reprint no 832

Yang H, Wang F, Miajima M (2015) Investigation of shallow landslides triggered by heavy rainfall during typhoon Wipha (2013), Izu Oshima Island, Japan. Geoenviron Disasters, 2: 15

Zhang L, Gao L, Zhou S, Cheung RWM, Lacasse S (2017) Stress testing framework for managing landslide risks under extreme storms. Proceedings of the 4th world landslide forum. Ljubljana

Part II

Sendai Landslide Partnerships, Kyoto Landslide Commitment, and International Programme on Landslides

Kyoto 2020 Commitment for Global Promotion of Understanding and Reducing Landslide Disaster Risk

WLF 2020 Kyoto Japan

Kyoji Sassa, Peter T. Bobrowsky, Kaoru Takara, and Badaoui Rouhban

Abstract

The Sendai Landslide Partnerships 2015–2025 was proposed and adopted as a voluntary commitment to the Sendai Framework for Disaster Risk Reduction. The Partnerships was signed by 22 global stakeholders committing to contribute to landslide disaster risk reduction. Under the Partnerships, ICL published a full color open access book "ISDR-ICL Sendai Partnerships 2015–2025" as Vol. 1 of Advancing Culture of Living with Landslides and two volumes of ISDR-ICL Interactive Landslide Teaching Tools. The Partnerships is effective for the promotion of landslide disaster risk reduction. However, it will be phased out by 2025. Landslide disaster risk reduction will remain a key necessity and become even more important after 2025 due to climate change and urban development. ICL and the partners of the Sendai Landslide Partnerships wish to develop this initiative further to 2025, 2030 and even beyond. Therefore, the Kyoto 2020 Commitment for Global Promotion of Understanding and Reducing Landslide Disaster Risk (KLC2020) was developed. Partners of KLC2020 are called upon to attend the meeting of the joint signatories and the declaration of the launching of KLC 2020 in November 2020.

Keywords

Sendai landslide partnerships • Sendai framework for disaster risk reduction • Sustainable development goals • Paris climate agreement • Interactive teaching tools

Preparation Process of the KLC2020

The concept of a Kyoto Landslide Commitment was initially proposed on 30 May 2017 at the high-level panel discussion of the Fourth World Landslide Forum held in Ljubljana, Slovenia. Based on that concept, a zero draft of the Kyoto 2020 Commitment for Global Promotion of Understanding and Reducing Landslide Disaster Risk (KLC 2020) was elaborated and examined in the 2017 ICL-IPL UNESCO Conference held in UNESCO Headquarters, Paris, on 29 November to 1 December 2017. Subsequently, a first draft of the Commitment was developed and considered during the 2018 ICL-IPL Conference—Planning of the Fifth World Landslide Forum (WLF5) and the Kyoto 2020 Commitment held at the National Kyoto International Conference Center (KICC) and the Disaster Prevention Research Institute, Kyoto University on 1–4 December 2018. Based on this draft, the final draft was prepared and submitted to the 2019 ICL-IPL UNESCO Conference held at UNESCO Headquarters on 16–19 September 2019. It was approved by the participants, and the representatives of 57 organizations signed it (some did so before the meeting and delivered their document to the secretariat). Further partners were since called upon, and were expected, to sign the final draft at a meeting initially due to take place at the Disaster Prevention Research Institute, Kyoto University, Kyoto, Japan on 19 April 2020. However, due to the Covid-19 pandemic, this meeting had to be cancelled, while the situation resulting from the pandemic got further worse in April and May 2020. Accordingly we decided to postpone the planned WLF5 by one year, that is to 2–6 November 2021, but at the same

K. Sassa (✉) · P. T. Bobrowsky · K. Takara · B. Rouhban
International Consortium on Landslides (ICL), Kyoto, Japan
e-mail: secretariat@iclhq.org

P. T. Bobrowsky
e-mail: peter.bobrowsky@canada.ca

K. Takara
e-mail: takara.kaoru.7v@kyoto-u.ac.jp

B. Rouhban
e-mail: b.rouhban@gmail.com

P. T. Bobrowsky
Geological Survey of Canada, Sidney, Canada

K. Takara
Kyoto University, Kyoto, Japan

K. Sassa et al. (eds.), *Understanding and Reducing Landslide Disaster Risk*,
ICL Contribution to Landslide Disaster Risk Reduction,
https://doi.org/10.1007/978-3-030-60196-6_7

venue (Kyoto International Conference Center) in Kyoto, Japan.

It has nevertheless been deemed inappropriate and detrimental to postpone both the publication of the full color books "Understanding and Reducing Landslide Disaster Risk" collecting the advances of landslide science and technology in the past three years as well as the launching of the Kyoto Landslide Commitment. In short, we cannot postpone the launching of the KLC2020.

Launching of the KLC2020

ICL decided to organize the 2020 ICL-IPL Conference on 2–6 November 2020 in a virtual mode. Because time difference widely varies from the west end of American continent to the east end of Asian continent, the meeting will daily be a 4-h one in different time for each continent. During this virtual Conference, a dedicated session for KLC 2020 will be scheduled on 5 November to host the KLC 2020 Joint signatory partners as well as the Declaration of the launching of the KLC 2020, This will coincide with the World Tsunami Awareness Day (Landslide induced tsunamis is a frontier of landslide disaster risk reduction). At this Session, Leaders of ICL supporting organizations and KLC 2020 signatories (past and as at that day) will deliver greeting messages to the event, recognize each other and discuss on the further action leading to the postponed WLF 5 now scheduled on 2–6 November 2021 in Kyoto, Japan.

Potential partners of the KLC2020 are the following:

1. ICL member organizations (full members, associate members and supporters)
2. ICL supporting organizations from the UN system, international or national organizations and programmes
3. Government ministries and offices in countries having more than 2 ICL on-going members
4. International associations/societies which contribute to the organization of WLF5 in 2021 and WLF6 in 2023
5. Other organizations having some aspects of activities related to understanding and reducing landslide disaster risk as their intrinsic missions.

Note:
KLC 2020 is a commitment to the Sendai Landslide Partnerships 2015–2025, the Sendai Framework for Disaster Risk Reduction 2015–2030, the 2030 Agenda Sustainable Development Goals, the New Urban Agenda and the Paris Climate Agreement. Its follow-up does not require any additional duties or responsibilities from the partners other than their own intrinsic works/mandates. All KLC2020 partners are those organizations who already have division(s)/entity(ies) whose intrinsic missions are related to understand and reduce landslide disaster risk. Accordingly, their success in carrying out their missions is in itself the successful commitment to KLC2020.

KLC 2020 is planned to continue further to 2025, 2030 and beyond. KLC 2020 will be examined at each World Landslide Forum every three years. Priority actions will be updated, so will Partners of KLC 2020, some will be added and some may leave due to change of society, landslide disaster, or partners themselves. As such, KLC2020 is a living and growing partnership.

ICL call upon potential partner organizations to consider the following full text of the Kyoto Landslide Commitment 2020, and indicate to ICL secretariat their willingness to participate in it inasmuch as such a participation will benefit their mission and work.

Full text of Kyoto Landslide Commitment 2020

Kyoto 2020 Commitment for Global Promotion of Understanding and Reducing Landslide Disaster Risk

A Commitment to the Sendai Landslide Partnerships 2015–2025, the Sendai Framework for Disaster Risk Reduction 2015–2030, the 2030 Agenda Sustainable Development Goals, the New Urban Agenda and the Paris Climate Agreement.

Preamble
Landslide disasters are the result of impacts of hazardous movement of soil and rocks that threaten vulnerable human settlements and infrastructure in mountains, cities, on coasts, and islands. An increase in the frequency and/or magnitude of heavy rainfall and shifts in the location, timing and periodicity of rainfall and permafrost/glacier degradation due to changing climate and global warming may significantly intensify the risk of landslides in many landslide prone areas.

Developments in mountains and coastal areas, including infrastructure construction such as roads, railways, energy and communication corridors, expansion of urban areas, including deforestation due to population growth and movement increase exposure to the hazards of landslides. Landslide disaster risk reduction is a globally important objective in all countries/regions where people living near mountains and on slopes are exposed to landslides.

The International Consortium on Landslides (ICL) proposed the 'ISDR-ICL Sendai Partnerships 2015–2025 for Global Promotion of Understanding and Reducing Landslide Disaster Risk' in the Working Session "Underlying Risk Factors" during the 3rd World Conference on Disaster Risk Reduction (WCDRR) in Sendai, Japan, 2015. The Sendai Partnerships was adopted and signed by 17 United Nations, international and national stakeholders. Joint efforts thereafter have been made and resulted in significant outcomes and materials including the edition and publication of the open access full color book "ISDR-ICL Sendai Partnerships 2015–2025", Vol. 1 of the Fourth World Landslide Forum (Ljubljana 2017), the edition of "Landslide Dynamics: ISDR-ICL Landslide Interactive Teaching Tools", as well as the enhanced publication of the monthly full-color journal Landslides: Journal of the International Consortium on Landslides.

The landslide risk to human settlements in mountainous and coastal areas in many countries will likely continue to rise including after the latter-half period of the Sendai Landslide Partnerships 2015–2025. In September 2015, the United Nations General Assembly adopted the 2030 Agenda for Sustainable Development and its 17 Sustainable Development Goals (SDG) including SDG 11 'Make cities and human settlements inclusive, safe, resilient and sustainable' and SDG 13 'Take urgent action to combat climate change and its impacts'. As a voluntary commitment to the 2030 Agenda, to the Sendai Framework for Disaster Risk Reduction 2015–2030 and to the Paris Agreement on Climate Change as well as to the Sendai Landslide Partnerships itself, participants in the Fourth World Landslide Forum considered and further endorsed the first outline of a commitment, the Kyoto Landslide Commitment 2020, as a stable framework to mobilize in the medium and long term a global alliance which will accelerate and incentivize action for landslide disaster risk reduction.

The High-Level panel discussion on "Strengthening Intergovernmental Network and the International Programme on Landslides (IPL) for ISDR-ICL Sendai Partnerships 2015–2025 for global promotion of understanding and reducing landslide disaster risk" was organized during the Fourth World Landslide Forum. The panelists were from the signatory organizations of Sendai Partnerships (ICL, UNESCO, WMO, FAO, UNU, ICSU, WFEO, IUGS, IUGG, Cabinet Office of Japan, Italian Civil Protection, Global Risk Forum, Davos) and new signatory organizations (the Indonesian National Agency for Disaster Management, the Administration of the Republic of Slovenia for Civil Protection and Disaster Relief, Ministry of Natural Resources and Environment, Vietnam, IRDR Science Committee, EuroGeoSurveys) as well as experts in this field.

The outcome of this High-Level panel discussion was reviewed by the Round Table Discussion to promote the Sendai Partnerships and the participants approved the 2017 Ljubljana Declaration on Landslide Risk Reduction. The Declaration endorsed the plan for the organization of the Fifth World Landslide Forum in Kyoto, Japan in November 2020 and the preparation of the Kyoto Landslide Commitment 2020 of a global alliance which aims, in the medium and long term, to accelerate and incentivize action for landslide disaster risk reduction to 2025, 2030 and beyond.

The Kyoto Landslide Commitment 2020 (KLC2020) is a framework aimed at providing key actors and stakeholders concerned with landslide risk at all levels and sectors with the tools, information, platforms, technical expertise and incentives to promote landslide risk reduction on a global scale. It supports the implementation, follow-up and review of the Sendai Framework, the 2030 Agenda for Sustainable Development, the New Urban Agenda and the Paris Climate Agreement as it addresses the adverse effects of climate change.

KLC2020 reaffirms the following resolution of the Sendai Landslide Partnerships, acknowledging that:

- Landslide disasters are caused by exposure to hazardous motions of soil and rock that threaten vulnerable human settlements in mountains, cities, on coasts, and islands.
- Climate change will intensify the risk of landslides in some landslide prone areas through an increase in the frequency and/or magnitude of heavy rainfall, and shifts in the location and periodicity of heavy rainfall.
- Global warming will intensify the risk of landslides in permafrost area and glacial lake outburst floods through snow and ice melting.
- Developments in mountains and coastal areas, including construction of roads and railways and expansion of urban areas due to population shifts, increase exposure to hazards of landslides.
- Although they are not frequent, strong earthquakes have potential to trigger rapid and long runout landslides and liquefaction. Earthquake-induced coastal or submarine large-scale landslides or megaslides (with depths on the order of hundreds of meters to one thousand meters) in the ocean floor can trigger large tsunami waves. These hazardous motions of soil and water impacting on exposed and vulnerable population can result into very damaging effects.
- The combined effects of triggering factors, including rainfall, earthquakes, and volcanic eruptions, can lead to greater impacts through disastrous landslides such as lahars, debris flows, rock falls, and megaslides.
- Understanding landslide disaster risk requires a multi-hazard approach and a focus on social and institutional vulnerability. The study of social and institutional as well as physical vulnerability is needed to assess the

extent and magnitude of landslide disasters and to guide formulation of effective policy responses.

- Human intervention can make a greater impact on exposure and vulnerability through, among other factors, land use and urban planning, building codes, risk assessments, early warning systems, legal and policy development, integrated research, insurance, and, above all, substantive educational and awareness-raising efforts by relevant stakeholders.
- The understanding of landslide disaster risk, including risk identification, vulnerability assessment, time prediction, and disaster assessment, using the most up-to-date and advanced knowledge, is a challenging task. The effectiveness of landslide disaster risk reduction measures depends on scientific and technological developments for understanding disaster risk (natural hazards or events and social vulnerability), political "buy-in", and on increased public awareness and education.
- At a higher level, social and financial investment is vital for understanding and reducing landslide disaster risk, in particular social and institutional vulnerability through coordination of policies, planning, research, capacity development, and the production of publications and tools that are accessible, available free of charge and are easy to use for everyone in both developing and developed countries.

We agree on the following priority actions of Kyoto 2020 Commitment for Global Promotion of Understanding and Reducing Landslide Disaster Risk in research and capacity building, coupled with social and financial investment:

Action 1 Promote the development of people-centered early warning technology for landslides with increased precision and reliable prediction both in time and location, especially in a changing climate context.
Action 2 Advance hazard and vulnerability mapping, including vulnerability and risk assessment with increased precision, as well as reliability as part of multi-hazard risk identification and management.
Action 3 Improve the technologies for monitoring, testing, analyzing, simulating, and effective early warning for landslides suitable for specific regions considering natural, cultural and financial aspects.
Action 4 Apply the ISDR-ICL Landslide Interactive Teaching Tools for landslide risk reduction in landslide prone areas and improve them with feedbacks from users in developed and less developed countries.
Action 5 Promote open communication with local governments and society through integrated research, capacity building, knowledge transfer, awareness-raising, training, and educational activities, to enable societies and local communities to develop effective policies and strategies for reducing landslide disaster risk, to strengthen their capacities for preventing hazards from developing into major disasters, and to enhance the effectiveness and efficiency of relief programs.
Action 6 Investigate the effect of climate change on rainfall-induced landslides and promote the development of effective rainfall forecasting models to provide earlier warning and evacuation especially in developing countries.
Action 7 Investigate the mechanism and dynamics of submarine landslides during earthquakes that may cause or enhance tsunamis, as well as develop and upgrade its hazard assessment and mitigation measures.
Action 8 Promote geotechnical studies of catastrophic megaslides and develop their prediction and hazard assessment.
Action 9 Foster new initiatives to study research frontiers in understanding and reducing landslide disaster risk by promoting joint efforts by researchers, policy makers and funding agencies.
Action 10 Facilitate and encourage monitoring, reporting on, and assessing progress made, through the organization of progress report meetings at the regional and national level, to take place in respective countries, in order to show delivery and performance on progress made towards achieving the Kyoto Landslide Commitment priority actions No.1–9. Participating parties and relevant stakeholders reporting on deliveries and achievements at these meetings are invited to report on this progress in the monthly full color journal "Landslides" so as to allow viewing progress in addressing landslide risk reduction. They are also encouraged to cooperate, as feasible and appropriate, with countries, the United Nations family, regional organizations, and all other partners and stakeholders concerned with landslide risk in their contribution to the Sendai Monitor System and the Voluntary National Reviews, and in their reporting on relevant key SDGs, notably on resilient and sustainable cities and climate action and on the Paris Agreement follow-up.

We submit that the above priority actions contribute to the four priority areas of the Sendai Framework and to the achievement of its seven global targets, in line with the "Words into Action" guidelines for Sendai Framework implementation, as well as of the SDG related targets. These actions also support landslide risk actors involved in building urban resilience so as to achieve coherence with the New Urban Agenda. Furthermore, they contribute to the discussion within the Global Platforms for Disaster Risk Reduction. Finally, they support the implementation of the Strategic Framework 2016–2021 of the United Nations Office for Disaster Risk Reduction (UNISDR).

We consider KLC2020 as a framework to enhance cooperation in landslide risk reduction internationally, but also as a platform aimed at providing support to regional, national and local efforts, triggering exchanges on good practices and twinning and building the capacity of institutions and professionals at the national and local levels.

Commitments by all participating parties are periodically reviewed and updated at the Triennial World Landslide Forum in which parties of KLC2020 are called upon to participate.

Fundamental Coordinating Commitments by the International Consortium on Landslides (ICL) and the Global Promotion Committee of the International Programme on Landslides (IPL) and others providing the common platform for the Kyoto Landslide Commitment 2020 include the following:

1. The Triennial Conference "World Landslide Forum" will be organized and the progress of Kyoto Landslide Commitment by all participating parties will be reported and examined for further development.
2. Landslides: Journal of International Consortium on Landslides will continue to be published monthly in full color and distributed to all participating parties.
3. Contribution fee and full color printing fee will continue to be waived to promote contribution from less developed countries and young researchers.
4. ICL provides that all parties of the Kyoto Landslide Commitment 2020 have the right to submit and publish news and reports of their activities in the "Landslides" journal. All parties will receive the digital access rights (tokens) to all issues of the journal (2002-present).
5. ICL will publish and update Landslide Dynamics: Landslide Interactive Teaching Tools (LITT) as a core activity for public education at each Forum. In early 2018, the first LITT (Vol.1 Fundamentals, Mapping and Monitoring, Vol.2 Testing, Risk Management and Country Practices) have been edited and published including PPT for lessons and PDF for reference in digital format.
6. ICL and the Global Promotion Committee of the International Programme on Landslides (GPC/IPL) will organize the annual IPL symposium and publish a series of books with ISBN numbers together with Research, Administrative and Strategic Review meetings at relevant venues such as UNESCO or elsewhere including the biennial Global Platform for Disaster Risk Reduction.
7. Other commitments by ICL and IPL groups will include:

- Landslide experts are called upon to gain trust and confidence from the local authorities and the communities facing the risk of landslides in order to effectively communicate the risk and urge local actions to help reduce the risk. Thus, ICL and IPL groups will promote a good dialogue at local levels throughout the activities of the Kyoto Landslide Commitment 2020.
- To promote cooperation between policy makers, national government authorities working on landslide risk reduction and landslide scientists and engineers, a joint round table discussion between ICL members and high-level Ministerial members will be organized at each triennial Forum.
- Community safeguard policy for the countries/areas which are affected by rain-induced rapid and long-travel landslides, earthquake-induced megaslides as well as coastal and submarine landslides will be examined in specific sessions at each Forum.
- To identify, whenever possible and appropriate, focal points at the national/regional level in as many countries/regions as possible for engagement with the Kyoto Landslide Commitment 2020 and for ensuring contact and coordination with the Secretariat.

We are conscious that KLC2020 will build and capitalize on the work and achievements of ICL and IPL notably the 2006 Tokyo Action Plan, the 2015 Sendai Partnerships and the outcomes of the World Landslide Fora.

We are committed to working together with Member States of the UN system, the UN family, regional organizations, and all other partners and stakeholders concerned with landslide risk, including civil society, academic, scientific and research entities, business, professional associates and private sector financial institutions, and the media.

We firmly believe that sustained cooperation and exchange between countries at a governmental level is needed if we are to promote in a sustainable way landslide risk reduction for resilience. Therefore, the timeliness and opportunity of having in place a platform or a mechanism at an intergovernmental level which would work to advise, strengthen and support decisions and initiatives on landslide risk reduction must be considered. This mechanism will enhance synergetic and concerted efforts not only among governmental entities but also between them and other sectors including the private sector and the civil society. We call upon stakeholders concerned to consider developing an Intergovernmental Panel on Landslide Risk Reduction in the framework of KLC2020 within the International Programme on Landslides (IPL) so as to raise the level of interest in this area and maintain it through support to a long term global alliance that will continue to 2025, 2030 and beyond. The Panel will help mobilize strong political interest and commitment of the international community as well as further scientific knowledge and technological know-how. It will advise on translating the objectives of ICL into meaningful programmes at the country level. The functions, form, governance and operation of the Panel will hopefully defined by relevant intergovernmental bodies.

A Call for Joining the Commitment

Competent global, regional, national, local institutions and entities participating in the Fifth World Landslide Forum are invited to support this initiative by joining and signing this Commitment through participation in clearly defined commitments for understanding and reducing landslide disaster risk. The potential parties are requested to make contact with the Secretariat of the host organization. Furthermore, ICL and competent national, regional and international institutions are encouraged to enter into bilateral arrangements including through a dedicated Agreement of Cooperation which will provide a framework of cooperation and facilitate collaboration in areas of common interest and which enable both parties to mutually benefit and develop their cooperation, for the benefit of landslide hazard-prone communities in the country concerned and worldwide.

Host Organization and Secretariat

The International Consortium on Landslides (ICL) hosts the Kyoto Landslide Commitment 2020 as a voluntary commitment to the Sendai Landslide Partnerships 2015–2025, the Sendai Framework for Disaster Risk Reduction 2015–2030 and the 2030 Agenda Sustainable Development Goals. The ICL Secretariat in Kyoto, Japan, serves as the Secretariat of the Kyoto Landslide Commitment 2020.

Signatories

Signature

Name

Position

Organization

Date

Appendix: Signatories of Kyoto Landslide Commitment 2020 as of 23 June 2020

No	Signatories	Position/Organization	Countries/Regions/International. Organizations	Date of signature
Host organization				
1	Peter T. Bobrowsky	President, International Consortium on Landslides	ICL	18/09/2019
ICL supporting organizations and other partners from Governmental and International organizations				
1	Miguel Clusener-Godt	Director, Division of Ecological and Earth Sciences, UNESCO	UNESCO	10/09/2019
2	Taikan Oki	Senior Vice-Rector, United Nations University	UNU	05/09/2019
3	Jacques de Méreuil	Executive Director, World Federation of Engineering Organizations	WFEO	18/09/2019
4	Qunli Han	Executive Director, Integrated Research on Disaster Risk	IRDR	18/09/2019
5	Qiuming Cheng	President, International Union of Geological Sciences	IUGS	02/02/2020
6	Juichi Yamagiwa	President, Kyoto University	Japan	18/09/2019
7	Akifumi Nakao	Director, International Cooperation Division, Disaster Management Bureau, Cabinet Office, Government of Japan	Japan	15/01/2020
8	Masaru Kunitomo	Director, Sabo Planning Coordination, Sabo Planning Division, Ministry of Land Infrastructure, Transport and Tourism	Japan	18/09/2019
9	Tomohiro Saeki	Head, Forest Disaster Prevention and Restoration Office, Forestry Agency, Ministry of Agriculture, Forestry and Fisheries	Japan	18/03/2020
10	Angelo Borrelli	Head, National Civil Protection Department, Italian Presidency of the Council of Ministers, Italy	Italy	18/09/2019
11	Darko But	Director General, the Administration for Civil Protection and Disaster Relief of the Republic of Slovenia	Slovenia	18/09/2019
12	Walter J. Ammann	CEO & President, Global Risk Forum GRF Davos	GRF Davos	18/09/2019
13	Mr. Rafig Azzam	President, International Association for Engineering Geology and the Environment	IAEG	01/10/2019
14	Nathalie Touze	Vice-President, International Geosynthetics Society	IGS	18/09/2019
ICL supporting organizations as well as ICL full members				
1	Julian SH Kwan	Geotechnical Engineering Office, Civil Engineering and Development Department, the Government of Hong Kong SAR	China	18/09/2019
2	Nicola Casagli	UNESCO Chair for the prevention and the sustainable management of geo-hydrological hazards, University of Firenze (UNIFI)	Italy	18/09/2019
3	Matjaž Mikoš	University of Ljubljana, Faculty of Civil and Geodetic Engineering	Slovenia	18/09/2019
ICL full members				
1	Renato Eugenio de Lima	Director, Center for Scientific Support on Disasters—Federal University of Paraná—Brazil (CENACID-UFPR)	Brazil	18/09/2019
2	Daniel Lebel	Director General, Geological Survey of Canada	Canada	18/09/2019
3	Michael T. Hendry	Associate Professor, University of Alberta	Canada	18/09/2019
4	Wei Shan	Dean of Cold Regions Science and Engineering, Northeast Forestry University	China	18/09/2019
5	Chang-Dong Li	China University of Geosciences (Wuhan)	China	18/09/2019
6	Lijun Su	Institute of Mountain Hazards & Environment, CAS	China	18/09/2019
7	Snježana Mihalić Arbanas and Željko Arbanas	Croatian Landslide Group	Croatia	18/09/2019

(continued)

No	Signatories	Position/Organization	Countries/Regions/International. Organizations	Date of signature
8	Guillermo Ávila	National University of Colombia	Colombia	18/09/2019
9	Vít Vilímek	Charles University	Czech Republic	18/09/2019
10	Josef Stemberk	Director, Institute of Rock Structure and Mechanics, the Czech Academy of Sciences	Czech Republic	18/09/2019
11	Hauke Zachert	Head, Institute and Laboratory of Geotechnics, Technical University Darmstadt	Germany	18/09/2019
12	Andro Aslanishvili	Head of LEPL National Environmental Agency of Georgia, Ministry of Environment Protection and Agriculture	Georgia	18/09/2019
13	Maneesha V Ramesh	Amrita Vishwa Vidyapeetham	India	18/09/2019
14	Dwikorita Karnawati	Head,Agency for Meteorology, Climatology, and Geophysics of the Republic of Indonesia (BMKG Indonesia)	Indonesia	18/09/2019
15	Teuku Faisal Fathani	Director,Center for Disaster Mitigation and Technological Innovation (GAMA-InaTEK), Universitas Gadjah Mada, Indonesia	Indonesia	18/09/2019
16	Mohammad Shekarchizadeh	President, Building & Housing Research Center	Iran	14/04/2019
17	Daniele Spizzichino	ISPRA-Italian Institute for Environmental Protection and Research	Italy	18/09/2019
18	Carlo Esposito	Centro di Ricerca CERI - Sapienza Università di Roma	Italy	18/09/2019
19	Giuseppe Mandrone	Dept. Earth Science, University of Torino	Italy	18/09/2019
20	Andrea Segalini	University of Parma, Dept. Of Engineering and Architecture	Italy	18/09/2019
21	Giovanna Capparelli	Camilab - Dimes Dept. University of Calabria	Italy	18/09/2019
22	Paola Reichenbach	Senior Researcher, Research Institute for Geo-Hydrological Protection, Italian National Research Council (IRPI-CNR)	Italy	22/06/2020
23	Hiroshi Yagi	President, Japan Landslide Society	Japan	18/09/2019
24	Fawu Wang	Director-General, International Consortium on Geo-disaster Reduction	Japan	18/09/2019
25	Zoran Gligorić	Dean, Faculty of Mining and Geology, University of Belgrade	Serbia	18/09/2019
26	Tomislav Popit	University of Ljubljana, Faculty of Natural Sciences and Engineering	Slovenia	18/09/2019
27	Miloš Bavec	Director, Geological Survey of Slovenia	Slovenia	18/09/2019
28	A. A. Virajh Dias	Central Engineering Consultancy Bureau	Sri Lanka	18/09/2019
29	Asiri Karunawardena	Director General, National Building Research Organisation	Sri Lanka	15/06/2020
30	Ray-Shyan Wu	Distinguished Professor, National Central University	Chinese Taipei	12/03/2020
31	Louis Ge	Department of Civil Engineering, National Taiwan University	Chinese Taipei	18/09/2019
32	Hans Guttman	Executive Director, Asian Disaster Preparedness Center	Thailand	18/09/2019
33	Oleksandr M. Trofymchuk	Institute of Telecommunications and Global Information Space, National Academy of Science of Ukraine	Ukraine	18/09/2019
34	Binod Tiwari	Associate Vice President, Office of Research and Sponsored Projects, Division of Academic Affairs, California State University, Fullerton	USA	18/09/2019
35	Nguyen Xuan Khang	Director General, Institute of Transport Science and Technology	Vietnam	18/09/2019
36	Tran Tan Van	Director, Vietnam Institute of Geosciences and Mineral Resources (VIGMR)	Vietnam	18/09/2019

(continued)

No	Signatories	Position/Organization	Countries/Regions/International. Organizations	Date of signature
ICL associate members				
1	Qiang Xu	Executive Deputy Director, State Key Laboratory of Geohazard Prevention and Geoenvironment Protection, Chengdu University of Technology	China	18/09/2019
2	Mario Parise	Department of Earth and Environmental Sciences, University Aldo Moro, Bari	Italy	18/09/2019
3	Michele Calvello	Geotechnical Engineering Group (GEG), University of Salerno	Italy	18/09/2019
4	Ryo Moriwaki	Director, Center for Disaster Management Informatics Research, Ehime University	Japan	18/09/2019
5	Nejan Huvaj	Middle East Technical University (METU)	Turkey	18/09/2019
6	Beena Ajmera	North Dakota State University	USA	18/09/2019
ICL supporters				
1	Shinro Abe	Adviser, Okuyama Boring Co., Ltd	Japan	18/09/2019
2	Nobuyuki Shibasaki	General Manager, Land Conservation Division, Nippon Koei Co., Ltd	Japan	18/09/2019
3	Ryosuke Tsunaki	Executive Engineer, Sabo & Landslide Technical Center	Japan	18/09/2019
4	Maki Yano	President, OSASI Technos, Inc	Japan	18/09/2019
5	Masaru Narita	President, Oyo Cooperation	Japan	22/04/2020
6	Taketoshi Marui	President, Marui & Co., Ltd	Japan	05/06/2020
7	Yamazaki Tsutomu	Director and General Manager of Technical Headquarters, Japan Conservation Engineer & Co. Ltd	Japan	16/06/2020

Note

KLC2020 partners are organizations having some aspects of activities related to understanding and reducing landslide disaster risk as their intrinsic missions. ICL members support the objectives of KLC2020, intellectually, practically, and financially. ICL supporting organizations support the objectives of ICL intellectually and practically within their capabilities.
ICL supporting organizations as well as ICL full members are «ICL supporting organizations having also the status of ICL full members» which support IPL activities financially in addition to providing ICL full membership dues. ICL associate members support the objectives of the Consortium and meet the financial obligations of 20% of ICL full membership due. ICL supporters support the objectives of the Consortium and provide funds for its activities.
Progress of activities of KLC2020 will be published under the category of News/Kyoto Commitment of the monthly Journal Landslides. The integral activities of KLC2020 will be a Commitment to the Sendai Landslide Partnerships 2015–2025, the Sendai Framework for Disaster Risk Reduction 2015–2030, the 2030 Agenda Sustainable Development Goals, the New Urban Agenda and the Paris Climate Agreement. Further KLC2020 partners are called upon to sign it at the virtual online joint signatory meeting on 5 November 2020.

International Consortium on Landslides (ICL): Proposing and Host Organization of SLP2015-2025 and KLC2020

Kyoji Sassa, Peter T. Bobrowsky, and Kaoru Takara

Abstract

The International Consortium on Landslides (ICL) was founded at the UNESCO-Kyoto University joint Symposium in January 2002. The ICL also decided to create the International Programme on Landslide (IPL) and a new international journal *Landslides* in the foundation meeting in 2002. The ICL was registered as a legal body under Japanese law in 2002 as a non-profit scientific organization. The ICL proposed the 2006 Tokyo Action Plan and exchanged a MoU to each of seven global stakeholders (ICL Supporting Organizations, ISO). Together with ISO, the ICL founded the triennial World Landslide Forum (WLF) in 2008, and in 2015 proposed the Sendai Landslide Partnerships 2015–2025. Based on partnerships, ICL and ISO are organizing the Fifth World Landslide Forum (WLF5) and will launch the Kyoto 2020 Commitment for Global Promotion of Understanding and Reducing Landslide Disaster Risk (KLC2020) on 5 November 2020.

Keywords

Sendai landslide partnerships (SLP) • Kyoto landslide commitment (KLC) • International programme on landsides (IPL)

Objectives

The principal objectives of the ICL are to:

(a) promote landslide research for the benefit of society and the environment, and capacity building, including education, notably in developing countries;
(b) integrate geosciences and technology within the appropriate cultural and social contexts in order to evaluate landslide risk in urban, rural and developing areas including cultural and natural heritage sites, as well as to contribute to the protection of the natural environment and sites of high societal value;
(c) combine and coordinate international expertise in landslide risk assessment and mitigation studies, thereby resulting in an effective international organization which will act as a partner in various international and national projects; and
(d) promote a global, multidisciplinary Programme on landslides, the International Programme on Landslides (IPL).

Histories

- The ICL history for each 3-year period, which is one term for ICL officers and the organization of triennial World Landslide Forum, is compiled from the foundation in 2002 to 2020 in Table 1.
- ICL was registered as a legal body (No.1300-05-005237) under the Japanese law in the Kyoto Prefectural Government, Japan in August 2002.
- ICL exchanged MoU with each of seven global stakeholders (UNESCO, UNDRR, UNU, FAO, WMO, ISC and WFEO) related to understanding and reducing landslide disaster risk in 2006
- ICL was approved in March 2007 as a scientific research organization (No. 94307) that can apply and receive scientific grants of the Ministry of Education, Culture, Sports, Science and Technology (MEXT), Japan and registered in the cross-ministerial research and development management system (e-Rad) of all ministries of Japan (No. 5010000002) in May 2008.

K. Sassa (✉) · P. T. Bobrowsky · K. Takara
International Consortium on Landslides (ICL), Kyoto, Japan
e-mail: secretariat@iclhq.org

K. Sassa et al. (eds.), *Understanding and Reducing Landslide Disaster Risk*, ICL Contribution to Landslide Disaster Risk Reduction,
https://doi.org/10.1007/978-3-030-60196-6_8

Table 1 ICL History-from foundation in 2002 to 2020

Date	Chronology of events
Foundation of the International Consortium on Landslide and its Interim Stage	
Interim President: Kyoji Sassa	
January 2002	The UNESCO-Kyoto University joint symposium "Landslide Risk Mitigation and Protection of Cultural and Natural Heritage" was organized in Kyoto, Japan. It was attended by six directors from Earth Science, Water Sciences, Cultural Heritage, Engineering Science of UNESCO, and deputy Secretary General of WMO, a representative from UNISDR, and also from the Ministry of Foreign Affairs and the Ministry of Education, Culture, Sports, Science and Technology (MEXT), as well as IGCP-425 members from around the world
January 2002	ICL was established by adopting the 2002 Kyoto Appeal "Establishment of a New International Consortium on Landslides" and the statutes of ICL and deciding the Interim President on 21 January 2002 during the same symposium. The main objective of ICL foundation is to establish a new International Programme on Landslides (IPL) after the end of IGCP-425. Most of the 31 subproject leaders of IGCP-425 received promotional benefits of their landslide research from authorization as the subproject leaders of the UNESCO/IUGS joint programme and wished to establish a new International Programme on Landslides (IPL)
August 2002	ICL was registered as a legal body under Japanese law in the Kyoto Prefectural Government, Japan in August 2002
November 2002	The First Session of the Board of Representatives (BOR) of ICL was organized at UNESCO Headquarters on 19–21 November 2002 The initial 33 ICL member organizations, mostly from IGCP-425 subproject leaders, attended, as well as the officers for the first three years. They decided to found an international journal "***Landslides***" as the initial IPL project Coordinating Project C100 as a core activity of ICL. ICL is basically a bottom up self-supporting organization with supports from UNESCO, WMO, FAO, UNISDR, Government of Japan and others
The First Period: 1 January 2003–31 December 2005	
President: Kyoji Sassa Vice Presidents: Peter T. Bobrowsky, Paolo Canuti, Romulo Mucho, Peter Lyttle Executive Director: Kaoru Takara; Treasurer: Claudio Margottini Number of ICL Member organizations: 33	
March 2003	ICL, Kyoto University and UNESCO established the UNITWIN (University Twining and Networking) Cooperation Programme on Landslide Risk Mitigation for Society and the Environment in March 2003. The establishment of UNITWIN Cooperation Programme was initially suggested by participants from UNESCO to authorize the International Programme on Landslides (IPL) as a UNESCO network activity
May 2003	The First ICL Steering Committee Meeting was organized at the headquarters of the Food and Agriculture Organization of the United Nations (FAO) in Rome, Italy. The steering committee meetings were organized annually or as necessary from then on
October 2003	The 2nd Session of the Board of Representatives (BOR) of ICL was organized at Simon Fraser University Harbour Centre, Vancouver, Canada on 28 October–1 November 2003
April 2004	ICL founded a new full color quarterly Journal "*Landslides*: Journal of the International Consortium on Landslides" in 2004, in cooperation with Springer after negotiating with several international publishers
September 2004	UNITWIN Headquarters Building was constructed by ICL and Kyoto University at the Kyoto University Uji Campus in September 2004
October 2004	The 3rd Session of the Board of Representatives (BOR) of ICL was organized at Druzba Hotel, Bratislava, Slovakia on 19–22 October 2004
January 2005	The World Conference on Disaster Reduction (WCDR) was held on 18–22 January 2005 in Kobe, Japan
January-June, 2005	ICL proposed and organized a session titled "New international Initiatives for Research and Risk Mitigation of Foods (IFI) and Landslides (IPL) in cooperation with the Flood group. ICL was well prepared for the session and proposed a "***Letter of Intent***" aiming to provide a platform for a holistic approach in research and learning on "Integrated Earth System Risk Analysis and Sustainable Disaster Management". A Letter of Intent is more flexible than agreements and Memorandums of Understanding, as suggested by Hans van Ginkel (Rector of United Nations University). The Letter of Intentwas agreed upon and signed by heads of seven global stakeholders of UNESCO, WMO, FAO, UNISDR, UNU, ICSU and WFEO from January to June 2005
October 2005	The First General Assembly of ICL, together with the 4th Session of the Board of Representatives (BOR) of ICL, was organized at Keck Center of the National Academy of Sciences, Washington D.C., USA on 12–14 October 2005

(continued)

Table 1 (continued)

Date	Chronology of events
The Second Period: 1 January 2006–31 December 2008	
President: Kyoji Sassa Vice Presidents: Peter T. Bobrowsky, Paolo Canuti, Oddvar Kjekstad, Peter Lyttle Executive Director: Kaoru Takara; Treasurer: Hiroshi Fukuoka Number of ICL Member organizations: 48	
January 2006	The Round Table Discussion "Strengthening Research and Learning on Earth System Risk Analysis and Sustainable Disaster Management within UN-ISDR as Regards "Landslides" was co-organized by ICL, UNESCO, WMO, FAO, UNISDR, UNEP, UNU and Kyoto University at United Nations University, Tokyo, Japan on 18–20 January 2006. This discussion aimed to implement the 2005 Letter of Intent
January 2006	The 2006 Tokyo Action Plan Strengthening Research and Learning on Landslides and Related Earth System Disasters for Global Risk Preparedness was adopted by the participants of the Round Table Discussion on 20 January 2006. The Tokyo Action Plan prosed a new stage of ***International Programme on Landslides (IPL)***, a programme of ICL for ISDR, which is managed by the IPL Global Promotion Committee consisting of all ICL member organizations and 7 global stakeholders (UNESCO, WMO, FAO, UNISDR, UNU, ICSU, and WFEO). New IPL activities include IPL projects, the World Landslide Forum (WLF) every three years, and the World Centres of Excellence on Landslide Risk Reduction (WCoEs) to be identified at each WLF. The First WLF succeeded the First General Assembly of ICL in 2005
April–December 2006	ICL exchanged Memorandums of Understanding to promote IPL with UNESCO, WMO, FAO, UNISDR, UNU, ICSU, WFEO within 2006
November 2006	The 5th Session of the Board of Representatives (BOR) of ICL was organized at UNESCO Headquarters on 23–24 November 2006
March 2007	ICL was approved in March 2007 as a scientific research organization (No. 94307) which can receive scientific grants of the Ministry of Education, Culture, Sports, Science and Technology (MEXT). Thereafter, ICL can apply for Scientific Grants and other Scientific programmes
April 2007	ICL was approved to be an NGO having operational relations with UNESCO in April 2007
November 2007	The 6th Session of the Board of Representatives (BOR) of ICL was organized at UNESCO Headquarters on 14–16 November 2007
May 2008	ICL was registered in the cross-ministerial research and development management system (e-Rad) of all ministries of Japan in May 2008
November 2008	The 7th Session of the Board of Representatives (BOR) of ICL was organized at the United Nations University, Tokyo, Japan on 17 November 2008
November 2008	The First World Landslide Forum (WLF1) was organized by ICL, IPL and Partners at the United Nations University, Tokyo, 18–21 November 2008. (The Second General Assembly of ICL to disseminate ICL-IPL activities was organized as the First World Landslide Forum from 2008.)
The Third Period: 1 January 2009–31 December 2011	
President: Paolo Canuti Vice Presidents: Oddvar Kjekstad, Peter Lyttle, Kaoru Takara Executive Director: Kyoji Sassa; Treasurer: Hiroshi Fukuoka Number of ICL Member organizations: 48	
November 2009	The 8th Session of the Board of Representatives (BOR) of ICL was organized at UNESCO Headquarters on 17–19 November 2009
November 2010	The UNITWIN programme was updated and developed to ***Landslide and water-related disaster risk management for society and the environment*** in a wider scope in November 2010
November 2010	The 9th Session of the Board of Representatives (BOR) of ICL was organized at UNESCO Headquarters on 16–19 November 2010
October 2011	The 10th Session of the Board of Representatives (BOR) of ICL was organized at FAO Headquarters, Rome, Italy on 3–5 October 2011
October 2011	The Second World Landslide Forum (WLF2) was organized by ICL, IPL and partners at the Food and Agriculture Organization Headquarters of the United Nations, Rome, 3–9 October 2011
The Fourth Period: 1 January 2012–31 December 2014	
President: Paolo Canuti Vice Presidents: Kaoru Takara, Yueping Yin, Claudio Margottini, Irasema Alcantara-Ayara Executive Director: Kyoji Sassa Treasurer: Hirotaka Ochiai Number of ICL Member organizations: 53	

(continued)

Table 1 (continued)

Date	Chronology of events
January 2012	ICL organized 10th anniversary meeting, Kyoto, Japan in January 2012 and adopted the ICL Strategic Organization of Plan 2012–2021
March 2012	ICL was reclassified as an NGO with a consultative partnership with UNESCO in March 2012
November 2012	The 11th Session of the Board of Representatives (BOR) of ICL was organized at UNESCO Headquarters, Paris, France on 20–23 November 2012
2013	"Landslides" Journal became a bimonthly journal from Vol. 10 in 2013
November 2013	The 12th Session of the Board of Representatives (BOR) of ICL was organized at Yamanouchi Hall, Shiran Kaikan, Kyoto University, Kyoto, Japan on 19–22 November 2013
June 2014	The 13th Session of the Board of Representatives (BOR) of ICL was organized at China National Convention Center, Beijing, China on 2 June 2014
June 2014	The Third World Landslide Forum (WLF3) was organized by ICL, IPL and partners at the National Convention Center of China, Beijing, in June 2014. ICL adopted the 2014 Beijing Declaration "Landslide Risk Mitigation: Toward a Safer Geoenvironment" at the WLF3 in Beijing
The Fifth Period: 1 January 2015–31 December 2017	
President: Yueping Yin Vice Presidents: Claudio Margottini, Irasema Alcantara-Ayara, Matjaž Mikoš, Dwikorita Karnawati Executive Director: Kyoji Sassa Treasurer: Kaoru Takara Number of ICL Member organizations: 57 (64 in 2016.10)	
March 2015	The 14th Session of the Board of Representatives (BOR) of ICL was organized at Tohoku Gakuin University, Sendai, Japan on 11–15 March 2015
March 2015	ICL proposed the ***"ISDR-ICL Sendai Partnerships 2015–2025 for global promotion of understanding and reducing landslide disaster risk"*** in the Third United Nations World Conference on Disaster Risk Reduction in Sendai, Japan, 2015
June 2015	Landslides: Journal of International Consortium on Landslides reached a 2015 Impact Factor of 3.049
March 2016	The 15th Session of the Board of Representatives (BOR) of ICL was organized at Kyoto University Uji Campus, Kyoto, Japan on 7–11 March 2016
January 2017	The number of pages for one issue of *Landslides*: Journal of the International Consortium on Landslides was increased from 200 pages/issue in 2016 to 300 pages/issue in 2017
May 2017	The Fourth World Landslide Forum (WLF4) was organized by ICL, IPL and partners at the Cultural and Congress Centre in Ljubljana, Slovenia, from 29 May to 2 June 2017 The concept of Kyoto 2020 Commitment was approved in WLF4. To promote Kyoto 2020 Commitment, ICL decided to create a new membership of ICL associates (20% membership fee of full members) at the ad-hoc BOR/ICL during WLF4. ICL members are defined to include full members, associate members and supporters
November–December 2017	The 17th Session of the Board of Representatives, 13th Session of the Global Promotion Committee, WLF5 organizing Committee as well as 2018 IPL Symposium were organized in Room 9 and Room 8, Fontenoy Building at UNESCO Headquarters on 29 November to 1 December 2017
The Sixth Period: 1 January 2018–31 December 2020	
President: Peter T. Bobrowsky Vice Presidents: Matjaž Mikoš, Dwikorita Karnawati, Nicola Casagli, Binod Tiwari, Željko Arbanas Executive Director: Kaoru Takara; Treasurer/Secretary General: Kyoji Sassa Number of ICL Member organizations: 99 in 2020.5. 69 full members: 16 associate members, 14 Supporters (ICL memberes are defined to include full members, associate members and supporters after WLF4)	
December 2018	The 18th Session of the Board of Representatives, 14th Session of the Global Promotion Committee, WLF5 organizing Committee as well as 2018 IPL Symposium were organized in Kyoto International Conference Center (KICC) and Disaster Prevention Research Institute, Kyoto University, in Kyoto, Japan on 1–4 December 2018
September 2019	The 19th Session of the Board of Representatives, 15th Session of the Global Promotion Committee, WLF5 organizing Committee as well as 2019 IPL Symposium were organized in Room IX, Fontenoy Building at UNESCO Headquarters on 16–19 September 2019
November 2020	The 20th Session of the Board of Representatives, 16th Session of the Global Promotion Committee, WLF5 organizing Committee as well as 2020 IPL Symposium will be organized in an online meetings on 2–6 November 2020. The Joint Online Signatories and Declaration of the Launching of KLC2020 is planned for 5 November 2020. Due to the time constraints over the globe, every day meetings will be organized for 4 h in the west end (PST) to east end (JPT) on the same day

- ICL was approved to be an NGO having operational relations with UNESCO in April 2007 and was reclassified as NGO with the consultative partnership with UNESCO in March 2012.
- ICL and the supporting organizations established the Sendai Landslide Partnerships (SLP) 2015–2025 in March 2015 in Sendai, Japan.
- ICL and partners of SLP 2015–2025 plan to further develop the initiative of the SLP2015-2025 after 2025, 2030 and even beyond, and will launch the Kyoto 2020 Commitment for global promotion of understanding and reducing landslide disaster risk on 5 November 2020.

Activities

- Edition and publication of *Landslides*: Journal of the International Consortium on Landslides every month. The online meeting of Screening editors is organized every week including American, European and Asian continents and decide to pass contributed papers to in-depth review or reject without in-depth review to manage the voluntary editors (173) and reviewers (1532) of Landslides as well as give authors a choice to either contribute to other journals or to different category of *Landslides* again.
- Publication of full colour books to disseminate the advance of science and technology for the latest three years (total of 23 books were published and are being published as shown in the Appendix of the Preface in this book).
- Publication of Landslide Interactive Teaching Tools (LITT) including text, PPT, PDF, and video tools for capacity development. Two volumes of LITT were published after WLF4. Some volumes of LITT will be published after WLF5.
- Identification and promotion of the World Centres of Excellence on Landslide Risk Reduction (WCoEs). The ICL called for and identified WCoEs at each Forum. Proposals from ICL members are evaluated by the Technical Evaluation Committee, and the results are further examined by the IPL Secretariat. Selected candidates are next submitted to the Independent Panel of Experts for additional evaluation and endorsement. Thc endorsed WCoE candidates are presented and approved by the Global Promotion Committee of IPL. The number of identified WCoEs was 35 from 2008–2020 in the past 4 periods.
- Identification and promotion of the IPL projects. IPL project proposals are invited for every year and the applications are orally presented in the IPL symposium every year. The application is evaluated by the IPL Evaluation committee and acceptance is decided by the Global Promotion Committee members who are ICL Board members and representatives of ICL Supporting organizations. The term of an IPL project is not fixed, but an annual report must be submitted to keep ongoing status. Currently 55 IPL projects are active.
- Activities of ICL regional and national networks. ICL Adriatic-Balkan network, ICL Italian Network, and ICL Japanese network have been established consisting of ICL full members, associate members, supporters, and cooperating organization in the area. Regional conferences and proceedings are published.
- Activities to promote international landslide initiatives: The Sendai Landslide Partnerships 2015–2025, which is the Sendai Framework Voluntary commitment, the planned Kyoto 2020 Commitment contribute to Sendai Framework 2015–2030, the 2030 Agenda Sustainable Development Goals, and the Paris Climate Agreement.

Memberships and Benefits

ICL is maintained by the membership fees from ICL members. The membership fee for full members is proportional to GNI/capita according to the World Bank and range from 500, 1000, 2000, to 3000 USD for full members. ICL members who contribute 5000 USD (2000 USD for IPL promotion fund in addition to 3000 USD for memberships) have the status of ICL supporting organization as well as ICL members. The membership fee for associate members is 20% of full members, namely 100, 200, 400, or 600 USD. The membership fee for an academic society is a half of each membership fee to promote academic activities.

The membership fee for supporters that are mainly companies working on landslide monitoring, testing, consulting, and landslide mitigation are 1000 USD or 2000 USD.

Benefits for ICL members are compiled in Table 2.

- Each full member is a Board member of ICL who has a voting right for all matters regarding ICL as part of the Board of Representatives of ICL (BOR/ICL) and a voting right for all matters of IPL in the Global Promotion Committee of IPL.
- Each full member is eligible to be an officer in ICL and IPL.
- Each Board member (3000–5000 USD) is recognized as member of the Editorial Board of the journal ***Landslides.***
- Each member organization can apply for travel support to attend an ICL-IPL meeting.
- ICL will accept one lower category of membership fee rank for developing countries such as 1000 USD instead of 2000

Table 2 ICL membership details and benefits

ICL members							
Categories of members	ICL supporting organizations and ICL full members		ICL full members	ICL society members	ICL supporters	ICL associates	
Countries and academic societies	Developed countries		Developing countries	All countries	All countries	Developed countries	Developing countries
Annual dues (USD)	5000	3000	2000, 1000, or 500	50% of annual dues	1000 or 2000	600	400 or 200 or 100
ICL board of representative status	O	O	O	O	×	×	×
ICL/IPL officers eligibility	O	O	O	O	×	×	×
IPL global promotion committee status	O	O	O	O	×	×	×
Voting rights for all ICL and IPL matters	O	O	O	O	×	×	×
Recognition in editorial board members of landslides journal	O	O	×	×	×	×	×
Recognition in ICL structure (last page of landslide journal and ICL books)	O	O	O	O	O	×	×
ICL/IPL meeting travel support access*	O	O	O	×	×	×	×
IPL project candidacy	O	O	O	O	×	×	×
World Centre of Excellence Candidacy (WCoE)	O	O	O	O	×	×	×
ICL/IPL web recognition	ICL/IPL	ICL/IPL	ICL/IPL	ICL/IPL	ICL/IPL	ICL	ICL
ICL/IPL business meeting participation	O	O	O	O	Observer	Observer	Observer
IPL symposium presentations	O	O	O	O	O	O	O
Discount registration for WLF participation	O	O	O	O	O	O	O
Submit to IPL/WCoE activities of landslides journal	O	O	O	O	×	×	×
Submit to Kyoto commitment of landslides journal	O	O	O	O	O	O	O
Landslides journal hard copies**	1–3	1–3	1–2	1	1–2	1	×
Landslides journal online tokens	3	3	1–2	1	1–2	1	1

*ICL will accept one rank lower category of membership fee for developing countries such as 1000 USD for 2000 USD and 500 USD for 1000 USD as ICL members. However, no travel support is allocated to those members who paid the one rank lower membership fee amount

**12 issues of landslides are published annually. Space is limited. Many members prefer only one hard copy. So number is written as 1–2 or 1–3

USD and 500 USD instead of 1000 USD as ICL members. However, no travel support is allocated to those members.

- All ICL members (full members, associate members and supporters) are encouraged to join the Kyoto Landslide Commitment. All KLC2020 partners are invited to contribute articles (new, reports, publication, and achievements that are less than 4 pages long) to the News/Kyoto Commitment category of Landslides.
- All full members are invited to propose IPL projects (every year) and apply for World Centre of Excellence status (every 3 years) as well as to report their activities in the IPL/WCoE category of Landslides.
- Each ICL member organization can receive 1–3 tokens that is accessible to all issues of Landslides from since its foundation, and 1–3 hard copies of the journal *Landslides* (see details in Table 2).
- The registration fee of the WLF5 is reduced for all ICL members.
- All ICL members are invited to orally present at annual IPL symposium and thier presentation is published in the CD proceedings.
- Other details of membership fee can be provided upon request to the ICL Secretariat.

Call for ICL Members

In order to invite wider partners to the planned Kyoto Landslide Commitment 2020, ICL has expanded ICL memberships by creating associate members that are 20% membership of full members (100–600 USD) and also creating academic society discount memberships that are 50% of each category since 2017. As a result, the number of ICL members has increased to 99 as of 2020.

ICL members (full members and associate members and supporters) are the core of KLC2020. All ICL members have some aspects of activities related to understanding and reducing landslide disaster risk as their intrinsic missions. It is clearly proven with the financial dues though the annual ICL membership. All activities of ICL members in their intrinsic mission lead to a contribution to the KLC2020. These activities should be shared with all other KLC2020 partners through publication in News/Kyoto Commitment of *Landslides* and presentation at the annual IPL Symposium during ICL-IPL annual conferences.

Those who are willing to join ICL and also KLC2020 signatory meeting on 5 November 2020 are requested to email ICL Secretariat (KLC2020 secretariat, secretariat@iclhq.org) using the subject "ICL membership and KLC2020 partner".

Note

1. Membership annual dues are based on the categories of the Gross national income per capita (GNI/Capita) in USD by World Bank; High income (3000), Upper middle income (2000), Lower middle income (1000), and Low income (500).
2. ICL full members who pay 5000 USD (3000 USD as full member + 2000 USD IPL supporting fund) are ICL supporting organizations.
3. Board members of category of 3000 USD and 5000 USD are listed in the Editorial Board (A4-A6 pages) of Landslide Journal.
4. Banners of Supporters are shown in the top page of ICL, IPL and ICL-JP webs.
   ICL Secretariat (Kyoto, Japan) accepts consultation from supporters and assists their initiatives for landslide disaster risk reduction activities through ICL-IPL Network.
5. ICL supporting organizations (UNESCO, WMO, FAO, UNISDR, UNU, ISC, WFEO, IUGS, IUGG) are members of IPL Global Promotion Committee with voting right.
6. Kyoto 2020 Commitment for Understanding and Reducing Landslide Disaster Risk is examined to be adopted at the World Landslide Forum 5 in 2020, Kyoto, Japan. All ICL members/supporters /associates are candidates of partners to sign the Kyoto 2020 Commitment.

2020 Kyoto Japan

The ICL Journal Landslides—16 Years of Capacity Development for Landslide Risk Reduction

Matjaž Mikoš, Kyoji Sassa, and Željko Arbanas

Abstract

Capacity building and capacity development for landslide risk reduction is an important pillar of the International Consortium on Landslides, Kyoto, Japan. This non-governmental organization with close to 100 full members, associates and supporters was established in 2001, and among many activities in this first two decades we may raise the latest overreaching one, namely the Sendai Partnerships 2015–2025 as the free commitment to Sendai Framework on Disaster Risk Reduction 2015–2030. The Kyoto Commitment 2020 to be discussed and accepted at 5th World Landslide Forum in Kyoto in November 2020, again stresses the importance for the ICL to raise awareness and enhance preparedness for landslide disasters as the ICL efforts for capacity building and capacity development in this field. The ICL stimulates landslide research that has to support capacity building for landslide risk reduction. Springer Nature publishes the journal *Landslides*: Journal of the International Consortium on Landslides since 2004. Being examined in the past by different authors from bibliometric and editorial point of view, this review paper focuses on the journal's 16 years of achievements (2004–2019). In these 16 years, 1313 papers were published on 16,286 pages, written by 5534 authors and with more than 1.1 million downloads and nearly 25,000 citations as in early 2020. The bibliometric analysis of *Landslides* and its comparison with a few selected similar journals of high reputation, among them *Engineering Geology* and *Bulletin of Engineering Geology and the Environment*, confirmed high rankings of *Landslides* in the research categories of geological & geotechnical engineering and engineering geology. Strong and weak points are discussed from the bibliometric point of view, stressing the need for higher internationality of co-authorship of published articles in order to be true international journal. Continuous publishing and the move to a monthly journal in 2018 has further increase journal's h-index and cited half-life of citations, but further editorial efforts should be directed to attract excellent review papers and focused technical notes to increase cites per paper and the number of Highly Cited papers. Until 2020, *Landslides* is the foremost journal in the field of landslide disaster risk reduction, and the top young international journal in the fields of geotechnical engineering and engineering geology.

Keywords

Capacity development • Citation analysis • Education • International collaboration • Journal metrics • Scientometric analysis

Introduction

The first issue of the international journal *Landslides*: Journal of International Consortium on Landslides was published in April 2004 (Springer 2020). Nowadays it is published by the international publishing company Springer Nature. The Editor-in-Chief is Professor Emeritus Kyoji Sassa, the Assistant Editor-in-Chief is Professor Željko Arbanas (both co-authors of this paper), there are eleven Associate Editors (one of them is first author of this paper) and ten Advisory Members. The journal is supported by over hundred editors and over thousand reviewers, and the whole

M. Mikoš (✉)
Faculty of Civil and Geodetic Engineering, University of Ljubljana, Jamova c. 2, 1000 Ljubljana, Slovenia
e-mail: matjaz.mikos@fgg.uni-lj.si

K. Sassa
International Consortium On Landslides (ICL), Kyoto, Japan

Ž. Arbanas
Faculty of Civil Engineering, University of Rijeka, Radmile Matejčić 3, 51000 Rijeka, Croatia
e-mail: zeljko.arbanas@gradri.uniri.hr

K. Sassa et al. (eds.), *Understanding and Reducing Landslide Disaster Risk*,
ICL Contribution to Landslide Disaster Risk Reduction,
https://doi.org/10.1007/978-3-030-60196-6_9

process is supported by a web-based manuscript submission and peer-review tracking system (e.g. Editorial Manager®).

The main aim of *Landslides* is to promote landslide science, technology, capacity building and capacity development, and to strengthen global cooperation for landslide risk reduction within the United Nations International Strategy for Disaster Risk Reduction (ISDR). *Landslides* is one of the main achievements of the International Consortium on Landslides, based in Kyoto, Japan. *Landslides* presents also an important ICL contribution to the world efforts within the Sendai Framework for Disaster Risk Reduction 2015–2030.

Critically analyzing a journal from a bibliometric and editorial point of view is an important task. Sassa et al. (2009) reviewed the achievements of *Landslides* in the first 5 years (2004–2008), and Sassa et al. (2015) did the same for the next 5 years (2009–2013). Sassa and Arbanas (2017) discussed the achievements of *Landslides* from 2004 until September 2016. They stressed the importance of the Journal Impact Factor (Clarivate Analytics) for visibility and reputation of the journal. They presented data on annual impact factors, 5-year impact factors, and citation data for most cited papers and for single journal volumes. Furthermore, they discussed and presented the view of the journal Editorial Board and some aspects of their editorial policy, such as article categories, classification of articles, editorial workflow, article downloading rates, and journal's best paper award.

Mikoš (2011) used data from ISI Web of Knowledge and Scopus for the first seven volumes of *Landslides* (2004–2010). He evaluated the current position of the journal in the landslide research community from a broader perspective than just using impact factor, journal ranking in a subject category and article downloading data. Other journal metrics, available at the time of the analysis, such as journal relatedness to other journals in the same field of science, citations half-life, immediacy impact, journal self-citations, Eigenfactors, Article Influence Score, and journal h-index, give a more detailed picture of Landslides.

Mikoš (2017) addressed the question from the paper's title—is *Landslides* a top international journal in the field of geological engineering and engineering geology? To answer this question, he performed a bibliometric (scientometric) analysis of the first 13 volumes of *Landslides* (2004–2016). He used different journal metrics (impact factor, 5-year impact factor, the number of highly cited papers, cited half-life of published papers, and Hirsch h-index) derived from the Web of Science (Clarivate Analytics 2017), Scopus (Elsevier 2017), and Google Scholar (Google 2017) for journals in the SCI category "geological engineering" and SCI category "multidisciplinary geosciences". Furthermore, he examined the journal's internationality in publishing articles using data on the domicile (country, institution) of authors and classification of articles published in Landslides.

This paper follows the approach of the paper Mikoš (2017) by using Web of Science, Journal Citation Reports, InCites and Essential Science Indicators database produced by Clarivate Analytics, and SCOPUS database by Elsevier in considering the first 16 years of *Landslides* (2004–2019) in the view of capacity development for landslide risk reduction.

Materials and Methods

Categories of Articles in *Landslides*

The journal *Landslides* publishes articles in six major categories (Sassa 2019b):

- *Original papers*: original research and investigation results.
- *Review papers*: review of current research and development of technology in a thematic area of landslide studies.
- *Recent Landslides*: reports of recent landslides including location.
- *Technical Notes*: research notes, reviews notes, case studies, progress of technology, and best practice in monitoring, testing, investigation and mitigation measures.
- *IPL/WCoE Activities*: Progress of IPL projects, World Centres of Excellence on Landslide Risk Reduction (WCoEs), and other IPL activities.
- *News/Kyoto Commitment*: news, reports, announcement of meetings, and all types of articles promoting the Kyoto Landslide Commitment 2020.

The description of categories changed during the period 2004–2019, some categories were omitted (e.g. Technical Development, ICL/IPL Activities), and some were introduced (*Thematic papers*) in a later publishing stage of *Landslides*. In this review paper, category *Thematic papers* was added to category *Original papers*, category *Technical Development* was added to category *Technical Notes*, category *ICL/IL Activities* was added to category *IPL/WCoE Activities*, category *News* was added to category *News/Kyoto Commitment*, all other published documents in *Landslides* were associated to the category *Other items*, including editorials, prefaces, discussions and replies, correction, erratum, retraction and book reviews.

Journal Metrics

The idea of the Impact Factor (IF) was introduced in 1951 by Eugen Garfield (Garfield 1955), and it has received support and criticism ever since (Garfield 2006). Today, the quality of a journal can be measured by numerous bibliometric parameters: impact factor (IF), 5-year impact factor, number of highly-cited papers, cited half-life, citing half-life, journal h-index, etc. Many more indices have been proposed, for a recent overview see Lando and Bertoli-Barsotti (2014).

The main databases now used for journal bibliometric analyses are the Web of Science (WoS) by Clarivate Analytics, and Elsevier's Scopus database. Recently, Google Scholar is frequently used, especially as it is free of charge and it yields higher bibliometric values due to its wide coverage of literature and documents. Therefore, when evaluating a journal for its reputation, it would be an advantage and less biased to use several databases and several journal metrics. In this review paper Google Scholar data were not used.

Clarivate Analytics' Web of Knowledge (WoK)

InCites Benchmarking & Analytics is a customized, web-based evaluation and research tool for bibliometric research among others (Clarivate Analytics, 2020). Journal Citation Reports (JCRs) annually show Journal Impact Factors (JIF) that are calculated by dividing the number of current year citations from all journals in the Web of Science (WoS) to the source items published in that journal during the previous two years. The journal *Landslides* is indexed in WoS in the SCI-Expanded categories: EG—Engineering, Geological (38 journals in 2020), and GM—Geosciences, Multidisciplinary (195 journals in 2020).

The WoS Core Collection covers peer-reviewed journals, dividing publication citations into several indices, those for journals are:

- Science Citation Index Expanded (SCI-Exp),
- Social Sciences Citation Index (SSCI),
- Arts & Humanities Citation Index (AHCI), and
- Emerging Sources Citation Index (ESCI).

It is clear from this development that a journal impact factor may rise annually not only because a journal reputation and impact is increasing but also because WoS database is expanding, covering more and more literature and therefore finding more and more citations to already published articles in journals. Furthermore, increasing the average number of references in published and citable items in WoS journals will increase the average impact factors. These facts should be kept in mind when analyzing a journal's reputation.

Clarivate Analytics Essential Science Indicators (ESI 2020) as a part of InCites offer analysis of top research output and is based on WoS journal article publication counts and citation data. As of March 24, 2020, we can get WoS data on the number of documents, the number of citations and cites per paper and the number of Highly Cited Papers for 96,782 authors, 6581 institutions, 9172 journals, and 149 countries/territories, as well as for 22 research fields. The highly cited papers must reach citations above the citation thresholds, i.e. the minimum number of citations obtained by ranking papers in a research field in descending order by citation count and then selecting the top fraction or percentage of papers. The Highly Cited Threshold reveals the minimum number of citations received by the top 1% of papers from each of 10 database years. For papers published in *Landslides*, the research field *Geosciences* is relevant.

Elsevier SCOPUS Database

Developing its own Scopus database, Elsevier offers different journal metrics, such as (Elsevier 2017):

- SJR—SCImago Journal Rank (SJR) considers both the number of citations received by a journal and the prestige of the journal based on where those citations come from.
- SNIP—Source Normalized Impact per Paper (SNIP) measures contextual citation impact by weighting citations based on the total number of citations in a subject field. It helps to compare a journal with competing journals in a subject area.
- IPP—The Impact per Publication (IPP) is based on citations in one year to articles, reviews, and conference papers published in the preceding three years, divided by the number of articles, reviews, and conference papers published in those three years.

New as of December 2016, the metric called CiteScore measures average citations received per document published in the serial—citations are considered that have been received in a given year for the documents published in the previous 3 years (note that a 2-year window is used for the Clarivate Analytics IF computation). A 3-year publication window is long enough to capture the citation peak of the majority of disciplines covered by Scopus. The new Scopus journal metric CiteScore could be a significant rival to the Journal Impact Factor (Van Noorden 2016). There are differences between IF and CiteScore rankings of journals.

It is also important that Scopus is a large database that covers more journals than WoS uses to compute IF. Scopus includes sources beyond journals such as books and conference proceedings (WoS covers these items by Book Citation Indices and Conference Proceedings Indices). CiteScore includes different document types from journals, not only research papers and review papers, but also editorials, prefaces, letters to the editor, corrections, news, and similar. CiteScore Rank indicates the rank position of the title in its subject area; therefore, it is very important for a journal to be properly classified into a journal category. Namely, across the various scientific domains, significant differences occur with respect to research publishing formats, frequencies and citing practices, the nature and organization of research and the number and impact of a given domain's academic journals. Cerovšek and Mikoš (2014) studied the relationships among citations, most-cited papers and h-indices across domains (Field of Science), confirming previously mentioned differences in citing practices.

Landslides is indexed in Scopus in the area of Physical Sciences, under Earth and Planetary Sciences subject area (including 2262 journals in 2020), and in the sub-subject area "Geotechnical Engineering and Engineering Geology" (including 176 journals in 2020).

Results and Discussion

Impact and Rankings of *Landslides*

Landslides started in 2004 with 4 issues per volume and 1 volume per year, was expanded to 6 issues per year in 2013 (bimonthly journal), and finally to a 12 issues per year in 2018 (monthly journal) (Table 1). The number of pages annually published increased from 305 pages in 2004 to over 2500 published pages in 2018 and 2019, and the number of annually published documents increased from 37 items in 2004 to 199 items in 2018 and 194 items in 2019.

The number of annual web downloads in SpringerLink increased from about 27,000 a year to over 170,000 a year in 2018 and over 100,000 in 2019, and the number of cited references in published articles increased from starting about 1000 references per year to 8982 references in 2018 and to 8465 references in 2019.

The journal exhibits a steady growth in two most important WoS-related journal parameters (Table 2 upper part): citable items in WoS (from ~30 to 174 items), and the total number of citations (over 4000 in 2018). The only troublesome year in this steady growth was 2009 (Vol. 6), when only 38 citable items and out of them only 21 research papers were published.

Looking at the WoS citations, *Landslides* received during this period 2004–2016 expressed by the Journal Impact Factor (JIF), the first IF was assigned in 2007. The fluctuations in IF and 5year IF observed in the period 2007–2018 can be partially ascribed to variations in the number of citable items (Fig. 1). The Pearson coefficients between IF and IF without Self Cites is 99.29% (2007–2018, n = 12), and between IF and 5-year IF is 95.20% (2009–2018, n = 10); both are very high and close to 100%.

Looking at IF without self-citations, from 2009 on the journal exhibits a steady increase (from 1.297 to 3.266). Looking at IF, from 2012 on the journal exhibits a steady increase (from 2.093 to 4.252). In this period the journal established itself as one of the leading journals in the SCI-Expanded category "Engineering, geological" (rank #1 since 2013), and as a member of Q1 journals in the SCI-Expanded category of "Geological Sciences, multidisciplinary" (top 20% since 2013).

The journal also exhibits a steady growth in two important Scopus-related journal parameters (Table 2 lower part): source documents in Scopus (from 37 in 2004 to 192 documents in 2018), and the total number of cites (as of March 11, 2020: 1980 citations to 37 documents published in 2004 and 1350 citations to 192 documents published in 2018).

Elsevier offers several journal metrics (https://journalmetrics.scopus.com/); looking at SNIP the journal achieved Q1 and was close to the top 10% in the Scopus category "Geotechnical Engineering and Engineering Geology". The same is true looking at Scimago Journal Rank (SJR) values in this period. The newly introduced Cite Score is even more in favor of the journal, ranking it into top 10% in this category since 2011, and ranking it #1 in 2013 and again in 2016. We compared WoS and Scopus journal metrics in Fig. 1.

Impact and Rankings of *Landslides*

Journal Impact Factor is determined using cited references in all journals screened by WoS. An increase in a journal's impact factor is a combined effect of a higher journal quality measured by its citation record due to a higher visibility, but also due to rising average number of references in all journals' articles in the database if this number overtakes the increasing number of published articles in journals.

Since the WoS database is continuously expanding its coverage (adding new journals) plus the journals covered already by the database are publishing more articles (with more references) that immediately give more citations to previously published articles, thus increasing journals' impact factors.

Therefore, we analyzed the average number of references per published article in three main article categories in *Landslides*: for Original papers (Fig. 2), Recent landslides (Fig. 3), and Technical notes (Fig. 4). We present the distribution of the number of references per article for all volumes of *Landslides* (left hand side on Figs. 2, 3 and 4), and then also the normalized number of references by dividing the total number of references by the total number of published articles of this category in each year.

In the category *Technical notes*, before 2009 (Vol. 6) there were not enough articles published in this category to present meaningful data. The total number of references per

Table 1 Overview of the published articles in *Landslides* in the period 2004–2019 for the first 16 volumes (Springer Nature 2020)—data collected early in January 2020

Article category	2004	2005	2006	2007	2008	2009	2010	2011	2012	2013	2014	2015	2016	2017	2018	2019	sum
Original paper	26	28	30	30	35	21	28	25	25	39	63	53	84	107	119	94*	807*
Review paper	–	–	–	–	–	1	–	–	–	–	2	2	2	1	2	1	11
Recent landslides	4	7	4	5	3	6	7	11	11	8	7	15	7	14	26	31	166
Technical note		1			1	5	4	7	5	11	8	14	16	23	20	37	152
IPL/WCoE activities	4	2	1	1	1	1	6	4	3	9	5	9	4	9	6	5	70
News/Kyoto commitment	–	–	–	–	–	–	–	–	–	–	–	–	–	3	9	12	24
Other items	3	3	6	1	2	6	6	2	3	1	1	5	8	5	17	14	83
Sum of published items	37	41	41	37	42	40	51	49	47	68	86	98	121	162	199	194	1313
Articles in web of science	34	39	41	37	40	38	45	46	44	64	86	97	120	161	199	194	1285
Articles in SCOPUS	37	39	41**	37	40	38	50	46	47	65	85	96	118	159	192	199	1289**
Landslides volume	1	2	3	4	5	6	7	8	9	10	11	12	13	14	15	16	
Number of issues	4	4	4	4	4	4	4	4	4	6	6	6	6	6	12	12	90
Number of pages published	305	352	358	381	440	335	485	534	544	819	1119	1203	1538	2108	2426	2419	15,386
Number of references used	680	988	849	2264	1176	925	1305	1438	1473	2378	4087	3825	6106	7091	8982	8465	52,032
Number of citations received in WoS	1645	1763	1094	1478	1578	1552	2053	1099	1512	1510	2634	1571	2040	1687	1092	270	24,578
Number of downloads in Springer	27,095	27,637	22,592	26,935	30,256	31,646	45,979	31,543	45,482	70,260	103,447	97,669	117,549	145,166	174,739	102,830	1,100,825

*including 5 thematic papers

**there is an error in SCOPUS database, in 2006 they are having 74 articles (i.e. preface and 32 papers from Evans et al. 2006), the total number of articles in SCOPUS is therefore 1322 articles

Table 2 Basic bibliometric parameters of *Landslides* in the period 2004—2019, using WoS (Clarivate Analytics 2020) and Scopus (Elsevier 2020) data

Journal metric	2007	2008	2009	2010	2011	2012	2013	2014	2015	2016	2017	2018
WoS total cites	155	231	460	461	535	760	1067	1310	1839	2388	3182	4187
WoS citable items	37	40	38	45	46	44	64	86	97	120	161	199
WoS cites used for IF	69	52	126	117	164	180	242	287	439	640	766	1131
IF	0.986	0.754	1.703	1.625	2.216	2.093	2.814	2.870	3.049	3.657	3.811	4.252
5 year IF			2.374	1.938	1.841	2.358	3.045	3.205	3.616	3.684	4.360	4.667
IF without self cites	0.728	0.637	1.297	1.472	1.702	1.790	2.313	2.430	2.548	3.120	3.049	3.266
Immediacy index	0.057	0.154	0.364	0.220	0.289	0.512	0.407	0.329	0.456	0.486	1.065	1.322
CiteScore					2.15	2.22	3.40	2.53	2.83	3.57	4.03	4.53
SJR	0.778	0.857	1.455	1.101	1.513	1.324	1.763	1.171	1.565	1.365	1.802	1.638
SNIP	1.066	0.966	1.269	1.414	1.942	1.859	1.873	1.596	1.789	2.047	2.259	1.928

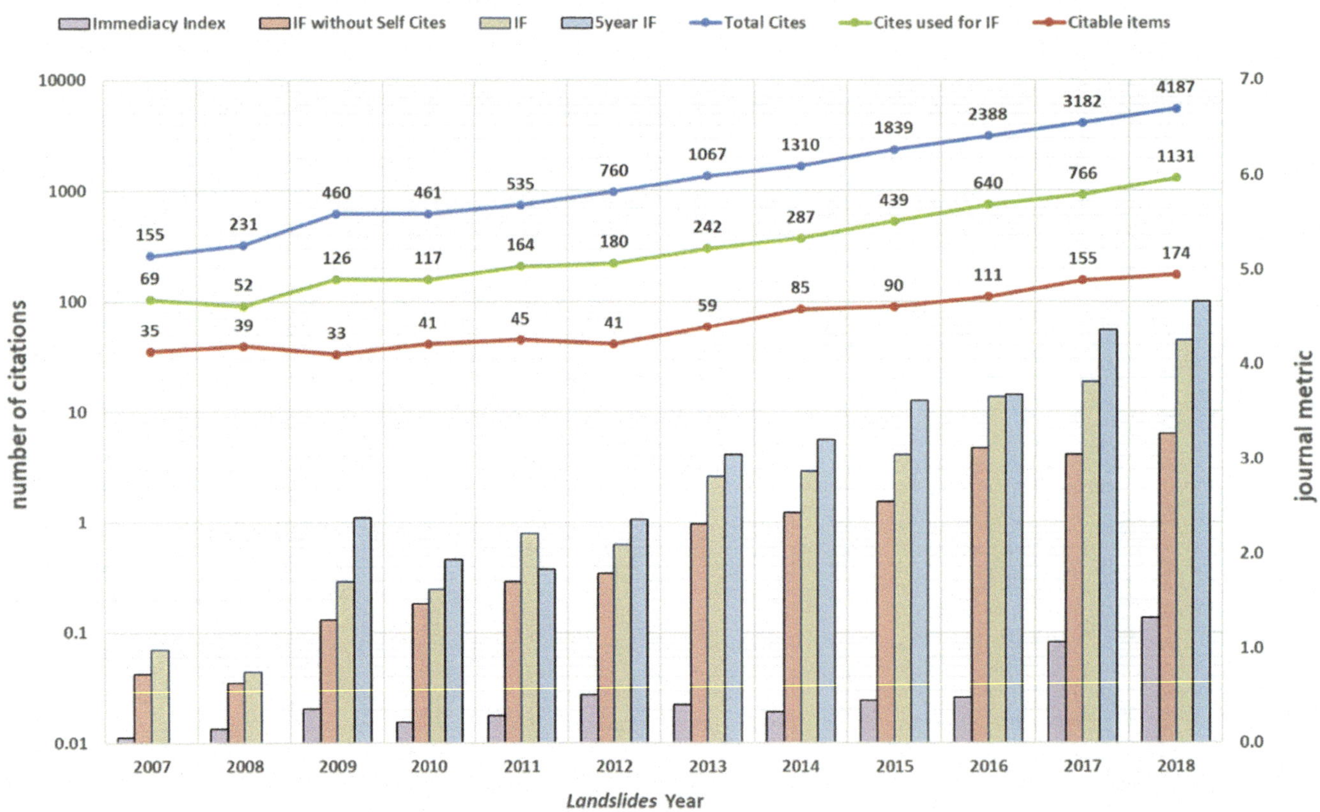

Fig. 1 Selected journal metrics for *Landslides* in the period 2007–2018

published article increases for all three article's categories. Considering the rising number of published articles that must receive citations in the next two years after the publication in order to maintain a journal's impact factor, the normalized number of references shows a decreasing trend in two categories: *Original papers* and *Technical notes*. Due to the fluctuating number of published articles in the category *Recent landslides*, there is no clear trend for this category.

Highly Cited Papers in *Landslides*

In *Landslides*, altogether there are 17 Highly Cited Papers (HCP) out of 1313 published papers in the period 2004–2019 (1.29%). The 10 most cited out of 17 HCP are given in Table 3.

The HCP are collected since 2009. Out of 17 HCP, 6 are open access papers (35.3%). Four HCPs are from 2019, three

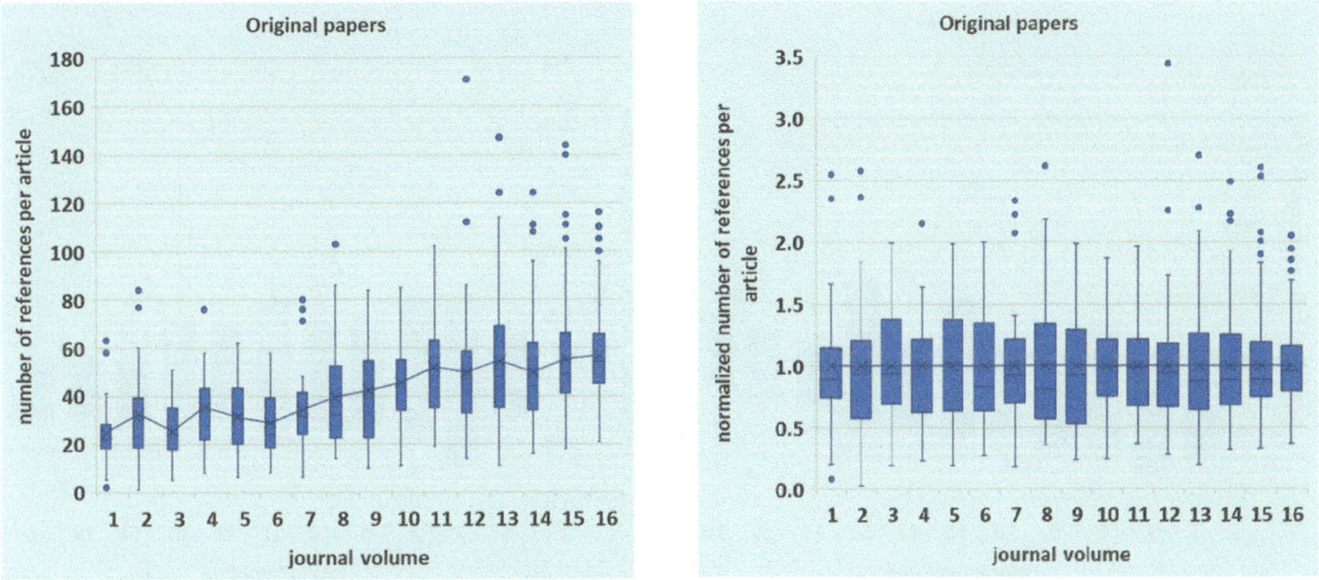

Fig. 2 Box plot of the number of references per article (left) and the normalized number of references per article by the total number of published articles (right) for 802 published *Original papers* in *Landslides*

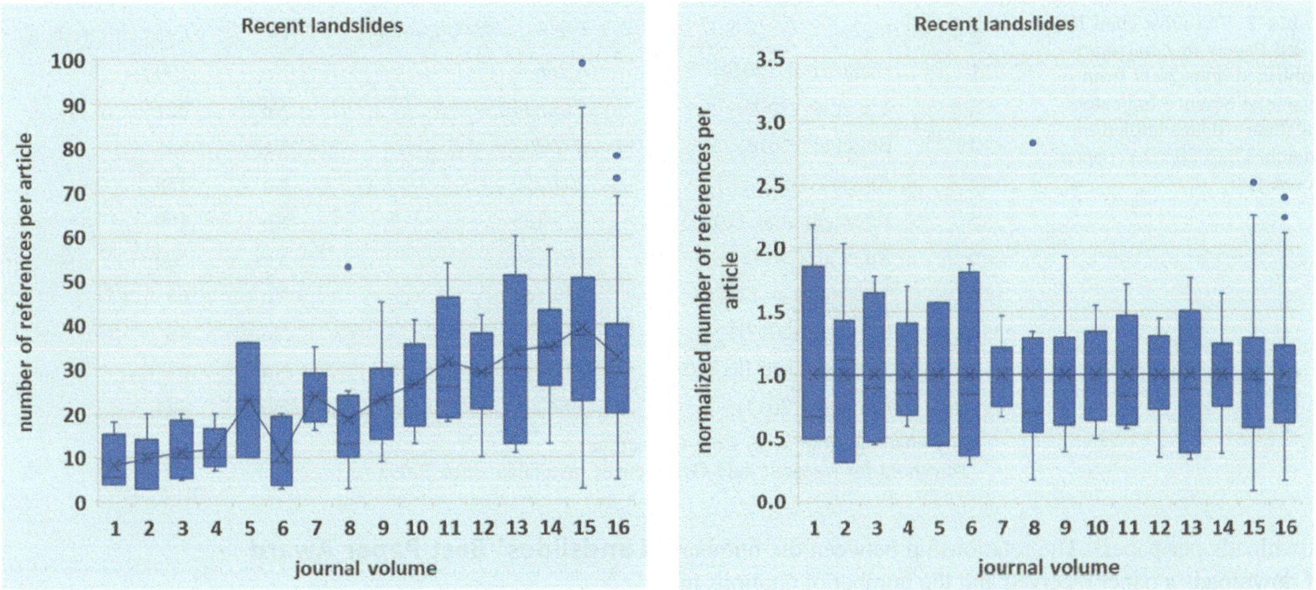

Fig. 3 Box plot of the number of references per article (left) and the normalized number of references per article by the total number of published articles (right) for 166 published *Recent landslides* in *Landslides*

HCPs are from 2018, 2016 and 2014, 2 HCPs are from 2017 and one HCP was published in 2012 and 2009. All papers in *Landslides* are put into research field of Geosciences, and citation thresholds for this research field for top 1% of papers published are also given in Table 3.

Most Downloaded Papers in *Landslides*

As of early January 2020, in Springer Link the number of downloads of 1313 published papers in *Landslides* in the period 2004–2019 exceeded 1.1 million (on average 838

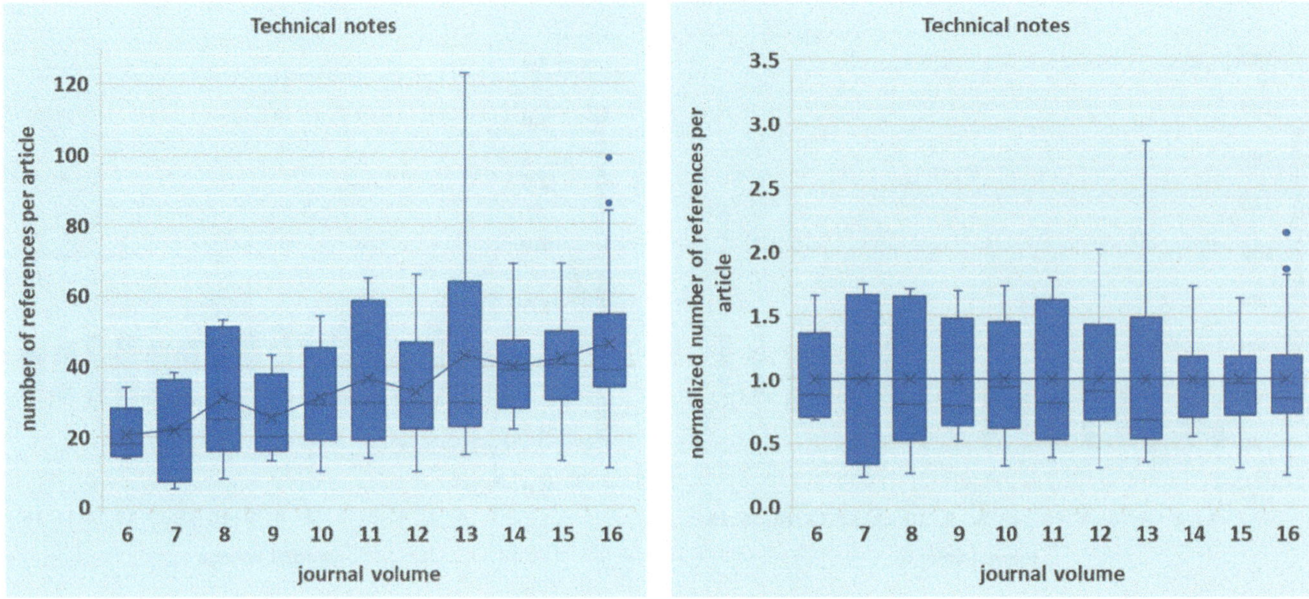

Fig. 4 Box plot of the number of references per article (left) and the normalized number of references per article by the total number of published articles (right) for 150 published *Technical notes* in *Landslides* in the period 2005–2019

Table 3 Ten most cited Highly Cited Papers in *Landslides* published after 2009 from Essential Science Indicators database—data sampled on March 27, 2020 (OA—open access)

#	Paper	Citations WoS*/Springer	OA	Citation threshold**
1	Hungr et al. (2014)	629/637	No	109
2	Yin et al. (2009)	313/334	No	221
3	Bui et al. (2016)	292/292	Yes	63
4	Akgun (2012)	204/219	No	158
5	Kavzoglu et al. (2014)	173/185	No	109
6	Xu et al. (2014)	165/170	Yes	109
7	Youssef et al. (2016)	139/164	No	63
8	Bui et al. (2017)	93/100	No	44
9	Tsangaros and Ilia (2016)	73/77	No	63
10	Fan et al. (2017)	59/62	No	444

*Only citations from WoS Core Collection
**Threshold for research field Geosciences (available since 2009)

downloads per paper). The relationship between the number of downloads a paper received and the number of citations in WoS of this paper is rather poor with a coefficient of linear correlation $R^2 = 0.3869$ (Fig. 5). The Pearson coefficient is 0.631, and when omitting papers without any citations the Pearson coefficient is comparable at 0.622.

The 10 most downloaded papers are given in Table 4. The numbers of downloads are well above the average of 838 downloads. The citations rate for these highly downloaded papers are quite different. Yin et al. (2009), Hungr et al. (2014), Xu et al. (2014) and Bui et al. (2016) are also Highly Cited Papers (see Table 4). The paper by Guzzetti et al. (2008) would certainly qualify, but HCPs can be only tracked since 2009.

Landslides' Best Paper Award

The Best Paper Award for the best paper published in Landslides has been given annually, beginning with the year 2004 for the first volume (Vol. 1) of the Journal. The selection of the Best Paper Award is carried out by the Best Paper Award Subcommittee. The judging and ranking of papers were based on a numerical grading system that involved three elements in the final score of the paper:

(i) Scientific and technical quality of the paper (up to 50%),
(ii) Impact on the profession and society (up to 35%), and
(iii) Quality of figures and tables (up to 15%).

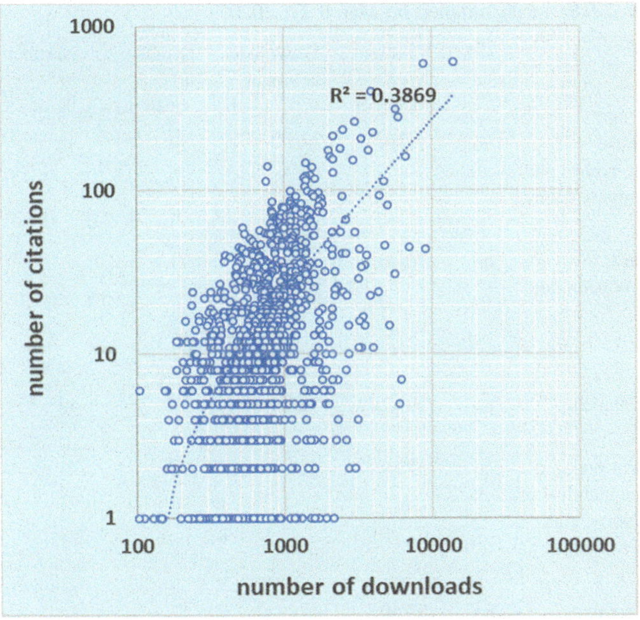

Fig. 5 Number of citations in Web of Science as a function of the number of downloads from Springer Nature web application Springer Link database—data sampled in early January 2020—only 1159 papers are shown with at least one citation

Table 4 Ten most downloaded papers in *Landslides* from Springer Link database—data sampled on March 27, 2020

#	paper	downloads	Open access	Citations WoS*/Springer
1	Hungr et al. (2014)	14,000	No	629/637
2	Ahmed (2015)	9190	Yes	41/49
3	Guzzetti et al. (2008)	8482	No	560/608
4	Intrieri et al. (2018)	7000	Yes	51/45
5	Xu et al. (2014)	6538	No	165/170
6	Tang et al. (2015)	6141	No	7/7
7	Carla et al. (2017)	6031	Yes	28/34
8	Liu et al. (2015)	6026	No	7/7
9	Bui et al. (2016)	5938	Yes	292/292
10	Yin et al. (2009)	5523	No	313/334

*only citations from WoS Core Collection

The proposal of Best Paper Award should be approved by the Board of Representatives of the International Consortium on Landslides. The Landslides Best Paper Awards from the Vol. 1 to Vol. 15 are listed in Table 5. None of these papers is an open-access paper, two papers are review papers (Hungr et al. 2014; Yavari-Ramshe and Ataie-Astiani 2016) and paper Hungr et al. (2014) is also Highly Cited Paper and at the top of the top 10 papers in Landslides regarding the number of downloads (Tables 3 and 4). The number of downloads is above the average of 838 downloads per paper in *Landslides*. The number of references is average (~44 per paper), but the number of figures is good with average of 18 ± 8 and median of 16.

Comparison Between SCI Journals: Landslides, Engineering Geology, Earth-Science Reviews, Geomorphology, and Bulletin of Engineering Geology and the Environment

Mikoš (2017) made a comparison among selected journals in the field of engineering geology and geotechnical engineering—we will follow this approach but will only compare *Landslides* to four SCI journals: *Engineering Geology (Eng Geol)*, *Earth-Science Reviews (ESR)*, *Geomorphology*, and *Bulletin of Engineering Geology and the Environment (Bull IAEG)*. The first three journals are highly estimated journals published by Elsevier and the fourth journal is the official journal of the International Association for Engineering Geology and the Environment (IAEG) that all cover topics from landslide research. We used selected journal bibliometric data from Web of Science, InCities and SCOPUS (Table 6).

The journals in comparison to *Landslides* are (much) older journals, two of them being published since mid-1960s—their h-index in WoS or SCOPUS database is above 100, with the only exception of *Bull IAEG* with h-index lower that of *Landslides*. Also Cited Half-life and Citing Half-Life for references in all journals under comparison is larger than that for *Landslides*.

Share of review articles in *Landslides* as well *Eng Geol* and *Geomorphology* is low (non-existent in *Bull IAEG*) when compared to *ESR* (i.e. the journal title says it), in which papers have very large average number of references per article (close to 200).

The share of open-access papers published in the period 2009–2019 in *Eng Geol* (6.6%) and in *Bull IAEG* (4.2%) is rather low if compared to *Geomorphology* (11.8%), *Landslides* (13.8%) and *ESR* (20.3%) that is reflected in a way also to the share (and number) of *Highly Cited Papers* in these journals, leading journal is definitely *ERS* with 8.27% HCP in the period 2009–2019, followed by *Bull IAEG* (1.74%) and *Landslides* (1.55%); below average of 1% are *Eng Geol* (0.86%) and *Geomorphology* (0.70%)—as *HCP*s are top 1% of published papers in a research field, *Eng Geol* and *Geomorphology* are well below average, *Landslides* and

Table 5 The *Landslides* Best Paper Awards from Vol. 1 to Vol. 15 (2004–2018)—data sampled on March 27, 2020

Year	Vol	No	Authors	Title	OA*	Citations WoS**/ Springer	Downloads Springer	Number of references	Number of figures
2004	1	1	Margottini C (2004)	Instability and geotechnical problems of the Buddha niches and surrounding cliff in Bamiyan Valley, central Afghanistan	no	14/18	397	24	16
2005	2	4	Baum RL, Coe JA, Godt JW, Harp EL, Reid ME, Savage WZ, Schulz WH, Brien DL, Chleborad AF, McKenna JP, Michael JA (2005)	Regional landslide-hazard assessment for Seattle, Washington, USA	no	76/76	1499	71	9
2006	3	2	Nadim F, Kjekstad O, Peduzzi P, Herold C, Jaedicke C (2006)	Global landslide and avalanche hotspots	no	197/209	3266	14	14
2007	4	4	Leynaud D, Sultan N, Mienert J (2007)	The role of sedimentation rate and permeability in the slope stability of the formerly glaciated Norwegian continental margin: the Storegga slide model	no	37/39	499	28	21
2008	5	4	Prochaska AB, Santi PM, Higgins JD, Cannon SH (2008)	A study of methods to estimate debris flow velocity	no	57/60	1179	62	16
2009	6	1	Lundström K, Larsson R, Dahlin T (2009)	Mapping of quick clay formations using geotechnical and geophysical methods	no	49/51	579	35	21
2010	7	3	Massey CI, Manville V, Hancox GH, Keys HJ, Lawrence C, McSaveney M (2010)	Out-burst flood (lahar) triggered by retrogressive landsliding, 18 March 2007 at Mt Ruapehu, New Zealand —a successful early warning	no	20/26	777	29	11
2011	8	2	Brideau MA, Pedrazzini A, Stead D, Froese C, Jaboyedoff M, van Zeyl D (2011)	Three-dimensional slope stability analysis of South Peak, Crowsnest Pass, Alberta, Canada	no	33/33	1085	46	25
2012	9	1	Pinyol NM, Alonso EE, Corominas J, Moya J (2012)	Canelles landslide: modelling rapid drawdown and fast potential sliding	no	44/44	858	17	33
2012	9	3	Sosio R, Crosta GB, Hungr O (2012)	Numerical modeling of debris avalanche propagation from collapse of volcanic edifices	no	38/38	818	55	12
2013	10	5	Staley DM, Kean JW, Cannon SH, Schmidt KM, Laber JL (2013)	Objective definition of rainfall intensity – duration thresholds for the initiation of post-fire debris flows in southern California	no	98/103	1471	62	11
2014	11	2	Hungr O, Leroueil S, Picarelli L (2014)	The Varnes classification of landslide types, an update	no	629/637	14,000	162	37
2015	12	5	Huang D, Cen D, Ma G, Huang R (2015)	Step-path failure of rock slopes with intermittent joints	no	34/35	1624	57	15
2016	13	6	Yavari-Ramshe S, Ataie-Astiani B (2016)	Numerical modelling of subaerial and submarine landslide-generated tsunami waves - recent advances and future challenges	no	44/44	2209	477	16
2017	14	3	Ruiz-Carulla R, Corominas J, Mavrouli O (2017)	A fractal fragmentation model for rockfalls	no	16/18	1109	53	17
2018	15	6	Frattini P, Crosta GB, Rossini M, Allievi J (2018)	Activity and kinematic behaviour of deep-seated landslides from PS-InSAR displacement rate measurements	N0	9/10	1484	64	12

*OA—open access

**only citations from WoS Core Collection

Table 6 A comparison between selected SCI journals:—data sampled on March 28, 2020

Indicator	Landslides	Engineering geology	Earth-science reviews	Geomorphology	Bulletin of engineering geology and the environment
Publisher	Springer Nature	Elsevier	Elsevier	Elsevier	Springer Nature
ISSN & eISSN	1612-510X 1612-5118	0013-7952 1872-6917	0012-8252 1872-6828	0169-555X 1872-695X	1435-9529 1435-9537
Published since	2004	1965	1966	1987	1970
Publication frequency	12 issues/year	12 issues/year	12 issues/year	24 issues/year	8 issues/year**
IF 2018	4.252	3.909	9.530	3.681	2.138
IF 2018 without self-cities (5 of IF 2018)	3.266 (76.8%)	3.156 (90.7%)	9.087 (95.4%)	3.295 (89.5%)	2.008 (93.9%)
IF 2017	3.811	3.100	7.491	3.308	1.825
IF 2016	3.657	2.569	7.051	2.985	1.901
IF 5 years	4.667	4.420	10.640	3.873	2.349
Immediacy Index	1.322	0.960	2.380	0.798	1.532
Cited Half-Life (years)	4.6	8.2	7.3	8.1	6.1
Citing Half-Life (years)	8.6	9.8	11.7	11.4	11.9
Articles in JCR 2018	172	297	142	329	126
Reviews in JCR 2018	2	6	63	12	-
Average references per article	49.2	44.3	186.3	70.2	41.3
Average references per review	139.0	72.8	193.3	144.4	-
Open access papers in WoS 2009–2019	151 out of 1,094 (13.8%)	161 out of 2,437 (6.6%)	282 out of 1,391 (20.3%)	490 out of 4,136 (11.8%)	58 out of 1,379 (4.2%)
open access HCP of total HCP—% HCP in WoS papers 2009–2019	6 out of 17—1.55% HCP in 1,094 papers	1 out of 21—0.86% HCP in 2,437 papers	31 out of 115—8.27% HCP in 1,391 papers	6 out of 29—0.70% HCP in 4,136 papers	1 out of 24—1.74% HCP in 1,379 papers
Citations of top 10 HCP (open access HCP in bold & italic)	629, 313, ***292***, 204, 173, ***165***, 139, 93, 73, 59	285, 152, 143, 131, 127, 113, 85, 83, 83, ***81***	1606, ***790***, ***711***, 664, ***553***, ***523***, ***508***, ***443***, 421, 421	1094, 341, 286, 250, 242, 238, 237, 222, 183, 177	***285***, 67, 61, 45, 40, 36, 36, 33, 24, 21,
WoS h-index (average citations per paper)	68 (18.6)	117 (22.59)	192 (63.79)	143 (30.2)	43 (8.29)
SCOPUS h-index (average citations per paper)	73 (20.76)	134 (26.64)	199 (65.70)	156 (33.51)	55 (10.85)
SCOPUS coverage	Since 2004-	Since 1965-	Since 1966-	Since 1987-	Since 1987-
Documents in SCOPUS: 2015–2019	86, 118, 159, 192, 199	268, 236, 249, 310, 304	137, 154, 147, 221, 251	476, 354, 389, 357, 351	93, 124, 120, 131, 508
CiteScore 2018	4.53	4.70	9.54	3.88	2.37
CiteScore rank (Category)	#6/176 (Geotechnical Engineering and Engineering Geology)	#4/176 (Geotechnical Engineering and Engineering Geology)	#3/181 (General Earth and Planetary Sciences)	#10/136 (Earth-Surface Processes)	#41/176 (Geotechnical Engineering and Engineering Geology)
CiteScore 2019*	5.47	5.62	10.81	4.12	2.97
SJR 2018	1.638	2.209	3.657	1.454	0.839
SNIP 2018	1.928	2.338	3.794	1.564	1.249

*as of April 9, 2020
**since 2019

Bull IAEG are well above average, and *ESR* excels in this regard.

These relations are to a certain extent influenced by the share of the published open-access papers, where this is visible when looking at the top *HCP*s and their citations—many such *HCP* in *ESR* are open-access papers, and much less in other journals. Also, journal bibliometrics from SCOPUS supports the excellent position of *ESR* when compared to the other journals, especially through their policy of publishing a lot of review papers (close to one third in 2018), and also other articles in *ESR* are citing a lot of references (close to 200 on average), and are therefore useful as state-of-the-art papers in specific fields of knowledge. A way forward for *Landslides* would be to publish more review papers, to popularize the policy of publishing open access, and to lower a bit the journal self-citation rate that is for *Landslides* well above 20%—still not a problem, but should be kept in mind.

International Cooperation as seen Through Multi-authorship of Published Articles in *Landslides*

Sassa et al. (2009) analyzed for the first 5-year period (2004–2009) of *Landslides* the number of individual authors from each country, counting them only once, regardless of the number of published papers, and each coauthor other than the first/corresponding author was counted as one. The number of authors found to be the greatest from Italy, then, China, Japan, the United States, Canada, India and Spain are following.

We looked at the period of 16 years (2004–2019) and analyzed the distribution of the number of authors per *Landslide* article category. The median number of authors for 1313 articles in the database is 4 (20.2% of all papers in this category). The highest median number of authors (5) is for the category *Recent landslides* (22.3% of all papers in this category), followed by 4 authors for the category *Original papers* (23.9% of all papers in this category), 3 authors for the category *Technical note* (18.7% of all articles in this category), 2 authors for the category News/KC2020 (27.3% of all papers in this category), and 1 author for the category *ICL/IPL Activities* (37.1% of all papers in this category). Differences between categories are not very apparent except for the category *ICL/IPL Activities* (Fig. 6), where close to 50% of the articles are single-author contributions.

Apart from the *News/KC* paper by Sassa et al. (2019b) on invited and accepted speakers of the Fifth World Landslide Forum in Kyoto, 2o2o, the multi-authorship "winning" articles are *Technical notes*: (i) about landslide databases in the Geological Surveys in Europe by Herrera et al. (2018) with 41 authors from 25 countries, (ii) about fatal landslides in Europe by Haque et al. (2016) with 22 authors coming from 18 different countries; and (iii) on impact of landslides on transportation routes during the 2016–2017 Central Italy seismic sequence by Martino et al. (2019) by 21 authors from two countries.

Furthermore, we looked at the period of 16 years (2004–2019) and have analyzed the distribution of the number of countries from which co-authors are coming, per *Landslide* article category. The majority of 1313 articles are published by author(s) coming from the same country (over 60% for all article categories in *Landslides*), the leading category is the Technical note category with 69.3% (Fig. 7).

Out of the total of 1313 articles in *Landslides*, 1305 articles has a corresponding author with a known country or territory, they are coming from 57 different entities, the order is (i.e. countries/territories with more than 10 records): People's Republic of China (287—22.0%), Italy (202—15.5%), Japan (146—11.2%), USA (68—5.2%), Canada (59

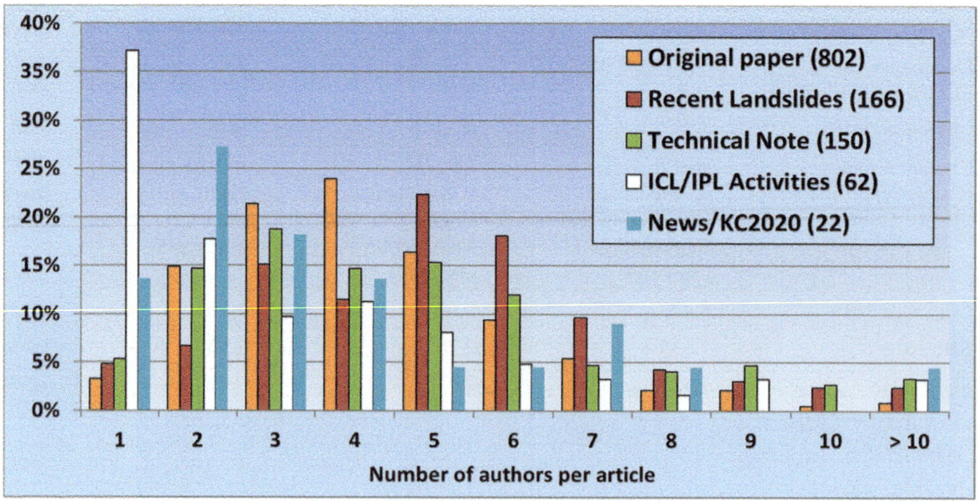

Fig. 6 Relative distribution of the number of authors given for 5 major article categories in *Landslides* (the number of articles per category is given in brackets)

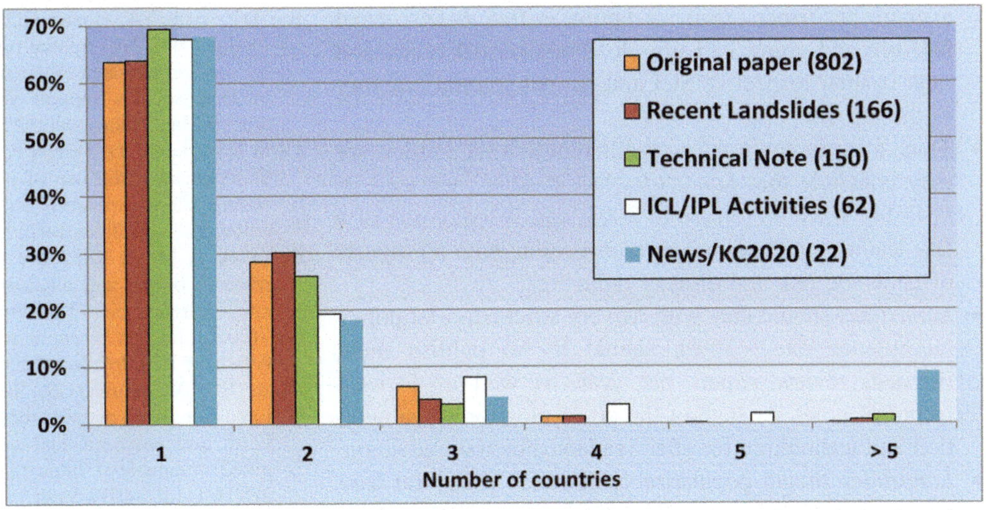

Fig. 7 Relative distribution of the number of countries of all contributing authors given for 5 major article categories in *Landslides* (the number of articles per category is given in brackets)

—4.5%), Spain (52), Switzerland (39), France (34), UK (30), Czech Republic (29), Taiwan (27), India (23), New Zealand (23), Slovenia (23), Hong Kong (20), Germany (18), Norway (16), Netherlands (10), South Korea (16), Austria (15), Australia (14), Iran (12), Mexico (11), Belgium (10), Greece (10), and Turkey (10).

Similar results are obtained from SCOPUS database with 1322 articles (1289 articles, reduced for papers in Evans et al. 2006) in *Landslides*, if all authors of all papers are considered, not only corresponding ones, the order is (minus in brackets are papers to be subtracted since published in the book by Evans et al. 2006): People's Republic of China (329), Italy (258–7), Japan (181), USA (120–4), Canada (91–9), Switzerland (82–2), UK (81–6), Spain (71–1), France (54–2), Hong Kong (49), Czech Republic (42), New Zealand (37–1), Taiwan (36), Germany (34–1), Australia (32), Norway (32–1), India (28), South Korea (27), Netherlands (26–1), Slovenia (25), Austria (22).

Sassa (2019b) presented an overview of the ICL members (full, associates, supporters)—ICL has 99 members from 37 countries/regions, mainly from Japan (17), Italy (14), and China (10), followed by South Korea (5), Czech Republic (4), Indonesia (4), Russian Federation (4), and Slovenia (4).

Looking at the authors' institution, the top 20 institutions with more than 20 published articles in *Landslides* in the period 2004–2019 published jointly 535 papers (more than one third of all published papers) and these institutions are: 6 from China (CAS—Chinese Academy of Sciences 95 papers, Chengdu University of Technology 65 papers, Institute of Mountain Hazards Environment of CAS 54 papers, China University of Geosciences 48 papers, University of CAS 35 papers, China Geological Survey 25 papers), 4 from Italy (CNR 53 papers, University of Florence 50 papers, IRPI CNR 31 papers, University of Salerno 22 papers), 2 from Czech Republic (Charles University Prague 24 papers, Czech Academy of Sciences 22 papers), Japan (Kyoto University 81 papers, ICL 27 papers) and USA (US Department of Interior 25 papers, USGS 25 papers), and 1 from France (CNRS 34 papers), Hong Kong (Hong Kong University of Science and Technology 37 papers), Slovenia (University of Ljubljana 20 papers), and Spain (Polytechnic University of Catalunya 22 papers). These institutions also contributed 8 Highly Cited Papers out of 17 published in *Landslides* so far. Unquestionably, the leading countries with contributions to *Landslides* are China and Italy that are also strongly present as members in the International Consortium on Landslides, Kyoto, Japan.

Conclusions

We analyzed all articles published in the first 16 years of the SCI journal *Landslides* using Clarivate Analytics' WoS, InCites, Journal Citation Reports and Essential Science Indicators, and Elsevier's SCOPUS databases, we can draw the following conclusions:

- *Landslides* was founded in 2004, got his first impact factor in 2007, and soon rose to Q1 among journals in several categories: SCI-Expanded "Engineering, Geological", SCI-Expanded "Geosciences, Multidisciplinary", and Scopus "Geological Engineering and Engineering Geology". In 2019, it is rank #4 in the last-mentioned SCOPUS category.
- *Landslides* is on a good way to compete with two well established journals that publish in the field of landslide research, namely *Engineering Geology*, *Bulletin of Engineering Geology and the Environment*, *Geomorphology* and *Earth-Science Reviews*. The other journals are several decades older than *Landslides* and therefore exhibit mature values in WoS and Scopus h-index over 100 (exception is *Bull IAEG*), well over 20 citations on

average per paper (again exception is *Bull IAEG*), cited half-life and citing half-life close or over 10 years, and rank in their respective SCI and Scopus journal category in the top 10.

- Thus, after an analysis using different journal metrics, we may conclude that *Landslides* has steadily increased its visibility since its release in 2004, and is today one of a few leading international journals dealing with all aspects of landslide risk and disaster reduction.
- *Landslides* should stay with its very strict editorial policy (acceptance rate ~36%), should try to publish more (invited) review papers per year, as well as focused technical notes on state-of-the art in monitoring, simulation and technology for effective landslide risk reduction.
- *Landslides* should popularize open-access policy (at least among ICL members) to be able to raise the share of Highly Cited Papers well above 1% of all published papers.

The ICL journal *Landslides* is explicitly mentioned in Action 10 of the Kyoto 2020 Commitment for Global Promotion of Understanding and Reducing Landslide Disaster Risk (Sassa 2019a) and is one of the most important pillars of the international activities of the International Consortium on Landslides and of its dedication to landslide disaster risk reduction in the world.

Acknowledgements This bibliometric analysis of *Landslides* was performed using databases that are available through the consortium to the University of Ljubljana staff and students. The analysis was partially financed by the Slovenian Research Agency (ARRS) and their financial support is greatly acknowledged—through the national Research Programme P2-0180 "Water Science and Technology, and Geotechnics".

References

Ahmed B (2015) Landslide susceptibility mapping using multi-criteria evaluation techniques in Chittagong metropolitan area, Bangladesh. Landslides 12(6):1077–1095

Akgun A (2012) A comparison of landslide susceptibility maps produced by logistic regression, multi-criteria decision, and likelihood ratio methods: a case study at Izmir, Turkey. Landslides 9 (1):93–106

Baum RL, Coe JA, Godt JW, Harp EL, Reid ME, Savage WZ, Schulz WH, Brien DL, Chleborad AF, McKenna JP, Michael JA (2005) Regional landslide-hazard assessment for Seattle, Washington, USA. Landslides 2(4):266–279

Brideau MA, Pedrazzini A, Stead D, Froese C, Jaboyedoff M, van Zeyl D (2011) Three-dimensional slope stability analysis of South Peak, Crowsnest Pass, Alberta, Canada. Landslides 8(2):139–158

Bui DT, Tuan TA, Klempe H, Pradhan B, Revhaug I (2016) Spatial prediction models for shallow landslide hazards: a comparative assessment of the efficacy of support vector machines, artificial neural networks, kernel logistic regression, and logistic model tree. Landslides 13(2):361–378

Bui DT, Tuan TA, Hoang ND, Thanh NQ, Nguyen DB, Liem NV, Pradhan B (2017) Spatial prediction of rainfall-induced landslides for the Lao Cai area (Vietnam) using a hybrid intelligent approach of least squares support vector machines inference model and artificial bee colony optimization. Landslides 14(2):447–458

Carlà T, Intrieri E, Di Traglia F, Nolesini T, Gigli G, Casagli N (2017) Guidelines on the use of inverse velocity method as a tool for setting alarm thresholds and forecasting landslides and structure collapses. Landslides 14(2):517–534

Cerovšek T, Mikoš M (2014) A comparative study of cross-domain research output and citations: Research impact cubes and binary citation frequencies. J Inform 8:147–161

ESI (2020) InCites essential science indicators. Clarivate analytics. Available at: https://esi.clarivate.com/IndicatorsAction.action

Evans SG, Mugnozza GS, Strom A, Hermanns RL (eds) (2006) Landslides from massive rock slope failure. Nato science series, IV. Earth and environmental sciences, vol 49. Springer, Dordrecht, p 662. Available at: https://link.springer.com/content/pdf/10.1007%2F978-1-4020-4037-5.pdf

Fan XM, Xu Q, Scaringi G, Dai LX, Li WL, Dong XJ, Zhu X, Pei XJ, Dai KR, Havenith HB (2017) Failure mechanism and kinematics of the deadly June 24th 2017 Xinmo landslide, Maoxian, Sichuan, China. Landslides 14(6):2129–2146

Frattini P, Crosta GB, Rossini M, Allievi J (2018) Activity and kinematic behaviour of deep-seated landslides from PS-InSAR displacement rate measurements. Landslides 15(6):1053–1070

Garfield E (1955) Citation indexes to science: a new dimension in documentation through association of ideas. Science 122 (3159):108–111

Garfield E (2006) The history and meaning of the journal impact factor. JAMA 295(1):90–93

Guzzetti F, Peruccacci S, Rossi M, Stark CP (2008) The rainfall intensity–duration control of shallow landslides and debris flows: an update. Landslides 5(1):3–17

Haque U, Blum P, da Silva PF, Andersen P, Pilz J, Chalov SR, Malet J-P, Jemec Auflič M, Andres N, Poyiadji E, Lamas PC, Zhang W, Peshevski I, Pétursson HG, Kurt T, Dobrev N, García-Davalillo JC, Halkia M, Ferri S, Gaprindashvili G, Engström J, Keellings D (2016) Fatal landslides in Europe. Landslides 13(6):1545–1554

Herrera G, Mateos RM, García-Davalillo JC, Grandjean G, Poyiadji E, Maftei R, Filipciuc T-C, Jemec Auflič M, Jež J, Podolszki L, Trigila A, Iadanza C, Raetzo H, Kociu A, Przyłucka M, Kułak M, Sheehy M, Pellicer XM, McKeown C, Ryan G, Kopačková V, Frei M, Kuhn D, Hermanns RL, Koulermou N, Smith CA, Engdahl M, Buxó P, Gonzalez M, Dashwood C, Reeves H, Cigna F, Liščák P, Pauditš P, Mikulėnas V, Demir V, Raha M, Quental L, Sandić C, Fusi B, Jensen OA (2018) Landslide databases in the geological surveys of Europe. Landslides 15(2):359–379

Huang D, Cen D, Ma G, Huang R (2015) Step-path failure of rock slopes with intermittent joints. Landslides 12(5):911–926

Hungr O, Leroueil S, Picarelli L (2014) The Varnes classification of landslide types, an update. Landslides 11(2):167–194

Intrieri E, Raspini F, Fumagalli A, Lu P, Del Conte S, Farina P, Allievi J, Ferretti A, Casagli N (2018) The Maoxian landslide as seen from space: detecting precursors of failure with Sentinel-1 data. Landslides 15(1):123–133

Kavzoglu T, Sahin EK, Colkesen I (2014) Landslide susceptibility mapping using GIS-based multi-criteria decision analysis, support vector machines, and logistic regression. Landslides 11(3):425–439

Lando T, Bertoli-Barsotti L (2014) A new bibliometric index based on the shape of the citation distribution. PLoS One 9(12):e115962. Available at: https://doi.org/10.1371/journal.pone.0115962

Leynaud D, Sultan N, Mienert J (2007) The role of sedimentation rate and permeability in the slope stability of the formerly glaciated

Norwegian continental margin: the Storegga slide model. Landslides 4(4):297–309
Liu J, You Y, Chen X, Chen X (2015) Mitigation planning based on the prediction of river blocking by a typical large-scale debris flow in the Wenchuan earthquake area. Landslides 13(5):1231–1242
Lundström K, Larsson R, Dahlin T (2009) Mapping of quick clay formations using geotechnical and geophysical methods. Landslides 6(1):1–15
Margottini C (2004) Instability and geotechnical problems of the Buddha niches and surrounding cliff in Bamiyan Valley, central Afghanistan. Landslides 1(1):41–51
Martino S, Bozzano F, Caporossi P, D'Angiò D, Della Seta M, Esposito C, Fantini A, Fiorucci M, Giannini LM, Iannucci R, Marmoni GM, Mazzanti P, Missori C, Moretto S, Piacentini D, Rivellino S, Romeo RW, Sarandrea P, Schilirò L, Troiani F, Varone C (2019) Impact of landslides on transportation routes during the 2016–2017 Central Italy seismic sequence. Landslides 16 (6):1221–1241
Massey C, Manville V, Hancox GH, Keys HJ, Lawrence C, McSaveney M (2010) Out-burst flood (lahar) triggered by retrogressive landsliding, 18 March 2007 at Mt Ruapehu, New Zealand —a successful early warning. Landslides 7(3):303–315
Mikoš M (2011) Landslides: a state-of-the art on the current position in the landslide research community. Landslides 8:541–551
Mikoš M (2017) Landslides: a top international journal in geological engineering and engineering geology? Landslides 14(5):1843–1854
Nadim F, Kjekstad O, Peduzzi P, Herold C, Jaedicke C (2006) Global landslide and avalanche hotspots. Landslides 3(2):159–173
Pinyol NM, Alonso EE, Corominas J, Moya J (2012) Canelles landslide: modelling rapid drawdown and fast potential sliding. Landslides 9(1):33–51
Prochaska AB, Santi PM, Higgins JD, Cannon SH (2008) A study of methods to estimate debris flow velocity. Landslides 5(4):413–444
Ruiz-Carulla R, Corominas J, Mavrouli O (2017) A fractal fragmentation model for rockfalls. Landslides 14(3):875–889
Sassa K (2019a) The fifth world landslide forum and the final draft of the Kyoto 2020 Commitment. Landslides 16(2):201–221
Sassa K (2019b) Journal landslides, the international consortium on landslides, and the Kyoto landslide commitment 2020. Landslides 16(9):1623–1628
Sassa K, Tsuchiya S, Ugai K, Wakai A, Uchimura T (2009) Landslides: a review of achievements in the first 5 years (2004–2009). Landslides 6:275–286
Sassa K, Tsuchiya S, Fukuoka H, Mikoš M, Doan L (2015) Landslides: review of achievements in the second 5-year period (2009–2013). Landslides 12(2):213–223
Sassa K, Arbanas Ž (2017) Landslides: journal of the international consortium on landslides. In: Sassa K, Mikoš M, Yin Y (eds) Advancing culture of living with landslides. WLF 2017. Springer, Cham, pp 257–267. Avalaible at: https://link.springer.com/chapter/10.1007/978-3-319-59469-9_21
Sassa K, Dang K, Guzzetti F, Casagli N, Tiwari B, Mikoš M, Vilimek V, Bobrowsky P, Konagai K, Arbanas Ž, Mihalić Arbanas S, Lu P, Sasahara K, Alcantara-Ayala I, Strom A, Hendry M, Yamagishi H, Tofani V, Cuomo S, Fathani F, Klimeš J, Wang F, Reichenbach P, Gokceoglu C, Higaki D, Koyama T (2019) Invited and accepted speakers of the fifth world landslide forum in Kyoto, 2020. Landslides 16(2):431–446
Sosio R, Crosta GB, Hungr O (2012) Numerical modeling of debris avalanche propagation from collapse of volcanic edifices. Landslides 9(3):315–334
Springer (2020) Landslides: journal of the international consortium on landslides—home page. https://www.springer.com/earth+sciences+and+geography/natural+hazards/journal/10346
Staley DM, Kean JW, Cannon SH, Schmidt KM, Laber JL (2013) Objective definition of rainfall intensity—duration thresholds for the initiation of post-fire debris flows in southern California. Landslides 10(5):547–562
Tang Y, Zhang Z, Wang C, Zhang H, Meng FW (2015) Characterization of the giant landslide at Wenjiagou by the insar technique using TSX-TDX CoSSC data. Landslides 12(5):1015–1021
Tsangaratos P, Ilia I (2016) Landslide susceptibility mapping using a modified decision tree classifier in the Xanthi Perfection, Greece. Landslides 13(2):305–320
Van Noorden R (2016) Impact factor gets heavyweight rival—CiteScore uses larger database and gets different results. Nature 540:325–326
Xu C, Xu XW, Yao X, Dai FC (2014) Three (nearly) complete inventories of landslides triggered by the May 12, 2008 Wenchuan Mw 7.9 earthquake of China and their spatial distribution statistical analysis. Landslides, 11(3):441–461
Yavari-Ramshe S, Ataie-Astiani B (2016) Numerical modelling of subaerial and submarine landslide-generated tsunami waves—recent advances and future challenges. Landslides 13(6):1325–1368
Yin YP, Wang FW, Sun P (2009) Landslide hazards triggered by the 2008 Wenchuan earthquake, Sichuan, China. Landslides 6(2):139–152
Youssef AM, Pourghasemi HR, Pourtaghi ZS, Al-Katheeri MM (2016) Landslide susceptibility mapping using random forest, boosted regression tree, classification and regression tree, and general linear models and comparison of their performance at Wadi Tayyah Basin, Asir Region, Saudi Arabia. Landslides 13(5):839–856

UNITWIN-UNESCO/KU/ICL Landslide, Earthquake and Water-related Disaster Risk Management for Society and the Environment Cooperation Programme

2020 Kyoto Japan

Kyoji Sassa, Ryosuke Uzuoka, and Kaoru Takara

Abstract

UNITWIN is the abbreviation for the university twinning and networking scheme. This UNESCO programme was established in 1992. During ICL foundation meeting in January 2002, participants from UNESCO advised to link the planned International Programme on Landslides (IPL) to one of UNESCO Programme for the promotion and the authorization. Then, ICL applied for UNITWIN programme to UNESCO soon after the foundation of ICL in 2002. UNITWIN-UNESCO/KU/ICL Landslides Mitigation for Society and the Environment Cooperation Programme was established in 2003 at Kyoto University, Kyoto, Japan. In 2010, the UNITWIN-UNESCO/KU/ICL Cooperation Programme was extended to "Landslide and Water-Related Disaster Risk Management" to deal with rainfall-induced landslides on slopes, as well as flood, sediment and debris flows in river systems. In 2019 the UNITWIN-UNESCO/KU/ICL UNITWIN Cooperation Programme was further extended to Landslide, Earthquake and Water-related Disaster Risk Management for Society and the Environment Cooperation Programme. This paper describes its progress and the activities of capacity development including the list of students and post-doctoral researchers within this programme.

Keywords

UNESCO • UNITWIN • International Programme on Landsides (IPL) • International Consortium on Landslides (ICL) • Kyoto university

Introduction

During the International Decade for Natural Disaster Reduction (IDNDR) (1990–1999), the Disaster Prevention Research Institute (DPRI) of Kyoto University conducted a research project on the Lishan Landslide in Xi'an, China, to assess the landslide hazard and protect a cultural heritage site, the Lishan Resort Palace from the Tan Dynasty (618–907), located at the foot of Lishan Slope. No landslide has occurred in this slope since the Tan Dynasty, but landslide experts in Japan and China indicated the necessity of detailed monitoring and testing, by mobilizing the most advanced technologies and establishing new landslide hazard assessment technology to protect this cultural heritage. This Japan–China joint research attracted the expert of UNESCO-IUGS Joint Programme IGCP (International Geological Correlation Programme). This project was invited to apply for IGCP project and approved as IGCP-425: "Landslide Hazard Assessment and Cultural Heritage" (1998–2004).

As a part of the activities of IGCP-425, the DPRI, Kyoto University (KU), organized an UNESCO-Kyoto University Joint Symposium in Kyoto, Japan in January 2002, which included participants from UNESCO, UNISDR, WMO, and Japanese ministries such as MOFA (Ministry of Foreign Affairs) and MEXT (Ministry of Education, Culture, Sports, Science and Technology). At this occasion, the International Consortium on Landslides was officially established. 6 participants from UNESCO proposed the establishment of a UNITWIN-UNESCO/KU/ICL Cooperation Programme Network to support the International Programme on Landslides (IPL) (Fig. 1). This manuscript report the development

K. Sassa (✉)
International Consortium on Landslides, Kyoto, Japan
e-mail: sassa@iclhq.org

R. Uzuoka
Disaster Prevention Research Institute, Kyoto University, Kyoto, Japan
e-mail: uzuoka.ryosuke.6z@kyoto-u.ac.jp

K. Takara
Graduate School of Advanced Integrated Studies in Human Survivability, Kyoto University, Kyoto, Japan
e-mail: takara.kaoru.7v@kyoto-u.ac.jp

K. Sassa et al. (eds.), *Understanding and Reducing Landslide Disaster Risk*, ICL Contribution to Landslide Disaster Risk Reduction,
https://doi.org/10.1007/978-3-030-60196-6_10

United Nations Educational, Scientific and Cultural Organization

Landslide and Water-related Disaster Risk Management

Fig. 1 Logo of UNITWIN-UNESCO/KU/ICL cooperation programme

of the UNITWIN-UNESCO/KU/ICL Network focusing the education aspect of doctoral students through the networks.

Products of UNITWIN-UNESCO/KU/ICL Cooperation Programme

The programme submits the activity report to UNESCO every 2 years. The last report submitted on November 2018 describes activities conducted between 2016–2018 (72 pages) including the list of publication, the list of Mater's degree, the list of Doctor's degree, the list of conference presentation, the list of seminar, the list of projects for two years in this network.

The number of ongoing IPL projects are 68 (refer the International Programme on Landslides in this book). The number of full color book publication at triennial World Landslide Forum to disseminate the latest advance of landslide research and technology is 22 including six books to be published in October 2020 (refer Preface in this book).

Table 1 presents the list of doctoral students trained by this network. Total 97 Ph.D. degree and 242 Mater's degree were awarded from 2013 to 2020.4.

One of the main teaching facilities in this network is the Dynamic-Loading Ring Shear Apparatus developed to study landslide dynamics and assessment of landslide initiation and motion. This apparatus is installed in UNITWIN laboratory in Kyoto University. And students invited from foreign countries to Kyoto University in this network published papers in Landslides or other journal. The those research papers are shown below.

Recently published papers in Kyoto University of the UNITWIN-UNESCO/KU/ICL Network

1. Do MD, Dang K et al. (2020) Analysis and modeling of a landslide-induced tsunami-like wave across the Truong river in Quang Nam province, Vietnam. Landslides, online published.
2. Nguyen DH, Sayama T, Sassa K, Takara K, Uzuoka R, Dang K, Pham VT (2020) A Coupled Hydrological—Geotechnical Framework for Forecasting Shallow Landslide Hazard—A Case Study in Halong City, Vietnam. Landslides, online published.
3. Tan Q, Sasa K, Dang K et al. (2020) Landslide hazard assessment around the 2016 Aranayake landslide, Sri Lanka based on soil testing and computer simulation. Landslides, online published.
4. Dang K, Sassa K, Konagai K, Karunawardena A, R. M. S. Bandara, Hirota K, Tan Q, Nguyen DH (2019) Recent rainfall-induced rapid and long-traveling landslide on 17 May 2016 in Aranayaka, Kagelle District, Sri Lanka. Landslides, Volume 16, Issue 1, pp 155–164.
5. Tien PV, Sassa K, Takara K, Fukuoka H, Dang K, Shibasaki T, Nguyen DH, Setiawan H, Doan HL (2018) Formation process of two massive dams following rainfall-induced deep-seated rapid landslide failures in the Kii Peninsula of Japan. Landslides, Volume 15, Issue 9, pp 1761–1778.
6. Lam HQ, Doan HL, Sassa K, Takara K, Ochiai H, Dang K, Abe S, Asano S, Do NH (2018) Susceptibility assessment of the precursor stage of a landslide threatening Haivan Railway Station, Vietnam. Landslides, Vol. 15 (2): 309–325.
7. Doan HL, Lam HQ, Sassa K, Takara K, Dang K, Nguyen KT, Pham VT (2017) The 28 July 2015 rapid landslide at Ha Long city, Quang Ninh, Vietnam. Landslides, Vol. 14 (3): 1207–1215.
8. Dang K, Sassa K, Fukuoka H et al. (2016) Mechanism of two rapid and long-runout landslides in the 16 April 2016 Kumamoto earthquake using a ring-shear apparatus and computer simulation (LS-RAPID). Landslides, Vol. 13: 1525–1534.
9. Sassa K, Dang K, Yanagisawa H et al. (2016) A new landslide-induced tsunami simulation model and its application to the 1792 Unzen-Mayuyama landslide-and-tsunami disaster. Landslides 13(6): 1405–1419.
10. Dang K, Doan HL, Sassa K, Do MD, Nguyen DH (2020) Hazard assessment of a rainfall-induced deep-seated landslide in Hakha city, Myanmar. Understanding and Reducing Landslide Disaster Risk. Springer, Vol. 4 Testing, Modeling and Risk Assessment. Accepted
11. Doan HL, Sassa K, Dang K, Miyagi T (2020) Simulation of Tsunami waves induced by coastal and submarine landslides in Japan. Understanding and Reducing Landslide Disaster Risk. Springer, Vol. 1 Sendai Landslide Partnerships and Kyoto Landslide Commitment. Accepted.

12. Doan HL, Sassa K, Dang K, Le HL (2020) Landslide hazard zoning based on the integrated simulation model (LS-Rapid). Understanding and Reducing Landslide Disaster Risk. Springer, Vol. 4 Testing, Modeling and Risk Assessment. Accepted.
13. Nguyen DH, Le QH, Sayama T, Sassa K, Takara K, Dang K (2020) An Integrated WebGIS System for Shallow Landslide Hazard Early Warning. Understanding and Reducing Landslide Disaster Risk. Springer, Vol. 2 Hazard and vulnerability mapping and zonation. Accepted
14. Sassa K, Dang K (2018) TXT-tool 0.081-1.1: Landslide Dynamics for Risk Assessment. Landslide Dynamics: ISDR-ICL Landslide interactive Teaching Tools. Springer, Vol.1 Fundamental, Mapping and Monitoring: pp 1–79
15. Lam HQ (2018) Video-tool 3.084-2.1: Manual for Undrained Dynamic-Loading Ring Shear Apparatus. Landslide Dynamics: ISDR-ICL Landslide interactive Teaching Tools. Springer
16. Sassa K, Setiawan H, He B, Gradiški K, Dang K (2018) TXT-tool 3.081-1.5: Manual for the LS-RAPID Software. Landslide Dynamics: ISDR-ICL Landslide interactive Teaching Tools. Springer, Vol.2 Testing, Risk Management and Country Practice: 191–224
17. Setiawan H, Sassa K, Dang K, Ostric M, Takara K, Vivoda M (2018) TXT-tool 3.081-1.6: Manual for the Undrained Dynamic-Loading Ring-Shear Apparatus. Landslide Dynamics: ISDR-ICL Landslide interactive Teaching Tools. Springer, Vol.2 Testing, Risk Management and Country Practice: 321–350
18. Setiawan H, Sassa K, Takara K, Miyagi T, Fukuoka H (2016) Initial Pore Pressure Ratio in the Earthquake Triggered Large-scale Landslide near Aratozawa Dam in Miyagi Prefecture, Japan. Procedia Earth and Planetary Science 16: 61–70

Appendix

See Table 1.

Table 1 Ph.D. Students and post-doctoral researchers participating in UNITWIN in the period of 2013–2020

Name	Years (Starting and completing year)	Status at UNITWIN (Ph.D. students and/or Post Doctorial students)	Current position	Country
Disaster Prevention Research Institute, Kyoto University Number of Ph.D. degree from 2013 to 2020: 19/Number of Master's degree from 2013 to 2020: 9				
Kenichiro Kobayashi	2009–2013 Ph.D.	P.D.	Assoc. Prof., Kobe U.	Japan
He Bin	2010–2013	P.D.	Prof., CAS	China
Apip	2008–2011 Ph.D. 2011–2013	Ph.D. Student P.D.	Researcher, LIPI	Indonesia
Luo Pingping	2012–2014	P.D. as JSPS Research Fellow	Prof., Chang'an University	China
Tomoko Teramoto	2009–2015 Ph.D.	Ph.D. student	Researcher, Foundation for River & Basin Integrated Communications (FRICS)	Japan
Maja Ostric	2010–2013 Ph.D.	Ph.D. student	Researcher, Croatia Water	Croatia
Nor Eliza binti Alias	2011–2014 Ph.D.	Ph.D. student	Asst. Prof., UTM	Malaysia
Duan Weili	2011–2014 Ph.D.	Ph.D. student	Asst. Prof., CAS	China
Hendy Setiawan	2014–2017 Ph.D.	Ph.D. student	Lecturer, Universitas Gadjah Mada (UGM)	Indonesia
Xue Han	2014–2017 Ph.D.	Ph.D. student	P.D., Sophia Univ., Japan	China
Vilaysane Bounhieng	2012–2015 Ph.D.	Ph.D. student	Safeguard Officer, Environmental Protection Fund Office, Laos	Lao PDR
Dang Quang Khang	2012–2015 Ph.D.	Ph.D. student	Research promotion officer, ICL	Vietnam
Josko Troselj	2012–2016 Ph.D.	Ph.D. student	Assistant Professor, Hiroshima Univ.	Croatia
Hu Maochuan	2013–2016 Ph.D.	Ph.D. student	P.D., Kyoto Univ.	China
Pham Hong Nga	2012–2019 Ph.D.	JSPS RONPAKU Fellow	Lecturer, Thuyloi University	Vietnam
Pham Van Tien	2015–2018 Ph.D.	Ph.D. student	Researcher, Institute of Transport Science and Technology, Vietnam	Vietnam

(continued)

Table 1 (continued)

Name	Years (Starting and completing year)	Status at UNITWIN (Ph.D. students and/or Post Doctorial students)	Current position	Country
Eva Mia Siska	2014–2018 Ph.D.	Ph.D. student	P.D. in Kyoto Univ.	Indonesia
Khai Lin Chong	2014–2017 Ph.D.	Ph.D. student	Lecturer, Univ. Utara Malaysia	Malaysia
Karlina	2015–2018 Ph.D.	Ph.D. student	Lecturer, Universitas Gadjah Mada (UGM)	Indonesia
Shi Yongxue	2016–2019 Ph.D.	Ph.D. student	Company worker in Tokyo	China
Adnan Artyunov	2016-	Ph.D. student	Student in Kyoto Univ.	U.S.A.
Toma Stoyanov	2016-	Ph.D. student	Student in Kyoto Univ.	Bulgaria
Nguyen Duc Ha	2016–2019 Ph.D.	Ph.D. student	Vietnam Institute of Geosciences and Mineral Resources	Vietnam
Saima Riaz	2016-	DPRI Research Fellow	Asst. Prof., Univ. of Eng. & Tech. Lahore	Pakistan
Sahare Anurag	2018–2021	Ph.D. student	Student in Kyoto Univ.	India
Adapa Gautham	2017–2020	Ph.D. student	Student in Kyoto Univ.	India
Yoshikazu Tanaka	2018–2020	P.D.	P.D. in Kyoto Univ.	Japan
Doan Huy Loi	2020-	Ph.D. student	Student in Kyoto Univ.	Vietnam
University of Florence, Italy Number of Ph.D. degree from 2013 to 2020: 41/Number of Master's degree from 2013 to 2020: 85				
Ascanio Rosi	2010–2013 2013–2019	Ph.D. student P.D.	Temporary Researcher, Univ of Florence	Italy
Federico Raspini	2010–2013 2013–2016	Ph.D. student P.D.	Researcher, Univ of Florence	Italy
Stefano Morelli	2010–2016 2019–2021	P.D. P.D.	Post Doc research fellow, Univ of Florence	Italy
William Frodella	2011–2014 2014–2019	Ph.D. student P.D.	Temporary Researcher, Univ of Florence	Italy
Luca Tanteri	2011–2014 2014–2018	Ph.D. student P.D.	Technician at the university civil protection center, Univ of Florence	Italy
Teresa Nolesini	2011–2014 2014–2018	Ph.D. student P.D.	Technician at the university civil protection center, Univ of Florence	Italy
Silvia Bianchini	2011–2014 2011–2016	Ph.D. student P.D.	Researcher, Univ of Florence	Italy
Federica Ferrigno	2011–2014 2014–2015	Ph.D. student P.D.	Post Doc Research fellow at the Higher Institute for Environmental Protection and Research (ISPRA)	Italy
Veronica Pazzi	2011–2020	P.D.	Research grant-holder, Univ of Florence	Italy
Federica Bardi	2012–2015 2015–2018	Ph.D. student P.D.	Technician, Univ of Florence	Italy
Sara Frangioni	2012–2015	Ph.D. student	Not available	Italy
Carlo Tacconi	2013–2016 2016–2020	Ph.D. student P.D.	Post Doc research fellow, Univ of Florence	Italy
Alessia Lotti	2013–2016 2016–2017	Ph.D. student P.D.	secondary school teacher	Italy
Anna Elisa Bandecchi	2014–2018	P.D.	Technician at the university civil protection center, Univ of Florence	Italy
Giulia Dotta	2015–2017 2017–2020	Ph.D. student P.D.	Research grant-holder	Italy
Teresa Salvatici	2015–2017 2017	Ph.D. student P.D.	Technician, Univ of Florence	Italy
Matteo Del Soldato	2014–2017	Ph.D. student	Post Doc research fellow, Univ. of Florence	Italy

(continued)

Table 1 (continued)

Name	Years (Starting and completing year)	Status at UNITWIN (Ph.D. students and/or Post Doctorial students)	Current position	Country
	2017–2020	P.D.		
Lorenzo Solari	2015–2018 2018–2019	Ph.D. student P.D.	supporting expert at EEA (European environmental agency)	Denmark
Tommaso Carlà	2015–2018 2018–2020	Ph.D. student P.D.	Post Doc research fellow	Italy
Michele D'ambrosio	2015–2018	Ph.D. student	Postdoctoral Research Fellow at the National Institute of Geophysics and Volcanology (INGV)	Italy
Laura Pastonchi	2016–2019	Ph.D. student	Postdoctoral Researcher at Cnr-IBIMET, National Research Council Institute of Biometeorology	Italy
Federico Marini	2016–2019	Ph.D. student	secondary school teacher	Italy
Liang Feng	2017–2020	Ph.D. student	student, Univ of Florence	Italy
Elena Benedetta Masi	2017–2020	Ph.D. student	student, Univ of Florence	Italy
Mattia Ceccatelli	2017–2020	Ph.D. student	student, Univ of Florence	Italy
Teresa Gracchi	2017–2020	Ph.D. student	student, Univ of Florence	Italy
Agnese Turchi	2018–2021	Ph.D. student	student, Univ of Florence	Italy
Tania Luti	2018–2021	Ph.D. student	student, Univ of Florence	Italy
Roberto Montalti	2018–2021	Ph.D. student	student, Univ of Florence	Spain
Monan Shan	2018–2021	Ph.D. student	student, Univ of Florence	China
Rachele Franceschini	2020–2023	Ph.D. student	student, Univ of Florence	Italy
Davide Festa	2020–2023	Ph.D. student	student, Univ of Florence	Italy
Artini Giada	2020–2023	Ph.D. student	student, Univ of Florence	Italy
Marco Lompi	2019–2022	Ph.D. student	student, Univ of Florence	Italy
Matteo Pampaloni	2019–2022	Ph.D. student	student, Univ of Florence	Italy
Genming Du	2019–2022 (interrupted)	Ph.D. student	Not available	China
Lorenzo Innocenti	2018–2021	Ph.D. student	student, Univ of Florence	Italy
Matteo Isola	2017–2020	Ph.D. student	student, Univ of Florence	Italy
Giulio Calvani	2016–2019 2019–2020	Ph.D. student P.D.	Grant holder, Univ. of Florence	Italy
Costanza Carbonari	2016–2019 2019–2020	Ph.D. student P.D.	Grant holder, Univ. of Florence	Italy
Tommaso Pacetti	2015–2018	Ph.D. student	Adjunct professor, Univ. of Florence	Italy
Chiara Arrighi	2013–2016	Ph.D. student	Temporary researcher, Univ. of Florence	Italy
Valentina Chiarello	2013–2016 2016–2020	Ph.D. student P.D.	Postdoctoral researcher, Univ. of Florence	Italy
Pina de Cicco	2013–2016 2016–2018	Ph.D. student P.D.	Private practice: Hydraulic Engineer, collaborator at "ATRE Ingegneria" (Florence)	Italy
University of Ljubljana, Slovenia Number of Ph.D. degree from 2013 to 2020: 9/Number of Master's degree from 2013 to 2020: 26				
Name	Years (Starting and completing year)	Status at UNITWIN (Ph.D. students and/or Post Doctorial students)	Current position	Country
Peter Lamovec	2009–2013 Ph.D.	Ph.D. student	Assist., Univ. Ljubljana	Slovenia
Martin Bombač	2010–2014 Ph.D.	Ph.D. student	Institute for Hydr. Research	Slovenia
Nejc Bezak	2012–2016 Ph.D.	Ph.D. student	Assist. Prof., Univ. Ljubljana	Slovenia
Zsuzsanna Engi	2012–2016 Ph.D.	Ph.D. student		Hungary

(continued)

Table 1 (continued)

Name	Years (Starting and completing year)	Status at UNITWIN (Ph.D. students and/or Post Doctorial students)	Current position	Country
			Researcher, West-Transdanubian Directorate for Water Management	
Gašper Rak	2013–2017 Ph.D.	Ph.D. student	Assist. Prof., Univ. Ljubljana	Slovenia
Jošt Sodnik	2014–2017 Ph.D.	Ph.D. student	Charted Eng., TEMPOS Ltd.	Slovenia
Tina Peternel	2013–2017 Ph.D.	Ph.D. student	Researcher, Geological Survey of Slovenia	Slovenia
Katarina Zabret	2014–2018 Ph.D.	Ph.D. student	Researcher, Institute for Waters of the Republic of Slovenia	Slovenia
Mateja Klun	2015–2020 Ph.D.	Ph.D. student	Assist., Univ. Ljubljana	Slovenia
Klaudija Sapač	2016–2020	Ph.D. student	Ph.D. Student, Univ. Ljubljana	Slovenia
Matej Radinja	2017–2021	Ph.D. student	Young Researcher, Univ. Ljubljana	Slovenia
Tamara Kuzmanić	2019–2023	Ph.D. student	Young Researcher, Univ. Ljubljana	Slovenia
Northeast Forestry University, China Number of Ph.D. degree from 2013 to 2020: 6/Number of Master's degree from 2013 to 2020: 8				
Wang Yuzhuo	2013–2019 Ph.D.	Ph.D. student	Suihua University	China
Hu Zhaoguang	2011–2017 Ph.D.	Ph.D. student	Zhengzhou Aerospace Vocational and Technical College	China
Zhang Kun	2013–2017 Ph.D.	Ph.D. student	Heilongjiang Institute of Architectural Technology	China
Guo Ying	2007–2013 Ph.D. 2014-2019	Ph.D. student P.D.	assistant professor Northeast Forestry University	China
Wang Chunjiao	2009–2015Ph.D.	Ph.D. student	Northeast Agricultural University	China
Jiang Hua	2009–2015 Ph.D.	Ph.D. student	Heilongjiang Institute of Technology	China
Hou Gui	2019-	P.D.	Northeast Forestry University	China
Yan Junxin	2018-	P.D.	Northeast Forestry University	China
Li Bing	2016-	P.D.	Northeast Forestry University	China
Wang Chunjiao	2017-	P.D.	Northeast Forestry University	China
Li Guodong	2016-	P.D.	Northeast Forestry University	China
Zhang Kun	2017-	P.D.	Northeast Forestry University	China
Li Yu	2019-	P.D.	Northeast Forestry University	China
Han Yangrui	2016-	P.D.	Northeast Forestry University	China
Du Wenxue	2016-	P.D.	Northeast Forestry University	China
Li Ping	2019-	P.D.	Northeast Forestry University	China
Wan Lijun	2008–2014 Ph.D. 2014–2019	Ph.D. student P.D.	Lecturer, Researcher Northeast Forestry University	China
Li Hongfeng	2016-	P.D.	Northeast Forestry University	China
Charles University, Prague, Czech Republic Number of Ph.D. degree from 2013 to 2020: 4/Number of Master's degree from 2013 to 2020: 13				
Jan Stejskal	2007–2013	Ph.D. student	Insurance Company	Czech Republic
Jan Burda	2008–2013	Ph.D. student	Research Institute	Czech Republic
Adam Emmer	2013–2017	Ph.D. student	Academy of Sciences	Czech Republic
Michal Kusák	2013–2017	Ph.D. student	Academy of Sciences	Czech Republic

(continued)

Table 1 (continued)

Name	Years (Starting and completing year)	Status at UNITWIN (Ph.D. students and/or Post Doctorial students)	Current position	Country
Amrita Vishwa Vidyapeetham, India Number of Ph.D. degree from 2013 to 2020: 4 / Number of Master's degree from 2013 to 2020: 6				
Rekha Prabha	2011–2018	Ph.D. student	Assistant Professor	India
Hemalatha T	2012–2020	Ph.D. student	Research Associate	India
Sangeeth Kumar	2015–2021	Ph.D. student	Research Associate	India
Balmukund	2019–2023	Ph.D. student	Junior research fellow	India
Landslide Group in National Central University, Chinese Taipei Number of Ph.D. degree from 2013 to 2020: 4/Number of Master's degree from 2013 to 2020: 30				
Truong-Nhat-Phuong, Pham	2016–2019	Ph.D. student	Lecturer in Vietnam, University of Architecture	Vietnam
Jun-Xue, Huang	2018-today	Ph.D. student	National Central University	Chinese Taipei
Che-Ming, Yang,	2011–2017	Ph.D. student	Post doctoral researcher, National Central University	Chinese Taipei
Wei-Ci, Li	2010–2017	Ph.D. student	Post doctoral researcher, National Central University	Chinese Taipei
Faculty of Mining, Geology and Petroleum Engineering, University of Zagreb (UNIZG-RGNF) Number of Ph.D. degree from 2013 to 2020: 4/Number of Master's degree from 2013 to 2020: 20				
Laszlo Podolszki	2008–2014 Ph.D.	Ph.D. student	Croatian Geological Survey	Croatia
Martin Krkač	2008–2015 Ph.D.	Ph.D. student	Assist. Prof., UNIZG-RGNF	Croatia
Petra Jagodnik	2010–2018 Ph.D.	Ph.D. student	Senior Lecturer, UNIRI-GF	Croatia
Sanja Bernat Gazibara	2015–2019 Ph.D.	Ph.D. student	Post Doctoral student, UNIZG-RGNF	Croatia
Faculty of Civil Engineering, University of Rijeka (UNIRI-GF) Number of Ph.D. degree from 2013 to 2020: 4/Number of Master's degree from 2013 to 2020: 37				
Sanja Dugonjić Jovančević	2007–2013 Ph.D.	Ph.D. student	Assist. Prof. UNIRI-GF	Croatia
Vedran Jagodnik	2008–2014 Ph.D.	Ph.D. student	Assist. Prof. UNIRI-GF	Croatia
Martina Vivoda Prodan	2010–2014 Ph.D.	Ph.D. student	Assist. Prof. UNIRI-GF	Croatia
Josip Peranić	2013–2019 Ph.D.	Ph.D. student	Post Doctoral student	Croatia
National Autonomous University of Mexico Number of Ph.D. degree from 2013 to 2020: 2/Number of Master's degree from 2013 to 2020: **8**				
Franny Giselle Murillo-García	2016–2020 Ph.D.	Ph.D. student	Student in UNAM	Mexico
Ricardo J. Garnica-Peña	2016–2020 Ph.D.	Ph.D. student	Technical researcher	Mexico

WLF 2020 Kyoto Japan

International Programme on Landslides (IPL): A Programme of the ICL for Landslide Disaster Risk Reduction

Qunli Han, Kyoji Sassa, and Matjaž Mikoš

Abstract

A new international scientific programme on Landslides was a major agenda in the foundation meeting of the International Consortium on Landslides in January 2002. The foundation of the International Programme on Landslides (IPL) was agreed at the first meeting of the Board of Representatives of the ICL at UNESCO in November 2002. This initial stage of IPL had coordinating projects proposed by the ICL and member projects proposed by each ICL member. The first IPL project was publication of International Journal of Landslides. The current second stage of IPL was defined by the 2006 Tokyo Action Programme on Landslides as an international programme managed by the IPL Global Promotion Committee consisting of ICL and ICL supporting organizations. The second stage of IPL includes IPL Projects conducted by ICL member organizations, the triannual World Landslide Forums and the World Centres of Excellence on Landslide Risk Reduction (WCoE). This paper describes those activities and presents the list of WCoE since 2008 and the list of IPL projects both in the initial stage of IPL projects (2002–2008) an in the second stage of IPL projects (2008–present).

Keywords

Sendai landslide partnerships 2015–2025 (SLP) • Kyoto landslide commitment 2020 (KLC) • International programme on landslides (IPL) • IPL projects • World center of excellence (WCoE)

Introduction

A new international scientific programme on Landslides was a major agenda in the foundation meeting of the International Consortium on Landsides in 21–25 January 2002. The foundation of the International Programme on Landslides (IPL) as a scientific programme of the ICL was agreed at the first Board of Representatives of the ICL at UNESCO in 19–21 November 2002. This initial stage of IPL included coordinating projects and member projects proposed by each ICL member. The first coordinating project was C100 Landslides: Journal of the International Consortium on Landslides. Then, ICL negotiated with several international publishers in order to publish an international full color journal including full color landslide photo and full color landslide map in 2003. It must be the core of the success of the planned International Programme on Landslides. The ICL agreed it with Springer in 2003 and the first issue of Landslides was published in April 2004. This journal provides all authors in developed countries and developing countries with free contribution fee and free full color printing fee.

Landslides have categories of Review papers, Original papers, Recent landslides, Technical notes, IPL/WCoE activities and News/Kyoto Commitment.

Q. Han · K. Sassa (✉) · M. Mikoš
GPC/IPL, International Consortium on Landslides, Kyoto, Japan
e-mail: secretariat@iclhq.org

Q. Han
IRDR-IPO, Beijing, China
e-mail: qunli.han@irdrinternational.org

K. Sassa
International Consortium on Landslides, Kyoto, Japan

M. Mikoš
University of Ljubljana, Ljubljana, Slovenia
e-mail: matjaz.mikos@fgg.uni-lj.si

K. Sassa et al. (eds.), *Understanding and Reducing Landslide Disaster Risk*,
ICL Contribution to Landslide Disaster Risk Reduction,
https://doi.org/10.1007/978-3-030-60196-6_11

Categories of Landslides (Core of All IPL Activities)

- Evaluation process for Review papers, Original papers, Recent landslides, Technical notes are same except the number of necessary reviewers, minimum 2 for Review papers and Original papers while 1 for Recent landslides and Technical note.
- To promote the contribution from developing countries, the category of Technical note includes case studies related to understanding and reducing landslide disaster risk in their countries or regions.
- Hot news (less than 3 months after occurrence) on landslides causing high-impact to society and environment in developing countries and developed countries will be published in News/Kyoto Commitment (less than 4 printed pages).
- Investigation report on recent landslides less than one or two years after occurrence will be published in Recent Landslides (less than 10 printed pages).
- Progress reports and achievement of IPL projects and World Centre of Excellence (WCoE) which was proposed and implemented by ICL members and cooperating members are published in the category of IPL/WCoE of Landslides.
- From 2018, we have started a new category of News/ Kyoto Commitment to disseminate news, reports and the information of each KLC 2020 partner.
- Organization of the Fifth World Landslide Forum and the promotion process of KLC2020, call participants and partners are published in this category.

Management of the International Programme on Landslides (IPL)

The current IPL was established by 2006 Tokyo Action Plan together with seven global stakeholders. The Global Promotion Committee of the International Programme on Landslides (GPC/IPL) has been established. Members of GPC/IPL is ICL board members (full members) and representatives of ICL supporting organizations. The ICL supporting organizations are initial seven global stakeholders (UNESCO, UNDRR, UNU, FAO, WMO, ISC, WFEO) and the International Union of Geological Sciences (IUGS), the International Union of Geodesy and Geophysics (IUGG), the Government of Japan (Cabinet Office, Ministry of Education, Sports, Science and Technology, Ministry of Agriculture, Forestry and Fisheries, Ministry of Land, Infrastructure and Tourism), and Kyoto University.

A Programme of the ICL for Landslide Disaster Risk Reduction

The GPC/IPL is organized at the annual ICL-IPL Conference together with the Board of Representatives of ICL members (BOR/ICL). BOR/ICL decides the ICL management matters. GPC/IPL decides the adoption of IPL projects based the technical evaluation committee, and approves the World Centre of Excellence (WCoE) in prior to each World Landslide Forum which are endorsed by the Independent Panel Experts, examined from secretariat aspect by the IPL secretariat, and evaluated by the technical evaluation team.

Salvano Briceño, Director of the UN Secretariat for Disaster Reduction accepted the role of the first chairperson of the Global Promotion Committee from 2008. Qunli Han, IRDR-IPO (Integrated Researech on Disaster Risk—International Programme Office) is the chairperson of the Global Promotion Committee after Salvano Briceñno. Current co-chairs are Taikan Oki (United Nations University, Senior Vice-Rector/United Nations Assistant Secretary General), Giuseppe Arduino (UNESCO, Chief, Section on Ecohydrology, Water Quality and Water Education), and Matjaž Mikoš (ICL Vice President, Dean of Faculty of Civil and Geodetic Engineering, University of Ljubljana).

The secretariat of the Global Promotion Committee is IPL World Centre hosted in the ICL Secretariat.

Secretary: Kyoji Sassa (IPL World Centre, Director).

Activities of WCoEs

The Landslide Sendai Partnerships 2015–2025, the Kyoto Landslide Commitment 2020, and World Landslide Forum are activities of the ICL and the IPL, a programme of ICL for landslide disaster risk reduction.

Other major activities of the IPL are the implementation of World Centers of Excellence on Landslide Risk Reduction (WCoE) and the IPL projects. The details of WCoEs is explained below.

Objectives of WCoE

To strengthen the International Programme on Landslides (IPL) including IPL projects in its area;

To create "A Global Network of entities contributing to landslide risk reduction";

To implement the Sendai Landslide Partnerships 2015–2025 for global promotion of understanding and reducing landslide disaster risk; and

To improve the global recognition of "Landslide Risk Reduction" and its social-economic relevance, and entities contributing to this field.

Criteria for WCoE Candidates

Governmental and non-governmental entities such as universities, agencies, and other institutions, and their subsidiary entities (faculties, departments, centres, divisions or others) which meet the following two conditions:

(1) Contributing to "Landslide Disaster Risk Reduction";
(2) Willing to support IPL intellectually, practically and financially by either joining ICL or/and contributing to IPL-GPC and promote "landslide research and disaster risk reduction" on a regional and/or global scale in a mutual beneficial manner.

Guidelines for WCoE

(1) Candidates of WCoEs must submit the application form to the Secretariat of the IPL Global Promotion Committee.
(2) Candidates will be evaluated from their achievements and current activities (scientific, technical and educational capacity, training courses, publications, dissemination of knowledge and information) and planned activities contributing to IPL.
(3) WCoEs will be identified at every World Landslide Forum (held every 3 years). The status as a WCoE will be given for 3 years until the next Forum.
(4) Each WCoE must submit an annual activity report each year which is uploaded in the IPL web and also report its activities in Landslides: Journal of the International Consortium on Landslides or/and a World Landslide Forum.
(5) The status as a WCoE may be extended for another 3 years in the same topics or in a revised topic by the IPL Global Promotion Committee, based on the activities carried out as WCoE.

Procedure for Identification of WCoEs

(1) The application form from an eligible entity will be submitted to the Secretariat of the IPL-GPC (IPL World Centre).
(2) Preliminary screening of received applications will be conducted according to the criteria by the secretariat of IPL World Centre. Feasible applications will be passed on for in-depth evaluation.
(3) Candidates of WCoEs passing to in-depth evaluation will be invited to orally present their proposals in the IPL Global Promotion Committee. Revision of application may be advised by the committee to avoid duplication of activities with other proposals or to improve the proposals when necessary.
(4) Final application forms will be evaluated by the technical evaluation committee.
(5) Summarizing all evaluation values, appropriate candidates will be selected, and a list of recommendations will be submitted from IPL World Center to the Independent Panel of Experts which consists of experts outside of ICL.
(6) The Independent Panel of Experts will review the recommendation, and endorse appropriate WCoEs.

(7) The Endorsement of WCoEs by the Independent Panel of Experts will be reported to the IPL Global Promotion Committee. WCoEs endorsed by the Panel must be approved by the vote of more than half of the participating members of the IPL Global Promotion Committee.

(8) The WCoE for the next 3 years will be announced at the World Landslide Forum.

Table 1 presents the chronological order of all World Centres of Excellence including both past and ongoing WCOEs.

Activities of IPL Projects

Objective of IPL Project is to identify and promote research, investigation, capacity development activities implemented by each ICL member.

The IPL project proposal is the right of ICL full members. Each ICL member can propose alone or toghert with associate members, supporters or any other cooperating organizations/individuals. IPL project leaders are requested to submit annul activity report by 31 December each year. A project with 2 years missing reports will be terminated.

Landslide research is not always well evaluated by experts of other field of scientists in national scientific grants or other funding agencies. Landslide researchers are often in minor group in each country. Aspect of Practice is important for landslide disaster risk reduction. But practice is not evaluated in scientific fields and advisers in scientific grant. Identification as an International Programme is some times to effectively support each project. It was a lesson from UNESCO-IUGS International Geoscience Programme (IGCP-425 Landslide Hazard Assessment of Cultural Heritage: 1998–2003). Initial 33 members of the ICL were the members of IGCP-425. All members were supported by group and sub-group activities of this international Project. All IGCP-425 members wish to establish the International Programme on Landslides which approve IPL Projects.

The initial IPL projects are C100 Landslides: Journal of the International Consortium on Landslides, and C101 Landslide risk evaluation and mitigation in cultural and natural heritage sites. And the subproject of C101 was C101-1 Landslide investigation and capacity building in Machu Pichu-Aguas Calientes area.

Table 2 presents the chronological order of the IPL Projects, including the initial IPL coordinating projects (C100, C101–C106 in 2002–2007) and IPL projects based on 2006 Tokyo Action Plan (2008–present).

Call for Cooperation to WCoEs and IPL Projects

Those who are willing to cooperate with WCoE and IPL Projects are requested to email ICL secretariat secretariat@iclhq.org and ICL Network Committee icl-network@iclhq.org.

Appendix

See Tables 1 and 2.

Table 1 The chronological order of all World Centres of Excellence including both past and ongoing WCOEs

No.	Title	Leader	Country/region	Organization	Status
World Centre of Excellence 2008–2011					
1	Scientific research for mitigation, preparedness and risk assessment of landslides	Yuepin Yin	China	China Geological Survey	Completed
2	Landslide field research and capacity building through international collaboration	Vit Vlimek	Czech Republic	Faculty of Science, Charles University in Prague	Completed
3	Earth observation advanced technologies for landslide monitoring, management and mitigation	Nicola Casagli	Italy	Department of Earth Science, University of Florence	Completed
4	Research and development of advanced technology for landslide hazard analysis	Alberto Presitininzi/Gabriele Scarascia-Mugnozza	Italy	Research Centre on Prediction Prevention and Control of Georisks of Rome University "La Sapienza"	Completed
5	Development of methodology for risk assessment of the earthquake-induced landslides	Hideaki Marui	Japan	The Japan Landslide Society	Completed
6	Implementation of National Slope Master Plan	Ashaari Mohamad/Che Hassandi bin Abdullah	Malaysia	Slope Engineering Branch, Public Works Department of Malaysia	Completed
7	Research on mitigation of landslide risk and training of specialists.	Farrokh Nadim	Norway	International Centre for Geohazards (ICG) at NGI	Completed
8	International Summer School on Rockslides and Related Phenomena in the Kokomeren River basin, Kyrgyzstan	Alexandar Strom	Russia and Kyrgyz	Institute of Geospheres Dynamics of Russian Academy of Science (IDG RAS) & Kyrgyz Institute of Seismology (KIS)	Completed
9	Mechanisms of landslides in over-consolidated clays and flysch	Bojan Majes/Matjaž Mikoš	Slovenia	University of Ljubljana, Faculty of Civil and Geodetic Engineering (UL FGG)	Completed
10	Landslide inventorization and susceptibility mapping in South Africa	S. Diop/SG Chiliza	South Africa	Engineering Geoscience Unit, Council for Geoscience	Completed
11	Promoting Knowledge sharing, Innovations and Institutions with South-South focus network on Landslide Risk Reduction in Asia	N. S. M. I. Arambepola	Thailand	Asian Disaster Preparedness Center	Completed
12	Conduct landslide hazard assessments and develop early warning systems	Peter Lyttle	USA	U.S. Geological Survey	Completed
World Centre of Excellence 2011–2014					
1	Canadian Landslide Loss Risk Reduction Strategy and Implementation	Peter T. Bobrowsky	Canada	Geological Survey of Canada	Completed
2	Risk Assessment and Disaster Mitigation Code for Long Run-out Landslides	Yueping Yin	China	China Geological Survey	Completed
3	Scientific research for landslide risk analysis and international education for mitigation and preparedness	Vit Vilimek	Czech Republic	Charles University, Faculty of Science	Completed
4	Research on landslide risk management harmonisation in support to European Union policy making	Javier Hervás	European Commission	Joint Research Centre, European Commission	Completed

(continued)

Table 1 (continued)

No.	Title	Leader	Country/region	Organization	Status
5	Training, Research and Documentation on Landslides Risk Management	Surya Parkash	India	National Institute of Disaster Management	Completed
6	Development of Community-based and Most Adaptive Technology for Landslide Risk Reduction	Dwikorita Karnawati	Indonesia	Universitas Gadjah Mada	Completed
7	Advanced Technologies for Landslides	N. Casagli, F. Catani	Italy	Department of Earth Science, University of Florence, Italy	Completed
8	Development of a methodology for risk reduction of earthquake-induced landslides	Daisuke Higaki	Japan	The Japan Landslide Society, Japan	Completed
9	Risk identification and land-use planning for disaster mitigation of landslides	Hideaki Marui	Japan	Niigata University, Institute for Natural Hazards and Disaster Recovery	Completed
10	Landslide monitoring and community based early warning systems	Irasema Alcántara-Ayala	Mexico	National Autonomous University of Mexico	Completed
11	Research on mitigation of landslide risk and training of specialists	Farrokh Nadim	Norway	International Centre for Geohazards	Completed
12	Annual Summer School on Rockslides and Related Phenomena in Kyrgyzstan	Alexander Strom	Russia and Kyrgyz	Inst. of Geospheres Dynamics of Russian Academy of Sciences & Kyrgyz Institute of Seismology	Completed
13	Mechanisms of landslides in over-consolidated clays and flysch	Bojan Majes	Slovenia	University of Ljubljana, Faculty of Civil and Geodetic Engineering	Completed
14	Promoting Knowledge, Innovations and Institutions with South-South focus through a Regional network of Landslide Risk Reduction	N. S. M. I. Arambepola	Thailand	Asian Disaster Preparedness Center	Completed
15	Scientific Research for Landslide Hazard Analysis, U.S. Geological survey	Peter Lyttle	USA	U.S. Geological Survey Landslide Programme	Completed
World Centre of Excellence 2014–2017					
1	Formation Mechanism Research, Disaster Warning and Universal Education of Cold Regions Landslide	Wei Shan	China	Research Center of Cold Regions Landslide, China	Completed
2	Scientific Research for Mitigation, Preparedness and Risk Assessment of Landslides	Wang Min	China	China Geological Survey, China	Completed
3	Scientific Research for Landslide Risk Analysis, Modeling, Mitigation and Education	Liang-Jenq Leu	Taiwan	Department of Civil Engineering, National Taiwan University, China, Taiwan	Completed
4	Landslide Risk Reduction in the Adriatic-Balkan Region through the Regional Cooperation	Željko Arbanas/Snježana Mihalić Arbanas	Croatia	Croatian Landslide Group, Croatia	Completed
5	Landslide Risk Assessment and Development Guidelines for Effective Risk Reduction	Josef Stemberk	Czech Republic	Institute of Rock Structure and Mechanics Czech Academy of Sciences & Charles University, Faculty of Science	Completed
6	Development of Community-based and Most Adaptive Technology for Landslide Risk Reduction	Dwikorita Karnawati	Indonesia	Universitas Gadjah Mada, Yogyakarta, Indonesia	Completed
7	Advanced Technologies for LandSlides (ATLaS)	Nicola Casagli	Italy	Department of Earth Science, University of Florence, Italy	Completed

(continued)

Table 1 (continued)

No.	Title	Leader	Country/region	Organization	Status
8	Emergency Response Support System for Large-scale Landslide Disasters	Satoshi Tsuchiya	Japan	The Japan Landslide Society (JLS), Japan	Completed
9	Risk Identification and Land-use Planning for Disaster Mitigation of Landslides	Hiroshi Fukuoka	Japan	Niigata University, Institute for Natural Hazards and Disaster Recovery, Japan	Completed
10	Implementation of National Slope Master Plan	Che Hassandi Abdullah	Malaysia	Slopes Engineering Branch, Public Works Department of Malaysia, Malaysia	Completed
11	Building Human Capacities and Expertise in Landslide Disaster Risk Managements	Ogbonnaya Igwe	Nigeria	Department of Geology, University of Nigeria, Nsukka, Nigeria	Completed
12	International Summer School on Rockslides and Related Phenomena in the Kokomeren River Valley, Tien Shan, Kyrgyzstan	Alexander Strom	Russia and Kyrgyz	Geodynamics Research Center—branch of JSC "Hydroproject Institute" & Institute of Seismology of National Academy of Sciences of Kyrgyz Republic	Completed
13	Mechanisms of Landslides and Creep in Over-consolidated Clays and Flysch	Ana Petkovšek	Slovenia	University of Ljubljana, Faculty of Civil and Geodetic Engineering (UL FGG), Ljubljana, Slovenia	Completed
14	Developing Model Policy Frameworks, Standards and Guidelines	Nihal Rupasinghe A. A. Virajh Dias	Sri Lanka	Central Engineering Consultancy Bureau, Colombo, Sri Lanka	Completed
15	Promoting Knowledge, Innovations and Institutions with South-South focus through a Regional network of Landslide Risk Reduction in Changing Climate Scenario in Asia	N. M. S. I. Arambepola	Thailand	Asian Disaster Preparedness Center (ADPC), Thailand	Completed
World Centre of Excellence 2017–2020					
1	Landslide Monitoring and Critical Infrastructure	Peter T. Bobrowsky	Canada	Geological Survey of Canada	On-going
2	Scientific research for mitigation, preparedness and risk assessment of Landslides	Yueping Yin	China	China Geological Survey	On-going
3	Formation mechanism research, disaster warning, and universal education of landslides in permafrost regions	Wei Shan	China	Institute of Cold Regions Science and Engineering, Northeast Forestry University	On-going
4	Center for Applied Landslide Research (CALaR)	Snjezana Mihalic Arbanas, Željko Arbanas	Croatia	Croatian Landslide Group from University of Zagreb and University of Rijeka	On-going
5	Landslide risk assessment and development guidelines for effective risk reduction—continuation	Vit Vilimek	Czech Republic	Charles University, Faculty of Science & Institute of Rock Structure and Mechanics Czech Academy of Sciences	On-going
6	Enhancement of the existing Real-time Landslide Monitoring and Early warning System in Western Ghats & Himalayas, India	Maneesha V Ramesh	India	Amrita University	On-going
7	Development of Community-based and Most Adaptive Technology for Landslide Risk Reduction	Dwikorita Karnawati	Indonesia	University of Gadjah Mada	On-going

(continued)

Table 1 (continued)

No.	Title	Leader	Country/region	Organization	Status
8	ATLaS: Advanced Technologies for LandSlides	Nicola Casagli	Italy	Department of Earth Sciences, University of Firenze (DST-UNIFI)	On-going
9	Methods and tools for landslide forecasting and risk mitigation and adaptation strategies	Fausto Guzzetti	Italy	Istituto di Ricerca per la Protezione Idrogeologica (IRPI), of the Italian National Research Council (CNR)	On-going
10	Landslide Hazards Mitigation Programs in the Korean Demilitarized Zone	Sangjun Im	Korea	Korean Society of Forest Engineering	On-going
11	Landslide Quantitative Risk Analysis Study for Malaysia	Che Hassandi Abdullah	Malaysia	Slope Engineering Branch, PublicWorks Department of Malaysia	On-going
12	Landslides Integrated Research for Disaster Risk Reduction	Irasema Alcántara Ayala	Mexico	National Autonomous University of Mexico (UNAM)	On-going
13	Characterizing past and planned activities: Klima 2050—Innovational methods for risk reduction associated to hydro-meteorologically induced landslides	José Cepeda	Norway	Norwegian Geotechnical Institute (NGI)	On-going
14	Central Asia rockslide inventory. Compilation and analysis	AlexanderStrom	Russia	JSC "Hydroproject Institute"	On-going
15	Harmonization of Landslide Data and Local Communities Capacity Building for Landslide Risk Reduction	Biljana Abolmasov	Serbia	University of Belgrade, Faculty of Mining and Geology	On-going
16	Landslides in Weathered Flysch: from activation to deposition	Ana Petkovšek	Slovenia	University of Ljubljana, Faculty of Civil and Geodetic Engineering (ULFGG)	On-going
17	Landslide risk reduction in Slovenia	Mateja Jemec Auflic	Slovenia	Geological Survey of Slovenia	On-going
18	Model Policy Frameworks, Standards, and Guidelines on Landslide Disaster Risk Reduction	A. A. Virajh Dias	Sri Lanka	Central Engineering Consultancy Bureau (CECB)	On-going
19	Characterizing past and planned activities: NBRO is the national focal point for landslide disaster risk management	Asiri Karunawardena	Sri Lanka	National Building Research Organization	On-going
20	Implementation of National Slope Master Plan	Oleksander Trofymchuk	Ukraine	The Institute of Telecommunication and Global Information Space (ITIGS) of the National Academy of Science of Ukraine (NASU)	On-going

Table 2 The chronological order of the IPL Projects, including the initial IPL coordinating projects (C100, C101–C106 in 2002–2007) and IPL projects based on 2006 Tokyo Action Plan (2008–present)

No.	Title	Organization	Country	Leader	Year	Status
The initial stage of the International Programme on Landslides (IPL: 2002–2008) under support from UNESCO						
C100	Landslides: Journal of the International Consortium on Landslides	Kyoto University, DPRI, RCL/ICL	Japan	Kyoji Sassa	2002 /2008–2010	Completed
C101	Landslide risk evaluation and mitigation in cultural and natural heritage sites	Kyoto University, DPRI, RCL/ICL	Japan	Kyoji Sassa and Paolo Canuti	2002/ 2008–2010	Completed
C101-1	Landslide investigation and capacity buildng in Machu Pichu-Aguas Calientes area	Kyoto University, DPRI, RCL/ICL	Japan	Kyoji Sassa	2002/ 2008–2011	Completed
C101-1-1	Low environmental impact technologies for slope monitoring by radar interferometry: application to Machu Picchu site	CIVITA Consortium/ENEA	Italy	Claudio Margottini	2002–2006	Completed
C101-1-2	Expressions of risky geomorphologic processes as well as paleogeographical evolution of the area of Machu Picchu	Charles University, Research Center of Earth Dynamic	Czech Republic	Vit Vilimek, Jiri Zvelebil	2002–2007	Completed
C101-1-3	Shallow geophysics and terrain stability mapping techniques applied to the Urubamba Valley, Peru: Landslide hazard evaluation	Instituto Geologico Minero y Metalurgico	Peru	Romulo Mucho, Peter T. Bobrowsky	2004–2006	Completed
C101-1-4	A proposal for an integrated geophysical study of the Cuzco region	Instituto Nazionale Di Oceanografia E Di Geofisica Sperimentale-OGS	Italy	Daniel Nieto Yabar	2004–2006	Completed
C101-1-5	UNESCO-Italian-ESA Satellite monitoring of Machu Picchu	University of Firenze, Earth Sciences Department	Italy	Paolo Canuti, Claudio Margottini,	2004–2006	Completed
C101-3	The geomorphological instability of the Buddha niches and surrounding cliff in Bamiyan valley (Central Afghanistan)	CIVITA Consortium/ENEA/ISPRA	Italy	Claudio Margottini	2002/ 2008–2015	Completed
C101-4	Stability assessment and prevention measurement of Lishan Landslide, Xian, China	Lishan Landslide Observatory, Xian Municipal Government	China	Qing Jin Yang	2002–2007	Completed
C101-5	Environment protection and disaster mitigation of rock avalanches landslides and debris flow in Tianchi Lake region and natural preservation area of Changbai Mountains, Northeast China	Jilin University, Environmental Geological Disaster Research Institute	China	Binglan Cao	2002–2007	Completed
C101-6	Conservation of Masouleh Town	Building & Housing Research Center	Iran	S. H. Tabatabaei	2002–2007	Completed
C101-7	Cultural and natural heritage threatened by landslides in the region of Iassy, Romania	Proexrom S.R.L.,Technical University, Civil Engineering Faculty	Romania	Nicolae Botu	2005–2007	Completed
C102	Assessment of global high-risk landslide disaster hotspots	International Centre for Geohazards, (ICG) in Oslo	Norway	Farrokh Nadim	2002–2004	Completed
C103	Global landslide observation strategy	Kyoto University, DPRI, Flood Section	Japan/Italy	Kaoru Takara, Nicola Casagli	2004/2008–2012	Completed

(continued)

Table 2 (continued)

No.	Title	Organization	Country	Leader	Year	Status
C104	World Landslide Database	Kyoto University, DPRI, RCL	Japan/Italy	Hiroshi Fukuoka, Nicola Casagli	2006/2008–2012	Completed
C105	Early Warning of Landslides	Kyoto University, DPRI, RCL/ICL	Japan	Kyoji Sassa	2007–2013	Completed
C106	Capacity building and outreach	ISPRA-Italian Institute for Environmental Protection and Research	Italy/Russia	Claudio Margottini, Alexander Strom	2008–2012	Completed
C106-1	Landslide museum in Civita di Bagnoregio	ENEA/ISPRA	Italy	Claudio Margottini	2006/2008–2012	Completed
International Programme on Landslides (IPL)—A programme of ICL for ISDR (2008–Present)—MoUs with UNESCO, WMO, FAO, UNISDR, UNU, ICSU and WFEO						
IPL-101-2	Landslides monitoring and slope stability at selected historic sites in Slovakia	Comenius University, Faculty of Natural Sciences	Slovakia	Jan Vlcko	2008–	Completed
IPL-106-2	International Summer School on Rockslides and Related Phenomena in the Kokomeren River Valley, Tien Shan, Kyrgyzstan	Institute of the Geospheres Dynamics, Russian Academy of Sciences/JSC "Hydroproject Institute"	Russia	Alexander Strom	2008–	Completed
IPL-112	Landslide mapping and risk mitigation planning in Thailand	Ministry of Agriculture and Cooperatives, Land Development Department	Thailand	Saowanee Prachansri	2008	On-going
IPL-139	Development of low-cost early warning system of slope instability for civilian use	University of Tokyo, Geotechnical Engineering Group	Japan	Ikuo Towhata/Taro Uchimura	2008	Completed
IPL-142	Seismic landslide hazards mapping in Sichuan	China Geological Survey	China	Yuepin Yin	2009–2011	Completed
IPL-143	Evaluation of sensitivity of the combined hydrological model (dynamic) for landslide susceptibility risk mapping in Sri Lanka	Central Engineering Consultancy Bureau (CECB)	Sri Lanka	A. A. Virajh Dias	2008–2012	Completed
IPL-144	SafeLand—Living with landslide risk in Europe: Assessment, effects of global change, and risk management strategies	International Centre for Geohazards, (ICG) in Oslo	Norway	Bjørn Kalsnes	2009–2012	Completed
IPL-145	Preparation of landslide risk map in Taleghan Area—Iran	Building & Housing Research Center	Iran	S. H. Tabatabaei	2009–2011	Completed
IPL-146	Spatial monitoring of joint influence of an atmospheric precipitation and seismic motions on formation of landslides in Uzbekistan (Central Asia)	Institute Hydroingeo, State Committee of Geology of Uzbekistan	Uzbekistan	Rustam Niyazov	2010–2012	Completed
IPL-147	Study on Debris Flow Controlling Factors and Triggering Mechanism in Peninsular Malaysia	Slope Engineering Branch, Public Works Department of Malaysia	Malaysia	Che Hassandi Abdullah	2010–2011	Completed
IPL-148	Geo-evaluation of the stability of slopes around crater lakes in Cameroon: The cases of lakes Nyos, Barombi, Mbo and Awing	University of Buea	Cameroon	Ntasin Edwin Bongsiysi	2010–2011	Completed
IPL-149	Canadian Landslide Best Practice Manual	Geological Survey of Canada	Canada	Peter T. Bobrowsky	2009–2015	Completed

(continued)

Table 2 (continued)

No.	Title	Organization	Country	Leader	Year	Status
IPL-150	Capacity building and the impact of climate-driven changes on regional landslide distribution, frequency and scale of catastrophe	Department of Geology, University of Nigeria, Nsukka	Nigeria	Ogbonnaya Igwe	2009–	Completed
IPL-151	Soil matrix suction in active landslides in flysch—the Slano Blato landslide case	University of Ljubljana, Faculty of Civil and Geodetic Engineering	Slovenia	Bojan Majes	2010–2012	Completed
IPL-152	Assessment of coastal landslides risk by innovative remote sensing techniques.	University of Roma "La Sapienza"	Italy	Gabriele Scarascia Mugnozza	2010–2011	Completed
IPL-153	Landslide hazard zonation in Kharkov region of Ukraine using GIS	Insutitute of Telecommunication and Global Information Space	Ukraine	Oleksandr M. Trofymchuk	2010–2011 2012–2014	Completed
IPL-154	Development of a methodology for risk assessment of the earthquake-induced landslides	Japan Landslide Society	Japan	Satoshi Tsuchiya	2009	Completed
IPL-155	Determination of soil parameters of subsurface to be used in slope stability analysis in two different precipitation zones of Sri Lanka	Central Engineering Consultancy Bureau	Sri Lanka	A. A. Virajh Dias	2009	On-going
IPL-156	Best practices for early warning of landslides in a changing climate scenarios	Asian Disaster Preparedness Center (ADPC)	Thailand	N. M. S. I. Arambepola	2009–2012	Completed
IPL-157	Dynamics of subaerial and submarine megaslides	ICL	Japan	Kyoji Sassa	2009–	On-going
IPL-158	Development of Community-based Landslide Early Warning System	Gadjah Mada University	Indonesia	Teuku Faisal Fathani	2009–	On-going
IPL-159	Development of Education Program for Sustainable Development in Landslide Vulnerable Area through Student Community Service.	Gadjah Mada University	Indonesia	Dwikorita Karnawati	2009–	On-going
IPL-160	Landslides and floods under extreme weather condition and resilient society	Kyoto University, DPRI, RCL	Japan	Hiroshi Fukuoka	2009–2012	Completed
IPL-161	Risk identification and land-use planning for disaster mitigation of landslides and floods in Croatia	Kyoto University, DPRI/Niigata University, R. Center for Natural Hazards and Disaster Recovery	Japan	Hiroshi Fukuoka/Hideaki Marui	2009–	Completed
IPL-163	Mechanical-mathematical modeling and monitoring for landslide processes	Russian Academy of Sciences, Institute of Environmental Geoscience	Russia	Svalova Valentina	2009–	Completed
IPL-165	Development of community-based landslide hazard mapping for landslide risk reduction at the village scale in Java, Indonesia	Gadjah Mada University	Indonesia	Dwikorita Karnawati	2010–	On-going
IPL-167	Landslides Mechanism and the Subgrade Stability Controlling Measures in Island Permafrost Area	Northeast Forestry University	China	Wei Shan	2010–	On-going

(continued)

Table 2 (continued)

No.	Title	Organization	Country	Leader	Year	Status
IPL-168	Engaging U.S. citizens in Landslide Science through the website, "Did You See It? Report a Landslide"	U. S. Geological Survey	USA	Rex Baum	2010–2013	Completed
IPL-169	Landslide hazard and risk assessment in Geyser Valley (Kamchatka)	Laboratory of Engineering Geodynamics, Geological Faculty, Moscow State University	Russia	Oleg V. Zerkal	2010–2012	Completed
IPL-170	Landslide susceptibility and landslide hazard zonation in volcanic terrains using Geographic Information System (GIS): A case study in the Río Chiquito-barranca Del Muerto watershed	Institute of Geography, UNAM	Mexico	Gabriel Legorreta Paulín	2010–2013	Completed
IPL-171	Study of the geotechnical characteristics of an unstable urban area of Barranquilla (Colombia) severely affected for slope instabilities and soil volume changes	Universidad Nacional de Colombia	Colombia	Guillermo Ávila	2010–2015	Completed
IPL-172	Documentation, Training and Capacity Building for Landslides Risk Management	National Institute of Disaster Management, New Delhi	India	Surya Parkash	2011–2014	Completed
IPL-173	Croatian virtual landslide data center	Faculty of Mining, Geology and Petroleum University of Zagreb	Croatia	Snjezana Mihalic	2011–	On-going
IPL-175	Development of landslide risk assessment technology and education in Vietnam and other areas in the Greater Mekong Sub-region	ICL/Institute of Transport Science and Technology	Japan, Vietnam	Kyoji Sassa/Nguen Xuan Khang	2011–	On-going
IPL-176	Slope Data Acquisition along Highways in Sabah State for hazard assessment and mapping	Slope Engineering Branch, Public Works Department of Malaysia	Malaysia	Che Hassandi Abdullah	2012–2013	Completed
IPL-177	Study on geological disasters focusing on landslides in and around Tegucigalpa City, Honduras	Universidad Politécnica de Ingeniería, UPI	Honduras	Aníbal Godoy	2012–2013	Completed
IPL-179	Database of Glacial Lake Outburst Floods (GLOFs)	Charles University, Research Center of Earth Dynamic	Czech Republic	Adam EMMER/Vit Vilimek	2012–	On-going
IPL-180	Introducing Community-based Early Warning System for Landslide Hazard Management in Cox's Bazaar Municipality, Bangladesh	Asian Disaster Preparedness Center (ADPC)	Thailand	N. M. S. I. Arambepola	2011–2013	Completed
IPL-181	Study of slow moving landslide Umka near Belgrade, Serbia	University of Belgrade, Faculty of Mining and Geology	Serbia	Biljana Abolmasov	2012–	On-going
IPL-182	Characterization of landslides mechanisms and impacts as a tool to fast risk analysis of landslides related disasters in Brazil	CENACID—UFPR (Center for Scientific Support in Disasters—Federal University of Parana)	Brazil	Renato Eugenio de Lima	2012–2014	Completed

(continued)

Table 2 (continued)

No.	Title	Organization	Country	Leader	Year	Status
IPL-183	Landslides in West Africa: impacts, mechanism and management	Department of Geology, University of Nigeria, Nsukka	Nigeria	Igwe Ogbonnaya	2012–	Completed
IPL-184	Study of landslides in flysch deposits of North Istria, Croatia: sliding mechanisms, geotechnical properties, landslide modeling and landslide susceptibility	Faculty of Civil Engineering University of Rijeka	Croatia	Željko Arbanas	2012–	On-going
IPL-185	Design and Validation of an Early Warning System for Landslides—DeVEL	Technische Universitat Darmstadt, Institute and Laboratory of Geotechnics	Germany	Rolf Katzenbach	2013–	Completed
IPL-186	Rock-fall hazard assessment and monitoring in the archaelogical site of Petra, Jordan	ISPRA-Italian Institute for Environmental Protection and Research	Italy	Claudio Margottini	2013–	On-going
IPL-187	Landslide hazards assessment and modeling and sediment yield	Institute of Geography, UNAM	Mexico	Gabriel L. Paulin	2013–	On-going
IPL-188	Study of slow-moving landslide Potoška Planina (Karavanke Mountain, NW Slovenia)	Geological Survey of SloveniaGeological Survey of Slovenia	Slovenia	Marko Komac	2013–	On-going
IPL-190	Landslide risk identification and resilience study in tectonically active mountains and sea floors	Niigata University, Research Institute for Natural Hazards and Disaster Recovery	Japan	Hiroshi Fukuoka	2015–	Completed
IPL-191	Landslide hazard zonation in Carpathian region of Ukraine using GIS	Institute of Telecommunication and Global Information Space	Ukraine	Yakovliev Yevhenii/Oleksandr Trofymchuk	2015–	On-going
IPL-192	Development of post-earthquake rainfall induced landslide (PERIL) hazard mitigation framework	California State University, Fullerton	USA and Nepal	Binod Tiwari	2015–	On-going
IPL-193	Integrated systems for landslides monitoring, early warning and risk mitigation along motorways	University of Calabria, DIMES, CAMILAB	Italy	Pasquale Versace	2015–	On-going
IPL-194	Public awareness and education programme for landslides management in Malaysia	Slope Engineering Branch, Public Works Department of Malaysia	Malaysia	Che Hassandi Abdullah	2015–	On-going
IPL-195	Study for mitigation and recovery of mud eruption disaster in East Java and modeling for risk reduction mudflow hazards	Parahyangan Catholic University	Indonesia	Paulus P. Rahardjo	2015–	On-going
IPL-196	Development and applications of a multi-sensors drone for geohazards monitoring and mapping	University of Firenze, Earth Sciences Department	Italy	Veronica Tofani	2015–	On-going
IPL-197	Low frequency, high damaging potential landslide events in "low risk" regions—challenges for hazard and risk management	Institute of Rock Structure and Mechanics Academy of Sciences	Czech Republic	Jan Klimeš	2015–	On-going
IPL-198	Multi-scale rainfall triggering models for Early Warning of Landslides (MUSE)	University of Firenze, Earth Sciences Department	Italy	Filippo Catani	2015–	On-going

(continued)

Table 2 (continued)

No.	Title	Organization	Country	Leader	Year	Status
IPL-199	The effect of root systems in natural slope erosion protection in the hill country of Sri Lanka	Central Engineering Consultancy Bureau	Sri Lanka	Pvip Perera	2015–	On-going
IPL-200	An assessment of the rock fall susceptibility based on cut slopes adjacent to highways and railways	Central Engineering Consultancy Bureau	Sri Lanka	H. M. J. M. K. Herath	2015–	On-going
IPL-201	Landslide inventory and Susceptibility map in Durres and Kavaja region	Albanian Geological Survey	Albania	Hasan Kulici	2016–	On-going
IPL-202	Ripley landslide monitoring project (Ashcroft, BC, Canada)	Geological Survey of Canada	Canada	Peter T. Bobrowsky/Claudio Margottini	2016–	On-going
IPL-203	Analysis and identify of landslides based on species distribution and surface temperature difference	Institute of Cold Regions Science and Engineering, Northeast Forestry University	China	Ying Guo	2016–	On-going
IPL-204	A study on socio-economic and environmental impacts of landslides	National Institute of Disaster Management, New Delhi	India	Surya Prakash	2016–	On-going
IPL-205	Integrated systems for landslides monitoring, early warning and risk mitigation along motorways	University of Calabria, DIMES, CAMILAB	Italy	Pasquale Versace/Giovanna Capparelli	2016–	On-going
IPL-206	Towards improved landslide mapping and forecasting	Istituto di Ricerca per la Protezione Idrogeologica of the Italian National Research Council	Italy	Fausto Guzzetti/Mario Parise	2016–	On-going
IPL-207	Evaluation on social research approach In determining "acceptable risk" and "tolerable risk" in landslide risk areas in Malaysia	Slope Engineering Branch, Public Works Department of Malaysia	Malaysia	Che Hassandi Bin Abdullah	2016–	On-going
IPL-208	Landslide disaster risk communication in mountain areas	Institute of Geography, UNAM	Mexico	Irasema Alcántara Ayala	2016–	On-going
IPL-209	Landslides and related sediment disaster project covering the entire South-East Nigeria, West Africa	Department of Geology, University of Nigeria, Nsukka	Nigeria	Igwe Ogbonnaya	2016–	Completed
IPL-210	Massive landsliding in Serbia following Cyclone Tamara in May 2014	University of Belgrade, Faculty of Mining and Geology	Serbia	Biljana Abolmasov	2016–	On-going
IPL-211	Development of wireless sensor network for monitoring and earlier warning of shallow and deep landslides (WISE-LAND)	Research Center for Geotechnology, Indonesian Institute of Sciences	Indonesia	Adrin Tohari	2016–	On-going
IPL-212	The construction of a global database of giant landslides on oceanic island volcanoes	Institute of Rock Structure and Mechanics, Czech Academy of Sciences	Czech Republic	Matt Rowberry	2016–	On-going
IPL-213	Real-time Landslide Monitoring and Early warning System in Western Ghats & Himalayas, India	Amrita Center for Wireless Networks & Applications	India	Maneesha Vinodini Ramesh	2016–	On-going

(continued)

Table 2 (continued)

No.	Title	Organization	Country	Leader	Year	Status
IPL-215	The development of paleo-landslides in the middle part of the Moskva River valley within the limits of the Moscow City	Engineering and Ecological Geology Department of the Geological Faculty of the Lomonosov Moscow State University	Russia	Oleg Zerkal	2016–	On-going
IPL-216	Diversity and hydrogeology of mass movements in the Vipava valley, SW Slovenia	University of Ljubljana, Faculty of Natural Sciences and Engineering, Department of Geology	Slovenia	Timotej Verbovšek	2016–	On-going
IPL-217	PROTHEGO—PROTection of European Cultural HEritage from GeO—Hazards	ISPRA—The Italian National Institute for Environmental Protection and Research—Researcher (Rome office)	Italy	Daniele Spizzichino/Claudio Margottini	2016–	On-going
IPL-218	Landslide rapid mapping from remote sensing	Tongji University	China	Ping LU	2017–	On-going
IPL-219	Rockfall hazard identification and rockfall protection in the coastal zone of Croatia	University of Rijeka, Faculty of Civil Engineering	Croatia	Željko Arbanas	2017–	On-going
IPL-220	Kostanjek landslide monitoring project (Zagreb, Croatia)	University of Zagreb, Faculty of Mining, Geology and Petroleum Engineering	Croatia	Martin Krkac	2017–	On-going
IPL-221	PS continuous streaming for landslide monitoring and mapping	Earth Sciences Department of the University of Firenze (DST-UNIFI)	Italy	Federico Raspini	2017–	On-going
IPL-222	Landslide risk analysis and mitigation in the ancient rock-cut city of Vardzia (Georgia)	Scientific and Technological Attaché Embassy of Italy in Egypt —UNESCO Chair at Florence University (Italy)	Italy	Claudio Margottini	2017–	On-going
IPL-223	Landslides in Africa: Understanding catastrophic failures and effective preventive measures in vulnerable regions of the continent	University of Nigeria, Nsukka	Nigeria	Igwe Ogbonnaya	2017–	On-going
IPL-225	Recognition of potentially hazardous torrential fans using geomorphometric methods and simulating fan formation	University of Ljubljana, Faculty of Civil and Geodetic Engineering	Slovenia	Matjaž Mikoš	2017–	On-going
IPL-226	Studying landslide movements from source areas to zone of deposition using a deterministic approach	Geological information Centre, Geological Survey of Slovenia	Slovenia	Mateja Jemec Aufli?	2017–	On-going
IPL-227	Development of a web based landslide information system for the landslides in Sri Lanka	Central Engineering Consultancy Bureau	Sri Lanka	K. M. Weerasinghe	2017–	On-going
IPL-228	General approach to landslide research and stabilization in Bosnia and Herzegovina	University of Tuzla—Faculty of Mining, Geology and Civil Engineering	Bosnia and Herzegovina	Sabid Zekan	2017–	On-going
IPL-230	Evolution-based key technology of landslide prevention in Three Gorges Reservoir region, China	China University of Geosciences (Wuhan)	China	Huiming Tang	2018–	On-going
IPL-231	Landslide mechanism considering Soil-Water-Vegetation coupling effects	Institute of Mountain Hazards and Environment, Chinese Academy of Sciences	China	Su Lijun	2018–	On-going

(continued)

Table 2 (continued)

No.	Title	Organization	Country	Leader	Year	Status
IPL-232	Investigations on landslides in Nilgiris, Tamil Nadu, India	Vellore Institute of Technology, Vellore, Tamil Nadu, India	India	S. S. Chandrasekaran	2018–	On-going
IPL-233	Archival Records and Documentation of Some Socio-economically Significant Landslides in India	National Institute of Disaster Management, New Delhi	India	Surya Parkash	2018–	On-going
IPL-234	Development of landslide detection system based on rainfall prediction and seismic aspect in Banjarnegara Region, Centre of Java, Indonesia	Agency for Meteorology, Climatology, and Geophysics of the Republic of Indonesia (BMKG Indonesia)	Indonesia	Dr. Munawar	2018–	On-going
IPL-235	EO4GEO—Towards an innovative strategy for skills development and capacity building in the space geo-information sector supporting Copernicus User Uptake	ISPRA-Italian Institute form Environmental Protection and Research	Italy	Luca Guerrieri and Daniele Spizzichino	2018–	On-going
IPL-236	A multiparametric field laboratory for the investigation on the relationship between material behavior and morphodynamic of landslides	DIA—Università degli Studi di Parma	Italy	Andrea Segalini	2018–	On-going
IPL-237	The role of time-dependent rock mass deformations and landscape evolution rates as predisposing factors for massive rock slope failures	Centro di Ricerca CERI—Sapienza Università di Roma	Italy	Carlo Esposito	2018–	On-going
IPL-238	Landslides Threatening Russian Cultural Heritage Sites	Russian State geological Prospecting University	Russia	D. N. Gobotsov	2018–	On-going
IPL-239	Detailed Interpretation and Evaluation of Dynamic Model Behavior of Pothupitiya Landslide Potential Area (Combined Ground Water and Slope Stability Dynamic Model under PC Raster environment)	Central Engineering Consultancy Bureau (CECB)	Sri Lanka	A. A. Virajh Dias	2018–	On-going
IPL-240	Global Lecture Series—Recent Advances on Landslide Analysis and Remediation	California State University, Fullerton & Tribhuvan University, Institute of Engineering	USA	Binod Tiwari	2018–	On-going
IPL-242	Studies of disasters related to natural and anthropogenic landslides in Brazil—Characterization of landslides triggers and impacts as a tool to rapid risk analysis	Center for Scientific Support in Disasters (CENACID)—Federal University of Paraná (UFPR)	Brazil	Renato Eugenio de Lima	2019–	On-going
IPL-243	Wildfire-related landslides in Italy: triggering mechanisms and propagation processes	Earth Sciences Department, University of Torino	Italy	Giuseppe Mandrone	2019–	On-going
IPL-244	Evolution mechanism and control of landslides induced by sudden rainstorm	China University of Geosciences (Wuhan)	China	Huiming Tang	2019–	On-going

(continued)

Table 2 (continued)

No.	Title	Organization	Country	Leader	Year	Status
IPL-245	Laboratory physical modeling of rainfall, slope deformation and landslides triggering	University of Calabria	Italy	Giovanna Capparelli	2019–	On-going
IPL-246	Classification and spatial distribution of landslides on dumps in brown coal basin in the Czech Republic	Brown coal research institute, Inc.; Charles University	Czech Republic	Martin Veselý	2019–	On-going
IPL-248	Innovation in slow-moving landslide risk assessment of roads and urban sites by combining multi-sensor multi-source monitoring data	Geotechnical Engineering Group (GEG), Dept. of Civil Engineering, University of Salerno	Italy	Dario Peduto	2019–	On-going
IPL-249	Development of early warning technology of rain-induced rapid and long-travelling landslides in Sri Lanka	ICL Headquarters & National Building Research Organization	Japan, Sri Lanka	Kazuo Konagai & Asiri Karunawardena	2019–	On-going

SATREPS Project for Sri Lanka with Regard to "Development of Early Warning Technology of Rain-Induced Rapid and Long-Travelling Landslides"

2020 Kyoto Japan

Kazuo Konagai, Asiri Karunawardena, and Kyoji Sassa

Abstract

Influenced by the recent global climate change, extreme rainfall events have become more frequent worldwide and resultant hydro-meteorological hazards are creating more deaths and devastations. One of the most remarkable disasters of rain-induced rapid long-travelling landslides (RRLL) in Sri Lanka took place at Aranayake, 70 km east of Colombo in 2016 (JICA Survey Team 2016) (JICA Survey Team in Survey results of Aranayake Disaster, 2016). The fluidized landslide mass ran over an about 2 km distance claiming the lives of 125 people. This tragic event has thus highlighted the importance of sophisticated early warning and disaster management mechanism even more than ever, because the presence of these hidden unstable soil masses as well as their run-out distances are very difficult to predict, and once they start sliding, it is almost impossible to stop them. Both the National Building Research Organisation, Sri Lanka (NBRO) and the International Consortium on Landslides (ICL) have jointly compiled a research proposal within the framework of SATREPS, standing for "Science and Technology Research Partnership for Sustainable Development," a Japanese government program that promotes international joint researches, and it passed the final round of selection on May 16, 2019. Thus, the new 5-years SATREPS project for Sri Lanka with regard to "Development of early warning technology of Rain-induced Rapid and Long-travelling Landslides (Project RRLL)", starts in 2020. This article reports on the outline of the project including its goals, plans of plots for developing individual technologies for the early warning system, etc.

Keywords

Rain-induced rapid long-travelling landslide • SATREPS • Sri Lanka • Early warning

Introduction

Sri Lanka, being an isolated island in the southern tip of India, usually experiences extreme weather patterns with two peaks of rainfalls in two monsoon seasons (Department of Meteorology, Sri Lanka 2020). Particularly the south-western monsoon from May to September brings rain to the southwest mountainous area of Sri Lanka. Out of 25 administrative districts in Sri Lanka, ten districts, approximately 30% of the total land area of the Island, have been mostly prone to landslides. Landslides had previously been perceived as isolated events, which occurred mainly due to natural causative factors with low vulnerability. However, these areas had long been the major area of tea and cinnamon plantations, being occupied by about 35% of the population of Sri Lanka. Moreover, the post-civil-war Sri Lanka have been attracting tourists with 5 of 7 UNESCO cultural and natural world heritages of Sri Lanka located in the landslide-prone areas. Studies have revealed that nearly 70% of the landslides in Sri Lanka are influenced by human induced interventions in prone areas such as rapid urbanization, population growth, inappropriate land management, development trends, deforestation in steep slopes and degradation of forests and natural resources. One of the most remarkable disasters to be mentioned took place at Aranayake in Kegalle District, 70 km east of Colombo (JICA Survey Team 2016). The fluidized landslide mass ran over a 2 km distance killing 125 people. This tragic event has thus highlighted the importance of sophisticated early warning

K. Konagai (✉) · K. Sassa
International Consortium on Landslides, Kyoto 606-8226, Japan
e-mail: kaz3776k@gmail.com

K. Sassa
e-mail: sassa@iclhq.org

A. Karunawardena
Ministry of Defence, National Building Research Organization, Colombo 05, Sri Lanka
e-mail: awardena@gmail.com

K. Sassa et al. (eds.), *Understanding and Reducing Landslide Disaster Risk*, ICL Contribution to Landslide Disaster Risk Reduction,
https://doi.org/10.1007/978-3-030-60196-6_12

and disaster management mechanism, because the presence of these hidden unstable soil masses as well as their long run-out distances are very difficult to predict, and once they start sliding, it is almost impossible to stop them.

Both the National Building Research Organisation, Sri Lanka (NBRO hereafter) and the International Consortium on Landslides (ICL hereafter) have jointly compiled a research proposal within the framework of SATREPS (Japan Science and Technology Agency 2020); SATREPS, "Science and Technology Research Partnership for Sustainable Development," is a Japanese government program, with two funding organizations Japan International Cooperation Agency (JICA hereafter) and the Japan Science and Technology Agency (JST hereafter), that promotes international joint researches, and it passed the final round of selection on May 16, 2019 that the new 5-years SATREPS project for Sri Lanka with regard to "Development of early warning technology of Rain-induced Rapid and Long-travelling Landslides (Project RRLL hereafter)", starts in 2020.

Outline of Project RRLL

The technologies to stabilize reactivated and creeping landslide masses have much progressed because their locations can be easily identified. However, recent tragic events in Sri Lanka, showing a soaring trend in general with some remarkable spikes in 2014, 2016 and 2017 as shown in Figs. 1 and 2, highlighted the difficulty in coping with devastations caused by RRLLs. Neither their locations nor early signs of movement can be easily identified in advance. Therefore, implementation of advanced and feasible technologies for early warnings of RRLLs are extremely crucial.

There has been a remarkable development reported in the international world of academy with the ICL as the core organization; they include forecasting of localized precipitation events, early detection of ground movements and relaying timely early warning to the last mile, namely, residents at landslide risk. Among them, the key technologies for the success of Project RRLL are:

1. Time prediction of heavy rainfalls and pore water pressure build-ups,
2. Site prediction of landslide initiations and motions, and
3. Effective risk communication and public education.

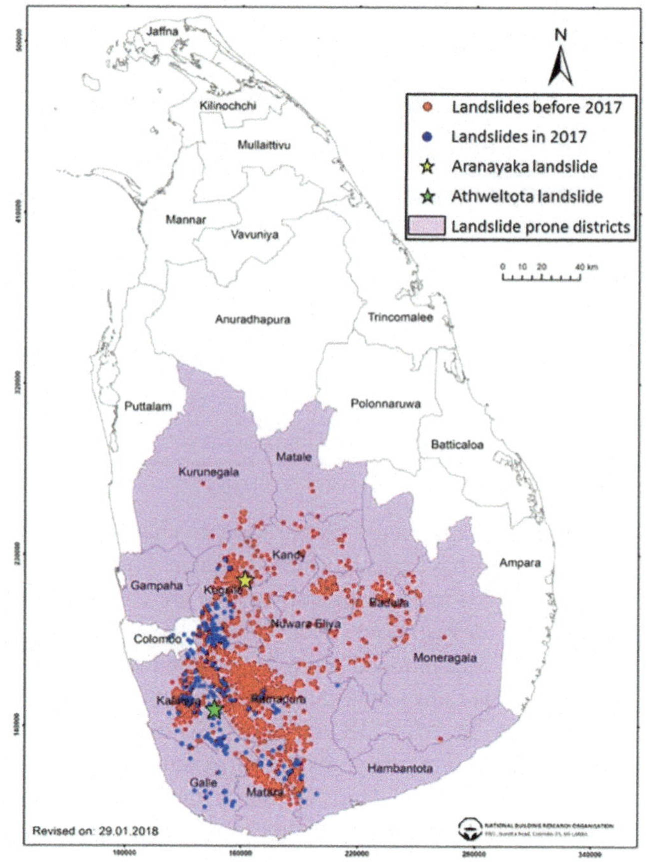

Fig. 2 Locations of landslides from 2003 to 2017 (credit: NBRO)

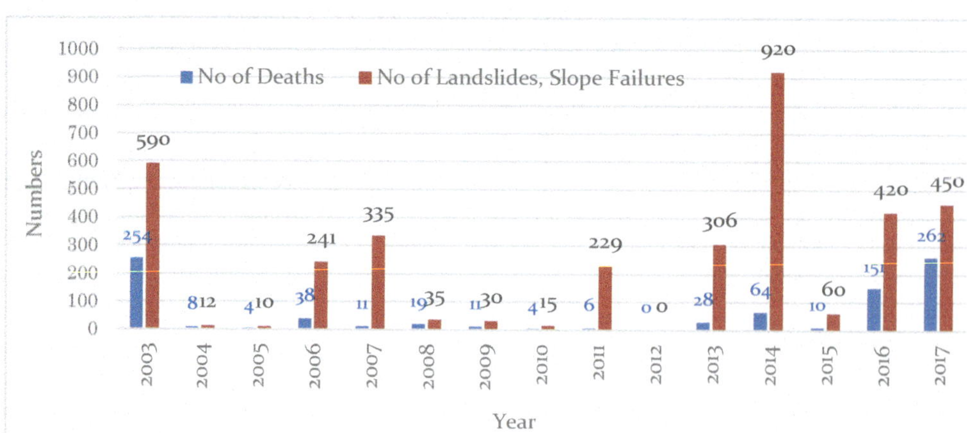

Fig. 1 Number of landslides and deaths from 2003 to 2017 (credit: NBRO)

This project is implemented by the NBRO of the Ministry of Defence (Formerly, Ministry of Irrigation and Water Resources Management and Disaster Management) with support from Department of Meteorology (DOM) and Disaster Management Centre (DMC), which are coming under the purview of Ministry of Defense and the Department of Irrigation (DOI) under the purview of Ministry of Mahaweli, Agriculture, Irrigation and Rural Development (Fig. 3). It is noteworthy that this project is complementary with the other ongoing JICA projects such as Project SABO for implementing some intangible measures including refining the current hazard maps and public educations.

Main Implementing Agency and Collaborating Entities in Sri Lanka

National Building Research Organization (NBRO), Ministry of Defence

Main Implementing Agency in Sri Lanka is National Building Research Organisation (NBRO), which has long been the national focal point for landslide risk management in Sri Lanka. NBRO is an autonomous institution established in 1984 and as assigned by the Government of Sri Lanka, NBRO has been carrying out landslide investigation

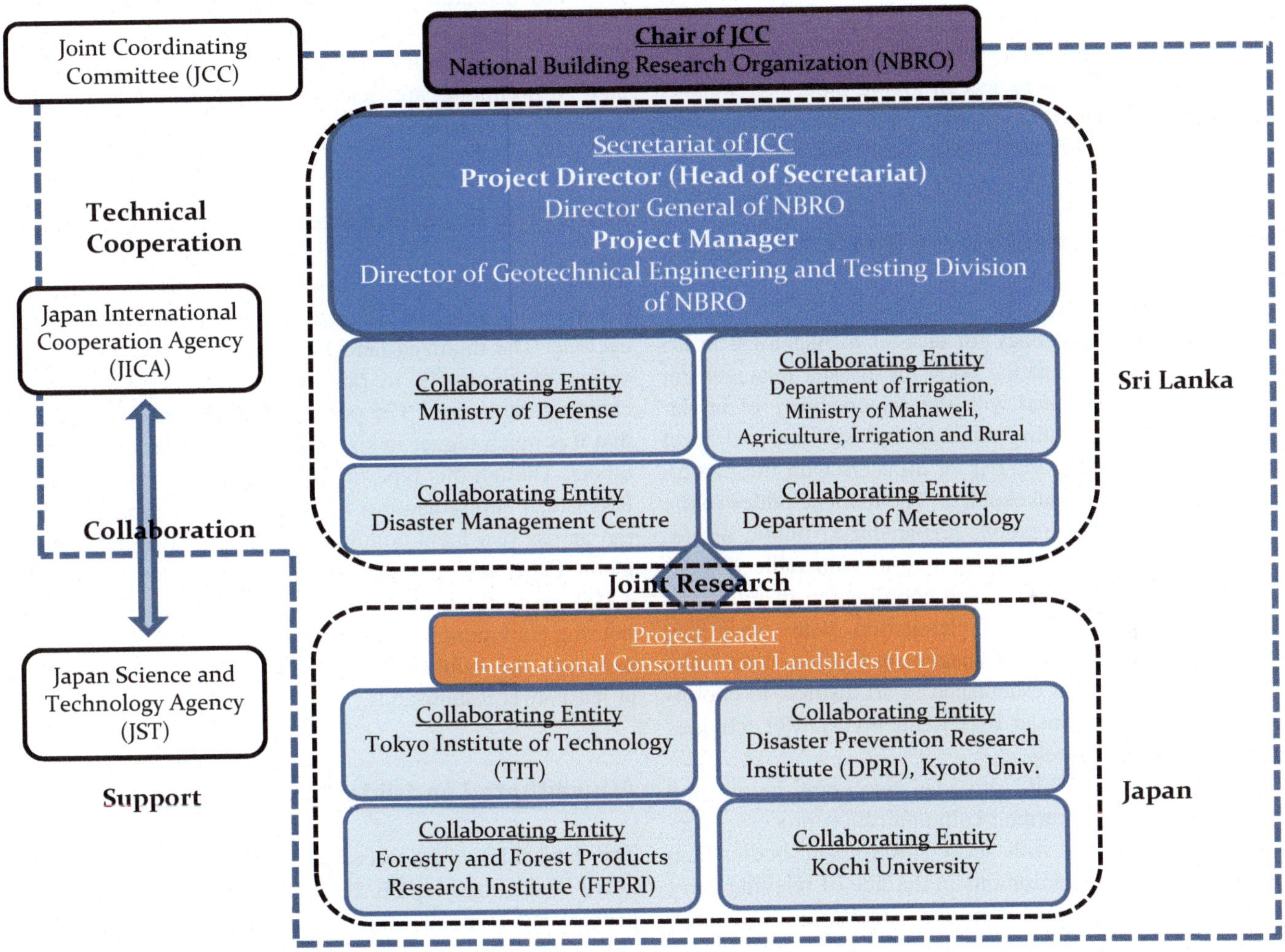

Fig. 3 Implementation structure for project RRLL (as of April 2020)

and studies since 1985. NBRO is a multi-disciplinary research and consultancy institution having its Head Office in Colombo and ten branch offices in landslide-prone districts of the country to implement landslide-disaster risk management activities. Real-time rainfall data from about 250 automated rain-gauges are currently being collected and used to issue early warning to vulnerable communities when rainfall intensities reach to certain threshold values.

Department of Meteorology (DOM), Ministry of Defence

DOM is the mandated national body for the provision of meteorological and climatological services as well as early warning services on weather related disasters and tsunami. It assists in the proposed project in giving inputs on weather forecasting, rainfall monitoring and prediction, thus strengthening the sediment disaster early warning process.

Disaster Management Centre (DMC), Ministry of Defence

DMC is the leading agency for disaster management in Sri Lanka. It is the coordinating body for disaster management in the country, mandated with the responsibility of implementing and coordinating national and sub-national level programs for reducing the risk of disasters with the participation of all relevant stakeholders. The main activities of the DMC are disaster mitigation, preparedness, public awareness, dissemination of early warning to vulnerable populations, emergency operations, and coordination of relief and post disaster activities in collaboration with other key agencies. District Disaster Management Coordination Units (DDMCUs) have been established in all districts to coordinate and carry out Disaster Risk Reduction (DRR) activities at the sub national levels.

In this project, DMC provides necessary inputs and support in the early warning communication process, especially in the liaison with local government bodies and community-based organisations in the line of relaying early warning communication from the Emergency Operation Centre of the DMC to communities at large, in the villages.

Department of Irrigation (DOI), Ministry of Mahaweli, Agriculture, Irrigation and Rural Development

DOI, with over a century of experience as a pioneer organization, takes responsivity for most of the development works in the irrigation sector. In this project, DOI plays an important role, as one of the most important stakeholders, to help develop a strategy for social implementation of the RRLL early warning system.

Pilot Study Sites

Geologically, the island of Sri Lanka is an extension of Peninsular India and forms part of the Indian Shield, one of the oldest and most stable parts of the earth's crust. Previous studies have suggested that the greater part of the landslide-prone areas is draped thick with weathered gneiss metamorphosed during Precambrian Era. The tropical climate has favoured deep weathering of these metamorphic rocks reaching tens of meters in these mountains with thick tropical vegetation drapes. Two pilot study sites, Aranayake and Athwelthota, are selected as representatives of two major types of RRLL (Fig. 4).

Aranayake Landslide Area

Aranayake landslide was triggered on May 17, 2016, by exceptional rainfall associated with a slow-moving tropical cyclone. The fluidized landslide mass from the relative elevation of about 600 m ran over an about 2 km distance claiming the lives of 125 people. This landslide is unique in that it is much bigger in size and its runout distance than the others. Though this type of landslides rarely occurs, a large RRLL can surely cause a big disaster. This landslide mass ran across two local communities, Elagipitiya and Debathgama Pallebage, having populations of about 1,500 and 1,100, respectively. Summing up populations of the neighbouring communities with similar risks of this type of RRLL expected, the number of beneficiaries of this project will be at least several thousands.

Athwelthota Landslide Area

This landslide, which occurred in Athwelthota area, Buduraliya District on May 26, 2017, destroyed 9 houses, killed 9 people and stopped traffic on a national highway. Each individual landslide of Athwelthota type will not cause surprisingly large disaster, but the number of landslides of this type can be very large causing extensive losses of human lives and properties.

During heavy rain of 2017, 37 RRLLs reportedly took place all at once claiming the lives of 262 people. In the above two pilot study sites, there remain unstable soil masses perching in and around atop of the exposed bare earth. Necessary pieces of equipment will be installed on/in these soil masses to measure causal factors of landslides and

Fig. 4 (Left) Alanayake landslide in 2016, and (right) Athwelthota landslide in 2017 (credit: NBRO)

creeping deformations of these soil masses; these pieces of equipment include pore pressure sensors, inclinometers, borehole extensometers, etc. Furthermore, the movements of these soil masses will be monitored with Persistent Scatterer Interferometric Synthetic Aperture Radar (PSInSAR), a remote sensing technique that uses radar signals from a satellite to accurately measure ground displacement. Using this technique, the motion of each scatterer structure can be very precisely tracked and ground deformation can be determined. These measurements will help develop infiltration models for the weathered gneiss in these areas.

Technologies to Be Developed

This project has the following three groups, G1, G2 and G3:

G1 works as a hub for this joint research, and integrates individual technologies developed at two pilot sites by Groups 2 and 3. Capacities of scientists/researchers of Sri Lanka are strengthened through this activity.

G2 develops technologies for (1) 24 h in-advance prediction of heavy rainfalls, and (2) assessing groundwater pressure build-up, initiation of an RRLL and its flowing dynamics.

G3 strengthens RRLL risk communication protocol, developing an augmented reality system for shearing predicted risk information and providing public education to develop capacities of the communities.

As said before, key technologies that will be developed in the above-mentioned three groups are for (1) precise weather forecast in mountain regions, (2) predicting groundwater pressure build-up, identifying locations of RRLLs and their moving areas, and (3) effective risk communication and public education. Here follow their details.

Precise Weather Forecast in Mountain Regions

Though the south-west region of Sri Lanka, where the south-western monsoon brings heavy rain between May to September, is our target region for precise weather forecast, it is desirable for the technology to be flexibly applied to wherever we need it, focusing on its future applications worldwide. From this perspective, we use MSSG as our generic platform for precise weather forecast. MSSG, standing for Multi-Scale Simulator for the Geo-Environment, is a coupled non-hydrostatic atmosphere–ocean-land model developed at the Japan Agency for Marine-Earth Science and Technology (JAMSTEC) (e.g., Takahashi et al. 2007, 2013; Onishi and Takahashi 2012). The developers of this MSSG join the research activity of G2 to help develop the rainfall prediction system suitable for Sri Lanka. The Yin-Yang grid system (Fig. 5) is used in both atmospheric and ocean components of MSSG. This grid model looks exactly like a tennis ball with two identical skins stitched together, thus the grid spacing becomes quasi-uniform allowing seamless zoom-in and zoom-out to be made anywhere on earth. On this MSSG, precise topographic effect and boundary-layer turbulence effect (Seifert and Onishi 2016) on the cumulonimbus clouds development particularly over upwind slopes can be taken into account for the better 24 h-in-advance prediction of heavy rainfalls in Sri Lanka.

Given the initial condition of weather variables such as winds, temperatures, atmospheric pressures, etc., three days before the Aranayake Landslide of May 2016 in Sri Lanka, MSSG simulated cumulated rain falls at Aranayake satisfactory in a decent manner as shown in Fig. 6. The blue line in this figure is the measured cumulative rainfall, while the red one shows what MSSG simulated. This simulation in

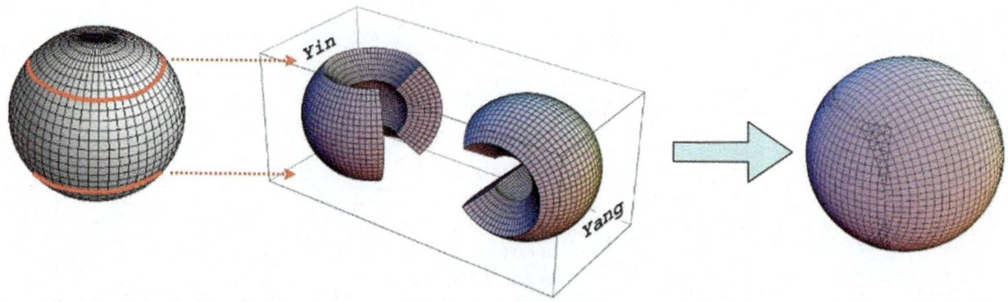

Fig. 5 Yin-Yang grid (https://www.research.kobe-u.ac.jp/csi-viz/members/kageyama/yin_yang_grid/index.html)

Fig. 6 requires huge computer resources, and thus was conducted on a super-computer at JAMSTEC with the upper-level computation nest covering a 200 km by 200 km area of the island and the lower-level nest at resolution of 500 m by 500 m. To make this simulation possible even on a workstation, parameters for the MSSG climate model are to be adjusted for better interpolation of weather at resolution of 2 km by 2 km. The obtained prediction at 2 km resolution will be spatially interpolated into 500 m resolution prediction with the aid of the super-resolution mapping technology based on the deep convolutional neural network (Onishi et al. 2019). This will realize the 24 h-in-advance reliable prediction of heavy rainfalls on a common workstation.

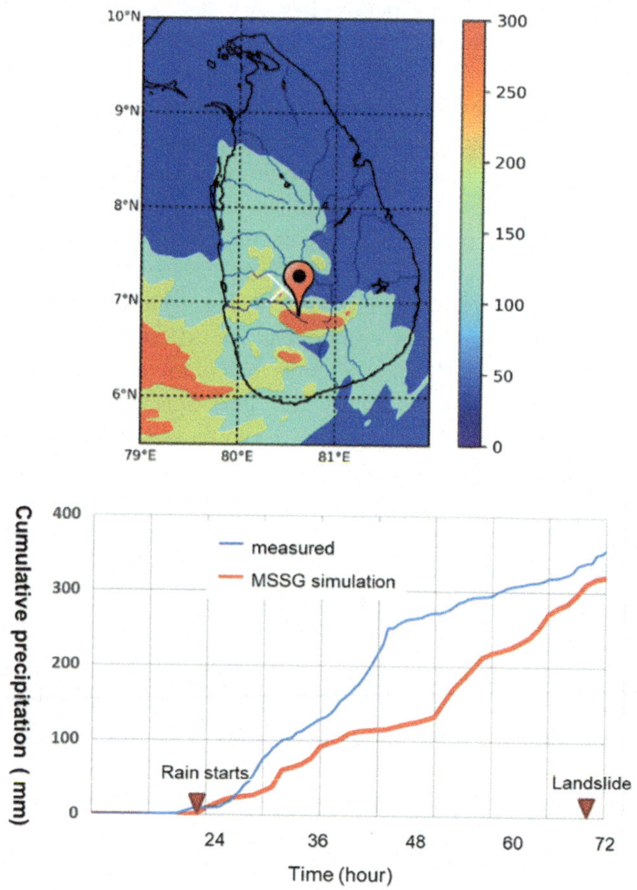

Fig. 6 Numerical simulation of 2016 Aranayake Rain on MSSG (by Onishi 2018)

Predicting Groundwater Pressure Build-Up, Identifying Locations of RRLLs and Their Moving Areas

This may not be an easy task given a basic question about how the surface cover-soil profile can vary over space in the vicinity of our pilot study sites. Though we are aware that this is a difficult task, we need at least to begin with the fundamental data that we will obtain from the two pilot sites. It is also worth verifying the feasibility of currently available numerical tools for evaluating how persistent rains would trigger a RRLL and how the fluidized soil mass of the RRLL would run out. Therefore, soil samples were taken from the source area of the Aranayake landslide (Fig. 7) for a series of undrained ring shear tests (Fig. 8). Given the parameters from the undrained ring shear test, a numerical tool, SLIDE (Liao et al. 2010) was first employed to reproduce the porewater-pressure buildup in the time domain, and then LS Rapid model (Sassa et al. 2010) was used to simulate how the landslide soil mass at Aranayake was detached and how the fluidized soil mass did run over that long distance of about 2 km (Fig. 9). These models for simulating infiltrations and fluidized soil dynamics need to be refined as we obtain much more pieces of information on site to be sure. However, the good agreement between the simulated and the observed runouts suggests that these tools will potentially be helpful for pursuing this project.

Effective Risk Communication and Public Education

It will take several hours to make one-day-in-advance forecast of the occurrence of RRLLs using the above-mentioned computer programs on high-performance workstations. The obtained results will then be relayed timely to the last mile as augmented reality dioramas of the predicted rains and

Fig. 7 Exposed slip surface immediately beneath the scar of Aranayake landslide. Soil samples were taken here. (N7.1554°, E80.4302°)

Fig. 8 Soil specimen mounted on high-speed high-stress ring-shear apparatus (Dang et al. 2019)

RRLLs with a real bird's-eye view of the terrain as its background on tablets' and/or PC screens (Fig. 10). The system physically allows for bi-directional communication between transmitter and receiver sides.

However, we need to take the followings into account for better implementation strategies:

1. In Sri Lanka, districts are the second level administrative divisions under provinces (Statoids 2015). Each district, administered under a District Secretary, is divided into several Divisional Secretary's (DS) divisions, which are further subdivided into extremely many local communities called "Grama Niladhari (GN)" divisions.
2. The number of local GN divisions in RRLL-prone areas is increasing year by year reflecting human interventions.
3. Though the Disaster Management Centre (DMC) of the Ministry of Défense has sole authority in terms of the issuance of evacuation alerts and evacuation orders, each Divisional Secretary (DS) can use his/her discretion in taking necessary actions in case of urgency.
4. NBRO has been helping their decisions providing them with alerting information of landslides.
5. In this regard, both NBRO and JICA has started a joint project "Project SABO" (NBRO and JICA 2020), for capacity strengthening on development of non-structural measures for landslide risk reduction. The goals include: (a) strengthening of hazard mapping and risk assessment capacity, and (b) revision of regional-level early warning issuance criteria, etc. Thus, Project RRLL is expected to play a complementary role vis-à-vis Project SABO, providing proactive information allowing people to take a precaution against RRLLs (Fig. 11).

How the Project has Come Up with Conclusions of Official Agreements

A signing ceremony for the Minutes of Meeting (MM) between NBRO, Sri Lanka and JICA, Japan was held on Oct. 15 at the auditorium of NBRO (Fig. 12). Mr. Satoshi Nakamura, Leader, Detailed Planning Survey Team, JICA, Japan, and Eng. (Dr.) Asiri Karunawardena, Director General, NBRO, Sri Lanka, signed the MM toward the implementation of Project RRLL.

In the same signing ceremony, Collaborative Research Agreement (CRA) between NBRO, Sri Lanka and ICL was also signed by Eng. (Dr.) Asiri Karunawardena, Director General, NBRO, Prof. Kazuo Konagai, Leader on the Japanese side of Project RRLL, Research Director at ICL, Prof. Kyoji Sassa, Secretary General, ICL and Dr. Kaoru Takara, Managing Director, ICL.

Record of Discussions of the Project RRLL was signed by the following officers: Mr. Fusato Tanaka, Chief Representative, JICA Sri Lanka Office, Eng. (Dr.) Asiri Karunawardena, Director General, NBRO, Major General (Retired). Kamal Gunaratne, Secretary, Ministry of Defence and Mr. Ajith Abeysekera, Director General, Department of

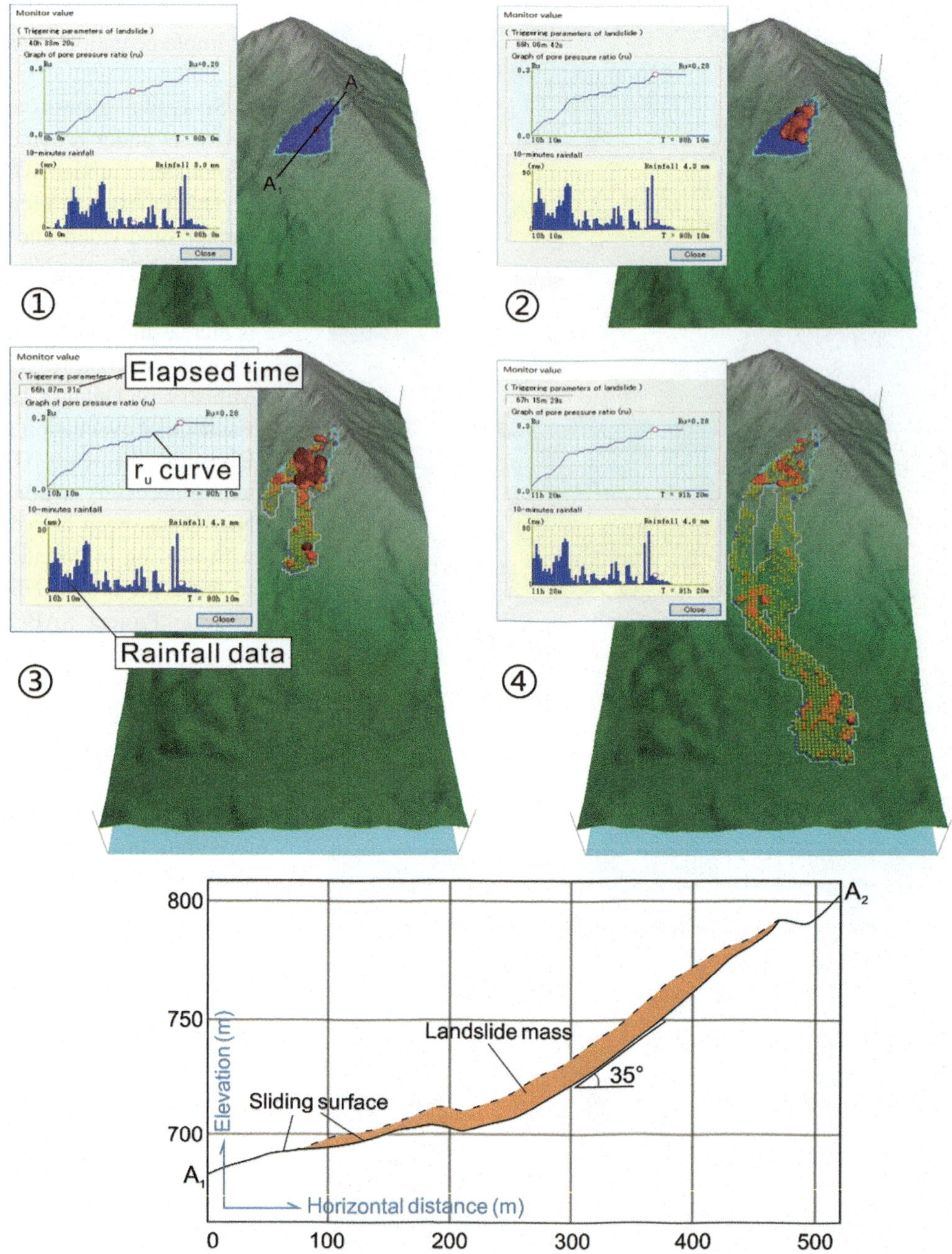

Fig. 9 Numerical simulation of the runout of the Aranayake landslide using LS-RAPID (by Tan et al. 2020)

External Resources, Ministry of Finance. This Record of Discussion is an official agreement between both the governments, to confirm implementation of the 05 years Project RRLL starting on Feb. 5, 2020.

Technical Cooperation Agreement for pursuing the Project RRLL over the 5-years period from March 1, 2020 to Feb. 28, 20,205 has been implemented between JICA and ICL.

JICA and ICL signed contract for the first year of the Project RRLL with the consent of the both parties. The first year starts on March 1, 2020 and ends up on May 31, 2021.

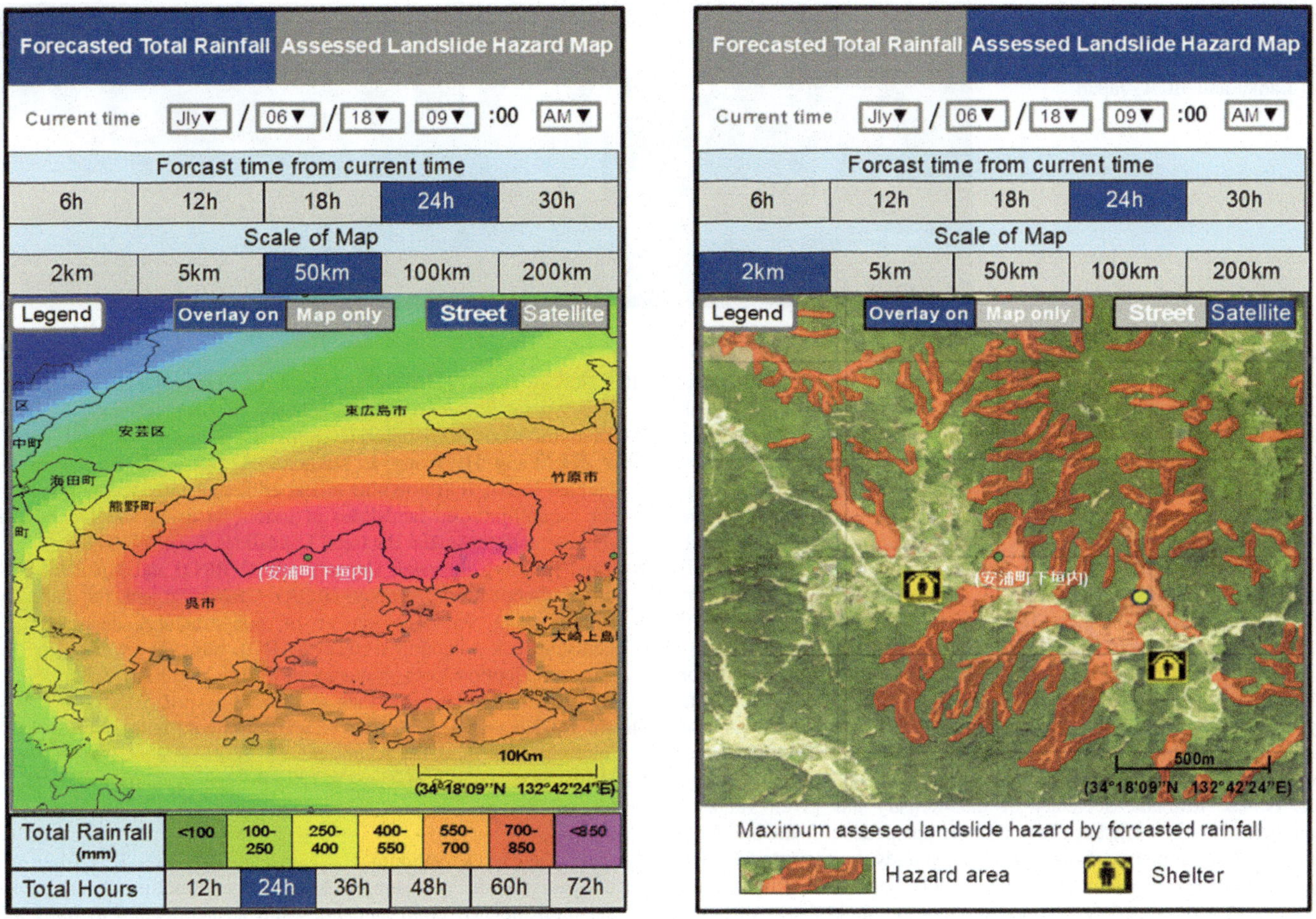

Fig. 10 Rendering image of augmented reality dioramas of the predicted rains and locations of RRLLs with a bird's-eye view of the area as their background on tablet's screen

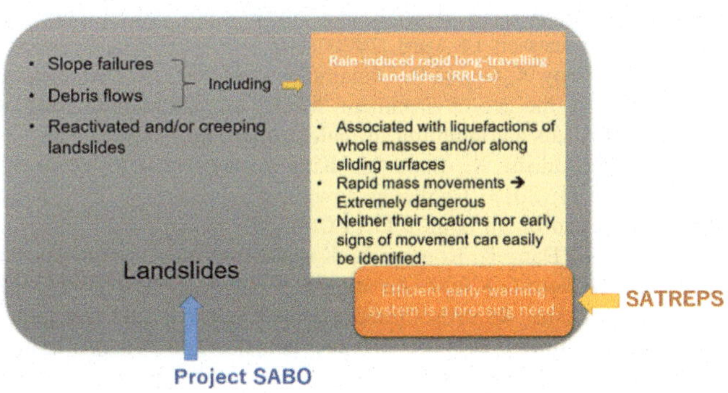

Fig. 11 Rain-induced landslides to be covered by both project SABO and project RRLL (SATREPS)

Summary

Influenced by the effects of global climate change, and more seriously, by human-induced interventions in landslide-prone areas, number of tragic RRLL events in Sri Lanka has been on a soaring trend in general with some remarkable spikes in 2014, 2016 and 2017. In this situation, ICL and NBRO are starting this project "Development of early warning technology of rain-induced rapid and long-travelling landslides in Sri Lanka" (Project RRLL). The project is in line with the activities by the Sri Lanka Comprehensive Disaster Management Programme, output 1.3 "National and community level landslide early warning system are in place". NBRO is currently running one more JICA technical cooperation project, Project SABO for capacity strengthening on development of non-structural measures for landslide risk reduction. Project RRLL is thus complementary with Project SABO providing extra lead time for emergency responses, evacuations, namely, one of the most important missing pieces of the jigsaw puzzle for landslide-hazard mitigation.

Acknowledgements In the midst of the currently ongoing coronavirus pandemic, we are now figuring out how we will firmly pursue Project "RRLL" in line with the strategy that we have developed. The good thing is that not only scientists but also all supporters from both governments are united in our common belief that the project will surely contribute to the Sustainable Development Goals (SDGs) of the United

Fig. 12 Signing ceremony on Oct. 15, 2019 for the Minutes of Meeting (MM) between NBRO, Sri Lanka and JICA, Japan

Nations, especially Goal 11 "Make cities and human settlements inclusive, safe, resilient and sustainable" through the landslide risk reduction for human settlements in mountainous areas and urban areas close to mountains. It would not have been possible for the authors to reach this start-up stage of the project without the support of Mr. Satoshi, Nakamura, Director, Mr. Naoya Orita, Assistant Director at the Disaster Risk Reduction Team, JICA, Prof. Takashi Asaeda, Research Supervisor, Mr. Kazuo Anazawa, Senior Associate Research Supervisor at JST, Mr. Takayuki Nagai, JICA expert, Mr. Kiyofumi Takashima, JICA Srilanka Office. The authors must also thank total 20 officers and scientists from NBRO, DOM, DMC, DMI of the ministry, Central Engineering Consultancy Bureau (CECB), and three major universities in Sri Lanka, who attended the RRLL Workshop at the Ministry of Irrigation, Water Resources and Disaster Management (supervisory authority of NBRO of that time) on June 20, 2018. The authors have largely been inspired by comments from them. Especially, Mr. N.A.S. Kumara, Secretary of the ministry has given valuable comments in this workshop from a broad perspective. Valuable comments and useful pieces of information to embody the project were also provided by Mr. Kenichi Sugawara, Ambassador Extraordinary and Plenipotentiary, successive Second Secretaries at the Embassy of Japan in Sri Lanka, and Mr. L.J.M.G. Chandrasiri Bandara, District Secretariat of Kegalle. Last but not least, the authors are greatly indebted to Dr. Kiyoharu Hirota, Dr. Khang Dang, Dr. Ms. Kumiko Fujita and Ms. Emi Ueda at ICL, for their ceaseless efforts to support the project.

References

Dang K, Sassa K, Konagai K, Karunawardena A, Bandara RMS, Hirota K, Tan Q, Ha ND (2019) Recent rainfall-induced rapid and long-traveling landslide on 17 May 2016 in Aranayaka, Kagelle district, Sri Lanka. Landslides 16:155–164

Department of Meteorology (2020) Sri Lanka, Climate of Sri Lanka. https://www.meteo.gov.lk/index.php?option=com_content&view=article&id=94&Itemid=310&lang=en. Last accessed 14 April, 2020

Japan Science and Technology Agency (2020) SATREPS for the earth, for the next generation. https://www.jst.go.jp/global/english/. Last accessed 14 April, 2020]

JICA Survey Team (2016) Survey results of Aranayake Disaster, JICA. https://www.jica.go.jp/srilanka/english/office/topics/c8h0vm00006ufwhl-att/160720.pdf. Last accessed 14 April, 2020

Liao ZH, Yang H, Wang J, Fukuoka H, Sassa K, Karnawati D, Fathani F (2010) Prototyping an experimental early warning system for rainfall-induced landslides in Indonesia using satellite remote sensing and geospatial datasets. Landslides 7(3):317–324

National Building Research Organization (NBRO) and Japan International Cooperation Agency (JICA) (2020) Project for capacity strengthening on development of non-structural measures for landslide risk reduction (Project SABO). https://www.nbro.gov.lk/index.php?option=com_content&view=article&id=197&catid=2&Itemid=101&lang=en. Last accessed 14 April, 2020]

Onishi R, Takahashi K (2012) A warm-bin-cold-bulk hybrid cloud microphysical model. J Atmos Sci 69:1474–1497

Onishi R, Sugiyama D, Matsuda K (2019) Super-resolution simulation for real-time prediction of urban micrometeorology. SOLA 15:178–182

Sassa K, Nagai O, Solidum R, Yamazaki Y, Ohta H (2010) An integrated model simulating the initiation and motion of earthquake and rain induced rapid landslides and its application to the 2006 Leyte landslide. Landslides 7(3):219–236

Seifert A, Onishi R (2016) Turbulence effects on warm-rain formation in precipitating shallow convection revisited. Atmos Chem Phys 16:12127–12141

Statoids (2015) Divisions of Sri Lanka. https://www.statoids.com/ylk.html. Last accessed 14 April, 2020

Takahashi K, Peng X, Onishi R, Ohdaira M, Goto K (2007) Multi-scale simulator for the geo-environment: MSSG and simulations. Use of high-performance computing in meteorology, pp 36–54

Takahashi K, Onishi R, Baba Y, Kida S, Matsuda K, Goto K, Fuchigami H (2013) Challenge toward the prediction of typhoon behaviour and down pour. J Phys Conf Ser 454:012072

Tan Q, Sassa K, Dang K, Konagai K, Karunawardena A, Bandara R M S, Tang H, Sato G (2020) An attempt to estimate the past and the future landslide hazards based on soil testing and computer simulation around 2016 Aranayake landslide, Sri Lanka, Landslides, Landslides 17:1727–1738

Central Asia—Rockslides' and Rock Avalanches' Treasury and Workbook

2020 Kyoto Japan

Alexander Strom and Kanatbek Abdrakhmatov

Abstract

More than 1000 large-scale rockslides, rock avalanches and DSGSDs exceeding ca. 1 million cubic meters in volume have been identified in the Central Asia region embracing the Pamir, Tien Shan, and Dzungaria mountain systems that belong to six states—Afghanistan, China, Kazakhstan, Kyrgyzstan, Tajikistan and Uzbekistan. Most of the catastrophic slope failures are prehistoric and quantitative parameters (area of the deposits, total area affected, volume, runout, height drop, etc.) of about 60% of them have been measured. Arid climate and lack of forestation provide preservation of landforms created by such slope failures of various types and good expressiveness of outcrops eroded in the landslide deposits. The case studies from this region are very didactic and the Kokomeren River basin in Central Tien Shan was selected for a 2 weeks long field training course—the Kokomeren Summer School on Rockslides and Related Phenomena that has been running annually since 2006.

Keywords

Central Asia • Rockslide • Rock avalanche • Field training • Pamir • Tien Shan • Dzungaria

A. Strom (✉)
Geodynamics Research Center, LCC, 3rd Novomikhailovsky passage, 9, Moscow, 125008, Russia
e-mail: strom.alexandr@yandex.ru

K. Abdrakhmatov
Institute of Seismology, National Academy of Science of Kyrgyz Republic, Asanbay 52/1, Bishkek, Kyrgyzstan
e-mail: kanab53@yandex.ru

Introduction

The Central Asian region embracing the Pamir, the Tien Shan, and the Dzungaria mountain systems located in Afghanistan, China, Kazakhstan, Kyrgyzstan, Tajikistan and Uzbekistan (Fig. 1), is one of the global "landslide hotspots" (Nadim et al. 2006). More than 1000 large-scale rockslides, rock avalanches and DSGSDs exceeding ca. 1 million cubic meters in volume occur in these mountainous systems, most of which are prehistoric, though several impressive slope failures occurred here during the last two centuries for which the historical records of natural disasters are available.

Rockslides and rock avalanches of almost all known types and subtypes (Hungr et al. 2014) are present—both long runout and forming compact landslide dams, those in confined and unconfined conditions, on the steep and rather gentle slopes composed of different sedimentary, igneous and metamorphic rocks (Strom and Abdrakhmatov 2018).

Due to the rugged terrain, arid climate, and lack of forestation, landforms created by such slope failures of various types as well as outcrops of the resultant landslide deposits are well preserved and several are extremely impressive. It makes this region a real rockslides' and rock avalanches' 'treasury' and, at the same time, an excellent 'workbook' for students and young landslide researchers.

The idea to arrange a field training course in this region was born during the post-conference field trip of the NATO advanced research workshop "Natural and artificial rockslide dams" that was held in Bishkek in June 2004. In 2005, the first full-color guidebook was prepared within the framework of the M111 Project of the International Program on Landslides (IPL) and, since 2006, this training course has been organized annually, and supported by the International Consortium on Landslides. Since 2017, it has been supported also by the UNESCO Almaty Cluster Office.

K. Sassa et al. (eds.), *Understanding and Reducing Landslide Disaster Risk*, ICL Contribution to Landslide Disaster Risk Reduction,
https://doi.org/10.1007/978-3-030-60196-6_13

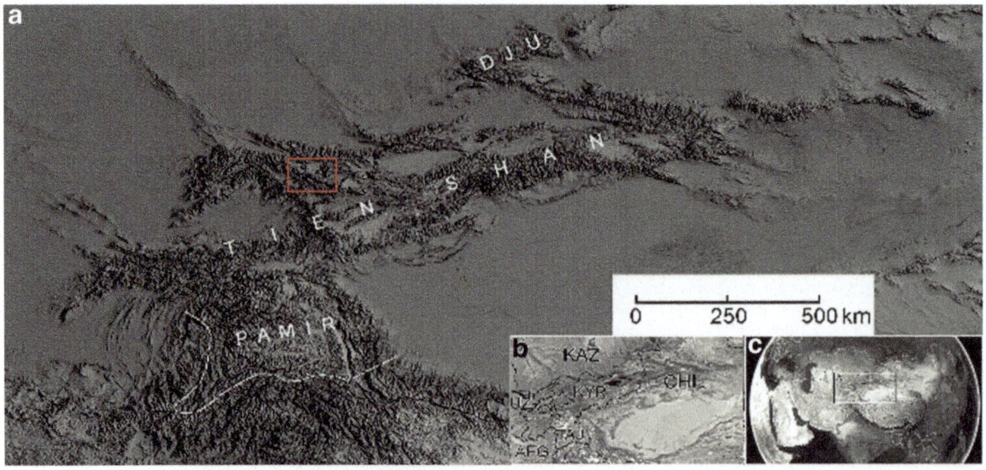

Fig. 1 **a** Study region; **b** political boundaries in the Central Asian region (from Google Earth); **c** location of the study region (outlined) in Asia. DJU, the Dzungarian Range; AFG, Afghanistan; CHI, China; KAZ, Kazakhstan; KYR, Kyrgyzstan; TAJ, Tajikistan; UZ, Uzbekistan. Dashed line indicates the conventional western and southern boundaries of the Pamir; red rectangle—area shown in Fig. 9. Modified after Strom and Abdrakhmatov (2018), with permission of Elsevier

Variability of Large-Scale Slope Failures Manifestations in Central Asia

Rock slope failures in the study region originate as rockslides of the rotational (Fig. 2), planar (Fig. 3), wedge (Fig. 4), and other types (Hungr et al. 2014).

Most of these failures converted into highly mobile rock avalanches, whose extreme mobility can be derived either from their long runout (Fig. 5) or from the significant runup (Fig. 6). Their morphologies strongly depend on the confinement conditions—a parameter that is critical for correct understanding of rock avalanche mobility (Shaller 1991; Strom and Abdrakhmatov 2018; Strom et al. 2019).

Besides overall morphological peculiarities allowing more detailed classification of rock avalanches based on the deposits' shape and debris distribution along the runout path (Strom 1996, 2006, 2010), many Central Asian case studies demonstrate impressive preservation of various minor landforms such as molards (Fig. 7) that also provide some information about processes acting during rock avalanche motion.

Bodies of numerous rockslides and rock avalanches have been deeply incised by erosion and good exposure of outcrops up to ca. 400 m high provide unique opportunity to investigate the internal structure and grain-size composition of the deposits. From these studies it was revealed that the vast majority of rock avalanches had moved as dry laminar granular flows retaining the same relationships between different lithologies involved in slope failure that characterize the internal structure of the host rock massifs (Fig. 8).

Fig. 2 The rotational Ornok rockslide, Central Tien Shan. Its headscarp is marked by yellow arrows. Frontal part of the rockslide converted into a rock avalanche, whose visible front is marked by orange arrows. Red arrows mark a thrust fault dividing Neogene and Palaeozoic deposits. Modified after Strom and Abdrakhmatov (2018), with permission of Elsevier

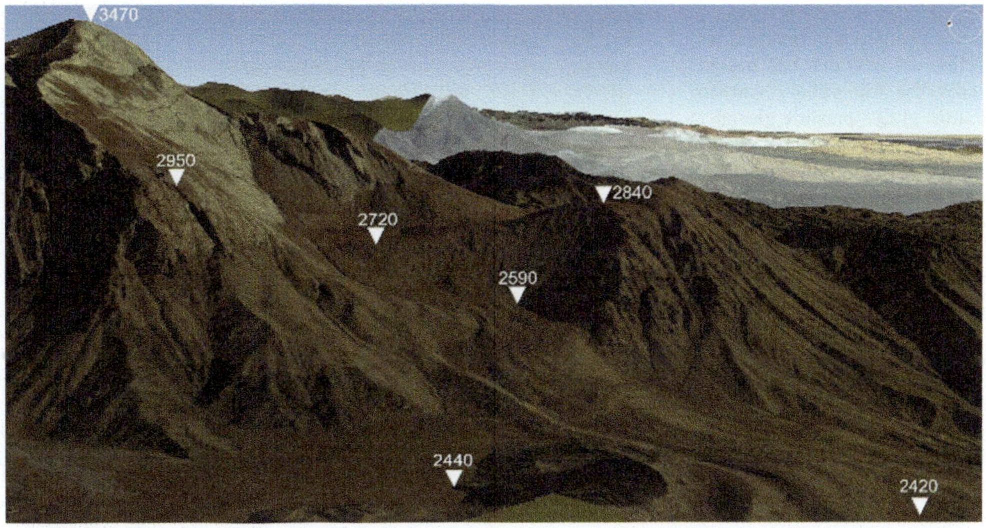

Fig. 3 The source and the proximal part of the Alasu planar rockslide deposition zone in Eastern Tien Shan that converted into long runout rock avalanche. 3D Google Earth eastward view. Varicoloured background is due to different space images available. After Strom and Abdrakhmatov (2018), with permission of Elsevier

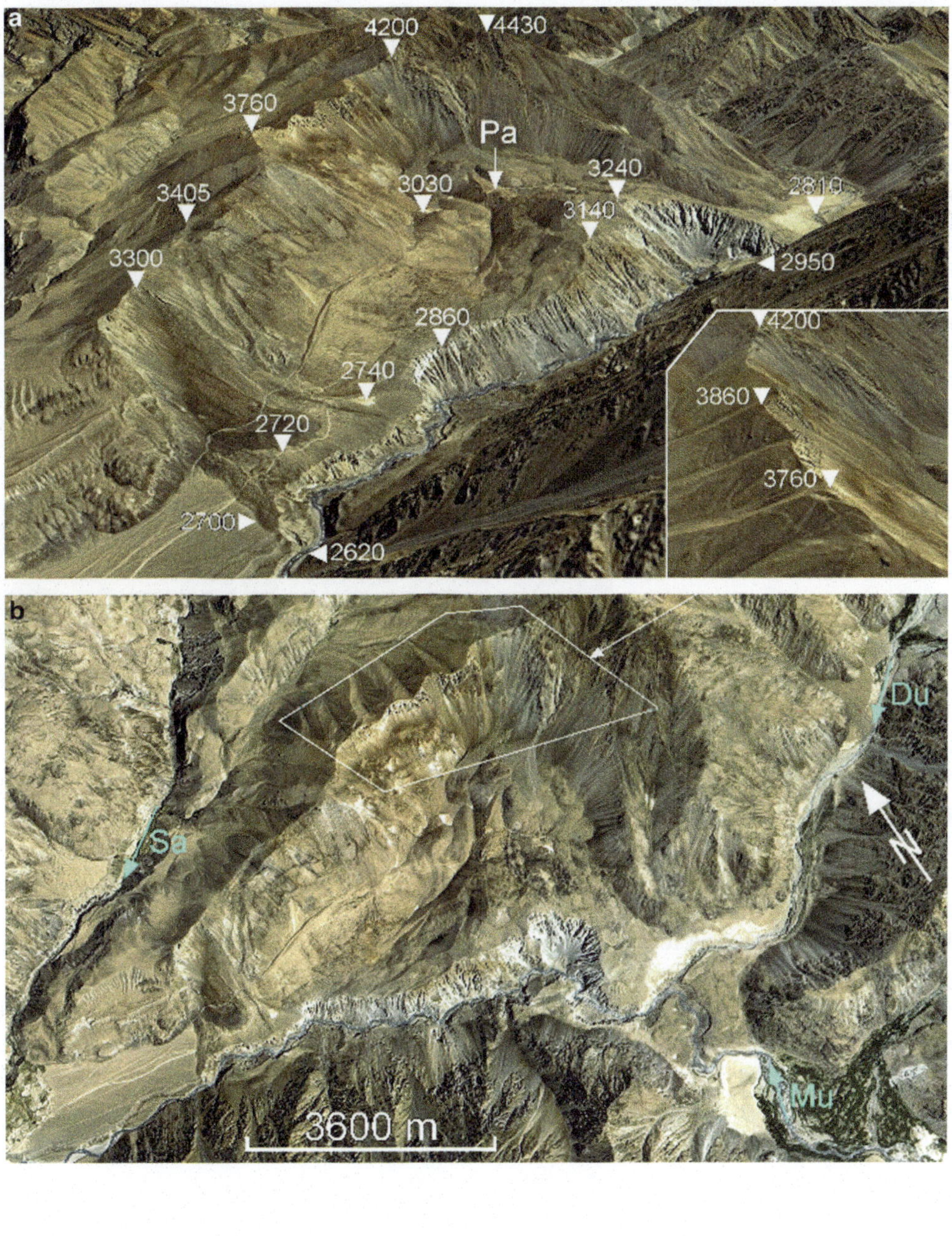

Fig. 4 The gigantic Pazhuk rockslide of the wedge type in Afghanistan. **a** 3D Google Earth view; **b** plan view. Beheaded gullies at the opposite side of the ridge between 4200 and 3760 m a.s.l. can be seen in the inset. Pa: the Pazhuk pass; Sa: the Sakhi River; Du: the Dura River; Mu: the Munjan River. After Strom and Abdrakhmatov (2018), with permission of Elsevier

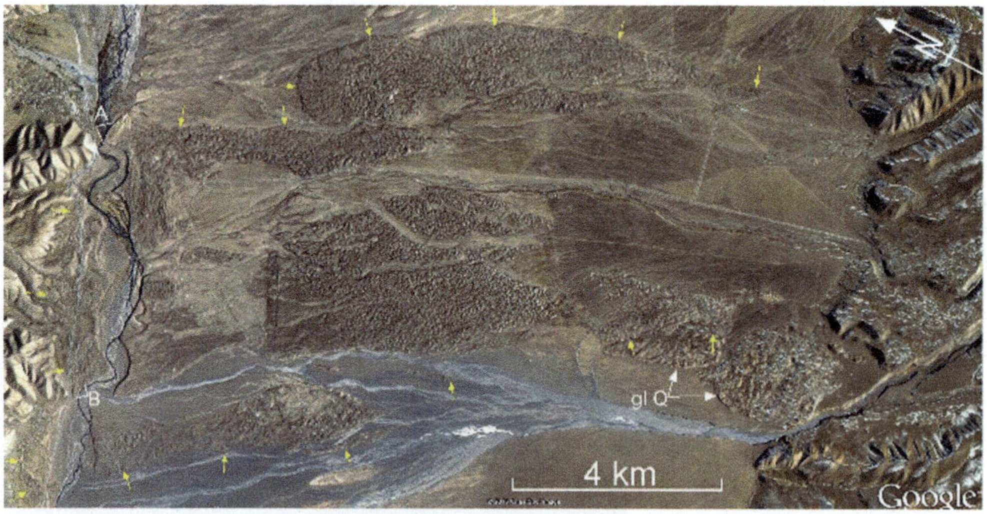

Fig. 5 The Komansu rock avalanche deposits (marked by yellow arrows) that had passed across the Alai valley, Northern Pamir. gl Q—preslide glacial deposits; **a** and **b**—sites where the internal structure of the deposits can be observed in details. Google Earth images. Modified after Strom and Abdrakhmatov (2018), with permission of Elsevier

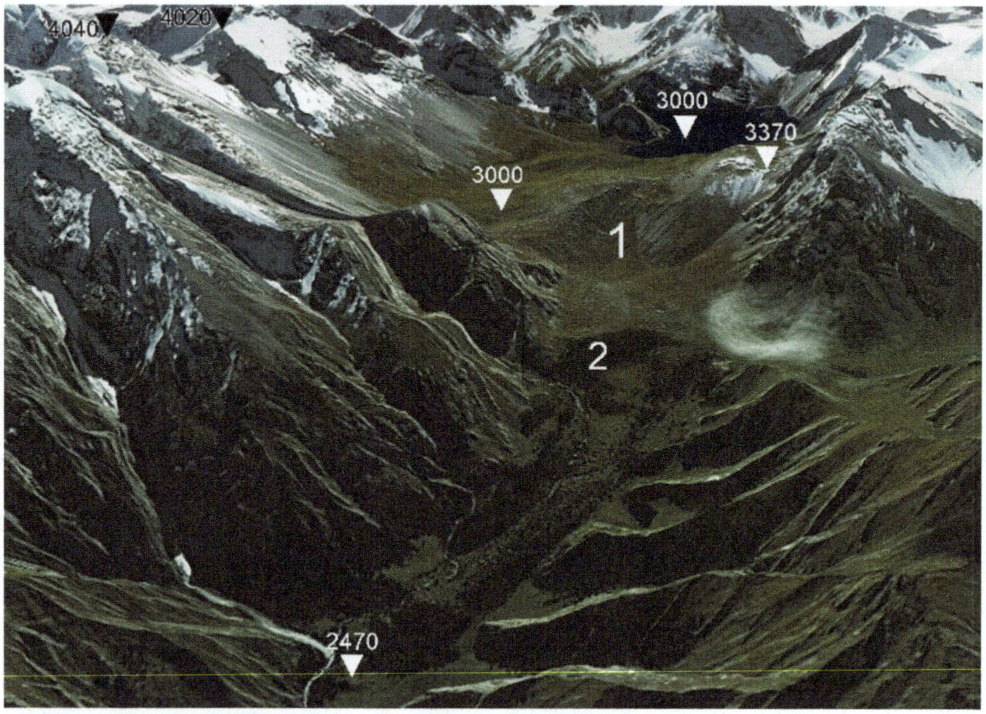

Fig. 6 The Primary Big Dragon lake rockslide about 2 km^3 in volume (Eastern Tien Shan) with 370 m run-up of its frontal part accompanied by the 4.3 km long deflected secondary rock avalanche that originated from the secondary scar 1 and moved up to 2470 m a.s.l. with an intermediate secondary scar 2. 3D Google Earth view. After Strom and Abdrakhmatov (2018), with permission of Elsevier

The Kokomeren Summer School on Rockslides and Related Phenomena

This two week long field training course aims to acquaint the attendees with the main morphological and structural peculiarities of rockslides and rock avalanches of various types, geological and geomorphic prerequisites of rock slope failures and geological evidence of river damming and of the disastrous outburst floods.

Since 2006, when the Summer School first started, 150 participants from Argentina, Austria, Belgium, China (including Hong Kong), Czech Republic, France, Germany, Great Britain, India, Italy, Japan, Kazakhstan, Korea, Kyrgyzstan, New Zealand, Norway, Poland, Russia, Slovakia, Slovenia, Switzerland, Spain, Taiwan, Tajikistan, USA and Uzbekistan have visited Tien Shan and participated in the daily field trips (Table 1).

The Kokomeren River basin was selected for a training course due to the unique concentration of very didactic rockslides and rock avalanches of different types ranging from a few million to more than 1 billion cubic meters in volume that originated on slopes composed of sedimentary, igneous and metamorphic rocks (Fig. 9). Some of them

Fig. 7 Molard on top of the Seit rock avalanche (Central Tien Shan). Its jigsaw puzzle structure reflects in-situ decomposition of large boulder ejected up from the deposits interior. Modified after Strom and Abdrakhmatov (2018), with permission of Elsevier

Fig. 8 Preservation of the internal structure in the deposits of the Kudara-Pasor rock avalanche (Western Pamir). Each layer can be traced within this body composed of heavily fragmented debris. After Strom and Abdrakhmatov (2018), with permission of Elsevier

Table 1 Number of participants of the Kokomeren summer school per year

Year	No. of attendees	Year	No. of attendees	Year	No. of attendees
2006	2	2011	10	2016	19
2007	4	2012	14	2017	16
2008	7	2013	8	2018	23
2009	10	2014	13	2019	20
2010	–[a]	2015	4		

[a]2010 was missed due to political crises in Kyrgyzstan

formed natural dams, both intact and completely dissected by erosion.

Besides numerous rockslides, rock avalanches, traces of valleys inundation and of powerful outburst floods, the study area provides impressive manifestations of the Neotectonics and Quaternary tectonics such as active faults, one of which was ruptured during the 1992 M7.3 Suusamyr earthquake, and numerous examples of tilted and folded pre-Neogene planation surfaces. One of the topics of the training course is to describe the paleoseismology of the region, paleoseismological interpretation of rockslides in particular.

All these features are located at a reasonably short distance from the base camp so that each of them can be visited in one day and require up to few hours of driving and hiking before reaching the site.

Unlike the LARAM and iRALL training courses that include both lectures in auditoriums and field visits to some sites (Casini et al. 2020; SKLGP 2016), the Kokomeren Summer School is a purely field training course—when organizing it we followed the proverb that exist in many languages: "it's better to see something once than hear about it a hundred times". Every day participants visit a new site guided by an experienced expert—Dr. Alexander Strom who started working in this region in 1990, found and described most of features demonstrated during Summer School (Strom and Abdrakhmatov 2006; Strom and Abdrakhmatov 2009, 2018).

First part of the training course is devoted to morphological peculiarities typical or rock avalanches of different types – primary, secondary and jumping.

Those of the primary type are exemplified by the Seit rock avalanche that originated as a rockslide of the irregular type and moved in laterally confined conditions. One more example is the Ak-Kiol rockslide dam originated as a wedge failure (collapse of the syncline core with hinge line dipping towards the river) that collided with the opposite valley slope. The Mini-Köfels rock avalanche, also of the primary type, is characterized by a very large—more than 200 m runup (see their location in Fig. 9).

Rock avalanches of the secondary type accompanied by the momentum transfer from the entire collapsing rock mass to its part retaining possibility of further motion are represented by several case studies (see Fig. 9). The Southern Karakungey rock avalanche collided with the opposite valley wall at an obtuse angle and its avalanche-like part turned almost 90° from the direction of its initial motion. At the Sarysu site, the collision that caused momentum transfer occurred with the slope base. The Snake-Head rock avalanche had passed through a funneled 'bottleneck'.

Jumping rock avalanches are represented by the Northern Karakungey and Kashkasu case studies. The latter formed a natural dam that remained intact and the dammed lake eventually silted completely.

Besides the case studies listed above, various morphological peculiarities of landslide bodies formed by the catastrophic rock slope failures and by rapid motion of the collapsing rock mass are demonstrated during field trips to other sites marked in Fig. 9.

The second part of the Summer School is focused mainly on the characteristic features of the internal structure of rockslide and rock avalanche deposits and on their grain-size

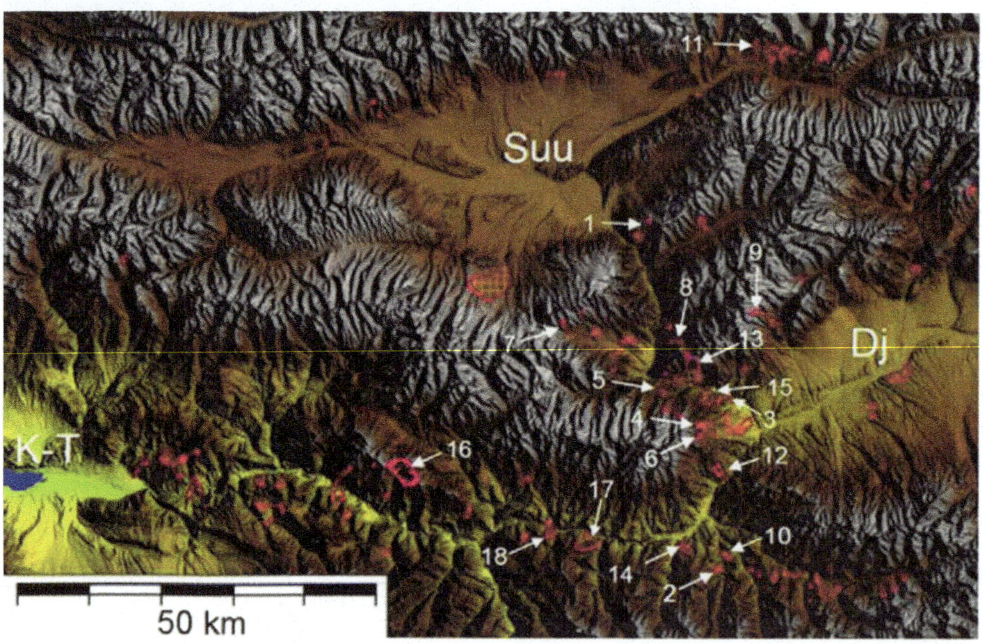

Fig. 9 Large landslides and rock avalanches in the Kokomeren River basin and adjacent part of the Naryn River basin. Suu, Dj and K-T—the Suusamyr, the Djumgal and the Ketmen-Tiube intermountain depressions. Selected features most of which are demonstrated during the training course: 1—Seit; 2—Ak-Kiol; 3—Mini-Kofels; 4—Northern Karakungey; 5—Kashkasu 6—Southern Karakungey; 7—Chongsu; 8—Sarysu; 9—Ming-Teke; 10—lower Ak-Kiol; 11—Snake-Head; 12—lower-Aral; 13—Kokomeren; 14—Ornok; 15—displaced Peneplain; 16—Kyzylkiol caldera-like cavity; 17—Karachauli; 18—lower Kokomeren

Fig. 10 Unmixed lithologies in the eastern part of the Lower Kokomeren rockslide body visible in the newly constructed road cut. MS—coarse metasediments' debris; FZ—tectonically comminuted rocks of the fault zone; N—debris of Neogene conglomerates; S—scree displaced from the base of the source zone; N-T—talus of Neogene conglomerates from the right bank of the Kokomeren River valley overlying rockslide deposits; RF—rockfall of Neogene conglomerates. After Strom and Abdrakhmatov (2018), with permission of Elsevier

Fig. 11 Participants of the 2019 Summer School together with local family in front of their yurt

composition. Several rockslides in the Kokomeren River valley, such as the Kokomeren, Ornok (see Fig. 2), Karachauli and Lower Kokomeren (Fig. 10) originated on high slopes composed of different types of rocks, whose debris can be easily recognized in rockslide bodies. It allows comparison of the mutual position of these rock types in situ and of debris after its emplacement that is extremely informative for a better understanding of rockslides and rock avalanches motion mechanism(s).

The detailed description of each case study visited during the field training course is presented in a full-color guidebook that is updated regularly and illustrated by numerous schemes, maps and aerial photographs. The Guidebook can be downloaded from the ICL website (https://icl.iplhq.org/).

During daily field trips we often meet local people staying in the summer shepherd's camps allowing students to be acquainted with the traditional life of these very kind and hospitable people (Fig. 11).

Fig. 12 Participants of the 2018 Summer school with certificates of attendance, our cooks and drivers. Photo taken at the base camp

Despite the Kokomeren River valley is located not very far from the Bishkek City—the Capital of Kyrgyzstan (see Fig. 1), it is a remote rural area without any hotels. That is why in our base camp located near the Aral village, on the Kokomeren River bank, all participants of the Summer School stay in tents with sleeping bags.

Electricity and mobile telephoning are available at the base camp. All participants receive a certificate of attendance (Fig. 12).

Future Plans and Conclusive Remarks

The 2020 Kokomerenm Summer School was cancel due to the Covid pandemia. In future, in the next years, we plan to continue this activity within the framework of the ongoing IPL-106-2 Project.

We also plan to rework the Summer School Guidebook adding detailed descriptions of the optimal way for each particular site with coordinates so that any interested researcher could arrange such a trip on their own if organizers will be unavailable.

Our experience gained during all the accomplished Summer Schools proves the efficiency of such field training. Explanation and discussion of what is visible at a particular site or outcrop help young landslide researchers to improve their knowledge of the phenomena in question significantly. A large variety of rockslide types and of the geological and geomorphic conditions of their occurrence learned in a short time provide a lot of information for comparative analysis—one of the most important method of studying and understanding natural phenomena.

Acknowledgements Authors are grateful to the executive director of ICL Prof. Kyoji Sassa for his permanent support of our activities. We also want to thank Kristine Tovmasyan and Natalya Kim from the UNESCO Almaty Cluster Office that cover all costs of participants from Kazakhstan, Kyrgyzstan, Tajikistan and Uzbekistan during last three years. We appreciative the anonymous reviewer for thorough check of the manuscript. And of course, we want to thank our drivers and cooks for their hard work during these field training courses.

References

Abdrakhmatov K, Strom A (2006) Dissected rockslide and rock avalanche deposits; Tien Shan Kyrgyzstan. In: Evans SG, Scarascia Mugnozza G, Strom A, Hermanns RL (eds) Landslides from massive rock slope failure. NATO Science Series: IV: Earth and Environmental Sciences, vol 49, pp 551–572. Springer, New York

Cascini L, Calvello M, Cuomo S (2020) LARAM school: an ongoing experience. This volume

Hungr O, Leroueil S, Picarelli L (2014) Varnes classification of landslide types, an update. Landslides 11:167–194

Nadim F, Kjekstad O, Peduzzi P, Herold C, Jaedicke C (2006) Global landslide and avalanche hotspots. Landslides 3:159–173

Shaller P J (1991) Analysis and implications of large Martian and terrestrial landslides, Ph.D. thesis. California Institute of Technology, Pasadena, CA, USA

SKLGP (2016) iRALL—International research association on large landslides. https://irall.sklgp.cdut.edu.cn/index.htm

Strom AL (1996) Some morphological types of long-runout rockslides: effect of the relief on their mechanism and on the rockslide deposits distribution. In: Senneset K (ed) Landslides. Proc. Of the Seventh International Symposium on Landslides, Balkema, Trondheim, Norway, Rotterdam, pp 1977–1982

Strom AL (2006) Morphology and internal structure of rockslides and rock avalanches: grounds and constraints for their modelling. In: Evans SG, Scarascia Mugnozza G, Strom A, Hermanns RL (eds) Landslides from massive rock slope failure, vol 49, pp 305–328. NATO Science Series: IV: Earth and Environmental Sciences

Strom AL (5–10 September, 2010) Evidence of momentum transfer during large-scale rockslides' motion. In: Williams AL, Pinches GM, Chin CY, McMorran TG, Massei CI (eds) Geologically Active. Proceedings of the 11th IAEG Congress, pp. 73–86. Auckland, New Zealand. Tailor & Francis Group, London

Strom A, Abdrakhmatov K (2009) International summer school on rockslides and related phenomena in the Kokomeren River Valley, Tien Shan, Kyrgyzstan. In: Sassa K, Canuti P (eds) Landslides. Disaster Risk Reduction, Springer, Berlin-Heidelberg, pp 223–227

Strom A, Abdrakhmatov K (2018) Rockslides and rock avalanches of Central Asia: distribution, morphology, and internal structure, p 449. Elsevier Netherlands, UK, USA (ISBN: 978-0-12-803204-6)

Strom A, Li L, Lan H (2019) Rock avalanche mobility: optimal characterization and the effects of confinement. Landslides 16:1437–1452

Results of Recent Monitoring Activities on Landslide Umka, Belgrade, Serbia—IPL 181

2020 Kyoto Japan

Biljana Abolmasov, Uroš Đurić, Jovan Popović, Marko Pejić, Mileva Samardžić Petrović, and Nenad Brodić

Abstract

Results of recent monitoring activities conducted from 2014 to 2019 are presented in the paper as a part of IPL 181 Project progress report. Recent monitoring activities are concentrated on several landslide monitoring techniques—automated GNSS monitoring system measurements, geodetic benchmark survey monitoring, UAV imaging, processing and analysis, and PSInSAR data processing and analysis. Results of all monitoring activities were analysed and used for cross-correlation and for verification of monitoring results obtained from different techniques. Displacement rates from GNSS measurements indicate that object point UmkaGNSS2 has moved 0.30 m towards the North and 0.50 m towards the West, while the vertical displacement was approximately −0.15 m for the 2014–2018 time span. Similar range of GNSS displacement rates were found in previously published results from monitoring activities realized from 2010–2014. PSInSAR data analysis also showed good correlation between nearest PS points and GNSS point for the same period of monitoring. Results from UAV and geodetic benchmarks survey showed very good correlation in displacement vectors' direction. According to the analyzed data it could be concluded that all monitoring results are in compliance with previous research results and confirm that the Umka is slow to very slow moving landslide with cyclic acceleration and deceleration phases.

Keywords

Landslide • Monitoring • GNSS • Geodetic survey benchmarks • UAV images • PSInSAR

Introduction

The IPL project No 181 titled "Study of slow moving landslide Umka near Belgrade" started in November 2012. Basic objective of the Project is to enable the analysis, correlation and synthesis of data obtained from various phases of investigation conducted on the Umka landslide after a few decades of research. Results received from geotechnical monitoring conducted during certain phases of research are compared with data from automated GNSS monitoring of last ten years and recent monitoring activities conducted in the last four years. Synthesis of research results help us define the mechanism and dynamics of this large, active, and slow landslide, with the final objective to propose adequate remedial measures. Project results would also help in better understanding of other landslides found on the right bank of the Sava river. More details about the project mission, objectives and goals can be found at Abolmasov et al. (2014, 2017).

Comprehensive analysis and results of previous geotechnical investigations and monitoring activities on Umka landslide from 2005–2014, were presented in Abolmasov et al. (2015). Recent monitoring activities are concentrated on several landslide monitoring techniques introduced after 2014—geodetic benchmark survey

B. Abolmasov (✉)
Faculty of Mining and Geology, University of Belgrade, Đusina 7, 11000 Belgrade, Serbia
e-mail: biljana.abolmasov@rgf.bg.ac.rs

U. Đurić · J. Popović · M. Pejić · M. Samardžić Petrović · N. Brodić
Faculty of Civil Engineering, University of Belgrade, Bul Kralja Aleksandra 84, 11000 Belgrade, Serbia
e-mail: udjuric@grf.bg.ac.rs

J. Popović
e-mail: jpopovic@grf.bg.ac.rs

M. Pejić
e-mail: mpejic@grf.bg.ac.rs

M. Samardžić Petrović
e-mail: mimas@grf.bg.ac.rs

N. Brodić
e-mail: nbrodic@grf.bg.ac.rs

K. Sassa et al. (eds.), *Understanding and Reducing Landslide Disaster Risk*, ICL Contribution to Landslide Disaster Risk Reduction,
https://doi.org/10.1007/978-3-030-60196-6_14

monitoring, UAV imaging, processing and analysis, and PSInSAR data processing and analysis, additionally to the existing GNNS monitoring system. The objective of this paper is to present the results of recent monitoring activities conducted from 2014 to 2019 as a part of IPL 181 Project report.

Study Area

The study area is located on the right bank of the Sava river, 25 km South-west of Belgrade, the capital of Serbia. Extensive geotechnical investigations and monitoring activities were conducted during several field campaigns in wider area during 1970–2006 (Vujanić et al. 1995; Mitrović and Jelisavac 2006). Most of the geotechnical investigations were performed for the Preliminary and Main Design for the Belgrade-Obrenovac Highway (E-763), and for the Umka urban plans and regulations. A summary of the geotechnical investigations results until 1995 can be found in Ćorić et al. (1996), while the summary of investigations and monitoring results until 2005 can be found in Mitrović and Jelisavac (2006).

Geometry, geological settings, mechanism and material properties of Umka landslide were well defined by previous geotechnical investigations. This landslide is fan-shaped, with the length along the slope of 900, 1650 m wide in the toe, reaching maximum depth of sliding surface at 26 m, and average slope gradient of 9°. Previous geotechnical research has shown that Umka landslide can be described as complex landslide within the stiff fissured Miocene (M_3^2) clayey marls. Landslide is active, with various phases of deceleration and acceleration, which are mostly in correlation with the Sava river level rise/drawdown, respectively, whereas landslide velocity is characterized as slow to very slow (Abolmasov et al. 2012, 2015).

The Umka landslide area is urbanized and populated with more than 490 inhabitants who are still living on the body of an active landslide. The state road IB 26 (from Belgrade to the state border with Bosnia and Herzegovina), is crossing landslide body and it is also constantly affected by slow displacement.

Previous Monitoring Activities (2010–2014)

Automated GNSS Monitoring

One of widely used system which is proven to be an effective and reliable tool for landslide monitoring is Global Navigation Satellite System (GNSS). Gili et al. (2000) give a general overview of the basic principles and discuss its applicability to landslide monitoring on Vallcebre landslide, in Spain. Since then, many published research papers presented successful landslide monitoring by GNSS and its integration with other observations (gained by other geodetic instruments such as automated total stations) across the world. GNSS landslide monitoring has proved its applicability especially for measuring surface deformations on large and slow-moving landslides (Mansour et al. 2011).

The first automated GNSS monitoring system in Serbia was established in March 2010, on Umka landslide (Abolmasov et al. 2012). The GNSS monitoring system consists of GNSS network and supporting software solution. The network is consisted of reference and object (monitoring) points on which GNSS stations (sensors) are mounted. Highly precise, multi-channel, multi-frequency systems (receivers and antennas) are used on all network points. Reference points are the integral part of the Active Geodetic Reference Network of Serbia (AGROS network), which is a permanent GNSS service of accurate satellite positioning over the Republic of Serbia.

The system is using two Leica Geosystems software solutions: GNSS Spider and GeoMoS (Geodetic Monitoring System). All observed GNSS measurements, with observation rate of 30 s, are collected by GNSS Spider and further forwarded, in a form of RINEX files, to GeoMoS Monitor and GeoMoS Analyzer on processing and further analysis, respectively.

The Umka landslide is represented by one object point (GNSS), which is located in the landslide body on the roof of a house (Abolmasov et al. 2012) (Fig. 1). This ten-years long project represents the longest continuous landslide monitoring in Serbia, and probably, one of the longest in the Balkan region. During these ten years of permanent monitoring, the GNSS network has changed one time due to the technical reasons, but the concept remained the same. The main change occurred due to the relocation of the Umka object point station (GNSS1), 25 m to Southwest—from one house to neighboring house in 2014 (Fig. 1). This change caused loss of more than 9 months of permanent monitoring, from the end of December 2013 until the September 2014 and the establishing new Umka monitoring point (GNSS2)(Fig. 1), already disscused and reported in Abolmasov et al. (2018).

During the first 45 months (March 2010–December 2013) the monitoring point Umka (GNSS1) has moved 0.46 m towards the North (Δx), and 0.70 m towards the West (Δy). Based on those results it can be concluded that the total 2D surface displacement was 0.84 m towards the Northwest, i.e. towards the Sava River. Furthermore, during the same period, the vertical displacement (Δz) of Umka GNSS1 sensor was nearly −0.30 m (Abolmasov et al. 2015).

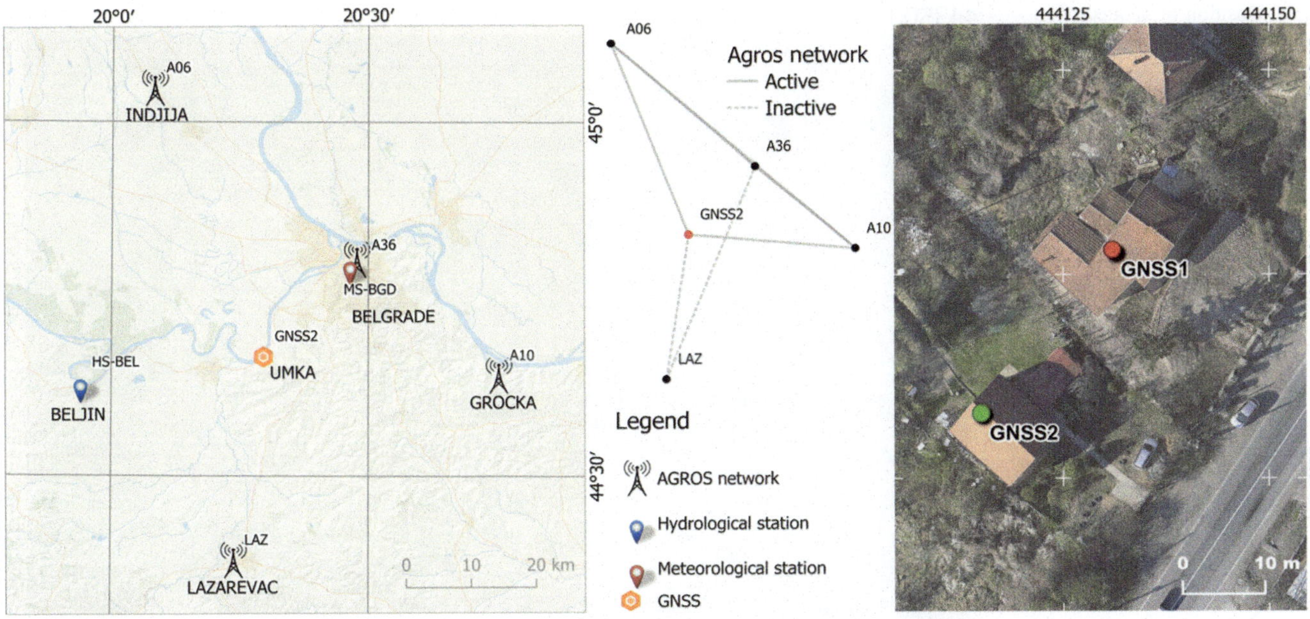

Fig. 1 Locations of Umka landslide area, GNSS stations, AGROS network stations, Belgrade Main meteorological station, and Beljin Sava river water level (hydrological) station

Recent Monitoring Activities (2014–2019)

In the past few years many authors integrated monitoring data from different sources to reduce uncertainities (Mateos et al. 2017, Casagli et al. 2017). In addition to the existing GNSS monitoring system on the Umka landslide, recent monitoring activities are composed of several newly introduced techniques for landslide monitoring: geodetic benchmark survey monitoring, UAV imaging with photogrammetric processing and analysis, and PSInSAR data processing and analysis. The main goals for introducing new monitoring techniques were: (1) to increase the number of surface monitoring points, (2) to test accuracy of existing and newly introduced monitoring techniques and (3) to compare monitoring data obtained from different techniques within same time span. Common to all implemented monitoring techniques is to measure displacement of the observed points (*dx*, *dy*, *dz*) on the landslide surface. Results of all monitoring data were analysed according to the measurments period and accuracy of monitoring techniques.

Data of climatological parameters and Sava River level are colected on daily basis from Hydrometerological Servise of Serbia from the begining of the monitoring project (2010), but correlation with monitoring results are not disscused and presented in this IPL181 Project report.

Geodetic Benchmarks Survey

In order to increse the number of surface monitoring points and to assess the reliability of photogrammetrically assessed displacements, conventional geodetic monitoring network was established during March 2018. The network initially consisted of 62 (1–62) object points, which were stabilized inside the landslide body and measured by RTK GNSS rover, as well as the four baseline points outside the landslide body in the stable ground. The high accuracy of the geodetic measurements in the research (positional <0.01 m and elevation accuracy <0.02 m), provides an accurate assessment of the displacements of object points. The location and distribution of the object points as well as baseline stable points are shown in Fig. 2. After processing, 59/62 points obtained from initial measurement were validated and they were used as “zero” measurement for further analysis. Two additional sequences of object points survey were conducted—in November 2018 and in March 2019. The main idea was to reconstruct surface displacement vectors from multiple sequences, as a represent of displacement within the landslide body. Displacement results were used for comparative analysis against photogrametric data obtained from UAV imaging covering the same period (2018–2019). Similar approach could be found in Peternel et al. (2017).

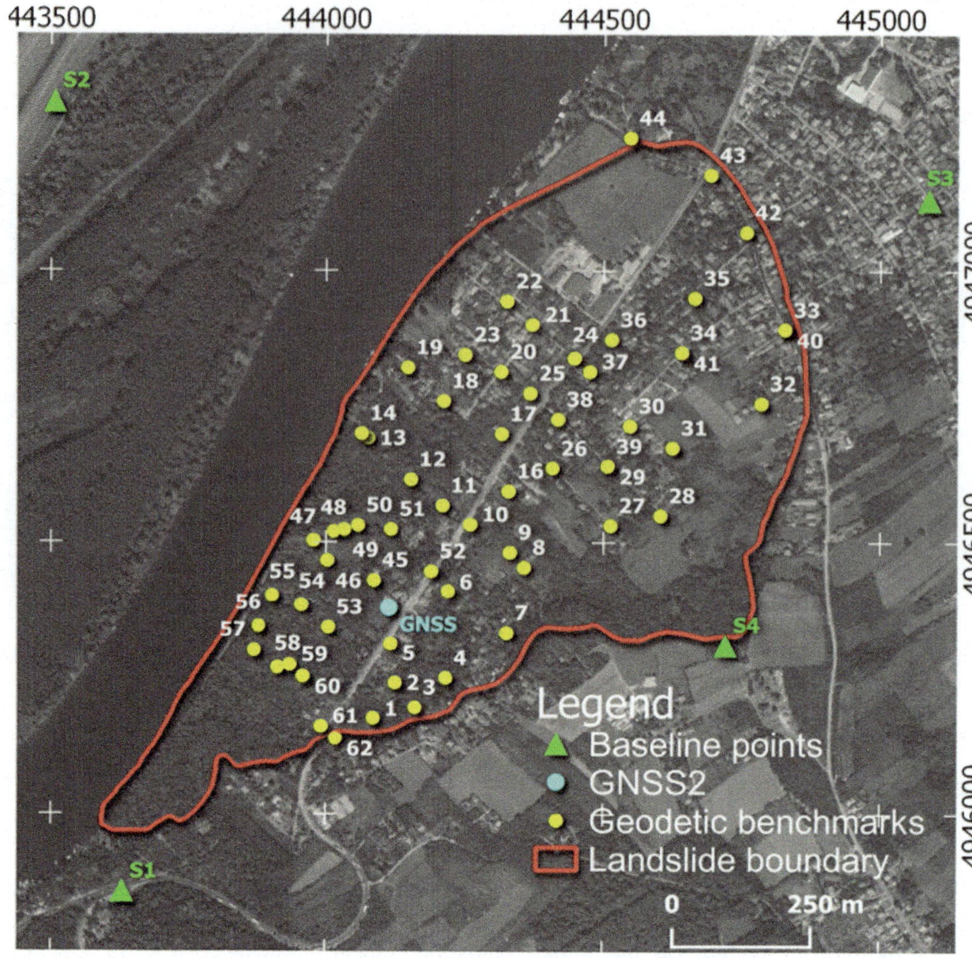

Fig. 2 Position of geodetic points and baseline points within Umka landslide area

UAV Imaging and Mapping

Using of small Unmanned Aerial Vechiles (UAV) systems and its implementation in Structure from Motion (SfM) photogrammetry is found very practical for landslide surface modeling and monitoring. Many advantages of UAV-based remote sensing for landslide characterization and monitoring were disccusesed in Colomina and Molina (2014), Balek and Blahut (2017), Peternel et al. (2017), Rossi et al. (2018). SfM method can provide accurate multitemporal and spatial surface products for landslide monitoring, primarily Digital Elevation Models (DEMs). Assuming the same procedure and same data quality, DEMs can be collected sequentially and compared. Change detection is used to reveal differences of the resulting DEMs, which are primarily caused by ground displacement.

Two aerial photogrammetric surveys were performed using aircraft DJI MATRICE 600 PRO industrial hexacopter with mounted DSLR camera Canon EOS 6D, resolution of 20.2 megapixels and focal length of 24 mm. More than 2000 images were taken by UAV during March 2018, and after manually removing blurred and oblique imagery, 1982 images were left for further processing. Forward image overlap was at least 90% and overlap between flight path rows was around 60%. UAV was flying at height of 80 m above take-off station achieving average pixel size of 2.2 cm. Seven flights were performed in order to cover the entire landslide area (Fig. 3).

"Zero" high resolution orthophoto and Digital Elevation Model (DEM) were principal SfM products from each sequence, as the same procedure was repeated in March 2019 and both obtained DEMs were analyzed and compared (DEM2018 vs DEM2019). These two sets of UAV aerial images (2018 and 2019) of the Umka landslide were used for generating the orthophotos, which were subsequently processed and analysed. Automated measurement and extraction of ground surface movements rate from these high resolution orthopohotos was conducted in Cosi-Corr software (https://www.tectonics.caltech.edu), with Statistical Correlator and kernel size 64 pix. Using Cosi-Corr software

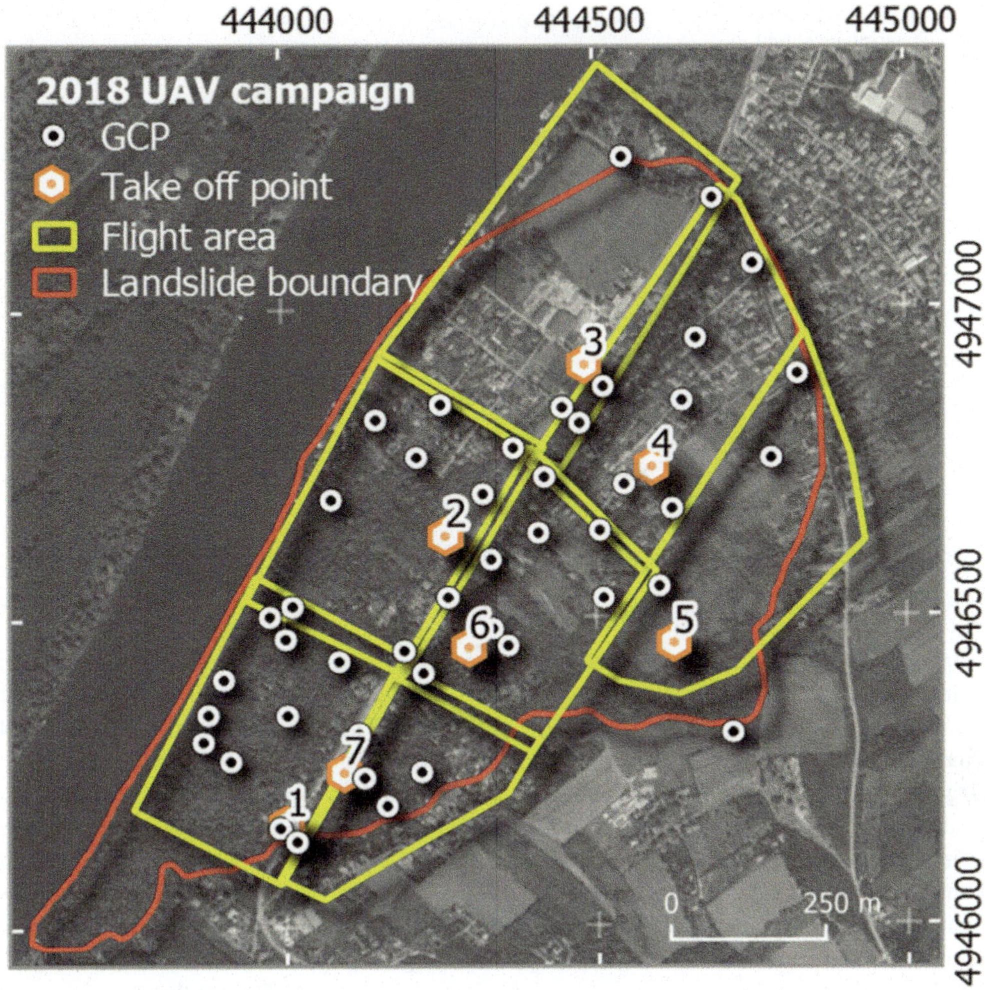

Fig. 3 UAV flight area (2018)

more than 25 k points were generated, and for each point 2D (x, y) vector and h (z) were calculated. According to the pixel size of processed images (2.2 cm) those generated points were filtered from noise, and only points with a higher displacement rate than 3 cm were selected and considered for further analysis and comparison. Those data were used for 2D points vector azimuth analysis and for comparison of points elevation (h) differences.

PSInSAR Data Analysis

Interferometric techniques named Persistent Scatter Interferometric Synthetic Aperture Radar (PSInSAR) have been developed and designed to generate time-series of ground deformations of individual coherent radar targets-Permanent Scatterers (PS). PSInSAR signal analysis allows estimating displacement of PS, acquisitions by acquisition. In the field of landslide monitoring the basic principles and possibility of using multi-temporal PSInSAR data were discussed in Tofani et al. (2013), Mateos et al. (2017), Casagli et al. (2017), Solari et al. (2019), Raspini et al. (2019).

In this research PSInSAR technique was used to analyze radar satellite images for displacement measurement from 2016 to 2018. ASF Earth Data search engine (https://vertex.daac.asf.alaska.edu) was used to select and download more than 140 radar images from ESA Sentinel 1a satellite mission over Umka landslide wider area (74 images from ascending and 69 images from descending acquisition mode) (Table 1).

PC compiled SarPROZ (Perissin 2016) software was used (https://www.sarproz.com) for the preliminary PSInSAR analysis. Standard procedure was applied for the extraction of master and slave images, co-registration, sparse point selection, APS estimation and Multi-temporal analysis of both radar datasets. After preliminary analysis, 4429 sparse

Table 1 PSInSAR images and their characteristics used in analysis

Sentinel 1a pass mode		
	Ascending	Descending
No of images	74	69
Level of processing	SLC	SLC
Acquisition level	IW	IW
Polarization	VV + VH	VV + VH
Subswath	2	2
Acquisition period	03.05.2016–31.12.2018	02.05.2016–30.12.2018
Heading angle	350°	190°
LOS	80°	280°
Swath	250 km	250 km
Spatial resolution	5 × 20 m	5 × 20 m
Incident angle	39,2°–40,3°	39.1°–39,5°
Track	175	153
Line	142	440–445

points were generated from ascending and 5465 sparse points from descending datasets. Fine tuning and filtering by location and coherence quality were also performed. The total of 71 stable radar targets from ascending and 84 radar targets from descending dataset were selected for detailed analysis. Finally, only 29 points from ascending and 15 points from descending dataset were found inside the active Umka landslide body, and only one from both datasets in vicinity of GNSS2. These two PS points were selected for correlation and validation versus GNSS2 permanent ground station data.

Results and Discussion

The analyzed results from recent GNSS measurements indicate that object point UmkaGNSS2 has moved 300 mm towards the North and 500 mm towards the West, i.e. the total cumulative 2D displacement was 650 mm towards the NW, while the vertical displacement was approximately 150 mm for the period September 2014–September 2018. Furthermore, it can be concluded that landslide displacement velocity of the target monitoring point GNSS2 varied during the observed time period (2014–2018) (Fig. 4).

The analyzed average annual 2D displacement was approximately 160 mm/year for analyzed time span (September 2014–December 2018), except in the period September 2014–March 2015 (Fig. 4), which was characterized by intensive fluctuation of the Sava River level (drawdown effect), which started dropping from high level for several meters in a short period. Fluctuation of the river level in 2014–2015 was followed by highest river discharge during floods in May 2014, particularly enhanced by discharge of the Djerdap lake accumulation downstream. This prominent level drop caused drawdown effect and quick redistribution of pore pressures. Similar correlation and the conclusion that followed from 2010–2014 monitoring data were reported in Abolmasov et al. (2015). According to the average annual displacement Umka landslide velocity can be classified as slow to very slow during measurements period of 50 months.

The 2D displacement vectors analysis was done for 59 stabilized geodetic survey points, i.e., dx, dy and dz were obtained from geodetic survey measurements. Zero measurements (March 2018) were compared with first sequence (November 2018). Directions and amount of superficial displacement were plotted for all points and spatial distribution of displacement vectors are presented on Fig. 5. Maximal displacement rates were recorded in central and SW part of the landslide body (95.8–139.1 mm) close to local secondary scarps. Taking into account time span between two survey sequences (eight months), displacement vectors rates were in line with obtained GNSS2 object point monitoring results for the same period.

UAV photogrammetric survey analysis from two series of imaging (2018 and 2019) was used for 2D displacement vector analysis and correlation with geodetic benchmarks survey data for the same period. Very good correlation was established between displacement vectors azimuth obtained from geodetic points survey and displacement vectors

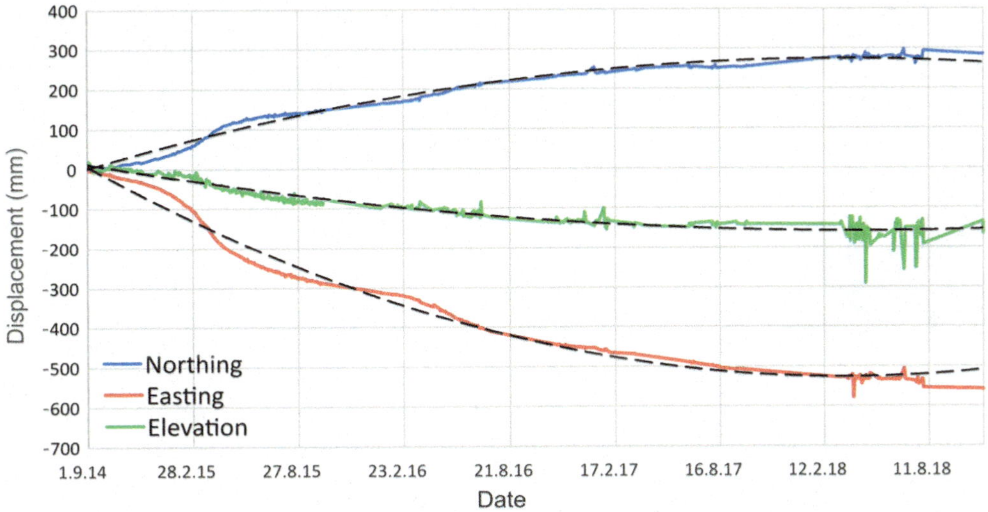

Fig. 4 Displacement of GNSS point from 2014 to 2018

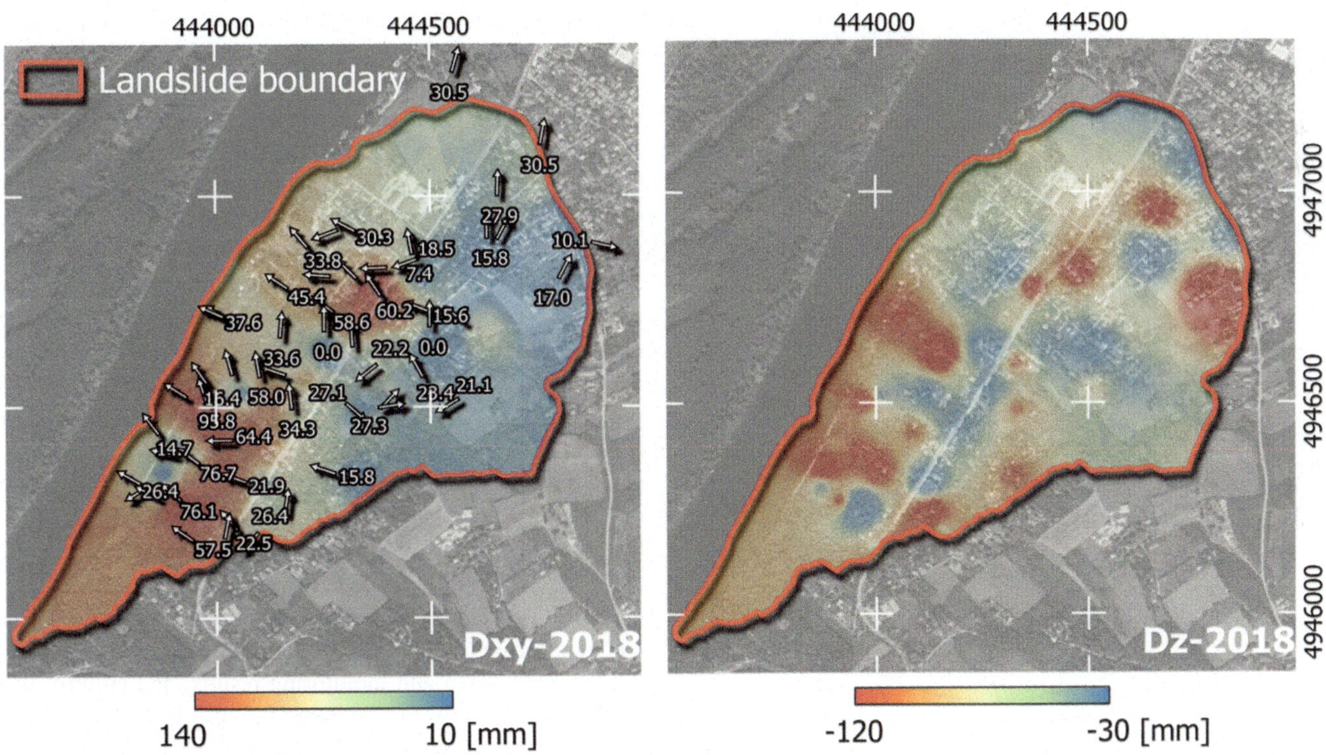

Fig. 5 Results of geodetic benchmarks survey; displacement vectors analysis; "zero" measurements (March 2018) versus November 2018 measurements

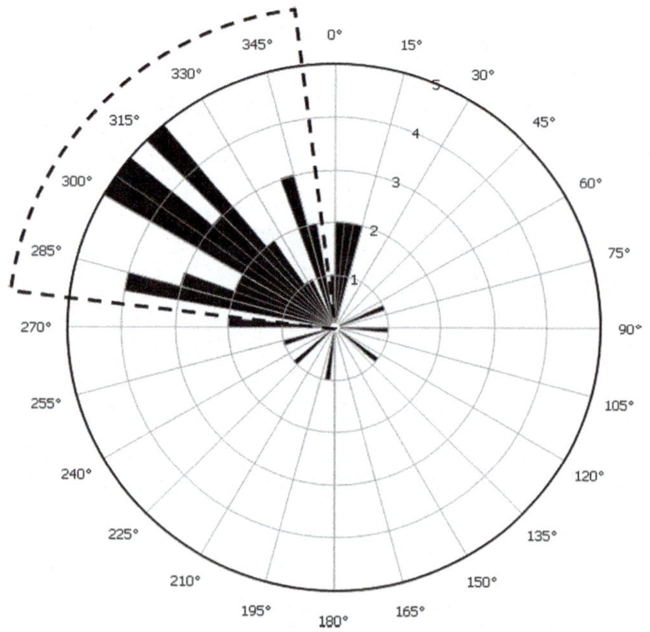

Fig. 6 Displacement vectors azimuth obtained from geodetic benchmarks survey campaigns 2018–2019

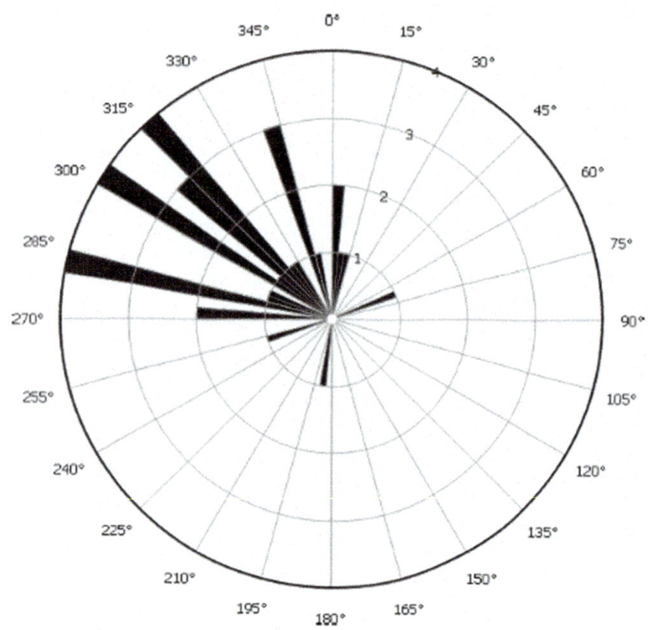

Fig. 7 Displacement vectors obtained from UAV imaging after 2018 and 2019 campaigns

obtained from UAV imaging after 2018 and 2019 campaigns (Figs. 6 and 7). Displacement vectors from both statistical analyses were within expected range of 285°–315°, which is also confirmed by Umka GNSS2 monitoring point.

The results of PSInSAR image analysis were correlated with GNSS2 object point monitoring. The analysis indicates that the selected point LOS displacements have a clear correlation and similar rates as *dz* GNSS2 permanent monitoring point (Figs. 4, 8),. Cumulative displacement for both GNSS2 and PS764/PS3954 were between 5 and 10 mm for the analyzed period of monitoring (March 2016–December 2018).

Spatial distribution and analysis of displacement rate of sparse points over the whole landslide body unfortunately are not evenly distributed due to small number of finally selected PS points. This is probably due to a fact that most of the landslide area is covered by vegetation.

Conclusion

Recent monitoring activities on the Umka landslide included several landslide monitoring techniques realized from 2014 to 2019. Results of all monitoring activities were analysed according to the longest common survey period and then used for cross-correlation and for verification of monitoring results obtained using different techniques.

Displacement rates from GNSS indicate that object point Umka GNSS2 has moved 0.30 m towards the North and 0.50 m towards the West, while the vertical displacement was approximately −0.15 m for the 2014–2018 time span. Similar range of GNSS displacement rates were found in previously published results from monitoring activities realized from 2010–2014. PSInSAR data analysis showed very good correlation between nearest PS points and GNSS point for the same period of monitoring (2016–2018). Results from geodetic survey benchmarks showed displacement rates in accordance to average displacement rates of GNSS2 object point. Results from UAV and geodetic benchmarks survey data analysis showed also very good correlation in vectors azimuth (for the period 2018–2019). According to the presented data it could be concluded that all monitoring results are in compliance with previous published research and monitoring results, and confirm that the Umka is a slow to very slow moving landslide.

Further research within IPL181 Project will be focused on continued monitoring and analysing: (1) spatial patterns and relationships between landslide and element at risk, (2) proposing quantitative risk model and (3) landslide risk management, according to the Project goals. Those research objectives will be also part of IPL248 joint Project research between Univeristy of Belgrade and Univeristy of Salerno.

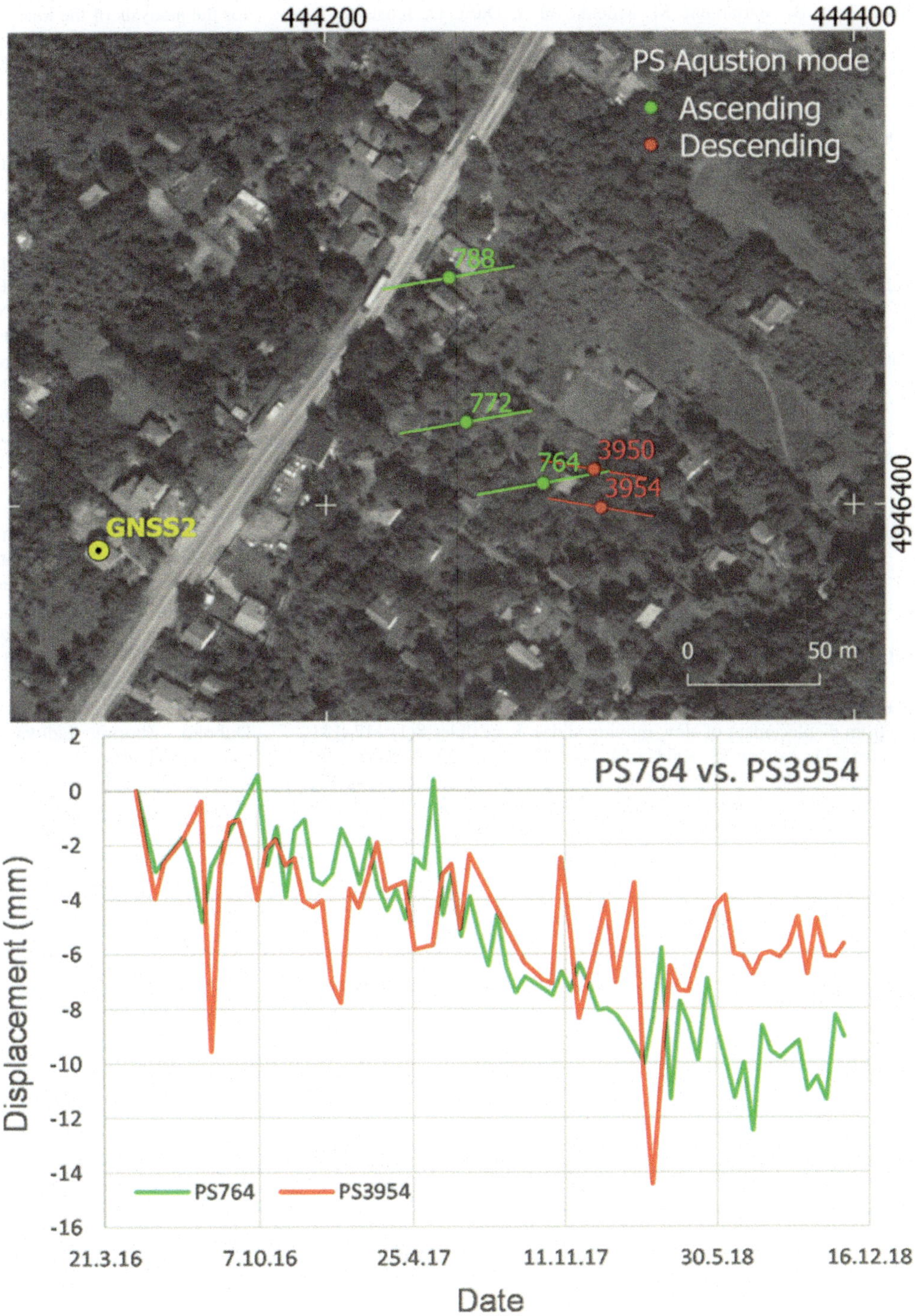

Fig. 8 Results of cumulative displacements of PS 764 (ascending mode-green) and PS3954 (descending mode-red) and their position within landslide Umka area; position of GNSS2 (2014–2019) (yellow)

References

Abolmasov B, Milenković S, Jelisavac B, Vujanić V, Pejić M, Pejović M (3–8 June, 2012) Using GNSS sensors in real time monitoring of slow moving landslides-a case study. In: Eberhardt E, Froese C, Turner K, Leroueil S (eds) Proceedings of the 11th international and 2nd American symposium on landslides and engineered Slopes, vol 2, pp 1381–1385. Banff, Canada. Taylor & Francis Group, London

Abolmasov B, Milenković S, Jelisavac B, Đurić U, Marjanović M (2014) IPL Project 181: study of slow moving landslide umka Near Belgrade, Serbia, In: Sassa K, Canuti P, Yin Y (eds) Landslide science for a safer geoenvironment, proceedings of 3rd World Landslide Forum, vol 1, pp 75–80. Springer International Publisher

Abolmasov B, Milenković S, Marjanović M, Đurić U, Jelisavac B (2015) A geotechnical model of the Umka landslide with reference to landslides in weathered Neogene marls in Serbia. Landslides 12 (4):689–702

Abolmasov B, Marjanović M, Milenković S, Đurić U, Jelisavac B, Pejić M (2017) Study of Slow Moving Landslide Umka near Belgrade, Serbia (IPL-181). In: Sassa et al. (eds) Advancing culture of living with landslides, proceedings of 4th world landslide forum, vol 1, pp 419–427. Springer International Publisher

Abolmasov B, Pejić M, Samardžić Petrović M, Đurić U, Milenković S (2018) Automated GNSS monitoring of Umka landslide—review of seven year's experience and results. In: Proceeding of the 3rd Regional symposium on landslides in the Adriatic-Balkan Region, Ljubljana 2017, pp 65–70 (11–13 October, 2017). Geological Survey of Slovenia, Ljubljana, Slovenia

Balek J, Blahut J (2017) A critical evaluation of the use of an inexpensive camera mounted on a recreational unmanned aerial vehicle as a tool for landslide research. Landslides 14:1217–1224

Casagli N, Frodella W, Morelli S, Tofani V, Ciampalini A, Intrieri E, Raspini F, Rossi G, Tanteri L (2017) Lu P (2017) Spaceborne, UAV and ground-based remote sensing techniques for landslide mapping, monitoring and early warning. Geoenvironmental Disasters 4:9

Colomina I, Molina P (2014) Unmanned aerial systems for photogrammetry and remote sensing: a review. ISPRS J Photogram Remote Sensing 92:79–97

Ćorić S, Božinović D, Vujanić V, Jotić M, Jelisavac B (17–21 June, 1996) Geotechnical characteristics of old landslides in Belgrade area. In: Senneset K (ed) Proceedings of the 7th international symposium on landslides, vol 2, pp 689–694. Trondheim, Norway, Balkema, Rotterdam

Gili JA, Corominas J, Rius J (2000) Using global positioning system techniques in landslide monitoring. Eng Geol 55(3):167–192

Mansour MF, Morgenstern NR, Martin CD (2011) Expected damage from displacement of slow-moving slides. Landslides 8(1):117–131

Mateos RM, Azañón JS, Roldán FJ, Notti D, Pérez-Peña V, Galve JP, Pérez-García JL, Colomo CM, Gómez-López JM, Montserrat O, Devantèry N, Lamas-Fernández F, Fernández-Chacón F (2017) The combined use of PSInSAR and UAV photogrammetry techniques for the analysis of the kinematics of a coastal landslide affecting an urban area (SE Spain). Landslides 14:743–754

Mitrović P, Jelisavac B (2006) Sanacija klizišta "Duboko." Materijali I Konstrukcije 49(1–2):46–59 ((in Serbian))

Peppa MV, Mills JP, Moore P, Miller PE, Chambers JE (2017) Brief communication: landslide motion from cross correlation of UAV-derived morphological attributes. Nat Hazards Earth Syst Sci 17:2143–2150

Perissin D (2016) Interferometric SAR multitemporal processing: techniques and applications. In: Multitemporal remote sensing, remote sensing and digital image processing, pp 145–176

Peternel T, Kumelj S, Ostir K, Komac M (2017) Monitoring the Potoška planina landslide (NW Slovenia) using UAV photogrammetry and tachymetric measurements. Landslides 14:395–406

Raspini F, Bianchini S, Ciampalini A, Soldato MD, Montalti R, Solari L, Tofani V, Casagli N (2019) Persistent Scatters continuous streaming for landslide monitoring and mapping: the case of the Tuscany region (Italy). Landslides 16:2033–2044

Rossi G, Tanteri L, Tofani V, Vannocci P, Moretti S, Casagli N (2018) Multitemporal UAV surveys for landslide mapping and characterization. Landslides 15:1045–1052

Solari L, Soldato MD, Montalti R, Raspini F, Thuegaz P (2019) A Sentinel-1 based hot-spot analysis: landslide mapping in north-western Italy. Int J Remote Sens 40(20):7898–7921

Tofani V, Raspini F, Catani F, Casagli N (2013) Persistent scatterer interferometry (PSI) technique for landslide characterization and monitoring. Remote Sensing 5:1045–1065

Vujanić V, Jotić M, Jelisavac B, Božinović D, Ćorić S (6–9 juni, 1995) Sinteza rezultata geotehničkih istraživanja klizištana Savi: Umka i Duboko. Zbornik radova Drugog simpozijuma istraživanje I sanacija klizišta, DonjiMilanovac, pp 335–351 (in Serbian). https://scihub.copernicus.eu (21.06.2019). https://vertex.daac.asf.alaska.edu (21.06.2019)

Landslides in Weathered Flysch: From Activation to Deposition (WCoE 2017–2020)

Matjaž Mikoš, Nejc Bezak, Janko Logar, Matej Maček, Ana Petkovšek, Dušan Petrovič, and Jošt Sodnik

Abstract

The Faculty of Civil and Geodetic Engineering of the University of Ljubljana (UL FGG), Slovenia, Europe, was voted in 2017 at the 4th World Landslide Forum in Ljubljana, Slovenia to be one of the 20 new World Centres of Excellence (WCoE) in Landslide Disaster Reduction for the period 2017–2020. This successful nomination followed the periods 2008–2011, 2011–2014, and 2014–2017 in which UL FGG successfully fulfilled the role as one of the WCoEs. We can divide the activities of the WCoE at UL FGG into international and national research activities. The international ones consisted of the ICL related activities, the international cooperation, the European research activities, and the bilateral cooperation. The national ones consisted of the national projects and the national research program "Water Science and Technology, and Geotechnics". In the paper, the before mentioned activities of the WCoE at UL FGG are elaborated into more detail, with a short list of top publications to show the dissemination and capacity building efforts.

Keywords

Capacity building • Debris flows • Flysch • Laboratory experiments • Landslides • Slovenia • WCoE

Introduction

The Faculty of Civil and Geodetic Engineering of the University of Ljubljana (UL FGG; https://www.en.fgg.uni-lj.si/), Slovenia, Europe, was proclaimed at the 4th World Landslide Forum in Ljubljana, Slovenia to be one of the 20 new World Centres of Excellence (WCoE) in Landslide Risk Reduction for the period 2017–2020 (https://iplhq.org/category/iplhq/world-centre-of-excellence-wcoe/).

The title of the WCoE at UL FGG was slightly changed from the WCoE 2014–2017 into a new title of WCoE 2017–2020 "Landslides in Weathered Flysch: from activation to deposition".

In this paper, we present the WCoE research activities in the domain of international and national activities, respectively. A more detailed review of past WCoE activities was published in a paper under News/Kyoto Commitment category in the journal Landslides (Mikoš and Petkovšek 2019).

International Research Activities

ICL Related Activities

Among the ICL/IPL related activities of the WCoE 2017–2020 at UL FGG we can list:

M. Mikoš (✉) · N. Bezak · J. Logar · M. Maček · A. Petkovšek · D. Petrovič · J. Sodnik
Faculty of Civil and Geodetic Engineering, University of Ljubljana, Jamova c. 2, 1000 Ljubljana, Slovenia
e-mail: matjaz.mikos@fgg.uni-lj.si

N. Bezak
e-mail: nejc.bezak@fgg.uni-lj.si

J. Logar
e-mail: janko.logar@fgg.uni-lj.si

M. Maček
e-mail: matej.macek@fgg.uni-lj.si

A. Petkovšek
e-mail: ana.petkovsek@fgg.uni-lj.si

D. Petrovič
e-mail: dusan.petrovic@fgg.uni-lj.si

J. Sodnik
e-mail: jost.sodnik@tempos.si; jost.sodnik@fgg.uni-lj.si

J. Sodnik
TEMPOS, Environmental Civil Engineering Ltd., Tehnološki park 21, 1000 Ljubljana, Slovenia

K. Sassa et al. (eds.), *Understanding and Reducing Landslide Disaster Risk*, ICL Contribution to Landslide Disaster Risk Reduction,
https://doi.org/10.1007/978-3-030-60196-6_15

- The cooperation in the ICL regional "Adriatic-Balkan Network", where we co-organized the 3rd Regional Symposium on Landslides in the Adriatic-Balkan Region (3rd ReSyLAB; https://www.geo-zs.si/ReSyLAB2017/) in October 2017 in Ljubljana, Slovenia, together with the Geological Survey of Slovenia (Jemec Auflič et al. 2018a; b). Furthermore, we were actively involved into the organization of the 4th ReSyLAB (https://www.geotehnika.ba/ReSyLAB_2019.htm) that was held in Sarajevo, Bosnia and Hercegovina in October 2019.
- The cooperation in the ICL thematic network "Landslide Monitoring and Warning Thematic Network—LaMaWaTheN" that is coordinated by the Croatian Landslide Group.
- We supported the ICL activities by serving in ICL/IPL bodies, i.e. as the elected ICL Vice President (Prof. Mikoš 2015–2017 and 2018–2020), and chairing the IPL Evaluation Committee (Prof. Mikoš 2012; https://iplhq.org/category/icl/structure-and-officers/), and as Editors for the ICL journal *Landslides* (Dr. Komac—Editor; Prof. Mikoš—Associate Editor); details to be found on the web: https://www.springer.com/journal/10346/editors).
- We supported the international visibility of the ICL by a research paper on the bibliometric impact of the ICL journal *Landslides* (Mikoš 2017) and professional books published by the ICL (Mikoš 2018a).
- Our experiences in the field of landslide disaster risk reduction have been submitted as so-called Text Teaching Tools (Bezak et al. 2018; Mikoš and Hübl 2018; Mikoš et al. 2018; Sodnik and Mikoš 2018a) to a new Springer reference work entitled "Landslide Dynamics: ISDR-ICL Landslide Interactive Teaching Tools".
- We reported on the successful organization of the 4th World Landslide Forum in Ljubljana, Slovenia in 2017, co-organized by UL FGG and Geological Survey of Slovenia (Alcántara-Ayala et al. 2017; Mikoš et al. 2017; Mikoš and Bezak 2018).
- The main ICL related activities in this period were dedicated to support joint preparation of the 5th World Landslide Forum in November 2020 in Kyoto, Japan (https://wlf5.iplhq.org/).

International Research Cooperation

The UNESCO Chair on Water-Related Disaster Risk Reduction at the University of Ljubljana (www.unesco-floods.eu) was established in 2016. We have reported about its activities at the 7th Asia-Europe Meeting on Sustainable Development Dialogue: Integrated Water Management in the Twenty First Century (Mikoš 2018b). Among Chair's activities, also landslide research and landslide risk and landslide disaster reduction can be found. We have started closer cooperation with the chairs around the world, among others:

- UNESCO WENDI Chair in Water, Energy and Disaster Management for Sustainable Development at the University of Kyoto, Japan;
- UNESCO Chair on Geoenvironmental Disaster Reduction at Shimane University, Japan;
- UNESCO Chair on Intersectoral Safety for Disaster Risk Reduction and Resilience of University of Udine, Italy.

In April 2019, we helped UL FGG to organize together with the Slovenian Chamber of Engineers (IZS) the World Construction Forum in Ljubljana, Slovenia (www.wcf2019.org) with the main theme on *Buildings and Infrastructure Resilience*, including a special theme on *Disaster Risk Management & Governance for Resilient Communities*. At the end of this event with more than 600 participants from 50 countries from 5 continents, the 2019 Ljubljana Declaration on Buildings and Infrastructure Resilience (WCF 2019) was adopted, as a contribution to the 2030 Agenda for Sustainable Development Goals.

Among different recommendations, the 2019 Ljubljana Declaration gives the following recommendations in the field of Disaster Risk Management & Governance for Resilient Communities: "*Participants recommend new policies, an adaptation of education and extensive research and development around these specific goals*:

(a) *Increasing the community awareness about the risks due to natural and other hazards by introducing concepts of short-term and long-term risk tolerance*;
(b) *Developing and implementing structural design codes and integrated tools for quantitative risk and resilience assessment for variety of natural and other hazards*;
(c) *Developing policies for comprehensive retrofitting and upgrading of the world's built environment with consideration of tolerable community resilience*;
(d) *Providing new insights into the behavior of buildings and infrastructures exposed to natural and other hazards in order to enhance the resilience of communities.*

In October 2019, we hosted the 2nd Meeting of the VISUS Experts (MUVEx); UNESCO promotes a multi-hazard school safety assessment methodology, called VISUS (Visual Inspection for defining Safety Upgrading Strategies) (VISUS 2020).

Bilateral Research Cooperation

As a results of a bilateral research project with the ICL member at the University of Rijeka, Croatia (2016–2017) on landslides in flysch, we published a paper on soil-water characteristic curves (SWCC) in flysch rock masses (Peranić et al. 2018a, b). During this study, a detailed examination of water retention characteristics was performed for a specific type of residual soil (CL) formed by the weathering of a flysch rock mass. Results presented in Peranić et al. (2018b) can be used to assess the influence of desaturation of the residual soil covering flysch slopes during dry summer periods by providing key-in material properties required to analyze the transient rainfall infiltration process.

In the scope of a bilateral research project with the University of Hannover entitled "*Stochastic rainfall models for rainfall erosivity evaluation*" we are working on the evaluation of different methods that can be used for stochastic rainfall generation. Different stochastic rainfall generation methods are evaluated from the rainfall erosivity perspective. Rainfall erosivity is tightly related to rainfall intensity and is therefore one of the variables that affects soil erosion and large mass movements such as debris flows.

In the scope of a bilateral research project with the Chongqing University of Technology, China (2018–2020) entitled "*Evaluation of intelligent learning techniques for prediction of hydrological data: useful case studies in China and Slovenia*" we are focusing on the evaluation of the different intelligent learning techniques for environmental modelling. So far, we have applied selected methods mostly for the streamflow modelling (i.e. modelling of the rainfall-runoff process) (Sezen et al. 2019; Bai et al. 2019).

National Research Projects

In December 2017, a Ph.D. thesis was finished with the title "*Debris-flow hazard assessment on torrential fans*", and the results were published at the INTERPRAEVENT 2018 Congress in Pacific Rim (Sodnik et al. 2018) and in a professional journal for civil engineers in Slovenia (Sodnik and Mikoš 2018b). Furthermore, we have studied the impact of forests on debris flows (Fidej et al. 2018).

In the scope of the IPL-225 research project "*Recognition of potentially hazardous torrential fans using geomorphometric methods and simulating fan formation*", we evaluated impact of a sequence of debris flows on the debris cone formation. For this purpose, we used RAMMS:DEBRIS FLOW software (RApid Mass MovementS—developed by WSL, Switzerland). Details about the study are available in Bezak et al. (2019a). We found that debris flow sequence has an impact on the fan characteristics but differences among different sequences are not statistically significant (Bezak et al. 2019a). Furthermore, Fig. 1 also shows part of the obtained results.

Moreover, the RAMMS:DEBRIS FLOW model was used for debris flood modelling—it was successfully applied for the May 2018 event that occurred near the Krvavec ski resort in N Slovenia (Bezak et al. 2020). Modelling was performed for two adjacent torrential catchments with similar characteristics. However, different parameter sets were found to be the most suitable for adjacent torrents, which could indicate that numerous small details impact on the fan formation (e.g., exact identification of the debris flow or flood source area).

In order to be able to recognize torrential fans that resulted from debris flow activity and deposition (i.e. debris-prone fans), and to be able to distinguish them from torrential fans, we developed different morphometric methods. A new method of 3D visibility network was developed and their statistical patterns were applied on different 3D surfaces, also on torrential fans (Babič et al. 2019a). Furthermore, two new methods for estimating fractal dimension in 3D space with an average space component of Hurst exponent H was developed, and applied to complex surfaces, also torrential fans (Babič et al. 2019b).

As a part of the IPL-226 research project entitled "*Studying landslide movements from source areas to zone of deposition using a deterministic approach*" we investigated if hydrological modelling could be included in early warning systems (EWS). These systems often include the empirical rainfall thresholds. However, this approach was recently questioned in literature [cited references in Bezak et al. (2019b)]. Therefore, we have proposed a new method for the prediction of the landslides using the results of the conceptual lumped hydrological model. Additional information and evaluation of the proposed approach can be found in Bezak et al. (2019b).

In 2019–2020, UL FGG together with the Geological Survey of Slovenia is working on the national project financed by the Ministry of the Environment and Spatial Planning (investment of the national Climate Fund) entitled "*Feasibility study for protection of the Koroška Bela village from landslides and debris flows*". The research team is evaluating which protection and/or mitigation measures could be realized in the area above and/or in the Koroška

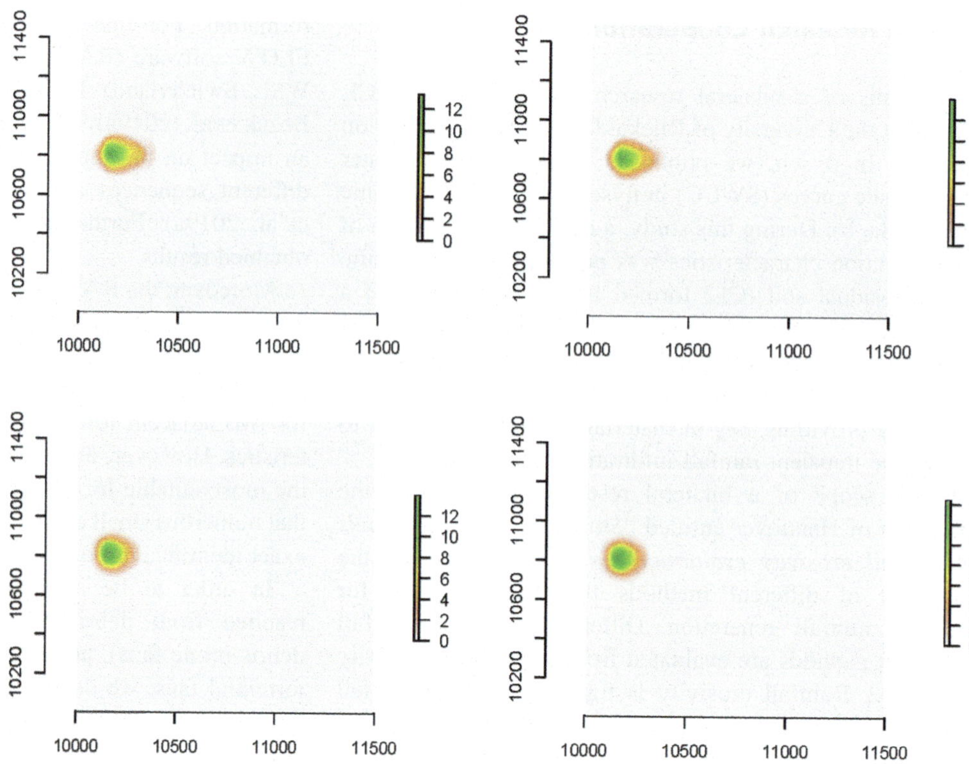

Fig. 1 Impact of the debris flow sequence on the debris flow fan after multiple simulations. Upper two panels show fan for a set of model parameters and lower two panels show fan using different set of parameters. Difference between left and right panel is in the debris flow sequence. Artificial terrain was used, x- and y-axis in Cartesian coordinates, in meters. Legend shows deposition in meters above terrain. Adopted after Bezak et al. (2019a)

Bela village in order to protect its close to 2000 inhabitants from potential debris flows triggered in unstable landslide masses in the torrential watersheds above the village.

As part of this feasibility study, we have evaluated climate change impact on the water cycle and IDF curves in the area that may be used for estimation of debris-flow magnitudes in the decades to come. The results are included in a separate paper in the WLF5 book series (Bezak et al. submitted).

National Research Program

Moreover, a large part of our landslide research activities is also related to the national research program "*Water Science and Technology, and Geotechnics*". Within the framework of this 5-year research program (2017–2021), we also focus on topics related to torrents and landslides.

In a small steep forested torrent, we tested the applicability of remote sensing techniques (e.g. Terrestrial Laser Scanner—TLS) to estimate morphological changes after flash floods, and developed semi-automatic algorithms for recognition of pieces of large woody debris (LWD) inside the torrential channel (Bezak et al. 2017).

We compared different methods for the determination of granulometric characteristics of coarse aggregates and sediments (Matič et al. 2019). Additionally, we have recently investigated trends and temporal changes in the intensity–duration-frequency (IDF) curves and extreme event clustering (Bezak and Mikoš 2019). Figure 2 shows a part of this investigation. Additional description of the main results can be found in Bezak and Mikoš (2019).

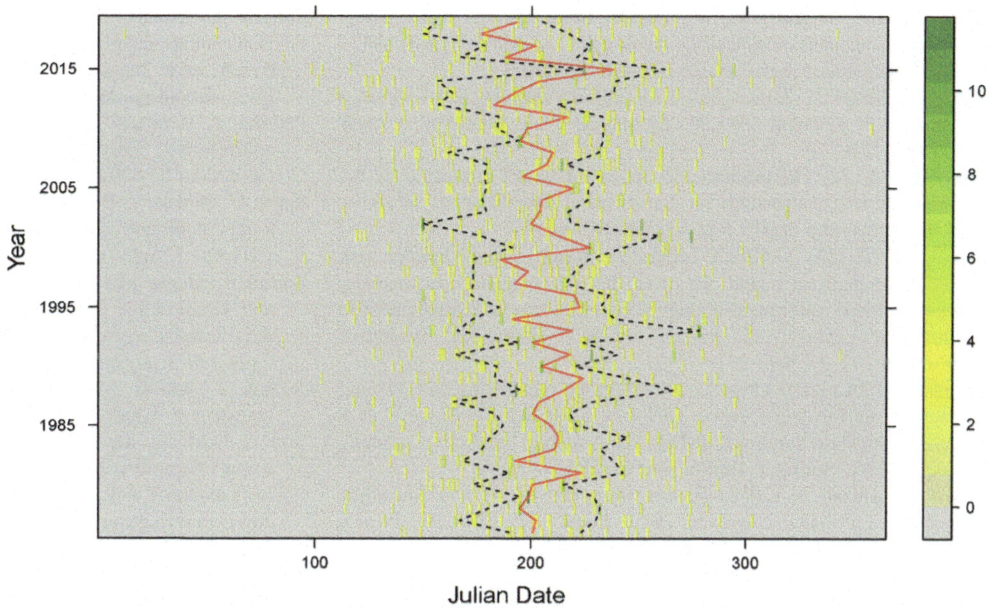

Fig. 2 Clustering of the five maximum 5-min rainfall intensities for 10 rainfall stations in Slovenia for the 1975–2018 period. Legend indicates number of events on specific day. Since 10 stations are considered the maximum number is 10 and this situation means that on a specific day on all stations maximum event was recorder. Red line shows the median Julian data and dotted black lines show 25 and 75% Julian date. Adopted after Bezak and Mikoš (2019)

Acknowledgements Our research activities in the field of landslide research have been mainly financed by the Slovenian Research Agency (ARRS) and their financial support is greatly acknowledged—through the national Research Programme P2-0180 "Water Science and Technology, and Geotechnics", and the research projects J7-8273 "Recognition of Potentially Hazardous Torrential Fans Using Geomorphometric Methods and Simulating Fan Formation", and J1-8153 "Studying Landslide Movements from Source Areas to Zone of Deposition using a Deterministic Approach", respectively.Both research projects were approved by the International Programme on Landslides (IPL) as the IPL-225 and IPL-226 Project, respectively (https://iplhq.org/category/iplhq/ipl-ongoing-project/). Moreover, our research on landslides in the Koroška Bela torrential watershed was financially supported by the Ministry of the Environment and Spatial Planning of the Republic of Slovenia.

References

Alcántara-Ayala I, Sassa K, Mikoš M, Han Q, Rhyner J, Takara K, Nishikawa S, Rouhban B, Briceño S (2017) The 4th World landslide forum: landslide research and risk reduction for advancing the culture of living with natural hazards. Int J Disaster Risk Sci 8 (4):498–502. https://doi.org/10.1007/s13753-017-0139-4

Babič M, Huber MA, Bielecka A, Soycan M, Przegon W, Gigović L, Drobnjak S, Sekulović D, Pogarčić I, Miliaresis G, Mikoš M, Komac M (2019) New method of visibility network and statistical pattern network recognition usage in terrain surfaces. RMZ—Mater Geoenviron Mater Geookolje 66(1):13–25. https://doi.org/10.2478/rmzmag-2019-0006

Babič M, Miliaresis G, Mikoš M, Ambu R, Calì M (2019) New method for estimating fractal dimension in 3D space and its application to complex surfaces. Int J Adv Sci Eng Inf Technol 9(6):2154–2159. https://doi.org/10.18517/ijaseit.9.6.9480

Bai Y, Bezak N, Sapač K, Klun M, Zhang J (2019) Short-term streamflow forecasting using the feature-enhanced regression model. Water Resour Manage 33(14):4783–4797

Bezak N, Mikoš M (2019) Investigation of trends, temporal changes in intensity-duration-frequency (IDF) curves and extreme rainfall events clustering at regional scale using 5 min rainfall data. Water 11(10/2167). https://doi.org/10.3390/w11102167

Bezak N, Grigillo D, Urbančič T, Mikoš M, Petrovič D, Rusjan S (2017) Geomorphic response detection and quantification in a steep forested torrent. Geomorphology 291:33–44. https://doi.org/10.1016/j.geomorph.2016.06.034

Bezak N, Brilly M, Šraj M, Mikoš M (2018) Intensity-duration-frequency curves for rainfall-induced shallow landslides and debris flows using copula functions: TXT-tool 2.386-1.1. In: Sassa K et al (eds) Landslide dynamics: ISDR-ICL landslide interactive teaching tools, vol 2. Springer, Berlin, pp 425–431

Bezak N, Sodnik J, Mikoš M (2019a) Impact of a random sequence of debris flows on torrential fan formation. Geosciences 9(2/64). https://doi.org/10.3390/geosciences9020064

Bezak N, Jemec Auflič M, Mikoš M (2019b) Application of hydrological modelling for temporal prediction of rainfall-induced shallow landslides. Landslides 16(7):1273–1283. https://doi.org/10.1007/s10346-019-01169-9

Bezak N, Jež J, Sodnik J, Jemec Auflič M, Mikoš M (2020) An extreme May 2018 debris flood case study in northern Slovenia: analysis, modelling, and mitigation. Landslides. https://doi.org/10.1007/s10346-019-01325-1

Bezak N, Peternel T, Medved A, Mikoš M, Submitted Climate change impact on the water balance of the Koroška Bela area, NW Slovenia. Submitted to WLF5 Springer book series

Fidej G, Mikoš M, Jež J, Kumelj Š, Diaci J (2018) Assessment of forest protective function against debris-flows. Gozdarski Vestnik 76 (4):167–180

Jemec Auflič M, Mikoš M, Verbovšek T, Arbanas Ž, Mihalić Arbanas S (2018) 3rd Regional symposium on landslides in the Adriatic-Balkan region (3rd ReSyLAB)—a final report. Landslides 15(2):381–384. https://doi.org/10.1007/s10346-017-0934-4

Jemec Auflič M, Mikoš M, Verbovšek T (eds) (2018b) Advances in landslide research: proceedings of the 3rd regional symposium on landslides in the Adriatic Balkan region, 11–13 Oct 2017, Ljubljana, Slovenia. Geological Survey of Slovenia, Ljubljana. https://doi.org/10.5474/9789616498593

Matič M, Bezak N, Mikoš M (2019) Optical granulometry of coarse aggregates and sediments. Acta Hydrotechnica 32(56):59–57

Mikoš M (2017) Landslides: a top international journal in geological engineering and engineering geology? Landslides 14(5):1843–1854. https://doi.org/10.1007/s10346-017-0869-9

Mikoš M, Bezak N, Jemec Auflič M (2017) Četrti Svetovni forum o zemeljskih plazovih—WLF4, Ljubljana, 2017. Gradbeni Vestnik 66:178–180

Mikoš M (2018a) The bibliometric impact of books published by the International Consortium on Landslides. Landslides 15(8):1459–1482. https://doi.org/10.1007/s10346-018-1019-8

Mikoš M (2018b) UNESCO chair on water-related disaster risk reduction and its recent activities. In: 7th Asia-Europe meeting, sustainable development dialogue, integrated water management in the 21st century: addressing imminent challenges, Budapest, Hungary, 11–12 Sept 2018

Mikoš M, Hübl J (2018) Practice guidelines on monitoring and warning technology for debris flows: TXT-tool 2.386-1.2. In: Sassa K et al (eds) Landslide dynamics: ISDR-ICL landslide interactive teaching tools, vol 1. Springer, Berlin, pp 567–585

Mikoš M, Bezak N (2018) Report from the 4th World Landslide Forum, Ljubljana, Slovenia, 2017. Ujma, 32, 284–292. https://www.sos112.si/slo/tdocs/ujma/2018/284-292.pdf

Mikoš M, Čarman M, Papež J, Jež J (2018) State-of-the-art overview on landslide disaster risk reduction in Slovenia: TXT-tool 4.386-1.1. In: Sassa K et al (eds) Landslide dynamics: ISDR-ICL landslide interactive teaching tools, vol 2. Springer, Berlin, pp 683–691

Mikoš M, Petkovšek A (2019) Faculty of civil and geodetic engineering, University of Ljubljana. Landslides 16(9):1815–1819. https://doi.org/10.1007/s10346-019-01231-6

Peranić J, Arbanas Ž, Foresta V, Cuomo S, Maček M (2018a) Determination of soil water retention curve of residual soil from a flysch rock mass. In: Ng CWW (ed) Unsaturated soils. 7th International conference on unsaturated soils, 3–5 Aug 2018, 6 p

Peranić J, Arbanas Ž, Cuomo S, Maček M (2018b) Soil-water characteristic curve of residual soil from a flysch rock mass. Geofluids, ID 6297819, 15 p. https://doi.org/10.1155/2018/6297819

Sezen C, Bezak N, Bai Y, Šraj M (2019) Hydrological modelling of karst catchment using lumped conceptual and data mining models. J Hydrol 576:98–110

Sodnik J, Mikoš M (2018a) Two-dimensional debris-flow modelling and topographic data: TXT-tool 3.386-1.1. In: Sassa K et al (eds) Landslide dynamics: ISDR-ICL landslide interactive teaching tools, vol 2. Springer, Berlin, pp 235–250

Sodnik J, Mikoš M (2018b) Landslides at debris flow hazard assessment. Gradbeni Vestnik 67:120–131

Sodnik J, Maček M, Mikoš, M (2018) Estimating landslide volumes using LS-rapid model—the 2000 Stože landslide in NW Slovenia. In: Yamada T (ed) Large scale sediment disasters in orogenic zones and countermeasures: symposium proceedings, INTERPRAEVENT 2018, pp 32–41

VISUS (2020) School Safety Assessment: VISUS Methodology. https://www.unesco.org/new/en/natural-sciences/special-themes/disaster-risk-reduction/school-safety/safety-assessment-method-visus/

WCF (2019) The 2019 Ljubljana Declaration on Buildings and Infrastructure Resilience—Contributing to the 2030 Agenda for SDGs. https://www.wcf2019.org/2019-ljubljana-declaration-on-buildings-and-infrastructure-resilience/index.html

Report of the Croatian WCoE 2017–2020: From Landslide Mapping to Risk Assessment

Snježana Mihalić Arbanas, Sanja Bernat Gazibara, Petra Jagodnik, Marin Sečanj, Vedran Jagodnik, Martin Krkač, and Željko Arbanas

Abstract

In this paper scientific activities of the Croatian Landslide Group (CLG), World Centre of Excellence on Landslide Risk Reduction (WCoE) of the International Consortium on Landslides (ICL) for the period 2017–2020, are shortly described. The results of scientific research are presented through the fields of landslide science related to landslide identification and mapping as well as landslide hazard and risk assessment. It is concluded that the resulting landslide inventory maps in detailed scale as well as landslide susceptibility and risk maps in large and small scale, provide necessary information for landslide risk management in Croatia at local and national level. Besides application of scientific results in Croatia, the general objectives of the ICL WCoE are reached by contributing to landslide risk reduction.

Keywords

Landslide maps • Landslide inventory • Susceptibility assessment • Risk assessment • Dinarides • Pannonian basin

Introduction

The Croatian Landslide Group (CLG) from University of Zagreb and the University of Rijeka encompass scientists from the field of engineering geology and geotechnics dealing with landslide science. Scientist from the CLG started joint research in 2009. The CLG become World Centre of Excellence on Landslide Risk Reduction (WCoE) of the International Consortium on Landslide (ICL) for the period 2014–2017 (Mihalić Arbanas et al. 2017a) and the period 2017–2020. General objectives of ICL WCoE are: (1) to strengthen the International Programme on Landslides (IPL) and IPL Global Promotion Committee; (2) to create "A Global Network of Entities Contributing to Landslide Risk Reduction"; and (3) to improve the global recognition of "Landslide Risk Reduction" and its social-economic relevance, and entities contributing to this field.

This article presents scientific research of the CLG in the period 2017–2020, related to the landslide identification and mapping as well as landslide susceptibility zonation and landslide risk assessment. From the summarised overview of the main results of the scientific research, it could be seen that the CLG is academic entity that has been contributing to landslide risk reduction in the area of the Dinarides and the Pannonian Basin in Central Europe. This contribution is in accordance to the role of a stakeholder of the "Sendai Framework for Disaster Risk Reduction 2015–2030".

S. Mihalić Arbanas (✉) · S. Bernat Gazibara · M. Sečanj · M. Krkač
Faculty of Mining, Geology and Petroleum Engineering, University of Zagreb, Pierottijeva 6, 10000 Zagreb, Croatia
e-mail: smihalic@rgn.hr

S. Bernat Gazibara
e-mail: sbernat@rgn.hr

M. Sečanj
e-mail: msecanj@rgn.hr

M. Krkač
e-mail: mkrkac@rgn.hr

P. Jagodnik · V. Jagodnik · Ž. Arbanas
Faculty of Civil Engineering, University of Rijeka, Radmile Matejčić 3, 51000 Rijeka, Croatia
e-mail: petra.jagodnik@gradri.uniri.hr

V. Jagodnik
e-mail: vedran.jagodnik@gradri.uniri.hr

Ž. Arbanas
e-mail: zeljko.arbanas@gradri.uniri.hr

K. Sassa et al. (eds.), *Understanding and Reducing Landslide Disaster Risk*, ICL Contribution to Landslide Disaster Risk Reduction, https://doi.org/10.1007/978-3-030-60196-6_16

Landslide Identification and Mapping

Research on landslide inventory mapping in Croatia encompass identification of shallow landslides using innovative techniques based on ALS (Airborne Laser Scanning). Ongoing research of development of mapping methodologies deals with landslides in the Dinarides and the Pannonian Basin (Mihalić Arbanas et al. 2017b).

Landslides in the Pannonian Basin

Area of hills and lowlands in the Pannonian Basin in Croatia are prone to very small-to-moderately small (<10^5 m^3) superficial-to-moderately shallow (<20 m) landslides. Development of the methodology for identification and mapping of landslides in the Pannonian Basin started using data derived by ALS performed in December 2013 for the Podsljeme area (size of 180 km^2) in Zagreb (capitol of Croatia). This data set is of crucial value because it is collected after MORLE (Multiple-Occurrence Regional Landslide Events) with more than 900 landslides (re)activated in the NW Croatia in winter 2012/2013 (Bernat Gazibara et al. 2018b). The topographic maps derived from DTM 0.3 × 0.3 m used for interpreting the landslide morphology were hillshade maps, slope maps, contour lines, and curvature and surface roughness maps. The landslide inventory map (Fig. 1) was produced for a pilot area (21 km^2) by visual intepretation to derive repesentative and complete data set for statistical analysis. In total, the landslide inventory consists of 702 landslides with a non-uniform distribution across the pilot area (Bernat Gazibara 2019a; Bernat Gazibara et al. 2019b). The prevailing types are shallow soil slides, and only 10 landslides were classified as earth flows. Nearly 65% of all identified landslides on the LiDAR DTM were evaluated as "high confidence" due to the clearly visible landslide features on the LiDAR derivatives. Based on the landslide inventory, the total landslide area is 0.5 km^2 or 2.43% of the pilot area (21 km^2). The mean landslide density is 33.3 slope failures per square kilometers, and the smallest landslide identified in the pilot area has an area of 43 m^2 (Bernat Gazibara et al. 2017). The most frequent landslides in the inventory have an area of approx. 400 m^2, and 90% of the landslide bodies showed a size between 100 and 2000 m^2 (Bernat Gazibara and Mihalić Arbanas 2018; Bernat Gazibara et al. 2018a). The frequency–size distribution of all mapped landslides in the pilot area (Fig. 2) shows "universal distribution" described

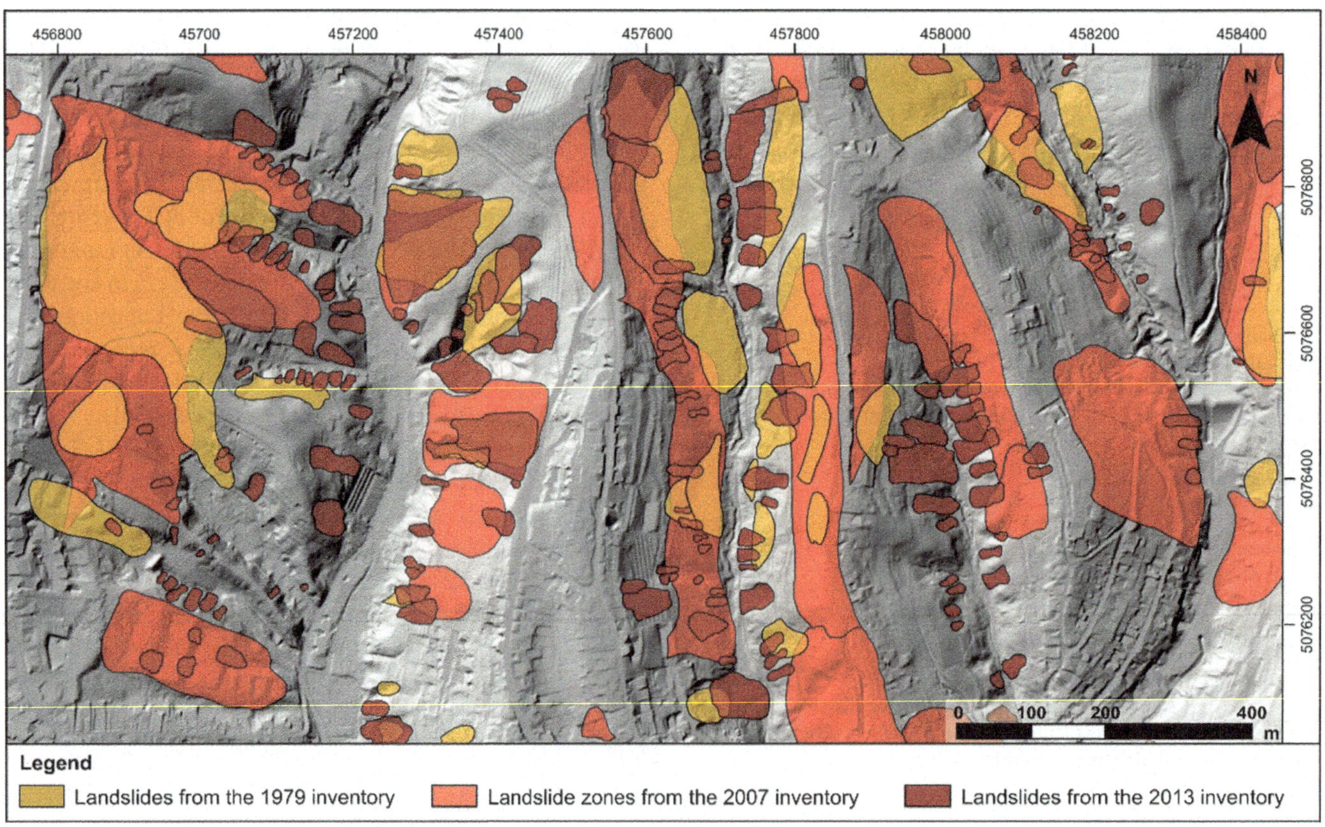

Fig. 1 Comparison of landslide contours from historical landslide inventories from 1979 and 2007 and LiDAR based landslide inventory from 2013 (Bernat Gazibara et al. 2019c)

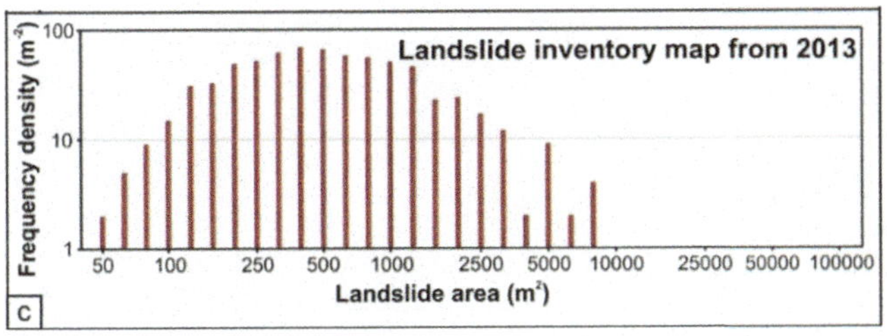

Fig. 2 Frequency-size distribution of LiDAR based landslide inventory (Bernat Gazibara et al. 2019a)

by Malamud et al. (2004), which implies that the occurrence of a landslides in the pilot area is a result of either natural triggering factors (e.g., rainfall or rapid snowmelt) or anthropogenic interventions with regional impacts (e.g., widespread deforestation and urbanization).

Based on the analysis performed by Bernat Gazibara et al. (2019b), more than 80% of the mapped landslides appear in forests, 10% in agricultural areas, and only 7% on artificial surfaces. The frequency-size distributions of mapped landslides show that there is no difference in the landslide size among the land use units, but the mean landslide density in forests (84 ls/km^2) is six times higher than that in agricultural areas and on artificial surfaces (14 ls/km^2). One of the possible explanations is that LiDAR-based landslide inventories can often be incomplete on settlements and arable lands due to frequent anthropogenic influences (Bell et al. 2012; Petschko et al. 2015; Steger et al. 2016). More probable is that the higher landslide density in forests is associated with prevailing steep slopes and gullies. In contrast, the majority of housing and infrastructure is constructed on relatively flat terrain, i.e., the prevailing slope angles (62%) in urban areas range from 0 to 12°. Based on the population density grid (1 × 1 km) for the Podsljeme area, approx. 43,000 people live within a zone of 50 m from 418 mapped landslides. It can be concluded that the landslides in the pilot area may affect further urbanization and that elements at risk are highly exposed to landslide hazards. The landslide density map obtained from the LiDAR-based inventory shows that 68% of the study area (14.3 km^2) is unstable or landslide prone. Based on the highest degree of landslide inventory completeness, it can be concluded that this density map gives the best proxy estimation of landslide susceptibility in the study area.

Bernat Gazibara et al. (2019c) compared LiDAR-based landslide inventory compiled for the pilot area (21 km^2) with two historical landslide inventories from the Podsljeme area based on geomorphological field mapping and historical records in 1979 and 2007. Landslide inventory from 2007 has status of formal landslide inventory of the City of Zagreb that is in common use for physical planning and construction, land-use planning and civil protection. Comparison of three landslide inventory maps (Fig. 2) was performed using the following analyses: frequency-area distribution, geographical discrepancy of landslides and landslide density maps. Results shows significant differences in the number of identified landslides as well as in the size and distribution of landslides. Cartographic error between analyzed landslide inventory maps is approximately 85–96% and it can be assigned to different geomorphological interpretations (1979 vs. 2007) and different techniques used for landslide mapping (1979 vs. 2013 and 2007 vs. 2013). Comparison of inventories showed the unreliability of the existing historical inventories and usefulness of LiDAR data for preparation of complete and reliable landslide inventory maps.

Landslides in the Vinodol Valley in Dinarides

Research on both landslides and erosion processes in the specific geological and morphological conditions in the tectonic valley in the Dinarides, composed of carbonate and flysch rocks, has been undertaken in the framework of the doctoral thesis by Đomlija (2018). In this study, nine different topographic datasets are derived from 1-m resolution LiDAR DTM that are visually analysed, in order to study in detail past and current hazardous processes, geological materials and geomorphological units in the study area. For each of these purposes, specific criteria for identification and mapping are developed on the basis of the preliminary visual interpretation of high resolution (HR) LiDAR imagery. In order to create historical landslide inventory of the Vinodol Valley by applying the modified Varnes classification of landslide types (Hungr et al. 2014), identification of materials involved in sliding processes was required. Thus, the first step in this research was identification of engineering geological units of the Vinodol Valley, on the basis of the type of material. In total, 14 engineering geological units were identified in the Vinodol Valley, which were classified to lithological complexes and lithological types.

The topography of the Vinodol Valley is characterized by diversity of soil erosion phenomena formed in the flysch bedrock (Đomlija 2018). The historical erosion inventory map of the Vinodol Valley is created (Fig. 3a) by identification and mapping of all past and current erosion phenomena (i.e., gully erosion, sheet erosion, and the combined effect of rill and sheet erosion) based on visual interpretation of HR LiDAR imagery. The methodology for identification and mapping of soil erosion processes is described in Đomlija et al. (2019a). This study reveals the two-step procedure in a visual interpretation of HR LiDAR imagery, including preliminary and detailed interpretation. The preliminary visual interpretation serves for a clear definition of identification and mapping criteria of the erosion phenomena, whose application during the detailed visual interpretation of LiDAR imagery considerably improves the

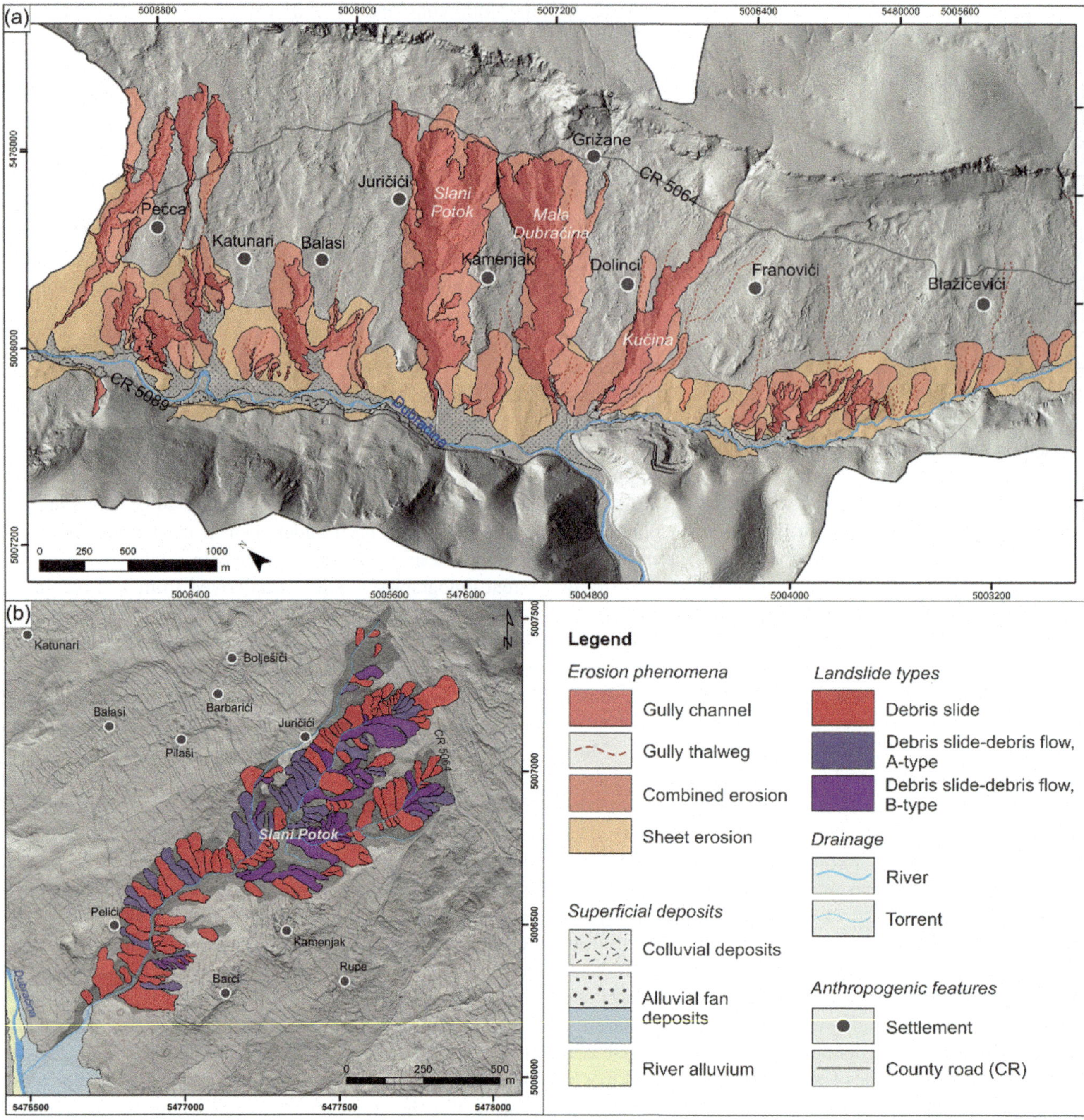

Fig. 3 **a** Detail from the historical erosion inventory map of the Vinodol Valley, for the central part of the Valley (Đomlija et al. 2019a), and **b** historical landslide inventory map of the Slani Potok gully (Jagodnik et al. 2020)

accuracy of mapping results and reduces the consumed time. The study of Đomlija et al. (2019a) determines the highest potential for direct visual identification and mapping of the gully erosion phenomena, while the method is proven to be inapplicable for direct mapping of the sheet erosion. However, sheet erosion processes could have been indirectly identified on certain LiDAR imagery, based on the recognition of colluvial deposits accumulated at the foot of the eroded slopes.

Most of the landslides identified in the Vinodol Valley are initiated within gully landforms (Đomlija 2018). Geomorphological historical landslide inventory is created for the Slani Potok gully (0.48 km^2), the largest gully phenomenon formed in the Vinodol Valley (Jagodnik et al. 2020). In total,

181 landslide phenomena are visually identified from seven HR LiDAR derivatives (Fig. 3b). Two landslide types are determined according to Hungr et al. (2014): (i) debris slides; and (ii) debris slide-debris flows. Landslides are mainly very small to moderate small (ICL 2018), with depths varying from one to several meters, and their topography is strongly modified by the erosion processes. However, the results of this study indicate that the visual interpretation of LiDAR derivatives is also an effective method for identification and mapping of relatively small and shallow landslides initiated within gullied topography. The use of as many LiDAR maps as possible is recommended, to correctly recognize and map landslide boundaries, especially if significant differences in the visibility of landslide topography on LiDAR maps exist.

There are five types of landslides (Hungr et al. 2014) identified along the carbonate cliffs (2.12 km^2) in the Vinodol Valley, based on the visual interpretation of seven different LiDAR imagery (Đomlija et al. 2019b): (i) rock fall; (ii) rock topple; (iii) rock irregular slide; (iv) rock slope deformation; and (v) rock avalanche. Only the rock irregular slides and rock slope deformations are possible to directly identify and map. Namely, talus deposits generally indicate rock falls and rock topples (e.g., Soeters and van Westen 1996). Moreover, distinct morphological features of talus deposits, as well as of certain accumulations of limestone olistoliths and boulders, recognized on the LiDAR imagery, indicated also the occurrence of rock avalanches in the Vinodol Valley (Đomlija et al. 2019b).

Landslide Hazard and Risk Assessment

In the period 2017–2020 scientist of the CLG started landslide susceptibility mapping in site-specific scale at the selected locations in the Dinarides. Susceptibility maps, that ranks slopes into categories according to its potential to become source area of rock falls or topples, are the most useful for use as preliminary information in engineering related to steep artificial or natural cuts in carbonate rocks. Contrary, at the national scale, heuristic analysis of landslide susceptibility was performed for the area of whole Croatia to indicate areas prone to sliding. Risk assessment was performed only in national scale, based on heuristic analysis, to get official preliminary information for risk evaluation.

Site Specific Rock-Fall Susceptibility

Site-specific rockfall susceptibility assessment was performed for the steep slopes with complex morphology above the historical Town of Omiš with total area of 6.5 ha (Sečanj et al. 2017, 2019a) and for the 900 m long rock cuts along the road in the Krka National Park (Sečanj et al. 2018, 2019b). Rockfall susceptibility at the slope scale is based on the analysis of the relation between discontinuities and slope morphology (Matasci et al. 2018), that can indicate if a certain instability mechanism is possible at a specific location on the rock slope. Remote sensing techniques, based on TLS and UAV photogrammetry, were employed to rapidly obtain orientation of rock faces, including in the field inaccessible areas, while discontinuity orientation data was obtained from the high-resolution point clouds and 3D models using Split-FX (Split Engineering LCC 2017) and CloudCompare software (CloudCompare 2017). Possible rockfall sources were identified by automated 3D spatial kinematic analysis, that analyze kinematic possibility of specific instability mechanism defined by the geometry of the slope and discontinuities. The input parameters for the 3D spatial kinematic analysis were:

(1) quasi-homogenous engineering geological zones (EGZ), defined by rock mass properties, representative discontinuity sets and the geometrical properties of the slope; (2) the orientation of the slope; (3) orientations of the discontinuities and discontinuity intersections; and (4) the average value of discontinuity friction angle. Also, for each model cell within a certain EGZ, Kinematic Hazard Index (KHI) defined by Casagli and Pini (1993), was calculated for all instability mechanisms. Spatial kinematic analysis and the calculation of the KHI were automated with the in-house MATLAB scripts and functions using vector analysis for each polygon of high-resolution 3D digital model. Result of the analysis were values of Maximal Kinematic Hazard Index (MKHI) for each polygon, which were interpolated to represent rockfall susceptibility map or 3D model where the probability of instabilities is expressed by the color scale, varying from green to red, as the MKHI increases.

Analyzed rock cut along the road in the Krka National Park showed high probability of kinematic instability mechanism occurrence with the MKHI up to 33%, related to wedge failure, while for steep slopes above the Town of Omiš MKHI was around 50%, related to wedge failure (Fig. 4). Location of the high values of MKHI at the slopes above the Town of Omiš (14–37%) were in good correspondence with potentially hazardous rock blocks and potential rockfall source areas identified by previous field investigations described in Arbanas et al. (2019).

National Landslide Susceptibility Zonation

First landslide susceptibility assessment for the territory of the Republic of Croatia was carried out by the CLG in 2019, for the purpose of development of the National Physical Plan. The heuristic analysis was used to produce map in the scale of 1:100,000 depicting zones of limitations related to

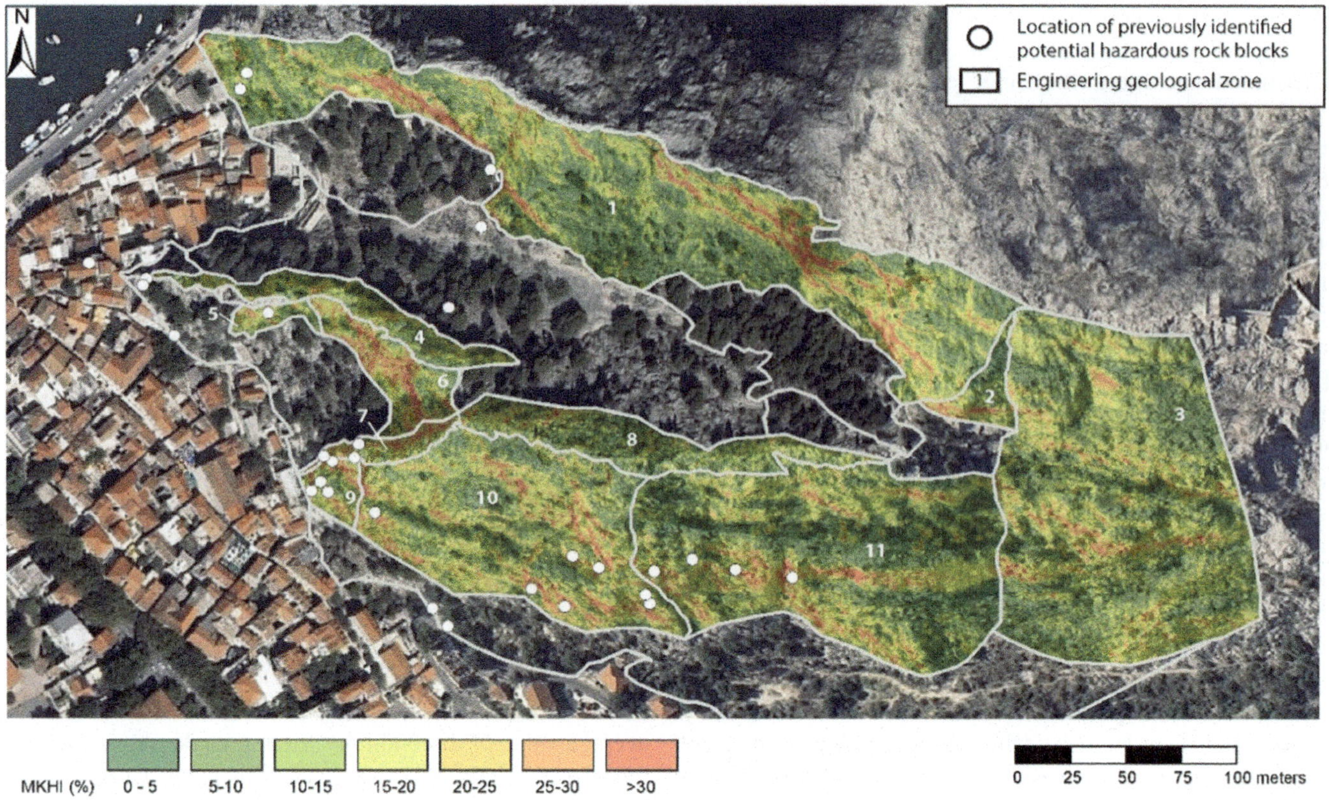

Fig. 4 Results of the automated kinematic analysis and MKHI calculation. The higher MKHI, the higher probability that the analyzed instability mechanism will take place. Potential hazardous rock blocks identified by previous investigations are depicted to show overlapping with rockfall susceptibility assessment (after Sečanj et al. 2019a)

sliding. Input data for the analysis were derived from Basic Geological Map in scale of 1:300,000 and from 20-m resolution DEM. Geological stratigraphic units has been empirically reclassified into six units based on its relationship to sliding. Factor slope map has been created according to geomorphological classification of slope inclination (IGU 1968) in order to define classes characterised by dominant geomorphological processes. Figure 5 depict resulting preliminary landslide susceptibility map of the Republic of Croatia derived from two casual factors, lithological units where sliding can occur and slopes inclination of 5–55°. The spatial distribution of landslide-prone areas shows that the highest frequency of geomorphological and geological conditions for landslide occurrence is present in the continental part of the Croatia, i.e., in the Pannonian Basin.

Given the relative proportion of landslide-prone areas in relation to the area of the county, the largest proportion is in the Krapina-Zagorje County, where 77.39% of the area is susceptible to sliding. The largest area of susceptible zones per county is in the Sisak-Moslavina County, where landslides can occur in zones of 1390.9 km^2. Based on performed landslide susceptibility analysis, approximately 28% of the territory of the Republic of Croatia is potentially prone to sliding with respect to geological (rock/soil type) and geomorphological conditions (slope inclination).

National Landslide Risk Assessment

Landslide risk assessment in Croatia was undertaken by the CLG in 2018 in the framework of the national risk assessment covering the geographic context of the whole country (56,594 km^2) and following recommendations from guidelines of the European Union (EC 2010) The EU guidelines on national risk assessments and mapping do not advocate any particular risk criteria, benchmarks or standards, but do encourage transparency in this area including use of nomenclature according to the standard ISO Guide 73:2009 (ISO 2009) and UNISDR terminology (UNISDR 2009) as well as defined procedures of risk identification, risk analysis and risk evaluation. These stages can be compared to the framework of landslide risk assessment and management developed by Fell et al. (2008a, b).

Risk identification is the process of finding, recognizing and describing risks. It serves as a preliminary step for the subsequent risk analysis stage. The methodology for applied

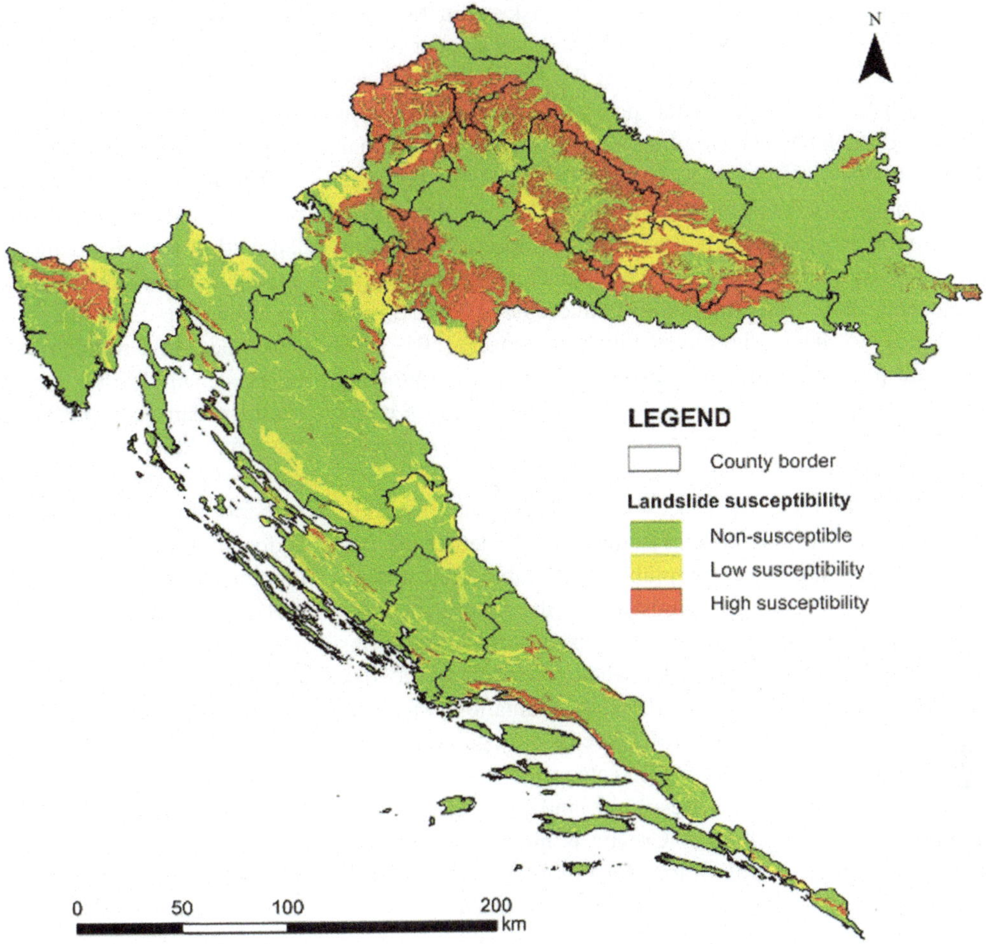

Fig. 5 Landslide susceptibility map of Croatia in the scale 1:100.000

landslide risk identification is described in Mihalić Arbanas et al. (2019). The outcome of the risk identification stage was a listing of two different identified risks and risk scenarios, followed by its descriptions. Both scenarios were based on a coherent and internally consistent set of assumptions about key relationships between hazardous event and consequences as well as driving forces in the form of landslide susceptibility, triggering event (precipitation) and its frequency. Like any other simplification of reality, the definition of a scenario entails subjective assumptions about the number of expected landslides and geographic coverage of scenarios. Namely, risk analysis for both scenarios was performed for the area of approx. 15,800 km^2 in NW Croatia that presents geographic coverage experienced MORLE in 2013 and 2018. For other parts of Croatia, MORLE events and impacts, which have so far not occurred, were also considered, assuming extreme hydrometeorological conditions (precipitation) that may happen as a consequence of climate changes and emerging landslide risk. It was essential that all information leading to the definition of scenarios was made explicit so that they can be reviewed and updated (Croatian Platform for Disaster Risk Reduction 2019).

Risk analysis is the process to comprehend the nature of risk and to determine the level of risk, as it is defined by standard ISO 31000:2018 (ISO 2018). Both risk scenarios were used in the risk analysis stage, aiming to establish quantitative estimates for impacts and probabilities. The assessment was performed to be as objective as possible, but due to lack of systematic records about historical hazards (i.e., number of landslide events) or consequences (i.e., losses), the main criteria was established based on expert opinion. Single-risk analysis estimated the risk of a singular hazard in isolation from other hazards or risk scenarios. Mihalić Arbanas et al. (2019) describes applied methodologies and results of the risk analysis, addressing the following subjects for MORLE from 2013 and 2018: hazard analysis; geographical analysis (location, extent); temporal analysis (frequency, duration, etc.); dimensional analysis (scale, intensity); probability of occurrence; vulnerability analysis; identification of elements and people potentially at risk (exposure); identification of impacts (physical, economic, environmental, social/political); assessment of likely impacts. An analysis of self-protection capabilities reducing exposure or vulnerability is planned to be performed in the next step of national risk assessment. Risk estimation was performed by classification of likelihood and impact of two hazardous scenarios. The impact of the Scenario 1 presenting MORLE with the worst possible consequences was estimated to have catastrophic human impact (more than 1500 affected people), significant economic and environmental impact (losses in range of 200–950,000,000 euros); and minor or significant political/social impacts (depending on observed elements at risk). The impact of the Scenario 2 presenting the most probable unfavorable MORLE was estimated to have significant human impact (501–1500 affected people), significant economic and environmental impact (losses in the range of 200–950,000,000 euros); and limited, minor or moderate political/social impacts (depending on observed elements at risk).

Risk evaluation is the process of comparing the results of risk analysis with risk criteria to determine whether the risk and/or its magnitude are acceptable or tolerable. Risk criteria are the terms of reference against which the significance of a risk is evaluated. Risk evaluation is used to make decisions about the significance of risks whether each specific risk should be accepted or treated. Risk evaluation was performed using the risk matrix recommended by guidelines EC (2010) classifying risk into four categories (low, medium, high and very high risk). Resulting landslide risk for both scenarios in the Republic of Croatia is high. The matrix is also used as a visualization tool for multiple risks that have been identified in the national risk assessment, to facilitate comparing the different risks.

Discussion and Conclusions

As a stakeholder of the "Sendai Framework for Disaster Risk Reduction 2015–2030", which belongs to academic, scientific and research entities, the CLG focuses on the disaster risk factors and scenarios, including emerging disaster risks, in the medium and long term; increases research for regional, national and local applications; supports action by local communities and authorities; and supports the interface between policy and science for decision-making.

The research of landslide disaster risk factors and scenarios, performed in the period 2017–2020 at the national level in Croatia, resulted in identification and description of two different risks and risk scenarios. The Scenario 1 presents MORLE with the worst possible consequences, estimated as: catastrophic human impact, significant economic and environmental impact, and minor or significant political/social impacts. The Scenario 2 presents the most probable unfavorable MORLE with consequences estimated as: significant human, economic and environmental impact and limited, minor or moderate political/social impacts. Based on quantitative estimates of impacts and probabilities of both scenarios showed that landslide risk in the Republic of Croatia is high. Results of landslide hazard and risk assessment are also important for long-term adaptation to climate changes and extreme weather conditions.

The CLG have been continuously increasing research for regional, national and local applications by development of landslide mapping methodologies as well as compilation of landslide maps in the Dinarides and the Pannonian Basin, including: (i) landslide inventories in the form of geomorphological inventories of shallow landslides and of different types of landslides associated with erosion phenomena (multi-hazard approach); (ii) landslide susceptibility mapping aimed at spatial prediction of initiation and runout zones of rock falls in local (site-specific) scale; (iii) landslide susceptibility mapping aimed at spatial prediction of landslides in regional and national scale. All landslide maps, together with data and information about registered and/or predicted landslides, supports action by local communities and authorities dealing with physical planning and construction, land-use planning and civil protection.

There are the two major results of the CLG, from the period 2017–2020, that supports the interface between policy and science for decision-making. First is official national landslide susceptibility map in the scale 1:100,000 of the Republic of Croatia, developed for the purpose of development of the National Physical Plan. The map will support decision making related to landslide risk prevention through the system of physical and land-use planning. The next strategically important Croatian document is national landslide risk assessment, that is crucial to supports decision making in the civil protection at the national level. Both national documents are basic documents for following-up development of a series of national and regional legislative documents that will regulate landslide risk management, from prevention and mitigation to landslide risk recovery.

References

Arbanas Ž, Vivoda Prodan M, Dugonjić Jovančević S, Peranić J, Udovič D, Bernat Gazibara S, Krkač M, Sečanj M, Mihalić Arbanas S (2019) Rockfall modelling and rockfall protection at the slopes above the city of Omiš, Croatia. In: Proceedings of the ISRM specialised conference "Geotechnical Challenges in Karst", 11–13 Apr 2019. Omiš, Croatia, pp 121–126

Bell R, Petschko H, Röhrs M, Dix A (2012) Assessment of landslide age, landslide persistence and human impact using airborne laser scanning digital terrain models. Geogr Ann Seri A, Phys Geogr 94:135–156

Bernat Gazibara S (2019) Methodology for landslide mapping using high resolution digital elevation model in the Podsljeme area (City of Zagreb). Ph.D. thesis, Faculty of Mining, Geology and Petroleum Engineering, University of Zagreb, Zagreb, Croatia (In Croatian)

Bernat Gazibara S, Mihalić Arbanas S (2018) Landslides in city of Zagreb urban area. In: Proceedings of the symposium "Urban utopia—approaches to urban space planning", 10–12 Nov 2017. Zagreb, Croatia, pp 17–17 (In Croatian)

Bernat Gazibara S, Krkač M, Sečanj M, Mihalić Arbanas S (2017) Identification and mapping of shallow landslides in the City of Zagreb (Croatia) using the LiDAR-based terrain model. In: Proceedings of the 4th World Landslide Forum, vol 2. May 29–June 2, 2017. Ljubljana, Slovenia, pp 1093–1100

Bernat Gazibara S, Krkač M, Sečanj M, Mihalić Arbanas S (2018a) Landslide inventory mapping based on LiDAR data. In: Proceeding of the VII conference of Croatian platform for disaster risk reduction, 11–12 Oct 2018. Zagreb, Croatia, pp 194–200 (In Croatian)

Bernat Gazibara S, Krkač M, Vlahek I, Pavlić K, Begić H, Zekan S, Sečanj M, Mihalić Arbanas S (2018b) Extreme precipitation events and landslides activation in Croatia and Bosnia and Herzegovina. In: Proceeding of the 3rd regional symposium on landslides in the Adriatic-Balkan region, 11–13 Oct 2017. Ljubljana, Slovenia, pp 19–24

Bernat Gazibara S, Cindrić Kalin K, Erak M, Krkač M, Đomlija P, Arbanas Ž, Mihalić Arbanas S (2019a) Landslide hazard analysis in national-scale for landslide risk assessment in Croatia. In: Proceeding of the 4th regional symposium on landslides in the Adriatic-Balkan region, 23–25 Oct 2019. Sarajevo, BIH, pp 175–182

Bernat Gazibara S, Krkač M, Mihalić Arbanas S (2019b) Landslide inventory mapping using LiDAR data in the City of Zagreb (Croatia). J Maps 15(2):773–779

Bernat Gazibara S, Krkač M, Mihalić Arbanas S (2019c) Verification of historical landslide inventory maps for the Podsljeme area in the City of Zagreb using LiDAR-based landslide inventory. Rudarsko-Geološko-Naftni Zbornik 34(1):45–58

Casagli N, Pini G (1993) Analisi cinematica della stabilità in versanti naturali e fronti di scavo in roccia. In: Proceedings of the 3rd Convegno Nazionale dei Giovani Ricercatori in Geologia Applicata, Potenza, 28–30 Oct 1993 (In Italian)

CloudCompare (v2.9.1) [GPL software], 2017

Croatian Platform for Disaster Risk Reduction Main Working Group (2019) Disaster risk assessment of the Republic of Croatia, App. 1 Scenario Development Report, Zagreb, 562 p

Đomlija P (2018) Identification and classification of landslides and erosion phenomena using the visual interpretation of the Vinodol Valley digital elevation model. Ph.D. thesis, Faculty of Mining, Geology and Petroleum Engineering, University of Zagreb, Zagreb, Croatia (In Croatian)

Đomlija P, Bernat Gazibara S, Arbanas Ž, Mihalić Arbanas S (2019) Identification and mapping of soil erosion processes using the visual interpretation of LiDAR imagery. ISPRS Int J Geo-Inf 8(10):438–465

Đomlija P, Jagodnik V, Arbanas Ž, Mihalić Arbanas S (2019b) Landslide types identified along carbonate cliffs using LiDAR-based DTM imagery—examples from the Vinodol Valley, Croatia. In: Proceedings of the 4th regional symposium on landslides in the Adriatic-Balkan region, 23–25 Oct 2019. Sarajevo, BIH, pp 91–96

European Commission, EC (2010) Risk assessment and mapping guidelines for disaster management. https://ec.europa.eu/echo/files/about/COMM_PDF_SEC_2010_1626_F_staff_working_document_en.pdf. Last accessed: 1st Sept 2019

Fell R, Corominas J, Bonnard C, Cascini L, Leroi E, Savage WZ (2008a) Guidelines for landslide susceptibility, hazard and risk zoning for land-use planning. Eng Geol 102(3–4):85–98

Fell R, Corominas J, Bonnard C, Cascini L, Leroi E, Savage WZ (2008b) Guidelines for landslide susceptibility, hazard and risk zoning for land-use planning. Eng Geol 102(3–4):99–111

Hungr O, Leroueil S, Picarelli L (2014) The Varnes classification of landslide types, an update. Landslides 11:167–194

ICL, International Consortium on Landslides (2018) Instructions for World Reports on Landslides. https://www.iplhq.org. Last accessed: 5th Mar 2020

IGU, International Geographical Union (1968) The unified key to the detailed geomorphologial map of the world, 1:25,000–1:50,000. Folia geografica, series geographica-physica 2, Krakow

ISO (2009) ISO GUIDE 73:2009, Risk management—Vocabulary, 17p

ISO (2018) ISO 31000:2018, Risk management—Guidelines, 18 p

Jagodnik P, Jagodnik V, Arbanas Ž, Mihalić Arbanas S (2020) Landslide types in the Slani Potok gully, Croatia. Geologia Croatica 73(1):13–28

Malamud BD, Turcotte DL, Guzzetti F, Reichenbach P (2004) Landslide inventories and their statistical properties. Earth Surf Proc Land 29:687–711

Matasci B, Stock GM, Jaboyedoff M, Carrea D, Collins BD, Guérin A, Matasci G, Ravanel L (2018) Assessing rockfall susceptibility in steep and overhanging slopes using three-dimensional analysis of failure mechanisms. Landslides 15:859–878

Mihalić Arbanas S, Arbanas Ž, Krkač M et al (2017a) Landslide risk reduction in Croatia: scientific research in the framework of the WCoE 2014–2017, IPL-173, IPL-184, ICL ABN. In: Proceedings of the 4th World Landslide Forum, vol 1. May 29–June 2, 2017. Ljubljana, Slovenia, pp 301–303

Mihalić Arbanas S, Sečanj M, Bernat Gazibara S et al (2017b) Landslides in the Dinarides and Pannonian Basin—from the largest historical and recent landslides in Croatia to catastrophic landslides caused by Cyclone Tamara (2014) in Bosnia and Herzegovina. Landslides 14:1861–1876

Mihalić Arbanas S, Bernat Gazibara S, Sečanj M et al. (2019) Landslide risk management in Croatia: Current state. In: Proceedings of the 4th regional symposium on landslides in the Adriatic-Balkan region, 23–25 Oct 2019. Sarajevo, BIH, pp 243–250

Petschko H, Bell R, Glade T (2015) Effectiveness of visually analyzing LiDAR DTM derivatives for earth and debris slide inventory mapping for statistical susceptibility modeling. Landslides 13 (5):857–872

Sečanj M, Mihalić Arbanas S, Kordić B, Krkač M, Bernat Gazibara S (2017) Identification of rock fall prone areas on the steep slopes above the town of Omiš, Croatia. In: Proceeding of the 4th World Landslide Forum, vol 5, May 29–June 2, 2017. Ljubljana, Slovenia, pp 481–487

Sečanj M, Bernat Gazibara S, Mihalić Arbanas S, Krkač M, Martinko M, Arbanas Ž (2018) Identification of potentially unstable rock blocks on the road cut in the Krka National park, Croatia. In: Proceedings of the 3rd regional symposium on landslides in the Adriatic-Balkan region, 11–13 Oct 2017. Ljubljana, Slovenia, pp 113–118

Sečanj M, Mihalić Arbanas S, Krkač M, Bernat Gazibara S, Arbanas Ž (2019a) Preliminary rockfall susceptibility assessment of the rock slopes above the town of Omiš (Croatia). In: Proceedings of the ISRM specialised conference "Geotechnical Challenges in Karst", 11–13 Apr 2019. Omiš, Croatia, pp 347–352

Sečanj M, Mihalić Arbanas S, Krkač M, Bernat Gazibara S, Arbanas Ž (2019b) Rockfall susceptibility assessment at the slope scale. Abstracts Book of the 6th Croatian Geological Congress, 9–12 Oct 2019. Zagreb, Croatia, pp 172–173

Soeters R, van Westen CJ (1996) Slope instability recognition, analysis, and zonation. In: Turner AK, Schuster RL (eds) Landslides,

investigation and mitigation. Transportation Research Board, Special Report 247, Washington DC, 129–177. ISBN 0-309-06151-2
Split Engineering, LCC. Split-FX V 2.4, Tucson, AZ (2017)
Steger S, Brenning A, Bell R, Petschko H, Glade T (2016) Exploring discrepancies between quantitative validation results and the geomorphic plausibility of statistical landslide susceptibility maps. Geomorphology 262:8–23
United Nations International Strategy for Disaster Reduction, UNISDR (2009) UNISDR terminology on disaster risk reduction. UNISDR, Geneva, 35p

LARAM School: An Ongoing Experience

Leonardo Cascini, Michele Calvello, and Sabatino Cuomo

Abstract

LARAM is an International School on "LAndslide Risk Assessment and Mitigation", born in 2005 as a pioneering didactic and research initiative of the Geotechnical Engineering Group of the University of Salerno. The main activity of LARAM is a yearly 2-week School for Ph.D. students and Young Doctors. Other initiatives include shorter courses, scientific sessions and workshops, research projects and other dissemination activities. This paper discusses the inspiration, the foundations and the main pillars of the School, and provides a brief overview of the main activities developed since 2006.

Keywords

Landslide • Risk • Assessment • Mitigation • Students

Introduction

LARAM is an International School on "LAndslide Risk Assessment and Mitigation" of the Geotechnical Engineering Group (GEG) of the University of Salerno.

As reported in the web page (www.laram.unisa.it), the goals of LARAM are: (i) to develop high educational interdisciplinary programs for assessing, forecasting and mitigating landslide risk over small and large areas; (ii) to promote the creation of "on the job" vocational training programs aimed at solving real landslide risk problems using the most advanced theories and methodologies in the fields of geotechnical engineering, geomechanics, geology, mathematical modelling, monitoring, GIS techniques.

LARAM was born in 2005 at the University of Salerno, as a pioneering didactic and research initiative. However, the origin and the basic ideas behind LARAM come from further away.

This paper summarizes the context where LARAM was conceived, as well as the fundamentals, the structure, the participation, the results achieved till now, and the School's perspective towards the future.

Inspiration

On 5–6th May 1998 more than 100 slope failures and several tens of catastrophic flow-like landslides impacted 5 Municipalities (Sarno, Quindici, Siano, Bracigliano, San Felice a Cancello) of the Campania region (Southern Italy), causing 160 victims and huge economic damage over about 60 km^2.

The day after, the Italian Department of National Civil Protection assigned the responsibility of the scientific management of the post-event emergency to the geotechnical engineering team of the University of Salerno, which at that time counted of only few persons. From the very beginning, the team was handed the needed technical and economic support by the National Government. Due to the magnitude of the landslide disaster, within only few days, tens of researchers and technicians arrived to Salerno as volunteers to contribute to the post-event emergency activities.

From that, an interdisciplinary working group of more than 100 persons was created and organized in specific Operative Units (O.U.), strictly interacting with each other. These O.U. included geotechnical, hydraulic and structural engineers, geologists, architectures, experts in geomatics and other technicians (Cascini 2004).

L. Cascini
LARAM School, Fisciano, Italy
e-mail: l.cascini@unisa.it

M. Calvello · S. Cuomo (✉)
Department of Civil Engineering, University of Salerno, 84084 Fisciano, Italy
e-mail: scuomo@unisa.it

M. Calvello
e-mail: macalvello@unisa.it

K. Sassa et al. (eds.), *Understanding and Reducing Landslide Disaster Risk*, ICL Contribution to Landslide Disaster Risk Reduction,
https://doi.org/10.1007/978-3-030-60196-6_17

The large number and the high-quality profile of the persons composing the working group convinced the Department of National Civil Protection to ask the University of Salerno to map the residual landslide risk, in the whole 60 km^2 area, by 18th May 1998, i.e. in only 12 days.

The deadline was fulfilled despite the challenge of such a strict schedule and the absence, at that time, of well-established procedures for field data acquisition and residual risk mapping, neither in Italy nor elsewhere. The presentation of the work developed in those days had a great impact on the public opinion in Italy, and was the starting point of the several future activities aimed to mitigate landslide risk. These comprised a number of specific tasks, the following 8 years, including the construction of engineering works inside and, to a lesser extent, outside the affected territory.

Based on the considerable success of the first 12-days phase of the disaster management, the National Government decided to definitively boost the development of landslide and flooding risk zoning in Italy. Thus, in the following 4 years, the Local Territorial Authorities have completed the task of the landslide risk zoning all over Italy. For an area of about 17,000 km^2 of Southern Italy, this activity was developed under the scientific coordination of the University of Salerno.

The large amount of work done by the Geotechnical Engineering Team of the University of Salerno from 1998 to 2002 was publicly presented in the International Conference on Fast Slope Movements Prediction and Prevention for Risk Mitigation, held in Naples in June 2003, which was attended by the top international experts in the field of landslide risk. An immediately later presentation was also done at the Environmental Ministry and to the Central Government, which funded at the University of Salerno a Centre of Excellence named CERIUS (Forecasting and prevention of hydrogeological risk over large areas), Principal Investigator, multidisciplinary approach, work principles and operative procedures of this centre were the same used to face the 1998 emergency management.

CERIUS lasted few years (2003–2005) but it was indeed the incubator of the LARAM School, which was ideally conceived as a long-term dissemination vehicle of landslide risk know-how for high-level technicians and researchers dealing with the analysis, assessment, and mitigation of landslide risk or related topics and tasks.

On 21st April 2005, the LARAM School was officially created at the University of Salerno with the specific goal, among others, to mutually bridge the gap between practice and theory, i.e. to turn the experience that followed the 1998 catastrophic landslides into a chance for disseminating innovative procedures, while also driving advanced theories into the standard practice applied in the field.

Foundation and Pillars

The structure of LARAM School has been simple yet effective so far. Starting from the foundation, it reads as follows.

The Management (M) of the School has been in charge of the Director of the School, Leonardo Cascini, until his retirement in 2017. Then, also considering the growing amount of initiatives, the LARAM management was entrusted to the President and to two Coordinators, to be individuated every three years among the Professors of Geotechnics of the University of Salerno. For the 2018–2020 term, Leonardo Cascini, Michele Calvello and Sabatino Cuomo were appointed, respectively, President and Coordinators.

The Technical Committee (TC) was chaired by our beloved colleague and friend Giuseppe (Peppe) Sorbino from the first edition of the School in 2006 until 2013, followed by Dario Peduto, who is currently in charge of the TC. Size and composition of TC have been changing remarkably over the years, starting from 2 members, in addition to the Chairperson, in the very beginning of the LARAM experience, and going up to 7 members in 2018, and back to 3 in 2020. This can be seen as a sign of a changing School in a changing world.

The Scientific Partners (SP) were fundamental at different stages of the LARAM experience. Without going into specific details, some of SPs were strong scientific supporters of the LARAM idea, other ones were austere reviewers. Both types of contributions helped LARAM in growing and maturing in these 15 years and in becoming what it has become. We are delighted to mention, among them, ALERT Geomaterials (The Alliance of Laboratories in Europe for Research and Technology), the International Centre for Geohazards (ICG) based in Norway, the Technical University of Athens (NTUA) and the Italian Geotechnical Society (Associazione Geotecnica Italiana, AGI).

The Sponsors (S) gave to LARAM the opportunity to take off and sail. The CMPA (Comunità Montana Penisola Amalfitana, Mountain Authority of Amalfi Peninsula), a local territorial authority for land management in the Amalfi-Sorrento district, was the sponsor of LARAM for the first three years and Dott. Raffaele Ferraioli, a true modern patron, deserves special thanks for having shared the feeling to institute LARAM. The above mentioned Centre of Excellence CERIUS, the Consortium CUGRI (Consorzio Inter-Universitario per la Previsione e Prevenzione dei Grandi Rischi), and the CUEB (Centro Universitario Europeo per i Beni Culturali) supported the LARAM initiatives for some years. Important it has also been the financial and logistical support of the Department of Civil Engineering of the University of Salerno, and the funds raised by the Geotechnical Engineering Group from Regione Campania and the EU-FP7 project SAFELAND (Living with landslide risk in Europe: a large-scale integrating collaborative research project).

However, the true pillars of LARAM are the School's Scientific Committee and the Students.

The Scientific Committee (SC) is the true pride of LARAM School, as it is composed of esteemed experts in the many fields that contribute to define, analyze, assess and manage Landslide Risk. The SC lasts three years. Every year, the current SC is in charge of setting the criteria for the selection of students, defining programme and contents of the Courses, choosing the lecturers, and evaluating the results of the School. Looking back at the 5 SCs set up so far, it is amazing to recognize 5 Dream Teams, composed of world-wide recognized experts and, above all, open-minded friends willing to invest their time in training young researchers.

The Students (St) can be considered the treasure of LARAM, and also its main goal. We will come back to them later.

LARAM Editions

The 2-Week Yearly School

The 2-week long school for Ph.D. students is the main yearly initiative of LARAM. For each edition of the school, 40 Ph.D. students are selected to attend the residential courses, selected among applicants working in civil engineering, environmental engineering, engineering geology, geology, geography or related fields. Since 2012, the School can be also attended by 5 (max.) Young Doctors who defended their Ph.D. thesis in the previous 5 years. The courses include formal lessons, tutorials and field training.

The first 3 Editions (2006–2008) of the LARAM School were organized in September, in Ravello, a beautiful village in the heart of the Amalfi Coast. This location was chosen for several reasons: (i) it is inside the territory of competence of the sponsor (CMPA), (ii) it is an area with significant hazard coming from shallow landslides, rockfalls and hyperconcentrated flows, (iii) it is a UNESCO Heritage site, and last but not least (iv) it lies in a marvellous touristic district, perfect to promote a newly born scientific initiative.

The 2 later Editions (2009–2010) of the School were held in downtown Salerno. This choice was inevitable for LARAM, being a creature of the University of Salerno but it was also a tribute to the city of "Schola Salerni", the first and most important "modern" medical institution created in Europe in the Middle Ages (IX century).

For 2 years (2011–2012) the LARAM School moved to Chengdu (China), renowned worldwide for the Giant Panda Park but also sadly famous for the 2008 Sichuan 7.9 M earthquake, after which extensive research have been performed on coseismic landslides and combined hazards. In Chengdu, the SKLGP (State Key Laboratory of Geohazard Prevention and Geoenvironment Protection) was the Host Institution and the Sponsor. The School was organized by Prof. Huang Runqiu, a true friend of LARAM, and it was held in November, to adjust to local climate. LARAM came back again in Chengdu in 2015.

LARAM landed in the classrooms of the University of Salerno in September 2013, and again in 2014 and from 2016 to 2018, thus granting the participating students to possibility to also live the daily life of the UniSa Campus.

The last edition of the School was held abroad again, in September 2019, hosted and sponsored by the University of Lausanne (Switzerland), thanks to the commitment and organizing skills of Prof. Michel Jaboyedoff.

And this year, which marks the 15th edition of the School, where will LARAM be? Good question. The UniSa Campus has been chosen again to host, in September, the LARAM School 2020.

What are the benefits of moving from one Country to another? It is a great opportunity to involve different teams of lecturers, attract different potential students and doing field work in completely different geoenvironmental contexts.

Special Editions and Other Initiatives

In these same years, something more also happened.

During the same days of the Schools, a series (2006–2014) of LARAM Workshops were organized, lasting one day, as forums to disseminate the latest developments in Landslide Risk Management methodologies among researchers, professionals and authorities in charge of territory governance in Italy and Europe. The first 3 Workshops were held in Ravello, and the others in the city of Salerno. For all these events, we moved in strict synergy with the local Italian associations of practitioners, including engineers, geologists, and architectures, but also taking advantage of the international pool of LARAM lectures present for the School. Special attention was also devoted to the non-English speaking attendants providing professional simultaneous translation. It is a matter of fact that knowledge transfer and true dissemination cannot avoid tackling also such issues.

Some special Editions of LARAM are also part of the LARAM story. A 1-week Course was held in September 2011 in Salerno, totally devoted to the outcomes of the EU-FP7 Safeland Project. 2011 was also the year of the 1st

LARAM School held in China. And this year, in February 2020, another 1-week Course has been held in India, hosted and sponsored by the Indian Institute of Technology Roorkee (IIT-Roorkee), with Prof. Mahendra Singh as Local Organizer. The Course was set up in cooperation with the Norwegian University of Science and Technology (NTNU), and the Indian Geotechnical Society (IGS). As novelty for LARAM, we re-casted the composition of the class to accommodate the needs of the potential Indian attendants. Thus, the call for participation to select the 40 students of the class was organized as follows: (i) 10 participants among Indian Ph.D. students, Young Ph.D. holders and Young Professors, (ii) 10 participants among non-Indian Ph.D. students, Young Ph.D. holders and Young Professors, (iii) and 20 Indian Professionals.

Other relevant initiatives are also worth to be mentioned, among the ones organized and listed in the section "Initiatives" of the LARAM web page.

In 2013, a Session titled "Geomechanics in Slopes" was organized within the ALERT Workshop 2013, as the 1st joint initiative within the framework of an Official Memorandum of Understanding signed between LARAM and ALERT in 2012.

In 2015, LARAM and SKLGP (State Key Laboratory of Geohazard Prevention and Geoenvironment Protection) of the Chengdu University of Technology (China) organized an academic trip for the LARAM Staff, including visits, presentations and discussions, in the following leading Chinese Institutions for landslide risk analysis and mitigation: SHANGHAI, Tongji University; WUHAN, China University of Geosciences; XI'AN, Changan University and Xian Geological Survey; BEIJING, Institute of geology and geophysics and Chinese Academy of Sciences; CHENGDU, Chengdu University of Technology.

In 2016, the 8th Olek Zienkiewicz ALERT Course was held at the University of Salerno in combination with the LARAM School.

In 2017, the first Edition of Meet the Experts: LARAM was launched, consisting in a series of short interviews, i.e. landslide chats, with experts in the field of landslide analysis and management. The first 8 interviews were recorded at the University of Salerno during the LARAM School 2017. In 2018, a second Edition was organized.

In October, 2020 a Session of the ALERT Workshop will be co-organized again, with the title "Forecasting landslide displacements". Topics will mainly include geomechanics-driven models, but also engineering-oriented models and procedures will be presented, within the context of risk theory, to provide information relevant for early-warning systems, urban planning maps, and land control tools operational, both at slope scale and over wide areas.

The Students

The selection of the Ph.D.s and Young Doctors who are granted the possibility to participate to the LARAM Schools and Courses has been always a priority for the Management team. Indeed, a very detailed and comprehensive Application Form has been set up since the first edition of the School and used, over the years, to evaluate the applicants and select the LARAM classes. Every year, during the selection process, we have always strived to select the most promising students as well as to compose a class that was widely international and heterogeneous in terms of field of expertise, scientific maturity and main topics addressed.

Number, geographical location and typology of applications received by students have been changing over the years. Clear trends cannot be recognised, however a correlation with the location of the School can be inferred. For instance, the three editions of the School that were held in Chengdu (China) recorded a high number of applications by Chinese and other Asian students and a significantly lower number of applications from European countries, if compared with the Schools held in Italy. Conversely, in the latter cases, the presence of applications from Italy and other European countries has been predominant. This is also related to the cost the selected applicants must sustain for their participation to the School. Besides the registration fees, free until 2015, travel and accommodation costs also play a major role in screening candidates potentially interested in the courses offered by the LARAM School.

Overall, more than 600 attendants participated to the 14 LARAM Schools organized over the past 14 years. In the 10 editions of the School held in Italy (Ravello or Salerno, from 2006 to 2018, excluding 2011, 2012 and 2015), the lectures have been attended by 473 students belonging to 148 universities from 54 different countries. The three editions organised in China (Chengdu, in 2011, 2012 and 2015) have been attended by 135 students belonging to 70 universities from 23 countries. The edition organized in Lausanne, Switzerland (in 2019) was again attended by an international class of students belonging to 21 universities from 11 countries. In summary, 192 universities provided Ph.D. students to the 14 editions of the LARAM School from 2006 to 2019.

To investigate how the students valued the LARAM experience, a questionnaire was set up, since the first year, addressing different didactic and logistic aspects of the School. At the end of every edition, during the last day, the questionnaires have been handed out to the students, filled, and analysed (in quasi real-time) by the organizing committee, in order to present the results to the class during the closing ceremony. Figure 1 shows, as an example, the results of the questionnaires filled by the students of the

LARAM School 2019. The feedback was very satisfactory for almost all the investigated aspects, and an absolutely positive evaluation (all answers were either excellent or good) was provided for the general programme and for the general organization. The only poor evaluation was recorded for the tutorials, with about 50% of the students judging them negatively.

A comparison with evaluations from students attending other LARAM Schools (Fig. 2) shows that this is not the first time such evaluations occur for this specific didactic aspect (Fig. 2c). The likely explanation for the fluctuating performance of tutorials is that the time devoted to these activities within the programme of the School has been, in most years, relatively short. The SC and the lecturers have always been aware that tutorial activities require a significant amount of hours to be properly conducted, yet typically they have chosen to "sacrifice" hands-on activities in favour of a wider exposure to many different renown lecturers addressing many different topics from many different points of view (each one dealt with only for a few hours). In that respect, this choice appears to be well perceived by the students, as the results of the questionnaire indicate that the vast majority of them (over 80% in all the editions of the School) appreciated the general programme and the quality of lectures (Fig. 2a, b), as well as the general organization of the School (Fig. 2e). Concerning the School location (Fig. 4d), the level of appreciation is generally good or very good, with the exception of 2016. It was not the first time, not it would have been the last, that the School was held at the Salerno Campus (from 2013 to 2018), where the teaching facilities were always well valued. Yet, it was the only time Campus dormitories, located in a suburban area like the rest of the Campus, were chosen for the students' accommodation. It appears that the students appreciate much more residing in the lively city centre of Salerno (where they were in 2010, 2014, and again in 2017 and 2018).

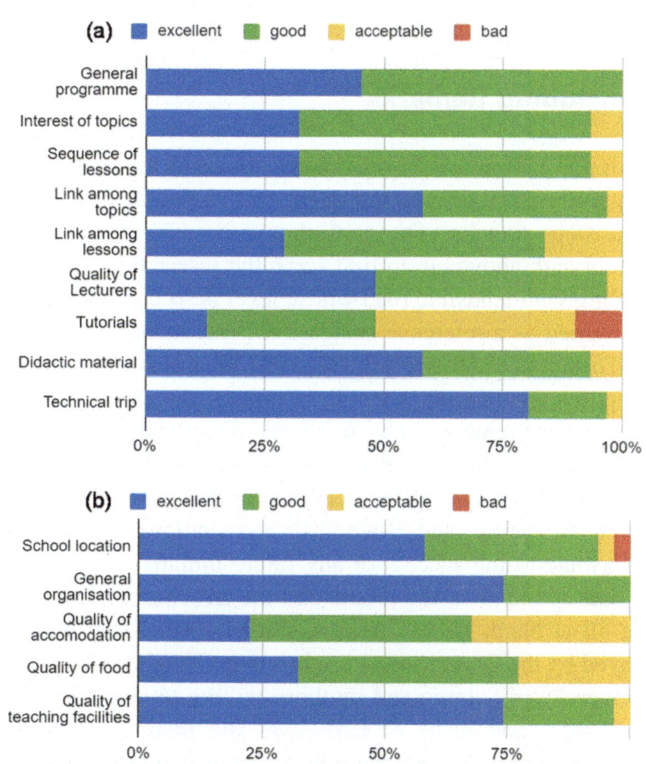

Fig. 1 Answers to questionnaires filled from students of LARAM School 2019: **a** didactic aspects, **b** logistic aspects

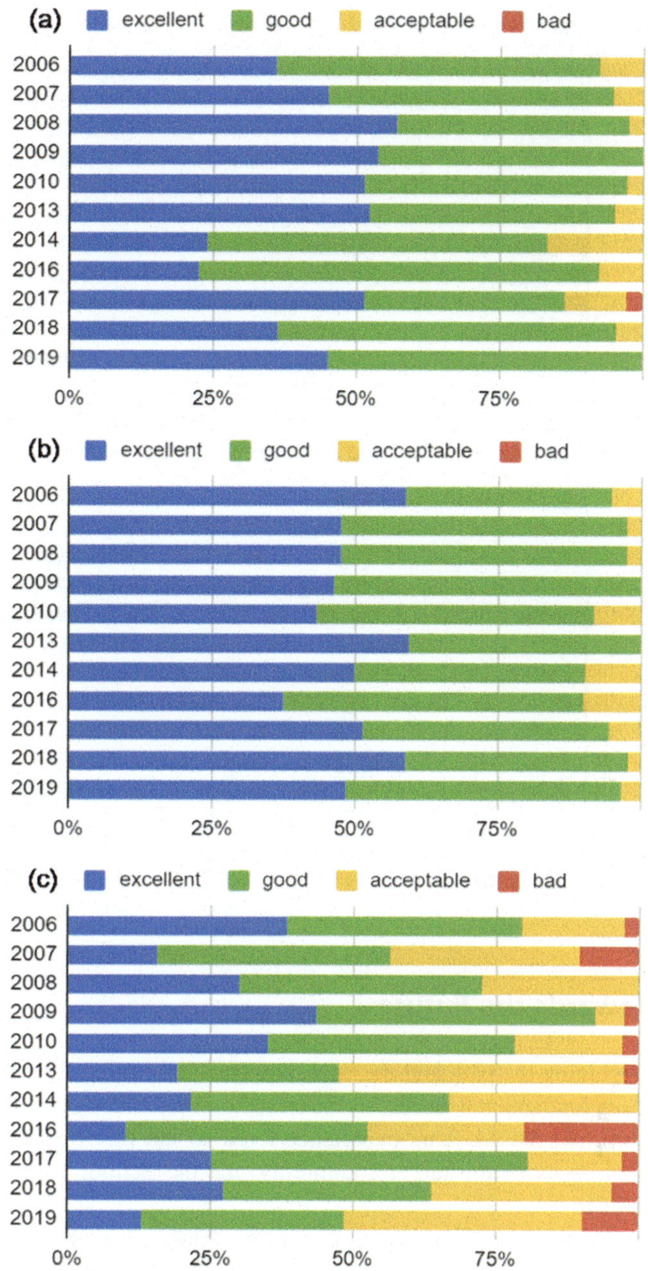

Fig. 2 Comparison of some questionnaire's answers from students of LARAM Schools held in Europe: **a** general programme, **b** quality of lectures, **c** tutorials, **d** school location, **e** general organization

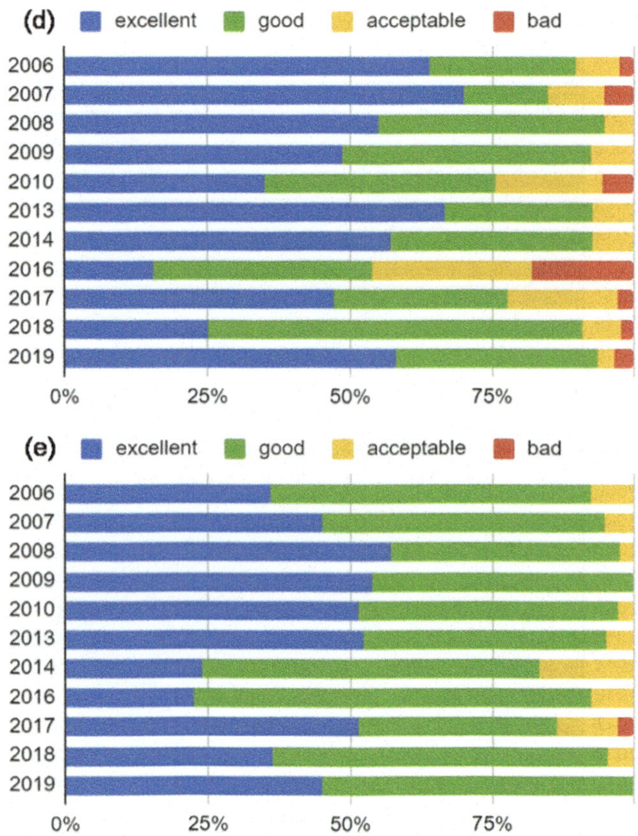

Fig. 2 (continued)

Towards the Future

Designing the path and the next steps must take into account three fundamental issues: (i) the context where LARAM was born, (ii) the 14 years experience lived so far, and (iii) the new needs of the target community that the School would refer to.

As for the first topic, the work done in a difficult post-emergency context, which required either high reliability or quick decisions, was highly instructive. The main lesson learnt was that multidisciplinary and multiscale approaches (Fell et al. 2008) are advisable to: (i) enhance the knowledge of the different disciplinary sectors involved in landslide risk assessment and mitigation and, at the same time, (ii) combine within a single process the needs of the territorial authorities and the technicians called to solve specific difficult problems. This same approach, applied to the LARAM School, has proven to be an attractor for students.

Concerning the 14 years experience, we already analyzed the flattering feedback from the students, but it is also worth commenting on the opinion of the prestigious scholars involved as members of the Scientific Committee. In fact, most of them regularly contributed to the courses and, in some cases, they lectured at almost every School's edition. The retention rate of the lecturers testifies that LARAM was a shared idea, born as a pioneering one and consolidating over time. Some meetings of the SC were also focused on the LARAM mission and one of the most shared opinions was to attribute to the school the goal of "long-term dissemination" (Cascini et al. 2012). We can state that we are succeeding in achieving that goal, as also testified by many students who have attended the School, continued to deal with landslide-related issue in their post-doctoral career and profitably used the concepts learned, albeit years later. It is generally recognized that the worldwide landslide risk management community has been paying a great attention to the LARAM experience over time. Indeed, LARAM was the vehicle for several Italian and international research and educative projects and initiatives, as already mentioned before.

From the past towards the future, LARAM has the need to renew itself while maintaining its inspiring principles. Involving brilliant past LARAM students as lecturers is a way to build and to renew over time the scientific community, which can grow in knowledge and size. Exploring new fields of knowledge, contiguous to landslide risk, is a way to export the high level of knowledge achieved so far. Employing new teaching methods, respectful of the fundamental principles of LARAM, could make the students deal with real cases to be analyzed and managed through advanced theories and helpful pragmatism.

Concluding Remarks

LARAM is now a consolidated School and the experience acquired so far suggests that great care should be paid to the next steps to take. The goal should be strengthening the school and, above all, to best serve the research community dealing with landslide risk, whose level is so high that could be termed as a reference standard for risk management.

To increase the strength of LARAM means, for sure, to keep the quality of Scientific Committee, applicants and lecturers extremely high. A strong interaction with other high-level Scientific Communities should also be seen as a new-old challenge. Indeed, involving Italian and International Scientific and Technical Associations can surely contribute to the success of any future initiative.

The foundation, the pillars and the structure of LARAM are all at hand. As in the past, a fundamental role will be played by the Geotechnical Engineering Group (GEG) of the University of Salerno and by those willing to share the new challenges and paths of LARAM. The final goal is promoting more and more the enhancement of know-how and procedures applied in the field of landslide risk, which is very significant in many areas of the world.

References

Cascini L (2004) The flowslides of May 1998 in the Campania region, Italy: the scientific emergency management. Riv Ital Geotecnica 2:11–44

Cascini L, Sorbino G, Calvello M, Cuomo S (2012) The LARAM school: teaching "LAndslide risk assessment and mitigation" to Ph. D. students. In: SFGE 2012, Proceedings of conference on shaking the foundations of geo-engineering education, Galway, Ireland, pp 211–218

Fell R, Corominas J, Bonnard C, Cascini L, Leroi E, Savage WZ (2008) Guidelines for landslide susceptibility, hazard and risk zoning for land use planning. Eng Geol 102:85–98

Advanced Technologies for Landslides (WCoE 2017–2020)

Nicola Casagli, Veronica Tofani, Filippo Catani, Sandro Moretti, Riccardo Fanti, Giovanni Gigli, Silvia Bianchini, and Federico Raspini

Abstract

The UNESCO Chair on Prevention and Sustainable Management of Geo-Hydrological Hazards, Department of earth Sciences, University of Florence has been a member of the International Consortium on Landslides (ICL) since 2002 and was designated as one of World Centres of Excellence (WCoE) for Landslide Risk Reduction four times for 2008–2011, 2011–2014, 2014–2017 and 2017–2020 with a project entitled "Advanced technologies for Landslides". In this paper we describe the activities carried out by the UNESCO Chair as a member of ICL and as WCoE and its contribution to the risk reduction policies promoted by the 2020 Kyoto Commitment.

Keywords

Italy • Landslides • Monitoring • Remote sensing • Risk reduction • Kyoto commitment

N. Casagli (✉) · V. Tofani · F. Catani · S. Moretti · R. Fanti · G. Gigli · S. Bianchini · F. Raspini
Department of Earth Sciences, University of Florence, Via la Pira 4, 50121 Florence, Italy
e-mail: nicola.casagli@unifi.it

V. Tofani
e-mail: veronica.tofani@unifi.it

F. Catani
e-mail: filippo.catani@unifi.it

S. Moretti
e-mail: sandro.moretti@unifi.it

R. Fanti
e-mail: riccardo.fanti@unifi.it

G. Gigli
e-mail: giovanni.gigli@unifi.it

S. Bianchini
e-mail: silvia.bianchini@unifi.it

F. Raspini
e-mail: federico.raspini@unifi.it

Introduction

The UNESCO Chair (UNESCO Chair) on Prevention and Sustainable Management of Geo-hydrological Hazards is based at the Department of Earth Sciences of the University of Florence (DST-UNIFI) that is an important center for research and higher training in Italy.

DST-UNIFI is a founding member of ICL since 2002 and in 2008 the DST-UNIFI was entitled as a World Centre of Excellence (WCoE) on Landslide Risk Reduction (2008–2011) by the Global Promotion Committee of International Programme on Landslides of UN-ISDR. This recognition was reaffirmed three times over for 2011–2014, 2014–2017 and 2017–2020.

The objective of the WCoE is to develop new methodologies and advanced technologies for landslide risk reduction. The UNESCO chair carry out research and development (R&D) for the prevention and management of landslides, in order to support policies and actions of risk reduction. In particular WCoE focuses on research activities concerning the landslide monitoring and early warning through innovative technologies, exploitation of EO (Earth Observation) data and technology to detect, map, monitor and forecast ground deformations, regional forecasting models and on activities related to education and training on landslides risk reduction.

The expertise of the research team includes remote-sensing techniques and application of space-borne and ground-based SAR interferometry, monitoring ground instabilities and development of early warning systems and GIS-based quantitative models for hazard and risk prediction. The group participates in research and technological development projects in several areas of the world, often in active collaboration with international, national and regional organizations and agencies.

Since 2002, the UNESCO Chair has coordinated or has been involved in several ICL/IPL projects and UNESCO

K. Sassa et al. (eds.), *Understanding and Reducing Landslide Disaster Risk*, ICL Contribution to Landslide Disaster Risk Reduction,
https://doi.org/10.1007/978-3-030-60196-6_18

Chair is the proposer of the ICL Italian network, that has been officially established in December 2018.

Based on the expertise gathered during the last years in the abovementioned activities, UNESCO Chair can provide its contribution to Kyoto 2020 Commitment for Global Promotion of Understanding and Reducing Landslide Disaster Risk that signed the Kyoto Commitment during the 2019 ICL-IPL Conference which at UNESCO Headquarters, Paris, on 16–19 September 2019 (Sassa 2018a, b, c, 2019).

Research Activity of WCoE

The WCoE project focuses on the three main activities structured in three packages:

WP1: Innovative technologies (Ground-based SAR interferometry, UAV, Laser Scanner) for landslide monitoring and early warning;

WP2: EO (Earth Observation) data and technology to detect, map, monitor and forecast ground deformations;

WP3: Regional landslide forecasting models.

WP1 performs monitoring activities of unstable slopes in order to estimate the deformational evolution of the landslide events (in space and time) and to implement the most suitable operational early warning systems (EWS) according to different critical situations (Carlà et al. 2019b; Frodella et al. 2018; Carlà et al. 2017; Bardi et al. 2017).

Several sites are monitored in Italy and all of them were monitored by using advanced sensors and portable instrumentations as ground-based synthetic aperture radar interferometers (GB-InSAR), terrestrial laser scanning (LIDAR), satellite interferometry (PS-InSAR), UAVs equipped with different sensors, GPS antennas, infrared thermography and traditional instrumentation (e.g. strain gauges, inclinometers, piezometers) (Del Soldato et al. 2018a; Di Traglia et al. 2018; Rossi et al. 2018; Casagli et al. 2017; Frodella et al. 2017; Lombardi et al. 2017; Carlà et al. 2019a, b).

The case study of Gallivaggio (Carlà et al. 2019b), for instance, describes a successful example of rockfall forecasting and risk management in proximity of the Gallivaggio sanctuary (San Giacomo Valley, Central Italian Alps) by means of ground-based radar interferometry. Since 2016, a GB-InSAR continuously monitored a roughly 500-m high, sub-vertical granitic slope. Monitoring data evidenced the presence and continued movement of a $\sim$ 5000 m^3 highly unstable mass, having a projected fall trajectory directly threatening the sanctuary and the adjacent segment of a national road. The rock mass eventually fell, and the failure-time was accurately predicted several hours in advance. Despite damage to both the road and buildings, the timely evacuation of the area prevented any loss of life. The case study provides a reference framework to better manage rockfall risk in areas where the installation of adequate protective barriers is not technically feasible.

Another example of integration of different monitoring techniques is the Bosmatto landslide located in Valle d'Aosta region in Northern Italy.

Bosmatto landslide (Carlà et al. 2019a) is a large phenomenon located in alpine environments with low accessibility. The characterization of such phenomenon is challenging, due to their complex nature as well as the limitations of monitoring techniques. The integration of different monitoring data from different sources can help reduce uncertainties and provide a characterization of the phenomenon.

The joint analysis of GNSS, satellite InSAR, and GBInSAR data (Fig. 1), collected over a time span of 16 years, provided crucial insights into the mechanism and behavior of the landslide. In particular, the fundamental deformation field of the slope was validated from the combination of

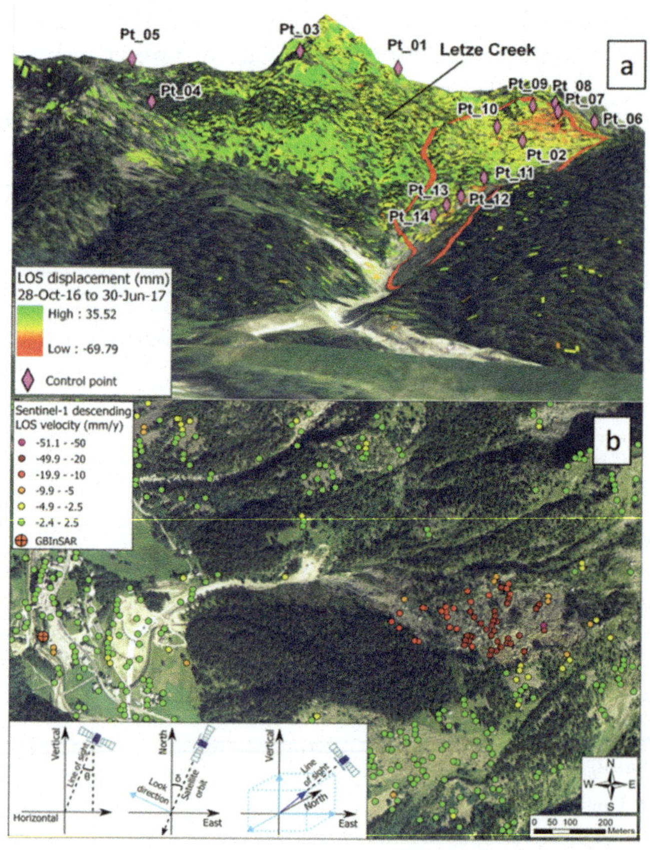

Fig. 1 Bosmatto landslide **a** Cumulative deformation map produced by the GBInSAR from 28 October 2016 to 30 June 2017. **b** Satellite InSAR data acquired in descending orbit over the study area. Modified from Carlà et al. (2019a)

GNSS and decomposed satellite InSAR records. It was suggested that the Bosmatto landslide is a large rockslide that moves in a generally coherent way according to a broadly roto-translational mechanism.

The work highlights challenges and implications related to slope monitoring in high alpine environments. Ku-band GBInSAR acquisitions were heavily affected by the presence of snow on the mountain slope. C-band Sentinel-1 acquisition also showed variation of the data points around the main trend during the winter seasons, although long-term trends could still be appreciated. C-band Sentinel-1 data are remarkably consistent with the GNSS measurements, especially concerning movements along the vertical component.

WP2 activities focus on the development of the satellite surveillance system exploiting the satellite data for the identification, mapping, monitoring and analysis of risk scenarios associated with landslides from local to regional scale (Intrieri et al. 2018; Pastonchi et al. 2018; Raspini et al. 2018; Rosi et al. 2014, 2016, 2018; Solari et al. 2019). The surveillance system is based on all the Earth Observation (EO) data (radar, multi- and hyperspectral) already available from several satellites (ERS, ENVISAT, RADARSAT, COSMO-SkyMed). Such integrated system is designed for the identification, rapid mapping, monitoring and analysis of risk scenarios. The launch of the SENTINEL-1 satellite (able to provide information in continuity with those from ERS and ENVISAT) offers new opportunities for monitoring the Earth's surface and for the evaluation of ground movements due to increased acquisition frequency and the regularity of acquisitions (Del Soldato et al. 2018a; Solari et al. 2018; Raspini et al. 2017).

A monitoring system based on the systematic processing of Sentinel-1 images processed by means of the SqueeSAR algorithm (Ferretti et al. 2011) has been firstly implemented for the Tuscany region in Italy and then for other two regions: Valle d'Aosta and Veneto. The methodology is the following: once a new Sentinel-1 acquisition is available, it is automatically merged with the existing SAR image archive, and the new data stack is processed to create continuously updated ground deformation data (Raspini et al. 2018, 2019).

Figure 2 related to Veneto region, measurement points are represented by dots coloured according to their mean velocity along the LOS of the satellite, expressed in millimetres per year, with positive values of mean velocity representing displacements towards the satellite, and negative value representing displacements away from the satellite. To each measurement point a deformation time series of the entire monitoring period is related.

Then, thanks to a specifically developed algorithm all the time series are analysed in order to detect anomalous points (i.e. points where a change in the dynamics of motion is occurring). Anomalous points are analysed with the support of thematic information to decide if an anomalous pattern can be considered worth to be reported to regional authorities.

WP3 activities focus on the optimization of the regional early warning system for landslide risk by means of meteorological nowcasting and real-time forecasting of slope movements that are characterized by rapid and very fast kinematic (Salvatici et al. 2018; Segoni et al. 2018a, b). The activities are related to the updating and application of a physically based model (HIRESSS model, Rossi et al. 2013) and to the improvement of statistical forecasting models (SIGMA and MACUMBA models). WP3 also deals with geotechnical characterization of soil cover in order to prepare reliable input data for the landslide forecasting physically based model (HIRESSS) (Bicocchi et al. 2019; Del Soldato et al. 2018b; Tofani et al. 2017).

In particular the activities of the WCoE has focused on the further improvement of the HIRESSS model developed to analyze shallow landslide triggering conditions on large scale at high spatial and temporal resolution using parallel calculation method (Salvatici et al. 2018). The HIRESSS model needs several geotechnical and hydrological input data and is able to compute the factor of safety at each selected time step (and not only at the end of the rainfall event) and at different depths within the soil layer and it can operate at any spatial resolution. Research activity has focused on the (i) analysis of the different approaches to spatialize the hydrological and geotechnical input data (Bicocchi et al. 2019), (ii) evaluation of the contribution of the vegetation root reinforcement in the regional stability modelling (Salvatici et al. 2018), (iii) validation of the model performance in several test sites in Italy (Salvatici et al. 2018; Tofani et al. 2017) (Fig. 3).

The activities of WP3 has been focused also on the development of a methodology to couple rainfall thresholds and susceptibility maps for dynamic landslide hazard assessment at regional scale (Segoni et al. 2018a, b). Both inputs are combined in a hazard matrix to get a spatially and temporally variable definition of landslide hazard: while statistical rainfall thresholds are used to obtain a temporal forecasting with very coarse spatial resolution, landslide susceptibility maps provide static spatial information about the probability of landslide occurrence at fine spatial resolution. The employ of the proposed procedure in a regional warning system brings two main advantages: (i) it is possible to better hypothesize when and where landslide are expected and with which hazard degree, thus fostering a more effective hazard and risk management (e.g., setting priorities of intervention); (ii) the spatial resolution of the regional scale warning system is markedly refined because from time to time the areas where landslides are expected represent only a fraction of the alert zone.

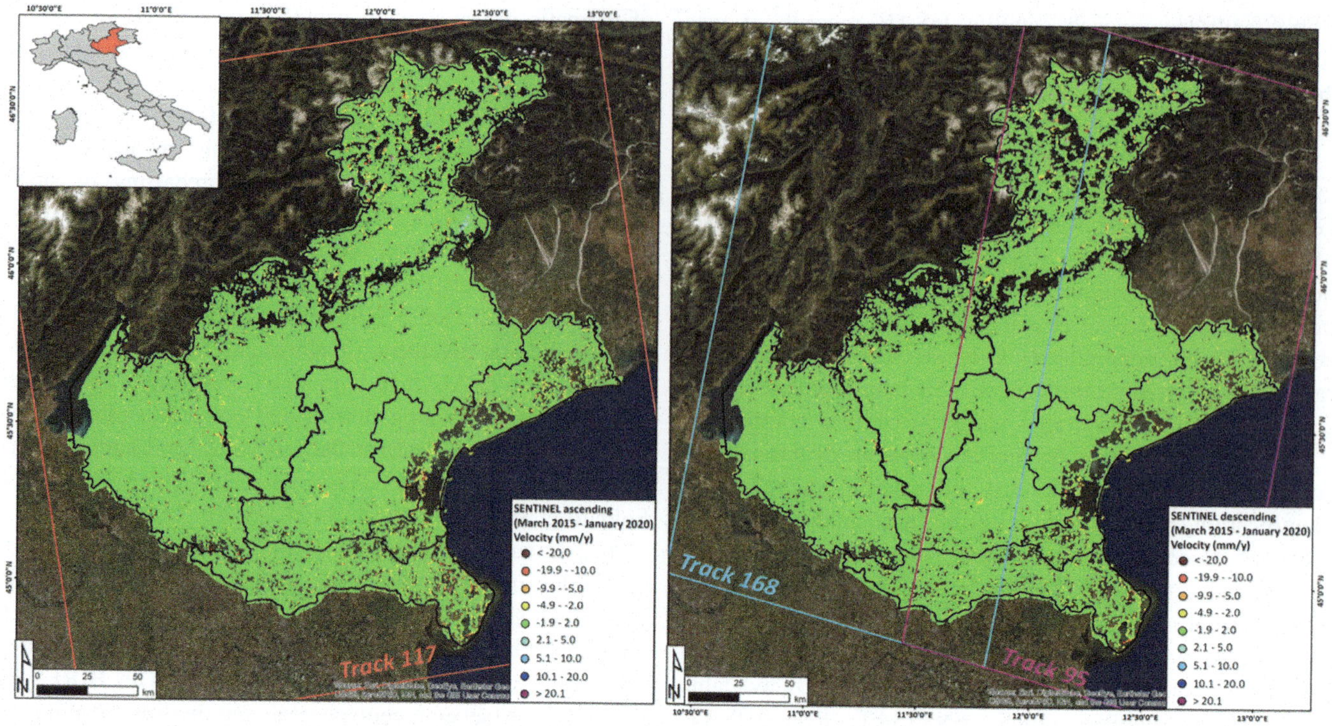

Fig. 2 Sentinel-1 ground deformation maps of the Veneto region related to the period March 2015–January 2020

Fig. 3 Distributed map of failure probability derived from HIRESSS model. The values of failure probability are aggregated at the first order level basins and inside each basis the colour identifies the number of instable pixels

IPL Projects

Since its involvement in ICL, UNESCO Chair has proposed several IPL projects. Currently the active projects are:

- IPL196: Development and applications of a multi-sensors drone for geohazards monitoring and mapping, Proposer: Veronica Tofani
- IPL198: Multi-scale rainfall triggering models for Early Warning of Landslides (MUSE), Proposer: Filippo Catani
- IPL 221: PS continuous streaming for landslide monitoring and mapping, Proposer: Federico Raspini and Silvia Bianchini.

IPL 196 has the objective to test the applicability of a multi-sensors drone for the mapping and monitoring of geohazards. In particular the project has two specific objectives: (i) development of the drone, sensors, safety and automation and (ii) application of the drone as a platform of integrated sensors (multispectral sensor, visible light camera, infrared camera and LIDAR) for the mapping and monitoring of geohazards.

IPL 198 aims at the enhancement of knowledge and methodologies related to the integration of landslide prediction models at different scales to build an effective operational multi-scale system for real-time early warning of rainfall triggered mass movements.

IPL 221 has the objective is to perform the transition from historical analysis of radar satellite image archives to real time monitoring of ground deformation at regional scale using radar satellite scenes. To accomplish this objective the short revisiting time and regularity of acquisitions of Sentinel-1 constellation of SAR (Synthetic Aperture Radar) satellite sensors were exploited.

The outcomes of these projects have been published in Landslides journal (IPL196, Rossi et al. 2018; IPL198, Tofani et al. 2017; IPL221, Raspini et al. 2019).

The activity reports are regularly published in the ICL IPL website.

ICL Italian Network

The ICL Italian network has been officially established in December 2018 during the ICL-IPL Conference, held in Kyoto on 1–4 December 2018. ICL Italian network, proposed and coordinated by the UNESCO Chair on Prevention and Sustainable management of Geo-hydrological hazards, currently counts 13 ICL members (7 Full members, 5 Associates and 1 Supporter) (Casagli and Tofani 2018). The general objective of the Italian ICL Network is to contribute at national level to the Sendai Partnership for Disaster Risk Reduction 2015–2025 for the national promotion of understanding, prevention and sustainable management of landslide risk disaster, for the safety of human life, society and the environment and to the Kyoto Commitment Commitment for Global Promotion of Understanding and Reducing Landslide Disaster Risk.

The Italian ICL network on landslides is formed by well-established scientific institutions, recognized both at national and international levels, with a long-dated expertise on landslide research for hydrogeological hazard assessment and landslide disaster risk reduction. The network partners have also developed strong and widespread synergies with national, regional and local administrations, technical stakeholders, and end-users for developing policies and procedures for landslide disaster prevention, management and mitigation.

Contribution to Kyoto 2020 Commitment

As a member of the International Consortium on Landslides (ICL) the UNESCO Chair on Prevention and Sustainable Management of Geo-hydrological Hazards is signatory of the ISDR-ICL Sendai Partnership 2015–2025 (Sassa 2015). The Sendai Partnership was accepted and signed during the 3rd United Nations World Conference on Disaster Risk Reduction (WCDRR) in Sendai (March 11–15th 2015). During the 4th World Landslide Forum (WLF4), held in Ljubljana from May 29–June 2 2017, the Chair has contributed to the "Ljubljana Declaration" on Landslide Risk Reduction, together with all of the participants to the forum, as a further commitment on the part of the global landslide community to the Sendai Framework for Disaster Risk Reduction 2015–2030 (Sassa 2017).

DST-UNIFI has signed Kyoto 2020 Commitment for global promotion of understanding and reducing landslide disaster risk during the 2019 ICL-IPL Conference at UNESCO Headquarters, Paris, on 16–19 September 2019. The Kyoto 2020 Commitment is a duty to the Sendai Landslide Partnerships 2015–2025, the Sendai Framework for Disaster Risk Reduction 2015–2030, the 2030 Agenda Sustainable Development Goals, the New Urban Agenda and the Paris Climate Agreement (Sassa 2019).

The UNESCO Chair, in line with the activity carried out in the framework of the WCoE and IPL projects will particularly contribute to some priority actions of Kyoto 2020 Commitment (Sassa 2018c):

Action 1 Promote the development of people-centered early warning technology for landslides with increased precision and reliable prediction both in time and location, especially in a changing climate context.

Action 2 Advance hazard and vulnerability mapping, including vulnerability and risk assessment with increased precision, as well as reliability as part of multi-hazard risk identification and management.

Action 3 Improve the technologies for monitoring, testing, analyzing, simulating, and effective early warning for landslides suitable for specific regions considering natural, cultural and financial aspects.

Action 6 Investigate the effect of climate change on rainfall-induced landslides and promote the development of effective rainfall forecasting models to provide earlier warning and evacuation especially in developing countries.

Action 9 Foster new initiatives to study research frontiers in understanding and reducing landslide disaster risk by promoting joint efforts by researchers, policy makers and funding agencies.

References

Bardi F, Raspini F, Frodella W, Lombardi L, Nocentini M, Gigli G, Morelli S, Corsini A, Casagli N (2017) Monitoring the rapid-moving reactivation of earth flows by means of GB-InSAR: the April 2013 Capriglio landslide (Northern Appennines, Italy). Remote Sens 9(2):165

Bicocchi G, Tofani V, D'Ambrosio M, Tacconi-Stefanelli C, Vannocci P, Casagli N, Lavorini G, Trevisani M, Catani F (2019) Geotechnical and hydrological characterization of hillslope deposits for regional landslide prediction modeling. Bull Eng Geol Env 78:4875–4891

Carlà T, Tofani V, Lombardi L, Raspini F, Bianchini S, Bertolo D, Thuegaz P, Casagli N (2019a) Combination of GNSS, satellite InSAR, and GBInSAR remote sensing monitoring to improve the understanding of a large landslide in high alpine environment. Geomorphology 335:62–75

Carlà T, Nolesini T, Solari L, Rivolta C, Dei Cas L, Casagli N (2019b) Rockfall forecasting and risk management along a major transportation corridor in the Alps through ground-based radar interferometry. Landslides 16:1425–1435

Carlà T, Intrieri E, Di Traglia F, Nolesini T, Gigli G, Casagli N (2017) Guidelines on the use of inverse velocity method as a tool for setting alarm thresholds and forecasting landslides and structure collapses. Landslides 14(2):517–534

Casagli N, Tofani V (2018) Establishment of ICL Italian network. Landslides 15:1907–1908

Casagli N, Frodella W, Morelli S, Tofani V, Ciampalini A, Intrieri E, Raspini F, Rossi G, Tanteri L, Lu P (2017) Spaceborne, UAV and ground-based remote sensing techniques for landslide mapping, monitoring and early warning. Geoenviron Disasters 4(9)

Del Soldato M, Riquelme A, Bianchini S, Tomàs R, Di Martire D, De Vita P, Moretti S, Calcaterra D (2018a) Multisource data integration to investigate one century of evolution for the Agnone landslide (Molise, southern Italy). Landslides 78:2387–2408

Del Soldato M, Pazzi V, Segoni S, De Vita P, Tofani V, Moretti S (2018b) Spatial modeling of pyroclastic cover deposit thickness (depth to bedrock) in peri-volcanic areas of Campania (southern Italy). Earth Surf Proc Land 43(1757):1767

Di Traglia F, Nolesini T, Ciampalini A, Solari L, Frodella W, Bellotti F, Fumagalli A, De Rosa G, Casagli N (2018) Tracking morphological changes and slope instability using spaceborne and ground-based SAR data. Geomorphology 300:95–112

Ferretti A, Fumagalli A, Novali F, Prati C, Rocca F, Rucci A (2011) A new algorithm for processing interferometric data-stacks: SqueeSAR. IEEE Trans Geosci Remote Sens 49(9):3460–3470

Frodella W, Ciampalini A, Bardi F, Salvatici T, Di Traglia F, Basile G, Casagli N (2018) A method for assessing and managing landslide residual hazard in urban areas. Landslides 15:183–197

Frodella W, Gigli G, Morelli S, Lombardi L, Casagli N (2017) Landslide mapping and characterization through Infrared Thermography (IRT): suggestions for a methodological approach from some case studies. Remote Sens 9(12)

Intrieri E, Raspini F, Fumagalli A, Lu P, Del Conte S, Farina P, Allievi J, Ferretti A, Casagli N (2018) The Maoxian landslide as seen from space: detecting precursors of failure with Sentinel-1 data. Landslides 15:123–133

Lombardi L, Nocentini M, Frodella W, Nolesini T, Bardi F, Intrieri E, Carlà T, Solari L, Dotta G, Ferrigno F, Casagli N (2017) The Calatabiano landslide (southern Italy): preliminary GB-InSAR monitoring data and remote 3D mapping. Landslides 14:685–696

Pastonchi L, Barra A, Monserrat O, Luzi G, Solari L, Tofani V (2018) Satellite data to improve the knowledge of geohazards in world heritage sites. Remote Sens 10(7):992

Raspini F, Bardi F, Bianchini S, Ciampalini A, Del Ventisette C, Farina P, Ferrigno F, Solari L, Casagli N (2017) The contribution of satellite SAR-derived displacement measurements in landslide risk management practices. Nat Hazards 86(1):327–351

Raspini F, Bianchini S, Ciampalini A, Del Soldato M, Solari L, Novali F, Del Conte S, Rucci A, Ferretti A, Casagli N (2018) Continuous, semi-automatic monitoring of ground deformation using Sentinel-1 satellites. Sci Rep 8(1):1–11

Raspini F, Bianchini S, Ciampalini A, Del Soldato M, Montalti R, Solari L, Tofani V, Casagli C (2019) Permanent Scatterers continuous streaming for landslide monitoring and mapping: the case of Tuscany Region (Italy). Landslides 16:2033–2044

Rosi A, Tofani V, Tanteri L, Tacconi SC, Agostini A, Catani F, Casagli N (2018) The new landslide inventory of Tuscany (Italy) updated with PS-InSAR: geomorphological features and landslide distribution. LANDSLIDES 15:5–19

Rosi A, Tofani V, Agostini A, Tanteri L, Tacconi Stefanelli C, Catani F, Casagli N (2016) Subsidence mapping at regional scale using persistent scatters interferometry (PSI): the case of Tuscany region (Italy). Int J Appl Earth Obs Geoinf 52:328–337

Rosi A, Agostini A, Tofani V, Casagli N (2014) A procedure to map subsidence at the regional scale using the persistent scatterer interferometry (PSI) technique. Remote Sens 6(11):10510–10522

Rossi G, Catani F, Leoni L, Segoni S, Tofani V (2013) HIRESSS: a physically based slope stability simulator for HPC applications. Nat Hazards Earth Syst Sci 13:151–166

Rossi G, Tanteri L, Tofani V, Vannocci P, Moretti S, Casagli N (2018) Multitemporal UAV surveys for landslide mapping and characterization. Landslides 15:1045–1052

Salvatici T, Tofani V, Rossi G, D'Ambrosio M, Tacconi Stefanelli C, Masi EB, Rosi A, Pazzi V, Vannocci P, Petrolo M, Catani F, Ratto S, Stevenin H, Casagli N (2018) Application of a physically based model to forecast shallow landslides at a regional scale. Nat Hazards Earth Syst Sci 18(7):1919–1935

Sassa K (2015) ISDR-ICL Sendai Partnerships 2015–2025 for global promotion of understanding and reducing landslide disaster risk. Landslides 12(4):631–640

Sassa K (2017) The 2017 Ljubljana declaration on landslide risk reduction and the Kyoto 2020 commitment for global promotion of understanding and reducing landslide disaster risk. Landaslides 14 (4):1289–1296

Sassa K (2018a) Zero draft of the Kyoto 2020 commitment for global promotion of understanding and reducing landslide disaster risk. Landslides 15(3):389–392

Sassa K (2018b) The Fifth World Landslide Forum—implementing and monitoring the ISDR-ICL Sendai Partnerships 2015–2025—organization plans, themes and sessions. Landslides 15(3):617–620

Sassa K (2018c) ICL Journal "Landslides" and the Kyoto 2020 commitment. Landslides 15(9):1705–1712

Sassa K (2019) The Fifth World Landslide Forum and the final draft of the Kyoto 2020 Commitment. Landslides 16(2):201–211

Segoni S, Tofani V, Rosi A, Catani F, Casagli N (2018a) Combination of rainfall thresholds and susceptibility maps for dynamic landslide hazard assessment at regional scale. Front Earth Sci 6:1–11

Segoni S, Rosi A, Fanti R, Gallucci A, Monni A, Casagli N (2018b) A regional-scale landslide warning system based on 20 years of operational experience. Water 10:1297

Solari L, Raspini F, Del Soldato M, Bianchini S, Ciampalini A, Ferrigno F, Tucci S, Casagli N (2018) Satellite radar data for back-analyzing a landslide event: the Ponzano (Central Italy) case study. Landslides 15(4):773–782

Solari L, Del Soldato M, Montalti R, Bianchini S, Raspini F, Thuegaz P, Tofani V, Casagli N (2019) A Sentinel-1 based hot-spot analysis: landslide mapping in north-western Italy. Int J Remote Sens 40(20):7898–7921

Tofani V, Bicocchi G, Rossi G, Segoni S, D'Ambrosio M, Casagli N, Catani F (2017) Soil characterization for shallow landslides modeling: a case study in the Northern Apennines (Central Italy). Landslides 14:755–770

Extreme Rainfall Event and Its Aftermath Analysis—IPL 210 Project Progress Report

2020 Kyoto Japan

Biljana Abolmasov, Mileva Samardžić Petrović, Ranka Stanković, Miloš Marjanović, Jelka Krušić, and Uroš Đurić

Abstract

The main results of recent activities on IPL210 Project summarize rainfall event, rainfall-induced landslides and their characteristics including type of movement, type of material involved and state of activity in two different regions in Serbia both affected by same extreme rainfall event in May 2014. In both regions dominant type of movement was sliding, but with different ratio between debris and soil as material involved. The analysis showed that there are no significant differences between first activated and reactivated landslides if two regions are compared.

Keywords

Landslides • Extreme precipitation • Data analysis • IPL project report

B. Abolmasov (✉) · R. Stanković · M. Marjanović · J. Krušić
Faculty of Mining and Geology, University of Belgrade, Đušina 7, 11000 Belgrade, Serbia
e-mail: biljana.abolmasov@rgf.bg.ac.rs

R. Stanković
e-mail: ranka.stankovic@rgf.bg.ac.rs

M. Marjanović
e-mail: milos.marjanovic@rgf.bg.ac.rs

J. Krušić
e-mail: jelka.krusic@rgf.bg.ac.rs

M. S. Petrović · U. Đurić
Faculty of Civil Engineering, University of Belgrade, Bul Kralja Aleksandra 84, 11000 Belgrade, Serbia
e-mail: mimas@grf.bg.ac.rs

U. Đurić
e-mail: udjuric@grf.bg.ac.rs

Introduction

The IPL project No. 210 titled “Massive landsliding in Serbia following Cyclone Tamara in May 2014” started in March 2016. The study area is located in the Western and Central part of the Republic of Serbia territory affected by Cyclone Tamara in May 2014. The project aims to summarize and analyse all relevant collected data, including historic/current rainfall, landslide records, aftermath reports, and environmental features datasets from the May 2014 event. In this respect, it was essential to produce unified large-scale inventories of the May 2014 event and use them for the state-of-the-art hazard analysis. Following these ideas, objectives of the proposed project include: (1) collecting all available (existing) and acquiring new landslides data, (2) analysing the trigger/landslide relation in affordable time span (past 15 years) and the May 2014 event, (3) relating the landslide mechanisms and magnitudes versus the trigger and its aftermath, (4) locating spatial patterns and relationships between landslides and geological and environmental controls, (5) proposing an overview susceptibility map of the event and (6) landslides numerical modeling on the site specific location/landslide mechanism. The Project is organized by University of Belgrade, Faculty of Mining and Geology and Faculty of Civil Engineering. Project beneficiaries are local communities, regional and national authorities. More detailed about the Project plan and program could be found in Abolmasov et al. (2017b, 2018).

This paper will present a progress report of the proposed project targets (1)–(3) performed by project participants in the last three years.

K. Sassa et al. (eds.), *Understanding and Reducing Landslide Disaster Risk*,
ICL Contribution to Landslide Disaster Risk Reduction,
https://doi.org/10.1007/978-3-030-60196-6_19

Previous Research

Study Area

The study area covered 11,840 km^2, i.e. 23 municipalities affected by different types of landslides in Western and Central parts of the Republic of Serbia. These municipalities were recognized as most vulnerable to floods, torrential floods and landslides by UNDP Office in Serbia during the post-disaster activities after May 2014 event (Abolmasov et al. 2017a). Geological and geomorphological settings are very complex as well as other environmental conditions in such a wide area. The types of movement and the types of material involved were depending on lithological type, local geomorphological characteristics, engineering geological properties, degree and depth of weathering substratum etc. as well as precipitation amount received during the May 2014 event.

Rainfall Event

In the third week of the May 2014, a massive low-pressure Cyclone Ivette (Tamara) swept through Western Balkan resulting in extreme precipitation for a short period that caused floods, torrential floods and massive landsliding in the Republic of Serbia (Marjanović et al. 2018). The highest precipitation for 72 h duration were recorded at several Main Meteorological Stations (MMS), Climatological (CS) and Precipitation Stations (PS) (Fig. 1). The flood event (14–15 May 2014) and landslides occurrences (15–18 May 2014) were caused simultaneously by extreme Cyclone activity (Prohaska et al. 2014), but massive landsliding was additionally initiated by antecedently introduced rainfall from April 15 to May 14 (Marjanović et al. 2018).

Landslides Data

The total number of 1751 landslides are mapped as open data file reports, according to the BEWARE Project deliverables (Abolmasov et al. 2017a). Different types of movement and types of involved material were registered during an extensive field campaign and analysis of remote sensing data (Đurić et al. 2017). It was a part of the objective (1) of the IPL 210 Project.

Progress Report—Recent Project Activities

Rainfall Event Data Sets

Recent research was concentrated on rainfall data distribution analysis in two different regions of the Republic of Serbia during May 2014 event. Namely, during BEWARE project activities and data analysis it was noticed that different types of movement and different types of material involved occurred depending on the cumulative and intensive precipitation amount (Aleotti 2004). Also, it was significant that the number of landslides occurrences were much larger in Western Serbia than in Central Serbia.

The recent analysis of rainfall data included available daily precipitation from the Hydro-meteorological Service of Serbia. Data collection was concentrated on the area already covered by Project BEWARE (Abolmasov et al. 2017a) and divided into two regions—Western Serbia and Central Serbia. Both regions were affected at the same time with May 2014 event—but with different antecedently (April–May 2014), and intense rainfall (14–16 May 2014) patterns.

A total number of 4 Main Meterological Stations (MMS—Valjevo, Loznica, Kraljevo, Kragujevac) and 48 Precipitation Stations (PS) from Western and Central part of Serbia were included in analysis, especially those where 72 h precipitation exceeded 100y maximums (Figs. 1 and 2).

Landslides Data Sets

Based on simple Cruden and VanDine (2013) landslides classification system, all data about type of mechanism were mapped on the field, during extensive field work. Five basic categories—slides, falls, topples, flows and lateral spreads were used for classification. If two different types of movements are involved it was suggested that dominant one should be chosen as the first type and the second mechanism should be after the slash, i.e. slides/flows. Complex movements are combination of two or more types of movements (but without visible dominant mechanism). The material involved categories were extended with the category anthropogenic, if artificial material was moved, in addition to existing ones (rock, debris, earth, complex). According to the Cruden and VanDine (2013) classification debris has 20–80% of the particles larger than 2 mm. Earth has 80% or more of the particles smaller than 2 mm. Earth is further

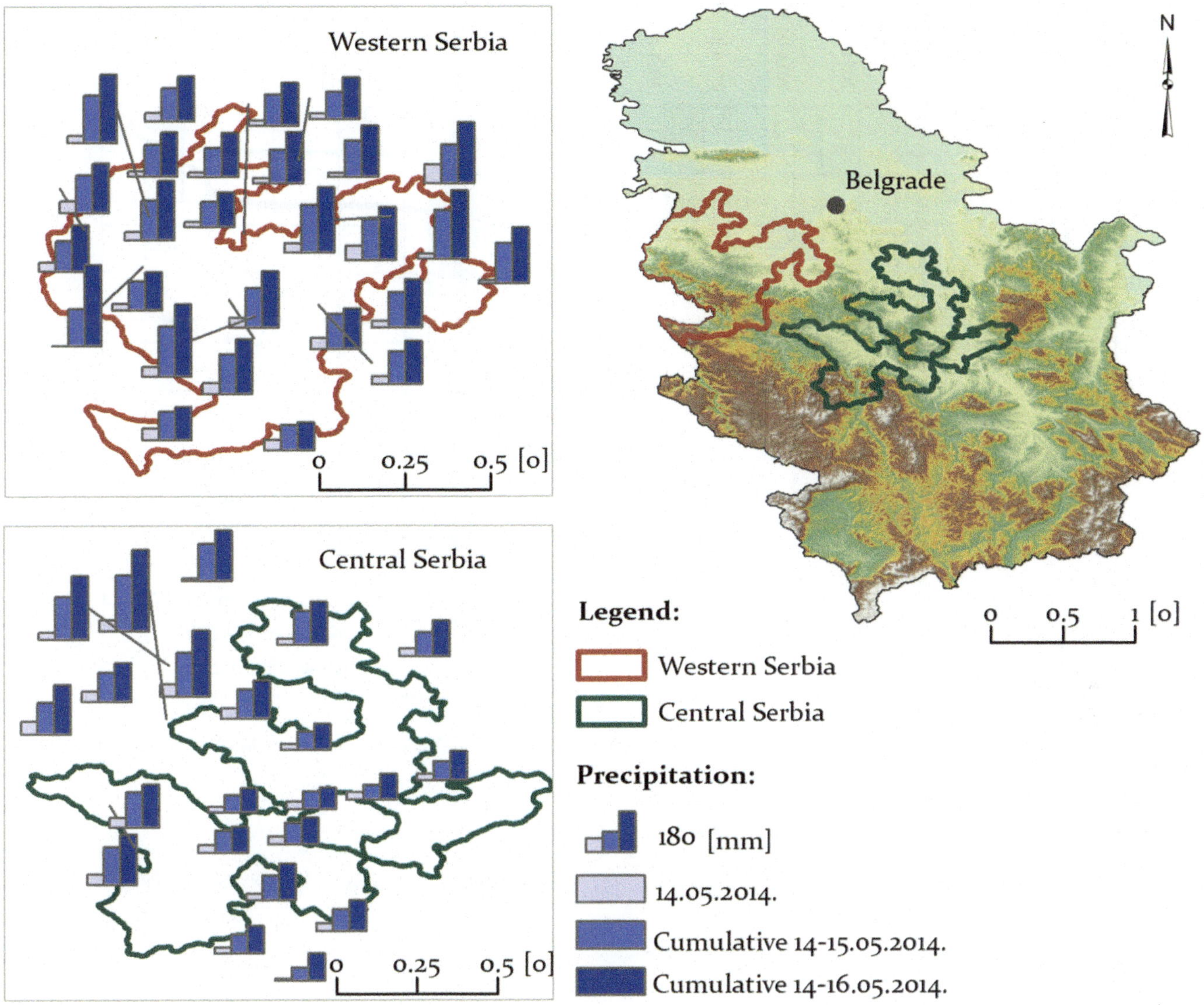

Fig. 1 Cumulative 72 h daily precipitation from 48 Precipitation Stations (PS) in both regions (14–16.05.2014)

subdivided into sand, silt and clay by grain size and plasticity, but those data are not used as a classifier in the analysis. If two or more materials are involved, it was suggested to select complex. Landslides data were unified and grouped in the same way as rainfall data, depending on the geographical position within Western Serbia or Central Serbia data sets (Table 1). Landslide data sets analysis also included type of mechanism versus state of activity in the selected regions (Table 2). The principal hypotesis was that differences in soil saturation caused by high antecedent precipitation and then intensive precipitation during 72 h could condition the ratio between first activated and reactivated landslides.

Results and Discussion

According to the rainfall data analysis, it is obvious that Western Serbia received more antecedent (15 April–13 May 2014), and intensive (14–16 May 2014) precipitation than Central Serbia. More saturation caused that the percent of flows and flow like landslides are several times higher in Western Serbia than in Central Serbia (Table 1; Fig. 3). The percent of slides is almost similar (80–85% of all mechanism). Geological composition and engineering geological conditions (state of weathering), determined types of material involved; namely debris and soil prevails in both regions

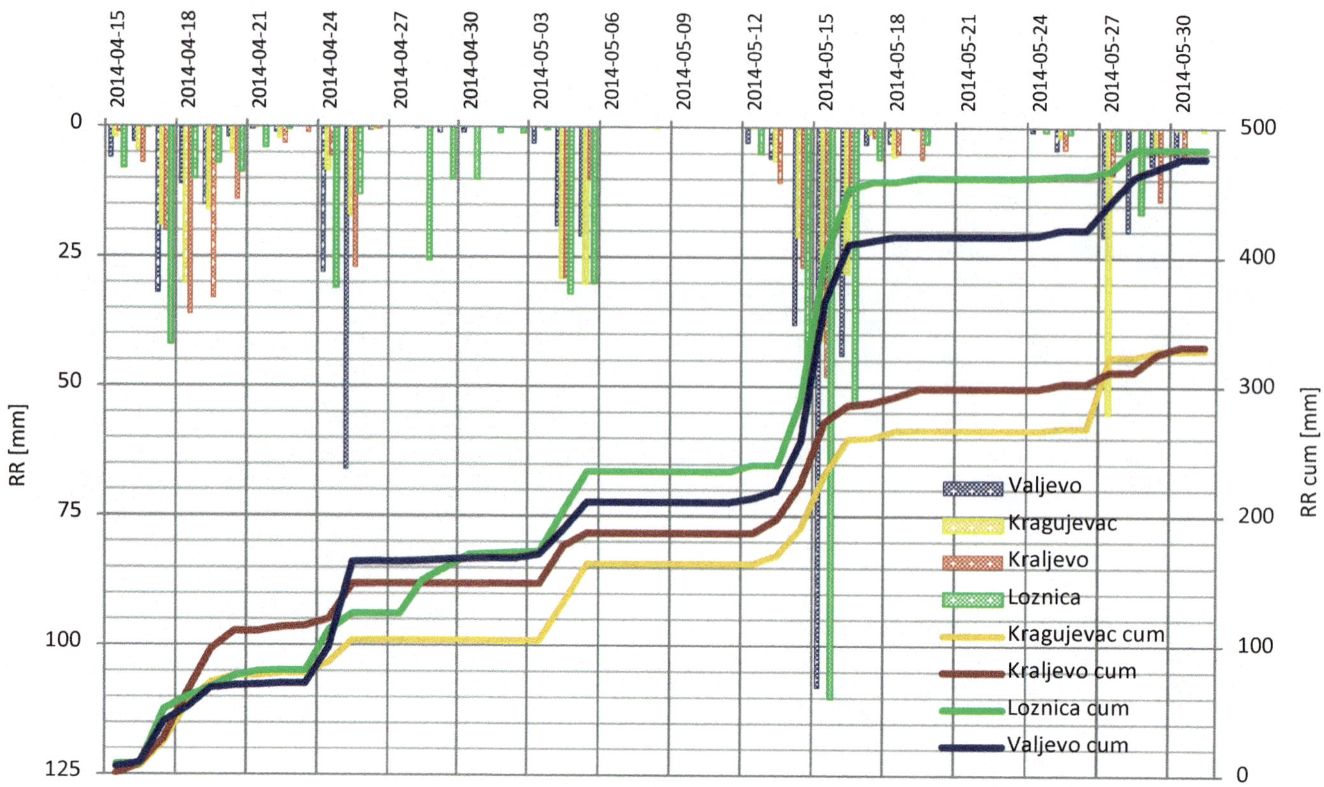

Fig. 2 Cumulative and daily precipitation from MMS (April–May 2014); Western Serbia-Valjevo and Loznica MMS; Central Serbia-Kraljevo and Kragujevac MMS

Table 1 Summary of landslides data from Western and Central Serbia according to the types of movement and types of material involved

Region	Type of movement	Type of material					
		Anthropogenic	Debris	Rock	Soil	Complex	Total
Western Serbia	Slides	6	549	2	354	21	932
	Slides/flows		83		23	2	108
	Rockfalls	1	7	3		6	17
	Rockfalls/slides		15			3	18
	Rockfalls/flows		3				3
	Flows	1	61		9	1	72
	Complex	2	5		7	1	15
	Total	10	732	5	393	34	1165
Central Serbia	Slides	4	132	1	356	5	498
	Slides/flows		9		17	1	27
	Rockfalls	1	11	6	1		19
	Rockfalls/slides	1	16	2	9		28
	Rockfalls/flows		2				2
	Flows		3		3		6
	Complex		1		5		6
	Total	6	174	9	391	6	586
Total		16	897	14	784	40	1751

Table 2 Summary of landslides data from Western and Central Serbia according to the types of movement and state of activity

Region	Type of movement	State of activity						
		Active	Dormant	Reactivated	Stabilized	Suspended	Abandoned	Total
Western Serbia	Slides	457	9	109	4	349	2	930
	Slides/flows	38		8	4	58		108
	Rockfalls	6		3	1	6		16
	Rockfalls/slides	10		3		5		18
	Rockfalls/flows					3		3
	Flows	20		2	4	46		72
	Complex	9				6	3	15
	Total	540	9	125	13	473	5	1165
Central Serbia	Slides	283	7	86	7	109	4	496
	Slides/flows	5		7	1	14		27
	Rockfalls	7		4	1	7		19
	Rockfalls/slides	15	7	2		3		27
	Rockfalls/flows					2		2
	Flows	2	1			3		6
	Complex	3	1				5	4
	Total	315	16	99	9	138	9	581
Total		855	25	224	22	611	14	1751

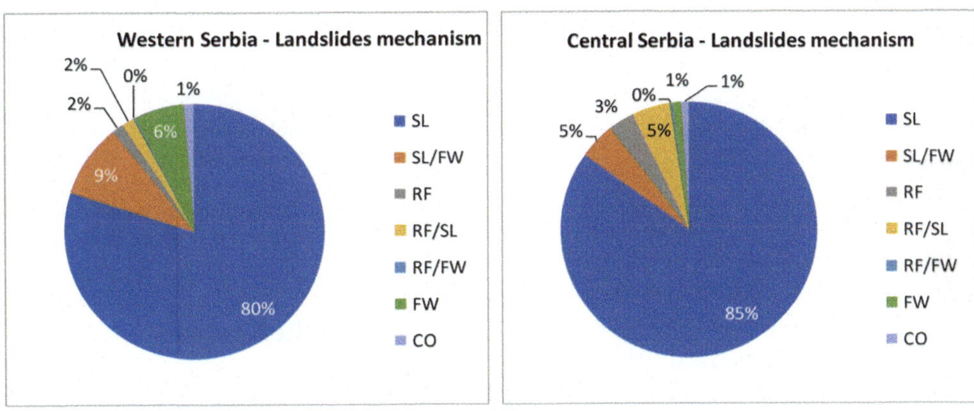

Fig. 3 Types of movement by region: SL-slides, SL/FW-slides/flows; RF-rockfalls; RF/SL-rockfalls/slides; RF/FW-rockfalls/flows; FW-flows; CO-complex mechanism

for all types of movement, but with different percent per region (Table 1). In the Western Serbia dominant type of material was debris (59%) whereas in Central Serbia was soil (71%) for slides for instance (Fig. 4). A similar ratio could be found in overall types of movement versus types of material involved analysis by region. Basic geological and engineering geological data were used from publicly available Basic Geological and Basic Engineering Geological Map of Serbia (sheets 1:25,000).

The mechanism of movement versus state of activity analysis showed that there are no significant differences between first activated and reactivated slides if two regions are compared (Fig. 5), unlike of flows and flow like landslides. The percent of active and suspended flows, flow like slides and complex movement are significantly higher in Western Serbia than in Central Serbia. The reason for that ratio could be found in extreme high antecedent and intense rainfall in Western Serbia compared to lower rates in Central Serbia.

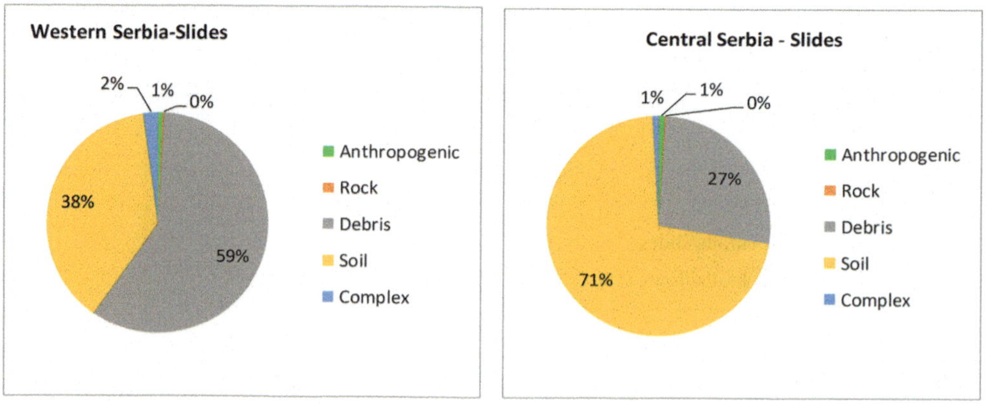

Fig. 4 Type of movement versus types of material involved by region (slides only)

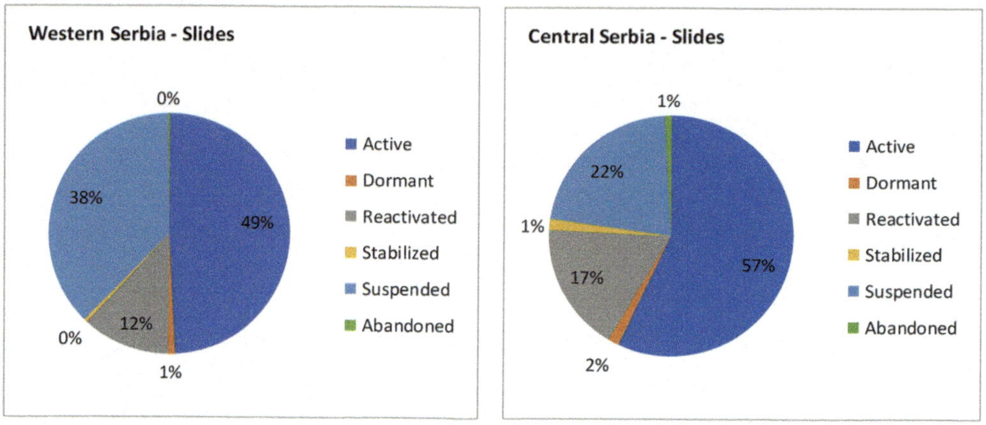

Fig. 5 Type of movement versus state of activity by region (slides only)

Conclusion

The main results of recent activities on IPL210 Project summarize rainfall event, rainfall-induced landslides and their main characteristics including types of movement, types of material involved and state of activity. The mutual relationships were analyzed with tempt to better define connections between the spatial distribution of rainfall pattern and mechanism of movement versus material involved, as well as relationship and influence on first activated/reactivated landslides in two different regions in Serbia affected by Cyclone Tamara in May 2014. Different distribution of landslides mechanism and types of material involved were found between Western and Central Serbia, but very slight connection was found between spatial rainfall distribution data versus landslides state of activity.

Further research within IPL210 Project will be focused on analysing: (1) spatial patterns and relationships between landslides and geological and environmental controls, (2) proposing overview susceptibility map of the May 2014 event for all affected area and (3) numerical modeling on the landslide sites accordingly to the landslides mechanism.

Acknowledgements IPL Project 210 will be not possible without Project BEWARE (BEyond landslide aWAREness) funded by the People of Japan and the UNDP Office in Serbia (grant No. 00094641). The project was implemented by the State Geological Survey of Serbia, and the University of Belgrade Faculty of Mining and Geology.

References

Abolmasov B, Damjanović D, Marjanović M, Stanković R, Nikolić V, Nedeljković S, Petrović Ž (2017a) Project BEWARE—landslide post-disaster relief activities for local communities in Serbia. In: Mikoš M et al (eds), Advancing culture of living with landslides, Proceedings of 4th world landslide forum, vol 3, Ljubljana 29 May–02 June 2017, pp 413–422

Abolmasov B, Marjanović M, Đurić U, Krušić J, Andrejev K (2017b) Massive landsliding in Serbia following cyclone Tamara in May 2014 (IPL-210). In: Sassa K et al (eds) Advancing culture of living with landslides, Proceedings of 4th world landslide forum, vol 1, Ljubljana 29 May–02 June 2017, pp 473–484

Aleotti P (2004) A warning system for rainfall-induced shallow failures. Eng Geol 73(3–4):247–265. https://doi.org/10.1016/j.enggeo.2004.01.007

Cruden D, VanDine DF (2013) Classification, description, causes and indirect effects-Canadian technical guidelines and best practices related to landslides: a national initiative for loss reduction. Geol Surv Canada Open File 7359

Đurić D, Mladenović A, Pešić-Georgiadis M, Marjanović M, Abolmasov B (2017) Using multiresolution and multi-temporal satellite data for post disaster landslide inventory in the Republic of Serbia. Landslides 14(4):1467–1482. https://doi.org/10.1007/s10346-017-0847-2

https://geoliss.mre.gov.rs/beware/. Accessed Jan 2020

https://www.hidmet.gov.rs/. Accessed Oct 2019

Marjanović M, Krautblatter M, Abolmasov B, Đurić U, Sandić C, Nikolić V (2018) The rainfall-induced landsliding in Western Serbia: a temporal prediction approach using decision tree technique. Eng Geol 232:147–159

Prohaska S, Đukić D, Bartoš Divac V, Božović N (2014) Statistical significance of the rainfall intensity that caused the May 2014 flood in Serbia. Water Res Manag 4(3):3–10

Complex Geomorphological and Engineering Geological Research of Landslides with Adverse Societal Impacts

Vít Vilímek, Jan Klimeš, Josef Stemberk, Jan Burda, Petr Kycl, and Jan Blahůt

Abstract

The Czech World Centre of Excellence (WCoE) on Landslide Risk Reduction 2017–2020 entitled as "Complex geomorphological and engineering geological research of landslides combined with assessment of adverse societal impacts" included research, education and awareness of landslide risks. Originally, the only participants were Charles University and the Czech Academy of Sciences; however, in the last two years we were joined by another two Czech institutions (Brown Coal Research Institute and Czech Geological Survey), strengthening possible impact of the work performed within the scope of the International Consortium on Landslides. In terms of landslide research, we focussed on various topics, i.e. national landslide susceptibility zonation, paleogeomorphological landslide research, rainfall triggering mechanism, a database of giant landslides, landslide dam analysis, glacial lake outburst floods, merging satellite-based landslide inventory data with on-site geomorphological investigations, and landslide susceptibility modelling. We performed our activities both at home and abroad. During the period starting from 2017, we performed various activities aimed at increasing the awareness of the general public as well as experts involved in landslide risk reduction. Student graduation works was performed in the framework of the WCoE at all levels (Bachelor's, Master's and Doctoral). In addition, we contributed to the overall ICL/IPL activities by performing editorial work, reviews and evaluations for the journal Landslides, book publications and IPL projects, as well as the WCoE.

Keywords

Landslide risk reduction • World centre of excellence • Capacity building • Community participation

Introduction

All activities performed in the framework of the World Centre of Excellence on Landslide Risk Reduction (WCoE 2017–2020) follow on from the former centre of excellence (Klimeš et al. 2017a; Stemberk et al. 2017). The current centre with the official title "Complex geomorphological and engineering geological research of landslides combined with assessment of adverse societal impacts" focuses on landslide risk reduction. The main tasks of the WCoE respect the basic principles of the United Nations International Strategy for Disaster Reduction (ISDR), with the following main objectives:

- To strengthen the International Program on Landslides (IPL) through analysis of long-term landslide monitoring and new approaches for hazard and risk mapping and assessment.

V. Vilímek (✉)
Faculty of Science, Department of Physical Geography and Geoecology, Charles University, Albertov 6, Prague 2, 128 43, Czech Republic
e-mail: vilimek@natur.cuni.cz

J. Klimeš · J. Stemberk · J. Blahůt
Institute of Rock Structure and Mechanics, The Czech Academy of Sciences, V Holešovičkách 41, Prague, 18209, Czech Republic
e-mail: klimes@irsm.cas.cz

J. Stemberk
e-mail: stemberk@irsm.cas.cz

J. Blahůt
e-mail: blahut@irsm.cas.cz

J. Burda
Brown Coal Research Institute, tř. Budovatelů 2830/3, 43401 Most, Czech Republic
e-mail: burda@vuhu.cz

P. Kycl
Czech Geological Survey, Klárov 3, Prague, 11821, Czech Republic
e-mail: petr.kycl@geology.cz

K. Sassa et al. (eds.), *Understanding and Reducing Landslide Disaster Risk*, ICL Contribution to Landslide Disaster Risk Reduction,
https://doi.org/10.1007/978-3-030-60196-6_20

- To meet the Sendai Partnerships 2015–2025 goals mainly in the analysis of the combined effects of landslide triggering factors, and with open communication with society through integrated research, capacity building as well as knowledge transfer.

The initial participating institutions, which also collaborated on the previous WCoE (2014–2017), were Charles University (Faculty of Science) and the Czech Academy of Sciences (Institute of Rock Structure and Mechanics). They were later joined by the Brown Coal Research Institute (BCRI, Department of Geotechnics and Hydrogeology) and the Czech Geological Survey (CGS). The BCRI provides research and practical solutions in slope stability assessment, focusing on areas heavily affected by open-pit coal mining and other industrial activities. The CGS has the statutory responsibility to gather, store and interpret geological information so that the state administration can take appropriate decisions about economic and environmental issues, including landslide risk reduction. These new competencies, combined with the strong research and education focus of the initial collaborating institutes, allow us to study landslides in a holistic context of landscape evolution and their societal impacts, to understand the complex interactions between all relevant geomorphological and social processes, which are often described individually. Many research results of the WCoE are being applied to landslide mitigation by local authorities or private companies. We also have long-time experience with field work in a variety of less developed countries, with more than 15 years of continuous research in Peru, as well as arctic regions (Spitzbergen Islands) and in systematic safety monitoring of unstable slopes in diverse geologies and environmental conditions (https://www.irsm.cas.cz/ext/tecnet/index_en.php).

All of the members of the WCoE participate in the education of university students through lecturing as well as leading their theses at Czech universities, as well as abroad (e.g. Argentina, USA).

The resources available for the activities of WCoE include indoor laboratories as well as geodynamic field laboratories located on landslides monitored for more than 10 years; worldwide geodynamic monitoring networks providing data from landslides/tectonic interactions (EU TecNet, https://www.irsm.cas.cz/ext/tecnet/index_en.php); field research facilities including two terrestrial laser scanners and high precision GPS stations; technical staff and Ph.D. students.

The detailed strategy of our research meets the following objectives:

(a) Landslide mapping based on field work or remote sensing data, collection of field samples for mineralogical analysis and granulometry as well as database construction.
(b) Analysis and interpretation of the results of the long-term landslide monitoring with respect to landslide hazard and risk assessment in selected regions.
(c) Landslide hazard and risk zonation with respect to risk mitigation and management considering the active participation of the affected communities.
(d) Capacity building—teaching undergraduate and Ph.D. students from developed and developing countries, field training during regular conferences "Slope movements and pseudokarst", implementation of advanced technologies (Africa, South America).

Methodology

We use a wide range methods suitable for different types of landslides being currently studied with respect to the regional settings (morphological-climatic zones) and the aims of the studies (e.g. scientific research or solution of practical, landslide-related problems). They include field sampling, state-of-art mapping deploying analyses of detailed digital elevation models (DEM) prepared with in-house technologies (e.g. unmanned aerial vehicles, UAV, ground-based laser scanner) or interpretation of publicly available DEMs and high-resolution optical images. Landslide movement monitoring is largely performed with an in-house developed 3D monitoring device (Klimeš et al. 2012) as well as the newest, commercially-available instruments for monitoring underground movements (Balek et al. 2019; Foglino et al. 2006). We also extensively use geophysical methods for site investigations (Chalupa et al. 2018) as well as monitoring of hydrological conditions on landslides (Tábořík et al. 2019; Hartvich et al. 2018) or their movements (Rinaldi-Montes et al. 2017). Targeted research is focussed on localities in the Czech Republic and other European countries (e.g. Italy, Spain, Norway, Switzerland, Germany, Poland, Slovenia, Greece and Slovakia). We also work in the Peruvian Andes and the western part of the USA.

Research Results

Studies in the Czech Republic

In order to meet the objectives of the WCoE, we focussed on landslides and their societal impacts in the Czech Republic (Klimeš et al. 2017b). We prepared the first national landslide susceptibility zonation map using GIS and multivariate statistical methods (Racek 2018). It precisely identified regions with known high landslide incidences and assigned others, which are so far unexplored in terms of detailed landslide inventory mapping and thus may represent areas of their increased occurrence, as being highly susceptible. We also maintain a national database of landslides reported in Google alerts (Klimeš et al. 2017b), which also collects information on landslide-related damage, showing that in the period between 2010 and 2018 two people died and thirteen were injured due to landslides. At a local scale, we presented several reports as an expert witness regarding landslide cases with high social or economic impact (e.g. interruption of the main highway connecting Prague with Germany). We also interpreted the movement of potentially dangerous landslides, mainly threatening local roads or railroads, showing that even during periods with below-average precipitation their movements may accelerate and negatively affect the transport infrastructure (Balek et al. 2019; Šilhán et al. 2019). We also initiated complex monitoring of dangerous rock slopes, which include rock deformations, surface temperatures and their attenuation with depth (up to 3 m) and other climatic variables (cf. precipitation, insolation).

Another topic was paleogeomorphological research in the Krušné hory Mts. (Burda et al. 2018). The aims of this study were to summarize the current knowledge of a large run-out, prehistoric landslide, and in particular, gather new data in order to estimate the age of the accumulation and reveal the movement mechanism. The reconstruction of a large rockslide-rock avalanche was supported by the analysis and interpretation of 216 boreholes and by GIS analysis of the original 1950s pre-mining surface using digitized, old military topographic maps. All three of the documented events occurred around the time of a significant climate change. The oldest event occurred due to the oldest Dryas warming, the largest event probably occurred at the end of the Younger Dryas (11,700 yBP), and the youngest was purely of a Holocene age, with the highest landslide frequency being during the Atlantic temperature fluctuations (approximately 8200 yBP).

Two significant rainfall episodes, which affected the eastern part of the Czech Republic in May and June 2010, caused the earthflow on the Lemešná Mt. in the Javorníky Range (Smolíková et al. in review). Antecedent rainfalls from the gauges situated close to the earthflow were analysed to identify the triggering conditions. The extreme values of all of the tested rainfall parameters only converge on the day of the earthflow. The rainfall analysis was also done in Krkonoše Mts. by Drábová (2018) which revealed that the local rainfall threshold for a debris flow is above 225 mm/day.

Research Performed Abroad

The first comprehensive and freely available global database of giant landslides on volcanic islands (https://www.irsm.cas.cz/ext/giantlandslides/) (Blahůt et al. 2018, 2019) contains georeferenced outlines (kml file) and detailed attribute information (xlsx file) on more than 180 documented cases. It illustrates how widespread these potentially extremely hazardous events are and demonstrates that more than half of them occurred during the last 0.5 Ma. The database will be continuously updated as new research is published.

Large attention of the WCoE was dedicated to research in the Peruvian Andes. Morphological analysis of landslide dams were performed in cooperation with Florence University (Tacconi et al. 2018). This paper covers a database of 51 cases of landslide dams in the Cordillera Blanca Mts. The main morphometrical parameters of the landslide have been determined through direct and indirect survey techniques. The data show a low variability in some of the main morphometrical parameter distributions, most likely due to an environmental control connected to the regional tectonic and glacial history. The results of the Cordillera Blanca analysis were compared to a large Italian landslide dam inventory, showing the differences as well as the similarities between the two datasets. Many of the stable dams are in disequilibrium with their surrounding environment.

The research topic of Glacial Lake Outburst Floods (GLOFs) is partly independent from the landslide research, but GLOFs are often triggered by slope movements and the resulting flood often has properties or debris flows due to high sediment load of a different origin (glacial, fluvial of from local landslides). The paper published by Emmer et al. (2018) helped to understand the effectiveness of remedial works.

Joint research with the University of Zurich and Gamma Remote Sensing (Switzerland, Strozzi et al. 2018) focused on merging satellite-based landslide movement data with on-site geomorphological investigation. It demonstrated a significant increase in the reliability of the landslide inventory maps and improvement in hazard assessment of individual landslides, which can be continuously updated through continuous satellite data collection. The used approach showed high applicability in semiarid environments of the Peruvian Andes, which at the same time proved to be highly challenging for regional-scale landslide susceptibility modelling (Bueechi et al. 2019). The applied models (cf. SINMAP, multiple regression and landslide density models) provided reasonable results despite the scarce input data, but they varied significantly depending on the used DEM resolution and the scale at which the landslide

Fig. 1 Sharing and evaluating the interim results of landslide risk reduction with the involved community is one of the basic requirements of equal partnership between experts and the intended beneficiaries of the project. It empowers community members to modify the project according their current needs and conditions (joint meeting of the community leaders and schoolchildren, Rampac Grande, Cordillera Negra, Peru)

inventory data were collected. A more local scale provided better susceptibility results. Work dedicated to the reduction of high landslide risk in the small community of Rampac Grande, Cordillera Negra, Peru, showed how important and mutually beneficial the involvement of the affected community is in all steps of landslide risk mitigation (e.g. goal definitions, data collection, evaluation and sharing, Fig. 1, Klimeš et al. 2019a). Information provided by the local inhabitants helped to improve the scientific understanding of landslide movement frequencies and magnitudes and thus improved hazard assessment. Their information also served as reliable field data for the validation and interpretation of the satellite observations.

Increasing Public Awareness in the Czech Republic

The WCoE performed several different activities to raise the awareness of the general public as well as experts involved in landslide risk reduction. Informing citizens about the possible negative impacts of landslides and information sources about landslide occurrences are the basic requirements for successful implementation of any risk reduction strategy. We presented the activities of the WCoE at a public workshop in Prague, which was intended for Czech experts from different disciplines (e.g. civil protection, architects, public administration). We also used this event to introduce a brochure dedicated mainly to local administrations and land-use planning departments. The brochure explains the best practice of landslide hazard reduction under the legal and environmental frameworks of the Czech Republic. Several activities focused on the general public while others were dedicated to motivated groups interested in outdoor activities (e.g. hiking or players of the Geocaching game). We prepared a series of articles for the general public published in the well-known Czech journal "Vesmír". They described e.g. losses caused by landslides in the Czech Republic or practical problems related to mitigation works as well as the environmental and cultural significance of landslides. We prepared a public hiking excursion through an area heavily affected by landslides in the Outer Western Carpathians, which was highly appreciated by its participants, represented mostly by elderly people with a strong interest in all of the environmental aspects of landscape development. We assumed that focusing on such highly motivated social groups could bring very good results in public awareness, since these people are very likely to share their experience with their families and friends. A similarly motivated group is made up of players of Earthcaching, which is a variation on the very popular Geocaching game. Therefore, we maintained already existing Earthcaching sites adding new ones focusing on landslide monitoring techniques. Members of our WCoE collaborated in the preparation of a series of short videos about a small community seriously affected by a landslide in 1967 (Carpathians, Czech Republic). The videos explain not only the landslide process responsible for destroying nine houses, but also focus on the social, cultural and economic damage caused by the event. They resulted in the serious marginalization of the village, which significantly affected families whose houses were not directly destroyed by the landslide.

Landslide Research Education

Several students prepared their graduation work in the framework of the WCoE at all levels (Bachelor's, Master's and Doctoral) thanks to the tutorial work of our researchers. We also took part in the landslide-oriented summer school—2nd Oltrepo Summer School (Pavia, Lombardy, Italy) in 2018, where one of our students initiated their Bachelor's thesis (Fig. 2). Our students also took part in the Kokomeren Summer School (supported by ICL).

Fig. 2 The 2nd Oltrepo Summer School (Pavia, Lombardy, Italy)

Conclusion

During the 2017–2020 period, our WCoE significantly enlarged the research team contributing to the worldwide network of institutions engaged in the International Consortium on Landslides (ICL). The new collaborators include an ICL full member, Brown Coal Research Institute, who provided hitherto missing expertise from research and practical solutions of slope stability problems in industrially heavily affected areas. The new ICL associate member, Czech Geological Survey, represents the main office dedicated to the collection and provision of landslide-related geological information to the state administration focussing on the effective reduction of landslide risk at the national level. We organized several activities for scientific as well as public audiences promoting our results gained within the IPL activities. Apart from the continuous landslide research, we also participated in the education of undergraduate as well as graduate students, leading several diplomas and doctoral theses. Members of the WCoE participated in all ICL activities, including editorial and reviewer work for the journal Landslides and proceedings of the World Landslide Forum conferences. One monothematic half-issue of the journal Landslides was also initiated and published (Klimeš et al. 2019b).

Acknowledgements The authors wish to acknowledge the financial support provided by the Ministry of Education, Youth and Sports of the Czech Republic (Program Inter-Excellence, Inter-Vector, LTV19014).

References

Balek J, Klimeš J, Blahut J, Štroner M, Urban R, Hartivch F (2019) Shallow landslide movements in clay rich rocks detected during subnormal precipitation period. Acta Geodyn Geomater 16:409–417

Blahůt J, Klimeš J, Rowberry M, Kusák M (2018) Database of giant landslides on volcanic islands—first results from the Atlantic Ocean. Landslides 15:1801–1814

Blahůt J, Balek J, Klimeš J, Rowberry M, Kusák M, Kalina J (2019) A comprehensive global database of giant landslides on volcanic islands. Landslides 16(10):2045–2052

Bueechi E, Klimeš J, Frey H, Huggel C, Strozzi T, Cochachin A (2019) Regional-scale landslide susceptibility modelling in the Cordillera Blanca, Peru—a comparison of different approaches. Landslides 16 (2):395–407

Burda J, Veselý M, Řehoř M, Vilímek V (2018) Reconstruction of a large run-out landslide in the Krušné hory Mts. (Czech Republic). Landslides 15(3):423–437

Chalupa V, Pánek T, Tábořík P, Klimeš J, Hartvich F, Grygar R (2018) Deep-seated gravitational slope deformations controlled by the structure of flysch nappe outlier: insights from large scale electrical resistivity tomography survey and LiDAR mapping. Geomorphology 321:174–187

Drábová Z (2018) Rainfall analysis of debris flows in the Obří důl Valley in the Krkonoše Mts., Czechia. AUC Geographica 53 (2):220–237

Emmer A, Vilímek V, Zapata ML (2018) Hazard mitigation of glacial lake outburst floods in the Cordillera Blanca (Peru): the effectiveness of remedial works. J Flood Risk Manag 11:5489–5501

Foglino L, Lovisolo M, Giusta D (2006) Contribution of DMS monitoring systems in the analysis of slide micro-movements for early warning management, risk assessment and evaluation of mitigating actions. Geophys Res Abstr 8:06122

Hartvich F, Táborík P, Blahůt J, Klimeš J, Balek J (2018) Complex DSGSD site Čeřeniště as a natural laboratory for landslide activity monitoring. Central European conference on geomorphology and quaternary sciences, 23–27.9.2018. Giessen, Germany

Klimeš J, Rowberry MD, Blahůt J, Briestenský M, Hartvich F, Košťák B, Rybář J, Stemberk J, Štěpančíkova P (2012) The monitoring of slow-moving landslides and assessment of stabilisation measures using an optical-mechanical crack gauge. Landslides 9:407–415

Klimeš J, Hartvich F, Tábořík P, Blahut J, Briestensky M, Stemberk J, Emmer A, Vargas R, Balek J (2017) Studies on selected landslides and their societal impacts: activity report of the Prague World Centre of Excellence, Czech Republic. Landslides 14:1547–1553

Klimeš J, Stemberk J, Blahut J, Krejčí V, Krejčí O, Hartvich F, Kycl P (2017) Challenges for landslide hazard and risk management in 'low-risk' regions, Czech Republic—landslide occurrences and related costs (IPL project no. 197). Landslides 14:771–780

Klimeš J, Rosario AM, Vargas R, Raška P, Vicuña L, Jurt C (2019) Community participation in landslide risk reduction: a case history from Central Andes, Peru. Landslides 16:1763–1777

Klimeš J, Calvello M, Auflič MJ (2019) Preface: thematic Issue "community participation for landslide disaster risk reduction." Landslides 16:1745–1746

Racek O (2018) Landslide susceptibility analysis of Czechia. Diploma Thesis, Charles University, Prague

Rinaldi-Montes, N, Rowberry M, Frontera C, Garcés J, Baroň I, Blahůt J, Pérez-López R, Pennos C, Martí X (2017) A contactless positioning system for monitoring discontinuities in three dimensions with geological and geotechnical applications. Rev Sci Instrum 88, Art. No. 074501

Smolíková J, Hrbáček F, Blahůt J, Klimeš J, Vilímek V, Loaiza Usuaga JC (in review) Analyses of the rainfall pattern triggering the Lemešná debris flow, Javorníky Range, the Czech Republic. Nat Hazards

Stemberk J, Vilímek V, Klimeš J, Blahůt J, Hartvich F, Balek J (2017) Landslide hazard and risk management (WCoE 2014–2017). In: Sassa J et al (eds) Advancing culture of living with landslides, vol 1. Springer, pp 373–377

Strozzi T, Klimeš J, Frey H, Caduff R, Huggel C, Wegmüller U, Rapre AC (2018) Satellite SAR interferometry for the improved assessment of the state of activity of landslides: a case study from the Cordilleras of Peru. Remote Sens Environ 217:111–125

Šilhán K, Klimeš J, Tichavský R (2019) The sensitivity of dendrogeomorphic approach to landslide movements. Geomorphology 347. 10.1016/j.geomorph.2019.1068690169-555X

Tábořík P, Hartvich F, Blahůt J, Balek J, Stemberk J, Kusák M (2019) Čeřeniště natural lab: a recent progress in the landslide monitoring instrumentation. In: Proceedings of Český ráj '19: state of geomorfological research in 2019. 3–5.4.2019. Czech Republic, p 57

Tacconi C, Vilímek V, Emmer A, Catani F (2018) Morphological analysis and features of the landslide dams in the Cordillera Blanca. Landslides 15(3):507–521

Report of the IPL-219, IPL-220 and Croatian WCoE 2017–2020: From Landslide Investigation to Landslide Prediction and Stabilization

Željko Arbanas, Josip Peranić, Martin Krkač, Vedran Jagodnik, Petra Jagodnik, and Snježana Mihalić Arbanas

Abstract

In this paper we described a part of scientific activities of the Croatian Landslide Group (CLG), World Centre of Excellence (WCoE) on Landslide Risk Reduction of the International Consortium on Landslide (ICL) for the period 2017–2020 including activities of IPL Projects IPL-2019 Rockfall Hazard Identification and Rockfall Protection in The Coastal Zone of Croatia and IPL-220 Kostanjek Landslide Monitoring Project (Zagreb, Croatia). The results of these scientific activities are presented through the fields of landslide science: landslide investigation and testing, landslide monitoring, landslide modelling and landslide stabilization and remedial measures. The manuscript presents the major results achieved in previously listed fields in the frame of Croatian Landslide Group WCoE activities and two IPL Projects.

Keywords

Landslide investigation • Laboratory testing • Monitoring • Landslide modelling • Landslide stabilization

Ž. Arbanas (✉) · J. Peranić · V. Jagodnik · P. Jagodnik
Faculty of Civil Engineering, University of Rijeka, 51000 Rijeka, Croatia
e-mail: zeljko.arbanas@gradri.uniri.hr

J. Peranić
e-mail: josip.peranic@gradri.uniri.hr

V. Jagodnik
e-mail: vedran.jagodnik@gradri.uniri.hr

P. Jagodnik
e-mail: petra.jagodnik@gradri.uniri.hr

M. Krkač · S. Mihalić Arbanas
Faculty of Mining, Geology and Petroleum Engineering, University of Zagreb, 10000 Zagreb, Croatia
e-mail: mkrkac@rgn.hr

S. Mihalić Arbanas
e-mail: smihalic@rgn.hr

Introduction

The Croatian Landslide Group (CLG) consisted of scientists from the University of Rijeka, Faculty of Civil Engineering and University of Zagreb, Faculty of Mining, Geology and Petroleum Engineering, Croatia, was promoted in the World Centre of Excellence (WCoE) on Landslide Risk Reduction of the International Consortium on Landslide (ICL) for the period 2017–2020. The Center for Applied Landslide Research (CALaR) of CLG set its main objective for development of scientific methods related to landslide identification, mapping, characterization, classification, laboratory testing, monitoring, prediction, modeling and stabilization with the application to landslide types characteristic for the Pannonian Basin and the Dinarides (Dinaric Alps). Through two International Programme on Landslides (IPL) Projects (IPL-219, IPL-220) as well as through the several national projects, CALaR was developing the wide spectra of activities within landslide science from landslide investigation to landslide prediction and stabilization. In this manuscript we will present the major results in the frame of Croatian Landslide Group WCoE activities and two IPL Projects achieved in fields of landslide investigation and testing, landslide monitoring, landslide modelling and landslide stabilization and remedial measures.

Landslide Investigation and Testing

Colluvial Materials from the Valići Lake Landslide

The role of the unsaturated zone for the occurrence of rainfall-induced landslides in flysch deposits was investigated in the framework of the doctoral research by Peranić (2019). Pilot area for the study was flysch material from the Valići Lake landslide reactivated in February 2014 (Mihalić Arbanas et al. 2017).

K. Sassa et al. (eds.), *Understanding and Reducing Landslide Disaster Risk*, ICL Contribution to Landslide Disaster Risk Reduction,
https://doi.org/10.1007/978-3-030-60196-6_21

Coupling between the transient process of rainfall infiltration and change of slope stability over time requires knowing the unsaturated property functions of soil materials. Providing a functional relationship between soil moisture, suction, and hydraulic conductivity, soil-water retention curve (SWRC) and the hydraulic conductivity function (HCF) are non-linear soil property functions governing the seepage process and build-up of the pore water pressures during a rainfall. The soil water retention properties of the residual soil from flysch rock mass were investigated by Peranić et al. (2018a, b). Measurements collected by different measurement techniques and devices (suction-controlled oedometers, standard and volumetric pressure plate extractors, HYPROP evaporation method device, WP4-T dew point potentiometer) were successfully combined to obtain a complete SWRC, both for intact and remolded samples undergoing drying and wetting paths. Peranić et al. (2020a, b) used observations about soil exhibiting normally- and over-consolidated behavior depending on the ongoing hydraulic path and applied suction, along with the obtained SWRC parameters and hydraulic conductivity measurement results obtained for saturated and unsaturated conditions to define the HCF relevant for analyzing changes in flysch slopes exposed to rainfall (Fig. 1).

Field measurements of the soil moisture as well as the laboratory measurements performed on intact samples collected at natural water content (w_n) (Peranić and Arbanas 2020) indicated that, during the dry period, surficial soil can undergo significant desaturation up to several meters of depth. Depending on the weather conditions in a period preceding to sampling, matric suction ($u_a - u_w$) values ranging from 30 up to 400 kPa were measured.

Peranić and Arbanas (2020) used modified direct shear apparatus to perform long-lasting wetting tests on unsaturated intact samples of residual soil from flysch rock mass to investigate hydro-mechanical response of the soil under the conditions of increasing water content (w) and reduction of $u_a - u_w$. Under imposed stress conditions, samples of low hydraulic conductivity were subjected to a prolonged wetting process simulating the process of rainfall infiltration in the field. Test results suggest that a gradual decrease of $u_a - u_w$ and an increase of w result in an increase of displacement rates under constant shear stress, which was interpreted as a failure of samples in partially saturated conditions (Fig. 2).

The relationship between the $u_a - u_w$ and shear strength of the residual soil from flysch rock mass was investigated by Peranić et al. (2020a, b). The obtained results are valuable for determination of how redistribution of pore-water

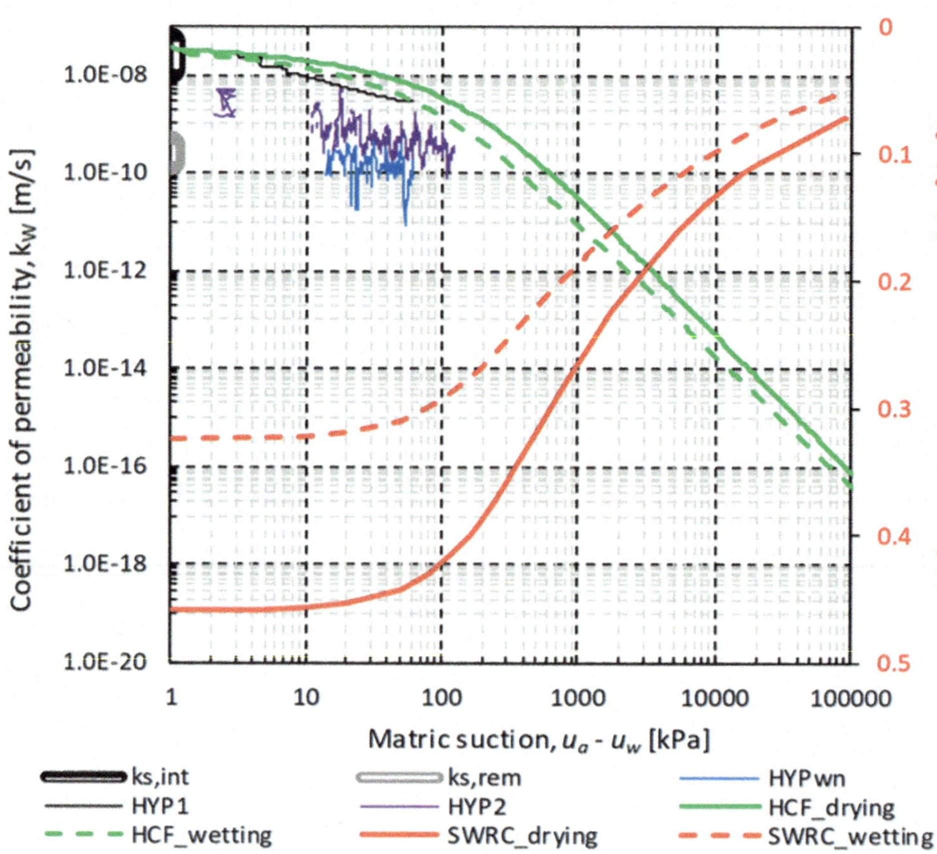

Fig. 1 Measured k_s, k_w ($u_a - u_w$) and the estimated HCF of residual soil from a flysch rock mass along with the SWRC (Peranić et al. 2020b)

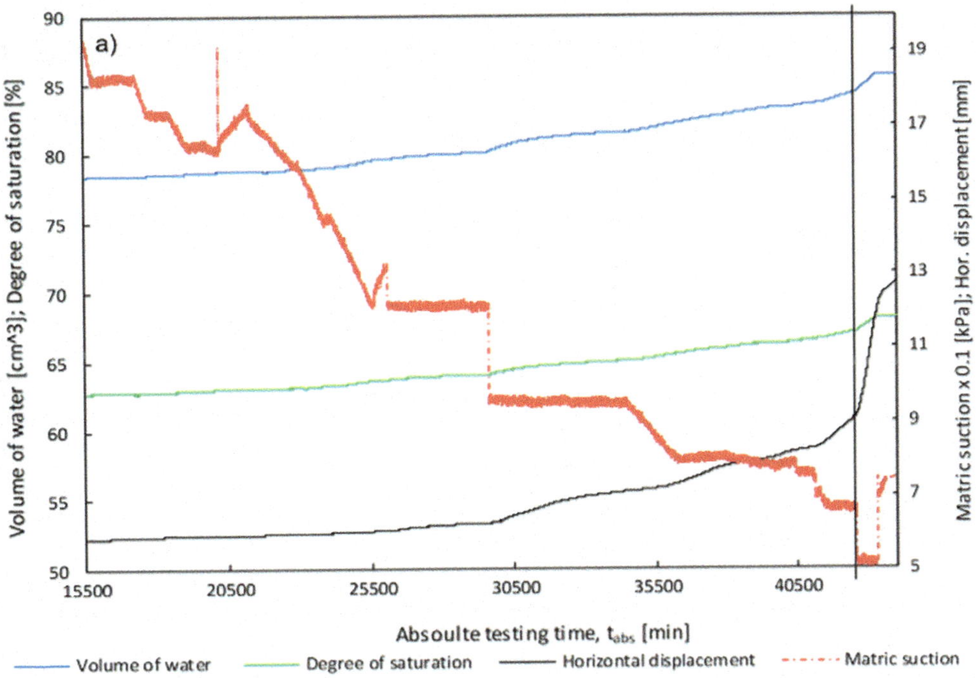

Fig. 2 Development of shear displacement, change of w, and $u_a - u_w$ with time during the wetting stage in UVSCDS11 test

pressures (pwp) (or water content) during rainfall affects shear strength of the soil.

Colluvial Materials in the Vinodol Valley

Limited number of debris slide materials from the Slani Potok gully has been preliminary tested to perform analysis of clay activity in the Vinodol Valley, and to establish relationship with sliding potential (Đomlija et al. 2019). Results of this study indicate that inactive and normally active clay minerals (i.e., kaolinite and illite) prevail in the colluvial deposits of the investigated landslides. The soils with lower activity generally have lower values of cohesion and can retain less quantity of water infiltration before a loss of the strength. According to these, soils with lower activity have consequently lower shear strength and are more sensitive to transition in weaker consistency caused by water content increasing. Those facts are very important for occurrences of shallow landslide which means that data about clay activity can be used for landslide susceptibility analyses in this study area.

As a part of regional neotectonic unit, the Vinodol Valley is considered to be seismically active zone with frequent small and even relatively high magnitude of earthquakes in the past (Herak et al. 2017). Beside rainfall, earthquake is considered to be one of the significant factors in landslide triggering (Keefer 2002; Xu 2015). Preliminary tests on cyclic behavior have been performed on colluvial deposits in the Vinodol Valley (Jagodnik et al. 2019a, b). Soil samples, taken from small debris slide, were subjected to laboratory tests in order to determine their behavior. Beside the soil classification tests, laboratory tests consisted of small strain tests using resonant column apparatus (Jagodnik et al. 2019a) and load control cyclic test using dynamic triaxial apparatus (Jagodnik et al. 2019b) under different cyclic stress ratios. The laboratory test gave the first insight of cyclic properties and cyclic behavior of such material. Figure 3 presents the proposed strength reduction curve, along with damping curve, for the colluvial deposits in the Vinodol Valley. Results are in good agreement with the one from literature (Vucetic and Dobry 1991; Kramer 1996). The performed tests showed that due to high content of clay fraction, cyclic stress ratio under 0.2 (Fig. 4) do not represent critical stress on cyclic behavior of colluvial deposits, which are related to lower magnitudes and ground accelerations.

The expected cyclic stress ratio in the Vinodol Valley for this type of material is between 0.2 and 0.25 which would present an issue only for high frequency seismic activity with the number of cycles larger than 100.

Landslide Monitoring

This section presents activities in the framework of the Project IPL-220 Kostanjek Landslide Monitoring Project (Zagreb, Croatia) and the Kostanjek Landslide Observatory established in 2011 (Mihalić Arbanas et al. 2018). The Kostanjek landslide is the deep-seated urban landslide with

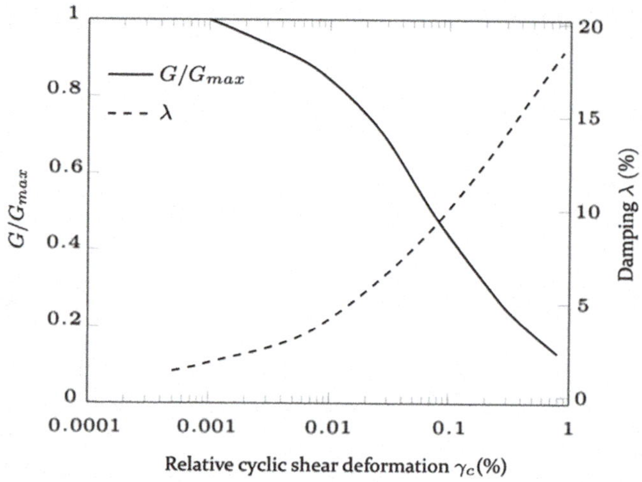

Fig. 3 Strength reduction damping curve for eluvial deposits in the Vinodol Valley, modified according to Jagodnik et al. (2019a)

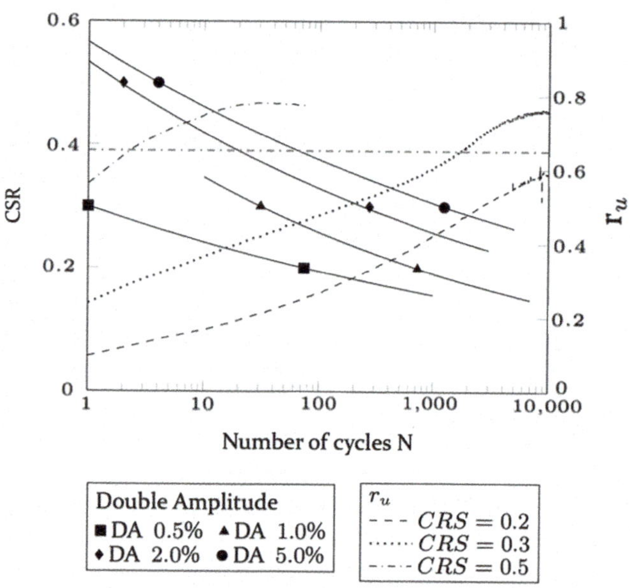

Fig. 4 Cyclic stress ratio and *ru* value for eluvial deposits from the Vinodol Valley, modified according to Jagodnik et al. (2019b)

the depth of approximately 90 m and the volume of the sliding mass evaluated to be 32 × 10^6 m^3. The sliding was developed in Miocene sedimentary rocks in the Pannonian Basin.

The Kostanjek landside monitoring system was set up in the period 2011–2014 and consists of multiple sensor networks for continuous observations of: (1) external triggers; (2) hydrological properties; (3) displacement/activity (Krkač et al. 2014). Analysis of monitoring data from 15 permanent GNSS stations showed similar patterns of landslide movements across the entire landslide area, but with different velocities (Krkač et al. 2019). The highest velocities were measured in the central part of the landslide, while the smallest are near the landslide boundaries. The total horizontal displacements in the central part of the landslide, during the period 2012–2018, were between 550 and 750 mm, and the vertical displacements were between −120 and −400 mm. Figure 5 presents the cumulative displacements measured with GNSS sensor at the central part of landside. From the GNSS measurements, totally seven periods of faster movement and seven periods of slower movement can be distinguished. The maximal observed velocity was 4.5 mm/day, which was reached in the first period of faster movement, during the first week of April 2013 (Krkač et al. 2019). The longest period of faster movement lasted from the end of January 2013 to the end of May 2013, during which the total horizontal displacement was 168 mm in central part of landslide. All seven periods of faster movement occurred as a consequence of groundwater level (GWL) rising periods. Periods of slower movement can be described as periods of rest or suspended state of activity according to Cruden and Varnes (1996). During periods of slower movement, the landslide velocities were up to 6 mm/month.

The data records from water level sensors in the central part of the landslide show GWL oscillations up to 8.5 m, between depths of 19–10.5 m, corresponding to variation of the pore pressures at the sliding surface, in the central part of landslide, from 425 to 510 kPa. Figure 5 presents GWL depths and pore pressures monitored at the central part of the landslide during the period 2013–2018. Altogether, 21 periods of groundwater level rise occurred, during which GWL relatively changed from 0.19 to 5.04 m (Krkač et al. 2019). The maximal observed groundwater level rise rate was 0.87 m/day, whereas the maximal observed groundwater level recession rate was 0.16 m/day. The average groundwater level recession rate was 0.05 m/day.

The cumulative precipitations that caused GWL rise in the central part of the landslide ranged 21–180 mm, depending on initial GWL depth. At the lower GWLs, for example in the summer months, higher amounts of precipitation are necessary to influence GWL rise, while at the higher GWL depths, during the autumn or winter months, smaller amounts of precipitation are necessary to influence GWL rise (Krkač et al. 2017). This phenomenon is related to the soil moisture conditions and different hydraulic conductivity above the GWL during the wet and dry periods (Krkač 2015).

Data measured with water level sensors located near the landslide boundaries, generally showed shallower GWL depths (3–7.5 m near the main scarp and 1–3.5-m at the right landslide flank) and higher susceptibility to short term and less intensive precipitations, as opposed to the depths of GWL in the central part of landslide. From the data comparison between GWL and movement, it was concluded that the threshold value of the GWL depth, for initiation of faster

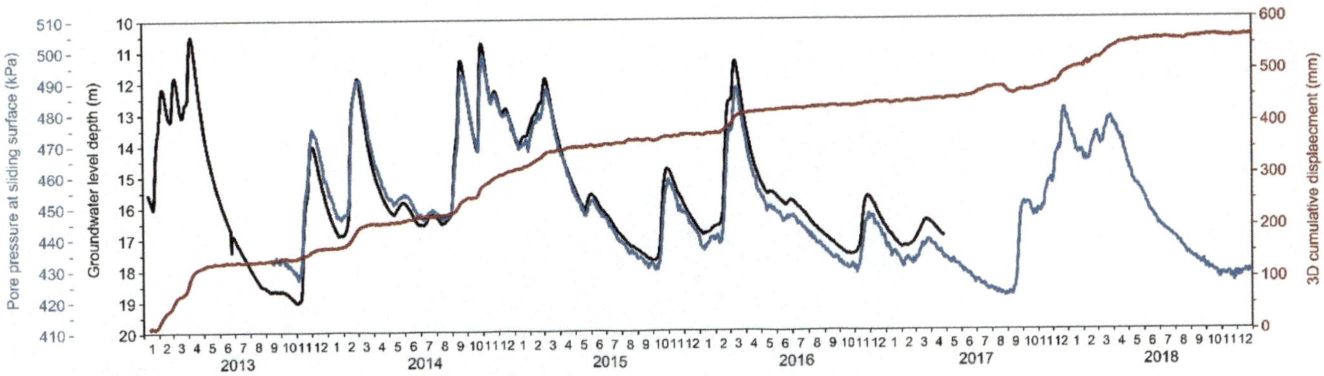

Fig. 5 Cumulative total (3D vector) displacements compared with groundwater level/pore pressure measured at the central part of Kostanjek landslide (Krkač et al. 2017)

landslide movement in the central part of Kostanjek landslide, is between 15 and 16 m (Krkač et al. 2019).

Analyses of precipitation, GWL and movement data of the Kostanjek landslide are aimed at landslide movement prediction. Some of the phenomenological models, tested for prediction of landslide movement are Random Forests (Krkač et al. 2017) and Multiple Linear Regression (Krkač et al. 2020). The models for prediction are divided into the models, for the prediction of GWLs from precipitation data and for the prediction of landslide velocity from GWL data. The GWLs and landslide velocities were predicted for 30 day periods (Fig. 5). The root mean square error analyses of k-fold cross-validation showed that the slightly better results were obtained from Random Forests than those from Multiple Linear Regression, indicating that both of these techniques can be used for practical prediction of landslide movement applied to important tool for decision making in crisis management that is necessary for risky landslide movements.

Landslide Modeling

Numerical Modelling of Valići Lake Landslide

Extensive field and laboratory measurement results were used to build a numerical model capable of analyzing the transient process of rainfall infiltration into flysch slope, simultaneously calculating the change of the factor of safety (FoS) of the slope. The obtained hydraulic (SWRC and HCF) and shear strength properties of the residual soil from flysch rock mass (Peranić 2020) were used to define material property functions required to perform infiltration and slope stability analysis in the GeoStudio (GEO-SLOPE) software suite. The 2D finite element software SEEP/W was used to solve a transient seepage equation.

Peranić et al. (2019) used the hourly rainfall measurements to define the flux boundary condition along the slope surface. Initial and boundary conditions were defined according to the collected in-situ measurements of the water content and groundwater level (Peranić 2019), laboratory measurements of the $(u_a - u_w)w_n$ on intact samples (Peranić et al. 2018a; Peranić and Arbanas 2020) and the average working level of the Valići Lake (Peranić et al. 2019). Initial pore-water pressure distribution with the corresponding FoS of the relevant sliding surface at the start of 167 days of simulation are shown in Fig. 6.

The results of rainfall infiltration analysis indicated slow dissipation of the $u_a - u_w$ during the first 130 days of simulation, following by the rapid increase of PWP from the 150th day of analysis (Fig. 7). During the last 17 days of analysis, the groundwater level increases for around 5 m, reaching almost the ground surface in the selected profile (Peranić et al. 2019).

PWP distribution profiles obtained for every 24 h of simulation in SEEP/W were introduced into the SLOPE/W module to calculate the FoS for 167 days of analysis. The obtained results indicate temporal change of the stability of flysch slope exposed to the triggering rainfall event. The results of numerical simulations showed that under defined conditions and applied rainfall, the slope remained stable during 167 days of infiltration, before the FoS dropped below unity. The shape and depth of the sliding surface, as well as the time of the failure onset corresponded to the field observations, previously performed slope stability back-analysis, and inclinometer surveys of the Valići Landslide.

Physical Modelling of Landslide Initiation

In October 2018, at the Faculty of Civil Engineering, University of Rijeka, started a four-year research Project “Physical modelling of landslide remediation constructions

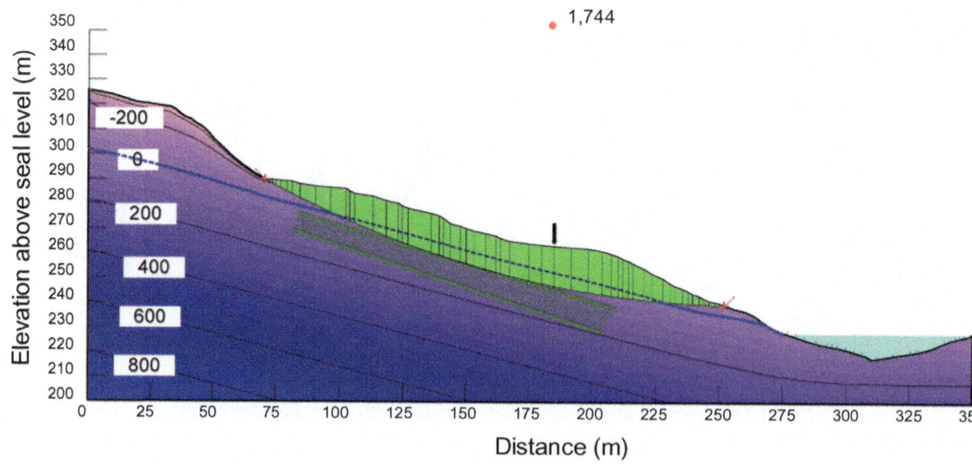

Fig. 6 Initial pore-water pressure distribution in the analyzed cross-section with the relevant sliding surface and corresponding FoS value

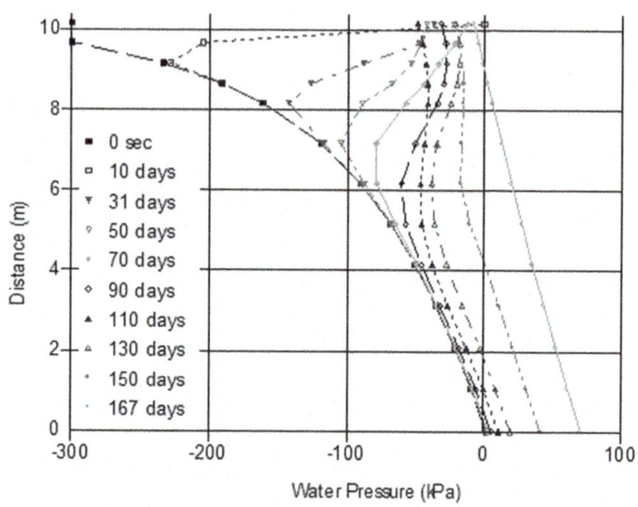

Fig. 7 Infiltration analysis results for the first 10 m of depth in the selected soil profile in terms of the change of $u_a - u_w$

behavior under static and seismic actions" funded by the Croatian Science Foundation. The Project aim is modelling of landslide remedial constructions' behavior in physical models of scaled landslides in static (rainfall triggered landslides) and seismic conditions (earthquake triggered landslides). Static landslide triggering conditions will be reached by artificial rainfall using a rainfall simulator. Seismic landslide triggering conditions will be induced by a pair of shaking platforms inducing seismic impulses (commonly used for structures testing under cyclic loading. Landslide movements will be observed with an innovative photogrammetric equipment and complex sensor network with ability to measure displacements, pore pressures and force pressures within a landslide body as well as at an applied remedial construction. Several different remediation constructions are planned to be embedded and their behavior tested in the small scale model: retaining walls, pile walls, buttressing and drainage systems. Results from both physical and numerical modelling will provide better insight into behavior of landslide remedial constructions in engineering practice (Arbanas et al. 2019a; Pajalić et al. 2019).

Methodology of the research Project consists of the following phases: (i) design, construction and testing of the platform for a landslide model; (ii) installation of the measuring equipment; (iii) construction of slope model; (iv) testing a model in static and seismic conditions (with and without remediation measures); (v) analysis of measured data and comparison with results of numerical simulations; vi) calibration of equipment and improvement of the model; and vii) reiteration of the same phases of testing. Approximate layout dimensions of the model platform are 1.5 × 2.0 m and a maximal depth of soil slope is 0.50 m. The platform will have possibility to adjust a slope angle and different depths of built-in slope material (Arbanas et al. 2019a; Pajalić et al. 2019).

The flume is constructed as a steel base construction with transparent Plexiglas sides in layout dimensions of 2.30 × 1.00 m that lies on the working table (Fig. 8). It was provided that the maximal height of the soil material in the slope model is to 30 cm. An inclination of the flume slope can be adjusted for a type of used soil material from 10° to 45°.

For the first test in the constructed flume, the fine sandy material was chosen to be built-in at an inclination of 30° (Arbanas et al. 2020). The strength properties of the Drava River Sand was determined in direct shear apparatus at the same relative density as it is in the flume of $D_r = 0.5$ and low normal stresses appropriate to the stresses present in the flume slope. The average friction angle is $\phi = 34.9°$. The test was conducted by exposing the slope to the artificial rain from three nozzles reaching an uniform artificial rainfall intensity of 58.23 mm/h. The first movements in the model slope were registered 126 min after the rainfall started. The landslide initiation was started after complete saturation of the slope foot and landslide occurrence was presented in

Fig. 8 Photo of physical model (Arbanas et al. 2020)

complete loss of soil strength in the slope toe and development of main scarp in the slope material (Fig. 9). The landslide initiation in the foot of the model slope, as well as further developing of landslide to the top of the model slope was explained based on results of the volumetric water content and soil suction observation during the test in Arbanas et al. (2020).

Landslide Stabilization and Remedial Measures

Field investigation, soil and rock laboratory testing and landslide remediation designs of several recent landslides were also conducted, as a part of CLG's activities in landslide risk reduction. The most of investigated landslides occurred as a sliding processes which affected roads in North-Wester part of Croatia, Istria and outback of the City of Rijeka, while the rock falls phenomena endangered cultural heritages and urban areas in coastal part of Croatia (Istria and Dalmatia). The investigated landslides and developed remediation measure design method in the period 2017–2020 are listed in Table 1. Some of designed landslide remediation measures, as well as rockfall protection measure are applied and completed, while some are in construction stage.

Fig. 9 Photo of landslide initiation and retrogressive landside development (Arbanas et al. 2020)

Rockfall Modelling and Rockfall Protection at the Slopes Above the City of Omiš

The old town of Omiš, is situated in the middle part of the Adriatic coast, at the mouth of the Cetina River in the toe of high limestone cliffs (Fig. 10) and it is very exposed to high rockfall hazards. The rockfall events along the limestone slopes were caused by unfavorable rock mass characteristics, rock mass weathering in combination with heavy rainfalls and the man-made influences.

Based on identified potential unstable block position and their approximate volumes, prevention or protection measures were chosen. The prevention of rockfalls by installing rock mass support systems including rock bolts and rock anchors in combination with steel ropes, steel wire fences and steel wire meshes are designed to hold the rock blocks on. To protect the building and infrastructure from running rock mass blocks, the rockfall protection barriers were designed to reduce the rockfall energy and suspend running blocks. The necessary energy absorption capacity of rockfall protection barriers was defined based on 2D rockfall analysis.

In total 22 potentially dangerous location were chosen and for each location prevention or protection measures or their combinations were designed. The main design included 8 locations of rockfall protection barriers with energy absorption capacity of 1000 and 2000 kJ (Fig. 11) as well as support measure for more than 30 potentially unstable blocks at the slopes above the City of Omiš (Arbanas et al. 2019b, c).

Investigation and Remediation of the Špičunak Landslide

The Špičunak Landslide is a deep-seated slow-moving landslide causing damage on the pavement of the state road D3 (Fig. 12) near the Lokve settlement in the Gorski Kotar region, Croatia, during the last 50 years (Vivoda Prodan et al. 2019). The landslide was formed at the contact of Permian shales and Jurrasic limestone and dolomites. The length of the landslide is 70 m, the width is 85 m and the depth of sliding surface is 20 m. Due to deformations on the

Table 1 Landslides and designed remedial measures in the period 2017–2020 by the CLG

Landslide name	Landslide type	Landslide area (m^2)	Remedial measures
Omiš	Rock falls	150,000	Rock anchoring, rock fall barriers
Radljevac	Rock avalanche	7500	Rock anchoring, rock fall barriers
Lidl	Rock falls	3000	Rock bolts, fences
Špičunak	Debris slide	6500	Anchored pile wall, drainage trenches, buttressing, bored drains
Pićan	Debris slide	1600	Gabion wall, bored drains
Štefanići	Debris slide	4000	Buttressing
Krnjak	Rock fall	3000	Gabion wall, fences
Brgat	Rock slide	45,000	Rockbolts, shotcreete

Fig. 10 A view at the City of Omiš and limestone slopes above the town

Fig. 11 A view at the rockfall protection barriers at limestone slopes above the City of Omiš

state and local roads, it was necessary to perform remediation measures. Based on the results of field investigations, the remedial measures were designed to stop further sliding and to increase general stability of the slope. The complex remedial measures design includes a pile wall anchored in the head with pre-stressed geotechnical anchors positioned below the state road, reinforced concrete grid construction and self-drilling bolts on the embankment stone linings, in combination with bored and trench drains below the road construction and in the landslide foot. Monitoring of the landslide during and after remediation using piezometers and inclinometers is also included in design. The remediation is planned to be conducted during the winter 2019 and spring 2020.

Fig. 12 A view at the Špičunak Landslide

Discussion and Conclusions

In the frame of the IPL Projects IPL-2019 Rockfall Hazard Identification and Rockfall Protection in The Coastal Zone of Croatia and IPL-220 Kostanjek Landslide Monitoring Project (Zagreb, Croatia), as well as through the activities of the Center for Applied Landslide Research (CALaR), WCoE on Landslide Risk Reduction of ICL, the scientists from the Croatian Landslide Group (CLG) were included the wide spectra of activities within landslide science from landslide investigation to landslide prediction and stabilization.

Investigation of landslide initiation is based on laboratory testing of soil material in combination with physical models testing in conditions of rain induced and earthquake induced landslides. Laboratory testing of soil samples taken from landslide bodies consist of different laboratory tests to determine soil properties in saturated and unsaturated conditions. A series of disturbed and undisturbed samples taken from colluvial material of landslide body in unsaturated condition were subjected to different tests to determine functional relationship between soil moisture, suction, and hydraulic conductivity, soil-water retention curve (SWRC) and the hydraulic conductivity function (HCF) in wetting and drying processes, (using different measurement techniques and devices such as suction-controlled oedometers, standard and volumetric pressure plate extractors, HYPROP evaporation method device, WP4-T dew point potentiometer), as well as unsaturated shear strength of soil in modified direct shear apparatus through long-lasting wetting tests. The obtained testing results enabled establishment of numerical models of a landslide initiation that can include different rainfall intensities in time, appropriate infiltration, coupled process of seepage of infiltrated water through unsaturated soil and consequential changes in hydraulic conductivity and unsaturated soil shear strength. Taken into account all these processes and changing soil properties, a real state of stability can be determined depending on simulation of long time rainfall intensities, duration and current soil condition, that enables establishing of an early warning system usable for deep seated landslides in slopes built in soil materials of low hydraulic conductivity.

The other type of soil testing is related to the dynamic (cyclic) behavior of a soil from sliding bodies using resonant column apparatus as well as dynamic triaxial apparatus. These testing are important for determining cyclic properties and cyclic behavior of such material through definition of strength reduction damping curves and cyclic stress ratio necessary for an estimation of landslide hazard at a certain probable earthquake intensity.

Physical modeling of landslides initiation under static (rainfall) and dynamic (earthquake) condition is one of the most important issue to be investigated from the CLG members using the knowledge acquired through sophisticate testing methods applied to soils from typical landslides bodies in flysch deposit environments. Physical modeling of landslides initiation would enable explanation of some of details that in real landslides can stay hidden, such are distribution of effective stresses in a slope and changes in suction that have an important role in landslide initiation. Another issue of landslide physical modeling is better understanding of different remediation measures' impact on slope stability. Although the remediation measures have an important role in landslide risk reduction, their real impact on slope stability was never deeply researched, and remained at the level of engineering analysis. Physical modeling of different remediation measures would enable better insight in their consequences and more efficient use in in remediation projects.

Prediction of landslide movements in case of deep seated slow moving landslides is one of the more demanding issues in an early warning system establishment. Development of any phenomenological model requires detailed knowledge of relations landslide triggering factors—internal forces—resistance and movements that can enable selection of parameters that impact on final values of a landslide movements. Although numerous prediction models are developed based on relatively low numbers of input parameters, the most of them are not proven and cannot adequately react in circumstances that were not established in so called learning period of the model. Models with more (independent and dependent) parameters are more robust and can adequately react in unexpected conditions. Developed prediction model based on 8 years of monitored data ensure stability of landslide movement prediction that can be used (with less adjustments) for any deep seated slow moving landslide.

Landslide and rockfall remediation designs, as a rule, are the task of the professional engineers, but some more complex and demanding cases, require much more knowledge in the field of landslide investigation, analyses, modeling and remediation. Protection of the City of Omiš (Middle Dalmatia, Croatia) from rockfall threats was a challenging issue from rockfall sources identification, rockfall runout analysis as well as rockfall protection measures design. The absence of landslides (including rockfalls) inventories, as well as landslides susceptibility and hazard assessment (Arbanas et al. 2018), required comprehensive approach in landslides susceptibility analysis, spatial 3D simulation of rockfall runouts, as well as structural analyses of protection measures. An innovative approach using recent experiences in rockfall hazard analysis and rockfall structural protection using remote sensing techniques was developed and applied in the project from field investigation, through rockfall protection design and construction.

All described activities implemented through two IPL Projects (IPL-2019 and IPL-220), as well as through the activities of the Center for Applied Landslide Research (CALaR), WCoE on Landslide Risk Reduction of ICL, represent significant contribution to the landslide science presented in journals and at conferences (listed below). The continuation of these scientific activities would result in practical contributions to landslide hazard and risk reduction.

Acknowledgements Some of researches presented in this paper were supported by Croatian Science Foundation under the Project IP-2018-01-1503 Physical modelling of landslide remediation constructions behaviour under static and seismic actions (ModLandRemSS), as well as University of Rijeka under the Projects UniRi-Tehnic-18-113 Laboratory research of static and cyclic behaviour at landslide activation and UniRi-Tehnic-18-276 Research of Rockfall Processes and Rockfall Hazard Assessment. These supports are gratefully acknowledged. The authors also want to thank to all Projects' members and their help in researches conducting.

References

Arbanas Ž, Udovič D, Sečanj M, Đomlija P, Mihalić Arbanas S (2018) Recent experiences in rockfall hazard and risk assessment. In: Proceedings of VII conference of Croatian platform for disaster risk reduction, 11–12 Oct 2018. National Protection and Rescue Directorate, Zagreb, pp 227–236 (in Croatian)

Arbanas Ž, Pajalić S, Jagodnik V, Peranić J, Vivoda Prodan M, Đomlija P, Dugonjić Jovančević S, Čeh N (2019a) Development of physical model of landslide remedial constructions' behaviour. In: Proceedings of 4th regional symposium on landslides in Adriatic Balkan Region, Sarajevo, Bosnia and Herzegovina, 23–25 Oct 2019. Geotechnical Society of Bosnia and Herzegovina, Sarajevo, pp 103–108

Arbanas Ž, Sečanj M, Vivoda Prodan M, Dugonjić Jovančević S, Peranić J, Bernat Gazibara S, Krkač M, Udovič D, Mihalić Arbanas S (2019b) Protection of the City of Omiš, Croatia, from rockfall threats. In: Proceedings of the 4th regional symposium on landslides in the Adriatic-Balkan region, 23–25 Oct 2019. Geotechnical Society of Bosnia and Herzegovina, Sarajevo, pp 251–255

Arbanas Ž, Vivoda Prodan M, Dugonjić Jovančević S, Peranić J, Udovič D, Bernat Gazibara S, Krkač M, Sečanj M, Mihalić Arbanas S (2019c) Rockfall modelling and rockfall protection at the slopes above the city of Omiš, Croatia. In: Proceedings of the ISRM specialised conference. "Geotechnical challenges in karst". Croatian Geotechnical Society, Zagreb, pp 121–126

Arbanas Ž, Jagodnik V, Peranić J, Pajalić S, Vivoda Prodan M, Čeh N (2020) Physical model of rainfall induced landslide in flume test: preliminary results. In: Proceedings of 4th European conference on physical modelling in geotechnics. Lulea University of Technology (in Press)

Cruden DM, Varnes DJ (1996) Landslide types and processes. In: Turner AK, Schuster RL (eds) Landslide investigation and mitigation (Special Report/Transportation Research Board, National Research Council; 247). National Academy Press, Washington DC, Chapter 3, pp 36–75

Đomlija P, Prša M, Jagodnik V, Arbanas Ž (2019) Preliminary testing of clay activity from landslide deposits in Dubračina River Basin, Croatia. In: Proceedings of 4th regional symposium on landslides in the Adriatic-Balkan region, 23–25 Oct 2019. Sarajevo, Bosnia and Herzegovina, pp 121–126

Herak D, Sović I, Cecić I et al (2017) Historical seismicity of the Rijeka region (northwest External Dinarides, Croatia)—Part I: earthquakes of 1750, 1838, and 1904 in the Bakar epicentral area. Seismol Res Lett 88(3):904–915

Jagodnik V, Đomlija P, Oštrić K, Arbanas Ž (2019a) Strength reduction curve of soils from eluvial deposits in Vinodol Valley, Croatia. In: Proceedings of the 4th regional symposium on landslides in the Adriatic—Balkan region. Društvo za geotehniku u Bosni i Hercegovini, pp 109–114

Jagodnik V, Đomlija P, Vorić L, Arbanas Ž (2019b) Cyclic stress ratio of landslide deposits in Vinodol Valley, Croatia. In: Proceedings of the 4th regional symposium on landslides in the Adriatic—Balkan region. Društvo za geotehniku u Bosni i Hercegovini, pp 79–84

Keefer DK (2002) Investigating landslides caused by earthquakes—a historical review. Surv Geophys 23:473–510

Kramer SL (1996) Geotechnical earthquake engineering. Prentice-Hall, London, p 653

Krkač M, Bernat Gazibara S, Arbanas Ž, Sečanj M, Mihalić Arbanas S (2020) A comparative study of random forests and multiple linear regression in the prediction of landslide velocity. Landslides 17:2515–2531

Krkač M (2015) A phenomenological model of the Kostanjek landslide movement based on the landslide monitoring parameters. Dissertation, University of Zagreb (In Croatian)

Krkač M, Bernat Gazibara S, Sečanj M, Arbanas Ž, Mihalić Arbanas S (2019) Continuous monitoring of the Kostanjek landslide. In: Proceedings of the 4th regional symposium on landslides in the Adriatic-Balkan Region, 23–25 Oct 2019. Sarajevo, Bosnia and Herzegovina, pp 43–48

Krkač M, Mihalić Arbanas S, Arbanas Ž, Bernat S, Špehar K, Watanabe N, Nagai O, Sassa K, Marui H, Furuya G, Wang C, Rubinić J, Matsunami K (2014) Review of monitoring parameters of the Kostanjek landslide (Zagreb, Croatia). In: Sassa K, Canuti P, Yin Y (eds) Proceedings of the 3rd world landslide forum 'Landslide science for a safer geoenvironment', vol 2. Springer, Netherland, pp 637–645

Krkač M, Špoljarić D, Bernat S, Mihalić Arbanas S (2017) Method for prediction of landslide movements based on random forests. Landslides 14(3):947–960

Mihalić Arbanas S, Sečanj M, Bernat Gazibara S, Krkač M, Arbanas Ž (2017) Identification and mapping of the Valići Lake Landslide (Primorsko-Goranska County, Croatia). In: Proceedings of the 2nd regional symposium on landslides in the Adriatic-Balkan region, 14–16 May 2015. Belgrade, Serbia, pp 197–202

Mihalić Arbanas S, Krkač M, Bernat Gazibara S, Komac, M, Sečanj M, Arbanas Ž (2018) TXT-tool 2.385-1.1 A comprehensive landslide monitoring system: the Kostanjek Landslide, Croatia. In: Landslide dynamics: ISDR-ICL landslide interactive teaching tools, volume 1: fundamentals, mapping and monitoring. Springer Publishing, Heidelberg, pp 449–464

Pajalić S, Jagodnik V, Arbanas Ž (2019) Physical model for testing the behavior of landslide remediation constructions. In: Proceedings of 7th conference Zajednički temelji 2019, Rijeka, 25–27 Sept 2019. Rijeka, Croatia, pp 37–42

Peranić J (2019) Importance of geotechnical cross-section unsaturated zone for landslide occurrence in flysch deposits. Dissertation, University of Rijeka, Faculty of Civil Engineering, Rijeka, Croatia

Peranić J, Arbanas Ž (2020) Impact of the wetting process on the hydro-mechanical behavior of unsaturated residual soil from flysch rock mass: preliminary results. Bull Eng Geol Env 79:985–998

Peranić J, Arbanas Ž, Cuomo S, Maček M (2018a) Soil-water characteristic curve of residual soil from a Flysch rock mass. Geofluids. Volume 2018, Article ID 6297819, p 15

Peranić J, Arbanas Ž, Foresta V, Cuomo S, Maček M (2018b) Determination of soil water retention curve of residual soil from a flysch rock mass. In: Proceedings of the 7th international conference on unsaturated soils, 3–5 Aug 2018. Hong Kong, China, pp 379–384

Peranić J, Jagodnik V, Arbanas Ž (2019) Rainfall infiltration and stability analysis of an unsaturated slope in residual soil from flysch rock mass. In: Proceedings of the XVII ECSMGE-2019, 1–6 Sept 2019. Reykjavik, Iceland, pp 4364–4371

Peranić J, Moscariello M, Cuomo S, Arbanas Ž (2020a) Hydro-mechanical properties of unsaturated residual soil from a flysch rock mass. Eng Geol 269, p 12

Peranić J, Moscariello M, Cuomo S, Arbanas Ž (2020b) Determination of hydraulic conductivity and shear strength properties of unsaturated residual soil from flysch rock mass. In: Proceedings of the 4th European conference on unsaturated soils, 24–26 June 2020. Lisbon, Portugal

Vivoda Prodan M, Krkač M, Mihalić Arbanas S, Arbanas Ž (2019) Remediation measures on a deep-seated slow-moving landslide. In: Proceedings of the 4th regional symposium on landslides in the Adriatic-Balkan Region, 23–25 Oct 2019. Geotechnical Society of Bosnia and Herzegovina, Sarajevo, pp 169–174

Vucetic M, Dobry R (1991) Effect of soil plasticity on cyclic response. J Geotech Eng 117:89–107

Xu C (2015) Preparation of earthquake-triggered landslide inventory maps using remote sensing and GIS technologies: Principles and case studies. Geosci Front 6:825–836

Part III

Landslide-Induced Tsunamis

Simulation of Tsunami Waves Induced by Coastal and Submarine Landslides in Japan

Doan Huy Loi, Kyoji Sassa, Khang Dang, and Toyohiko Miyagi

Abstract

Members of the International Consortium on Landslides (ICL) together with Kyoto University have developed the undrained dynamic-loading ring-shear apparatus (UD-RSA) and the integrated simulation model (LS-Rapid) for the evaluation of the initiation and motion of landslides using landslide dynamic parameters for landslide hazard assessment. In order to develop the landslide-induced tsunami hazard assessment technology based on underwater landslide motion simulated by UD-RSA and LS-Rapid, a new landslide induced tsunami model (LS-Tsunami) has been developed. It is verified using the world's largest well-documented landslide induced-tsunami disaster with 15,153 deaths in Unzen, Nagasaki, Japan in 1972. The LS-Tsunami simulation results agreed with the observations in the historical records. The study of the hypothetical Senoumi (stone flower sea) submarine landside in Suruga Bay, Japan based on the UD-RSA testing of the sample (volcanic ash) cored from 200 m deep under the sea floor (by the International Ocean Discovery Program-IODP) was implemented. As there is no historical record and the date has not yet been identified, this submarine landslide is called "hypothetical." The initiation and motion of the landslide was simulated by LS-Rapid. LS-Tsunami model was applied to the Senoumi submarine landslide and used to estimate the tsunami height at the excavation site along Ota River in Shizuoka Prefecture where three layers of tsunami deposits have been found by Fujiwara et al. (Quat Sci Rev, 2019). LS-Tsunami analysis presented that tsunami wave caused by the Senoumi submarine landslide had an inundation depth of around 1.5 m (3.5 m tsunami height above sea level) at the excavation site. In 2009, a M_w 6.4 Suruga Bay earthquake occurred triggering a small submarine landslide within the Senoumi area. The submarine landslide induced a tsunami (Baba et al. in Advances in Natural and Technological Hazards Research. Springer, pp 485–494, 2012). Using LS-Rapid and LS-Tsunami, three conditions for triggering the tsunami wave were considered: (1) caused by fault motion only, (2) caused by landslide motion only, and (3) wave caused by both landslide motion and fault motion. Each mode was compared to monitored tidal wave forms along Suruga Bay. LS-Tsunami simulation result agreed with the monitored results including a very big tsunami wave monitored by the Yaizu tidal gauge. Application of LS-Tsunami for three cases in Japan: the 1792 Unzen coastal landslide, the Senoumi submarine landslide, and the 2009 Suruga Bay submarine landslide has proved the precision and reliability in both the maximum tsunami height and time variation of the tsunami wave. Accordingly, this technology is applied to hazard assessment for the potential retrogressive landslides of the Senoumi landslide in Suruga bay which are likely triggered by earthquakes along the Nankai Trough in the future. The result showed 20–50 m inundation depths of Tsunami in Yaizu city.

D. H. Loi (✉) · K. Sassa · K. Dang
International Consortium on Landslides, 138-1, Tanaka Asukaicho, Sakyo-ku, Kyoto, 606-8226, Japan
e-mail: doanhuyloidkt@gmail.com

K. Sassa
e-mail: sassa@iclhq.org

K. Dang
e-mail: khangdq@gmail.com

D. H. Loi
Graduate School of Engineering, Kyoto University, Kyoto, Japan

K. Dang
University of Science, Vietnam National University, Hanoi, Thanh Xuan, 334 Nguyen Trai, Hanoi, Vietnam

T. Miyagi
Advantechnology Co. Ltd., 1-4-8 Kakyoin Aobaku, Sendai, 980-0013, Japan
e-mail: c1934009@mail.tohoku-gakuin.ac.jp

K. Sassa et al. (eds.), *Understanding and Reducing Landslide Disaster Risk*, ICL Contribution to Landslide Disaster Risk Reduction,
https://doi.org/10.1007/978-3-030-60196-6_22

Keywords

LS-tsunami • Suruga Bay • Ring shear test • Unzen–Mayuyama megaslide • Submarine landslide

Introduction

This paper aims to establish landslide-induced tsunami hazard assessment technology based on the assessment of the initiation and motion of landslides. It consists of three parts and 8 sub-parts.

Part 1 describes the 1972 earthquake-induced coastal megaslide in Unzen volcano and the landslide-induced.

tsunami disaster over Ariake sea in Japan. It is the world's largest landslide induced tsunami disaster causing 15,153 deaths.

- Development of undrained dynamic loading ring shear testing for landslide-induced tsunamis: Fig. 1
- Development of tsunami simulation model (LS-Tsunami) based on landslide simulation model (LS-Rapid): Figs. 2 and 3
- Application of LS-Rapid and LS-Tsunami for the Unzen landslide and the associated tsunami disaster along with comparisons between the simulation results and historical records: Figs. 4, 5, 6, 7, 8, 9 and 10

Part 2 describes the study of the hypothetical Senoumi submarine landslide in Suruga bay, Japan, verification of tsunami height estimated by LS-Rapid and LS-Tsunami, and tsunami monitoring data in 2009 Suruga Bay earthquake.

- Study of the hypothetical Senoumi submarine landslide in Suruga Bay, Japan based on IODP drilling, seismic-loading ring shear testing, and LS-Rapid simulation: Figs. 11, 12, 13, 14, 15, 16, 17, 18 and 19.
- Application of LS-Tsunami to this Senoumi submarine landslide and the comparison between the simulation results and the tsunami deposits found along Ota River, Shizuoka Prefecture: Figs. 20, 21, 22, 23 and 24.
- Verification of the simulation results from LS-Rapid and LS-Tsunami with the tidal gauge monitoring data for the tsunami height in 2009 M_w 6.4 Suruga Bay earthquake: Figs. 25, 26, 27, 28, 29 and 30.

Part 3 describes the tsunami hazard assessment induced by a potential retrogressive landslide in the Senoumi landslide topography in Suruga Bay, Japan.

- Tsunami hazard assessment triggered by retrogressive landslides of Senoumi submarine landslide: Figs. 31, 32 and 33.
- Landslide-induced tsunami hazard risk in Yaizu City, which is close to the Senoumi submarine landslide, was visualized in an inundation depth map (100 m mesh) Fig. 34.

Development of Undrained Dynamic Loading Ring Shear Testing for Landslide-Induced Tsunamis

Figure 1a, b, c illustrate the location to sample of soil within a sliding surface. Figure 1a is for landslide on land. The sample is taken by (1) drilling in unstable mass which has not yet moved or (2) from the material in the source area of the already moved landside. Both cases need evaluation of the material in which a sliding surface was formed and is going to be formed. Normally 2–3 different samples will be tested and analysed. Figure 1b is for submarine landslide. In the study of Senoumi hypothetical landslide, we tested volcanic ash taken from around 200 m below the sea floor by IODP deep sea drilling. Figure 1c illustrates the sampling from a moving deposit. When a landslide mass is moving from the source area onto a depositional area, a certain depth of deposit will be scraped forming the shear surface within the deposit. In this case, a sample is taken from the deposit to measure the landslide dynamic parameters which are necessary for the integrated simulation model (LS-Rapid). Figure 1d shows the initial stress by gravity (mg) and seismic stress (ma), where m is the mass, g is gravitation acceleration and a is the acceleration. Figure 1e is the sample within the shear box. The sample is consolidated by initial stresses. The stress changes caused by earthquake, rainfall or both are, then, reproduced in the UD-RSA using a stress controlled system. When a failure occurs, the lower ring starts to rotate while the upper ring is held fixed by the shear stress load cells. Figure 1f presents the initial stress (I) and stress path due to stress change by earthquake or pore pressure change by rainfalls. If the stress reaches the failure line, a landslide is initiated. Figure 1g, h present the apparatus along with the loading and monitoring systems, respectively. The precise shear stress acting on the sliding surface is monitored by S1 and S2, the normal stress on the shear surface is monitored by the vertical load cell N in Fig. 1h. Pore pressure during shearing is monitored by P in Fig. 1g, h. Additional details of the undrained dynamic loading ring shear apparatus is explained by Sassa et al. (2014a, b), and Sassa and Dang (2018).

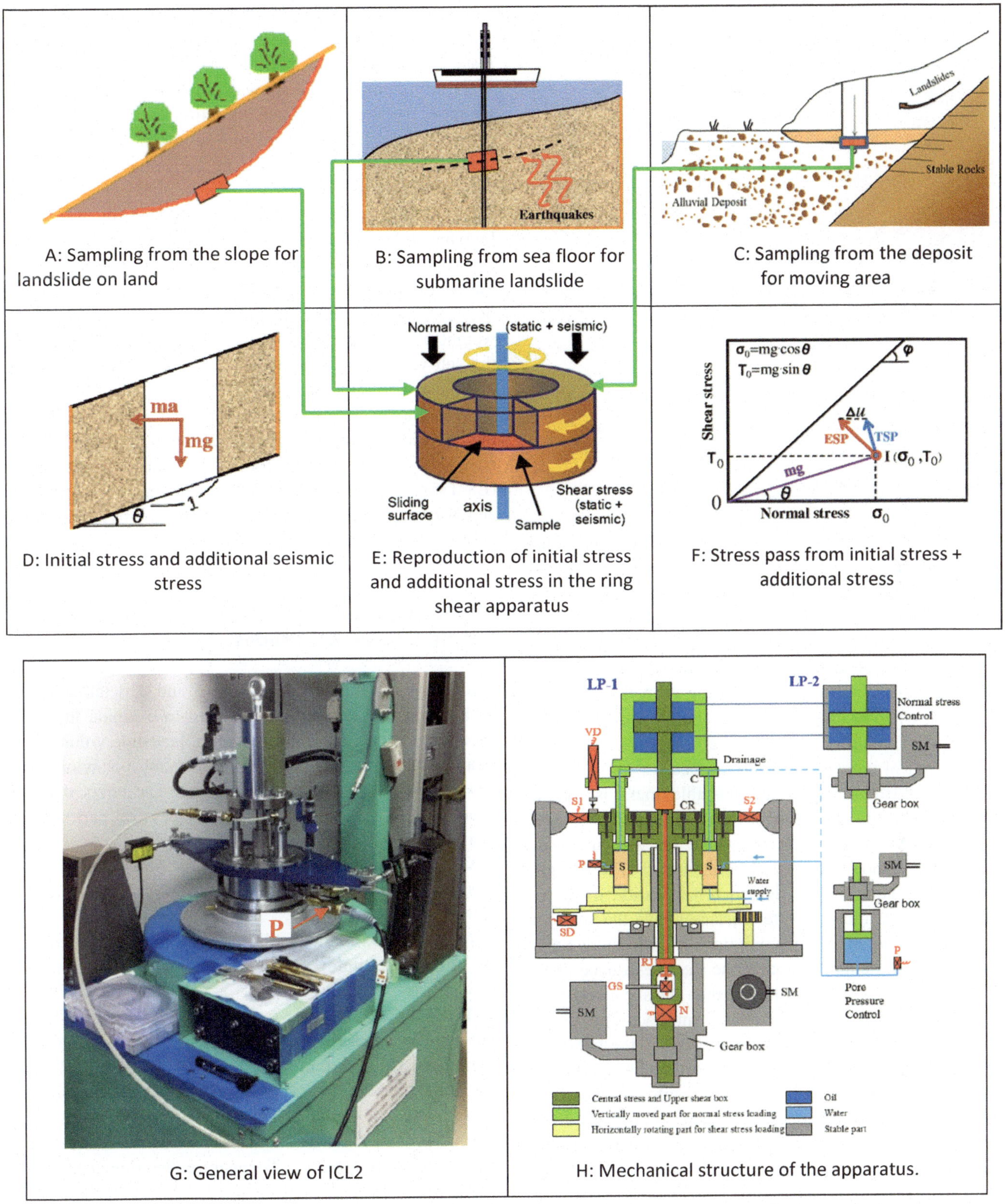

Fig. 1 Undrained dynamic-loading ring shear testing for landslide hazard assessment

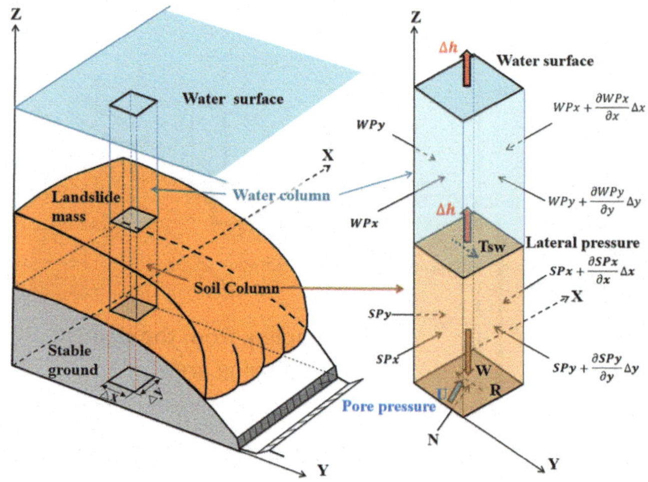

Fig. 2 Basic principle of the landslide-induced-tsunami simulation model (Sassa et al. 2016)

Development of LS-Tsunami Model

Figure 2 presents the basic principles of the landslide-induced tsunami model. The left part of this figure is same concept as the integrated landslide simulation model (LS-Rapid), which considers a vertical imaginary column containing the forces acting within a moving landslide mass. The only difference is that the moving mass is below a water surface as shown in the right part of Fig. 3. A tsunami wave is caused by elevating the water body caused by the movement of landslide mass on the sea floor, namely the upward vertical movement of the sea bed.

When landslides occur, the unstable mass is moving down and enters or travels across the sea floor. The altitude of the submarine ground level (where the soil and water columns meet and interact) will be increased. The rising of submarine ground surface due to the landslide mass will increase the elevation of water surface above the sea level datum by an amount, Δh, in a unit time, Δt. The fundamental assumptions of the model are: (1) the shear resistance within the soil (R) should be much higher than the shear resistance (τ_{sw}) between soil and water at this interface and (2) the effective stress acting on the surface between the bottom of the soil column and the top of the sea floor is regarded as unaffected by the increase in the water height. Assumption 2 is based on the following two reasons: (1) the surficial stratigraphy on the ocean floor is made up of loose particles, which are expected to be sufficiently permeable to transmit the changes in water pressure due to fluctuations in the water height to the bottom of the soil column above the sea floor and (2) the changes in the water pressure are quickly transmitted from the top to the bottom of soil column because the soil column of the landslide mass is both unconsolidated and water saturated (elastic wave velocity is around 1.5 km/s for water) and thus, can be regarded as instantaneous.

From assumption 1, τ_{sw} was neglected in the model. In addition, the shear resistance mobilized in the shear surface of large-scale landslides should be much higher the tangential force by water. Hence, no tangential force between soil and water is considered.

When a landslide mass is submerged in water, the buoyant force of the water will act on the soil mass. Buoyancy reduces the apparent weight of soil mass to the buoyant unit weight of the soil mass which is the difference between unit weight of the soil mass in onshore γ_t) and unit weight of sea water (γ_w). In the case of partial submersion, the self-weight of column will only change for the part of soil mass below the water level.

While many parameters are inputted in LS-RAPID for the landslide motion (Sassa et al. 2010), only two additional parameters are used for the tsunami simulation (LS-Tsunami). These parameters are the unit weight of water and Manning's roughness coefficient for the basal resistance between water and ground.

Application of the Model to Two Hypothetical Simple Coastal Landslides

Two hypothetical landslides (Cases A and B, detailed below) on an imaginary slope were selected to examine the influence of each parameter (water depth, landslide volume, and steady state shear strength) on the tsunami behavior in the new landslide-induced tsunami model. The effect of water depth can be found discussed in our previous paper (Sassa et al. 2016).

Case A: Maximum depth = 100 m, Maximum length = 500 m, Maximum width = 300 m, Volume = 5.8 × 106 m^3.

Case B: Maximum depth = 313 m, Maximum length = 1230 m, Maximum width = 1000 m, Volume = 170 × 106 m^3.

The simple slope was produced by LS-Rapid. The top and bottom of the slope was horizontal, while the slope angle was 35°. For both landslides, the heads and the toes are located at the top of the slope and the water surface, respectively. The distance between the left end and the coast is 2000 m, and the distance from the coast to the bottom of the figure is 100,000 m. To analyze the influence of the steady state shear strength, a fixed value the peak friction angle, ϕ_p, of 42° and a friction angle during motion, ϕ_m, of 40° were selected.

We tested the influence of steady-state shear resistance parameter (τ_{ss}) by changing the τ_{ss} values with following

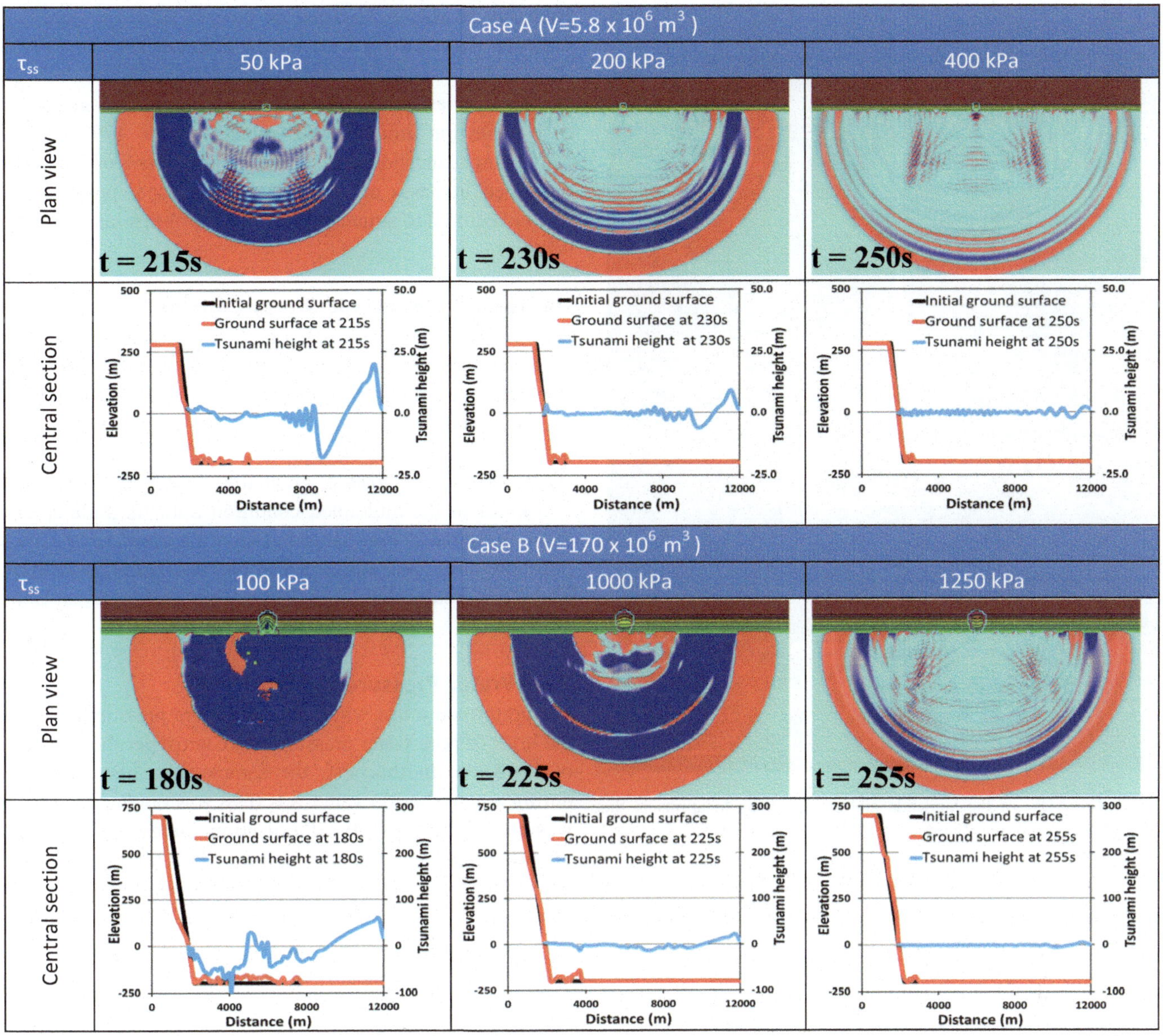

Fig. 3 Tsunami simulation results for two different volume of landslides (5.8 × 106 and 170 × 106 m^3) with 6 different value of steady state shear resistance (τss = 50, 100, 200, 400, 1000 and 1250 kPa). Tsunami (plan view) and landslide profiles (at the central section) when the tsunami front reached 12,000 m are presented. The bright red color presents waves more than 3 m above sea level and the dark blue color presents waves less than 3 m below sea level. Modified from Sassa et al. (2016)

steps 50, 200, and 400 kPa for landslide A and 100, 1000, and 1250 kPa for landslide B. The water depth was 200 m. The motion of the water wave from the LS-Tsunamis calculated at each 0.05 s. The results at the central section were plotted at each 5 s as shown in Fig. 3. Figure 3 and Table 1 show the relationship between the steady-state shear strength and the tsunami behavior. The results show that higher steady-state shear resistances could lead to the smaller maximum velocities (v_p) and maximum tsunami heights. Increasing τ_{ss} values result in less unstable areas being simulated.

Application of the Undrained Dynamic-Loading Ring-Shear Testing and LS-Tsunami Model to the Unzen-Mayuyama Landslide and Its Resulting Tsunami Wave

The first application of this simulation tools was the reproduction of the failure and movement behaviour of the largest landslide disaster named the 1792 Unzen-Mayuyama megaslide (volume: 3.4×10^8 m^3; maximum depth:

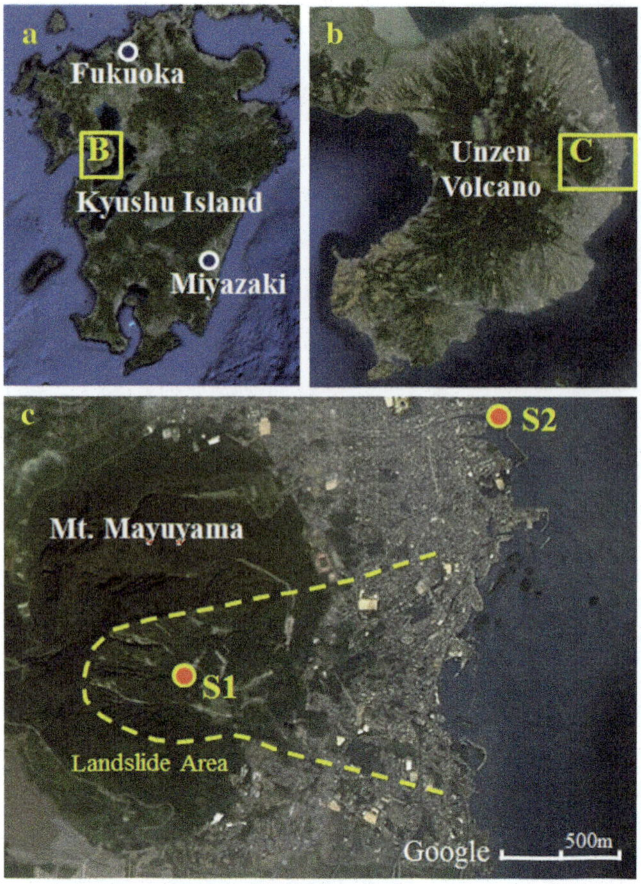

Fig. 4 Overview of the landslide and the sampling locations in the Unzen Mayuyama area (Sassa et al. 2014a)

400 m) triggered by an earthquake. This landslide and the landslide-induced tsunami wave killed a total of 15,153 persons (Usami 1996). The landslide disaster was well recorded by many researchers and the government. The main data from the 1792 Unzen-Mayuyama disaster including the pre- and post-failure topography and the height of tsunami wave on the Shimabara Peninsula and the Kumamoto and Akusa coasts.

Sample from the 1792 Unzen–Mayuyama Landslide Area

The 1792 Unzen–Mayuyama landslide in Shimabara City, Kyushu Island, Japan was selected to examine the capacity of ICL-2. Figure 4 shows overview of the landslide and the location of sampling. A sample (S1) was taken from the sandy zone exposed along a torrent side slope in the source area where the initial sliding surface was formed. Sample 2 (S2) was taken along the coastal area outside the landslide moving area as a representative soil in which a sliding surface would be formed during the motion of the initial landslide.

Results of ICL-2 Testing of the 1792 Unzen-Mayuyama Landslide Samples

A series of tests were conducted on sample S1 and S2 to investigate the initiation and motion of the 1792 Unzen-Mayuyama landslide, respectively. The detail of results is described in Sassa et al. (2014a). This paper presents test results required to simulate the initiation of this landslide (S1 sample).

Undrained Monotonic Stress Control

As a basic test, a series of monotonic undrained increasing shear stress tests that were conducted at various consolidation pressures ranging from 300 to almost 3000 kPa. For each test, the normal stress was first increased to the desired stress in the drained condition. During the shearing state, shear stress was loaded gradually at a rate of $\Delta\tau = 1 - 5$ kPa/s in the undrained condition until the failure. The result is shown in Fig. 5a, b. In two of the tests, large excess pore pressure generation was observed during shearing. These test results suggested the possibility of rapid landslide motion.

Pore-Water Pressure Control Test

We performed a test with increasing pore pressure to check the level of pore water pressure that can trigger the initiation of landslide. In this test, the sample was first saturated, during which the BD value was 0.98 was obtained. Then, it was consolidated to 3.0 MPa of normal stress and 1.5 MPa of shear stress in a drained condition. This preparation stage aims to reproduce the initial stress in the slope as shown by a black line in Fig. 5c. These initial stresses correspond to the slope of arctan (1.5/3.0) = 26.5°, similar angle to the landslide block. Additional water pressure was then applied to the shear box at a rate of $\Delta u = 1$ kPa/s to reproduce the pore-pressure-induced landslide process. Sample was in the failure state where the pore-water pressure reached more than 1.2 MPa or a pore-water pressure ratio, r_u of 1.2/3.0 = 0.4. The friction angle at failure was 39.4°.

Examination of Earthquake Loading at the Unzen-Mayuyama Landslide

There was no record of 1792 earthquake, so we investigated recent earthquake-induced landslides to find a similar one to apply for the landslide simulation. According to Usami (1996), the greatest earthquake hit on 21 May 1792 has estimated magnitude of M = 6.4 ± 0.2.

Based on the effect of earthquake, the Japanese seismic intensities (Usami 1996) are as follows:

V: 80–250 cm/s^2 (where walls and fences are cracked, and Japanese gravestones fall down).

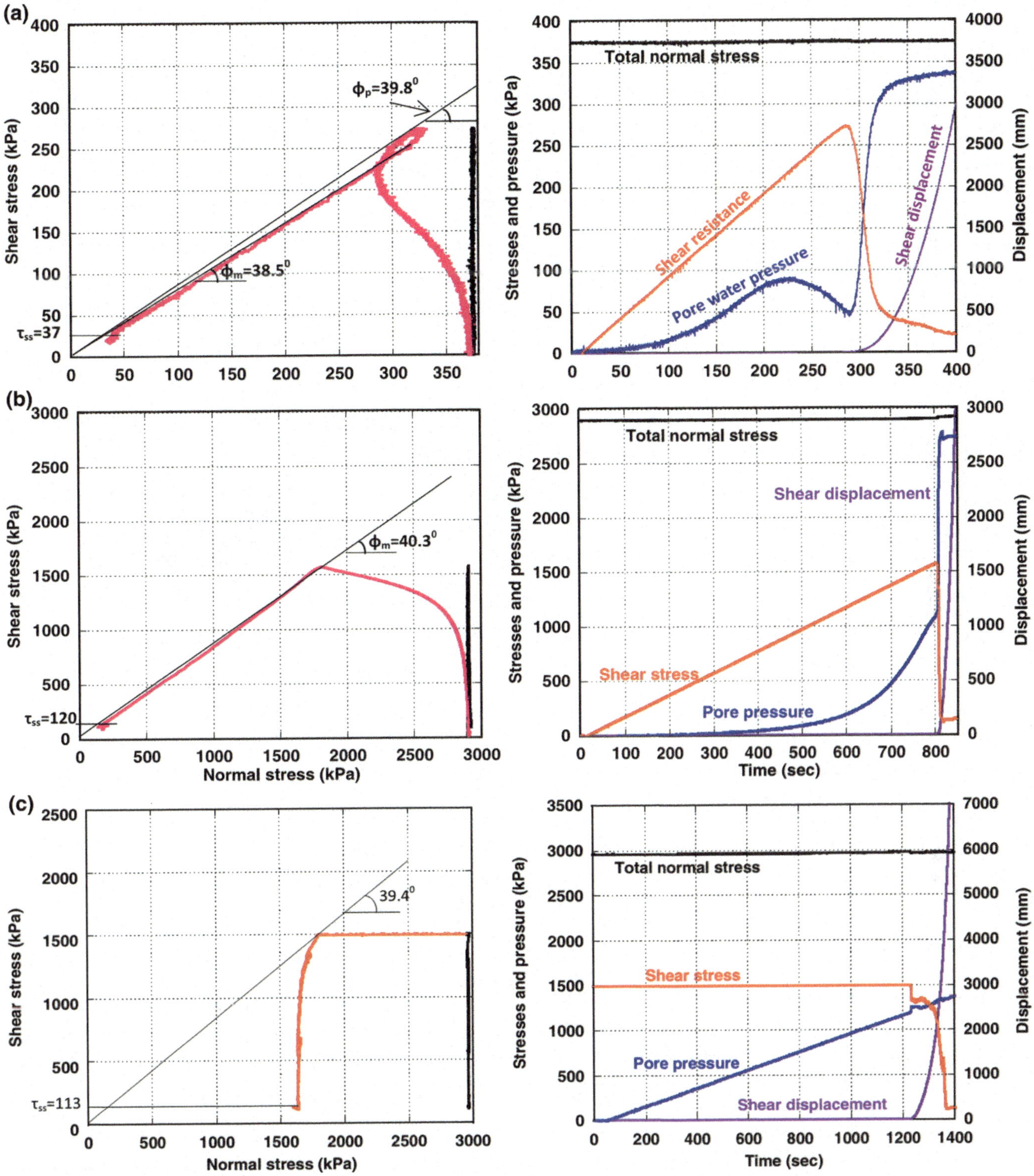

Fig. 5 Ring shear test results. Sample: S1 from Unzen–Mayuyama. **a**, **b** Undrained monotonic stress control tests at different normal stresses: **a** Normal stress = 375 kPa, BD = 0.93, $\Delta\tau/s = 1$ kPa/s. **b** Normal stress = 2900 kPa, BD = 0.96, $\Delta\tau/s = 2$ kPa/s. **c** Pore pressure control test: Normal stress = 3000, BD = 0.98, Shear stress = 1500 kPa, $\Delta u/s = 1$ kPa/s. Modified from Sassa et al. (2014a)

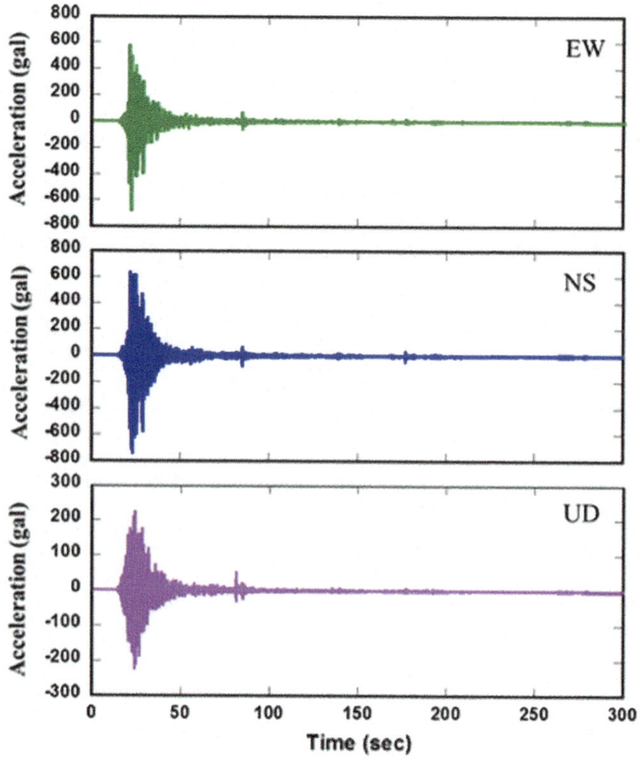

Fig. 6 The 2008 Iwate–Miyagi earthquake record (MYG004) from K-net of National Research Institute for Earth Science and Disaster Prevention (NIED)

VI: 250–400 cm/s^2 (where less than 30% of Japanese wooden houses are destroyed).

VII: More than 400 cm/s^2 (where more than 30% of the houses are destroyed, landslides are triggered and surface fault rupture is seen).

Based on investigations, more than 30% of the houses were destroyed in the Shimabara area and the Unzen Restoration Office estimated that the seismic intensity which triggered the Unzen–Mayuyama landslide was VII. Exact seismic accelerations may never be known, but were probably at least 400 cm/s^2. The estimate of an acceleration of more than 400 cm/s^2 by the Restoration Office is probably correct, as it is based on a detailed study of house damages.

In recent years, the 2008 Iwate-Miyagi Earthquake (M = 7.2) triggered the Aratozawa landslide (1300 m in length, 900 m in width and 67 million cubic meters) in Miyagi Prefecture. Figure 6 shows the seismic record at MYG004 (National Research Institute for Earth Science and Disaster, Prevention, NIED). The maximum recorded acceleration was 739.9 cm/s^2 in the NS component. As it is of a similar scale and both triggered megaslides, it was decided to use the Iwate-Miyagi earthquake wave form recorded in Miyagi Prefecture (MYG004) for the ring-shear simulation test and also the computer simulation for 1792 Unzen-Mayuyama landslide.

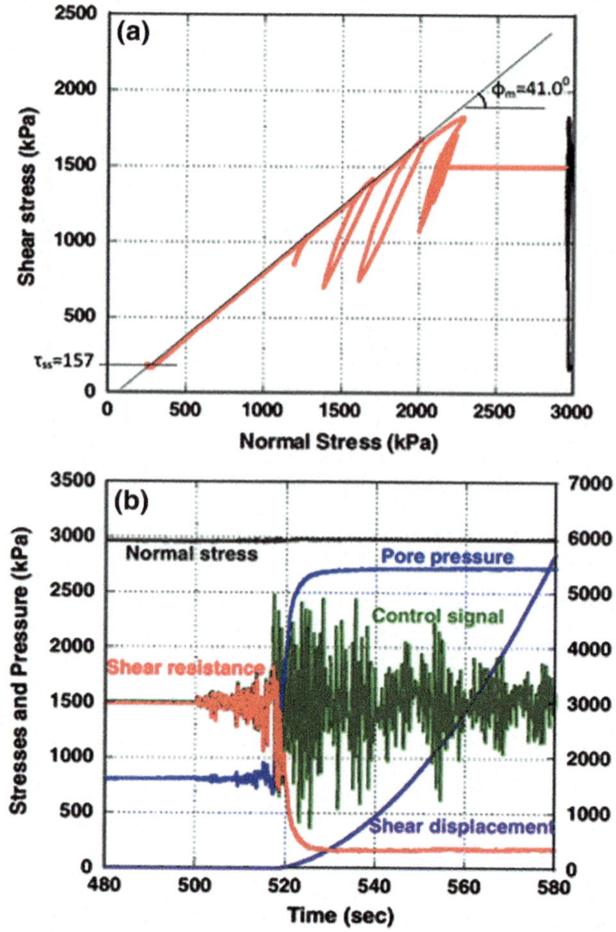

Fig. 7 Undrained pore water pressure and seismic loading test. Sample: S1 from Mayuyama *Source* Area. BD = 0.94 (Sassa et al. 2014a)

Seismic Loading Test

The most advanced and complicated test of the new ring-shear apparatus is the seismic loading test. It can be used to simulate earthquake-induced landslide in which effects of pore-water pressure and earthquake shaking can be combined.

Seismic loading test was performed on sample S1. The sample was initially saturated (BD = 0.94) and consolidated to a normal stress of 3 MPa and shear stress of 1.5 MPa. The corresponding slope angle is arctan (1.5/3.0) = 26.6°. It is not possible to estimate the exact value of the initial pore water ratio before disaster. However, it must be smaller than 0.4 which is the critical pore water pressure which causes a landslide without an earthquake (Fig. 5c). Pore water pressure was then increased to 800 kPa, namely a pore water pressure ratio of r_u = 800/3000 = 0.27 to the initial slope condition. The additional shear stress was given by loading the NS component of the 2008 Iwate-Miyagi earthquake record (maximum acceleration of which was 739.9 cm/s^2) at MYG004. For precise pore pressure monitoring as well as

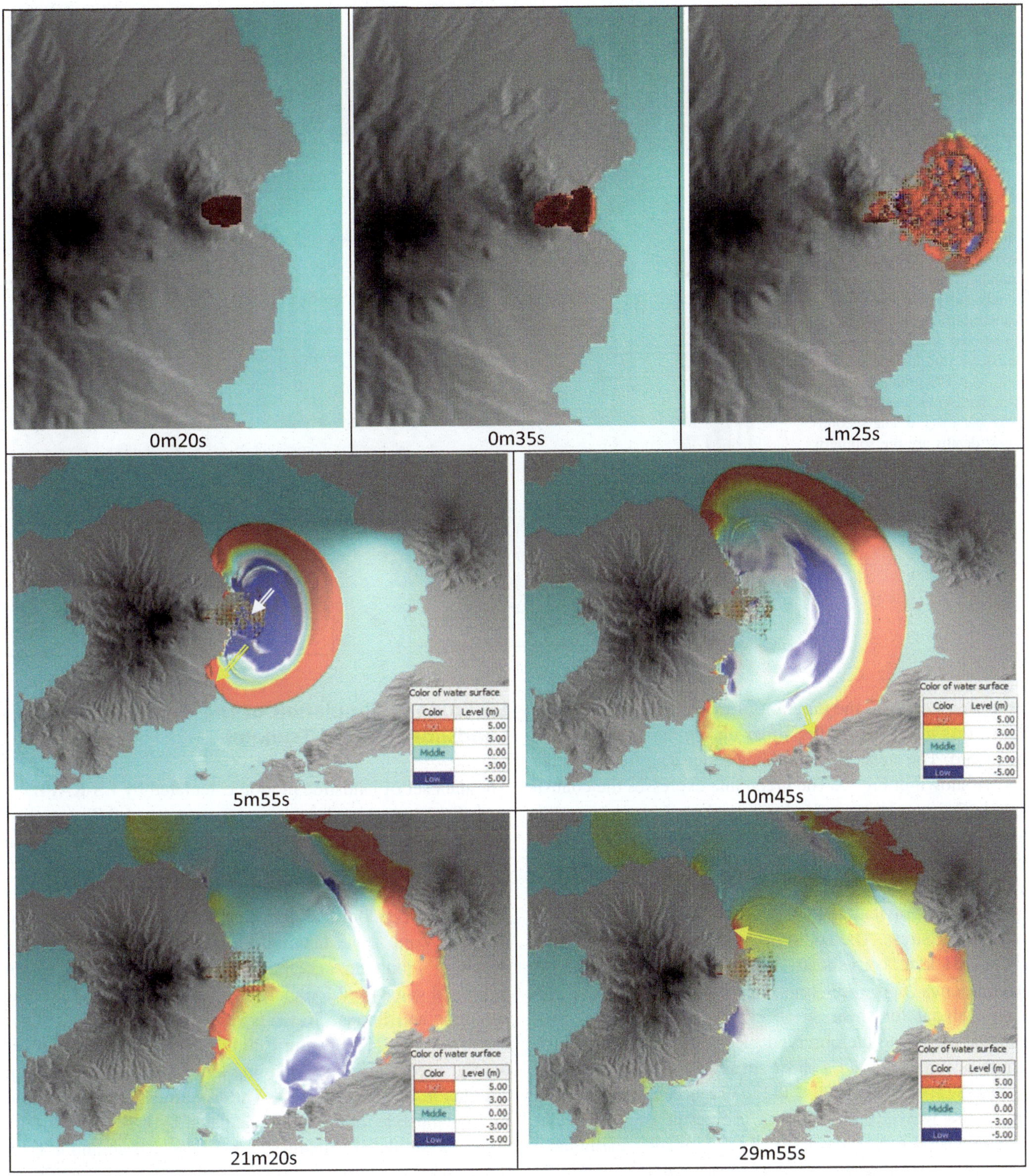

Fig. 8 LS-Tsunami simulation result for the 1792 Unzen-Mayuyama landslide-induced tsunami disaster. In this figure, the red, yellow, aqua, and white present the wave heights: more than 5 m, 3–5 m, 3 to −3 m, −3 to −5 m and less than −5 m, respectively. Modified from Sassa et al. (2016)

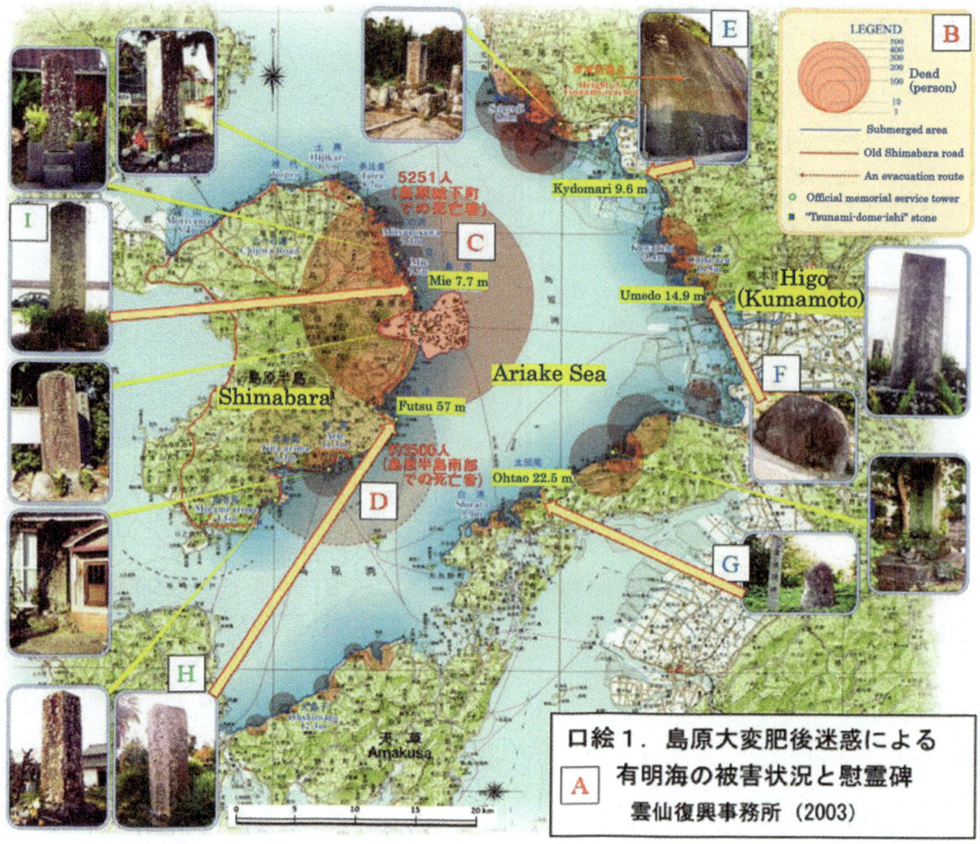

Fig. 9 Records of disaster by the Unzen landslide-induced tsunami (by Unzen Restauration Office 2003) Total number of deaths is 15,153 persons. The size of circles is proportional to the number of human fatalities in the area. The legend for the number of deaths is show in the right-top corner. **a** Disasters around Ariake Sea and monuments by the Shimabara-Taihen, Higo-Meiwaku. "Catastrophe" in Shimabara area and "Annoyance" or "adversely affected" in Higo (Kumamoto) region. **b** The numbers of deaths are shown in the circles (the largest is 500 persons) **c** The greatest deaths are Shimabara town around the castle (5251 persons). **d** The second largest deaths are the southern part of Shimabara Peninsula (around 3500 persons)

servo-stress control, the duration of the seismic acceleration record was increased by five times during the application of the record. The stress path and time series data from seismic test are shown in Fig. 7. The green line expresses the control signal. The maximum and minimum shear stress values are 2469 kPa (1500 + 969 kPa) and 369 kPa (1500–1131 kPa), respectively. The additional shear stress was calculated as $\Delta\tau = ma = 1500{*}a/g$, where, a is the earthquake acceleration, g is the gravitational acceleration, and 1500 kPa is the initial shear stress. The additional shear stresses of 969 kPa and −1131 kPa correspond to earthquake accelerations of +633 cm/s^2 and −739 cm/s^2, respectively. Therefore, the control signal for shear stress given to the sample corresponded exactly to the observed acceleration data. In Fig. 8, failure occurred around a shear stress of 1825 kPa, namely, a/g = (1825 − 1500)/1500 = 0.22. The acceleration needed for failure was 216 cm/s^2. This test result suggested that with initial pore pressure ratio of 0.27, smaller earthquake shaking (around 216 cm/s^2/633 cm/s^2 = 0.34) than the Iwate-Miyagi earthquake could have triggered Unzen Mayuyama landslide. The steady-state shear strength was 157 kPa.

LS-Tsunami Simulation of the Unzen-Mayuyama Landslide-Induced Tsunami

The data set of landslide motion calculated by LS-RAPID at each 0.1 s was saved, and used as input to LS-Tsunami as the tsunami trigger. The LS-RAPID result is reported in Sassa et al. (2014a).

To calculate basal resistance in the sea, LS-Tsunamis required two additional parameters as the unit weight of seawater and the Manning's roughness coefficient. The Manning's roughness coefficient is not a tested value, but an empirical value used for analysis of water movement such as tsunamis and floods. Previous research (MLIT 2012; Tinti and Tonini 2013; Imamura 2009) showed the most commonly used value of Manning's coefficient for tsunami simulation is 0.025 m 1/3 s. The tsunami motion generated by the landslide motion was calculated by LS-Tsunami. The result of simulated tsunami height at each mesh and at each 0.05 s was saved. The simulated result is presented in Fig. 8.

At 0 min 20 s, the initial landslide motion was shown in the first figure on the top. The second figure on the top at 35 s presents the landslide mass continuously moving and

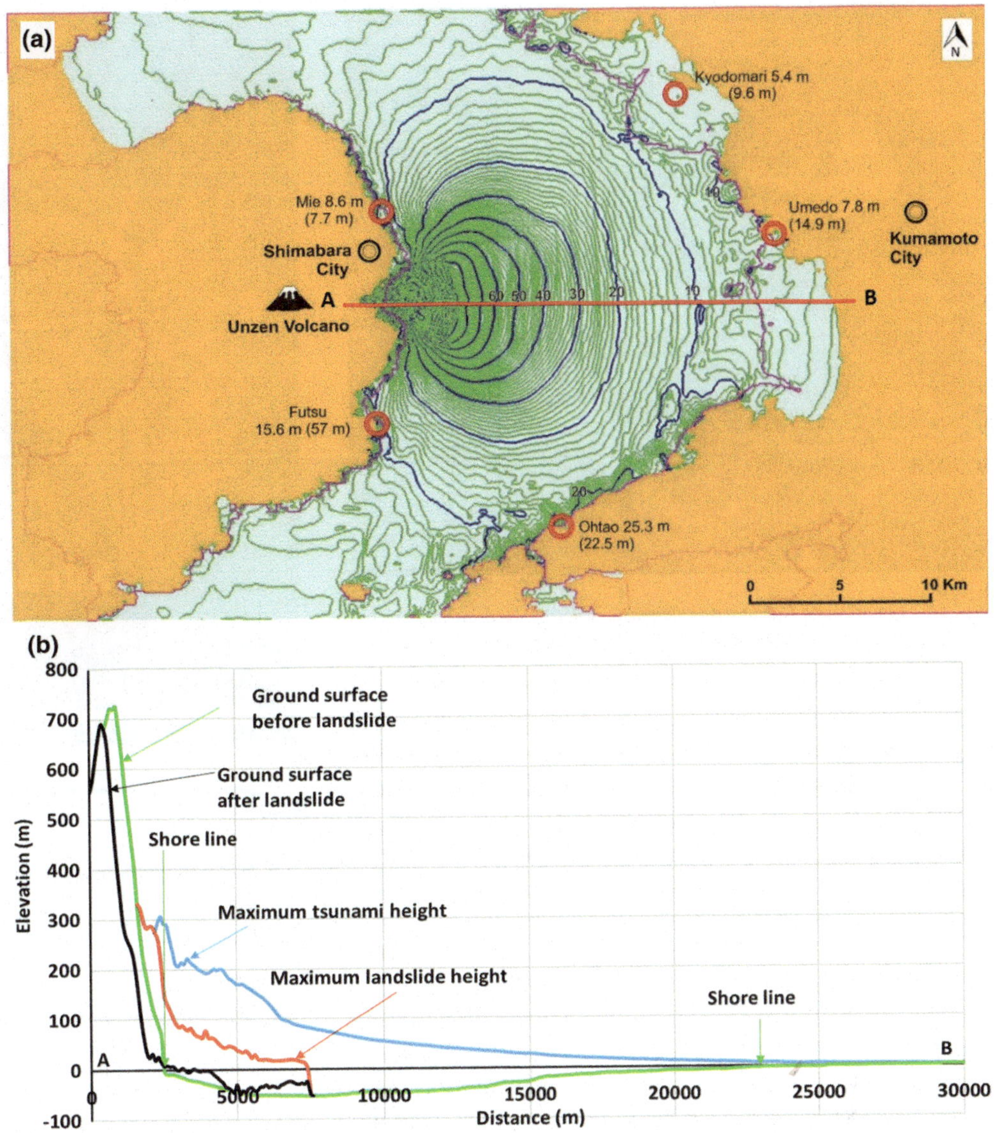

Fig. 10 Maximum tsunami height at each mesh obtained by LS-Tsunami: **a** presents contours of the maximum tsunami height at each mesh (one meter contours in green and 10 m contours in blue) and a comparison between the tsunami height by LS-Tsunami and the tsunami height in the historical record (within brackets) at five locations along the Kumamoto coast and the Shimabara coast. **b** Presents the profile of the maximum tsunami height at each mesh along the central section A–B

reaching the shoreline. Then, landslide mass entered into the sea and triggering the tsunami wave which was shown in the third figure on the top (1 min 25 s). The left figure (5 min 55 s) in the center shows that the tsunami wave was expanded into the Ariake Sea and landslide blocks (dark brown color dots) was viewed above sea level, as shown by a white arrow. The tsunami wave reached Kumamoto Prefecture on the opposite bank right at 10 min 45 s. The left figure (21 min 20 s) on the bottom illustrates that the tsunami wave was given back from the opposite bank to attack the southern part of Shimabara peninsula where around 3500 persons were killed by the Tsunami (Fig. 9). The right figure on the bottom at 29 min 55 s shows another returned wave attacking the northern side of Mayuyama landslide area. One red arrow at 5 m 55 s shows that the first direct tsunami wave (15–16 m) reached the Futsu town in the southern part of Shimabara Peninsula. Two yellow arrows at 21 m 20 s and 29 m 55 s show the returned waves reaching the Shimabara coast.

Comparison Between the LS-Tsunami Simulation Result and the Historical Record

The tsunami disaster reported by the Unzen Restoration Office (2002, 2003) is shown in Fig. 9. The highest points reached by the tsunami were marked by Tsunami-Dome-Ishi (tsunami stopping stones). Tsunami-Dome-Ishi are found in Kyodomari, in Umedo and in Ohtao along the coast of Kumamoto Prefecture, which were carefully investigated by Tsuji and Hino (1993). Tsuji and Murakami (1997) investigated tsunami heights by finding stone pillars which were built in the Shimabara Peninsula as a memory of the tsunami disaster. This figure was based on those investigations and

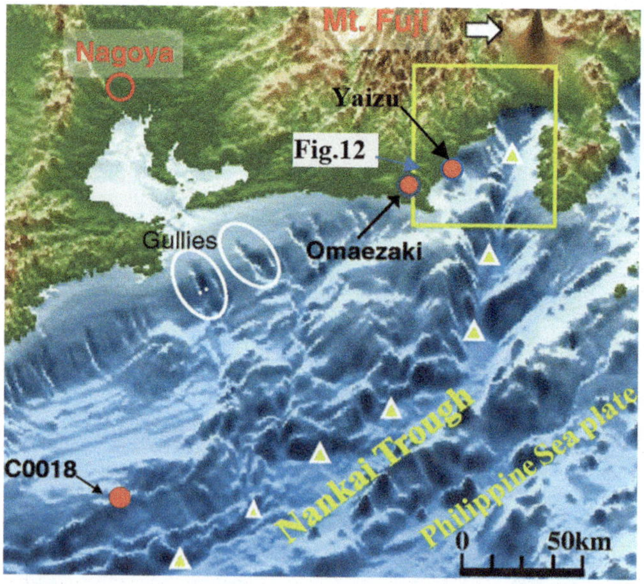

Fig. 11 Location of Suruga Bay (Yellow Square). Red circles are two sampling points. Modified from Sassa et al. (2012)

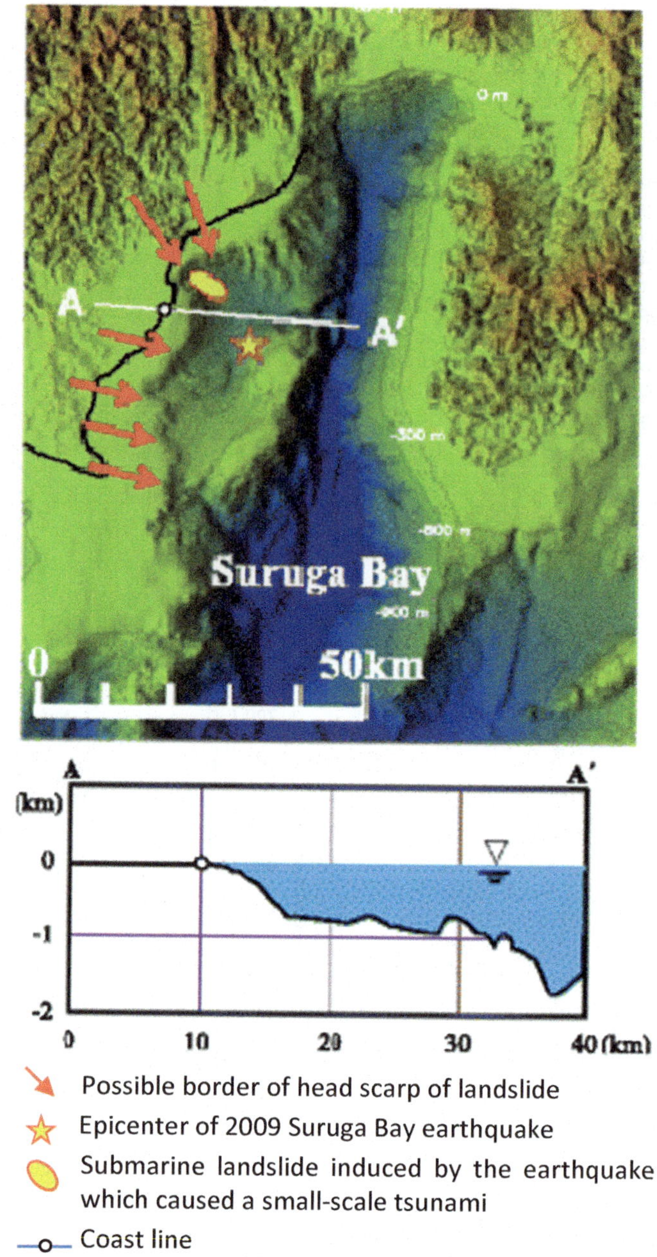

Fig. 12 Characteristic bathymetric of Suruga Bay area. Modified from Sassa et al. (2012)

includes the historical data as well as the stone pillars. The original image for Fig. 9 is a Japanese printed figure, which cannot be changed. English explanations have been added to the figure caption by adding A–D. The word of Shimabara-Taihen, Higo-Meiwaku in A means the "Catastrophe" in Shimabara area and "Annoyance" or "adversely affected" in Higo (Kumamoto) region in the opposite coast. This word "Shimabara-Taihen, Higo-Meiwaku" is often used in Japan for the cases in which something happens in the opposite half of the globe and the adverse effects strike Japan.

Futsu is a special case where the maximum tsunami height had a large difference between the simulation results and the observations. Due to the gulf-shaped coast line, the first direct tsunami wave attacking the cliff of the Futsu terrace probably was amplified. The tsunami wave flowed up to the terrace (Tsuji and Murakami 1997). Figure 9 at 5 m 55 s shows the instant the tsunami wave attacked Futsu. A return wave also struck Futsu from a different direction at 21 m 20 s. Around 3500 people were killed in this area. The bathymetric and topographic data at 100 m resolution used in the simulation of LS-Tsunami cannot include the effect of local amplification of tsunami height due to the sharp topography along this part of the coast.

The elevation of tsunami stopping stone on the slope is not always close to the average height of calculated mesh. In the 100 m resolution grip, if the coastal area has a 5–10° slope, the elevation difference over a horizontal difference of 50 m will be 4.3–8.8 m. Figure 10a includes the comparison of tsunami height by LS-Tsunami and the historical record by the local people except at Futsu. Considering slopes of 5–10°, it can be reasonably said that the estimated tsunami hazard area and the tsunami heights along the coast by LS-Tsunami are well matched with the historical records.

Figure 10b presents the ground surface before (green) and after (black) the landslide, the maximum landslide height during the motion (red) and the maximum tsunami height at each mesh (blue). The maximum tsunami height is around 300 m. It is the same level with the case of Senoumi hypothetical landslide (Fig. 21) and the potential retrogressive landslide A (Fig. 32).

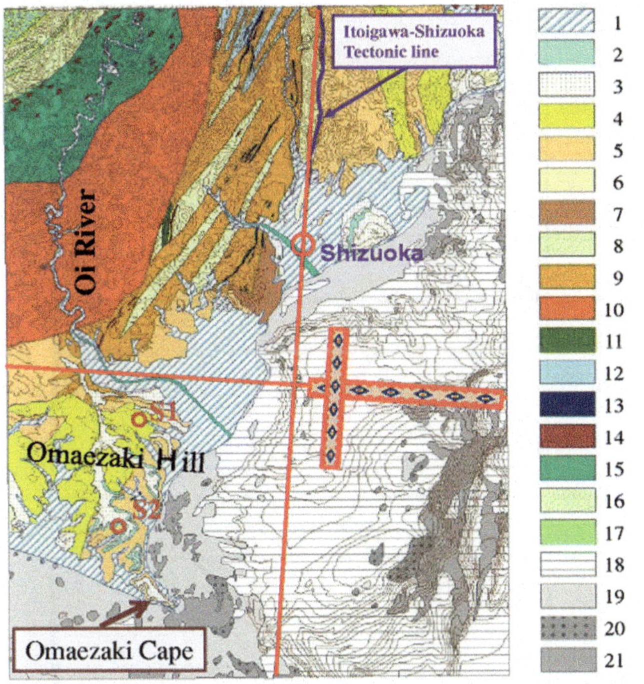

1: Holocene (Alluvial deposit); 2: Pleistocene silty deposits; 3: Pleistocene terrace gravels; 4:Pliocene (Kakegawa groups); 5: Upper Miocene (Sagara groups); 6: Lower Miocene; 7: Paleogene alkari basalt; 8: Oligocene; 9: Eocene to Ologocene; 10: Eocene; 11: Diabase; 12: Turbidity & Shale; 13: Limestone; 14: Green stone & Red chert; 15: Latest Cretaceous - Paleogene; 16: Late Cretaceous (Sandstone-rich alternation); 17: Late Cretaceous (Mudstone-rich alternation); 18: Silt; 19: Sand; 20: Gravel; 21: Rock.

Fig. 13 Geological Map in the Senoumi area and assumed shear bands [This map was redrawn based on the geological map (Tsuchi 2001), the bathymetric mesh data (Kishimoto 2000), possible extension of the Itoigawa–Shizuoka tectonic line and assumed two lines of shear bands and sampling points (S1 and S2)]

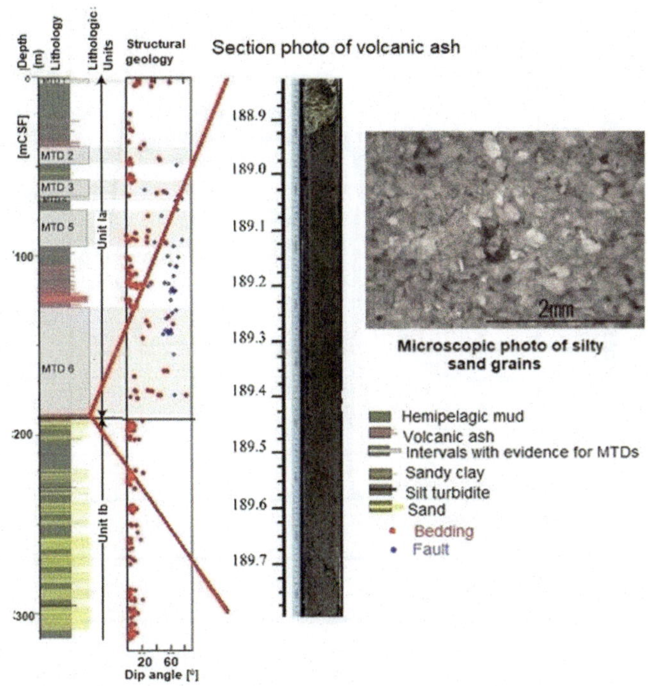

Fig. 14 IODP drilling and sample (Strasser et al. 2012)

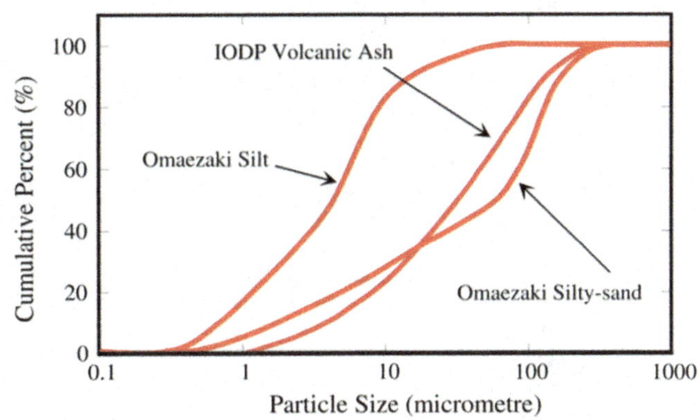

Fig. 15 Grain size distribution curves for three samples from the Senoumi submarine landslide (Sassa et al. 2012)

Ring-Shear-Testing and LS-Rapid Simulation of a Hypothetical Senoumi Submarine Landslide

Studying submarine mass movements and their consequences started in 1952 with the work of Heezen and Ewing on the Grand Banks slide and tsunami (Locat and Lee 2009). To date, many catastrophic submarine landslide events that have generated tsunamis have been reported including a submarine landslide at the Nice International airport in 1979 that caused the loss of human lives and important construction damages (Dan et al. 2007), the dramatic landslide-generated Papua New Guinea tsunami 1998, which devastated three villages with the loss of over 2200 lives (Tappin et al. 2008) and the landslide-induced tsunami by the Great Japan earthquake in March 2011. A more recent example is the tsunami that resulted from the magnitude Mw 7.5 2018 Sulawesi earthquake. The huge earthquake and tsunami hit property and life with more than 2000 fatalities. It surprised researchers as it followed an earthquake generated from a strike-slip rupture mechanism, which was not expected to generate catastrophic tsunamis (Omira et al. 2019). After the earthquake-induced submarine mass movement which generated the Papua New Gunea Tsunami, a renewed interest has grown in a topic now called tsunamigenic landslides (Bardet et al. 2003).

The 2011 Tohoku earthquake (Mw 9.0) triggered a powerful tsunami with a maximum tsunami height of 40 m. The destructive tsunamis caused major disasters along the north eastern coast of Japan. For preparing a similar mega-earthquake in the future, the tectonic motion in the Nankai trough has attracted attention. Some submarine landslides were found in the Nankai Trough (Cochonat et al.

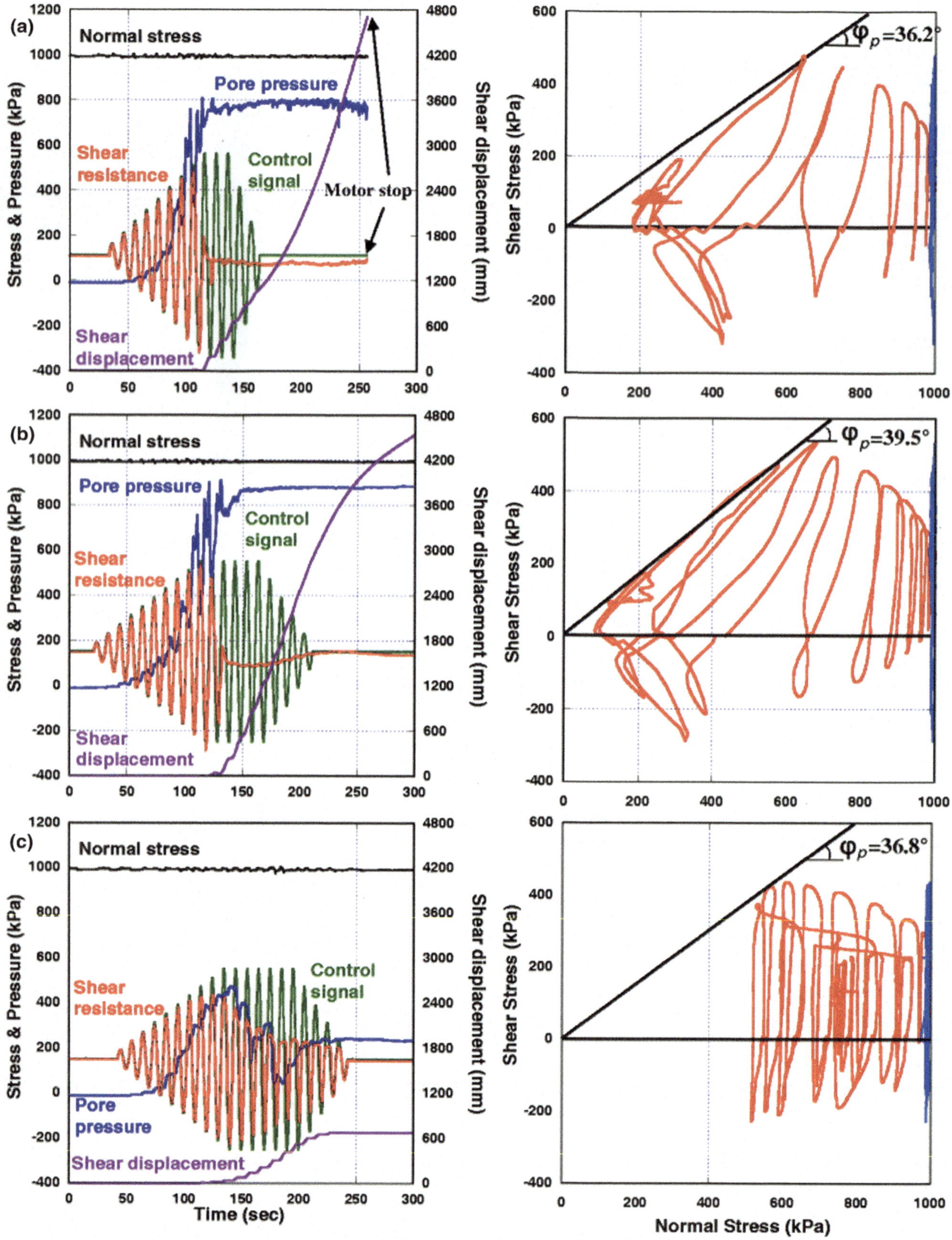

Fig. 16 Results from undrained ring shear tests on **a** IODP volcanic ash, **b** Omaezai Neogene sand, and **c** Omaezaki Neogene silt (Sassa et al. 2012)

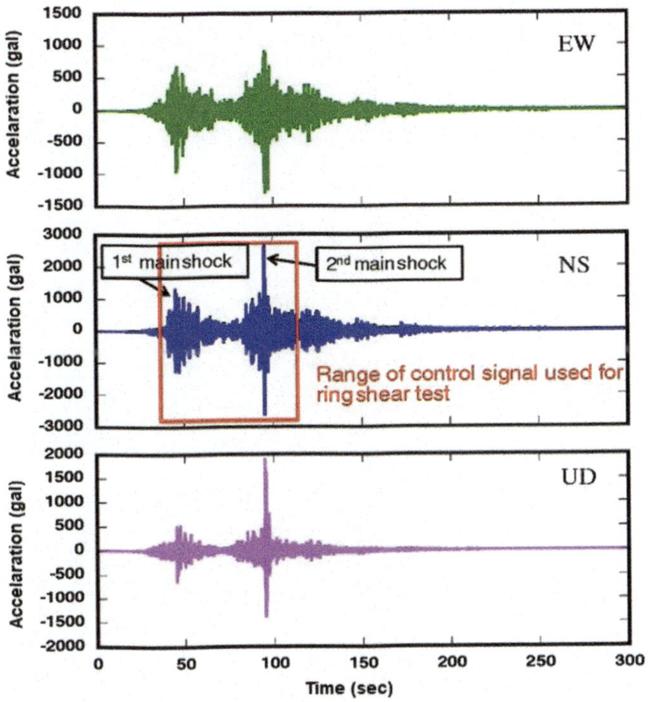

Fig. 17 2011 Tohoku earthquake record (MYG004)

2002; Strasser et al. 2011), which have not been well studied. The purpose of this section is to investigate the detailed features of Senoumi area in Suruga Bay that may be prone to submarine landslides. The Nankai Trough (Fig. 11) is a submarine trough, which is recognized as a plate boundary megathrust where the Philippine Sea Plate subducts under SW Japan (Ando 1975). Many submarine slope deformation features were found along the Nankai Trough between Kii Peninsula and Izu Peninsula from a 250 m mesh bathymetry of the sea floor. The authors selected the Senoumi area to study due to its close proximity to the densely populated Yaizu city which contains important transportation facilities (the Shinkansen line and the Tokyo–Nagoya highway).

Red arrows in Fig. 12 illustrate the possible head scarp of the Senoumi megaslide. A small submarine landslide triggered by the Mw 6.4 earthquake in 2009 and its epicentre is plotted in this figure. This small submarine landslide generated a small tsunami in Suruga Bay, as described in the next section. The lower part of Fig. 12 shows cross-section A–A′ that passes through a gully. The slopes from the head scarp to the toe range between 80 and 120. The shape of Senoumi submarine landslide was established under the following assumptions. Since the exit of the Suruga Trough is very narrow, the Senoumi features might be formed when the landslide mass is similar to a liquid and flows over a long distance with very small shear resistance before a retrogressive landslide occurs. In Fig. 13, the red line represents

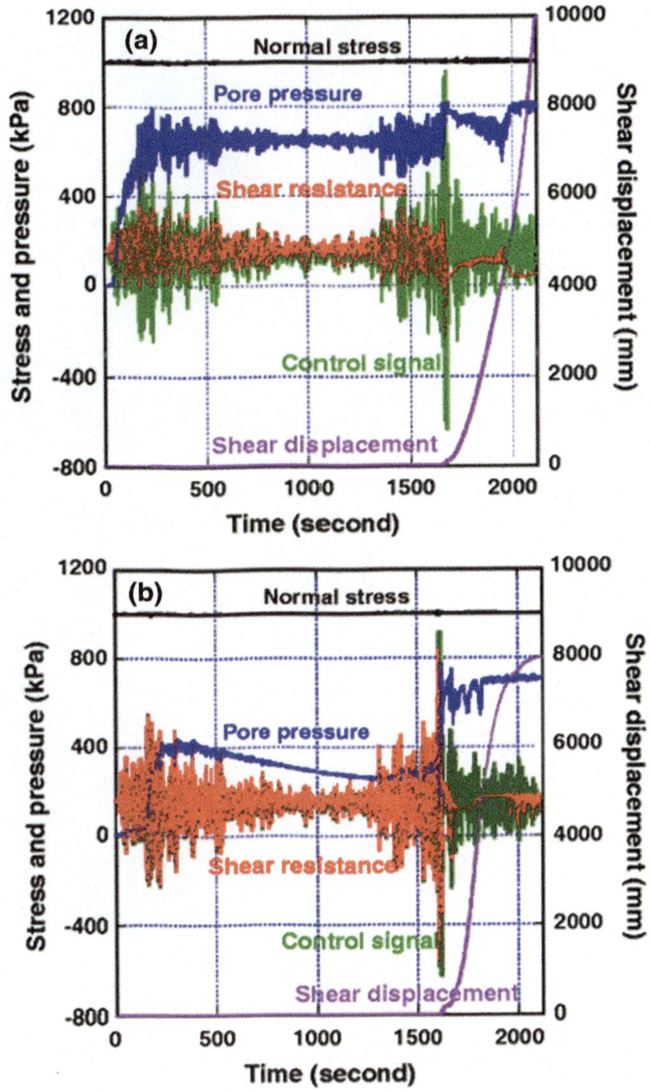

Fig. 18 Undrained seismic loading test results using the 2011 Tohoku earthquake record for **a** the volcanic ash and **b** Neogene sand. Modified from Sassa et al. (2012)

the Itoigawa–Shizuoka tectonic line. Shear bands in Senoumi area may develop parallel to this line. In addition, the Oi River flows in the same direction with this line but its shape was altered from the NS to the EW direction. Therefore, we believe that shear bands may exist along perpendicular to the Itoigawa–Shizuoka tectonic line. The geological map (Fig. 13) shows that the offshore area is primarily composed of silts, but also of sands, gravels, and rocks. Neogene sediments (yellow to dark yellow) covered the Omaezaki Hill, Izu Peninsula and eastern side of Itoigawa-Shizuoka tectonic region. Therefore, we assumed the soil layers of the Senoumi area before landslide are likely formed by Neogene deposits from Oi River, including volcanic ash.

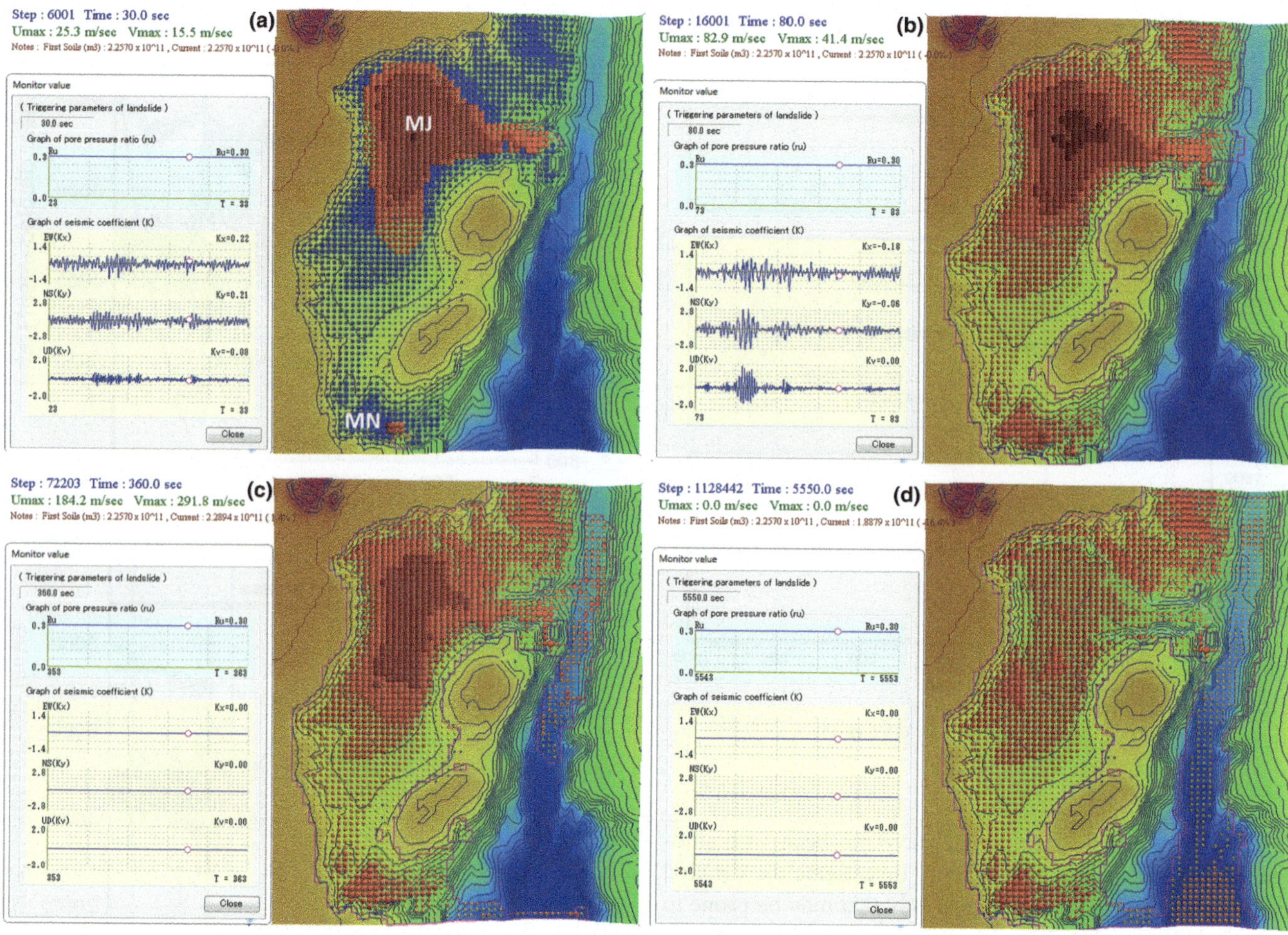

Fig. 19 Landslide simulation result for IODP volcanic ash when the 1.0 × Tohoku earthquake record (MYG004) was applied

Transportable Undrained Loading Ring Shear Apparatus ICL-1

This research was conducted before the development of the new ring shear apparatus ICL-2. Therefore, the ring shear apparatus, ICL-1, was used for this study. ICL-1 is a transportable and compact high stress undrained loading ring shear apparatus that is a part of Science and Technology Research Partnership for Sustainable Development (SATREPS) program for Croatia. The maximum capacity of ICL-1 is 1000 kPa.

Testing Samples

Three samples collected in this study included submarine volcanic ash, inland Neogene silty-sand, and a sample from the silt deposit. Additional details are provided below.

Submarine Deposits in Nankai Trough

Submarine samples were taken from core samples at Site C0018 (Fig. 11). A Pleistocene to Holocene succession of a slope basin on the accretionary prism of Nankai Trough was investigated and cored during the Expedition 333 of The Integrated Ocean Drilling Program (IODP) in December 2010 (Strasser et al. 2012). Figure 14 presents six mass transport deposits (MTDs) with the thicknesses up to 60 m. The volcanic layer, which is located about 188–189 m below sea floor, was used for the Senoumi submarine landslide. A section and microscopic photo of this layer are shown in Fig. 14.

Neogene Deposits in the Omaezaki Hill

According to Japan Landslide Society (1996), landslides often take place in Neogene deposits, which are found in the Omaezaki hill (Fig. 13). The 2004 Mid-Niigata Prefecture Earthquake that triggered landslides in Neogene sediments

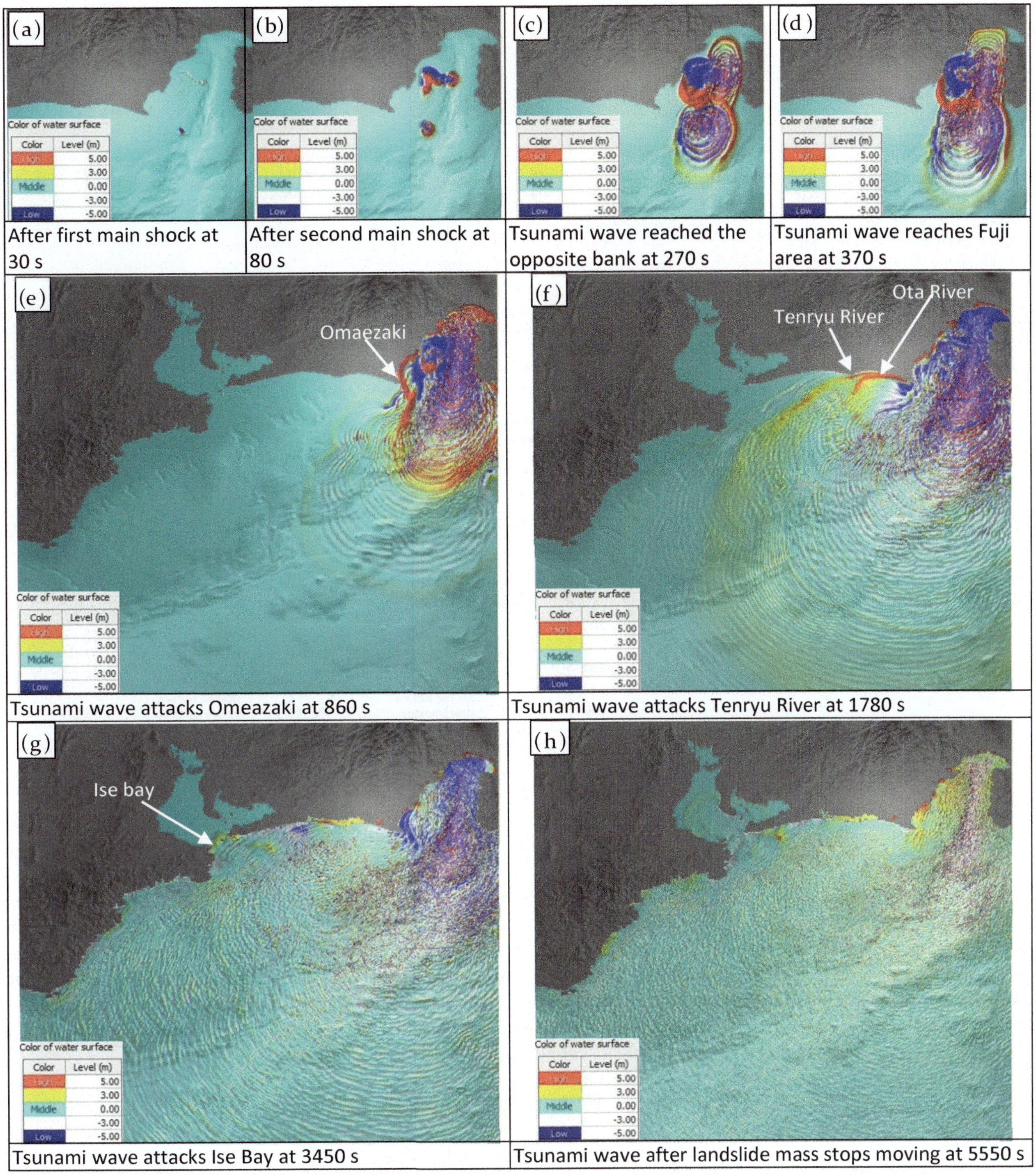

Fig. 20 LS-Tsunami simulation results when considering the hypothetical Senoumi landslide in the IODP volcanic ash triggered during the 1.0 Tohoku earthquake record (MYG004). In this figure, the red, yellow, light blue, white and dark blue represent the wave heights corresponding to more than 5 m, 3 to 5 m, 3 to −3 m, −3 to −5 m and less than −5 m, respectively

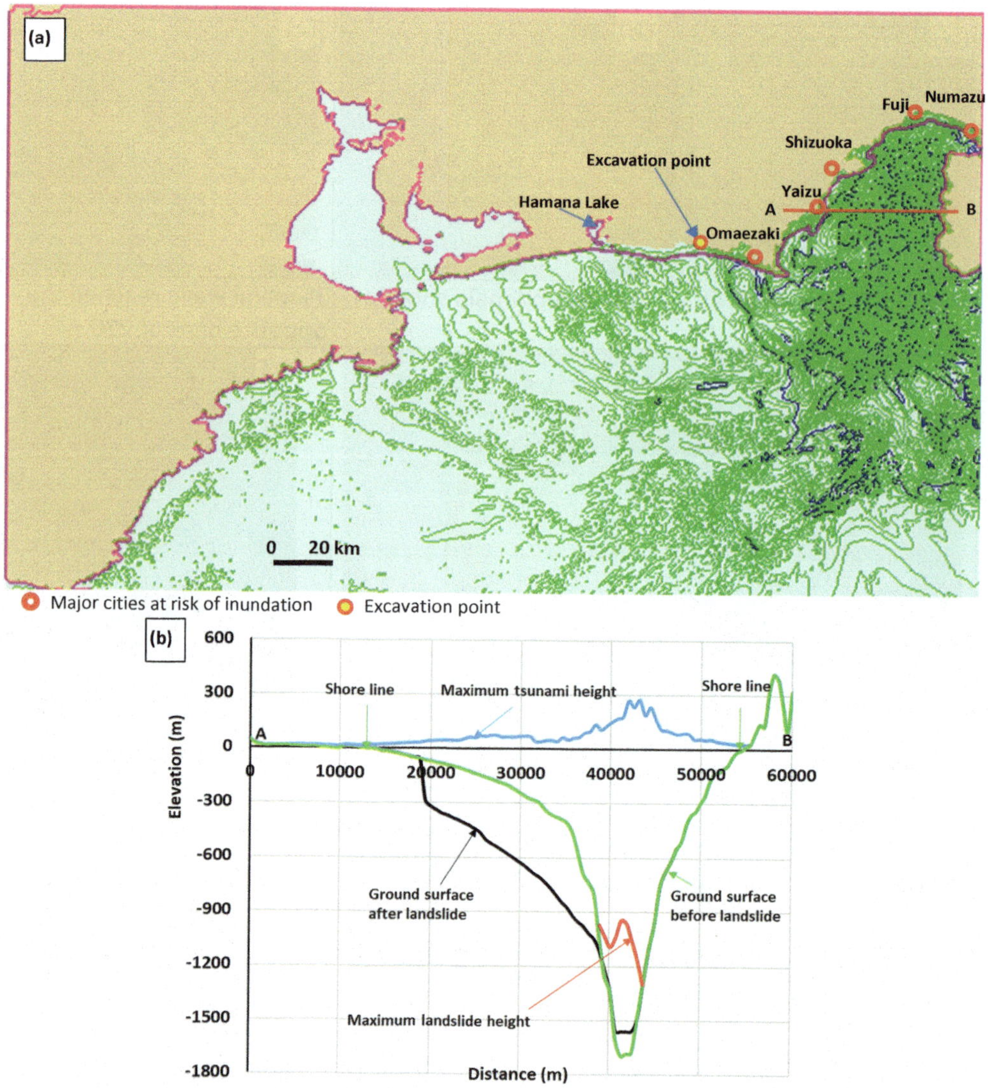

Fig. 21 Maximum tsunami height caused by the hypothetical Senoumi landslide triggered in the IODP volcanic ash by 1.0 Tohoku earthquake record (MYG004) at each mesh. **a** presents the contour of the maximum tsunami height at each mesh (2 m contours are in green, while 10 m contours are in blue). **b** presents the profile of the maximum tsunami height at each mesh along the central section A–B

was reported by Sato et al. (2005) and Sassa et al. (2005). A large shear strength reduction in the Neogene sand sample was found from their results, which can indicate the occurrence of rapid moving, earthquake-induced landslides. Neogenen-sand (S1) and silt (S2) were taken from Omaezaki Hill for the Senoumi submarine landslide study area. This was based on the assumption that a continental shelf will be formed with the same materials which are found in the onshore area.

Undrained Ring Shear Tests

Initial Condition

The total unit weight of submarine soil layer was 18.6 kN/m^3 and the unit weight of sea water was 10.1 kN/m^3 (Expedition 333 Scientists, 2012), resulting in a normal stress of 1000 kPa (the maximum capacity of the ICL-1) at a burial depth of 117 m. Although this depth was not close to the depth of volcanic ash sample (around 189 m), we decided to apply 1000 kPa as initial normal stress since it is the maximum capacity of this apparatus. The initial shear stress was chosen as 160 kPa to reproduce a slope of $\tan^{-1}(160/1000) = 9.09°$ (corresponding to expected slope angles from 8° to 12°).

Undrained Cyclic Loading Tests

Figure 16 shows the result of cyclic tests on IODP volcanic ash and Neogene sand and silt. The shear stress was gradually varied at each cycle until it reached 400 kPa (large enough to cause failure). The cyclic load was applied at a frequency of 0.1 Hz. The green line shows the control signal of shear stress, while the black, red and blue colored lines present the normal stress, mobilized shear resistance and generated pore water pressure, respectively.

The results demonstrate several findings as follows:

Fig. 22 **a** Location of tsunami deposit investigation conducted by Fujiwara and **b** the main survey line. Modified from Fujiwara et al. (2019)

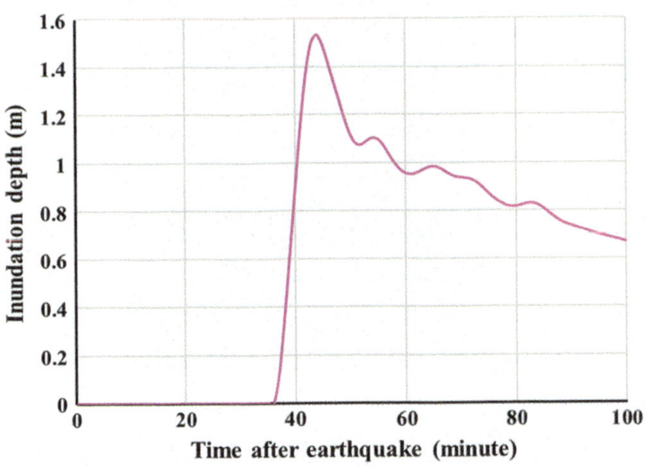

Fig. 23 Simulated tsunami wave form at the excavation point caused by the hypothetical Senoumi landslide in IODP volcanic ash during the 1.0 Tohoku earthquake record (MYG004)

Figure 16a shows high pore water pressures were generated during cyclic loading in the volcanic ash. The steady state resistance was 72 kPa and is the lowest value of the three samples. The peak friction angle reached 36.2° and the pore pressure ratio was 0.80.

Although a high pore water pressure ratio of 0.85 was observed in the Neogene sand, the shear speed was increased only during cyclic loading, but it slowed down after the stop of cyclic loading. The steady state resistance was close to the initial shear stress of 160 kPa. Therefore, no acceleration at the final stage was found. The peak friction angle was 39.5°, which is the highest value among these three samples.

During cyclic loading the pore water pressure ratio in the Neogene silt was small when compared to the other two samples. The peak friction angle was 36.8°. After failure, shear displacement occurred with increasing cycles and was stopped at around 600 mm.

Undrained Seismic Loading Ring Shear Test

The Cabinet Office of the Government of Japan is investigating the risk of a mega-earthquake in the Suruga trough in the future. We used the seismic wave at station MYG004 at Tsukidate in Miyagi Prefecture (Fig. 17), which is the largest recorded earthquake acceleration, to examine the mega-earthquake triggered submarine landslide on the volcanic ash and Neogene sand samples.

From the cyclic loading test, all samples failed before the additional shear stress reached 400 kPa. Therefore, we decided to apply 0.3 times of MYG004 (2699 gal in NS component wave), which corresponded to a maximum additional shear stress of 810 kPa (2699 × 0.3 × 1000) in the seismic loading ring shear test. Since the capacity of servo-shear control motor of the ICL-1 is limited, the seismic acceleration record that was 30 times slower in speed was applied. Figure 18 shows the seismic loading test results for the IODP volcanic ash sample (top figure) and Neogene sand sample (bottom figure). During the first main shock, high pore water pressures was generated in the IODP ash while the Neogene sand generated smaller pore pressures. The shear displacement rapidly increased in the second main shock which reproduced the occurrence of rapid landslide on 9.09° slope. The shear displacement was seen to decelerate when additional smaller shear stresses were applied on the Neogene sand. In Fig. 18b, the control signal line and shear resistance line overlapped until the second main shock suggesting that the Neogene sand did not fail during the first main shock.

LS-RAPID Simulation Result for Senoumi Area

The soil parameters used in this computer simulation were obtained from ring shear tests, as detailed in Sassa et al.

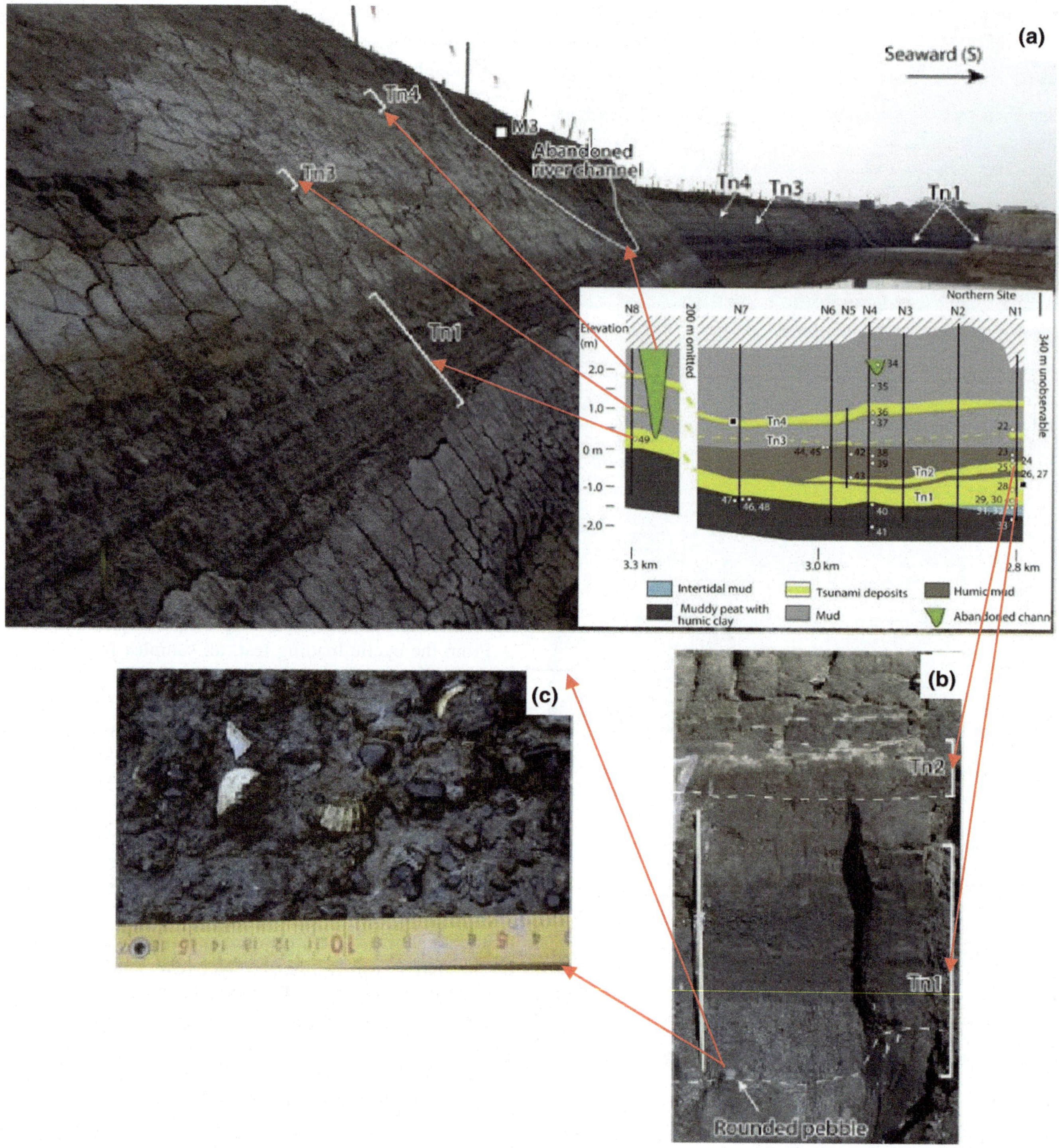

Fig. 24 Photograph of tsunami deposits at the northern site. Modified from Fujiwara et al. (2019). **a** Overview of the excavated walls with excavation heights of about 4 m at site N8. Unconsolidated relatively coarse-grained beds are exposed as grooves along the excavated walls over hundreds of meters. Bed Tn2 was not identified at this location due to its distance from the coast. **b** At Site N1, photograph showing beds Tn1 and Tn2. **c** Close up of lower part of bed Tn1 at Site N1 showing rounded pebbles and marine shell fragments

(2012). Table 2 present the parameters for the LS-RAPID computer simulations.

During an earthquake, a seismic load is applied as detailed in Sassa et al. (2010). The three components of the 2011 Tohoku Earthquake ground motion (Fig. 17) were inputted to LS-RAPID. Although the distances from the epicenter are almost the same, the recorded accelerations at MYG004 (2933 gal) and at FKS009 (810 gal) are very

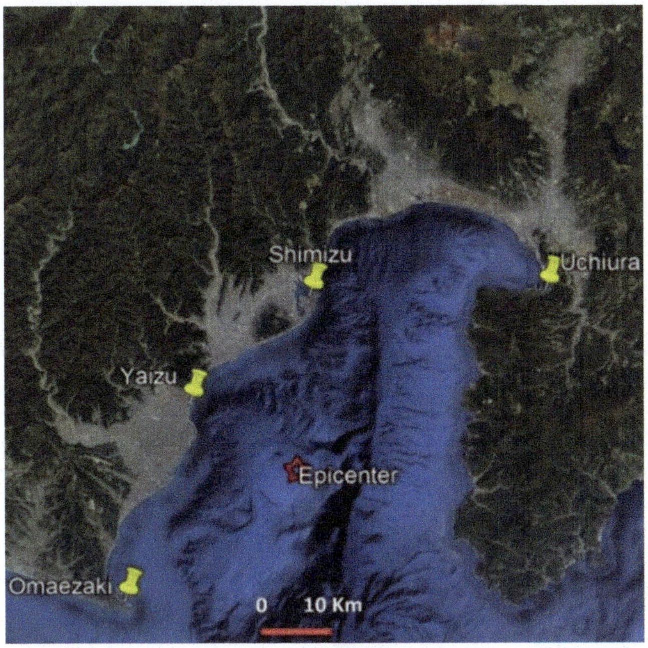

Fig. 25 Location of tide gauge stations

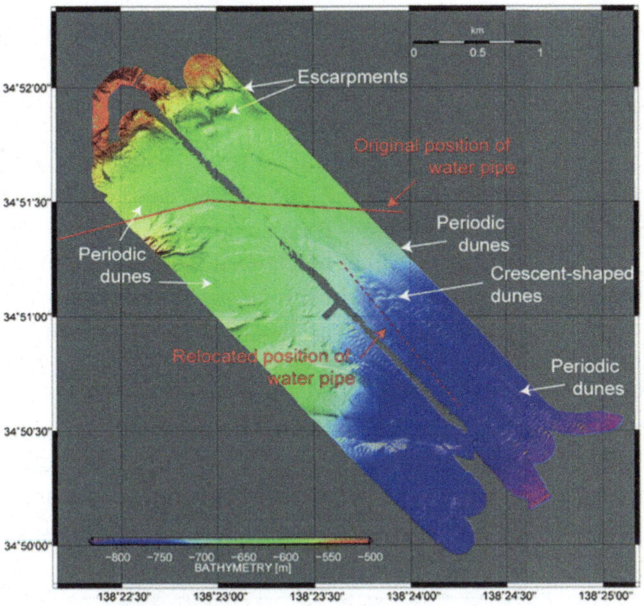

Fig. 26 Micro-bathymetric map obtained from data recorded by the AUV Urashima showing escarpments and bed forms (Baba et al. 2012)

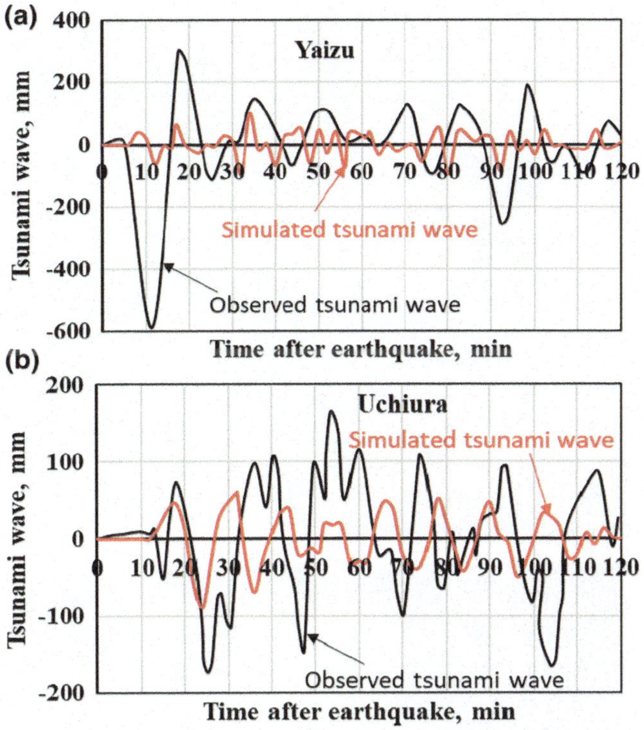

Fig. 27 Comparison of the sunami wave resulting from only the fault motion and the observed tsunami wave at: **a** Yaizu station and **b** Uchiura station

different. The wave form of MYG004 was used, but acceleration amplitudes were multiplied by 1.0, 0.7, 0.4, and 0.3. Another triggering factor applied in this simulation was the increase in the excess pore pressure ratio to 0.3. Table 3 shows the results related to the triggering of the landslide. In the volcanic ash, the record corresponding to 0.3 × MYG004 could cause a landslide while no landslides were triggered in the Neogene silt. Only a minor landslide (MN) was triggered by 0.3 × MYG004 record for IODP Neogene sand. Figure 19 illustrates the simulation results for volcanic ash using the record corresponding to 1.0 × MYG004. A major landslide (MJ) occurred soon after the first shock (Fig. 19a). Soon after the second shock, both MJ and MN were subjected to flow with the velocities more than 0.5 m/s. The landslide mass, then, moved down along the Suruga trough (Fig. 19c). "D" corresponods to a time of 5550 s, at which point the entire landslide mass had stopped moving.

Tsunami Simulation Result of the Senoumi Hypothetical Landslide for IODP Volcanic Ash with the Application of the 1.0 × Tohoku Earthquake Record (MYG004)

The most dangerous scenarios corresponded to the IODP volcanic ash sample when using the 1.0 × Tohoku earthquake record. Thus, we decided to use the landslide motion from this scenario for the tsunami simulation. Manning's roughness coefficient of 0.025 was used in the LS-Tsunami model as an additional parameter (Sassa et al. 2016). The simulated result is presented in Fig. 20. Figure 20a at 30 s shows the tsunami soon after the first shock. It appears to be only a local tsunami occurring in Suruga Bay. Figure 20b at

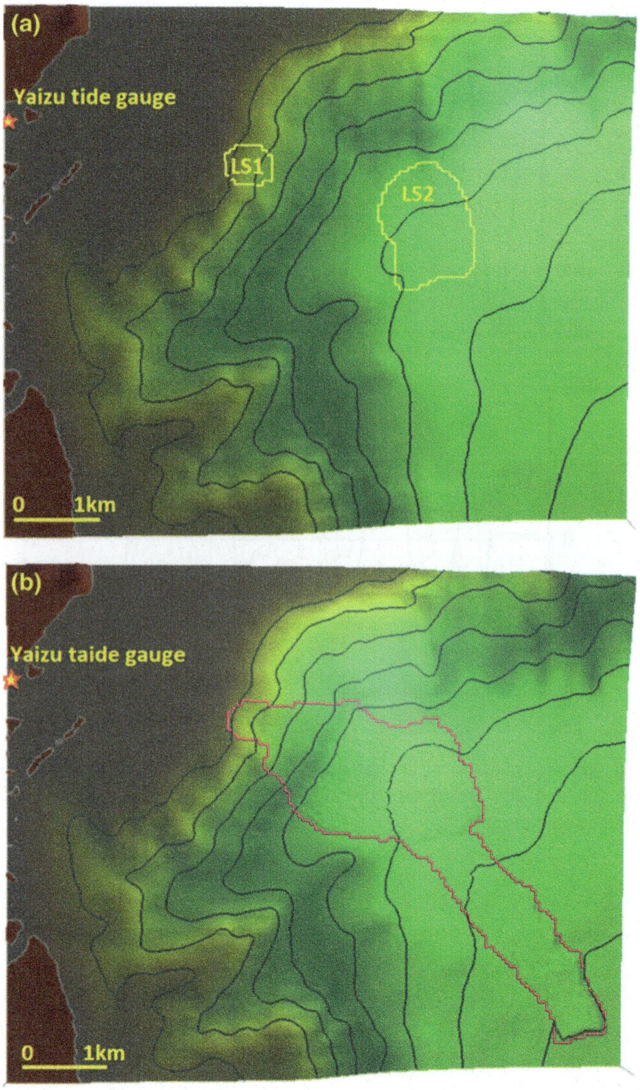

Fig. 28 LS-rapid simulation area (100 m contour). **a** presents the location of two landslides (LS1 and LS2). The yellow lines are the periphery of the landslide source areas. **b** contains the border of the landslide moving area after the termination of movement (red line). Two landslide masses were combined during motion

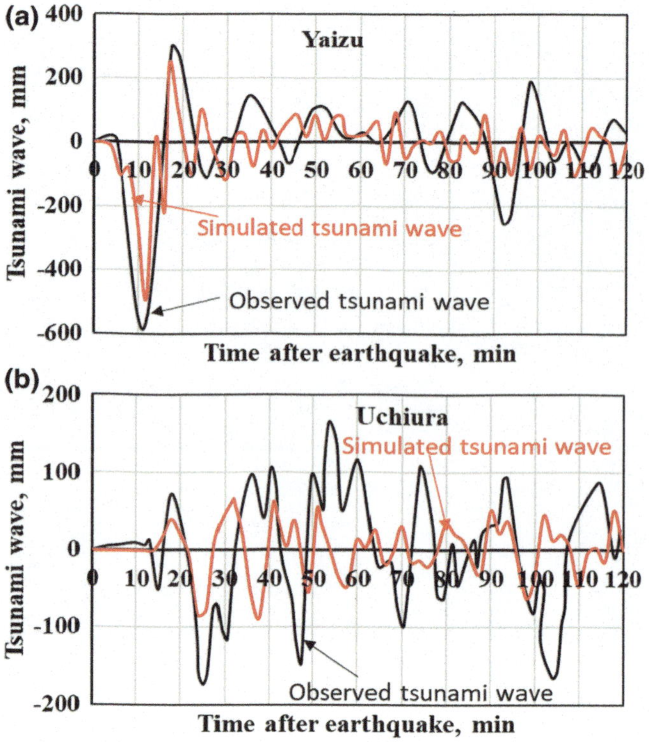

Fig. 29 Comparison of the simulated tsunami wave (red line) induced by the landslide motion at: **a** Yaizu station and **b** Uchiura station and observed tsunami wave (black line)

80 s showed the tsunami soon after the second shock. The negative tsunami heights are shown in blue and positive tsunami heights are in red.

Tsunami waves expanded onto the opposite bank (Fig. 20c). Figure 20d illustrates that the tsunami wave will hit the Fuji area, which is the third largest city in terms of population in Shizuoka Prefecture. Omeazaki city was attacked by tsunami wave at 860 s (Fig. 20e). Figure 20f (1780s) shows that the tsunami wave moved downward to attack Tenryu River, which is Japan's ninth longest river. The right figure on the bottom at 5550 s shows the tsunami wave when the landslide stopped moving.

Figure 21a shows the maximum tsunami height at each mesh. The contour lines at 2 m are very dense inside the Suruga Bay. Cross-section AB illustrates the tsunami height (blue line) and the landslide height (red line) at each mesh. The maximum tsunami height was close to 300 m, while the maximum landslide height at the bottom of the Suruga trough was around 750 m. All cities (Yaizu, Shizuoka, Fuji, Numazu and Omaezaki) around the Suruga Bay should have been attacked by tsunami wave generated from the hypothetical Senoumi landslide. However, no historical records or tsunami deposits were found. This was one of the reasons we named the Senoumi submarine landslide as hypothetical.

Finding Tsunami Deposits Along Ota River, Shizuoka Prefecture

Fujiwara and his colleagues conducted a tsunami deposit investigation during the river improvement project. Figure 22a presents the Otagawa Lowland, which is located to the east of the Tenryu River. The excavation site is about 3 km from the shoreline. Due to its location, fine-grained soils were deposited in this area. Figure 22b shows the main survey lines in the northern site (N1–N8).

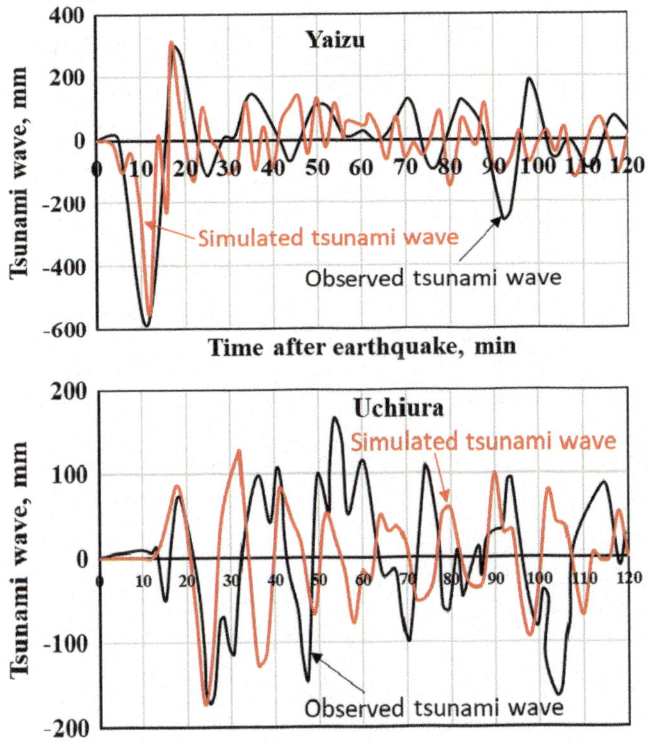

Fig. 30 Comparisons between the simulated tsunami wave (red line) induced by motions from both the landslide and the fault at: **a** Yaizu station and **b** Uchiura station and the observed tsunami wave (black line)

Figure 23 shows the inundation depth as a function of time at the excavation point in Ota River as calculated by the LS-Tsunami simulations. The inundation depth is calculated by subtracting ground elevation from tsunami height. The maximum inundation depth at this point was close to 1.5 m, suggesting that the tsunami deposits found in the lowland of the Ota River may have been deposited by the tsunami triggered by the Senoumi landslide. Tsunami waves are caused by motion of both of the fault and the landslide during big earthquakes. The example of combined effects was examined for the case of 2009 Surga Bay earthquake (Figs. 25, 26, 27, 28, 29 and 30).

Nine mega-earthquakes occurred in the late seventh century and in 887, 1096, 1361, 1498, 1614, 1707, 1854 and 1944 CE in Nankai trough. The last recurrence interval is only 90 years. It was reported in Fujiwara et al. (2019) that four tsunami deposits were clearly found in Ota River and the two younger deposits are related to the 1096 and 1498 earthquakes. From northern site cross sections (Fig. 24), four potential deposit layers named Tn1, Tn2, Tn3, and Tn4 were found, of which Tn1 is the thickest layer. In the southern site, three tsunami deposit layers (Ts1, Ts2, and Ts3) were also detected with Ts1 being the thickest. The deposit layers in the northern site were more visual than those in the southern site.

Fig. 31 Shapes and cross sections of hypothetical Senoumi landslide (HSL) and potential retrogressive landslides A and B (PRLA and PRLB, respectively)

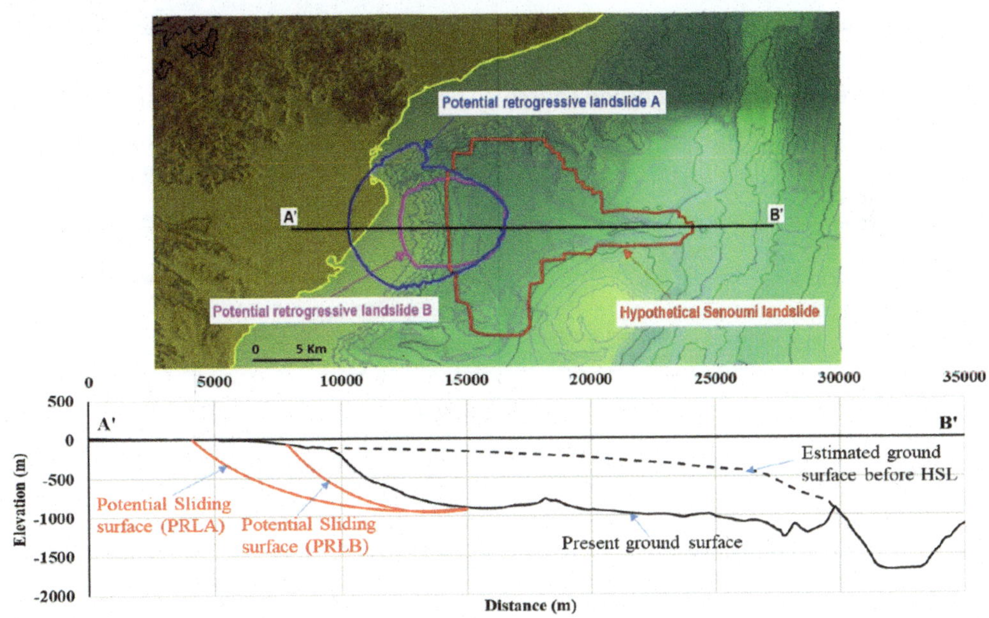

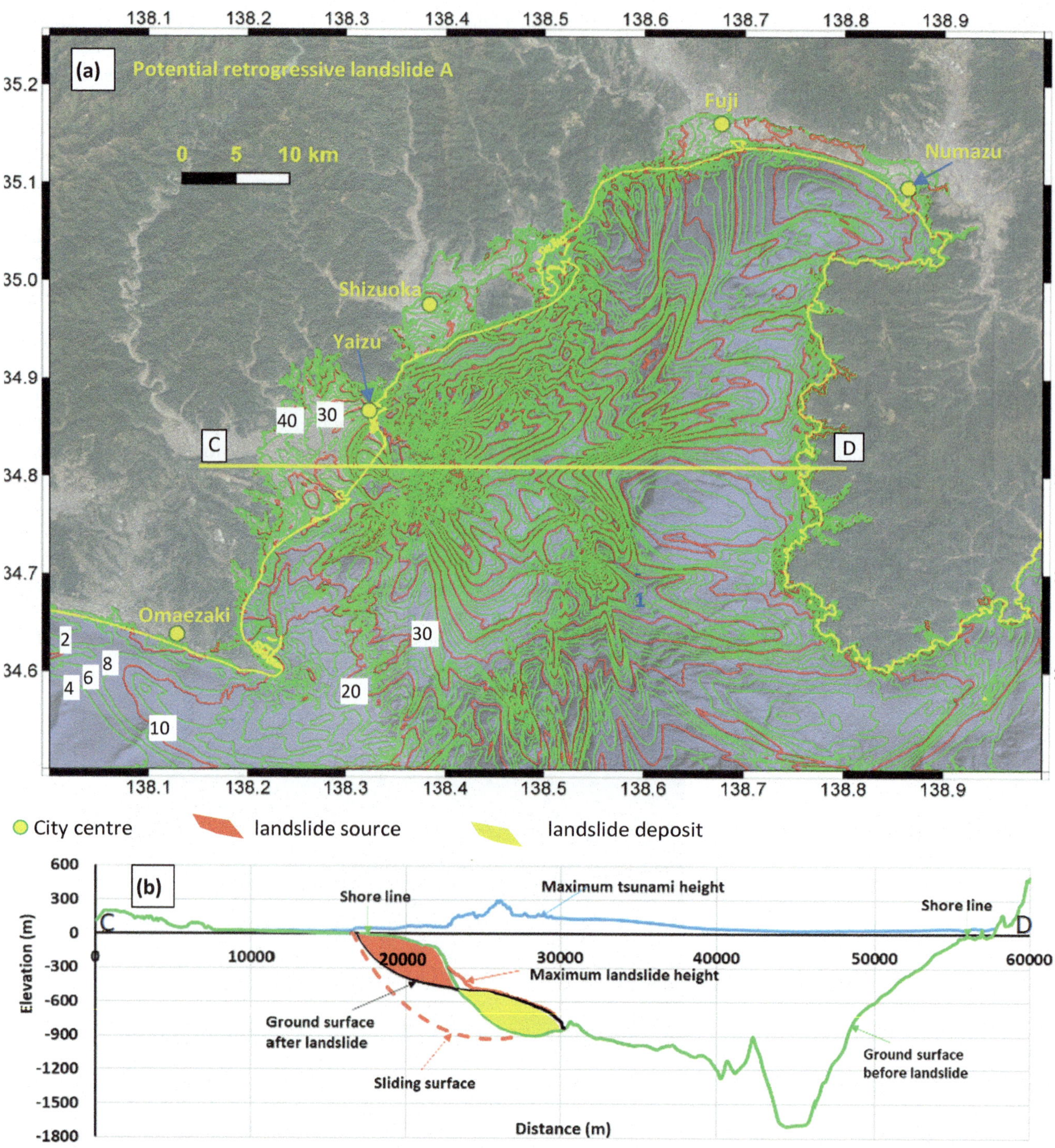

Fig. 32 Maximum tsunami height caused by potential retrogressive landslide A in Neogene sand triggered by 0.7 Tohoku earthquake record (MYG004). In this figure, tsunami height are in meters above sea level. **a** presents the contour of the maximum tsunami height at each mesh (with 2 m contours shown in green and 10 m contours in blue). **b** presents the profile of the maximum tsunami height at each mesh along section C–D

Detailed stratigraphy is presented in Fig. 24. Four potential sandy tsunami deposits were recognized from the excavation site. An overview of the excavated walls from site N8 is illustrated in Fig. 24a. At this site, Tn2 was not found. Two tsunami deposit layers (Tn1, Tn2) were identified at site N1 (Fig. 24b). Close examination of the lower part of the Tn1 bed at site N1 revealed rounded pebbles and marine shell fragments as shown in Fig. 24c. Tn1–Tn4 tsunami deposits were possibly deposited between the late seventh century and 887, 1096, 1361, 1498, 1614,

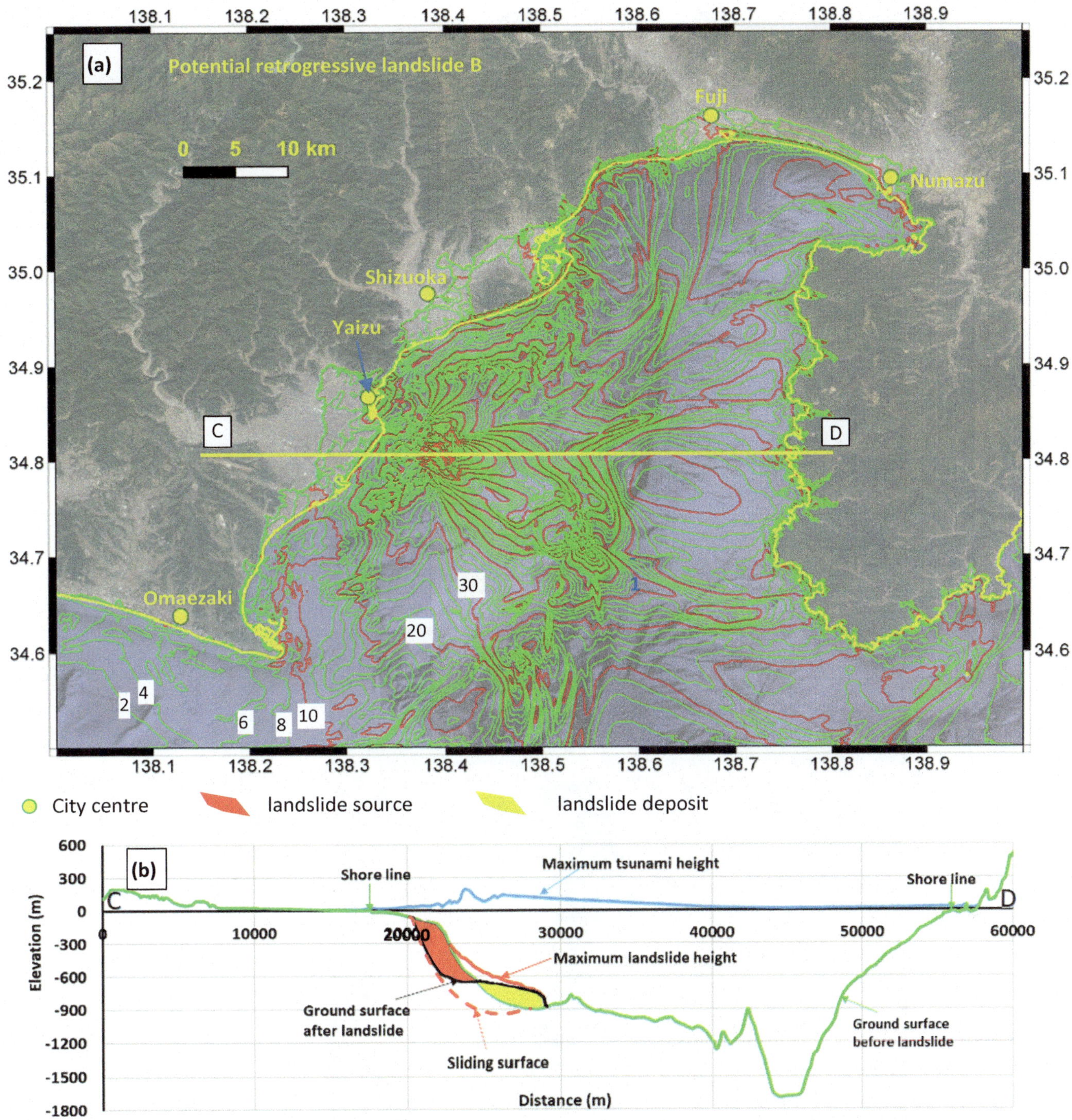

Fig. 33 Maximum tsunami height caused by potential retrogressive landslide B in Neogene sand triggered by 0.7 Tohoku earthquake record (MYG004). In this figure, tsunami height are in meters above seal level. **a** presents the contour of the maximum tsunami height at each mesh (with 2 m contours shown in green and 10 m contours in blue). **b** presents the profile of the maximum tsunami height at each mesh along section C–D

1707, 1854 and 1944 (Fujiwara 2019) earthquakes. A certain tsunami hazard in the coastal area should have occured in this area and at those times. However, there is no historifcal record of tsunami disaster similar to the 1972 Unzen disaster case, probably because the tsunami wave was not high and it impacted agricultural land or coastal forests without many houses. Whereas the actual origin of the tsunami deposits is not always known, submarine landslides are often considered as the main trigger causing a large tsunami. The enormous submarine landslides with volumes of several

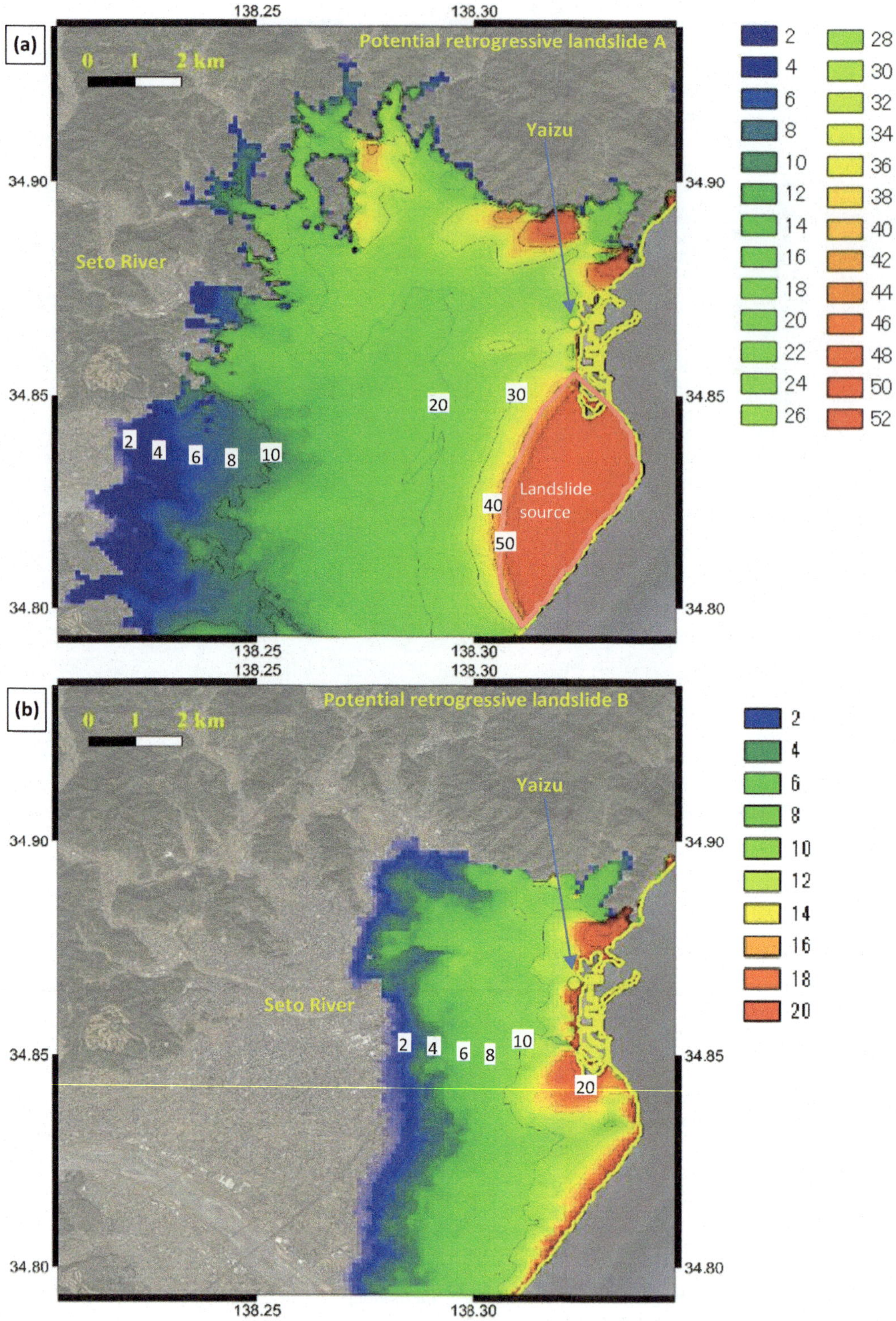

Fig. 34 Inundation map at Yaizu City center caused by **a** potential retrogressive landslide A and **b** potential retrogressive landslide B in Neogene sand triggered by 0.7 Tohoku earthquake record (MYG004). Inudation depth (color scale in meter) = maximum tsunami height—ground elevation

thousand to tens of thousands of cubic kilometers may cause tsunamis with large run-up heights. For example, the giant Storregga slide generated a tsunami that hit the west coast of Norway (with run-up heights of 10–12 m), Scotland (4–6 m), Shetland (20–30 m) and the Faroes (>10 m) (Bryn et al. 2005). The hypothetical Senoumi landslide also can

Table 1 Relation between the steady-state shear resistance and tsunami behavior. Modified from Sassa et al. (2016)

Landslide Volume (m^3)	τ_{ss} (kPa)	v_p (m/s)	Tsunami traveling time (s) to 10 km from the coast	Maximum tsunami height (m) in Fig. 3
A: 5.8×10^6	50	42.8	215	20.0
	200	38.0	230	9.2
	400	13.3	250	2.2
B: 170×10^6	100	55.7	180	62.0
	1000	44.7	225	23.4
	1250	30.2	255	5.9

cause a tsunami that could cause widespread effects in the Suruga Bay could resulting in the deposition of materials at the excavation point.

Reproduction of the 2009 Earthquake-Triggered Submarine Landslide-Induced Tsunami in Suruga Bay, Japan as an Evidence of the Reliability of LS-Tsunami

The M_w 6.4 Suruga Bay earthquake took place at 05:07 JST, 11 August 2009. The earthquake triggered the embankment failure of the TOMEI expressway caused 5 days of traffic jams. Some routes were also affected due to other slope failures. The Yaizu deep-sea water facility was subjected to extensive damage. A deep-sea water pipe extending to the depth of 687 m was ruptured (Baba et al. 2012) and moved 2 km downslope from its original location by the submarine mass movement. This small landslide was investigated by the Autonomous Underwater Vehicle Urashima of JAMSTEC (Baba et al. 2012). A tsunami generated by the 2009 Suruga Bay Earthquake was observed at the Yaizu tidal gauge station, and the Shimizu, Uchiura and Omaezaki wave-meter stations (Fig. 25). Observed tsunami wave forms from the 2009 Suruga Bay at these four stations were corrected by Satake et al. (2010). From these results, the tsunami wave at Shimizu is much smaller than the tsunami waves recorded at the Yaizu station. This data may be have affected by ocean wave. We selected two stations: Yaizu station (close to the landslide source area) and Uchiura station (on the opposite bank) to check the reliability of the LS-Tsunami simulations. Baba et al. (2012), based on the studies by Satake (2001) and Abe et al. (2008), simplified the real landslide movement as the vertical downward movement of a plate in the source area and the vertical upward movement of another plate in the deposition area. From this research, the recorded tsunami wave height was much higher than the one obtained in the simulation. The height in the simulation was estimated using only the fault model, while the wave form was simulated considering the effects of both the fault movement and the submarine landslide. These results matched fairly well with the observed tsunami wave heights. A limitation of the previous research was not using a landslide dynamic model to simulate the submarine landslide- induced tsunami. In order to rectify this problem, we decided to use LS-Rapid to reproduce the submarine landslide motion, using which LS-Tsunami was utilized to simulate the tsunami wave.

Evidence of Submarine Landslide

Bathymetric data were acquired from the remotely controlled autonomous underwater vehicle (AUV) Urashima, which is the world's largest autonomous deep-sea exploration robot (Aoki et al. 2008). The bathymetric data at about 1 m intervals can be obtained from Urashima using a multi-beam echo sounder. Baba and his colleuges at Japan Agency for Marine-Earth Science and Technology (JAMSTEC) conducted two investigations in Suruga Bay using Urashima on February 27 and March 1, 2010. Figure 26 shows the micro-bathymetric map of Suruga Bay. The red line is the water pipe which was moved about 2 km down slope from its original location to the red dashed line by the submarine landslide. The upperpart of the figure presents two escarpments with a width of 450 and 10–15 m deeper than the adjacent seafloor (Baba et al 2012). It is morphological evidence of submarine landslides. The landslides took place on a slope which is about 1 km north of the original position of the water pipe. The landslide material flowed at least 3–4 km and resulted in periodic dunes.

Tsunami Simulation

The tsunami triggered by only the fault motion was simulated using the crustal deformation data. Table 4 presents the input data for the fault motion model. The tsunami wave results at Yaizu and Uchiura stations are shown in Fig. 27.

Table 2 Parameters used in LS-RAPID computer simulation of Senoumi landslide

Parameters	Value	Source
Unit weight of the mass and sea water		
Total unit weight of the mass γ_τ γ_τ	18.6 kN/m^3	Data from the IODP-333
Unit weight of sea water (γ_ω)	10.1 kN/m^3	Average sea water density (1,026 kg/m^3)
Triggering factors		
Excess pore pressure ratio in the fractured zone (r_u)	0.3	Assumption
Acceleration of the 2011 Tohoku earthquake: 1.0 times	2,699 gal	Wave form of the ground motion record at MYG004
Shear strength parameters		
Shear strength parameters of the IODP volcanic ash		
Steady-state shear resistance (τ_{ss}) in the source area	72 kPa	Test data
Lateral pressure ratio ($k = \sigma_h/\sigma_v$)	0.85	Estimated from test data
Friction angle at peak (ϕ_p)	36.2°	Test data
Friction angle during motion (ϕ_m)	34.5°	Test data
Shear displacement at the start of strength reduction (D_L)	10 mm	Test data
Shear displacement at the end of strength reduction (D_U)	72 mm	Test data
Shear strength parameters of the Omaezaki Neogene silty-sand		
Steady-state shear resistance (τ_{ss}) in the source area	135 kPa	Test data
Lateral pressure ratio ($k = \sigma_h/\sigma_v$)	0.90	Estimated from test data
Friction angle at peak (ϕ_p)	39.5°	Test data
Friction angle during motion (ϕ_m)	35.8°	Test data
Shear displacement at the start of strength reduction (D_L)	8 mm	Test data
Shear displacement at the end of strength reduction (D_U)	100 mm	Test data
Shear strength parameters of the Omaezaki Neogene silt		
Steady-state shear resistance (τ_{ss}) in the source area	201 kPa	Test data
Lateral pressure ratio ($k = \sigma_h/\sigma_v$)	0.70	Estimated from test data
Friction angle at peak (ϕ_p)	36.8°	Test data
Friction angle during motion (ϕ_m)	23.5°	Test data
Shear displacement at the start of strength reduction (D_L)	20 mm	Test data
Shear displacement at the end of strength reduction (D_U)	400 mm	Test data
Moving area		
Steady sate shear resistance	10 kPa	See Sassa et al. (2004)
Parameters of the function for nonfictional energy consumption		
Energy loss coefficient of the function	1	See Sassa et al. (2010)
Threshold value of velocity	100 m/s	Impossible velocity

Table 3 Simulation results for landslide triggering (Sassa et al. 2012)

Earthquake record	0.3 × MYG004	0.4 × MYG004	0.7 × MYG004	1.0 × MYG004
IODP volcanic ash	*Landslide*	Landslide	Landslide	Landslide
Omaezaki Neogene silty-sand	No (only MN)	*Landslide*	Landslide	Landslide
Omaezaki Neogene silt	No	*Landslide*	Landslide	Landslide

Table 4 Seismic parameters of the main shock (NIED 2009)

Items	Value
Time	05:07, 11/8/2009
Slip (m)	0.8
Strike (degree)	307
Slip angle (degree)	119
Fault length (km)	15
Fault width	15
Depth	20

From the results, it is clear that the simulated tsunami waves by fault motion do not agree with observed tsunami waves at these stations.

The soil parameters for landslide simulation were estimated from previous UD-RSA data for shallow landslides. The data associated with the landslide motion calculated in LS-Rapid at each 1 s was saved, and used as an input to the LS-Tsunami model.

Landslide and tsunami simulation need the bathymetric and topographic data before the landslide occurred. The pre-collapse bathymetry and topography at 250 m resolution were obtained from Kishimoto (2000). The 50 m resolution bathymetric and topographic data after 2009 earthquake was created by JAMSTEC in 2010. In order to simulate the small scale of the 2009 submarine landslide (LS1), 50 m mesh data of the slope surface before sliding was estimated from 250 m mesh data for the LS-Rapid model. We edited the shape of the contour lines inside the landslide area of the 50 m mesh data to be similar to the shape of those in 250 m data to derive the elevation before sliding.

Two landslides (LS1, LS2) were created using ellipsoidal parameters. The location of LS2 is based on the investigation performed by Baba et al. (2012), shown in Fig. 26. Interpreting the 50 m mesh DEM and the corresponding contours, we estimated a small landslide LS1, which is a part of the headscarp of the Senoumi landslide and close to Yaizu station. In Fig. 28a, the yellow lines are the boundaries of the landslides. The volumes of LS1 and LS2 used for the simulation are 2662 m^3 and 17,213 m^3, respectively. The soil parameters for the landslide simulation were estimated by back analyses dependent on the location, size and movement of the submarine landslide.

Figure 29 illustrates the tsunami wave induced by the landslide motion at the Yaizu and Uchiura stations. The tsunami heights from the models at each 2 min interval were calculated and compared with observed data. The compound tsunami wave (Fig. 30) was calculated by summing the tsunami wave induced by the fault motion with that induced by the landslide motion. The compound tsunami wave gives a clearly better prediction of the tsunami height than the model that considers only the fault motion. The nearest observation location is Yaizu port with an amplitude of about −60 cm, which could not be explained by the tsunami wave induced by the fault motion only, as seen in Fig. 27. The maximum tsunami height simulated by the compound model is—55 cm at Yaizu port as seen in Fig. 30, better matching the observations at the site. The maximum positive tsunami heights in both the simulation results and the recorded data are 30 cm. For Uchiura station, we find only small differences between the obseved data and the simulation results until 24 min after the earthquake. In summary, there is good agreenment between our simulations and the observed data, especially in the early stages of the time series. In the late stages, the tsunami wave is not reproduced well and might be affected by the rebound from coastal protection structures. These constructions are complex and not shown within a 250 m resolution mesh.

Hazard Assessment of Tsunami Induced by a Potential Retrogressive Landslide in the Senoumi Landslide Topography in Suruga Bay, Japan

Excavation of the toe of the slope for road and/or railway construction or houses cause landslides in the upper slope. River erosion in the toe of slope can also cause landslides. A landslide occurring in the lower slope is equivalent to excavations of the toe of the upper slope. Landslides in the upper slope of previous landslides are called retrogressive landslides.

Often, more than one retrogressive landslide can occur. A small initial failure may develop into a large retrogressive slide and result in huge loss of lives (L'Heureux 2012). Quinn et al. (2011) presented a new mechanical model for the development of some "retrogressive" landslides in sensitive clay.

Hampton et al. (1996), Kvalstad et al. (2005), and Gue (2012) indicated that retrogressive failure is one of the causes of submarine landslides. One characteristic of submarine landslides are their enormous sizes. The largest submarine landslide, so far described, is named Agulhas slum. It is located off of south east Africa with a volume of 20,331 km^3. The maximum depth of this slide is about 800 m (Dinggle 1977). The giant Storregga slide that occurred 8200 years ago is also well known as one of the largest submarine slides, which generated a great tsunami (Bryn et al. 2005; Locat and Lee 2009). The slide volume was about 3,000 km^3 with its deepest part more than 700 m below the present seabed (Kvalstad et al. 2005). On the other hand, the Usoi landslide triggered by a M_s 7.6 earthquake in Tajikistan on 18 February 1911 created the highest landslide dam in the world. The rock slide with volume of approximately 2.0 km^3 dammed the Murghab River to the height of 600 m and impounded Lake Sarez to the depth of 550 m (Schuster and Alford 2004). Other deep-seated submarine landslides are listed in Hampton et al. (1996) including slides in Cape Fear (700 m deep) and Orkdalsfjord (500 m deep).

Sassa et al. (2012) presented research on the Senoumi submarine megaslide in Suruga Bay in Japan. Their results were based on undrained dynamic-loading ring shear tests on a volcanic ash sample cored from the mass transport deposit (MTD No.6: Landslide deposit) found around 200 m deep under the sea floor (by the International Ocean Discovery Program-IODP). The results were presented in Fig. 14.

Figure 19 presented the LS-RAPID simulation results based on the 2011 Tohoku earthquake record (MYG004). The first main shock triggered a major landslide (MJ) in Fig. 19a. The MJ landslide area retrogressively expanded as shown in Fig. 19b–d, forming the current very wide Senoumi landslide topography. The retrogressive landslide reached close to the Yaizu coast. The border of the MJ is shown in with the red line in the upper figure in Fig. 31.

The year the Senoumi landslide occurred is unknown. However, excavation of the slope due to the Senoumi Landslide should accelerate the ground water flow in the upper slope, while the unloading causes a stress reduction and gradual slope deformations in the upper slope. This will result in weathering and increase the risk of landslide occurrence in the upper slope. Though the upper slope (potential retrogressive landslide A) and B did not move at the time of Senoumi landslide, it is possible for it to fail as a result of future strong seismic shaking. Thus, it is necessary to prepare for a future retrogressive landslide and the resulting tsunami coming at the Nankai Trough similar to what happened after the 2011 Tohoku Earthquake.

A potential smaller retrogressive landslide B (PRLB in Fig. 31) is a minimally sized landslide that includes the current head scarp of the Senoumi landslide topography. A potential greater retrogressive landslide A (PRLA) is a landslide that includes the coast of Yaizu. It may either occur independently or concurrently with the potential retrogressive landslide B. We examined the tsunami hazard risk by both potential landslides (A and B).

To simulate landslide initiation, motion and tsunami generation, we assumed both of them were slump landslides. Two potential sliding surfaces were created using ellipsoidal parameters. We assume that the sliding surfaces pass to the toe of steepest region of the Senoumi area. The maximum vertical depths of the sliding surface of PRLA and PRLB are approximately 700 m and 500 m, respectively. The PRLA has a volume of 22.6 km^3, which is 3.3 times bigger than the volume of PRLB (6.8 km^3). The yellow line in Fig. 31 is the coast line, while the purple (PRLB), blue (PRLA) and red lines (the initial block of the hypothetical Senoumi landslide) are landslide boundaries. The main scarp of PRLA is located in the Yaizu City area.

Conditions of LS-RAPID Simulation

Sassa et al. (2012) tested three sample related to Senoumi landslide: Volcanic ash which was cored by drilling, Omaezaki Neogene sand and Omaezaki Neogene silt which are taken at the ground surface as representative of the widely distributed soils in this area. The sliding surface of Senoumi landslide is rather flat. Volcanic ash may deposit in the flat areas. However, the sliding surface of PRLA and PRLB are steep, especially the surface for PRLB. Some non-plastic silts can fail or liquefy easily. We, then, used the landslide dynamic parameters of Neogene sand. We used the

0.7 × 2011 Tohoku earthquake wave (MYG 0004) as the triggering factor for both landslides (PRLA and PRLB). Bathymetric and topographic data for 50 m mesh resolution around Suruga Bay was obtained from Baba et al. (2012) (JAMSTEC). Due to capacity of computer, the map is used as a 100 m mesh resolution for the LS-RAPID and LS-Tsunami simulations.

Tsunami Hazard Assessment in Coasts Along Suruga Bay by LS-Tsunami

The initiation and motion of PRLA and PRLB landslides were simulated in LS-RAPID under the above conditions. The simulated landslide movement is, then, entered into LS-Tsunami and the tsunami wave triggered was estimated. The results associated with the landslide motion and tsunami wave for PRLA are presented in Fig. 32. Figure 32a presents the maximum tsunami height at each mesh on the map. Each 2 m contour line is shown in green while each 10 m contour is shown in red. The coastline is shown in yellow along with the central cross section C–D. Figure 32b presents the section of ground surface before (green line) and after (black line) the landslide, the maximum landslide height during motion at each mesh (red line), and the maximum tsunami height at each mesh (blue line). The initial landslide block is shown with a red block, while the landslide deposit after movement is shown with a yellow block. The maximum tsunami height along section C–D is 300 m. The tsunami wave is caused by a vertical (upward or downward) movement of the sea floor. The 300 m tsunami wave is comparable to the vertical movement of sea floor induced by the landslide movement as shown in Fig. 32b.

Figure 33 presents the simulation results from LS-RAPD and LS-Tsunami for potential retrogressive landslide PRLB. As the landslide size is smaller, the resulting tsunami height is also smaller, but it is still around 160 m. The overall distribution characteristics of tsunami height change significantly between the two scenarios. It is clear that Yaizu city is at the highest tsunami risk when compared to the other cities along Suruga Bay.

In Figs. 32 and 33, the tsunami height is presented above the sea level. However, to focus on the tsunami risk on land, especially in the urban areas, it is better to determine the inundation depth, namely (tsunami height above sea level)—(ground elevation above sea level). Therefore, we present the inundation depth on land and the tsunami height on the sea in Fig. 34. This allows for the visualization of the tsunami hazard risk in Yaizu City, which is closest to the Senoumi landslide and retrogressive landslide A and B. The inundation depth is shown at a 2 m contour scale. Yaizu City center experiences inundation depths of about 30 m. Potential retrogressive landslide A includes the land submerging the coastal area under the sea.

Figure 34b presents the results for potential retrogressive landslide B. The inundation area and depth are smaller than those observed for PRLA. However, the Yaizu City center has inundation depths of 20 m. The inundation area of PRLA is ~ 91.0 km^2, which is 2.05 times bigger than the area of ~ 44.5 km^2 in PRLB.

Conclusions

A 400 m deep landslide in in Unzen Volcano and the resulting landslide induced tsunami wave in 1972 killed 15,153 people. In 2018, 4,340 people were killed by tsunami waves triggered by earthquake-induced coastal submarine landslides (Sassa et al. 2019). Landslide-induced tsunami disasters are not so frequent. However, coastal and submarine slopes are subjected to landslides by earthquakes. As shown in Fig. 14, a total of six landslide deposit layers (Mass Transport Deposit) are found in the IODP drilling (C0018) along the Nankai Trough, suggesting that submarine landslides occurred six times in this area. Landslides in coastal and submarine slopes along the Nankai Trough and the Suruga Trough are possible during a strong earthquake. One of the high-risk slopes is a retrogressive landslide in Senoumi Bay since the toe of slope is already excavated by previous landslides. This region provides one of the best sites to investigate potential sliding surfaces by drilling (around 700 m) on the coast. It is much easier to drill under the sea than IODP drilling such as C0018. This paper can be an important reference for the planning of coastal and submarine landslide investigation including drilling 700 m on the coast.

Tsunamis present serious hazards in coastal areas, including Japan. Hazard assessment of landslide-induced tsunamis should be incorporated in the coastal development planning process in both developing and developed countries. A reliable model of the landslide motion among the most important as it serves as the trigger of the landslide-induced tsunami hazards. The integrated model simulating the initiation and the motion of landslide (LS-RAPID) and the measurement of the landslide dynamic parameters by the undrained dynamic-loading ring-shear apparatus enabled estimation of the landslide motion as the trigger of a tsunami wave. LS-RAPID and the landslide-induced tsunami model (LS-Tsunami) could provide simulation results matching well with the recorded data. The simulation results of tsunami heights in the Unzen Mayuyama landslide and tsunami disasters and the tsunami wave in the 2009 Suruga Bay earthquake were presented as evidence of the reliability and precision of the new landslide-induced tsunami model described in this paper.

Accordingly, the combination of LS-Rapid and LS-Tsunami can be used for the hazard assessment of landslide-induced tsunamis triggered by strong earthquakes such as a potential retrogressive landslide in the Senoumi landslide topography in Suruga Bay. It will provide information to prepare for tsunami hazards by coastal or submarine landslides triggered by possible strong earthquakes along the Nankai Trough.

Further application to other cases of landslide-induced tsunamis by the participants and their colleagues of the Special Event for the World Tsunami Awareness Day of the Fifth World Landslide Forum on 5 November 2020 in Kyoto, Japan can continue to define the reliability and precision of this new model from different aspects.

Acknowledgements H. Yanagisawa of Tohoku Gakuin University has contributed to the development of LS-Tsunami (Sassa et al. 2016), and provided advice related to the tsunami simulations in this paper. S. Sassa of Port and Airport Research Institute, National Institute of Maritime, Port and Aviation Technology is appreciated for his comments and discussion on this paper. The bathymetric and topographic mesh data: Japan 250 m mesh was used for wide area analyses (Kishimoto 2000). The bathymetric and topographic mesh data in Suruga Bay area (50 m mesh) was provided by courtesy of T. Baba (Japan Agency for Marine-Earth Science and Technology: JAMSTEC). It was used for the LS-RAPID and LS-Tsunami simulation in Suruga Bay area. They are appreciated for providing this data.

References

Abe I, Goto K, Imamura F, Shimizu K (2008) Numerical simulation of the tsunami generated by the 2007 Noto Hanto earthquake and implications for unusual tidal surges observed in Toyama Bay. Earth Planets Space 60:133–138

Ando M (1975) Source mechanisms and tectonic significance of historical earthquakes along the Nankai Trough, Japan. Tectonophysics 27:119–140

Aoi S, Enescu B, Suzuki W, Asano Y, Obara K, Kunugi T, Shiomi K (2010) Stress transfer in the Tokai subduction zone from the 2009 Suruga Bay earthquake in Japan. Nat Geosci 3:496–500

Aoki T, Tsukioka S, Yoshida H, Hyakudome T, Ishibashi S, Sawa T, Ishikawa A, Tahara J, Yamamoto I, Ohkusu M (2008) Advanced technologies for cruising AUV URASHIMA. Int J Offshore Polar Eng 18(2):81–90

Ataie-Ashtiani B, Yavari-Ramshe S (2011) Numerical simulation of wave generated by landslide incidents in dam reservoirs. Landslides 8(4):417–432

Baba T, Matsumoto H, Kashiwase K, Hyakudome T, Kaneda Y, Sano M (2012) Microbathymetric evidence for the effect of submarine mass movement on tsunami generation during the 2009 Suruga bay earthquake, Japan. In: Yamada et al. (eds) Submarine mass movements and their consequences. Advances in Natural and Technological Hazards Research, vol 31. Springer, pp 485–494

Bardet JP, Synolakis CE, Davies HL, Imamura F, Okal EA (2003) Landslide tsunamis: recent findings and research directions. Pure Appl Geophys 160(2003):1793–1809

Biscarini C (2010) Computational fluid dynamics modelling of landslide generated water waves. Landslides 7(2):117–124

Bryn P, Berga K, Forsberg CF, Solheim A, Kvalstad TJ (2005) Explaining the storegga slide. Mar Pet Geol 22:11–19

Cabinet Office of the Government of Japan (Committee on megaquake model in Nankai Trough) (2011) Intermediate report of the study team for the Mega-quake model in Nankai trough. https://www.bousai.go.jp/jishin/chubou/nankai_trough/chukan_matome.pdf

Cabinet office of the Government of Japan (Committee on Megaquake Model in Nankai Trough) (2012) The tsunami-fault model in the 2011 off the Pacific Coast of 2011 Tohoku Earthquake, Reference No. 1 of the 12th Meeting of Megaquake Model in Nankai Trough. https://www.bousai.go.jp/jishin/chubou/nankai_trough/12/sub_1.pdf

Cochonat P, Cadet J, Lallemant S, Mazzotti S, Nouze H, Fouchet C, Foucher J (2002) Slope instabilities and gravity processes in fluid migration and tectonically active environment in the eastern Nankai accretionary wedge (KAIKO-Tokai' 96 cruise). Mar Geol 187:193–202

Dan G, Sultan N, Savoye B (2007) The 1979 Nice harbour catastrophe revisited: trigger mechanism inferred from geotechnical measurement and numerical modelling. Mar Geol 245:40–64

Dingle RV (2015) The anatomy of a large submarine slump on a sheared continental margin (southeast Africa). J Geol Soc London 134:293–310

Expedition 333 Scientists (2012) Site C0018. In: Henry P, Kanamatsu T, Moe K (eds) Proceedings IODP and the Expedition 333 Scientists, 333. Integrated Ocean Drilling Program Management International, Inc., Tokyo. https://doi.org/10.2204/iodp.proc.333.103.2012

Fine I, Rabinovich A, Bornhold B, Thomson R, Kulikov E (2005) The Grand Banks landslide-generated tsunami of November 18, 1929: preliminary analysis and numerical modelling. Mar Geol 215:45–57

Fujiwara O, Aoshima A, Irizuki T, Ono E, Obrochta S, Sampei Y, Sato Y, Takahashi A (2019) Tsunami deposits refine great earthquake rupture extent and recurrence over the past 1300 years along the Nankai and Tokai fault segments of the Nankai Trough, Japan. Quat Sci Rev

Fraser SA, Power WL, Wang X, Wallace LM, Mueler C, Johnston (2014) Tsunami inundation in Napier, New Zealand, due to local earthquake sources. Nat Hazard 70:415–445

Geological Survey of Japan. GSJ open-file report no. 353

Glimsdal S, L'Heureus JS, Harbitz CB, Pedersen GK (2013) Modelling of the 1888 landslide Tsunami, Trondheim, Norway. In: Proceeding the second world landslide forum landslide science and practice, vol 5. Springer, pp 73–79

Gue CS (2002) Submarine landslide flows through centrifuge modelling. Ph.D. Thesis. University of Cambridge

Hampton MA, Lee HJ, Locat J (1996) Submarine landslides. Rev Geophys 34(1):33–59

Imamura F (2009) Chapter 10. Tsunami modeling: calculating inundation and hazard maps. In: Bernard EN, Robinson AR (eds) THE SEA Tsunamis. Harvard University Press, London, pp 321–332

Inoue K (1999) Shimabara-Shigatsusaku earthquake and topographic changes by Shimabara catastrophe. J Jpn Soc Erosion Control Eng 52(4):45–54 (in Japanese)

Intergovernmental Oceanographic Commission (IOC) (1997) Numerical method of tsunami simulation with the leap-frog scheme. IUGG/IOC Time Project IOC Manuals and Guides, No. 3, UNESCO, 126 p

Japan Landslide Society (1996) Landslides in Japan, 5th edn

Kishimoto K (2000) Combined bathymetric and topographic mesh data: Japan 250m.grd

Kvalstad TJ, Andresen L, Forsberg CF, Bryn P, Wangen M (2005) The Storegga slide: evaluation of triggering sources and slide mechanics. Mar Petrol Geol 22:245–256

L'Heureux JS (2012) A study of the retrogressive behavior and mobility of Norwegian quick clay landslides. In: Proceedings of the 11th international symposium on landslides (ISL) 2nd North American symposium on landslides, Banff, Canada, 3–8 June 2012

Linsley RK, Franzini JB (1979) Water resource engineering, 3rd edn. McGraw-Hill

Locat J, Lee H (2009) Submarine mass movement and their consequences: an over view. In: Sassa K, Canuti P (eds) Landslides—disaster risk reduction. Springer, New York, pp 115–142

Ministry of Land, Infrastructure, Transport and Tourism (MLIT), Water and Disaster Management Bureau, Coastal management office (2012) Guideline for the assessment for Tsunami inundation area. Ver.1.00. https://www.mlit.go.jp/river/shishin_guideline/bousai/saigai/tsunami/shinsui_settei_ver100.pdf

NIED (National Research Institute for Earth Science and Disaster Prevention) (2009) Special page on 2009 Aug. 11 Suruga Bay Earthquake (in Japanese). https://www.hinet.bosai.go.jp/topics/suruga-bay090811/

Omira R, Dogan G, Hidayat R, Husrin S, Prasetya G, Annunziato A, Proietti C, Probst P, Paparo MA, Wronna M, Zaytsev A, Pronin P, Giniyatullin A, Putra PS, Hartanto D, Ginanjar G, Kongko W, Pelinovsky E, Yalciner AC (2019) The September 28th, 2018, Tsunamin Palu-Sulawesi, Indonesia: a post-event field survey. Pure Appl Geophys 176(4):1379–1395

Quinn PE, Diederichs MS, Rowe RK, Hutchinson DJ (2011) A new model for large landslides in sensitive clay using a fracture mechanics approach. Can Geotech J 48:1151–1162

Sassa K, Fukuoka H, Wang G, Ishikawa N (2004) Undrained dynamic-loading ring-shear apparatus and its application to landslide dynamics. Landslides 1(1):7–19

Sassa K, Fukuoka H, Wang F, Wang G (2005) Dynamic properties of earthquake-induced large-scale rapid landslides within past landslide mass. Landslides 2(2):125–134

Sassa K, Nagai O, Solidum R, Yamazaki Y, Ohta H (2010) An integrated model simulating the initiation and motion of earthquake and rain induced rapid landslides and its application to the 2006 Leyte landslide. Landslides 7(3):219–236

Sassa K, He B, Miyagi T, Strasser M, Konagai K, Ostric M, Setiawan H, Takara K, Nagai O, Yamashiki Y, Tutumi S (2012) A hypothesis of the Senoumi submarine megaslide in Suruga Bay in Japan—based on the undrained dynamic-loading ring shear tests and computer simulation. Landslides 9(4):439–455

Sassa K, Dang K, He B, Takara K, Inoue K, Nagai O (2014a) A new high-stress undrained ring-shear apparatus and its application to the 1792 Unzen–Mayuyama megaslide in Japan. Landslides 11(5):827–842

Sassa K, Bin H, Dang K, Nagai O, Takara K (2014b) Plenary: progress in landslide dynamics. Landslide science for a safer geoenvironment, vol 1. Springer, pp 37–67

Sassa K, Dang K, Yanagisawa H, Bin H (2016) A new landslide-induced tsunami simulation model and its application to the 1792 Unzen-Mayuyama landslide-and-tsunami disaster. Landslides 13(6):1405–1419

Sassa K, Dang K (2018) Txt-tool 0.081-1.1: landslide dynamics for risk assessment. In: Sassa K, Tiwari B, Liu K-F, McSaveney M, Strom A, Setiawan H (eds) Landslide dynamics: Isdr-icl landslide interactive teaching tools: volume 2: testing, risk management and country practices. Springer International Publishing, Cham, pp 1–79

Sassa S et al (2019) Liquefied gravity flow-induced tsunami: first evidence and comparison from the 2018 Indonesia Sulawesi earthquake and tsunami disasters. Landslides 16(1):195–200

Satake K (2001) Tsunami modeling from submarine landslides. In: Proceedings of the international tsunami symposium, pp 665–674

Satake K, Namegaya Y, Fujii Y, Okada M, Abe K, Imai K, Ueno T, Yamaghichi K, Miwa K, Yamamoto H (2010) In situ measurement of tide gauges response and corrections of tsunami waveforms from the 2009 Suruga Bay earthquake. Bull Earthq 85:1–14

Sato H, Sekiguchi T, Kojiroi R, Suzuki Y, Iida M (2005) Overlying landslides distribution on the earthquake source, geological and topographical data: the Mid Niigata prefecture earthquake in 2004. Landslides 2(2):143–152

Schuster RL, Alford D (2004) Usoi landslide dam and Lake Sarez, Pamir Mountains, Tajikistan. Environ Eng Geosci 10:151–168

Strasser M, Moore, G, Kimura G, Kopf A, Underwwd M, Guo J, Screaton E (2011) Slumping and mass transport deposition in the Nankai fore arc: evidence from IODP drilling and 3-D reflection seismic data. Geochem Geophys Geosyst 12, Q0AD13. https://doi.org/10.1029/2010GC003431

Strasser M, Henry P, Kanamatsu T, Thu M, Moore G, IODP Expedition Scientists (2012) Scientific drilling of mass-transport deposits in the Nankai accretionary wedge: first result of from IODP Expedition 333. Submarine mass movements and their consequences. In: Yamada et al. (eds) Advances in natural and technological hazards research, vol 31. Springer, New York, pp 671–681

Tappin DR, Watts P, Grilli ST (2008) The Papua New Guinea tsunami of 17 July 1998: anatomy of a catastrophic event. Nat Hazards Earth Syst Sci 8:243–266

Tinti S, Tonini R (2013) The UBO-TSUFD Tsunami inundation model: validation and application to a tsunami case study focused on the city of Catania, Italy. Nat Hazards Earth Syst Sci 13:1795–1816

Tsuchi R (2001) Geomorphology and geology of Shizuoka Prefecture —geology of Shizuoka Prefecture (2001 Revision version) with explanation. Naigai Map Co. Ltd. One map, pp 92

Tsuji Y, Hino T (1993) Damage and inundation height of the 1792 Shimabara landslide tsunami along the coast of Kumamoto Prefecture. Bull Earthq Res Inst Univ Tokyo 68:91–176 (in Japanese)

Tsuji Y, Murakami Y (1997) Inundation height of the 1792 Mayuyama landslide Tsunami the Shimabara Peninsula side. Historical Earthquake, No. 13, Society of Historical Earthquake Studies, pp 135–197

Unzen Restoration Office of the Ministry of Land, Infrastructure and Transport of Japan (2002) The Catastrophe in Shimabara—1791–92 eruption of Unzen–Fugendake and the sector collapse of Mayu-Yama. An English leaflet, p 23

Unzen Restoration Office of the Ministry of Land, Infrastructure and Transport of Japan (2003) The Catastrophe in Shimabara—1791–92 eruption of Unzen–Fugendake and the sector collapse of Mayu-Yama. A Japanese leaflet, p 44

Usami T (1996) Materials for comprehensive list of destructive earthquakes in Japan. University of Tokyo Press (in Japanese)

On the Use of Statistical Analysis to Understand Submarine Landslide Processes and Assess Their Hazard

Uri S. ten Brink and Eric L. Geist

Abstract

Because of their inaccessibility, submarine landslides are typically studied individually and at great effort and expense to provide knowledge of the specific site conditions where these landslides occur. Statistical analysis of submarine landslide scars can offer generalized perspectives on the processes that initiate submarine landslides and can help toward hazard assessment in areas that have not been studied in detail. The following review discusses more than a decade of development of statistical approaches to studying submarine landslides. Landslides were previously viewed together with other natural hazards, such as earthquakes and fires, as a phenomenon whose size distribution obeys an inverse power law. Inverse power law distributions are the result of self-organized avalanche processes, in which the final hazard size cannot be predicted at the onset of the disturbance. We find that volume and area distributions of submarine landslides along the U.S. Atlantic continental slope and along nine other margins worldwide do not follow an inverse power law. Rigorous statistical tests of several different probability distribution models indicate that the lognormal model is most appropriate for these siliciclastic environments. Lognormal distributions can be simulated by assuming that the area of slope failure depends on earthquake magnitude, in other words, failure occurs simultaneously over the area affected by horizontal ground shaking and does not cascade from nucleating sources. Therefore, the maximum landslide size can be predicted from the earthquake magnitude and the distance from the rupturing fault. Moreover, earthquakes $<\sim$M4.5 cannot generate significant submarine landslides. We further demonstrate that empirical, offshore landslide hazard curves can be developed from these lognormal landslide size distributions, if the duration of mapped landslide activity is known. In addition to hazard estimation, scaling relationships can yield insights on the physical processes associated with landslide failure. For example, the log-log relationship between volume and area of landslide scars in siliciclastic margins is observed to be almost linear implying that most landslides are translational. Carbonate margins, in contrast, show a power-law distribution of scar volumes and their volume to area relationship is $\sim$1.3. These results suggest that landslides in carbonate margins are governed by the random distributions of existing fissures, and they act like rock falls on land. Although earthquakes are the principal trigger of submarine landslides, the effects of earthquake frequency on slope stability can be counterintuitive. The average size of landslide scars decreases non-linearly with increasing frequency of earthquakes and increases with increasing sedimentation rate. The effect is interpreted as evidence for densification and shear strength increase of margin sediment, induced by repeated seismic shaking.

Keywords

Landslide area distribution • Landslide volume distribution • Area-volume relationship • Translational landslides • Lognormal distribution • Landslide triggering • Soil densification by shaking • Landslides and sedimentation rate • Landslide size and morphology • Failure of carbonates and sand and clay

U. S. ten Brink (✉)
U.S. Geological Survey, Woods Hole, MA 02540, USA
e-mail: utenbrink@usgs.gov

E. L. Geist
U.S. Geological Survey, Moffett Field, Falmouth, CA 94035, USA
e-mail: egeist@usgs.gov

K. Sassa et al. (eds.), *Understanding and Reducing Landslide Disaster Risk*, ICL Contribution to Landslide Disaster Risk Reduction,
https://doi.org/10.1007/978-3-030-60196-6_23

Introduction

Offshore landslides are a significant agent of sediment transport to the deep sea. These landslides have also been known to cause devastating tsunamis, and they also present direct hazards to wind farms, oil and gas platforms, pipelines, communication cables, and other offshore infrastructure. Landslide hazards onshore can be estimated from direct geotechnical sampling, detailed information on the site response to seismic ground shaking, the topographic slope, and subsurface physical properties. For offshore landslides, however, geotechnical sampling, information on site response and subsurface physical properties are rarely available and topographic slope is poorly correlated with the extent of slope failure (ten Brink et al. 2016). In this regard, offshore landslide studies have similar limitations to studies of other planets where physical processes are deduced from remote sensing coupled with modeling. Remote sensing of the offshore environment is carried out primarily by multibeam sonar, whose resolution is water depth dependent.

Landslide evidence in the bathymetry is preserved by scars on the ocean floor left behind after slope failures. The scar areas are defined by an excavation region, which is partly surrounded by scarps that break the local slope. There are uncertainties in estimating the size of the slope failure from the scar area. First, scar area often cannot be measured accurately because it often transitions into an area of debris deposition, and the debris may cover part of the excavation or evacuation area. Second, the multibeam sonar has finite horizontal and vertical resolutions. Third, the original volume of the evacuated material is unknown. This is important because landslide volume, not area, is the important parameter in understanding sediment transport and tsunami hazard.

Systematic mapping of sufficiently large contiguous regions and statistical analysis of the scars in these regions can alleviate these problems. The uncertainty in measuring the exact scar area can be estimated by the measurement of a large number of scars within a region, which as long as the measurement error is not systematic, provides a pattern of landslide area distribution. As will be discussed later, this pattern is lognormal in margins covered by siliciclatic sediments (Geist and ten Brink 2019) and may be power law in carbonate margins (ten Brink et al. 2006a). The error involved in measurement resolution can be overcome by adhering to a uniform grid resolution of the bathymetry and by mapping at a uniform scale. Finally, landslide volume was shown to be almost linearly correlated with landslide area in siliciclatic margins (Chaytor et al. 2009; ten Brink et al. 2014), hence we can use area as a proxy for landslide volume.

In addition to identifying a pattern in landslide area distribution, the statistical relationship between area and volume of landslides and the probability distribution of landslide area provide insights into landslide processes and into the underlying triggers, which are difficult to observe or measure directly. They can further be combined with other readily available parameters such as earthquake recurrence interval, sedimentation rate, and average slope to help design hazard curves for submarine landslides in different regions.

Inferring the Style of Slope Failure from Area-Volume Relationship

The compilation of bathymetric data along the U.S. Atlantic margin at a grid interval of 100 m (Andrews et al. 2013) allowed us to identify all but the very small (>0.1 km^2) landslide scars and calculate their areas and volumes. Landslide volumes were calculated by interpolating smooth surfaces through polygons that define the scarps of each slide, then gridding these smooth surfaces, and finally, subtracting these grids from the gridded topography of each scar (Fig. 1; ten Brink et al. 2006a; Chaytor et al. 2009).

The resulting datasets of landslide areas and volumes show several interesting relationships. Landslide volume, V, increases almost linearly with area, A: $V = kA^d$, where $d \sim 1.1$ (Fig. 2a). This relationship implies that the thickness of most slides is independent of the landslide area. This in turn leads to the inference that most landslides on the margin must be translational, failing along depositional layer boundaries with low basal shear strength, which separate units of similar thickness (Fig. 2b). Bathymetry and high-resolution seismic reflection data along the margin indeed indicate failure of thin (typically 20–100 m thick) broad layers (Twichell et al. 2009). The almost linear relationship is in contrast to volume–area relationships of subaerial landslides (d = 1.5, Hovius et al. 1997) and submarine landslides in carbonate margins, discussed in the section "Hazard Assessment in Carbonate Margins".

Lognormal Distributions of Landslide Scars

The cumulative distribution of both volume and area for submarine landslide scars along the U.S. Atlantic margin follows a lognormal distribution (Fig. 3; Chaytor et al. 2009). In other words, there are finite numbers of small landslides, very few large landslides, and about half of the landslide scar volumes are between 0.1 km^3 and 5 km^3. A similar distribution was derived from analysis of higher

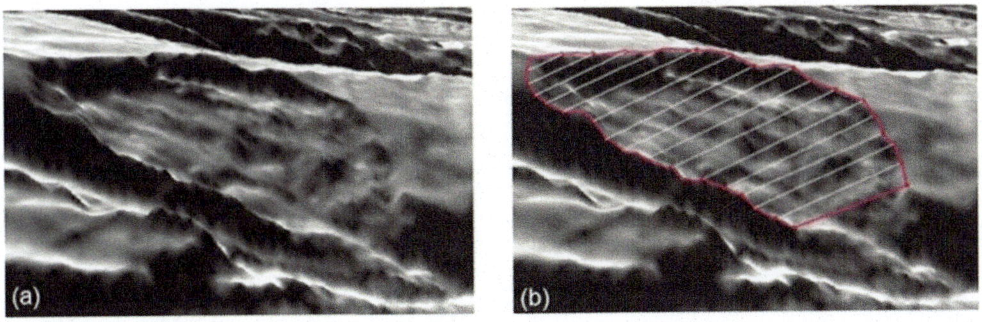

Fig. 1 **a** Perspective view of a failure scar area (grey shaded bathymetry; ~10x vertical exaggeration) **b** schematic example of a smooth-surface fit to the scar perimeter. The scar bathymetry is subtracted from the smooth surface to estimate the excavated volume (Chaytor et al. 2009)

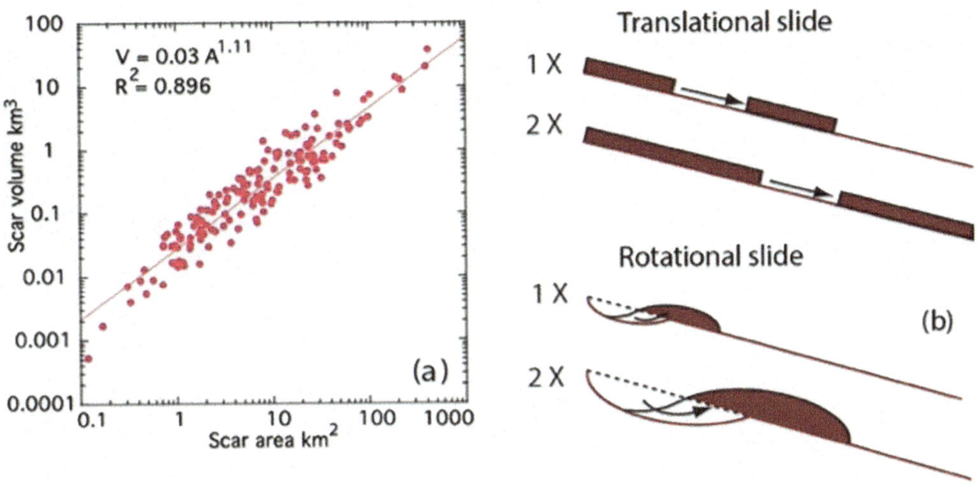

Fig. 2 **a** Area–volume relationship for 160 landslide scars in the mid-Atlantic margin. **b** Schematic drawings of translational and rotational slides each with one- and two-times scar length. Doubling the scar length increases the slide thickness for rotational slides but does not for translational slides. The change in slide thickness with slide area is expressed by the value of the exponent in landslide volume to area relationship: ~1 for translational slides and ≤1.5 for rotational slides (ten Brink et al. 2014)

resolution bathymetry data from the mid-Atlantic region of the margin (ten Brink et al. 2014; Fig. 3). Similar area and volume distributions were also obtained by considering a subset of the data that exclude poorly defined scars and a subset which groups adjacent scars that might have failed together (ten Brink et al. 2014).

The probability density function (pdf) for the lognormal distribution of landslide scar areas *f(a)* is given by

$$f(a) = \frac{1}{\sqrt{2\pi}\sigma a} e^{-\frac{[\ln(a)-\mu]^2}{2\sigma^2}} \tag{1}$$

where μ is the mean of the natural log of the distribution and σ is a shape parameter. The expected value (i.e., mean) for the lognormal distribution is given by $\exp(\mu + \sigma^2/2)$, the variance by $[\exp(\sigma^2) - 1]\exp(2\mu + \sigma^2)$, and the median by $\exp(\mu)$.

Estimates of lognormal distribution parameters for scar areas along the 10 margins in Fig. 4 show that the parameter μ varies significantly, with the U.S. Atlantic and Queen Charlotte margins being the high and low endmembers, respectively. These margins have the largest and smallest density of scars per unit area of the margin. In contrast, there is much less variation in the σ parameter among each of the margins. The lognormal distribution model is a natural distribution to model positive and skewed sizes but is not the only one. We examine the data fit to two other statistical distributions: the Weibull distribution and the tapered Pareto distribution previously used to model the size distribution of terrestrial landslides (Malamud et al. 2004). The lognormal distribution is a better fit for the data than the Weibull and tapered Pareto distributions based on the Akaike information criterion (Burnham and Anderson 2010) and the Maximum Likelihood criteria (Fig. 5).

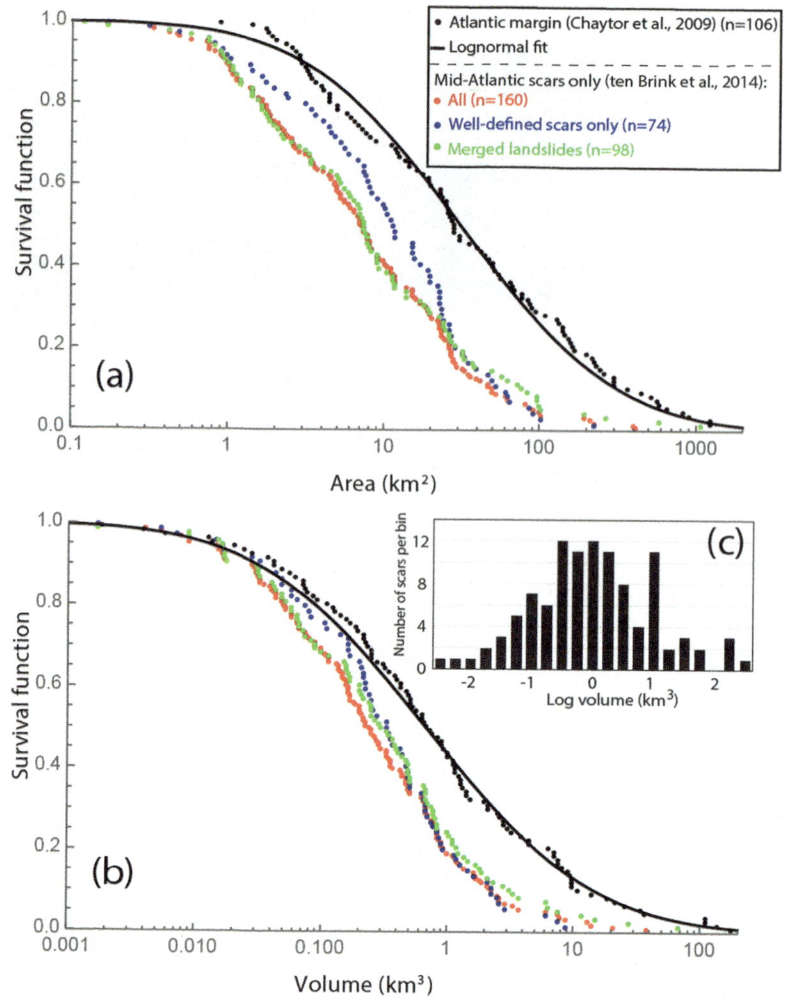

Fig. 3 Cumulative area (**a**) and volume (**b**) distributions of landslide scars along the U.S. Atlantic margin showing a lognormal-like distribution. Black dots—106 mapped landslides from the entire margin by Chaytor et al. (2009) and a lognormal fit to the data. Red dots—160 landslides in the U.S. mid-Atlantic region identified by ten Brink et al. (2014) in newer high-resolution bathymetry data. Blue dots—Subset of the best defined defined scars in the mid-Atlantic, (74 landslides). Green dots—98 interpreted landslide clusters from the mid-Atlantic region. **c** Frequency distribution of landslide volumes in Chaytor et al. (2009) binned at 0.25 log volume

Inferring Landslide Triggering Mechanisms from Scar Area Distribution

Lognormal Distribution Is Indicative of Triggering by Earthquakes

We next show that a simple earthquake-triggered landslide mechanism can produce the observed lognormal area distribution, by carrying out Monte Carlo simulations of landslide failure and examining their area distributions (Fig. 6). The simulations use two simple assumptions: (1) Infinite slope stability analysis is used to calculate landslide area, based on the acceleration needed to displace seafloor sediments, and (2) Horizontal acceleration whose amplitude and attenuation with distance from the fault are functions of the earthquake magnitude. A sample of earthquakes with magnitudes drawn from the Gutenberg-Richter relation is randomly distributed across a profile, which represents the U.S. continental margin.

Slope failure of sediments is assumed to initiate when the pseudo-static stress, which includes the downslope gravitational stress plus horizontal earthquake loading, exceeds the undrained shear strength. However, to cause finite displacement of the material, the peak earthquake acceleration must exceed sediment shear strength by a factor of 6.7 or more (Hynes-Griffin and Franklin 1984; Lee et al. 2000). In engineering literature, finite distance is taken to be at least 1 m (e.g., Newmark 1965; Hynes-Griffin and Franklin 1984), which is assumed to be sufficient for continued downhill movement.

The area susceptible to failure is calculated in two steps: First we calculate the maximum distance from the fault at which finite failure of the sea floor is possible, and second, we define the geometrical area surrounding the fault from the maximum distance to failure. We use typical sediment shear strength, published peak horizontal spectral acceleration curves for the eastern U.S. (Campbell 2003; Tavakoli and Pezeshk 2005), and we assume a commonly observed landslide thickness of 20–100 m for the margin (ten Brink

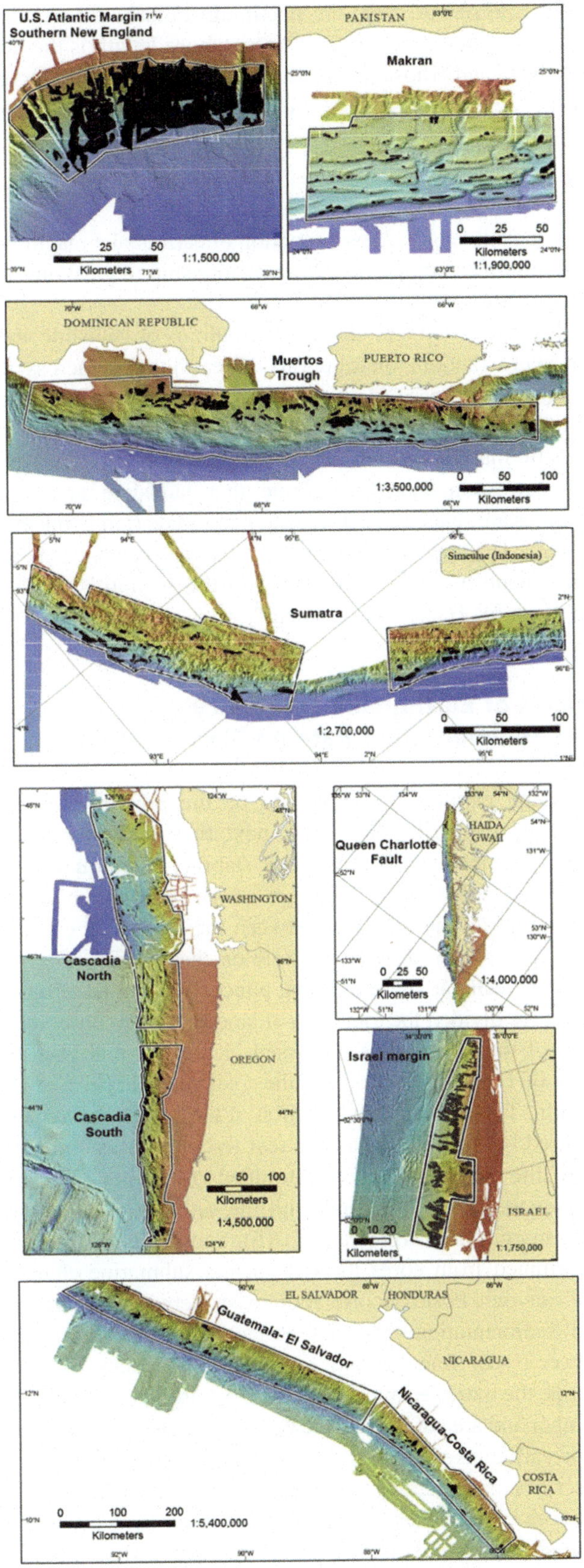

Fig. 4 Bathymetry maps and mapped scar areas (in black) of ten margins analyzed in this study (ten Brink et al. 2016)

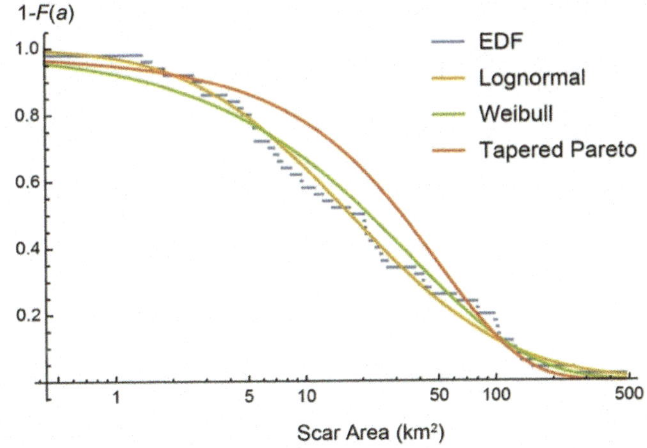

Fig. 5 Survival functions for three probability models (lognormal, Weibull, and tapered Pareto) in comparison to the empirical distribution function (EDF) for the U.S. Atlantic margin (Geist and ten Brink 2019)

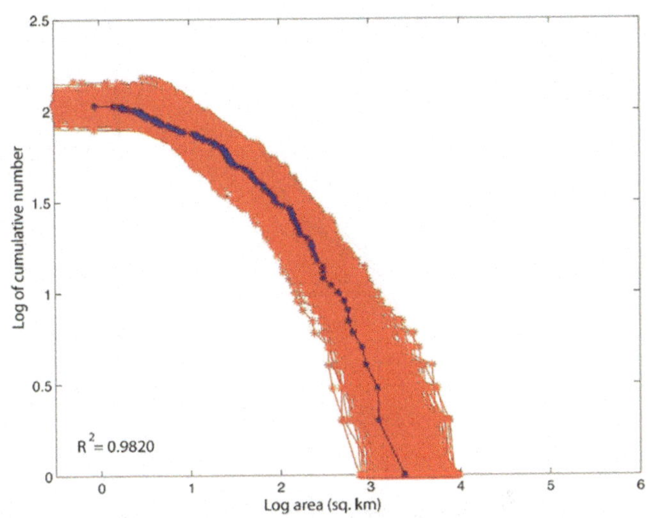

Fig. 6 Cumulative area distribution of landslide scars (blue curve) from the Atlantic margin (Chaytor et al. 2009) and 100 Monte Carlo distribution simulations (red curves) calculated using the relationship between slope stability and earthquake magnitude, and the Gutenberg–Richter relation of earthquake magnitudes (ten Brink et al. 2009a)

et al. 2009b). The calculated maximum distance, r_{max} from a rupturing fault to sites where submarine slope failure is expected, is that distance at which the modified peak spectral acceleration of the earthquake, k_{PSA} for a seismic wave period T, is equal or smaller than the critical acceleration, necessary to cause catastrophic displacement, k_y,

$$0.15 \times 3.5 \times k_{PSA(T=0.75)} \leq ky \qquad (2)$$

This distance increases with seabed slope angle and with earthquake magnitude and depends on the choice of soil strength, expressed as the ratio of undrained shear strength to overburden pressure, c/p (Fig. 7a).

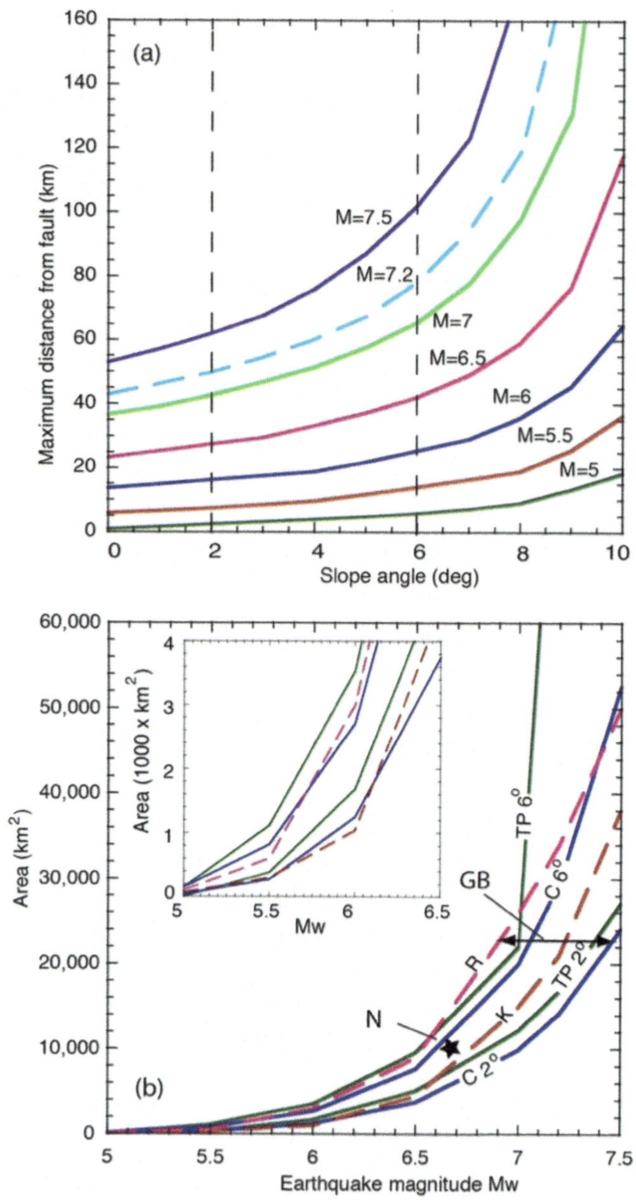

Fig. 7 **a** Maximum distance from a fault to sites where failure is expected, as a function of slope angle and earthquake moment magnitude (ten Brink et al. 2009b). Calculations used $c/p = 0.2$ and K_{PSA} of Campbell (2003). Dashed lines mark typical seafloor gradient range for the U.S. Atlantic continental margin. **b** Comparison between two methods used to derive the relationship between earthquakes and landslide areas. Solid curves were calculated by the slope stability analysis method for seabed gradients of 2° and 6° using two published peak spectral acceleration relationships for the eastern U.S, (Campbell 2003; Tavakoli and Pezeshk 2005 marked C and TP). Red dashed curves are observed maximum total area affected by landslides on land due to earthquakes from Keefer (1984; K) and Rodriguez et al. (1999, *R*). Inset is an enlargement for smaller earthquake magnitudes. Star N—total area affected by landslides due to the M6.7 Northridge earthquake. Double-arrowed line GB—maximum estimated failure area of the M7.2 ± 0.3 Grand Banks 1929 earthquake (Mosher and Piper 2007)

For the second step, the maximum failure area, A_L, is a rectangle containing the fault trace, whose length is the fault length, L, and whose half-width is the maximum distance to failure, r_{max}, plus two half circles with a radius r_{max} at either end of the fault,

$$A_L = \pi r_{\max^2} + 2Lr_{\max} \tag{3}$$

The fault length, L, as a function of earthquake magnitude is given by an empirical scaling relationship (e.g., Wells and Coppersmith 1994).

The calculated cumulative distributions of landslide areas for many random distributions of earthquakes across the margin, encompass (within one order of magnitude uncertainty) the observed landslide area distribution of the U.S. Atlantic margin and mimics its shape (Fig. 6). We interpret this correspondence to indicate that landslide areas and volumes are related to earthquake magnitude (ten Brink et al. 2009a). This relationship can be used for landslide hazard assessment, as will be discussed in the section "Statistical Analysis for Hazard Assessment".

Effects of Earthquake Frequency and Sedimentation Rate on Slope Stability

Although earthquakes appear to be the primary cause of landslide triggering, we show next that earthquakes also have the opposite effect on slope stability. Earthquakes can serve to stabilize the margin slope and limit failure depending on the interplay between the frequency of earthquakes and the sedimentation rate on the margin.

To investigate the stabilizing effect of earthquakes on the slope, we divided the total area of landslide scars observed in each of the 10 margins discussed in "Lognormal Distributions of Landslide Scars" by the total mapped area of that margin. This quantity, the fraction of the margin bathymetry covered by landslide scars or "scar fraction", was correlated with the independently determined frequency of large earthquakes ($\sim \geq$M7) and with the sedimentation rate for each margin (ten Brink et al. 2016). Sedimentation rate was determined from cores located outside submarine channels and canyons. Positive correlation exists between scar fraction and sedimentation rate, but a negative correlation exists between scar fraction and earthquake frequency. In other words, the more seismically active the margin, the smaller the number and area of landslides. Combining sedimentation rate with earthquake frequency into a single parameter, ψ, that describes the average thickness of sediment accumulation between earthquakes, provides the best positive correlation with scar fraction (Fig. 8a; ten Brink et al. 2016). The U.S.

Atlantic margin has had high sedimentation rate during the last glacial maximum, but low frequency of seismic activity. It shows abundant and large landslide scars. The seismically active margins of Makran and Sumatra (Fig. 4) have equal or higher sedimentation rate, but the frequency of their seismic activity is perhaps two orders of magnitude higher. Therefore, the average thickness of sediment that can accumulate between earthquakes in these active margins is almost two orders of magnitude smaller and the scar fraction is correspondingly a lot smaller (Fig. 8a).

This correlation suggests that slope stability increases with increasing frequency of earthquakes and with decreasing sedimentation rate. This increase in stability (likely associated with increased sediment shear strength;

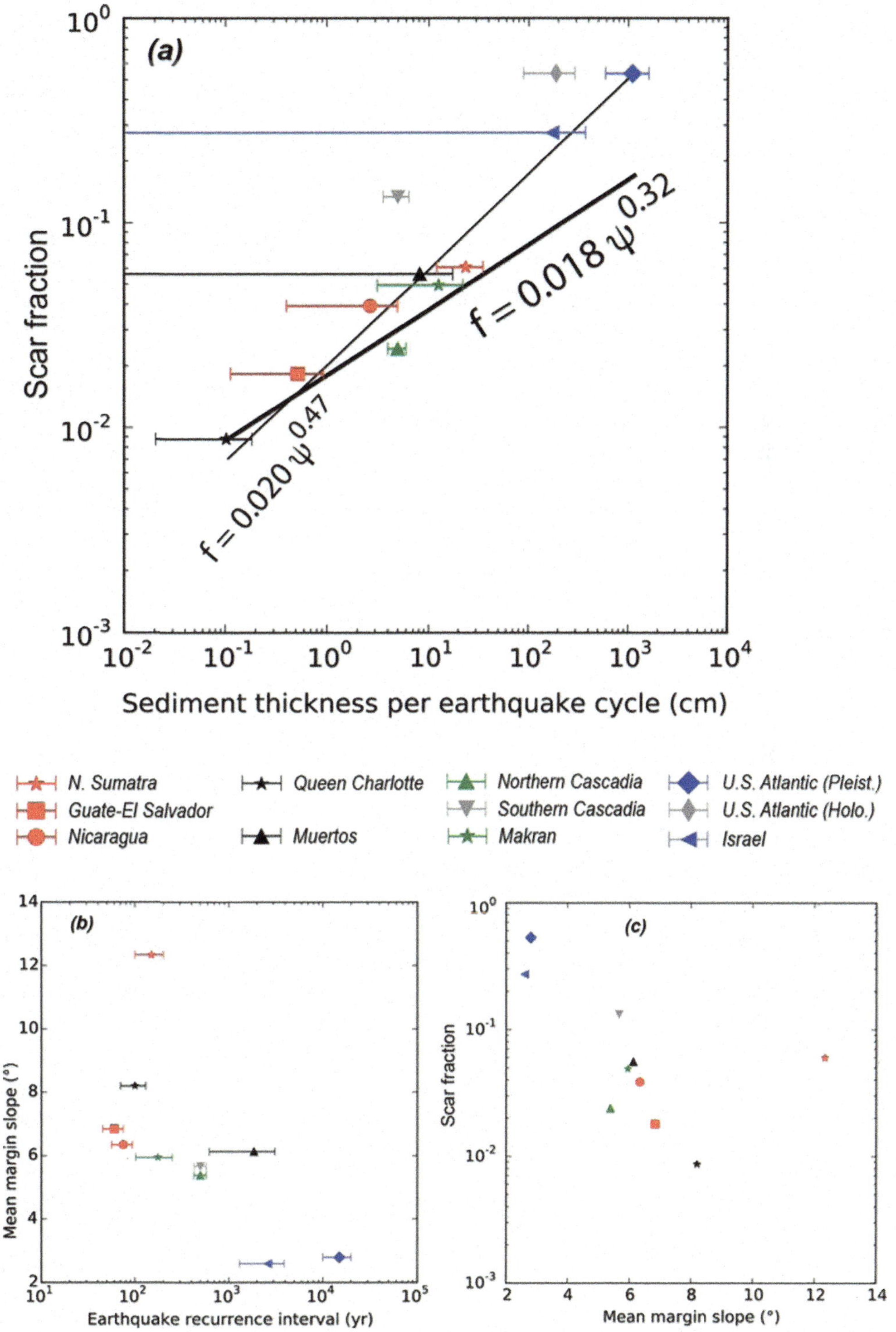

Fig. 8 **a** Fraction of total margin area covered by scars (scar fraction, f) for different margins shown in Fig. 3 as a function of sediment thickness that accumulates during an average interval between earthquakes, ψ. Estimated range of sedimentation rate is marked by error bars (log scale makes error bars longer on left side of each data point). Heavy solid line is best-fit power law. Thin black line indicates best-fit correlation, ignoring uncertainty in data, which was used as starting models for nonlinear least-squares solution with uncertainties. **b** Mean margin slope gradient as a function of earthquake recurrence interval. **c** Fraction of total margin area covered by scars as a function of mean gradient of margin (ten Brink et al. 2016)

Morgenstern 1967) is not linear, accelerating with decreasing interseismic sediment accumulation (ten Brink et al. 2016). There can be outliers to this correlation, such as Southern Cascadia (Fig. 8a), where tectonic activity or salt diapirism are expected to lead to relatively rapid oversteepening of the slope, and to an eventual failure regardless of seismic activity and sediment supply.

Cyclic loading in the absence of drainage causes reduction in shear strength and even liquefaction, because of increased pore pressure, leading to a decrease in effective stress (Lee et al. 2004). Positive correlation between increased failure and sedimentation rate indicates that rapid sedimentation is causing underconsolidation: i.e., excess pore pressures generated because the sediment rate is too high to allow full drainage. However, the stronger correlation with earthquake recurrence and with sediment accumulation per earthquake cycle (Fig. 8a) suggests that the excess pore pressure generated by the sedimentation process typically has time to dissipate, so that at least normal consolidation prevails. In sites of frequent earthquakes, the sediment may become overconsolidated by grain re-arrangement leading to tighter packing, and by excess fluid drainage.

Undrained shear strength values measured in Ocean Drilling Program (ODP) cores from the seismically active Japan, Costa Rica, and Cascadia margins are indeed significantly higher than those from the U.S. Atlantic and Amazon margins, which rarely exhibit seismic activity (Sawyer and DeVore 2015). Increased slope stability, and related shear strength of continental slope sediments, is also evident in the correlation between higher mean margin slope and increasing earthquake frequency (ten Brink et al. 2016; Fig. 8b). A general relationship between higher mean margin slope and smaller scar fraction (Fig. 8c) has also been previously reported by McAdoo et al. (2000).

Deducing Failure Processes of Submarine Landslides from Scar Area Distribution

Subaerial landslides have previously been viewed together with other natural hazards, such as earthquakes and fires, as phenomena whose size distributions follow power law relationship (Densmore et al. 1998; Guzzetti et al. 2002; Malamud and Turcotte 2006). These power law distributions have often been explained by self-organized avalanche processes, in which the final size cannot be predicted at the onset of the disturbance. In terms of landslides, this process assumes that failure nucleates in one or more locations, spreads to surrounding regions, and can coalesce to generate large failures.

We have documented lognormal distributions of landslide areas in submarine environment, not a power law distribution (Section "Lognormal Distributions of Landslide Scars"), and simulated the lognormal distribution using horizontal earthquake acceleration and its effect on slope stability (Fig. 6). Because slope destabilization is triggered by seismic shaking, it is expected to occur simultaneously within a finite affected area, which can be quite large (Fig. 7b). Earthquake-triggered terrestrial landslides, in fact, display this phenomenon, where a single earthquake causes multiple separate slope failures (Fig. 9a). A landslide does not propagate in an avalanche fashion from one or a few nucleating sources as expected in a power law distribution (Fig. 9b). Propagation of slope failure may be generally limited to no more than several hundreds of meters (e.g., Laberg and Vorren 2000), whereas scar areas can cover thousands of square kilometers. How much of an area, which is subjected to ground shaking, will actually fail depends on local variations in slope gradient, material strength, pore pressure, and the presence of pre-existing fractures.

Several observations support our hypothesis of many nucleating sources. First, the total seafloor area affected by the tsunamigenic 1929 Grand Banks landslide is estimated at 22,700 km^2, yet detailed seafloor maps show complete failure in only 1/3 of that area, whereas in the remaining 2/3 of the area patches of failures are interspersed with seafloor patches where failure was not detected (Mosher and Piper 2007). Second, the total area encompassing all the terrestrial landslides from a particular earthquake is a function of that earthquake magnitude (Keefer 1984; Rodriguez et al. 1999) and is comparable to the area calculated for the Atlantic margin from slope stability analysis, (Fig. 7b; ten Brink et al. 2009a).

We argue that the observed power law distribution of landslide areas on land is determined by the size distribution of contingent hill slopes and is not indicative of the initiation process. For example, the cumulative distribution of hill slope areas with unique compass directions ("aspect" in Geographic Information System terminology) in the region affected by the M6.7 1994 Northridge earthquake has a similar inverse power law distribution with a similar exponent to the observed landslide distribution in the region (Fig. 10, ten Brink et al. 2009a). Landslide distribution in southern California appears, therefore, to be controlled by the availability of slope areas to fail, not by the landslide process itself. In contrast, continental margins, and especially their lower slope regions, have fewer morphological constraints on the size of contiguous scars (ten Brink et al. 2009a). Their area distribution is therefore more representative of the initiation process.

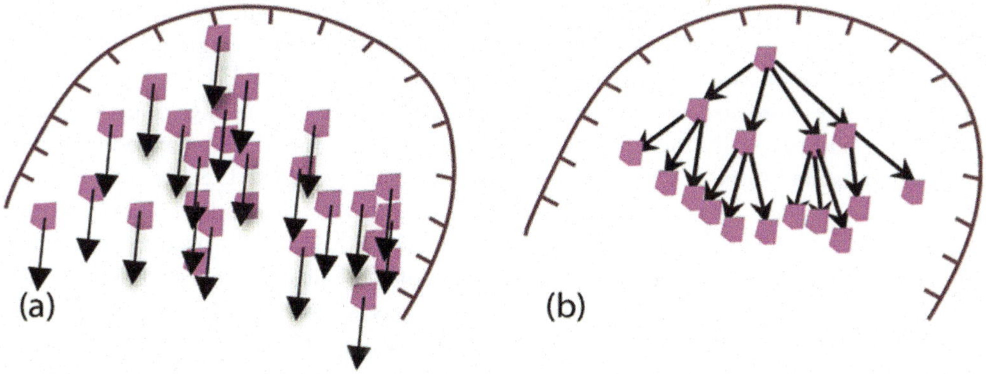

Fig. 9 Alternative modes of landslide initiation. **a** The landslide is an aggregate of many small slope failures that initiate almost simultaneously within an area affected by ground shaking. **b** The landslide is a cascading avalanche that propagates from one or a few sources, similar to the rupture of an earthquake, or to the spread of a forest fire (ten Brink et al. 2014)

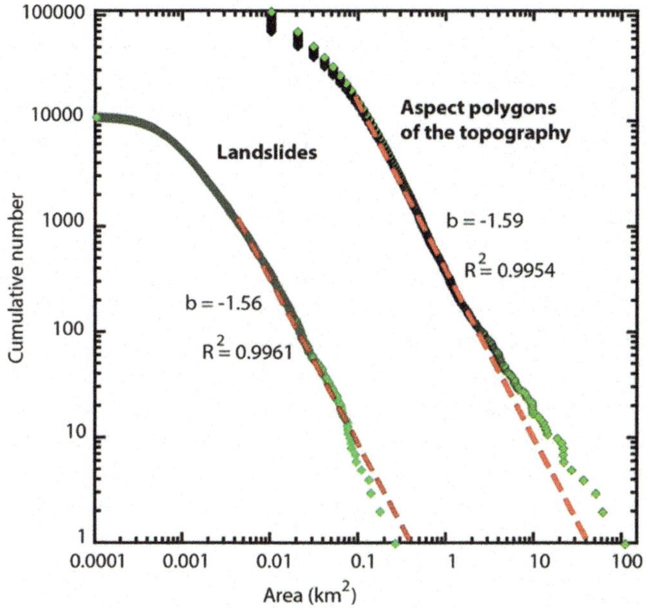

Fig. 10 Comparison between cumulative area distribution of aspect polygons in the topography of the Northridge area and the 1994 M6.7 Northridge earthquake landslides areas measured by Harp and Jibson (1995) (ten Brink et al. 2009a)

Statistical Analysis for Hazard Assessment

Warnings Based on Earthquake Location and Magnitude

The procedure to calculate maximum distance, r_{max} to slope failure from a rupturing fault, discussed in the section "Lognormal Distribution is Indicative of Triggering by Earthquakes", allows us to determine which earthquakes have the potential to generate submarine landslides. For example, the largest potential earthquake along the U.S. Atlantic margin is estimated to be M7.5 (Frankel et al. 1996). An M7.5 must be located within 100–150 km of the continental slope and outer rise to generate submarine landslides on a 2°–6° slope (Fig. 7a; ten Brink et al. 2009b). Therefore, historical earthquakes such as the 1888 Charleston and 1755 Cape Ann earthquakes would not have caused a significant slope failure on the continental slope. The calculated maximum failure distance from the fault, provides a conservative guideline for issuing tsunami warnings following an earthquake offshore the eastern U.S. If the earthquake magnitude is smaller, then it has to be located closer to the continental slope and rise to cause slope failure. Figure 7a also show that accelerations from small earthquakes (<M4.5) are too small to cause significant landslides even if the earthquakes are located under the continental slope. Keefer (1984) made similar empirical observations for terrestrial landslides in the continental U.S. (Fig. 7b), namely, earthquakes with local magnitude $M_L < 4$ do not generate landslides, and $M_L < 5$ earthquakes may not generate rock slumps, block slides, rapid soil flows, and subaqueous landslides.

If landslides are mainly triggered by earthquakes, then shouldn't the landslide rate model mirror the earthquake rate model? Unfortunately, however, slope failures depend on the availability of sediments ready to fail. As discussed in Sect. 4.2, if sedimentation rate between earthquakes is low, then sediments which previously did not fail will in fact increase their shear strength with additional shaking (Sawyer and Devore 2015; ten Brink et al. 2016). Slope strengthening with successive shaking explains the much smaller number of failure events encountered in drilling into the Hellenic forearc than was predicted by the earthquake frequency (Strozyk et al. 2010). It also explains the paucity of landslide scars following the giant M 9.1 Sumatra (Indonesia) and the M 8.8 Maule (Chile) earthquakes (Henstock et al. 2006; Völker et al. 2011).

On the other hand, as long as detailed geotechnical knowledge from the world's margins is lacking, we should assume for warning purposes a worse-case scenario in which landslide areas and volumes are correlated with earthquake magnitude (Fig. 7b) and modified by the distance of the rupturing fault from the continental slope.

Construction of Landslide Hazard Curves

An offshore landslide hazard curve that plots the exceedance probability as a function of scar area (or other landslide size metric) can be determined from the landslide area distribution derived above if the duration of mapped (T_D) activity that formed the landslides is known. We assume that landslides occur in time according to a stationary Poisson process (Urlaub et al. 2013), such that the probability of one or more landslides over T years follows an exponential distribution

$$P = 1 - \exp(-\lambda T) \quad (4)$$

where λ is termed the rate or intensity parameter. This is a time-independent distribution, meaning that the probability is constant with time. For comparison, time-dependent distributions have been previously evaluated at an Integrated Ocean Drilling Program site in the Gulf of Mexico with repeated mass transport deposits (Geist et al. 2013). The hazard curve combines the survival function for the landslide size distribution with the overall rate of occurrence (α):

$$\lambda(a) = \alpha[1 - F(a)] \quad (5)$$

The overall rate can be simply thought of as $\alpha = n/T_D$, where n is the number of landslides above the resolution threshold. T_D can be estimated using the age of the oldest landslide, although as described below, this age is frequently unknown. The most complete calculation of α includes uncertainty related to the open time intervals before the oldest slide and after the youngest slide (cf. Geist et al. 2013; Parsons 2008). The estimate of α also assumes that the rate is stationary since the age of the oldest landslide (cf. Lee 2009; Urlaub et al. 2013), which may not be valid in regions where the sedimentation rate varies quickly in geologic time, such as for glaciated margins.

The empirical determination of a landslide hazard curve as outlined above is distinct from a geotechnical probabilistic approach that is primarily used for subaerial landslides. The latter relies on estimates for expected ground shaking from nearby earthquakes, slope, and physical properties including soil shear strength (e.g., Rathje et al. 2014, Saygili and Rathje 2009, Wang and Rathje 2015). Unlike terrestrial environments, the size a landslide within a mapped region the marine environment is not correlated with its local slope. Direct measurement of physical properties such as shear strength and in situ pore pressure are rarely available for offshore regions, which is why an empirical approach is considered.

In many margins, the age of the oldest slide, as a proxy to determine the duration of mapped landslide activity, has yet to be determined. The U.S. Atlantic is one margin where the age of the oldest landslides ($\sim$20 kyr) can be estimated (ten Brink et al. 2014) and is used to demonstrate the calculation of an offshore landslide hazard curve. A hazard curve for this margin is shown in Fig. 11 (red), using the parameters estimated for the lognormal distribution and Eqs. (4) and (5) above. For comparison, a hazard curve for the Queen Charlotte margin is also shown in Fig. 11 (blue). In this case, we make the assumption that T_D is constrained by the time that the Cordilleran Ice Sheet began to retreat, approximately 14 kyr ago (Clague and James 2002). The validity of the stationary rate assumption for this margin depends on the length of time during which sedimentation changed from a glacial to interglacial rate. The hazard curves for the two data sets are truncated to the left at the minimum observed scar area determined from the minimum resolution afforded by the multibeam data, which varies between the data sets. The two hazard curves, representing end-member margins of the ones analyzed, are different by orders of magnitude for design probabilities of 0.001 or less (i.e., mean return times of 1 kyr or greater), primarily because of differences in the size distributions for the two margins. One could argue that the U.S. Atlantic and Queen Charlotte margins are outliers based on the pairwise comparison tests (Geist and ten Brink 2019) and that the typical range is smaller than shown in Fig. 11. The 10 margins discussed in the section "Lognormal Distributions of Landslide Scars", however, represent only a subset of all margins that have had landslide activity, making it difficult to determine whether the U.S. Atlantic and Queen Charlotte margins are true outliers.

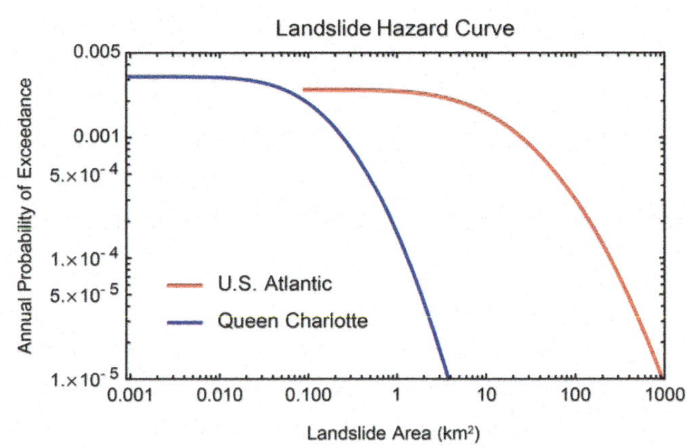

Fig. 11 Landslide hazard curves for the Queen Charlotte and U.S. Atlantic margins (Geist and ten Brink 2019)

It is useful to understand how different estimates of T_D affect the calculated landslide hazard curve. For this, we can take advantage of the fact the lognormal shape parameter σ is similar among most of the margins, as discussed in Sect. 3. The similarity of the shape parameter among margins results in hazard curves having a similar shape when plotted with a logarithmic landslide area axis (Geist and ten Brink 2019). Shown in Fig. 12 is a schematic landslide hazard curve using an average value of σ. Changes in T_D result in vertical shifts in the hazard curve (±10 kyr shown as an example in Fig. 12a), whereas changes in μ results in horizontal shifts in the hazard curve (range using max/min values of μ (0.74–1.98; Fig. 12b). This shift is also prominent in comparing the U.S. Atlantic and Queen Charlotte hazard curves in Fig. 11.

To assess the total landslide hazard for a margin, one can calculate what we term the "landslide release rate," analogous to seismic moment release rate along a fault (Frohlich and Wetzel 2007; Kagan 2002; Peterson and Seno 1984). Landslide release rate $\dot{A}$ (km^2/kyr) for a given mapped area

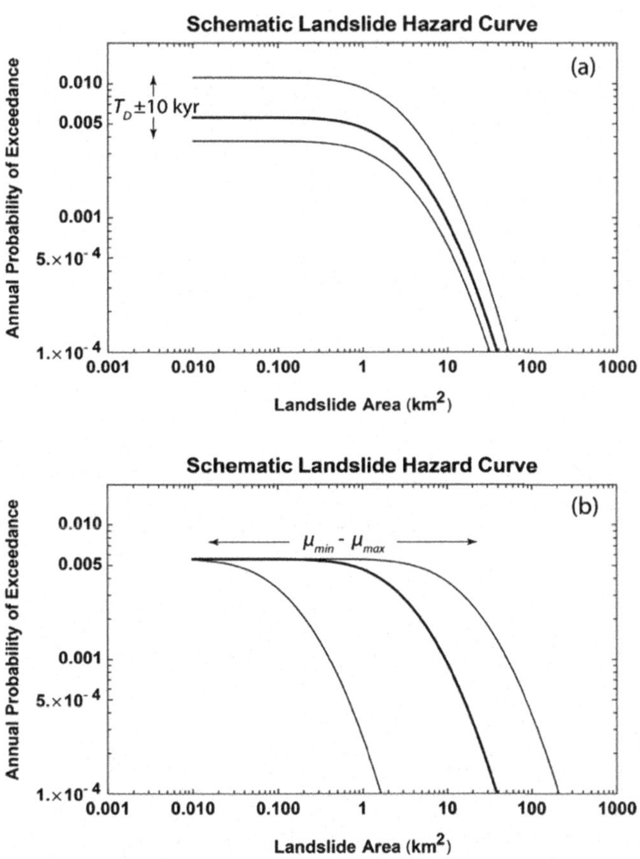

Fig. 12 Effect of changes in T_D (a) and μ (b) on calculated hazard curves, assuming constant σ. Schematic curves (heavy lines) calculated from average values of μ, σ, and n for the 10 margins in Fig. 4 and T_D = 20 kyr. Range is shown by thin curves (Geist and ten Brink 2019)

is related to the total scar area by considering the time period when landslide activity takes place

$$\dot{A} = \frac{A}{T_D} = \alpha \exp\left(\mu + \frac{\sigma^2}{2}\right) \quad (6)$$

where α is the overall rate of landslides above the resolution threshold. As with calculating the hazard curves, the duration of landslide activity is needed to calculate landslide release rate. In contrast to the hazard curves, landslide release rate varies with the area of the margin being considered. Although the estimated landslide release rate applies only to the mapped portion of the margin, the estimate can be extrapolated to unmapped portions of the margin as long as there is confidence that the size distribution applies to the unmapped portions. Landslide release rate can be normalized with respect to the mapped area, similar to how seismic moment release rate is normalized with respect to length of subduction zones in Peterson and Seno (1984).

Normalized landslide release rate based on the temporal rate of landslide scar formation yield the areal fraction of the margin that fails in landslides over time (kyr^{-1}). For the Queen Charlotte and U.S. Atlantic margins, the normalized landslide release rates are 4.6×10^{-4} and 2.9×10^{-2} kyr^{-1}, respectively, using Eq. (6) normalized with respect to the mapped areas and the same ages as for the hazard curve calculations described above. The total normalized landslide hazard for the U.S. Atlantic margin, therefore, is approximately 60 times greater than that for the Queen Charlotte margin.

Hazard Assessment in Carbonate Margins

A volume–area relationship of d = 1.3 was measured for 160 landslide scars measured along the edge of the carbonate margin north of Puerto Rico (Fig. 13b; ten Brink et al. 2006a). This relationship is different from the quasi-linear relationship in siliciclastic margins, discussed in the section "Inferring the Style of Slope Failure from Area-Volume Relationship" and is similar to some terrestrial measurements. The seafloor gradient in which these landslides occur often exceeds 20° and may reaches 45° reflecting the strong cohesion of carbonate rocks. The increase of slide volume as the square (d = 1.5) or cube (d = 1.25) root of the area indicate that these slides are mostly rotational or rock falls.

Translational slides of relatively non-cohesive material, such as along the U.S. Atlantic margin, may undergo simultaneous slope failure over a large area during ground shaking. However, areas and volumes of submarine landslides north of Puerto Rico obey an inverse power law (Fig. 13a; ten Brink et al. 2006a). An inverse power law distribution of landslide areas may be controlled by the

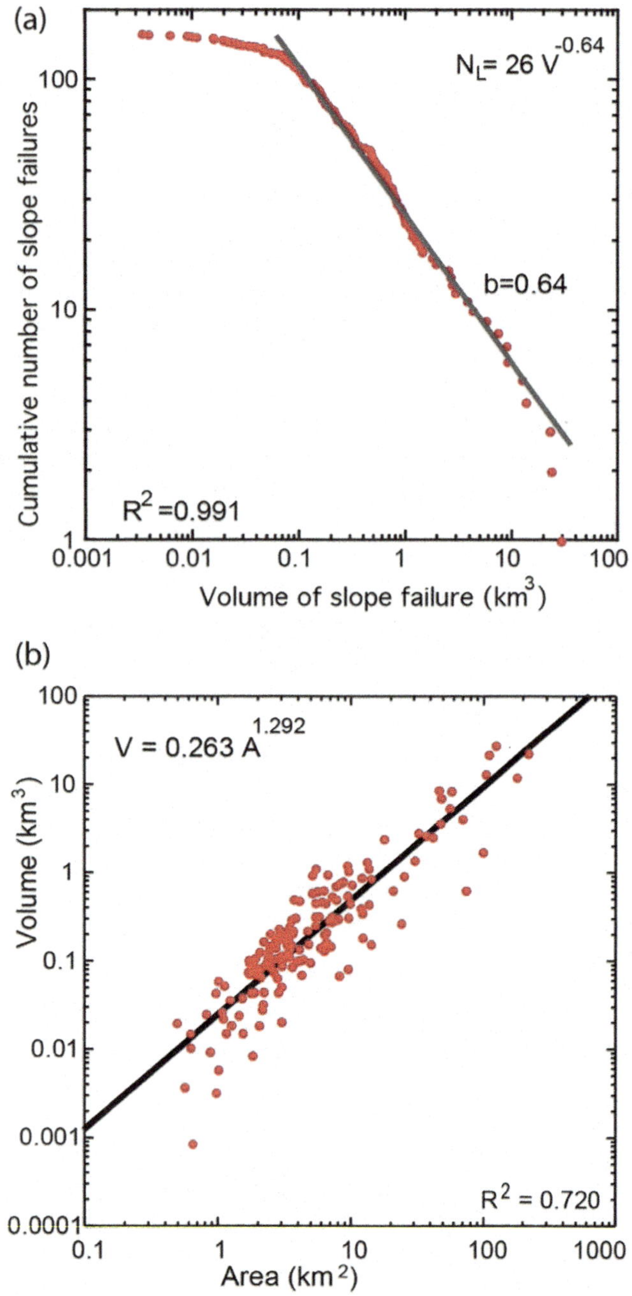

Fig. 13 **a** Cumulative volume distribution of submarine slope failures north of Puerto Rico. Dots—observations. Line—best fit regression line on a log–log plot. **b** area-volume distribution for 160 landslide scars in the carbonate margin north of Puerto Rico

distribution of fractures, and not by earthquake magnitude. Fissures and fractures observed in multibeam bathymetry images at the edge of the platform (ten Brink et al. 2006b) probably develop by tensile stresses and grow by carbonate dissolution in a system of seawater circulation. Slope failures during earthquakes may utilize these fractures. Tensile fracture systems (Katz and Aharonov 2006) and faults (Scholz and Cowie 1990) have been shown to obey an inverse power law distribution in the lab and in the field, attesting to their progressive development with time. It is, therefore, difficult to issue tsunami warnings from landslides in carbonate margins based on earthquake magnitude, because landslide area and volume depend on the availability and the size of unstable blocks, not on the earthquake magnitude as in siliciclastic margins.

Concluding Remarks

We have shown here that statistical approach to studying submarine landslide processes is capable of revealing many aspects of the landslide triggering process and of the associated hazard assessment of this phenomenon, which are difficult to achieve by in-depth studies of individual landslides. We hope that the increasing availability of high-resolution bathymetry over large margin areas, the improvement in GIS analysis methods, and advanced statistical methods will pave the way for additional insights into the workings of submarine landslides.

Acknowledgements The research was supported by the U.S. Nuclear Regulatory Commission Job V6166 and Job N6480 and the U.S. Geological Survey Coastal and Marine Hazards and Resources Program. We thank Tom Parsons for a helpful review and acknowledge the contributions of our coauthors of the papers cited in this summary.

References

Andrews BD, Chaytor JD, ten Brink US, Brothers DS, Gardner JV (2013) Bathymetric terrain model of the Atlantic Margin for marine geological investigations. U.S. Geolo Surv Open-File Rep 2012–1266

Burnham KP, Anderson DR (2010) Model selection and multimodel inference, 2nd edn, vol 488. Springer, New York

Campbell KW (2003) Prediction of strong ground motion using the hybrid empirical method and its use in development of ground-motion (attenuation) relations in Eastern North America. Bull Seismol Soc Am 93:1012–1033

Chaytor J, ten Brink US, Solow AR, Andrews BD (2009) Size distribution of submarine landslides along the U.S. Atlantic Margin. Mar Geol 264(1–2):16–27

Clague JJ, James TS (2002) History and isostatic effects of the last ice sheet in southern British Columbia. Quatern Sci Rev 21(1–3):71–87

Densmore A, Ellis M, Anderson R (1998) Landsliding and the evolution of normal-fault-bounded mountains. J Geophys Res 103:15203–15219

Frankel AD, Mueller CS, Barnhard TP, Perkins DM, Leyendecker EV, Dickman N, Hanson SL, Hopper MG (1996) National seismic-hazard maps; documentation June 1996. U. S. Geolo Surv Open-File Rep 96–532

Frohlich C, Wetzel LR (2007) Comparison of seismic moment release rates along different types of plate boundaries. Geophys J Int 171 (2):909–920

Geist EL, ten Brink US (2019) Offshore landslide hazard curves from mapped landslide size distributions. J Geophys Res: Solid Earth 124:3320–3334

Geist EL, Chaytor JD, Parsons T, ten Brink U (2013) Estimation of submarine mass failure probability from a sequence of deposits with age dates. Geosphere 9(2):287–298

Guzzetti F, Malamud BD, Turcotte DL, Reichenbach P (2002) Power-law correlations of landslide areas in central Italy. Earth Planet Sci Lett 195:169–183

Harp EL, Jibson RW (1995) Inventory of landslides triggered by the 1994 Northridge, California earthquake. U. S. Geol Surv Open-File Rep 95-0213

Henstock TJ, McNeill LC, Tappin DR (2006) Seafloor morphology of the Sumatran subduction zone: surface rupture during megathrust earthquakes? Geology 34:485–488

Hovius N, Stark CP, Allen PA (1997) Sediment flux from a mountain belt derived by landslide mapping. Geology 25:231–234

Hynes-Griffin ME, Franklin AG (1984) Rationalizing the seismic coefficient method. U.S. Army Eng Waterways Exp Stat Mis papers, GL-84–13, 21 pp

Kagan YY (2002) Seismic moment distribution revisited: II. Moment conservation principle. Geophysi J Int 149(3):731–754

Katz O, Aharonov E (2006) Landslides in a vibrating sandbox: what controls types of slope-failure and frequency magnitude relations? Earth Planet Sci Lett 247:280–294

Keefer DK (1984) Landslides caused by earthquakes. Geol Soc Am Bull 95(4):406–421

Laberg J, Vorren T (2000) The Trænadjupet slide, offshore Norway—morphology, evacuation and triggering mechanisms. Mar Geol 171:95–114

Lee HJ (2009) Timing of occurrence of large submarine landslides on the Atlantic Ocean margin. Mar Geol 264(1–2):53–64

Lee HJ, Locat J, Dartnell P, Minasian D, Wong F (2000) A GIS-based regional analysis of the potential for shallow-seated submarine slope failure. In: Proceedings 8th International symposium on Landslides, Cardiff, Wales, June 26–30, pp 917–922

Lee HJ, Orzech K, Locat J, Boulanger E, Konrad J-M (2004) Seismic strengthening, a conditioning factor influencing submarine landslide development. In: Proceedings of the 57th Canadian geotechnical conference (Session 7G): Québec, Canadian Geotechnical Society, pp 8–14

Malamud BD, Turcotte DL (2006) An inverse cascade explanation for the power-law frequency-area statistics of earthquakes, landslides and wildfires. In: Cello G, Malamud BD (eds) Fractal analysis for natural hazards. Geological Society Special Publications, London, pp 1–9

Malamud BD, Turcotte DL, Guzzetti F, Reichenbach P (2004) Landslide inventories and their statistical properties. Earth Surf Processes Landforms 29:687–711

McAdoo B, Pratson L, Orange D (2000) Submarine landslide geomorphology, U.S. continental slope. Mar Geol 169(1–2):103–136

Morgenstern NR (1967) Submarine slumping and the initiation of turbidity currents. In: Richards AF (ed) Marine Geotechnique. University of Illinois Press, Urbana, pp 189–210

Mosher DC, Piper DJW (2007) Analysis of multibeam seafloor imagery of the Laurentide Fan and the 1929 Grand Banks landslide area. In: Sakellariou D, Locat J, Lykousis V (eds) Submarine mass movements and their consequences. Springer, pp 77–88

Newmark NM (1965) Effects of earthquakes on dams and embankments. Geotechnique 15:139–159

Parsons T (2008) Monte Carlo method for determining earthquake recurrence parameters from short paleoseismic catalogs: example calculations for California. J Geophys Res 113:B03302

Peterson ET, Seno T (1984) Factors affecting seismic moment release rates in subduction zones. J Geophys Res 89(B12):10233–10248

Rathje EM, Wang Y, Stafford PJ, Antonakos G, Saygili G (2014) Probabilistic assessment of the seismic performance of earth slopes. Bull Earthq Eng 12(3):1071–1090

Rodriguez CE, Bommer JJ, Chandler RJ (1999) Earthquake-induced landslides: 1980–1997. Soil Dyn Earthquake Eng 18(5):325–346

Sawyer DE, DeVore JR (2015) Elevated shear strength of sediments on active margins: evidence for seismic strengthening. Geophys Res Lett 42:10216–10221

Saygili G, Rathje EM (2009) Probabilistically based seismic landslide hazard maps: an application in Southern California. Eng Geol 109 (3–4):183–194

Scholz CH, Cowie PA (1990) Determination of total strain from faulting using slip measurements. Nature 346:837–839

Strozyk F, Strasser M, Förster A, Kopf A, Huhn K (2010) Slope failure repetition in active margin environments: constraints from submarine landslides in the Hellenic forearc, eastern Mediterranean. J Geophys Res 115:B08103. https://doi.org/10.1029/2009JB006841

Tavakoli B, Pezeshk S (2005) Empirical-stochastic ground-motion prediction for eastern North America. Bull Seismol Soc Am 95:2283–2296

ten Brink US, Geist EL, Andrews BD (2006a) Size distribution of submarine landslides and its implication to tsunami hazard in Puerto Rico. Geophys Res Lett 33:L11307. https://doi.org/10.1029/2006GL026125

ten Brink US, Geist EL, Lynett P, Andrews B (2006b) Submarine slides north of Puerto Rico and their tsunami potential. In: Mercado A, Liu PL-F (eds) Caribbean tsunami hazard. World Scientific, Singapore, pp 67–90

ten Brink US, Barkan R, Andrews BD, Chaytor JD (2009a) Size distributions and failure initiation of submarine and subaerial landslides. Earth Planet Sci Lett 287(1–2):31–42

ten Brink US, Lee HJ, Geist EL, Twichell DC (2009b) Assessment of tsunami hazard to the U.S. Atlantic Coast using relationships between submarine landslides and earthquakes. Mar Geol 264:65–73

ten Brink US, Chaytor JD, Geist EL, Brothers DS, Andrews BD (2014) Assessment of tsunami hazard to the U.S. Atlantic margin. Mar Geol 353:31–54. https://doi.org/10.1016/j.margeo.2014.02.011

ten Brink US, Andrews B, Miller N (2016) Seismicity and sedimentation rate effects on submarine slope stability. Geology 44(7):563–566

Twichell D, Chaytor JD, ten Brink US, Buczkowski B (2009) Morphology of late quaternary submarine landslides along the U. S. Atlantic margin. Mar Geol 264:4–15

Urlaub M, Talling PJ, Masson DG (2013) Timing and frequency of large submarine landslides: implications for understanding triggers and future geohazard. Quatern Sci Rev 72:63–82

Völker D, Scholz F, Geersen J (2011) Analysis of submarine landsliding in the rupture area of the 27 February 2010 Maule earthquake. Mar Geol 288:79–89

Wang Y, Rathje EM (2015) Probabilistic seismic landslide hazard maps including epistemic uncertainty. Eng Geol 196:313–324

Wells DL, Coppersmith KJ (1994) New empirical relationships among magnitude, rupture length, rupture width, rupture area, and surface displacement. Bull Seismol Soc Am 84:974–1002

2020 Kyoto Japan

The Continuing Underestimated Tsunami Hazard from Submarine Landslides

David R. Tappin and Stephan T. Grilli

Abstract

Tsunamis generated by submarine landslides are, relatively, a recently identified hazard, which resulted from the Papua New Guinea event of 1998, when 2200 people died. Recognition of the tsunami hazard from submarine landslides has been possible mainly because of the recent development of advanced technology, such as multibeam echosounders, now available to image the seabed to high resolution. In addition, the architecture of submarine landslides developed from the marine mapping has been the basis for new numerical models of tsunami generation from seabed sediment movement. Tsunamis, post-dating PNG, where an earthquake mechanism was not realistic, have been considered in this recent context, and new relationships identified. These relationships have been particularly with strike-slip and large magnitude (great) earthquakes, as well as with small magnitude earthquake events where a dual mechanism (earthquake and landslide) is most likely. As a result, submarine landslide tsunamis are now recognised from all geological environments; passive, convergent and strike-slip margins as well as volcanoes. Despite these new advances in understanding, however, recognition of the hazard from submarine landslide tsunamis is still limited.

Keywords

Submarine landslide • Tsunami • Hazard • Earthquake

D. R. Tappin (✉)
British Geological Survey, Nottingham, NG12 5GG, UK
e-mail: drta@bgs.ac.uk

S. T. Grilli
University of Rhode Island, Narragansett, RI 02882, USA
e-mail: grilli@uri.edu

Introduction

Over the past 20–30 years, there have been a number of catastrophic tsunamis, which resulted in over 300,000 fatalities (Tappin 2017a), and it seems, that we are living in the 'Age of Tsunamis'. Most of these tsunamis (e.g. 2004, Indian Ocean) were generated from earthquakes but, the development of sophisticated seabed mapping capabilities, together with high resolution position fixing and the manipulation of large data volumes have resulted in the identification of numerous submarine landslides (SL) offshore of most continental margins (Tappin 2017c). These landslides pose a significant potential hazard to local coastal communities in the generation of tsunami.

The wake-up call for landslide tsunamis was in July, 1998, when a small magnitude, 7.1 earthquake triggered a tsunami which devastated local coastlines on northern Papua New Guinea (PNG), causing 2200 fatalities (Tappin et al. 2008) (Figs. 1 and 2). For the first time, an offshore marine survey programme, funded by Japan, identified a small volume (6 km^3) slump as the tsunami mechanism (Tappin et al. 2001). The data acquired during the marine surveys included multibeam echosounder and seismic data, sediment cores and seabed photographs and videos (Figs. 1 and 2). Based on these data, new numerical models for tsunami generation from submarine landslides were developed.

In this context, consideration of previous events, including those of the Grand Banks in 1929, and Storegga, 8200 years BP, resulted in a reappraisal of the previously unappreciated tsunami hazard from seabed sediment movement (Tappin 2017b).

On September 28th, 2018, an earthquake, together with subaerial and coastal landslides generated a tsunami in Palu Bay, Sulawesi, Indonesia (Arikawa et al. 2018; Gusman et al. 2019; Jamelot et al. 2019). There were 4100 fatalities associated with these events. Perhaps over 1000 perished in the tsunami. But as yet, the tsunami mechanism is still uncertain (Gusman et al. 2019; Jamelot et al. 2019).

K. Sassa et al. (eds.), *Understanding and Reducing Landslide Disaster Risk*,
ICL Contribution to Landslide Disaster Risk Reduction,
https://doi.org/10.1007/978-3-030-60196-6_24

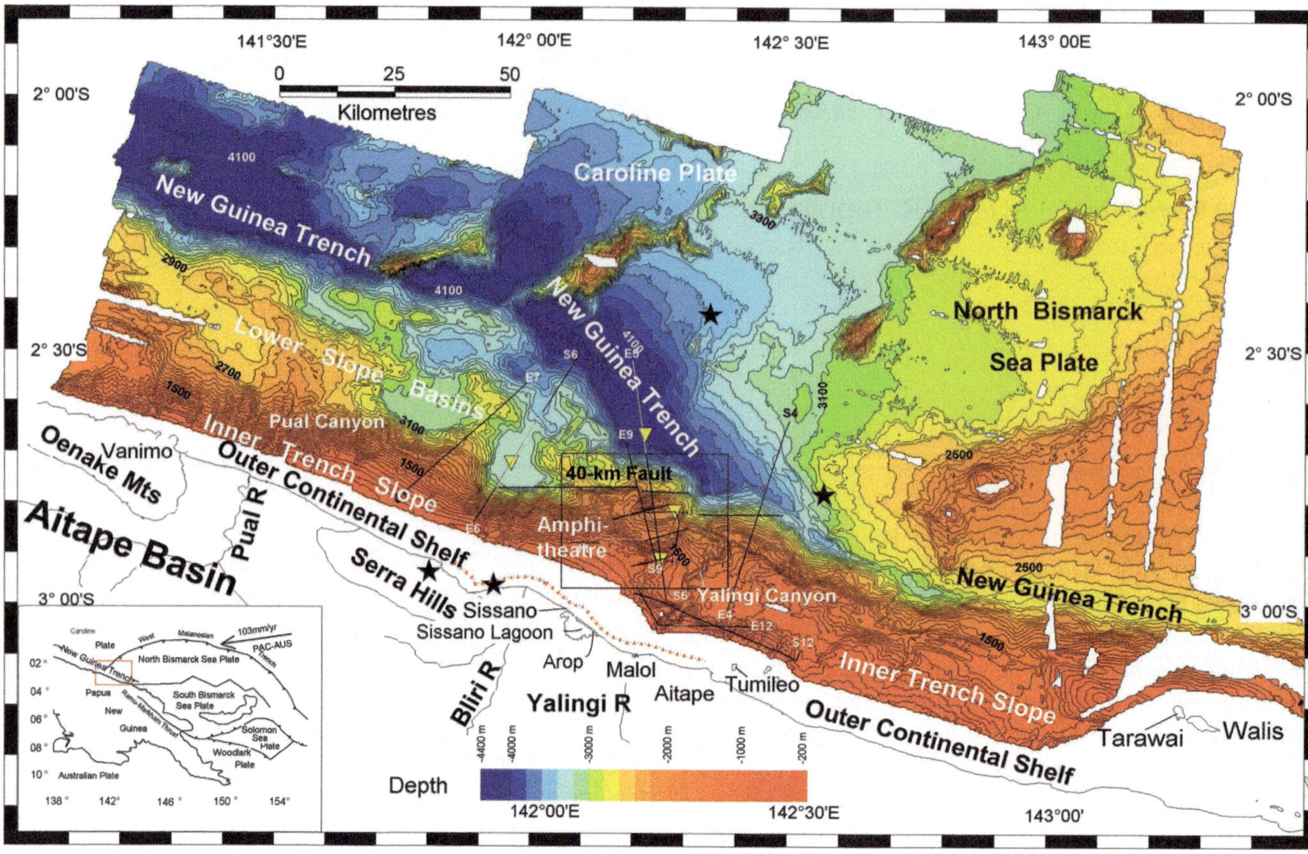

Fig. 1 Bathymetry and main morphologic elements offshore of northern Papua New Guinea together with the main coastal locations and features (Box is the area of map in Fig. 2). Red triangles = main area devastated by the 17th July 1998 tsunami. (Figure reproduced from Tappin et al. 2001)

Three months after the Palu event, an eruption of Anak Krakatau, in the Sunda Strait, resulted in a flank collapse, which generated a tsunami onto the adjacent coasts of Java and Sumatra that resulted in 437 fatalities (Grilli et al. 2019). Here we consider the present state of knowledge on submarine landslide tsunamis, their potential hazard, their impact and mitigation.

Evidence for Submarine Landslides

Mapping the seabed to the resolution we have available today resulted from three main technological developments: the multibeam echosounder (MBES), satellite global positioning systems and computers able to store and manipulate large (enormous) data volumes. All of these technologies have become readily and increasingly available over the past 30–40 years; MBES since 1990, GPS since the mid-1990s (Theberge and Cherkis 2013), and computing power over the past several decades. One of major advances resulting from these developments has been the acquisition seabed bathymetry over large area of the ocean seabed (Fig. 3), and the identification along many, if not most, continental and island margins, of submarine landslides, and the recognition of their common occurrenc (Fig. 4). Not all margins are mapped (Fig. 4), with major gaps off Africa, but good coverage along the margins of the Pacific and North Atlantic oceans. The seabed coverage, together with the history of historical tsunami events, allows us to present a new review of the hazard from submarine landslides.

Passive Margins

Submarine landslides are now mapped along many passive margins, although few are associated with identified tsunamis (Fig. 4). This may seem surprising, until it is realised that the identification of tsunamis, especially prehistoric events, is mainly from sediments deposited as the tsunamis flood the land. The preservation potential of these sediments is very low because of intense erosion in these environments.

Although few landslides are dated, many failed during glacial periods when sea levels were much lower and continental shelves more exposed, so sediments laid down are

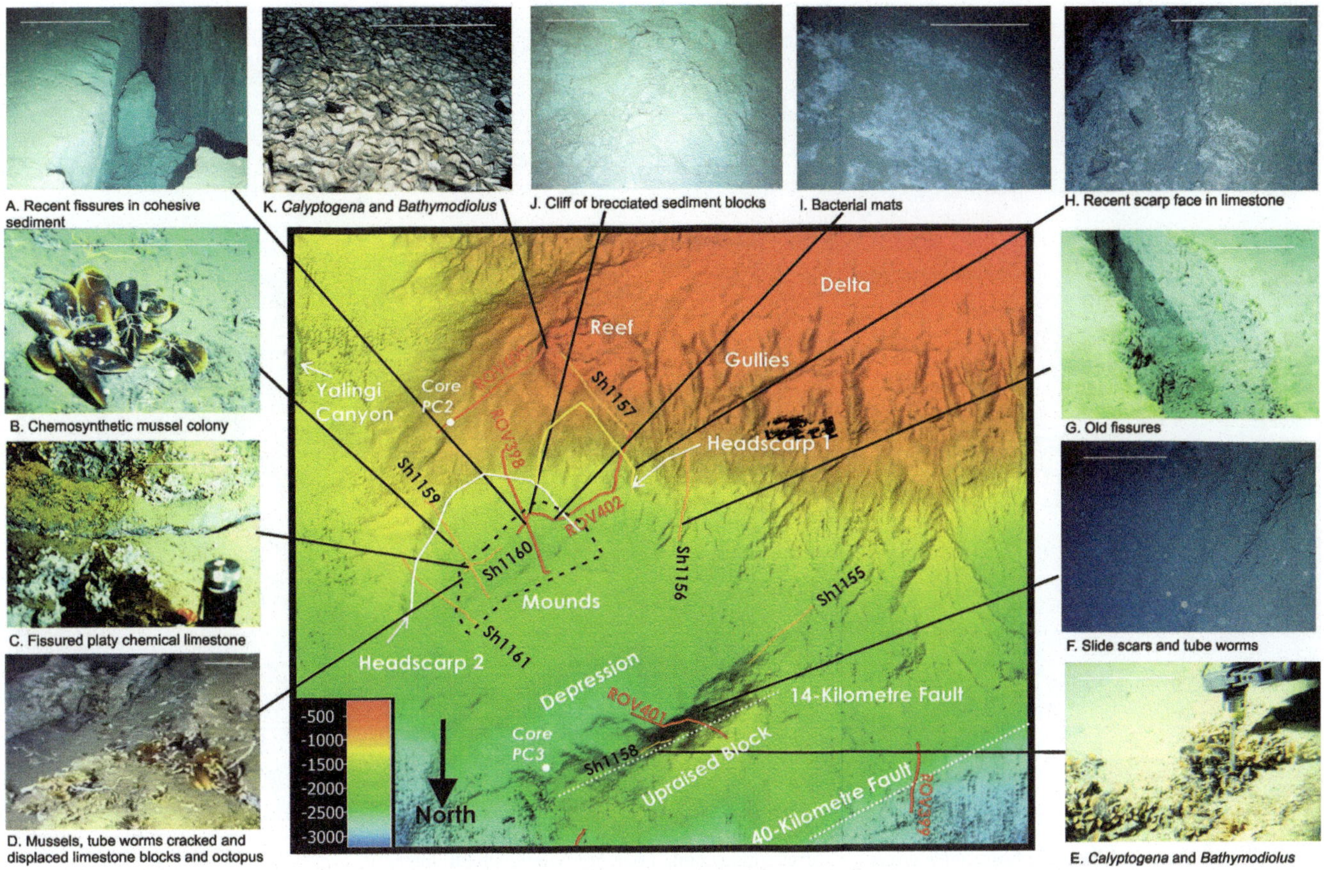

Fig. 2 Amphitheater area off Sissano lagoon (for location see Fig. 1) with main morphologic features labelled and photographs of significant seabed features (*Scales* solid white line = 1 m)

now below sea level and probably eroded during post-glacial marine transgression.

The best studied passive margins are those of the North Atlantic along the coasts of Europe and North America, where there is an extensive database of submarine landslides, their depositional character, age and triggering mechanisms (Fig. 4) (Tappin 2017c). From these studies, it is apparent that there is a strong climate control on submarine landslides, which is both geographical and temporal (Bryn et al. 2005). In the Atlantic, continental margins may be defined by latitudinal variations that may geographically be classified as 'glaciated' north of 56° N (southern tip of Norway), 'glacially-influenced' between 26° N to 56° N, and non-glaciated south of 26° N (Fig. 4). In these areas, the only major tsunamis recorded are from the prehistoric landslide event of Storegga off Norway dated at 8200 years BP and Grand Banks, 1929 off Canada (Bryn et al. 2005; Heezen and Drake 1964; Ruffman 1997). Both of these landslides are located in the 'glaciated' or 'glacially influenced' regions, and reflect the strong temporal climate control from the changing sedimentation regimes during the glacial and interglacial periods operating over the past 400,000 years. Although there is general recognition of the climate control on submarine landslide failure the relationships are complex, and as yet have to be established in detail, especially landslide triggering. The main constraint is the age dating of the landslides. Whereas there is now high-resolution age dating of climate change over the late Quaternary and Holocene, few landslides, especially in the Atlantic, are dated to high resolution.

Convergent Margins

There are number of submarine landslide tsunamis along convergent margins, with Papua New Guinea, 1998, perhaps the best known and researched. Other convergent margin landslide tsunamis include those of Messina (1908), Makran (1945) Aleutians (1946), Alaska (1964) Puerto Rico (1918), Flores Islands (1992) and Java (2006) (Fig. 4) (Fritz 2007; Fryer et al. 2004; Hornbach et al. 2008; Imamura et al. 1995; Parsons et al. 2014; Rastgoftar and Soltanpour 2016; Schambach et al. 2020; von Huene et al. 2014). There were 2200 fatalities from the PNG event, 600 people died in the

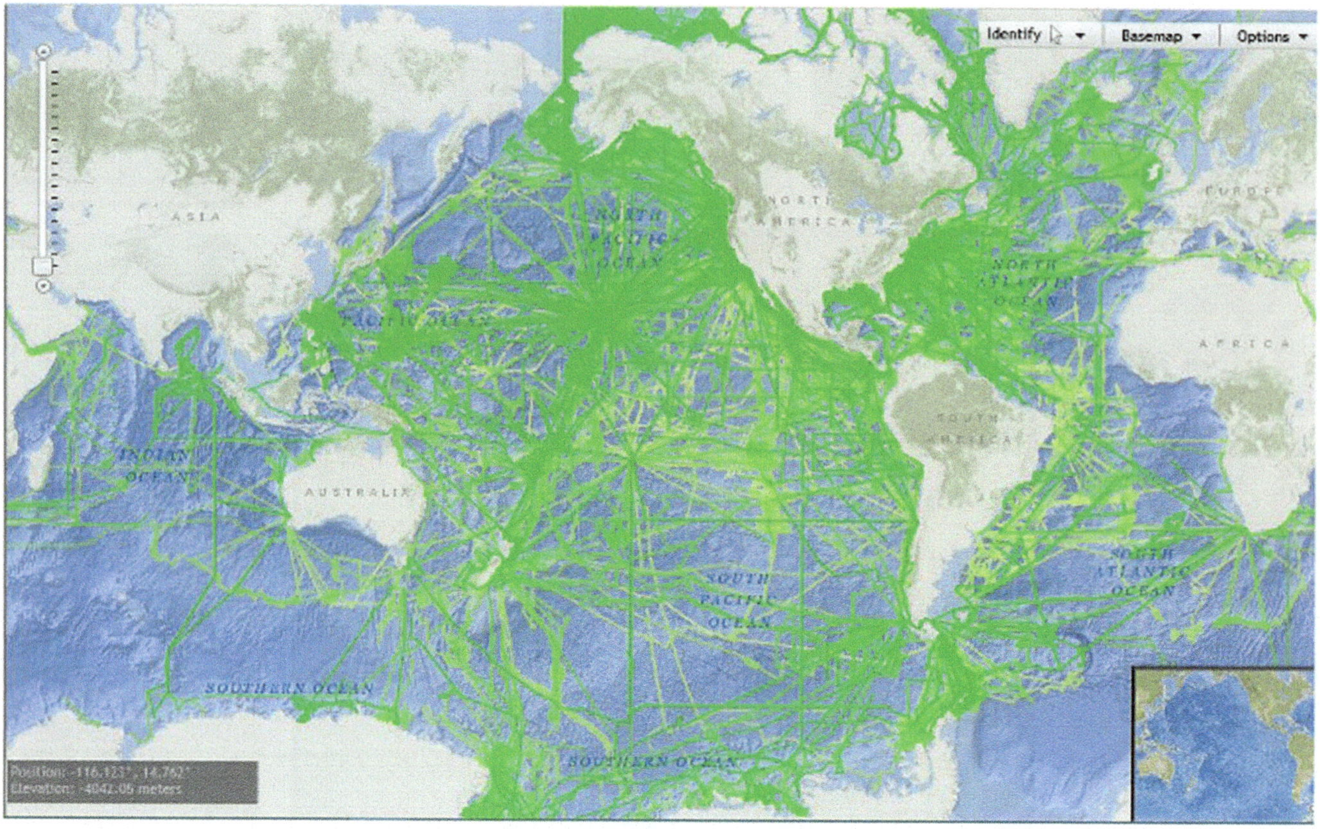

Fig. 3 World ocan multibeam echosounder coverage (https://maps.ngdc.noaa.gov/viewers/bathymetry/)

Java tsunami and 1000 at Flores. Although suspected, for most of these events the submarine landslide contribution remains uncertain, because there are no marine data to evaluate the seabed for landslides, that can be used to underpin numerical tsunami models. Hydroacoustic data has been used to evaluate the tsunami mechanisms of the Aleutians, Alaska, Messina, Puerto Rico and the Aleutians.

The great Alaska earthquake of 1964 directly generated a tsunami, but also triggered sediment failures in fjords, along the nearby coast, which in turn generated local tsunamis (Fig. 4) (Plafker et al. 1969). At Seward, a one kilometre section of the waterfront collapsed into the sea, creating a 10 m high tsunami, with the destruction compounded by the earthquake-generated tsunami, 10 m high, which arrived 30 min later. Most of the 13 people who died, were drowned in the tsunamis. At Valdez, the waterfront collapsed as the seabed failed, and generated a tsunami of up to 52 m in which 32 people died.

The most devastating convergent margin tsunami after PNG was Messina, in 1908, in the Ionian Sea (Fig. 4) where 2000 people are estimated to have drowned (Comerci et al. 2008). The earthquake magnitude (M_S = 7.1) located in the Messina Strait is incompatible with both the height and extent of local tsunami runups which extended southward along the east coast of Sicily (Schambach et al. 2020). There were seabed cable breaks in the southern Ionian Sea, so the most likely mechanism for most of the recorded tsunamis is a submarine landslide (or landslides) off the east coast of Sicily. There have been numerous attempts to identify their locations, and the recent paper by Schambach et al. (2020) identifies a submarine block slide off of Etna (Fig. 5), which explains the tsunami elevations along the central and southern coasts of Sicily, but requires additional landslides to be present farther north. The local tsunami from the Aleutians, 1946 earthquake is now recognised as the result of a submarine landslide. There is still controversy over the tsunami from the Puerto Rico earthquake of 1918, with the landslide contribution still uncertain (Hornbach et al. 2008; López-Venegas et al. 2008).

The most recent devastating tsunami, which again has proved seminal, was on March 11th 2011, on the east coast of Honshu Island, Japan (Lay et al. 2018; Saito et al. 2011). Yet again, the mechanism is controversial. The Mw 9.1 earthquake, was at first considered to be the only tsunami mechanism, however, this cannot explain the elevated (40 m) focussed runups along the Sanriku coast of northern Honshu Island, because these are too far north of the main earthquake rupture (Grilli et al. 2013; MacInnes et al. 2013). Numerical modelling, based only on the earthquake, cannot reproduce tsunami elevations along the Sanriku coast, so an

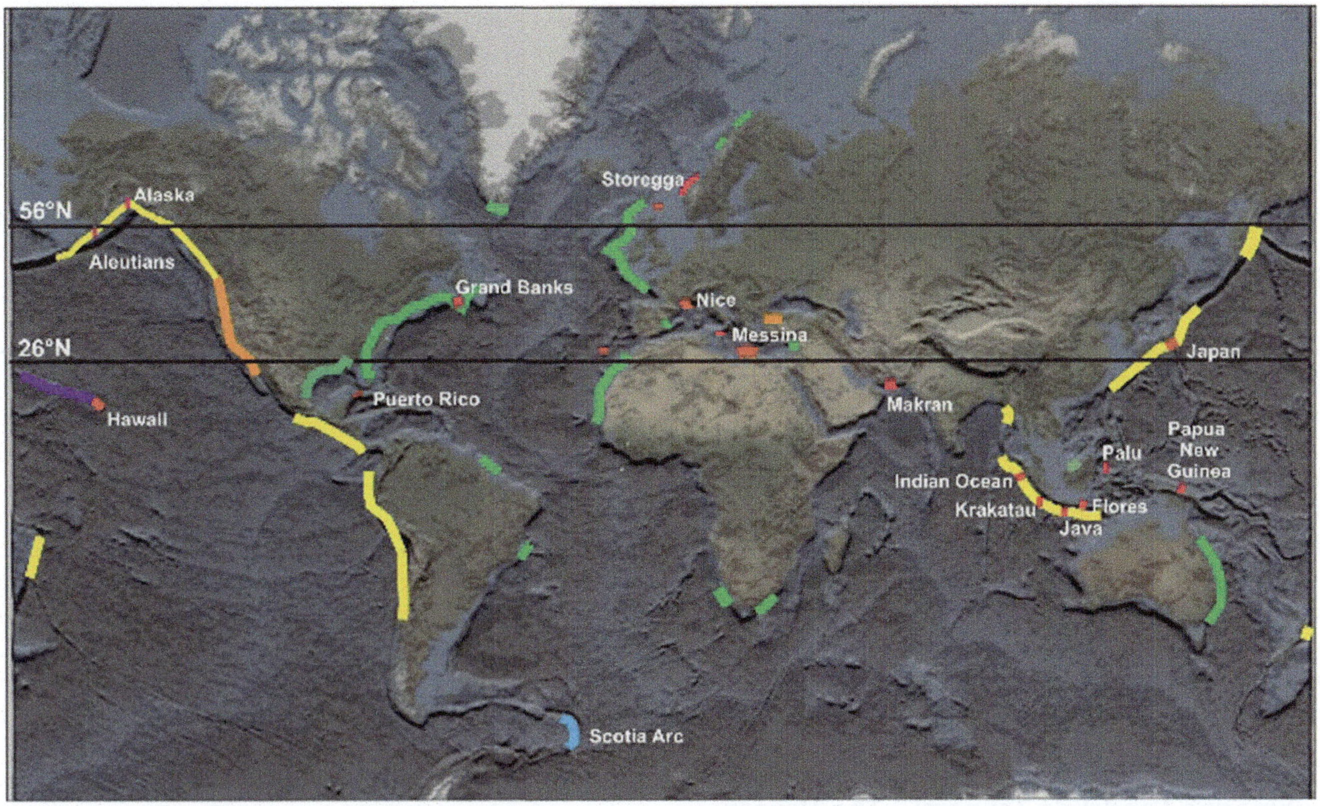

Fig. 4 Global distribution of mapped submarine landslides (SLs). Green, SLs on passive margins: Yellow, SLs located along convergent margins: Orange, SLs on strike slip margins; purple, volcanoes; Red, tsunamis associated with SLs

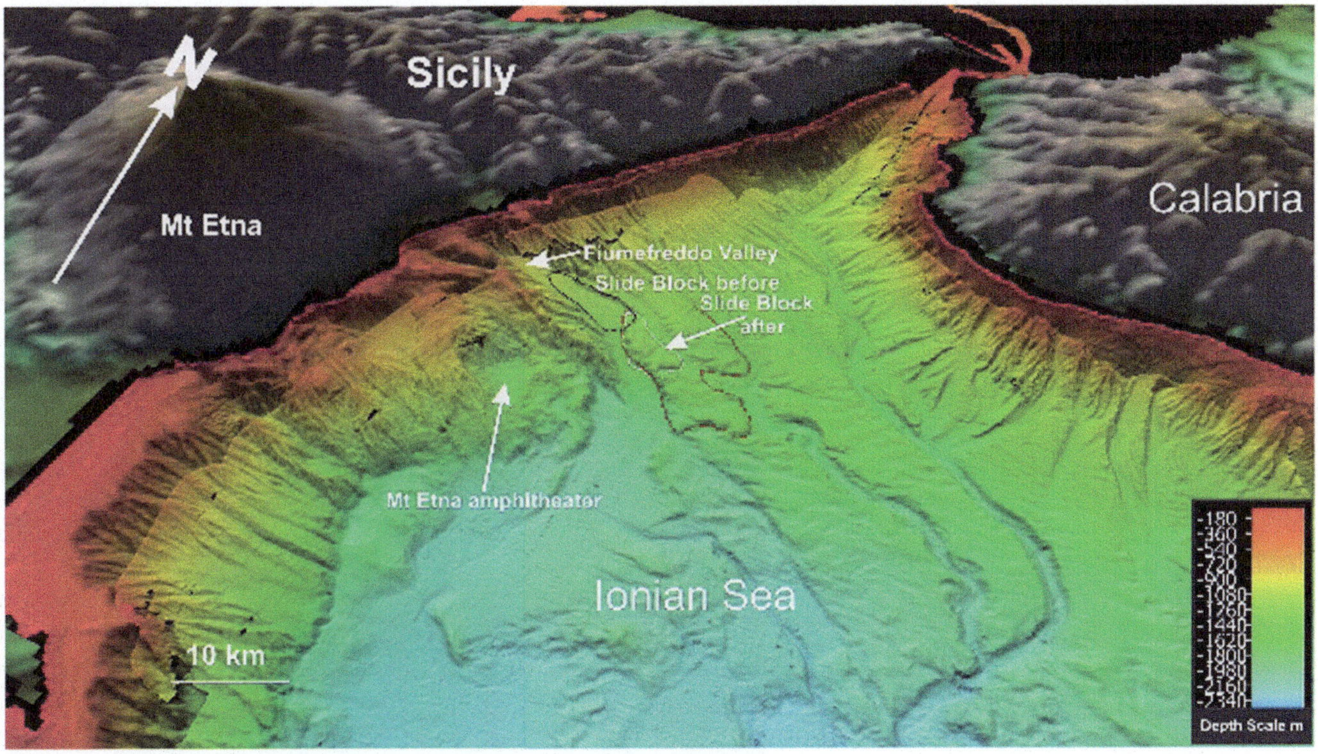

Fig. 5 3D image of the landslide block that contributed to the 1908 Messina tsunami (reproduced from Schambach et al. 2020)

additional submarine landslide mechanism is proposed. Numerical tsunami modelling of a dual mechanism, however, earthquake and landslide, identified from multibeam bathymetry from before and after the earthquake, located off Sanriku, (Fig. 4), best reproduces the tsunami elevations from the 2011 event (Tappin et al. 2012) (Fig. 6).

Strike-Slip Margins

Tsunamis from submarine landslides associated with strike slip earthquakes are more common, than might be realised. Events include those in the Gulf of Izmit (1999) and Haiti (2010) (Hornbach et al. 2010; Tinti et al. 2006). The tsunami from the Kaikoura event in New Zealand, in 2016 has been proposed as from a submarine landslide (Heidarzadeh and Satake 2017), and the earthquake that triggered the Grand Banks landslide in 1929, was strike slip (Bent 1995). Most recently, the tsunami generated in September, 2018 at Palu on Sulawesi, Indonesia, is likely a combination of coastal landsides, and the triggering strike-slip earthquake (Carvajal et al. 2019; Jamelot et al. 2019; Sassa and Takagawa 2019).

Submarine Landslide Tsunami—The Hazard Remains Undefined

The 1998 PNG tsunami was transformative in identifying the hazard from submarine landslides. The initial basis of the identification was the small earthquake and the field surveys, which identified the tsunami elevations, but it was the marine surveys, especially the MBES data, that proved the presence of the submarine landslide. Notwithstanding, it took many years for this to be accepted, and the reasons for this have been addressed in Tappin (2017b).

Extensive mapping of continental margins now shows the common presence of submarine landslides (Fig. 4). Along passive margins, there is a strong climate control on sediment failure, and research in the north Atlantic suggests that failure is related to sedimentation regimes controlled in part by the 120,000-year interglacial/glacial cycles, with triggering mainly from earthquakes (Bryn et al. 2005). The dominant controls on landslides along convergent margins are not as well established, but are more likely to be local sedimentation regimes, and triggering from earthquakes, although earthquake shaking is also shown to both weaken and strengthen sediment (Tappin 2009).

Only three major submarine landslide tsunamis have been identified, researched and validated; Storegga, Grand Banks, Papua New Guinea although with two others, Messina and Japan, there is strong evidence for a dual, earthquake and landslide, mechanism (Schambach et al. 2020; Tappin et al. 2014). Two of these, PNG and Japan, resulted in significant loss of life. Dual source, earthquake/landslide tsunamis are more common than previously recognised, with the earthquake mechanisms, either thrust or strike slip faults. Events such as Messina (1908), Makran (1945) Aleutians (1946), Alaska (1964) Puerto Rico (1918), Flores Islands (1992) and Java (2006), confirm the potential hazard from these dual mechanisms, some of which still require further research to understand their specifics.

Over the past year, large volume submarine landslides have been identified in the Scotia Arc in the south Atlantic which, although millions of years old, possibly generated

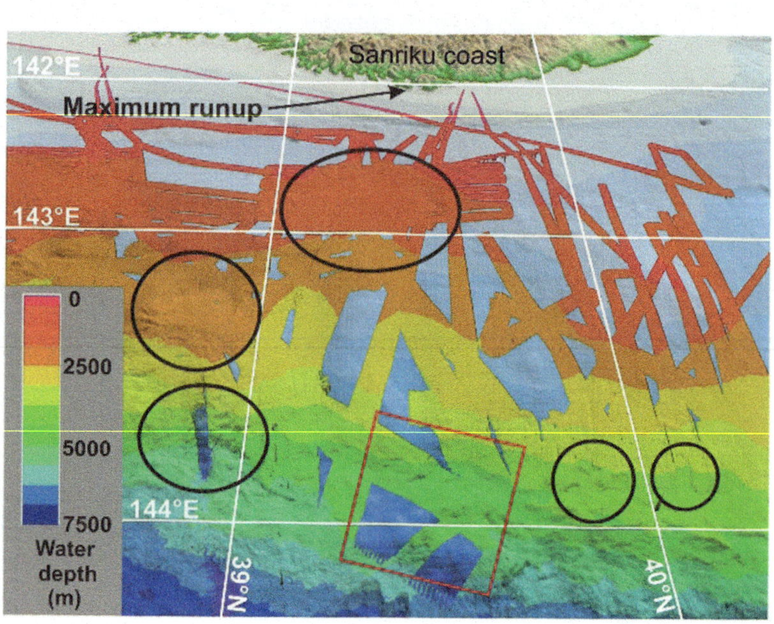

Fig. 6 Multibeam bathymetry off the east coast of Japan, showing submarine landslides (SLs). Black ellipses/circle are SLs: Red square is the location of the SMF triggered by the March 2011 earthquake (reproduced from Tappin et al. 2014)

significant tsunamis (Nicholson et al. 2020). Recent submarine landslide events have also been recognised in Indonesia in the Makassar Straight, between Sulawesi and Kalimantan (Brackenridge et al. 2020). A submarine landslide, tsunami hazard that has received little attention is that associated with strike slip faults. The earliest historical example was the Grand Banks tsunami of 1929, but more recent events include the 1999, Gulf of Izmit event, 2010, Haiti and, in late 2018, Palu, Sulawesi, Indonesia. It is likely that as more areas of seabed are mapped with hydroacoustics, more landslides (and possibly tsunamis) will be identified. For example, an unusual association of low angle, normal faulting triggering submarine landslides in the Banda Sea, west Indonesia has recently been identified as the cause of a major tsunami in 1852 (Cummins et al. 2020).

Acknowledgements D. R. Tappin publishes with the permission of the CEO of the British Geological Survey (United Kingdom Research and Innovation).

References

Arikawa T, Muhari A, Okumura Y, Dohi Y, Afriyanto B, Sujatmiko KA, Imamura F (2018) Coastal subsidence induced several tsunamis during the 2018 Sulawesi Earthquake. J Disaster Res sc20181204–sc20181204. https://doi.org/10.20965/jdr.2018.sc20181204

Bent AL (1995) A complex double-couple source mechanism for the Ms 7.2 1929 Grand Banks earthquake. Bull Seismol Soc Am 85:1003–1020

Brackenridge R, Nicholson U, Sapiie B, Stow D, Tappin DR (2020) Indonesian Throughflow as a preconditioning mechanism for submarine landslides in the Makassar Strait. Geol Soc. London, Special Publications 500:SP500-2019-2171 https://doi.org/10.1144/sp500-2019-171

Bryn P, Berg K, Forsberg CF, Solheim A, Lien R (2005) Explaining the storegga slide. Mar Pet Geol 22:11–19. https://doi.org/10.1016/j.marpetgeo.2004.12.003

Carvajal M, Araya-Cornejo C, Sepúlveda I, Melnick D, Haase JS (2019) Nearly instantaneous tsunamis following the Mw 7.5 2018 Palu Earthquake. Geophys Res Lett 46:5117–5126 https://doi.org/10.1029/2019gl082578

Comerci V et al (2008) One century after the 1908 Southern Calabria—Messina earthquake (southern Italy): a review of the geological effects. Geophys Res Abs 10

Cummins PR, Pranantyo IR, Pownall JM, Griffin JD, Meilano I, Zhao S (2020) Earthquakes and tsunamis caused by low-angle normal faulting in the Banda Sea. Indonesia Nat Geosci. https://doi.org/10.1038/s41561-020-0545-x

Fritz HM (2007) Extreme runup from the 17 July 2006 Java tsunami. Geophys Res Lett 34

Fryer GJ, Watts P, Pratson LF (2004) Source of the great tsunami of 1 April 1946: a landslide in the upper Aleutian forearc. Mar Geol 203:201–218

Grilli ST, Harris JC, Bakhsh TST, Masterlark TL, Kyriakopoulos C, Kirby JT, Shi F (2013) Numerical simulation of the 2011 Tohoku Tsunami based on a new transient FEM Co-seismic source: comparison to far- and near-field observations. Pure Appl Geophys. https://doi.org/10.1007/s00024-012-0528-y

Grilli ST et al (2019) Modelling of the tsunami from the December 22, 2018 lateral collapse of Anak Krakatau volcano in the Sunda Straits. Indonesia Sci Rep 9:11946. https://doi.org/10.1038/s41598-019-48327-6

Gusman AR et al (2019) Source model for the Tsunami Inside Palu Bay Following the 2018 Palu Earthquake. Indonesia Geophys Res Lett 46:8721–8730. https://doi.org/10.1029/2019gl082717

Heezen BC, Drake CL (1964) Grand Banks slump. AAPG Bull 48:221–225

Heidarzadeh M, Satake K (2017) Possible dual earthquake-landslide source of the 13 November 2016 Kaikoura, New Zealand Tsunami. Pure Appl Geophys 174:3737–3749. https://doi.org/10.1007/s00024-017-1637-4

Hornbach MJ et al (2010) High tsunami frequency as a result of combined strike-slip faulting and coastal landslides. Nature Geosci. Advance online publication. http://www.nature.com/ngeo/journal/vaop/ncurrent/abs/ngeo975.html#supplementary-information

Hornbach MJ, Mondziel SA, Grindlay NR, Frohlich C, Mann P (2008) Did a submarine slide triger the 1918 Puerto Rico tsunami? Sci Tsunami Hazards 27(2):1–31

Imamura F, Gica E, Takashi T, Shuto N (1995) Numerical Simulation of the 1992 Flores Tsunami: interpretation of Tsunami Phenomena in Northeastern Flores Island and Damage at Babi Island. Pure Appl Geophys 144:555–568

Jamelot A, Gailler A, Heinrich P, Vallage A, Champenois J (2019) Tsunami simulations of the Sulawesi Mw 7.5 event: comparison of seismic sources issued from a Tsunami warning context versus post-event finite source. Pure Appl Geophys. https://doi.org/10.1007/s00024-019-02274-5

Lay T (2018) A review of the rupture characteristics of the 2011 Tohoku-oki Mw 9.1 earthquake. Tectonophysics 733:4–36. https://doi.org/10.1016/j.tecto.2017.09.022

López-Venegas AM, Brink USt, Geist EL (2008) Submarine landslide as the source for the October 11, 1918 Mona Passage tsunami: observations and modeling. Mar Geol 254:35–46

MacInnes BT, Gusman AR, LeVeque RJ, Tanioka Y (2013) Comparison of earthquake source models for the 2011 Tohoku event using tsunami simulations and near-field observations. Bull Seismol Soc Am 103:1256–1274. https://doi.org/10.1785/0120120121

Nicholson U, Libby S, Tappin DR, McCarthy D (2020) The subantarctic front as a sedimentary conveyor belt for tsunamigenic submarine landslides. Mar Geol 106161. https://doi.org/10.1016/j.margeo.2020.106161

Parsons T et al (2014) Source and progression of a submarine landslide and tsunami: the 1964 Great Alaska earthquake at Valdez. J Geophys Res: Solid Earth 119:8502–8516. https://doi.org/10.1002/2014JB011514

Plafker G, Kachadoorian R, Eckel EB, Mayo LR (1969) Effects of the earthquake of March 27, 1964 on various communities vol 542–G, 43 p., 2 sheets, scales 1:2,500,000 and 1:250,000

Rastgoftar E, Soltanpour M (2016) Study and numerical modeling of 1945 Makran tsunami due to a probable submarine landslide. Nat Hazards 1–17. https://doi.org/10.1007/s11069-016-2356-3

Ruffman A (1997) Tsunami runup mapping as an emergency preparedness planning tool: the 1929 Tsunami in St. Lawrence, Newfoundland, vol. 1. Geomarine Associates, Contract Report for Emergency Preparedness Canada, Ottawa, Ontario

Saito T, Ito Y, Inazu D, Hino R (2011) Tsunami source of the 2011 Tohoku-Oki earthquake, Japan: Inversion analysis based on dispersive tsunami simulations. Geophys Res Lett 38:L00G19. https://doi.org/10.1029/2011gl049089

Sassa S, Takagawa T (2019) Liquefied gravity flow-induced tsunami: first evidence and comparison from the 2018 Indonesia Sulawesi earthquake and tsunami disasters. Landslides 16:195–200. https://doi.org/10.1007/s10346-018-1114-x

Schambach L, Grilli ST, Tappin DR, Gangemi MD, Barbaro G (2020) New simulations and understanding of the 1908 Messina tsunami for a dual seismic and deep submarine mass failure source. Mar Geol 421:106093. https://doi.org/10.1016/j.margeo.2019.106093

Tappin DR (2009) Mass transport events and their Tsunami hazard. In: Mosher DC, Shipp RC, Moscardilli L, Chaytor JD, Baxter CDP, Lee HJ, Urgeles R (eds) Submarine mass movements and their consequences. Springer Science + Business Media, pp 667–684

Tappin DR (2017a) The generation of Tsunamis. In: Encyclopedia of maritime and offshore engineering. Wiley. https://doi.org/10.1002/9781118476406.emoe523

Tappin DR (2017b) The importance of geologists and geology in tsunami science and tsunami hazard. Geological Society, London, Special Publications 456 https://doi.org/10.1144/sp456.11

Tappin DR (2017c) Tsunamis from submarine landslides. Geol Today 33:190–200

Tappin DR, Evans HM, Jordan CJ, Richmond B, Sugawara D, Goto K (2012) Coastal changes in the Sendai area from the impact of the 2011 Tōhoku-oki tsunami: Interpretations of time series satellite images, helicopter-borne video footage and field observations. Sediment Geol 282:151–174. http://doi.org/10.1016/j.sedgeo.2012.09.011

Tappin DR et al (2014) Did a submarine landslide contribute to the 2011 Tohoku tsunami? Mar Geol 357:344–361. https://doi.org/10.1016/j.margeo.2014.09.043

Tappin DR, Watts P, Grilli ST (2008) The Papua New Guinea tsunami of 17 July 1998: anatomy of a catastrophic event Nat Hazards Earth Syst Sci 8:243–266

Tappin DR, Watts P, McMurtry GM, Lafoy Y, Matsumoto T (2001) The Sissano Papua New Guinea tsunami of July 1998—offshore evidence on the source mechanism. Mar Geol 175:1–23

Theberge AE, Cherkis NZ (2013) A note on fifty years of multi-beam. Hydro Int

Tinti S, Armigliato A, Manucci A, Pagnoni G, Zaniboni F, Yalçiner AC, Altinok Y (2006) The generating mechanisms of the August 17, 1999 İzmit bay (Turkey) tsunami: Regional (tectonic) and local (mass instabilities) causes. Mar Geol 225:311–330. http://dx.doi.org/10.1016/j.margeo.2005.09.010

von Huene R, Kirby S, Miller J, Dartnell P (2014) The destructive 1946 Unimak near-field tsunami: new evidence for a submarine slide source from reprocessed marine geophysical data. Geophys Research Lett 41:2014 GL061759. https://doi.org/10.1002/2014gl061759

2020 Kyoto Japan

December 11, 2018 Landslide and 90-m Icy Tsunami in the Bureya Water Reservoir

Viacheslav Gusiakov and Alexey Makhinov

Abstract

On December 11, 2018, a large (with an estimated volume up to 25 million cubic meters) landslide occurred in the middle part of the Bureya water reservoir (Russian Far East). The landslide has completely blocked the river bed and prevented the replenishment of the Bureya HPP reservoir from a large part of its catchment area. In the deep (down to 70 m) reservoir, the landslide generated a destructive tsunami-like wave whose impact on the shore was emphasized by a thick (up to 20 cm) ice cover. The maximum run-up that was reached at the point 2.8 km away from the landslide on the right bank of the Sredniy Sandar river which flows into the reservoir just against the landslide, turned out to be equal to 90 m. The maximum penetration of the wave along the Srendiy Sandar valley was 3.75 km, the water stopped at an altitude of 78 m above the initial water level leaving a pile of tree trunks up to 2 m in height. The maximum run-up reached on the gentle slope of the right (northern) bank, faced directly to the landslide, was up to 62 m, with an inundation distance up to 650 m from the water edge. All the trees and soil cover in this area were completely removed. The height of the mound of broken trees left at the inundation limit in this area reached 5 m. The run-ups on the left (southern) slopes, at a distance of up to one kilometer from the landslide, were within 35–45 m in upstream direction, and within 25–35 m in the downstream direction. Field surveys found that the actual run-ups were by 5–10 (sometimes up to 15) meters higher than the trim line of complete forest and soil destruction visible in air photos and satellite images of the affected area.

Keywords

Tsunami • Landslide • Run-up • Inundation • HPP water reservoir • Tsunami hazard

Introduction

On December 11, 2018, at 14:48 of local time, a large (with an estimated volume up to 25 million cubic meters) landslide occurred on the left (southern) slope of the Bureya water reservoir (Russian Far East) (Fig. 1). The landslide occurred in the middle part of the reservoir where the river valley is oriented almost strictly from east to west. The Bureya river is the third-longest (after the Zeya and the Usury rivers) tributary of the Amur river, the largest river in the Russian Far East. The Bureya HPP built in 2009 with its dam of 140 m high and 20.9 km^3 water reservoir plays an important role in the Russian Far East electric supply system. It has installed capacity of 2.1 MW and produces up to 17% of electricity in the regional electric system. The landslide completely blocked the river bed and prevented the replenishment of the Bureya reservoir from a large part of its catchment area that created a federal-level emergency situation. In the deep (down to 70 m) water reservoir, the landslide generated a destructive tsunami-like wave whose impact on the banks was emphasized by a thick (up to 20 cm) ice cover.

The climate of the area is sharply continental, with the long cold winter (average January temperature is about −28 °C) and relatively short warm summer (average July temperature is about +18 °C). Vegetation is dominated by forests consisting mostly of larch-trees and birch-trees with minor inclusions of pine-trees and alders. One of important features of the territory is the wide presence of permafrost

V. Gusiakov (✉)
Siberian Division, Institute of Computational Mathematics and Mathematical Geophysics, Russian Academy of Sciences, Novosibirsk, Russia
e-mail: gvk@sscc.ru

A. Makhinov
Institute of Water and Ecological Problems, Far East Division, Russian Academy of Sciences, Khabarovsk, Russia
e-mail: amakhinov@mail.ru

K. Sassa et al. (eds.), *Understanding and Reducing Landslide Disaster Risk*,
ICL Contribution to Landslide Disaster Risk Reduction,
https://doi.org/10.1007/978-3-030-60196-6_25

Fig. 1 General view of the landslide scar on the southern slope of the left bank and the landslide body blocking the river bed. Note the trees, remaining in inclined but standing position on the top of the rear part of the landslide. Photo by A. Makhinov taken on 25.12.2018 from a helicopter MI-8

rocks with thickness from 50 to 100 m (Kondratieva 2019). The islands of permafrost are associated with the river valleys and the northern slopes of the terrain.

The landslide did not have any discernible trigger. It occurred in winter in very stable weather conditions when the average daily temperatures were about −30 °C. Despite the Byreya reservoir is located in the earthquake-prone region, no seismic quakes were recorded in this particular area for that day.

Due to the remoteness of the area and the fact that the region is sparsely populated, the event went unnoticed and become publicly known only on December 22, 2018 when two local hunters came to the site and took photos of the affected area. The first reconnaissance helicopter flight to the site was made on December 25, 2018. This quick survey, however, confirmed that the landslide had completely blocked the river bed and generated a destructive tsunami-like wave that removed the forest and the soil in the area of most severe impact (Fig. 2). The first on-site field survey was completed in the middle of January 2019, when the initial instrumental measurements of the landslide body were made. This was followed by several other short-time helicopter visits made in February, March and April 2019, but a detailed on-site survey of the landslide and tsunami run-up measurements were only performed in June 2019.

In mid-February 2019, the Russian Ministry of Defense carried out a special operation to make a water passage in the body of the landslide and restore the integrity of the basin to provide a free flow from the upper to the lower part of the reservoir. This was done to ensure the normal functioning of the Bureya HPP and to prevent flooding of the settlements and roads upstream the landslide during spring flood. The initial 50 m wide passage made in February was widened up to 150 m in April-May by spring waters flow that eventually eliminated the risk of flooding.

For the Russian territory, the occurrence of such a significant tsunami-like phenomenon generated by a coastal slope failure, is a rare event but on a global scale such phenomena occur quite regularly. Only for the last two decades, the global tsunami catalog has recorded five large (more than 50 m in height) landslide-generated tsunamis that occurred on the banks of reservoirs, lakes, fjords and bays: November 21, 2000 Greenland (51 m); April 21, 2007 Aysen Fjord, Chile (60 m); April 4, 2007 Chehalis Lake, Canada (38 m); October 18, 2015 Icy Bay, Alaska (190 m); June 17, 2017 Greenland (90 m). Historically, a record-high

Fig. 2 The area on the right (northern) bank of the Bureya reservoir near the mouth of the Sredniy Sandar river. Note the evident trim line, below which vegetation and soil cover have been completely removed by the upslope water flow and the gullies produced on the steep coastal slope by backwash curents. Photo by A. Makhinov taken on 25.12.2018 from a helicopter MI-8

(525 m) wave was generated on July 10, 1958 in Lituya Bay in Alaska after a collapse of its steep eastern slope, triggered by a strong local earthquake (Miller 1960). The largest hydraulic disaster caused by a slope failure occurred on October 9, 1963 on the Vajont Dam in northern Italy. A collapse of 270 million cubic meters of rocks from the steep slope of Monte Toc located in the immediate proximity to the dam, caused a 250 m wave that overtopped the dam and rushed down a narrow gorge, destroying the downstream villages of Langarone, Piave and Rivalta where 1910 people were killed (Panizzo et al. 2005).

Landslide

The landslide occurred in the middle part of the Bureya reservoir about 90 km from the dam, at the point of 50.5589° N, 131.4719° E. In this area, the Bureya river flows almost strictly from east to west; its valley forms a deep gorge with a high (about 400 m) and steep (slopes up to 30–35°) left bank and a relatively flat, terrace-like right bank of about 1 km wide that goes up to 50 m above the present water level. After construction of the dam, the valley was flooded to form a water reservoir of about 500–550 m wide and down to 70 m in depth. The tributaries' estuarine sections are also filled with water and form narrow and deep bays of 1–3 km long. In this part of the river channel, the water flow is pressed to the left bank and cuts the base of the slope, constantly increasing the steepness and reducing its stability.

The bedrocks composing the steep left slope of the valley are represented by Proterozoic igneous rocks penetrated by younger dikes (Makhinov et al. 2019). In this area, the rocks are severely fractured by inclusions of stronger monolithic blocks up to several meters in size. The fracturing essentially weakened the stability of the slopes, preparing the conditions for a massive slope failure. After the beginning of reservoir filling, groundwater began to penetrate under the base of the slope, increasing the water content in the fractured rock in its lower part and reducing slope stability. The stability is further reduced due to significant (15–20 m in range) fluctuations in the water level caused by seasonal variations of the flow rate and uneven HPP water consumption that led to accelerated erosion of the lower part of the slopes.

The failure of the southern slope occurred at the elevation of ~550 m above the initial water level at the time of the event. The landslide's separation bowl is deeply cut into the slope and, being rectangular in plan, has a considerable size of about 600 m in length and 500 m in width (Fig. 3). Along the upper edge of the amphitheater there are several small blocks of rocks that are marked by gaping cracks and ready for displacement but remain in place. In the lower part of the amphitheater formed a thin cover of rock fragments of different sizes that continuously roll down from the exposed slope.

Fig. 3 General view of the landslide scar on the southern bank of the Bureya water reservoir and the body of the landslide with a passage, initially made in February 2019 and then extended by the spring flood in April-May 2019. Photo by A. Ostroukhov taken from quadrocopter "Fantom4" on June 19, 2019

The landslide completely blocked the reservoir with a high rocky dam of 800 m in length and from 150 m (at its base) up to 620 m (in the middle part) in width. The total surface area of the landslide body is 270 thousand m^2. The landslide's total volume calculated based on the results of geodetic measurements, is 24.5 million m^3, with about 4.5 million m^3 of rocks in its surface part. In plan, the landslide body has a fan-shaped form and a complex relief (Fig. 4). The main body of the landslide is shifted to the right bank, which indicates that the slide was of high speed and significant kinetic energy. The wide rear part of the landslide is represented by weakly deformed strata of displaced bedrock, as well as by chaotic piles of huge monolithic blocks and smaller debris. The frontal part of the landslide, which was stopped by the ledge of the terrace-shaped slope of the right bank, underwent the most significant deformation. As a result, several short and high mounds were formed in its upper part, separated by rather deep linear depressions in the form of steep slope ditches (Makhinov 2019).

The rock mass entered the water reservoir and produced a destructive tsunami-like wave that traveled upstream and downstream at a distance exceeding 10 km. The water wave stripped vegetation up to tens of meters above the shoreline and eroded soil layer down to the depth of freezing (70 cm) along extended parts of the shore.

A prominent trim line is evident in these areas, below which vegetation has been almost completely removed. Along the shoreline, the geomorphic impacts of tsunami differ over short distances due to changes in shoreline configuration and slope steepness. The erosion is most severe near the landslide, especially in the areas of the north coast that lie directly against the collapsed slope. Erosion traces are less noticeable further from the landslide and on the bank slopes, protected from direct exposure to waves. The degree of erosion also decreases upslope, away from the shoreline where the soil cover remains largely undamaged due to reduced velocity of currents.

Tsunami Impact

Due to severe winter conditions in the area (thick snow cover, low daytime temperatures), the detailed instrumental measurements of the run-up heights are possible only in the

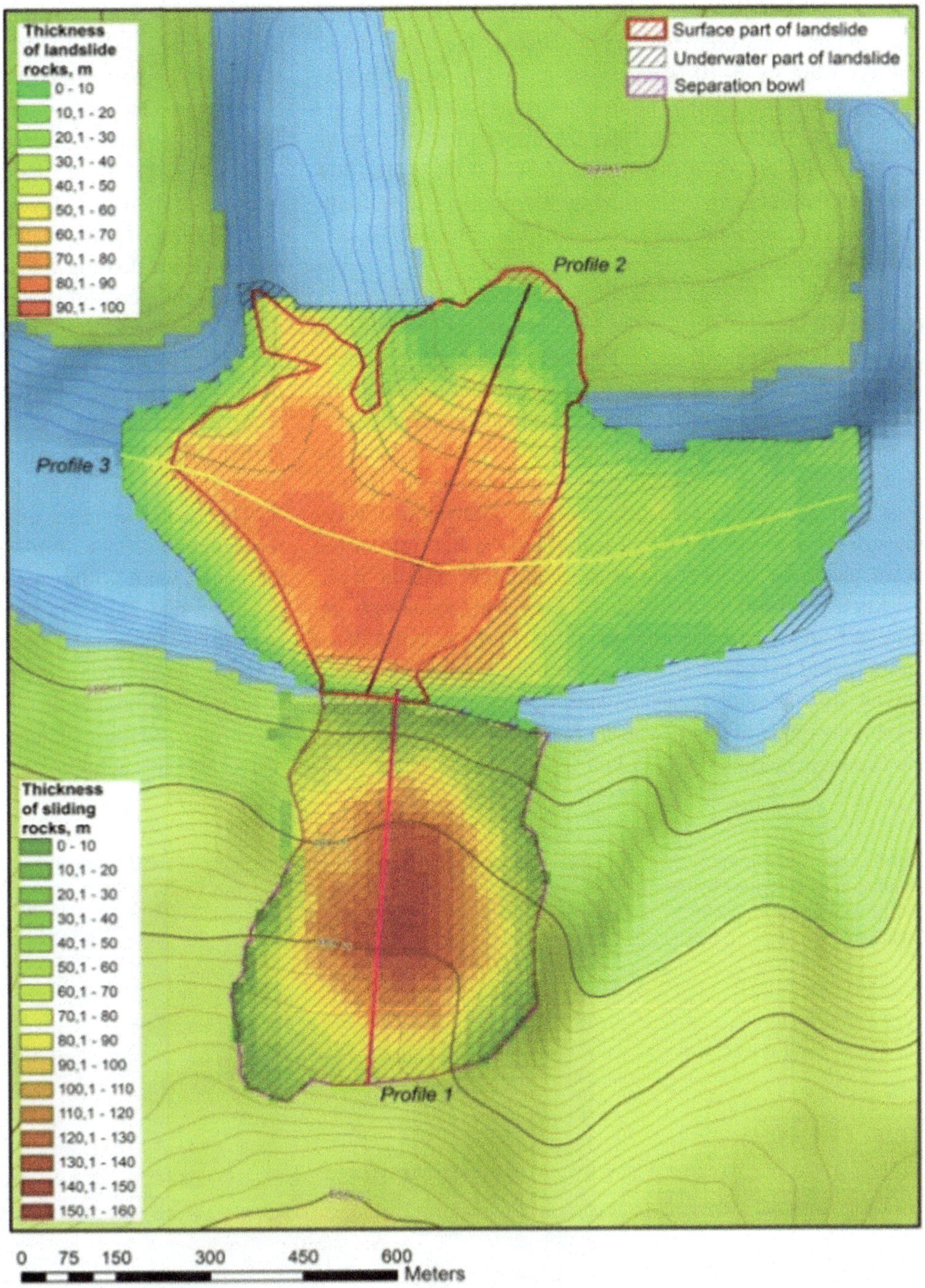

Fig. 4 Model of the landslide based on the results of field geodetic measurements made in January of 2018

summer season after the ice cover melts and the annual spring flash flood passes. These measurements were made during a field survey of the affected area that was conducted in the second half of June 2019.

During the field survey, the inundation boundaries and tsunami run-up heights were substantially clarified as compared to their initial estimates obtained from the analysis of air photos and satellite images. In total, about 50 run-up heights were measured using a barometric altimeter built into the Garmin portable GPS receivers. Before each measurement or a series of measurements at nearby points, the receiver was zeroed to the reservoir's water level on the day of measurement. The absolute water level in the landslide area was obtained by interpolating the elevation marks at the hydrological posts of the Bureya HPP dam and in the Chekunda village that have a long continuous history of level monitoring.

In June 2019, six months after the event, the limit of the wave run-up along the slopes was clearly marked by the presence of tree branches and pieces of wood on the upslope side of tree trunks, which were flooded but remained intact. The in-site measurements demonstrated that the actual run-ups were by 5–10 (sometimes up to 15) meters higher than the trim line visible in air photos and satellite images made soon after the event. Near the upper limit of inundation, partially uprooted tree trunks indicated the water flow direction that had generally gone up the slope away from the reservoir, but in some cases—toward the reservoir (due to backwash).

Along the large part of the trim line clearly visible on air photos and satellite images, the wave left the piles of broken and damaged trees (Fig. 5). Backwash currents produced deep (up to 1.0–1.5 m) gullies on the steep parts of the right (north) bank lying directly against the landslide (Fig. 6).

The most severe impact of the water wave fell on the banks of the Sredniy Sandar river that flows in the Bureya river just across the landslide (Fig. 7). The highest wave run-up, 90 m above the initial water level at the time of the event, occurred at the right bank of Sredniy Sandar river, 2.8 km away from the landslide (Fig. 8). This value puts the Bureya tsunami of December 11, 2018, in the seventh place on the world list of the strongest tsunamis of landslide origin having instrumental measurements of this parameter. The maximum penetration of a water wave along the valley of the Sredniy Sandar was 3.75 km, its progress stopped at an altitude of 78 m above the initial water level at the point of 50.5844° N, 131.4629° E, where the wave left a pile of tree trunks up to 2 m in height. The backwash current from the Sredniy Sandar valley carried all the cut trees back in the Bureya reservoir to cover the water surface with an almost continuous layer of damaged tree trunks with stripped branches. This thick layer of wood remained frozen in ice until the end of May when the spring flood cleared the water surface from floating trees.

The gentle coastal slope on the right (northern) bank lying directly opposite the landslide suffered the strongest wave impact. The high velocity water flow completely removed both vegetation and soil leaving only damaged stumps and exposed tree roots on the place of dense forest that grew here earlier (Fig. 9) The maximum inundation level reached here 62 m with maximum horizontal flooding up to 650 m.

The run-up heights on the left (southern) slopes of the reservoir, located within a distance of up to 1 km from the landslide, were within a range of 36–48 m in the upstream direction and 25–33 m in the downstream direction. At the same time, the run-up decreased (down to 15–20 m) at the parts of the left coast directly adjacent to the landslide. On several sharp breaks of the reservoir's coastline (capes) the wave had overlapped the capes and the water stream had rolled into the neighboring bays.

Fig. 5 Debris along the inundation limit formed by the cut and broken trees. Photo by V. Gusiakov

Fig. 6 Gullies on the opposite (right) bank of the Bureya river made by the backup wash currents. Photo by V.Gusiakov

Fig. 7 Right bank of the Sredniy Sandar river. The inundation reached here 70 m above the initial water level at the distance about 2 km from the landslide. Photo by V. Gusiakov

Fig. 8 Left bank of the Sredniy Sandar river. The place of the maximum run-up height (90 m above the initial water level) reached here at the distance about 2.8 km from the landslide. Photo by V. Gusiakov

Fig. 9 Damaged stumps and exposed tree roots on the gentle coastal slope located directly opposite the landslide on the right (northern) bank of the Bureya River. Note the clearly marked trim line on the left (southern) bank of the reservoir. Photo by A. Makhinov taken on 25.12.2018

The overview map of all the measured run-ups of the Bureya tsunami plotted in the PDM/TSU graphic shell (PDM/TSU 2020) is shown in Fig. 10. In this map, run-ups are plotted as the scaled vertical lines positioned according to the geographical coordinates of run-up measurement point. The run-up distribution demonstrates a strong directivity of the water wave produced by slope failure with its energy maximum radiated in the direction of mass movement and the leading role of narrow river valleys in increasing the wave height.

Conclusion

The coastal landslide that occurred on the left bank of the Bureya water reservoir on December 11, 2018 is the largest event of this type that happened in the Russian Federation for the last decades. The landslide did not have any discernable trigger. It occurred in winter in very stable weather conditions with the average daily temperature of about −30 °C. No seismic activity was recorded in this area for that day. The

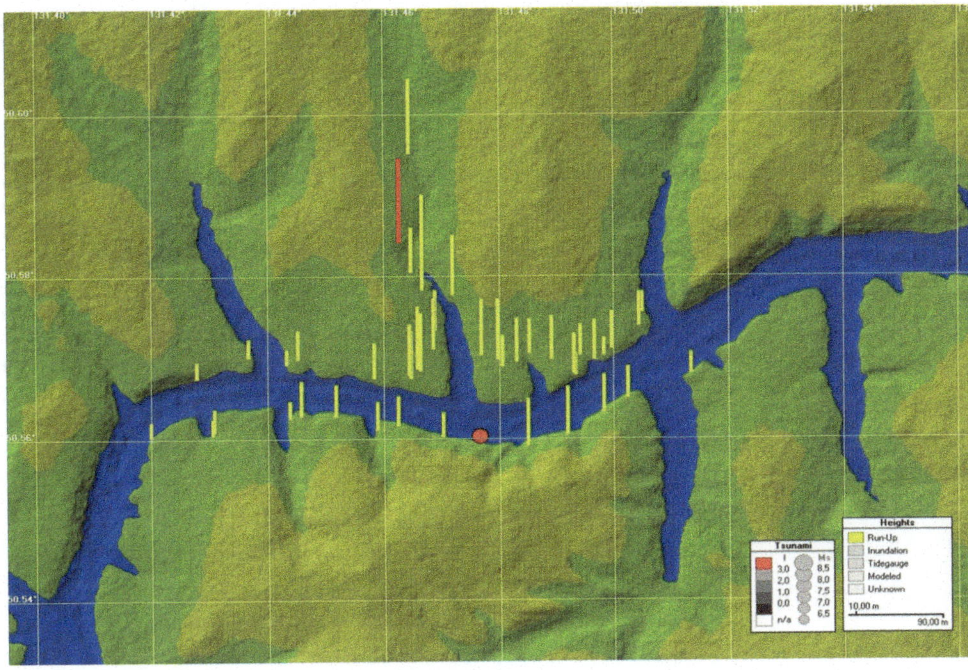

Fig. 10 Map of the measured run-up heights of the Bureya tsunami plotted in the PDM/TSU graphic shell. The red dot marks the landslide. The highest run-up (90 m) is highlighted in red

landslide completely blocked the Bureya River bed and prevented water supply from the 2/3 of river's catchment area that was critical for the normal functioning of the Bureya HPP in winter conditions when the water supply is low.

The event occurred in the very remote, sparsely populated area of the Russian Far East and remained unnoticed for more than a week. The first reconnaissance helicopter flight to the site was made only on December 25, 2018. Due to severe winter conditions in the area (thick snow cover, daytime temperatures down to −35 °C), the detailed field survey was conducted only in June 2019 after the annual spring flash flood was over. During the survey, the detailed instrumental measurements of the landslide body and tsunami run-up heights along the river banks were made.

The landslide generated a destructive tsunami-like wave whose impact on the banks was emphasized by a thick (up to 20 cm) ice cover. The maximum run-up at the right bank of the Sredniy Sandar River reached 90 m above the initial water level at the time of the event. The maximum penetration of the water wave along the valley of the Srendiy Sandar River was 3.75 km, the water stopped at the altitude of 78 m above the initial water level.

In the area of large run-up, high-velocity water currents loaded with thick ice cut or uprooted all the trees (mostly larches and birches) and completely removed the soil down to freezing depth (about 70 cm). Backwash currents produced deep (up to 1.0–1.5 m) gullies on the steep parts of the right (north) bank lying directly against the landslide.

In June of 2019, six months after the event, the upper limit of water inundation along both river banks could be still clearly identified by the tree remnants and the piles of tree trunks left by water along the line of maximum in-land penetration. In the field survey, it was found that the actual run-ups were by 5–10 m (sometimes up to 15 m) above the trim line identified on the air photos and satellite images.

The wave energy decay and run-up heights are determined by a distance to landslide, bathymetry and shoreline configuration relative to a wave propagation direction. The observed pattern of run-up heights suggests that the landslide mass and velocity drive the initial wave size, while the energy decay and inundation heights are defined by a combination of distance to the landslide, bathymetry, wave propagation direction and shoreline configuration.

Propagation of the water wave along the river bed both in upstream and downstream directions was limited by a distance of 10–12 km, although soon after the event the traces of water disturbance was clearly visible (as the lines of broken ice along the coastline) up to the distance of 20–25 km.

The Bureya landslide occurred far away (90 km) from the Bureya HPP and did not directly affect the station's dam. However, the event has demonstrated the potential threat of slope instability for the safety of HPP dams, especially for those having their water reservoirs in a mountain region. In Russia, the two largest HPPs (Sayano-Shushenskaya and Krasnoyarskaya HPPs) are located in such regions. Engineering surveys of slope stability within at least 10 km of HPP dams should be a high priority task to ensure the safety of both HPPs.

Acknowledgements The authors wish to thank Andrey Ostroukhov for providing the high-resolution quadrocopter photo of the landslide, Natalia Reznichenko—for her help in polishing-up the English version, and Tamara Kalasnikova - for her assistance in formatting of the manuscript. The authors also thank the Khabarovsk Regional Headquarter of the Ministry for Civil Defence and Emergencies (EMERCOM) for providing helicopter transportation to the site of the event. The results presented in this paper were supported by the RAS Project no. 0315-2019-0005.

References

Kondratieva L (2019) Bureysky landslide and ecological risks. Vestn Far East Branch Russ Acad Sci 2:45–55. https://doi.org/10.25808/08697698.2019.204.2.005 (In Russian)

Makhinov A, Kim V, Ostroukhov A, Matveenko D (2019) Large landslide in the valley of the Bureya River and tsunami in the reservoir of the Bureya hydroelectric power station. Vestn Far East Branch Russ Acad Sci 2:35–44. https://doi.org/10.25808/08697698.2019.204.2.004 (In Russian)

Makhinov A (2019) Landslide and Tsunami on the Bureya River on December 11, 2018. Priroda 4:12–22. https://doi.org/10.7868/S0032874X19040021 (In Russian)

Miller DJ (1960) Giant waves in Lituya Bay, Bull Seis Soc Am 50:253–266

Panizzo A, De Girolamo P, Di Risio M, Maistri A, Petaccia A (2005) Great landslide events in Italian artificial reservoirs. Nat Hazards Earth Syst Sci 5:733–740

PDM/TSU (Parametric Data Manager for Tsunami data management) graphic shell (2020) NTL/ICMMG SD RAS, https://tsun.sscc.ru/PDM.htm, Accessed 15 Apr 2020

The Link Between Upper-Slope Submarine Landslides and Mass Transport Deposits in the Hadal Trenchs

2020 Kyoto Japan

Kazuko Usami, Ken Ikehara, Toshiya Kanamatsu, Arata Kioka, Tobias Schwestermann, and Michael Strasser

Abstract

Sediment gravity flow is a density current that moves down a slope due to the gravitational force. Submarine landslides are a trigger mechanism for sediment gravity flows, and the movement of a large amount of sediment mass has the potential for generating tsunami. An oceanic trench is a terminal depression in the hadal environment that receives mass transport deposits (MTDs) from the slope. Thick, acoustically transparent layers were found in small basins along the northern Japan Trench. Sediment cores obtained from these basins were composed of thick, homogeneous mud, suggesting the presence of distal MTDs. Radiocarbon dates obtained using bulk organic carbon from the sediment indicated that these MTDs were formed around 2000 years ago and were correlated with each other. Benthic foraminiferal assemblages in the MTDs contained upper-slope species. There are many submarine landslides distributed along the upper slope of the Hidaka Trough; thus, the upper slope of this trough is important not only as a sediment source but also as a source of carbon for the Japan Trench, which is an effective sink of organic carbon. However, the exact locations of the submarine landslide on the upper slope have not yet been determined. Furthermore, from the geohazard point of view, the upper slope is an area of large sediment movements, which should be considered in the context of tsunami hazard mitigation.

Keywords

Submarine landslide • Mass transport deposit • Foraminifera • Japan Trench

Introduction

Sediment gravity flow is a density current that moves down a slope due to the gravitational force (e.g. Pickering and Hiscott 2015). Submarine landslides are a trigger for the generation of sediment gravity flows. Because surface sediments with a high water content, together with sea water, intrude into the sliding mass, the initial landslide develops into a high-density sediment gravity flow (Pickering and Hiscott 2015), resulting in mass transport deposits composed of the slide, slump, subaqueous debris-flow deposits, and turbidites (e.g. Pickering and Corregidor 2005). For transformation from subaqueous debris flow to turbidity current, hydroplaning is an important process (Mohrig et al. 1998). For example, upper-slope submarine landslides were triggered by the 1929 Grand Banks earthquake; these transformed into submarine debris flows, and subsequently into turbidity currents (Heezen and Ewing 1952; Piper et al.

K. Usami · K. Ikehara (✉)
National Institute of Advanced Industrial Science and Technology, Geological Survey of Japan, Tsukuba Central 7, 1-1-1 Higashi, Tsukuba, 305-8567, Japan
e-mail: k-ikehara@aist.go.jp

K. Usami
e-mail: usami-kzk@janus.co.jp

K. Usami
Energy Consulting Department, Safety and Environmental Analysis Unit, Japan NUS Co. Ltd, Nishi-Shinjuku Prime Square 5F, 7-5-25 Nishi-Shinjuku, Shinjuku-Ku, Tokyo, 160 0023, Japan

T. Kanamatsu
Japan Agency for Marine-Earth Science and Technology, Research Institute of Marine Geodynamics, 2-15 Natsushima-Cho, Yokosuka, 237-0061, Japan
e-mail: toshiyak@jamstec.go.jp

A. Kioka
Department of Earth Resources Engineering, Kyushu University, 744 Motooka, Nishi-Ku, Fukuoka, 819-0395, Japan
e-mail: kioka@mine.kyushu-u.ac.jp

T. Schwestermann · M. Strasser
University of Innsbruck, Institute of Geology, Innrain 52f, Innsbruck, 6020, Austria
e-mail: Tobias.Schwestermann@uibk.ac.at

M. Strasser
e-mail: Michael.Strasser@uibk.ac.at

K. Sassa et al. (eds.), *Understanding and Reducing Landslide Disaster Risk*,
ICL Contribution to Landslide Disaster Risk Reduction,
https://doi.org/10.1007/978-3-030-60196-6_26

1999; Schulten et al. 2018). These turbidity currents broke submarine cables that were lying on the slope and formed turbidite beds on the basin floor (Heezen et al. 1954). A tsunami caused by this earthquake that had a height of 13 m hit the coast of Newfoundland and damaged houses along the coast. Recent tsunami simulations indicate that submarine landslides caused the height of the tsunami to be greater locally (Lovholt et al. 2018). Submarine landslides can be observed from bathymetric maps and seismic reflection records around the Japanese islands (e.g. Ikehara et al. 1990; Noda et al. 2013; Moriki et al. 2017). However, little sedimentological evidence on the transformation of flows from landslide to turbidity current through to submarine debris-flow has been reported.

Giant earthquakes drive large amounts of sediment and organic carbon from shallower water to the deeper hadal zone (e.g. Kioka et al. 2019a, b; Mountjoy et al. 2018). The 2016 Kaikoura earthquake produced submarine landslides on the uppermost slope of the Kaikoura Canyon. The flow, which resulted in eroding the channel floor and rims, transported around 850 Mt of sediment and 7.2 ± 3.1 Mt of carbon to the deep Hikurangi Channel (Mountjoy et al. 2018). Kioka et al. (2019a) calculated the volume of sediments transported by the 2011 Tohoku-oki earthquake to the small basins in the Japan Trench, and concluded that the largest amount of sediment (54×10^6 m^3) was deposited in the southernmost basin, where the Nakaminato Submarine Canyon is located. In contrast, small amounts of sediment ($3–5 \times 10^6$ m^3) were deposited in the basins of the central Japan Trench, where no trench-connected submarine canyon exists. These facts indicate the importance of submarine canyons to the transport of sediments to the deep hadal zone.

A thick, acoustically transparent layer was observed in sub-bottom profiler records of the northern Japan Trench; the Ogawara Submarine Canyon is located on the canyon slope here (Kioka et al. 2019b). This layer extended to at least four small basins along the trench axis, and the total volume of the layer was calculated as $413\ (+85/-26) \times 10^6$ m^3 (Kioka et al. 2019b). Submarine landslides cause tsunami to become locally bigger; it is, therefore, important to clarify the relationship between such large volume sediment-transport events and large sediment failures in slope areas in order to evaluate the largest class of tsunami along the Pacific coast of northern Tohoku and south–eastern Hokkaido.

Several large ($M > 7.6$) earthquakes, which have the potential to trigger submarine landslides and sediment gravity flows, have occurred near the northern Japan Trench (Toda 2016). These were the 1994 Sanriku-oki (M 7.6) earthquake, the 1968 Tokachi-oki (M 7.9) earthquake, the 1933 Sanriku-oki (M 8.3) outer rise earthquake, the 1931 (M 7.6) earthquake, the 1987 (M 7.7) earthquake, and the 1896 Sanriku-oki (M 8.2) tsunami earthquake. The 2011 Tohoku-oki (Mw 9.1) earthquake occurred south of the study area.

In this paper, we describe the sediment lithology of four cores recovered from different basins in the northern Japan Trench, and examine the age of mass transport deposits (MTDs) from three of the cores. Further, we report preliminary results on benthic foraminiferal assemblages in the MTD taken from one of the cores, and show that the upper slope is the origin of the MTD. This result suggests that submarine landslides at the upper-slope influence the hadal environment.

Materials, Data and Methods

We collected four sediment cores from different basins along the axis of the Japan Trench (Fig. 1). Cores KS14-16 PC08 (Lat.: 39°45.942'N, Lon.: 144°16.229'E, water depth: 7368 m, core length: 843.7 cm) and KS18-10 PC01 (Lat.: 40°05.63'N, Lon: 144°19.57'E, water depth: 7541 m, core length: 639.3 cm) were collected from basins NO3 and NO5 (Kioka et al. 2019b), respectively, during cruises KS-14-16 and KS18-10 of the R/V Shinsei–Maru. Cores GeoB21817 (Lat.: 40°23.743'N, Lon.: 144°25.226'E, water depth: 7608 m, core length: 971 cm) and MR17-06 PC01 (Lat.: 40°27.341'N, Lon: 144°26.930'E, water depth: 7590 m, core length: 560.1 cm) were collected from basin NO8 (Kioka et al. 2019b), the northernmost basin of the Japan Trench, during cruises SO251a of the R/V Sonne and MR17-06 of the R/V Mirai, respectively. The cores were vertically split, and logged visually. Radiocarbon age determinations using bulk organic carbon from the sediment were conducted for seven samples from core KS18-10 PC01 and nine samples from MR17-06 PC01. Bulk sediment samples with volumes of around 5–10 cm^3 were used for the measurements. Before the measurements were made, the samples were washed by acid to remove carbonates. The radiocarbon dates were determined at the AMS radiocarbon facility of Beta Analytics Co. Ltd. To examine the benthic foraminiferal assemblages, four samples from core GeoB21817 were analyzed. The samples were washed gently and filtered using a 63 μm mesh. After splitting the residues to obtain the desired number of benthic individuals, we identified and counted the number of individual benthic foraminifera.

Lithological columns for cores KS14-16 PC08, KS18-10 PC01, and GeoB21817 were shown in Ikehara et al. (2018) and Kioka et al. (2019b). The sub-bottom profiler records of

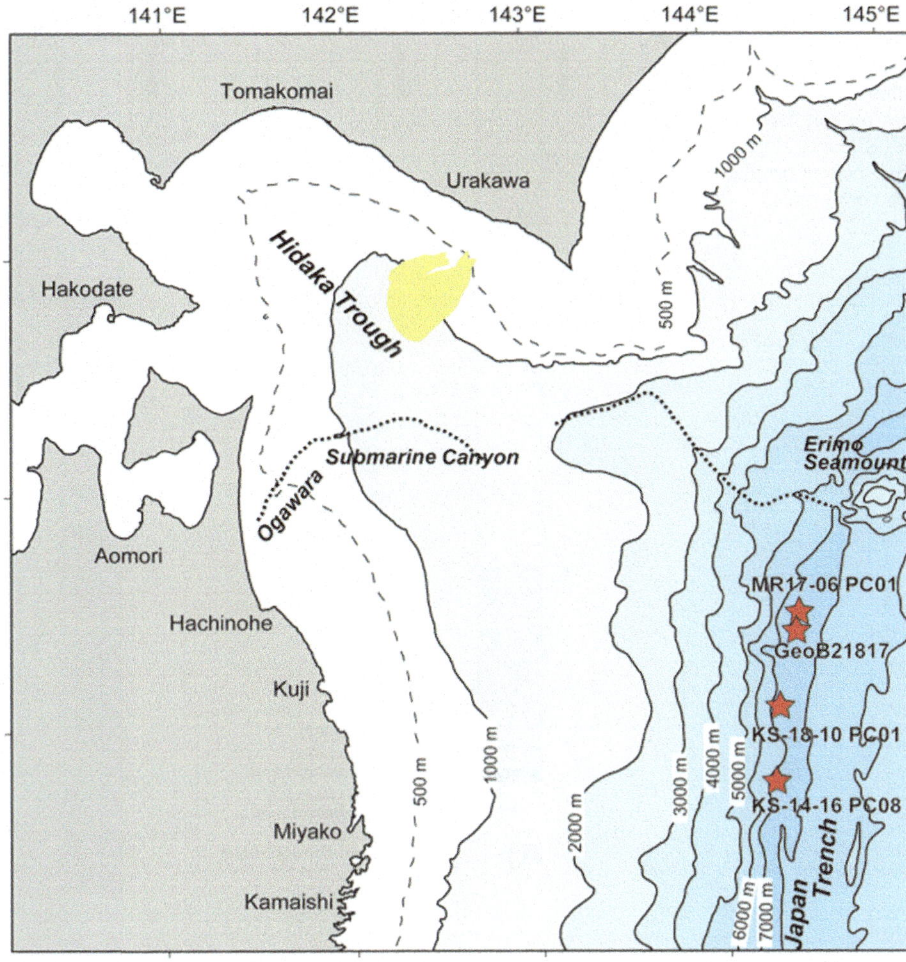

Fig. 1 Bathymetry of the study area and locations of the cores. The areas of the upper slope of the Hidaka Trough marked in yellow are the youngest submarine landslides studied by Noda and Katayama (2013)

each basin and the radiocarbon dates of bulk organic carbon from 12 horizons of core GeoB21817 were given in Kioka et al. (2019b).

Results and Discussion

Characteristics of the MTDs in the Observed Cores, Their Ages, and the Correlation Between Them

Apart from core KS14-16 PC08, the other cores (cores KS18-10 PC01, GeoB21817, and MR17-06 PC01) had similar lithologies (Fig. 2a). The upper 1.2–2.7 m of the cores (the upper unit) were composed of bioturbated diatomaceous mud, a type of hemipelagic mud. No calcareous remains were observed in this unit. In core KS14-16 PC08, two fine-grained turbidites which had thicknesses of several tens of centimeters were intercalated in the upper unit. Below the upper unit, homogeneous diatomaceous mud without bioturbation (the lower unit) was present. Calcareous foraminiferal tests were found in the lower unit of core GeoB21817. The thickness of the lower unit was more than 2.5 m in the case of core KS14-16 PC08, more than 5 m for core KS18-10 PC01, more than 7 m for core GeoB21817, and more than 4.5 m for core MR17-06 PC01. Thin sand layers were observed in the lower part of core KS14-16 PC08 and in the middle to lower part of core MR17-06 PC01 (Fig. 2a). A gentle fining upward trend was recognized in core MR17-06 PC01. The lower unit had a homogeneous structure without bioturbation and a sand layer intercalation similar to those found by Ikehara et al. (2016) in the thick turbidite layers of the central Japan Trench; it was also similar to divisions T3 to T7 of the ideal fine-grained turbidite unit described by Stow and Shanmugam (1980). Thus, because the basins of the Japan Trench are terminal basins, the lower unit was thought to consist of fine-grained turbidite and to be the distal facies of the MTD. Although Kioka et al. (2019b) indicated the occurrence of a thin ($\sim$80 cm), acoustically transparent layer (U1-NO8) within the upper unit in the northernmost basin (NO8), no fine-grained turbidite was observed in the upper unit of cores GeoB21817 and MR17-06 PC01 collected from basin NO8.

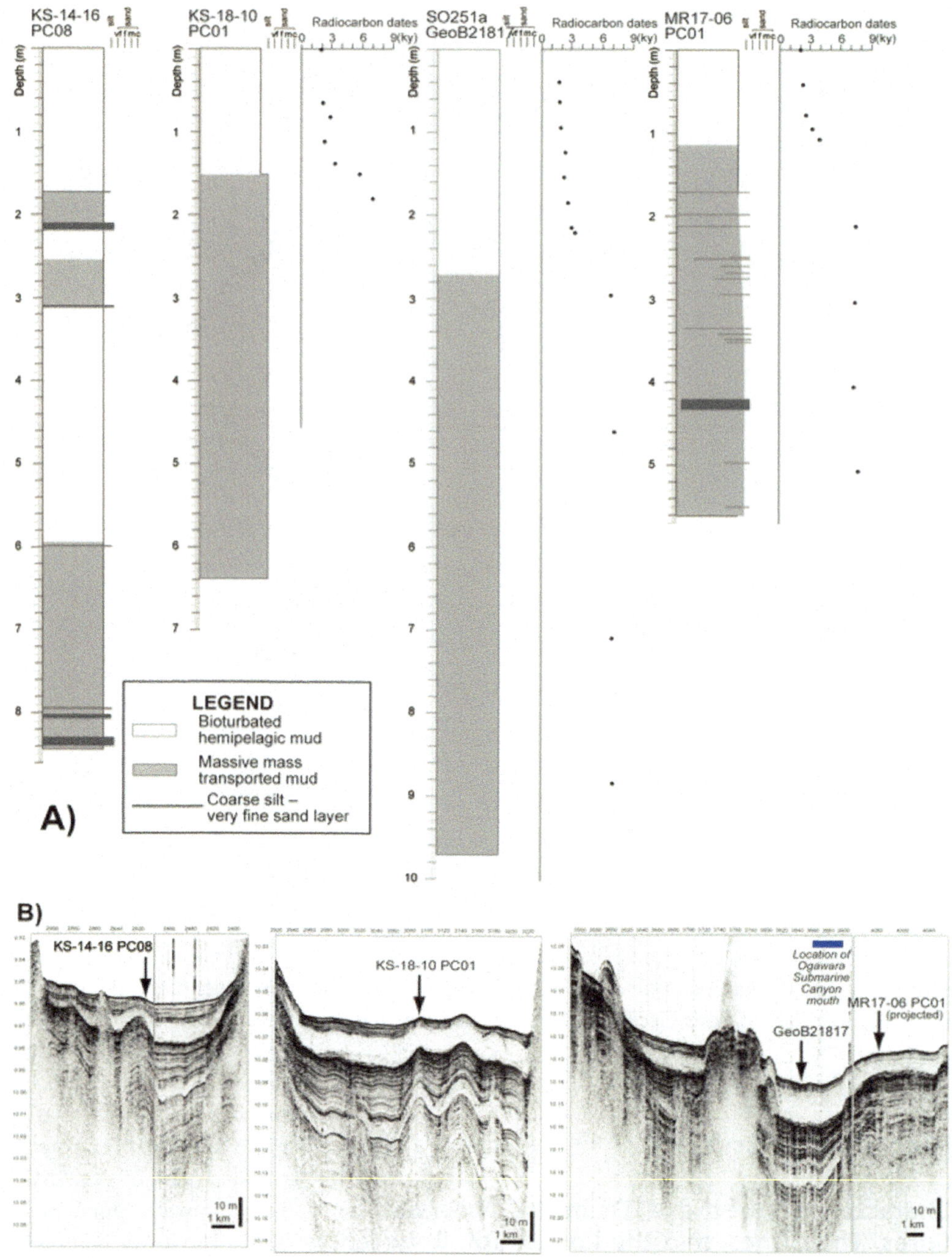

Fig. 2 **a** Columnar sections and radiocarbon dates of the bulk organic carbon in the cores taken in this study. The radiocarbon dates for core GeoB21817 are after Kioka et al. (2019b). **b** Sub-bottom profiler record at each coring sites

The radiocarbon dates of the upper units of cores KS18-10 PC01, MR17-06 PC01, and GeoB21817 (Kioka et al. 2019b) showed a linear relationship with the sub-bottom depths of the measured horizons (Fig. 2a and Table 1). The radiocarbon dates of the MTDs were in the range 5710–7630 years BP. In the case of the mixture of young hemipelagic mud and older MTD mud near the lithological boundary in core KS-18-10 PC01, the radiocarbon dates ranged from 6767 to 7630 years BP. The dates obtained near the bottom of the upper units of the three cores ranged from 3280 to 3800 years BP. This similarity in ages obtained using bulk sedimentary organic carbon suggests that the three distal MTDs are correlated with each other, representing a single depositional event. The radiocarbon dates of bulk sedimentary organic carbon are older than those of planktonic foraminifera (Ikehara 2000), with the age offset in the central Japan Trench estimated at around 1600 years (Ikehara et al. 2016; Bao et al. 2018). Thus, simple regression of the offset indicates the distal MTDs were deposited approximately 2000 years BP. Kioka et al. (2019b) also correlated thick, acoustically transparent layers in the SBP records (Fig. 2b), which were correlated with our

Table 1 Results of radiocarbon age determinations

Core	Depth (cm)	Conventional radiocarbon age
KS-18-10 PC01	0–2	1910 ± 30
	70.3–72.3	2070 ± 30
	84.3–86.3	2810 ± 30
	114.3–116.3	2220 ± 30
	141.3–143.3	3280 ± 30
	153.3–155.3	5710 ± 30
	181.3–183.3	6960 ± 30
MR17-06 PC01	0.8–2.8	1920 ± 20
	47.8–49.8	2200 ± 30
	83.9–85.9	2490 ± 30
	99.9–101.9	3110 ± 30
	111.9–113.9	3800 ± 30
	218.1–220.1	7340 ± 30
	309.1–311.1	7280 ± 30
	409.1–411.1	7210 ± 30
	510.8–512.8	7630 ± 30

distal MTDs, based on acoustic stratigraphy. They also calculated the volume of the acoustically transparent layers in each trench basin from the area and morphology of the basin and the layer thickness. The total volume of the layers in the four northern Japan Trench basins (NO3, NO5, NO7 and NO8) was given as 413 (+85/−26) × 10^6 m^3.

Origin of Distal MTD in Core GeoB21917

The characteristic feature of the composition of the sediment in core GeoB21817 was the occurrence of well-preserved calcareous foraminiferal tests even in the hadal environment below the calcium carbonate compensation depth (CCD). In contrast, no foraminifera occurred within the upper unit. The occurrence of calcareous foraminifera in the distal MTD suggests that the MTD material originated above the CCD, which is located at around 4000 m in the northwest Pacific (Berger et al. 1976).

Furthermore, benthic foraminiferal assemblages included the characteristic species of the outer shelf to upper slope. The benthic foraminiferal assemblages of the four MTD samples from depths 578–580 cm, 675–677 cm, 777–797 cm and 886–888 cm that were examined were similar, and composed mainly of *Stainforthia fusiformis* (abundance: 65–82%), accompanied by *Chiostomellina fimbriata* (3–6%), *Brizalina pacifica* (3–5%), *Stainforthia spinosa* (0–3%), and *Nonionellina labradorica* (0–6%) with calcareous test. Agglutinated species were very rare. The composition of the uppermost (578–580 cm) sample was relatively diverse, and contained relatively large species such as *Islandiella japonica* (3%), *N. labradorica* (6%), and *Uvigerina akitaensis* (3%). The depth of the habitat of benthic foraminiferal assemblages in the northwest Pacific is remarkable (e.g. Matoba 1976; Keller 1980). Keller (1980) indicated that agglutinated species of benthic foraminifera are dominant below a depth of 3000 m. The most dominant species, *S. fusiformis,* is abundant on outer shelves to upper slopes (Gooday and Alve 2001). Although *S. fusiformis* was not found off Sendai, the main habitat of other species off Sendai can be summarized as (Matoba 1976): *I. japonica* (abundant at 300–600 m), *N. labradorica* (> 300 m and abundant at 500–700 m), *S. spinosa* (< 1000 m and abundant at < 100 m), *B. pacifica* (400–1200 m), *C. fimbriata* (> 500 m and abundant at ~1000 m), and *U. akitaensis* (> 800 m). Based on the modern habitats of these species, it was concluded that the origin of the MTD mud is the upper slope at a water depth of around 600 m.

Another reason for interest in benthic foraminifers is their good state of preservation even in the ultra-deep environment and after long-distance transport from the upper slope. Quick deposition of the distal MTDs prevents the reaction of calcareous tests of benthic foraminifera with cold and HCO_3^--unsaturated water near the ocean floor, thus preserving the foraminiferal tests in the distal MTDs. It is surprising that well-preserved examples of benthic foraminiferal species with fragile, thin-walled tests, such as *S. fusiformis*, *S. spinosa,* and *B. pacifica,* were observed. Usami et al. (2017) found that fragile, thin-walled benthic foraminiferal species could be transported down slopes by turbidity currents to depths several hundred meters greater. In fact, the results of this study indicate that many individuals of such fragile species can be well-preserved despite being transported to depths around 7 km deeper by turbidity currents.

Possible Link Between Submarine Landslides on the Upper Slope and MTDs in the Hadal Japan Trench Floor

Many surface and buried submarine landslides have been observed on the upper slope of the Hidaka Trough (Morita et al. 2011a, b; Noda et al. 2013). Gas hydrate instability and related fluid migration is a possible trigger for these slides (Morita et al. 2011a, b; Noda et al. 2013). Most of the headwall domain of the sliding masses along the Hidaka coast, on the northern and northeastern slopes of the Hidaka Trough (Fig. 1), has occurred at the shelf edge of the uppermost slope, and the translational domain stops at the foot of the slope of the Hidaka Trough (Noda et al. 2013). The youngest slide along the Hidaka coast occurred to the southwest of Urakawa, which occurred between 17,000 and 6000 years ago due to a high sediment supply both from the land and the shelf, which in turn resulted from post-glacial transgression and from surface water with a high primary productivity (Noda et al. 2013). This age does not agree with the date of the deposition of the distal MTDs in the Japan Trench basins ($\sim$2000 years BP). The western slope of the Hidaka Trough, where the head of Ogawara Submarine Canyon is located, is another candidate for recent submarine slope failure. Morita et al. (2011a, b) indicated the presence of many buried submarine landslide masses at the foot of the slope and on the floor of the western slope. However, no study on recent submarine landslides at this slope has been conducted. Although it has not yet been possible to identify the precise location of the source of the distal MTDs, the benthic foraminiferal assemblages and the condition of the upper slope suggest that a submarine landslide on the upper slope was the most likely source of the large amount of sediments forming the distal MTDs in the hadal Japan Trench. Thus, the Hidaka Trough upper slope has a connection with the hadal Japan Trench floor. Because the hadal oceanic trench is a sink of organic carbon–as suggested by Kioka et al. (2019a, b)–it is also important that the upper slope is a source not only of sediments but also of organic carbon due to high primary productivity in the surface water along the southern Hokkaido coast. Furthermore, from the geohazard point of view, the upper slope is a trigger area for large sediment movements, which should be considered for tsunami hazard mitigation. Although post-failure dynamics is important for geohazard evaluation (Sassa 2016), sediment dynamics analysis using sedimentary structures and successions (e.g. Talling et al. 2004) has not been applied for this case because no complete MTD sequence has been recovered from slope to trench areas.

Conclusions

Large sediment movements have occurred in relation to submarine landslides and have the potential to generate tsunami. Distal MTDs are possible records of submarine landslides. In the small basins of the northern Japan Trench, the distal MTDs are found as thick, fine-grained turbidites. The distal MTDs cover at least four small basins; the maximum thickness and total volume of the MTDs has been calculated as 11.1 m and $> 400 \times 10^6$ m^3, respectively (Kioka et al. 2019b). The date of the deposition of the distal MTDs was estimated as $\sim$2 ky BP based on radiocarbon dates from three cores recovered from the Japan Trench basins. Well-preserved benthic foraminifera were contained in the distal MTD, even in the ultra-deep hadal environment below the CCD. Benthic foraminiferal assemblages suggested that the distal MTDs originated at the upper slope. Many submarine landslides have been reported at the upper slope of the Hidaka Trough (Morita et al. 2011a, b; Noda et al. 2013). Therefore, submarine landslides at the upper slope most likely provide the majority of sediments to the hadal Japan Trench basins. Thus, the upper slope of the Hidaka Trough is an important area as sediment and carbon source to the Japan Trech. Furthermore, the upper slope is important area of large sediment movements, which are able to generate tsunami.

Acknowledgements We express our thanks to the captains, crews, technicians and scientists on board cruises KS-14-16, KS18-10, SO251a, and MR17-06 for their help. We also thank Prof. S. Hasegawa of Tohoku University for his comments on benthic foraminifera taxonomy. This work was supported by KAKENHI's "Japan Trench Deep-sea Research Project for Assessing Shallow Seismic Slips and Their History (No. 26000002)" and the MEXT research project "Research Project for Compound Disaster Mitigation on the Great Earthquakes and Tsunamis around the Nankai Trough region." T.S. and M.S. were supported by the Austrian Science Fund (FWF; Project Nr. 29678).

References

Bao R, Strasser M, McNichol AP, Haghipour N, McIntyre C, Wefer G, Eglinton TI (2018) Tectonically-triggered sediment and carbon export to the Hadal zone. Nat Commun 9:121. https://doi.org/10.1038/s41467-017-02504-12018

Berger WH, Adelseck CG Jr, Mayer LA (1976) Distribution of carbonate in surface sediments of the Pacific Ocean. J Geophys Res 81:2617–2627

Gooday AJ, Alve E (2001) Morphological and ecological parallels between sublittoral and abyssal foraminiferal species in the NE Atlantic: a comparison of *Stainforthia fusiformis* and *Stainforthia* sp. Prog Oceanogr 50:261–283

Heezen BC, Ewing M (1952) Turbidity currents and submarine slumps, and the 1929 Grand Banks Earthquake. Am J Sci 250:849–873

Heezen BC, Ericson DB, Ewing M (1954) Further evidence for a turbidity current following the 1929 Grand Banks Earthquake. Deep-Sea Res 1:193–202

Ikehara K (2000) Comparison of radiocarbon ages of planktonic foraminifera and bulk organic carbon in marine sediments. Bull Geol Surv Japan 51:299–307

Ikehara K, Sato M, Yamamoto H (1990) Sedimantation in the Oki Trough, southern Japan Sea, as revealed by high resolution seismic records (3.5 kHz echograms). J Geol Soc Japan 96:37–49

Ikehara K, Kanamatsu T, Nagahashi Y, Strasser M, Fink H, Usami K, Irino T, Wefer G (2016) Documenting large earthquakes similar to the 2011 Tohoku-oki earthquake from sediments deposited in the Japan Trench over the past 1500 years. Earth Planet Sci Lett 445:48–56

Ikehara K, Usami K, Kanamatsu T, Arai K, Yamaguchi A, Fukuchi R (2018) Spatial variability in sediment lithology and sedimentary processes along the Japan Trench: Use of deep-sea turbidite records to reconstruct past large earthquakes. In Tsunami: Geology, Hazards and Risks, Geol. Soc. London. Spec Publ 456:75–89. https://doi.org/10.1144/SP456.9

Keller G (1980) Benthic foraminifers and paleobathymetry of the Japan Trench area, Leg 57, Deep Sea Drilling Project. Init Rep DSDP 56/57(2):835–865

Kioka A, Schwestermann T, Moernaut J, Ikehara K, Kanamatsu T, McHugh CM, dos Santos Ferreira C, Wiermer G, Haghipour N, Kopf AJ, Eglinton TI, Strasser M (2019a) Megathrust earthquake drives drastic organic carbon supply to the hadal trench. Sci Rep 9:1553, https://doi.org/10.1038/s41598-019-38834-x

Kioka A, Schwestermann T, Moernaut J, Ikehara K, Kanamatsu T, Eglinton TI, Strasser M (2019b) Event stratigraphy in a hadal oceanic trench: The Japan Trench as sedimentary archive recording recurrent giant subduction zone earthquakes and their role in organic carbon export to the deep sea. Front Earth Sci 7:319. https://doi.org/10.3389/feart.2019.00319

Lovholt F, Schulten I, Mosher D, Harbitz C, Krastel S (2018) Modelling the 1929 Grand Banks slump and landslide tsunami. In: Lintern DG et al (eds.) Subaqueous mass movements. Geological Society, London, Special Publications, 477, https://doi.org/10.1144/SP477.28

Matoba Y (1976) Recent foraminiferal assemblages off Sendai, northeast Japan. In: Schafer CT, Pelletier BR (eds) First international symposium on Benthic Foraminifera of Continental Margins, Part A, Ecology and Biology, Maritime Sediments, vol 1. Special Publication, pp 205–220

Mohrig D, Whipple KX, Hondzo M, Ellis C, Parker G (1998) Hydroplaning of subaqueous debris flows. Geol Soc Am Bull 110:387–394

Moriki H, Kumamoto T, Nakata T, Goto H, Izumi N, Nishizawa A (2017) Identification of landslide and its characteristics on the seafloor around Japan using anaglyph images. Rep Hydrogr Oceanogr Res 54:1–16

Morita S, Nakajima T, Hanamura Y (2011a) Submarine slump sediments and related dewatering structures: Observations of 3D seismic data obtained for the continental slope off Shimokita Peninsula. NE Japan J Geol Soc Japan 117:95–98

Morita S, Nakajima T, Hanamura Y (2011b) Possible graound instability factor implied by slumping and dewatering structures in high-mehane-flux continental slope. In Yamada Y et al (eds) Submarine mass movements and their consequences, Advances in natural and technological hazards research, Springer, vol 31, pp 311–320

Mountjoy JJ, Howath JD, Orpin AR, Barnes PM, Bowden DA, Rowden AA, Schimel ACG, Holden C, Horgan HJ, Nodder SD, Patton JR, Lamarche G, Gerstenberger M, Micallef A, Pallentin A, Kane T (2018) Earthquakes drive large-scale submarine canyon development and sediment supply to deep-ocean basins. Sci Adv 4, eaar3748, https://doi.org/10.1126/sciadv.aar3748

Noda A, Katayama H (2013) Sedimentological Map of Hidaka Trough. Mar Geol Map Ser, no. 81(CD), Geological Survey of Japan, AIST

Noda A, TuZino T, Joshima M, Goto S (2013) Mass transport-dominated sedimentation in a foreland basin, the Hidaka Trough, northern Japan. Geochem. Geophys Geosys 14:2638–2660. https://doi.org/10.1002/ggge.20169

Pickering KT, Corregidor J (2005) Maa-transport complexes (MTCS) and tectonic control on basin-floor submarine fans, Middle Eocene, south Spanich Pyrenees. J Sed Res 75:761–783

Pickering KT, Hiscott RN (2015) Deep marine systems: processes, deposits, environments, tectonics and sedimentation. AGU and Wiley, p 657

Piper DJW, Cochonat P, Morrison ML (1999) The sequence of events around the epicentre of the 1929 Grand Banks earthquake: initiation of debris flows and turbidity current inferred from sidescan sonar. Sedimentology 46:79–97

Sassa S (2016) Submarine liquefied flow dynamics and their analytical framework with experimental and field validations. Rep Port Airport Res Inst 55:75–91

Schulten I, Mosher DC, Krastel S, Piper DJW, Kienast M (2018) Surficial sediment failures due to the 1929 Grand Banks Earthquake, St. Pierre Slope. In: Lintern DG et al (eds) Subaqueous mass movements, geological society, London, Special Publications, p 477, https://doi.org/10.1144/SP477.25

Stow DAV, Shanmugam G (1980) Sequence of structures in fine-grained turbidites: comparison of recent deep-sea and ancient flysch sediments. Sediment Geol 25:23–42

Talling PJ, Amy LA, Wynn RB, Peakall J, Robinson M (2004) Beds comprising debrite sandwiched with co-genetic turbidite: origin and widespread occurrence in distal depositional environments. Sedimentology 51:163–194

Toda S (2016) Crustal earthquakes. In: The geology of Japan. Geol Soc London, pp 371–408

Usami K, Ikehara K, Jenkins RG, Ashi J (2017) Benthic foraminiferal evidence of deep-sea sediment transport by the 2011 Tohoku-oki earthquake and Tsunami. Mar Geol 384:214–224

Tsunami from the San Andrés Landslide on El Hierro, Canary Islands: First Attempt Using Simple Scenario

Jan Blahůt and Byron Quan Luna

Abstract

This paper presents the first attempt to model a tsunami genesis and propagation from an incipient volcano slope failure termed San Andrés Landslide located on the El Hierro Island, Canary Islands, Spain. A rather conservative landslide scenario compared to other studies is proposed. The scenario comprises a subaerial failure of a block more than 2.5 km long and 7.5 km wide with volume of almost 6 km^3. The initial wave from this landslide reaches 80 m and its propagation through Atlantic Ocean has been modelled using DELFT 3D model. Results show that even a conservative scenario can have very severe consequences, especially in the adjacent islands. High to moderate waves are expected to affect also European SW and African NW coasts. As in any tsunami simulation however, the maximum slide speed is crucial for generating a tsunami wave. For that reason, future attempts should focus on more accurate landslide dynamic modelling to obtain realistic behaviour of the sliding mass to assess possible tsunami scenarios.

Keywords

Volcanic flank collapse • Tsunami modelling • Hazard scenario • San andrés landslide • El hierro • Canary islands

Introduction

Tsunamis generated from volcanic flank collapses are considered as one of the major geological hazards as it can be seen from the recent Anak Krakatau example in Indonesia (Walter et al. 2019). On the other hand, they can be also observed as an important agent in inter-island biological colonisation (García-Olivares et al. 2017).

Canary Islands are among areas where several giant landslides occurred in the past (Blahůt et al. 2018a, 2019). Thus, several studies were focused on tsunamis in the past years, mostly on the La Palma Island. Ward and Day (2001) published a study focused on the possible consequences of a catastrophic failure of the western flank of the Cumbre Vieja volcano. They assumed a landslide with a volume from 150 to 500 km^3. The energy generated by the largest volume slide block sliding at 100 m/s may produce waves 10-25 m high in the Atlantic coast of the North America. Their study was highly criticized for unrealistic assumptions and overestimation of both the catastrophic landslide scenario and the wave model used. Mader (2001) used SWAN code to model the Cumbre Vieja tsunami and he concludes, that the expected tsunami would result in an intermediate wave in distant areas rather than shallow water tsunami. He assumes that the wave on the US east coast would be less than 3 m high. Pararas-Carayannis (2002) examined the assumptions of the tsunami modelling made by Ward and Day (2001) and also investigated flank failures of Mauna Loa and Kilauea in Hawaii in 1868 and 1975 and caldera collapses and large slope failures associated with volcanic explosions of Krakatau in 1883 and of Santorin in 1490 B.C. He states that mega tsunami generation, even from the larger slope failures of island stratovolcanoes, is extremely unlikely to occur and concludes that the threat of mega tsunami generation from massive flank failures of island stratovolcanoes has been greatly overstated. This was confirmed by other studies of Løvholt et al. (2008), Zhou et al. (2011), Abadie et al. (2012)

J. Blahůt (✉)
The Czech Academy of Sciences, Institute of Rock Structure and Mechanics, V Holešovičkách 94/41, Prague, 18209, Czechia, Czech Republic
e-mail: blahut@irsm.cas.cz

B. Quan Luna
DNV GL, Digital Technology Center, Veritasveien 1, Høvik, 1363, Norway
e-mail: Byron.Quan.Luna@dnvgl.com

K. Sassa et al. (eds.), *Understanding and Reducing Landslide Disaster Risk*, ICL Contribution to Landslide Disaster Risk Reduction,
https://doi.org/10.1007/978-3-030-60196-6_27

and Tehranirad et al. (2015) who used depth averaged Boussinesq, THETIS 3D Navier-Stokes and FUNWAVE-TVD models respectively.

Giachetti et al. (2011) modelled collapse of the Güìmar debris avalanche on Tenerife Island. They used two scenarios of 44 km^3 falling as one block and as five retrogressive failures of equal volume of 8.8 km^3. They used VolcFlow code (Kelfoun et al. 2010) and concluded that the initial wave reached up to 500 m and that the tsunami deposits found on the nearby Gran Canaria Island were deposited from this debris avalanche. Paris et al. (2017) used also VolcFlow model to simulate failures from the northern flank of Tenerife Island (Icod collapse), which occurred approx. 170 ka. They state, that this failure occurred together with large explosive eruption, which represents a new type of volcano-tectonic event on oceanic volcanoes.

This brief summary of examples only from Canary Islands shows the devastating potential of volcanic flank collapses able to generate tsunamis. In recent years however, it has been shown that large debris avalanches may have collapsed as much smaller events. Hunt et al. (2013) analysed data from around western Canary Islands showing that debris avalanches occurred rather as multistage collapses reducing the individual volumes of landslides from up to 350 km^3 to less than 100 km^3. This is in accordance with the previously mentioned work of Abadie et al. (2012) and also a recent work of León et al. (2017). León et al. (2017) analysed in detail the El Golfo debris avalanche on the El Hierro Island and distinguished two different events with volumes of 84 and 234 km^3 respectively.

As volcanic flank collapses may also occur at persistent slumps, which may have been active over periods as long as 10^4–10^5 years prior to catastrophic failure (McGuire 2006), we focused this analysis on a creeping San Andrés Landslide on the El Hierro Island (Canary Islands) in order to assess its tsunami hazard potential. In our scenario, we used a rather conservative approach using a relatively small landslide and investigated its tsunami potential effects on El Hierro and the neighbouring islands.

Study Area

El Hierro is the smallest and westernmost island of the Canary Islands (Fig. 1a). The characteristic three-point star morphology of the island is a result of a number of gravitational slope failures (Fig. 1c). Until now, seven debris avalanches have been identified: Tiñor (<880 ka), Las Playas I (545–176 ka), Las Playas II (176–145 ka), El Julan (>158 ka), El Golfo A (176–133 ka), El Golfo B (87–39 ka), and Punta del Norte with unknown age (Masson 1996; Urgeles et al. 1996, 1997; Carracedo et al. 1999, 2001; Masson et al. 2002; Longpré et al. 2011; Becerril et al. 2016; León et al. 2017).

The San Andrés Landslide (Klimeš et al. 2016) is located on the southeast flank of the El Hierro volcanic edifice and it is defined by a group of pronounced faults which represent landslide detachment planes. It is a large slump, sensu Moscardelli and Wood (2008), or a deep-seated gravitational slope deformation, sensu Sorriso-Valvo et al. (1999) or Agliardi et al. (2001). For detailed description of the landslide and its geology, please refer to Blahůt et al. (2018b, 2020a).

San Andrés Landslide has been continuously monitored since 2013 and shows a creep movement reaching 0.5 mm. a^{-1} (Blahůt et al. 2017, 2018b, 2020b). Recently, a study has been made by Blahůt et al. (2020c), which analyses its current stability. It used limit equilibrium analysis on two predefined slip surfaces. The outcomes show, that the landslide is currently stable. However, a critical factor of safety equal to one can be reached by seismic loading of PGA = 0.29 g in case of a slip plane, which is subaerial and is between 200–400 m deep. For that purpose a scenario has been developed, which includes sudden slip of a rather small, failure-prone part of the San Andrés Landslide laying above the sea water level.

Methodology

The dynamic numerical modelling of the tsunami generated by the San Andrés Landslide was divided into two parts. The first part consisted of modelling the tsunami genesis based on the defined landslide scenario characteristics. The second part was the simulation of the tsunami propagation that included the results of the tsunami genesis and the behaviour of the resulting waves.

Tsunami Genesis

The genesis of the tsunami was assessed based on the geometrical characteristics and morphology of the San Andrés Landslide. The height drop (H) of the landslide was computed at 772 m; having a length (L) of 2655 m and a width (W) of 7490 m (Fig. 1b, c). These characteristics results in a run-out H/L ratio of 0.29. In agreement with the critical factor of safety slip plane, the depth of the landslide (D) was demarcated at 300 m as a continuous depth block that moves parallel to the shear plane. In this context, the maximum slope is computed having 37.4 degrees.

The landslide final volume was estimated as 5.965 km^3 and the deposit area that comprised 160.4 km^2. The

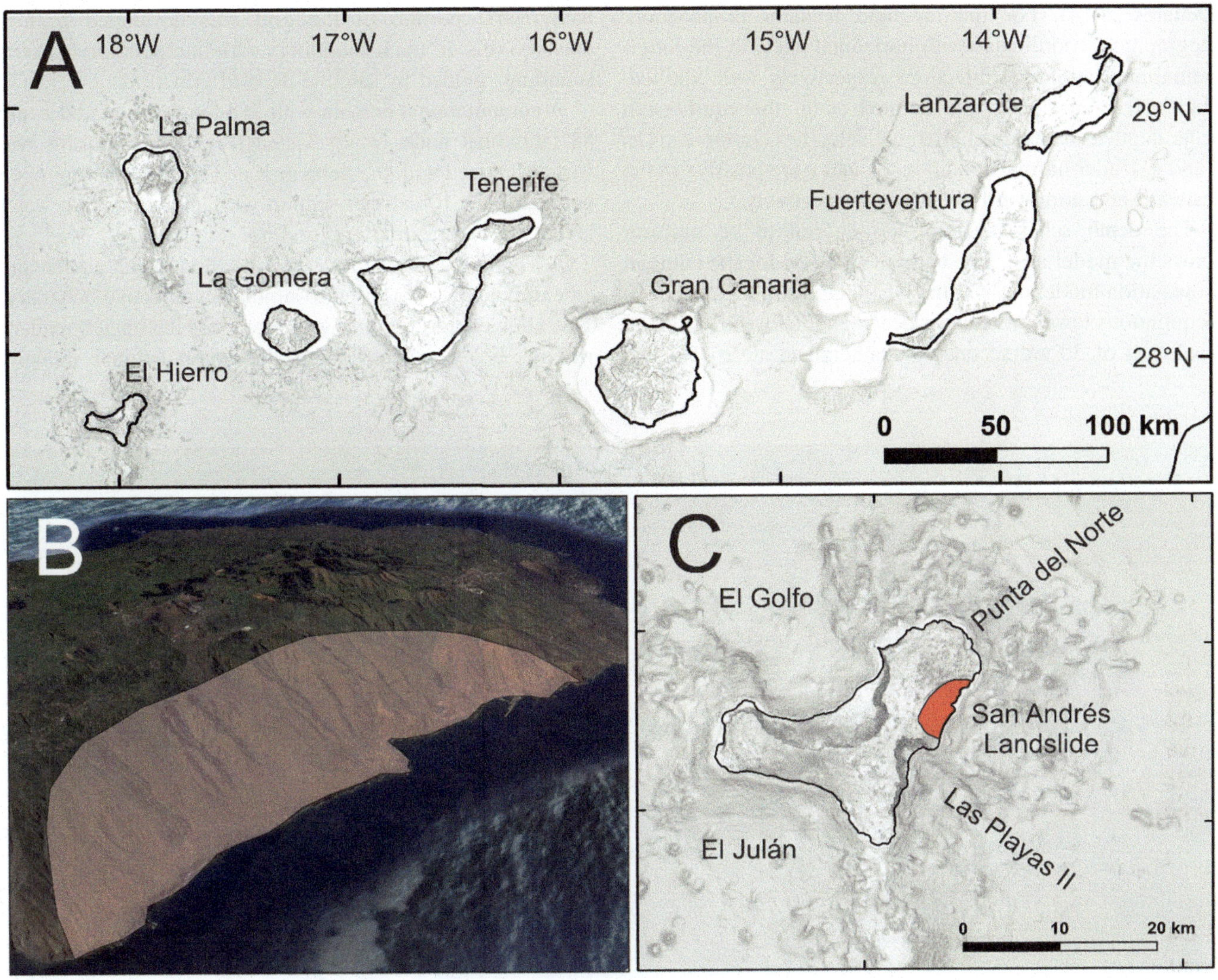

Fig. 1 **a** Location of the study area within Canary Islands Archipelago. **b** Oblique Google Earth image of the tsunami source landslide. **c** Detailed map of El Hierro with historically known slope failures. The tsunami source landslide is highlighted in red

bulk specific density of the failed mass selected was 1900 kg.m^{-3}. The computed velocity of the landslide when it reached the water was 94.5 m.s^{-1}.

Two empirical approaches were used for the calculations regarding the behaviour of the landslide in order to create a critical initial wave height that will trigger a tsunami. According to Noda (1970), the wave height can be modelled as:

$$\eta = F\lambda \quad (1)$$

where η is wave height in m; F is the Froude number and λ is max thickness of the slide. This resulted in an initial wave height of 80.28 m.

Based on Huber and Hager (1997) model, the initial wave height can be obtained by:

$$H = 0.88 \sin \alpha \left[\left(\sqrt{\frac{\rho s}{\rho}} \right) \left(\frac{Vs}{b} \right) \left(\sqrt{\frac{d}{x}} \right) \right] \quad (2)$$

where α is the slope angle at the impact site, ρs is the density of the flow, ρ is the density of water, Vs is the volume of the sliding mass, b is the finite shore distance equal to the width of the slide at impact with water, d is the water depth and x is the distance from the impact site. The wave height computed using this formula was 80.55 m.

Tsunami Propagation

For the wave propagation, we used depth-averaged quantities from DELFT 3D as initial conditions for tsunami simulations in two horizontal dimensions (2HD), applying a finite difference wave model including the Coriolis terms

(Deltares 2014). For the far field tsunami propagation, geographical coordinates with horizontal axes, in the longitudinal and latitudinal directions respectively were applied. The surface elevation was defined as η, the equilibrium water depth as h, and the horizontal velocities components as u and v. Under normal conditions, u and v are positive in the eastward and northward directions, respectively.

The depth schematization was chosen to be uniform across the model area. The bathymetry used for the tsunami propagation modelling was the GEBCO_2014 grid, which is a continuous terrain model for ocean and land with a spatial resolution of 30 arc seconds (Weatherall et al. 2015). The bathymetric portion of the grid was developed from a database of ship track soundings with interpolation between soundings guided by satellite-derived gravity data.

A computational domain with 218 longitudinal nodes and 88 latitudinal nodes was selected (Fig. 2). A uniform grid was discretized with 0.2 delta in X and Y direction. For open boundaries, a 10 cell per section was applied using a water level type with astronomic forcing.

A TPXO 7.2 Global Inverse Tide Model with a Riemann type of boundary was used to simulate a reflective boundary. This allowed for outgoing waves to cross the open boundary without being reflected into the computational domain.

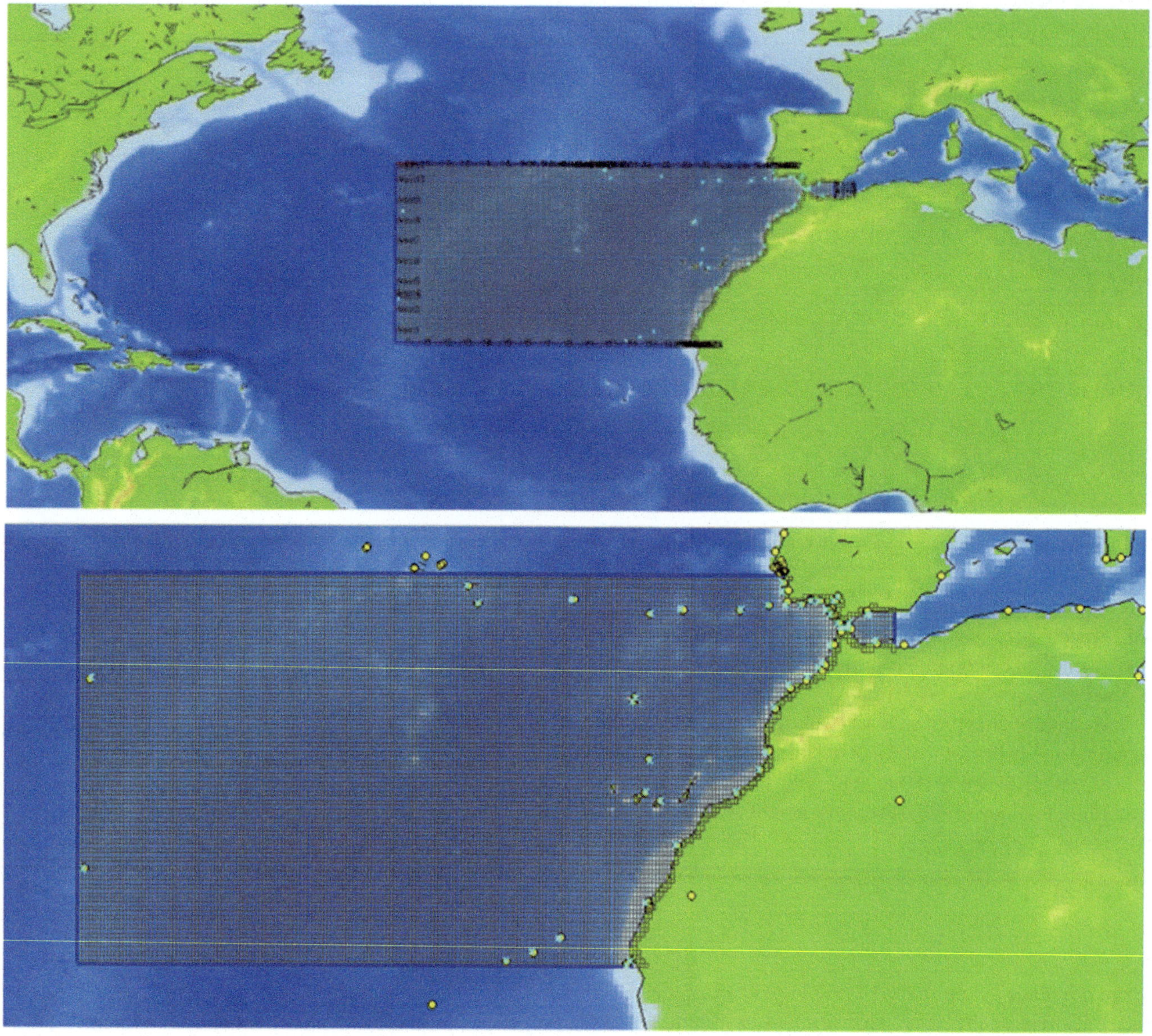

Fig. 2 Gridded computational domain that was selected for modelling the tsunami propagation. The light blue dots represent observation points used for monitoring the propagation inside the computational domain

A cyclic numerical scheme was selected for the advective terms in the advection-diffusion equation.

For observation points, IHO Tidal stations were selected to monitor the time-dependent behaviour of the computed quantities as a function of time at a specific location (i.e. water elevations, velocities). Observation points represent an Eulerian viewpoint at the results with amplitude and phase as parameters. Observation points are located at cell centres (i.e. at water level points) (Fig. 2).

The initial crest amplitude generally depends strongly upon the Froude number (ratio of slide speed to wave celerity), as well as the frontal slide area. The wave generation was defined as critical when the Froude number is close to unity, and the slide is then efficient in generating waves. The landslide material was defined as a fluid with the density of basalt (representing granules), and it begins to flow under the influence of gravity as soon as the calculation begins. The material flow first pushes the water above the lower part of the slide up. The progression of the slide along the bottom continues to pump energy into the water wave that is driven ahead of it, until the hydrodynamic drag and friction with the bottom slow and eventually stop the slide run-out.

Results

Using the maximum speed of the slide is 94.5 m/s gives a Froude number of 0.84, and is thus effective in generating a high leading wave localized just ahead of the slide front. As mentioned in the methodology section, the initial wave height of 80.5 m was selected. A reference time of 1 day (24 h) with a computational time using a batch output time step of 1 h was used. Figure 3 shows graph of wave oscillations at three places in the Canary Islands: Los Cristianos on Tenerife, Santa Cruz on La Palma and Las Palmas on Gran Canaria. It can be noted the generally uniform wave behaviour during first 4–5 h of simulation, which turns into dispersive mode in the remaining time. The simulated waves reach more than 70 m in case of Tenerife and La Palma islands. Las Palmas on Gran Canaria shows much lower, yet devastating, waves due to non-direct exposure (eastward orientation) of the port to the tsunami source.

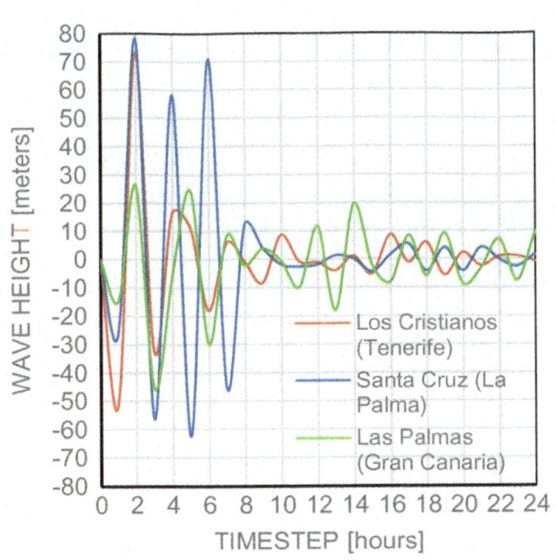

Fig. 3 Graph of wave oscillations on three places on the Canary Islands. Note that graph is built using data from the batch timestep of 1 h only

Figure 4 shows spatial results of the tsunami propagation model. The progression of the waves during first 24 h of the simulation is visible at selected timesteps.

The different wave heights for each time step can be noted in the map's legend, which defines spatio-temporal aspects of the propagation. The results show that waves higher than 2 m are likely along the SW European coast in Portugal and Spain and more than 3.5 m high waves are likely on Moroccan coast. Generally dispersive nature of the waves in the second half of the simulation is also notable.

Discussion and Conclusions

Simulation shows that the slide speed is close to critical, effectively generating an initial wave of approx. 80 m height and the wave propagation is genuinely dispersive. In the far-field, propagation becomes increasingly complex due to the combined effects of dispersion, refraction, and interference in the direction of propagation. Interference of the trailing waves are found to decrease the decay of the maximum amplitude.

Results show that this scenario considering only approx. 6 km^3 can cause very severe consequences, comparable with the large failure scenarios of more than 100 km^3 as in the La Palma case. Consequences of the El Hierro scenario would be largest on the Canary Islands, but the model outputs also suggest that the South-eastern Europe (Portugal and Spain) and African Morocco´s coast would also face consequences.

In any tsunami simulation, the maximum slide speed is crucial for generation of the tsunami wave. Sliding mass of slower velocities (slower than 94.5 $m.s^{-1}$ in our case) would barely have enough power to generate a tsunami wave. For that reason, we believe that future studies of potential San Andrés Landslide failure should focus on a realistic landslide dynamic modelling, including different rheological properties of the falling mass. This will allow to accurately define the dynamic properties of the sliding mass and lead to more realistic tsunami scenarios.

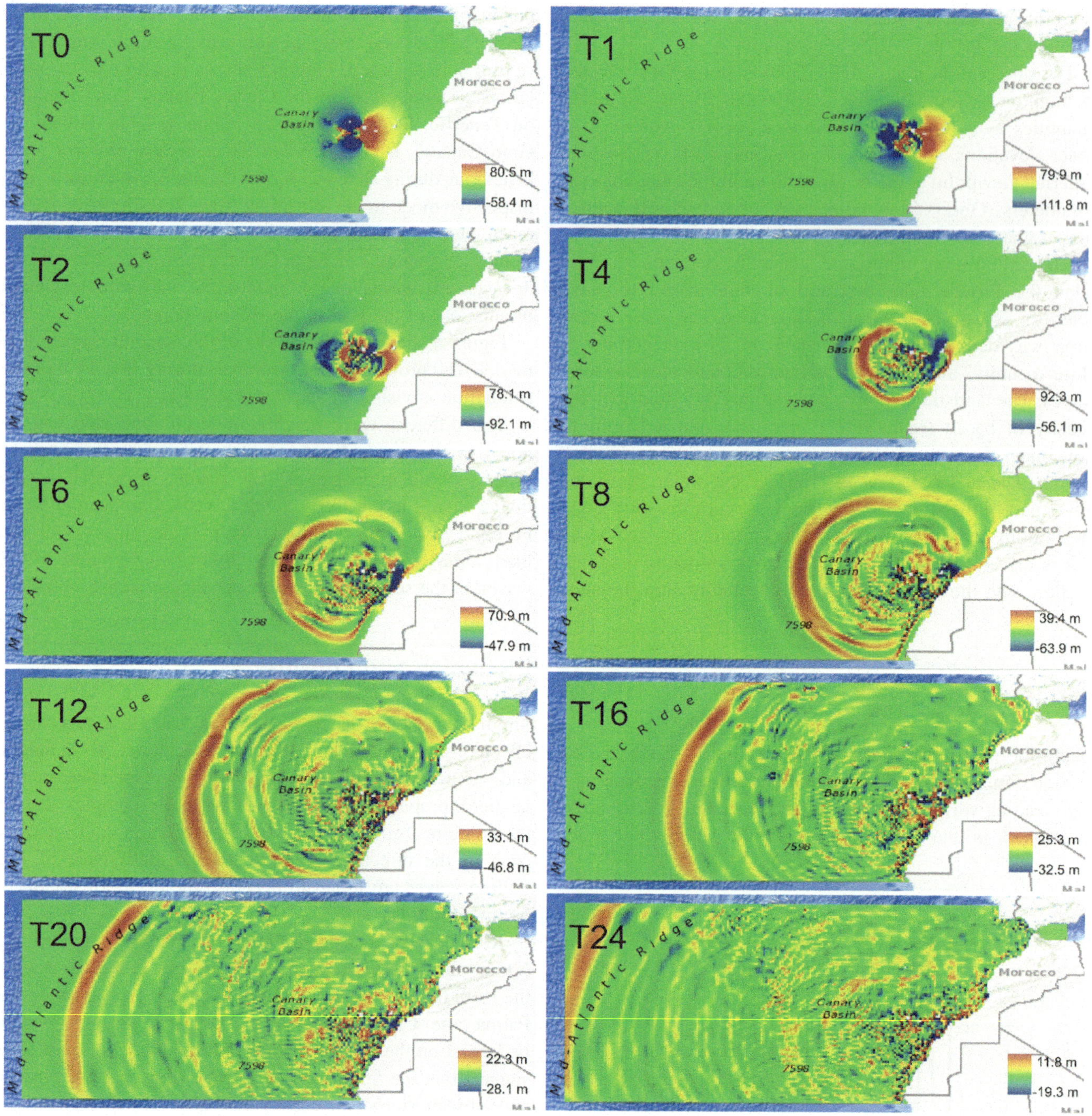

Fig. 4 Results of the tsunami propagation model from the San Andrés Landslides. Different timesteps are shown starting from the initial wave at T0 and ending after 24 h of simulation

Acknowledgements This work was supported by the Czech Science Foundation (GJ16-12227Y) and by the long-term conceptual development research organisation RVO: 67985891.

References

Abadie SM, Harris JC, Grilli ST, Fabre R (2012) Numerical modelling of tsunami waves generated by flank collapse of the Cumbre Vieja Volcano (La Palma, Canary Islands): Tsunami source and near field effects. J Geophys Res 117:C05030. https://doi.org/10.1029/2011JC007646

Agliardi F, Crosta G, Zanchi A (2001) Structural constraints on deep seated slope deformation kinematics. Eng Geol 59:83–102. https://doi.org/10.1016/S0013-7952(00)00066-1

Becerril L, Galve J, Morales J, Romero C, Sánchez N, Martí J, Galindo I (2016) Volcanostructure of El Hierro (Canary Islands). J Maps 12:43–52. https://doi.org/10.1080/17445647.2016.1157767

Blahůt J, Rowberry M, Balek J, Klimeš J, Baroň I, Meletlidis S, Martí X (2017) Monitoring giant landslide detachment planes in the era of big data analytics. In: Mikoš M, Arbanas Ž, Yin Y, Sassa K (eds). Advancing culture of living with landslides, vol 3. Springer, Cham, pp. 333–340. https://doi.org/10.1007/978-3-319-53487-9_38

Blahůt J, Klimeš J, Rowberry M, Kusák M (2018a) Database of giant landslides on volcanic islands - first results from the Atlantic Ocean. Landslides 15:823–827. https://doi.org/10.1007/s10346-018-0967-3

Blahůt J, Baroň I, Sokol Ľ, Meletlidis S, Klimeš J, Rowberry M, Melichar R, García-Cañada L, Martí X (2018b) Large landslide stress states calculated during extreme climatic and tectonic events on El Hierro, Canary Islands. Landslides 15:1801–1814. https://doi.org/10.1007/s10346-018-0993-1

Blahůt J, Balek J, Klimeš J, Rowberry M, Kusák M, Kalina J (2019) A comprehensive global database of giant landslides on volcanic islands. Landslides 16:2045–2052. https://doi.org/10.1007/s10346-019-01275-8

Blahůt J, Mitrovic-Woodell I, Baroň I, René M, Rowberry M, Blard PH, Hartvich F, Meletlidis S (2020a) Volcanic edifice slip events recorded on the fault plane of the San Andrés Landslide, El Hierro, Canary Islands. Tectonophysics 776:228317. https://doi.org/10.1016/j.tecto.2019.228317

Blahůt J, Balek J, Eliaš M, Meletlidis S (2020b) 3D Dilatometer time-series analysis for a better understanding of the dynamics of a giant slow-moving landslide. Appl Sci 10(16):5469. https://doi.org/10.3390/app10165469

Blahůt J, Olejár F, Rott J, Petružálek M (2020c) Current stability modelling of an incipient San Andrés giant landslide on el Hierro Island, Canaries, Spain - first attempt using limited input data. Acta Geodyn Geomater 17, 1(197):89–99. https://doi.org/10.13168/AGG.2020.0006

Carracedo JC, Day S, Guillou H, Pérez Torrado F (1999) Giant quaternary landslides in the evolution of La Palma and El Hierro, Canary Islands. J Volcanol Geoth Res 94:169–190. https://doi.org/10.1016/S0377-0273(99)00102-X

Carracedo JC, Rodríguez Badiola E, Guillou H, De la Nuez H, Pérez Torrado F (2001) Geology and volcanology of the western Canaries: La Palma and El Hierro. Estud Geol 57:171–295

Deltares (2014) Delft3D-FLOW: simulation of multi-dimensional hydrodynamic flows and transport phenomena, including sediments. Deltares, Netherlands

García-Olivares V, López H, Patiño J, Alvarez N, Machado A, Carracedo JC, Soler V, Emerson BC (2017) Evidence for mega-landslides as driver of island colonization. J Biogeogr 44(5):1053–1064. https://doi.org/10.1111/jbi.12961

Giachetti T, Paris R, Kelfoun K, Pérez-Torrado FJ (2011) Numerical modelling of the tsunami triggered by Güìmar debris avalanche, Tenerife (Canary Islands): comparison with field-based data. Marine Geol 284:189–202. https://doi.org/10.1016/j.margeo.2011.03.018

Huber A, Hager WH (1997) Forecasting impulse waves in reservoirs. Commission Internationale des Grands Barrages, Dix-neuvieme Congres des Grands Barrages, Florence, pp 993–1005

Hunt JE, Wynn RB, Talling PJ, Masson DG (2013) Multistage collapse of eight western Canary Island landslides in the last 1.5 Ma: Sedimentological and geochemical evidence from subunits in submarine flow deposits. Geochem Geophys Geosyst 14(7):2159–2181. https://doi.org/https://doi.org/10.1002/ggge.20138

Kelfoun K, Giachetti T, Labazuy P (2010) Landslide-generated tsunamis at Reunion Island. J Geophys Res 115:F04012. https://doi.org/10.1029/2009JF001381

Klimeš J, Yepes J, Becerril L, Kusák M, Galindo I, Blahůt J (2016) Development and recent activity of the San Andrés Landslide on El Hierro, Canary Islands, Spain. Geomorphology 261:119–131. https://doi.org/10.1016/j.geomorph.2016.02.018

León R, Somoza L, Urgeles R, Medialdea T, Ferrer M, Biain A, García-Crespo J, Mediato JF, Galindo I, Yepes J, Gonzáles FJ, Gimenez-Moreno J (2017) Multi-event oceanic island landslides: New onshore-offshore insights from El Hierro Island, Canary Archip elago. Marine Geol 393:156–175. https://doi.org/10.1016/j.margeo.2016.07.001

Longpré M, Chadwick J, Wijbrans J, Iping R (2011) Age of the El Golfo debris avalanche, El Hierro (Canary Islands): new constraints from laser and furnace 40Ar/39Ar dating. J Volcanol Geoth Res 203:76–80. https://doi.org/10.1016/j.jvolgeores.2011.04.002

Løvholt F, Pedersen G, Gisler G (2008) Oceanic propagation of a potential tsunami from the La Palma Island. J Geophys Res 113: C09026. https://doi.org/10.1029/2007JC004603

Mader CL (2001) Modeling the La Palma landslide tsunami. Science of Tsunami Hazards 19:150–170

Masson D (1996) Catastrophic collapse of the volcanic island of Hierro 15 ka ago and the history of landslides in the Canary Islands. Geology 24:231–234. https://doi.org/10.1130/0091-7613(1996)024%3c0231:CCOTVI%3e2.3.CO;2

Masson D, Watts A, Gee M, Urgeles R, Mitchell N, Le Bas T, Canals M (2002) Slope failures on the flanks of the western Canary Islands. Earth Sci Rev 57:1–35. https://doi.org/10.1016/S0012-8252(01)00069-1

McGuire WJ (2006) Lateral collapse and tsunamigenic potential of marine volcanoes. Geological Society, London, Special Publications 269(1):121. https://doi.org/10.1144/GSL.SP.2006.269.01.08

Moscardelli L, Wood L (2008) New classification system for mass transport complexes in offshore Trinidad. Basin Res 20:73–98. https://doi.org/10.1111/j.1365-2117.2007.00340.x

Noda E (1970) Water waves generated by landslide. Proceedings of the American society of civil engineers. J Waterw Harbors Div 96(4):835–855

Pararas-Carayannis G (2002) Evaluation of the threat of mega tsunamis generation from postulated massive slope failures of island stratovolcanoes on La Palma, Canary Islands, and on the island of Hawaii. Science of Tsunami Hazards 20:251–277

Paris R, Coello Bravo JJ, Martín Gonzáles ME, Kelfoun K, Naure F (2017) Explosive eruption, flank collapse and megatsunami at Tenerife ca. 170 ka. Nature Communications 8:15246. https://doi.org/10.1038/ncomms15246

Sorriso-Valvo M, Gullà G, Antronico L, Tansi C, Amelio M (1999) Mass movement, geologic structure, and morphologic evolution of the Pizzotto-Greci Slope (Calabria, Italy). Geomorphology 30:147–163. https://doi.org/10.1016/S0169-555X(99)00051-3

Tehranirad B, Harris JC, Grilli AR, Grilli ST, Abadie S, Kirby JT, Shi F (2015) Far-Field Tsunami impact in the North Atlantic basin from large scale flank collapses of the Cumbre Vieja Volcano. La Palma. Pure and Appl Geophys 172:3589. https://doi.org/10.1007/s00024-015-1135-5

Urgeles R, Canals M, Baraza J, Alonso B (1996) The submarine El Golfo debris avalanche and the Canary debris flow, west Hierro Island: the last major slides in the Canary Archipelago. Geogaceta 20:390–393

Urgeles R, Canals M, Baraza J, Alonso B, Masson D (1997) The most recent megalandslides on the Canary Islands: the El Golfo debris avalanche and the Canary debris flow, west El Hierro Island. J Geophys Res Solid Earth 102:20305–20323. https://doi.org/10.1029/97JB00649

Walter TR, Haghighi MH, Schneider FM, Coppoloa D, Motagh M, Saul J, Babeyko A, Dahm T, Troll VR, Tilmann F, Heimann S, Valade S, Triyono R, Khomarudin R, Kartadinata N, Laiolo M, Massimetti F, Gaebler P (2019) Complex hazard cascade culminating in the Anak Krakatau sector collapse. Nat Commun 10:4339. https://doi.org/10.1038/s41467-019-12284-5

Ward SN, Day S (2001) Cumbre Vieja Volcano potential collapse at La Palma, Canary Islands. Geophys Res Lett 28:3397–3400. https://doi.org/10.1029/2001GL013110

Weatherall P, Marks KM, Jakobsson M, Schmitt T, Tani S, Arndt JE, Rovere M, Chayes D, Ferrini V, Wigley R (2015) A new bathymetric model for world´s oceans. Earth and Space Science 2(8):331–345. https://doi.org/10.1002/2015EA000107

Zhou H, Moore CW, Wei Y, Titov VV (2011) A nested-grid Boussinesq type approach to modelling dispersive propagation and runup of landslide generated tsunamis. Nat Hazards Earth Syst Sci 11(10):2677–2697. https://doi.org/10.5194/nhess-11-2677-2011

A Sedimentological Study of Turbidite Layers on a Deep–Sea Terrace in the Japan Trench

2020 Kyoto Japan

Kiichiro Kawamura

Abstract

I report turbidite layers collected from a deep-sea terrace. The sample was recovered at 1555 m in water depth using a 4 m piston core system during the cruise YK14-21 by *R/V Yokosuka*. The sediments are composed of silty clay interbedded with five turbidite layers (T1–T5). Wood skins were observed at the base of the T4 and T5 indicating the ^{14}C age being 41,720 ± 520 year BP and an age older than 43,500 being the Last Glacier Period. I analyzed the paleocurrent directions of these turbidite layers using magnetic methods. As a result, the turbidite layers were mostly transported repeatedly from east as NE Japan arc, and deposited on a slope in the deep-sea terrace. On the basis of the ^{14}C ages, these turbidite layers might not be related to the recent large earthquakes. However, we need further to do a basic sedimentological study to identify the earthquake-and/or tsunami-generated turbidite layers in the Tohoku region.

Keywords

Turbidite • Japan trench • Deep-sea terrace • ^{14}C age determination

Introduction

The 2011 off the Pacific coast of Tohoku Earthquake (hereafter the 2011 Tohoku Earthquake) being the largest earthquake mechanically recorded in Japan occurred at 14:46 JST on March 11th, 2011. The Japan Meteorological Agency opened detailed information (Mw 9.0; epicenter location 38°6.2'N, 142°51.6'E and the hypocenter depth 24 km, Fig. 1). This largest earthquake excited the large tsunami (hereafter the 2011 Tohoku Tsunami), which devastated the Pacific coast off NE Japan. The runup height of ~ 40 m was recorded at Miyako city, easternmost NE Japan (e.g. Tsuji et al. 2011, Fig. 1).

The 2011 Tohoku Tsunami induced a strong turbidity current to transport rapidly from terrestrial to deep-sea. Arai et al. (2013) discussed the tsunami-generated turbidity current that were detected by an ocean bottom pressure gauge deployed before the 2011 Tohoku Earthquake. They reported an abrupt temperature increasing after about one and half hours of the main shock. The gauge flew down simultaneously ~1 km east from the original position. The turbidity current was deposited widely a turbidite silt—sand layer from ~200 m to ~3000 m in water depth. These turbidites should be a key to understanding recurrence of such tsunamis.

In this study, I described in detail a piston core sample collected from the landward trench slope of the Japan Trench. "Four turbidite layers were observed including plant materials indicating geological ages of 41,720 ± 520 year BP and an age older than 43,500 year BP. These layers indicate that the paleocurrent directions are mostly from east to west. I conclude that the turbidites would be supplied from NE Japan, but these turbidites might not be related to recent earthquake events. However, we need further to do a basic sedimentological study to identify the earthquake-and/or tsunami-generated turbidite layers in the Tohoku region.

Materials

The studied specimens were collected from a deep–sea terrace in the Japan Trench during *R/V Yokosuka* cruise YK14–21 of Japan Agency for Marine Science and Technology (JAMSTEC).

K. Kawamura (✉)
Yamaguchi University, 1677-1 Yoshida, Yamaguchi City, Yamaguchi 753–8512, Japan
e-mail: kiichiro@yamaguchi-u.ac.jp

K. Sassa et al. (eds.), *Understanding and Reducing Landslide Disaster Risk*,
ICL Contribution to Landslide Disaster Risk Reduction,
https://doi.org/10.1007/978-3-030-60196-6_28

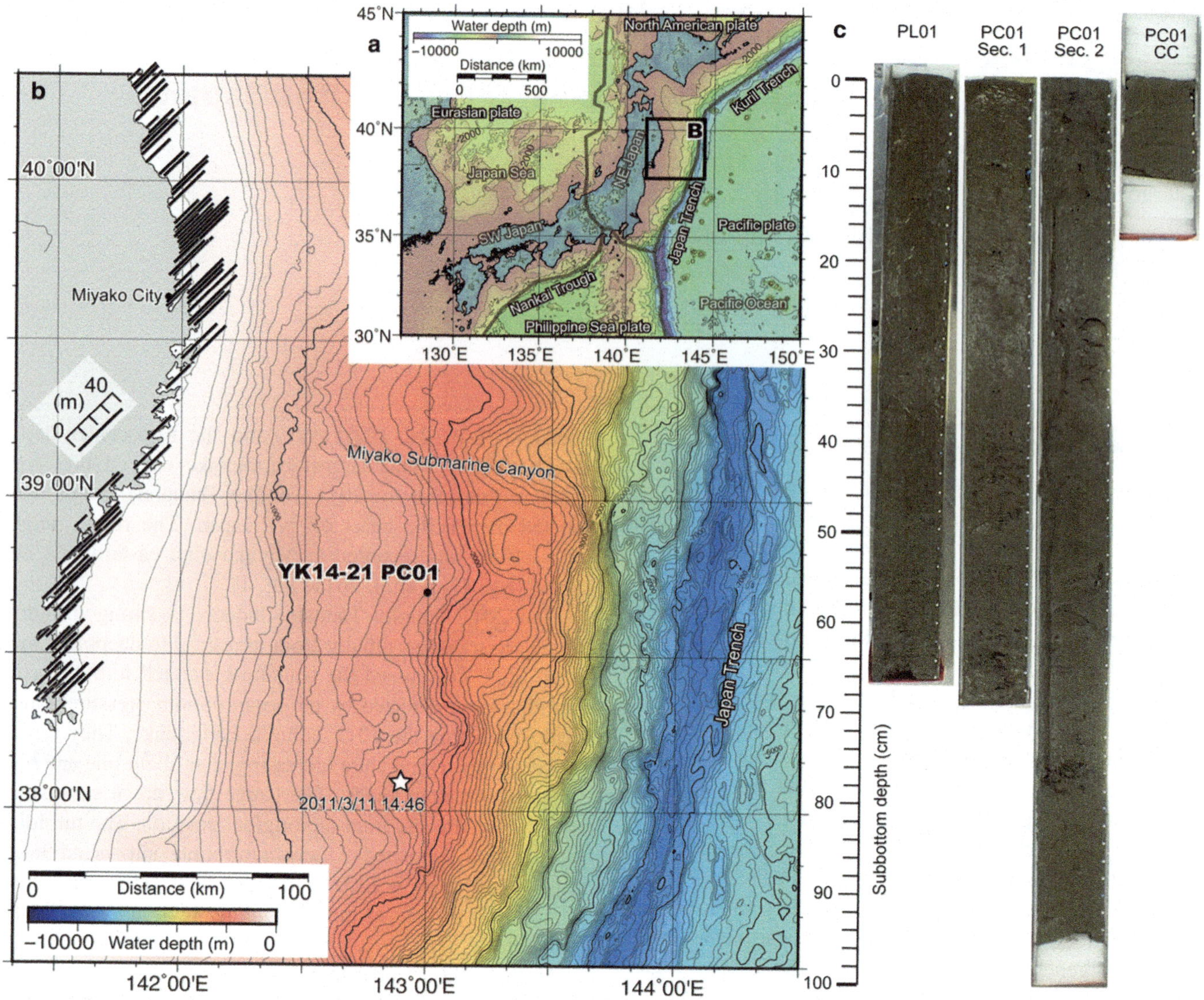

Fig. 1 Locality of the study area **a**, the coring site **b** and photographs of the core samples. **a**: The study area is located on a deep–sea terrace in the Japan Trench. **b**: The coring site of YK14-21 PC01 is a gentle slope at the top of a spur, where is located between amphitheater–shaped depressions of ~ 30 km wide. At the north of the spur, there are several submarine canyons through the depression. The run-up heights of the 2011 Tohoku Tsunami are shown as black bars along the Pacific coast off NE Japan (Tsuji et al. 2011). A star indicates the epicenter of the 2011 Tohoku Earthquake. **c**: Photographs of the core sample YK14-21 PC01 and PL01. The white dots in the core samples are in 2 cm intervals

The coring site of the piston core (PC01) and the pilot core (PL01) is at 143° 01.200’ E, 38° 46.000’ N and water depth is 1555 m (Fig. 1).

I recovered the specimens using a 4 m piston corer with a 1 m pilot corer. The core samples of PC01 and PL01 are 175 cm in length with 7.5 cm in diameter and 63.5 cm in length with 7.4 cm in diameter, respectively. The core samples were cut by 1 m long and split into working and archive halves onboard.

For description of the lithofacies, I conducted smear slide observation and grain–size analyses. The grain–size distribution was obtained using a Master Sizer 2000 (Malvern instruments ltd.) laser diffraction grain–size analyzer. A sediment of weight approximately 0.1 g wet for each sample was dipped into water and dispersed by an ultrasonic vibrator for 30–60 s just before measurement.

PC01 is composed mainly of sandy silt at 0–144 cm and clayey silt at 144–175 cm (Figs. 1 and 2). The sandy silt layer is interbedded with three pumiceous turbidite units at 30–72 cm, 82–99 cm and 129–144 cm (Fig. 2). These units are characterized by grading structure with coarse pumiceous sands at the base.

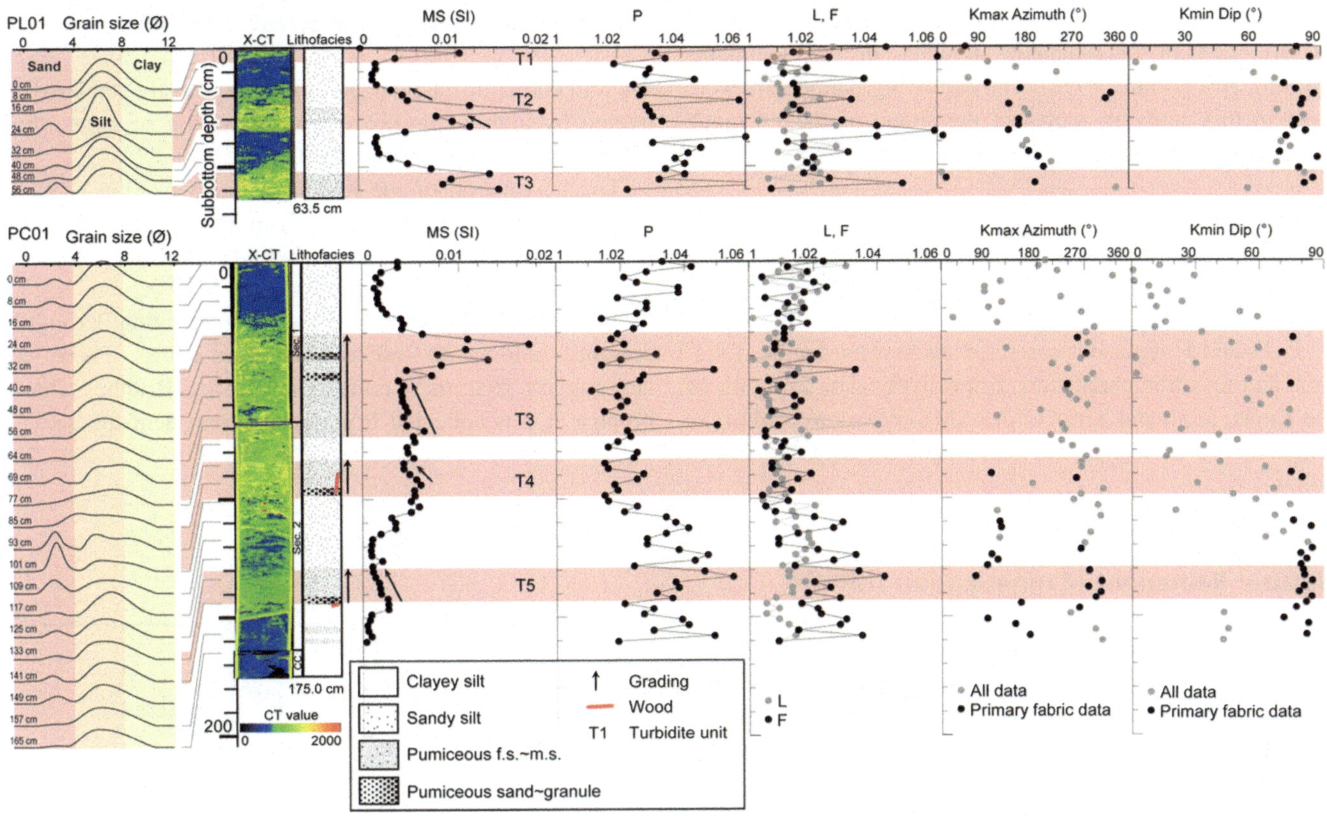

Fig. 2 Grain size distribution, X–ray CT image, lithofacies and magnetic properties of YK14-21 PL01 and PC01. Magnetic susceptibility (MS) decreases gradually with increasing of sub-bottom depth in the turbidite units T2, T3, T4 and T5. This corresponds to grading structure in the units

There are wood skins at the bases of the lower two turbidite units, T4 and T5 (Fig. 2). ^{14}C ages of these wood skins were 41,720 ± 520 year BP and an age older than 43,500 year BP. These ages correspond to the Last Glacier Period.

Lithofacies of PL01 is mainly composed of sandy silt with three pumiceous sand layers, which is similar to the upper part of the PC01. But the sediments in PL01 do not critically correspond to those in PC01 (Fig. 2).

Based on comparison of lithofacies, I concluded that the lowermost turbidite unit in PL01 corresponds to the uppermost turbidite unit in PC01. The surface sediments (probably ~ 40 cm thick) of PC01 blew off during the coring operation. This conclusion is supported by characteristics of magnetic properties as shown in later discussion.

Thus, a total of five pumiceous turbidite units (T1–T5 in PL01 and PC01) were observed as follows.

PL01: T1 = 0–6 cm, T2 = 16–23 cm and T3 = 52–63.5 cm
PC01: T3 = 30–73 cm, T4 = 83–99 cm and T5 = 129–142 cm

Methods

I used two measurements being magnetic susceptibility, its anisotropy and natural remanent magnetization to analyze paleocurrent directions in the core samples as follows.

For these measurements, the core samples were packed into plastic cubes of 7 cm^3 with right orientation. The remains of the samples were used for smear slide observations and grain–size analysis.

Magnetic Susceptibility (MS and Anisotropy of Magnetic Susceptibility (AMS)

The MS and AMS was measured using a KappaBridge KLY–4 magnetic susceptometer (AGICO) at a setting of 0.4 G low magnetic fields.

The AMS is geometrically represented by a magnetic susceptibility ellipsoid with three principal axes: the maximum (Kmax), intermediate (Kint), and minimum (Kmin)

magnetic susceptibility. In general, the ellipsoid is controlled by arrangement of magnetic particles in the sediments (Tarling and Hrouda 1993). The following parameters were used in this study to represent the shape of the magnetic susceptibility ellipsoid:

P = Kmax/Kmin
F = Kint/Kmin (Stacey et al. 1960), and
L = Kmax/Kint (Balsley and Buddington 1960).

P, F, and L represent corrected anisotropy degree, foliation, and lineation parameters, respectively. The azimuths of the Kmax, Kint, and Kmin of the AMS were corrected to the magnetic north using the paleomagnetic data as shown below.

Natural Remanent Magnetization (NRM)

The NRM was measured using a pass–through superconducting magnetometer Model 760R (2G Enterprises): increasing direction of X axis is perpendicular to the split face of the archive half of the core; increasing direction of Y axis is toward left on the split surface looking upcore; increasing direction of Z axis is perpendicular to both axes. This XYZ coordinates are the same as for the AMS measurement.

Stepwise alternating–field demagnetization up to 800 G was performed, and the magnetic direction (paleo–north) was plotted in a vector endpoint diagram.

Results

Magnetic Susceptibility (MS and Anisotropy of Magnetic Susceptibility (AMS)

The MS ranges from 2.0×10^{-3} to 5.0×10^{-3} SI at the sandy silt layer and from 4.0×10^{-4} to 9.0×10^{-4} SI at the clayey silt layer (Fig. 2). The MS decreases gradually in the turbidite units as shown by arrows in Fig. 2.

There are three peaks of 1.0–2.0 $\times 10^{-2}$ SI at 4 cm, 26 cm and 56 cm in PL01. On the basis of the MS curve, the lowermost peak in PL01 would correspond to the uppermost peak at 34 cm in PC01 (Fig. 2).

P, F and L values range generally from 1.01 to 1.03 throughout the core samples (Fig. 2). There are several peaks larger than 1.04 in the turbidite units (Fig. 2).

The dip angles of the Kmin range mostly higher than 70° in PL01, whereas those are scattered in PC01. Particularly there are no high angles at the uppermost sandy silt layer from 0 to 30 cm in PC01. This might result from a secondary fluidization with falling sidelong of the core on deck.

Natural Remanent Magnetization (NRM)

The directions of magnetization become mostly stable when treated by alternating–field demagnetization of larger than 200 G and/or smaller than 600 G (Fig. 3).

The declination of the remanent magnetization changes downward from 80° toward 60° in PC01. This variation suggests that the piston corer was twisted anticlockwise when it penetrated into the sediments.

The inclination of the remanent magnetization is roughly constant, holding 30°–60° of downward values. These values suggest that they correspond to the Brunhes chron, because it was collected from the northern hemisphere.

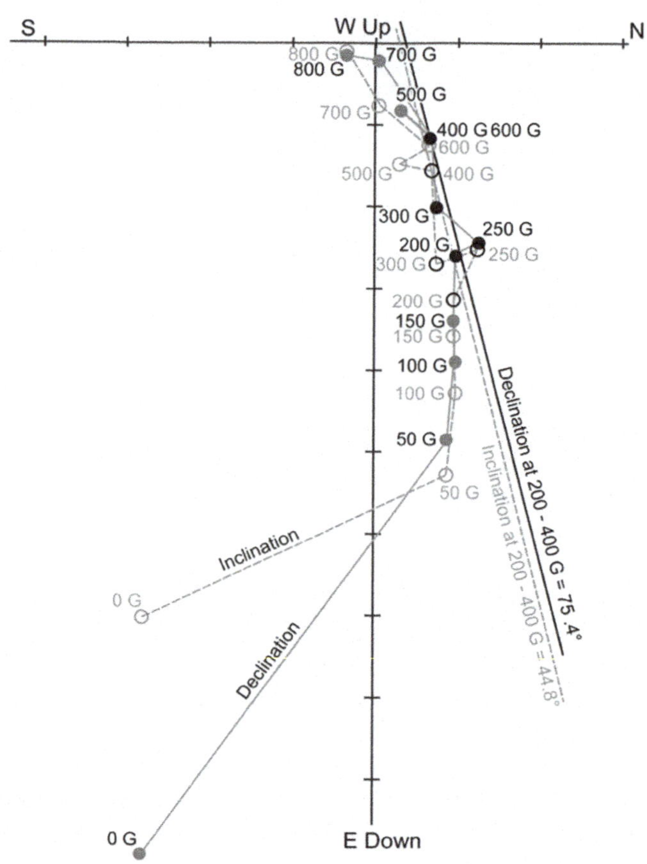

Fig. 3 Vector endpoint diagram by the stepwise alternating–field demagnetizations (PC01 52.3–54.4 cm in depth). The dots and the circles show the declination and the inclination respectively. Each long of the line between the node and dots (circles) show the intensities of the remanent magnetizations. The intensities are decreasing according the alternating field becomes strongly step by step. Each direction of the line shows the degrees of the declinations and inclinations of the remanent magnetizations. The degrees become constantly according the alternating field becomes strongly step by step. 0°, 90°, 180° and 270° in this diagram show the degree of declination, 0° corresponds to the direction of X axis. 90° up and 90° down show the degree of inclination

Paleocurrent Analysis

Because the core sample is rotated during coring, I have to correct the original azimuth of Kmax (int, min) by using the NRM as magnetic north of the samples. The calculations are simple as follows.

$$\text{cKmax (int, min)} = \text{oKmax(int, min)} - \text{dNRM}$$

cKmax (int, min): corrected azimuth of Kmax (int, min) with magnetic north
oKmax (int, min): original azimuth of Kmax (int, min)
dNRM: declination of NRM.

The primary fabric indicator in Tarling and Hrouda (1993) was used to remove the secondary fabric data (e.g. secondary deformation by coring disturbance, bioturbation, tectonics and so on) from raw data. In this study, I used the parameters to select automatically the primary fabrics as follows; dip of Kmax > 20°, dip of Kmin < 80° and F > L.

The azimuths of cKmax and the dip of cKmin are shown in Fig. 2. There are primary fabric data in all the turbidite units. The azimuths of cKmax concentrate mostly 270° in the turbidite units in PC01 and 180° in the turbidite units in PL01 (Fig. 2).

The data of the cKmax, cKint and cKmin corrected by magnetic north are shown in stereographic diagrams (lower hemisphere) (Fig. 4). The directions of cKmax, cKint and cKmin are shown by solid squares, solid triangle and circles, respectively. All the correlated data do not show any preferred orientation.

The primary fabric data selected from all the correlated data show clearly preferred orientations in the turbidite units T3 and T5 (Fig. 4). The paleocurrent directions shown by arrows are judged with the dip direction of the Kmin as discussed in Kawamura et al. (2002). The paleocurrents in T3 and T5 show mostly from west to east, indicating transportation from NE Japan to the Japan Trench, although that in T1 is from north to south.

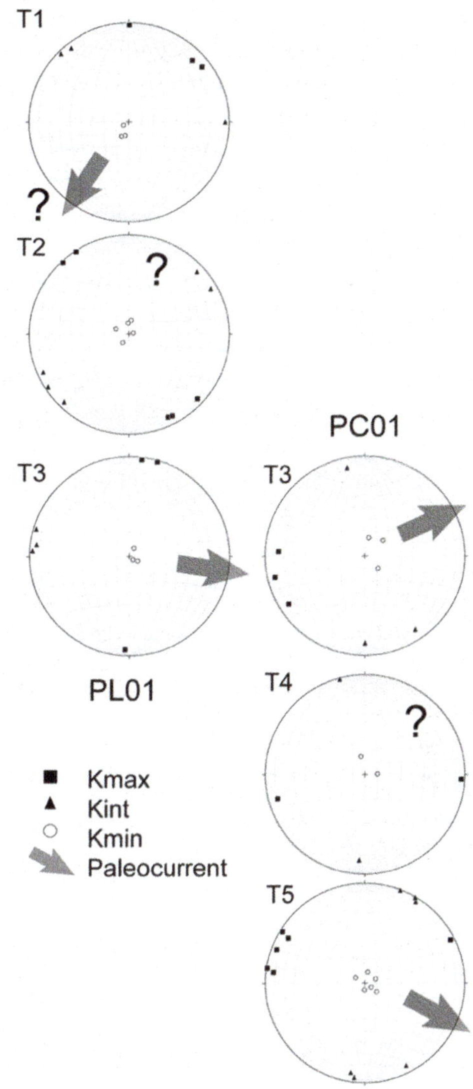

Fig. 4 Paleocurrent direction analyses in the turbidite units T1–T5. Kmax, Kint and Kmin = maximum, intermediate and minimum magnetic susceptibility directions as cKmax, cKint and cKmin (see text). These directions are projected in the stereographic diagrams of the lower hemisphere. The north in these diagrams is upward. The paleocurrent directions shown by arrows are judged with the dip direction of the Kmin as discussed in Tarling and Hrouda (1993) and Kawamura et al. (2002). The paleocurrents in T3 and T5 show mostly from west to east, indicating transportation from NE Japan to the Japan Trench

Concluding Remarks

A sediment core sample was collected from a deep-sea terrace at 1555 m in water depth. The sediments are composed of silty clay interbedded with five turbidite layers (T1–T5). Wood skins were observed at the base of the T4 and T5 indicating the ^{14}C age being 41,720 ± 520 year BP and an age older than 43,500 year BP. These indicate that these turbidite layers were transported with wood skins, which should come from coastal regions.

The wood skins in the turbidite were mixed when a turbidity current occurred or flowed down. It is known that the ^{14}C ages of these constituent grains in the turbidites do not coincide with the ages in generation and/or sedimentation of the turbidite, but rather are older. However, at least I can say that the depositional ages of the turbidite layers are younger than the ^{14}C age of the wood skins.

The paleocurrent directions of these turbidite layers were analyzed using magnetic methods. As a result, the turbidite layers were mostly transported from east as NE Japan arc. These results imply that turbidity currents have come repeatedly on a slope in the deep-sea terrace.

According to Arai et al. (2013), a giant turbidity current occurred at 2011 Tohoku-Oki tsunami, and the turbidite layer was deposited widely on trench slopes along the Japan Trench. These widely deposited turbidite layer in the Japan Trench are recorded repeatedly in the deep-sea sediments. Ikehara et al. (2016) documented historical three turbidite layers by 2011 Tohoku-Oki, 1611 Keicho and/or 1454 Kyotoku and 869 Jogan.

This study result indicates clearly that the turbidity currents occurred repeatedly in this region, but the ^{14}C ages of the wood skins correspond to the Last Glacier Period. This means that these turbidite layers might not be related to the recent large earthquakes. However, we need further to do a basic sedimentological study to identify the earthquake- and/or tsunami-generated turbidite layers in the Tohoku region.

Acknowledgements I thank chief scientist Makoto Yamano and captain Takafumi Aoki and crew members of *R/V Yokosuka* also supported the cruise and recovering the core sample. Professor Masafumi Murayama, Professor. Yuhji Yamamoto, Dr. Toshiya Kanamatsu and Mr. Takuya Matsuzaki supported the X-ray CT, grain size paleomagnetism and AMS analyse. Mr. Keita Yoshimoto and Ms. Mebae Kuranaga helped for core sampling and laboratory measurements. I thank Dr. Khang Dang and one anonymous reviewer for their constructive suggestions. This study was performed under the cooperative research program of the Centre for Advanced Marine Core Research (CMCR), Kochi University (14B049).

References

Arai K, Naruse H, Miura R, Kawamura K, Hino R, Ito Y, Inazu D, Yokokawa M, Izumi N, Murayama M, Kasaya T (2013) Tsunami-generated turbidity current of the 2011 Tohoku-Oki earthquake. Geology 41:1195–1198. https://doi.org/10.1130/G34777.1

Balsley JR, Buddington AF (1960) Magnetic susceptibility anisotropy and fabric of some Adirondack granites and orthogneisses. Am J Sci 258A:6–20

Ikehara K, Kanamatsu T, Nagahashi Y, Strasser M, Fink H, Usami K, Irino T, Wefer G (2016) Documenting large earthquakes similar to the 2011 Tohoku-oki earthquake from sediments deposited in the Japan Trench over the past 1500 years. Earth Planet Sci Lett 445:48–56

Kawamura K, Ikehara K, Kanamatsu T, Fujioka K (2002) Paleocurrent analysis of turbidites in Parece Vela Basin using anisotropy of magnetic susceptibility. J Geol Soc Jpn 108:207–218

Stacey FD, Jophin G, Lindsay J (1960) Magnetic anisotropy and fabric of some foliated rocks from S.E. Australia. Geophys Pure Apply 47:30–40

Tarling DH, Hrouda F (1993) The magnetic anisotropy of rocks. Chapman & Hall, London, p 217p

Tsuji Y, Satake K, Ishibe T, Kusumoto S, Harada T, Nishiyama A, Kim HY, Ueno T, Murotani S, Oki S, Sugimoto M, Tamari J, Heidarzadeh M, Watada S, Imai K, Choi BH, Bae JS, Kim KO, Kim HW (2011) Field surveys of tsunami heights from the 2011 off the Pacific coast of Tohoku, Japan Earthquake. Bull Earthq Res Inst University of Tokyo 86:29–279

Flank Failure of the Volcanic Turtle Island and the Submarine Landslide in the Southernmost Okinawa Trough

2020 Kyoto Japan

Pi-Chun Huang, Shu-Kun Hsu, Song-Chuen Chen, and Ching-Hui Tsai

Abstract

The Turtle Island (Kueishantao) is situated in the southernmost Okinawa Trough, a back-arc basin to the north of the Ryukyu Arc. Based on bathymetry around the Turtle Island, obviously hummocky seafloor indicating slope failures have occurred. The Turtle Island could have erupted four times in the last 7000 years. Previous studies have pointed out that volcanic eruptions may be related to island flank collapses. The bathymetric hummocky feature reveals submarine debris avalanches in the north, south and east side. In order to understand the landslide mechanism, we have conducted detailed marine geophysical surveys in the north part of the debris avalanches. The multi-beam bathymetric data, sub-bottom profiler, sidescan sonar, sparker seismic reflection data and remotely operated vehicle (ROV) dives data are used to identify the landslide features, including the depositional and erosional characters. Our results show that the debris avalanche is probably related to the horseshoe scar of the subaerial flank collapsed event. The debris avalanches display a runout distance of up to 4 km northeastward and form a ~ 4 km^2 hummocky terrain with displaced blocks. The biggest block could be ~ 50 m high and 50 m in diameter. Identified by the sparker seismic data, the chaotic facies of the landslide deposits can be divided into three MTD (Mass Transport Deposit) units (MTD1, MTD2 and MTD3). The main volcanic debris avalanche deposits are identified as MTD3 with a characteristic of broken and discontinuous lateral reflections. Correlated with seismic data and sub-bottom profiles, multiple landslide events are recognized on the basis of different MTD units. The last landslide event is probably related to a faulting and a slope failure of the island and could be highly associated with a climatic event.

Keywords

Flank failure • Debris avalanche • Okinawa trough • Submarine landslide

Introduction and Geological Setting

The Okinawa Trough backarc basin extends from Japan in the northeast to Taiwan in the southwest. The Okinawa Trough is formed due to the subduction of the Philippine Sea Plate beneath the eastern margin of the Eurasian Plate (Sibuet et al. 1987; 1995, 1998). However, the opening of the southern Okinawa Trough (SOT) is closely related to the post-collision of the Taiwan orogeny (Wang et al. 1999). The SOT is ~ 100 km wide in the east and becomes narrower westward and vanishes in the Ilan Plain (Fig. 1). Sibuet et al. (1998) suggested that volcanic intrusions were emplaced during Pleistocene along the active NE-SW normal faults. According to the interpretation of geophysical data, extensive magmatic activities and volcanic intrusions were found in SOT (Sibuet et al. 1998; Deffontaines et al. 2001). Lee (2005) has pointed out 12 submarine volcanoes and hydrothermal vents distributed in SOT, especially concentrated close to Kueishantao volcanic island (KST) which is located in the westernmost part of the Okinawa Trough (Fig. 1). Most of the submarine volcanoes have as high as

P.-C. Huang · S.-K. Hsu (✉)
Department of Earth Sciences, National Central University, Taoyuan City, Taiwan
e-mail: hsu@ncu.edu.tw

P.-C. Huang
e-mail: bat2266@gmail.com

S.-C. Chen
Central Geological Survey, New Taipei City, Taiwan
e-mail: songlin@moeacgs.gov.tw

C.-H. Tsai
Center of Environment Studies, National Central University, Taoyuan City, Taiwan
e-mail: chinghuitsai.s0642002@gmail.com

K. Sassa et al. (eds.), *Understanding and Reducing Landslide Disaster Risk*, ICL Contribution to Landslide Disaster Risk Reduction,
https://doi.org/10.1007/978-3-030-60196-6_29

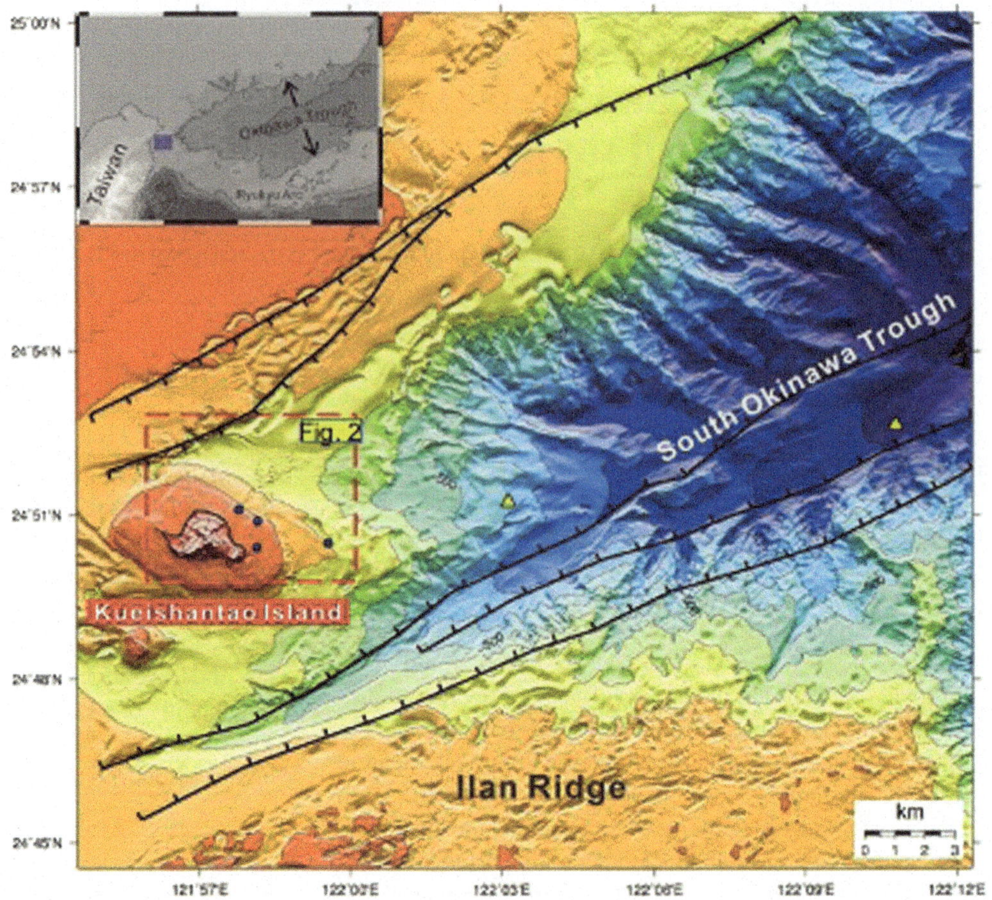

Fig. 1 Topography of the Kueishantao island and the surrounding area

200 m above seafloor, and are distributed at water depths between 1500 and 2000 m.

KST covers an area of ~2.8 km^2, having a distance of 3.3 km from east to west and a distance of 1.7 km from north to south. By analyzing K–Ar isotope, the formation age of the KST is estimated to be 20 ± 10 ka (Juang and Chen 1985). The KST magma chamber is thought to be migrated from the deep portion beneath the Ilan plan (Lin et al. 2004). KST is mainly composed of basaltic andesite lava flow and pyroclastic deposits, belonging to a typical composite volcano (Chiu et al. 2010). Both K–Ar dating (Juang and Chen 1985) and TL dating (Chen et al. 2001) show that the latest eruption age of KST is around 7 ± 0.7 ka from xenolith sediments within the volcanic rocks. Besides, there are 2 sequences of lava flow and 2 sequences of pyroclastic flow accumulated above the xenolith indicating high possibility of at least 4 times eruption after 7000 years BP (Chen et al. 2001; Song 2013). KST is similar to a sea turtle as its appearance, therefore it can mainly be divided into three part according to the morphology: Kueitou (turtle-head), Kueichia (turtle-shell) and Kueiwei (turtle-tail). Pyroclastic flow deposits are mainly distributed at Kueitou and interbedded between lava flows at Kueichia. Kueitou has a shape of half cone that could have collapsed due to a poor cementation.

Methods and Data

To study the flank failure of the KST, we have collected and analyzed multi-beam bathymetry, high-resolution sub-bottom profiler, sidescan sonar, sparker seismic-reflection data to understand the offshore landslide deposits distribution, characteristic and the possible landslide mechanisms. We have also used camera images from a remotely operated underwater vehicle (ROV) in some specific sites.

Bathymetric multi-beam data were collected with Kongsberg EM710. The EM710 system operates with the frequency range from 65 to 100 kHz and has 400 beams and a track coverage up to 5.5 times of the water depth. Bathymetric data are processed by CARIS™ software for further analysis. Bathymetric data had been corrected such as for seawater depth, pressure, temperature, salinity and a resolution up to 1 m.

High-resolution deep-towed sonar data were acquired with EdgeTech 2000-CSS. The sub-bottom profiler operated with the frequency range from 0.5 to 12 kHz, which has resolutions from 8 to 19 cm. The sidescan sonar has a resolution of 6.3 cm at 100 kHz and 1.8 cm at 400 kHz, respectively.

In total, we have collected 67 lines of sparker seismic-reflection profiles, which contain 60 lines in NW–SE direction and 7 lines in NE-SW direction. The sparker system was conducted with ELP1250 source model of 6000 J and with 3 s shooting rate and 0.125 ms sampling rate. Vertical resolution can be up to 4 ms. Data acquisition was performed using a SIG high-resolution streamer with 2 channels.

Results

A flank collapse event is generally considered as one of the important submarine landslide hazards. Our high-resolution bathymetric data around the KST has revealed the submarine hummocky morphology located north of the island. Although a previous study on land has pointed out the flank collapse scar in the northern flank, the associated submarine landslide of the KST was never studied. After analyzing our data, we conclude the following findings.

1. Based on the high-resolution bathymetric data, we have found the hummocky deposits in the northeastern part off the island. Diverse blocks range from ~10 m high

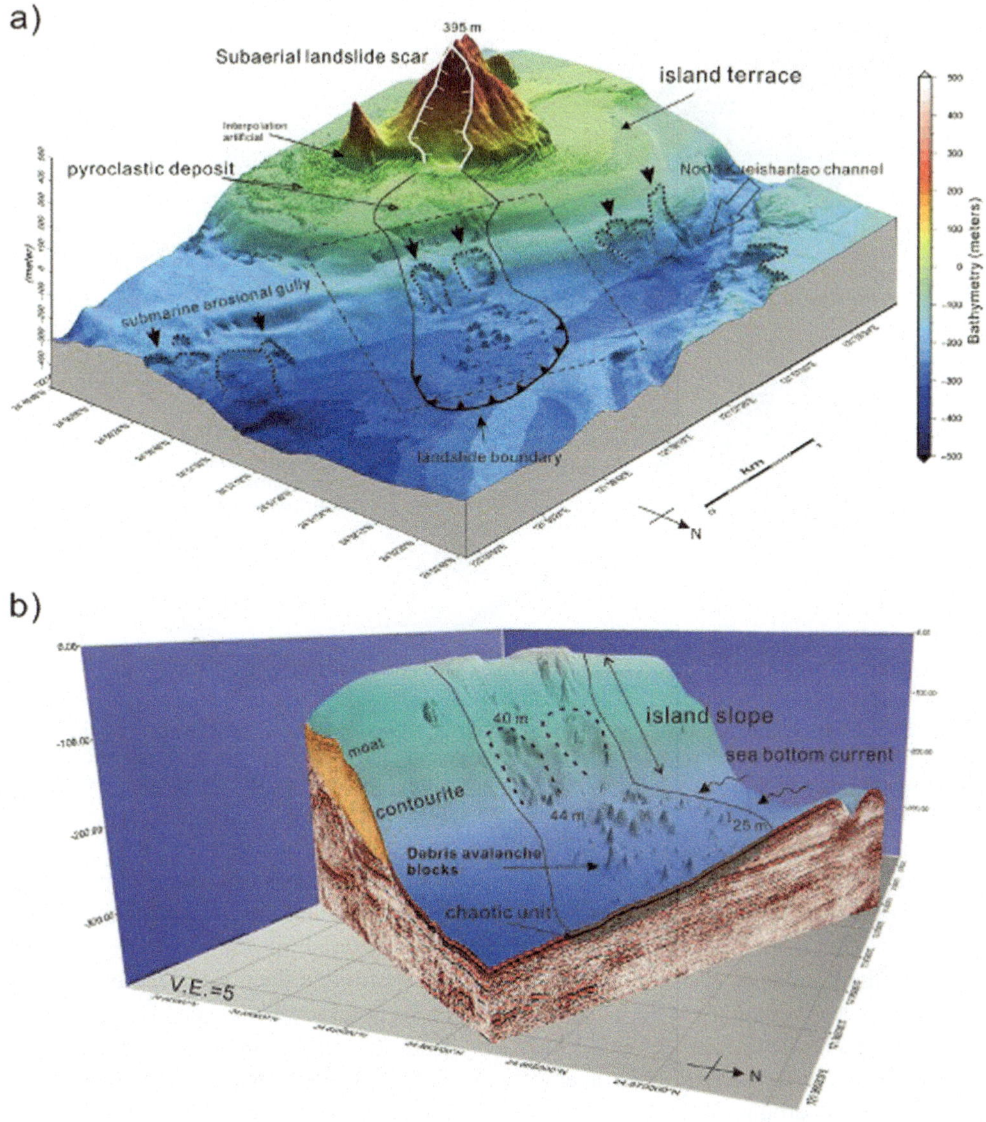

Fig. 2 **a** 3D image of the morphology showing the landslide scar and the debris avalanche. The flank collapse scar is marked by white line and the erosional gully along island slope is marked by dashed line. Black solid line indicated the main distribution of the landslide boundary. **b** Pseduo3D seismic data depicting the location of the landslide area

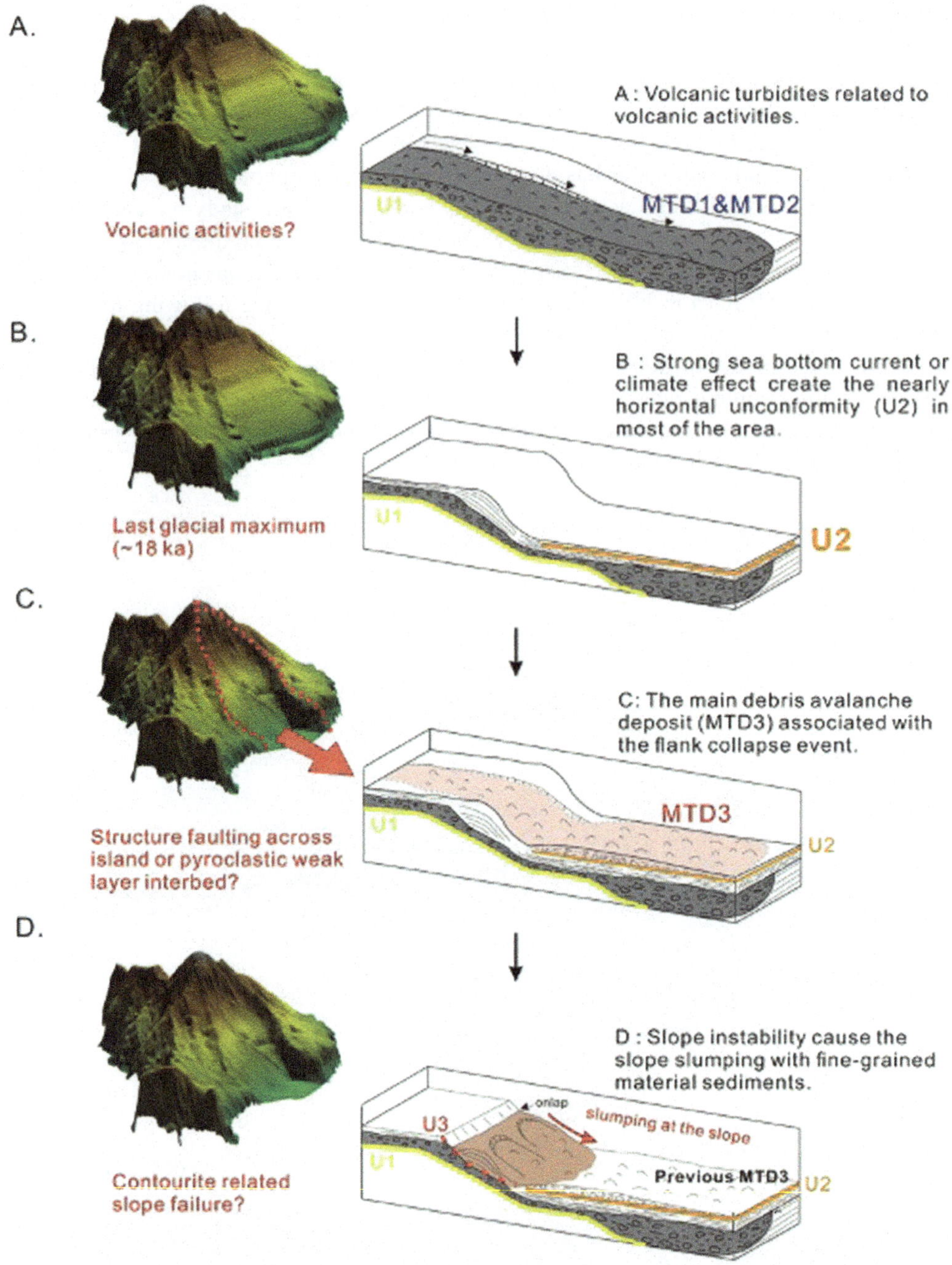

Fig. 3 Schematic illustration of the submarine failure events and possible mechanisms. **a** The volcanic-activities related deposit (MTD1 & MTD2) first occurred and deeply buried in the strata. **b** Horizontal unconformity form during the last glacial period due to erosion or depositional changes. **c** Main debris avalanche event triggered and sourced from the volcanic flank collapse. **d** Slope instability cause the slope slumping with mixed sediments of fine-grained sediments

to ~50 m high. Total area of the landslide deposits is around 10 km^2 and the run-out distance is ~3 km long (Fig. 2).

2. The horseshoe-shaped flank collapse scar can be recognized from the DTM data and is thought to be the source of submarine debris avalanche. Total scar area is 0.3 km^2 and the estimated volume is 0.012 km^3. The average slope gradient of the scarp is around 30°.
3. Previous analysis of sediments particles in the study area has shown that the main deposits are mud, fine-grained, clay material of the seafloor. However, SBP profiles show strong reflections of the blocky deposits, indicating these blocks are high density material and are different from the surrounding environment.
4. Sidescan sonar images show high backscatter intensities on the island shelf with scattered blocks deposits and the

lineation at the foot of island slope. Those linear structures are related to the seafloor erosion by the sea bottom current.

5. ROV images have provided a good constraint on acoustic images showing the blocky deposits on the seafloor. Characteristics of these images indicate that the scattered blocks are different from the muddy seafloor environment. Besides that, scattered shell debris can be found on the seafloor showing the carbonate deposits. The high backscatter signals from sidescan sonar images are related to the scattered blocks deposits.
6. Based on the sparker seismic profiles, four types of the seismic facies can be distinguished (Fig. 3): volcanic facie, chaotic facie I, chaotic facie II, sedimentary facie. Volcanic facie is assumed to a source directly from the island and is distributed mainly on the island shelf. The main landslide deposits are recognized as MTD3 (Mass Transport Deposit 3) from the chaotic facie distribution. However, MTD1 and MTD2 are deposited at the deeper part and may related to some volcanic activities.
7. Base on the seismic stratigraphy, we are able to distinguish unconformities U1, U2 and U3. The unconformity U1 is located at the deepest part with a strong acoustic reflection. Unconformity U2 is distributed at the foot of the slope area and is almost horizontally onlap unconformity U1. Unconformity U3 is only located in the eastern part of the seismic profiles, corresponding to the island slope gliding plane.

Summary

The turtle island is an active volcanic island located in the westernmost Okinawa Trough back-arc basin. Several MTDs identified around the Turtle Island are related to geological or seismological events. Particularly, a flank failure occurred in the northeast side of the Turtle island. This failure could be triggered by a structural faulting due to an earthquake. This flank failure has also caused a debris avalanche and a patch of hummocky seafloor to the northeast of the island.

References

Chen Y-G, Wu W-S, Chen C-H, Liu T-K (2001) A date for volcanic eruption inferred from a siltstone xenolith. Quatern Sci Rev 20:869–873

Chiu C-L, Song S-R, Hsieh Y-C, Chen C-X (2010) Volcanic characteristics of Kueishantao in northeast Taiwan and their implications. Terr Atmos Ocean Sci 21:575–585

Deffontaines B, Liu C-S, Jacques A, Lee C-T, Sibuet JC, Tsai Y-B, Lailemand S, Lu C-Y, Lee C-S, Hsu S-K, Chu H-T (2001) Preliminary neotectonic map of onshore-offshore Taiwan. Terr Atmos Oceanic Sci 339–349

Juang W-S, Chen J-C (1985) Geochronology and geochemistry of volcanic rocks in northern Taiwan. Bull Cent Geol Surv 31–66

Lee I-L (2005) The study of active submarine volcanoes and hydrothermal vents in the Southernmost Part of Okinawa Trough [Master thesis in Chinese]. National Taiwan Ocean University, Taiwan, p 55

Lin JY, Hsu SK, Sibuet JC (2004) Melting features along the western Ryukyu slab edge (northeast Taiwan): tomographic evidence. J Geophys Res: Solid Earth 109

Sibuet JC, Deffontaines B, Hsu SK, Thareau N, Formal L, Liu CS (1998) Okinawa trough backarc basin: early tectonic and magmatic evolution. J Geophys Res Solid Earth 103:30245–30267

Sibuet JC, Hsu S-K, Shyu C-T, Liu C-S (1995) Structural and kinematic evolutions of the Okinawa Trough backarc basin. Backarc basins, Springer, pp 343–379

Sibuet JC, Letouzey J, Barbier F, Charvet J, Foucher JP, Hilde TW, Kimura M, Chiao LY, Marsset B, Muller C (1987) Back arc extension in the Okinawa Trough. J Geophys Res Solid Earth 92:14041–14063

Song S-R (2013) The kueishantau—an active volcanic Island of Taiwan

Wang KL, Chung SL, Chen CH, Shinjo R, Yang TF, Chen CH (1999) Post-collisional magmatism around northern Taiwan and its relation with opening of the Okinawa Trough. Tectonophysics 308(3):363–376

Numerical Simulation for Tsunami Generation Due to a Landslide

2020 Kyoto Japan

Taro Kakinuma

Abstract

Tsunamis generated by falling rigid bodies, or a falling fluid, have been numerically simulated using the MPS model, in the vertical two dimensions. The tsunami height, immediately after the large circles enter the water, does not depend much on the offshore still water depth, while the tsunami-height reduction is suppressed, when the offshore still water depth is shallower. Conversely, the tsunami height, immediately after the small circles enter the water, increases as the offshore still water depth is shallower. Both the tsunami height, immediately after the falling bodies enter the water, and the reduction rate of tsunami height, are larger for the large circles than for the small circles. Furthermore, in the cases where the falling bodies include both the large and small circles, the reduction rate of the water level near the wave source is larger, when the large circles are also stacked on the offshore side at the initial condition. The tsunami height of the first wave, immediately after the rigid masses enter the water, is almost the same regardless of the rigid-mass shape. The tsunami height of the second wave is larger than that of the first wave in the cases of the right-angled isosceles triangles and the right triangles. The reduction rate of tsunami height for the rigid masses is larger than that for the circles. When the falling body is any of the three rigid masses, a tsunami component traveling toward the shore and running up the slope is confirmed. When the slope is milder underwater, the tsunami height due to the falling fluid is larger than that for the uniform-slope case.

Keywords

Tsunami • Landslide • Rigid body • MPS method

Introduction

Tsunamis can be triggered by not only submarine earthquakes but also by landslides. In 1792, the tsunamis due to a landslide, or a sector collapse, at Mt. Mayu, Japan, traveled over the Ariake Sea, resulting in a runup on the opposite shore (Togashi and Hirayama 1993). Such tsunamis are not necessarily generated only by soil or rocks: an excursion ship could be hit by tsunamis due to a partial collapse of glacier near a coast on Svalbard Islands, Norway (Marchenko et al. 2012). These tsunamis are generated through an interaction between water motion and falling bodies, such that the tsunami generation process is rather complicated.

In the present study, we have investigated several fundamental characteristics of tsunami generation caused by a landslide, or a sector collapse, on the basis of vertically two-dimensional results, obtained through a numerical simulation based on a Lagrangian method, where the falling body is assumed to be a rigid mass, a group of rigid bodies, or a fluid, which moves down a slope.

Numerical Method and Conditions

Tsunami generation due to a landslide or a sector collapse has been studied numerically with various methods, including a moving particle semi-implicit (MPS) model designed by Koshizuka and Oka (1996) (e.g. Gotoh et al. 2011). We use the numerical model developed by Iribe and Nakaza (2011), based on the MPS method, where the water surface level is determined using the spatial gradient of particle-number density, to inhibit pressure disturbance at the water surface. The interparticle distance is 0.005 m in the

T. Kakinuma (✉)
Graduate School of Science and Engineering, Kagoshima University, 1-21-40 Korimoto, Kagoshima, 890-0065, Japan
e-mail: taro@oce.kagoshima-u.ac.jp

K. Sassa et al. (eds.), *Understanding and Reducing Landslide Disaster Risk*,
ICL Contribution to Landslide Disaster Risk Reduction,
https://doi.org/10.1007/978-3-030-60196-6_30

present cases. No turbulence model is utilized for fluid motion, and both fluid viscosity and surface tension are neglected for simplicity.

Figure 1 shows the target domain inside a water basin with a slope, where the gradient of the slope is 45°. The distance between the starting point of the slope and the offshore wall is 3.00 m, and the still water depth off the slope is uniformly 0.09 m or 0.245 m. The origin of the coordinate axes is at the shoreline in the still water condition. The positive direction of the x-axis is the horizontal offshore direction, while the positive direction of the z-axis is the upward direction.

The release of objects, stacked on the slope, is a sector collapse to generate tsunamis, where the falling objects are assumed to be rigid bodies as follows:

1. Large circle: The diameter of large circles is 2.0 cm. A large circle consists of 21 rigid particles.
2. Small circle: The diameter of small circles is 0.5 cm. A small circle consists of 4 rigid particles.
3. Right triangle: A right triangle consists of 276 rigid particles.
4. Rectangle: A rectangle consists of 276 rigid particles.

   We have also conducted hydraulic experiments, using a group of cylinders with a diameter of 2.0 cm, although the experimental results cannot be shown in this paper owing to space limitations. The fluid density is 1000 kg/m^3, while the density of the rigid particles is 2600 kg/m^3. Both the elasticity and the plasticity of the falling bodies are neglected for simplicity.
   The tsunamis due to the falling object, assumed to be the following fluid, are also investigated, where the slope gradient is different between above and below the still water level:

5. Fluid, the density of which is 2600 kg/m^3.

Tsunamis Caused by Falling Large Circles

As shown in Fig. 2, 16 large circles are arranged on the slope at the initial time t = 0.0 s.

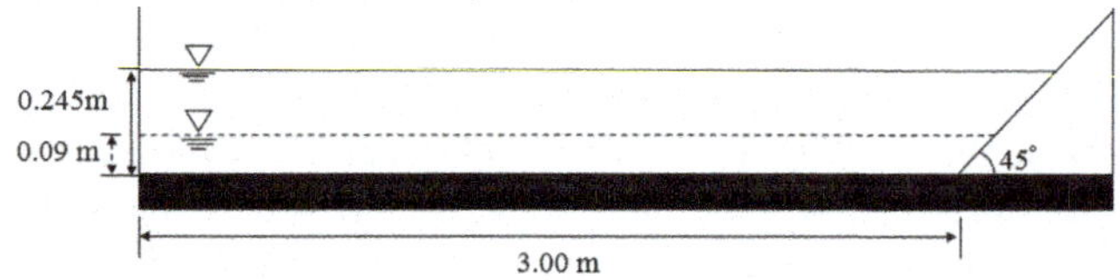

Fig. 1 The target domain inside a water basin with a slope

Fig. 2 A sketch of the large circles at the initial condition

Figure 3 shows the numerical simulation result, where the uniform still water depth off the slope, h, is 0.09 m.

The water surface displacements at x = 0.6 m and 2.15 m are shown in Fig. 4, where x denotes the distance from the shoreline in the still water condition. In this paper, the results for water surface displacements are moving average values with a sampling interval of 0.01 s.

The tsunami height is defined as the maximum value of water surface displacement at each location. According to Fig. 4a, the tsunami height at x = 0.6 m is 0.072 m when h = 0.09 m, and 0.068 m when h = 0.245 m, such that the difference between them is not large, that is, the tsunami height, immediately after the large circles enter the water, does not depend much on the still water depth. Conversely, according to Fig. 4b, the tsunami height at x = 2.15 m is 0.066 m when h = 0.09 m, and 0.032 m when h = 0.245 m, such that the tsunami-height reduction rate differs greatly for the two cases, for the reduction in the tsunami height is suppressed owing to shallowing, showing the steeper wave front, in the former with the shallower offshore still water depth.

Tsunamis Caused by Falling Small Circles

As indicated in Fig. 5, 66 rigid small circles are placed on the slope at the initial time t = 0.0 s. Both the total mass, and the vertical position of the center of gravity at t = 0.0 s, of all the small circles are equal to those of all the large circles shown in Fig. 2.

Shown in Fig. 6 are the water surface displacements at x = 0.6 m and 2.5 m.

According to Fig. 6a, the tsunami height at x = 0.6 m is 0.061 m when h = 0.09 m, and 0.046 m when h = 0.245 m, such that there is a difference in the tsunami height, immediately after the small circles enter the water, where the tsunami height increases as the offshore still water depth is shallower. It should be noted that the second wave, reflected

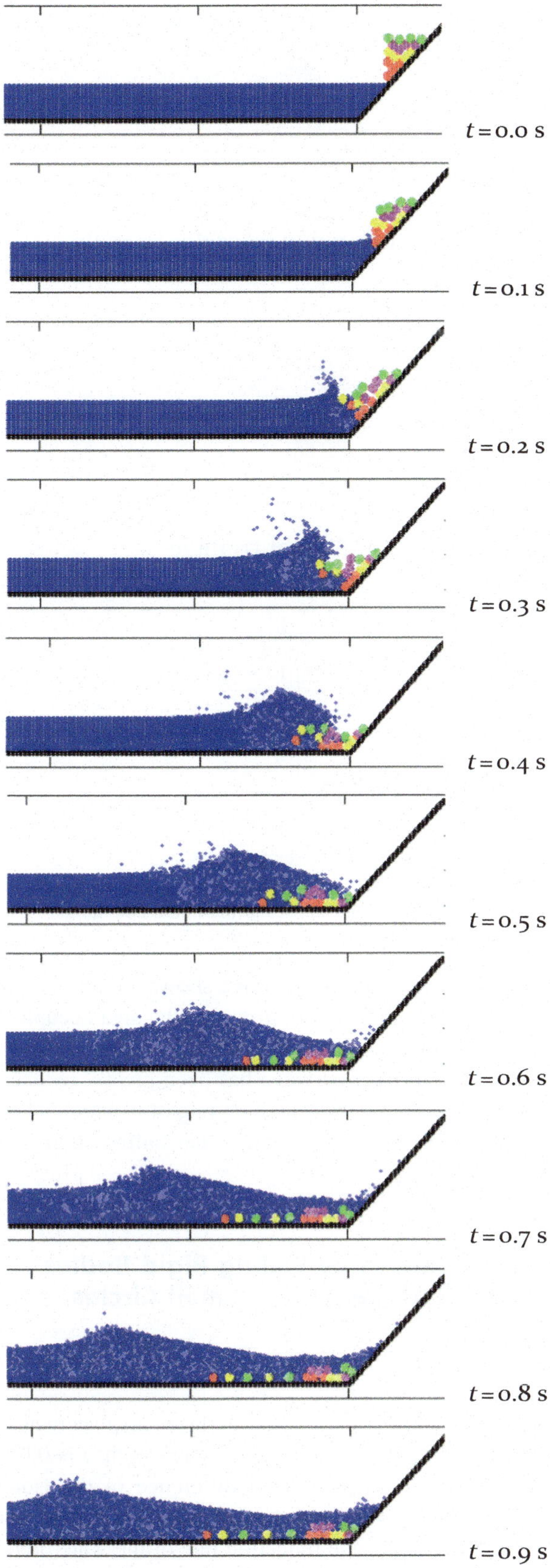

Fig. 3 The numerical result of the tsunamis caused by the falling large circles, where $h = 0.09$ m

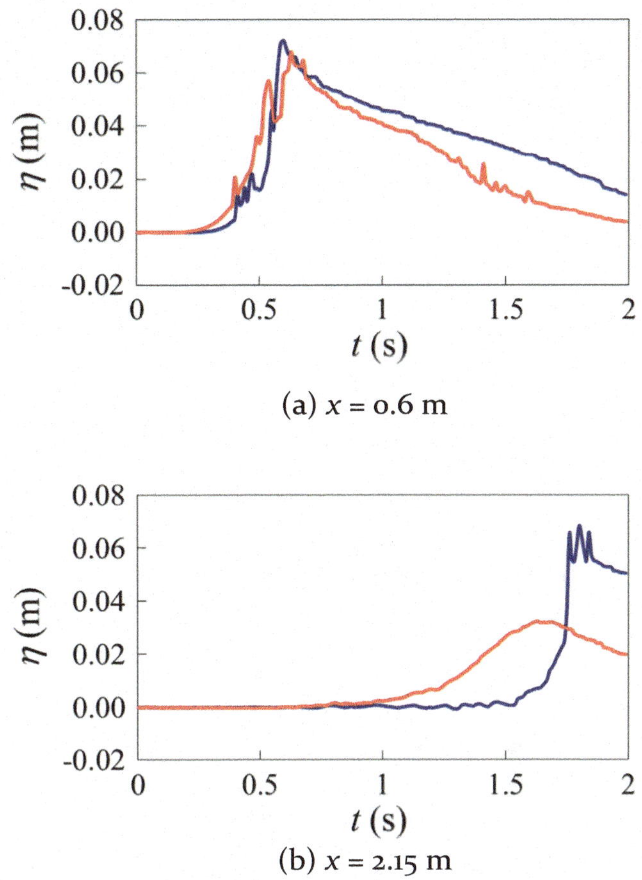

Fig. 4 The water surface displacements due to the falling large circles. The blue lines represent the water surface displacements when $h = 0.09$ m, and the red lines when $h = 0.245$ m

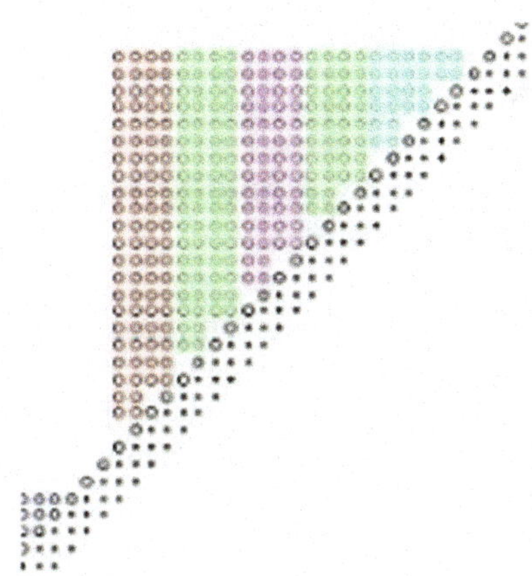

Fig. 5 A sketch of the small circles at the initial condition

on the slope, remarkably appears in the deeper case, at $t \simeq 1.4$ s.

According to Figs. 4a and 6a, the tsunami height, immediately after the rigid bodies enter the water, is larger when the falling bodies are the large circles than when those

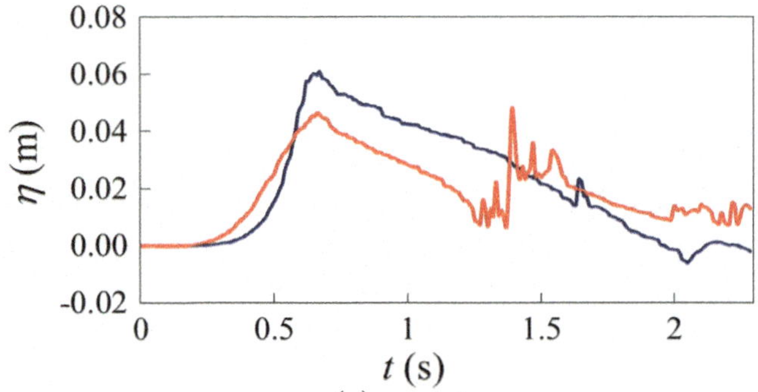

(a) $x = 0.6$ m

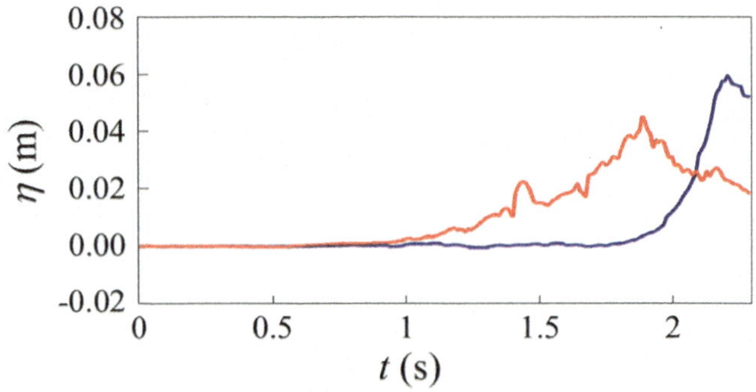

(b) $x = 2.5$ m

Fig. 6 The water surface displacements due to the falling small circles. The blue lines represent the water surface displacements when $h = 0.09$ m, and the red lines when $h = 0.245$ m

are the small circles, for more water can enter between the small circles. Conversely, according to Fig. 6b, the tsunami height at $x = 2.5$ m is 0.060 m when $h = 0.09$ m, and 0.045 m when $h = 0.245$ m, such that the reduction rate of tsunami height is lower for the falling small circles than for the falling large circles.

Tsunamis Caused by a Falling Right Triangle or a Falling Rectangle

As illustrated in Fig. 7, the falling body is a right-angled isosceles triangle with an equilateral length of 0.11 m, a right triangle, or a rectangle, where both their area, and their vertical position of the center of gravity at $t = 0.0$ s, are equal.

Figure 8 shows the water surface displacements at $x = 0.6$ m and 2.5 m, where the offshore still water depth h is 0.245 m. The tsunami height of the first wave, immediately after the rigid mass enters the water, is almost the same regardless of the rigid-mass shape. The tsunami height of the second wave, however, is lower than that of the first wave only in the case of the rectangle.

(a) A right-angled isosceles triangle

(b) A right triangle

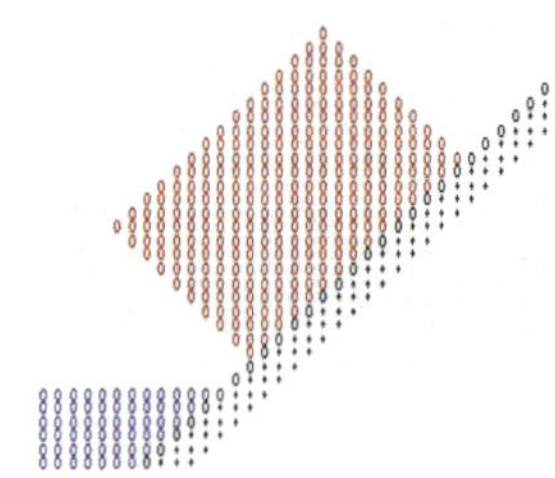
(c) A rectangle

Fig. 7 Sketches of the rigid masses at the initial conditions

In all the cases, the second wave shows a relatively large wave height, for water cannot flow into the area occupied by the rigid mass after landing. At the location $x = 2.5$ m, the tsunami height of the first wave has been greatly reduced in all the cases, such that the reduction rate of tsunami height is larger than that for the cases where the falling rigid bodies are the circles.

Tsunamis Caused by Falling Rigid Bodies Including Both Large and Small Circles

As shown in Fig. 9, both the large and small circles are loaded at the initial time.

Figure 10 shows the water surface displacements at $x = 0.6$ m, where the offshore still water depth h is 0.09 m. Although there is no significant difference in the tsunami height for these cases, immediately after the falling bodies plunge into the water, the reduction rate of the water level at

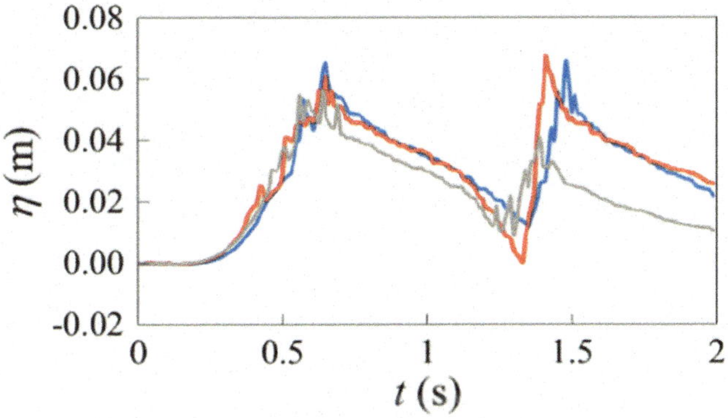

(a) $x = 0.6$ m

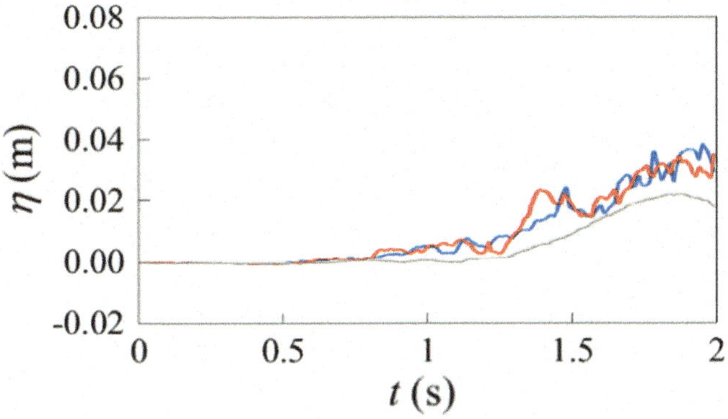

(b) $x = 2.5$ m

Fig. 8 The water surface displacements due to the falling rigid masses, where $h = 0.245$ m. The blue lines represent the water surface displacements due to the right-angled isosceles triangle; the red lines due to the right triangle; the gray lines due to the rectangle

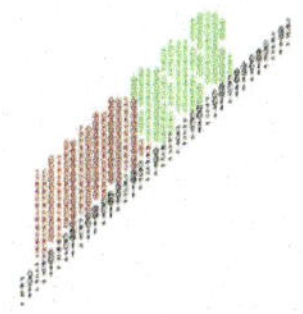

(a) The large circles are on the onshore side.

(b) The large circles are on the offshore side.

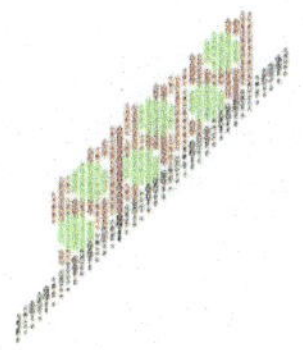

(c) The large and small circles are mixed.

Fig. 9 Sketches of the rigid bodies, including both the large and small circles, at the initial conditions

this location is larger, when the large circles are also stacked on the offshore side at the initial condition.

Tsunamis Running Up the Slope Where the Landslide has Occurred

In general, when a landslide occurs in water, the influence of the density ratio is lower than when a landslide occurs on land, for the surrounding of the collapsed body is water, such that the falling speed of the collapsed body is slow, and the tsunami is unlikely to grow (Kakinuma 2016).

It is necessary, however, to consider the tsunami component that travels toward the shore and runs up on land, when tsunamis are caused by a nearshore landslide whether the landslide begins above or below the sea level.

Figure 11 shows the numerical simulation results at $t = 0.8$ s, i.e., when the height of the upstream end becomes large on the slope, where the falling mass is initially placed on the slope above the still water level, as shown in Fig. 7. The offshore still water depth h is 0.245 m. When the rigid

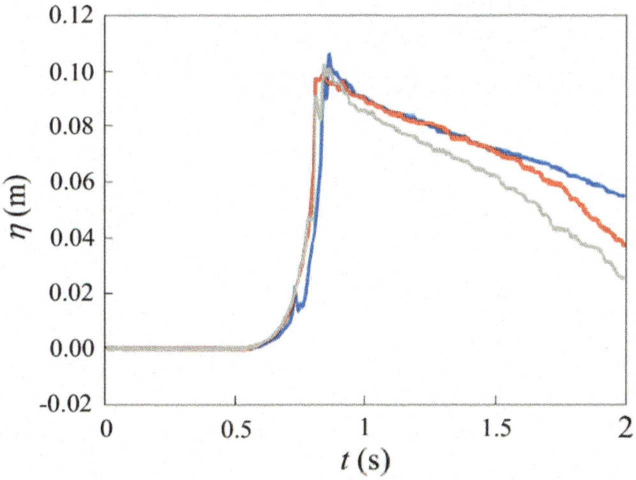

Fig. 10 The water surface displacements at $x = 0.6$ m, due to the falling rigid bodies including both the large and small circles, where $h = 0.09$ m. The blue line represents the water surface displacement when the large circles are initially arranged on the onshore side; the red line when the large circles are initially stacked on the offshore side; the gray linc when the large and small circles are mixed at the initial condition

mass is the right-angled isosceles triangle, the right triangle, and the rectangle, as shown in Fig. 7, the runup height z_{up} is 0.11 m, 0.094 m, and 0.047 m, respectively, such that if the offshore still water depth h is 10.0 m, with a length scale of

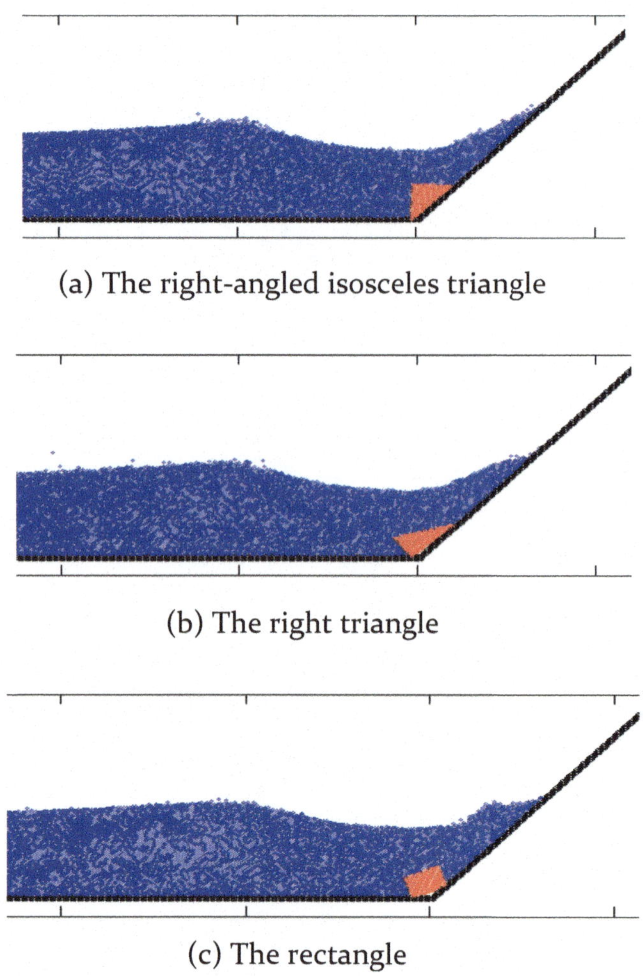

Fig. 11 The numerical simulation results at $t = 0.8$ s, where the falling rigid masses at the initial time $t = 0.0$ s are shown in Fig. 7; $h = 0.245$ m

0.0245, the runup height $z_{\rm up}$ is around 4.5 m, 3.8 m, and 1.9 m, respectively. These values are the runup heights at which wooden structures can be damaged.

Tsunamis Due to a Falling Fluid, Where the Slope Gradient is Different Between Above and Below the Still Water Level

As shown in Fig. 12, the fluid, the density of which is 2600 kg/m^3, is loaded on the slope with a gradient of 45°, at the initial time, where the bottom of the fluid is located at $z = 0.1$ m in Cases 2, 4, and 6; the slope gradient is different between above and below the still water level. In Cases 1 and 2, the slope gradient is milder below the still water level than above it, owing to sediments supplied from a river. In Cases 3 and 4, the coast shows a coastal terrace. In Cases 5 and 6, the coast was eroded below the sea level. The offshore still water depth h is 0.09 m, where the water density is 1000 kg/m^3.

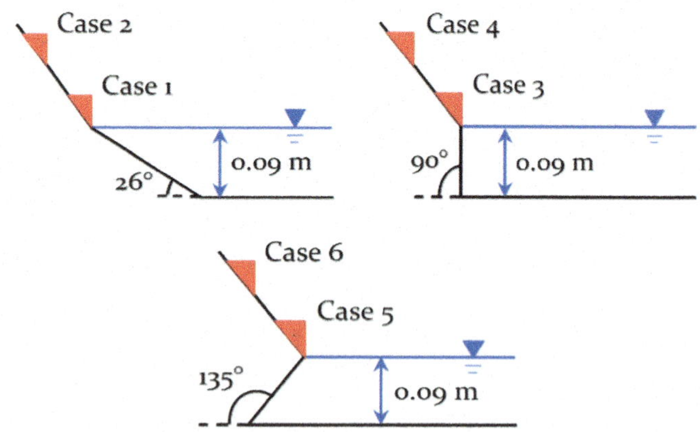

Fig. 12 Sketches of the falling fluids at the initial conditions in Cases 1–6

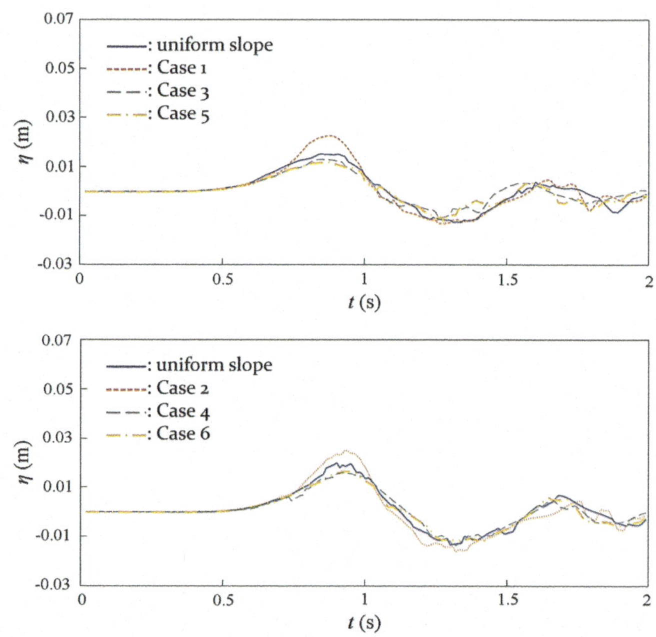

Fig. 13 The water surface displacements at $x = 0.7$ m in Cases 1–6 shown in Fig. 12, as well as that for the case where the slope gradient is uniform

Shown in Fig. 13 are the water surface displacements at $x = 0.7$ m in Cases 1–6 shown in Fig. 12, as well as that for the case where the slope gradient is the same above the water level and underwater.

Only in Cases 1 and 2, the tsunami height is larger than in the uniform-slope case, for the slower movement of the falling fluid, on the milder slope, makes the tsunami grow in the shallower water.

Conclusions

The tsunamis generated by the falling bodies were numerically simulated using the MPS model, in the vertical two dimensions. The tsunami height, immediately after the small circles entered the water, increased as the offshore still water depth was shallower. Both the tsunami height, immediately after the falling bodies entered the water, and the reduction rate of tsunami height, were larger for the large circles than for the small circles. In the cases where the falling bodies included both the large and small circles, the reduction rate of the water level near the wave source was larger, when the large circles are also stacked on the offshore side at the initial condition.

The tsunami height of the second wave was larger than that of the first wave in the cases of the right-angled isosceles triangles and the right triangles. The reduction rate of tsunami height for the rigid masses was larger than that for the circles. When the falling body is any of the three rigid masses, a tsunami component traveling toward the shore and running up the slope was confirmed.

When the slope is milder underwater, the tsunami height due to the falling fluid was larger than that for the uniform-slope case.

Acknowledgements Sincere gratitude is extended to Dr. Tsunakiyo Iribe, University of the Ryukyus, who allowed me to apply his computational program based on the MPS method. I also express my gratitude to Ms. Manami Higashi, Ms. Aya Oyama, Mr. Ryo Sawada, and Mr. Mitsuru Yanagihara, who contributed to the numerical simulation, when they were student members of our laboratory.

References

Gotoh H, Ikari H, Matsubara T, Ito T (2011) Numerical simulation on tsunami due to sector collapse by solid-liquid two-phase flow model based on accurate particle method. J JSCE B2 (Coastal Eng) 67(2): I_196-I_200 (in Japanese with an English abstract)

Iribe T, Nakaza E (2011) An improvement of accuracy of the MPS method with a new gradient calculation model. J JSCE B2 (Coastal Eng) 67(1):36–48 (in Japanese with an English abstract)

Kakinuma T (2016) Tsunami generation due to a landslide or a submarine eruption. In: Tsunami. Mokhtari M (ed). InTech. ISBN 978-953-51-2677-5. pp 35–58

Koshizuka S, Oka Y (1996) Moving-particle semi-implicit method for fragmentation of incompressible fluid. Nucl Sci Eng 123:421–434

Marchenko AV, Morozov EG, Muzylev SV (2012) A tsunami wave recorded near a glacier front. Nat Hazards Earth Syst Sci 12: 415–419

Togashi H, Hirayama Y (1993) Hydraulic experiment on reappearance of the Ariake-kai tsunami in 1792. Proc IUGG/IOC Int Tsunami Symp (Tsunami '93):741–754

Dealing with Mass Flow-Induced Tsunamis at Stromboli Volcano: Monitoring Strategies Through Multi-Platform Remote Sensing

Federico Di Traglia, Teresa Nolesini, and Nicola Casagli

Abstract

Volcano landslides or explosions-induced mass flows constitute an important trigger of tsunamis. Even if landslide-induced tsunamis can produce more local impacts comparable earthquake-induced tsunamis, large volume failure of volcanic edifice may cause tsunamis with widespread effects. Considering this, successful strategies for volcano slope instability detection must involve the integration of different methodologies for mapping, monitoring, and automated approaches for early warning, integrating field-based studies, geomorphological mapping, remote sensing data, geophysical and geochemical investigations, and/or numerical modelling. In this contribution, the applications of different remote sensing techniques products for the identification, mapping, and forecasting mass movements in the island of Stromboli are presented. The integration of space-borne and ground-based Synthetic Aperture Radar displacement data with the analysis of (topographic- and SAR amplitude images based) change detection allowed the identification the evolution of the slope instability phenomena and the geomorphological processes affecting the Stromboli unstable slopes. Ground based SAR devices are the key-instruments for the operational approach to mitigating landslide risks, being used to monitor the slope instability and to detect the inflation/deflation of the crater area.It is crucial to emphasize the importance of smart integration of space borne-derived hazard information with permanent-sited, operational monitoring by GBInSAR devices to detect areas impacted by mass wasting and volcanic activity.

Keywords

Volcano slope instability • Landslide-induced tsunami • InSAR • GBInSAR • Topographic change detection • SAR amplitude • Stromboli

Introduction

Landslides, including volcano flank collapses or explosions-induced mass flows, constitute the second-most important trigger of tsunamis after earthquakes (Harbitz et al. 2014). Landslide-induced tsunamis produced a great variety of effects (Fig. 1), resulting in more local impacts comparable to earthquake-induced tsunamis. However, large volume landslides (volume >1000 km^3) may cause tsunamis with more widespread effects (i.e. Sassa et al. 2016; Williams et al. 2019). Large-scale volcanic edifice collapse has been identified at >400 Quaternary volcanoes worldwide (Siebert et al. 2006), and several studies have also revealed the tendency for volcanoes to collapse repeatedly in their lifetime (Schaefer et al. 2019). Repetitive failure is possible at volcanoes with persistent magmatic activity, in active tectonic settings, or high magma extrusion rates that rapidly reconstruct the edifice (Schaefer et al. 2019). The first global database of giant landslides on volcanic islands comprises a total of one hundred and eighty-two entries (Blahůt et al. 2019): seventy-five are located in the Atlantic Ocean (most of them located in Canary islands, Madeira, Cabo Verde archipelago, and Azores islands); sixty-seven giant landslides derived from volcanic edifices in the Pacific Ocean; and forty giant are in the Indian Ocean, but these have all taken place on the island of Réunion. In the last 150 years, there have been 12 large-volume collapse events (0.05 km^3 or larger)

F. Di Traglia (✉) · N. Casagli
Dipartimento di Scienze Della Terra, Università Degli Studi di Firenze, Via La Pira 4, 50121 Firenze, Italy
e-mail: federico.ditraglia@unifi.it

N. Casagli
e-mail: nicola.casagli@unifi.it

T. Nolesini
Centro Per La Protezione Civile, Università Degli Studi di Firenze, Piazza San Marco 4, 50121 Firenze, Italy
e-mail: teresa.nolesini@unifi.it

K. Sassa et al. (eds.), *Understanding and Reducing Landslide Disaster Risk*, ICL Contribution to Landslide Disaster Risk Reduction,
https://doi.org/10.1007/978-3-030-60196-6_31

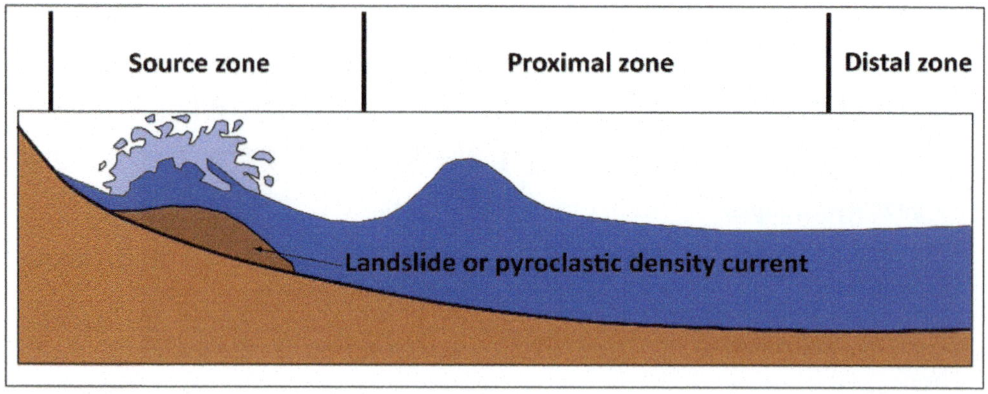

Fig. 1 Conceptual model of mass-flow induced tsunamis

that have resulted in debris avalanches, debris flows and/or tsunami, emphasizing the widespread occurrence of volcanic mass movements (Schaefer et al. 2019).

Due to their complex nature, successful strategies for volcano slope instability hazard assessments, especially in coastal and island environments, must involve the integration of different methodologies for detection, mapping, and monitoring, possibly with the most automated approaches possible for early warning. This often requires the combined use of field-based studies and geomorphological mapping with remote sensing, geophysical and geochemical investigations, and/or numerical modeling.

In this paper, the applications of different remote sensing techniques products for the identification, mapping, and forecasting mass movements in island volcanoes are presented. To emphasize the potential utility of the integrated method, the case-study of Stromboli (Italy) volcano (Fig. 2) is presented, as (Di Traglia et al. 2014; Schaefer et al. 2019): (1) it recently experienced moderate to major instability events; (2) its slopes are prone to landslides; (3) it is persistently active; (4) landslide-induced tsunamis could affect populated areas, and; (5) a variety of remote sensing techniques were applied.

Materials and Methods

Stromboli Island

Stromboli island (Fig. 2) is the 916 m-high emerged part of a $\approx$ 3000 m high active volcano in the southern Tyrrhenian sea. Stromboli's persistent activity consists of frequent, small scale, explosions (Rosi et al. 2013). The most hazardous phenomena at Stromboli Island are mass flow-induced tsunamis (Maramai et al. 2005), which make it one of the tsunami sources in the Mediterranean Sea (Cerase et al. 2019). This activity, showing intensity and frequency fluctuations over time, is often punctuated by higher intensity explosions (major or paroxysmal explosions), lava overflows from the crater, and/or by flank eruptions, with the outpouring of lava flows from ephemeral vents, as occurred in 2002–03, 2007 and 2014 (Rosi et al. 2013; Di Traglia et al. 2018a).

Mass flows can be of two types:

- Intrusion-related landslides from the NW unstable flank of the volcano (Sciara del Fuoco; SdF), as occurred on 30th December 2002;
- By the entry into the sea of pyroclastic density currents (PDCs) produced during paroxysmal, as occurred on 3rd July 2019 and 28th August 2019 (Turchi et al. 2020).

Tsunamis occurred in recent times as in 1879, 1916, 1919, 1930, 1944, and 1954, 2002, and 2019 (Fig. 3) accounting for an average of 1 tsunami every 20 years (Maramai et al. 2005), and affecting the coast of Stromboli (Nave et al. 2010), and secondly, the coasts of the other Aeolian islands and the Tyrrhenian coasts of southern Italy (Fornaciai et al. 2019).

Deposits of potentially tsunamigenic landslides have been discovered in marine sediments offshore of the island of Stromboli (Di Roberto et al. 2010). Recent findings revealed the occurrence of three tsunamis likely related to repeated flank collapses struck during the Late Middle Ages (Rosi et al. 2019).

Methods

Optical and Synthetic Aperture Radar (SAR) sensors have often been used to maps areas affected by lithological and morphological changes, i.e. to identify areas impacted by eruptive and post-eruptive (landslides or floods) phenomena (i.e. Di Traglia et al. 2018b), whereas multi-temporal Digital Elevation Models (DEMs) allowed the quantification of topographic changes (i.e. Di Traglia et al. 2020).

Geodetic techniques are particularly useful for determining the extent and current state of volcano slope

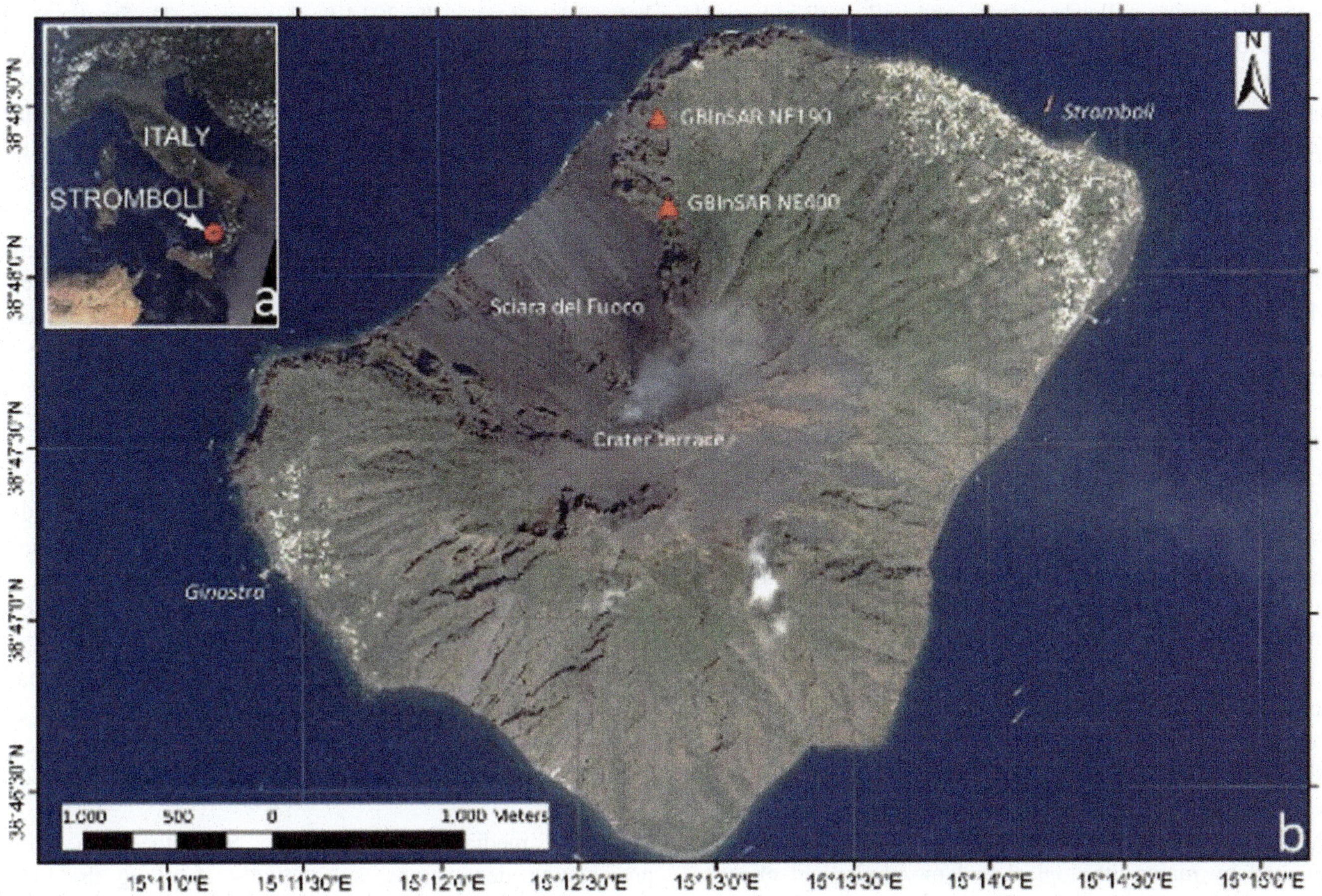

Fig. 2 **a** Geographic location of the Aeolian Islands (image collected by SENTINEL-3 on 19 July 2018); **b** PLÉIADES-1B image (28 August 2017) of Stromboli Island

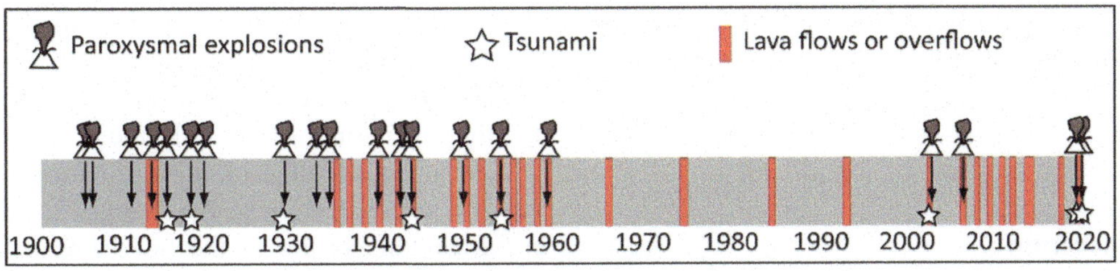

Fig. 3 Volcanic activity and tsunamis at Stromboli volcano (modified after Maramai et al. 2005)

instability, with many benefits arising from the use of synthetic aperture radar (SAR) data (Schaefer et al. 2019 and references therein). Measuring surface deformation using the phase difference between two space-born SAR images (differential interferometric SAR, DInSAR; Massonnet and Feigl 1998) allows for the recognition of centimeter-scale displacements of the ground along the satellite line of sight (LOS) direction. Processing a long stack of images using multi temporal (MT) InSAR techniques allows for the detection of millimeter-scale displacements over long time frames through the reduction of error sources (i.e. Ferretti et al. 2001; Berardino et al. 2002).

Ground-Based InSAR (GBInSAR) has the additional advantage of producing frequent SAR images (on the order of seconds to minutes), resulting in very high frequency slope maps and time series. With these fast repeat time intervals, GBInSAR has led InSAR from mapping to surveillance and early-warning applications (Di Traglia et al. 2014). Moreover, the use of GBInSAR in the Ku-band (17–17.1 mm radar) can penetrate dust clouds (abundant especially during collapse events), and will work in variable light and atmospheric conditions (Calvari et al. 2016).

The NE portion of the summit crater terrace and the northern portion of the SdF are monitored by two GBInSAR devices, located in a stable area N of the SdF (Fig. 4). The first GBInSAR (GBInSAR NE400; Model: GB-InSAR LiSALab; Revisiting time; 11 min; Antonello et al. 2004) was installed in February 2003, during the 2002–03 flank eruption, while the second device (GBInSAR NE190; Model: GB-InSAR LiSAmobile k09; Revisiting time; 2 min) was installed on 14 December 2014, after the 2014 flank eruption.

Moreover, a total of 85 (22 February 2010-18th December 2014; descending orbit) COSMO-SkyMed (CSK; X-band) and 47 (23 February 2015–15 October 2016; descending orbit) SENTINEL-1A (SNT; C-band; Fig. 4c) SAR images, were exploited and processed using the SqueeSAR algorithm (Ferretti et al. 2011). To reduce the atmospheric disturbance Atmospheric Phase Screen (APS) filtering was performed (Ferretti et al. 2001).

A large set of multi-temporal data of the SdF was acquired, including LiDAR data, tri-stereo PLEIADES-1 imagery, high-spatial-resolution (HSR) optical imagery (QUICKBIRD and PLEIADES-1), and moderate spatial resolution (MSR) SENTINEL-2 Multi-Spectral Instruments (MSI) imagery. Multi-temporal data permits to maps areas affected by major lithological and morphological changes, and the volumes of deposited/eroded material (Fig. 5b). The reflectivity (amplitude) evolution of CSK SAR images, were analysed in order to constrain the geomorphologic evolution of the SdF (Fig. 5b). The results lead to the identification of topographical variations and slope processes that occurred in response to the variation in eruptive intensity.

Results

Ground Displacement

The analysis of the entire dataset of GBInSAR measurements allowed the assessment of the deformation field of the northern part of the summit crater area and the Sciara del Fuoco depression (Di Traglia et al. 2014). In detail, the main displacements recognized can be related to different factors (Fig. 6): (1) the inflation/deflation respectively immediately before and after each new effusive event; (2) the bulging of localized sectors of the volcano involved in the vent opening; (3) the gravitational sliding of the Sciara del Fuoco infill; (4) the movement of lava flows. During the period from 1st January 2010—6th August 2014, the GBInSAR recorded slow ground displacement (<0.05 mm/h; see Di Traglia et al. 2018a, b). Inflation of the summit plumbing system has been recorded since 30th May 2014, 2 months before the onset of the 2014 flank eruption (Di Traglia et al. 2018a). Moreover, an increase in the displacement rate (up to 230 mm/h) associated with the breaching of the summit a rea starting approximately 15 h before the occurrence of the collapse that occurred at the eruption onset. After the eruption, there were recorded displacements in the SdF related to three types of movement: (1) gravitational re-adjustment of the lava breccia, sometimes evolving in rockfalls (Di Traglia et al. 2018b), (2) thermal contraction of the lava field (Schaefer et al. 2019), and (3) persistent flank motion (Schaefer et al. 2019).

Change Detection

The integration of the TCD and the analysis based of the SAR images were the basis for the definition of the geomorphological evolution of the SdF steep-slope, which is frequently affected by an accumulation of volcanic material, erosion, and mass-wasting phenomena (Di Traglia et al. 2018b). During the period between May 2012 and May 2017, large topographic changes occurred (Di Traglia et al. 2020). The largest variations were related to the emplacement of the 2014 lava flow field in the NE part of the SdF (Fig. 5).

In the median portion of the SdF, the lava filled pre-existing depressions, substantially changing the morphology of the SdF. In the central part of the SdF, from 2012 to 2017, a huge volcanoclastic wedge accumulated. During the same time period, the SdF has been affected by massive erosive phenomena. The older lava has been affected by headward erosion carried on, to lose material. Besides this, two large areas have been affected by erosion: one constrained between the 2014 lava and the volcaniclastic wedge and the other south of the volcaniclastic wedge (Fig. 5a).

CSK images, collected during low-intensity Strombolian activity, are characterized by changes in the backscattering in the central portion of the SdF, with a loss of the SAR amplitude between the two consecutive images. During periods characterized by more frequent/intense Strombolian activity, lava overflows and the coarse-grained deposits emplaced by means of gravel flows produced lobes mainly located in the central part of the SdF. Gravel flow deposits are marked by a strong increase in backscattering. Areas characterized by a decrease in SAR amplitude correspond to the deposition of fine-grained material produced during gravel flow emplacement. Long-term analysis highlights areas characterized by morphological/lithological variations at a longer time scale (Fig. 5b). The area that suffered major changes is the central SdF, with evidence of erosion up-slope and near-shoreline deposition. Larger variations are also recognizable near the 2014 vent, with the marked low amplitude ratio in the NEC talus area.

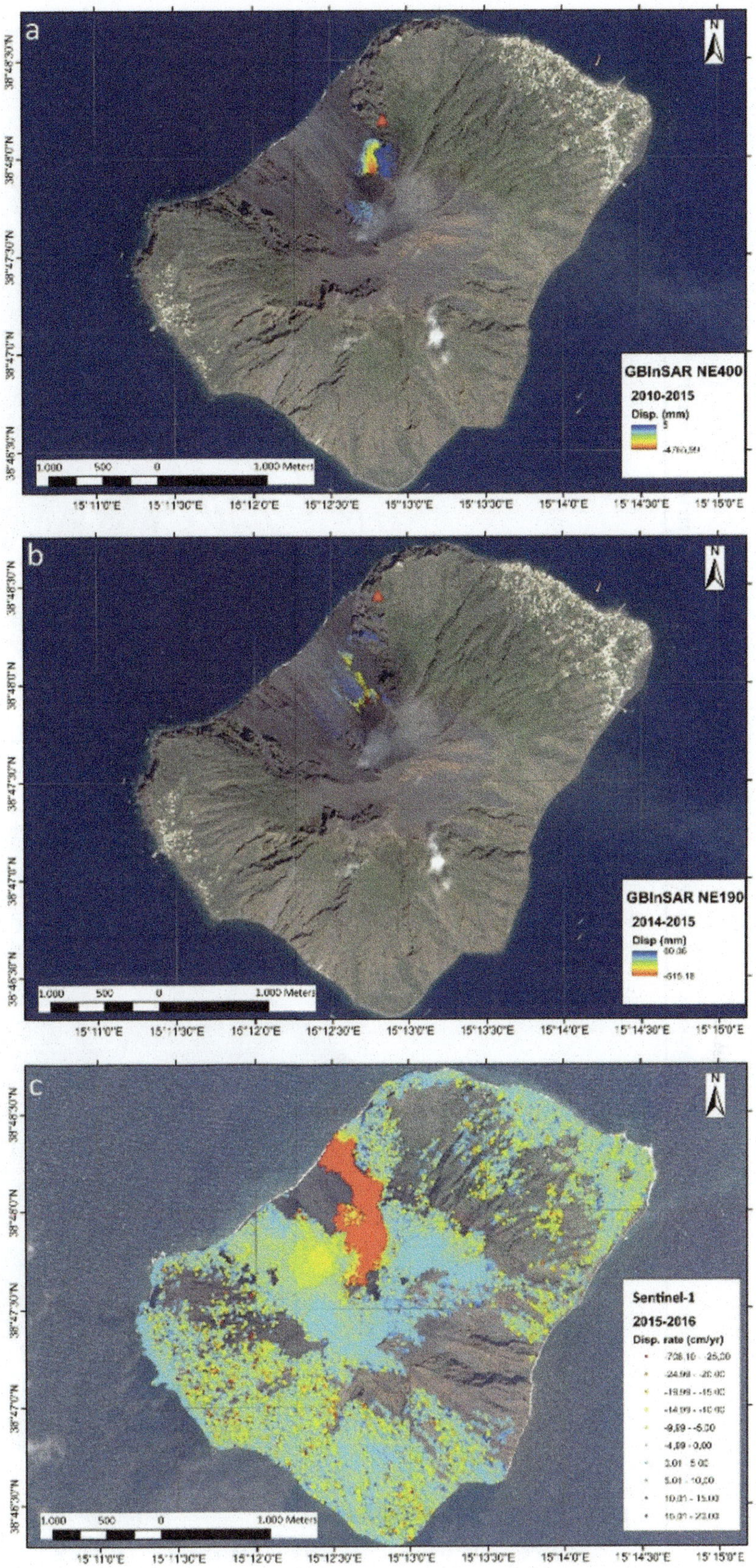

Fig. 4 Displacement data from **a** GBInSAR NE 400; **b** GBInSAR NE 190; **c** Sentinel-1

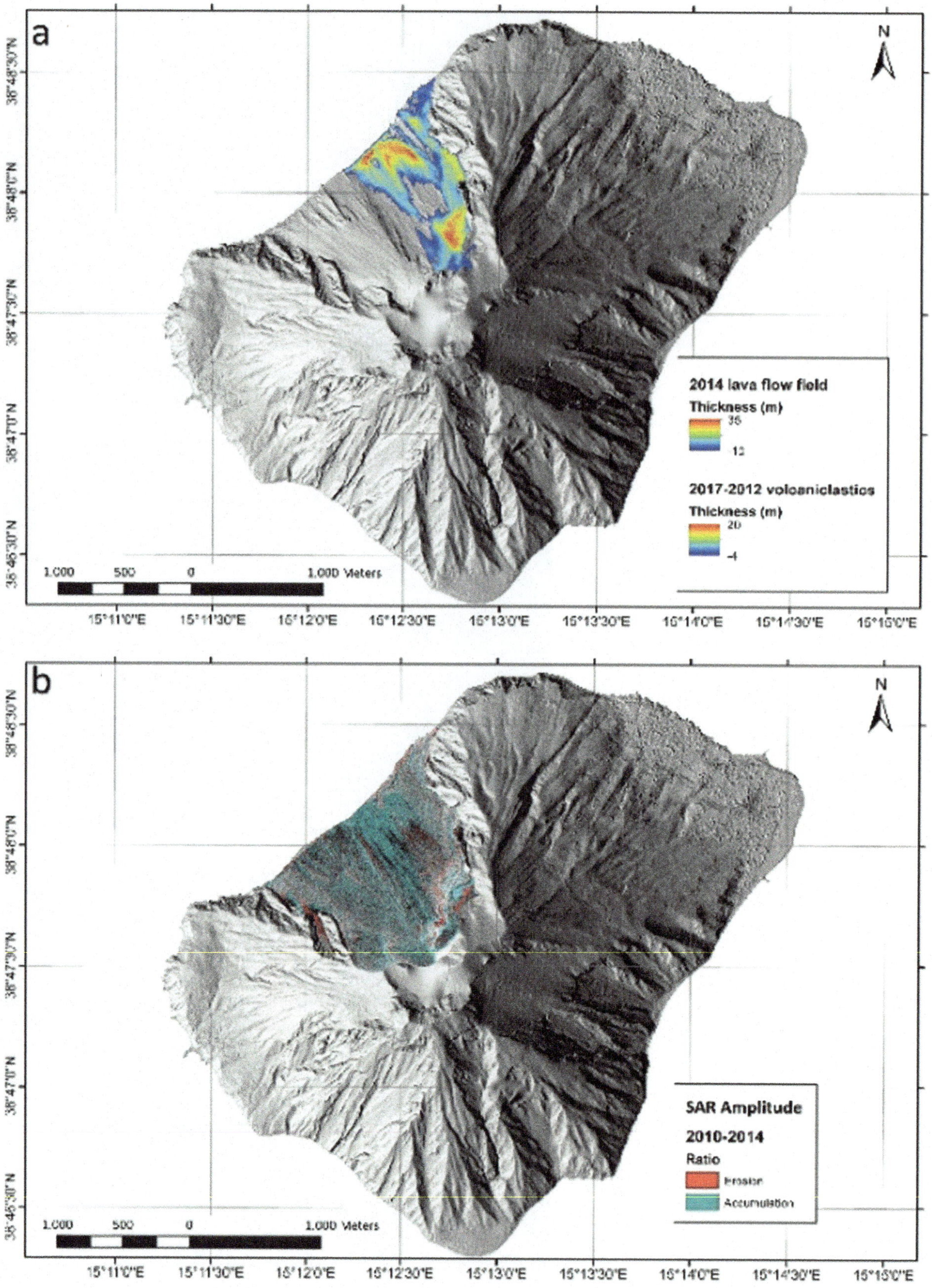

Fig. 5 Change detection from **a** topographic data (2012–2017); **b** SAR amplitude data (2010–2014)

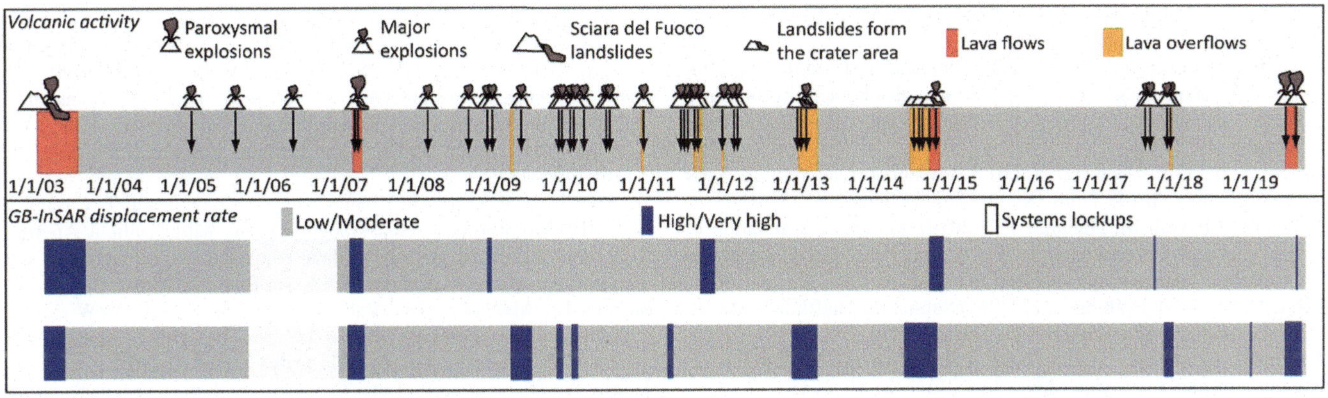

Fig. 6 Volcanic activity, mass-flows, and GBInSAR data at Stromboli volcano (modified after Di Traglia et al. 2014)

Discussion and Conclusive Remarks

At Stromboli, volcanic activity fluctuated in the intensity and eruptive style from mild-explosive (Strombolian) to effusive. This produced the emplacement of volcaniclastic material, of a lava flow filed, and several lava overflows. During periods characterized by a low intensity/frequency of Strombolian activity, the production of materials ejected from the crater terrace towards the SdF was generally low, and erosion was the prevailing process, mainly affecting the central sector of the SdF. After the 2014 flank eruption, the gravitational re-adjustment of the lava breccia sometimes produced rockfalls.

While space-borne allowed the identification of low displacements (up to 1.5 mm/h) in the SdF, the GBInSAR data are able to catch fast movement, even in a relatively small area (the northern part of the SdF). The integration of displacement data and the analysis of (topographic- and SAR amplitude images based) change detection allowed the identification the evolution of the slope instability phenomena and the geomorphological processes affecting the SdF slope. During periods characterized by low-intensity Strombolian activity, displacements within the SdF were low, and the CSK images showed backscattering and homogeneity values coherent with a phase of erosion of the volcanoclastic deposits of the central portion of the SdF.

GBInSAR devices resulted in an operational approach to mitigating landslide risks with SAR by defining a relationship between measured displacement rates and the associated instability and eruptive hazards. The GBInSAR system has been used to monitor the slope instability in the northern sector of the SdF and also to detect the inflation and deflation of the N sector of the crater area, allowing us to define the unstable area within the SdF and also to outline the structural framework of the summit area. This approach can be applied to other volcanic areas that are characterized by similar phenomena, but at the same time it can be generalized and applied in other geological contexts that are characterized by steep slopes.

Acknowledgements This work has been financially supported by the "Presidenza del Consiglio dei Ministri—Dipartimento della Protezione Civile" (Presidency of the Council of Ministers—Department of Civil Protection). This publication, however, does not reflect the position and the official policies of the Department.

References

Berardino P, Fornaro G, Lanari R, Sansosti E (2002) A new algorithm for surface deformation monitoring based on small baseline differential SAR interferograms. IEEE Trans Geosci Remote Sens 40(11):2375–2383

Blahůt J, Balek J, Klimeš J, Rowberry M, Kusák M, Kalina J (2019) A comprehensive global database of giant landslides on volcanic islands. Landslides 16(10):2045–2052

Calvari S, Intrieri E, Di Traglia F, Bonaccorso A, Casagli N, Cristaldi A (2016) Monitoring crater-wall collapse at active volcanoes: a study of the 12 January 2013 event at Stromboli. Bull Volcanol 78(5):39

Cerase A, Crescimbene M, La Longa F, Amato A (2019) Tsunami risk perception in southern Italy: first evidence from a sample survey. Nat Hazards Earth Syst Sci 19(12)

Di Roberto A, Rosi M, Bertagnini A, Marani MP, Gamberi F (2010) Distal turbidites and tsunamigenic landslides of Stromboli volcano (Aeolian Islands, Italy). In: Submarine mass movements and their consequences. Springer, Dordrecht, pp 719–731

Di Traglia F, Nolesini T, Intrieri E, Mugnai F, Leva D, Rosi M, Casagli N (2014) Review of ten years of volcano deformations recorded by the ground-based InSAR monitoring system at Stromboli volcano: a tool to mitigate volcano flank dynamics and intense volcanic activity. Earth Sci Rev 139:317–335

Di Traglia F, Calvari S, D'Auria L, Nolesini T, Bonaccorso A, Fornaciai A, Esposito A, Cristaldi A, Favalli M, Casagli N (2018a) The 2014 effusive eruption at Stromboli: New insights from in situ and remote-sensing measurements. Remote Sens 10(12):2035

Di Traglia F, Nolesini T, Ciampalini A, Solari L, Frodella W, Bellotti F, Fumagalli A, De Rosa G, Casagli N (2018) Tracking morphological

changes and slope instability using spaceborne and ground-based SAR data. Geomorphology 300:95–112
Di Traglia F, Fornaciai A, Favalli M, Nolesini T, Casagli N (2020) Catching geomorphological response to volcanic activity on steep slope volcanoes using multi-platform remote sensing. Remote Sensing. 12(3):438
Ferretti A, Prati C, Rocca F (2001) Permanent scatterers in SAR interferometry. IEEE Trans Geosci Remote Sens 39(1):8–20
Fornaciai A, Favalli M, Nannipieri L (2019) Numerical simulation of the tsunamis generated by the Sciara del Fuoco landslides (Stromboli Island, Italy). Sci Rep 9(1):1–12
Harbitz CB, Løvholt F, Bungum H (2014) Submarine landslide tsunamis: how extreme and how likely? Nat Hazards 72(3):1341–1374
Maramai A, Graziani L, Tinti S (2005) Tsunamis in the Aeolian Islands (southern Italy): a review. Mar Geol 215(1–2):11–21
Massonnet D, Feigl KL (1998) Radar interferometry and its application to changes in the Earth's surface. Rev Geophys 36(4):441–500
Ferretti A, Fumagalli A, Novali F, Prati C, Rocca F, Rucci A (2011) A new algorithm for processing interferometric data-stacks: squeeSAR. IEEE Trans Geosci Remote Sens 49(9):3460–3470
Rosi M, Pistolesi M, Bertagnini A, Landi P, Pompilio M, Di Roberto A (2013) Stromboli volcano, Aeolian Islands (Italy): present eruptive activity and hazards. Geol Soc London, Mem 37(1):473–490
Sassa K, Dang K, Yanagisawa H, He B (2016) A new landslide-induced tsunami simulation model and its application to the 1792 Unzen-Mayuyama landslide-and-tsunami disaster. Landslides 13 (6):1405–1419
Schaefer LN, Di Traglia F, Chaussard E, Lu Z, Nolesini T, Casagli N (2019) Monitoring volcano slope instability with Synthetic Aperture Radar: A review and new data from Pacaya (Guatemala) and Stromboli (Italy) volcanoes. Earth Sci Rev 192:236–257
Siebert L, Alvarado GE, Vallance JW, de Vries BVW (2006) Large-volume volcanic edifice failures in Central America and associated hazards. Volcanic Hazards Central Am 412:1
Turchi A, Di Traglia F, Luti T, Olori D, Zetti I, Fanti R (2020) Environmental aftermath of the 2019 stromboli eruption. Remote Sens 12(6):994
Williams R, Rowley P, Garthwaite MC (2019) Reconstructing the Anak Krakatau flank collapse that caused the December 2018 Indonesian tsunami. Geology 47(10):973–976

Detailed Seafloor Observations on a Deep-Sea Terrace Along the Japan Trench After the 2011 Tohoku Earthquake

Kiichiro Kawamura, Ayaka Wada, Miriam Römer, Michael Strasser, Hiske G. Fink, Yoshihiro Ito, and Ryota Hino

Abstract

We describe in detail seabed features on a deep-sea terrace of the Japan Trench caused by the 2011 Tohoku Earthquake based on visual seafloor observations. Data of this study was acquired using towed camera surveys and a remotely operated vehicle (ROV) dive in water depths between about 1300 and 2300 m and was conducted three, six and twelve months after the 2011. Damaged benthic organisms and disseminated bio-fragments might record the flow directions of the seismically or tsunami induced turbidity current. These records would be preserved into the geological record, if they remain undisturbed and getting buried before being reworked.

Keywords

Japan trench • Seabed video observations • Bacterial mat • Tsunami-generated turbidity current

K. Kawamura (✉)
Graduate School of Sciences and Technology for Innovation, Yamaguchi University, 1677-1 Yoshida, Yamaguchi city, 753-8512, Yamaguchi, Japan
e-mail: kiichiro@yamaguchi-u.ac.jp

A. Wada
Yamaguchi University, 1677-1 Yoshida, Yamaguchi city, 753-8512, Yamaguchi, Japan
e-mail: kayaadawa0201@gmail.com

Present Address:
A. Wada
Zoshindo, Juken-Kenkyu-Sha, 3-3-6 Shinmachi, Nishiku, Osaka-City, 550-0013, Osaka, Japan

M. Römer · H. G. Fink
MARUM—Center for Marine Environmental Sciences, University of Bremen, Leobener Strasse, 28359 Bremen, Germany
e-mail: mroemer@marum.de

H. G. Fink
e-mail: hfink@marum.de

M. Strasser
Institute of Geology, University of Innsbruck, Innrain 52, Raum, 6O 305, 6020 Innsbruck, Austria
e-mail: Michael.Strasser@uibk.ac.at

Y. Ito
Kyoto University, Disaster Prevention Research Insitute, Gokasho, Uji-City, Kyoto, 611-0011, Japan
e-mail: ito.yoshihiro.4w@kyoto-u.ac.jp

R. Hino
Research Center for Prediction of Earthquakes and Volcanic Eruptions, Graduate School of Science, Tohoku University, 6-6 Aza-Aoba, Aramaki, Aoba-ku, Sendai, 980-8578, Miyagi, Japan
e-mail: hino@m.tohoku.ac.jp

Introduction

On 11 March 2011, the 2011 off the Pacific Coast of Tohoku Earthquake (hereafter 2011 Tohoku Earthquake) triggered a devastating tsunami, which struck the Pacific coast of Honshu with run-up heights of up to 40 m (Satake et al. 2013). This region has experienced repeatedly large earthquake/tsunami events, as for example in 1933 the Showa Sanriku, in 1896 the Meiji Sanriku and in 869 the Jogan events (e.g. Goto et al. 2012). These events have been recorded as mud and sand layers on coastal plains, providing information on past-earthquake/tsunami events and their recurrence intervals (Goto et al. 2012). In addition to onshore deposits, it has been documented that past-earthquake/tsunami events are recorded in deep-sea sediments (e.g. Ikehara 2012; Arai et al. 2013; Ikehara et al. 2016; Kioka et al. 2019a).

On the basis of various comparison studies before and after the 2011 Tohoku Earthquake and Tsunami, seabed disturbances resulting in deep-sea turbidity (Noguchi et al. 2012; Kioka et al. 2019b; Molenaar et al. 2019), deep-sea tsunami-generated turbidity currents (Arai et al. 2013), appearance of new fissures and changes of cold seep assemblages (Tsuji et al. 2013), submarine landslides and related phenomena (Kawamura et al. 2012; Strasser et al. 2013) have been described. Some of these biological, chemical, sedimentary and deformational processes have the

K. Sassa et al. (eds.), *Understanding and Reducing Landslide Disaster Risk*, ICL Contribution to Landslide Disaster Risk Reduction,
https://doi.org/10.1007/978-3-030-60196-6_32

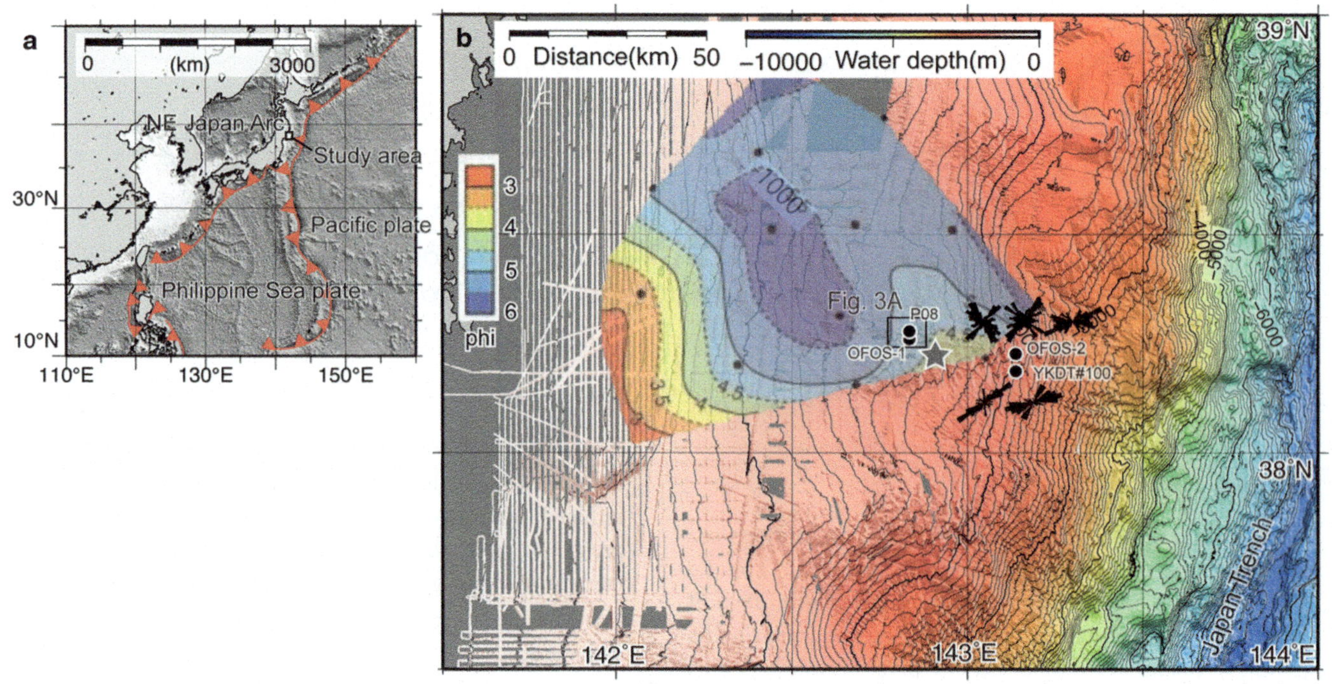

Fig. 1 **a** Map of the study area. The Japan Trench formed along a convergent margin between the Pacific plate and the Okhotsk plate. **b** Locality of the studied dive surveys (for details of surveys see Table 1). Rose diagrams show the direction of bio-fragments in the sites OFOS-2 (upper) and YKDT#100 (lower) (see text). Contour area by rainbow color shows a mean grain size distribution of the tsunami-generated turbidite as shown in Arai et al. (2013). The mean grain size is represented by phi scale: > 3 phi = > fine sand, 3–4 phi = very fine sand, 4–5 phi = coarse silt, 5–6 phi = medium silt. A gray star indicates the epicenter of the 2011 Tohoku Earthquake. A black square indicates the location of Fig. 3a.

potential to be recorded in the sediments as fossil evidence of this event. The lasting impact of a Tsunami event on seafloor stratum/features as well as benthic organisms has been reported for shallow seas after the Tohoku Earthquake and Tsunami (e.g. Seike et al. 2018). However, to decipher the tsunami/earthquake history not only from shallow seas, but also from deep-sea sediments, the recordable evidence has to be described by examining the datasets obtained after the 2011 Tohoku Earthquake.

In this paper, we describe in detail seabed features on the deep-sea terrace of the Japan Trench using camera observations (Fig. 1). We report on the effects of the tsunami regarding the biota and stratum of the seafloor in the region affected by the 2011 event. Surveys on the gentle slopes from ~1300 to ~2300 m water depths were conducted in 2011 and 2012: (i) three months, (ii) six months, and (iii) one year after the earthquake. Based on this comparison study including four seabed video surveys conducted during three different research cruises, we found both transient and permanent seabed phenomena related to the 2011 Tohoku Earthquake and Tsunami at the seafloor.

Deep-Sea Sedimentary Processes Triggered by the 2011 Tohoku Earthquake

The fore-arc basin in the Japan Trench developed from the Oligocene to the Holocene (e.g. von Huene and Lallemand 1990) and shows continuing subsidence of ~320 m due to tectonic erosion (von Huene and Lallemand 1990). This subsidence resulted in deposition of a thick (>1 km) fore-arc sedimentary section since the late Oligocene (von Huene and Lallemand 1990). The ongoing subsidence might be the fundamental mechanism for tsunami earthquake generation in this region as suggested by Tsuji et al. (2013).

A strong thick turbidity anomaly was detected above the seafloor in water depths of ~2000 to ~6000 m within one month after the 2011 earthquake (Noguchi et al. 2012). The suspended particles were composed of sediment materials, similar to the surface materials of the sediment cores, and were generated locally by a disturbance due to slope failure (Noguchi et al. 2012). Such turbidity events were still observed locally on a landward trench slope in places on

June 3rd-14th 2011 during the research cruise YK11-E04 Leg 1 by *R/V YOKOSUKA* (Shipboard Scientists 2011).

According to Arai et al. (2013), a tsunami-generated turbidity current attacked the ocean bottom pressure gauges (hereafter OBPs) and ocean bottom seismometers (OBSs), which were deployed in water depths from 300 to 5700 m along the landward slope of the Japan Trench. The average current speed was calculated as 2.4–7.1 m/s, whereas topographic effects (e.g. in valleys) could have caused accelerated current speeds. In fact, one of the OBPs moved ~1 km east from the original position at ~0.3 m/s (Arai et al. 2013).

Sediments deposited by the turbidity current were widely distributed at the seabed. All OBSs and OBPs were buried by turbidite layers as thick as ~10 cm (Arai et al. 2013; Fig. 1b).

After the depositions of the turbidite, diatoms and coccoliths from a spring plankton bloom on April 2011 settled onto the deep-sea bed (Oguri et al. 2013). Plankton carried radionuclide diffused from the Fukushima Dai-ichi nuclear disaster, and dispersed widely on the deep-sea bed as greenish diatomaceous and coccolith deposits (Oguri et al. 2013).

Description of the Survey Results

In total, three camera sled deployments and one ROV dive where undertaken in water depths between ~1300 and ~2300 m on the deep-sea terrace in the Japan Trench during four research cruises conducted three, six and twelve months after the 2011 Tohoku Earthquake and Tsunami (Table 1):

Survey Three Months After the Earthquake (June 2011)

Visual seabed observations of the deep-sea terrace were obtained using the deep towed TV camera system of YKDT#100 during research cruise YK11-E04 Leg 2 with *R/V YOKOSUKA* (Table 1). The water depths ranged between ~1500 and ~2000 m (Fig. 1b).

During this survey, we observed NE-SW trending cliffs of ~5 m height (Fig. 2a). The cliffs dipped westwards. Blocks of several centimeters to meters in diameter were colonized by sponges and sea anemones disseminated widely on gentle slopes, and covered by a mud layer. We found alignments of bio-fragments oriented in a NE-SW direction (Fig. 1a).

Survey Six Months After the Earthquake (September 2011)

The remotely operated vehicle (ROV) *HAKUYO3000* of SNK Ocean Co. Ltd. dived on the deep-sea terrace (Table 1). The aim of the dive was to recover an OBP. The survey area was around site P08 in a water depth of ~1450 m (Table 1; Figs. 1b and 3b).

During the dive survey, we observed numerous dead benthic organisms at the seafloor (mainly brittle stars), which were covered with white bacterial mats, some of them with extents of ~1 m (Fig. 2b). The dead organisms formed assemblages several meters in diameter. The bacterial mats were spatially related to these organic accumulations. The OBP was buried by mud and sand layers of 10–20 cm thickness (Fig. 2c). We interpret that these features would result from the tsunami-generated turbidity currents as described by Arai et al. (2013) at this location.

Surveys One Year After the Earthquake (March 2012)

We surveyed again around site P08 and the transect of previous dive YKDT#100 using a deep-towed camera mounted on the Ocean Floor Observation System (OFOS) during research cruise SO219A with *R/V SONNE* (Table 1). The first survey, OFOS-1, was performed in water depths

Table 1 Dive survey list

Vessel	Device	Survey-name	Date	Start-coodinates		End-coordinates		Water depth (m)
R/V YOKOSUKA YK11E-04 Leg 2	Deep-tow camera	YKDT #100	21 June, 2011	38° 13.980′N	143° 09.507′E	38°13.993′N	143° 04.637′E	1500–2000
A chartered ship	*HAKUYO3000* ROV	P08	25 Sep., 2011	38° 16.977 N	142° 49.918E	38°16.977′N	142° 49.918′E	1450
R/V SONNE SO219A	*OFOS* Video sled	OFOS-1	18 Mar., 2012	38° 17.065 N	142° 49.892E	38°15.733′N	142° 49.872′E	1450
R/V SONNE SO219A	*OFOS* Video sled	OFOS-2	18 Mar., 2012	38° 13.98 N	143° 06.66E	38°14.214′N	143° 10.883′E	1300–2300

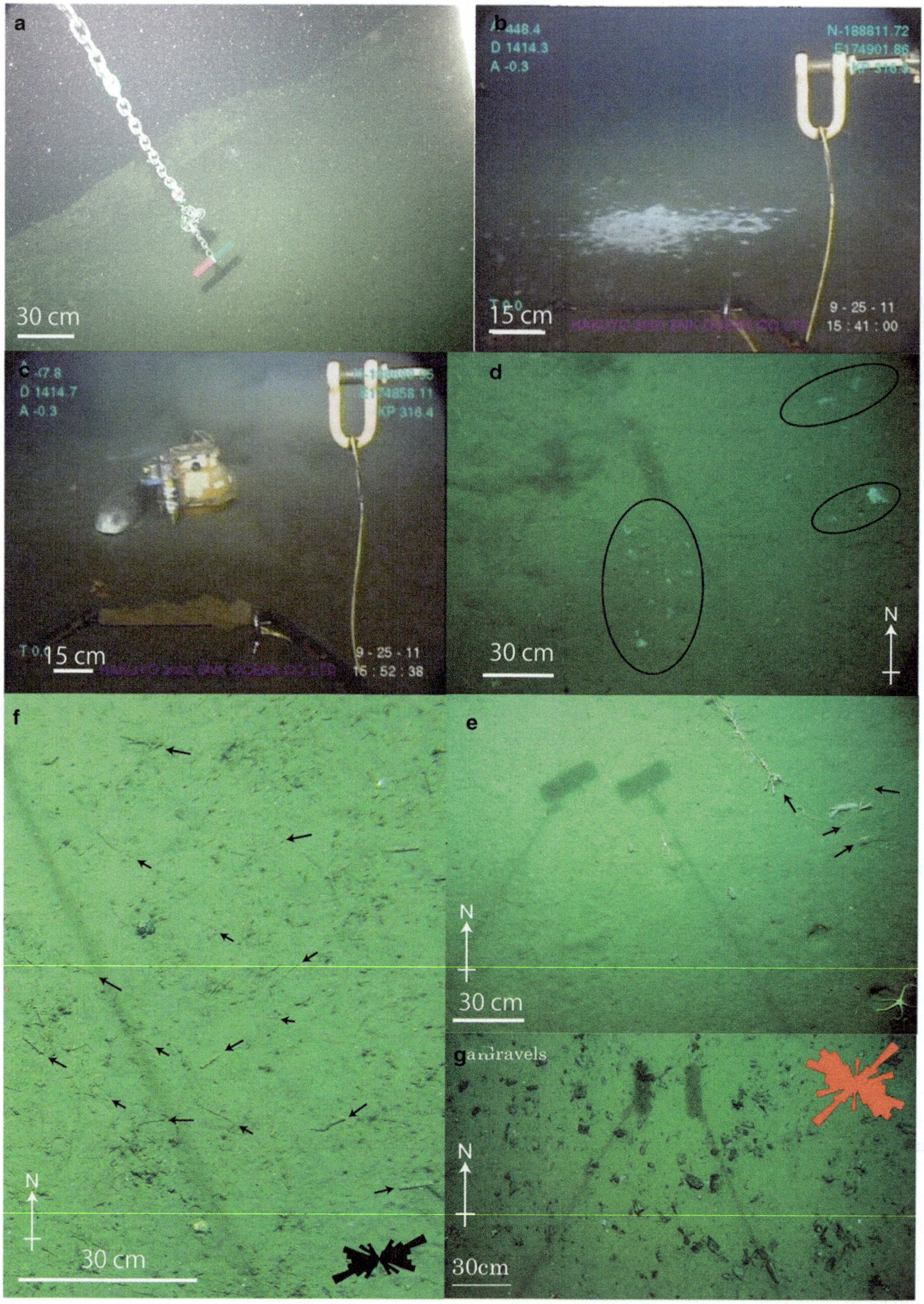

Fig. 2 Seabed features after the 2011 Tsunami. **a** NE-SW trending cliffs of ~5 m height, YKDT#100. **b** Bacterial mats of ~1 m in diameter, ROV *HAKUYO 3000* survey at P08. **c** Buried OBP in mud and sand layers of 10–20 cm thickness, *ROV HAKUYO3000* survey at P08. **d** Small bacterial mats indicated by black circles, OFOS-1. **e** Sea pens tilted to NE direction. These sea pens presumably indicate the paleo-current direction from SW to NE, OFOS-1. **f** Bio-fragments aligned to NE–SW and/or W–E directions, OFOS-2. **g** The longest axes of gravels aligned to NE–SW and NW–SE directions, OFOS-2

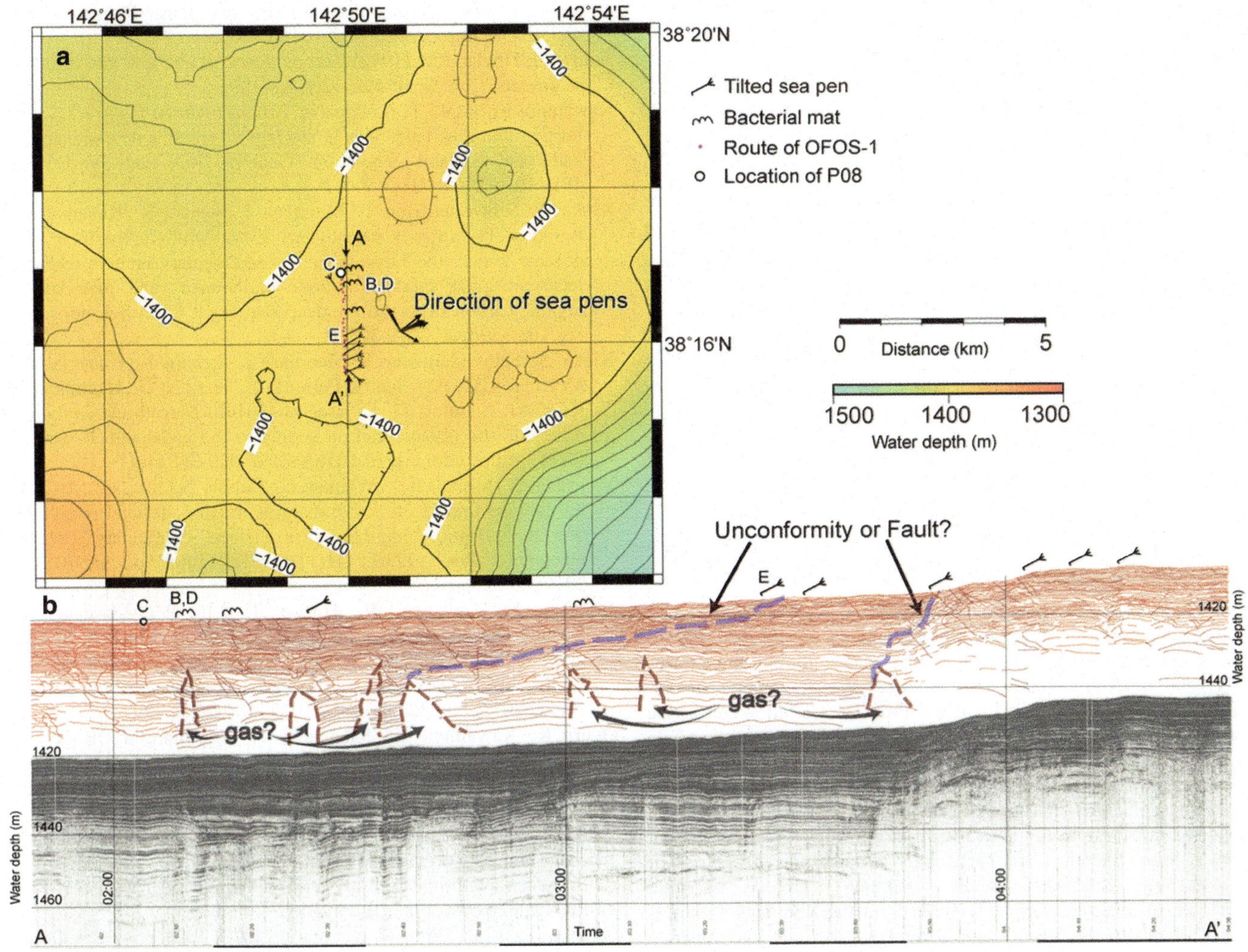

Fig. 3 **a** Detailed bathymetric map and the location of dive surveys OFOS-1 and P08 survey. The location is indicated in Fig. 1a. Direction of sea pens indicates a paleo-current direction as shown in Fig. 2e. Alphabets A, C, D and E in this figure indicate locations of Fig. 2b, c, d and e, respectively. Arrows along the OFOS-1 survey show the start and end points of the survey. The parasound survey in Fig. 3b was done between A' and A during the OFOS-1 survey. **b** Original parasound echogram (lower part) and its interpretation (upper part). There are seven transparent zones interpreted as gas conduits. The occurrences of bacterial mats seem not to be directly related to the transparent zones

ranging from 1300 to 2300 m (Fig. 1b, 3a). Dive OFOS-2 was conducted at a water depth of ~1400 m (Fig. 1b). Along both OFOS dive tracks, sub-seafloor imaging was performed by using the ship parametric sediment echo-sounder system (hereafter parasound) onboard *R/V SONNE* (Fig. 3b). During the surveys, the parasound recorded echograms using a primary frequency of 18 kHz for water column imaging and a second frequency of ~4 kHz for sediment profiling.

During the survey OFOS-1, we observed small bacterial mats of only cm-scale diameter on site P08 (Fig. 2d). The parasound data show several acoustically transparent zones in the subbottom echogram (Fig. 3b). This acoustic blanking may indicate increased fluid content in the sediment pore space, possibly including light hydrocarbon gases (Judd and Hovland 1992). However, it might be more probable that the mats are related to the decomposition of dead benthic organisms, as also supported by our observations during the ROV survey at site P08 in September 2011. Based on our data, we cannot certainly evaluate the origin of the bacterial mats. Further seabed observations along the Japan Trench are needed.

During the survey OFOS-2 in the same area as dive YKDT#100, we found alignments of sea pens, bio-fragments, and gravels. The sea pens were tilted in a northeast direction (Fig. 2e). Most of the longest axes of the bio-fragments align NE-SW and/or W-E (Fig. 2f) and most of the longest axes of the gravels align NE-SW and NW–SE (Fig. 2g). These directions are about the same as those observed during dive YKDT#100.

Concluding remarks

We compared visual seabed observations acquired during seabed surveys performed three, six and twelve months after the 2011 Tohoku Earthquake and Tsunami. Three and six months after the earthquake and tsunami, we observed many dead benthic organisms (mostly brittle stars), which were covered with white bacterial mats and disseminated bio-fragments. One year after the earthquake and tsunami, we could identify only few and comparably smaller bacterial mats. The short-term existence of bacterial mats indicates a relation to the fault seeps and/or rapid consumption of benthic organic matter. The bacterial mats grew rapidly within three months and disappeared within the following months.

Damaged benthic organisms and disseminated bio-fragments might record the flow directions of the seismically or tsunami induced turbidity current. These records would be preserved into the geological record, if they remain undisturbed and getting buried before being reworked.

Acknowledgements We thank captains and crews of *R/V YOKOSUKA* and *R/V SONNE* for onboard assistance during the cruises YK11E-04 in 2011 and SO219A in 2012. We also appreciate Dr. Katsunori Fujikura, Dr. Ken Takai and Prof. Dr. Gerold Wefer as the chief scientists of YK11-E04 Leg 1, 2 and SO219A for their great help during the cruise. Cruise SO219A was funded by the German Bundesministerium für Bildung und Forschung (BMBF) and the Deutsche Forschungsgemeinschaft (DFG-project WE 992/58-1).This study was supported in part by a Grant-in-Aid from the Ministry of Education, Culture, Sports, Science and Technology of Japan to the first author (20,340,139; KANAME), the research project Tohoku Ecosystem-Associated Marine Sciences from the Ministry of Education, Culture, Sports, Science and Technology to K. F. and Austrian Science Fund (project P29678 to M.S). We further acknowledge Dr. Khang Dang and one anonymous reviewer for their helpful comments, which improved the manuscript considerably.

References

Arai K, Naruse H, Miura R, Kawamura K, Hino R, Ito Y, Inazu D, Yokokawa M, Izumi N, Murayama M, Kasaya T (2013) Tsunami-generated turbidity current of the 2011 Tohoku-Oki earthquake. Geology 41:1195–1198. https://doi.org/10.1130/G34777.1

Goto K, Chague-Goff C, Goff J, Jaffe B (2012) The future of tsunami research following the 2011 Tohoku-oki event. Sed Geol 282:1–13. https://doi.org/10.1016/j.sedgeo.2012.08.003

Ikehara K (2012) Offshore earthquake- and/or tsunami-induced sediment transports and their deposits: Importance of marine sediment study for understanding past earthquakes and tsunami. Sedimentological Soc Jpn 71 (2):141–147 (Japanese with English abstract)

Ikehara K, Kanamatsu T, Nagahashi Y, Strasser M, Fink H, Usami K, Irino T, Wefer G (2016) Docomenting large earthquakes similar to the 2011 Tohoku-oki earthquake from sediments deposited in the Japan Trench over the past 1500 years. Earth Planet Sci Lett 445:48–56. https://doi.org/10.1016/j.epsl.2016.04.009

Judd AG, Hovland M (1992) The evidence of shallow gas in marine sediments. Cont Shelf Res 12:1081–1095

Kawamura K, Sasaki T, Kanamatsu T, Sakaguchi A, Ogawa Y (2012) Large submarine landslides in the Japan Trench: a new scenario for additional tsunami generation. Geophys Res Lett 39: L05308, 5PP. 10.1029/2011GL050661

Kioka A, Schwestermann T, Moernaut J, Ikehara K, Kanamatsu T, Eglinton TI, Strasser M (2019a) Event stratigraphy in a hadal oceanic trench: the Japan trench as sedimentary archive recording recurrent giant subduction zone earthquakes and their role in organic carbon export to the deep sea. Front Earth Sci. https://doi.org/10.3389/feart.2019.00319

Kioka A, Schwestermann T, Moernaut J, Ikehara K, Kanamatsu T, McHugh CM, dos Santos Ferreira C, Wiemer G, Haghipour N, Kopf AJ, Eglinton TI, Strasser M (2019b) Megathrust earthquake drives drastic organic carbon supply to the hadal trench. Sci Rep 9:1553. https://doi.org/10.1038/s41598-019-38834-x

Molenaar A, Moernaut J, Wiemer G, Dubois N, Strasser M (2019) Earthquake impact on active margins: tracing surficial remobilization and seismic strengthening in a slope sedimentary sequence. Geophys Res Lett 41(11):1195. https://doi.org/10.1029/2019GL082350

Noguchi T, Tanikawa W, Hirose T, Lin W, Kawagucci S, Yoshida-Takashima Y, Honda MC, Takai K, Kitazato H, Okamura K (2012) Dynamic process of turbidity generation triggered by the 2011 Tohoku-Oki earthquake. Geochem Geophys Geosyst 13(11):2012. https://doi.org/10.1029/2012GC004360

Oguri K, Kawamura K, Sakaguchi A, Toyofuku T, Kasaya T, Murayama M, Fujikura F, Glud RN, Kitazato H (2013) Hadal disturbance in the Japan Trench induced by the 2011 tohoku-oki earthquake. Sci Rep 3:1915. https://doi.org/10.1038/srep01915

Satake K, Fujii Y, Harada T, Namegaya Y (2013) Space distribution of coseismic slip of the 2011 Tohoku Earthquake as inferred from tsunami waveform data. Bull Seismol Soc Am 103(2B):1473–1492. https://doi.org/10.1785/0120120122

Seike K, Sassa S, Shirai K, Kubota K (2018) J Geophys Res Oceans 123(2):1376–1392. https://doi.org/10.1002/2017JC013746

Shipboard Scientists (2011) R/V Yokosuka Cruise Report YK11-E04 Leg 1—Impact by the huge earthquake on marine ecosystem including environments, chemical, geology and geophysics in the Off Sanriku area. JAMSTEC, Yokosuka, p 25. (https://www.godac.jamstec.go.jp/catalog/data/doc_catalog/media/YK11-E04_leg1_all.pdf)

Strasser M, Kolling M, Ferreira S, Fink HG, Fujiwara T, Henkel S, Ikehara K, Kanamatsu T, Kawamura K, Kodaira S, Römer M, Wefer G, R/V/ Sonne Cruise SO219A and JAMSTEC Cruise MR12-E01 scientists (2013) A slump in the trench: tracking the impact of the 2011 Tohoku-Oki earthquake. Geology 41:935–938. https://doi.org/10.1130/G34477.1

Tsuji T, Kawamura K, Kanamatsu T, Kasaya T, Fujikura K, Ito Y, Tsuru T, Kinoshita M (2013) Extension of continental crust by anelastic deformation during the 2011 Tohoku-oki earthquake: The role of extensional faulting in the generation of a great tsunami. Earth Planet Sci Lett 364:44–58. https://doi.org/10.1016/j.epsl2012.12.038

von Huene R, Lallemand S (1990) Tectonic erosion along the Japan and Peru convergent margins. Geol Soc Am Bull 102:704–720. https://doi.org/10.1130/0016-7660

Part IV

Landslides at UNESCO Designates Sites and Contribution from WMO, FAO, IRDR

Landslides at UNESCO-Designated Sites

Irina Pavlova, Soichiro Yasukawa, Aurélien Dousseron, and Vincent Jomelli

Abstract

UNESCO-designated sites (World Heritage sites, Biosphere Reserves and UNESCO Global Geoparks) promote sustainable development, and focus on the protection of natural and cultural heritage or the conservation and sustainable use of biodiversity and geological resources. More than 2000 UNESCO-designated sites may be partly or entirely exposed to various natural hazards, including landslides, with potential impacts on the communities living in or near the sites, and on their livelihoods. Because of their high cultural and symbolic value, the impact of the loss or damage of a UNESCO-designated site can resonate across the world. In the domain of landslides risk reduction, UNESCO cooperates with International Consortium on Landslides to evaluate landslide risks and regional at local levels and promote good practices in landslide risk management.

Keywords

UNESCO designated sites • Landslides • Global exposure assessment • Natural heritage • Cultural heritage

I. Pavlova (✉) · S. Yasukawa
UNESCO, Natural Science Sector, Paris, France
e-mail: i.pavlova@unesco.org

S. Yasukawa
e-mail: s.yasukawa@unesco.org

A. Dousseron · V. Jomelli
Laboratory of Physical Geography, CNRS, Meudon, France
e-mail: aurelien.dousseron@gmail.com

V. Jomelli
e-mail: Vincent.JOMELLI@lgp.cnrs.fr

K. Sassa et al. (eds.), *Understanding and Reducing Landslide Disaster Risk*,
ICL Contribution to Landslide Disaster Risk Reduction,
https://doi.org/10.1007/978-3-030-60196-6_33

Introduction: UNESCO Designated Sites

UNESCO Global Geoparks

UNESCO Global Geoparks are single, unified, geographical areas where sites and landscapes of international geological significance are managed through the holistic concept of protection, education, and sustainable development (UNESCO 2020a). Their bottom-up approach consists of combining conservation with sustainable development while involving local communities. As of April 2020, there are 140 UNESCO Global Geoparks in 38 countries.

Biosphere Reserves

Biosphere Reserves are areas comprising terrestrial, marine, and coastal ecosystems. Each Biosphere Reserve promotes solutions reconciling the conservation of biodiversity with its sustainable use (UNESCO 2020b). Biosphere Reserves are 'Science for Sustainability support sites'—special places for testing interdisciplinary approaches to understanding and managing changes and interactions between social and ecological systems, including conflict prevention and management of biodiversity. As of April 2020, there are 686 Biosphere Reserves in 122 countries, including 20 transboundary sites.

World Heritage Sites

To be on the World Heritage (WH) List, sites must be of 'Outstanding Universal Value' and meet at least one out of ten selection criteria (UNESCO 2020c). A list of WH properties is maintained and up-dated every year by an inter-governmental Committee, also known as the WH Committee. Within UNESCO, the WH Centre is the focal point and coordinator for all matters related to WH, it coordinates both reporting on the condition of sites and the emergency actions undertaken when a site is threatened. From the initial 12 properties recognized in 1978, the List grew to reach 981 properties from 161 States Parties at the beginning of 2020.

Natural Hazards at UNESCO Designated Sites

UNESCO designated sites are located in geographical settings which may be partly or entirely exposed to natural hazards and extreme weather events which can potentially impact the communities living in or near the sites and their livelihoods. Because of their high cultural and symbolic value, the impact of the loss or damage of a UNESCO designated site can resonate across the world.

In 2017, UNESCO DRR experts undertook a global assessment, aiming to overview disaster risk reduction at UNESCO designated sites, in particular to provide qualitative information concerning the global exposure of UNESCO-designated sites to natural hazards (UNESCO 2020d).

According to the results, 92% of UNESCO designated sites could be potentially exposed to at least one out of the main natural hazards (Pavlova et al. 2019). In particular, 94% of Biosphere Reserves, 96% of UNESCO Global Geoparks and 85% of World Heritage sites.

Overall, landslides and earthquakes are the most frequent geohazards (33 and 18% of all sites, respectively) while wildfires and floods (41 and 39% of all sites, respectively) are the most frequent among hydrometeorological hazards. Such hazards as cyclones, storms, extreme temperature and droughts episodes have a high probability to be triggered at around 10% of all UNESCO sites. 5% of sites are potentially exposed to volcanic eruptions, tsunamis, desertification issues and snow and ice related hazards.

The regional distribution varies as illustrated on the Fig. 1. In Europe and Asia, floods appear to be the most frequent hazard, followed by wildfires in Europe and landslides in Asia. Wildfires are at the top of the hazard list for Africa, Latin and Northern Americas and Oceania; while floods stay the second most frequent hazard in Africa, Oceania and Northern America. Landslides stay on the second position in Latin America and Asia.

Exposure of UNESCO Designated Sites to Landslides

Past Landslides at UNESCO Designated Sites

In recent years, landslides have already caused extensive damage to UNESCO designated sites. Different types of landslides frequently occur on the slopes surrounding such sites as the Machu Picchu World Heritage site or Nanda Devi Biosphere Reserve, damaging access roads and tourist paths. In February 2010, Monarch Butterfly Biosphere Reserve in Mexico Several UNESCO Global Geoparks in Europe, such as Adamello Brenta, Swabian Alb, Karavanke, Beigua and Apuan Alpes UNESCO Global Geoparks, are exposed to landslides and avalanches. For instance, historical earthquakes with magnitudes greater than 6, as well as recurring events generated by heavy rainfalls and debris-flows shook the area several times claiming many victims and causing severe damage at Appuan Alpes UNESCO Global Geoparks (Bartelletti et al. 2017) (Fig. 2). Multiple designated sites are no exception. For instance, two existing triple designated sites, Mont-Perdu, France and Spain, and Jeju, South Korea, are exposed to landslides and avalanches.

Global Assessment: Methodology

This paper present detailed analyzes of the exposure to landslides at 1914 UNESCO designated sites worldwide. Analyses were performed by UNESCO DRR experts together with collaborators from the Laboratory of Physical Geography, CNRS.

First, the exposure of UNESCO designated sites was estimated using open-access data on physical exposure to geological hazards, coming from Global Risk Data Platform and Global Volcanism Program. Second, the Periodic Reporting registered data, which is part of the World Heritage monitoring system and includes questions on the geological hazards that threaten each site, was examined to estimate World Heritage site managers' risk awareness. Finally, estimated and registered datasets for each World Heritage site were compared.

The preparatory step was the creation of a united geospatial database of all 1901 UNESCO designated sites, provided by respective Secretariats of each UNESCO site Programme. The database consolidated the information on each site's general description (name, country, criteria of inscription, etc.) as well as its spatial location. Further work was then undertaken to add to the database all available information on disaster risk reduction issues, including exposure and vulnerability to risks, current experience on prevention and mitigation measures, awareness raising activities and site managers' needs.

An open-source database coming from the Global Risk Data Platform (GAR 2013), a multi-agency effort to share spatial information on global risk from natural hazards, was used to estimate exposure to landslides. The annual physical exposure to landslides triggered by earthquakes and precipitation events datasets were used in the study. These

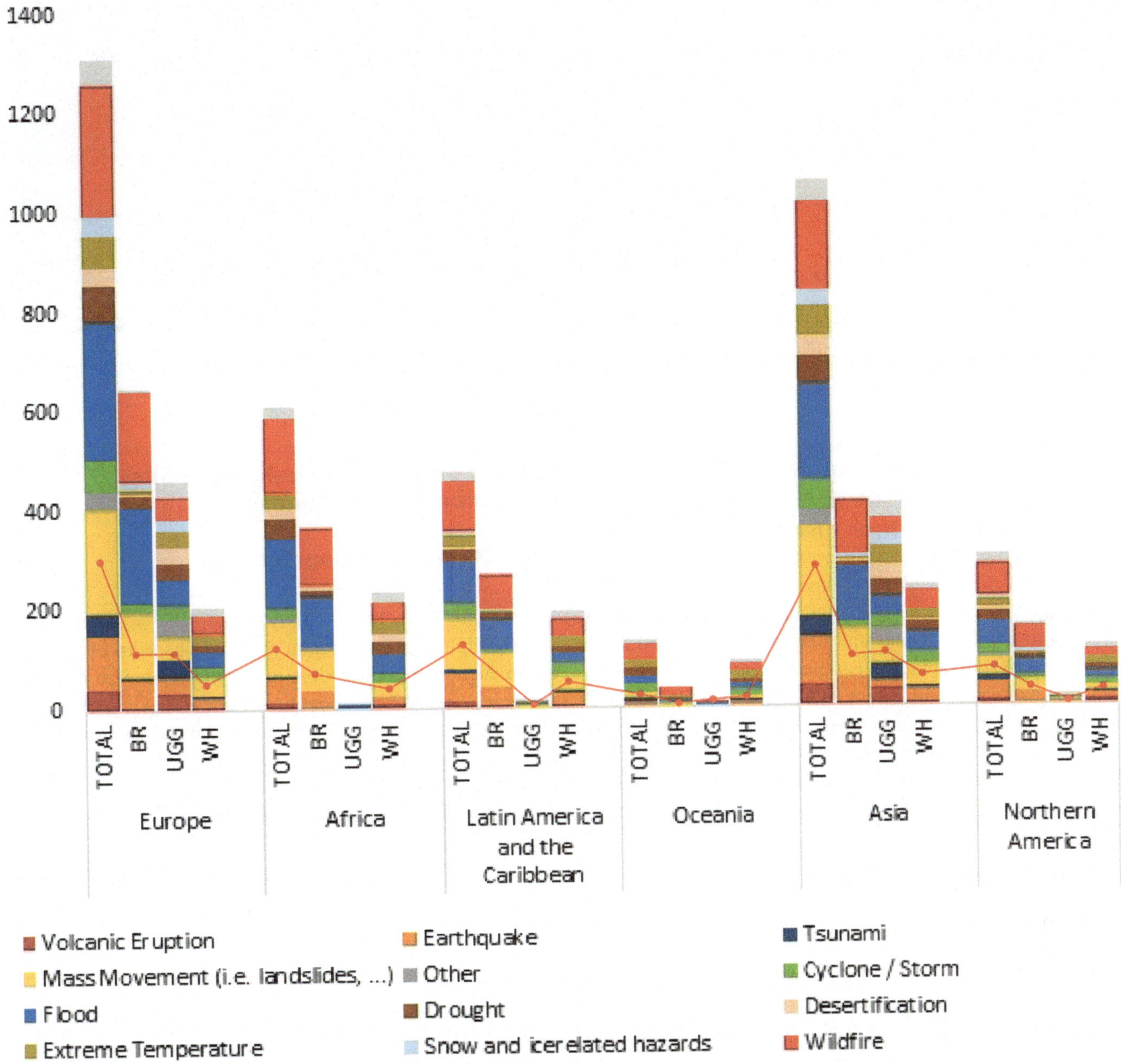

Fig. 1 Number of hazards at UNESCO designated sites per type of hazard (Biosphere reserve (BR); UNESCO Global Geopark (UGG); UNESCO World Heritage (WH)), per region and per site's type (Pavlova et al. 2019)

datasets were modelled using global data and aggregated to ~5 km resolution for distribution.

The spatial database of UNESCO designated sites was associated with datasets containing information about landslide risk exposure (Pavlova et al. 2017). If the site location intersected with a positive raster value of the hazard, this site was assumed to be potentially exposed to landslide hazard.

Global Assessment: Results

Results based on the GRPD dataset shows that 24.5% of UNESCO designated sites could be potentially exposed to landslides. Of these sites, 7.9% are Biosphere Reserves, 3% UNESCO Global Geoparks and 13.6% are World Heritage sites.

Fig. 2 Historical mudflows that damaged villages in Apuan Alpes UNESCO Global Geopark in 1902 (left) and in 1996 (right) (Bartelletti et al. 2017)

It is interesting to have a closer look at a regional distribution of sites between UNESCO regions (Fig. 3 and Table 1). Asian region is the most exposed to landslides, followed by Europe and North America, followed by number of exposed sites in Latin America. Least regional number of UNESCO designated sites exposed to landslides is in Africa and Arab States regions. Such regional distribution is due to both number of sites and orographic characteristics.

Regarding geographical distribution, most of landslides risks are associated with major mountainous areas. Table 2 illustrates major mountain areas and selected UNESCO designated sites potentially exposed to landslides. This list is indicative but illustrates well that all three types of UNESCO designated sites in all UNESCO regions are exposed to different types of landslides.

Actions on UNESCO Designated Sites Protection Against Landslides

UNESCO and ICL Cooperation

UNESCO has been a catalyst for international, inter-disciplinary cooperation in many aspects of disaster risk reduction and mitigation, in collaboration with both UN Agencies and other scientific entities (UNESCO 2007). Through the interface of education, natural and social sciences, culture and communication and information, UNESCO supports Member States to implement the Sendai Framework for Disaster Risk Reduction, which is in line

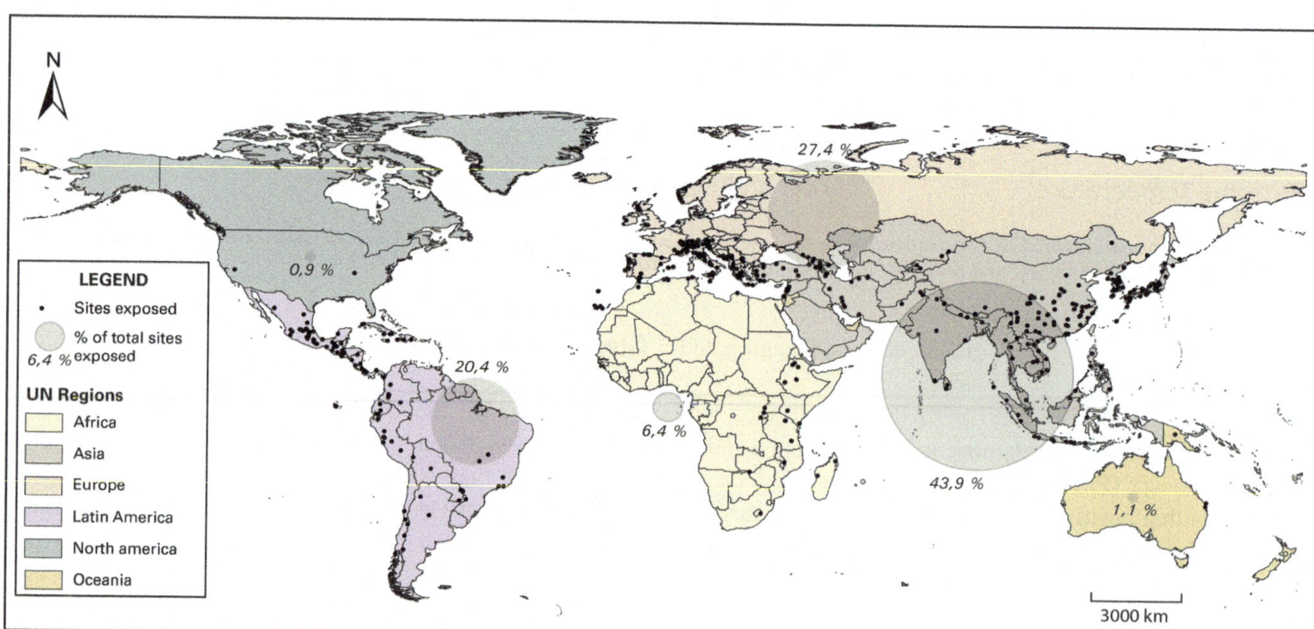

Fig. 3 Global landslides distribution according UN regions

Table 1 UNESCO regions and number of sites potentially exposed to landslides

UNESCO regions	Number of sites potentially exposed to landslides
Afrique	30
Asia	206
Europe	129
Latin America and the Caribbean	96
North America	4
Oceania	5

Table 2 Major mountain areas and selected UNESCO designated sites potentially exposed to landslides

UNESCO region	Major mountain area	Selected UNESCO designated sites exposed to landslides BR = Biosphere Reserve, UGGP = UNESCO Global Geopark, WHS = World Heritage site
Africa	Mitumba Mounts	Kahuzi-Biega National Park WHS
Africa	Ethiopian Highlands	Rock-Hewn Churches, Lalibela WHS
Africa	Great Rift Valley	Ngorongoro UGGP
Africa	Tsaratanana mounts	Rainforests of the Atsinanana WHS
Africa	Drakensberg	Maloti-Drakensberg Park WHS
Africa	Ankaratra Massif	Tsingy de Bemaraha Strict Nature Reserve WHS
Arab states	Atlas	Gouraya BR
Arab states	Atlas	Tipasa WHS
Arab states	Atlas	Medina of Tétouan (formerly known as Titawin) WHS
Arab states	Atlas	Ichkeul BR
Arab states	Atlas	Archaeological Site of Volubilis WHS
Arab states	Western desert	Archaeological Site of Cyrene WHS
Asia and Oceania	Himalayas	Khangchendzonga National Park WHS
Asia and Oceania	Plateau of Tibet	Huanglong BR
Asia and Oceania	Siwalik Hills	Chitwan National Park WHS
Asia and Oceania	Fergana Valley	Sulaiman-Too Sacred Mountain WHS
Asia and Oceania	Khasi Hills	Nokrek BR
Asia and Oceania	Alataw mounts	Altyn-Emel BR
Asia and Oceania	Sikhote-Alin Range	Sikhote-Alin BR
Asia and Oceania	Qinling mountains	Funiushan UGGP
Asia and Oceania	Taihang mounts	Mount Wutai WHS
Asia and Oceania	New Guinea Highlands	Kuk Early Agricultural Site WHS

(continued)

Table 2 (continued)

UNESCO region	Major mountain area	Selected UNESCO designated sites exposed to landslides BR = Biosphere Reserve, UGGP = UNESCO Global Geopark, WHS = World Heritage site
Europe and N. America	Cordiera Cantabrica	Muniellos BR
Europe and N. America	Appennini	Historic Centre of Urbino WHS
Europe and N. America	Dinaric Alps	Golija-Studenica BR
Europe and N. America	Kjølen Mountains	West Norwegian Fjords – Geirangerfjord and Nærøyfjord WHS
Europe and N. America	Caucasus Mounts	Upper Svaneti WHS
Europe and N. America	Balkan mounts	Boyana Church WHS
Europe and N. America	Pyrenees	Catalan Romanesque Churches of the Vall de Boí WHS
Europe and N. America	Alps	Monte Viso BR and WHS
Europe and N. America	Anatolia	Hattusha: the Hittite Capital WHS
Europe and N. America	Lesser Caucasus	Monasteries of Haghpat and Sanahin WHS
Europe and N. America	Les Laurentides	Charlevoix BR
Europe and N. America	Sierra Nevada	Yosemite National Park UGGP and WHS
Latin America	Patagonia	Andino Norpatagonica BR
Latin America	Brazilian Highlands	Bosque Mbaracayú BR
Latin America	Cordiera Oriental	Fuerte de Samaipata WHS

with the 2030 Agenda for Sustainable Development and the Paris Climate Agreement.

In the domain of landslide hazard, UNESCO actively supports The International Consortium on Landslides (ICL) that has made steady progress on cooperative research and capacity building in landslide risk mitigation. The knowledge and experience acquired over the past years through the Consortium represents a success story of international scientific cooperation (Arduino et al. 2017).

Since the establishment of the Consortium in 2002, UNESCO actively participates in its various activities. UNESCO is committed to the promotion and implementation of the "Sendai Partnerships 2015–2025 for Global Promotion of Understanding and Reducing Landslide Disaster Risk" (Partnership) (Sassa 2015). according to its mandate in DRR, UNESCO will continue to support the Patnership, especially on; development of global, regional and national multi-hazard early warning systems to natural hazards, including landslides; improve scientific basis to develop technologies and tools to cope with landslide, as part of multi-risk identification and management; strengthening capacity for floods and landslide monitoring and forecasting; enhance the capacity of schools and local communities to prepare for and respond to environmental and other threats and disasters; provide policy support and technical assistance for capacity enforcement in landslide disaster risk reduction; enhance research, partnerships and international scientific cooperation on landslide disaster risk reduction; collaborate with international partners and cross sectors, UNESCO field offices, UNESCO Category Two Centers on the topic of landslide disaster risk reduction (UNESCO 2020e).

UNESCO assists Member States and its designated sites in strengthening livelihood capacities in Disaster Risk Reduction. UNESCO encourages the identification of risks, protection from different hazards, as well as fostering climate change resilience, and the preservation of UNESCO designated sites and their communities. UNESCO continuously contributes to building capacity in DRR (UNESCO 2020e), developing innovative policy (UNESCO 2015), tailoring management strategies (UNESCO et al. 2010; UNESCO 2017), and recognizing the value of resilient protected area systems (Fassoulas et al. 2018; UNESCO 2016).

In the 5th World Landslide Forum in November 2020 in Japan, UNESCO hosts a session on "Landslides at UNESCO designated sites", where international experts present existing practices of risks assessment and discuss how to move leverage on existing global ICL expertise. Further cooperation should include joint international projects, targeting detailed local and regional assessment of landslides and solutions for local landslide management at various UNESCO designated sites at landslide risk.

References

Arduino G, Badaoui R, Yasukawa S, Makarigakis A, Pavlova I, Shirai H, Han Q (2017) UNESCO's contribution to the implementation of UNISDR's global initiative and international consortium on landslides. 4 landslide forum

Bartelletti A, Amorfini A, Ottria G (2017) Rains & ruins in the apuan alps. arco Regionale delle Alpi Apuane—Apuan Alps Unesco Global Geopark. p 81

Despotaki V, Silva V, Lagomarsino S, Pavlova I, Torres J (2018) Evaluation of seismic risk on UNESCO cultural heritage sites in Europe. Int J Architectural Heritage 12(7–8):1–14

Fassoulas C, Watanabe M, Pavlova I, Amorfini A, Dellarole E, Dierickx F (2018) UNESCO Global Geoparks: living laboratories to mitigate natural induced disasters and strengthen communities' resilience. In: Il Sileno Edizioni G (eds) Natural hazards and disaster risk reduction policies. p. 290

GAR (2013) The global exposure database for GAR 2013. Background Paper Prepared for the 2013 Global Assessment Report on Disaster Risk Reduction, UnitedNations, Geneva, Switzerland. p 22

Pavlova I, Makarigakis A, Depret T, Jomelli V (2017) Global overview of the geological hazard exposure and disaster risk awareness at world heritage sites. J Cultural Heritage 28

Pavlova I, Fassoulas C, Watanabe M, Cupa P, Ikhram R, Yasukawa S (2019) UNESCO designated sites—natural and cultural heritage sites as platforms for awareness raising. Contributing Paper to GAR 2019

Sassa K (2015) ISDR-ICL sendai partnerships 2015–2025 for global promotion of understanding and reducing landslide disaster risk. Landslides 12:631–640

UNESCO (2007) Disaster preparedness and mitigation: UNESCO's role. UNESCO, France, Paris

UNESCO, ICCROM, ICOMOS, IUCN (2010) Managing disaster risks for world heritage. Paris, France

UNESCO (2015) Policy for the integration of a sustainable development perspective into the processes of the World Heritage Convention. France, Paris

UNESCO (2016) Lima action plan for UNESCO's MAB programme and its world network of biosphere reserves (2016–2025). Lima, Peru

UNESCO (2017) Strategy for reducing disaster risks at World Heritage properties, WHC-07/31.COM/7.2. Paris, France

UNESCO (2020a) International geoscience and geoparks programme (IGGP). https://www.unesco.org/new/en/natural-sciences/environment/earth-sciences/international-geoscience-and-geoparks-programme/. Last Accessed 14 Apr 2020

UNESCO (2020b) Man and the biosphere programme. https://www.unesco.org/new/en/natural-sciences/environment/ecological-sciences/man-and-biosphere-programme/. Last Accessed: 14 Apr 2020

UNESCO (2020c) World heritage centre. https://whc.unesco.org/. Last Accessed: 14 Apr 2020

UNESCO (2020d) Disaster risk reduction in UNESCO designated sites. https://en.unesco.org/drr-sites. Last Accessed 14 Mar 2020

UNESCO (2020e) Disaster risk reduction. https://www.unesco.org/new/en/natural-sciences/special-themes/disaster-risk-reduction/. Last Accessed: 14. Apr 2020

Traditional Knowledge and Local Expertise in Landslide Risk Mitigation of World Heritages Sites

Claudio Margottini and Daniele Spizzichino

Abstract

Landslide mitigation in UNESCO and other heritage sites requires an approach that safeguards the site. This approach is put in place in order to avoid any modifications of the original status in terms of authenticity and integrity. To do so, any mitigation project has to provide effective solutions—very often innovative—to be combined with original architecture and environment. The harmonisation of advanced solutions with historical heritage then requires a mitigation project (sometimes pioneering), with a design that minimises the impact of new intervention. This possibly will aid in recovering traditional materials and/or solutions or the re-use of original construction techniques. In low income Countries, the usage of original construction techniques ensures maintenance by local workers, hence enhancing sustainability. The present research investigates 10 UNESCO and other heritage sites affected by landslides, in an attempt to understand whether traditional techniques and local expertise may play a role in mitigating the expected damage and, at the same time, maintaining the original authenticity and integrity of the site. The investigation demonstrates that the mitigation of landslide risk through traditional techniques can help in the preservation of the integrity and authenticity of the heritage site, enhancing local job opportunities and ensuring long term maintenance. A similar result can also be achieved when making recourse to modern low impact measures (e.g. Nature-Based Solutions). Evidently, it is not always possible to apply the cumulative body of knowledge, know-how, practices and representations maintained and developed by peoples with extended histories, to all typologies of mass movements. The best focus is found in the application to shallow landslide phenomena, with limited involved volumes.

Keywords

UNESCO world heritage site • Landslide • Mitigation • Traditional knowledge • Local expertise

Introduction

In order to prevent or mitigate landslide disasters, several strategies have been developed, structural and non-structural. Prerequisite to all that works is an appropriate understanding of landslide hazards and their consequences are implemented in terms of the spatial extent, temporal occurrence, magnitude (size), and intensity (velocity). More in detail, understanding landslide hazard is mainly dealing with the deep scientific knowledge of mechanical parameters for involved materials (soil/rock/debris), triggering factors, involved volumes, and evolution (including run-out). Many high technologies, monitoring and modelling approaches have been developed in recent years but still some uncertainties remain.

According to the above it is possible to say that, in addressing the goals of landslide mitigation, the role of science is crucial; scientific knowledge and appropriate technologies are central to resolving the safety of potential affected people and infrastructures, mainly allowing mitigation solutions that are not heavily impacting on the environment, as well as ensuring long term efficacy. In other words, sustainable. However, science does not constitute the only form of knowledge, and closer links need to be

C. Margottini (✉)
Embassy of Italy in Egypt, Cairo, Egypt
e-mail: claudio.margottini@gmail.com

C. Margottini
UNESCO Chair for Prevention and Sustainable Management of Geo-Hydrological Hazards, Florence University, Florence, Italy

D. Spizzichino
Institute for Environmental Protection and Research (ISPRA), Via V. Brancati 48, 00144 Roma, Italy
e-mail: daniele.spizzichino@isprambiente.it

K. Sassa et al. (eds.), *Understanding and Reducing Landslide Disaster Risk*,
ICL Contribution to Landslide Disaster Risk Reduction,
https://doi.org/10.1007/978-3-030-60196-6_34

established between science and other forms and systems of knowledge in addressing sustainable landslide mitigation issues and problems at the local level, such as traditions and natural resources management and ecosystem conservation.

Traditional societies, usually with strong cultural roots, have nurtured and refined systems of knowledge of their own, relating to such diverse domains as, meteorology, geology, ecology, botany, agriculture, and many others.

Such knowledge systems represent enormous wealth. Not only do they represent other approaches to the acquisition and construction of knowledge and harbour information often as yet unknown to science, but they are also expressions of other relationships between society and nature in general, and of sustainable ways of managing natural resources in particular (Rosswall 2002).

However, the research community, with the exception of some disciplines specifically focused on studying traditional societies and traditional knowledge, such as ethnology, ethnobotany and ethnoscience, has not yet engaged in ways of better linking science to other knowledge systems. To do so would bring important advantages to both sides, and provide, to those in need of knowledge for pursuing sustainable development goals. a broader range of empirical information (Rosswall 2002).

This article investigates some case studies related to landslide hazard mitigation in UNESCO and other heritage sites, trying to identify the added value provided by traditional knowledge and local expertise. For the purposes of this article, traditional knowledge refers to the combination of all the techniques that are designed according to the experience of local populations and applied with the use of local materials and local workers. This view is quite similar to the nature-based approach, that is now progressively rising in many fields of engineering geology.

Materials and Methods

Protection of cultural heritage from landslides is a science at the border between Engineering Geology and Conservation of Cultural Heritage. To maintain the authenticity and integrity of the site, any mitigation project has to provide effective solutions, very often innovative, to be combined with original architecture and environment. The harmonisation of advanced solutions with historical heritage further require a mitigation project (sometimes pioneering), with a design that is minimising the impact of new intervention, possibly recovering traditional materials and/or solutions or the re-use of original construction techniques. In low income Countries, the usage of original construction techniques also ensures maintenance by local workers, thus enhancing sustainability.

Clearly, not all landslides can be stabilised with traditional techniques. Large volume phenomena or complex landslides may require a more invasive intervention. However, in these cases it is possible to reduce the visual impact of consolidation work (Margottini 2013a, 2013c, 2015, 2018; Margottini and Spizzichino 2014).

Finally, when dealing with oral traditions, not based on a real comprehension of physical processes, the efficacy on time of adopted solutions must be properly evaluated. Furthermore, such long-term safety has to be established, taking into consideration the current climate change effects and the impact on landslides recurrence interval.

Thus, not many experiences are currently available about mitigation of landslides with traditional techniques. An investigation in the two major peer review databases has revealed a few publications that were recovered by the indicator "landslide" + "traditional knowledge". Specifically, 15 articles in Scopus (https://www.scopus.com) and 3 in ISI Web of Knowledge (https://www.webofknowledge.com). Furthermore, most of these articles are only marginally related to landslides.

A different case of an indication of "climate change" + "traditional knowledge" where Scopus reports 547 peer reviewed articles and ISI Web of Knowledge 313 papers. This demonstrates that, in current scientific literature, traditional knowledge was playing a significant role in the adaptation to climate change, while still being marginal in landslide science.

Maitra (2009) reports how local communities in Darjeeling Hills (northeastern India) own a range of traditional measures to mitigate landslides. Such as: agroforestry practices; improving terrace riser; fencing (dry wall fencing and bio fencing), mixed and inter cropping. A set of indicators was also reported by Maitra (2009) as signs of landslide early warning such as unusual animal behavior, displacement of doors and windows, and other signs.

Dasanayaka and Matsuda (2019) identified 8 local knowledge-based practices/actions in Sri Lanka. A questionnaire survey was distributed to 80 representatives of indigenous population, in an attempt to identify their experience of landslide disaster, as reported in Table 1.

More specific traditional mitigation works are reported in Margottini (2015), Margottini (2018) and Margottini and Spizzichino (2014).

In order to evaluate the contribution of traditional mitigation measures to landslide risk, a comparison between experiences in heritage sites and current state of art in the field was performed. UNESCO and other heritage sites were selected because, in these case studies, any intervention has to avoid modifications of the original status, in terms of authenticity and integrity (ICOMOS 1964; UNESCO 2019). Landslide mitigation was finally here, considering the need

Table 1 Level of awareness to mitigation actions in a selected indigenous community in Sri Lanka (Dasanayaka and Matsuda 2019)

Local knowledge based practices/actions	Answer (80)	
	N	%
Text basic awareness on Landslides	55	69
Text unusual earth cracks	27	34
Text continuous heavy rain	70	87
Warning system based on temple bell	74	93
Practicing "Aththam Kramaya"	38	47
Terrace paddy fields/Forest reservation	46	57
Live fencing	64	80
Stone walls	66	83

to recover as much as possible, obtaining traditional knowledge and the application of the state of art low impact techniques, or "to invent" pioneering solutions. Most likely utilising local expertise and local materials.

Landslide mitigation generally considers structural and non-structural measures. Then, contributes to reduce (or mitigate) hazard, vulnerability of exposed elements, risk or to share residual risk (Vaciago 2012). Table 2 provides a classification of current mitigation strategies (Lacasse et al. 2013).

From a physical point of view, structural measures consider extensive works aimed at reducing the unstable forces. Intensive stabilization works by applying external forces; intensive stabilization works by increasing the internal resistance of the soil.

For the purpose of the present paper, in order to compare historical traditional techniques with current landslide mitigation works, an inventory of detail measures was required. The web site https://www.larimit.com/, developed a toolbox updating the information of Vaciago (2012) and Lacasse et al. (2013) from Safeland project (SafeLand 2011), was considered a useful reference point.

The toolbox includes approximately 70 structural mitigation measures, divided into eleven categories, which belong to class of measures either reducing hazard or reducing consequences (Table 3).

Following are the 10 UNESCO and other heritage sites affected by landslides. Here, 72 mitigation projects were conducted mainly by the authors, in an attempt to enhance the contribution of traditional knowledge and local expertise (whether possible).

Shahr-e-Zohak, Afghanistan (Rock Fall, Mud Flow, Solifluction, Soil Erosion, Gully Erosion)

The archaeological remains of Shahr-e Zohak (Fig. 1) are placed on a steep hill at the confluence of the Bamiyan and Kalu rivers (Margottini et al. 2015). The fortress is protected by ramparts, built along the steep cliffs bounding the site, equipped with several watchtowers and protected by three more orders of walls. All the structures are made of mud bricks placed on top of stony foundations.

The site is affected by incised erosional phenomena on the eastern side, while the hilltop is mainly hit by diffuse erosion and soil mobilization during snowmelt. Monument deterioration due to weathering is coupled with the lack of

Table 2 Classification of mitigation measures (structural measures on top, non-structural measures on bottom) (Lacasse et al. 2013)

Class	Component of risk addressed	Brief description	Notes and other terms used
Stabilisation	Hazard (H)	Eng'g works to reduce the landslide probability of occurrence	Preventive, remedial, hard, soft, active stabilization
Control	Vulnerability (V) (consequence)	Eng'g works to protect, reinforce, isolate the elements at risk from the landslide area	Preventive, hard, soft, passive stabilization
Avoidance	Elements (E) (consequence)	Temporary or permanent reduction of exposure w/warning systems, evacuation/-relocation, sheltering, land-use planning	Reduction of the exposure of the elements at risk. Monitoring, early warning systems and civil protection procedures, often temporary, selective avoidance measures
Tolerance	Elements (E) (consequence)	Awareness, acceptance and/or sharing of risk	Indirect reduction of the exposure of the elements at risk

Table 3 Structural mitigation measures from https://www.larimit.com/

Category 1: nature based solution (NBS) for erosion control—living approach	Share Zohak (soil)	Tiwanaku (soil)	Civita (soil and rock)	Machu Picchu (soil, debris and rock)	Antananarivo (soil, debris and rock)	Vardzia (rock)	Maaloula (rock)	Petra (rock)	Orongo (rock)	Bamiyan (rock)
Hydroseeding			+							
Turfing					**					
Tree bushes direct/pit planting (Live transplanting)	**				**					
Live/inert fascines and straw wattles	**		**		**					
Brush mattresses										
Brush layering					**					
Live stakes (Live poles)					**					
Live smiles										
Category 2: nature based solution (NBS) for erosion control—combined living/not living approach	**Share Zohak (soil)**	**Tiwanaku (soil)**	**Civita (soil and rock)**	**Machu Picchu (soil, debris and rock)**	**Antananarivo (soil, debris and rock)**	**Vardzia (rock)**	**Maaloula (rock)**	**Petra (rock)**	**Orongo (rock)**	**Bamiyan (rock)**
Geotextiles (Rolled erosion control products)					**					
Drainage blankets					**					
Beach replenishment/nourishment										
Rip-rap					**					
Rock dentition										
Category 3: Modifying the slope geometry—mass distribution	**Share Zohak (soil)**	**Tiwanaku (soil)**	**Civita (soil and rock)**	**Machu Picchu (soil, debris and rock)**	**Antananarivo (soil, debris and rock)**	**Vardzia (rock)**	**Maaloula (rock)**	**Petra (rock)**	**Orongo (rock)**	**Bamiyan (rock)**
Removal of (actual or potentially) unstable soil/rock mass										
Removal of loose or potentially unstable blocks/boulders (trimming and scaling)						###		###		**
Removal of material from driving area										
Substitution of material in driving area with light weight fill										
Addition of material to the area maintaining stability										
Terracing (NBS)		###								
Category 4: modifying the surface water regime—surface drainage	**Share Zohak (soil)**	**Tiwanaku (soil)**	**Civita (soil and rock)**	**Machu Picchu (soil, debris and rock)**	**Antananarivo (soil, debris and rock)**	**Vardzia (rock)**	**Maaloula (rock)**	**Petra (rock)**	**Orongo (rock)**	**Bamiyan (rock)**
Surface drainage works (Ditches, Channels, Pipework)	###	**		###	###	**		###		###
Local regrading to facilitate runoff										

(continued)

Table 3 (continued)

Category 4: modifying the surface water regime—surface drainage	Share Zohak (soil)	Tiwanaku (soil)	Civita (soil and rock)	Machu Picchu (soil, debris and rock)	Antananarivo (soil, debris and rock)	Vardzia (rock)	Maaloula (rock)	Petra (rock)	Orongo (rock)	Bamiyan (rock)
Sealing tension cracks					**	**				**
Impermeabilization (Geo Membranes, Impervious facing *-penka*)		###								
Vegetation—Hydrological effects (NBS)					**					
Hydraulic Control Works (Channel Lining and Check Dams)			**	###	**	+		###		
Diversion channels			**			+				
Live pole drains (NBS)					**					
Live/rock check dams (NBS)										
Category 5: modifying the groundwater regime—deep drainage	**Share Zohak (soil)**	**Tiwanaku (soil)**	**Civita (soil and rock)**	**Machu Picchu (soil, debris and rock)**	**Antananarivo (soil, debris and rock)**	**Vardzia (rock)**	**Maaloula (rock)**	**Petra (rock)**	**Orongo (rock)**	**Bamiyan (rock)**
Shallow trenches filled with free-draining material		**								
Deep trenches filled with free-draining material										
Sub-horizontal drains (conventional drilling)										
Sub-horizontal drains (directional drilling)										
Small and medium diameter vertical wells										
Drainage tunnels, Adits, Galleries with secondary drains or as outlet for wells										
Category 6: modifying the mechanical characteristics of the unstable mass	**Share Zohak (soil)**	**Tiwanaku (soil)**	**Civita (soil and rock)**	**Machu Picchu (soil, debris and rock)**	**Antananarivo (soil, debris and rock)**	**Vardzia (rock)**	**Maaloula (rock)**	**Petra (rock)**	**Orongo (rock)**	**Bamiyan (rock)**
Vegetation—mechanical effects (NBS)										
Substitution										
Compaction from surface		###								
Deep compaction (vibrocompaction—vibroreplacement—vibrodisplacement),										
Mechanical deep mixing with lime and/or cement										
Low pressure grouting with cementitious or chemical binder										

(continued)

Table 3 (continued)

Category 6: modifying the mechanical characteristics of the unstable mass	Share Zohak (soil)	Tiwanaku (soil)	Civita (soil and rock)	Machu Picchu (soil, debris and rock)	Antananarivo (soil, debris and rock)	Vardzia (rock)	Maaloula (rock)	Petra (rock)	Orongo (rock)	Bamiyan (rock)
Jet grouting										
Modification of groundwater chemistry (e.g. lime piles)										
Category 7: transfer of loads to more competent strata	Share Zohak (soil)	Tiwanaku (soil)	Civita (soil and rock)	Machu Picchu (soil, debris and rock)	Antananarivo (soil, debris and rock)	Vardzia (rock)	Maaloula (rock)	Petra (rock)	Orongo (rock)	Bamiyan (rock)
Counterfort drains (trench drains intersecting basal shear plane)										
Piles										
Barrettes (diaphragm walls)										
Caissons—mechanical effects										
Soil nailing										
Dowels and harnessing						+		+	+	**
Rock bolting			+	**	+	**		**	**	**
Strand anchors			+		+		+	**	**	**
Soil nail and root technology —SNART (Hybrid)										
Category 8: Retaining structures to improve the slope stability	Share Zohak (soil)	Tiwanaku (soil)	Civita (soil and rock)	Machu Picchu (soil, debris and rock)	Antananarivo (soil, debris and rock)	Vardzia (rock)	Maaloula (rock)	Petra (rock)	Orongo (rock)	Bamiyan (rock)
Reinforced soil structures										
Gabion walls		**	**		**			**		
Crib walls	**				**					
Drystack masonry walls										
Mass concrete or masonry walls	###	**		###	**			###		###
Reinforced concrete stem walls					+					
Vegetated gabions (Hybrid)										
Live crib walls (NBS)			**		**					
Category 9: passive control works for deviating the path of a landslide	Share Zohak (soil)	Tiwanaku (soil)	Civita (soil and rock)	Machu Picchu (soil, debris and rock)	Antananarivo (soil, debris and rock)	Vardzia (rock)	Maaloula (rock)	Petra (rock)	Orongo (rock)	Bamiyan (rock)
Deflection structures (Berms)										
Rock sheds										
Channelization structures (Lateral walls)										
Category 10: passive control works for dissipating the energy of a landslide	Share Zohak (soil)	Tiwanaku (soil)	Civita (soil and rock)	Machu Picchu (soil, debris and rock)	Antananarivo (soil, debris and rock)	Vardzia (rock)	Maaloula (rock)	Petra (rock)	Orongo (rock)	Bamiyan (rock)
Debris racks										
Baffles (Impediments)										
Check dams					**					

(continued)

Table 3 (continued)

Category 10: passive control works for dissipating the energy of a landslide	Share Zohak (soil)	Tiwanaku (soil)	Civita (soil and rock)	Machu Picchu (soil, debris and rock)	Antananarivo (soil, debris and rock)	Vardzia (rock)	Maaloula (rock)	Petra (rock)	Orongo (rock)	Bamiyan (rock)
Attenuator systems										
Afforestation (NBS),										
Live gully breaks (NBS)										
Category 11: passive control works for arresting and containing landslides	**Share Zohak (soil)**	**Tiwanaku (soil)**	**Civita (soil and rock)**	**Machu Picchu (soil, debris and rock)**	**Antananarivo (soil, debris and rock)**	**Vardzia (rock)**	**Maaloula (rock)**	**Petra (rock)**	**Orongo (rock)**	**Bamiyan (rock)**
Rigid barrier										
Flexible barriers						+				
Ditches and embankments										
Debris retention basin										

Traditional techniques; ** Modern techniques with low impact; + Modern techniques

Fig. 1 Shahr e Zohak

an adequate drainage system to collect runoff. Many constructions collapsed or are prone to collapse. Ramparts located on the steep hillslopes are severely affected by gully erosion and siphoning, which causes depressions infilled with eroded and weathered building material. The access path is locally eroded or buried by debris cones. The western margin of the plateau has been rapidly retreating due to collapses, while the citadel is in danger due to diffuse or gully erosional processes developed on all its sides.

A conservation plan for the Citadel of Shahr-e Zohak was prepared by (Margottini et al. 2015), to identify mitigation measures against soil erosion, in the upper part of the hill and to reduce water runoff, in the middle/lower flanks. The mitigation strategy comprises a nature-based approach, making recourse to traditional techniques and local knowledge (bushes, fascine, check dam, channels, masonry walls and more) (Fig. 2). Among traditional techniques, the recovery of ancient masonry walls was proposed, as well as the use of local bushes, to be used for the mitigation of soil erosion. Since this bush is locally used as biomass for heating in winter, a special nursery was agreed with the local population, with plant production to be jointly used to contrast soil erosion, and offered to the population in order to avoid destroying the soil erosion mitigation fields.

Tiwanaku, Bolivia (Silt Rotational Slide, Earth Flow, Solifluction, Rill Erosion)

Tiwanaku is an Archaeological World Heritage Site, located at an elevation of 3885 m.a.s.l., south of Lake Titicaca, in the altiplano or high plateau region of the Republic of Bolivia. It is in the province of Ingavi, department of La Paz, 60 km from the city of La Paz. A major highlight of the site is the Akapana pyramid, an extensive earth mound hill, with only few remains of the original coverage of stone and columns (Fig. 3).

The pyramid is suffering from extensive erosional phenomena such as: rill erosion, mass movement and creeping. The majority of the phenomena mentioned above are caused by the uncontrolled flow of the rainwater, which flows freely through the depressions (natural or cultural), but also infiltrates and deforms the materials of the deposits, and appears to be affecting the stability of the internal structures. Judging from site inspection, it is possible to note the large toe deposit originated by the erosion/dismantling of the earth mound. Some potential surficial mass movement, creeping, and the evidence of deformed walls, suggests internal deformation to the mound. (Margottini 2013b).

For these particular morphological processes affecting the pyramid, the most suitable mitigation measure is to control the deepening of the small watersheds, due to accelerated linear erosion. Surely, one of the main problems is the uncontrolled flow of the rain water coupled with the snowmelt. In order to reduce the above free flow and to avoid linear and superficial erosion along the slope of the pyramid, a drainage system (able to collect and evacuate the water) must be properly designed. It must be implemented and coupled with the consolidation. This can be obtained with the diffused use along the selected channel of:

- Implementation of a system of small earthen/rock-and-wood channels at the upper part of the catchments. This would start from the top to collect the surface water;
- Final collection and extraction of water outside the area using pipes;
- Small gabions (filled with local stone) for the control of the deepening eroded channels.

Possible alternatives to the use of gabions may be provided by:

- Drywall (always realized with local stone);
- Tiwanaku style walls;
- Small check dams (stone and woods).

Any adopted solution must be clearly identified as new construction and can never be confused with ancient works. The use of gabions should be preferred due to their

Fig. 2 Ancient masonry wall to be restored (left) and a local bush to be used against soil erosion (center and right)

Fig. 3 Digital model of Akapana, from laser scanning and drone survey, integrated with orthophoto

flexibility, cheapness, and simplicity of production and installation. Due to possible limits and constraints and, after several meetings with UNESCO consultants and local managers, the final adopted measures were the use of small gabions (filled with local stone—sand stone—Arenisca). In order to reduce and mitigate the impact of the measures (sustainable mitigations) and to recreate the original shape of the Akapana Pyramid, it was also decided that any adopted solutions must be located following the ancient terraced wall path (rocky staircase), see Fig. 4.

Earth terraces were also realized to maintain the original tradition. The material was compacted by hand and by foot, with the help of an old system based on a flat panel where to stand over (Fig. 5).

Impermeabilization of surfaces and terraces was implemented by recovering an old Inca technique, which is still in place. This is made by using a combination of straw, mud and cactus-juice (penka) (Fig. 6).

Civita di Bagnoregio, Italy (Rock Fall, Rock Block Topple, Rock Rotational Slide, Rock Wedge Slide, Silt Rotational Slide, Debris Flow, Mud Flow, Earth Flow, Complex)

Civita di Bagnoregio (Margottini 2013c) lies in a hilly area surrounded by steep valleys intensively eroded by two torrents presenting an E-W direction (Fig. 7). The cliff is composed of a 20–25 m top layer of jointed ignimbrite, underlain by intensively stratified pyroclastic formation. The Quaternary tuffs rest on a bedrock of a Plio-Pleistocene clayey succession, in which forms the valley in the major area.

Due to the particular geological and geomorphological features, large and frequent slope instability phenomena

Fig. 4 Gabions with exposed surfaces composed by the same typology of local stones, originally covering most of the pyramid

Fig. 5 Rill terracing and compaction of soil in Tiwanaku

Fig. 6 From cactus to cactus-juice, to penka

occur. In the clayey formation, landslides are represented by mudflows, debris-flows and rotational slides. In the upper portion of the volcanic cliff, rock-falls, toppling, and block-slides are the common landslide typologies, as a result of a retrogressive mechanism of erosion.

Bolt and anchors have been discovered in the upper rocky part of the cliff, by means of hand-excavated shafts located along the border of the cliff. This typology realized in the external portion of the cliff (unstable tuff blocks) has the scope to consolidate the most hazardous and risk sector of the town with the internal and most stable tuff rock through nails, anchors, grouting and injections (Fig. 8). The principal aim is to transfer, through deep anchors, the stress of the external unstable portion to a stable portion of the cliff (internal massive tuff). The structural passive anchors provide a lateral restraint of the cliff preventing toppling, and rock fall phenomena of unstable block portions (Margottini 2013c).

In the lower clay flanks, only minor interventions were executed. They include water diversion through modification of river drainage at the bottom of the valley, live crib walls to protect a country road from mudslides (Fig. 9), check dams and, finally, live/inert fascines and hydroseed in the upper part. Beyond traditional, the site was continuously experiencing new solutions and technologies, with a nature-based approach in clay formation.

Machu Picchu, Peru (Rock Fall, Rock Planar Slide, Debris Slide, Debris Avalanche)

Machu Picchu Inca citadel (Canuti et al. 2009), designated by UNESCO as a World Heritage site since 1983 (Fig. 10), is affected by frequent landslides that threaten security of inhabitants and tourist exploitation (i.e. landslides of March 2004, February 2008, January 2010).

The Geology of the area is characterized by granitoid bodies that were emplaced in the axial zones of the main rift system. Deformation of granite caused by cooling and tectonic phases, originated in 4 main joint sets, regularly spaced (few decimeters to meters). Several slope instability phenomena have been identified and classified according to mechanism, material involved and state of activity. They are mainly related to rock falls, debris flows, rock slides and debris slides. Origin of phenomena is kinematically controlled by the structural setting and relationship with slope

Fig. 7 Civita di Bagnoregio

Fig. 8 Low impact anchors and nails through shaft excavated inside the rocky cliff of Civita di Bagnoregio

face (rock falls, rock slide and debris slides). The accumulated materials are very often the source for debris flow. Geomorphological evidences of deeper deformations are still under investigation.

Traditional techniques have been applied in the conservation/reconstruction of ancient terraces (andenas), as well as in the watering collection and distribution, originally for both agricultural and human needs (Figs. 11 and 12).

Fig. 9 Live crib walls in the clay flanks of Civita di Bagnoregio, protecting a local road (from left 1993, 1993 and 1995)

Fig. 10 Machu Picchu and the andenas in the Eastern side

Antananarivo, Madagascar (Rock Fall, Sand Slides, Debris Slide, Complex)

The city of Antananarivo (Ciampalini et al. 2019) is located in the central highlands of Madagascar. A capital city of 1.5 million people, and the largest urban centre of the highlands. From a geological point of view the Analamanga hill area consists of a granite ridge bounded by steep slopes and rock walls, mantled by soils and debris slope cover. Considering the frequent heavy cyclonic rains, frequently hitting the highlands during the rainy season, the geomorphological context is particularly prone to strong erosion and geo-hydrological hazards. These include various types of landslides on hillslopes (from rock falls to shallow phenomena), and flash flooding in the plain area, as recently proved during the winters of 2015 and 2018 natural disastrous events (Fig. 13).

After some field work and desk study to mitigate landslides hazard and to properly manage the sewer system of the town, the following actions were suggested (Ciampalini et al. 2019):

1. Implementation and proper maintenance of the sewer system. There is a real necessity of a comprehensive plan guiding the local administration in the management of landslide and flash flood hazard. This will consider: (i) Development of a detailed hazard and risk map, also taking in account multiple hazards (e.g. seismicity); (ii) Mitigation of landslide and flash flood hazards in a changing climate; (iii) Rainfall water collection and discharge. A project for managing the run-off of free water is necessary. This may also include waste-water from houses. The proper management of surficial water may help in reducing the shallow landslides that are affecting the Antananarivo hills.
2. Mitigation of soil erosion areas, which can potentially trigger shallow landslides in case of heavy rains. The use

Fig. 11 Recovering damaged andenas (terraces) with traditional technique

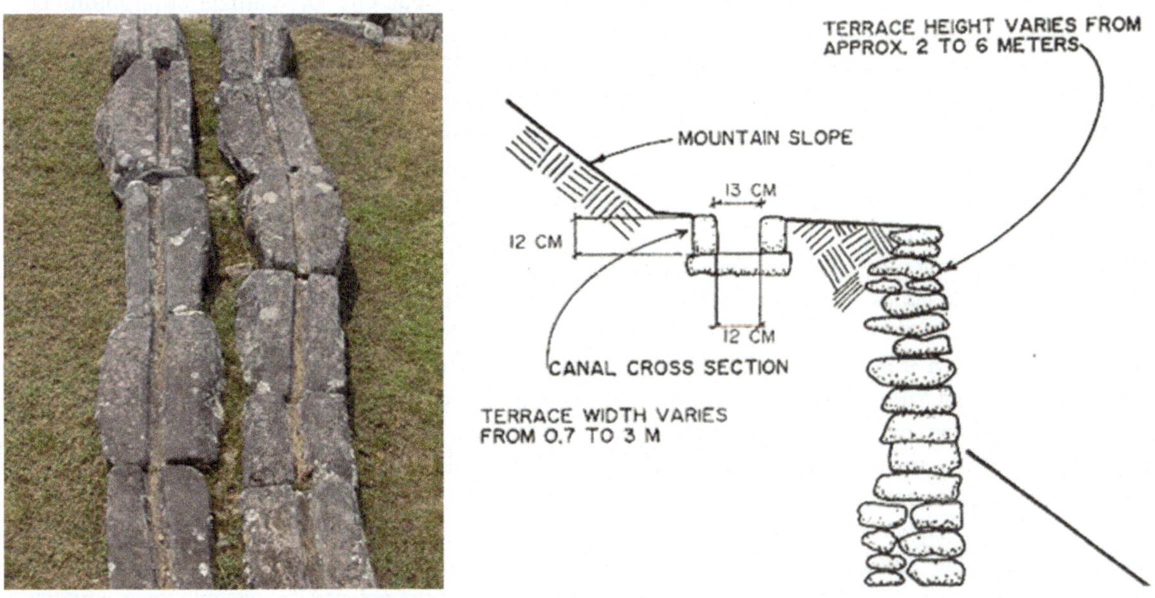

Fig. 12 Ancient watering system in Machu Picchu (drawing from Wright et al. 1997)

of nature-based solutions (NBS) will be an effective solution, in order to recreate a more natural environment in the Analamanga hill area (biofilter fabric and grassing, viminates, wooden drainage channels). The most appropriate solution and type of vegetation must be investigated by a botanic/green engineering expert. Local

Fig. 13 View of the flank of Antananarivo with soil slip, after 2015 winter (courtesy Bureau National de Gestion des Risques et des Catastrophes)

constructions already use tradition masonry walls to face soil slip (Fig. 14).

3. Stabilization-aggradation of steep slopes affected by gully erosion. The use of granite stone gabions can be an effective solution, and can be easily managed by local workers. A proper design is required in any site, to investigate the soil conditions and the most suitable foundation.
4. Avoiding construction of new buildings, unless it does not affect the occurrence of new landslides, as per the DRM plan of the UNESCO site.
5. Management of waste materials and avoidance of burning garbage along the hillslopes. There is a need to avoid the production of cohesionless material on top of the soil that may flow down, even if it may be light rainfall.
6. Mapping and monitoring cultural heritage – scaling up and zooming in. It is crucial to develop and validate innovative cost-effective methods and non-destructive diagnostic technologies, to map and monitor the state of preservation, and the impact of landslide hazard on movable and immovable cultural heritage. A map of heritage sites and exposure to risk must be prepared. Some advanced monitoring technologies (e.g. ground based radar interferometry, wireless sensor networks, laser scanning, etc.) aimed at understanding processes before the conservation can be implemented in relevant or high risky sites.

Vardzia, Georgia (Rock Fall, Rock Block Topple, Rock Rotational Slide, Rock Planar Slide, Rock Wedge Slide, Boulder Fall, Debris Flow)

The rock-cut city of Vardzia (Margottini et al. 2015; Boldini et al. 2018) is a cave monastery site in south-western Georgia (Fig. 15). It is excavated from the slopes of the Erusheti mountain on the left bank of the Mtkvari River. The main period of construction was the second half of the twelfth century. The caves stretch along the cliff for some five hundred metres and up to forty meters inside. The monastery consists of more than six hundred hidden rooms spread over thirteen floors, in which it was possible to be protected from the Mongol domination.

The site was affected by frequent slope instability processes, mainly rock slides and falls, along the main discontinuities systems that are identified on the entire vulcanite tuff façade.

Many modifications occurred to the site in the recent years. Removal and cleaning from unstable blocks, implementation of water drainage, harnessing by metallic net of adequate stiffness and resistance, anchored by dowels along the edges of the potentially unstable mass, and passive bolts. Low impact and recourse to traditional knowledge, when possible, was implemented. A local company was trained to operate in such a field, since there were none available in Georgia (Fig. 16).

Maaloula, Syria (Rock Planar Slide)

Fig. 14 Traditional masonry walls to face soil slip in Antananarivo favelas

The Maaloula village (Margottini 2013d) is situated in the Palmyrides Mountains (Syria), close to the anti-Lebanon orogenic zone. The morphology of the village is the result of important tectonic activity, which resulted in an impressive narrow valley intersecting an anticline (Fig. 17). Maaloula is well known as the place of the first example of Christianity sepulture, such as the Santa Tekla tomb (IV c. A.D.). Local population still speak in the traditional Aramaic language.

From the geological point of view, the cliff is composed of hard, massive Oligocene limestone and forms nearly vertical walls 50–70 m in height, fractured by subvertical small faults. In the lower part of the flanks a general cover of

Fig. 15 The site of Vardzia

rock debris is present. In the area, many rockfalls and rock slides have been detected. They exhibit volumes from meter to thousands of cubic meters, ranging from fall and small rock slides, to tens of thousands of cubic meters for few giant rock slides.

The area was also affected by destructive earthquakes as in 1759, November 25th producing a ground motion ranging between VII and VIII EMS-92 (Sbeinati et al. 2005).

A consolidation project was provided by local experts, consisting of anchors in the upper part of the cliff (from top to down) and a small buttress integrated with a nail at the bottom of the large block. Large use of concrete was applied to limit the surface water drainage. This project was suggested to be revised in order to limit the landscape impact and to find a more suitable mitigation technique.

Petra, Jordan (Rock Fall, Rock Block Topple, Rock Planar Slide, Rock Wedge Slide, Debris Flow)

The rock cut city of Petra (SW Jordan) is a world-wide famous UNESCO (United Nations Educational, Scientific and Cultural Organization) cultural heritage site known for the beauty and uniqueness of its monumental area.

The hand-carved rock monuments of Petra are entirely cut in the sandstone of Umm Ishrin and Disi formations (Fig. 18). Lithologic and petrographic characteristics of the two formations are controlling weathering related conservation problems, affecting, as well, shear strength parameters of the rocks.

The sandstone is generally brittle in the whole Petra park, thus with a large variability of strength, and a high frequency of discontinuities of various origin (i.e. faults, joints, weathering), which promotes a block composition of rocks. Structural condition determines a rock-fall potential activity that may involve unstable volumes, from 0.1 m3 up to over 500 m3 (Margottini and Spizzichino 2017). Also, small lateral valleys occasionally produce destructive flash floods.

A global master plan for the site was prepared by the authors, mainly suggesting the following:

- Clean, grub and scale the slope of all loose rock. (Mechanical or manual scaling) (Fig. 19, left);
- Fix unstable blocks (medium and large size) where scaling is not possible by use of anchors, bolts and dowels. Occasionally if not visible and for small size blocks, harnessing (wire mesh);
- Construction of small check dams along lateral channels at the end of the Siq (where they are not yet present);
- Recover original Nabatean water drainage system;
- Regular maintenance of check dams, by emptying the dams from the loose material accumulated during past events (Fig. 19, right).

Orongo, Easter Island, Chile (Rock Planar Slide)

The village of Orongo, part of Rapa Nui National Park, is located in Easter Island, above the inner wall of the crater of

Fig. 16 Scaling of unstable blocks with local climbers (left) and cover-up the head of passive rock bolt (right)

Fig. 17 View of Santa Tekla Monastery in Maaloula

Fig. 18 The narrow canyon (Siq) going to the treasury tomb and the loose blocks on the upper part of the rocky cliff

Rano Kau's volcano at an altitude of ca. 300 m a.s.l. Prominent historical remains are the numerous petroglyphs that represent the ancient ceremonial of the birdman cult (Fig. 20).

Recent topographical and geomorphological analysis of the area have provided clear evidence of slope instability, along the southern external flank of the caldera. The increased stress has promoted rock sliding of weathered material located in the upper part of the slope until the limit of the crater edge. A multi-temporal analysis of landslide activity through a 3D-GIS modelling coupled with a back-analysis was implemented, in order to define present slope stability conditions, and possible future evolution of the slope.

In order to reduce the possibility of triggering new slope instability phenomena, along the Rano Kau cliff beneath the Orongo village, a Master Plan of mitigation measures was proposed. The measures, which adopt a low visual impact restoration, include the following:

Waterproofing of ground surface, avoiding infiltration and pore pressure excess:

- Surface water collection and channeling;
- Drainage trench for subsurface water flowing.

Protection of material potentially prone to slide:

- Iron net with green reforestation;
- Strengthening of the rock blocks with grouting.

General slope consolidation:

- Scaling of unstable blocks (small to medium size);
- Deep anchor with steel cable on top;
- Stabilisation of unstable individual blocks.

This is to provide useful tools for local authorities to implement preventive landslide risk mitigation measures, for the protection of the cultural heritage in Orongo.

Bamiyan, Afghanistan (Rock Fall, Rock Planar Slide, Sheet Erosion, Gully Erosion)

The historical site of Bamiyan, Fig. 21, is affected by geomorphological deformation processes. And, were worsened by the explosion of the Buddhas in March 2001, this destroyed the statues that date back to the sixth century AD (Margottini 2014).

Not only was invaluable cultural heritage irremediably lost, but the consequences of the explosions, as well as the collapse of the giant statues also added greatly to the geological instability of the area. Traces of rocks which recently slid and fell are relevant proof of the deterioration of its stability conditions, and most parts of the site now appear prone to collapse in the near future.

Under the coordination of UNESCO, a global project to assess the feasibility conditions for the site's restoration was developed (Margottini 2004 and 2014); field data was collected, and a mechanism for the potential cliff and niches' evolution was provided. In the meantime, some consolidation works were carried out in the most critical rock

Fig. 19 Climber scaling unstable blocks (left) and one of the ancient dams on the northern side of Al-Khubtha Mountain (right)

Fig. 20 View of the Orongo village and of the global landscape of Easter Island, with Rano Kau's volcano

fall-prone areas, to avoid any further collapse in the coming winter season. But also, to enable archaeologists to safely catalogue and recover the Buddha statues' remains, from the floor of the niches. The emergency activities started in October 2003 and finished in 2012. These included:

1. The installation of a monitoring system, to evaluate any possible deformation of the cliff.
2. The realization of temporary protection with steel ropes, and two iron beams suitable to avoid lateral deformation, and, inside the niche, from blocks destabilized by the explosion. Among the temporary work, just after the consolidation of the niche's wall, a wire net was installed over the back wall of both niches, in order to enable archaeologists to work on the ground floor in safe conditions.
3. The final stabilization of the East niche:
   In this area anchors, nails and grouting were introduced (Fig. 20). To reduce the risk of rock fall and collapse, particular care was given to the problem of grouting material due to the extreme high slaking capability of siltstone.
4. Minimization of intervention (anchor/nail head finishing) completes the execution of work. Anchor and nail heads were designed to be placed slightly inside the rock and then covered by a mortar, allowing a total camouflage of the work. A number of tests for the best mixture of cement, local clay/silt and water, used to cover the anchor/bolt heads, were also developed in 2003, in cooperation with ICOMOS experts. The results provided the best chromatic stability and robustness of the mixture.

Some minor consolidation and impermeabilization works were implemented in the Western Buddha niche in 2009. Consolidation works were mainly implemented by professional climbers, directly operating on the cliff as well as working on the inner niche scaffolding (Fig. 22).

Finally, after the stabilization of the external part of the Eastern Buddha niche, in 2009–2012 the inner back wall (shear zone of explosion) was stabilized. This was done by means of small anchors and limited grouting. Also aimed at fixing the still existing original part of the Statue plaster, jointly executed by Engineering Geologists and Conservators.

Results: Traditional Knowledge and Local Expertise in Landslide Mitigation

The selected case studies allowed, for the first time, an overview about the application of traditional knowledge to mitigate landslide risk. If traditional knowledge has been defined as a cumulative body of knowledge, know-how, practices and representations maintained and developed by peoples with extended histories of interaction with the natural environment, the application to landslide mitigation allows to relate human experience and local materials to safeguard the built environment. These sophisticated sets of understandings, interpretations and meanings are part and parcel of a cultural complex that encompasses language, naming and classification systems, resource use practices, ritual, spirituality and worldviews (International Council for Science 2002).

Fig. 21 The Western niche of Bamiyan after the destruction of Buddha statue

To understand the efficacy of traditional knowledge a comparison of 72 available mitigation measures, adopted in the 10 UNESCO and other heritages sites, have been used to compile Table 3. In such sites, the application of traditional techniques is clearly the best way to maintain the authenticity and integrity of the heritage, as requested by the UN heritage convention of 1972 (UNESCO 1972, 2019) and the charter of Venice, 1964 (ICOMOS 1964).

Table 3 is organised in terms of structural mitigation measures (https://www.larimit.com/), against solutions adopted in the 10 case studies described in the previous paragraph. Following that, all adopted measures are ranked according to the quality of the chosen approach, such as traditional technique, modern technique, and low impact modern technique.

The first view of the table clearly shows that most of the sites have been mainly stabilised against superficial processes. Such processes mainly consider soil erosion control and drain and control of surface water regime. In case of soil erosion, mainly the application of Nature-Based -Solution (NBS) was implemented, a modern approach with low impact techniques (https://www.larimit.com/). A special case is the use of "penka", in Bolivia, a mixture of mud, straw and cactus juice producing an outstanding material for soil impermeabilization and reinforcement.

A wide recourse to the traditional system was implemented for surface water control. For example, channels, ditches and masonry walls, as in Bamiyan (Fig. 23) and, mainly, Shar-e-Zohak. In recent years the use of NBS approach was widely implemented in heritage sites, allowing the mitigation of surface processes with low impact techniques.

Stabilisation of debris material with stone masonry walls was also implemented in many sites (Fig. 24), such as Machu Picchu (Peru). The used stones are generally from the same debris material.

Unstable blocks were often removed using local workers, as in Petra (Jordan).

In the case of rock settlements, e.g. rupestrian sites, some drainage works that were in continuity with the past have been discovered, using traditional techniques and materials. Sometimes complementing with modern materials, e.g. cement, to ensure longer efficacy. In addition, the removal of unstable and endangered blocks, as in Petra (Jordan) was proposed in sites where the local population usually climb high slopes.

Finally, in such rupestrian sites, there was mainly the need to transfer the load to more competent strata. A wide recourse to anchor/bolt/dowel was necessary. it is mainly passive, and applying a modern technique, possibly with low

Fig. 22 Local workers, properly trained, consolidated the back side of the niche and bonded the remaining small pieces of Buddha Statue to the consolidated surface (left). The scaffolding inside the niche (right)

impact (e.g. covering the bolt head with the same material as the cliff).

The traditional techniques described in previous chapters and summarised in Table 3 are not only preserving the heritage site, but also based on the skill of the local population, ensuring the long-term maintenance of mitigation works.

Discussion

The present paper investigates 10 UNESCO and other heritage sites affected by landslides in an attempt to understand whether traditional techniques and local expertise may play a fundamental role in mitigating the expected damage and, in the meantime, maintaining the original authenticity and integrity of the site. Such two concepts are of paramount importance for the inclusion in the World Heritage List (UNESCO 1972), as clearly stated by the Operational Guidelines for the Implementation of the World Heritage Convention (UNESCO 2019).

In this guideline (UNESCO 2019), integrity is a measure of the wholeness and intactness of the natural and/or cultural heritage and its attributes, while authenticity is linked to the value of the heritage, that is the degree to which information sources about these values may be understood as credible or truthful. Evidently, any mitigation work to preserve from landslide risk, must be envisioned in this contest.

Factually, mitigating landslide risk in UNESCO and other heritage sites, constrains the adoption of an approach that requires the protection of authenticity and integrity of the site. This is where traditional knowledge may provide a very effective compendium of practical measures, by using local materials and where the local population is capable of realizing such works and finally, to perpetually maintain them.

Clearly, the above statement is not valid elsewhere. Traditional knowledge seems to be applied to shallow landslide phenomena, involving a reduced volume of materials. In case traditional expertise cannot be used for the complexity of phenomena or for the volume of the mass movements, a low impact solution must be envisaged. Tables 4 and 5 report the various case studies versus depth of phenomena (Table 4) and versus displaced mass. It must be also noticed that in some cases, such as Machu Picchu, ancient engineered works have been conducted to drain surface water, avoid infiltration and, consequently, the triggering of medium to deep landslides.

Fig. 23 An old channel for water retainment and injection for sealing of an open discontinuity, on top of Western Buddha niche (Afghanistan)

Fig. 24 The famous stone masonry walls of Machu Picchu (andenas) with indication of soil creeping

Noticeably, the reported case studies are not exhaustive of natural variability affecting UNESCO heritage sites. Nevertheless, various landslide typologies, different materials, and diverse climate regions, rheology, etc. are considered. The resulting picture, even if not complete, may propose an outcome having a general validity.

The case studies also introduce another relevant issue, in relation to the adopted measures for mitigating landslide risk. Traditional knowledge is quite effective against landslide threats but, to become very effective, they should be obtained with extensive knowledge of geomorphological and hydrological processes that affect the heritage site. This was not always the case in the past. Identifying the different approaches and ranking the various typologies of investigation in the available case studies, clearly shows the variety of geomorphological studies in the many sites, which are clearly depending on local natural processes. Table 6 reports such variability in the methodological approach and the level of investigation in the various steps. From this table, it is quite evident how any proposed mitigation measure, especially when dealing with delicate traditional solutions, needs a considerable amount of scientific knowledge when it comes to acting processes. Sometimes this level of knowledge requires the consideration of the social and economic contest, where the heritage site is located.

The reported Table 6 shows how relevant is an approach that is not only limited to the site of the monument but includes an area of interest where it is possible to better understand the environmental flow chart. This is evident in the first steps of the study process (field, aerial, map, GIS) that are evidently related to a large investigated area. The need to investigate an area wider than the site is a mandatory requirement for rupestrian sites, where the monument and the surrounding environment are affected by the same geomorphological processes.

An attempt to synthesize all the available information, is reported in the following Fig. 25. It is possible to notice that, with reference to the 10 UNESCO and other heritages case studies and the 72 specific measures analysed, in both soil and rock geological conditions, we can remark:

- Heritage sites are mainly affected by shallow phenomena, concentrated in the first meters in depth;
- Volumes involved, in average, are of few tens of m^3, generally lower than 5 m^3;
- Traditional knowledges are playing a role mainly in the management of surface water regime, retaining structures (masonry walls, check dams, etc.), measures in cleaning and securing rocky cliff, minor terracing and landscaping;
- Most of the intervention in mitigating landslide hazard and risk in UNESCO and other heritage sites is developed by means of modern measures with low impact;
- Invasive modern technologies are applied only in deep phenomena or where the involved volumes are relevant. This category is significant but not prevailing.

Considering that heritage sites have survived for many centuries, it is evident that their location was mainly selected to be safe against major natural hazard, according to the knowledge of that time. This is probably justifying the missing of large landslides affecting the sites, but only accentuating minor and surficial phenomena.

In conclusion, mitigation projects of UNESCO and other heritage sites, especially in developing Countries, are advanced laboratories for sustainable mitigation of landslide risk. The use of traditional techniques and/or low impact measures, as well as mainly nature-based solutions, are the milestone of such approach.

This creates local job opportunities (e.g. local climbers in Petra or Vardzia; production and installation of penka in Tiwanaku; drilling activities in Vardzia), contributing to the local economy, as well as raising awareness against landslides.

Furthermore, since such measures are managed by local workers belonging to local communities, the long-term maintenance of the mitigation work is mainly ensured by them. Of course, without having to recourse to external experts and workforces coming from abroad.

Without disregarding that, in some cases and jointly with the application of traditional knowledge, the involvement of local communities and local materials may contribute to realize modern effective mitigation measures with low impact. Nature-based solutions, such as those proposed in Shahr-e-Zohak (Afghanistan) mixing traditional surface drainage systems with the use of vegetation, are an example of such approach.

Conclusion

The paper investigates some UNESCO and other heritage sites affected by landslides, in an attempt to understand whether traditional techniques and local expertise may play a role in mitigating the expected damage and, in the meantime, not degrading the site with invasive measures.

Undoubtedly, mitigating landslide risk in heritage places restricts to an approach that, by definition, it needs to preserve the authenticity and integrity of the site. Traditional knowledge may provide a very effective compendium of practical measures, also by using local materials, where the local population is capable to realize such works and to maintain them again, on a regular basis.

The research demonstrates that the mitigation of landslide risk through traditional techniques, can help in the

Table 4 Depth of main geomorphological processes and main typologies of mitigation

	Share Zohak (soil)	Tiwanaku (soil)	Civita (soil and rock)	Machu Picchu (soil, debris and rock)	Antananarivo (soil, debris and rock)	Vardzia (rock)	Maaloula (rock)	Petra (rock)	Orongo (rock)	Bamiyan (rock)
Shallow (0–3 m)	###	###	**/ +	###	###	###		###	**	###
Medium (3–10 m)			+	**	**		+	**	**	**
Deep (10–25 m)			**							
Extremely deep (>25 m)										

Traditional; ** Not traditional but with low impact; + Modern

Table 5 Volume of displaced mass and main typologies of mitigation

	Share Zohak (soil)	Tiwanaku (soil)	Civita (soil and rock)	Machu Picchu (soil, debris and rock)	Antananarivo (soil, debris and rock)	Vardzia (rock)	Maaloula (rock)	Petra (rock)	Orongo (rock)	Bamiyan (rock)
Small: up to a few m^3;	###	###	**	###	###	###/**		###/ +		###/**
Medium: 5 to 50 m^3	**		**	**	**	**		**/ +		**
Large: > 50 m^3			**/ +	**	**/ +		+		+	**

Traditional; ** Not traditional but with low impact; + Modern

Table 6 Methodological approach and level of investigation in the different steps

	Share Zohak (soil)	Tiwanaku (soil)	Civita (soil and rock)	Machu Picchu (soil, debris and rock)	Antananarivo (soil, debris and rock)	Vardzia (rock)	Maaloula (rock)	Petra (rock)	Orongo (rock)	Bamiyan (rock)
Field reconnaissance	**C**	**C**	**C**	**C**	**C**	**C**	**C**	**C**	**C**	**C**
Aerial reconnaissance	**C**	**C**	**B**	**C**	**C**	**C**	**B**	**B**	**B**	**C**
Map analysis	**C**	**C**	**C**	**C**	**B**	**C**	**B**	**C**	**B**	**C**
Computerised landslides terrain analysis and GIS	**C**	**C**	**B**	**C**	**B**	**C**	**B**	**C**	**B**	**B**
Historical analysis		**B**	**C**	**B**	**A**	**B**		**A**		**A**
Meteo-climatic seting	**A**	**B**	**C**	**C**	**B**	**C**	**A**	**C**	**C**	**A**
Geophysical studies		**B**	**A**	**B**		**C**			**B**	**A**
Drilling			**B**	**B**		**C**	**B**			
Laboratory and in-situ tests	**A**	**B**	**C**	**C**	**A**	**C**	**A**	**C**	**B**	**C**
Investigation of groundwater conditions			**A**	**A**	**A**	**A**	**B**			**A**
Instrumentation and monitoring			**B**	**C**		**C**		**C**		**B**
Soil/rock stability analysis (2 and 3D)	**A**	**C**	**B**	**B**		**C**	**B**	**C**	**A**	**B**

A Basic; *B* Intermediate; *C* Advanced

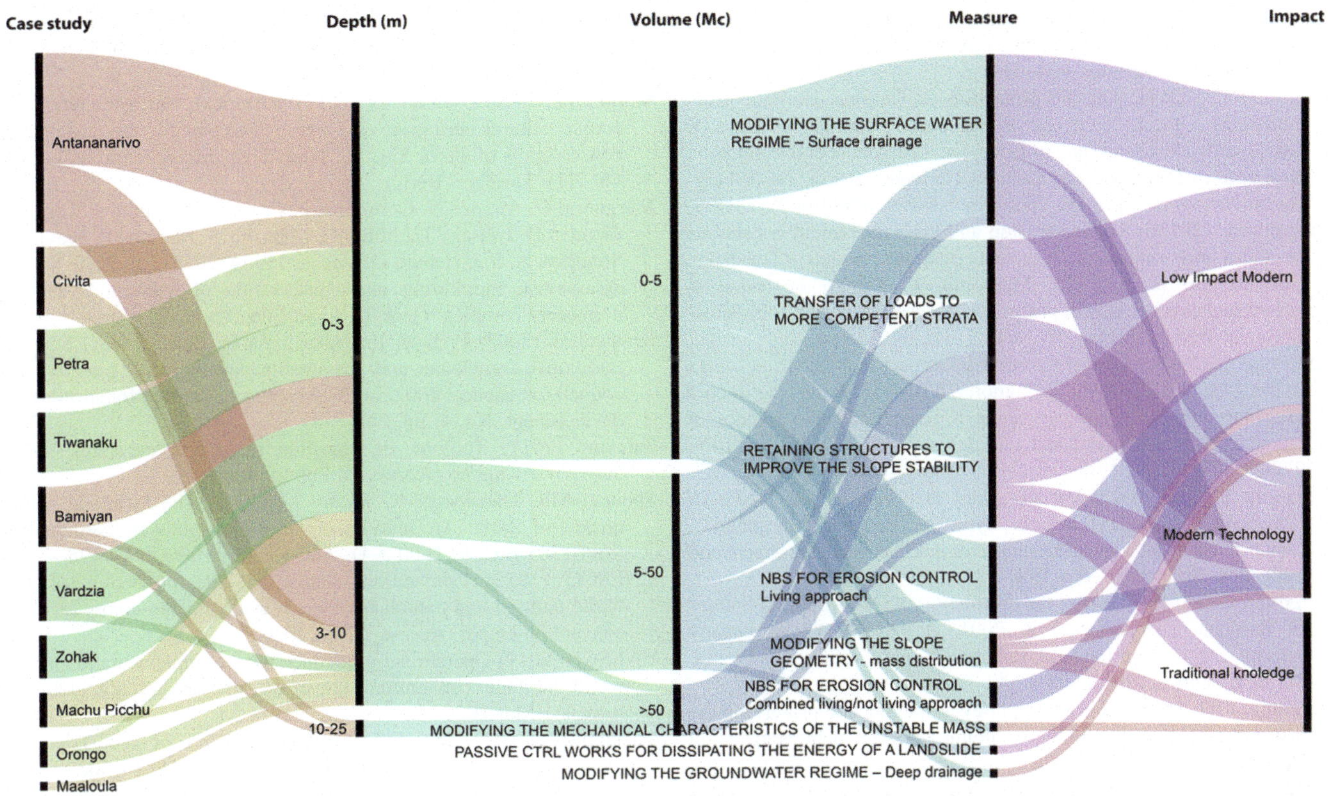

Fig. 25 Alluvial diagram showing the flows between case studies, typologies of mass movements in terms of depth and involved volume, adopted mitigation measures and related impact (traditional knowledge, modern technologies with low impact, modern technologies)

preservation of the integrity and authenticity of the heritage site. Besides, it enhances local job opportunities and ensures long term maintenance. A similar result can be achieved when making recourse to modern low impact measures (e.g. NBS).

Clearly, it is impossible to apply the cumulative body of knowledge, know-how, practices and representations maintained and developed by peoples, with extended histories of interaction with the natural environment, everywhere. In few cases with large landslides, the need to preserve the heritage site is prevailing with respect to the impact of the chosen solutions. But also, in these cases, it is possible to invent new solutions, mitigating hazard and risk but, in the meantime, safeguarding the landscape and the site. The case study of Civita di Bagnoregio (Italy) is a milestone in this direction.

Acknowledgements The authors are deeply grateful to the many institutions and agencies supporting the researchers in this paper. UNESCO central headquarter and many local UNESCO offices greatly contributed to the obtained results. Similarly, we deeply thank ICOMOS, the National Authority for Heritage Conservation of Georgia and the Région Ile-de-France à Madagascar.Last but not least, most of the activities have been conducted within the UNESCO Chair on Prevention and Sustainable Management of geo-hydrological hazards, in Florence University (Italy).

References

Boldini D, Guido GL, Margottini C, Spizzichino D (2018) Stability analysis of a large-volume block in the historical rock-cut city of Vardzia (Georgia). Rock Mech Rock Eng 51(1):341–349

Canuti P, Margottini C, Casagli N, Delmonaco G, Falconi L, Fanti R, Ferretti A, Lollino G, Puglisi C, Spizzichino D, Tarchi D (2009) Monitoring, geomorphological evolution and slope stability of Inca citadel of Machu Picchu: results from Italian INTERFRASI project. In: (a cura di): Sassa K, Canuti P, Landslides: disaster risk reduction. Springer, Berlin. ISBN: 978-3-540-69966-8

Ciampalini A, Frodella W, Margottini C, Casagli N (2019) Rapid assessment of geohydrological hazards in antananarivo (Madagascar) historical centre for damage prevention. Geomatics, Nat Hazards Risk 10(1):1102–1124. https://doi.org/10.1080/19475705.2018.1564375

Dasanayaka U, Matsuda Y (2019) A study on local knowledge in adaptation to landslide disasters in Sri Lanka. Eng J 23:6

ICOMOS (1964) International charter for the conservation and restoration of monuments and sites (the Venice charter 1964. https://www.icomos.org/charters/venice_e.pdf. Access 1 Mar 2020

International Council for Science (2002) Science, traditional knowledge and sustainable development. In: ICSU series on science for sustainable development, No. 4: p 24

Lacasse S, Kalsnes B, Vaciago G, Choi YJ, Lam A (2013) A web-based tool for ranking landslide mitigation measures. In: Proceedings of the 18th ICSMGE, Paris. France

Maitra H (2009) Integration of Indigenous methods of landslide risk reduction into national disaster management framework. Disaster Dev 3(2):54 (2009)

Margottini C (2004) Instability and geotechnical problems of the Buddha niches and surrounding cliff in Bamiyan valley, Central Afghanistan. Landslides 1:41–51

Margottini C (2013a) On the protection of Cultural Heritage from landslides . In: (a cura di): Margottini C, Canuti P, Sassa K (eds) Landslide science and practice: risk assessment and mitigation. Key note in Landslide and Cultural Heritage session. Heidelberg-Berlin-New york, Springer Verlag Inc.. ISBN: 978-3-642-31318-9

Margottini C (2013b) Surface erosion and mass movement constraints in the conservation of Akapana pyramid mound (Tiwanaku, Bolivia). In: (a cura di): Margottini C, Canuti P, Sassa K (eds) Landslide science in practice: risk assessment and mitigation. Springer, Berlin. ISBN: 978-3-642-31318-9

Margottini C (2013c) Low environmental impact consolidation works in the rock cliff of Civita di Bagnoregio (Central Italy). In: (a cura di): Margottini C, Canuti P, Sassa K (eds) Landslide science in practice: risk assessment and mitigation. Springer, Berlin. ISBN: 978-3-642-31318-9

Margottini C (2013d) Large rock slides and falls in the cradle of Christianity: Maaloula (Damascus, Syria). In: (a cura di): Margottini C, Canuti P, Sassa K, Landslide science in practice: risk assessment and mitigation. Springer, Berlin. ISBN: 978-3-642-31318-9

Margottini C (2014) (ed) After the destruction of Giant Buddha Statues in Bamiyan (Afghanistan) in 2001. A UNESCO's emergency activity for Recovering and Rehabilitation of Cliff and Niches. Springer, Berlin. ISBN: 978-3-642-30050-9

Margottini C (2015) Engineering geology in shaping and preserving the historic urban landscapes and cultural heritage: achievements in UNESCO World Heritage Sites. In: Lollino et al. (eds) Engineering geology for society and territory—Vol. 8—Preservation of Cultural Heritage. Proceedings of IAEG XII CONGRESS. Springer, Torino

Margottini C (2018) TXT-tool 4.039-2.1 On the protection of cultural heritages from landslides. In: Sassa et al. 2018, Landslide dynamics: ISDR-ICL landslide interactive teaching tools, vol. 2, pp 647–667

Margottini C, Spizzichino D (2014) The management of cultural heritage in sites prone to natural hazard. Mem Descr Carta Geol d'It XCVI (2014)

Margottini C, Spizzichino D (2017) Historical "infrastructures" to access cultural heritages: engineering geology for the sustainable conservation of Petra Siq. In: Innovative infrastructure solutions, vol 2(1). Springer, Berlin

Margottini C, Antidze N, Corominas J, Crosta GB, Frattini P, Gigli G, Giordan D, Iwasaky I, Lollino G, Manconi A, Marinos P, Scavia C, Sonnessa A, Spizzichino D, Vacheisvili N (2015) Landslide hazard assessment, monitoring and conservation of Vardzia Byzantine monastery complex. Georgia. Landslides 12(1):193–204

Rosswall T (2002) Preface. In: Nakashima D, Elias D (eds) Science traditional knowledge and sustainable development. international council for science. 2002. In: ICSU series on science for sustainable development. No. 4: pp. 24

Safeland (2011) Toolbox of landslide mitigation measures—NGI. https://www.ngi.no (Access 14 Feb 2020)

Sbeinati MR, Darawsheh R, Mouty M (2005) The historical earthquakes of Syria: An analysis of large and moderate earthquakes from 1365 B.C. to 1900 A.D, Ann Geophys 48(3):347–435

UNESCO (1972) Convention concerning the protection of the world cultural and natural heritage. https://whc.unesco.org/archive/convention-en.pdf. Access 1 Mar 2020

UNESCO (2019) Operational guidelines for the implementation of the world heritage convention. https://whc.unesco.org/en/guidelines/. Access 1 Mar 2020

Vaciago G (2012) The safeLand compendium of landslide risk mitigation measures. In: Margottini C et al. (eds) Landslide science and practice, vol. 6.# Springer, Berlin

Wright KR, Kelly JM, Zegarra AV (1997) Machu Picchu: ancient hydraulic engineering. Journal of hydraulic engineering 123 (10):839–843. https://www.scopus.com. Access 14 Feb 2020. https://www.webofknowledge.com. Access 14 Feb 2020

Reconstruction of the Slope Instability Conditions Before the 2016 Failure in an Urbanized District of Florence (Italy), a UNESCO World Heritage Site

Stefano Morelli, Veronica Pazzi, Veronica Tofani, Federico Raspini, Silvia Bianchini, and Nicola Casagli

Abstract

On May 25th, 2016, following a massive water leakage from the city water supply pipeline a portion of the urbanized riverbank collapsed along the Arno River in the city centre of Florence (Italy), UNESCO World Heritage site. Understand the dynamics driving the landscape evolution (from the slope up to the riverbank) is useful to know the historical growth of such a territory and to plan a farsighted and sustainable urban development. Therefore, a historical characterization of the area surrounding the Arno riverbank collapse was made considering the interaction between the anthropic pressure and the local geological hazards or those factors that activate them. Based on this reconstruction, some advanced technologies (like satellite imaging), integrated with more traditional techniques, have been employed (a) to analyse the landscape condition almost up to the trigger moment, (b) to study the dynamics of the event, (c) to monitor the possible spatial evolution of the 2016 landslide, and (d) to characterize the damage condition of the involved structures. The results show very low deformation rates, indicating relatively stable ground conditions before the event.

S. Morelli (✉) · V. Pazzi · V. Tofani · F. Raspini · S. Bianchini · N. Casagli
Department of Earth Sciences, University of Firenze, Via La Pira, 4, 50121 Florence, Italy
e-mail: stefano.morelli@unifi.it

V. Pazzi
e-mail: veronica.pazzi@unifi.it

V. Tofani
e-mail: veronica.tofani@unifi.it

F. Raspini
e-mail: federico.raspini@unifi.it

S. Bianchini
e-mail: silvia.bianchini@unifi.it

N. Casagli
e-mail: nicola.casagli@unifi.it

Keywords

Riverbank landslide • Satellite imaging • Cultural heritage monitoring • Hazard assessment • Landslide awareness

Introduction

The southern extremity of the historic centre of Florence, on the hydrographical left side of the Arno River, is bordered by a belt of rolling hills which yield an evocative landscape of inestimable cultural, historic and artistic value where the beauty of nature blends harmoniously with the fervour of the human activity. However, because of materials nature below the walking surface this area is subjected to a certain geological fragility that, in history, has contrasted such great beauty with recurring damaging events. Earthquakes, floods, and mass movements are here the main natural hazards affecting the city, showing respectively a high relevance for the entire built-up area, a high impact for the plain (including the reshaped sub-flat areas close to the river), and confined heavy consequences for the urbanized hillsides up to the riverbank slopes. The latter type of events, which for the riverbanks occurred even during floods due to the action of strong flows, is analysed in this paper. Starting from a historical reconstruction of the city evolution, we initially report the failures, the collapses, and the landslides occurred in the past up to describe the collapse of the artificial structure of the modern riverbank occurred in 2016 just upstream the Oldest Bridge (Ponte Vecchio in Italian) of Firenze.

Without the contextualization of the complex historical evolution of the slope and its alterations, it is difficult to understand the last instability event and the risks that can be incurred during this kind of events. This aspect, here described, was taken into consideration during the 2016 event for the emergency management and the design choices in the subsequent securing and final restoration.

K. Sassa et al. (eds.), *Understanding and Reducing Landslide Disaster Risk*, ICL Contribution to Landslide Disaster Risk Reduction,
https://doi.org/10.1007/978-3-030-60196-6_35

The Study Area

The area affected by the 25th May landslide (red area in Fig. 1, named Lungarno Torrigiani) involves a portion of the left riverbank of the Arno River just upstream of the medieval stone closed-spandrel segmental arch bridge called Old Bridge (Ponte Vecchio in Italian). It is located in the naturally narrower section of the river and for this reason it was chosen as the first crossing point in history (Mariotti Lippi et al. 2013). This area also corresponds to the hill foot called "Poggi dei Magnoli" (slope inclination between 20 and 45%) and commonly recognized by the population as "Poggio delle Rovinate" (Rovinate hill) because of the succession of landslides in past times ("Rovinate" means landslides in ancient Italian language), involving sometimes even the riverside.

From a geological point of view (Fig. 1) this area is set on the formations belonging to the Monte Morello Unit that represents the solid bedrock on which most of the buildings in this part of the city are anchored. Upward there are the Villafranchian fluvio-lacustrine deposits and finally the Holocene fluvial deposits. In detail, the Unit of Monte Morello is represented by the Pietraforte Formation in the lower portion, with lithofacies consisting of calcareous sandstones, shaly limestone with clay and marl (Cretaceous late), and by the Sillano Formation in the upper portion with lithofacies consisting of shales with limestone and sandy-limestone layers (Upper Cret.—Lower Eoc.) (Fanti 2006). The fluvial-lacustrine sediments in sandy and gravelly facies with abundant clayey matrix are higher up, followed on the top by the Holocene fluvial deposits constituted by pebbles and sands, with silts and clays.

The study area does not have a real aquifer. However, a degree of permeability is locally present such as to allow meteoric waters a relevant underground circulation. It is caused by altered and disrupted (also for the past gravitational events) clay layers and by intense fractured arenaceous-calcareous interlayers. At the end of June 2016, the water levels in local piezometers were found to be between -7.0 m and −7.5 m from the street level (Asioli et al. 2017).

The current riverbank morphology (Fig. 2) is the product of a major urban restructuring and reorganization of the public spaces approved in 1866 and completed in 1872 in the overall framework of the modernization necessary to let Florence be the capital of Italy (1865–1871). Before these works, which have maintained almost the same resulting structure until today, the area still belonged to the river's meandering actions. It was in morphological continuity with the doorstep of the more proximal buildings whose main facades were facing the hill. Only the secondary facades were oriented towards the river with a direct access to it through vegetable or botanical gardens, exclusive spaces or private moorings. According to the project conserved in the Historical Archive of the Florence Municipality, the new riverbank (Fig. 2) was created by constructing a road several meters raised above the water level. In practical, it interrupted the direct usability of the river spaces and established a physical barrier between the elements that have lived together for centuries: the population and the watercourse.

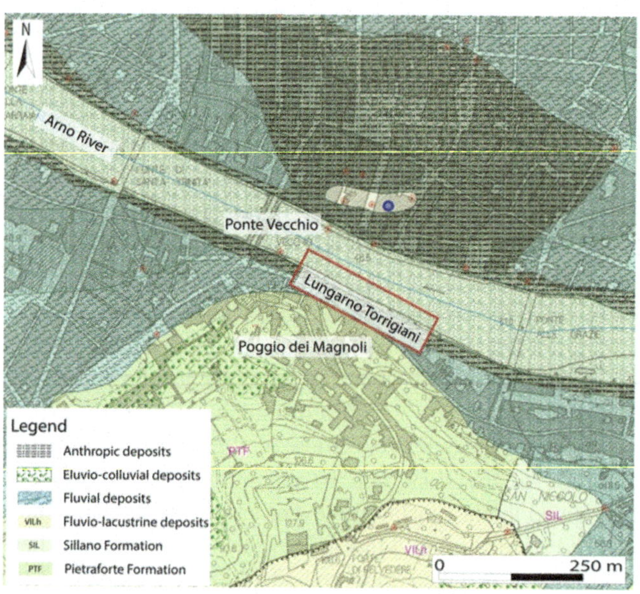

Fig. 1 Geological map of the zone surrounding the study area (red rectangle)

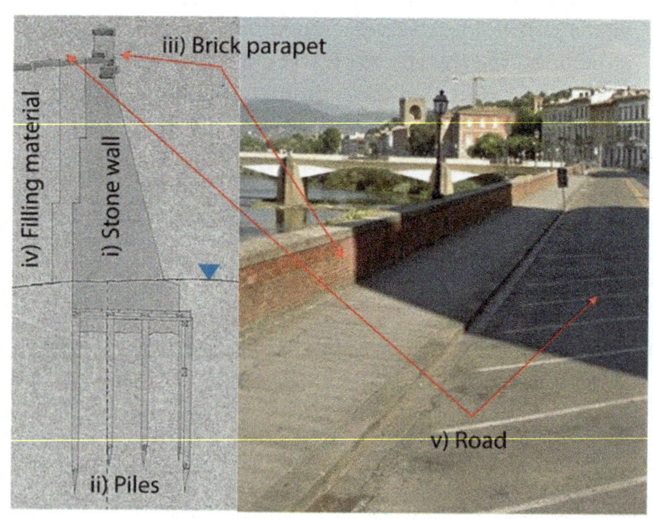

Fig. 2 The current riverbank morphology. (i) the vertical stone retaining wall; (ii) four rows of piles that anchor the wall directly to the substrate; (iii) the brick parapet; (iv) filling material with buried subservices; and (v) the paved road. Blue triangle is the water level

The riverbank new structure (Fig. 2) can be summarized with the following components: (i) a vertical stone retaining wall anchored directly to the substrate of the riverbed with four rows of piles; (ii) a brick parapet (about 1.20 m high); (iii) filling material and buried subservices between the stone wall and the original riverbank; (v) a paved road at the top. The filling operation was mainly managed with compacted landfill. Moreover, the riverbank was completed with the insertion of an arched vault culvert of metric dimensions just adjacent to the buildings' foundations, operating in junction with a lifting water station of a coeval aqueduct system that was built almost 600 m upstream.

In the first decades of the '900, a more modern and functional aqueduct pipe was added few centimetres below the surface in connection to the whole city network. During the nineteenth century restyling also the facades of the aligned facing buildings were rearranged and embellished to become real main facades such as for "Palazzo Capponi alle Rovinate", whose name, thanks to the appellative "Rovinate", remembers once again the local land instability.

Space–Time Reconstruction of the Instability Framework

Past Riverbanks Failures

In 1175, the fifth circle of city walls was completed to encompass and defend some hamlets built on the Arno river left bank, tripling the urbanized area with a secure and stable fortification (Day Jr 2002). By means of this urban reorganization, the city structure migrated its geographical barycentre closer to the Arno riverbed from a raised and ancient river terrace, which was occupied by the first Roman settlements (in 59 B.C.), and permanently inhabited for more than 11 centuries (Fig. 3). The two main consequences were: (a) the river was completely included in its natural narrowest stretch (Morelli et al. 2014) and (b) the inhabited area grew from a floodplain, less exposed to flooding and swamping, to a more floodable area. Thanks to renovations, integrations, and reconstructions of the old Roman walls in the north area of the city, the transition was very quick, and it was completed in just three years (Guidoboni and Ferrari 1995). From this moment onwards, all the natural riverbanks were progressively transformed into artificial urban morphologies thanks to slope reprofiling and accumulations of imported backfill (Schenk 2006).

Consequently, in this confined arrangement, the watercourse started to be a very active subject in the history of the city, both as indispensable support for the maintenance of the daily social-economic activities (especially those arisen to exploit directly the hydraulic power during ordinary flow conditions) and as a negative actor, during the most considerable flood events. In fact, as it is possible to read in many works published in Italian, (and here not cited) several damaging inundations were repeatedly observed over the centuries, and in some cases, riverbank collapses and/or damages to structures and buildings directly facing the riversides were induced by turbulent flows.

However, tangible and nonphysical consequences were commensurate with the different urban development during

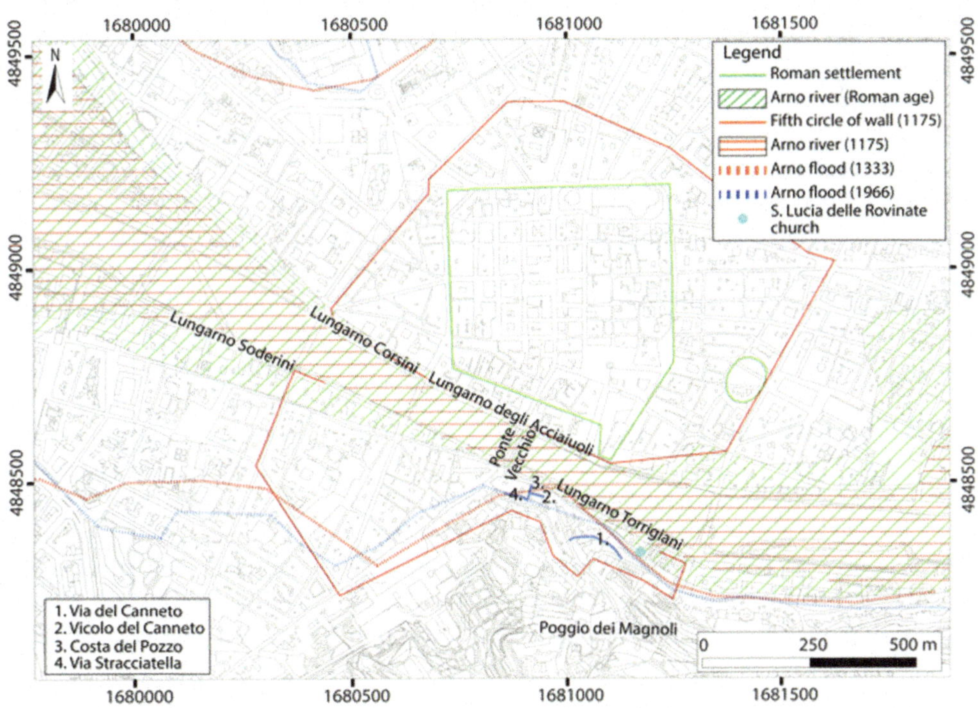

Fig. 3 The Florence city centre over the years

historical periods. The exposure to the river-related risks increased, from time to time, according to the gradually change of the general structure of the city until today. However, the memory of the most significant events has been lost caused by the age-old habit of the population to these episodes and the wide and systemic omission in the chronicles compared with the other types of calamitous events. During the history even the resonance of the events varied because of the momentary economic consequences and the fluctuation of attention and problems perception relating to river dynamics, with substantial and consolidated reduction in more recent times. In the contemporary age, the grow of the emotional emphasis in people's consciousness, resulting from an increasingly deep awareness of the high artistic and historical heritage value of the historical assets of the city (Guidoboni and Ferrari 1995), has been added to all this, in addition to the value of the human life quality.

In this context, as it is possible to read in the historical cornicles, on July 4th, 1965 a portion of the left riverbank, downstream out of the study area, rapidly collapsed within the Arno River after a leakage from the subterranean water supply network that hugely saturated the fill soil. Such occurrence engulfed the retaining wall, which was vertically built to protect the anthropic riverbank from high flood currents, and to sustain the above levee wall represented by a low masonry parapet. The resulting landslide was characterized by a volume of about 4000–5000 m^3, a length and a width of 40 m and 6 m, respectively. Despite the central location of the event site, one person was killed and no one else was involved. The reconstruction works started immediately after the due checks and were so effectively executed that the disastrous flood of the following year (November 4th, 1966) did not cause any damages in this stretch. In the morning of the 1966 devastating event, also the studied riverbank did not suffer any failures or significant damages to the main structure (only to the street furniture or minor artefacts) from the rapid rise of the water level and the consequent flood spreading that began nearby around 7.00 am. At the contrary, a portion of the right riverbank (Lungarno degli Acciaiuoli and Lungarno Corsini embankments) suffered significant collapses for long stretches because of the erosive strength of the river flows.

Past Slope Instabilities

First reports of significant landslides on the behind hillside, south the study area ("Poggio dei Magnoli", or Magnoli hill), date back to 1284. At that time, the urbanization had gone far beyond the plain next to Ponte Vecchio and it was consolidating on the surrounding hills. On April 2nd, 1284 after a very intense rainfall, a sudden mass movement collapsed into the riverbed. As mentioned in the historical chronicles, in that occasion, more than 50 buildings erected on the steepest hill collapsed killing many inhabitants. It was hypothesized that houses had been poorly built, without a proper terracing and retaining walls.

In 1373, 1490, and 1547 similar injurious events occurred in this area after prolonged rainfalls (Asioli et al. 2017). The latter landslide, composed by a big mixture of rocks and earth, had highly impacted the urban layout, and eighteen houses were completely devastated. The consequent human vicissitudes of this tragedy struck a chord in the whole population that it was necessary to find a practical solution for the future. Consequently, in 1565, the Duke of Florence tried to solve once for all the problem of the slope instability. Realizing that the structural works would not only be sufficient, he prohibited constructing new buildings in all the affected area and a plate with such ban was affixed on site for future memory. Substantially this ban was respected for a very long period and only some buildings in the most distal part were recovered with appropriate protection works like the old church of Santa Lucia dei Magnoli which was later popularly renamed church of "Santa Lucia delle Rovinate" (light blue dot in Fig. 3). The last damages to the remaining houses were documented in 1651. They were caused by incessant rains that, after a persistence of about four months, triggered local slope deformations and movements. This indicates that a large part of the problem was resolved (i.e., the human influence), but the issue of surface and groundwater circulation still existed.

Problems at the hydraulic draining system existed for a long time and with substantial evidences (continuous water rising, moisture surfaces, and stagnation along some impluvia) that some streets were renamed over time reflecting these characteristics. For example, "Via del Canneto" (assigned from the fifteenth century, 1. in Fig. 3) and "Vicolo del Canneto" (2. in Fig. 3) got their name from marsh with natural thickets of canes that once could be found in wetlands nearby the river. In the same area, also the abundant presence of not too deep underground waters is testified by the streets' toponyms, which are here dedicated to the existence of some public wells in past centuries. For example, there is "Costa del Pozzo" (meaning "Little Street - that runs along the slope—of the Well") which was included in the past adjacent to the road currently renamed as "via Stracciatella" (3 and 4 in Fig. 3).

From the great landslide of 1547 to the beginning of the twentieth century no further news of slope instabilities were recorded. Nevertheless, as it is possible to read in the chronicles, modest gravitational movements were reactivated in the area of the Bardini Garden in 1916, and in

1925–1926 some slope deformations caused the shift of the Bardini Garden road by about 3 m towards the river. From 1926 to today no relevant movements occurred.

The 2016 Pre-collapse Condition

Following the failure of 25th May 2016, four datasets of very high-resolution X-band (wavelength of 3.1 cm) satellite images were acquired and processed to retrieve information about pre-event surface displacement phenomena. The purpose of the radar data analysis was to detect and record any pre-event deformation along and around the study area, through the use of advanced InSAR (Interferometric Synthetic Aperture Radar) techniques, suitable for mapping slow deformation phenomena (from mm up to a few cm per year).

Ground displacement data were obtained using the SqueeSAR technique (Ferretti et al. 2011), a second-generation PSI (Persistent Scatterer Interferometry) algorithm. Methods belonging to this family are based on the processing of long temporal series of co-registered, multi-temporal SAR images acquired over the same target area from the same acquisition geometry. The main idea behind these methods is to identify pixels and groups of pixels sharing similar radar returns during the entire observation period, via proper statistical analyses. Using this sparse grid of radar targets, PSI techniques are able to statistically identify, quantify and remove atmospheric noise from SAR phase data and to create a history of motion for each target with millimetre accuracy. Images acquired, along ascending and descending orbits, by the Italian Constellation COSMO-SkyMed and by the German satellite TerraSAR-X (both with 3×3 m of ground resolution) were processed with the SqueeSAR approach. Obtained results provided valuable information on ground movements before the event and are reported in Fig. 4, where measurements points are classified according to their average velocities. Positive values correspond to motion toward the satellite; negative values correspond to motion away from the satellite (in mm/year). Both COSMO-SkyMed and TerraSAR-X datasets show, for the area of the Lungarno Torrigiani, very low deformation rates, ranging between −1.5 and 1.5 mm/yr, indicating relatively stable ground conditions (green points). All the information included in Fig. 4 agree that no significant movements before the 25th May 2016 event can be observed. High-resolution multi-interferometric results highlight the absence of movements in this sector and exclude the presence of active slope instability before the 25th May 2016 failure.

The 2016 Riverbank Failure

At 6:00 a.m. of May 25th, 2016 part of the Lungarno Torrigiani road collapsed with a partial slide of the terrigenous layer towards the riverbed. This failure was approximately 4 m in high and 150 m in length (cavity volume: about 1300 m^3) and the retaining wall was deformed but did not collapse. Late in the night of May 24th, 2016 a large amount of water observed over the street, despite the absence of rainfalls, and the decrease in the water pressure within a ND 600 pipeline gave advance of this event. Remote sensing technologies like laser scanner, total station, camera for close-range photogrammetry together with in situ analysis (boreholes, electrical resistivity topographies, downhole, and seismic noise) and laboratory tests were employed to characterize the riverbank landslide and the wall cracks (Morelli et al. 2020). To forecast any possible extension of the failure, a monitoring system was set up. It involved a high-resolution ground-based interferometric radar, located on the right riverbank, three seismic stations, placed on the riverbank wall, four crackmeters and nine biaxial tiltmeters laid on the facades of the buildings and remotely controlled. A detailed analysis of the failure was performed to identify the damage condition of the involved structures, to define the causes of the failure, and to mitigate and preserve the cultural heritage site. This analysis followed the strategy of developing a reasoned incremental observational capacity, progressively focusing on the most significant points on a smaller scale with increasingly specific devices after understanding the investigated scenarios. The management of these instruments, especially the more sophisticated ones, has produced a large amount of data that have been used only partially for the contingent purposes of rapid monitoring. Many of the unused stored data were subsequently reorganized for characterizing the event and, thanks to small additions of appropriate data acquisitions, understanding the residual risk in the area. Such analyses are not the focus of this paper.

Discussion and Conclusion

The Duke of Florence's ban in 1565 was the first example of a farsighted and sustainable urban development. In the following centuries, the manipulation of the territory was concentrated on the riverbanks, reaching its maximum pressure in the last decades. The maintenance of these lowland areas however did not sufficiently take into account the natural features of the territory and its relations with the

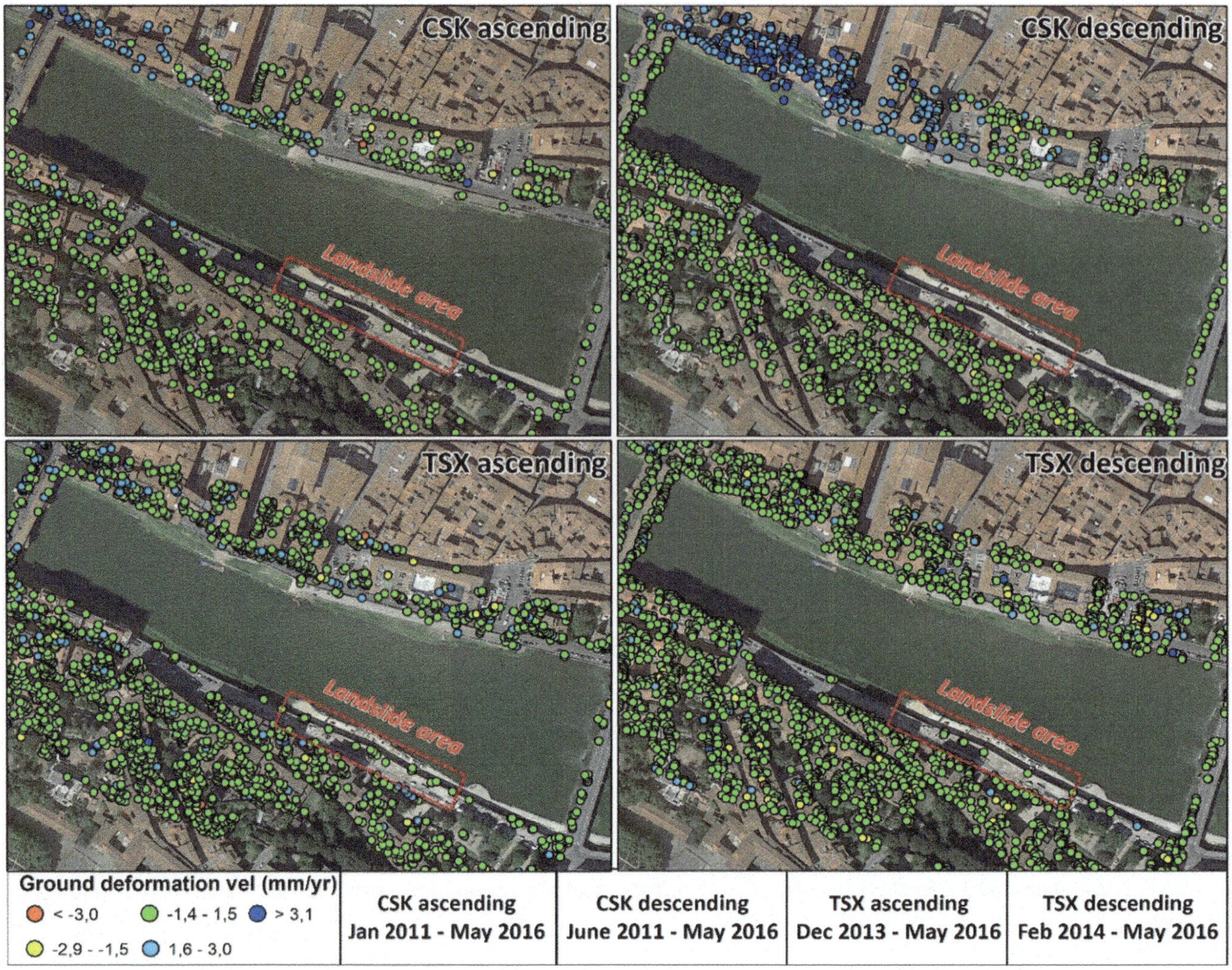

Fig. 4 Pre-event ground deformation for the area affected by the May 25th riverbank landslide obtained processing very high-resolution COSMO-SkyMed (CSK) and TerraSAR-x (TSX) satellite acquisitions

slope instabilities. Then, the last disastrous flood of 1966 completely shifted the focus on the riverbank preservation due to high waters levels in the river, developing specific plans only for inundations. The 2016 collapse, due to an atypical circulation of underground water, hit unexpectedly, but with few damages that it was possible to study, preparing the community to face again these ancient problems.

Acknowledgements Many thanks to all the people of the Earth Sciences Department of the Florence University involved during the first emergency hours, in the monitoring system setup and to all of them that collaborated in the characterization and monitoring activities.

References

Asioli C, Bertero A, Cribari F, Agostini C, Chiarugi M, Galli O, Tognotti M, Spinelli P, Aiello E, Micheloni M (2017) Intervento di messa in sicurezza in somma urgenza del tratto interessato da dissesto sul Lungarno Torrigiani a Firenze: progetto ed esecuzione delle opere di consolidamento, conservazione e ripristino del muro spondale dislocato. In: La geotecnica nella conservazione e tutela del patrimonio costruito. XXVI convegno nazionale di geotecnica, Roma, 20–22 giugno 2017. 2:413–423 (in Italian)

Day WR Jr (2002) The population of florence before the black death: survey and synthesis. J Medieval Hist 28(2):93–129

Fanti R (2006) Slope instability of San Miniato hill (Florence, Italy): possible deformation patterns. Landslides 3(4):323–330. https://doi.org/10.1007/s10346-006-0060-1

Ferretti A, Fumagalli A, Novali F, Prati C, Rocca F, Rucci A (2011) A new algorithm for processing interferometric data-stacks: SqueeSAR. IEEE Trans Geosci Remote Sens 49(9):3460–3470

Guidoboni E, Ferrari G (1995) Historical cities and earthquakes: florence during the last nine centuries and evaluations of seismic hazard. Ann Geophys 38(5–6):617–647

Mariotti Lippi M, Bellini C, Mori Secci M, Gonnelli T, Pallecchi P (2015) Archaeobotany in Florence (Italy): landscape and urban development from the late Roman to the Middle Ages. Plant Biosyst Int J Dealing All Aspects Plant Biol 149(1):216–227. https://doi.org/10.1080/11263504.2013.822433

Morelli S, Battistini A, Catani F (2014) Rapid assessment of flood susceptibility in urbanized rivers using digital terrain data: application to the Arno river case study (Firenze, northern Italy). Appl Geogr 54:35–53. https://doi.org/10.1016/j.apgeog.2014.06.032

Morelli S, Pazzi V, Tanteri L, Nocentini M, Lombardi L, Gigli G, Tofani V, Casagli N (2020) Characterization and geotechnical investigation of a riverbank failure in Florence, Italy, an UNESCO World Heritage Site. J Geotech Geoenvironmental Eng. https://doi.org/10.1061/(ASCE)GT.1943-5606.0002305

Schenk GJ (2006) '...prima ci fu la cagione de la mala provedenza de' Fiorentini...' Disaster and "Life World"—reactions in the commune of florence to the flood of November 1333. Medieval Hist J 10(1–2):355–386. https://doi.org/10.1177/097194580701000213

WLF
UNESCO · ICL · ISC · WFEO · FAO · UNDRR · WMO · UNU
2020 Kyoto Japan

Integrating Kinematic Analysis and Infrared Thermography for Instability Processes Assessment in the Rupestrian Monastery Complex of David Gareja (Georgia)

William Frodella, Daniele Spizzichino, Giovanni Gigli, Mikheil Elashvili, Claudio Margottini, Alberto Villa, Paolo Frattini, Giovanni Crosta, and Nicola Casagli

Abstract

The present paper describes the recent support activities implemented during the fall of 2018 in the David Gareja monastery complex area (South-eastern Georgia). During last decades, the growing cultural interest in the Georgian rock hewn monasteries founded in the sixth century has been constantly accompanied by conservation and management problems, due mainly to weathering and rocks collapses. In order to verify instability processes affecting the several monasteries in the area, detailed topographic surveys (Terrestrial Laser Scanner—TLS and Unmanned Aerial Vehicle-based Digital Photogrammetry—UAV DP) were performed jointly with Infrared Thermographic (IRT) analysis, geological and geomorphological field surveys. Geo-structural analysis was implemented in order to analyse the landslide types and processes, while kinematic analyses were performed in order to define the most frequent instability mechanism affecting the sites. IRT revealed thermal anomalies potentially connected to erosion, weathering and instability. The outcomes of the performed analyses highlighted that geo-structural setting, joint and stress released promotes rock instability processes in all the David Gareja monastery complex, representing a starting point for the implementation of a preliminary master plan based on low impact mitigation measures and monitoring systems.

Keywords

Cultural heritage • Landslides • Kinematic analysis • Infrared thermography

Introduction

Landslides and more generally slope instabilities are an important factor endangering cultural heritage sites, especially prehistoric sites, earth/rock monuments, and archaeological sites affected by environmental processes and degradation (Margottini and Vilimek 2014). The David Gareja monastery complex area is located in Kakheti semi-desert region (Eastern Georgia), about 25 km southeast of Tbilisi (Fig. 1). It includes the complex of 19 Medieval monasteries with approximately 5000 cells for monks and is part of the UNESCO Tentative List since 2007.

The dozens of cave monasteries decorated with unique frescoes are the best examples of harmonious interaction of man-made structures with the dramatic landscape. They bear the traditional principles of sustainable living and are considered as the masterpieces of Georgian Medieval art (Vv. 2018). From a geological point of view the area is constituted by a Miocene-Pliocene shallow and coastal succession formed by sandy clays, brownish-gray clays with inter-bedded sandstones, and rare motley continental conglomerate inter-layers and thin layers of volcanic ash (Gamkrelidze et al. 2017).

In detail, the analysed monasteries are constituted by soft sedimentary rocks, from very coarse-grained, pebbly thick-bedded sandstones, to coarse-grained soft sandstones, siltstones and clays. From a structural point of view the whole stratigraphic sequence is characterized by a monocline structure, with strata dipping SW with low-gentle angles. For all the monasteries, the distribution, typology and frequency of the potential instabilities have been identified, highlighting the possible triggering factors and/or

W. Frodella (✉) · G. Gigli · C. Margottini · N. Casagli
UNESCO Chair on Prevention and Sustainable Management of Geo-Hydrological Hazards of the University of Florence, Largo Fermi 1, 50142 Florence, Italy
e-mail: william.frodella@unifi.it

D. Spizzichino
ISPRA, Via V. Brancati 48, 00144 Rome, Italy

M. Elashvili
Faculty of Natural Sciences and Engineering, Ilia State University, Kakutsa Ave 3/5, Tblisi, Georgia

A. Villa · P. Frattini · G. Crosta
University of Milano-Bicocca, Piazza dell'Ateneo Nuovo, 1, 20126 Milan, Italy

K. Sassa et al. (eds.), *Understanding and Reducing Landslide Disaster Risk*, ICL Contribution to Landslide Disaster Risk Reduction,
https://doi.org/10.1007/978-3-030-60196-6_36

predisposing factors. Only after the definition of the potential instability processes it will be possible to assess a suitable monitoring plan for these causes and to implement a general master plan intervention, in order to take into account the specificity of the individual phenomena.

Materials and Methods

During the second half of 2018, two separate field survey were carried out in the David Gareja monastic complex area, in order to investigate the sites of Lavra, Natlismcemeli, Dodo Gareji, and Sabereebi (Fig. 1).

For each site the following analysis were carried out: TLS and DP surveys; Infrared thermographic surveys; structural and geo-mechanical surveys; and global kinematic analyses. During the field surveys structural data (discontinuity and bedding planes orientations, were collected, in order to obtain input data for the kinematic analysis (Figs. 2a, 3).

The topographic surveys of the analysed monasteries were carried out by means of a long-range 3D laser scanner (RIEGL VZ-400) (Riegl 2017). UAV-DP images were processed using Agisoft Photoscan (Agisoft 2017). The kinematic analysis was performed by means of the DiAna-k Matlab tool (Gigli et al. 2012), while the obtained 3D models were visualized by means of CLOUDCompare software (DGM 2015). The thermographic surveys were performed by using a hand-held thermal camera (FLIR SC620 model; FLIR 2009), while the data processed with FLIR Tools + software (FLIR 2015).

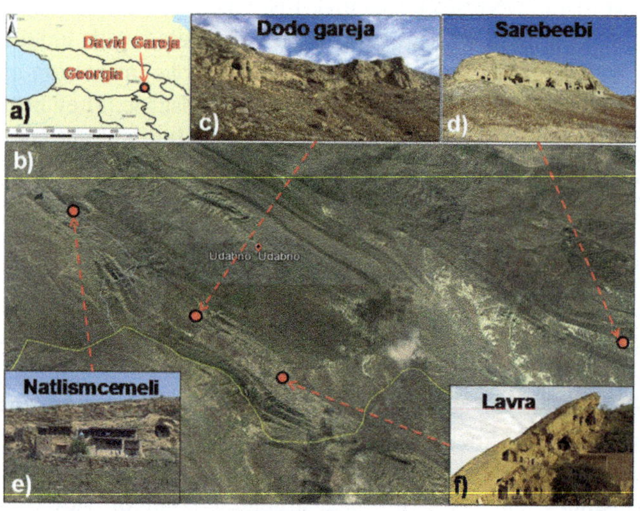

Fig. 1 aLocation of the study area of David Gareja (**a**, **b**). The investigated monasteries (**c–f**)

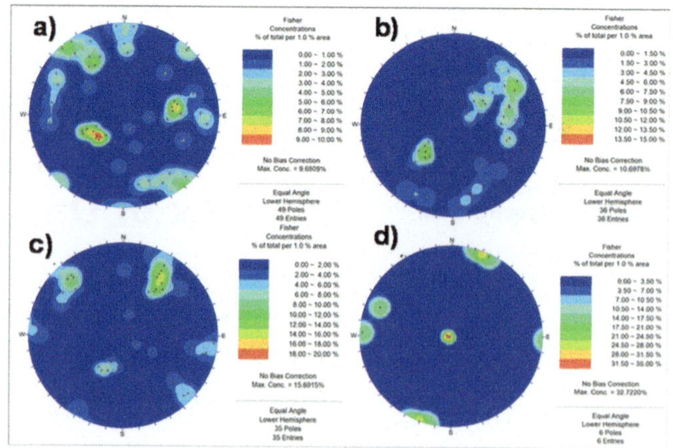

Fig. 2 Stereoplot of discontinuity orientations taken at: **a** Lavra; **b** Natlismcemeli; **c** Dodo Gareja; **d** Sabereebi

Topographic Surveys

The TLS employed device, by measuring with high accuracy (millimeter or centimeter) the laser time-of-flight between the sensor and the reflecting targets, is capable of obtaining the exact position of a mesh of points (point cloud), characterized by (x, y, z) cartesian coordinates and reflectance value R (x, y, z), providing in a short time a high resolution 3D surface digital model of the scanned object (Oppikofer et al. 2009). DP is a well-established and low-cost technique for acquiring dense 3D geometric information in slopes and from stereoscopic overlaps of photo sequences captured by a calibrated digital camera (Chandler 1999). TLS and UAV-DP are more and more used for the analysis of slopes characterized by instability processes, as safely allows in a short time a high detailed and accurate 3D representation of the investigated rock mass plano-altimetric morphological and geostructural setting, and is therefore particularly suitable for geostructural characterization and unstable rock cliffs monitoring applications (Remondino et al. 2011; Crosta et al. 2015).

Landslide Kinematics Global Analysis

In order to define the main instability processes, a spatial kinematic analysis can be performed by using discontinuity orientation data, semi-automatically and manually extracted from the obtained slope 3D surface models.

Given the slope geometry and the discontinuity orientation, this kind of analysis is capable of establishing the location and the probability of kinematically feasibility of the following mechanisms: (i) plane failure (PF) (Hoek and

Table 1 IRT survey parameters of the analysed sites

Case study	Slope aspect	Sensor-target mean distance (m)	Image res. (cm)	Avg. air T (C°)/ Rh (%)
Natlismcemeli	S	110/20	7.1/1.2	13.4/65.9
Lavra	S/NW	40	2.6	14.4/63.4
Dodo Gareji	S	90/20	5.8/1.2	16.9/52.6
Sabereebi	SW	100	6.5	22.6/39.8

Bray 1981); (ii) wedge failure (WF) (Hoek and Bray 1981); (iii) block toppling (BT) (Goodman and Bray, 1976); (iv) flexural toppling (FT) (Goodman and Bray 1976; Hudson and Harrison 1997). Casagli and Pini (1993) introduced a kinematic hazard index for each instability mechanism. These values are calculated by counting poles and discontinuities falling in critical areas of the stereographic projection. By using specific software, such as KARS (Casagli and Pini 1993), Rock Slope Stability (Lombardi 2007), or specifically designed Matlab tools such as DiAna-k (Gigli and Casagli 2011; Gigli et al. 2012, 2014) it is possible to load a great number of discontinuities with different friction angles. Intersection lines are calculated automatically, together with the equivalent friction angle, based on the friction angles of the intersecting planes and the shape of the wedge (Casagli and Pini 1993). The analysis can be performed for specific slope orientations, or for each cell of a 3D surface (true 3D kinematic analysis). This method overcomes many limitations of the traditional approaches, as it is possible to employ true 3D surfaces, and the kinematic conditions leading to the investigated instability mechanisms have been extended to overhanging slopes. Moreover, a global kinematic index (GKI) has been introduced, to quantitatively define the rock instability hazard for each sector of the slope, whatever the effective instability mechanisms. The input data of this method are the high-resolution 3D meshes and the discontinuities extracted with the manual and semiautomatic methods. As the orientation of fractures is related to the tectonic processes that have been acting in the investigated areas, a kinematic analysis can be useful to highlight the rock wall sectors which are more prone to instability processes. One of the limits of the graphic kinematic analysis is that of considering an infinite persistence, zero friction angle, not taking into account the variability of the parameters that determine the friction angle. In fact, the roughness, the uniaxial compression strength and the eventual filling of the discontinuities differ according to the fractures. These are identified by combining fracture dip and dip directions with local slope orientations.

Infrared Thermography (IRT)

Thermography is a remote sensing imaging technique accomplished by using Infrared calibrated thermal cameras, which sensors are capable of detecting the thermal radiation of the analysed scenario. The product of an IRT survey is a digital image acquired by the thermal camera array detector, which following the correction of the sensitive parameters (such as object emissivity, path length, air temperature and humidity) is converted by the built-in processor in a surface temperature map.

In the analyses of slope instability phenomena mapping surface temperature can lead to the detection of thermal anomalies which can reveal the presence of potential criticalities such as: (i) structural discontinuities and open fractures; (ii) moisture or a seepage zones; (iii) ledge-niche systems Local weather data (air temperature and relative humidity) was acquired during the survey by means of a pocket thermo-hygrometer for the image correction (Table 1). IRT provides non-contact, wide area detection of subsurface defects, to be used as an alternative or complement to conventional inspection technologies in a wide variety of geological applications (Gigli et al. 2014; Frodella et al. 2017; Di Traglia et al. 2018).

Preliminary Assessment of Slope Instabilities

Lavra and Natlismcemeli Monasteries

Considering the complex morphology of the area (the Lavra monastery complex is built within the interception of two rock slabs) there is not a predominant instability mechanism (Fig. 3a, b), but depending on the local slope face orientation these are represented by plane failure, wedge failure, flexural toppling and free fall in correspondence of the overhangs.

The Natlismcemeni monastery area is affected by different landslide typologies in terms of type, size, magnitude and activities. The performed analyses allowed to assess the main instability mechanisms affecting the rock cliff, which is represented by plane failure, followed by wedge failure, and subordinately by flexural toppling and free fall (Fig. 4a). The latter can be locally predominant in overhanging sectors, such as the cave niches. In general, the rock slopes are highly prone to kinematic instability mechanisms (GKI max >60%). The thermographic analysis assessed dry conditions for the both the monasteries' rock cliffs, while warm thermal anomalies were detected in correspondence of the topmost sector sectors of the potentially cave ledge-niche systems, were free fall mechanisms are relevant (Fig. 4c, d and 5).

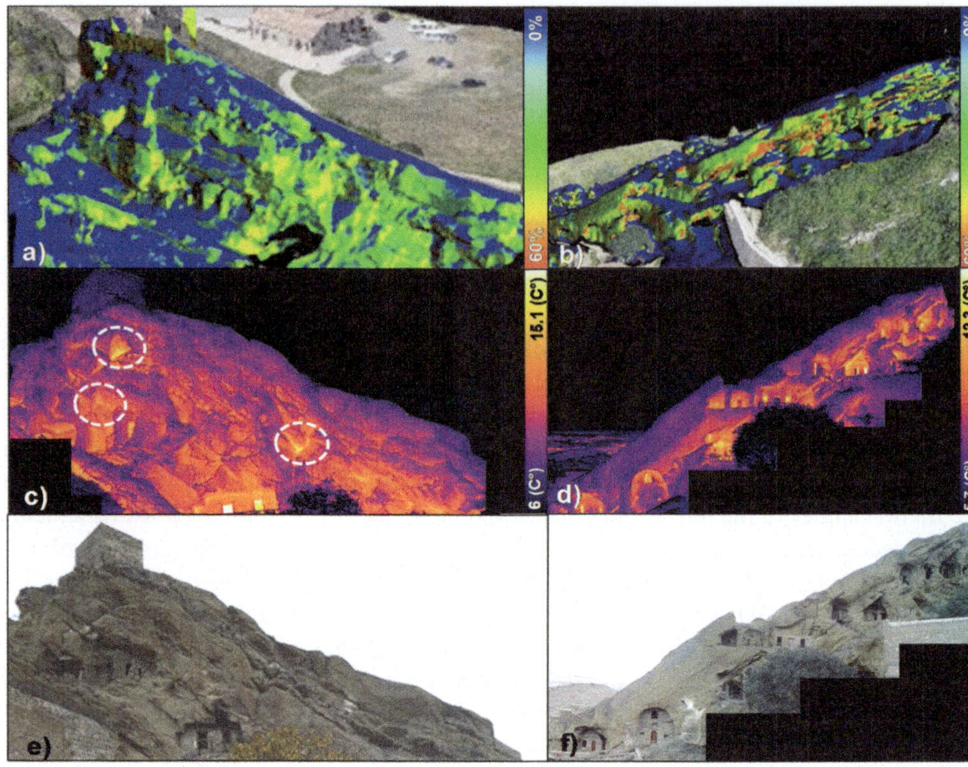

Fig. 3 Lavra monastery Global Kinematic index (GKI%) (**a**: north slab sector, **b**: east slab sector); mosaicked surface temperature maps (**c**: north slab sector, **d**: east slab sector): warm thermal anomalies (white dashed ovals) highlight ledge-niche systems; corresponding optical images (**e**, **f**)

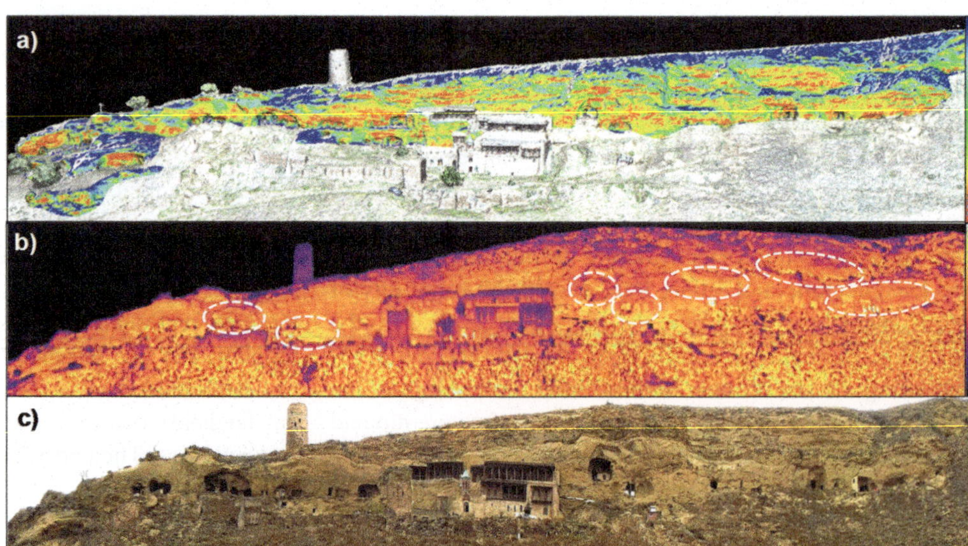

Fig. 4 Kinematic analysis—Global Kinematic index (GKI%) (**a**), mosaicked surface temperature map (white ovals enhance the warm thermal anomalies; (**b**) and corresponding optical image of Natlismcemeli (**c**)

Sabereebi and Dodo Gareji Monasteries

Regarding both monasteries a 3D surface model was not available, therefore the kynematic analysis was performed by graphical stereographic projections considering a slope mean orientation. The material in which this Sabereebi cave complex was carved is represented by a transition between a weak-sandstones and partially cemented sediments (sands and clays) (Fig. 5a, b).

For this reason, the discontinuities are not particularly evident nor frequent; nevertheless, a recent collapse of a sand block. seem to be locally deeply influenced by the structural setting. The graphical kinematic analysis confirms high plane and wedge failure indexes (Fig. 5c, d). IRT analysis showed warm thermal anomalies in correspondence of the slope face affected by the recent collapse and on the underlying slope talus sectors, where erosion is exposing bare soil (Fig. 5b, d). In Dodo Gareji monastery the main detected instability processes are wedge and plane failures, displaying a probability of occurrence of 51% and 31%, respectively (Fig. 6a–d). The infrared thermographic analysis revealed linear cold thermal anomalies in correspondence of the slope talus underlying the monastery, where to rills are eroding the accumulated loose excavation material (Fig. 6e).

Discussions and Concluding Remarks

The present paper describes recent support activities implemented during the scientific mission of June and November 2018. During the mission periods, different monasteries were

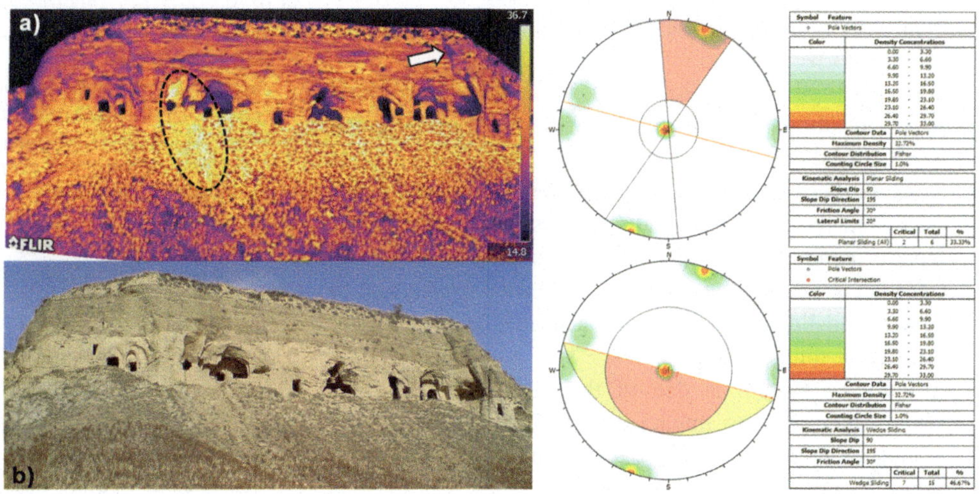

Fig. 5 Mosaicked surface temperature map of Sabereebi (black oval highlights detachment-erosion area, white arrow points out cold thermal anomaly in open fracture; **a**) and corresponding optical image (**b**); kinematic analysis (**c**: planar failure, **d**: wedge failure)

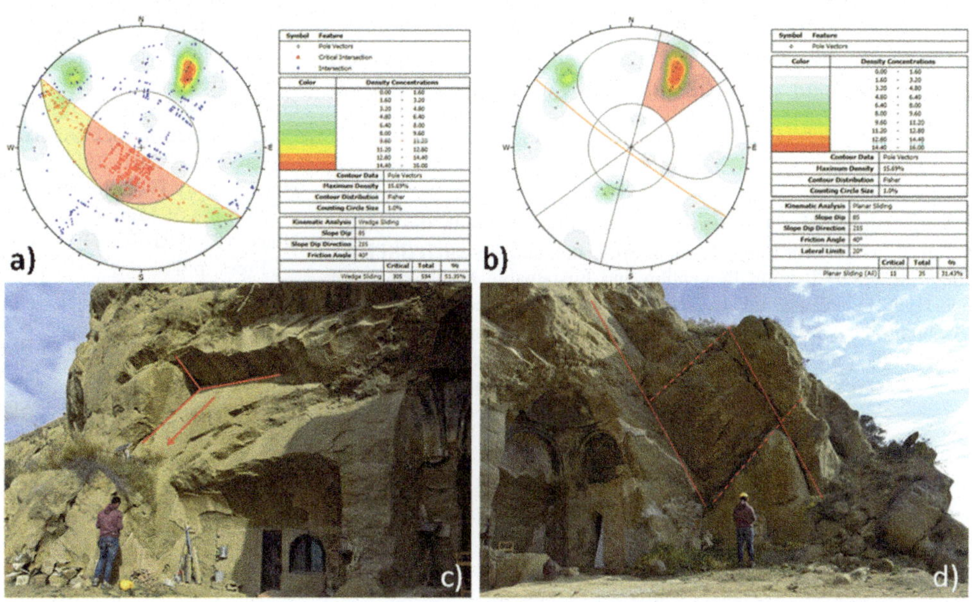

Fig. 6 Graphical kinematic analysis: Wedge failure mechanism (**a**); Planar failure mechanism (**b**). Field evidences of kinematic mechanisms: wedge failure (**c**) and plane failure (**d**)

surveyed in order to collect geomechanical parameters, verify and calibrate preliminary stability models, define preliminary mitigation measures and a monitoring system implementation.

In general, it can be assumed that there is not a predominant instability process: in fact, they can all be reconducted to rock collapses directly dependent on local structural setting differently interacting with the slope face. The performed Kinematic analysis and IRT surveys proved to be effective tools to define preliminary landslide and erosional processes affecting the sites: in fact, sectors characterized by the highest GKI index match very well with the detected thermal anomalies in enhancing potentially unstable ledge-niche systems. As preliminary conclusion, the following main instability predisposing factors were recognised:

- The David Gareja monastery complex area is constituted mainly by soft sedimentary rock creating instability processes and weathering;
- Geo-structural setting, joint and stress release promote rock instability processes in all the investigated monasteries;
- Rock samples collection and laboratory tests are in progress and they will define the main strength and deformation parameters useful for future stability models;
- Geological and geomechnical models are a useful tool to define landslide mechanism and activities as well as the priority of mitigation measures;
- Monitoring system is one of the main non-structural, sustainable and low impact mitigation measure for the management of the tourist exploitation of the sites and future sustainable exploitation polices.

  Only a multi-disciplinary approach can define a new paradigm for the conservation and mitigation measures. Understanding the instability processes is the main target in order to define general master plan of mitigation measures and most suitable and sustainable mitigation measures.

Acknowledgements The activities were carried out with the coordination of Institute for Environmental Protection and Research (ISPRA). The UNESCO Chair on Prevention and Sustainable Management of Geo-Hydrological Hazards of the University of Florence performed the Kynematic analysis, Infrared Thermographic survey and the field surveys (jointly with ISPRA), the University of Milano-Bicocca performed the TLS surveys, ILIA University the UAV-DP surveys.

References

Agisoft (2017) Agisoft Photoscan Datasheet; https://www.agisoft.com/pdf/photoscanpro_1_4_en.pdf. Last Access Dec 2019

Casagli N, Pini G (1993) Analisi cinematica della stabilità di versanti naturali e fronti di scavo in roccia. Geologia Appl e Idrogeologia 28:223–232 (in Italian)

Chandler J (1999) Effective application of automated digital photogrammetry for geomorphological research. Earth Surf Proc Land 24:51–63

Crosta GB, Lollino G, Paolo F, Giordan D, Andrea T, Carlo R, Davide B (2015) Rockslide monitoring through multi-temporal LiDAR DEM and TLS data analysis. In: Engineering geology for society and territory, vol. 2. Springer, Cham, pp 613–617

DGM ABRM (2015) CloudCompare-user's manual for version 2.6.1 https://www.danielgm.net/cc/doc/qCC/CloudCompare%20v2.6.1%20-%20User%20manual.pdf

Di Traglia F, Nolesini T, Ciampalini A, Solari L, Frodella W, Bellotti F, Fumagalli A, De Rosa G, Casagli N (2018) Tracking morphological changes and slope instability using spaceborne and ground-based SAR data. Geomorphology 300:95–112

FLIR Systems Inc. (2009) ThermaCAM SC620 technical specifications. http://www.flir.com/cs/emea/en/view/?id=41965

FLIR Systems Inc. (2015) FLIR tools+datasheet. https://www.infraredcamerawarehouse.com/content/FLIR%20Datasheets/FLIR%20ToolsPlus%20Datasheet.pdf

Frodella W, Gigli G, Morelli S, Lombardi L, Casagli N (2017) Landslide mapping and characterization through infrared thermography (IRT): suggestions for a methodological approach from some case studies. Remote Sens 9(12):1281

Gamkrelidze I, Okrostsvaridze A, Gagnidze N (2017) Field trip guide of the fourth plenary conference. In: IGCP 610 "from the caspian to mediterranean: environmental change and human response during the quaternary" (2013–2017). Fourth Plenary Conference and Field Trip, Tbilisi, Georgia, 2-9 Oct 2016

Gigli G, Casagli N (2011) Semi-automatic extraction of rock mass structural data from high resolution LIDAR point clouds. Int J Rock Mech Min Sci 48(2):187–198

Gigli G, Frodella W, Mugnai F, Tapete D, Cigna F, Fanti R, Intrieri E, Lombardi L (2012) Instability mechanisms affecting cultural heritage sites in the Maltese Archipelago. Nat Hazards Earth Syst Sci 12:1–21

Gigli G, Frodella W, Garfagnoli F, Morelli S, Mugnai F, Menna F. Casagli N (2014) 3-D geomechanical rock mass characterization for the evaluation of rockslide susceptibility scenarios. Landslides 11 (1):131–140

Goodman RE, Bray JW (1976) Toppling of rock slopes ASCE specialty conference on rock engineering for foundations and slopes. Boulder Colorado 2:201–234

Hoek E, Bray JW (1981) Rock slope engineering. Revised third edition. Institute of Mining and Metallurgy, London

Hudson JA, Harrison JP (1997) Engineering rock mechanics. Pergamon Press, p 444

Lombardi L (2007) Nuove tecnologie di rilevamento e di analisi di dati goemeccanici per la valutazione della sicurezza. Ph.D. Thesis, Università degli studi di Firenze (in Italian)

Margottini C, Vilímek V (2014) The ICL network on "landslides and cultural & natural heritage (LACUNHEN)". Landslides 11(5):933–938

Oppikofer T, Jaboyedoff M, Blikra L, Derron MH, Metzer R (2009) Characterization and monitoring of the Åknes rockslide using terrestrial laser scanning. Nat Hazards Earth Syst Sci 2009(9):1003–1019

Remondino F, Barazzetti L, Nex F, Scaioni M, Sarazzi D (2011) UAV photogrammetry for mapping and 3d modeling–current status and future perspectives. Int Arch Photogrammetry Remote Sens Spat Inf Sci 38(1):C22

Riegl (2017) www.riegl.com/uploads/tx_pxpriegldownloads/10_DataSheet_VZ-400_2017-06-14.pdf

Vv Aa (2018) David gareji monasteries and hermitages, georgia—technical report. The 7 Most Endangered 2018, Europa Nostra. Campbell Thomson EIB-Institute, Luxembourg 15 Mar 2019

Shallow Landslide Susceptibility Assessment in the High City of Antananarivo (Madagascar)

William Frodella, Daniele Spizzichino, Andrea Ciampalini, Rosi Ascanio, Claudio Margottini, and Nicola Casagli

Abstract

The urban area of Antananarivo hosts the most important built Cultural Heritage sites, recently inscribed in the UNESCO World Heritage tentative list: the High City of Antananarivo, built on top of the Analamanga hill and encompassing the Rova royal complex together with several important Churches and Cathedrals. During the first months of 2015 the city was severely affected by geo-hydrological hazards due to heavy cyclonic rain, resulting in flooding in the Ikopa river plain area and widespread shallow landslides along the Analamanga hill slopes. This event caused thousands of evacuees and casualties, showing the vulnerability of the site to geo-hydrological hazards. Field data and remote sensing data interpretation were combined in order to produce a detailed geological-geomorphological map in order to understand the processes acting in the Analamanga hill area. With the aim of analyzing the landslide-prone areas with respect to the High City Cultural Heritage and structures a shallow landslide susceptibility map was also created. The obtained maps will provide management-planning tools to be used as a first step towards a risk reduction strategy in the High City UNESCO Core Zone and the surrounding Buffer Zone.

Keywords

Geomorphological mapping • GIS • Remote sensing • Cultural heritage • Landslides

Introduction

Antananarivo is located in the central highland region (18.55′ South; 47.32′ East) of Madagascar, on top of the Analamanga hill, a 3 km long ridge (1436 m a.s.l. at its highest point) rising to about 200 m above the Ikopa River plain area (Ciampalini et al. 2019). The High City represents the first core of the city, which developed at the end of the seventeenth century from the hilltop to the hillslopes (Middle City) and reaching the river valley during the colonial period (Low city). Currently the High City represents an important cultural heritage site of Madagascar, encompassing the new stone brick-built Rova and Chapel, the high dignitaries' buildings, such as the baroque-style architecture Andafiavaratra palace, the Cathedrals of Andohalo and Ambohipotsy, built in the second half of nineteenth century (Fig. 1) (Frodella et al. 2020). For these reasons the High City is included since 2016 in the UNESCO World Heritage tentative list.

Between January and March 2015 Antananarivo was hit by the twin cyclones Bansi and Chedza, resulting in severe flooding in the river plain area, and in widespread shallow landslides along the hillslopes of the Middle Town (Frodella et al. 2020; Fig. 2). This event caused diffuse structural damage affecting several urban areas along the hillslopes and the plain area, causing 20 casualties and an estimate of 36,000 evacuees.

Following this event detailed field mapping campaigns were carried out in October 2017 and May–June 2019. The outcomes were integrated in a GIS environment with high resolution DEMs, in order to understand the geological, geomorphological and landslide features of the Analamanga

W. Frodella (✉) · R. Ascanio · C. Margottini · N. Casagli
UNESCO Chair on Prevention and Sustainable Management of Geo-Hydrological Hazards of the University of Florence, Largo Fermi 1, 50142 Florence, Italy
e-mail: william.frodella@unifi.it

D. Spizzichino
ISPRA, Via V. Brancati 48, 00144 Roma, Italy

A. Ciampalini
Earth Sciences Department, University of Pisa, Via Santa Maria, 53, 56126 Pisa, PI, Italy

K. Sassa et al. (eds.), *Understanding and Reducing Landslide Disaster Risk*, ICL Contribution to Landslide Disaster Risk Reduction,
https://doi.org/10.1007/978-3-030-60196-6_37

Fig. 1 The High City cultural heritage: **a** The Rova Palace; **b** Andafiavaratra Palace; **c** The Royal Chapel; **d** Ambohipotsy Cathedral; **e** Andohalo Cathedral

Fig. 2 Twin cyclones Bansi and Chedza hitting Madagascar East coast (**a**) (available at: www.eumetsat.int); aerial pictures of the triggered shallow landslides within the city on March 5th event (**b**, **c**) (courtesy of BNGRC; https://www.bngrc.mg)

hill area (Frodella et al. 2020). The surveys covered the area represented by the UNESCO Buffer zone, which includes the High City area (UNESCO Core zone). The final purpose was to improve the detailed geological and geo-morphological knowledge of an area characterized by limited scientific data, in order to understand the interaction between the natural landforms, the urban evolution and the acting instability processes, as a first step towards a risk management and a conservation strategy of the Antananarivo High City.

Geomorphological-Geological Features

The Analamanga hilltop is represented by a flat area rising gradually from north to south, and is delimited on the western hillside from NW to SW by a sub-vertical rock wall (47°–84° slope angle class) class) which links with the plain area with low angle surfaces. The eastern hillside has a minor slope angle (around 30°). From a geological point the view the Analamanga hill is composed by a migmatitic granitoid batholite, covered discontinuously by the various products of the progressive stages of weathering of the crystalline bedrock. These latter are formed by an irregular sequence reported here, from bottom to top (Figs. 3 and 4):

1. Granite fresh rock (=F), in which various fracture network develop, widely outcropping in the steep rock walls at the top of the western slopes, and in the south-eastern slope toe (in correspondence with the abandoned quarries) (Fig. 3a, b);
2. Laminated and slightly-moderately weathered granite (saprock = SW–MW), cropping out on top of the rock walls (Fig. 3a, b);
3. From highly to completely weathered granite (saprolite = HW–CW), discontinuously cropping out especially in the hilltop and on the eastern hillside (Fig. 3a);
4. Completely decomposed residual soils (laterite = RS), formed by mainly loamy sands, completely encircling both the of hillslope toes (Fig. 3a, c). The higher thickness of these deposits (up to a few tens of meters) is located in correspondence with the low energy surfaces at the bottom of the western slope;
5. Loose sandy eluvial/colluvial cover (=SC), formed mainly by slightly pebbly sands, irregularly outcropping especially in the hilltop and the eastern hillside (Fig. 3a, d).

Slope deposits mantle the western and eastern slopes in fan-like shapes, and are constituted by cobbles, blocks and scattered boulders, deposited by slope gravitational processes, in a coarse sandy matrix formed by the weathering of the bedrock (Fig. 3e). Furthermore, the intense anthropic activity in the last centuries has created detrital deposits composed by heterogeneous residual building materials, from fine to coarse-grained (anthropic deposits = AD; Fig. 3d). These may cover the above-mentioned sequence in correspondence with the inhabited areas. Alluvial deposits (=AL), formed by organic-rich clays and silts, mantle the river plain areas in correspondence with the rice fields.

Slope Instability Processes

The landslides recognized in the field involve the first one-two meters of bedrock alteration products, therefore are represented by shallow phenomena. Following the classification of Cruden and Varnes (1996) these are reported as follows: (1) Debris rotational/translational slides occurring in HW-CW-RS-AD; (2) Earth-debris rotational/translational slides involving RS and AD. The first type of phenomena involves the slope cover and generally occurs on the hilltop and on the south-eastern hillside; it is also characterized by small volumes and reduced mobility, usually not reaching the hydrographic network. Nevertheless, in case of cyclonic rain these can evolve in unchanneled debris flows, as in the March 2015 event. The second phenomena occur along the slopes of the main creek channels deeply cutting the slope toe at the foot of the western hillside (lavakas = gullies) and may evolve into channeled earth-debris flows following intense rainfall events during the cyclones. Their related level of risk is high, due to the interaction with man-made structures, such as buildings located directly inside the channels, which often restrict their section, and road paving often causing their culverting in the downstream sector.

Shallow Landslide Susceptibility Map

Landslide susceptibility maps (LSM) are a graphical representation of the relative probability of the occurrence of landslides in a given area, without taking into consideration the probability of occurrence in time (Brabb 1984). LSMs can be obtained in a variety of ways and a very ample literature is available on the subject, relying on at least 20 yr of

Fig. 3 Geological features of the study area: road cut showing an example of different granite weathering products (**a**); fresh granite (on the bottom) and limited granite (on the top) (**b**); laterite terrace slope cut for housing (**c**); eluvial-colluvial cover mixed with rubbish on vegetated slope (**d**); detail of heterogeneous slope deposits (**e**)

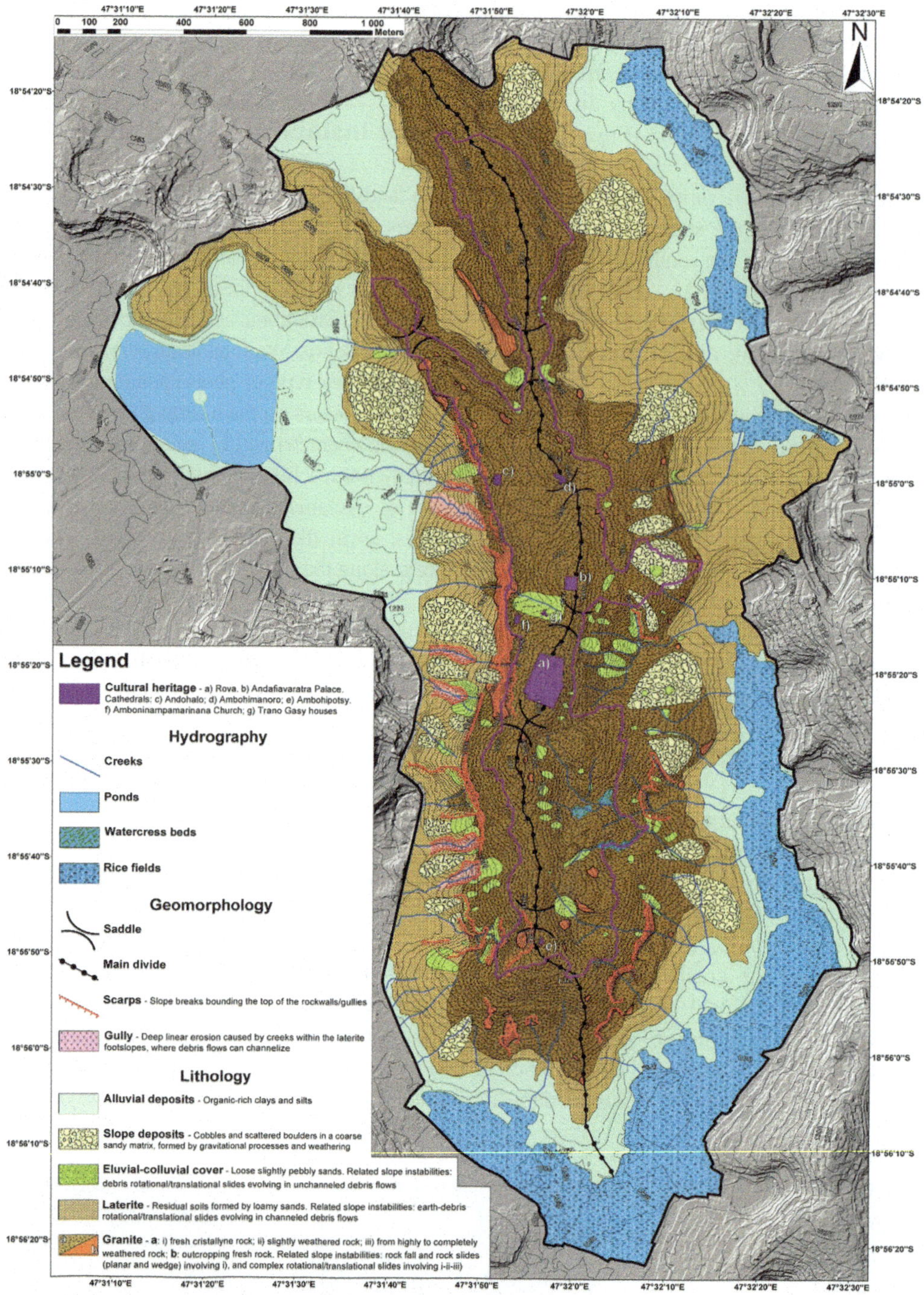

Fig. 4 Geomorphological-Geological map of the Analamanga hill area (after Frodella et al. 2020)

history of susceptibility assessment for mass movements. The first adopted method is probably the so called "heuristic mapping", carried out by a team of expert geomorphologists through the definition of a set of conditioning factors. These latter lead to landslide development in a given area on the basis of field surveys and aerial photograph interpretation, supported by ancillary map data such as geological maps (Casagli et al. 2004). On the other hand, it has the drawback of being subjective (Ardizzone et al. 2002). For these reasons many authors started to propose quantitative assessment methods, based on a set of uniquely defined conditioning factors to increase LSM reproducibility and on a series of weighting techniques to improve accuracy and robustness. To map the shallow landslide susceptibility, we used the 'Random forest', a machine-learning algorithm for non-parametric multivariate classification (Breiman 2001).

Among its advantages, the random forest technique allows the employment of both categorical and numerical variables, it accounts for interactions and nonlinearities between variables, it allows exploration of a large number of explanatory variables (as it intrinsically emphasizes only those variables of high explanatory power), and no assumption is required about the distribution of the data (Catani et al. 2013). We therefore fed the machine-learning algorithm with a large number of input parameters: flow direction, flow accumulation, topographic wetness index, elevation, profile curvature, planar curvature, slope gradient, aspect, lithology and NDVI map. Morphometric attributes were derived from a digital elevation model (DEM) with 2 m pixel size. NDVI was used to differentiate forest area from bare soil or rock and urban areas, since a land use map was not available for the country. Lithological and landslide data were retrieved from Ciampalini et al. (2019) and Frodella et al. (2020) and consist of 72 landslides. To map landslide susceptibility, we

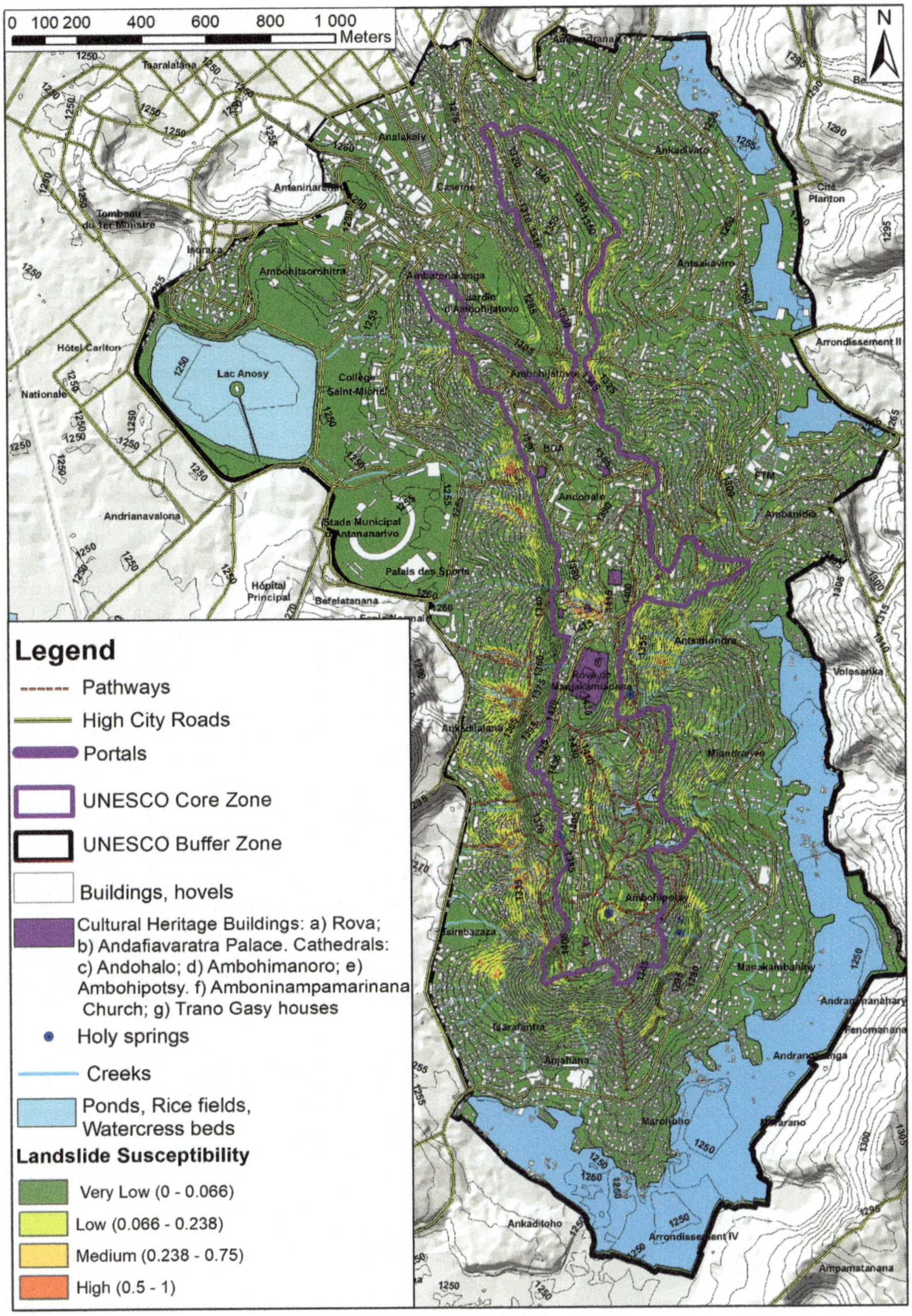

Fig. 5 Shallow landslide susceptibility map of the Analamanga Hill area

randomly sampled the study area to select 70% of the pixels for training and 30% for testing and in each data the percentages of landslide and non-landslide pixels was equally distributed (50–50%). The final output of the methodology is a raster with a 2 × 2 m cell size, where each pixel has a percentage value expressing the probability of landslide potential. According to our analysis, the susceptibility values range from 0 to 1 (Fig. 5). Cultural heritage elements such as sacred springs, portals, palaces and churches shown in Fig. 1 are also reported, together with buildings, and linear infrastructures (roads and pathways).

Concluding Remarks

The rapid urban development of Antananarivo, together with the lack of a proper urban planning, have caused several environmental problems such as intense deforestation and quarrying, unauthorized slope cutting and terracing for the construction of illegal hovels. Moreover, there is a general lack of proper drainage and sewer systems. This urban pressure, together with the geological-geomorphological features of the Analamanga hill makes the area particularly prone to shallow landslide phenomena, especially during the frequent heavy cyclonic rain events often affecting the central highlands. The steep slope sectors, the widespread weathering and fracturing of the granitoid bedrock, the abovementioned problems related to urbanization, together with the heavy cyclonic rainfall, represent natural predisposing/triggering factors for geo-hydrological hazards such as shallow landslides, as testified by the March 2015 event. The identification of landslide processes on the field was not an easy task, due to rapid modification of the landforms and slope cover caused by the intense hovel urbanization. Nevertheless, the adopted heuristic approach has provided the detection of landslide processes with a satisfactory resolution. The shallow landslides connected to the 2015 event slightly affected the UNESCO Core Zone (Figs. 4 and 5), except for its southern area, along the pathway connecting the Rova palace with Ambohipotsy Cathedral. Within the UNESCO Buffer zone, the most exposed areas are the depressed houses/hovels of the Middle City area and the mid-slope pathways, particularly the foot of the western hillslope, where shallow debris rotational/translational slides develop. Here earth-debris rotational/translational slides involving residual soil were identified along the slopes of the main creek channels deeply cutting the slope toe at the foot of the western hillside. Here intense linear creek erosion of the soft lateritic soil cover creates large gullies, which are rapidly expanding (as shown by the presence of slope instabilities in correspondence of their apex) and damaging the road pavement and buildings. The middle-southern eastern slope is less prone to shallow landslides, except for the area located in correspondence of the creek basin east of the Rova (Fig. 5). The slope north hill sector shows a general stability due to the widespread urban cover and the low steepness of the slope, while the southern sector shows some prone areas at the top of the abandoned quarry niches (Figs. 4 and 5). The produced maps represent a first step for land-use planning and represent a contribution to a general master plan of mitigation measures for the High City and its cultural heritage protection from landslide hazard.

Acknowledgements The authors would like to thank: Paris Region Expertise-Madagascar (PRX)and RC Heritage/RCh consultants for providing topographic data, logistic support and suggestions.

References

Ardizzone F, Cardinali M, Carrara A, Guzzetti F, Reichenbach P (2002) Impact of mapping errors on the reliability of landslide hazard maps. Nat Hazards Earth Syst Sci 2:3–14

Brabb EE (1984) Innovative approaches to landslide hazard mapping. In: Proceedings 4th international symposium on landslides. vol 1, Toronto, pp 307–324

Breiman L (2001) Random forests. Mach Learn 45:5–32

Casagli N, Catani F, Puglisi C, Delmonaco G, Ermini L, Margottini C (2004) An inventory-based approach to landslide susceptibility assessment and its application to Virginio river basin, Italy. Environ Eng Geosci 10:203–216

Catani F, Lagomarsino D, Segoni S, Tofani V (2013) Landslide susceptibility estimation by random forests technique: sensitivity and scaling issues. Nat Hazards Earth Syst Sci 13:2815–2831

Ciampalini A, Frodella W, Margottini C, Casagli N (2019) Rapid assessment of geo-hydrological hazards in Antananarivo (Madagascar) historical centre for damage prevention. Geomatics, Nat Hazards Risk 10(1):1102–1124

Cruden DM, Varnes DJ (1996) Landslides: investigation and mitigation. Chapter 3-Landslide types and processes. Trans Res Board Spec Rep (247)

Frodella W, Spizzichino D, Ciampalini A, Claudio, Margottini C, Casagli N (2020) Hydrography and geomorphology of antananarivo high city (Madagascar). J Maps 1–12

Thermo-Mechanical Cliff Stability at Tomb KV42 in the Valley of the Kings, Egypt

2020 Kyoto Japan

Rodrigo Alcaíno-Olivares, Matthew A. Perras, Martin Ziegler, and Kerry Leith

Abstract

The Valley of the Kings (or Kings' Valley, KV) in Egypt is surrounded by tall, subvertical cliffs of marl and limestones which are close to archaeological sites and visitor pathways. A previous analysis of the slope above the tomb KV42 entrance suggested that vertical rock discontinuities (i.e., tectonic joints and a lateral tension cracks) have a large impact on the stability of the cliff. The site at KV42 is being monitored with a weather station, a crack meter and a seismometer. Preliminary data analysis of temperature cycles suggests a correlation with cyclic crack aperture changes of a prominent tension crack in the cliff above KV42. Weather data from April 2018 to present helped in understanding the environmental conditions of the site. During a field campaign in April 2019, an infrared sensor was installed to systematically capture thermal infrared images of the slope above KV42. The thermal response of the rock slope measured by this sensor was used in correlation with the environmental conditions and the geomechanical response of tension crack. The in-situ data were used in this study to develop a thermo-mechanical numerical model in FLAC®, by inputting the thermal boundary conditions and geometry of the rock column, and to differentiate gravitational and thermally induced stress acting in the rock mass. Results showed that peak total displacements values are delayed with respect to peak values of rock temperatures, similar to the observations made in real measurements. The modelled rock temperatures are higher than the measured data by 3% in warm months and by 68% in cooler periods. The total displacement trend is similar to the measured data, however, the model underestimates the peak displacement by 0.2 mm, which is 34% lower than the measured data. The research forms the basis of an approach to incorporate weather conditions into long-term stability modelling of rock masses.

Keywords

Environmental monitoring • Thermo-mechanical modelling • Infrared thermographic survey

Introduction

Coupled thermo-mechanical (TM) models have been used to quantify the influence impact of thermally induced stress in rock masses such as nuclear waste disposal or geothermal systems (Huang et al. 2017; Eggerston et al. 2018). The lack of direct observations under natural conditions makes it difficult to decouple the influence of fluctuating climatic variables on near-surface rock mass damage evolution and behaviour, which is an emerging subject of interest to the scientific community (Bakun-Mazor et al. 2013; Collins and Stock 2016; Collins et al. 2018; Alcaíno-Olivares et al. 2018; Guerin et al. 2019). Decoupling the individual processes leading to fracturing events and understanding the driving mechanisms at both the laboratory and field scales requires further research.

In this context, recent measurements from the Valley of the Kings, or Kings' Valley (KV), Luxor, Egypt, indicate a

R. Alcaíno-Olivares (✉)
Department of Civil Engineering, York University, 4700 Keele St, Toronto, M3J 1PR, Canada
e-mail: ralcaino@yorku.ca

M. A. Perras
Department of Civil Engineering, York University, Toronto, M3J 1PR, Canada
e-mail: mperras@yorku.ca

M. Ziegler
Department of Earth Science, Swiss Federal Institute of Technology Zurich—ETH, 8006 Zurich, Switzerland
e-mail: martin.ziegler@erdw.ethz.ch

K. Leith
Swiss Federal Institute of Technology Zurich—ETH, Zurich, Switzerland
e-mail: kerry.leith@erdw.ethz.ch

K. Sassa et al. (eds.), *Understanding and Reducing Landslide Disaster Risk*, ICL Contribution to Landslide Disaster Risk Reduction,
https://doi.org/10.1007/978-3-030-60196-6_38

potential mechanism of elastic opening and closing cycles of cracks, which was previously introduced in Alcaíno-Olivares et al. (2019). The KV contains 64 tombs of ancient Egyptian pharaohs, excavated into the base of or close to steep cliffs of the valley (Fig. 1a). The marlstone cliffs are exposed to sun radiation and daily temperature fluctuations, between 25 °C and 50 °C in summer and between 5 °C and 20 °C in winter. A rock slabs forming above the tomb 42 (KV42) on the cliff face are being monitored to record the response of the rock mass displacements and the local weather conditions (Fig. 1b). Preliminary data suggest that cyclic variations of temperature and relative humidity lead to opening and closing of the lateral tension crack measured by a crack meter placed in between the cliff and the rock column (Alcaíno-Olivares et al. 2019). This implies that cyclic fatigue from thermal strain could be leading to a state where crack growth is possible, although it has not been observed directly to date at the site.

Previous numerical modelling, which focused on the mechanical influence of rock mass structures on the column's stability, suggested that tectonic and gravitational fractures at the site have a large effect on the stability of the area (Alcaíno-Olivares et al. 2019; Maissen 2019). For this reason, the aim of this investigation is to scrutinize the thermal processes in the rock column to differentiate between the thermally induced stresses and gravitational stresses driving the deformation of the rock slabs above KV42. Therefore, a field campaign was carried out to install an infrared thermographic (IRT) sensor to monitor the rock column (Fig. 1c). This sensor complements the existing system that records the environmental parameters at the site and the rock mass response to these fluctuations. The field campaign and numerical simulation are detailed in this paper.

Site Investigation

Infrared Thermographic (IRT) Surveying

An IRT system was installed at the KV42 site to capture thermal contrasts of the rock's surface due to rock mass emittance of infrared waves, and then to determine the changes and distribution of rock temperatures within and around the rock column at shallow depths (i.e. less than 30 cm). Of particular interest are the effects of direct sunlight exposure and temperature fluctuations caused by radiation changes, relative humidity and wind. The conceptual framework of these influences are sketched in Fig. 2a. The IRT survey does not damage or alter the rock surface following the inherent work restrictions in KV (which ensures preservation of the archaeological site). Similar surveys have been carried out in different geological environments to capture the temperature profiles of a wide range of features, from geothermal studies to landslide monitoring (Chen 2017; Frodella et al. 2017; Coggan et al. 2016), rock mass (Fiorucci et al. 2018) and integrity assessment of ancient buildings (Gil et al. 2017).

The mounted IRT station was added to the existing power supply and communication systems and consists of an infrared camera (FLIR Lepton, Longwave Infrared—LWIR) and a microcontroller (Raspberry-PI). The LWIR camera

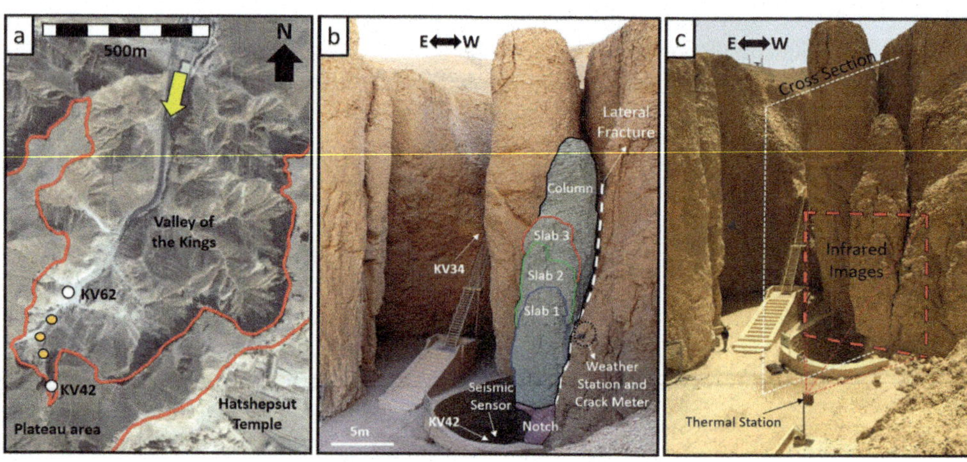

Fig. 1 Overview of the area around KV42. a: Satellite image of the Valley of the Kings. The red lines indicate the crest of the cliffs (Google Earth 2015). The yellow arrow points to the main entrance to the archaeological sites. Yellow dots mark the pedestrian path between KV62 and KV42 b: Photograph of the cliffs surrounding KV42 with view towards South. The rock column under investigation has been shaded in green. At its foot a notch occurs (purple). The lateral fracture (white dashed line) and the individual slabs (solid coloured lines) are highlighted. c: The image shows the location of the infrared sensor and the maximum extension of the area being captured (red dashed line). The white dashed frame indicates the location of the geological cross section that was used for the numerical model in this study

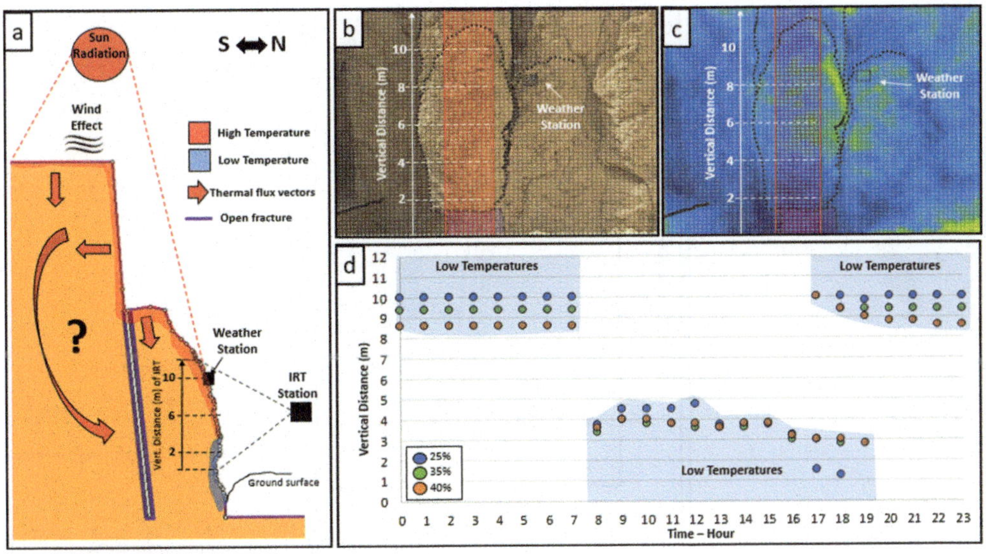

Fig. 2 Infrared thermographic (IRT) surveying above KV42. **a** A south-north sketch (seen in Fig. 1c) to show different features of the rock mass and equipment at the site. The purple line shows a prominent open fracture between the back of the rock column and the cliff. **b** Photograph taken from the IRT station. Pixels have been added for reference purposes. **c** Thermal image of the rock mass with pixels to show IRT's coverage. **d** Graph with typical distribution of the lowest temperature regions across a full day for the rock mass. The vertical axis is the distance indicated in Fig. 2b. 20 pm to 6 am) the lowest temperatures are concentrated on the top of the measured column

captures the thermal response of the rock mass over a wavelength range of 8 to 14 μm (common thermal infrared spectrum) with a thermal sensitivity lower than 0.05 °C, to deliver a raw value associated with the temperature of each pixel. The LWIR image has 80 (horizontal) by 60 (vertical) active pixels. The IRT system was installed in a sealed box within a wooden case, with the focus centred on the rock slab above KV42 (Fig. 2b), at a mean distance of about 11 m from the rock, which implies that each pixel has a dimension of approximately 0.2 × 0.2 m (Fig. 2b).

The portion of the rock mass contained within the limits of the outlined pixels was used to obtain an average thermal behaviour of the rock column (Fig. 2c, red squares). The average raw value per row of pixels within this red enclosure limit (60 rows) were calculated hourly to determine the vertical position of the hottest pixel at each hour, from an imagery acquired every 15 min. Then, other pixels with percentages of 25, 35 and 40 of the maximum temperature were allocated vertically to define the fluctuation of the cool and warm regions of the column. This is summarized in Fig. 2d. Images were acquired from the 20th April to 20th August, 2019. The results show that the bottom 4 m of the column remains cool during daytime (8am–19 pm), while it becomes the warmest region (comparatively) during night times (20–7am). These findings were used to determine different thermal regions along the rock column, by using the rock surface temperature information from IRT survey data.

Characterization of Local Weather Conditions and Heat Fluxes

The local weather conditions in the KV have been gathered by recording data from a weather station (WS) placed above the KV42 (WSKV42). The data have been compared with data from another weather station located at the top of the tomb KV62 (Tutankhamun's tomb), owned by Helio Scientific Egypt (WSKV62). The locations of these stations are given in Fig. 1a. Figure 3a shows average hourly values of air temperature (AT), wind speed (WS) and solar radiation (SR) for the summer and winter seasons at both stations. Data show AT average values are usually 20 °C higher in the summer months (June to August) wit an average AT of 35 °C compared to the winter months (December to February) with an average of 15 °C. AT average values for both WSKV42 and WSKV62 are similar along the measured months. Differences between SR for WSKV42 and WSKV62 are the direct result of topographic differences between the sites, where WSKV42 has less sun exposure due to the vicinity of surrounding cliffs. WS values are different during the day and the night, which is related to the wind direction. During the day the wind blows from the entrance of the KV to the plateau, whilst during the night the wind direction usually changes to blow from the plateau to the entrance (Fig. 1a). It was also found that speeds at KV42 are usually lower than at KV62 during the day, potentially

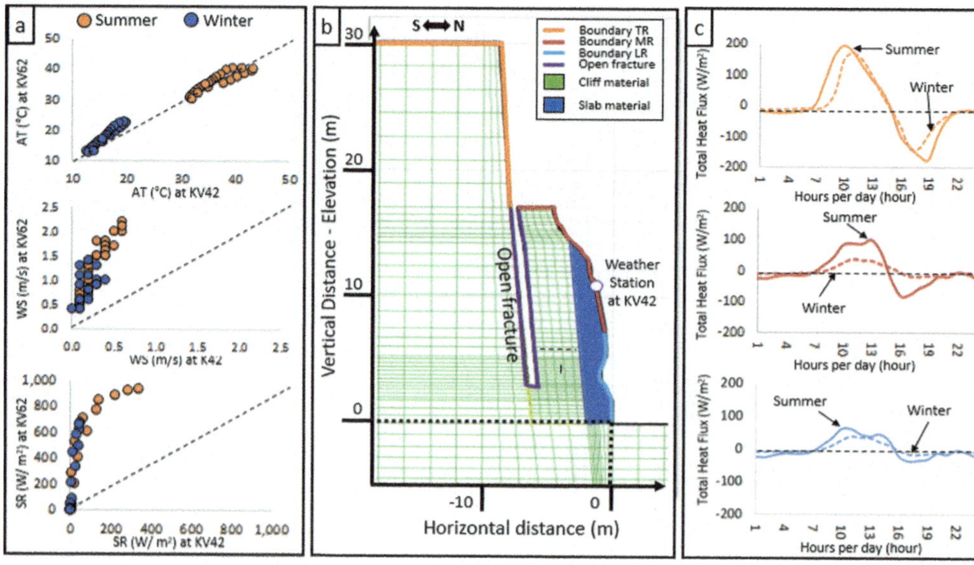

Fig. 3 Definitions of thermal response of the rock mass under environmental conditions. **a** AT, WS and SR measured at KV42 and KV62 for summer and winter months. **b** cross section of numerical model and boundary conditions. TR: top region. MR: middle region. LR: low region. c: examples of heat fluxes for winter (dashed lines) and summer (solid lines) months were determined using the approach of Bentz (2000). Legend from Fig. 3b applies to curves in Fig. 3c

due to losses along the pedestrian path (yellow dots in Fig. 1a) where the wind deviates around hills and divides along different paths locally between KV62 and KV42.

Based on this environmental characterization and the results from the IRT survey, three regions have been defined in terms of the daily rock mass thermal response (Fig. 3b). The top region (TR, orange line) is fully irradiated by the sun and defined in the model based on the SR measured at KV62, due to a very low influence of the local topography. A middle region (MR, red line) is essentially influenced by the local morphology and is irradiated for a few hours of the day only. Finally, a lower region (LR, blue line) is the coldest part of the rock mass and its height was determined from the IRT analysis. MR and LR are also affected by local changes in KV42 for WS and wind direction (WD) during the day, whilst TR is an open surface exposed to global variations in KV for WS and WD.

A thermal balance calculation was conducted to estimate the heat fluxes acting at the boundaries of the TR, MR, and LR, using the approach of Bentz (2000). The radiant flux from the rock mass to the atmosphere was calculated using empirical relationships (Algarni and Nutter 2015). Fluxes were calculated by using the average value per hour per month for weather conditions recorded in the KV. Thirty-six hourly curves (12 per region) were estimated, however, only two curves are plotted in Fig. 3c (summer and winter) to show the contrast in fluxes throughout the day for the outer boundary of the rock mass. During the night the heat flux conditions were all calculated from a thermal balance with no input radiation. These fluxes were used to simulate the thermo-mechanical response of the cliff across different seasons in FLAC®.

Numerical Model

Setup

A two-dimensional numerical finite difference model was developed using Itasca's FLAC® software. The model aimed to replicate the mechanical displacements using thermal fluctuation inputs and the geometry from Fig. 3b. Geotechnical and thermal input properties are listed in Table 1. External forces from seismic events and the influence of water on the rock mass behaviour was not incorporated in the simulations. The model's ground level (y = 0) was set at the toe of the slope (x = 0). The level of the debris was not incorporated in the model as it has been shown to not have a large influence on the stability of the column (Maissen 2019). The southern and northern boundaries are 40 and 20 m away from the origin respectively, and both can move freely in the y-axis only. The bottom boundary is 20 m below the origin, and it has been restricted from both lateral (x-axis) and vertical (y-axis) movement.

The back fracture was modelled as an open area between the back slope and the column. The mesh was refined in the column region to give higher resolution. The slab fractures were not included at this stage of the simulation, although their elastic properties were set differently than the column and cliffs as calculated by Maissen (2019), indicated in

Table 1 Geotechnical and thermal properties of the intact rock and rock mass for the KV42 model input. Rock Mass serefers to cliffs and column

Property	Intact rock	Slabs	Rock Mass	References
Density (kg/m^3)	2100	–	–	Maissen (2019)
Young's modulus (GPa)	8.0	4.2	7.0	Maissen (2019)
Poisson's ratio	0.21	–	–	Maissen (2019)
Thermal Cond. (W/m/°K)	2.0	–	–	Davis et al. (2007)
Heat capacity (J/kg/°K)	900	–	–	Kappelmeyer and Haenel (1974)
Thermal expansion (1/°K)	8.1×10^{-6}	–	–	Harvey (1967)

Table 1. Thermal boundary conditions were applied to the different regions of the model, by incorporating heat fluxes from Fig. 3c as monthly curves. The rock mass was modelled as thermally isotropic with an initial rock temperature (RT) of 25 °C. The simulation time started on the 1st of April 2017 and lasted for 3 years. The null displacement value has been set to April 2018, and all total displacements (TD, total length of the resulting 2D displacement vectors) values are relative to the one reached at that simulation time. Therefore, solutions from April 2017 to March 2018 are meant to prepare conditions for analysis of the results after April 2018.

Preliminary results

Simulated results for RT and TD are given in Fig. 4. The mechanical elastic solution is shown in Fig. 4a, with a maximum TD of 3 mm. After adding thermally induced stress, the maximum RT was observed at the top vertex of the cliff (x = -10 m, y = 30 m). The rock column showed lower RTs than the cliff. The RT contours showed that the low temperature acquired by the bottom region of the column generates a gradient of heat flow towards the base of the column, therefore the thermo-mechanical solution (Fig. 4c) showed larger increments of TD at the cliff's near surface (slope face) and the upper part of the column, with a TD up to 4.0 mm. The open fracture at the back of the column was not set as an active thermal boundary. Then it combines thermal processes from the front part of the column to provoke inclined isothermal curves dipping to the right side of the column (North). As it can be seen in Fig. 4b, at the same horizontal level within the rock column, we can have a gradient of temperature from 30 °C to 40 °C at a distance of 5 m (thickness of the column at its central height). This can be observed in the time series of RT fluctuations at points close to the open fracture.

In fact, survey transects (S1, S2, and S3) and survey points (P1, P2, and P3) were taken to illustrate the time-dependant fluctuations of RT and TD conditions of the rock mass (Fig. 4b, c). The time series data for S1, S2, and S3 are plotted in Fig. 5 from April 2018 to October 2019, which matches with the timeline that the system has been recording in-situ data. The fluctuation of RT across the year reflects the cyclic behaviour of the boundary conditions not only per hour as it was shown in Fig. 3c, but also in terms of warmer and cooler months as it is seen in Fig. 5. It is relevant to note that all surveys show a remarkable delay in the change of RT at different depths. For instance, peak values of the RT for S2 are obtained in June–July for the shallow layers (at surface), whilst at 6 m depth the peak temperature is reached in September–October.

In the same manner, the RTs at depth fluctuate less than the surface RTs (blue and orange lines). While the RT can vary by 35 °C between summer and winter at the surface, at 6 m below ground the variation is on the order of 10 °C. Points located at 20 and 30 m only show a small increase over time (purple and green line). However, for those points surveyed at the rock column, the fluctuations extend down to 13 m at least due to the boundary conditions at the side of the column and its narrow geometry (5 m thick). These climatic-dependent variations also influence the displacements throughout the year.

Trends of the TD show cumulative displacement increases throughout the year, particularly for the rock surface with a TD of 1.5 mm at S1 and 2.5 mm at S2 (presented in Fig. 5 by the blue dashed lines). Nevertheless, the increase in TD is smaller at greater depth. This is in contrast to the column behaviour, which shows cyclic trends of displacements with values of TD up to 1.4 mm, although returning to null displacement after a yearly cycle (i.e. from April to April). Similar to the fluctuation of temperatures, the displacements vary less with depth along the column at least to 13 m underneath the top of the column. The model indicates a variation of Δ0.5 mm between the cold and warm period at that depth.

Analysis

Results from nodal queries are plotted in Fig. 6 for the average values of both RT and TD with time. For easier comparison, we added the measured data of RT at KV42 and TD of the lateral fracture which strikes N-S (see Fig. 1b,

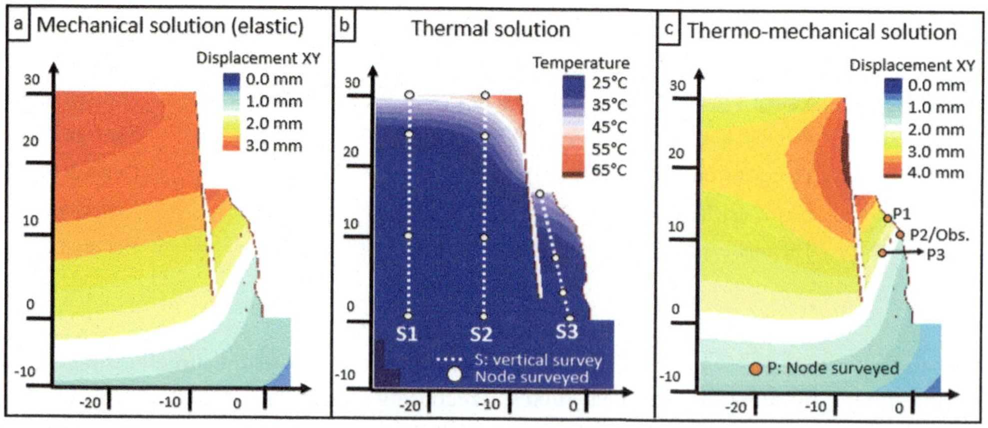

Fig. 4 Contoured numerical model results, showing **a** the mechanical elastic solution prior to applying the thermal loading, **b** the rock temperature variation within the rock mass, and **c** the total displacement for the thermo-mechanical solution for May 2018. The white dots in Fig. 5b are queried nodes to build the vertical surveys. Orange dots in Fig. 5c are queried nodes close to in-situ measured points. P1 locates on the surface at the top of slab 1 (see Fig. 1b). P2 is at the same elevation as WSKV42 (see Fig. 2a). P3 is in the middle vertical point of the column behind slab 3, at 7 m depth from the top of the rock column and along the vertical survey S3

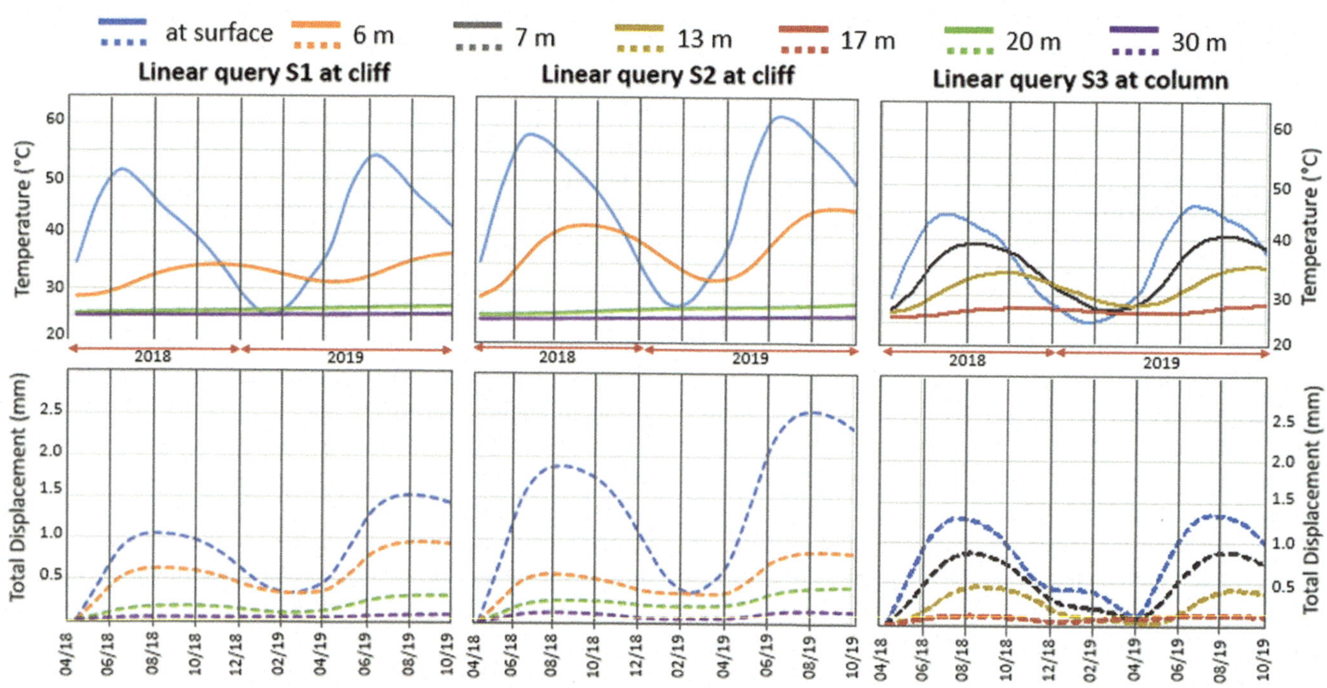

Fig. 5 Numerical results of temperature and displacement for 3 survey transects (see Fig. 4b). The colour of the lines indicates the depth of the point from the top part of the survey line. Note that S3 has different depths than S1 and S2

measured displacement is perpendicular to the modelled section). The data recorded contained some data gaps between September and December 2018, which were filled in by correlating existing trends of RT in KV42 with weather values from KV62. The estimated values in these data gaps are indicated by dashed black lines. Details of instrumentation and data from the site can be found in Alcaíno-Olivares et al. (2019).

Regarding trends of the RT, it is worth mentioning that P3 is the only point located in the inner section of the column, and therefore shows a delay in the heating/cooling process (see preliminary results) as well as the accumulated TD from the upper layers, which would explain its peak of TD during August–September, whilst the RT still increases. In general, the modelled RTs from P1 and P2 match the trend and magnitude of the measured records at KV42 for the summer, but the model does not capture the low temperatures in the winter (observed data are 68% lower than modelled data). During the colder months, the simulation is not able to dissipate the heat properly to cool the rock mass

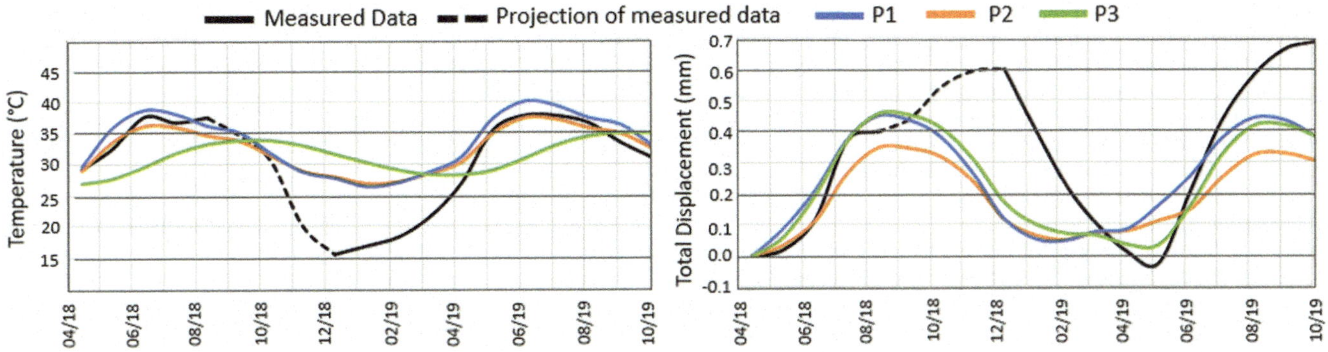

Fig. 6 Comparison between measured and simulated RT and TD results obtained from three points located close to WS (see Fig. 4c)

down to 15–25 °C as it has been recorded at the site. This may be explained due to (1) the initial temperature of the rock mass, which may hinder RT decreases below 25 °C, (2) an insufficient model volume to dissipate the heat, (3) sensor errors to measure RT, which may be influenced by the air temperature, as the sensor is placed within the fracture behind slab 1 (see Fig. 1b), and/or (4) air circulation within the rock mass fracture network, which would locally change the RT.

Modelled TD magnitudes are smaller than the measured TD for the warmest (−18%) and coolest period (−70%). This could possibly be caused by the different measurement direction (~W–E; out of plane) compared to the model section (~N–S) used and the differences in RT for the winter season. In order to numerically model the correct orientation of the measurement direction a 3D model will be developed for further stages in this investigation. Nevertheless, the measured trend and response to temperature cycles indicated a delay between peak RT and peak TD, which were captured in the 2D simulation. In detail, positive peak RT values occur during June–August, whilst maximum TD peaks were obtained for September–November for P1, P2, and the measured data. This out-of-phase behaviour has been also reported in other geological environments, e.g., by Collins et al. (2016) and Alcaíno-Olivares et al. (2018).

Conclusions

A methodology to analyse the thermo-mechanical behaviour of a rock slope has been designed for this work and implemented numerically in FLAC®. The impact of using local environmental values (from local weather stations) and the visualization of the rock mass surface temperatures (from an IRT system) helped to develop the model boundary conditions and heat fluxes and the mechanical response of the rock mass. Results showed that simulated values of RT are 3% lower in summer and 68% higher in winter than measured data. Furthermore, the thermo-mechanical model reached annually 34% of the observed displacements. Some environmental elements have not been considered at this stage of the research. For instance, the thermal effect of the open fracture and of minor fractures and their exposure to wind and fluctuations in air temperature, rainfall effects, or the presence of swellable clay minerals. In addition, the built model has a simplified geometry (i.e. a smaller surface area for heat addition and dissipation then in reality) which might lead to larger fluctuations of TD. Future research will consider these three-dimensional effects, refine heat flux functions along with the addition of structural elements to represent the slab fractures. Finally, rock plastic behaviour will be implemented to quantify the thermo-mechanical impact on the rock mass stability under environmental influences.

Acknowledgements The authors would like to express their gratitude to the University of Basel and the King's Valley and Life Histories of Theban Tombs projects (SNF grant number 162967), We particularly thank Susanne Bickle, Andrea Loprieno-Gneirs and Elina Paulin-Grothe for their great support and allowing us to join their team, and the Permanent Committee of the Supreme Council of Antiquities in Cairo for permission to carry out the works at the site. We also acknowledge Helio Scientific Egypt for allowing us to use their environmental data captured above the tomb KV62. Funding has been provided by the National Science and Engineering Research Council of Canada (NSERC) through the Discovery Grant program (grant reference RGPIN-2018-05918).

References

Alcaíno-Olivares RD, Perras MA, Ziegler M, Maissen J (2019) Cliff stability at tomb KV42 in the Valley of the Kings, Egypt: A first approach to numerical modelling and site investigation. In: 53rd ARMA Symposium. New York, United States. June 2019

Alcaíno-Olivares R, Hayden D, Perras MA, Leith K, Loew S (2018) Fracture growth on långören Island, Finland, on the hottest day on record in 2014. In: 71st Canadian geotechnical conference. edmonton, Canada. Sept 2018

Algarni S, Nutter D (2015) Survey of sky effective temperature models applicable to building envelope radiant heat transfer. ASHRAE Trans 121(2):351–364

Bakun-Mazor D, Hatzor YH, Glaser SD, Santamarina JC (2013) Thermally versus seismically induced block displacements in Masada rock slopes. Int J Rock Mech Min Sci 61:196–211

Bentz DP (2000) A computer model to predict the surface temperature and time-of-wetness of concrete pavements and bridge decks. No. 6551. NIST Interagency. United States.

Chen C-Y (2017) Applications of thermal images for monitoring surficial temperature changes of naked slope. Proc Eng Technol Innovation 5:1–37

Collins BD, Stock GM, Eppes MC, Lexiw SW, Corbett SC, Smith JB (2018) Thermal influences on spontaneous rock dome exfoliation. Nat Commun 9:1–11

Collins BD, Stock GM (2016) Rockfall triggering by cyclic thermal stressing of exfoliation fractures. Nat Commun 9:395–400

Coggan JS, Pascoe DM, Eyre ML, Howe JH (2016) Integrated use of terrestrial laser scanning and thermal imagery for characterization of hydrothermally altered granites. Aust Geomech Soc Sydney Aust 5:855–860

Davis MG, Chapman DS, Van Wagoner TM, Armstrong PA (2007) Thermal conductivity anisotropy of metasedimentary and igneous rocks. J Geoph Res Solid Earth 112(B5):1–7

Eggertsson GH, Lavallée Y, Kendrick JE, Markússon S (2018) Impact of thermo-mechanical stimulation on the reservoir rocks of the geothermal system at Krafla, Iceland. In: 42nd Workshop on Geot Res Engineering. Standford University, California. Feb 2017

Fiorucci M, Marmoni G, Martino S, Mazzanti P (2018) Thermal response of jointed rock masses inferred from infrared thermographic surveying (Acuto, Italy). Sensors 18(7):1–25

Frodella W, Gigli G, Morelli S, Lombardi L, Casagli N (2017) Landslide mapping and characterization through Infrared Thermography (IRT): suggestions for a methodological approach from some case studies. Remote Sens 9(12):1–25

Gil E, Lerma C, Vercher J, Mas Á (2017) methodology for thermal behaviour assessment of homogeneous Façades in heritage buildings. J Sens 2017:1–13

Guerin A, Jaboyedoff M, Collins BD, Derron MH, Stock GM, Matasci B, Boesiger M, Lefeuvre C, Podladchikov YY (2019) Detection of rock bridges by infrared thermal imaging and modelling. Sci Rep Nat Res 9(1):1–19

Harvey RD (1967) Thermal expansion of certain Illinois limestones and dolomites. No. 415. Illinois State Geological Survey. United States

Huang X, Tang CA, Tang SB, Xiw LM (2017) Numerical test of thermal cracking behaviour in a nuclear waste disposal. In: 4th ISRM young scholars symposium on rock mechanics, international society for rock mechanics and rock engineering. Jeju, South Korea. May 2017

Kappelmeyer O, Haenel R (1974) Geothermic with special reference to application. Berlin Gebrueder Borntraeger Geoexploration Monographs Series, 4

Maissen J (2019) The fractured rock cliff above KV42, Valley of the Kings, Egypt. MS thesis, Swiss Federal Institute of Technology Zurich, Switzerland

Collaboration in MHEWS Through an Integrated Way

The Great Efforts Contributed by Multi-stakeholder Partnership at National, Regional and International Levels

Xu Tang, Kyoji Sassa, Guy P. Brasseur, Johannes Cullmann, and Zheqing Fang

Abstract

In recent years, great progress has been made in the development of knowledge and practices related to the multi-hazard early warnings and in strengthening the related multi-stakeholder partnership. Global initiatives are gaining momentum to improve multi-hazard early warning systems (MHEWS) and so boost the resilience of the most vulnerable countries to extreme weather/water / geological, environment and health-related events and its impacts to the sustainability of social and economic development. However, understanding its interconnectivity and interoperability and developing impact-based and risk-informed methodologies in its integration in the all relevant hazard aspects are still needed to be strengthened. This article presents an overview of advances and challenges in multi-stakeholder partnership for improving MHEWS in an integrated way in order to better achieve the Target-G of the Sendai Framework on DRR, Paris Agreement on Climate Change and SDGs. We focus on major international cooperation and partnerships on MHEWS and its applications, but not limited to, the International Multi-Hazard Early Warning System Network (IN-MHEWS), Climate Risk and Early Warning Systems (CREWS), Regional Integrated MHEWS in Africa and Asia (RIMES); and many specific MHEWSs interfaces, such as Global Disaster Alerting Coordination System (GDACS), Global Multi-hazard Alert System (GMAS), MeteoAlarm/Alert Systems, All Risk Integrated System TOwards Trans-boundary hoListic Early-warning—European Natural Hazards Scientific Partnership (ARISTOTLE-ENHSP), etc. as well as relevant multi-stakeholder partnership platforms to connect related MHEWS with its application networks between international organizations, governments, non-governmental organizations (NGOs), private sectors at regional and national levels, such as United Nations Disaster Assessment and Coordination (UNDAC), Global Water Partnership(GWP), Environment and Humanitarian Action (EHA) Network, Global network on Monitoring, Analysis, and Prediction of Air Quality (MAP-AQ), Public Health Emergency Operations Centre Network (EOC-NET), Humanitarian Networks and Partnerships Week (HNPW), and Forecast-based Financing mechanism and programme (FbF) etc. The further actions to facilitate UN Member States for improving its capacity in MHEWS partnership at national and local levels through strengthening multi-stakeholder partnership at international and regional levels are discussed and recommended in the conclusion and discussion.

Keywords

Impact-based forecasting and integrated risk management • Interactivity • Interoperability • Multi-hazard early warning systems •

X. Tang
Department of Atmospheric and Oceanic Sciences, Fudan University, 200433 Shanghai, China
e-mail: xutang2016@gmail.com

ISC/UNDRR Integrated Research on Disaster Risk, International Programme Office, 100094 Beijing, China

K. Sassa (✉)
International Consortium on Landslides, Kyoto, 606-8226, Japan
e-mail: sassa@iclhq.org

G. P. Brasseur
Max Planck Institute for Meteorology in Hamburg, 20146 Hamburg, Germany
e-mail: guy.brasseur@mpimet.mpg.de

National Center for Atmospheric Research in Boulder, PO Box 3000, Boulder, CO 80307-3000, USA

J. Cullmann
WMO Water, Ice and Snow Coordination, World Meteorological Organization, CH-1211 Geneva, Switzerland
e-mail: jcullmann@wmo.int

Z. Fang
School of Atmospheric Sciences, Nanjing University, 210023 Nanjing, China
e-mail: fangzheq@163.com

K. Sassa et al. (eds.), *Understanding and Reducing Landslide Disaster Risk*, ICL Contribution to Landslide Disaster Risk Reduction,
https://doi.org/10.1007/978-3-030-60196-6_39

Multi-stakeholder partnerships governance • Risk interconnectivity • Public and private partnership (PPP) • And public and public partnership (PUP)

Introduction

What is new in today's increasingly interconnected society is the diversity of threats and hazards, and the complex interaction among them, which result in "an unprecedented global creation of risks" (The Global Risk Report 2020/WEF). While facing humanitarian emergencies with multiple hazards, particularly in the context of hydrometeorological, geological and environment related phenomena and increasingly in disease outbreaks, more efforts are needed to facilitate Members and Member States to improve its capacity on Multi-Hazard Early Warning Systems (MHEWS) (UN 2006; WMO 2017, 2019). The overview on MHEWS, 4th WLF 2017 illustrated the concerted international efforts for advancing MHEWSs, specifically in addressing mainly in (1) context from EWSs to MHEWSs, (2) what's a (MH)EWSs, (3) recent advances in EWSs, (3) gaps/challenges related to EWSs, (4) tracking progress at international level and (5) introduction of IN-MHEWS while this paper mainly focus on multi-stakeholder partnership in MHEWS and its applications (Luther et al. 2017).

To "substantially increase the availability of and access to MHEWS" is one of the seven global Disaster Risk Reduction (DRR) targets of the Sendai Framework (target-G). Over the past decades, DRR was far from being the purview of a single agency or programme, as the "concept and practice of reducing disaster risks through the application of analysis to reduce causal factors of disasters" (UNDRR 2009). It consists of representatives from across areas such as government bodies, civil society, private sectors, research institutes and non-government organizations (NGOs) (UNDRR 2016).

Globally, in 2019, about 9000 people lost their lives in natural catastrophes compared with 15,000 in 2018 while the weather extremes, as an important part of natural disasters and consequence of climate change impact, are remaining at top 1 among top 5 global risks in terms of likelihood and at top 4 risks in terms of impact in 2019. This confirms the overall trend towards lower numbers of victims thanks to better DRR, including MHEWS (GRR/WEF 2020). However, increasing diseases outbreaks have shown its strong negative interconnectivity which undermines stability and sustainable development of social and economic development (Corona Virus (COVID-19) "Infodemic" and Emerging Issues through a Data Lens: The Case of China, Hua and Shaw 2020; Changing Rapid Weather Variability Increases Influenza Epidemic Risk in a Warming Climate, Liu et. al. 2020).

Despite progress in strengthening MHEWS across the world, significant gaps remain, especially in continuously building multi-stakeholder partnership among the actors and agencies concerned, lack of integrated standardized operating procedures/protocols to seamless all levels and interoperable information systems across all relevant disciplines and full chains in MHEWS, limited public awareness and participation in risk management, insufficient actionable political commitment, and limited public/private financial support for the implementation of these systems (UNDRR 2006a, b; UNEP 2012; Clinton et al. 2016). Lack of institutional harmonization of the bottom-up approach achieved at national and local levels and top-down approach facilitated to Members and Member States by international and regional organizations and its relevant programmes//projects on sharing knowledge, capacity development and early actions in the major application areas in MHEWS is a critical issue to be addressed.

There are distinct characteristics that need to be understood and addressed—aspects of the interconnectivity of disaster risks, its impacts and interoperability in implementing and delivering MHEWS—in addition to effective and acceptable multi-stakeholder governance mechanism. Answering these challenges calls for a more integrated approach to acknowledge the complexity of threats, risks and address such multiple interactive features. Current health crisis on Covid-19 is calling on that early warnings and its effective awareness for early actions are critical important to be reinforced (UNCG 2020). Best Practice has been made by the meteorological community through development and implementation of Impact based Forecasting and Warning Services (WMO No. 1150 2015) for the advisory services on hydromet hazard triggered MHEWs and its emergency response (Luther et al. 2019).

The complexity of the interaction between different spheres of the earth system that human in situated and the interconnections of the risks and impacts created from weather, water, climate, ocean and related environmental processes are increasingly challenging to the sustainability of social and economic development, especially in building a pathway on risk-informed and sustained social and economic development (UNDRR 2015). Risks generated by the interaction of complex human and natural systems, amplified by changes in the climate, are increasing the propensity for systems reverberations, setting up feedback loops with cascading consequences that are larger, more complex and more difficult to foresee—ultimately reversing efforts to achieve the 2030 Agendas (GAR 2019; Keys et al. 2019). No single government or agency has the necessary resources and capacities to address all these challenges on its own. Therefore, forging and working in partnerships with stakeholders, such as international agencies, national and local governments (GNDR 2013), non-governmental organizations (NGOs), academia, the private sector and the media, and engaging in networks and other collaborative efforts is essential for meeting the goals/targets of SDGs, Paris Agreement on Climate Change and Sendai Framework on DRR through addressing impact-based and risk-informed integrated MHEWS at local, national, regional and international levels (UNFCC 2015).

While rapidly growing best practices on MHEWSs at national and local levels such as best practices documented in *Implementing Hazard Early Warning Systems* (Rogers and Tsirkunov 2011), *Institutional Partnership in MHEWSs* (Golnaraghi 2012), *Five Approaches to Build Functional Early Warning Systems* (UNDP 2018), *Proceedings of MHEWC-I* (INMHEWS 2017), and *Proceedings of MHEWC-II* (INMHEWS 2019, to be published) are gaining momentum to improve multi-hazard early warning systems and so boost the resilience of the most vulnerable countries to extreme weather and the impacts of climate change. The importance of multi-stakeholder partnership for disaster risk reduction (DRR), sustainable development, adaptation of climate change has been repeatedly highlighted in major international agendas (UN 2015).

For example, CREWS was proposed by France Governments and related international organizations especially WMO and IN-MHEWS was initiated by 11 international organizations and Member States as major outcomes of the Working Session on Early Warning at the Third United Nations World Conference on Disaster Risk Reduction (WCDRR-III) in Sendai, Japan, in 2015. "Achieving a more integrated approach to multi-hazard early warning systems requires new ways of thinking about the intergovernmental and non-governmental cross-sectoral working arrangements and partnerships to deliver end-to-end and people-centred systems" is one of the outcomes of the 2nd Multi-hazard Early Warning Conference (MHEWC-II) in 2019.

MHEWS should engage all relevant actors to increase the effectiveness, efficiency, consistency, interoperability and utilization of impact-based forecasting and risk informed warning services. Against this background, the article presents the recently established multi-stakeholder partnership at national, regional and international levels in an integrated way with recent advances and remaining gaps and challenges for discussion and consideration (UNDRR 2019a,b).

Multi-stakeholder Partnership on MHEWS at International Level

Efforts to strengthen multi-stakeholder partnerships have been achieved in many ways through several global initiatives, including how to improve interconnectivity through MHEWS interfaces, and how to build Interoperability across multiple international organizations. The examples include, but not limited to, United Nations Disaster Assessment and Coordination (UNDAC), Humanitarian Networks and Partnerships Week (HNPW), Climate Risk and Early Warning System (CREWS), IN-MHEWS, GWP and EHAN.

Partnerships of Global MHEWSs Interfaces with Humanitarian and Crisis Management Networks

The interoperability of MHEWS interfaces with response networks for actions are critical. GDACS (as a MHEW interactive information system/partnership), Global Water Partnership (GWP) (as a multi-stakeholder action network and intergovernmental collaboration Interface) and GMAS (as a MHEW integrated services framework for partnership) with the United Nations Disaster Assessment and Coordination (UNDAC), Inter-agency Steering Committee (IASC) on Humanitarian Coordination, Global Crisis Centres Network (GCCN) and HNPW, as a value-added Partnership/Network in support of services for Humanitarian activities, and with other special emergency response networks in thematic areas, such as EOC-NET, EHAN, and UN Operation and Crisis Centre (UNOCC) etc.

The United Nations Disaster Assessment and Coordination (UNDAC) is part of the international emergency response system for the first phase of a sudden-onset emergency by rapid deployment of specialized teams. UNDAC teams, consisting of experienced humanitarians and disaster management experts, can deploy at short notice (12–48 h) anywhere in the world. They are provided free of charge to the disaster-affected country, and deployed upon the request of the United Nations Resident or Humanitarian Coordinator and/or the affected Government.

The GCCN provides a community of practice for national and regional crisis centres as well as other international actors (e.g. scientific organisations, NGOs, humanitarian agencies) that deal with disaster information analysis and disaster response after major sudden-onset disasters.

The GCCN is an informal coordination mechanism to support information exchange and analysis among international actors. The GCCN has developed procedures for alerting its members, and to establish an initial coordination process. As such, the GCCN aims at leveraging the collective knowledge, expertise and experience of its members, which include (but are not limited to) disaster managers, logisticians, scientists (e.g. meteorological services), humanitarian agencies, as well as specialized organisations and NGOs. The GCCN coordination process is expected to contribute to the development of a Common Operational Picture (COP) and to enable comprehensive information analysis in support of decision making in the Affected State and among responders with regard to international assistance and international response coordination. The GCCN also provide products to the affected Member States as follow:

- Product catalogue: the creation of a standard product catalogue of information products that support situation analysis and decision making among international responders in the first phase after major disasters.
- The web-based STN Knowledge Base (prototype), an interactive tool with comprehensive information about mechanisms and actors in international disaster and humanitarian response.
- Impact Mapping: the Impact Mapping is produced based on scientific modelling within a short time frame after the disaster event and made available on the GDACS website (GDACS 2020).
- Predictability in satellite-based mapping: feedback from Copernicus EMS (Annett Wania, EC/JRC). COPERNICUS is the European Union's Earth Observation Programme, which provides since 2012 on-demand rapid satellite mapping products on a 24/7 basis. The products can be delivered within 24 h from the request (through the EC/ECHO-ERCC), and range from reference situation maps, the first impact estimates to detailed impact assessment maps.

The Inter-agency Steering Committee (IASC): Created by the United Nations (UN) General Assembly resolution 46/182 in 1991 IASC is the longest-standing and highest-level humanitarian coordination forum of the UN system, bringing together the executive heads of 18 UN and non-UN organizations to ensure the coherence of preparedness and response efforts, formulate policy, and agree on priorities for strengthened humanitarian action The IASC is chaired by the Emergency Relief Coordinator (ERC) and facilitates the leadership role of the UN Secretary-General by regularly convening to ensure better preparation for, as well as a rapid and coherent response to, humanitarian crises.

The responsibilities of the IASC include: making strategic and policy decisions with system-wide implications; endorsing major operational decisions; arbitration where no consensus can be reached by other IASC structures; advocating common principles, collectively or individually on behalf of the IASC; approving the work plans of the IASC structures; bringing issues to the attention of the Secretary-General and Security Council through the ERC; and, designating Humanitarian Coordinators and selecting coordination arrangements.

The Committee is supported by subsidiary bodies; groups of decision-makers and experts who inform and carry out the priorities set by the IASC.

The IASC community extends beyond the Committee itself—it is a network of humanitarian actors and experts dedicated to delivering timely assistance to people in need. In addition to IASC-endorsed guidance, Accountability to Affected Populations, Protection from of Sexual Exploitation and Abuse, Persons with Disabilities, Gender and Age, and Mental Health and Psychosocial Support, as well as resources that draw linkages between these areas.

The cluster approach (Fig. 1), designed by IASC, is to strengthen system-wide preparedness and technical capacity to respond to humanitarian emergencies, and provide clear leadership and accountability in the main areas of humanitarian response. At country level, it aims to strengthen partnerships, and the predictability and accountability of international humanitarian action, by improving prioritization and clearly defining the roles and responsibilities of humanitarian organizations.

Clusters are groups of humanitarian organizations, both UN and non-UN, in each of the main sectors of humanitarian action, e.g. water, health and logistics. They are designated by IASC and have clear responsibilities for coordination (OCHA 2020a).

Great efforts have been taken to strengthen such interconnectivity, One of the examples to show how such multi-stakeholder partnership has been enhanced is GDACS actively engaging in the Humanitarian Networks and Partnerships Week (HNPW) through addressing GDACS, as technical advisory platforms on MHEWS being embedded into real actions on crisis containment and mitigation through the provision of risk-informed MHEWS advisory to a large audience of the network. GDACS consultations have been regularly made through its web-based interaction and regular consultation workshops with its users from the Member States, international organizations, NGOs and private sectors. Supported by UNITAR-UNOSAT, the GDACS-Satellite Mapping Coordination System platform (SMCS) has constantly contributed by different satellite mapping groups (e.g. UNOSAT, Copernicus Emergency Management Service, Space Charter), which allows GDACS stakeholders and the wider humanitarian community to determine at real-time which satellite images are collected where and which entity is working on what type of analysis. GDACS-Satellite Mapping Coordination System platform (SMCS) (2018) has been activated and utilized in 14 major disaster events. The activation of the SMCS in the Sept 2018 Indonesia earthquake was used as a case study to demonstrate the platform.

HNPW, co-chaired by the United Nations Office for Coordination of Humanitarian Affairs (OCHA) and the Swiss Agency for Development and Cooperation (SDC), provides a unique forum for humanitarian networks and partnerships to meet and address key humanitarian issues. One of the largest humanitarian events of its kind, it gathers

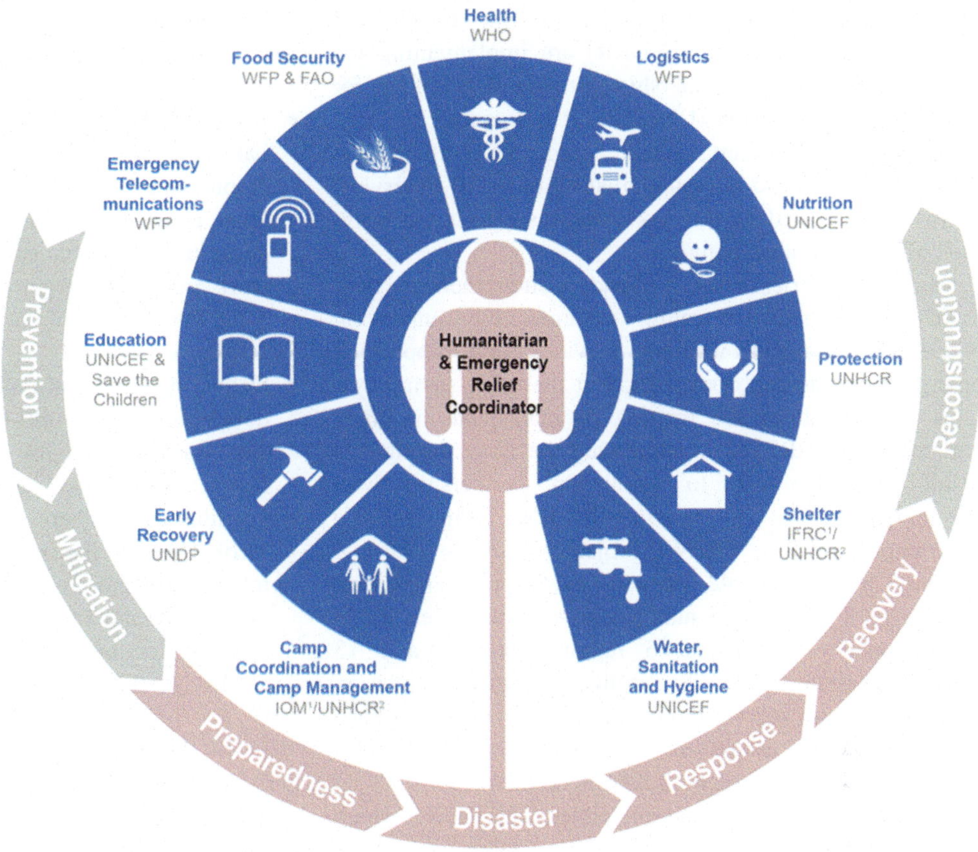

Fig. 1 Demonstration of the cluster approach. *Source* https://www.humanitarianresponse.info

participants from the UN, NGOs, Member States, the private sector, the military, academia and beyond to discuss and solve common challenges in humanitarian affairs. Among them, MHEWS related initiatives have actively engaged in.

During HNPW 2020, the technical session on tropical cyclone impact estimation for humanitarian preparedness and response has been held. This 1.5-day meeting was the first workshop jointly organized by the Joint Research Centre of the European Commission (EC-JRC) and OCHA in the context of the Global Disaster Alert and Coordination System (GDACS), and by the World Meteorological Organization (WMO). It brought together operational meteorologists and hydrologists and satellite experts with practitioners from disaster management and humanitarian agencies to discuss innovative solutions for early warning advice on tropical cyclone-related hazards and for estimating their impacts on vulnerable and exposed populations and humanitarian operations. It catalysed positive dialogue between the scientific and humanitarian communities, resulting in an increased understanding of each other's work, challenges and constraints and showcased examples of what is already working well as well as areas where to further improve and collaborate (OCHA 2020b).

Partnership on Funding Mechanism for MHEWS

CREWS, initiated by French Government and relevant international organizations, such as WMO during WCDRR-III in 2015, is a mechanism that funds Least Developed Countries (LDC) and Small Island Developing States (SIDS) for risk-informed early warning services, CREWS is an excellent example for strengthening the partnership between donor countries, development agencies, international organizations and the countries to be supported for enhancing its capacity on MHEWS. CREWS works directly with countries to increase the availability of, and access to, early warning systems. CREWS focuses on ensuring that early warnings, related both to the weather and climate events are risk-informed.

Country and regional projects are implemented by the countries with the support of implementing partners who provide technical assistance and capacity development. This includes the twining of two or more National Meteorological and Hydrological Services and by leveraging the expertise of regional and international institutions. The CREWS Steering Committee with 17 members who have committed to support CREWS, regularly reviews information on capacity gaps, demands and leveraging potential across LDCs and SIDS to prioritize its investments.

WMO and the World Bank's Global Facility for Disaster Reduction and Recovery (GFDRR) are implementing the CREWS initiative in partnership with the UN Office for Disaster Risk Reduction (UNDRR). It is financed by Australia, France, Germany, Luxembourg, the Netherlands, Switzerland, UK and also supported by other 10 observers. A trust fund hosted by the World Bank supports the activities of implementing partners.

Many countries have already strengthened their multi-hazard early warning systems by enhancing hydro-meteorological warning services and improving emergency plans and operations. But these life-saving systems and structures are missing or inadequate in many countries. The CREWS is addressing the issue through its result-based approach to strengthen partnership with life-saving systems and structure in those countries who have implemented its CREWS Projects.

The CREWS has 42 Partners for its implementation including 13 NMHSs, 5 national government agencies, 5 international organizations, 6 regional bodies, and 13 research institutes. CREWS already launched initiatives in 18 countries and sub-regions including Mali, Burkina Faso and several Pacific islands to strengthen forecast capabilities and ensure warnings reach all who need them.

The general standards and guidelines for CREWS which include partnerships on the Disaster Risk Knowledge/Risk-informed system design, warning dissemination and communication, preparedness and response capabilities/ability to respond can be found at https://www.crews-initiative.org/en.

IN-MHEWS Partnership for Strengthening Coordination

11 international and national agencies established the International Network for Multi-Hazard Early Warning Systems (IN-MHEWS) as a major outcome of the Working Session on Early Warning at the Third United Nations World Conference on Disaster Risk Reduction (WCDRR-III) in Sendai, Japan, in 2015 and 22 international organizations have actively participated in the network until the 2nd MHEW Conference in Geneva, May 2019.

IN-MHEWS under the Sendai Framework is to foster coordination, cooperation, collaboration, and networking to facilitate the sharing of expertise and good practice for multi-hazard early warning systems as a national strategy for disaster risk reduction, climate change adaptation, and building resilience. Besides, it aims to guide and advocate the implementation and/or improvement of multi-hazard early warning systems, share lessons learnt regarding early warning and increases the efficiency of investments in such systems for enhanced societal resilience.

The major contributions of IN-MHEWS and its members are with the following examples:

- Regular MHEW Conference organized prior to the Global Platform on DRR as its preparatory session bringing its outcomes and recommendations to the relevant higher level consultation during the GP, The MHEWC is a consultation mechanism between members of the network and the platform for sharing best practices;
- Joint Publication of the Checklist on MHEWS (WMO 2018a) and others;
- Delivering Guidelines by members of IN-MHEWS on the development of Multi-hazard impact-based forecasting and warning services (WMO), IFRC Forecast Based Financing for Early Action as examples;
- Facilitating members to implement Common Alert Protocol (CAP) to the transmission of multi-hazard warning information and promoting cataloguing and identification of hazards and utilization of Unique Identify to record and transfer data on hazards and its associated hazard clusters.

IN-MHEWS through its first and second Multi-Hazard Early Warning Conference (MHEWC-I 2017, II 2019) has addressed the importance of multi-stakeholder partnership and its major issues related are as follows:

- Partnerships between the scientific and research communities, early warning information providers and humanitarian and development practitioners are essential.
- Need for partnership with impact domain experts and data holders.
- Collaboration between Climate Risk and Early Warning Systems Initiative (CREWS) and World Bank Global Facility for Disaster Reduction and Recovery (WB-GFDRR) is important in measuring the effectiveness of MHEWS.
- Partnerships with the public and private sector are essential to creating MHEWS that usable, useful and used.

Concerning its next strategic step for IN-MHEWS, sustained multi-stakeholder partnership governance has been recognized. The critical issues should be raised for further actions as follows:

- Improvement of multi-stakeholder partnership governance model through establishing an Alliance on MHEWS to leverage current IN-MHEWS;
- Encouragement of engagement of members of IN-MHEWS through improving its rotation mechanism to manage its practical coordination between major

international organizations; Whether it should be steered by a steering committee to advise a consulted planning for collaboration/partnership for the network/alliance and co-chaired by representatives from international organizations, development agencies, academia institutes, NGOs and private sectors;
- Expert advisory teams on different thematic areas of MHEWS, such as climate extremes and public health (Rogers 2011), atmospheric emergency response for nuclear power plant accidents etc. should be established according to the features of cascading impacts of risk inter-connection between original hazard and derived hazards and the relevant authorized international organizations, associated with academia institutes, private sectors should lead these teams;
- Effective connections of the mechanism of IN-MHEWS with the frameworks for UN humanitarian assistance and emergency response, post-disaster assessment for actions and investment;
- Effective connection with national, regional and international MHEWS related networks, programmes/projects and activities, such as IN-MHEWS and CREWS, IAEA Nuclear Accident Incident and Emergency System, WHO Emergency Response Programme, IHP/UNESCO and regional flood programmes, Weather Ready Nation Project (WRN), etc.;
- Effective inter-connectivity between highly relevant centres established and authorized by inter-governmental organizations, such as European Centre for Medium-range Weather Forecast (ECMWF), International Center for El Niño Research (CIIFEN), WMO World Meteorological Centres (WMCs) and its Regional Specialized Meteorological Centres (RSMCs), UNOCHA/EU-JRC Global Disaster Alerting, Coordination Centre (GDACS) and Incidental Emergency Response Centre of IAEA, etc.;
- Strengthening PPE through the special arrangement to facilitate engagement of private sectors in IN-MHEWS;
- Taking actions to promote "no one left behind" through enhancing capacity development for LDCs and SIDS on MHEWS.

MHEWS Partnership in Thematic Areas, Such as Cascading Impact Chain Relate to Landslide

The aims of MHEWS are not only to address the interoperability issue for identifying integrated efforts contributed by multi-stakeholder through establishing capacities on scientific/technological advisories and concrete supporting facilities in support of building resilient society in a cost/effective way but also to address its interconnectivity issue to identify cascading impact processes triggered by the original hazard events. It is important for us to take a smart science-based approach in a more precise manner. Typically, landslide can be triggered by both hydrometeorological factors (e.g. heavy rainfall) and geological factors (e.g. earthquake). Overlapping with vulnerable areas, it can trigger tremendous loss and damage in a critical location. Great efforts to improve international cooperation on MHEWS for landslide have been made in recent years. At the 2nd United Nations World Conference on Disaster Reduction, which was held in Kobe, Japan, on 18–22 January 2005, the International Consortium on Landslides (ICL) co-organized a session which resulted in a global partnership and platform taking a holistic approach to research and learning on 'Integrated Earth system risk analysis and sustainable disaster management'. This partnership was forged through a "Letter of Intent", which was signed by UNESCO, UNDRR, WMO, FAO, UNU, ICSU, and WFEO who have committed to support MHEWS and its response for landslide (Sassa 2019).

In addition, the participating scientific and academic institutions and governmental and non-governmental organizations proposed that the Sendai Partnerships 2015–2025 for Global Promotion of Understanding and Reducing Landslide Disaster Risk in the Third UN World Conference on Disaster Risk Reduction in Sendai, Japan in March 2015. The partnership was signed by 22 global stakeholders including WMO and ICL and developed through the Fourth World Landslide Forum in Ljubljana, Slovenia in 2017. This sound global platform will be mobilized thr0ugh the Fifth World Landslide Forum in Kyoto, Japan in 2020 and in the coming decade to pursue prevention, to provide practical solutions, education, communication, and public outreach to reduce landslide disaster risk.

An example at national level: National landslide early warning services operated by National Building Research Organization in Sri Lanka (Fig. 2). A landslide hazard zonation mapping programme within the 10 landslide prone districts has been developed. The maps which display the distribution of the severity of landslide hazard potential in a given area were intended to be used with associated guidelines as a decision-making tool for development of central highlands of the country. It is also used for identification of elements at landslide risk and can be utilized in relocation, rehabilitation, allocation of relief funds and insurance purposes also. Mapping is carried out at 1:50,000 scale and at 1:10,000 scale. An early warning service has been operated based on such risk mapping and impact-based forecasting.

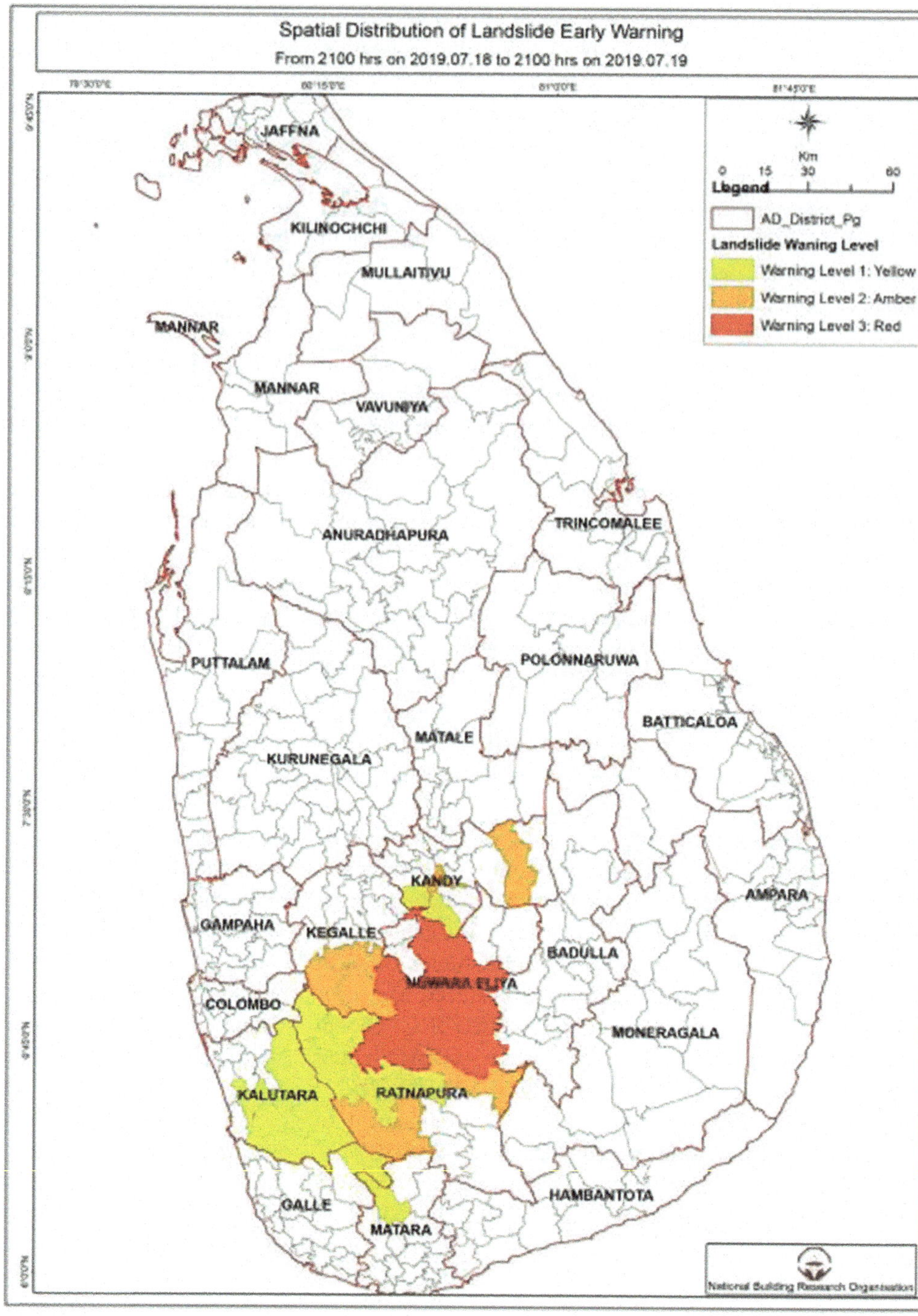

Fig. 2 Landslide Early Warning Issued 21.00 h on 18–07-2019 to 21:00 h on 2019–07-19. *Source* https://nbro.gov.lk/index.php?option=com_content&view=article&layout=edit&id=250&lang=en

Partnership on Environment and Humanitarian Action (EHA) Network

The Environment and Humanitarian Action (EHA) Network is an informal network aiming to avoid, minimize, or mitigate environmental impacts of humanitarian action and to promote environmentally responsible humanitarian programming through collaboration and cooperation. It was established in 2013 with the objective of mainstream environmental considerations in humanitarian action (UNEP 2020b).

The EHA network seeks to mitigate environmental impacts during humanitarian response and to promote environmentally responsible humanitarian programming. Network members jointly work to advance humanitarian policy, strengthen knowledge on EHA, conduct advocacy and provide technical support to humanitarian operations.

People can join the network on an individual or institutional basis. The network holds meetings approximately every three months as well as one annual face-to-face meeting in connection with the Humanitarian Networks and Partnerships Week of the Leading Edge Programme.

Global Network on Monitoring, Analysis, and Prediction of Air Quality (MAP-AQ) and its Support of the Frontiers and Professional Partners

The Global Network MAP-AQ (Fig. 3) is a specialized multi-stakeholder partnership that has been endorsed as a sponsored activity of the International Global Atmospheric Chemistry (IGAC) Project and is directly contributing to the objectives of the Global Atmosphere Watch (GAW) at WMO. The overarching goal of MAP-AQ is to constitute and develop a consortium of expert groups to coordinate and enhance research and services that will help mitigate air pollution, specifically in regions of the world where high concentrations of pollutants are observed. It aims to develop and implement a global air pollution monitoring, analysis and prediction system and alliance with downscaling capability in regions of the world that are affected by high levels of atmospheric pollutants, in particular in Asia, Latin America and Africa (NCAR 2020).

Fig. 3 International system/network for monitoring, analysis, and prediction of air quality (MAP-AQ). *Source* https://www2.acom.ucar.edu/map-aq

The partners of MAP-AQ network includes (1) expert groups in charge of the modelling system development and validation, space data analysis and assimilation; (2) regional and local representatives contributing to the development, dissemination and validation of the products; (3) members of international projects in support of programmes, such as the EU/ESA Copernicus programme in particular the Copernicus Atmospheric Monitoring Service (CAMS).

MAP-AQ expects to develop partnerships with other international and humanitarian frontier programs and their networks including UNEP (Climate and Clean Air (CCA) Coalition), WHO (capacity building to tackle air pollution), World Bank (WB projects for nations/regions in which MAP-AQ has initiated activities), and the EHA network etc.

Global Water Partnership

The Global Water Partnership (GWP) is a global action network with over 3,000 Partner organisations in 179 countries. The network has 68 accredited Country Water Partnerships and 13 Regional Water Partnerships (GWP 2018).

The network is open to all organisations involved in water resources management: developed and developing country government institutions, agencies of the United Nations, bi- and multilateral development banks, professional associations, research institutions, non-governmental organisations, and the private sector.

GWP's action network provides knowledge and builds capacity to improve water management at all levels: global, regional, national and local. GWP does not operate alone. Its networking approach provides a mechanism for coordinated action and adds value to the work of many other key development partners.

GWP is gearing up for its continued support to countries on climate change adaptation, leading towards the implementation of the Paris Agreement. GWP activities under the Global Water, Climate, and Development Programme (WACDEP) aim to strengthen the resilience of countries to climate change.

The WACDEP Coordination Unit (CU) in Africa is strengthened through relevant strategic alliances and partnerships with Multilateral Development Banks, UN agencies, and others. The CU provides global/regional thematic leadership on strengthening regions and countries. Cross-regional technical support functions are also being established for GWP Asian regions in collaboration with established relevant strategic allies such as UNEP, UNDP, ADB, IWMI, ASEAN, and others.

Multi-Stakeholder Partnership at Regional Level

There is a growing number of regional MHEWS initiatives to actively address joint efforts on trans-boundary issues. Such as, but not limited to SSE-MHEWS-A for South-East Europe, CDEMA for Caribbean region, RIMES for Africa and Asia and GMAS-A, as well as many regional multi-hazard interfaces has been established by ARISTOTLE and GMAS etc. (WMO 2020).

ARISTOTLE in Europe and SSE-MHEWS-A in Southeast Europe

ARISTOTLE-ENHSP (also known as ARISTOTLE-2. being the continuation of the previous ARISTOTLE 2016–2018, Fig. 4) is a European Natural Hazard Scientific Partnership Project financed by the Directorate-General for European Civil Protection and Humanitarian Aid Operations (DG-ECHO) that delivers world-leading multi-hazard advice capability to the Emergency Response Coordination Centre (ERCC) (ARISTOTLE 2020).

ARISTOTLE-ENHSP has been designed to offer a flexible and scalable system that can provide new hazard-related services to the EU Emergency Response Coordination Centre (ERCC) and to create a pool of experts in the field of Meteorology and Geophysics of Europe that can support the ERCC concerning situation assessments in crises.

Within ARISTOTLE, a multi-hazard scientific partnership and its governance have been established across Europe to exploit the available scientific and technological expertise and to assure mutual learning and improved coordination based on a multi-hazard approach, including the definition and implementation of the required prevention countermeasures, enhancing the information provided to the ERCC and, through the governance structure, ensuring a collaborative dialogue with the national mandated civil protection authorities.

The 24*7 services system is set up in a way that it builds upon and adds value to existing information systems and sources, and that it fully respects national and regional responsibilities.

South-East European Multi-Hazard Early Warning Advisory System (SSE-MHEWS-A)

South-East Europe has experienced a significant number of severe meteorological and hydrological events in recent years. This will, in turn, increase demand for improved early warning for communities under threat from such natural hazard as well as a need for more preparedness in those communities to improve their resilience. WMO initiated the South-East European Multi-Hazard Early Warning Advisory System (SEE-MHEWS-A) project in 2016 to assist Members in the region to achieve these objectives (WMO 2018b). This project builds on the outcomes of several recent projects in the region related to disaster risk reduction that were implemented with funding from the European Union, United Nation agencies, the World Bank and several other international and national organizations such as U.S. Agency for International Development (USAID).

The previous projects demonstrated that there is a need to strengthen regional partnership to address gaps in forecasting and warning provision at the national and regional levels, particularly in transboundary areas. To achieve this, the development of a regional multi-hazard early warning advisory system—consisting of information and tools for forecasters at National Meteorological and Hydrological

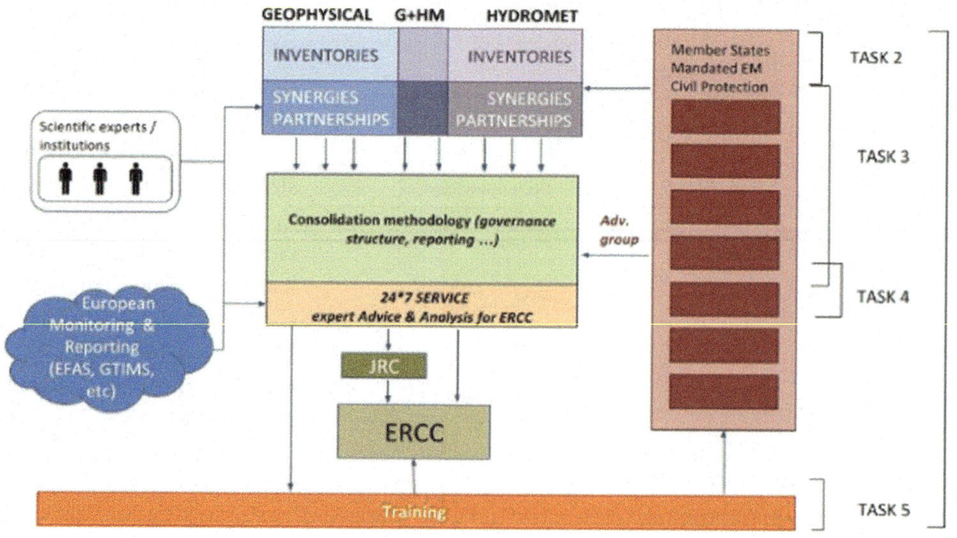

Fig. 4 Demonstration of ARISTOTLE partnership among the participants and of a network of scientific and operational centres run by a governance body. *Source* https://pilot.aristotle.ingv.it/

Services (NMHSs) and harmonized national early warning systems—is essential. The development of the SEE-MHEWS-A will support the NMHSs in fulfilling their mandate to provide timely and accurate warnings to minimize the impacts on people, infrastructure and industry of hazardous weather events and to protect the lives and livelihoods of the people.

SEE-MHEWS-A is providing operational forecasters with effective and tested tools for forecasting hazardous weather events and their possible impacts to improve the accuracy of warnings and their relevance to stakeholders and users. On a single virtual platform, the system will collect existing information, products and tools for the provision of accurate forecasts and warnings to support hazard-related decision-making by national authorities. Furthermore, the system will function as a cooperative platform where forecasters from different countries can work together on the identification of potential hazards and their impacts, especially when impending weather hazards may have potential impacts in many countries. Development, implementation and operation of the "cloud-based" Common Information (and communication) Platform (CIP) for SSE-MHEWS-A to facilitate access to-, and dissemination of model outputs, post-processing tools and post-processed products, such as nowcasting, dissemination of warnings, such as via its partnership with MeteoAlarm, and communication among forecasters to coordinate advisories and warnings especially in transboundary areas.

During the first phase of the SEE-MHEWS-A project, which was supported by the USAID in 2016–2017, a detailed Implementation Plan was developed in cooperation with the NMHSs of the region, WMO Secretariat and various stakeholders in the fields of meteorology, hydrology and disaster risk reduction. The Implementation Plan outlines the overall technical and governance structure of the SEE-MHEWS-A system to be realized until 2023. The Directors of the NMHS of Albania, Bosnia and Herzegovina, Croatia, Cyprus, Greece, Hungary, Israel, Kosovo (UNSCR 1244/99), the former Yugoslav Republic of Macedonia, Republic of Moldova, Montenegro, Romania, Slovenia, Turkey, and Ukraine declared in June 2017 their intention to collaborate towards the implementation of activities and projects leading to full operation of the SEE-MHEWS-A, within the scope broadly portrayed in the Implementation Plan.

Regional Integrated MHEWS (RIMES) in Africa and Asia

The Regional Integrated Multi-Hazard Early Warning System for Africa and Asia (RIMES) is an international and intergovernmental institution aiming to provide regional warning services related to the tsunami and associated hydrometeorological hazards to its member states, which include many countries in South and Southeast Asia (RIMES 2020). RIMES evolved from the efforts of countries in Africa and Asia, in the aftermath of the 2004 Indian Ocean tsunami, to establish a regional early warning system within a multi-hazard framework for the generation and communication of early warning information and capacity building for preparedness and response to trans-boundary hazards. Tamil Nadu System for Multi-hazard potential impact assessment, Alert, emergency Response planning and Tracking (TN-SMART)—A web-GIS based Decision Support System to strengthen preparedness, response, recovery and mitigation measures during multi hazards of flood, cyclone and tsunami (Fig. 5).

RIMES has 18 core member countries, supported by 14 collaborating countries, 15 collaborating partners including WMO, ECMWF, IOC/UNESCO, 3 universities and 9 research institutes. There are 6 development partners including UNDP, UN Environment, FAO, ESCAPE, DANIDA and USAID. RIMES operates from its regional early warning centre located at the campus of the Asian Institute of Technology in Pathumthani, Thailand. In addition, RIMES partner with research organizations, such as Deltares, on projects implementing early warnings systems in-country, such as early flood warning in Bangladesh (Cumiskey et al. 2015).

MHEWS in the Caribbean: Partnership Through Caribbean Disaster Emergency Management Agency (CDEMA) and the Application of the Early Warning Systems Checklist in the region

In 2017, after the devastating effect that Irma and Maria left in their wake, the American, British, French and Dutch Caribbean territories, Dominica and Antigua were awaiting relief from America, Britain, France, the Netherlands and Venezuela—geographically distant countries, emphasizing the need for a region of no borders, that is, to create one singular emergency unit, specifically for the Caribbean region, equipped with the necessary resources (boats, helicopters, planes, etc.) to provide rapid relief in the event of disasters. CDEMA has established and implemented the necessary protocol in dealing with border issues (CDEMA 2020).

'Strengthen integrated early warning systems for more effective disaster risk reduction in the Caribbean through knowledge and tool transfer' is an initiative aimed strengthening integrated Early Warning Systems (EWS) in Antigua and Barbuda, Saint Vincent and the Grenadines (SVG), Dominica, Dominican Republic, Saint Lucia and Cuba through the effective leveraging of tools and knowledge. The Project is being implemented by UNDP in close collaboration with IFRC, CDEMA, DIPECHO partners and national counterparts.

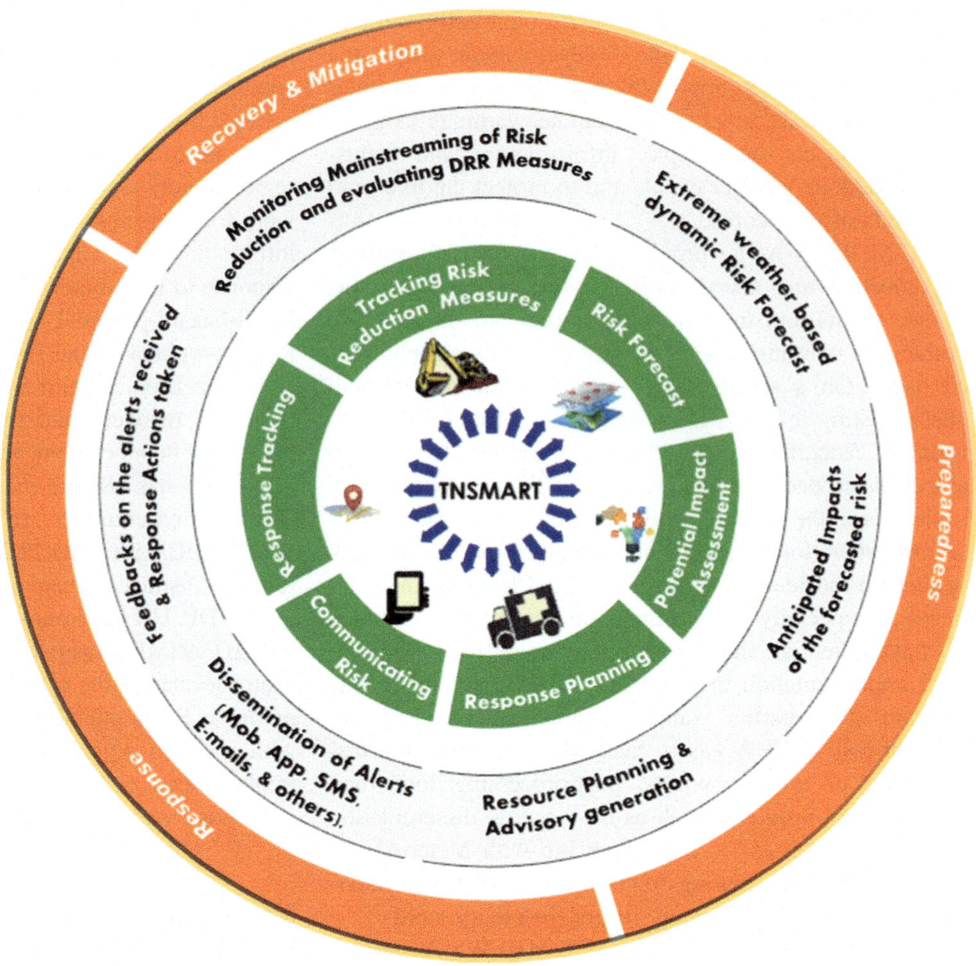

Fig. 5 TNSMART—a web-GIS based Decision Support System to strengthen partnership in preparedness, response, recovery and mitigation measures during multi hazards of flood, cyclone and tsunami. *Source* https://beta-tnsmart.rimes.int/index.php/login/login_form.

The objective of the Project is to improve EWS for more effective Disaster Risk Reduction (DRR) in the Caribbean and to move toward the realization of a more integrated system, through concrete actions addressing existing gaps. This initiative seeks to emphasize the 4 components of EWS—and close priority gaps—at a national level, contributing to the integration of national and community EWS and addressing sustainability and national ownership of EWS through 4 following efforts:

- Increase access to existing tools and knowledge of EWS at a national and regional level;
- Provide integrated EWS solutions in five target countries through knowledge sharing;
- Increase EWS effectiveness in five target countries through concrete priority actions;
- Ensure EWS knowledge transfer, documentation and communication.

MHEWS in the Caribbean brought together representatives from CDEMA, UNDP, ECHO, national governments, governmental organizations, non-governmental organisation (NGOs), donors, and regional institutions. It created a space for sharing perspectives and promoting the harmonization of actions towards enhancing Early Warning Systems (EWS) in the Region.

Many partnerships were also formed with disaster management ministries and national disaster systems in a shared ambition to grow and strengthen the organization's delivery capacity with stakeholder engagement, including the political directorate, and the necessary their commitment to implement DRR strategies, consider gender equality, establish mechanisms for technical and financial sustainability and build the existing capacity incorporating new productive capacities.

The policy director should enable individuals, governments and others to take timely action to reduce disaster risk in advance of hazardous events, therefore suggesting that effective EWS are really about partnerships and strengthening capabilities through cooperation and collaboration.

Looking for opportunities to establish partnerships with the private sector around closing the deficit the government has in providing adequate resources for the most vulnerable. The public/private partnerships are critical for enhancing preparedness through the creative financing scheme. For example, consideration of the impacts on the bottom line of the cruise ships industry to get the private sector to also invest in this broader agenda for preparing.

There was also the recognition of the roles of the two consortiums facilitated through the partnership with OXFAM and IFRC. The work led by OXFAM further detailed as it relates to the gaps that were identified at the local level, for example, in Cuba work was done on the tool for the farm level looking at how that information can be shared with and managed by the national level and sharing it with the region.

In the Caribbean region, the National Risk Network launches, with Saint Lucia and Haiti, the public/private partnership action plans at country level.

Multi-stakeholder Partnership at National Level

The tremendous efforts have been made at the national level to strength national MHEWS partnership with multi-stakeholder. Successful MHEWS have been implemented in many countries such as Cuba, Bangladesh, France, Germany, Japan, China and the United States. Many examples of MHEWS span a broad spectrum of geographic and climatic conditions in both developed and developing countries and address a variety of hydrometeorological and other hazards.

The benefits have been gained for building DRR resilient society through strengthening multi-stakeholder partnership, such as the Vigilance system in France, an example of joint efforts of multi-agency collaboration for making an early warning system with a multi-Hazard approach; the Weather Ready Nation partnership in U.S. through its Weather-Ready Nation Ambassador™ (www.weather.gov/wrn/ambassadors); strategic partnership with private sectors for scaling-up multi-hazard early warning in Indonesia.

Multi-sector and Multi-level Participation in Indonesia

The 2004 tsunami disaster and the HFA provided the driving incentives for institutional change in disaster risk management (UNDRR 2005). A Presidential Regulation legitimized the establishment of the multi-level Disaster Management Agency (BNPB, BPBD at the national and sub-national levels) for improved links and coordination from the national to the local. Furthermore, local authorities are legitimizing the establishment and operations of Multi-level Emergency Operations Centres as another critical and important architecture component and structure for improved decision-making and governance throughout the country.

Institutionalizing and embedding the Indonesian Tsunami Early Warning System (INATEWS) within the Disaster Management Agency (BNPB) as a larger architecture is a key step towards a multi-hazard approach and improved inter-institutional coordination and performance. Furthermore, the global-regional governance framework for tsunami hazard and risks under UNESCO-IOC coordination, and the developing multi-level architectures and structures synchronized within the existing decentralization are ideal polycentric multi-layered architectures for optimum interlink between levels and improved hazard and sustainable risk governance in Indonesia.

In terms of decision-making links, the national disaster platform supporting the Hyogo Framework for Action and the new steering committee consisting of the state and professional citizens of the multi-level Disaster Management Agency gives legitimacy to multi-sector, multi-level participation and decision-making in Indonesia (Poterie and Baudoin 2015).

MHEWS Partnership in Urban Areas

Today, 55% of the world's population lives in urban areas. By 2050 another 2.5 billion people will be added to urban areas, which will be 68% of the world's population, mostly in Africa and Asia. Urban areas are inherently more vulnerable to risks and stresses, as set out in the Sendai Framework for Disaster Risk Reduction, brought about by climate change and natural hazards owing to their high concentrations of population and economic activities (Nuha et al. 2018; CEB 2019). This is exacerbated by the fact that cities are frequently located in low-lying coastal areas, with particularly vulnerable populations often living on outright hazardous land. Enormous progress has been made by integrated efforts provided by city governments, local communities, private sectors and NGOs through strengthening its partnership in MHEWS, emergency response, contamination and mitigation although its political structures varying with different assignments of roles and responsibilities in relation to disaster risk reduction. For example, the densely populated urban region of Shanghai, China has provided corresponding illustrations of the emphasis on ensuring clarity regarding stakeholders' roles and responsibilities that is a feature of all successful MHEWS (Tang 2006, 2008).

The China Meteorological Administration (CMA) and Shanghai Municipal Government (SMG) jointly support the Shanghai Multi-Hazard Early Warning System as a WMO demonstration project with 'Multi-agency Response' as the core. The project integrates diversified advanced technologies into a multi-hazard warning process, advancing improved multi-agency coordination and cooperation through a multi-link communication platform with responsible emergency response and rescue agencies. The MHEWS is organized around its "4 + 1" technical platforms and three-level standard system on multi-agency coordination and cooperation. The technical platforms are: Multi-Hazard Detection and Monitoring, Forecast and Warning Information Generation, Multi-agency Coordination and Cooperation Support, Dissemination and User Application Platforms, and the Multi-Hazard Information Database. The three-level standard systems comprise: a Multi-agency Coordination and Cooperation Standard System, Safety Community Standard System and Regional Joint Defence Standard System. The MHEWS provides technical support to the Shanghai Emergency Response Platform and has been introduced into the Emergency Response Headquarters of the SMG. It provides forecast and warning services to the SMG's emergency response command centre, which is responsible for public emergency response actions and the delivery of emergency-related information. The network has been fully operational to provide emergency response services to Shanghai Expo2010. Following Expo2010, WMO has conducted an assessment of the MHEWS. Remarkable progress has been made. The dissemination platform has entered the testing stage; the forecast and warning information generation and multi-agency coordination and cooperation platforms are entering the development stages, with some modules already in operation. Warning subsystems for city traffic safety, heat wave and human health, power and energy security, and bacterial food poisoning are operational. There has been significant progress with grassroots level delivery of warning messages and with the integration of information into the city grid management. Breakthroughs have been made in multi-agency coordination and cooperation. "The emergency response plan of Shanghai Municipality for rain, snow and freezing weather disasters" and "The emergency response plan of Shanghai Municipality for heavy fog disasters" have been officially distributed by the general office of SMG for actions.

Multi-stakeholder Partnership with Private Sector and NGOs

Scenario-Based Risk Insurance for Multi-Hazard Impacts

Climate change risk is intensifying and is a serious threat to the insurability of communities and economies around the world. In the insurance market, the need to understand the uncertainty posed by concurrent hazards has already been recognized (e.g., Lloyd's 2016). Insurance plays a role in reducing vulnerability to financial losses and risk sharing, either through formal insurance, micro-insurance, or crop insurance, and can be a mechanism for vulnerability reduction in the face of extreme weather events (e.g. IPCC 2012; Kunreuther 1996; Surminski and Hudson 2017).

In order to ensure an insurable, resilient and sustainable world, UNEP Finance Initiative announced a partnership with 16 of the world's largest insurers to develop a new generation of risk assessment tools designed to enable the insurance industry to better understand the impacts of climate change through its weather/climate extremes on their business (UNEP 2020a).

To make use of the latest climate science, including some of the most advanced, forward-looking climate scenarios available. The tools and indicators were jointly developed and piloted by the Insurer Group will incorporate the latest scenario analysis to assess climate-related physical and transition risks in insurance portfolios.

The leading insurers that will work together with the UN are all signatories to UNEP FI's Principles for Sustainable Insurance (PSI), a global best-practice sustainability framework and the largest collaborative initiative between the UN and the insurance industry.

In 2013, the Zurich Insurance Company made a monetary commitment of up to USD 22.7 million over five years to support the creation of a flood resilience model together with the International Federation of Red Cross and Red Crescent Societies. The model is based on innovative pre-event mitigation measures and targets poor communities around the world. As part of this effort, a successful activity is the implementation of the mobile application Z-alert in Indonesia. Today, Z-alert provides notifications on various hazards such as fire, typhoons, and tsunamis with the ability for private citizens to add and verify warnings.

At the regional and national level, Data necessary for catastrophe risk quantification in Asia and in SE Asia, is in general poor in terms of availability, accessibility and quality. Singapore launched the Natural Catastrophe Data Analytics Exchange (NatCatDAX) in 2016, to address the lack of holistic and good quality data for natural catastrophe risks in the region, which has led to a growing protection gap (NatCatDAX 2020). The outcome is a high resolution, objective and widely accepted data and analytics platform which would enhance the analytics and understanding of catastrophe risks. It consists of two cat risk models developed by the Institute for Catastrophe Risk Management (ICRM), specifically a Jakarta flood risk (JKT FL) and a Singapore Earthquake (SIN EQ) risk assessment model (Su et al. 2018). The models quantify the relevant hazards, assess the exposures and vulnerabilities from inputs provided by end-users, and compute the relevant loss risk metrics (ICRM 2020).

This is a Public–Private Partnership, supported by the Monetary Authority of Singapore (MAS), and led by the ICRM of the Nanyang Technological University, Singapore (NTU), in collaboration with the insurance industry. Other successful examples are the Caribbean Catastrophe Risk

Insurance Facility (CCRIF), Pacific Catastrophe Risk Assessment and Financing Initiative (PCRAFI) and the UK flood insurance scheme.

Partnership with NGOs, an Example from Implementing the Early Action Protocol (EPR) for Delivering the Disaster Relief Emergency Fund (DREF) in a Forecast Based Early Action Manner

With the new Forecast-based Financing mechanism and programme (FbF), the International Red Cross and Red Crescent Movement is reshaping the future of the global humanitarian system. Based on forecast information and risk analysis, FbF releases humanitarian funding for pre-agreed activities (German Red Cross 2020). For early actions to be performed quickly and efficiently before disaster strikes, funds through DREF are allocated automatically to be the first on the ground providing help to those in need when a specific threshold is reached. A dedicated financing mechanism (EPR) is the key for taking fast and effective action before disaster strikes. This is why Forecast-based Action (FbA) was set up. To establish FbA, experts analyse the relevant natural hazards, assess the impacts of previous disasters and look at vulnerability data.

FbF programme has supports of numerous partners in the chain of partnership, including a large network of renowned scientists to provide advisory services, business/Foundations, humanitarian agencies such as WFP, IFRC, UNOCHA and local actors with 16 Red Cross and Red Crescent national societies in Africa, the Americas and Asia–Pacific.

The German Red Cross coordinates the development of FbF with the support of the Federal Foreign Office and important institutional partners, such as the Red Cross Red Crescent Climate Centre.

With the FbF methodology, forecasts have successfully triggered early action by National Societies in Peru, Togo, Uganda, Bangladesh and Mongolia.

Public and Private Partnership (PPP) on Delivering Warnings and Emergency Alerts

The Common Alerting Protocol (CAP), achieved by multi-stakeholder partnership, is an international standard format for emergency alerting and public warning. It is designed for "all-hazards", related to weather events, earthquakes, tsunami, volcanoes, public health, power outages, and many other emergencies. Today, approximately 75% of the world's population lives in a nation that has already implemented, or is currently in the process of implementing, a national-level, official source of CAP alerts. It is important to note that this percentage is growing steadily.

CAP has been strongly supported and adopted by multi-stakeholder partnership between International organizations (e.g. International Telecommunication Union (ITU), WMO and OASIS standard organization etc.) the emergency management community (such as the International Association of Emergency Managers (IAEM) and national emergency response agencies), humanitarian organizations (such as UNOCHA etc.) and other NGOs (e.g. IFRC etc.), and a broad range of commercial organizations (such as IBM, Google etc.).

The Wireless Emergency Alerts (WEA) system in U.S. implemented by using CAP. WEA is an essential part of national wide emergency preparedness. Since its launch in 2012, the WEA system been used more than 33,000 times to warn the public about dangerous weather, missing children, and other critical situations—all through alerts on cell phones. WEA is a public safety system that allows customers who own certain wireless phones and other enabled mobile devices to receive geographically-targeted, text-like messages alerting them of imminent threats to safety in their area. WEA enables government officials to target emergency alerts to specific geographic areas—lower Manhattan, for example.

The Google Public Alerts is implemented also based on CAP. Google Public Alerts is Google's help platform to show relevant official weather, public safety and earthquake alerts around the world and emergency messages such as evacuation notices for hurricanes, and everyday alerts such as storm warnings. While Google Public Alerts can't guarantee that you'll see every alert when using Google services, the platform is doing its best to show what's important when you need it as a useful additional source of information. Currently, the platform publishes content from its partners in some of countries, such as U.S., Australia, Canada, Colombia, Japan, Indonesia, Mexico, the Philippines, India, New Zealand, and Brazil etc.

PPP on delivering official warnings and emergency alerts is also important for the maritime safety at sea and off-shore zones. Multi-hazard, such as gust, extreme wave, storm surge, sea ice and dense fog has significant influences on maritime safety on Marine Shipping and Coastal Social and Economic Development. The Global Maritime Distress and Safety System (GMDSS), coordinated by 19 countries that invest human, material and financial resources to issue Marine Safety Information (MSI) bulletins to the entire maritime community at no charge. This IMO (International Maritime Organization)/ WMO Worldwide Met-Ocean Information and Warning Service (WWMIWS) is to undergo a significant change with a new service to support the Global Maritime Distress and Safety System (GMDSS) in the early 2020s according to IMO approved INMARSAT's 'Fleet Safety' solution, Iridium (May 2018) and BeiDou BDS (May 2019) (Fig. 6) Navigation Satellite Networks solutions for GMDSS services respectively.

Facing the rapidly rising costs and workload, it is hoped that such a partnership will be best to reduce the increasing burden on the 19 METAREA Issuing Services that have long shouldered the costs and responsibility for warning mariners of hazardous weather at sea. In addition, consultation process is on the way to find solution to reduce the cost through grouping users' applications on the utilization of such new satellite communication for the IMO's Navigation, Communications and Search and Rescue (NCSR).

PUP-PPP Multiple Disasters' Damage and Loss Data Recording

Collecting, achieving, standardizing and processing loss and damage data are essential for multi-hazard risk mapping, integrated assessment and developing impact-based forecasting and warning services for MHEWS (EC-JRC 2015). Several best practices have been obtained through PPP.

For example, JRC/EU provided guidance for sharing loss data across organisations, among the EU Member States and with EU and international institutions, proposes a minimum set of loss indicators that should be part of any operational disaster loss database.

The cross-cutting role and utility of loss data should be discussed across government departments, including emergency management, urban planning, and government budget and across all government scales and participative governance fora (local to national). High-level requirements should be identified based on the public and private needs analyses across sectors. Implementation of the data recording should be embedded in a Public-Public Partnership (PUP) and/or Public Private Partnership (PPP) modes to ensure participation and ownership of all stakeholders.

Conclusion and Discussion

Collective experiences have been derived from the successful multi-stakeholder partnerships at all levels that have proven effective in reducing losses of life and property in the face of disasters caused by multiple natural hazards. In all

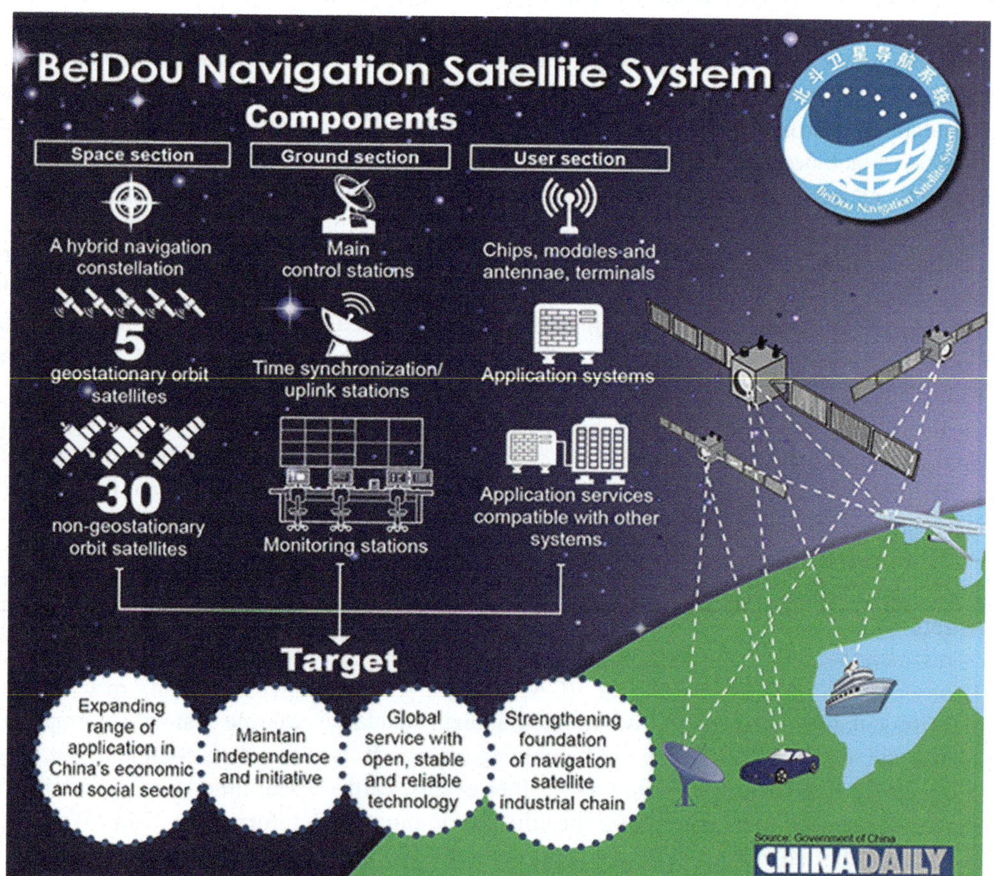

Fig. 6 BEIDOU satellite navigation system launched. *Source* China Daily

these cases, MHEWS with an integrated approach is a critical component of comprehensive disaster risk reduction management for risk informed sustainable development. Regarding to response the major challenges and existing gaps on strengthening multi-stake-holder partnership in MHEWSs mentioned above, greater efforts need to be reinforced in multi-stakeholder partnership to effectively harmonize the bottom-up approach achieved based on best practices at national and local levels and the top-down approach facilitated at international and regional levels for improving countries' capacity in MHEWSs (Gaillard and Mercer 2013). Therefore, it is important to set up platform to bring them together regularly and more effectively to address the issues on MHEWS partnership.

To achieve the Sendai Framework (target-G), great efforts should still be made by multi-stakeholder MHEWS partnerships through integrated and holistic modes. There already have had different kind of partnership mechanisms with inter-governmental feature (e.g. CREWS, ARISTOTLE, GMAS), official authorized mandates' feature (e.g. UNDAC, IASC, GWP), volunteer contribution feature (e.g. HNPW, IN-MHEWS) and PPP feature (e.g. UNEP FI's Principles for Sustainable Insurance (PSI)). To strengthen and harmonize a high-level official multi-stakeholder MHEWS partnership governance is our desire and long-term goal, but currently, we need to go through a practical way based on mutual interests and commitments in the community. There are many successful experiences in building such alliances, such as GWP, HNPW, Global Alliance for Urban Crisis, International Business Alliance for Corporate Ocean Responsibility, and they have shown its strength and weakness in managing its work.

Alternatively, a joint commitment alliance is more achievable. To ensure that MHEWS multi-stakeholders' partnership governance could be agreed by the community. The following issues should be considered:

- Initiate a partnership framework and Forum to unite all MHEWS related bodies in an open manner (e.g. similar to HNPW),
- Thereby establish an MHEWS Partnership Alliance committed to working effectively with its partners through agreed governance arrangements (e.g. Alliance Charter), which chaired by leading international organizations in the thematic areas, and co-chaired by principle research institutes, private sectors, and NGOs in a rotation base,
- Within the alliance, expert advisory teams on different thematic areas of MHEWS, such as climate extremes and public health, atmospheric emergency response for nuclear power plant accidents etc., should be established according to the features of cascading impacts of risk inter-connection between original hazard and derived hazards and the relevant authorized international organizations, associated with academia institutes, private sectors should lead these teams.
- Providing a sustainable platform to share knowledge, technologies, trainings, and even simulation exercises, for future collaboration and partnership. Specific Offices for special purpose or specific regions for the Alliance, which could also be aligned with Knowledge Centres or Portals could be considered. As an example. UN-SPIDER is a programme of the United Nations Office for Outer Space Affairs (UNOOSA), with offices in Vienna, Beijing and Bonn. The Bonn office systematically compiles relevant information on how to use Earth Observation, satellite communication and satellite navigation for disaster risk management and emergency response. This information is made available on UN-SPIDER's Knowledge Portal. Similarly, IRDR has also an International Programme Office in Beijing. Both of them are strongly supported by the German Government and the Chinese Government through the donation of human and financial resources for establishing and maintaining the offices.
- To prioritize its activities on MHEWSs and its supporting governance mechanism on partnership through planning and developing common understanding and recognition with committed actions.
- Facilitating UN Member States in MHEWSs through scaling up its best practices of countries, international organizations and private sectors to establish a fit-for-purpose, user-oriented, and people-centred platform.
- Consultation Groups on financing, risk insurance and crisis and humanitarian management should be established under the Alliance to develop strong joint efforts with these two "End" in order to develop 3 End approach for the Alliance.

Acknowledgements The authors are grateful to the information sharing by various international organizations, government agencies, private sectors and academics. These include, but not limited to, the CREWS, World Bank Group, UNDRR, UNEP, UNDP, UNOCHA, UNOOSA, UNESCO, WHO, WMO, IMO, IAEA, RIMES, IRDR, IFRC, UNDAC, IASC, GWP, WEF, ICL, GCCN, CDEMA, DG-ECHO-ERCC, EU-JRC, IN-MHEWS and Member States and private sectors who shared its best practices mentioned in the paper. We would also like to show our special gratitude to Dr. Jian Liu, UN Environment Chief Scientist and Director of the Science Division, for shari‹ng his pearls of wisdom with us during the course of writing this paper.

References

All Risk Integrated System TOwards Trans-boundary hoListic Early-warning (ARISTOTLE) (2020) European natural hazard scientific partnership. https://aristotle.ingv.it/tiki-index.php. Last Accessed 22 Mar 2020

Caribbean Disaster Emergency Management Agency (CDEMA) (2020) About us, URL: https://www.cdema.org/about-us. Last Accessed 22 Mar 2020

Climate Risk and Early Warning System (CREWS), CREWS Initiatives. https://www.crews-initiative.org/en. Last Accessed 22 Mar 2020s

Clinton J, Zollo A, Marmureano A, Zulficar C, Parolai S (2016) State-of-the art and future of earthquake early warning in the European region. Bull Earthq Eng 14:2441–2458

Cumiskey L, Hoang T, Suzuki S et al (2015) Youth participation at the third UN world conference on disaster risk reduction. Int J Disaster Risk Sci 6:150–163. https://doi.org/10.1007/s13753-015-0054-5

De la Poterie AT, Baudoin MA (2015) From Yokohama to Sendai: approaches to participation in international disaster risk reduction frameworks . Int J Disaster Risk Sci 6(2):128–139

European Commission Joint Research Center (EC-JRC) (2015) Guidance for recording and sharing disaster damage and loss data. EC-JRC, Luxembourg

Field CB, Barros V, Stocker TF, Dahe Q (2012) Changes in climate extremes and their impacts on the natural physical environment. In: Managing the risks of extreme events and disasters to advance climate change adaptation (special report of the intergovernmental panel on climate change), pp 109–230. https://doi.org/10.1017/CBO9781139177245 (Chapter 3)

Gaillard JC, Mercer J (2013) From knowledge to action: bridging gaps in disaster risk reduction. Prog Hum Geogr 37(1):93–114

German Red Cross (2020) Manual—forecast-based financing. https://fbf.drk.de/. Last Accessed 22 Mar 2020

Global Disaster Alert and Coordination System (GDACS) (2020) GDACS overviews and partners. https://www.gdacs.org/About/overview.aspx. Last Accessed 22 Mar 2020

Global Network of Civil Society Organisations for Disaster Reduction (2013) View from the frontline: beyond 2015. https://gndr.org. Last Accessed 22 Mar 2020

Global Water Partnership (GWP) (2018) Climate insurance and water-related disaster risk management. GWP. https://www.gwp.org. Last Accessed 22 Mar 2020

Global Water Partnership (GWP) (2019) Mobilising for a water secure world. GWP, Stockholm

Golnaraghi M (ed) (2012) Institutional partnerships in multi-hazard early warning systems: a compilation of seven national good practices and guiding principles. Springer, Berlin

Hua J, Shaw R (2020) Corona virus (COVID-19) "infodemic" and emerging issues through a data lens: the case of China. Int J Environ Res Public Health 2020(17):2309

Institute of Catastrophe Risk Management (ICRM) (2020) Research projects. https://icrm.ntu.edu.sg/Research/Pages/default.aspx. Last Accessed 22 03 2020

International Network for Multi-Hazard Early Warning Systems (INMHEWS) (2019) Governance document. INMHEWS, Geneva

Keys P, Galaz V, Dyer M, Matthews N, Folke C, Nyström M, Cornell SE (2019) Anthropocene risk. https://doi.org/10.1038/s41893-019-0327-x

Liu Q, Tan ZM, Sun J, Hou Y, Fu C, Wu Z (2020) Changing rapid weather variability increases influenza epidemic risk in a warming climate. Environ Res Lett 15(4)

Luther J, Hainsworth A, Tang X, Harding J, Torres J, Fanchiotti M (2017) World meteorological organization (WMO)-concerted international efforts for advancing multi-hazard early warning systems. Adv Culture Living Landslides. WLF 2017. Springer, Cham

Luther J, Andrioli M, Honoré C, Tang X (2019) Contributing to building a weather-ready, resilient and sustainable society. World Meteorol Organ Bull Geneva 68(1)

National Center for Atmospheric Research (NCAR) (2020) Monitoring, analysis, and prediction of air quality (MAP-AQ). https://www2.acom.ucar.edu/map-aq. Last Accessed 22 Mar 2020

Natural Catastrophe Data Analytics Exchange (NatCatDAX) (2020) https://www.natcatdax.org/Aboutus.aspx. Last Accessed 22 Mar 2020

Nuha E, Charles E, Virginia M (2018) Building urban resilience for disaster risk management and disaster risk reduction. Procedia Eng 212(2018):575–582

Regional Integrated Multi-Hazard Early Warning System for Africa and Asia (RIMES) (2020) Projects sorted by program area. https://www.rimes.int/?q=projects. Last Accessed 22 Mar 2020

Rogers D (2011) Partnering for health early warning systems. World Meteorological Organization Bulletin, Geneva, 60 (1).

Rogers D, Tsirkunov V (2011) Implementing hazard early warning systems 2011

Sassa K (2019) The Kyoto landslide commitment 2020: first signatories. Landslides 16:2053–2057

Su H, Cheung S, Lo E (2018) Multi-objective optimal design for flood risk management with resilience objectives. Stoch Env Res Risk Assess 2018:32

Tang X (2006) Managing disaster risk in a mega-city. World Meteorol Organ Bull Geneva 55(4)

Tang X (2008) New challenges for weather services in changing urban environments. World Meteorol Organ Bull Geneva 57(4)

United Nations (UN) (2006) Global survey on early warning systems. An assessment of capacities, gaps and opportunities towards building a comprehensive global early warning system for all natural hazards. UN, New York

United Nations (UN) (2015) Transforming our world: the 2030 Agenda for sustainable development. UN, New York

United Nations Communications Group (UNCG) (2020) United Nations system response to COVID-19 core messages. UNCG, Geneva

United Nations Environment Programme (UNEP) (2012) Early warning systems: state-of-art analysis and future directions. UNEP, Nairobi

United Nations Development Programme (UNDP) (2018) Five approaches to build functional early warning systems. UNDP, Geneva

United Nations Environment Programme (UNEP) (2020a) Finance Initiative. https://www.unepfi.org/. Last Accessed 22 Mar 2020

United Nations Environment Programme (UNEP) (2020b) EHA connect. https://ehaconnect.org/. Last Accessed: 22 Mar 2020

United Nations Framework Convention on Climate Change (UNFCCC) (2015) Adoption of the Paris agreement. In: Conference of the parties. Twenty-first session. 30 November–11 December 2015. UNFCC, Paris

United Nations Office for Disaster Risk Reduction (UNDRR) (2005) Hyogo framework for action 2005–2015: building the resilience of nations and communities (HFA). UNDRR, Geneva

United Nations Office for Disaster Risk Reduction (UNDRR) (2006a) Platform for the promotion of early warning. Basics of early warning. http://www.unisdr.org/2006/ppew/whats-ew/basicsew.htm . Accessed 12 Nov 2020

United Nations Office for Disaster Risk Reduction (UNDRR) (2006b) Global survey on early warning systems. An assessment of capacities, gaps and opportunities towards building a

comprehensive global early warning system for all natural hazards. UNISDR, Genev
United Nations Office for Disaster Risk Reduction (UNDRR) (2009) UNDRR terminology on disaster risk reduction. UNDRR, Geneva
United Nations Office for Disaster Risk Reduction (UNDRR) (2010) German committee for disaster reduction, emerging challenges for early warning systems in context of climate change and urbanization. UNDRR, Geneva
United Nations Office for Disaster Risk Reduction (UNDRR) (2015) Making development sustainable: the future of disaster risk management. Global assessment report on disaster risk reduction. UNDRR, Geneva
United Nations Office for Disaster Risk Reduction (UNDRR) (2016) United nations plan of action on disaster risk reduction for resilience. UNDRR, Geneva
United Nations Office for Disaster Risk Reduction (UNDRR) (2019a) Global Assessment Report on Disaster Risk Reduction. UNDRR, Geneva
United Nations Office for Disaster Risk Reduction (UNDRR) (2019b) global platform for disaster risk reduction. https://www.UNDRR.org/conference/2019/globalplatform/home. Last Accessed 22 03 2020
United Nations Office for the Coordination of Humanitarian Affairs (OCHA) 2020a What is the cluster approach?. https://www.humanitarianresponse.info/en/about-clusters/what-is-the-cluster-approach. Last Accessed 22 Mar 2020
United Nations Office for the Coordination of Humanitarian Affairs (OCHA) (2020b) Humanitarian networks and partnerships week (HNPW). https://www.unocha.org/humanitarian-networks-and-partnerships-week-hnpw. Last Accessed 22 Mar 2020
United Nations System Chief Executives Board for Coordination (CEB) (2019) United Nations system-wide strategy on sustainable urban development. CEB, Geneva
World Economic Forum (WEF) (2020) The global risks report 2020. WEF, Geneva
World Meteorological Organization (WMO) (2015) WMO Guidelines on multi-hazard impact-based forecast and warning services. WMO, Geneva
World Meteorological Organization (WMO) (2017) A disaster risk reduction roadmap for the world meteorological organization. WMO, Geneva
World Meteorological Organization (WMO) (2018a) Multi-hazard early warning systems: a checklist. WMO, Geneva
World Meteorological Organization (WMO) (2018b) South-East European multi-hazard early warning advisory system (SEE-MHEWS-A). WMO, Geneva
World Meteorological Organization (WMO) (2019) World meteorological congress abridged final report of the eighteenth session. WMO, Geneva
World Meteorological Organization (WMO) (2020) WMO pilot project to enhance the capability of meteorological disaster risk reduction in RAII (Asia) (GMAS-A). https://gmas.asia/. Last Accessed 22 Mar 2020

2020 Kyoto Japan

Resilient Watershed Management: Landscape Approach to Climate Change and Disaster Risk Reduction

Yuka Makino, Thomas Hofer, Mustapha Azdad, and Faizul Bari

Abstract

Resilient Watershed Management differs from traditional approaches in that it incorporates *climate change* and *natural disaster risk* into the overall watershed management planning process. Underpinning these two attributes is its use of a *landscape approach* to natural resource management and risk mitigation, which factors in all aspects of land use and societal needs, from the highest snow mountains, to the middle hills, to the coastal plains. Most importantly, the umbrella under which all three concepts fall is the *sustainable development* of a watershed's inhabitants. Resilient Watershed Management also incorporates an *integration of the physical and social sciences* as fundamental to its operations, from the first analysis of stakeholder vulnerabilities to the baseline data required for the most meaningful monitoring and evaluation plan. The use of *local knowledge* within the planning process is also stressed. Case studies from FAO projects in Morocco and Pakistan provide evidence of the benefits of this approach. In both cases, the investment and objectives addressed land degradation while engaging the local communities and income generation. In Morocco, the project's objective was combating soil erosion and reducing the risk of flooding in overgrazed and eroded area. The investments included the improvement of irrigation systems, apple orchards, promoting income-generation activities, and enhancing the capacity of local institutions. In Pakistan, landslide stabilization was conducted through integrated and participatory watershed management. Activities included bioengineering and soil conservation measures, capacity development of stakeholders, strengthening natural resource management-based enterprises, and the improvement of local livelihoods. In both cases, the involvement of governmental institutions was key in ensuring the follow-up and replication of the projects in similarly affected areas. It is through adaptive management, the integration of key stakeholders and institutions, and the recognition of the landscape as the unit of action, that these communities successfully enhanced their risk management.

Keywords

Resilience • Watershed • Risk • Landscape • Integrated management

Y. Makino (✉)
FAO Forestry Department, Rome, Italy
e-mail: Yuka.Makino@fao.org

T. Hofer
FAO Regional Office of Asia and the Pacific, Bangkok, Thailand
e-mail: Thomas.Hofer@fao.org

M. Azdad
FAO Office in Maroc, Rabat, Maroc
e-mail: mustaphazdad@gmail.com

F. Bari
FAO Office in Pakistan, Islamabad, Pakistan
e-mail: faizul.bari@fao.org

Introduction

Building resilient communities and landscapes is an urgent global imperative, never more so than now as the impacts of climate change increase. Communities worldwide, especially those living in watersheds and whose livelihoods, well-being and identity depend on these ecosystems, can be particularly vulnerable to hazards of increasing magnitude and frequency. Population growth, unsustainable resource management and ecosystem degradation exacerbate the situation and reduce the ability of watersheds to provide ecosystem services in a changing climate. To address these issues, the Food and Agriculture Organization of the United Nations (FAO) has developed the Resilient Watershed Management (RWM) approach (FAO 2017). This landscape management strategy includes disaster and climate risk in watershed

K. Sassa et al. (eds.), *Understanding and Reducing Landslide Disaster Risk*,
ICL Contribution to Landslide Disaster Risk Reduction,
https://doi.org/10.1007/978-3-030-60196-6_40

management and addresses the environmental and socio-economic dimensions of risk and watershed management.

The RWM approach responds to global needs. For example, it is aligned with the targets and goals of the Sendai Framework for Disaster Risk Reduction, the Sustainable Development Goals (SDGs) and the Paris Agreement from the UN Framework Convention on Climate Change. Furthermore, according to the 2019 Global Assessment Report on Disaster Risk Reduction (UNDRR 2019), landscape management approaches need to evolve so as to integrate disaster risk reduction (DRR) with climate change adaptation and mitigation. RWM is one such approach. Adopting such strategies is critical for countries as they move towards risk-informed socio-economic development in a rapidly changing climate. As shown by the two case studies in Morocco and Pakistan presented, landscape management mechanisms that integrate DRR, climate change adaptation and mitigation can increase the resilience of communities. These case studies offer valuable lessons for the future of watershed management and provide evidence of the importance of moving towards RWM.

Principles of Resilient Watershed Management

The earliest records of watershed management date back to 800 BC (Wang et al. 2016). More recently, until the 1970s and 1980s, this approach focused primarily on hydrological processes, but changed once water scarcity was recognized as a human problem, driven by industrial development, population growth and climate change (FAO 2006). To address these issues, watershed management evolved to include biological, physical and socio-economic elements of the landscape, as well as a focus on ecosystems as a whole rather than on their isolated features. The results of implementing this more holistic approach showed that adaptive management techniques and the involvement of all relevant communities and stakeholders were key to achieving sustainable management. In its publication *Watershed management in action: Lessons learned from FAO field projects*, FAO presents the steps required to implement modern watershed management (FAO 2017).

At the 1992 Earth Summit, countries recognized the benefits of holistic management of water resources and adopted the concept of Integrated Water Resource Management (IWRM). The United Nations Environment Programme (UNEP) defines IWRM as a "*process which promotes the coordinated development and management of water, land and related resources to maximize the resultant economic and social welfare in an equitable and sustainable manner.*" IWRM and Integrated Watershed Management (IWM) are often used interchangeably, or may also be used to address different scales where IWRM happens at national level and IWM is conducted at watershed level. IWRM, and consequently IWM, are now recognized and integrated widely into policies and national planning and development programmes. IWRM (and IWM) has been identified as a key contributor to achieving the 2030 Sustainable Development Agenda (UNEP 2018). Indeed, SDG Indicator 6.4.1 particularly measures the degree of integrated water resource management implementation. The uptake of this approach has been so extensive that 80% of countries that have provided information on IWRM (172) as part of SDG 6 reporting, have been shown to have the foundations for IWRM implementation (UNEP 2018).

In recent years, IWRM and IWM have benefited from advancing technology in the fields of geographic information system, remote sensing and other monitoring techniques. However, gaps remain in implementation, and the scope of work is changing. This is particularly the case with climate change risk and disaster risk management. As a result, there are many opportunities for work to improve monitoring processes and streamline risk management into landscape management approaches. However, the challenge is how these management approaches can adapt and cope in a rapidly changing climate, where risk is increasing faster than resilience (WWAP 2019). This particular challenge is attracting attention in the international arena. For example, the 2019 Global Assessment Report on Disaster Risk Reduction (UNDRR 2019) highlights the fact that the holistic nature of IWRM/IWM has allowed for policy developments to be integrated with the water, food, energy, climate change and human health nexus, enabling multidisciplinary approaches to be adopted that bring together different sectors to address risk management. It is in the context of this space and these needs that the RWM approach has been developed to include risk in watershed management (Fig. 1).

Landscape and Integrated, a Risk-Based Approach to Watershed Management

The Resilient Watershed Management approach integrates a risk perspective into the various stages of integrated watershed management, where risk refers to both disaster and climate. Resilient Watershed Management applies the landscape approach, and a landscape may consist of one or more watersheds. According to FAO, "*A landscape approach deals with large-scale processes in an integrated and multidisciplinary manner, combining natural resource management with environmental and livelihood considerations. The landscape approach also factors in human activities and their institutions, viewing them as an integral part of the system rather than as external agents*" (FAO 2019).

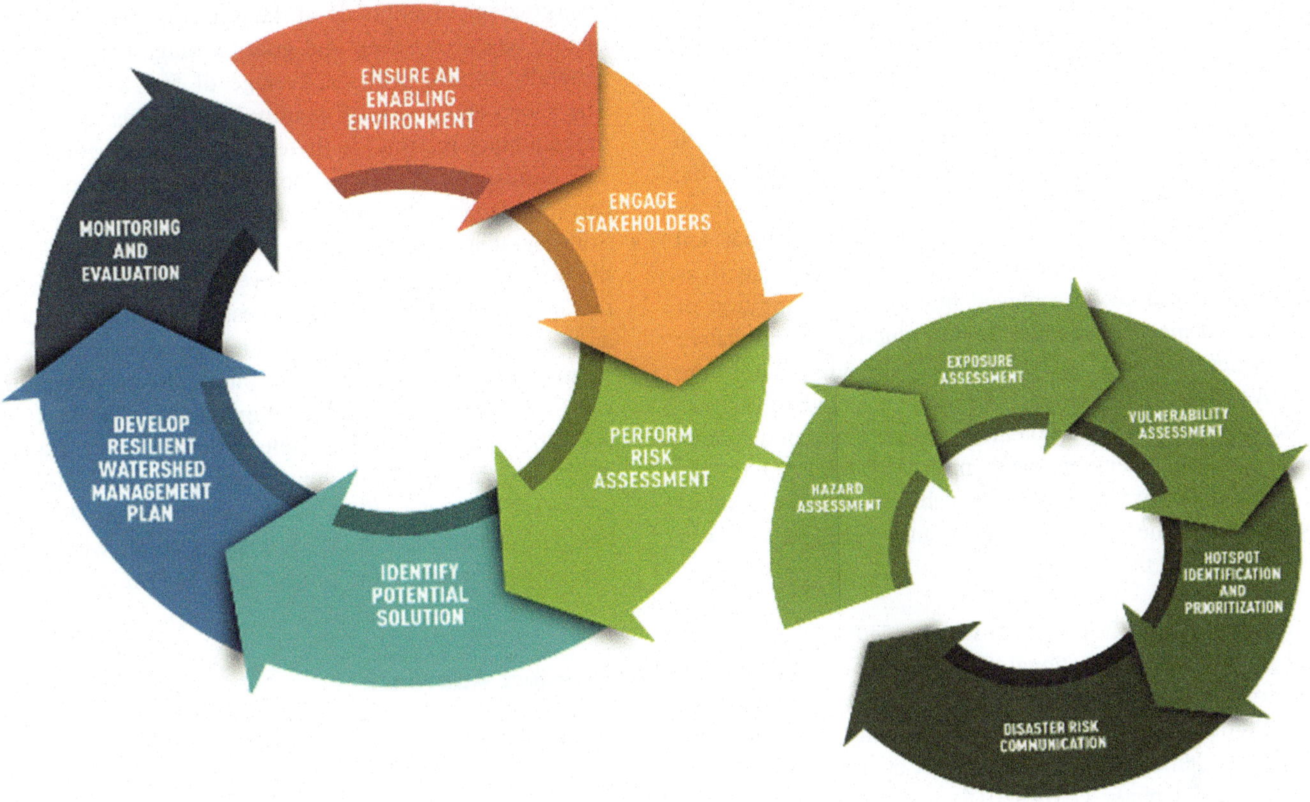

Fig. 1 The resilient watershed management cycle (FAQ 2021)

By adopting an integrated and landscape approach, RWM ensures that risk is considered in a holistic way as it takes into account various land uses and activities within the landscape. This allows for a better understanding of the different interests of stakeholders and the related natural resource use (forestry, agriculture, fisheries, fodder, fuelwood, non-timber forest products, ecotourism services, flood control, water storage, filtration, biodiversity conservation and carbon sequestration, among others). Furthermore, RWM recognizes the social dimension (such as gender dynamics, equity, customary management practices), livelihoods, natural resource management (NRM) and ecosystem services provided by the watershed.

In summary, Resilient Watershed Management implies:

- a landscape approach to risk-based integrated watershed management;
- an integrated approach to risk and vulnerability reduction;
- recognizing multiple hazards and their different cascading effects;
- combining short-term planning in priority hotspots with long-term planning to account for the changing climate;
- disaster risk management and climate change adaptation and mitigation at watershed level;
- ensuring that upstream solutions do not have negative impacts on downstream areas and vice versa; and
- flexibility to adapt to potential disturbances, largely as a result of climate change.

The RWM approach aims to provide a mechanism for countries to improve the resilience of communities by bringing together sectors and actors at different scales of governance. In addition, it provides a platform to combine national and subnational policies with local actions in key high-risk areas or hotspots. This ensures a better understanding of risk and the uptake of measures. It is hoped that countries will adopt this approach and adapt it to their local situations in order to help communities grow and develop sustainably, while fostering the wise use of their natural resources.

Case Study 1—Morocco

General and Local Context

The Resilient Watershed Management project is situated in the upstream area of the Moulouya river basin, which originates at the junction of the Middle and High Atlas in the Province of Midelt and empties into the Mediterranean Sea in northern Morocco. The southern border of the project area is made up by the Mouaskar and Ayachi mountain range. The project covers a total area of 1600 km^2, of which 1100 km^2 are located in the Anseguemir and 500 km^2 in the Outat watershed.

The overall objective of the project was to combat soil erosion and reduce the risk of natural hazards, such as floods. This was to be achieved through the promotion of institutional mechanisms and the introduction of innovative approaches to soil and water conservation, and the improvement of livelihoods. The project aimed to demonstrate the feasibility and potential impact of the new watershed co-management approach promoted by FAO (2006, 2017) and to explore the applicability of this 'model' approach to other parts of the country. The project was implemented in the following phases:

- The first phase (2010–2013) was funded by the Spanish Agency for International Development Cooperation. During this phase, the project area was selected, local institutional capacities were strengthened and a pilot co-management plan for the project area was developed.
- The second phase (2014–2015) was financed partly by FAO and partly by the Swiss Agency for Development and Cooperation (SDC). During this interim period, the methodology and techniques used in Phase I were strengthened.
- The third phase (2016–2019) was fully funded by SDC. It focused on facilitating and promoting the participation of local stakeholders in project implementation and the development process.

Morocco experiences recurring droughts that accelerate the process of natural resources degradation. Soils are highly vulnerable and subject to erosion. The density of forests and overall forest cover is declining and, as a result of overgrazing, nomadic pastoral areas are threatened by desertification. Natural disasters, particularly floods, are increasing.

The rural population of the two watersheds is approximately 37,000, spread over 54 douars (villages). The population density in the two watersheds is 23 people/km^2. Both watersheds are affected by degradation, with generally low vegetation density. The project area faces various challenges related to degradation of the natural environment, the most important of which are:

- Overexploitation of wood: The poorest household use between 2 and 3 tonnes of fuelwood during the winter season. Affluent households consume 2.5 times more—between 4 and 8 tonnes/inhabitant/year.
- Overgrazing: Livelihoods of the nomadic pastoralists are mainly based on sheep and goat farming. The herds are generally mixed, with an average of 500 heads/herder. The sheep and goats are mainly local breeds adapted to mountainous regions, with relatively low performance. The number of animals in the pastoral areas exceeds the carrying capacity by three to four times.
- Soil degradation: Between 80 and 90% of the surface of the watersheds is impacted by severe soil erosion, at a rate that significantly exceeds the threshold tolerated for arid and semi-arid areas (2 tonnes/hectare (ha)/year). The estimated average soil loss amounts to 40–50 tonnes/ha/year, a value that is considered extremely high. In 1992 and 2008, the city of Midelt suffered heavy floods, originating in the Oued Outat watershed. Some 10% of the vulnerable sites identified by the National Plan for Flood Protection are located in the Greater Moulouya watershed (including the project intervention zone).

The Project Approach

The challenge was to identify and test innovative and locally adapted approaches to protect infrastructure and the renewable natural resources of the watersheds against surface erosion and floods. This was to be done by strengthening institutional tools, as well as by introducing innovative techniques for the conservation and development of soils and local species of vegetation, with the ultimate objective of contributing to improved resilience and well-being for communities. The Government of Morocco has drawn on the experience of FAO to assist in developing an innovative participatory and integrated approach to natural resource management.

Political and public institutions play a leading role in terms of territorial development. They identify the challenges faced by different regions and develop participatory action plans to resolve them. Within this context, two specific institutional structures have been set up to manage the project. These are the Interinstitutional Steering Committee (ISC) and the Interinstitutional Technical Committee (ITC):

- The ISC is chaired by the Governor of the Province of Midelt. It brings together the decision-makers of the province and the presidents of the communes concerned in order to make decisions regarding the regionalization and implementation of the plans.
- The ITC is chaired by the Chief of the Circle of Midelt and operates under the supervision of the Governor of the Province of Midelt. It brings together specialists from the technical services of the provincial institutions, the

communes and local bodies, such as cooperatives and associations. The ITC decides on the technical, social and economic feasibility of the proposed actions, as well as of the implementation mechanisms.

Facilitation of communication between these bodies is provided by the Project Management Unit, composed of consultants and facilitators recruited by FAO. One tool that has proved highly effective has been to include rural communities in project management through consultation workshops. This approach has enabled the indigenous and empirical knwoldege of local people to be captured, and integrated into project implementation. It has led to the launch of a number of microprojects, which specifically addressed the problems identified by communities. As a result, participants at community level have gradually acquired a sense of responsibility for implementing project activities. Similarly, local authorities have gained confidence to entrust these tasks to the communities.

Main Project Achievements

The project's income-generating activities focused on youth and women. It supported communities to invest in local processing facilities, which allowed them to transform local products into value added items, and to manage these facilities in a collective way. Such pilot processing units were established in the project's second phase and are run by youth and women's cooperatives—an important feature for sustainability.

In the context of soil and vegetation conservation, the management of microwatersheds serves a practical purpose. Based on short periods of intervention, it enables good practices to be highlighted, as well as participatory management with communities. In addition, it facilitates the clarification of certain aspects of data collection, interpretation and extrapolation. Overall, it allows the positive impact and replicability of small interventions to be demonstrated, especially in situations where budgets are modest. In line with this approach, project interventions focused on three microwatersheds, with a total area of 14,000 ha. They were based on a comprehensive plan for each site, developed through close consultation with the relevant communities and technical institutions. Risk factors present in the community were identified during this process (Table 1). The provincial institutions are committed to replicating the successful practices in their regular programmes.

The interventions mainly consisted of:

- Promoting income-generating activities for the benefit of 841 young people and women in the microwatersheds. Distribution of 36,000 fruit trees, 7414 chickens of local breeds and 360 beehives—total investment USD 234,695.
- Constructing and rehabilitating irrigation water diversion channels over 6320 m. 1946 farmers benefited from these interventions—total cost of engineering works USD 223,500.
- Soil and water conservation activities were carried out to protect 103 ha of cultivated fields, and important infrastructure installed against soil erosion and floods. For that purpose, 6670 m^3 of gabions were fixed along the slopes of prioritized river creeks. 1694 farmers directly or indirectly benefited, with investment totalling USD 491,000. In addition, with an investment of USD 166,547, 127.6 ha of pilot and experimental plots were stabilized with anti-erosion plantations, benefiting 1663 inhabitants. The good practices are replicable in the entire area, which would significantly increase the number of beneficiaries.
- Capacity-building of institutions, partners and communities involved 16 training sessions for 510 beneficiaries and 5 exchange visits for 86 beneficiaries—total investment USD 85,000.

The total project investment for implementation of these pilot activities was USD 1,200,742.

A consistent approach throughout the project involved ensuring the integration of project activities into the provincial authorities' regular programmes over the entire territory of the two basins. Accordingly, a significant number of diverse activities implemented in the project area were financed directly by the provincial institutions. Some of these acivities benefited from and duplicated the good practices promoted within this project. The following lists government-led activities related to the project's objectives:

- Improvement of the hydro-agricultural system through the maintenance and rehabilitation of 124 km of irrigation water diversion channels and the creation of 3 drinking water points for livestock investment USD 8,717,000.
- Contribution to the value increase of the small-scale farming sector through the distribution of 35,000 fruit tree seedlings—investment USD 51,000.
- Water and soil conservation and combating erosion by installing 13,500 m^3 of gabions—investment USD 420,000.
- Opening and maintenance of 44 km of village access roads—investment USD 4,330,000.
- Rehabilitation and maintenance of local vegetation and enrichment of 5700 ha with aromatic, medicinal and pastoral plants—investment USD 530,000.
- Flood protection works on 10 km of river banks—investment USD 670,000.

Table 1 Identified risk factors and mitigation activities

Risk factors	Mitigation activities
Nomadic pastoralists do not respect the protection measures established for the forest plantations	Silvopastoral planning study proposed
Excessive fuelwood exploitation	Launch of a study on the search for alternatives to fuelwood consumption
Poor management of runoff	Dialogue with specialized agencies to identify a more sustainable approach to water management
Fragmentation of private irrigated lands and mismanagement of collective lands	Establishment of farmers' groups to identify more rational and planned management of collective and private lands
Structure of the associations is often weak	Limit associations' interventions to tasks of coaching the population and implementing cooperatives' activities
Illiteracy of women	Combat illiteracy through dialogue with the primary education sector and by promoting education campaigns for women
Spatial and temporal variability of rainfall	Collaboration with projects working on climate change adaptation
Challenge to integrate all institutions into the fight against environmental degradation, of forests in particular	Promote awareness of institutions and encourage them to take ownership of the watershed co-management plan established in the context of the project

In total, the provincial institutions invested USD 14,718,000 in implementing these complementary activities in the project areas.

For implementation of project activities, the Project Management Unit team used all forms and types of social structures (modern and traditional), which are specific to each village and locality. At the same time, executives from the provincial institutions were closely involved in field operations and follow-up through the ISC and ITC.

Follow-Up and Exit Strategy

During the project's final workshop, participants and key stakeholders agreed that governance of the Co-Management Plans will be ensured by the Governor of the Midelt Province through the Provincial Technical Committee, which meets periodically to coordinate and steer all institutional activities.

Case Study 2—Pakistan

General and Local Context

A project was implemented from January 2007 to December 2009 to assist the Earthquake Rehabilitation and Reconstruction Authority (ERRA) and partners in restoring livelihoods in an earthquake affected area of Pakistan. The watershed management component involved landslip stabilization through integrated and participatory watershed management. The project was located in the nine earthquake affected districts of Khyber Pakhtunkhwa (KP) and Pakistan Administered Kashmir (PAK). The total affected area of KP was 16,000 km^2, while an 8900 km^2 area was affected in PAK.

On 8 October 2005, a devastating earthquake measuring 7.6 on the Richter scale struck part of North West Frontier Province, now renamed Khyber Pakhtunkhwa, and Pakistan Administered Kashmir in the northwest of Pakistan. In total, 3–4 million people from nine districts were affected, with an estimated death toll exceeding 80,000. A FAO damage assessment survey (November 2006) estimated the total cost of damage and losses to the agriculture and livestock sector at USD 409 million. In addition, a large number of landslides occurred in various catchment areas of the earthquake-affected districts. A significant number of irrigation channels were destroyed and damaged, with repercussions on food security and local livelihoods.

Under the early recovery programme, FAO implemented a project to assist the ERRA and its partners in restoring livelihoods in the earthquake affected area of Pakistan. A major component was collaborative watershed management, with the objective of stabilizing the landslides through a combination of measures, under the leadership of watershed management committees. In addition, a number of income-generating activities for local people were introduced. These contributed to the sustainable management of forests, while also helping to improve local livelihoods.

The project addressed the major issue of landslide stabilization, damaged water channels and rehabilitation of degraded lands through planting and soil conservation measures, capacity development of the project implementers, and the creation of an enabling environment for NRM-based enterprises, contributing to the improvement of local livelihoods. Awareness raising of sustainable watershed management was also a key activity. All activities were based on a holistic and integrated planning approach, with the overall objective of creating resilience for the environment and communities to future shocks.

The Project Approach

For the enabling environment, it was important to ensure that key stakeholders could play a role. For this purpose, local communities were grouped into watershed management committees, responsible for identifying needs and subsequently implementing interventions. These committees received technical support from FAO, while the Forest Department committed to assisting local communities in implementing integrated watershed management plans. Technical support for bioengineering structures was also secured from the International Centre for Integrated Mountain Development.

Local stakeholders included members of local community organizations, forest departments and international organizations. The private sector was also involved, with a focus on developing various NRM-based enterprises. The watershed management committees were the vehicles for implementation of the project. Local non-governmental organizations were also engaged to support local communities.

Prior to preparation of the watershed management plans, the project conducted a number of surveys to identify risks and vulnerabilities. Hazard mapping of the project sites was also implemented. Based on the findings of these surveys, 17 Resilient Watershed Management Plans were formulated, spread over the nine earthquake affected districts, to be implemented by the watershed management committees. These plans followed the principles of collaborative watershed management, prepared in a participatory manner, involving all key stakeholders. Particular attention was paid to the priorities of local communities.

Based on the needs assessments and surveys and on the specifics of the watershed management plans, a number of activities and solutions were identified, including: delineation of the selected watersheds; Participatory Rural Appraisal including upstream and downstream relations; implementation of priority activities; and capacity building of beneficiaries and line department staff. The main field interventions were landslide stabilization through planting and bioengineering structures; rehabilitation of irrigation channels; establishment of private nurseries by women; kitchen gardening and backyard poultry; establishment of watershed plantations; and NRM-based income-generating activities.

Main Project Achievements

The project was highly effective in reducing landslide risks and increasing the resilience of local communities. The bioengineering practices were particularly successful and were widely replicated in the relevant districts. The independent evaluation report confirmed that 18 months after the project closure, the beneficiary groups continued their prevention and mitigation measures, ensuring project sustainability. The 17 watersheds have significant potential to serve as learning sites for the dissemination of knowledge, information and best practices, deliver on-the-job training and expand the integrated watershed management approach to build resilient landscapes in the country.

The post-project evaluation shows that local people involved in the project are continuing the NRM-based enterprise activities, which provide them with additional income through private nurseries, kitchen gardens, collection and processing of medicinal plants, and backyard poultry. Capacity-building prepared nursery staff to engage with the 10 billion trees afforestation programme. They earn much higher revenues than other nursery growers who were not exposed to the innovative practices. The protection of the catchments also contributed to enhancing their functions, services and products, with a direct positive impact on local livelihoods.

Follow-Up and Exit Strategy

Both the quantity and quality of the implemented interventions were evaluated, using a participatory monitoring and evaluation methodology, and progress of project implementation was reviewed on a monthly basis. In addition, FAO conducted an independent evaluation, reviewing impacts one and a half years after the project closure. The project was rated very high in terms of the creation of resilient landscapes through watershed management. The evaluation report stated that the forest department at province and national levels institutionalized key processes. The forest department is applying the project methodology and strategy in other catchment areas, as well as to the 10 billion Trees Afforestation Programme in Khyber Pakhtunkhwa.

Conclusion

People's livelihoods and well-being have been increasingly affected by the negative impacts of climate change. As the magnitude and frequency of natural hazards increase, countries and communities are now in a position where actions need to take place to strengthen the resilience of people and systems in order to reduce existing disaster risk while adapting to a changing climate. Resilient Watershed Management provides an efficient and effective tool to address these challenges that are global in nature, but which require actions at different scales of governance and collaborative approaches across sectors.

References

FAO (2006) The new generation of watershed management programmes and projects. FAO forestry paper 150. FAO, Rome

FAO (2017) Watershed management in action—lessons learned from FAO field projects. FAO, Rome

FAO (2019) Cited 4 Feb 2020. www.fao.org/land-water/overview/integrated-landscape-management/en/

FAO (2021) Resilient watershed management handbook. FAO Forestry Paper 204. Under draft. FAO, Rome

UNESCO World Water Assessment Programme (2019) The United Nations world water development report 2019: leaving no one behind. Paris

United Nations Environment Programme (2018) Progress on integrated water resources management. Global baseline for SDG 6 indicator 6.5.1: degree of IWRM implementation

United Nations Office for Disaster Risk Reduction (2019) Global assessment report on disaster risk reduction. Geneva

Wang G, Mang S, Cai H et al (2016) J For Res 27:967. Cited 4 Feb 2020. https://doi.org/10.1007/s11676-016-0293-3

2020 Kyoto Japan

Integrating DRR into the Conservation and Management Mechanisms of the Internationally Designated Sites—View of IRDR

Qunli Han and Fang Lian

Abstract

Internationally designate sites (IDAs) refers to Biosphere Reserves (BR), World Heritage (WH) properties and Global Geoparks designed by UNESCO for global conservation of biological, cultural and geological diversities and sustainable use. The relevance of DRR in IDAs has been seen increasing due to the concern over climate change impacts and earthquake, landslides and volcano eruptions in recent years. What at stake are firstly about the conservation values of IDAs. It is also about safety of people, both living near IDAs and visitors, as most IDAs are also important tourism destinations. IDAs are thus highly relevant to SFDRR Priorities and Sustainable Development Targets. WLF4 in 2017 provided an important occasion to underscore landslide disaster risks in IDAs and the needs for ICL's global support to IDAs. Given disaster response in practice remains more a common approach in IDAs, 2018 Huangshan Dialogue, co-sponsored by UNESCO-HIST and IRDR, produced an initial set of recommendations for improvement. For better integration of DRR in IDAs practices, one needs also to look into the statutory instruments used in IDAs. In parallel, greater effort is required to further develop an international DRR initiative for IDAs, in order to facilitate research cooperation, data and knowledge sharing, multi-early warning and fast assessment, access to expertise and capacity building. All these are of interest of IRDR in pursuing a new partnership with UNESCO, ICL and others in its planned new DRR research agenda for 2030.

Keywords

Internationally designate sites • Disaster risk reduction • Partnership • IRDR • UNESCO • ICL

The Question of Relevance

Internationally designate sites (IDAs) is a relatively new term people use to refer to Biosphere Reserves (BR), World Heritage (WH) properties and UNESCO Global Geoparks (UGG) designed by UNESCO.[1] Biosphere reserves are terrestrial and coastal ecosystems designed by Man and the Biosphere (MAB) Programme of UNESCO to promote solutions to reconcile the conservation of biodiversity with its sustainable use (UNESCO 2018b). Together, these sites constitute a World Network of Biosphere Reserves (UNESCO 2018a). World Heritage sites are those places recognized of having 'Outstanding Universal Values' (OUVs) in accordance the criteria (6 cultural and 4 natural criteria) of the Operational Guidelines of the WH Convention (UNESCO 2018d). Each WH site is unique and irreplaceable. A new IDA category is UNESCO Global Geoparks, referring to single, unified geographical areas where sites and landscapes of international geological significance are managed with a holistic concept of protection, education and sustainable development (UNESCO 2018c).

As of 2019, there are 701 Biosphere Reserves in 124 countries, 1121 sites inscribed in the World Heritage List, among which 869 cultural, 213 natural and 39 mixed, and 114 Global Geoparks in 41 countries. The total terrestrial coverage of IDAs are already over 1 billion ha with an estimated human population over 200 million. Although there is no proper statistic data regarding the overall

[1]There is also a category of wetland of international importance under Ramsar Convention, which is regarded also as IDAs but not so much relevant to the topic of this paper.

Q. Han (✉)
IRDR International Programme Office, B705, Dengzhuangnan Rd., Beijing, 100094, China
e-mail: qunli.han@irdrinternational.org

F. Lian
IRDR International Programme Office, B713, Dengzhuangnan Rd., Beijing, 100094, China
e-mail: fang.lian@irdrinternational.org

K. Sassa et al. (eds.), *Understanding and Reducing Landslide Disaster Risk*, ICL Contribution to Landslide Disaster Risk Reduction,
https://doi.org/10.1007/978-3-030-60196-6_41

economic activities of IDAs at present, the fact that IDAs provide bases for livelihoods of this large population is already a strong evidence. Tourism sector has been in rapid growth over three decades, resulting not only in direct economic revenues, but also job opportunities as well as the fast development of transportation infrastructures and recreational facilities.

The geographic locations and environmental conditions of many IDAs, especially those of natural areas, make many of them highly exposed and vulnerable to various natural, and increasingly, man-made hazards. As shown in Fig. 1, 92% of UNESCO designated sites could be potentially exposed to at least one out of the main natural hazards, and overall landslides are the most frequent geohazards (Pavlova et al. 2018). Jiuzhaigou, one of the most scenic sites in Qionglai Mountain range in China, which is both BR and WH, suffered first in 2008 earthquake of Wenchuan, and the 2017 earthquake with epicentre right at the site. This later earthquake and the earthquake-induced landslides (Fig. 2) caused enormous damage to the site, and heavy blow to the tourism industry (Zhao et al. 2018), which is central in local economy. Earthquake in Nepal in 2015, also produced devastating damage to the cultural WH Kathmandu Valley. A part from loss of human life, the economic damage is hard to bear, given the weight of tourism sector in Nepal (Fallahi 2015). Different types of landslides frequently occur on the slopes surrounding such sites as Machu Picchu WH site in Peru (Klimeš 2013), damaging the accessing roads and tourist paths. Climate-induced weather extreme events cause growing concerns to conservation communities. 2019 extensive forest fire in Australia has caused huge loss to Australia's natural heritage. The impacts and consequences of this disaster remain to be assessed.

The relevance of DRR in IDAs is clear and high. In fact, the four priorities of SFDRR in terms of reducing losses, improving governance, investing in DRR and building back better are all relevant to IDAs. Further, if one considers the overall purposes of IDAs in support sustainability of the planet, these sites can be considered as the entry points to connect SFDRR with SDGs and Paris Agreement on Climate Change and the related international conventions.

Recent International Effort to Connect IDAs with DRR

Integration of DRR into IDAs is a process of transformative change. It takes time and require leading institutions especially UNESCO to push for it. During 2006–2007 period, the WH Committee has developed a document entitled 'Issues related to the state of conservation of World Heritage properties: Strategy for Reducing Risks from Disasters at World Heritage properties' (UNESCO 2007) which was adopted at the 31st session in Christchurch of New Zealand in 2007. A series of new efforts since SFDRR establishment should be mentioned and underlined:

Sendai Process. The establishment of SFDRR in 2015 certainly generated a new momentum for all to look at the risk perspective instead of disaster response in IDAs. In fact, UNESCO was actively involved in the discussions and preparations of SFDRR from all its sectors but especially from Sectors of Natural Sciences that responsible for BRs and UGGs and its scientific programmes in ecology (MAB), hydrology (IHP) and geology (IGCP), and from Culture Sector on WH. In this process toward the 2015 Sendai Conference, a specific support of UNESCO was made toward the preparation and launching the *Sendai Partnerships 2015–2025 for Global Promotion of Understanding and Reducing Landslide Disaster Risk* (ICL et al. 2015), which was initiated by International Consortium of Landslide (ICL) on landslide research cooperation. The Sendai Partnership later became a multi-agency commitment to which IRDR also signed up as partner in 2017.

UNESCO 2015 Survey. UNESCO intersectoral group on DRR carried out a survey (https://www.soscisurvey.de/naturlhazardsunescosites/) in IDAs in 2015 addressed to site managers. The survey aimed to provide an overview of natural hazard distribution at UNESCO-designated sites. In Asia's IDAs, nearly 80 indicated being affected or exposure by landslide and mud-rock flow (Fig. 3). Yet, the sites that have made comprehensive risk assessments are less about 10 percent worldwide, showing a huge need as well as potential roles for DRR research communities to make contribution.

WLF4 in Ljubljana 2017. The 4th World Landslide Forum provided an occasion to present UNESCO DRR survey in IDAs and to build connection with landslide research communities, especially in relation to the implementation of the Sendai Partnership 2015–2025. In the panel discussions at WLF4, the risk assessment and disaster management in the cultural heritage at the imperial resort palace of Lishan, Xian, China (Fukuoka et al. 2005; Sassa et al. 1996) was referred as a pioneering case in addressing landslide risk at heritage areas. The discussion followed outlined the potential needs for scientific and technological advisory service for many IDAs concerned with landslide risk, and this later was included in the agenda of WLF5 2020.

The Huangshan Dialogue 2018. A further discussion was made at the 2018 Huangshan Dialogue in November 2018, co-sponsored by UNESCO-HIST and IRDR (HIST 2018). The Huangshan Dialogue, a biennium platform started by UNESCO and its category II centre HIST on the use of space technologies for IDAs in 2014, is the only platform bringing together all three IDA mechanisms outside UNESCO Headquarters. Understood the importance of the Huangshan Dialogue, IRDR took initiative to convene a

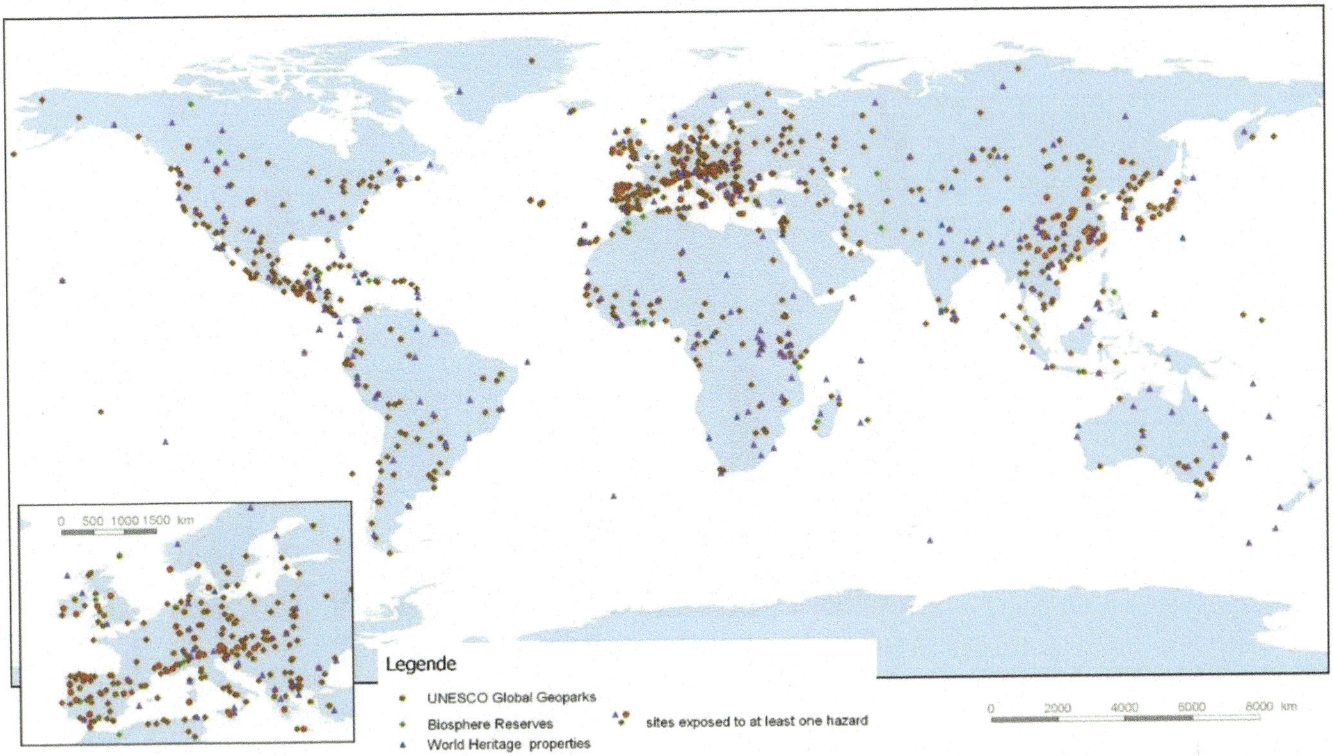

Fig. 1 UNESCO designated sites (BR, UGG, WH) potentially exposed to at least one natural hazard

Fig. 2 Landslides induced by 2017 earthquake caused serious damage to part of the landscape of Jiuzhaigou. Photo by the authors in 2018

session entitled 'Disaster Risk Assessment and Mitigation for UNESCO Sites'. The session examined a number of field case studies and further clarified the needs for IDAs to work in DRR in line with the SFDRR. Huangshan Dialogue and produced an initial but specific set of recommendations (Table 1) which made clear connection of IDAs to SFDRR, as well as to other post 2015 agenda such as SDGs and Paris Agreement.

The outcome of Huangshan on DRR for IDAs is important. While some recommendations have been implemented as projects, such as the case in Jiuzhaigou site as well as input toward GP2019, most of them remain to be done.

Integration Within Respective IDAs Statutory Mechanisms

To ensure DRR become integrated in the IDA systems which obligate the countries to take expected actions, there is a need to reflect SFDRR in the statutory mechanisms of IDAs. This is particularly concerning the nomination process, monitoring and assessment missions and periodic reporting, at both state and site levels. Currently, the integration status in this regard differs in IDAs.

Among three categories of IDAs, WH is best positioned in terms of integrating disaster risk issues into the conservation regime. The Strategy for Reducing Risks from Disasters at World Heritage properties' which was adopted in 2007 set five broad objectives that are fairly consistent with SFDRR:

1. Strengthen support within relevant global, regional, national and local institutions for reducing risks at World Heritage properties
2. Use knowledge, innovation and education to build a culture of disaster prevention at WH properties

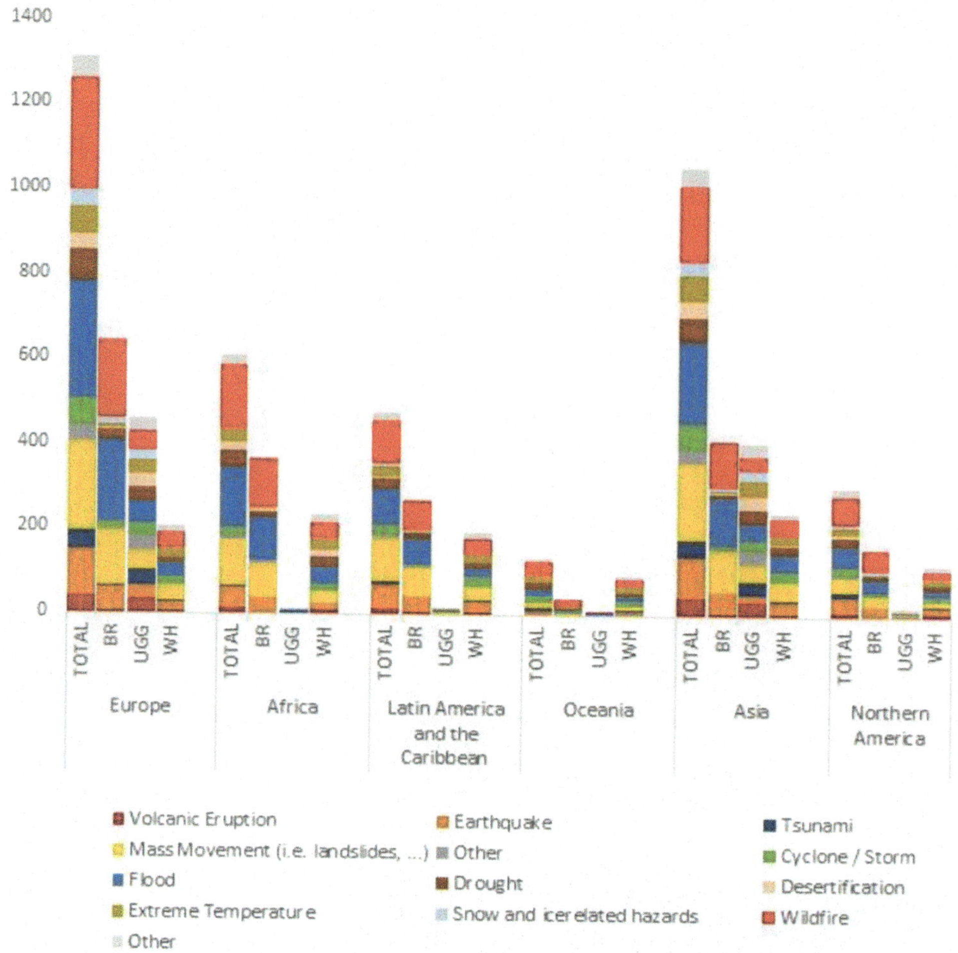

Fig. 3 Number of hazards at UNESCO designated sites per type of hazard (Pavlova et al. 2018)

Table 1 Recommended actions for disaster risk reduction for UNESCO sites (IRDR 2019)

The following recommendations and key discussion points were made during the session:
1. The use of remote sensing and related technologies for hazard risk assessment and early warning has significant potential for further application
2. An integrated, comprehensive global database on the application of remote sensing and related technologies for disaster risk reduction would be of considerable value
3. Engineering solutions to mitigate disaster risk must be designed so as to be appropriate in the context of local conditions, with standards aligned with trends in hazard intensities
4. The comprehensive, recently completed DRR survey among UNESCO-designated sites has yielded important data and results with the potential to serve as the basis for decision-and policy-making
5. Existing levels of site-to-site and other modalities of international cooperation do not reach their full potential—considerable benefits could be derived from increased and intensified cooperation
6. Local and traditional knowledge of cultural and natural heritage—ranging from knowledge of techniques, materials, landscape ecology, agriculture and more—are essential components in reducing and mitigating disaster risk and should be given full consideration
Towards the implementation of DRR at UNESCO-designated sites, the following actions were recommended by the participants of the session
1. Consider the establishment of an international task group on DRR for IDAs. This task group could be composed of committed DRR organizations such as IRDR, IDMR and ICL, under the overall guidance of UNESCO. First understanding on the modality of such a group would be discussed between UNESCO and IRDR in the upcoming year
2. Continue the discussion on DRR for IDAs started at the Huangshan Dialogue, with particular attention to relevant indicators under international frameworks (Sendai, 2030 Agenda and the SDGs, Paris Agreements, New Urban Agenda, SAMOA Pathway, etc.)
3. Identify and promote concrete DRR actions at IDAs through the design and implementation of field projects and research cooperation, with focus on sites affected by recent major natural hazards such as Jiuzhaigou Biosphere Reserve/World Heritage and Kathmandu Valley as demonstration cases
4. Through UNESCO and IRDR, bring the issue of DRR for IDAs to the attention of the 2019 Global Platform for Disaster Risk Reduction

3. Identify, assess and monitor disaster risks at WH properties
4. Reduce underlying risk factors at WH properties
5. Strengthen disaster preparedness at World Heritage properties for effective response at all levels.

In 2008, the World Heritage Committee further adopted a standard list of factors (14 factors) affecting the Outstanding Universal Value of World Heritage properties as part of the questionnaire of the Periodic Reporting exercise (Section II) (UNESCO 2008). Among the 14 factors, the factor on climate change and sever weather events covers storms, flooding, drought, desertification, and the factor of sudden ecological or geological events includes volcanic eruption, earthquake, tsunami/tidal wave, avalanche/landslide, erosion, siltation/deposition and fire. Other factors cover infrastructure development that may affect the OUV of WH properties.

In BR nomination and periodic review, there is only one general reference under field monitoring concerning natural hazards. There is neither requirement nor mechanism on risk assessment and measures for risk reduction for sites to be functioning as BR. The MAB Council in 2016 made decision to develop a new Operational Guidelines for the World Network of Biosphere Reserves,[2] and this offers a good opportunity of introducing DRR in the BR system both in the nomination and periodic review process, as well as the site management plans. This Operational Guidelines is still in the process of elaboration.

In the criteria ii of UGG, it articulates that UNESCO Global Geoparks should '… promote awareness of key issues facing society in the context of the dynamic planet we all live on, including but not limited to increasing knowledge and understanding of: geoprocesses; geohazards; climate change' (UNESCO 2015). This indicates the roles in science education related to geo-hazards and related risk. However, it is not specific yet in the operational guidelines how this role on education on geohazards can be fulfilled.

Conclusion: IRDR's View on Key Actions Required for Further Integration

From above discussion, IRDR has considered the following are the key actions or priority actions toward further DRR integration in IDAs.

1. There is a need to have better coherence in the statutory mechanisms for IDAs in DRR. This will have to be driven primarily by UNESCO itself with support of the Member States, which can also be complemented by the use of international centers of excellence associated with IDAs.
2. A much greater partnership development on DRR for IDAs is essential. The gaps in terms of bringing on board research and technical institutions, especially in risk assessment and support to preventive actions remain wide and need to be narrowed soon. In the area of landslide, for instance, there should be long-term cooperation agreement established with those that lead landslide research, such as ICL/IPL which possesses the largest network of landslide expertise. The Sendai Commitment to be signed at WLF5 in 2020, is a good opportunity for such partnership engagement.
3. Risk assessment in IDAs need to be planned and organized systematically for all IDAs, with particular focus on vulnerability and exposure. In this regard, the risk assessment should lead to the formulation long-standing assessment mechanisms for IDAs, and the formulation of multi-early warning systems for IDAs. There should be long-term expert teams per hazard category such as landslides, flood, wildfire and biological hazards established and engaged in IDAs.
4. Fast disaster response and assessment and post-disaster monitoring in the restoration process is another area for cooperation. To this end, IRDR stresses the importance of cooperation across policy, technical and engineering as well as social and economic dimensions. In relation to restoration, there is also a special question to many IDAs, and in particular regarding WH properties. Priority 4 of SFDRR on calls on building back better but whether it would be possible or appropriate to 'build back' when part of the core conservation values of IDAs, especially OUVs, is damaged remain unclear. Such dilemma has been seen in a number of IDAs in post-disaster recovery work. For instance, in the recent case in Jiuzhaigou, when the earthquake has damaged some scenic spots, experts are very divided in response actions.
5. A multi-sectors joint report of DRR on IDAs is needed. This report could indicate progress made, gaps identified, challenges encountered, and lessons learned in DRR practices on IDAs. The experiences should be shared by science and academia community, the public sector, private sectors and other related stakeholders. This publication could be complimentary to global DRR reporting such as GAR series.
6. Capacity building in terms of training and education, young researchers and managers, database development, use of new technologies especially digital technologies should be enhanced. In this regard, connecting

[2]The debates on this matter were included in the section XVII of the final MAB Council Report: https://www.unesco.org/new/fileadmin/MULTIMEDIA/HQ/SC/pdf/FINAL_REPORT_27_MAB-ICC_en-v2.pdf.

DRR-related Centres of Excellence in different international organisations for coherent programmes on IDAs should be pursued.

With these actions undertaken, IRDR anticipates an international initiative gradually established for DRR in IDAs within the new DRR research agenda toward 2030.

References

Fallahi A (2015) Post disaster needs assessment sector reports. National Planning Commission, Government of Nepal, Kathmandu

Fukuoka H, Sassa K, Wang G, Wang F, Wang Y, Tian Y (2005) Landslide risk assessment and disaster management in the Imperial Resort Palace of Lishan, Xian, China (C101-4). In: Landslides. Springer, pp 81–89

HIST (2018) The 3rd Huangshan Dialogue on UNESCO sites and sustainable development. https://huangshandialogue3.csp.escience.cn/dct/page/1. Accessed 12 Mar 2020

ICL et al. (2015) Sendai partnerships 2015–2025 for global promotion of understanding and reducing landslide disaster risk. Sendai

IRDR (2019) Outcome document of DRR session in 2018 Huangshan Dialogue. https://www.irdrinternational.org/2019/01/04/outcome-document-of-drr-session-in-2018-huangshan-dialogue/. Accessed 12 Mar 2020

Klimeš J (2013) Landslide temporal analysis and susceptibility assessment as bases for landslide mitigation, Machu Picchu, Peru. Environ Earth Sci 70:913–925

Pavlova I, Fassoulas C, Watanabe M, Canet C, Cupa P (2018) UNESCO designated sites–natural and cultural heritage sites as platforms for awareness raising

Sassa K, Fukuoka H, Scarascia-Mugnozza G, Evans S (1996) Earthquake-induced-landslides: distribution, motion and mechanisms. Soils Found 36:53–64

UNESCO (2007) Issues related to the state of conservation of world heritage properties: strategy for reducing risks from disasters at world heritage properties. Paris

UNESCO (2008) List of factors affecting the properties. https://whc.unesco.org/en/factors/. Accessed 12 Mar 2020

UNESCO (2015) Operational guidelines for UNESCO global geoparks

UNESCO (2018a) Biosphere reserves. https://www.unesco.org/new/en/natural-sciences/environment/ecological-sciences/biosphere-reserves/. Accessed 12 Mar 2020

UNESCO (2018b) Man and the biosphere programme. https://www.unesco.org/new/en/natural-sciences/environment/ecological-sciences/man-and-biosphere-programme/. Accessed 12 Mar 2020

UNESCO (2018c) UNESCO global geoparks. https://www.unesco.org/new/en/natural-sciences/environment/earth-sciences/unesco-global-geoparks/. Accessed 12 Mar 2020

UNESCO (2018d) World heritage properties. https://whc.unesco.org/en/criteria/. Accessed 12 Mar 2020

Zhao B, Wang Y-S, Luo Y-H, Li J, Zhang X, Shen T (2018) Landslides and dam damage resulting from the Jiuzhaigou earthquake (8 August 2017), Sichuan, China. R Soc Open Sci 5:171418

Landslide Hazard and Risk Assessment for Civil Protection Early Response

2020 Kyoto Japan

Giuseppe Esposito and Olga Petrucci

Abstract

This paper presents a series of severe landslide disasters occurred in Italy, during which the local scientific community supported civil protection authorities in the management of the emergency responses. Depending on the event characteristics, scientific support focused on landslide mapping, damage assessment, monitoring, early warning, and designing of countermeasures. Relevant studies, published after the disasters and describing these activities, highlight that, in case of major events, the scientific community can provide a significant support in decision-making processes and intervention strategies, by means of multi-disciplinary skills, experience and resources. The Italian example of cooperation between the scientific community and civil protection authorities here described highlights as a knowledge transfer from theoretical frameworks to practical applications can optimize the disaster response operations. Even if this study focuses on the Italian situation, it may represents a starting point to evaluate the real contribution provided by landslide experts in disaster responses worldwide, for understanding weaknesses and strengths. In many countries, in fact, local authorities are not able to provide timely and effective responses also because of a lacking or insufficient support of scientists.

Keywords

Landslide disaster • Emergency response • Scientific support • Italy

G. Esposito (✉) · O. Petrucci
National Research Council, Research Institute for Geo-Hydrological Protection (CNR-IRPI), Via Cavour 4/6, 87036 Rende, CS, Italy
e-mail: giuseppe.esposito@irpi.cnr.it

O. Petrucci
e-mail: o.petrucci@irpi.cnr.it

K. Sassa et al. (eds.), *Understanding and Reducing Landslide Disaster Risk*, ICL Contribution to Landslide Disaster Risk Reduction,
https://doi.org/10.1007/978-3-030-60196-6_42

Introduction

Landslides are ubiquitous in any terrestrial environment with slopes, driven by natural factors and/or human activities (Froude and Petley 2018). Hazard and risk assessments are essential to prevent landslide losses due to the impact on urbanized areas, as extensively documented in the scientific literature (e.g., Glade et al. 2005; Hungr et al. 2005; Fell et al. 2008; van Westen et al. 2008), as well as to provide an appropriate post-event response. In the latter case, experts may support the civil protection authorities with a series of scientific activities depending on specific requirements and conditions, like characteristics of the event (e.g., spatial extent, triggering mechanism, geomorphic and geological conditions), and the presence of settlements and population. For instance, the spatial extent of a landslide event may influence time, resources, and approaches for the assessments; landslide properties, as typology and geometry, may influence the selection of risk mitigation measures.

In this paper, we describe scientific activities carried out within the response to major landslide disasters occurred in Italy (Fig. 1), after which the Italian Government declared the "state of emergency". The several examples show how the scientific community supported the civil protection authorities in decision-making processes and intervention strategies. Outcomes may represent a starting point to perform a global evaluation of the effective role played by scientists in landslide disaster responses, in order to facilitate the transfer of knowledge from theoretical frameworks to operational tools, and improving the current situation.

Post-landslide Scientific Activities

In the following paragraphs, we provide a description of the selected case studies and related scientific articles (Table 1) focusing on the supporting activities, such as landslide mapping, damage assessment, landslide monitoring/early

Fig. 1 Location of the selected landslide disasters occurred in Italy. The numbers indicate the year of occurrence of each event

warning, and designing of countermeasures. The last case describes the most damaging among the considered landslide events, requiring multiple activities.

Event Landslides Inventory Mapping

A landslide inventory map is the simplest form of landslide mapping (Hansen 1984; Guzzetti et al. 2000). Landslide inventory maps show information that can be used to investigate the distribution, types, pattern, recurrence and statistics of slope failures, and to study the evolution of landscapes dominated by mass-wasting processes (Guzzetti et al. 2012; Reichenbach et al. 2018). Depending on the scope, the extent of the study area, the scale of the base maps, the quality and detail of the existing information, an inventory map can be prepared using different techniques, such as field mapping, visual interpretation of optical data and analysis of satellite imagery (Guzzetti et al. 2000, 2012; Mondini et al. 2011). An event inventory map is produced when multiple landslide processes are associated with a given rainstorm, earthquake, or snowmelt triggering event (Reichenbach et al. 2018). The objective of the detection and mapping of landslides associated to an event is the determination of the location, size, boundaries, type, and other related information of the mass movements, including the source area and the volume of displaced mass (Zhong et al. 2020; Hervás and Bobrowsky 2009).

In Italy, landslide inventory maps have been produced after several landslide disasters. As an example, we report the mapping activity carried out after the event that occurred on 1 October 2009, along the Ionian coast of Sicily (southern Italy), where an area of 19.5 km^2 was affected by a series of shallow landslides triggered by an intense storm. Landslides and flooding resulted in 37 fatalities and innumerable injured people. After the event, Ardizzone et al. (2012) prepared a detailed landslide inventory map at 1:10,000 scale for the Briga and Giampilieri catchments. The inventory was obtained by means of field surveys carried out in the period from October to November 2009, and visual interpretation of pre-event and post-event stereoscopic and pseudo-stereoscopic aerial photographs. The produced inventory map showed: (i) the distribution and types of the event landslides triggered by the 1 October 2009 rainstorm, (ii) the distribution and types of the pre-existing landslides, and (iii) other geomorphological features related to fluvial processes and slope movements.

Table 1 List of the selected landslide disasters and scientific activities, with related references

Scientific activity	Disaster	References
Event landslides inventory mapping	Giampilieri (2009)	Ardizzone et al. (2012)
Damage assessment	Cavallerizzo (2005)	Iovine et al. (2006), Petrucci and Gullà (2009, 2010)
Monitoring and early warning	Montaguto (2010)	Giordan et al. (2013), Allasia et al. (2013), Ferrigno et al. (2017)
Designing of countermeasures	Cinque Terre (2011)	Cevasco et al. (2013, 2014), Galve et al. (2016)
Multiple activities in major disasters	Sarno (1998)	Cascini (2004), Versace et al. (2005), Capparelli and Versace (2014)

Damage Assessment

The analysis of the damage distribution induced by landslides is essential to understand the exposure of the infrastructures and the population, and to take appropriate remedial measures and effectively respond to the events (Tomás et al. 2018). A comprehensive approach for the surveying and classification of landslide-induced damage plays a key role in the strategy to better delineate mass movement boundaries by categorizing its detectable impacts on the ground, as well as to improve knowledge of the instability or to avoid repeated occurrences (Del Soldato et al. 2017).

One approach is the Support Analysis Framework (SAF) developed by Petrucci and Gullà (2009) that can be used to assess damage indices related to mass movements already occurred, or the potential outcomes of dormant phenomena reactivations. The authors started from damage schematizations available in the literature, for developing a tool-form aimed to convert descriptions of landslide effects into numerical indices expressing direct, indirect and intangible damage.

Petrucci and Gullà (2010) applied the SAF tool to the reactivation of a complex landslide occurred on 7 March 2005 in the Calabria Region (southern Italy). The major collapse was preceded by the opening of cracks within the urban center of Cavallerizzo di Cerzeto, allowing the complete evacuation of the population (Iovine et al. 2006). After the event, the Cavallerizzo community was delocalized in a new permanent settlement, realized in a safe place. By means of the SAF analysis, all the three types of damage were classified as high, in a rank formed by four levels (low, medium, high, very high). The simulation of different damage scenarios, related to pre-disaster conditions, highlighted that a preventive damage assessment before the 2005 landslide reactivation would have pointed out that some structural measures, even if not perfectly successful, would have limited damage to the county road and buildings, maintaining all of the indices under the medium level, and avoiding the high indirect consequences.

Monitoring and Early Warning

Monitoring is essential to identify the ongoing spatial and temporal evolution of a landslide, to evaluate the triggering mechanisms, to set up and validate numerical models required to forecast the possible evolution of the phenomenon, to identify early warning indicators, and to plan active and passive mitigation measures (Corsini et al. 2005). A landslide monitoring system can be based on different types of methods, such as geotechnical, hydrogeological, geophysical and topographic. A redundancy of the monitored parameters and of the monitoring methods should be considered to design a system for early warning purposes, as recommended by Pecoraro et al. (2019).

As an example of monitoring strategy implemented within a local early warning system, we report the case of the Montaguto complex earthflow, occurred in April 2010 in southern Italy. Due to the mass movement, provincial and national roads together with the railway were interrupted for three months, with severe consequences for the residents, local economy, and national railway system of southern and central Italy. According to Giordan et al. (2013), the earthflow was characterized by a total length of 3.1×10^3 m, and an area of 6.6×10^5 m^2. Due to the relevant dimensions, topographic changes of the landslide surface were monitored by means of an integrated monitoring system formed by three robotized total stations (RTSs), and a GB-DInSAR. A meteorological station was also installed close to the unstable area.

The three RTSs repeated the topographic measurements of benchmarks (optical prisms placed in the active zone) in the framework of the ADVanced dIsplaCement monitoring system for Early warning (ADVICE) implemented by Allasia et al. (2013). ADVICE consists in a set of tools that passes automatically, and in near-real-time, from the acquisition of displacements data to the divulgation of the monitoring results via the Internet. During the early emergency response, the system was used to assure safety conditions for workers involved in the removal of landslide material from both the roads and railway. After this phase, surface displacement data were used to analyze the medium/long-term evolution of the landslide, as well as to design protection and mitigation measures. Currently, the monitoring network has the key role to control the stability and effectiveness of remedial measures, and to detect eventual landslide reactivations.

The GB-DInSAR was set to detect both short- and long-term movements, developing interferograms with a temporal baseline spanning from 3.5 min to 1 month (Ferrigno et al. 2017). During the first weeks of the monitoring activity, the GB-DInSAR was coupled with a webcam and an infrared thermal camera (IRT). A comparison between the optical images and the interferograms was crucial for the interpretation of the radar images and therefore the detection of unstable areas, whereas the IRT allowed a very accurate investigation of the water flow paths, wet areas and drainage directions (Ferrigno et al. 2017). The GB-DInSAR was also used for early warning purposes.

Besides the early warning use, the integrated monitoring system was utilized to quantify the surface movements and velocity trends in the several landslide sectors, as well as for assessing the related kinematical characteristics.

Designing of Countermeasures

Structural and non-structural countermeasures can be realized after landslide disasters to mitigate the risk that may still affect urban settlements. Generally, this is a difficult task given that a wide range of issues, such as hazard analysis, potential loss estimation and cost-effectiveness evaluation have to be considered. From a technical point of view, scientific activities should be firstly aimed at selecting the mitigation strategy and deciding, for example, if reducing the likelihood or the consequences of potential landslides (Crozier and Glade 2006). Besides technical aspects, the selection of countermeasure also depends on economic issues. As highlighted by Agliardi et al. (2009), in fact, the residual risk and the costs and benefits associated to different protection scenarios should be estimated, in order to assess both the technical performance and the cost efficiency of different mitigation options.

An interesting case study related to this topic is the "Cinque Terre" geo-hydrological disaster, occurred in October 2011. The Cinque Terre National Park, a "World Heritage Site" declared by UNESCO in 1997, represents a famous touristic destination in north-west Italy, due to the characteristic terraced landscape overlooking sea. During the 2011 disaster, the Vernazza catchment was affected by a very intense rainfall event that triggered thousands of shallow landslides, widespread erosive and depositional processes and floods, which caused three fatalities. In the post-event phase, civil protection authorities realized structural interventions locally on the unstable slopes and along the stream channels, for a short-term risk mitigation. Planning of these interventions was supported by an inventory map of the occurred geo-hydrological processes produced during the emergency phase by Cevasco et al. (2013), as well as by an engineering geological zoning based on laboratory and in-situ investigations (Cevasco et al. 2014). Analyses of Cevasco et al. (2013, 2014) were also used by Galve et al. (2016) to identify long-term measures for reducing the shallow landslide-related risk in the Vernazza catchment. By means of a quantitative risk assessment, these authors verified that the most cost-effective measure to stabilize the terraced slopes on a decadal scale were not structural interventions but reforesting. In fact, the considered structural measures for restoring terraces, such as (1) a combined structure formed by a dry stone wall reinforced with a live crib wall and (2) vegetated rock gabions resulted too expensive for being cost-effective solutions.

Findings described in this case study underline hence the importance of field surveys and quantitative risk assessments to design cost-effective short- or long-term risk mitigation measures, at local and wide scale.

Multiple Activities in Major Disasters

The response phase of a disaster management cycle includes the environmental analysis and the determination of emergency priorities (Zhou et al. 2018). In this phase, scientists can be involved in decision-making processes to support several activities like evacuation, warning issue and emergency strategies development. A landslide emergency response is always different because strictly dependent from the type of processes, the extension of the affected area, and damage severity.

In 1998, the Italian Civil Protection faced one of the most severe landslide disasters ever occurred in Italy. On 5–6 May, after a period of prolonged rainfall, multiple mud/debris flows triggered by shallow landslides invaded five towns, involving an area of about 60 km^2 and displacing a volume of 2,000,000 m^3 of pyroclastic and carbonate rocks (40% derived from the eroded materials along the channels) (Capparelli and Versace 2014). The flow processes caused 160 fatalities and huge damage to the urban settlements. In the early emergency phase, the civil protection authorities asked a support to the national scientific community concerning the following operations (Cascini 2004): (i) to map the urban areas still exposed to the mud/debris flow risk within the five towns; (ii) to implement an early warning system based on rainfall thresholds and aimed at the temporary evacuation of people; (iii) to predispose the guidelines for realizing control works aimed at the recovery and safeguarding of the risk zones; (iv) to start up a geo-hydrological Territorial Survey, consisting of a technical team of geologists and engineers which had to evaluate and monitor critical situations in the field, also during weather alerts (Versace et al. 2005). All these requests were to assure safety conditions to thousands of people that were evacuated far from the mountains. As stated by Cascini (2004), major difficulties in answering such requests were due to the absence of adequate studies related to the occurred phenomena in the area until then, and the short time available to respond to the relevant questions. For these reasons, a temporary research unit coordinated by the University of Salerno, formed by dozens of researchers coming from all over Italy and specialized in different scientific sectors was created. Such multidisciplinary unit operated until 15 October 1998, accomplishing with success all the scientific activities necessary for the emergency response.

With respect to the other case studies described in this contribution, this event required a very complex scientific support focused on different types of activities. The lacking of appropriate scientific knowledge required relevant efforts to fill the research gap timely. This purpose was reached mostly with data acquired in the field by the geo-hydrological Territorial Survey, proving the key role played by multi-disciplinary teams of experts in the management of early response to major landslide disasters.

Concluding Remarks

Slope instability processes occur all over the world, often with similar mechanisms but with different consequences, depending also by emergency preparedness and response capacities of communities. The reported case studies refer to landslide disasters in which Italian experts supported civil protection authorities in the response phase. Such examples would remark as the landslide scientific community may provide a valuable support in post-event decision-making processes and interventions by means of different types of activities. The Italian Civil Protection, in fact, can take advantage of a network of multi-disciplinary research teams located throughout the country. In case of disaster, one or more teams can be activated according to the characteristics of the landslide emergency, in order to provide a scientific support for the response operations. In countries where, for different reasons, the scientific community cannot be directly involved in supporting civil protection authorities, relevant difficulties can occur in both prevention and emergency operations. A lack of economic and institutional capacity for prevention efforts can often be found, in fact, as significant factors in explaining the different impact of disasters on developed and developing countries (Fiala 2017). The latters suffer 95% of disaster deaths worldwide (Linnerooth-Bayer et al. 2011). Therefore, more efforts should be done by local governments together with the landslide research community to fill this gap, as pointed out during the High-Level Panel Discussion at the 4th World Landslide Forum in Ljubljana, Slovenia (Sassa 2017). In the light of this, analyses focusing on post-event activities like the one here presented should highlight weaknesses and strengths of local situations to improve them.

Looking at the Italian experiences, besides the positive aspects here described, it is worth underlining that a series of improvements may be necessary also in this country. The geo-hydrological Territorial Survey created during the 1998 disaster, for example, represents a valuable tool to acquire scientific data in the field, as well as to monitor the areas at risk. Although this non-structural measure was adopted definitely in the official Italian legislation, currently, this is operational in few towns only. Besides this, few research centers have remote sensing technologies to support monitoring operations in the field, or personnel with specific expertise to manage the acquired data. As highlighted in the monitoring experience of the Montaguto landslide, advanced monitoring instruments available to research centers can be exploited not only for pure research aims (e.g., understanding a landslide geomorphic evolution) but also for early warning purposes. However, the financial support to equip more teams with such technologies is still insufficient.

The described activities highlight hence that hazard and risk assessments for disaster management purposes are difficult tasks and require proper expertise and skills. Specific competences are necessary to perform field surveys devoted to mapping and damage evaluation, as well as remote sensing experts can be essential to survey wide or inaccessible areas affected by mass movements. Further competences may be required in engineering geological investigations, emergency planning and design of countermeasures (e.g., Caruso et al. 2015). In the light of this and considering that landslide risk conditions are increasing in some areas worldwide (Gariano and Guzzetti 2016), we believe that civil protection authorities can no longer respond to disasters without an adequate scientific support of landslide experts.

References

Agliardi F, Crosta GB, Frattini F (2009) Integrating rockfall risk assessment and countermeasure design by 3D modelling techniques. Nat Hazards Earth Syst Sci 9:1059–1073

Allasia P, Manconi A, Giordan D, Baldo M, Lollino G (2013) ADVICE: a new approach for near-real-time monitoring of surface displacements in landslide hazard scenarios. Sensors 13:8285–8302

Ardizzone F, Basile G, Cardinali M, Casagli N, Del Conte S, Del Ventisette C, Terranova O (2012) Landslide inventory map for the Briga and the Giampilieri catchments, NE Sicily, Italy. J Maps 8 (2):176–180

Capparelli G, Versace P (2014) Analysis of landslide triggering conditions in the Sarno area using a physically based model. Hydrol Earth Syst Sci 18:3225–3237

Caruso A, Esposito G, Scarpino S, Versace P (2015) Geo-hydrological risk management on the Amalfi Coast: the village of Atrani emergency plan. In Lollino G, Manconi A, Guzzetti F, Culshaw M, Bobrowsky PT, Luino F (eds) Engineering geology for society and territory, vol 5. Springer International Publishing, Switzerland, pp 815–819. ISBN 978-3-319-09048-1

Cascini L (2004) The flowslides of May 1998 in the Campania Region, Italy: the scientific emergency management. Riv Ital Geotec 2:11–44

Cevasco A, Brandolini P, Scopesi C, Rellini I (2013) Relationships between geo-hydrological processes induced by heavy rainfall and land-use: the case of 25 October 2011 in the Vernazza catchment (Cinque Terre. NW Italy). J Maps 9(2):289–298

Cevasco A, Pepe G, Brandolini P (2014) The influences of geological and land use settings on shallow landslides triggered by an intense rainfall event in a coastal terraced environment. Bull Eng Geol Environ 73:859–875

Corsini A, Pasuto A, Soldati M, Zannoni A (2005) Field monitoring of the Corvara landslide (Dolomites, Italy) and its relevance for hazard assessment. Geomorphology 66:149–165

Crozier MJ, Glade T (2006) Landslide hazard and risk: issues, concepts and approach. In Glade T, Anderson MG, Crozier MJ (eds) Landslide hazard and risk. Wiley, Chichester, pp 1–40. ISBN 0-471-48663-9

Del Soldato M, Bianchini S, Calcaterra D, De Vita P, Di Martire D, Tomás R, Casagli N (2017) A new approach for landslide-induced damage assessment. Geomatics Nat Hazards Risk 8(2):1524–1537

Fell R, Corominas J, Bonnard Ch, Cascini L, Leroi E, Savage WZ (2008) Guidelines for landslide susceptibility, hazard and risk zoning for land use planning. Eng Geol 102(3–4):85–98

Ferrigno F, Gigli G, Fanti R, Intrieri E, Casagli N (2017) GB-InSAR monitoring and observational method for landslide emergency management: the Montaguto earthflow (AV, Italy). Nat Hazards Earth Syst Sci 17:845–860

Fiala O (2017) Natural disasters and individual behaviour in developing countries. Springer International Publishing, Switzerland, 195 p. ISBN 978-3-319-53903-4

Froude MJ, Petley DN (2018) Global fatal landslide occurrence from 2004 to 2016. Nat Hazards Earth Syst Sci 18:2161–2181

Galve JP, Cevasco A, Brandolini P, Piacentini D, Azañon JM, Notti D, Soldati M (2016) Cost-based analysis of mitigation measures for shallow-landslide risk reduction strategies. Eng Geol 213:142–157

Gariano SL, Guzzetti F (2016) Landslides in a changing climate. Earth Sci Rev 162:227–252

Giordan D, Allasia P, Manconi A, Baldo M, Santangelo M, Cardinali M, Corazza A, Albanese V, Lollino G, Guzzetti F (2013) Morphological and kinematic evolution of a large earthflow: the Montaguto landslide, southern Italy. Geomorphology 187:61–79

Glade T, Anderson M, Crozier MJ (2005) Landslide hazard and risk. Wiley, Chichester, 802 p. ISBN 0-471-48663-9

Guzzetti F, Cardinali M, Reichenbach P, Carrara A (2000) Comparing landslide maps: a case study in the upper Tiber River Basin, Central Italy. Environ Manag 25(3):247–363

Guzzetti F, Mondini AC, Cardinali M, Fiorucci F, Santangelo M, Chang KT (2012) Landslide inventory maps: new tools for an old problem. Earth Sci Rev 112(1):42–66

Hansen A (1984) Landslide hazard analysis. In: Brunsden D, Prior DB (eds) Slope instability. Wiley, New York, pp 523–602

Hervás J, Bobrowsky P (2009) Mapping: inventories, susceptibility, hazard and risk. In: Sassa K, Canuti P (eds) Landslides—disaster risk reduction. Springer-Verlag, Berlin, pp 321–349. ISBN 978-3-540-69970-5

Hungr O, Fell R, Couture R, Eberhardt E (2005) Landslide risk management. Balkema, Amsterdam, 774 p. ISBN 9780415380430

Iovine G, Petrucci O, Rizzo V, Tansi C (2006) The March 7th 2005 Cavallerizzo (Cerzeto) landslide in Calabria-Southern Italy. In: Culshaw MG, Reeves HJ, Jefferson I, Spink T (eds) Engineering geology for tomorrow's cities. Paper number 785. The Geological Society of London, London, pp 1–12. ISBN 978-1-86239-290-8

Linnerooth-Bayer J, Lofstedt RE, Sjostedt G (2011) Transboundary risk management. CRC Press, London, 352 p. ISBN 9781853835377

Mondini AC, Guzzetti F, Reichenbach P, Rossi M, Cardinali M, Ardizzone F (2011) Semi-automatic recognition and mapping of rainfall induced shallow landslides using satellite optical images. Remote Sens Environ 115:1743–1757

Pecoraro G, Calvello M, Piciullo L (2019) Monitoring strategies for local landslide early warning systems. Landslides 16(2):213–231

Petrucci O, Gullà G (2009) A support analysis framework for mass movement damage assessment: applications to case studies in Calabria (Italy). Nat Hazards Earth Syst Sci 9:315–326

Petrucci O, Gullà G (2010) A simplified method for landslide damage scenario assessment based on historical data. Nat Hazards Earth Syst Sci 52:539–560

Reichenbach P, Rossi M, Malamud B, Mihir M, Guzzetti F (2018) A review of statistically-based landslide susceptibility models. Earth Sci Rev 180:60–91

Sassa K (2017) The 2017 Ljubljana Declaration on landslide risk reduction and the Kyoto 2020 Commitment for global promotion of understanding and reducing landslide disaster risk. Landslides 14 (4):1289–1296

Tomás R, Abellán A, Cano M, Riquelme A, Tenza-Abril AJ, Baeza-Brotons F, Saval JM, Jaboyedoff M (2018) A multidisciplinary approach for the investigation of a rock spreading on an urban slope. Landslide 15:199–217

van Westen CJ, Castellanos Abella EA, Sekhar LK (2008) Spatial data for landslide susceptibility, hazards and vulnerability assessment: an overview. Eng Geol 102(3–4):112–131

Versace P, Caruso A, Cassetti M, Capparelli G (2005) Il Presidio Territoriale e la Gestione dell'Emergenza a Sarno. Quaderni Del Camilab 3(3):55–84

Zhong C, Liu Y, Gao P, Chen W, Li H, Hou Y, Nuremanguli T, Ma H (2020) Landslide mapping with remote sensing: challenges and opportunities. Int J Remote Sens 41(4):1555–1581

Zhou L, Wu XH, Xu ZS, Fujita H (2018) Emergency decision making for natural disasters: an overview. Int J Disaster Risk Reduct 27:567–576

Size Matters: The Impact of Small, Medium and Large Landslide Disasters

2020 Kyoto Japan

Irasema Alcántara-Ayala

Abstract

Owing to the nature of landslide triggering mechanisms, lack of realistic documentation on the impact of landslide disasters at global, national and subnational scales has existed for many years. Data from two sources was used to examine the discrepancies about the impact of landslide disasters by considering both, high magnitude-low frequency and high frequency-low magnitude events. Analysis of this landslide disaster data for thirteen countries (Argentina, Bolivia, Chile, Colombia, Costa Rica, Ecuador, Guatemala, Mexico, Nepal, Nicaragua, Perú, Sri Lanka, and Venezuela) revealed larger differences than those previously reported. Variation among number of landslide disasters between databases was expressed in three orders of magnitude, whereas number of human losses differed in two orders of magnitude, and people affected in one.

Keywords

Landslides • Disasters • Database • Impact • DesInventar • EM-DAT

Introduction

It has been difficult to estimate the real impact of landslides on society and the environment. Owing to its nature, a complete documentation of the impact of landslide disasters is difficult to attain as the occurrence of other hazards directly linked to their triggering mechanisms, such as precipitation, earthquakes and volcanic activity, mask landslide consequences.

As already noted in earlier studies (Guha-Sapir and Below 2002; Marulanda et al. 2010; Kron et al. 2012), existing world-wide disaster databases have been developed to meet specific requirements and objectives, therefore, their use and applicability carry both, advantages and limitations. This is of particular relevance for research on landslide disasters.

Previous studies have reported a series of discrepancies among available datasets (Petley 2012; Kirschbaum et al. 2015; Froude and Petley 2018), and although several assays have recognised the general impact of high magnitude-low frequency landslide disasters, research has yet to systematically investigate the causality and effect of high frequency-low magnitude disasters associated with landsliding.

Notwithstanding the major drawbacks of landslide disaster databases, few studies have attempted to compare quantifications of the impact of landslide disasters. As reported by Froude and Petley (2018) underestimation of number of landslide disaster events and fatalities derived from EM-DAT database has been recognised. To this regard, Petley's comparative study (Petley 2012) found for the period 2004–2010 an underestimation of the number of fatal landslide events by ~2000% and of the death toll by 430%, whereas Kirschbaum and collaborators detected that between 2007 and 2013, underestimations accounted for ~1400% of fatal landslide events, and 331% of life losses (Kirschbaum et al. 2015).

In an attempt to further contribute to the analysis of the estimations of landslide disaster impact, records from EM-DAT[1] (CRED: EM-DAT 2019) repository were considered for the different regions of the world, and data for some countries included in both EM-DAT and DesInventar[2] (DesInventar 2019) databases were also used to identify additional disparities.

I. Alcántara-Ayala (✉)
Institute of Geography, National Autonomous University of Mexico (UNAM), 04510 Mexico City, Mexico
e-mail: ialcantara@igg.unam.mx

[1]www.emdat.be.
[2]www.desinventar.org.

K. Sassa et al. (eds.), *Understanding and Reducing Landslide Disaster Risk*, ICL Contribution to Landslide Disaster Risk Reduction,
https://doi.org/10.1007/978-3-030-60196-6_43

Methodology

Data for this paper were produced by CRED EM-DAT and DesInventar databases. For the former, an initial period between 1900 and 2019, was considered, whereas for the latter it varied from country to country.

A map of the spatial distribution of high frequency-low magnitude landslide disaster events, including information regarding number of landslide disaster events, associated human losses and people affected, for the different regions of the world, between 1900 and 2019, was created by using the information available in the EM-DAT database (Fig. 1).

Considering that this paper aimed at documenting some uncertainties associated with the information regarding the impact of landslide disasters, it was beyond its scope to undertake an assessment of landslide disasters at global scale, but rather to compare the results obtained from the two sources for the selected countries in terms of number of landslide disasters, human losses and total people affected.

Selection of countries for a further comparative analysis was carried out according to data availability for each of them in both data repositories, as follows: Argentina, Bolivia, Chile, Colombia, Costa Rica, Ecuador, Guatemala, Mexico, Nepal, Nicaragua, Perú, Sri Lanka, and Venezuela.

Results

EM-DAT: High Magnitude-Low Frequency Disaster Events

Notwithstanding the limited evidence of a consistent documentation of the impact of landslide disasters around the globe, results derived from EM-DAT database indicated that between the period 1900 and 2019, a total of 788 high magnitude-low frequency landslide disasters involved 70,884 human losses worldwide and affected a population of circa 14.5 million (Fig. 1).

Regarding the regional distribution of landslide disaster occurrence for the same period, Asia accounted for more than half of the number of all landslide reported disasters (56%). The Americas had a 25% share of the number of global landslide disasters, Europe and Africa both took less than a tenth of landslide disasters occurrence (9 and 7% correspondingly) and Oceania 3% (Fig. 1).

The number of human losses in Europe, Africa and Oceania showed values of 6%, 5% and 1%, respectively. Additionally, as much as 55% of global landslide disaster casualties were from Asia, compared to 33% from the Americas. On the other hand, more than a third of the total number of people reported affected was concentrated in Asia (59%), while in the Americas was as high as 39%. Africa, Europe and Oceania accounted for figures of 1.46, 0.26 and 0.15% accordingly (Fig. 1).

Discrepancies Between Databases on Selected Countries

Larger differences than those previously reported have been obtained when comparing landslide disaster data for thirteen countries on national basis between EM-DAT database and DesInventar. Although EM-DAT offers figures from 1900 to 2019, periods of time of analysed records presented here varied according to time frames of national disaster data bases comprised within DesInventar. Considering that contrasting the two databases by using percent points was skyrocketing, comparisons were made in orders of magnitude.

Costa Rica was the country with highest disparities regarding number of disasters, as in EM-DAT only one event was registered, whilst in DesInventar 3902 disasters were included between 1968 and 2019. Similar results were found for Colombia, Bolivia, Ecuador and Venezuela where records from EM-DAT showed 45, 8, 14 and 3 disasters respectively, while numbers in DesInventar were three order of magnitude higher with 53,605, 5498, 7899 and 1509 disasters, correspondingly. Nicaragua was the country with the lowest disparities; one disaster was included in EM-DAT and 47 in DesInventar. Annual average of landslide disasters based on the information provided by DesInventar was highest in Colombia with 515 events, and the lowest in Nicaragua, with 2 disaster events. When information for all countries was summed up, discrepancies between data represented three orders of magnitude. Total number of disasters estimated by EM-DAT was of 147, whereas in DesInventar it was as high as 82,944 (Table 1).

The contrast concerning human losses was highest in Nicaragua, and the lowest in Guatemala, where 29 and 716 disasters were included in EM-DAT correspondingly, whereas 2044 and 1210 were registered in DesInventar, respectively (Table 2).

Total human losses for all selected countries differed in one order of magnitude. EM-DAT records included a death toll of 11,304, while in DesInventar was as high as 84,901 (Table 2).

In terms of the number of people affected by landslide disasters, the highest annual average of 289,316 was registered in Colombia. Guatemala and Mexico followed with 128,419 and 71,944 persons respectively. Results of the analysis yielded the highest difference of four orders of magnitude between DesInventar and EM-DAT for Mexico. Data collected for Sri Lanka, Costa Rica and Colombia involved three orders of magnitude, whereas in the case of

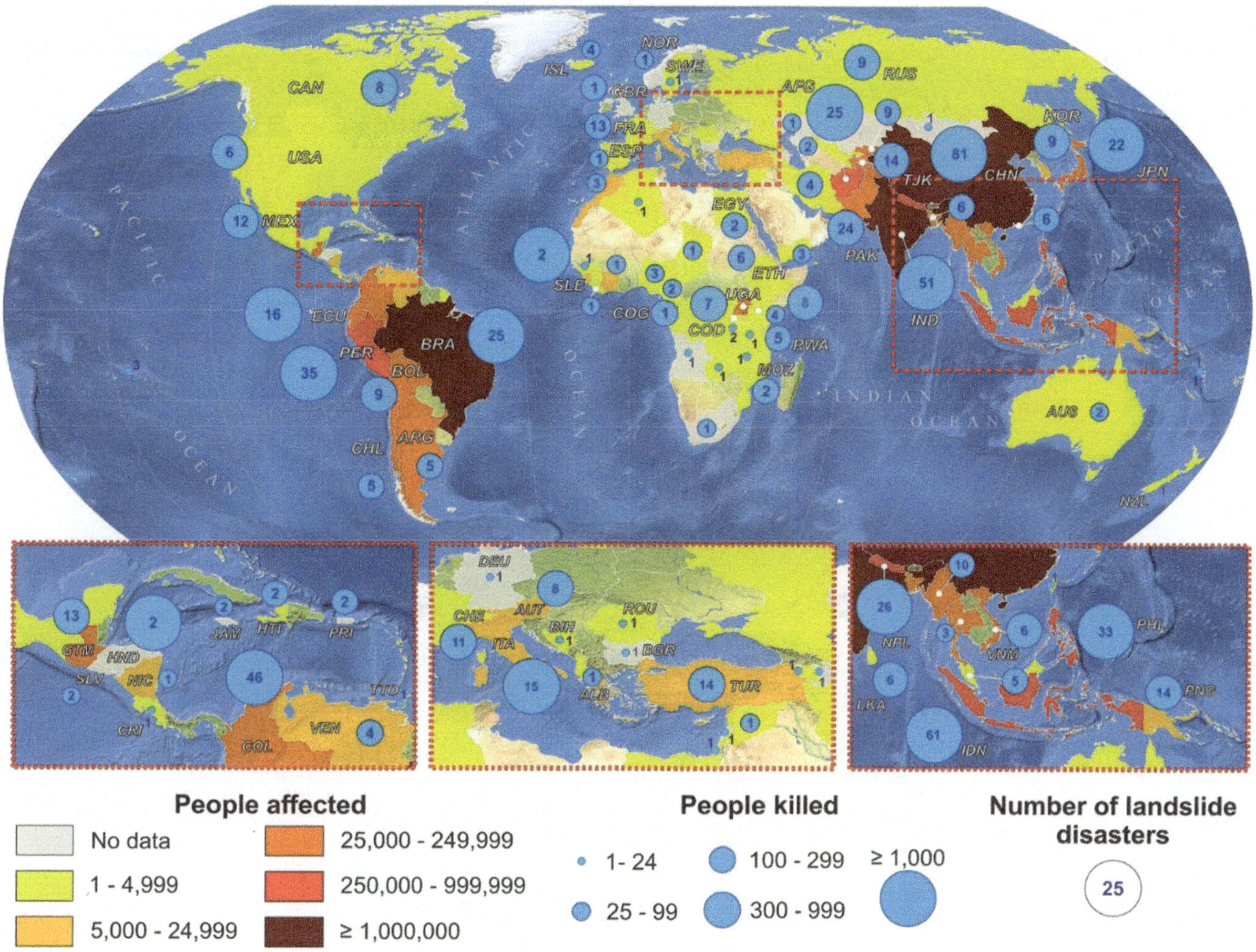

Fig. 1 Spatial distribution of high frequency-low magnitude landslide disasters in terms of occurrence, associated human losses and people affected, between 1900 and 2019. *Data source* EM-DAT database

Guatemala was of two orders of magnitude. The variation for Venezuela, Chile, Bolivia, Nicaragua and Argentina came about an order of magnitude, whilst in the case of Ecuador, Nepal and Perú was the same order of magnitude. Combined data for all countries produced a discrepancy of an order of magnitude between both databases (Table 3).

Discrepancies in Latin-American Countries Between Databases (1970–2013)

Owing to availability of data, the period 1970–2013 was considered for further analyse nine Latin-American countries: Argentina, Bolivia, Colombia, Costa Rica, Chile, Ecuador, Mexico, Peru and Venezuela. According to DesInventar, highest number of disaster landslides occurred in Colombia (8484), Costa Rica (3407), and Peru (3382). Yet the differences of data between DesInventar and EM-DAT databases was of four, and three orders of magnitude for Costa Rica, and Venezuela, respectively, whereas for all other countries was two. Whereas in EM-DAT, a total of 108 disasters were recorder for all the countries, 22,373 disasters were included in DesInventar. This means a difference of two orders of magnitude when data for all countries was added on each database (Fig. 2).

Annual average of number of disasters was highest in Colombia (N = 193), followed by Costa Rica (N = 77) and Peru (N = 77), and Ecuador (N = 61). Lowest values were estimated for Argentina and Chile with annual averages of six and nine disasters, respectively (Fig. 2).

Data concerning fatal victims showed a disparity of an order of magnitude between both databases for Peru, Mexico, Venezuela, Argentina and Costa Rica, whereas records included for Colombia, Ecuador, Bolivia and Chile were in

Table 1 Comparison between the number of landslide disasters registered in EM-DAT and DesInventar databases for selected countries

Number of landslide disasters				
Country	Time frame	EM-DAT	DesInventar	Annual average DesInventar
Costa Rica	1968–2019	1	3902	77
Colombia	1914–2018	45	53,605	515
Bolivia	1970–2015	8	5498	122
Ecuador	1970–2019	14	7899	161
Venezuela	1970–2015	3	1509	34
Mexico	1970–2013	10	1852	43
Nepal	1971–2007	15	2249	62
Guatemala	1988–2015	11	1629	96
Perú	1970–2013	30	3382	79
Sri Lanka	1965–2007	3	687	16
Chile	1970–2014	3	388	9
Argentina	1970–2015	3	297	7
Nicaragua	1992–2013	1	47	2
Total		147	82,944	N/A

Table 2 Comparison between human losses associated with landslide disasters registered in EM-DAT and DesInventar databases for selected countries

Human losses				
Country	Time frame	EM-DAT	DesInventar	Annual average DesInv
Nicaragua	1992–2013	29	2044	93
Costa Rica	1968–2019	7	221	4
Mexico	1970–2013	214	2701	61
Colombia	1914–2018	3675	45,726	435
Bolivia	1970–2015	311	2848	62
Sri Lanka	1965–2007	119	927	22
Perú	1970–2013	3363	21,799	495
Venezuela	1970–2015	134	854	19
Argentina	1970–2015	58	280	6
Nepal	1971–2007	1575	4211	114
Ecuador	1970–2019	900	1723	34
Chile	1970–2014	203	357	8
Guatemala	1988–2015	716	1210	43
Total		11,304	84,901	N/A

the same order of magnitude. Joint records for all countries showed a difference of one order of magnitude. According to DesInventar, highest figures of human losses occurred in Peru, Colombia and Mexico, amounting 21,799, 5784 and 2701, with yearly averages of 495, 131 and 61, respectively.

Lowest number of fatal victims occurred in Costa Rica (N = 207), Argentina (N = 266) and Bolivia (N = 409). Total human losses estimated by EM-DAT for all countries were 8006, whereas in DesInventar were as high as 33,924 (Fig. 3).

Resulting estimates in terms of affected people between DesInventar and EM-DAT indicated four, three and two orders of magnitudes for Mexico, Costa Rica and Colombia, correspondingly. Likewise, in the case of Chile, Venezuela, and Argentina a difference of an order of magnitude was identified, whilst, the difference found for Peru and Ecuador did not reach one order of magnitude. In contrast to all above mentioned data, records for Bolivia showed a difference of one order of magnitude between EM-DAT and DesInventar given the fact that records included in the former were of

Table 3 Comparison between total people affected associated with landslide disasters registered in EM-DAT and DesInventar databases for selected countries

Total people affected				
Country	Time frame	EM-DAT	DesInventar	Annual average DesInventar
Mexico	1970–2013	320	3,165,546	71,944
Sri Lanka	1965–2007	130	150,980	3511
Costa Rica	1968–2019	200	120,501	2317
Colombia	1914–2018	78,493	30,378,225	289,316
Guatemala	1988–2015	58,073	3,595,737	128,419
Venezuela	1970–2015	21,518	251,563	5469
Chile	1970–2014	82,841	652,071	14,490
Bolivia	1970–2015	182,053	1,123,451	24,423
Nicaragua	1992–2013	5769	29,652	1348
Argentina	1970–2015	32,364	121,424	2640
Ecuador	1970–2019	81,606	225,640	4513
Nepal	1971–2007	442,618	481,214	13,006
Perú	1970–2013	790,678	603,743	13,721
Total		1,776,663	40,899,747	N/A

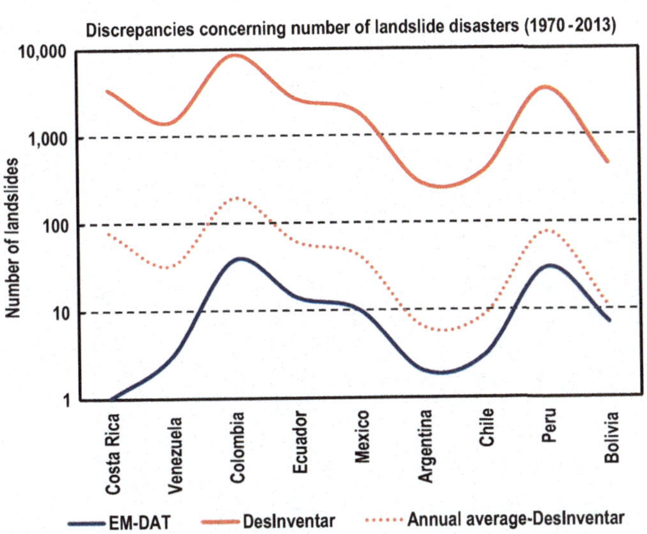

Fig. 2 Discrepancies regarding the number of landslide disasters for selected Latin-American countries between databases

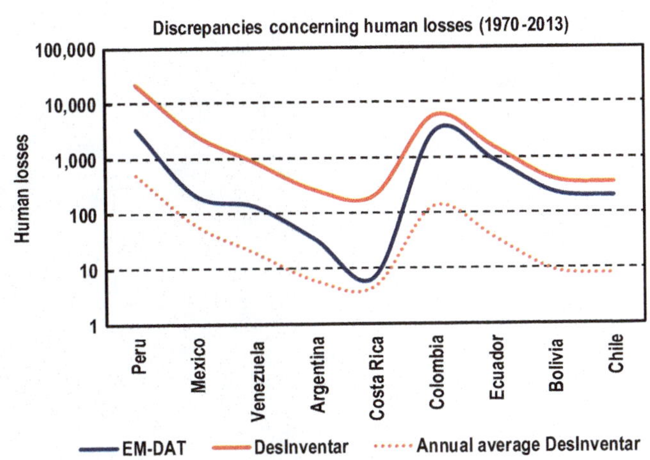

Fig. 3 Discrepancies regarding the number of human losses associated with landslide disasters for selected Latin-American countries between databases

170,653, whereas in the later corresponded to 30,237. The main difference of such data was the inclusion in EM-DAT of 165,000 people affected on February 1st, 1994 by landslides that were not accounted for in DesInventar. A difference of an order of magnitude between DesInventar and EM-DAT databases was estimated for overall data for all countries analysed. On the other hand, based on DesInventar, highest records of affected persons were recognised for Mexico, Colombia and Chile with totals of 3,165,546, 2,656,600 and 651,521, besides annual averages of 71,944, 60,377 and 14,807, correspondingly (Fig. 4).

Discussion and Concluding Remarks

The research data in this work was drawn from two main sources, which provide information regarding both, high magnitude-low frequency, and high frequency-low magnitude events. It has often proved difficult to separate out the landslide disaster impact from those caused by its triggering mechanisms. Due to information availability constraints, this paper cannot provide a thorough world-wide comparison of landslide disaster databases. Owing to disparities and lack of systematic information on the impact of landslide disasters at global, national, subnational and local scales in-depth

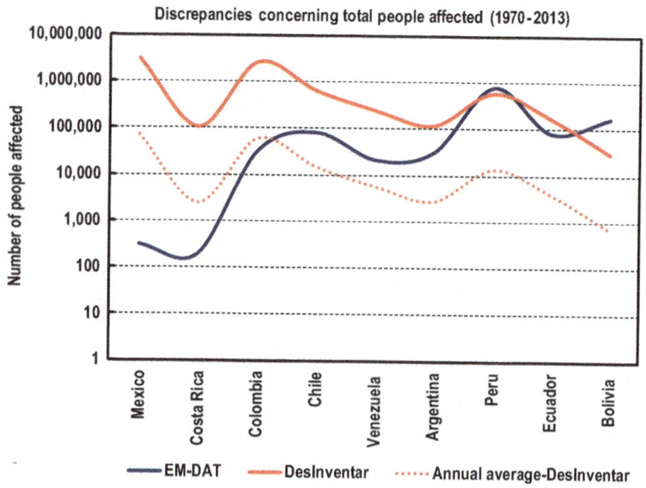

Fig. 4 Discrepancies regarding the number of total people affected associated with landslide disasters for selected Latin-American countries between databases

landslide disaster risk assessment is impeded. It can be argued, however that engineers and Earth scientists have been keen to understand the dynamics of landsliding, through instrumentation, monitoring and mapping, as well as contributed to a great extent to the establishment of Landslide Early Warning Systems. Nonetheless, specific landslide disaster risk drivers have not been identified and hence, appropriately addressed. Current research and strategies regarding landslide disaster risk reduction and management reflect such weaknesses. Consequently, at a more everyday level, it would be important to systematically document, understand, and analyse the complex dynamics of the increasing vulnerability conditions of communities exposed to landsliding, particularly, but not exclusively, those strongly intertwined to the occurrence of high frequency-low magnitude events (Fig. 5).

A considerable amount of care needs to be taken to accurately document landslide risk and landslide disaster

Fig. 5 Increasing vulnerability and exposure to landsliding, Teziutlán, Puebla, México

occurrence, so that a comprehensive understanding of the social construction of landslide disaster risk by all relevant stakeholders can be encouraged (Blaikie et al. 1994; Oliver-Smith et al. 2016, 2017). This would allow to recognise and deal with dynamic pressures or drivers of landslide disaster risk, in addition to enhancing the availability of consistent landslide hazard, vulnerability and exposure data, to fill the gaps of landslide knowledge (Alcántara-Ayala et al. 2017) that can be 'useful, usable and used' (Boaz and Hayden 2002).

At a general level, the analysis presented here raises a series of questions in need of detailed exploration from an integrated disaster risk research perspective (IRDR 2013) in regard not only of the temporal and spatial distribution of landslide disasters, but concerning vulnerability and exposure to landslides, as the main ingredients for the creation of landslide disaster risk. This type of research could provide useful insights into the ways in which landslide disaster risk research can be used for policy making and practice to promote disaster risk reduction and management (Alcántara-Ayala et al. 2015; Cutter et al. 2015).

Finally, when properly documented, and owing to the mounting vulnerability of communities exposed to landslides due to lack of adequate territorial management, from which main landslide disaster risk drivers derive from, coupled with climate change, sooner than later the impact of landslide disasters worldwide could exceed that of the disasters triggered by floods.

Acknowledgements Thanks are due to Ricardo J. Garnica-Peña, for his help to produce the final version of the map and to the anonymous reviewer. Financial support was kindly provided by UNAM-DGAPA, through the project PAPIIT IN300818.

References

Alcántara-Ayala I, Altan O, Baker D, Briceño S, Cutter S, Gupta H, Holloway A, Ismail-Zadeh A, Jiménez Díaz V, Johnston D, McBean G, Ogawa Y, Paton D, Porio E, Silbereisen R, Takeuchi K, Valsecchi G, Vogel C, Wu G, Zhai P (2015) Disaster risks research and assessment to promote risk reduction and management. In: Ismail-Zadeh A, Cutter S (eds) ICSU-ISSC ad hoc group on disaster risk assessment. ICSU, Paris

Alcántara Ayala I, Murray V, Daniels P, McBean G (2017) On the future challenges for the integration of science into international policy development for landslide disaster risk reduction. In: Sassa K, Mikoš M, Yin Y (eds) Advancing culture of living with landslides, vol 1. ISDR-ICL Sendai partnerships 2015–2025. Springer, pp 143–154

Blaikie P, Cannon T, Davis I, Wisner B (1994) At risk: natural hazards, people's vulnerability and disasters. Routledge, New York

Boaz A, Hayden C (2002) Pro-active evaluators: enabling research to be useful, usable and used. Evaluation 8(4):44053

CRED: EM-DAT (2019) The International Disaster Database, Centre for Research on the Epidemiology of Disasters—CRED. Université Catholique de Louvain, Brussels. Available at: https://www.emdat.be. Accessed 22 Dec 2019

Cutter SL, Ismail-Zadeh A, Alcántara-Ayala I, Altan O, Baker DN, Briceño S, Gupta H, Holloway A, Johnston D, McBean GA, Ogawa Y, Paton D, Porio E, Silbereisen RK, Takeuchi K, Valsecchi GB, Vogel C, Wu G (2015) Global risks: pool knowledge to stem losses from disasters. Nature 522:277–279

DesInventar (2019) DesInventar—inventory system of the effects of disasters. Corporación OSSA, Cali. Available at: https://desinventar.org. Accessed 22 Dec 2019

Froude MJ, Petley DN (2018) Global fatal landslide occurrence from 2004 to 2016. Nat Hazards Earth Syst Sci 18:2161–2181

Guha-Sapir D, Below R (2002) The quality and accuracy of disaster data a comparative analysis of three global data sets. Technical document. The ProVention Consortium and the Disaster Management Facility, The World Bank, WHO Centre for Research on the Epidemiology of Disasters, University of Louvain School of Medicine

IRDR (2013) Integrated research on disaster risk strategic plan 2013–2017. Beijing

Kirschbaum D, Stanley T, Zhou Y (2015) Spatial and temporal analysis of a global landslide catalog. Geomorphology 249:4–15

Kron W, Steuer M, Löw P, Wirtz A (2012) How to deal properly with a natural catastrophe database – Analysis of flood losses. Nat Hazards Earth Syst Sci 12:535–550

Marulanda MC, Cardona OD, Barbat AH (2010) Revealing the socioeconomic impact of small disasters in Colombia using the DesInventar database. Disasters 34:552–570

Oliver-Smith A, Alcántara-Ayala I, Burton I, Lavell A (2016) Forensic investigations of disasters (FORIN): a conceptual framework and guide to research. IRDR FORIN publication no. 2. Integrated research on disaster risk. ICSU, Beijing, 56 pp

Oliver-Smith A, Alcántara-Ayala I, Burton I, Lavell A (2017) The social construction of disaster risk: seeking root causes. Int J Disaster Risk Reduct 22:469–474

Petley D (2012) Global patterns of loss of life from landslides. Geology 40:927–930

Practices of Public Participation Early Warning System for Geological Hazards in China

Shengnan Wu, Yu Lei, Pihua Yin, Peng Cui, and Zhengtao Zhang

Abstract

Early warning systems (EWSs) are the essential tools for disaster risk reduction and have been implemented in practices all around the world. The Chinese government, in consideration of its physical and social condition, developed the Public Participation Monitoring and Warning (PPMW) System as an affordable solution to reduce disaster risk. Governments steer the whole system, and residents are trained on how to identify, monitor, and escape from hazards to establish a PPMW network for early warning and emergency response. This system has been implemented in practice for more than twenty years and proved its efficiency, especially in terms of disaster mortality reduction in China. This article introduced its function and took Liangshan Prefecture to explain how the PPWM was set up and operated during the "719" Boli Landslide in 2018 at the community level to avoid casualties. The PPWM is one of the practices of the Chinese government to engage the public in disaster risk reduction, which can be an excellent experience to share with other less developed and populated countries facing similar disaster situations as China.

Keywords

Early warning • Public participation • Geological hazard • China

Introduction

The United Nations Office for Disaster Risk Reduction (UNDRR) defined early warning systems (EWSs) as "The set of capacities needed to generate and disseminate timely and meaningful warning information to enable individuals, communities and organizations threatened by a hazard to prepare and to act appropriately and insufficient time to reduce the possibility of harm or loss." (UNISDR 2009, p. 5). EWSs of natural hazards are increasingly advocated by scientists and stakeholders and play essential roles in disaster risk reduction and management.

Complex physical conditions make China prone to geological hazards, among which landslides, rockfalls, and debris flows are the most frequent types. According to the annual report of the Ministry of Natural Resources of the People's Republic of China (available at: https://www.mnr.gov.cn/) (Fig. 1), from the year of 2008 to 2018, China had suffered 152,807 geological hazards in total, including 31,055 rockfalls, 10,039 debris flows and 102,992 other types of landslides as per the classification by Hungr et al. (2014). Although the number of geological hazards events varied from year to year, they strike each year and had significant impacts on Chinese society. Within the decade (2008–2018), official statistics from the Ministry of Natural Resources (available at: https://www.mnr.gov.cn/) showed that these geological hazards had caused 7037 people to die or to miss and a direct economic loss of 46.995 billion RMB in total. Therefore, China, like many other countries, is encountered with many threats from geohazards and has high demands to seek solutions to disaster risk reduction. This article will explain how the Chinese government

S. Wu · P. Cui · Z. Zhang
Institute of Geographic Sciences and Natural Resources Research, Chinese Academy of Sciences, Beijing, 100101, China
e-mail: shengnan@imde.ac.cn

P. Cui
e-mail: pengcui@imde.ac.cn

Z. Zhang
e-mail: zhangzt@igsnrr.ac.cn

Y. Lei (✉)
Institute of Mountain Hazards and Environment, Chinese Academy of Sciences, Chengdu, 610041, China
e-mail: leiyu@imde.ac.cn

P. Yin
Liangshan Prefecture Geological Environmental Center, Liangshan, 615000, China
e-mail: 398126050@qq.com

K. Sassa et al. (eds.), *Understanding and Reducing Landslide Disaster Risk*, ICL Contribution to Landslide Disaster Risk Reduction,
https://doi.org/10.1007/978-3-030-60196-6_44

concerns its physical conditions and socio-economic development levels and organize relevant stakeholders to monitor and manage geological hazards; and, hope to provide an option for other countries facing similar disaster situations with China.

Due to the limited length of this paper, we have selected one recent landslide event occurred at Boli village which is located in remote mountainous areas to explain how the public participation system is operated to save people in a real disaster event.

Early Warnings and Chinese PPMW System

The EWS is a vital tool to reduce disaster risks, as to provide residents with more time to flee from hazards and enables local authorities to evacuate or shelter people in advance (Rogers and Tsirkunov 2011). Different parts of the world have implemented EWSs to manage geological hazards in various ways. When new technologies and scientific advances are playing increasing roles in monitoring and warning, many developed areas integrated InSAR, UAV, and other new tools with the EWSs to increase their accuracy and efficiency (Fang et al. 2011; Niethammer et al. 2011; Peppa et al. 2016; Thiebes and Glade 2016; Mantovani et al. 2016; Mateos et al. 2017) and utilized EWSs at different scales in many cases around the world (Alcántar-a-Ayala and Garnica 2013). This type of EWSs is more reliant on the instruments, technology, well-developed networks (Angeli et al. 2000; Tarchi et al. 2003; Osanai et al. 2010), and to require more investment and a higher level of science and techniques. However, for some less developed countries, such as China, with a large number of disadvantaged populations in remote mountainous areas, at this stage of development, it is less preferential to adopt this type of EWSs due to a considerable investment in finance and personnel to relevant stakeholders. Therefore, for China, developing an affordable solution to manage geological hazards is imperative.

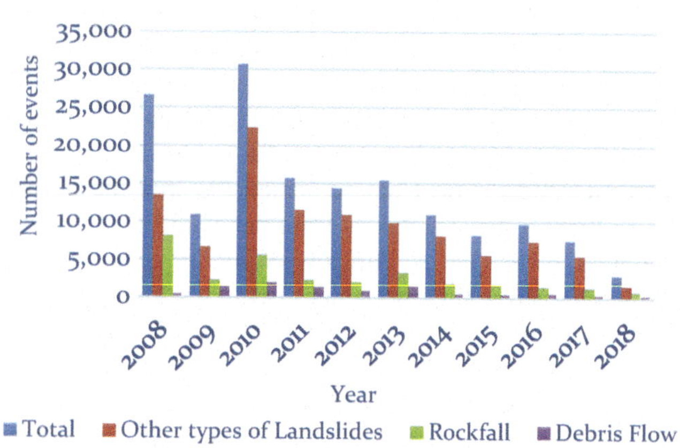

Fig. 1 Geological hazards in China (2008–2018)

China is a mountainous country with 70% of the total land area, where lives about 600 million population (Chen et al. 2010). Due to the complex geological conditions, geological hazards stroked mountainous areas every year and brought adverse impacts on societies. For reducing disaster risks, China needs to set up EWSs to provide more time for people to evacuate before disasters. However, full coverage of all potential hazard sites requires an enormous amount of resources in terms of personnel and finance (Chen et al. 2012). Therefore, with a large number of populated mountainous areas, China needs to find its way to set up EWSs for geological hazards into consideration with its physical and social conditions.

The Chinese government created the PPMW system, which is to engage professionals, officials, and residents in the disaster risk management process. In the PPWM system, relevant stakeholders are educated on how to identify, monitor, and escape from hazards and then establish a network for early warning and emergency response. It is a people-centered and low-cost EWS, which means that the system relies much on people to monitor, analyze and investigate hazards, communicate and disseminate alerts and warnings, and respond to emergencies at a local scale.

On March 1st, 2004, the National Regulation for Geological Hazards Mitigation was initiated (State Council of China 2003), in which implantation of PPMW was legislated in two articles. Since the enactment in 2004, 39,888 townships in China have implemented the PPMW to manage geological hazards.

Like other disaster-prone areas, the Liangshan prefecture government implemented the PPMW system Liangshan, Sichuan Province, is located in the mountainous areas with complex geological conditions in the southwest of China. It covers an area of 60,400 km^2 and has a population of 5.3 million; meanwhile, there endured many geological hazards.

In the next part of this paper, we took the Liangshan Prefecture government as an example to explain how to set up the PPMW system and how it works during emergency response.

Implementing PPMW System: Top-Down

Institutional Capacity Building

Chinese administrative structure is composed of the Central government and locals in five levels, including the provincial government, city government, county government, township government, and village committee. In disaster-prone areas, local governments are legislatively to implement PPMW from the top- to the bottom-level to manage disasters.

Local governments can make small adjustments by integrated with its own conditions, when they set up PPMW systems. The Liangshan Prefecture government set up the five-level PPMW system in its jurisdiction by six components (Fig. 2): Liangshan Prefecture (city-level), 17 counties, 404 townships, 3737 villages, and 4404 appointed responsible persons, and residents. The governments organize residents living near the monitoring sites to form a monitoring group who is responsible for routinely monitoring and reporting to the village committee.

Operating the whole system requires a well-designed institutional structure. The Liangshan Prefecture government played the leading role in the whole system. Each year, the provincial government, with local governments and professionals, always formulates emergency response plans and organizes meetings on geological hazard prevention and mitigation before the rainy season (high risk in geohazards), during which the Prefecture government and 17 county governments shall agree on the duties and obligations of their roles in disaster management. During the raining season, officials of Liangshan Prefecture conduct everyday meetings according to the latest information and released alerts of geological hazards.

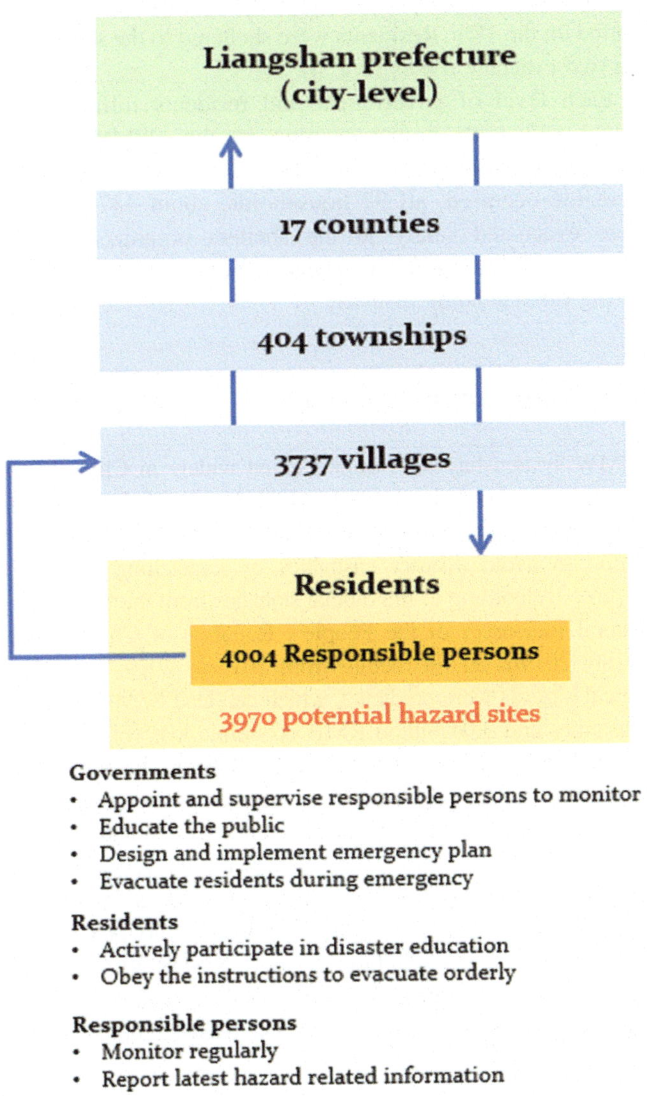

Fig. 2 The PPMW system in Liangshan Prefecture

Further, for monitoring, the Liangshan Prefecture government shall investigate potential hazard sites and appoint responsible persons from residents to monitor geohazards with support from professionals. During the rainy season, they also set up a '24-h duty system' to monitor disasters and a checking system to make sure there is no communication breakdown from responsible persons to report emergencies.

Meanwhile, the County government is a vital link between the Prefecture and the township governments for delivering information and lead its affiliations. The township government, at the frontier, is in charge of educating residents and organizing them to evacuate during emergencies.

Public Education for Disaster Risk Reduction

Public education for disaster risk reduction can mobilize people through clear messages, supported with detailed information. Residents have the advantage of familiarity with the local environment and close to the potential hazard sites. With proper training, engaging the public in disaster management is expected to make full use of their grassroots knowledge in the process of disaster management.

In the PPMW system, governments take responsibility to increase public awareness and public education for disaster risk reduction in order to turn available knowledge into proper actions in various ways (Liu et al. 2006). Like other local governments, the Liangshan government provide public disaster education to potentially affected households in multiple ways, such as giving public lectures, organizing open-air movies, and broadcasting videos.

In the PPMW, one compulsory way of local governments is to issue 'two cards': The Hazard Mitigation Card and the Hazard Avoidance Card. The 'two-cards' is called 'Life Card (MNR 2018a), which is a metaphor to Life Vest that save lives during emergencies. Hazard Mitigation Card includes necessary information on geological hazards, evacuation instruction, and emergency contacts. Different from the Hazard Mitigation Card, the Hazard Avoidance Card is customized for each family and considers the particular needs of each family, including the involvement of the elderly and the disabled.

As the highest local government in its PPMW system, the Liangshan government prepared and issued 'two cards' together with the residents to each household at high risks and let its residents to know more about disaster risk knowledge.

Emergency Response: Bottom-Up

Emergency Response Process

The effective emergency response relies on the well-designed institutional structure of PPMW (Fig. 2). In the PPMW system, usually, communities make bottom-up emergency responses and evacuate under the instructions from superior governments.

Responsible persons, appointed by governments, are required to check potential hazards sites routinely. During the raining season, they shall increase the frequency of checking when government release alerts. When they investigate any emergency, they will immediately report to the village committee. Then, the village will report to the township.

Correspondingly, together with professionals, governments shall soon afterward reach the site to check and evaluate risks. If there were possibilities to damage the village, either people or properties, local governments, village committees, and professionals shall together formulate emergency plans and evacuate residents nearby.

Response to the "719" Landslide in Boli Village

The Boli village (E 101°01′, N 27°29′) is a disaster-prone village under the administration of the Liangshan Prefecture. The PPMW system has been adopted there for many years. It is a five-level system guided by the instruction from the Liangshan Prefecture (Fig. 3).

On the 19th of July 2018, a mega landslide struck the village, with a volume of about 18 million m^3 (2000 m 300 m $\times$ 30 m) (Fig. 3). This landslide collapsed 186 houses and damaged 200 acres farmlands, which caused direct economic loss at 13 million yuan (Wu et al. 2018). However, there were no casualties during the landslide.

According to the official document of the Liangshan Province, the successful avoidance of mortalities in this landslide disaster, much of the credits should go to the PPMW system. Thus, the article used the '719' Landslide to illustrate how the PPMW system operated to respond to emergencies and reduce casualties from the community level.

Before the "719" landslide on 13th, soon after a responsible person found a few large cracks at one of the monitoring sites, residents and the Boli village committee immediately made PPMW bottom-up responses. When the responsible person investigated cracks, he straight away realized that it was a forewarning sign of geological hazards and reported his findings to the township. The township reported to the county and the prefecture governments. Meantime, the Prefecture, county, township government, and village committee immediately took actions to respond according to the emergency plan formulated before the raining season.

After receiving the warning from the township, the steering groups of the Liangshan Prefecture government forthwith sent professionals to check on sites and confirmed that the area is subject to a potential landslide and set up warning signs (Fig. 4). The head of each village gathered to discuss emergency plans with support from professionals, including the steering groups at Prefecture. The responsible persons were 24-h on duty and monitored any further development of cracks. Residents who lived nearby the cracked area had been evacuated right away on the 14th. A few days later, with the report of the continuous development of cracks, evacuation started on the 18th. Residents were sheltered to the safety sites via two evacuate routes (Fig. 4).

Each level of government and residents fulfilled their duties and took proper actions in the PPMW system accordingly. By the evening of the 18th, the day before the landslide occurred, all 97 households, about 437 villagers were evacuated safely to the shelters organized by the governments and village committee (Fig. 4). No people died during this landslide disaster.

PPMW System Outcomes

PPMW system has been implemented widely in China in the last two decades and achieved remarkable outcomes. The data in practices have proved that the PPMW system make contribution to avoid disaster casualties by evacuating people in advance. According to the official statistics from the Ministry of Natural Resources of the People's Republic of China (MNR 2018b, 2019b, 2020), merely from 2017 to 2019, 2460 geological hazards were predicted, which can lead 87,907 potential causalities and 3.04 billion RMB economic loss (Fig. 5). During the same period, only 322 geohazards killed people, accounting for 2.13% of the total number. Some studies (e.g. Liu et al. 2006; Li et al. 2012; Chen et al. 2012; Wu et al. 2018) stated its effectiveness to reduce disaster mortality, especially in developing countries or regions with limited resources. The PPMW system has made contributions to achieve the global goals of the Sendai Framework (UNISDR 2015) and can be a reference for other less developed areas facing similar disaster situations with China to involve residents in the process of disaster risk reduction.

Fig. 3 The '719' Boli landslide and evacuate routes

Fig. 4 Warning signs (Translation of text on the board: landside is dangerous. Traffic and pedestrians are prohibited.) and safety sites

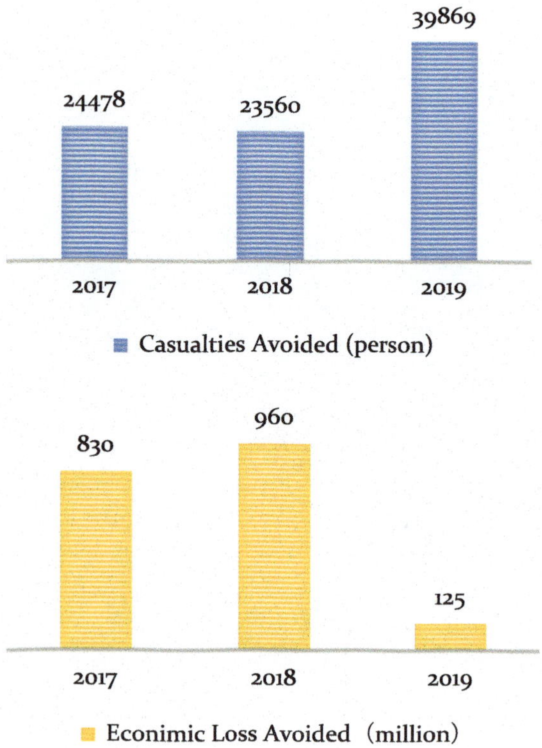

Fig. 5 Casualties and economic loss avoided (2017–2019)

Conclusion

EWSs are essentials in disaster risk reduction by providing more time for people to respond during a disastrous event. Many countries adopted EWSs in practices in different ways. The Chinese government, concerning the socio-economic development level in its current phase, set up the PPMW system in an affordable way. The primary function of this system is to organize residents to evacuate timely before disasters and reduce casualties. In this PPMW, governments steer the whole system and educate the public to monitor and implement for early warning and emergency response. Governments, responsible persons, and residents fulfill their roles accordingly at each level in the system.

The 20-year implementations of PPMW in China have been conclusively proved to be efficient in reducing disaster mortalities, which devoted to the goals of the Sendai Framework in reducing disaster mortalities. Thus, PPMW can be an option for some less developed but populated countries facing similar disaster situations as China, to engage the public in the EWSs to bring down both financial and technical investments and to decrease disaster-related mortalities.

Acknowledgements This work is supported by the International Partnership Program (131551KYSB20160002) and NSFC (Y9K1070070). The Liangshan Prefecture Geological Environmental Center is especially crediting with providing the case study of the "7.19" Landslide.

References

Alcántara-Ayala I, Garnica RJ (2013) Landslide monitoring and warning systems in Mexico. In: Landslides: global risk preparedness. Springer, Berlin, Heidelberg, pp 299–314

Angeli MG, Pasuto A, Silvano S (2000) A critical review of landslide monitoring experiences. Eng Geol 55(3):133–147

Chen G, Fang Y, Gao Y (2010) Report on the development of mountain areas in China. Commercial Press, Beijing. ISBN 9787100071420

Chen R, Chen S, Wu C et al (2012) Debris flow disaster mitigation and prevention countermeasure in Taiwan, Shuili Xuebao. 0559-9350 (2012) 2-0186-07

Fang M, Qi Y, Zhang J (2011) Research of Lanzhou city geological disaster mass observation and mass prevention informationization based on WebGIS. Remote Sens Technol Appl 26(2):137–146

Hungr O, Leroueil S, Picarelli L (2014) The Varnes classification of landslide types, an update. Landslides 11(2):167–194

Li Q, Huang L, Li D (2012) Important role of monitoring and prevention in geological disaster prevention and mitigation in Hubei Province. Resour Environ Eng 26(1):92–94

Liu C, Zhang M, Meng H (2006) Study on the geo-hazards mitigation system by residents' self-understanding and self-monitoring. J Disaster Prev Mitig Eng 02:175–179

Mantovani M, Devoto S, Piacentini D, Prampolini M, Soldati M, Pasuto A (2016) Advanced SAR interferometric analysis to support the geomorphological interpretation of slow-moving coastal landslides (Malta, Mediterranean Sea). Remote Sens 8(6):443

Mateos RM, Azañón JM, Roldán FJ, Notti D, Pérez-Peña V, Galve JP, Devantèry N (2017) The combined use of PSInSAR and UAV photogrammetry techniques for the analysis of the kinematics of a coastal landslide affecting an urban area (SE Spain). Landslides 14 (2):743–754

MNR (Ministry of Natural Resources of the People's Republic of China) (2018a) Summary of the achievements of geological hazard prevention and reduction in the recent 10 years in China. [Online]. Available at: https://www.mlr.gov.cn/xwdt/jrxw/201805/t20180516_1776256.htm. Accessed 19 Sept 2018

MNR (Ministry of Natural Resources of the People's Republic of China) (2018b) Summary and forecast of geological hazards in China in 2017. [Online]. Available at: https://mini.eastday.com/bdmip/180108095143885.html. Accessed 08 May 2019

MNR (Ministry of Natural Resources of the People's Republic of China). (2019a) Geological hazard annual report in 2018. [Online]. Available at: https://www.cigem.cgs.gov.cn/gzdt_4839/dwdt_4861/201904/t20190417_479382.html. Accessed 08 May 2019

MNR (Ministry of Natural Resources of the People's Republic of China) (2019b) Summary and forecast of geological hazards in China in 2018. [Online]. Available at: https://www.sohu.com/a/287710844_120032925. Accessed 08 May 2019

MNR (Ministry of Natural Resources of the People's Republic of China) (2020) Summary and forecast of geological hazards in China in 2019. [Online]. Available at: https://www.jinchuan.gov.cn/xxgkml/zfbm/gtfj/gzxx_1963/ywgzgtj/202001/t20200113_134022.html. Accessed 08 May 2019

Niethammer U, Rothmund S, Schwaderer U et al (2011) Open source image-processing tools for low-cost UAV-based landslide investigations. Int Arch Photogramm Remote Sens Spat Inf Sci 38(1):C22

Osanai N, Shimizu T, Kuramoto K, Kojima S, Noro T (2010) Japanese early-warning for debris flows and slope failures using rainfall indices with radial basis function network. Landslides 7(3):325–338

Peppa MV, Mills JP, Moore P et al (2016) Accuracy assessment of a UAV-based landslide monitoring system. ISPRS-Int Arch Photogramm Remote Sens Spat Inf Sci 41:895–902

Rogers D, Tsirkunov V (2011) Costs and benefits of early warning systems. Global assessment rep

State Council of China (2003) National rule for geological hazards mitigation. Available at: https://www.gov.cn/gongbao/content/2004/content_63064.htm. Accessed 15 Sept 2018

Tarchi D, Casagli N, Fanti R, Leva DD, Luzi G, Pasuto A, Silvano S (2003) Landslide monitoring by using ground-based SAR interferometry: an example of application to the Tessina landslide in Italy. Eng Geol 68(1–2):15–30

Thiebes B, Glade T (2016) Landslide early warning systems—fundamental concepts and innovative applications. In: Aversa S, Cascini L, Picarelli L, Scavia C (eds) Landslides and engineered slopes: experience, theory and practice. Proceedings of the 12th international symposium on landslides, Napoli, June 2016, pp 12–19

UNISDR (2009) Terminology on disaster risk reduction. [Online]. Available at: https://www.unisdr.org/files/7817_UNISDRTerminologyEnglish.pdf. Assessed 13 Sept 2019

UNISDR (2015) Sendai framework for disaster risk reduction. [Online]. Available at: https://www.preventionweb.net/files/43291_sendaiframeworkfordrren.pdf. Assessed 13 Sept 2019

Wu K, Chen N, Hu G, Huang N, Zhang Y (2018) Emergency investigation to 719 landslide disaster in Boli Village, Yanyuan County, Sichuan, China. Mount Res 36(05):806–812

Part V

Education and Capacity Development for Risk Management and Risk Governance

Early Warning Systems in Italy: State-of-the-Art and Future Trends

2020 Kyoto Japan

Emanuele Intrieri, Giulia Dotta, Federico Raspini, Ascanio Rosi, Samuele Segoni, and Nicola Casagli

Abstract

Landslide risk in Italy is one of the highest worldwide. The strategy to face this problem largely relies on early warning systems. A national early warning system based on weather forecast is regulated by a national law, although each of the twenty regions composing Italy has a high degree of autonomy. For example, in Emilia-Romagna Region, rainfall data are statistically analysed and multiple values of standard deviation are used to define warning thresholds. On the other hand, a correct risk communication toward the population using appropriate means and content is necessary to make the warnings effective. Recently, a new possibility for regional scale, displacement-based early warning systems came to existence and is here presented.

Keywords

Early warning systems • Landslides • Sentinel-1 • Rainfall thresholds • Risk communication

E. Intrieri (✉) · G. Dotta · F. Raspini · A. Rosi · S. Segoni · N. Casagli
Department of Earth Science, Università degli Studi di Firenze, via G. La Pira 4, 50121 Firenze, Italy
e-mail: emanuele.intrieri@unifi.it

G. Dotta
e-mail: giulia.dotta@unifi.it

F. Raspini
e-mail: Federico.raspini@unifi.it

A. Rosi
e-mail: ascanio.rosi@unifi.it

S. Segoni
e-mail: samuele.segoni@unifi.it

N. Casagli
e-mail: nicola.casagli@unifi.it

Introduction

Landslide risk in Italy is the highest among European countries while worldwide is second only to China, Japan and countries of Central and South America (Canuti et al. 2001). Italy has the highest cumulative number of deaths or missing people because of landslides and the highest expected yearly loss of life in Europe (Forli and Guida 2009).

In the period 1900–1999, 5939 fatalities (60 per year) due to landslides have been registered (Guzzetti 2000), of which, more than 2000 were caused by Vajont landslide alone (Barla and Paronuzzi 2013). Not including this catastrophic event, in the decade 1990–1999, 263 casualties were recorded (Guzzetti 2000). Between 1900 and 2002 the number of displaced or homeless persons is 175,000 and landslides with significant damages to the population occurred in 1328 municipalities (Salvati et al. 2003). The IFFI project (Inventory of Landslide Phenomena in Italy) produced a map of the landslides in the whole national territory and identified 485,000 landslides, covering an area of 20,721 km^2, i.e. 6.9% of the country (APAT 2007). According to ISPRA (2018), almost 1.3 million people live in high or very high landslide risk areas in Italy. Concerning the economic impact, if floods are included, the total cost of works to reduce geo-hydrological risk in Italy is estimated around 40 billion euros (MATTM 2003), but the yearly government funding rarely reaches 400 million euros (Trezzini et al. 2013). Funds are instead regularly spent to face emergency in much less cost-effective solution.

When structural countermeasures are not a viable option, early warning systems (EWS) assume great importance. These are defined as "monitoring devices designed to avoid, or at least to minimize, the impact imposed by a threat on humans, damage to property, the environment, or/and to more basic elements like livelihoods" (Medina-Cetina and Nadim 2008). EWS are effective when they manage to balance their instrumental/technical element (necessary for

K. Sassa et al. (eds.), *Understanding and Reducing Landslide Disaster Risk*, ICL Contribution to Landslide Disaster Risk Reduction,
https://doi.org/10.1007/978-3-030-60196-6_45

monitoring and forecasting) with their social/communication part (Intrieri et al. 2013).

In Italy a national warning system based on weather forecasts is regulated by law and implemented in every region, down to each municipality. In the following section we discuss how this system is used together with rainfall thresholds to provide landslide early warning at regional level. EWS also require a relevant effort in social strategies, because an accurate prediction is useless if it is not properly communicated to the population. Therefore, we present the guidelines for a correct communication that are currently employed in a group of municipalities in Italy and that are compatible to be used in the whole country. Finally we describe a novel, alternative regional early warning system, based on the use of Sentinel-1 constellation to provide continuous displacement monitoring aimed at detecting anomalous accelerations and possibly anticipating landslide failures.

Rainfall-Based EWS

The National Warning System

Italian civil protection is a complex system entrusted to all administrative levels of the State. A "central" national department (under the direct control of the Prime Minister) coordinates a network of regional centres, each one in charge of forecasting and managing hazards inside its administrative territory, while at the local level the most important authority of civil protection is the mayor of each municipality. Other bodies (e.g., provinces) constitute an intermediate level between the municipality and the regions, ensuring an effective management of complex hazard scenarios.

According to this multi-level organization, each of the twenty Italian regions has a so-called "functional centre" in charge of meteorological monitoring and forecasting. This activity includes a warning system for landslide hazard covering the whole region and issuing, at least daily, a criticality state among four possible levels (absent, ordinary, medium, and high). To improve the forecasting effectiveness and the emergency response, regions are subdivided into alert zones (158 in total), each monitored and alerted independently. Every criticality level is coded with a colour (green, yellow, orange and red) and is characterized by a standardized description of the typical scenario expected on the alert zone.

In the past decades, every regional functional centre has developed its own landslide warning system based on statistical rainfall thresholds, which, according to the current state-of-the-art, is the most effective technique to get reliable landslide forecasts at the regional scale (Piciullo et al. 2018; Segoni et al. 2018). This technique is particularly widespread because it is easily implemented, operate and communicated. For instance, the input data for a threshold definition consists only in a landslide inventory (with known time of triggering) and the corresponding rainfall measurements. Statistical techniques are then used to find a relationship between some rainfall parameters (e.g. rainfall intensity and duration) and the minimum rainfall conditions associated to landslide triggering. This relationship is expressed by a simple mathematical law, which can be visualized as a line in a diagram, and landslide hazard can be monitored indirectly by monitoring the triggering factor (rainfall): when the monitored (or forecasted) rainfall exceeds the threshold an alert is issued.

The alerts issued by the regional centre are addressed to the mayors of the municipalities of the alert zones, which are in charge of activating all due countermeasures.

Such countermeasures should be carefully planned and written in a civil protection plan, which is compulsory for every municipality. The plan contains a detailed description of the scenarios expected at the municipality level, a careful planning of the emergency response through objectives, strategies and personnel and means to be activated. The plan also describes the management procedures and how to convey the necessary information to the citizens.

SIGMA Model

In this section an example of a statistical model used to implement an operative EWS is presented.

The model is named SIGMA, it is based on rainfall thresholds and it is in use in the Emilia Romagna (ER hereafter) Civil Protection Agency.

The development and the refinement of the model took place in a time span of ca. 20 years, by the strict collaboration between scientist and public authorities (Segoni et al. 2018); in this time the model was continuously updated and refined, using new data and considering the requirement of the final user (the ER Civil Protection Agency).

The model is based on the statistical analysis of long rainfall time series and on the hypothesis that landslide triggering is related to anomalous or extreme values of accumulated rainfall (Martelloni et al. 2012).

To define the thresholds, rain data series have been analysed and multiple values of standard deviation (σ) have been used to discriminate between ordinary and extraordinary rainfall events.

For each rain accumulation period (ranging from 1 to 365 days) several σ values, from 1 to 3 (with step of 0.05 σ) have been calculated, so that a series of σ curves were defined.

The results of the analyses have been used to define a decisional algorithm, where measured rainfalls are combined with forecasted data to define four warning levels (Fig. 1). Shallow landslides have been associated to very intense (high σ values) and short rainfall (1–3 days), while deep landslides have been associated to longer rainy periods, variable on seasonal base (4–60 days in dry season and up to 245 in the wet season).

To select the proper σ value for each kind of landslide a fine calibration, based on comparison of landslide events and accumulated rainfall, was carried out.

Since ER is geomorphologically heterogeneous, with mountains in the southern part and plains in the northern part, the whole territory was divided into 8 zones (corresponding to the alert zones) and each of them into several homogeneous territorial units (TU).

The thresholds were then calibrated for each TU and the results were aggregated to define a unique decisional algorithm, used to implement the EWS.

Risk Communication

Being able to forecast is a fundamental part of an EWS but it is useless if the last step is not properly accomplished: risk communication, that is the communication toward the population employed before and during an anticipated event. In Italy the city mayor is responsible for transmitting alerts to the population, however, since there is no national regulation or indication for such specific task, we present here some recommendations from the guidelines derived from the Interreg European project Proterina and already adopted in several Italian municipalities (Intrieri et al. 2020).

The warning message should be issued to the population once the regional functional centre transmits to the municipality the colour-code assigned to the connected alert zone. The communication to the citizens can be entrusted to the management of the municipal civil protection office (if present in the municipality), the press office, or the mayor secretariat, with the contribution of all the municipal personnel competent in these matters and the whole civil protection system (for example volunteering or first responders such as fire fighters).

The information to be included in the message should include:

- Colour-code of the criticality level.
- Municipality where the alert takes place.
- Risk type (in this case landslide risk): municipalities are often exposed to different risks and risk scenarios, which are considered in municipal civil protection planning and must be clearly specified in the alert messages.
- Alert duration: the duration of the alert should always be provided; the exact hour of start and end of duration of alerts is an information not precisely relatable to the actual period of risk. A qualitative description of the alert duration is more advisable for communication purposes (for example "in the evening of 4 November").
- "Simplified" and updated event description: weather forecasts and regional warnings are provided at a larger scale than municipalities. Furthermore, the actual ground effect of weather conditions (event scenarios) can considerably vary depending on the specific features of the territory and structures. Therefore, in the message to the

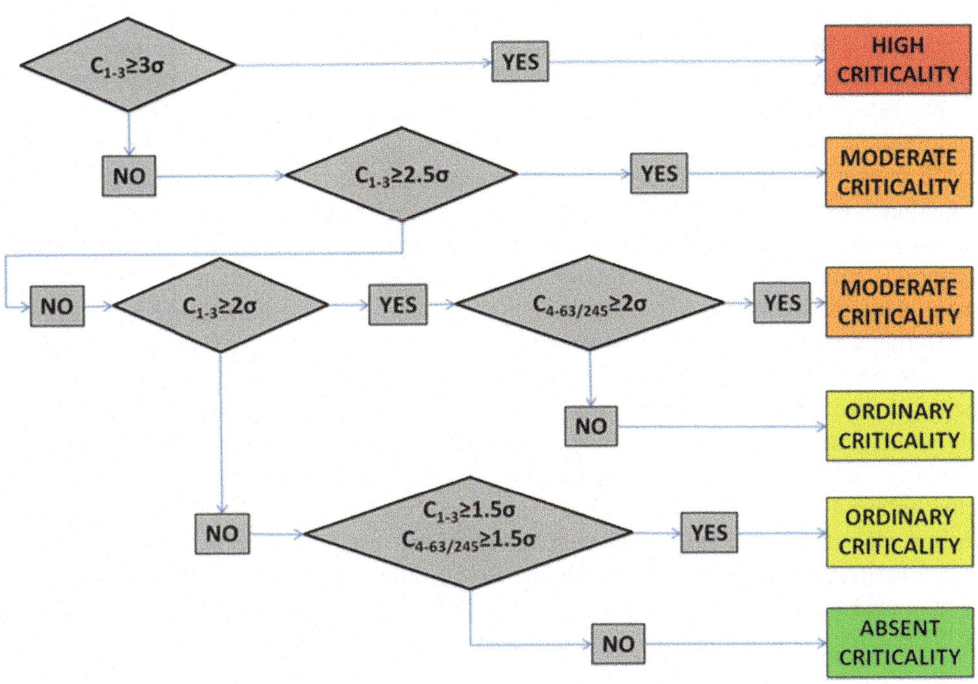

Fig. 1 The decisional algorithm of SIGMA (from Martelloni et al. 2012), where C_{i-j} is the accumulated rainfall from day i to day j

people, the description of the weather forecast and criticality level must be contextualized and made relevant to the territory, referring to some local issues and risk scenarios, using terms understandable to the general public.
- Safe behaviours to employ before, during and after an event; the transmission of self-protective rules to the population, whenever possible, is best done using infographics, images, photos or links.

The communication should be institutional (formalized by the municipality), efficient, rapid, reliable and it should use multiple communication channels in order to reach a higher number of citizens. For this reason, the guidelines include a hierarchy of the communication channels depending on the colour-code with the purpose of activating specific channels only for the most critical communications (red alert). This allows a municipality to activate and supervise the channels with high priority and to leave behind those for which the human and economic resources are not enough. In fact, once a channel is activated (for example the alert messages are disseminated through a post on the Facebook page of the municipality), people will expect to always receive communications through that medium. Moreover, if some channels are activated only during the highest levels of alert this contributes to generate a different risk perception.

The alert message should adopt a shared glossary (in particular concerning hashtags when messages are spread through social networks), so that the same terminology is used by all the municipalities during the alert communication phase, and it should preferentially use infographics, to convey the safe behaviours to be adopted in case of emergency.

For all alert levels, the municipality website and variable-message signs (VMS) have been placed higher up in the list of communication channels since each municipality should have its own institutional site, which should be made known to the citizens and always be kept updated; regarding VMS, they are present along the main places and infrastructures of a municipality (bus stops, along important communication roads, at the entrance of the city) and are the main means to reach also non-resident people.

During a yellow alert, additional communication channels that can be activated are specific mobile apps and the social networks Facebook and Twitter, in priority order. The apps allow the citizens to receive notifications or push messages whenever an alert is issued, in some cases, they provide other relevant information such as risk maps and civil protection plan. Facebook and Twitter are the two most used social networks in Italy that are also suited to real-time and institutional communication (differently from Instagram or YouTube, for example); the main difference between the two social networks is that Twitter has a maximum limit of characters (280 characters) and is less popular in Italy and worldwide (We Are Social 2019).

During the orange and red alerts, e-mails and instant messaging applications are recommended over social networks. Specific apps are still preferred because, in the case of e-mails and instant messaging applications (like WhatsApp), the citizens have to register to a service and provide personal information, which is less likely action than just downloading an app.

Concerning the red alert, a further means of communication to the citizen can be the phone call, that is the information service that transmits a pre-recorded message of the mayor, which must be concise to be effective. This channel has been placed after Municipality website and VMS in terms of priority and is recommended only for the red alert because it has the strongest impact on the people risk perception and it should be reserved when the risk is at its highest and the possibility of a false alarm is low; furthermore, this service can be expensive for a municipality with little economic resources.

All these indications are exemplified in Fig. 2.

Displacement-Based EWS

Landslide Forecasting Using Kinematic Parameters

So far, EWSs based on displacement monitoring have been implemented only at the slope scale (Intrieri et al. 2012; Michoud et al. 2013) and relied on the availability of detailed, high-frequency data from sensors installed in situ. While the regional scale EWSs like the one described in the previous section are based on indirect indicators of slope instabilities (such as accumulated rainfall) displacement data are a direct sign of the dynamic conditions of a landslide (Intrieri and Gigli 2016).

In recent years, advances in satellite sensors, increase of computing capacity, refinement of data screening tools contributed to the design of a new paradigm in satellite-based monitoring systems. The launch of Sentinel-1 mission opened a new opportunity for large-scale monitoring applications thanks to the increased acquisition frequency, the regularity of acquisitions and the policy on data access. Sentinel-1 data, continuously processed and analysed, can be used as a tool for a systematic tracking of ground deformation at regional scale and as a key information layer for risk mitigation.

The rationale behind this application is that anomalous accelerations can indicate that the landslide has entered the tertiary creep and it is approaching collapse (Saito 1969).

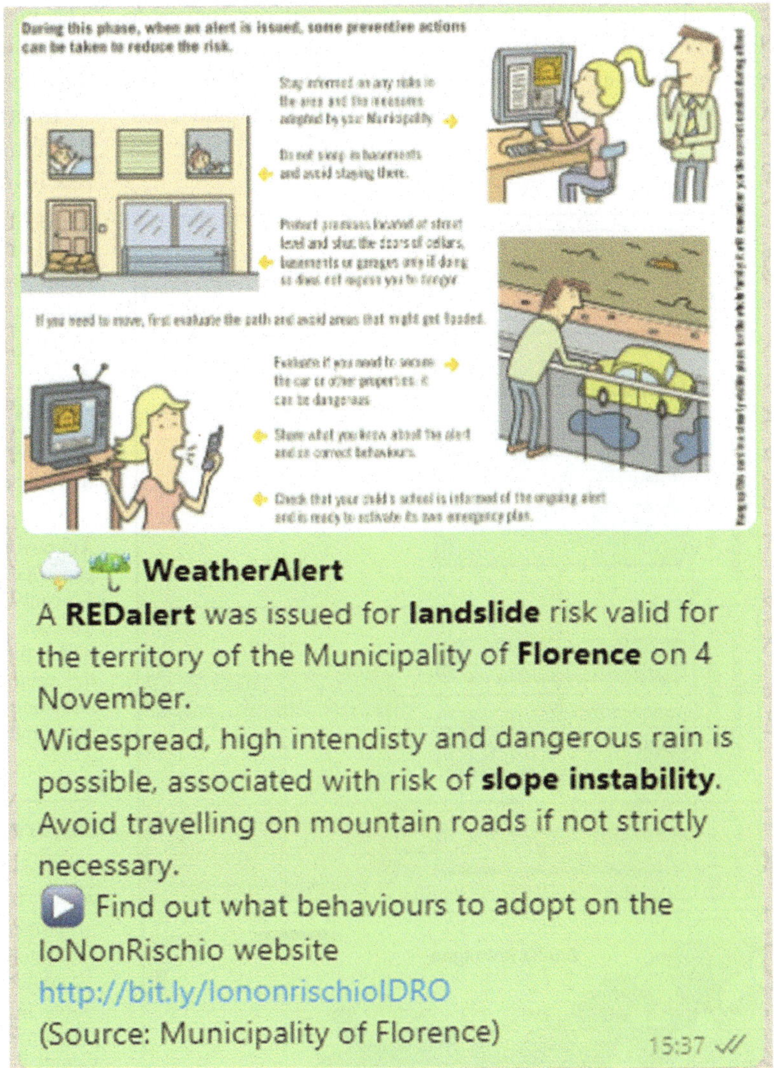

Fig. 2 Example of a warning message compiled using the guidelines and conveyed via WhatsApp

At this point specific forecasting methods exist to determine the probable time of failure (Fukuzono 1985; Intrieri et al. 2019). Recent experiences have proven that landslide forecasting using Sentinel-1 is feasible, although they were all cases applied in retrospect (Carlà et al. 2019; Intrieri et al. 2018). Even if forecasting methods are not implemented, the detection of deviations form a linear trend of the displacements can still be used as a reliable warning threshold.

The Satellite Monitoring System

Starting from October 2016, the Tuscany Region adopted, first region in the world, a regional monitoring system based on the systematic processing of Sentinel-1 images. Operatively, a specific processing chain for both ascending and descending geometry has been implemented. Once a new Sentinel-1 acquisition is available, it is automatically downloaded and added to the existing archive. The new data stack is then entirely reprocessed using the SqueeSAR algorithm to generate new ground deformation maps and updated displacement time series (Ferretti et al. 2011). This monitoring system was conceived to mark the transition from historical satellite analysis of radar imagery to a dynamic streaming of displacement information at regional scale.

Displacement time series, systematically updated with the most recent available Sentinel-1 acquisition, are analysed to identify anomalous points (i.e., points where a change in the dynamics of motion is occurring) (Raspini et al. 2018). When detected, changes in the deformation pattern are analysed and interpreted, with the support of thematic information, optical images and in situ data, to assign a triggering factor and providing possible warning (Fig. 3). The presence of groups of anomalies along with their temporal persistency (i.e., anomalies repeated in at least two consecutive updates) are the most important parameters to link registered trend changes to ground instabilities. Reports also the Tuscan municipalities classified according to the total number of persistent anomalies related to landslides.

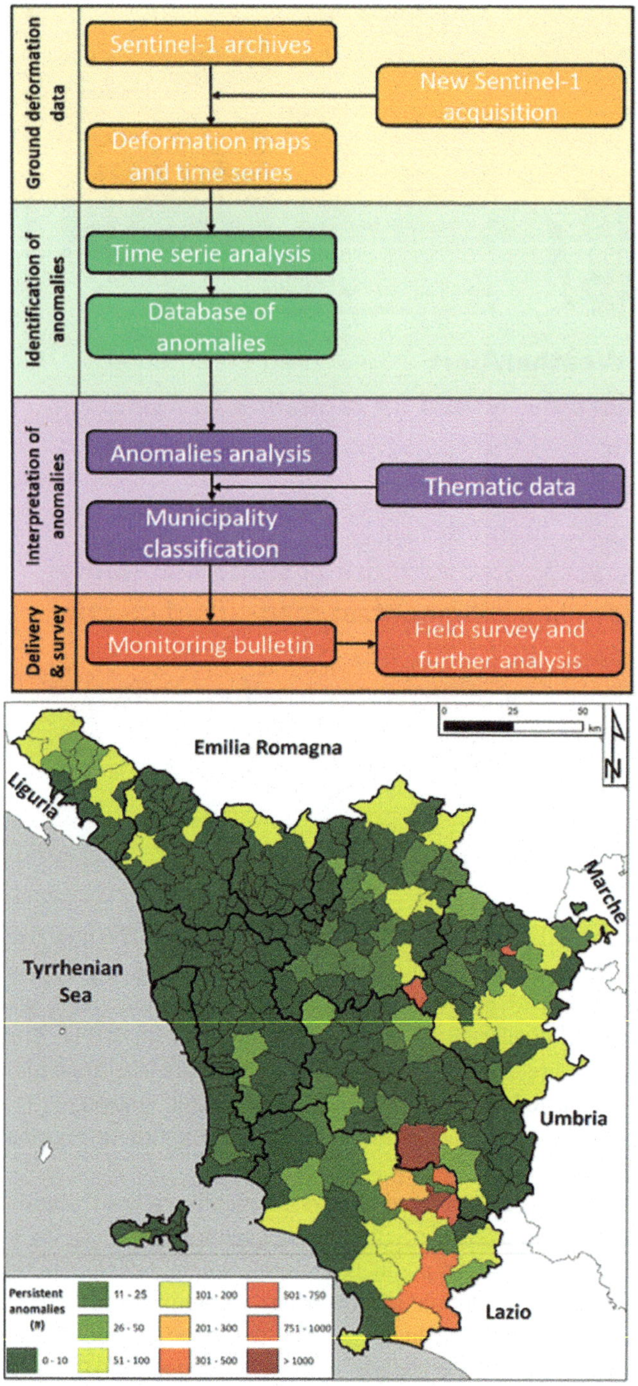

Fig. 3 *Upper* Flow diagram showing the adopted procedures within the monitoring system. *Lower* Classification of the municipalities in Tuscany Region according to the number of persistent anomalies of movement related to slope instability. Modified from Raspini et al. (2019)

The presence of a cluster of persistent anomalies affecting elements at risk determines a significant level of risk, with the necessity of further analyses.

Information on anomalous areas (areas where a change in the dynamics of motion is occurring) affecting elements at risk is routinely delivered to regional authorities in charge of the geohazard management practices in the form of monitoring bulletins. Whenever needed, field investigations are then performed to determine the severity of the hazard, initiate management of the risk and decide, together with local authorities, the most appropriate actions to mitigate the threats.

This project proved successful and spread also to Valle d'Aosta Region and Veneto Region, where monitoring is now active.

Acknowledgements The satellite monitoring system presented in this paper has been founded and supported by the Regional government of Tuscany. The communication guidelines have been financed by ANCI Toscana through the European project "PROTERINA-3Évolution" within the Interreg Italia-Francia Marittimo 2014–2020 program and by the project "Realizzazione di linee guida per riduzione del rischio alluvione tramite un approccio non strutturale" coordinated by Emanuele Intrieri and financed by the University of Florence, Italy.

References

APAT Rapporti 78/2007 (2007) Rapporto sulle frane in Italia. Il progetto IFFI (Inventario dei Fenomeni Franosi in Italia)—Metodologia, risultati e rapporti regionali. Dipartimento difesa del suolo, Roma, 681 p. ISBN 978-88-448-0310-0

Barla G, Paronuzzi P (2013) The 1963 Vajont landslide: 50th anniversary. Rock Mech Rock Eng 46:1267–1270

Canuti P, Casagli N, Pellegrini M, Tosatti G (2001) Geo-hydrological hazards. In: Vai M (ed) Anatomy of an orogen: the Apennines and adjacent Mediterranean basins. Kluwer Academic, Dordrecht/Boston, pp 513–532 (Chap 28)

Carlà T, Intrieri E, Raspini F, Bardi F, Farina P, Ferretti A, Colombo D, Novali F, Casagli N (2019) Perspectives on the prediction of catastrophic slope failures from satellite InSAR. Sci Rep 9(1):1–9

Ferretti A, Fumagalli A, Novali F, Prati C, Rocca F, Rucci A (2011) A new algorithm for processing interferometric data-stacks: SqueeSAR. IEEE Trans Geosci Remote Sens 49(9):3460–3470

Forli A, Guida T (2009) Il rischio idrogeologico in Italia. Adempimenti e tecniche operative d'intervento. Sistemi Editoriali, Roma, 412 p. ISBN 9788851305727

Fukuzono T (1985) A method to predict the time of slope failure caused by rainfall using the inverse number of velocity of surface displacement. Landslides 22:8–13

Guzzetti F (2000) Landslide fatalities and the evaluation of landslide risk in Italy. Eng Geol 58(2):89–107

Intrieri E, Gigli G (2016) Landslide forecasting and factors influencing predictability. Nat Hazards Earth Syst Sci 16:2501–2510

Intrieri E, Gigli G, Mugnai F, Fanti R, Casagli N (2012) Design and implementation of a landslide early warning system. Eng Geol 147:124–136

Intrieri E, Gigli G, Casagli N, Nadim F (2013) Brief communication. Landslide early warning system: toolbox and general concept. Nat Hazards Earth Syst Sci 13:85–90

Intrieri E, Raspini F, Fumagalli A, Lu P, Del Conte S, Farina P, Allievi J, Ferretti A, Casagli N (2018) The Maoxian landslide as seen from space: detecting precursors of failure with Sentinel-1 data. Landslides 15(1):123–133

Intrieri E, Carlà T, Gigli G (2019) Forecasting the time of failure of landslides at slope-scale: a literature review. Earth Sci Rev 193:333–349

Intrieri E, Dotta G, Fontanelli K, Bianchini C, Bardi F, Campatelli F, Casagli N (2020) Operational framework for flood risk communication. Int J Disaster Risk Reduct. https://doi.org/10.1016/j.ijdrr.2020.101510

ISPRA (2018) Data yearbook in numbers. In: Environmental data yearbook 2018. 86/2019, 164 p. ISBN 978-88-448-0941-6

Martelloni G, Segoni S, Fanti R, Catani F (2012) Rainfall thresholds for the forecasting of landslide occurrence at regional scale. Landslides 9:485–495

MATTM (Ministero dell'Ambiente e della Tutela del Territorio e del Mare) (2003) Direzione Generale per la difesa del suolo—Pianificazione territoriale provinciale e rischio idrogeologico: rapporto 2003. Roma

Medina-Cetina Z, Nadim F (2008) Stochastic design of an early warning system. Georisk 2:223–236

Michoud C, Bazin S, Blikra LH, Derron MH, Jaboyedoff M (2013) Experiences from site-specific landslide early warning systems. Nat Hazards Earth Syst Sci 13:2659–2673

Piciullo L, Calvello M, Cepeda JM (2018) Territorial early warning systems for rainfall-induced landslides. Earth Sci Rev 179:228–247

Raspini F, Bianchini S, Ciampalini A, Del Soldato M, Solari L, Novali F, Del Conte S, Rucci A, Ferretti A, Casagli N (2018) Continuous, semi-automatic monitoring of ground deformation using Sentinel-1 satellites. Sci Rep 8:7253

Raspini F, Bianchini S, Ciampalini A, Del Soldato M, Montalti R, Solari L, Tofani V, Casagli N (2019) Persistent scatterers continuous streaming for landslide monitoring and mapping: the case of the Tuscany region (Italy). Landslides 16(10):2033–2044

Saito M (1969) Forecasting time of slope failure by tertiary creep. In: Proceedings of the 7th international conference on soil mechanics and foundation engineering, Mexico City, vol 2, pp 677–683

Salvati P, Guzzetti F, Reichenbach P, Cardinali M, Stark CP (2003) Map of landslides and floods with human consequences in Italy. Publication CNR GNDCI n. 2822, Scale 1:1,200,000

Segoni S, Rosi A, Fanti R, Gallucci A, Monni A, Casagli N (2018) A regional-scale landslide warning system based on 20 years of operational experience. Water 10(10):1297

Trezzini F, Giannella G, Guida T (2013) Landslide and flood: economic and social impacts in Italy. In: Margottini C, Canuti P, Sassa K (eds) Landslide science and practice. Springer, Berlin, Heidelberg, pp 171–176

We Are Social (2019) https://wearesocial.com/global-digital-report-2019. Accessed 28 Jan 2020

Community-Based Landslide Risk Management in Contrasting Social Environments, Cases from the Czech Republic

Jan Klimeš and Ping Lu

Abstract

The contribution presents four cases illustrating advantages as well as flaws of community-based LDRR approaches under various environmental and social conditions. It shows that the individualization along with preferred competing interests (e.g. housing development) negatively affect LDRR at the community level. Nevertheless, even under unfavourable community conditions, its individual members may still effectively protect their interests involving the local state administrations into mitigation landslide risk, which they perceived as high. LDRR on the community level could be further hindered by legal fragmentation and institutional diversification, which could prevent implementation of desired mitigation measures due to unavailability of funds or missing definition of responsibilities. Nevertheless communities can overcome even such institutional obstacle, although it requires much larger collaboration involving other external actors (e.g. non-governmental organizations).

Keywords

Community • Landslide risk reduction • Citizen science

J. Klimeš (✉)
Institute of Rock Structure and Mechanics, The Czech Academy of Sciences, V Holešovičkách 41, 18209 Prague, Czech Republic
e-mail: klimes@irsm.cas.cz

P. Lu
College of Surveying and Geo-Informatics, Tongji University, Shanghai, 200092, China
e-mail: luping@tongji.edu.cn

Community-Based Landslide Risk Management

Communities should be in the focal point in search for effective landslide disaster risk reduction (LDRR) strategies (Klimeš et al. 2019a) as evidences point out that the widely applied top-down approaches have not resulted into desired reduction of landslide related loses (Froude and Petley 2018). Community responses to landslide hazard depend to a different degree on external (e.g. administrative, financial or legal) conditions and inputs, which are set by regional or state policies. Therefore any community-based LDRR strategy has to fit into this broader framework, or the external actors should consider and include the community views into their policies. This requires intensive communication and collaboration between various local, regional or national actors especially with respect to their needs and preferred mitigation measures (Maes et al. 2019). Nevertheless, there are increasing evidences that the effectiveness of community-based actions is being challenged by individualization (Beck 2012). It has been observed in regions with long history of tied community bounds (Vincent 2018) as well as in areas with challenging historical development with respect to communities (Raška 2019).

The contribution presents four cases from the Czech Republic (Fig. 1) illustrating potential flaws of community-based LDRR approaches as well as examples of policy-driven community actions searching for approaches of effective collaboration among distinct actors at various spatial scales.

Landslide Risk Reduction and Individualization at the Community Level or Institutional Diversification

Allotment Gardens

Gardening is very popular leisure time activity in the Czech Republic where a large number of allotment gardens were

K. Sassa et al. (eds.), *Understanding and Reducing Landslide Disaster Risk*, ICL Contribution to Landslide Disaster Risk Reduction,
https://doi.org/10.1007/978-3-030-60196-6_46

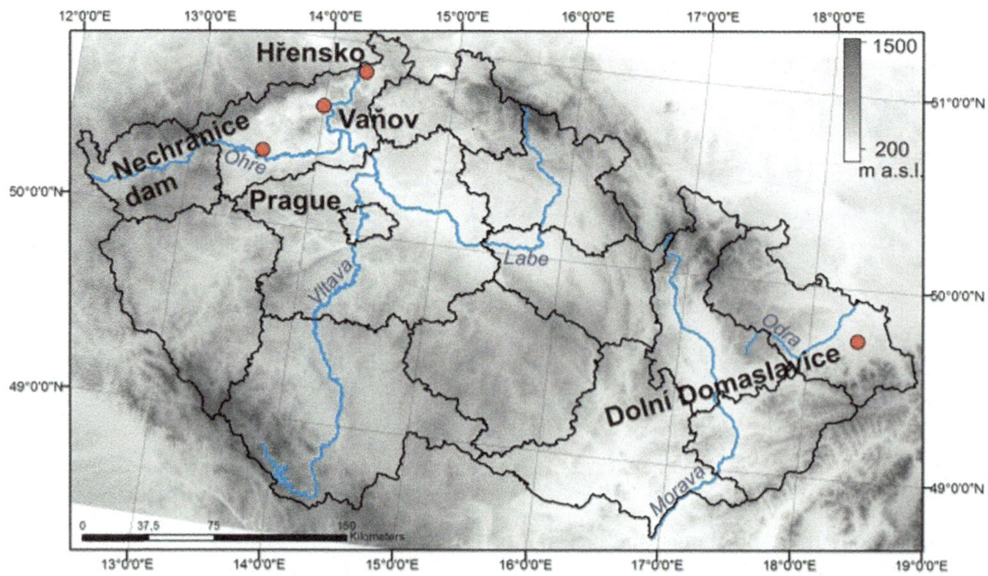

Fig. 1 The Czech Republic with sites discussed in the text (black lines are country and department limits, the capital city is marked by black colour)

established. They are usually located close to the cities on sites which are not suitable for housing constructions. It is very often because of they are on slopes prone to landslide occurrences, which is also the case of the allotment gardens on the south rim of the Nechranice dam (Fig. 1) located in Tertiary basin filled with clays overlying volcanic rocks. Poor stability conditions of the slope were identified and described in detail before the dam construction (1960s) resulting into abandonment of railroad construction project and prohibition of any housing development (Müllerová et al. 2017), which was abolished from unknown reasons in 1980s (Raška 2019; Rybář 1991). Since then, private owners build number of huts designated for second housing within the high landslide hazard zone. Three landslides gradually developed between December 2014 and February 2015 (compare with reported historical landslide occurrence on 14.2.1958 in Špůrek 1972) causing serious damage to the buildings and gardens. One of the landslides (with dimensions 90 m × 90 m) resulted into damage or destruction of 6 huts (Fig. 2). Their owners suffered considerable financial as well as emotional loss as they were totally unaware of any landslide hazard at the site and had no strategy to coop with landslide driven damage. The other landslide (240 m long and 90 m wide) developed just 250 m away affecting two gardens. Their owners were adapted to such events since they observed occasional shallow landsliding on their land and owned the gardens since 1970s thus were well informed about the landslide investigations in the 1960s. They used shipping container to store garden furniture and other implements and have not planted any fruit trees within their gardens. This adaptation limited the use of the gardens, but ensured that their owners suffered no lose due to landslide activity.

Despite the proximity of these two landslides, landowners within the single allotment garden failed to share potentially important information about the high landslide hazard of the site. Also the local administration office failed to provide this information to the new landowners who then had very little chance to prepare them for possible landslides. This can be explained by high individualization at the community level and missing legal obligations at the institutional level of administration offices responsible for land development.

Fig. 2 Hut seriously damaged by the landslide movement during December 2014–February 2015 near the Nechranice dam (for location see Fig. 1). The hut was located on the accumulation lobe and was latter demolished

Individualization in Landslide Emergency Response

Emergency responses to natural disasters or immanent landslide hazard are in the Czech Republic well defined and organized with a key role of local (e.g. village, town) administrations (Krejčí et al. 2002). It could be illustrated by remediation of recurrent, actively moving landslide in neovolcanic rocks in the Labe River valley (Vaňov village) near Ústí nad Labem town (Fig. 1, Raška et al. 2014). Its repeated activity in 1994 and 1995 upset the local community, which demanded quick and effective safety measures. Based on the geological evaluation and monitoring results, it had been stabilized with drainages and retaining wall and local inhabitants were informed about the performed works and their results (Raška 2019). The local community as a whole was satisfied with this solution. Nevertheless almost 20 years later, landowners directly bordered with the landslide toe noticed damage of the retaining wall. They repeatedly plead for its repair at the city administration, which finally resolved this problem and organized public meeting of the community members with private company responsible for the repairs as well as experts of the Czech Geological Survey representing state policies in the LDRR.

This example illustrates failure of the local administration in regular inspections of costly technical landslide mitigation measures, which was replaced by the active participation of the local inhabitants possibly directly affected by failure of the safety measures. Their action represented rather individual than communal involvement (Raška 2019). This individualization was largely caused by uneven spatial distribution of landslide hazard, which did not affect the major part of the community. Such conditions naturally lead to individualization of emergency responses, which in this case was limited to few families who were actively defending their properties by attracting attention of the local administration.

The direct financial help of local administrations or state agencies during landslide emergency responses is not always possible or timely due to legal fragmentation and institutional diversification. Therefore, in some cases, the local administrations search alternative ways how to support community members affected by landslides. Such an alternative response followed landslide which seriously damaged three houses in rural area of the Outer Western Carpathians in 2010 (Dolní Domaslavice, Fig. 1). It occurred during intensive precipitations in highly weathered Tertiary flysch rocks (Pánek et al. 2011). Geological survey and underground drainage performed by local private company was financed by regional branch of the Catholic Church charitable organization, which included the landslide remediation into its help to the victims of the regional floods.

Institutional Diversification

Hazardous landslides may originate outside the community lands, which involves external actors (e.g. institutions or other communities) into the LDRR process. An example is a small village of Hřensko (288 inhabitants in 2019, CSO 2019) located at the mouth of local stream to the Elbe River in the N Bohemia (Fig. 1). It extends along both sides of the stream in about 150 m deep and narrow valley, which was carved into the sub-horizontally layered, thick bedded Cretaceous sandstones. Their outcrops form dominant and unstable rock cliffs above the village, which experienced number of historical damaging rockfalls (Blahůt et al. 2013; Müllerová et al. 2017). The historical records and modelling results show, that both valley slopes pose significant rockfall hazard (Blahůt et al. 2013). Complex approach integrating monitoring with field, expert-based hazard assessment and continuous mitigation works had been applied to reduce the rockfall risk (Zvelebil et al. 2005). Despite of this effort, rockfall seriously affected important international road connection in 2009 (Blahůt et al. 2013). It stimulated demand for more reliable risk reduction measures resulting into construction of dynamic rockfall barriers on the south slopes of the valley requiring governmental investments of at least 14 million USD during 2011–2015 (the cost is based on freely available state expenditures data described in detail in Klimeš et al. 2017). Nevertheless, the opposite slopes remained without any protection. The south, protected slopes belong to the national park (The National Park Bohemian Switzerland), while the opposite slopes are owned by the forest public company (Lesy ČR, a.s.). Although both institutes belong to the state, they differ in function and are managed based on distinct laws. In the case of the forest public company, valid norms are ambiguous with respect to its responsibility to mitigate landslide risk originating on its property (Müllerová et al. 2017) and therefore, no barriers were constructed there so far.

This example well illustrates difficult situation in practical reduction of landslide risk due to institutional diversification caused by fragmentation of legal norms dealing with landslide hazard and risk mitigation. Historical perspective on the institutional diversification with respect to the community-based LDRR puts its onset to the foundation of the democratic republic of the Czechoslovakia after the 1st World War (Raška 2019). The high degree of institutional diversification remained during the totalitarian regimes (1939–1989) and further continued since the early 1990 with the renewal of democracy and free-market economy, which contributed to this trend by introduction of private contractors as new and important actors in the LDRR process (Raška 2019). The private companies not only provide technical services and solutions for landslide mitigation measures implemented at the community level, but often actively initiate LDRR projects funded through public budgets managed by local or regional administration offices.

Role of Governmental Policies in LDRR

State involvement in various stages of LDRR is necessary due to associated high costs or large scale of mitigation works exceeding community spatial limits and technical abilities. At the same time, governmental regulations have often been related with ineffective, largely top-down approaches which do not respect enough community needs and local conditions. Land-use planning and development regulations are often perceived in this way. Nevertheless, careful revision of recently valid laws, regulations and praxis in land-use planning in the Czech Republic, contradicts this narrative of "top-down, restrictive approach". It shows that the legal framework defines obligation to consider landslide hazard as limiting factor of development only in general way without specifying data sources for, and specific procedures of the hazard or risk assessment (Müllerová et al. 2017). It leaves large space for initiative of local authorities which may adjust the land-use planning to the local, community needs (Klimeš et al. 2020). Nevertheless, in some cases, the local authorities try to use this vague definition to minimize development regulations preferring land development (e.g. constructions of new houses) over minimizing landslide risk as this attitude has often positive political feedback. Some observations (compare with sub-chapter Allotment gardens above) suggest that such attitude may result in increasing landslide risk although there are no qualitative or quantitative data to support this hypothesis at the national scale.

Discussion and Conclusions

Synergies of State-Wide Policies with Community-Based Actions for Landslide Risk Reduction

Ideally, government defined policies would stimulate the community-based actions of LDRR best suited for the current state of the given community and its local natural and social conditions. Despite number of problems which obstacle this collaboration (e.g. different prioritization of the preferred landslide mitigation measures, Maes et al. 2019) there are examples suggesting that this type of collaboration leads to effective reduction of landslide risks outside the Czech Republic. Peruvian government launched in 2015 intensive campaign to prepare the country for sever weather caused by expected strong El Niño phenomena (UNISDR 2016), which is usually associated with high frequency of damaging landslides (Vilímek et al. 2013). Although no extreme precipitations occurred (Ramírez and Briones 2017), this campaign raised awareness at the community level and provided additional finances to implement measures mitigating landslide hazard leaving the responsibility of their implementation largely on local governments and communities. One of the rural communities in the Andes (Rampac Grande, Cordillera Negra, Ancash, Klimeš et al. 2019b) did not spend the provided funds during 2015/2016 season as the rains were only moderate, but due to intensive precipitations they decided to use them in January 2017. They enlarged the river trough of the stream with frequent debris flow occurrence which run through their village. Several days after the work was done, major debris flow occurred on the stream causing no damage as the trough was large enough to conduct its material. Although, the rationale behind the decision to postpone the implementation of the mitigation measure from 2015/2016 rainy season to the next one (2016/2017) is not entirely clear and includes probably high degree of subjectivity, it turned to be a correct decision preventing damage to the community during the severe 2017 El Niño event which was not predicted with sufficient anticipation to prepare corresponding mitigation plans on the state level (Ramírez and Briones 2017).

Systematic collaboration between state administration and local communities has been applied to lower the landslide risk in remote regions of China. It represents complex process of LDRR including expert identification of high hazard sites which require further attention; training of local people to perform regular, visual monitoring of the landslide movement activities; and defining warning procedures in the case of dangerous landslide movement requiring warning the local population. Several cases of sucessful landslide warnings and prevention of damage during testing phase (Guo et al. 2002) suggest that it represents low cost and applicable approach of LDRR. Such a close collaboration of community members with landslide experts on the landslide hazard assessment and early warning has all signs of citizen science approach (Trojan et al. 2019).

The presented examples show that the individualization is an important negative factor of effective LDRR at the community level. But, it does not prevent entirely the community members to actively participate in risk reduction procedures as shows the case when local administration implemented repairs of landslide mitigation structures based on information and inquires of the individual community members. In cases of missing regulations defining specific responsibilities of administration offices (e.g. regular check and maintenance of landslide mitigation measures) communities may act for them, when they feel that the inaction could increase the landslide risk directly affecting them. On the other hand, several examples of land-use planning praxis suggest that transferring more power and responsibilities from central government to the local administrations may not always results into landslide risk reduction. The local administrations may prefer other interests (e.g. housing development and increasing number of the community inhabitants) over the risk reduction especially when they expect positive community feedback of such a policy

(Müllerová et al. 2017). Legal fragmentation of the landslide hazard issues along with functional diversification of governmental institutions may effectively prevent to perform desired LDRR measures preventing the risk reduction. In such cases, the possibility of communities to act for the missing state action is very limited since the desired measures are very often too costly or too complex to be performed by the communities. Nevertheless, the example of the involvement of the non-governmental charity organization in landslide risk mitigation suggest, that alternatives still exists. We suggest that crowd-funding or citizen science approaches (Trojan et al. 2019) represent an alternative or complementary method to governmental role in natural disasters mitigation. We think, it is well illustrated with the Chinese approach of including local inhabitants into the landslide movement monitoring process (Guo et al. 2002) as low-cost and applicable alternative to instrumental monitoring and early warning, which otherwise still requires significant expert involvement and large funds allocation.

Presented cases show that effective, long-term sustainable solution of LDRR requires fair collaboration across various scales of governance starting with individual citizens or communities reaching to the governmental institutions. Such collaboration may be successful only if the participants will understand and consider that failure of any of the involved actors may causes failure of the entire process resulting into failure of the main aim—reducing risk of landslide disasters. As illustrate the described cases from the Czech Republic, there are number of flaws of the institutional chain required for effective LDRR.

Acknowledgements This research was supported by the Ministry of Education, Youth and Sports of the Czech Republic (Program Inter-Excellence, Inter-Vector, LTV19014). It is also part of the ongoing International Program on Landslides project no. 197 'Low frequency, high damaging potential landslide events in "low-risk" regions – challenges for hazard and risk management' and it was carried out thanks to the support of the long-term conceptual development research organization RVO: 67985891.

References

Beck U (2012) Individualism. In: Ritzer G (ed) The Wiley-Blackwell encyclopedia of globalization. https://doi.org/10.1002/9780470670590.wbeog292

Blahůt J, Klimeš J, Vařilová Z (2013) Quantitative rockfall hazard and risk analysis in the selected municipalities of the České Švýcarsko National Park, northwestern Czechia. Geografie 118:205–220

CSO (2019) https://www.czso.cz/csu/czso/pocet-obyvatel-v-obcich-za0wri436p. Accessed 29 Jan 2020

Froude MJ, Petley DN (2018) Global fatal landslide occurrence from 2004 to 2016. Nat Hazards Earth Syst Sci 18:2161–2181. https://doi.org/10.5194/nhess-18-2161-2018

Guo LY, Wei Z, Guo H (2002) Achievement of mass monitoring and preventing on landslide and debris flow in experimental counties of south Gansu and Shaanxi provinces. Bull Soil Water Conserv 22:38–39

Klimeš J, Stemberk J, Blahut J, Krejčí V, Krejčí O, Hartvich F, Kycl P (2017) Challenges for landslide hazard and risk management in 'low-risk' regions, Czech Republic—landslide occurrences and related costs (IPL project no. 197). Landslides 14:771–780

Klimeš J, Calvello M, Auflič MJ (2019a) Preface: thematic issue "community participation for landslide disaster risk reduction". Landslides 16:1745–1746

Klimeš J, Rosario AM, Vargas R, Raška P, Vicuña L, Jurt C (2019b) Community participation in landslide risk reduction: a case history from Central Andes, Peru. Landslides 16:1763–1777

Klimeš J, Müllerová H, Woitsch J, Bíl M, Křížová B (2020) Century-long history of rural community landslide risk reduction. Int J Disaster Risk Re 51:101756. https://doi.org/10.1016/j.ijdrr.2020.101756

Krejčí O, Baroň I, Bíl M, Jurová Z, Hubatka F, Kirchner K (2002) Slope movements in the flysch Carpathians of Eastern Czech Republic triggered by extreme rainfalls in 1997: a case study. Phys Chem Earth 27:1567–1576

Maes J, Mertens K, Jacobs L et al (2019) Social multi-criteria evaluation to identify appropriate disaster risk reduction measures: application to landslides in the Rwenzori Mountains, Uganda. Landslides 16:1793–1807

Müllerová H, Klimeš J, Hálová M, Blahůt J, Gibas P, Woitsch J, Pauknerová K (2017) Landslides—underestimated thread (in Czech). Czech Academy of Sciences, Prague, p 80

Pánek T, Brázdil R, Klimeš J, Smolková V, Hradecký J, Zahradníček P (2011) Rainfall-induced landslide event of May 2010 in the eastern part of the Czech Republic. Landslides 8:507–516

Ramírez IJ, Briones F (2017) Understanding the El Niño Costero of 2017: the definition problem and challenges of climate forecasting and disaster responses. Int J Disaster Risk Sci 8:489–492

Raška P (2019) Contextualizing community-based landslide risk reduction: an evolutionary perspective. Landslides 16:1747–1762

Raška P, Hartvich F, Cajz V, Adamovič J (2014) Structural setting of the Čertovka landslide (Ústí nad Labem, Czech Republic): morphostructural analysis and electrical resistivity tomography. Geol Q 58:85–98

Rybář J (1991) Untersuchung der Hangbewegungen in der ČSFR. Felsbau 9:178–181

Špůrek M (1972) Historical catalogue of slide henomena. Stud Geogr 19. Geografický ústav ČSAV, Akademia, Brno, 178 p

Trojan J, Schade S, Lemmens R, Frantál B (2019) Citizen science as a new approach in geography and beyond: review and reflections. Morav Geogr Rep 27:241–253

UNISDR (2016) https://www.unisdr.org/archive/50577. Accessed 10.12.2018. https://www.czso.cz/csu/czso/pocet-obyvatel-v-obcich-za0wri436p

Vilímek V, Hanzlík J, Sládek I, Šandov M, Santillán N (2013) The share of landslides in the occurrence of natural hazards and the significance of El Niño in the Cordillera Blanca and Cordillera Negra Mountains, Peru. In: Sassa K, Rouhban B, Briceño S, McSaveney M, He B (eds) Landslides: global risk preparedness. Springer, Berlin, Heidelberg, pp 133–148

Vincent S (2018) Transformations of collectivism and individualism in the Peruvian Central Andes: a comunidad over three decades. Ethnography 19:63–83

Zvelebil J, Vařilova Z, Paluš M (2005) Tools for rock fall risk integrated management in sandstone landscape of the Bohemian Switzerland National Park, Czech Republic (M121). In: Sassa K, Fukuoka H, Wang F, Wang G (eds) Landslides—risk analysis and sustainable disaster management. Springer, pp 119–126

Refinement Progresses on Freeway Slope Maintenance After a Huge Landslide Disaster

2020 Kyoto Japan

Wen-I. Wu, Tsai-Ming Yu, Chia-Yun Wei, Lee-Ping Shi, San-Shyan Lin, and Jen Cheng Liao

Abstract

After a huge landslide disaster, the maintenance refinement progresses on freeway such as emergency treatment, manual revision, establishment of management system and overall inspection are emphasized. The completely blocked traffic was opened after 55 days ongoing excavation and construction. The slope maintenance manual was totally revised based on the actual operation action. There are patrolling, periodical and special inspections. Some of the slope and the anchor inspection results are shown in the paper. All of grade A or B slopes have been upgraded to at least grade C (no obvious signs of instability) by the reinforcement method.The system is composed by lifecycle-based maintenance and management system, slope inspection operation system, slope information sharing platform and slope action management platform. All of the inspection and monitoring data are stored in the system. The data management time can be saved more than 30% and the inspection efficiency can increase about 50% by using the system. The freeway slopes are safeguarded from the results of overall inspection and restoration. The slope sorting results can help to control the maintenance sequence and budget planning.

Keywords

Slope maintenance management • Slope safety • Management system • Overall inspection

W.-I. Wu (✉) · T.-M. Yu · C.-Y. Wei
Freeway Bureau, No. 70, Banshanya, Liming Village, Taishan Dist., New Taipei City, 24303, Taiwan
e-mail: wyiwu@freeway.gov.tw

T.-M. Yu
e-mail: tm_yu@freeway.gov.tw

C.-Y. Wei
e-mail: jubby923@freeway.gov.tw

L.-P. Shi · J. C. Liao
Taiwan Construction Research Institute, 11F, No. 190, Sec. 2 Zhongxing Rd., Xindian Dist., New Taipei City, 23146, Taiwan
e-mail: lepnhsu@tcri.org.tw

J. C. Liao
e-mail: jcliao@tcri.org.tw

S.-S. Lin
Department of Harbour and River Engineering, National Taiwan Ocean University, No. 2, Pei-Ning Rd., Keelung, 20224, Taiwan
e-mail: sslin@mail.ntou.edu.tw

Introduction

The freeway network is the most important land transportation system in Taiwan. There are 3 longitudinal line (No. 1, 3 and 5) on north–south direction and 5 transverse line (No. 2, 4, 6, 8 and 10) on east–west direction. It was very shocked that all the lanes of the No. 3 Freeway at 3.1 k were blocked due to huge landslide disaster on the clear weather condition. It occurred suddenly on April 25, 2010. Three cars were buried, caused 4 casualties. One cross over bridge was thrusted to collapse at south down lane. The covered area was approximately 200 × 60 m^2 and total volume of landslide was estimated over 210,000 m^3. The landslide overview is shown in Fig. 1. Its failure mechanism was studied by Ching and Liao (2013). The refinement progresses after this tragic event can be divided into emergency treatment, revision of maintenance manual, establishment of management system and overall inspection which are enunciated as follows.

Emergency Treatment

The collapsed debris was cleared away comprehensively 6 days after landslide. Two lanes of north up and south down were opened after 37 days clearing and construction. Then 3 lanes of both directions were opened after 55 days

K. Sassa et al. (eds.), *Understanding and Reducing Landslide Disaster Risk*, ICL Contribution to Landslide Disaster Risk Reduction,
https://doi.org/10.1007/978-3-030-60196-6_47

Fig. 1 The landslide overview at 3.1 k of No. 3 Freeway

ongoing construction. In order to understand the sliding risks of all dip slopes along No. 3 Freeway, Chinese Taipei Geotechnical Society (CTGS) was requested to do the inspection on them by Ministry Of Transportation and Communications (MOTC). After sight inspection for 32 dip slopes, it is suggested that there were 17 slopes needed to do further detail safe evaluation (Chinese Taipei Geotechnical Society 2011).

The suggestions for the Freeway Bureau are as follows:

1. Establishing a standard slope inspection system for freeway which include slope referred information, slope safety evaluation and maintenance priority, slope safety inspection and monitoring operation regulation.
2. Reviewing the referred regulation in maintenance manual which include inspection objects, methods and items, inspection frequency and inspector qualification, etc.
3. Studying out slope monitoring operation procedures which include monitoring standard, method and installation plan, instrumental maintenance and criteria used, etc.
4. Establishing junior engineer education and training programs and regular courses.
5. Establishing long turn slope management system which include plan, design, construction and operation information, disaster records, maintenance and reinforcement data, etc.

Revision of Maintenance Manual

There were 7 chapters in the slope maintenance manual of Freeway Bureau before 2010. It was totally revised after 3.1 k disaster occurred. The contents were increased into 12 aspects which included explanation, slope grading, slope inspection, slope monitoring, ground anchor inspection, slope safety evaluation, slope grading method, slope maintenance, slope reinforcement and restoration, freeway slope full lifecycle maintenance and management system, education and training, and slope management meeting.

Slope Inspection

The freeway slopes can be divided into A, B, C and D grades which are defined as follow:

1. Grade A: obvious signs of instability can be observed, it needs to be announced immediately and take some effective actions collocated with intimate inspection and monitoring.
2. Grade B: some suspicious signs of instability can be observed, some maintenance, reinforcement and repairing need to be taken with increasing inspection and monitoring.

3. Grade C: no obvious signs of instability, it needs to take some inspections or regular maintenances, and monitoring done as needed.
4. Grade D: stable condition, it still needs to take casual inspections.

There are 3 kinds of slope inspections which are patrolling, periodical and special. The patrolling inspection need do at least once a day in daytime by car to see if there is any obvious influence to the safety of the user; The periodical inspection is the comprehensive investigation to find deterioration of facility and slope safety condition beforehand; The frequency from grade A to grade D is monthly, quarterly, yearly, and every 3 years, respectively; The special inspection need to be done after typhoon (in the area covered by 7 level storm circle), heavy rain (in the area of 24 h cumulative rainfall more than 200 mm or in the area of 3 h cumulative rainfall more than 100 mm), earthquake (in the area of earthquake intensity more than 4) or depending on situation of man-made destruction (e.g., fire, damaged by vehicle bumping, etc.). The periodical and special inspection scope should cover the slope sliding area where may be outside the freeway right-of-way.

There are 3 main items for periodical and special inspections which include slope surface, stability facility and drainage. Each contains several items which need to be filled with the degree of damage condition (high, medium and low). All of photos with high, medium and low degree of damage conditions of each item have been sorted out in the data bank of maintenance system (Taiwan Construction Research Institute 2017). Some of them are shown in Fig. 2.

In Fig. 3, it can be seen that the percentage shortage of drainage increased from 36.8 (2014) to 50.3% (2017). On the other hand, the percentage shortage of stability facility decreased from 26.5 (2014) to 15.9% (2017). It means that more than half of the shortages are drainages after the slope maintained. Therefore, it was decided that the drainage inspection must be finished before the rainy season (from May to November) to increase the safety of slope.

Anchor Inspection

The anchorage function of slope anchor lost is the reason causing the sliding disaster of 3.1 k. Therefore the inspections of anchor conditions are very important. There are periodical and special inspections for slope anchors. The frequency of periodical inspection is half a year at least for slope grade A, every 2 years at least for slope grade B and every 4 years for slope grade C or D which can be done partially every year. The special inspections start when anchor device functions are dying away or lost from the slope inspections or monitoring results and it may threat the safety of the slopes. The inspection items include protective cover appearance of anchor head, component of anchor head, wire corrosion condition inspected by endoscope and existing load examined by lift-off test.

For each slope, every anchor head cover appearance should be inspected and some percentage of all anchors for the anchor head component, wire corrosion inspection and lift-off test. The results of appearance inspections can be used to determine which anchor heads are appropriate to be opened. The choice principals of component inspections and lift-off tests are evenly distributed on anchors in the slope. Each inspected condition can be classified into loss of function (X), very poor (A), poor (B), passably (C), and normal (D). The details of anchor function evaluation and grading method are defined in the manual (Freeway Bureau 2018).

The photo of grade for component of anchor head is shown in Fig. 4. The anchor inspection results from 2011 to 2013 are shown in Fig. 5 (Taiwan Construction Research Institute 2017). It is observed that the protective cover appearance inspection results cannot get the actual anchorage condition, the proportions of passably and normal anchors are about 86% and only 0.3% for loss of function. But for lift-off test results, the proportions of passably and normal anchors are about 67 and 10% for loss of function, as shown in Fig. 5.

Slope Safety Improvement

The slope grade adjusted time is when slope is inspected (step 1), monitored (step 2) or done on anchor inspection (step 2) and done by reinforcement construction (step 3). In step 1, the slope grading is based on the damage condition from the result of slope inspection and disaster potential factor evaluation, i.e., attitude, disaster history and active fault influence. The grade gained from step 1 is called the initial slope grade which can be used as the index of the priority of maintenance. In step 2, if monitoring value is more than warning, or the total anchorage function grade is A or B, and the risk scale assessing is more than medium, the slope safety analysis need to be done. The slope grade is determined by the analysis results. The grading method is illustrated in the manual (Freeway Bureau 2018).

All of the free ends of the existing anchors have been grouted to protect the tendons before 2013, and their anchor heads have been covered with galvanized plate and the internal spaces have been filled specified grease. It can be seen that the anchors have been intensified for their corrosion resistant ability. After 2013, all the free end tendons of the new installed anchors needed to be set on polyethylene (PE). Due to the importance of the freeway kept on smooth flow condition, all grade A and B slopes have been upgraded

Fig. 2 Photos of degrees of damage conditions for some slope inspection items

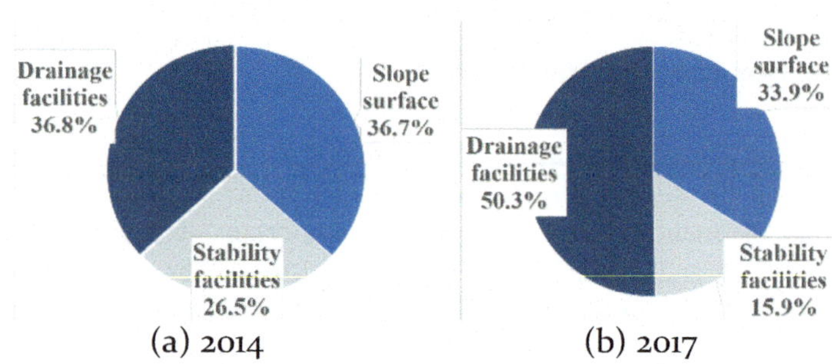

Fig. 3 The shortage statistics of main items of freeway slope inspections in **a** 2014 and **b** 2017

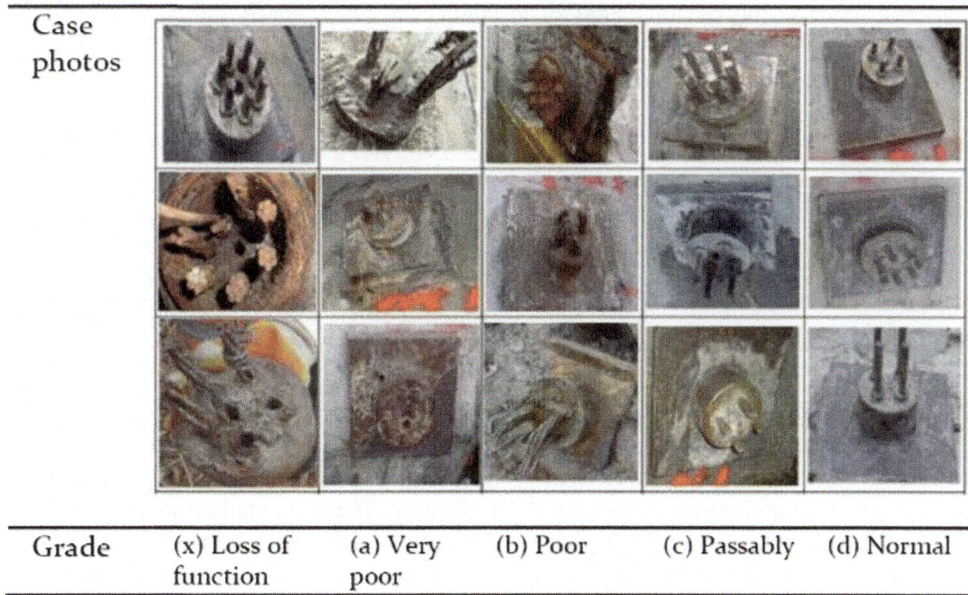

Fig. 4 Grade for component of anchor head

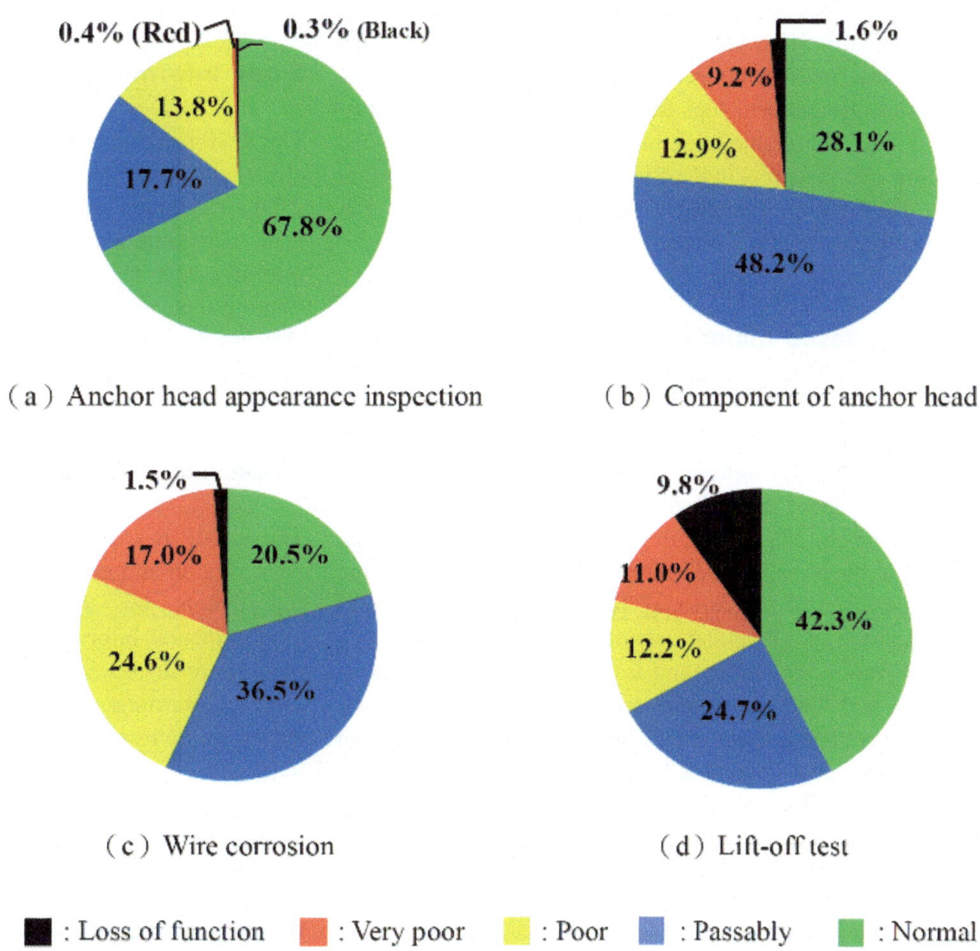

Fig. 5 Inspection results of anchors from 2011 to 2013

to at least grade C (no obvious signs of instability) by the reinforcement method.

The slope maintenance procedures are regulated in the manual. They are written down based on the actual operation action. The maintenance effects are reviewed in the slope management meetings and some improved procedures are proposed for inefficient or insufficient cases. Therefore it is a real and continuously revised manual.

Establishment of Management System

The data from planning, construction and operation service is very important for the slope maintenance. Therefore, it is decided to build up a full lifecycle maintenance and management system for freeway slope after 3.1 k disaster. The system is developed for the purpose of increasing the slope facility life and safe condition. The system is composed by lifecycle-based maintenance and management system (LMMS), slope inspection operation system (SIOS), slope information sharing platform (SISP) and slope action management platform (SAMP). It supports the functions that include information collection and integration platform, electronic management tool for maintenance manual, information and communication integration for feedback and assisting decision making, service of customized management, information exchanged and shared platform, etc. (Taiwan Construction Research Institute 2017).

Lifecycle-Based Maintenance and Management System (LMMS)

The LMMS is the core of the management system. It is used for the database by building up the referred page and column. It also gives a platform for the user of Bureau or its Region Branch Office to fill in the necessary data. There are functional block and shared block. The former contains the slope data area of fundamental, plan, design and construction, monitoring, slope and anchor inspection, maintenance and renovation, education and training, management meeting, etc., and the management area of fax/e-mail, statistical analysis, system control, etc. The latter contains common area of latest announcement, to-do list, message board etc., for the use of the public service. Each slope unit has its identification number based on the coding principle. All of the instrumentation and anchor serial number is also generated from this ID. The slope curriculum vitae can be searched by this number.

Except the referred data was collected in the LMMS, it can show the function of inquiry, prompt notice, statistical chart or table, alert transmission, report production, etc.; it can also display the slope and external map information with space information platform for overlay comparison (Lin et al. 2017).

Slope Inspection Operation System (SIOS)

The SIOS is the auxiliary system for the slope inspection with tablet operation. All the inspection forms are built up in the SIOS. It is connected with the LMMS and the data can be checked and inputted on line. The slope inspection record can be directly returned to the database if network connected or temporary stored by off line method if network disconnected.

Android or iOS operation system are supported for 3G/4G smartphone or tablet. The location can be shown on the action device. There is a checking mechanism, i.e., the track of the inspector should 100% pass through the check points. The position and photo of shortage can be displayed and checked on the screen. Photo samples of degree of damage conditions are established to help judging on site.

Slope Information Sharing Platform (SISP)

The SISP is a platform for the users to upload and download electronic data files of various documents. It includes 9 databases which are fundamental, inspected, maintained data, detection and monitoring data, safety analysis and reinforcement results, disaster case history, education and training data, specification and manual, and the others. It is connected with the LMMS to provide the necessary data. The data can be searched by key word or file category.

Slope Action Management Platform (SAMP)

The SAMP is a decision making platform. It contains ordinary version and emergency response version. The reminding events on the home page, slope information, environmental information and engineering/technical information are displayed in the ordinary version. The weather forecast, warning/message and monitoring information are shown in the response version. The functions of hierarchical management, situation (formal and emergency) management, synchronous and immediate information and integrated external resource are provided in this system.

Effects of Management System

There are totally 2567 slopes along freeway including 927 excavation slopes and 1640 embankments. There are 148 slopes installed with 1998 monitoring sets and 160 slopes stabilized with 30,608 ground anchors. All of the monitoring and inspection data have been stored in the system. And the 62,361 inspection data (including slope condition photos) can be inquired by the system. Meanwhile, more than 300 engineering and technical service cases were provided. The different management and statistic tables and synthetic working results can be generated. It is evaluated that the data management time can be saved more than 30%. The inspections by action devices can increase about 50% working efficiency, automatic check processing can decrease the time and human resource.

Overall Inspection

According to the regulation of the manual, it is necessary to do overall inspection after 4 or 5 years slope maintenance. The objectives are to confirm the slope maintenance effectiveness and for the more advanced utilization. The overall inspection working items include review of slope data, checking on all slope inspected results and slope safety evaluation and sorting of maintenance sequence (Taiwan Construction Research Institute 2019).

Review of Slope Data

The reviews include fundamental, monitoring and detection, safety analysis, maintenance and reinforcement data. It was found that some construction drawing or topographic map did not accord with actual situation, some slope width or height did not conform to reality or some layer attitudes were unclear in fundamental data. Some slope initial grading with B_i had not been maintained or reinforced before the end of that year in inspection data. There were 24 inclinometers shown the obvious sliding layers or trends of sliding in the monitoring data. All of the above problems have been improved or continuously paid attention to or followed. The slope grading results meet with manual regulations.

Checking on All Slope Inspected Results

The old inspection results were checked on sites by the professional engineers who should attend the training course. It was wanted to find the following results:

1. Whether the slope mileage was correct?
2. Whether did the medium or high degree of damage condition accord with actual situation compared with 2018 inspection item?
3. Was there any new abnormal phenomenon generated?
4. Whether did the slope facility need to be maintained or repaired?
5. What is the risk scale of the slope by fundamental data evaluation and on site checking?
6. What is the potential sliding mode evaluated by the fundamental data and on site checking?

The problem percentages of checking results of the above 1, 2, 3 and 4 items are quite small, and the improvement have already been done by each Region Branch Office. For the 927 excavation slopes, there are 13 slopes with high risk scale (sliding event can cause passer-by safety and traffic problem) and 363 slopes with medium risk scale (sliding scope only to shoulder and small safety and traffic influence); For 1640 embankments, there are none with high risk scale and 3 with medium risk scale. The sliding mode can be divided into 3 kinds, i.e., rock fall, rock sliding and earth sliding. For the excavation slopes, 32 belong to rock fall, 144 belong to rock sliding, 730 belong to earth sliding (most of them belong to surface erosions). All of the embankments are earth sliding slopes and most of them belong to surface erosions.

Slope Safety Evaluation

After reviewing of slope data and checking on all slopes on sites, the maintenance performance results proposed by professional engineer have been examined by geotechnical, geological, soil and water conservation, instrumentation etc. experts. If the results were still not clarified, the site surveys were done by geological experts and necessary drilling boreholes were suggested. There are 93 slopes with 2520 m boring depth. Soil physical property tests, soil direct shear tests rock direct shear tests and rock uniaxial tests from borehole samplings have been done to determine the parameters for analyses. There are 103 slopes selected for slope safety evaluation. In general, the slope safeties of them are in accordance with the specifications.

Sorting of Maintenance Sequence

The influence factors considered for slope maintenance sorting are as followings: slope grading, existing sliding trend, layer attitude, slope height, installed instrumentation, gradient, initial slope grading and risk scale, etc. The disaster potential factors are given to different weight scores for sorting evaluation shown as Table 1. The sorting results have been evaluated for each Region Branch Office. They are beneficial to the maintenance sequence and budget planning.

Table 1 Weight scores of disaster potential factors for sorting evaluation

	Attitude (3)	Disaster history (2)	Active fault (1)
High (3)	Dip slope with daylight of sliding layer, dip > 20° (9)	Disaster or reinforcement history within 5 year (6)	Close to active fault < 100 m (3)
Medium (2)	Dip slope with daylight of sliding layer, dip < 20° (6)	Disaster or reinforcement history more than 5 year (4)	Close to active fault < 200 m (2)
Low (1)	Oblique slope, escarpment, dip slope but no daylight (3)	No disaster or reinforcement history (2)	Not close to active fault (1)

Conclusion

It is concluded that the maintenance refinement progresses on freeway include emergency treatment, manual revision, establishment of management system and overall inspection. The slope maintenance manual was totally revised based on the actual operation action. All of the existing anchors have been intensified for its corrosion resistant ability. All of the grade A or B slopes have been upgraded to at least grade C (no obvious signs of instability) by the reinforcement method. All of the inspection and monitoring data are stored in the system. The data management time can be saved more than 30% and the inspection efficiency can increase about 50% by using the system. The freeway slopes are safeguarded from the results of overall inspection and restoration. The slope sorting results can help to control the maintenance sequence and budget planning.

Acknowledgements The work was supported by the Freeway Bureau, MOTC. Thanks to all people providing helps on this work.

References

Chinese Taipei Geotechnical Society (2011) Forensic study on the dip slope failure of No. 3 Freeway at 3.1 k. Investigation report. MOTC, Taiwan, 174 p (in Chinese)

Ching J, Liao HJ (2013) A spatial variability view of Freeway-3 dip slope failure in Taiwan. J GeoEng 8(1):1–10

Freeway Bureau, MOTC (2018) Maintenance manual of Freeway, Chap 3. Taiwan, 86 p (in Chinese)

Lin BS, Wei CY, Shi LP, Liao JC, Lin XH (2017) Installation plans of rainfall stations for the maintenance management of Freeway slopes. In: Proceedings of the 5th international conference on geotechnical engineering for disaster mitigation and rehabilitation. CTGS, Taipei, pp 385–390

Taiwan Construction Research Institute (2017) Technical consultation for Freeway slope maintenance and management. General report. Taiwan, 277 p (in Chinese)

Taiwan Construction Research Institute (2019) Overall inspection of Freeway slope. Work result report. Taiwan, 218 p (in Chinese)

Landslide Exposure Community-Based Mapping: A First Encounter in a Small Rural Locality of Mexico

2020 Kyoto Japan

Ricardo J. Garnica-Peña, Gerardo Cardón-Idelfonso, and Irasema Alcántara-Ayala

Abstract

This research aimed to analyse the exposure of households and other buildings to landslides by means of community-based mapping in Huehueymico, a small rural locality of 884 inhabitants located in Teziutlán, Puebla, Mexico. Despite the severe impact of the disaster of October 1999 associated with rainfall induced landslides, there has been a lack of interest of local authorities to promote landslide disaster risk preparedness in rural isolated areas. Under such account, it is suggested that community-based mapping can be regarded as a very valuable alternative to enhance disaster risk education as a first encounter between residents and landslide disaster risk awareness. The methodology undertaken comprised the use of an unmanned aerial vehicle (UAV), in order to generate high resolution base maps, field evaluation of the conditions and state of conservation of the buildings, and a couple of workshops directed to undertake the community-based mapping. Results regarding the landslide exposure community-based map are presented in this investigation.

Keywords

Community-based mapping • Landslide exposure • Disaster risk • Uavs

R. J. Garnica-Peña (✉) · I. Alcántara-Ayala
Institute of Geography, National Autonomous University of Mexico (UNAM), 04510 Mexico City, Mexico
e-mail: garnica@igg.unam.mx

I. Alcántara-Ayala
e-mail: ialcantara@igg.unam.mx

G. Cardón-Idelfonso
Faculty of Philosophy and Letters, National Autonomous University of Mexico (UNAM), 04510 Mexico City, Mexico
e-mail: gerardo101295@gmail.com

Introduction

Owing to the extreme rainfall event of October 1999, hundreds of landslides occurred in the Sierra Norte de Puebla (SNP), causing considerable negative effects in the localities of the region (Lugo et al. 2005). Teziutlán, Puebla was one of the municipalities where the greatest human and economic losses were concentrated, including 263 human losses and 1.5 million affected people. Damages were especially severe in the neighbourhoods of La Gloria, Juárez, Atoluca, France, Siete Sabios, La Aurora, Huehueymico, among others (Bitrán 2001).

After the 1999 disaster, diverse investigations have been carried out in Teziutlán at municipal and local scales to understand landslide dynamics and the construction of landslide disaster risk associated to high levels of vulnerability and exposure (Alcántara-Ayala et al. 2017). However, there has been an absence of attention to rural isolated localities. Therefore, in this research, a landslide community-based approach was undertaken to create a participative landslide exposure map, as a first step into the enhancement of landslide disaster risk awareness for the residents of the locality of Huehueymico, a rural area, where 12 people lost their lives and 24 houses were affected during the disaster of 1999.

Study Area

Huehueymico is one of the 33 rural towns of Teziutlán and is located approximately 4 km northeast of the municipal capital (Fig. 1). It is comprised within the Trans-Mexican Volcanic Belt in the SNP, in the province of Puebla, Mexico, at 1720 ma.s.l.

Existing data on the population of the rural town of Huehueymico are scarce. Most updated information indicates that in 2010, there was a population of 884 inhabitants,

K. Sassa et al. (eds.), *Understanding and Reducing Landslide Disaster Risk*,
ICL Contribution to Landslide Disaster Risk Reduction,
https://doi.org/10.1007/978-3-030-60196-6_48

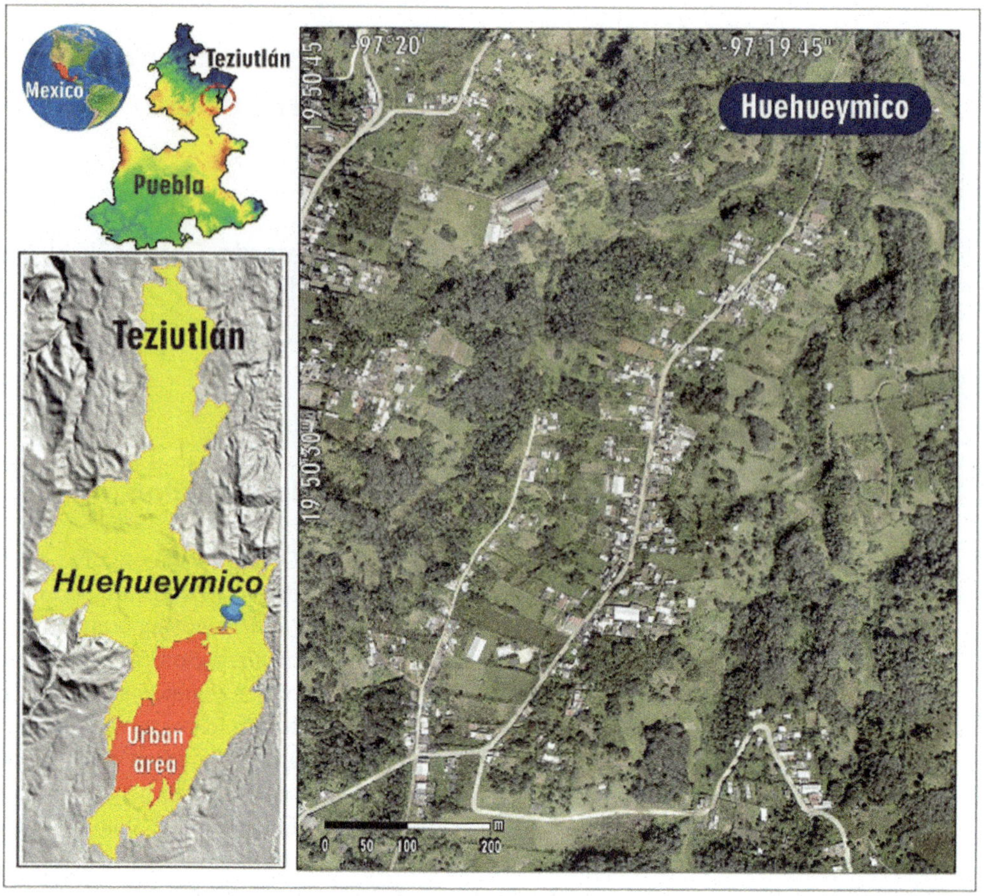

Fig. 1 Location of Huehueymico, in the municipality of Teziutlán, Puebla, Mexico

414 males and 470 females, and a total of 178 inhabited households.

The lithological character of the town of Huehueymico is associated entirely with volcanic activity. According to the Mexican Geological Service (SGM 2011), Huehueymico is situated on the highly landslide susceptible pyroclastic ramp formed by deposits of the Caldera de los Humeros Volcano, located to the south of Teziutlán. According to García (2004), the climate present in the study area is temperate humid with rainfall all year round and average annual temperature that varies between 12 and 18 °C. Both, lithological and the climatic factors are of great relevance for the occurrence of landslides.

Community-Based Mapping

Participatory or community-based mapping was originated within the field of cultural geography in the 1990s, at the time when it was decided to bring back the development of indigenous mapping (Diez and Escudero 2012). The greatest advantage of this type of cartography is the involvement of communities contributing to the generation of information regarding their own environment, culture, traditions and problems (Mora and Jaramillo 2003). In other words, community mapping takes advantage of local knowledge for research, and problem solving.

Samodra et al. (2018) pointed out that community mapping is an effective method for generating information in areas where research has not been carried out or with scarce or no data. Several studies have been undertaken to use community-based mapping to involve people into different aspects linked to disaster risk reduction (DRR). Cadag and Gaillard (2012) carried out Participatory 3-Dimensional Mapping in the Philippines pointing out its significance for risk assessment and DRR planning. The work of Sullivan-Wiley and collaborators (2019) included participatory mapping related to floods, droughts, landslides, among other hazards in Uganda.

Although very often community mapping is not sustainable due to lack of funding, there are successful cases in both developed and developing countries, such as Jushibo in

Japan or Guardianas de la Ladera (guardians of the hill-slopes) in Colombia, on which community participation plays an essential role for disaster risk reduction.

According to Barragán-Giraldo (2016) four criteria should be met for the elaboration of participatory mapping: (1) production process: refers to the strategy and the objective with which the community mapping will be carried out, which must be inclusive and open; (2) community ideology: community mapping must be related to a topic of importance for the communities, concerning their needs and potentially usable by them; (3) local knowledge: maps should show local information and knowledge; (4) cartographic convention: it is not necessary that participatory maps resemble official maps; however, the closer, the more likely they could be considered as appropriate communication tools.

By involving communities in participatory mapping, it is possible to identify and understand the different ingredients of disaster risk and to engage people in diverse activities

Table 1 Criteria used for the grid surveys

Criteria	North–South survey	East–West survey
Area (ha)	88.13	71.87
Flight height above ground (m)	150	150
Images	701	590
Mean pixel resolution (cm)	4.1	4.1
Overlapping (%)	80 × 80	80 × 80
Total flight time (minutes)	38	30

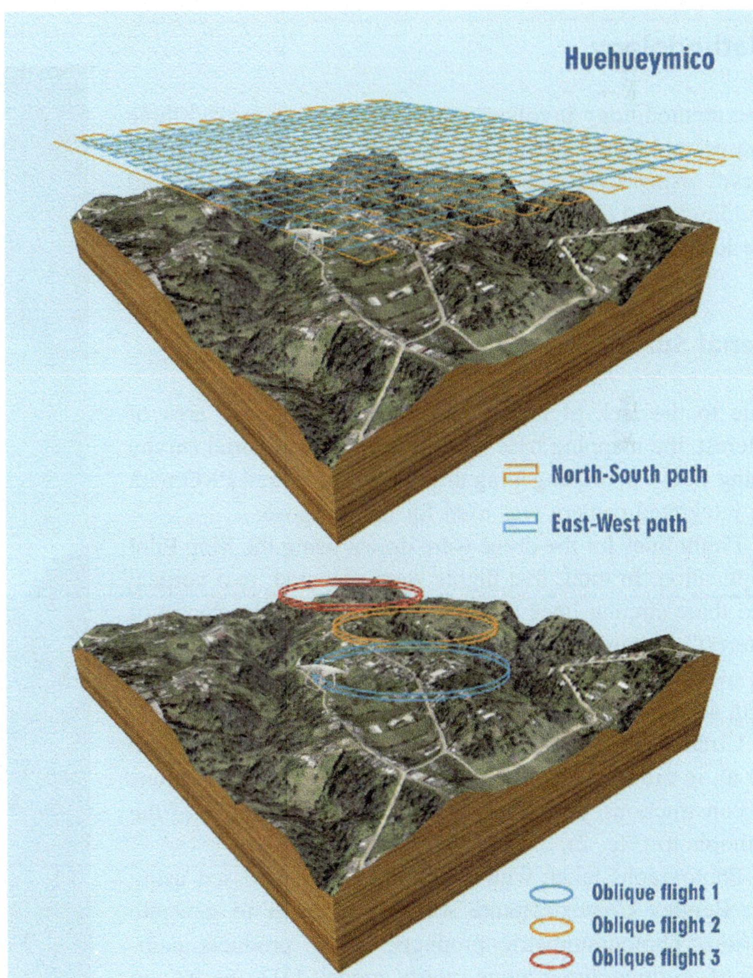

Fig. 2 Vertical and oblique surveys

aimed at reducing existing vulnerability (Canevari-Luzardo et al. 2015) and exposure, including strategies for raising awareness and preparedness at local level.

The significance of community-based mapping provides the possibility to generate spatial information, that can be used not only by researchers, but by the communities, and other type of stakeholders to manage their territory. The motivated participation of residents is essential to obtain valuable information from their environment based on scientific and traditional knowledge to get them involved into decision making and practice. Through mapping people get in contact with their own space (Mora and Jaramillo 2003). Recognise needs and can be committed to a series of challenges to improve living conditions.

Recent literature has shown the significance of community-based approaches for landslide research and disaster risk reduction strategies at local levels in different mountain regions of the world (Karnawati et al. 2018; Fukuoka and Dok 2013; Coutinho et al. 2016; Alcántar-a-Ayala et al. 2018).

Methodology

The methodology involved in the research comprised three phases developed during the summer of 2019: (1) Aerial survey using UAV; (2) Field evaluation of households and buildings; and (3) Community-based mapping workshops in the locality of Huehueymico, Teziutlán, Puebla.

Aerial Survey Using UAV

Due to the lack of high-resolution images of the area of interest, the mapping base was created from an aerial survey using UAV. A rotating wing drone DJI Phantom 4 PRO with an integrated camera was used for this purpose.

Flight lines for the drone were drawn using the Map Pilot application. In total, five flights were prepared: two vertical and three circular lines for oblique shots using the "point of interest" capture option. Linear flights were performed with an overlap of 80% at the top and bottom of the image, and with 40% overlap for the sides of the photographs (Table 1).

Circular flights were made at two different heights, 50 and 60 m, in order to obtain greater detail of the lateral sections of constructions, aimed at getting a better resolution of the orthophoto (Fig. 2).

Photographs taken with the drone were processed using the Bentley ContextCapture software in order to generate various photogrammetric products. These products comprised a three-dimensional mesh model (3D mesh), an orthophoto, a digital surface model (DSM) and point cloud in LAS format. The orthophoto and the DSM were generated at 5 cm pixel size and were used to create a pair of maps, 6.5 × 4.5 m in size, which were printed out to perform the community-based mapping workshop.

Field Evaluation of Buildings

Owing to its condition of rural locality, no official cartography was available to start the research. Based on field evaluation and information provided by the residents of Huehueymico, it was possible to identify and map the boundaries of the locality.

By using Map it GIS and a GPS, buildings were evaluated in the field in terms of conservation state, land use and available basic services. Assessed elements included location relative to the slope, existence or conditions of sewer systems, lighting, water intake, foundations, in addition to land use, number of house storeys, type of floor, presence of vegetation and cracks, and level of deterioration of households or buildings.

Fig. 3 Community-based mapping workshop

Fig. 4 Rainfall induced landslide triggered in 2018

The information resulted from field evaluation was used in combination with terrain attributes derived from the DSM produced through the aerial survey using UAV, including slope shape, topographic wetness index, slope inclination, distance to river currents, along with proximity to existing landslides, were used to define a landslide exposure index. In this context, exposure was defined as "the situation of people, infrastructure, housing, production capacities and other tangible human assets located in hazard-prone areas" (UNISDR 2017). Buildings were classified into three categories: high, moderate and low exposure to landslide.

Community-Based Workshop

A couple of community-based workshops were carried out at the middle school of Huehueymico; only members of the student community participated (Fig. 3). The first stage of the workshop consisted of a series of introductory lectures to provide information and insights concerning landslides, risk, vulnerability, exposure, disasters, GIS, UAVs and basic cartography. Examples of cartographic inputs from previous work developed for the Sierra Norte de Puebla elaborated with UAVs were also provided (Garnica-Peña and Alcántara-Ayala 2017).

Fig. 5 Community-based mapping workshop

An orthophoto and an altimetric map generated from the aerial survey using UAV were printed on canvases (Fig. 3). Relevant landslide symbology was created to be used in the community mapping workshops. It included a series of elements to identify landslide symptoms and controlling factors. Additionally, iconography was used to identify meeting points, potential evacuation routes and roads in good and poor conditions. The iconography was printed out on 9 by 6 cm^2 cardboard for better manageability.

Maps were extended on the yard of the middle school and symbols placed on two tables located near the maps. Participants were invited to walk on the large size maps to recognise territorial features of Huehueymico. They were later asked to locate particular features of its locality, such as the church, schools and their own houses in both maps. Once the participants were familiarised with their locality through both maps, recognition of landslide symptoms, controlling factors, existing landslides (Fig. 4), meeting points, and households was undertaken.

Discussions regarding factors influencing the stability of the slopes were held. After a consensus was reached with regards to the elements that needed to be included in the map and their location, the information was combined with the field evaluation of households and building regarding exposure by using GIS.

Validation of the resulted maps was undertaking by walking around the locality aided by a high-accuracy GPS receiver, and with the information provided by the civil protection authorities (Fig. 5).

Concluding Remarks

Owing to its high resolution, drone-generated cartographic inputs allowed efficient identification of different components of landslides exposure at the local level through field evaluation and community-based mapping. Although much remain to be done, the performed landslide mapping was highly valued for raising awareness by participants and was considered by them as a significant first step into landslide disaster risk awareness (Fig. 6).

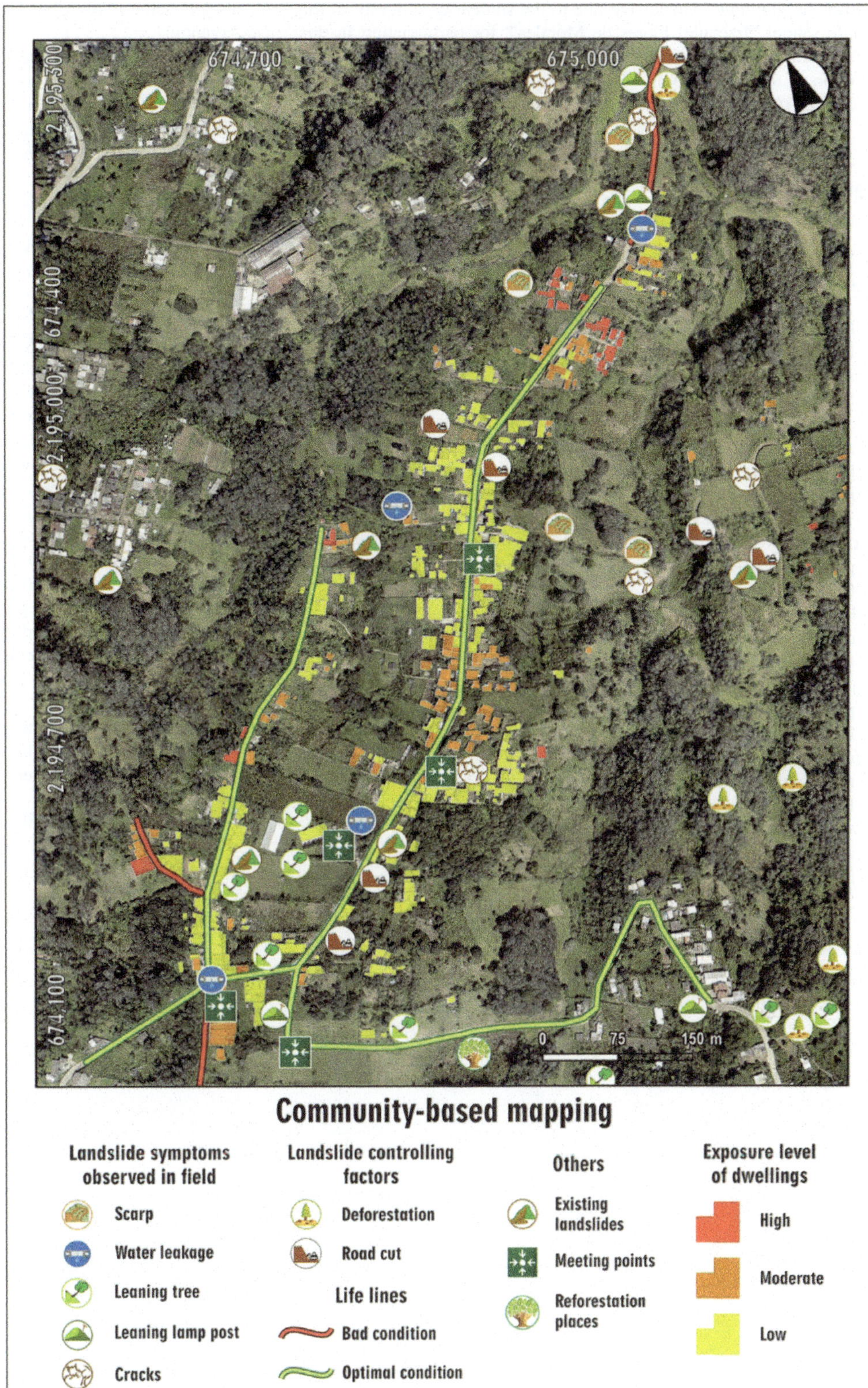

Fig. 6 Landslide exposure community-based map of Hueyhueymico, Teziutlán, Puebla, México

Acknowledgements Thanks are due to the student community of the middle school "Ezequiel Flores de Morales", for participating in this research during the community-based mapping workshops and opening the doors of their classrooms. Financial support was kindly provided by UNAM-DGAPA, through the project PAPIIT IN300818.

References

Alcántara-Ayala I, Garnica-Peña RJ, Coll-Hurtado A, Gutiérrez de MacGregor MT (Coords.) (2017) Inestabilidad de laderas en Teziutlán, Puebla. Factores inductores del riesgo de desastres, Instituto de Geografía, Universidad Nacional Autónoma de México, In Spanish p 223

Alcántara-Ayala I, Garnica-Peña RJ, Murillo-García FG, Salazar-Oropeza MO, Méndez-Martínez A, Coll-Hurtado A (2018) Landslide disaster risk awareness in Mexico: community access to mapping at local scale. Landslides 15(8):1691–1704. https://doi.org/10.1007/s10346-018-1010-4

Barragán-Giraldo DF (2016) Cartografía social pedagógica: entre teoría y metodología. Revista Colomb de Educ 70:247–285

Bitrán D (2001) Características del impacto socioeconómico de los principales desastres ocurridos en México en el periodo 1980–1999. CENAPRED

Cadag JRD, Gaillard JC (2012) Integrating knowledge and actions in disaster risk reduction: the contribution of participatory mapping. Roy Geogr Soc 44(1):100–109

Canevari-Luzardo L, Bastide J, Choutet I, Liverman D (2015) Using partial participatory GIS in vulnerability and disaster risk reduction in Grenada. Clim Dev 9(2):95–109

Coutinho RQ, Henrique HM, Duarte CC, Nascimento DM (2016) Risk mapping for landslides and erosion in the municipality of Ipojuca-PE-Rurópolis community. Landslides and engineered slopes. Experience, theory and practice pp 699–706

Diez J, Escudero H (2012) Cartografía social. Investigación e intervención desde las ciencias sociales, métodos y experiencias de aplicación. Colección Extensión. Universidad de la Patagonia. Argentina

Fukuoka H, Dok A (2013) Precursor process and triggering mechanism of rapid landslides under extreme weather conditions, and an attempt of ICT-based participatory joint mapping of past landslides with experts in developing countries. J Disaster Res 1:165–166

García E (2004) Modificaciones al sistema de clasificación climática de Köppen. Insitutio de geografía, serie libros num. 6. Universidad Nacional autónoma de México, In Spanish. p 91

Garnica-Peña, RJ, Alcantara-Ayala, I (2017) Multi-temporal landslide evaluation by using UAV: Some insights on disaster risk in Teziutlan, Puebla Mexico. Advancing culture of living with landslides, advances in landslide science vol 2, pp 209–218. https://doi.org/10.1007/978-3-319-53498-5-24

Karnawati D, Fathani TF, Wilopo W, Andayani B (2018) Landslide dynamics: ISDR-ICL landslide interactive teaching tools. TXT-tool 4.062-1.1 community hazard maps for landslide risk reduction. pp 599–606. https://doi.org/10.1007/978-3-319-57777-7-36

Lugo J, Zamorano J, Capra L, Inbar M, Alcántara-Ayala I (2005) Los procesos de remoción en masa en la sierra norte de Puebla, octubre de 1999: Causa y efectos. Rev Mex De Cienc Geol 22:212–228

Mora H, Jaramillo M (2003) Aproximación a la construcción de Cartografía social a través de la geomática. Ventana informática 11 (enero-junio) pp 129–146

Samodra G, Chen G, Sartohadi J, Kasama K (2018) Generating landslide inventory by participatory mapping: an example in Purwosari area, Yogyakarta, Java. Geomorphology 306:306–313

SGM (2011) Carta geológico-minera: Teziutlán E14–B15 (geological-mining map: Teziutlán E14–B15). Servicio Geológico Mexicano, In Spanish

Sullivan-Wiley KA, Gianotti AGS, Connors JPC (2019) Mapping vulnerability: opportunities and limitations of participatory community mapping. Appl Geogr 105:45–57. https://doi.org/10.1016/j.apgeog.2019.02.008

UNISDR (2017) Report of the open-ended intergovernmental expert working group on indicators and terminology relating to disaster risk reduction. UNISDR, Geneva

Co-Producing Data and Decision Support Tools to Reduce Landslide Risk in the Humid Tropics

Elizabeth A. Holcombe, Rose Hen-Jones, Paul J. Vardanega, Mair E. W. Beesley, Charlotte E. L. Gilder, and Elisa Bozzolan

Abstract

Rainfall-triggered landslides are increasing in the humid tropics, and Small Island Developing States are disproportionately affected. Frequent shallow slides in hillside cuttings along roads and in communities hinder sustainable development. Larger, less frequent storms cause hundreds of landslides that block lifeline roads, impede disaster response and reverse economic growth. Top-down Disaster Risk Reduction (DRR) policies and approaches aiming to transfer conventional landslide assessment science and engineering practices are not always suitable in these data- and resource-limited contexts. This paper recognises the emergence of co-production approaches as part of the resilience paradigm response to DRR science-policy-practice gaps. We present a case study from Saint Lucia, Eastern Caribbean, in which government engineers and policy-makers have partnered with the authors to co-produce landslide hazard assessment data and prototype decision support tools to strengthen landslide hazard management along lifeline roads.

Keywords

Rainfall-triggered landslides • Small Island Developing States (SIDS) • Knowledge co-production • Lifeline roads • Geotechnical data • Stochastic slope stability modelling

E. A. Holcombe (✉) · R. Hen-Jones · P. J. Vardanega · C. E. L. Gilder · E. Bozzolan
Department of Civil Engineering, University of Bristol, Queen's Building, University Walk, Bristol, BS8 1TR, U.K.
e-mail: liz.holcombe@bristol.ac.uk

M. E. W. Beesley
Formerly Affiliated to Department of Civil Engineering, University of Bristol, Queen's Building, University Walk, Bristol, BS8 1TR, U.K.

Introduction

The occurrence and impact of rainfall-triggered landslides is increasing in the humid tropics due to deforestation, urbanisation and road construction (Froude and Petley 2018). Slope failures in hillside cuttings such as those along roads and in informal communities are becoming more common (ibid). High frequency shallow soil slides in cut slopes have a localised impact on lives and livelihoods, and these 'everyday' (extensive) disasters can be part of a pattern of risk accumulation that hinders sustainable development (Bull-Kamanga et al. 2003). Extreme rainfall events can trigger hundreds of slope failures over wide areas causing fatalities, blocking and damaging roads, and impeding disaster response. For countries with low economic resilience these intensive disaster events can divert public expenditure, reverse economic growth and increase national debt (IMF 2016; World Bank 2017).

The impact of disaster risks on development, and vice-versa, is recognised in the UN's Sendai Framework for Disaster Risk Reduction, SFDRR, (UNISDR 2015) and the Sustainable Development Goals. It is widely acknowledged that top-down development policies struggle to deliver practical hazard mitigation measures for DRR on the ground (Wamsler 2007; Aitsi-Selmi et al. 2015). Bottom-up community-based approaches associated with the development paradigm (e.g. Wisner et al. 2012) deliver local DRR, often with an emphasis on community capacity development and vulnerability reduction. However, there is a knowledge-action gap between these two spatial and organisational scales (Gaillard and Mercer 2013). To address this gap the SFDRR calls for "actionable knowledge" that translates new DRR science into policy and practice whilst ensuring research is informed by end-user needs (UNISDR 2015). Knowledge production and 'co-production' are often cited as key components to any such endeavour (e.g. Weichselgartner and Pigeon 2015; Scolobig and Pelling 2016).

K. Sassa et al. (eds.), *Understanding and Reducing Landslide Disaster Risk*, ICL Contribution to Landslide Disaster Risk Reduction,
https://doi.org/10.1007/978-3-030-60196-6_49

The interconnected human-physical causes and impacts of rainfall-triggered landslide risks provide a valuable context for exploring DRR science-policy interactions, gaps and the potential utility of co-production approaches (Scolobig and Pelling 2016). This paper draws on our experience of longstanding partnerships with community residents, government civil engineers and policymakers to deliver landslide hazard knowledge and mitigation measures in Saint Lucia, Eastern Caribbean. We first outline typical scientific and technical challenges to landslide hazard assessment, and the science-policy interactions that may influence translation of information into actions in this context. Then, to illustrate a potential approach to addressing these challenges, we present a case study on the co-production of data on road cut slopes and prototype landslide risk management decision-support tools with our partners in Saint Lucia.

Landslide Risk Reduction Data, Knowledge and Action Gaps in Small Island Developing States (SIDS)

To identify typical landslide risk reduction knowledge-action gaps in the humid tropics it is helpful to consider the conventional scientific methods used for hazard assessment and how this informs policy and practice. It should be noted that the policy-practice (and funding) context can determine the scientific knowledge production process adopted. We organise this overview in terms of spatial scales (see Fig. 1).

Slope stability is determined by local topographic, geotechnical, hydrological and surface-cover factors. Constructing roads or informal communities on slopes can reduce stability where vegetation is removed, slopes cut, and drainage altered (Holcombe et al. 2016).

Top-down DRR policies typically call for wide-area Geographical Information Systems (GIS)-based hazard assessments at national or city scales to inform funding priorities and land use planning. These conventional landslide susceptibility maps cannot represent localised or dynamic landslide factors or processes, or the effects of risk mitigation and climate change. More advanced empirical-statistical and physics-based GIS hazard assessments require landslide inventories and high-resolution spatial data (van Westen et al. 2008). Yet, despite rapid progress in topographic remote sensing, it is difficult to generate detailed enough data for GIS-based cut slope stability assessment (van Westen 2016).

Civil engineers need data from geotechnical investigations and slope surveys to analyse instability mechanisms in individual slopes and design slope stabilising measures. Practitioners typically assume a static water table in such analyses, which fails to capture the dynamic hydrological processes that often trigger landslides in soils (c.f. Anderson et al. 1997). Physics-based modelling is also too data- and time-intensive for hazard assessment at road network scales. Instead rapid field reconnaissance of slopes and qualitative scoring of hazard and vulnerability by experts, can be used to develop databases for road asset and risk management (e.g. GEO 2013).

The thinly spread institutional capacity of small, low-to middle-income countries, such as many SIDS, means that municipal engineers have limited resources for data collection or adopting the off-the-shelf technologies often transferred under top-down programmes. Without an adequate inventory of past landslides or man-made slopes, or data on slope profiles and soil properties, many conventional landslide hazard assessment methods are not suitable. As Fig. 1 illustrates, these data gaps translate into knowledge and

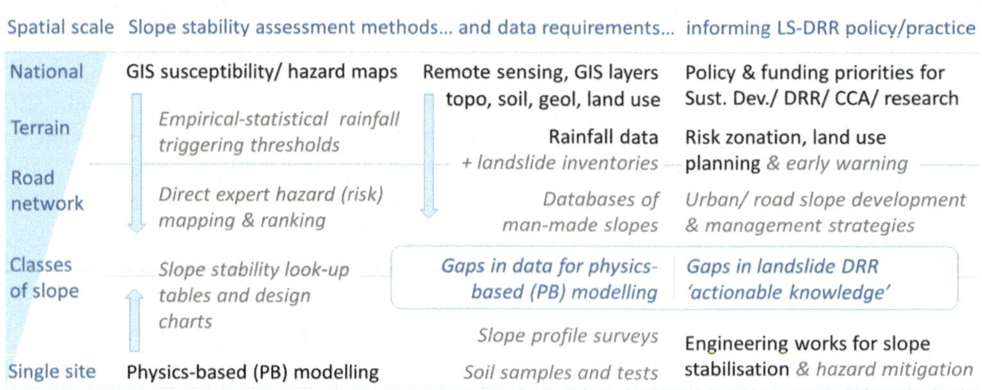

Fig. 1 Conventional landslide hazard assessment methods, data and policy/practice, and typical gaps (*italics*) in actionable knowledge for landslide hazard management in data and resource-limited countries such as SIDS

action gaps between top-down DRR policy and site-specific slope engineering works. Therefore, civil engineers and disaster risk management agencies in the humid tropics struggle to anticipate and mitigate rainfall-triggered landslides effectively in communities and along lifeline road networks.

Partnerships for Landslide Risk Reduction: Saint Lucia

Community-Based Landslide Risk Reduction

Between 2004 and 2011 the Government of Saint Lucia (GoSL) partnered with Anderson and Holcombe to develop the 'Management of Slope Stability in Communities' (Mossaic) methodology and deliver landslide hazard mitigation projects in ten informal urban communities (Anderson and Holcombe 2013). The local actors were community residents, the social development fund, and several government ministries. Funding came from GoSL, the World Bank and other development banks. The Mossaic approach has three foundations that help to overcome knowledge-action gaps at community scales, and enable co-production of effective mitigation works:

1. *A community-base*: Residents of hillside communities are not just seen as those "at risk", but as people with a practical knowledge of local slope features affecting landslide hazard (e.g. drainage routes, soil depths, signs of instability), and who can actively contribute to delivering landslide mitigation works.
2. *A scientific base*: Community-based mapping and elicitation of local engineers' expert knowledge on soils allows local slope processes to be diagnosed using the Combined Hydrology And Stability Model (CHASM). If water infiltration is found to be a dominant destabilising factor, then rainwater capture and surface water drainage measures are designed and built.
3. *An evidence base*: Delivering cost-effective measures that improve slope stability can influence risk management practice and policy (World Bank 2013).

Landslide Hazards Along Lifeline Roads at National Scales

The longstanding Mossaic partnership with GoSL engineers provided the basis for a recent national scale initiative to start to address knowledge-action gaps hindering effective management of landslides along roads in Saint Lucia. Roads are important for economic development and are the primary lifeline infrastructure for disaster response and recovery (World Bank 2017). Yet the slope cuttings along road corridors in mountainous tropical regions are many times more landslide-prone than natural slopes (Larsen and Parks 1997), and there is often low network redundancy due to limited physical space and financial resources in SIDS. Landslides along road networks account for a large share of disaster losses. Hurricanes Tomas in St Lucia in 2010, and Maria in Dominica in 2017, caused losses exceeding 40% and 220% GDP respectively. Both events triggered hundreds of landslides in man-made cut slopes along roads, hindering disaster response, recovery and longer-term economic activity.

Between 2016 and 2018 GoSL engineers and University of Bristol researchers co-produced a prototype Platform for Road and Infrastructure Slope Management (PRISM) that builds on the three Mossaic foundations—combining *local knowledge* with landslide hazard modelling *science* to deliver an *evidence base* for LS-DRR practice and policy.

In this case the local knowledge is from GoSL civil engineers and based on their professional training and experiences (such as knowledge of Saint Lucia's soil types, past landslides, and working with governmental processes and policies, for example). To this we add and digitise data that can be collated from disparate sources, such as paper copies of geotechnical laboratory tests results; and new data from field reconnaissance of road cut slopes. The methods of data co-production, processing and curation are designed to align with existing working practice and institutional capacity, rather than being a one-off project or product. They address some of the missing data sources identified in italics in Fig. 1. and provide the basis for building up national databases of soils properties and cut slopes. They are also designed to be mapped in GIS and connected with ongoing road asset and risk management initiatives.

To increase the information obtainable from these new, but still limited, geotechnical, slope profile and rainfall data we use stochastic CHASM modelling to assess the stability of typical ranges of road cut slope geometries and soil types. Sensitivity analysis is used to identify influential slope properties and patterns of stability response to rainfall. Finally, decision-support tools are co-produced that translate the simulation results into rainfall thresholds, look-up tables for rapid slope assessment, and recommendations for targeted data acquisition. The methods for co-producing the prototype data and decision support tools are described in the remainder of this paper.

Co-producing the Prototype Platform for Road and Infrastructure Slope Management (PRISM) in Saint Lucia

Prototype National Cut-Slope Database

The primary purposes of the PRISM cut-slope database are to: (i) provide GoSL engineers with a georeferenced inventory of the slopes along roads, which can be added to over time; (ii) facilitate basic hazard assessment using a simplified and adapted version of the Priority Ranking System for Man-made Slopes (GEO 2013); (iii) provide the slope geometry parameter ranges for stochastic modelling in CHASM; and (iv) link to slope stability look-up tables generated by the stochastic modelling methodology. After reviewing previous experiences of IT transfer with GoSL engineers, it was agreed that the prototype database should be an MS Excel workbook (to be managed by a nominated engineer) with macros for data-entry, calculating the hazard scores and reading look-up table results.

A review of available GIS data and reports from previous landslide and road mapping projects confirmed that road cut-slopes had previously been mapped on the primary road network (Mott MacDonald 2013). However, slope characteristics were not in the GIS attribute table (just a single hazard score). Fieldwork was undertaken with GoSL engineers to check whether these slopes matched our own criteria for cut-slope identification, namely: consistent geometry and angle of cut and the surrounding topography (shape of wider slope); and consistent material type and strata along its length (see Fig. 2). Agreement was generally good, but in some cases, the Mott MacDonald (2013) slopes needed to be subdivided into two or more individual slopes. Challenges to slope identification included vegetation obscuring the slope, cut slope crests being indistinct through erosion, and GPS handsets (or mobile phone GPS) losing signal near steep hills so coordinates became inaccurate.

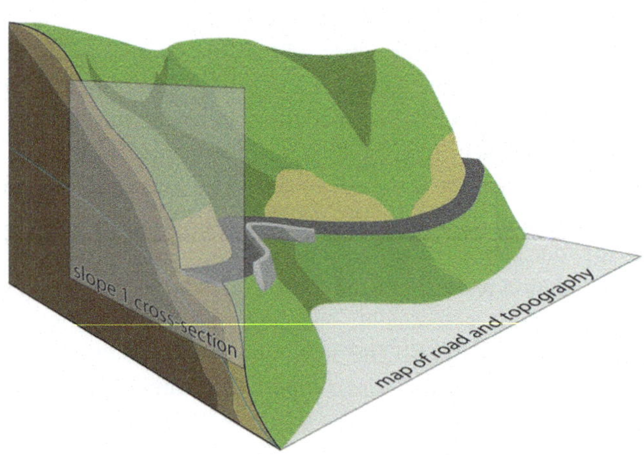

Fig. 2 Cut slopes along roads (yellow–illustrative) and an estimated profile (labeled 'slope 1 cross-section') for initial stability assessment

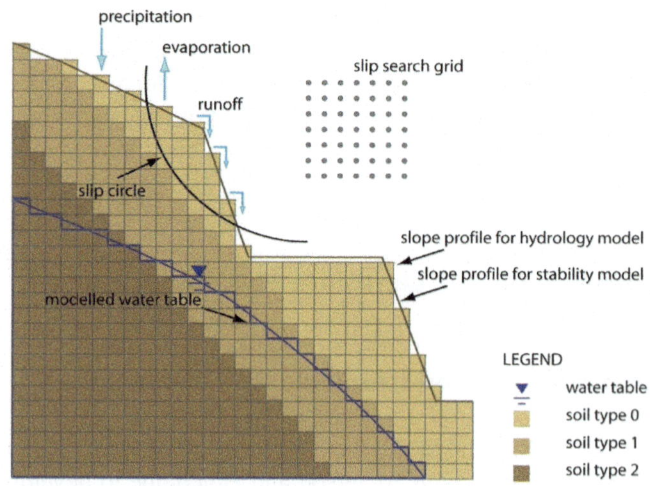

Fig. 3 Slope profile for analysis in CHASM (adapted from Anderson and Holcombe 2013)

The next stage was to design a method for recording slope features for stability ranking and/or as inputs to CHASM (Fig. 3). Cut slope heights and angles were measured using an Abney level and a laser rangefinder (Leica DISTO S910). The instruments gave cut slope heights within 0.5 m of each other; however, slope angles were much more easily and accurately obtained with the laser rangefinder. Natural slope angles ~80 m upslope and downslope of the cut and road (after Larsen and Parks 1997) were obtained from a Digital Elevation Model.

The most complex aspect of the cut-slope survey design was the characterisation of the soil in enough detail to allow an appropriate statistical distribution of soil strength parameters to be assigned for stochastic modelling and for cross-referencing with the resulting stability look-up tables. We combined local engineering knowledge of soils; data from geology and soil maps and surveys; and soil test results from known locations, to define three soil 'families' that can be recognised in the field: (A) tropical residual soils formed in-situ through weathering (often clayey and red/brown); (B) volcanic deposited soils of unsorted agglomerates and fine-grained matrix; (C) volcanic deposited soils of pumice/tuff and ash-derived lithosols (grey silty sandy soils) (Shepheard et al. 2019). The degree of weathering of the slope materials is assessed on site and assigned a weathering grade from VI—all rock material converted to soil, to I—fresh rock (Fookes 1997).

Finally, factors are recorded for slope risk scoring:

- Facilities or features up/down-slope of cut slope.
- Drainage on the slope; evidence of water ingress or seepage; presence of water-carrying services.

- Surface protection or vegetation cover.
- Evidence or knowledge of past instability.
- Engineering judgement of stability and risk.

Prototype National Soil Geotechnical Database

The soil data required for CHASM analysis are the effective cohesion and effective friction angle, unit weight, saturated and residual moisture contents, saturated hydraulic conductivity and soil moisture characteristic curve. Figure 3 shows an idealised tropical residual soil profile with weathering parallel to the natural slope ('soil' type 2 is unweathered volcanic bedrock).

The initial reason for compiling a soil geotechnical database for Saint Lucia was to give the basis for statistical analysis to provide parameter distributions for stochastic CHASM modelling, or for estimating distributions (such as the hydraulic parameters) from index properties such as soil texture. GoSL engineers routinely carry out basic soil tests for construction projects or after a landslide, but data from separate projects are rarely combined for use in wider analysis. Other reasons for creating the database include digitisation of paper records of soil test results; enabling verification of test results against each other or related parameters; and creation of a searchable and mappable geotechnical dataset that can be analysed with respect to geological and soil maps and used to estimate soil properties for preliminary project studies.

Shepheard et al. (2019) provides full details of the soil A, B, C classification approach, digitisation of 91 laboratory test records, benchmarking of shear box test apparatus, and the statistical analyses of soil property correlations and parameter distributions. The prototype soil database is currently in the form of an MS Excel workbook, like the prototype slope database. Over 50 additional soil test results have recently been digitised from archived records in the GoSL materials laboratory; and the database continues to be updated. A separate prototype MS Excel workbook for geotechnical test data entry was also co-designed with technical officers in the GoSL materials laboratory, so that results could be digitised during testing (with automated calculations and error-checking).

Prototype Decision-Support Tools and Information

The development of soil and cut slope data collection methods and prototype databases in Saint Lucia has started to address some of the data gaps at the road network and 'slope class' scales identified in our conceptual framework (Fig. 1). Translation of this data into slope hazard assessment information, knowledge and decision-support tools for LS-DRR practice and policy can also be mapped onto this framework. We provide a brief overview of these prototype outputs here (details of the methods and results are the subject of a paper in preparation):

Slope hazard (and vulnerability) ranking at road network scales is done using a macro in the cut slope database. Scores are calculated from slope geometry, instability and vulnerability factors recorded in the field (after GEO 2013). This information will allow engineers to identify and prioritise high risk slopes for further investigation and landslide mitigation works. The scores of 36 cut slopes in the initial database, which were surveyed during development of the fieldwork methodology, show good agreement with observations of previous landslides in these locations.

Rainfall-triggering thresholds (national scale), priorities for *targeted data acquisition*, and *slope stability look-up tables* ('slope class' and site scales) are generated using a stochastic physics-based modelling approach described in detail by Almeida et al. (2017). The first step is to translate soil and slope information from the new databases, statistical analyses, local knowledge and literature review into CHASM input parameter distributions (slope geometry, strata, initial water table, geotechnical properties, and rainfall intensity and duration). CHASM simulates rainfall infiltration, groundwater seepage, negative and positive pore water pressures, and slope stability over time (see Anderson et al. 1997, and Wilkinson et al. 2002, for equations and numerical scheme). CHASM is run tens-of-thousands of times using parameter combinations generated from the Saint Lucia data. When predicted landslides are plotted with respect to the associated rainfall, we can estimate the threshold above which landslides are likely to be triggered in road cut slopes. Rainfall thresholds can be used for emergency planning and early warning at national scales. The experimental threshold in Fig. 4 was estimated using Pareto

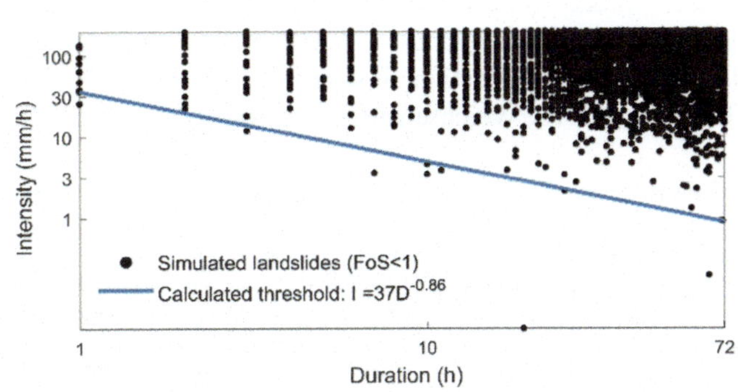

Fig. 4 Prototype landslide-triggering rainfall threshold, I = intensity (mm/h), D = duration (h), for residual soil cut slopes in Saint Lucia (>9000 simulated landslides)

optimisation to capture 99.8% of simulated landslides (minimise area and maximise points above the line).

Classification And Regression Tree (CART) analysis of simulation input–output matrices is used to identify influential slope parameters (after Almeida et al. 2017) and provide engineers with information for prioritising data collection. To create physics-based look-up tables for rapid slope assessment a second set of simulations is run with fixed design storms. The CART analysis is constrained using six 'observable parameters' that engineers can estimate in the field to within specified ranges (cut slope height and angle, low to very high effective friction angle and cohesion, upslope angle, top strata depth). This analysis produces experimental decision-trees based on the six observable parameters which, once tested and verified, should provide look-up tables for physics-based assessment of cut slopes in the field, or as a macro in the slope database.

Conclusions

The PRISM data and decision-support tools described here are still at a proof of concept stage. Further work on the prototype PRISM is needed to test and refine the data, methods and tools; continue to integrate with practice and policy in Saint Lucia; and define an 'adaptable blueprint' for use in similar locations. Adaptations could include road loading (negligible with respect to soil unit weight for this study's narrow, 6.5 m, pavements) and, with a suitable model, seismic loading (excluded here due to the low frequency of earthquake-triggered landslides versus frequent rainfall triggering).

This case study shows a possible way of addressing some of the scale, data, knowledge and action gaps that hinder effective management of landslide hazards along lifeline roads in the tropics (i.e. creating both horizontal and vertical connections in Fig. 1). Co-production has combined local knowledge with science and created appropriate tools as a basis for practice and policy. This has been a reflexive and social process involving technical and political decision-makers throughout; rather than being linear from data to knowledge to action, or science to practice and policy (Kasperson 2010; Weichselgartner and Pigeon 2015).

Acknowledgements The prototype PRISM methods described in this paper were co-produced with our partners in the Government of Saint Lucia Ministry of Infrastructure, Ports, Energy and Labour. We gratefully acknowledge their inputs of technical knowledge and experience, and their time and practical support during field visits. Funding came from two EPSRC-GCRF University of Bristol institutional sponsorship pump-priming awards (2016, 2018); and from The World Bank for additional Research Associate time (2017, 2018), and for two GoSL engineers to visit University of Bristol (2017).

References

Aitsi-Selmi A, Blanchard K, Al-Khudhairy D, Ammann W, Basabe P, Johnston D et al (2015) UNISDR STAG report: science is used for disaster risk reduction. https://preventionweb.net/go/42848 Accessed 31 Jan 2020

Almeida S, Holcombe EA, Pianosi F, Wagener T (2017) Dealing with deep uncertainties in landslide modelling for disaster risk reduction under climate change. Nat Hazard Earth Sys Sci 17(2):225–241

Anderson MG, Lloyd DM, Kemp MJ (1997) Overseas road note 14: hydrological design manual for slope stability in the tropics. Transport research laboratory, Crowthorne, U.K. (ISSN 0951-8797). 64 p

Anderson MG, Holcombe EA (2013) Community-based landslide risk reduction: managing disasters in small steps. The world bank, Washington D.C. (ISBN 978-0-8213-9456-4). p 404

Bull-Kamanga L, Diagne K, Lavell A, Leon E, Lerise F, MacGregor H et al (2003) From everyday hazards to disasters: the accumulation of risk in urban areas. Environ Urban 15(1):193–204

Fookes PJ (1997) Tropical residual soils. Geological Society, London

Froude MJ, Petley DN (2018) Global fatal landslide occurrence from 2004 to 2016. Nat Hazard Earth Sys Sci 18:2161–2181

Gaillard JC, Mercer J (2013) From knowledge to action: bridging gaps in disaster risk reduction. Prog Human Geog 37(1):93–114

GEO (2013) The new priority ranking systems for man-made slopes and retaining walls. GEO report no. 284. P F K Cheng. Geotechnical engineering office, CEDD, The government of Hong Kong, special administrative region. p 179

Holcombe EA, Beesley M, Vardanega PJ, Sorbie R (2016) Urbanisation and landslides: hazard drivers and better practices. Proc ICE Civ Eng 169:137–144

IMF (2016) Small states' resilience to natural disasters and climate change. IMF Policy Paper, International Monetary Fund, Washington D.C, 124p

Kasperson RE (2010) Science and disaster reduction. Int J Disaster Risk Sci 1(1):3–9

Larsen MC, Parks JE (1997) How wide is a road? The association of roads and mass-wasting in a forested montane environment. Earth Surf Proc 22(9):835–848

MacDonald M (2013) Landslide risk assessment for saint lucia's primary road network. Hurricane Tomas rehabilitation and reconstruction. Final feasibility report. Sept 2013. Government of Saint Lucia. p 258

Scolobig A, Pelling M (2016) The co-production of risk from a natural hazards perspective: science and policy interaction for landslide risk management in Italy. Nat Hazards 81(S1):7–25

Shepheard CJ, Vardanega PJ, Holcombe EA, Hen-Jones R, de Luca F (2019) Minding the geotechnical data gap: appraisal of the variability of key soil parameters for slope stability modelling in Saint Lucia. Bull Eng Geol Environ 78(7):4851–4864

UNISDR (2015) Sendai framework for disaster risk reduction, 2015–2030. United Nations International Strategy for Disaster Reduction, Geneva, Swiss, 37p

van Westen CJ, Castellanos Abella EA, Sekha LK (2008) Spatial data for landslide susceptibility, hazards and vulnerability assessment: an overview. Eng Geol 102(3–4):112–131

van Westen CJ (2016) National scale landslide susceptibility assessment for Saint Lucia. CHARIM caribbean handbook on risk information management, world bank GFDRR, ACP-EU natural disaster risk reduction program. https://www.charim.net/sites/default/files/handbook/maps/SAINT_LUCIA/SLULandslideReport.pdf Accessed 30 Jan 2020

Wamsler C (2007) Bridging the gaps: stakeholder-based strategies for risk reduction and financing for the urban poor. Environ Urban 19 (1):115–142

Weichselgartner J, Pigeon P (2015) The role of knowledge in disaster risk reduction. Int J Disaster Risk Sci 6:107–116

Wilkinson PL, Anderson MG, Lloyd DM (2002) An integrated hydrological model for rain-induced landslide prediction. Earth Surf Proc Land 27:1285–1297

Wisner B, Gaillard JC, Kelman I (2012) Handbook of hazards and disaster risk reduction. Routledge, London

World Bank (2013) Saint Lucia: leaders in reducing landslide risk. News feature story. https://www.worldbank.org/en/news/feature/2013/06/11/saint-lucia-oecs-leader-reducing-landslide-risk Accessed 30 Jan 2020

World Bank (2017) Climate and disaster resilient transport in small Island developing states: a call for action. The World Bank, Washington D.C, 131p

Effective Global Communication on Disaster Mitigation of Landslides Through E-Conferencing

2020 Kyoto Japan

A. A. Virajh Dias, N. N. Katuwala, and S. S. I. Kodagoda

Abstract

E-conference is one of the latest additions that has been promising by a collective approach which is so far not used enough among earth science researchers and students whom dedicated on landslide disaster risk reduction to a great extent. The usefulness of the E-conference is where people participate in a conference through an electronic medium unlike in a traditional conference which is held in a room. It strongly contributes to the democratization of Landslide science, represent modern approach to education and information sharing in a way that has less impact on the environment by saving CO_2 emissions related to transportation in traditional conferences. The E-conference 2015 of the World Centre of Excellence (WCoE), Central Engineering Consultancy Bureau (CECB) was an amalgamation of 300 participants representing almost all the continents in the World and 20 keynote experts and facilitators, supporting for an effective communication platform. The E-conference 2017, linking 120 new participants and eight experts representing different countries and opening three social media–Facebook portals was conducted to address community engagement through UN volunteers, professionals, environmentalist, readers, entertainers, researches, students and society at large. Keynote authors uploaded papers, technical notes and these plus comments were somewhat interesting due to open tract of e-flow throughout the conference and linking emails which added multiple information over the web portal where participants can further be sharing comments through email streaming volunteer groups. A Local Forum was also being carried out in parallel to the International E-conference in order to deeply confer the pressing matters concerned on landslides mitigation in national significance. E-conference addresses issues in different perspective and mobility which receive quite important observations on advocacy, neutrality, debriefing and interpretation on landslide disaster risk reduction and management.

Keywords

E-conference • Disaster • Risk • Mitigation • Landslides • International programme of landslides (IPL)

Introduction

The world is experiencing rapid increases of landslides disaster with a significant increase of climate change (Bhandari 2015). Social awareness on disaster events is fairly accessible in various territories due to technological advances of telecommunication, e-learning, and mass-media facility. An excessive and unplanned urban growth leads to various physical, social and economic vulnerabilities and impacts of disasters are highly detrimental when they occur in urban environments (Amaratunga 2015a). LearnWorlds, Open-edX, LearnDash, WizIQ and many more e-learning platforms are available today which discuss on media preferences but access by disaster proven educational groups are limited. It is therefore important in developing awareness on community specific approaches (Dias and Wijewardana 2002) on disaster reduction with free and direct professional interaction

A. A.V. Dias (✉)
Integrated Watershed and Water Resources Management Project (IWWRMP), Ministry of Mahaweli, Agriculture, Irrigation and Rural Development, No. 500, T. B. Jaya Mawatha, Colombo, 10, Sri Lanka
e-mail: aavirajhd@yahoo.com

A. A.V. Dias · N. N. Katuwala · S. S. I. Kodagoda
Natural Resources Management and Laboratory Services, Central Engineering Consultancy Bureau (CECB), No. 415, Bauddhaloka Mawatha, Colombo, 7, Sri Lanka
e-mail: nkatuwala@gmail.com

S. S. I. Kodagoda
e-mail: sudheeraik@gmail.com

K. Sassa et al. (eds.), *Understanding and Reducing Landslide Disaster Risk*, ICL Contribution to Landslide Disaster Risk Reduction,
https://doi.org/10.1007/978-3-030-60196-6_50

to educate the resilience on disasters. Web based research interaction of professional teams are very much attractive nowadays and mobility of such approaches also seems to be valid with the use of smart mobile phone. The e-conferencing is one of the effective tools proposed by the Central Engineering Consultancy Bureau (CECB) and well-practiced and materialised according to the theme of the World Centre of Excellence (WCoE) on Developing Model Policy Framework, Standards and Guidelines promoted by the International Programme of Landslides (IPL) and the Global Promotion Committee (GPC).

World Centre of Excellence (WCoE)

It was indeed a pleasure for the Central Engineering Consultancy Bureau (CECB) to obtain the prestigious opportunity, to be entitled as one of the "World Centre of Excellence on Landslide Disaster Reduction 2014–2017" on 'Model Policy Frameworks, Standards and Guidelines on Landslide Disaster Reduction' at the World Landslide Forum 3 in Beijing, China (Fig. 1) and subsequent extending and awarding the same during the World Landslide Forum 4 in Ljubljana, Slovenia (Fig. 2) for the period of 2017–2020. As an internationally recognized institution, CECB believes that it is a duty to establish an e-global cooperation platform for better disseminating and understanding of Disaster Risk Management (Dias et al. 2017) and appreciate the global image of the IPL and the Global Promotion Committee (GPC) on landslide disaster risk reduction.

Fig. 1 The CECB was awarded WCoE 2014–2017 by Ms. Irina Bokowa, UNESCO director general at world landslide forum 3, June 2014, China national convention centre, Beijing, China

Fig. 2 The CECB was re-awarded WCoE 2017–2020 by Mr. Qunli HAN, UNESCO DG representative at WLF4 held in Cankarjev dom—cultural and congress centre in Ljubljana, Slovenia, June, 2017

First International E-Conference in 2015

Working towards a successive approach on WCoE theme, the Natural Resources Management and Laboratory Services (NRM&LS) of CECB decided to launch the first E-conference in 2015 and was preceded under the theme of Developing Model Policy Frameworks, Standards and Guidelines on Landslide Disaster Reduction. The conference 2015 was an amalgamation of over 300 participants from 18 countries, 2 international organizations, and 2 United Nations volunteer groups, community representatives, scientists, engineers, researchers, students, experts, politicians, and policy makers working in the area of landslide risk reduction. Participants from almost all the continents and 20 keynote experts/facilitators, supported for an effective communication platform throughout the project. The main topic was sectionalized in to three subordinated topics stated below for the ease of contextualizing the imperative facts and figures such as the technical issues, social and ethical aspects, required recommendations and regulations related with landslide disaster phenomena. Related topics were:

1. Theme A: Developing Conceptual Model Policy Frameworks to Understand Causes, Effects and Mitigations of Landslide Occurrences (Bhandari 2015).
2. Theme B: Implementation of Applicable Guidelines/Teaching Tools to Establish Essential Synergies in Landslide Disaster Phenomena (Amaratunga 2015b).
3. Theme C: Originate Pertinent Standards for Humanitarian Activities in support of Effective Risk Reduction and Mitigation on Landslide Occurrences (Rupasinghe 2015).

The paper filters the most significant facts discussed throughout the conference and summarizes important findings which suits maximum usage towards a safer society living in landslide disaster prone areas.

A Local Forum (LF) was also being carried in parallel to the International E-conference in order to deeply confer the pressing matters occurred due to landslides in national significance. The LF activity encourages cohesion and sustained dialogue within local research communities, and was targeted at a specific disciplinary or interdisciplinary landslide disaster management groups and offering add questions typically with changing themes, at regular intervals. In addition, a consultation workshop was conducted through the volunteer groups to identify the thematic areas to come up with feasible solutions for the issues faced by vulnerable communities from the grass root level (Rupasinghe et al. 2014). A Question Session was also instigated in order to clarify all the issues that have emerged through the conference deliberations utilizing the expertise knowledge in finding the best solution. The experts reviewed the globally available imperative information, identified research needs and gaps in existing recommendations directly relevant to the respective problems. The conference was carried out through an official website, Fig. 3 (https://e-conference.crdcecbsl.lk/). Similarly, the 2nd E- Conference (Fig. 4) on Landslide Disaster Risk Reduction was held during 30th April–30th July 2017.

It was a continuation of some related contents of the 2015 and more elaboration on youth empowerment on disaster risk reduction, disaster resilient buildings and communication of the indigenous knowledge aspects in landslide disaster mitigation. In addition, 60 participants from 10 countries, 2 international organizations, and 2 United Nations organizations, scientists, engineers, researchers, students, experts participated for the event. Related topics were:

1. Theme A-Youth Empowerment in Landslide Disaster Risk Reduction through Knowledge Sharing (Nupearachchi 2017).
2. Theme B-Disaster Resilient Housing, Building, Land Management and Agricultural Development (Amaratunga 2017).
3. Theme C-Indigenous Knowledge in Landslide Disaster Management, (Katuwala 2017).

It was our pleasure to work in partnership with International Consortium of Landslides (ICL), IPL, UNDP, UN Volunteers, National Building Research Organisation (NBRO) and other participants from all over the world (Fig. 5) towards the landslide disaster risk reduction e-conferences approach.

Key Stakeholder Expressions

At the very first round of conference, many explanations were broadly discussed through expert's opinions and therefore, landslide risk is defined as "the combination of the probability of an event and its negative consequences". Hazard is defined as "a dangerous phenomenon, human activity or condition that may cause loss of life, injury or other health impacts, property damage, loss of livelihood and services, social and economic disruption, or environmental damage" (Bhandari 2015). The stakeholders were much interested on projection of landslide scenarios, definitions and estimation of the associated risk which reflects the need on development of model policy frameworks and appropriate community level standards. However, the mobility of e-conference received quite high contents of observations of advocacy, neutrality, debriefing and interpretations as listed below.

1. Landside is a natural process of degradation. However, unmanageable urbanization or illegal constructions in foot hills or inappropriate land use patterns in agriculture and deforestation are major societal and indirect contribution on landslides.
2. Risk is defined as "the combination of the probability of an event and its negative consequences". Hazard is mostly related to "a dangerous phenomenon, human activity or condition that may cause loss of life, injury or other health impacts, property damage, loss of livelihoods and services, social and economic disruption, or environmental damage".
3. Less attention on practical approach on dissemination of knowledge on disaster management among landside vulnerable communities.
4. Practicing of model policy approach disaster risk reduction and continuous engagement of the model approach is limited.
5. Early warning is one of the major discussion points among many stakeholders. This will ensure early stage of assurance to avoid the risk of vulnerable communities. Wide promotion on simple and economical instrumentation foresees and avoids the danger of landslides.
6. Many early warning centers are usually empowered with the institutional based entities but it is advising or time to think in terms of community-centric early warning systems which is now practicing in many developing and law income countries.
7. Late response to react on emergency is an attitudinal risk of a person during implementation of disaster risk management. The community response is usually high if the history of event was a closer record.

Fig. 3 Web portal 2015; https://e-conference.crdcecbsl.lk/

8. The research community should be encouraged to re-visit many case studies and develop scientifically proven educational tools for management of wide spectrum of falsifying facts and correction.
9. Simplicity and community attracting methodologies are useful for dissemination among the wide range of people and such mechanisms can be published through a social media networks.
10. Undertaking collaborative research and having frequent brainstorming sessions and advisory on new policy formulations and amendments are required.

Facebook Interaction and Societal Interest

Advancement of information technology and reduction of the price of communication devices compared to the per capita income had made the World Wide Web (WWW) a powerful tool of knowledge dissemination. Several Social media networks that function through this medium have become popular. Since Facebook which is a strong social media network with a high degree of user friendliness has linked all stages of life of averagely literate community, it

Fig. 4 Website provisions for the e-conferences 2017; www.econference2017.crdcecbsl.lk

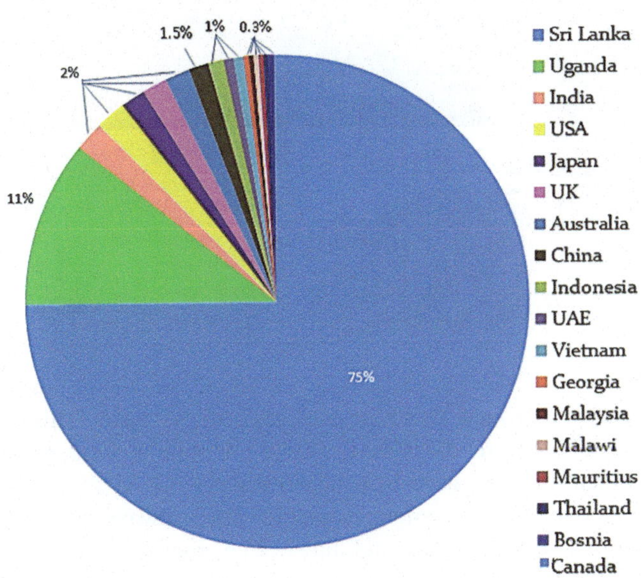

Fig. 5 Participants of first International E-conference on landslide disaster risk reduction—2015; Total number of participants, 313 and total number of countries participated, 18

was decided to utilize this facility for knowledge dissemination and data collection. Three Facebook pages were created to achieve the objectives of the e-conference as shown in the Figs. 6, 7 and 8.

The Facebook domains were made public and steps were taken to use graphics and cartoons to attract the affected and interested community.

English was selected as the Main medium used for communication as it is the most familiar out of the living languages, to the authors, and general society. Nevertheless, posting in any other language was not restricted. Also, the level of complexity of language or the descriptions were maintained low to increase its accessibility. Simple queries were forwarded to the community aiming at assessing the level of awareness and also to grab relevant and useful information that existed worldwide. Responses were used to build up the proceedings of the e-conference.

Indigenous Knowledge Aspects

Incorporating the traditional knowledge of local tribes and indigenous communities, with the existing scientific knowledge will facilitate the process of landslide disaster prevention and preparedness in a sustainable way. The indigenous knowledge transmitted either orally or experientially among the local communities and also the best practices that have been used by the indigenous communities to survive in their habitats integrating their rich knowledge in nature and natural phenomena may be unknown with

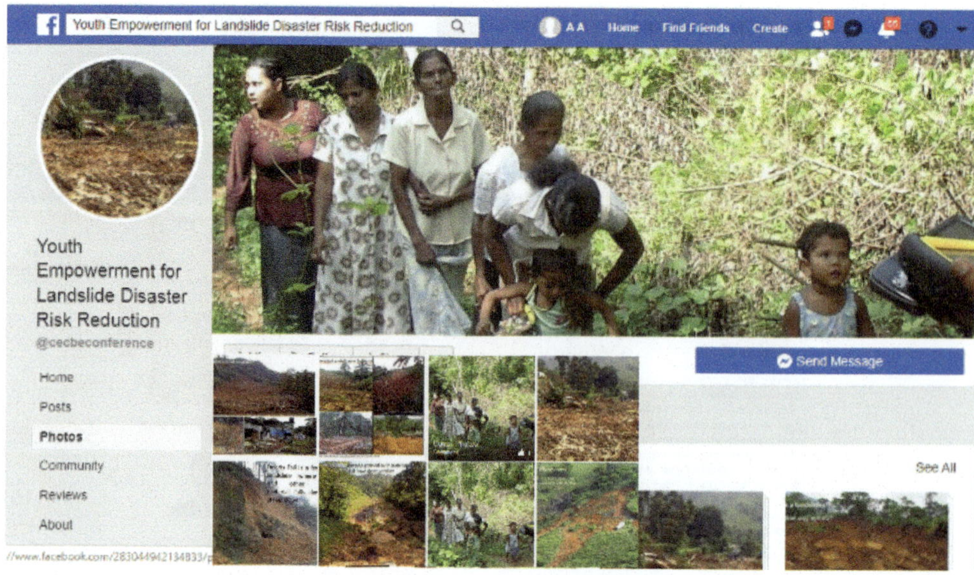

Fig. 6 Youth empowerment in landslide disaster risk reduction

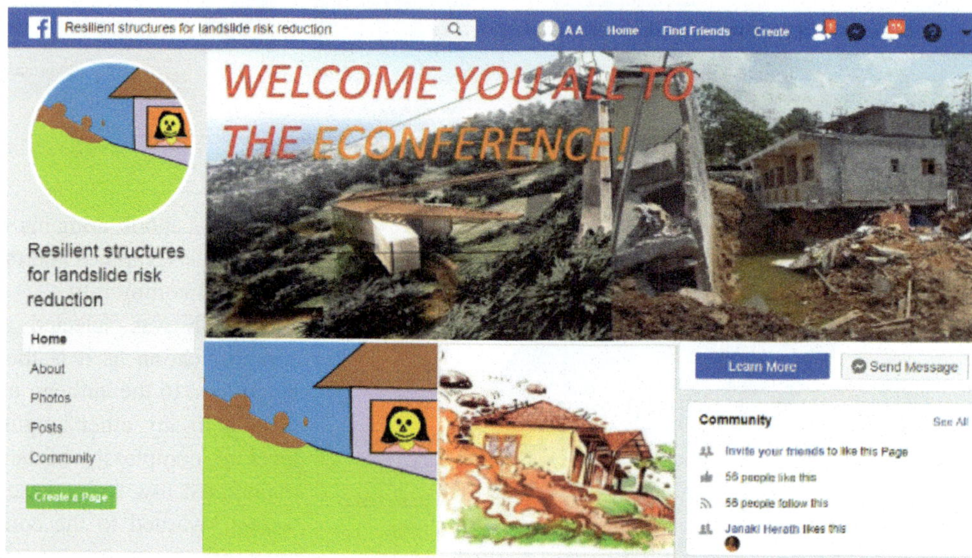

Fig. 7 Disaster resilient housing, building, land management and agricultural development

future generations. With regarding the landslides, especially the behavioural changes of the animals and insects are closely monitored by the village communities as they believe that the animals, insects and birds are more sensitive to the natural changes of the environment and their behavioural changes and calling patterns indicate an upcoming catastrophic phenomenon. Therefore these pre-disaster signals led them to seek for precautions and save their lives. In addition, the indigenous knowledge on certain geological changes in their vicinity is also remarkable. Even without any scientific analysis or testing method, the indigenous communities can identify that a disaster may occur due to the change in the colour of ground water or a change of its taste. This is probably due to the strong interaction with natural environment in every aspect of their life style.

Above all, indigenous societies have always given a spiritual value to nature and thus even a structural change or any development had been done utilizing a natural element such as caves or using local material with minimum disturbance to the higher grounds. The natural caves used as monasteries in the past indicates the methods and practices developed by local communities from their advanced understanding of the local environment for long term habitation in highland territories.

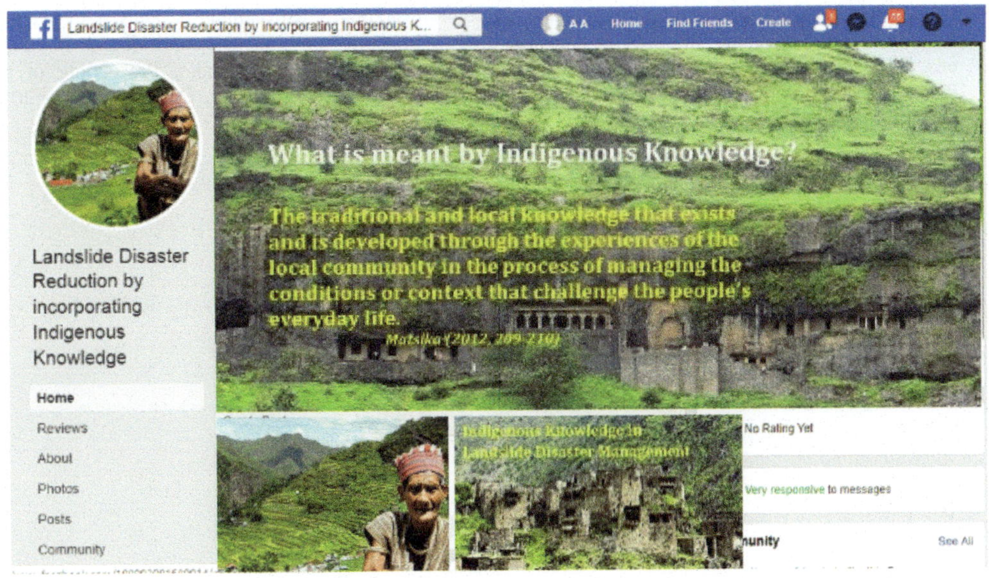

Fig. 8 Indigenous knowledge in landslide disaster management

Conclusions

The final outcome of the e-proceedings is to filter the most significant facts discussed throughout the conference in order to increase the readability and obtaining the maximum usage of it towards the society. The institutional target is focused more towards the social aspects such as to enrich the landslide risk awareness among citizens (including school students, workmen and laymen) and ensure the effective transmission of this message among public to enhance the adaptability with the mechanisms involving landslide mitigation and slope protection measures discussed throughout the conference.

It was certainly a fruitful global discussion in supporting to bring the sustainable humanitarian future to fruition. Risks and findings must be systematically integrated into policies, plans in disaster risk reduction. At last, e-conference has indicated one milestone of establishing an e-framework for IPL intellectually, practically, economically by contributing to IPL-GPC which promotes "landslide research and risk reduction" on a regional and/or global scale in a mutually beneficial manner.

Acknowledgements This paper forms an integral part of the "World Centre of Excellence (WoCE)" being recognised by the Central Engineering Consultancy Bureau (CECB) of the Ministry of Mahaweli, Agriculture, Irrigation and Rural Development. It is published with their permissions. The views expressed in the paper are however those of the authors only. Our grateful thanks are due to Eng. U.S. Karunarathna, Chairman and Eng. T.D. Wickramarathna, General Manager, Central Engineering Consultancy Bureau for the permission. Eng. Aravinda Kalugaldeniya, Addl. General Manager, Natural Resources Management and Laboratory Services is appreciated for his continuous encouragement and guidance towards research and publications.

References

Amaratunga D (2015a) A disaster resilient built environment in urban cities: the need to empower local governments. Proceedings of the first international E-conference on landslide disaster risk reduction, https://e-conference.crdcecbsl.lk, WCoE, CECB, 24th March–24th April 2015

Amaratunga D (2015b) Originate pertinent standards for humanitarian activities in support of effective risk reduction and mitigations on landslide occurrences. Proceedings of the first international E-conference on landslide disaster risk reduction, https://e-conference.crdcecbsl.lk, WCoE, CECB, 24th March–24th April 2015

Amaratunga D (2017) Disaster resilient housing, building, land management and agricultural development proceedings of second international E-conference on landslide disaster risk reduction, https://e-conference.crdcecbsl.lk, WCoE, CECB, 30th April–30th May 2017

Bhandari RK (2015) Developing conceptual model policy frameworks to understand causes, effects and mitigations of landslide occurrences proceedings of the first international E-conference on landslide disaster risk reduction, https://e-conference.crdcecbsl.lk, WCoE, CECB, 24th March–24th April 2015

Dias AAV, Wijewardana PR (2002) Community base participatory model in natural disaster preparedness—landslides, Proceedings 9th world emergency management conference, TIEMS, Toronto, Canada, 14–17 May, 2002

Dias AAV, Katuwala N, Herath HMJMK, Perera PVIP, Sahabandu KLS, Rupasinghe N (2017) Model policy frameworks, standards and guidelines on disaster reduction (WCoE 2014–2017)"; Advancing culture of living with landslides; vol 1 ISRD-ICL sendai partnerships 2015–2025, pp 365–374

Katuwala NN (2017) Indigenous Knowledge in landslide disaster management; Proceedings of second international E-conference on landslide disaster risk reduction, https://e-conference.crdcecbsl.lk, WCoE, CECB, 30th April–30th May, 2017

Nupearachchi CN (2017) Youth empowerment in landslide disaster risk reduction through knowledge sharing. Proceedings of second international E-conference on landslide disaster risk reduction, https://e-conference.crdcecbsl.lk, WCoE, CECB, 30th April–30th May, 2017

Rupasinghe N (2015) Implementation of applicable guidelines/teaching tools to establish essential synergies in landslide disaster phenomena. Proceedings of the first international E-conference on landslide disaster risk reduction, https://e-conference.crdcecbsl.lk, WCoE, CECB, 24th March–24th April, 2015

Rupasinghe N, Dias AAV, Hennayake SK (2014) Role of intervening agencies and officials in emergency risk management of landslides, Sri Lanka; Proceeding of the world landslide forum3 (WLF3), Beijing, China, 2–6 June 2014; vol. 4, pp 700–705

2020 Kyoto Japan

ICT-Based Landslide Disaster Simulation Drill: Road to Achieve 2030 Global Commitment

Mohamad Fazli Sardi, Ahmad Fairuz Mohd Yusof, Khamarrul Azahari Razak, Rudzidatul Akmam Dziyauddin, Siti Hajar Othman, and Munirah Zulkaple

Abstract

Information and Communication Technology (ICT) is an important driver for socio-economic innovation, and a promising tool for better managing future risk in a changing climate. ICT technology has been widely applied to support disaster management cycles (prevention/mitigation, preparedness, response, and recovery/rehabilitation). This paper presents a new approach based on inclusion of multi-stakeholders in strengthening community-based disaster risk reduction (DRR) in the urbanized region. We developed a functional framework to promote an ICT driven bottom-up DRR approach and increase coping with resilience capacity of vulnerable communities. This study was carried out based on the landslide that occurred on 28 November 2016 at Taman Idaman (Serendah, Selangor) with a potential effected 83,000 residents in the vicinity. We designed and implemented a disaster management training scheme based on a simulation drill involving multi-stakeholders in the most urbanized state in Malaysia. The training scheme was established based on the metamodel based geospatial multi-disaster prototype system, which was later tested in the Full Scale Exercise (FSX) simulation drill. Remarkably the FSX drill was the first recorded landslide disaster simulation drill carried out at a district level characterized by ICT-Geospatial data in Malaysia. It was successfully organized with aims not only to improve the efficiency of the rescue effort, but also to enhance the community's participation in disseminating early warnings and to utilize an ICT enabler for building disaster resilient communities. We depicted 10 critical factors to design a landslide disaster simulation drill for supporting a local DRR resilience strategy. As a conclusion, an ICT-based landslide disaster simulation drill has potential to be up-scaled and replicated by empowering community knowledge, building capacity of local champions and promoting digital inclusivity towards assessing disaster risk and building resilient societies, in line with the 2030 global commitment in DRR.

Keywords

Disaster risk management • Landslide • ICT • Simulation drill • Community-based disaster risk reduction

M. F. Sardi (✉) · K. A. Razak · R. A. Dziyauddin
UTM Razak Faculty of Technology and Informatics, Universiti Teknologi Malaysia (UTM), 54100 UTM Kuala Lumpur, Malaysia
e-mail: mohamadfazlisardi@graduate.utm.my; fazli@civildefence.gov.my

K. A. Razak
e-mail: khamarrul.kl@utm.my

R. A. Dziyauddin
e-mail: rudzidatul.kl@utm.my

M. F. Sardi
Malaysia Civil Defence Force, Prime Minister's Department, Jalan Padang Tembak, 50556 Kuala Lumpur, Malaysia

A. F. M. Yusof
Selangor Disaster Management Unit, Selangor State Secretary Office, 7th Floor, Sultan Salahuddin Abdul Aziz Shah Building, 40503 Shah Alam, Selangor, Malaysia
e-mail: ahmad_fairuz@selangor.gov.my

K. A. Razak
Disaster Preparedness and Prevention Center, Malaysia-Japan International Institute of Technology (MJIIT), University Teknologi Malaysia (UTM), 54100 UTM Kuala Lumpur, Malaysia

S. H. Othman
Department of Computer Science, Faculty of Engineering, Universiti Teknologi Malaysia (UTM), 81310 Johor Bahru, Johor, Malaysia
e-mail: hajar@utm.my

M. Zulkaple
Prime Minister's Department, National Disaster Management Agency (NADMA) Malaysia, Block D, Pusat Pentadbiran Kerajaan, Putrajaya, Malaysia
e-mail: munirah@nadma.gov.my

K. Sassa et al. (eds.), *Understanding and Reducing Landslide Disaster Risk*, ICL Contribution to Landslide Disaster Risk Reduction,
https://doi.org/10.1007/978-3-030-60196-6_51

Introduction

Landslides remained the main natural hazard and disaster causing high number of human losses in Malaysia. Despite many landslides studies have been carried out and largely reported in 1999–2019 by Jaapar (2006), Sew and Chin (2006), Gue See-Sew (2009), Komoo et al. (2011), Kazmi et al. (2017), PWD (2017), Sardi et al. (2019), reducing landslide risk and its cascading impacts is a challenging task. Malaysia is ranked the 10th highest in the the world in term of landslide frequency in 2007–2019 as reported by The National Aeronautics and Space Administration (NASA) in The Global Landslide Catalogue (Update 2019) (Kostis, 2019). In addition, the Slope Engineering Branch, Public Work Department of Malaysia stated about 42,500 landslide hotspots that are closely monitored nationwide in 2019 (PWD 2019). The landslide disaster also triggered the development of the national policy in disaster management as stated in the National Security Council (NSC) Directives No. 20 Policy and Mechanism of National Disaster Management and Relief. This NSC Directives No. 20 was established after the occurrence of the Highland Towers collapsed in 1993, claimed 48 lives. It is worth to mention that no special guideline is available for the rescue teams and also local authorities before the 1993s tragedy. Despite the amendment was done on the NSC Directive No. 20 in 2012, the standard operating procedures (SOPs) for landslide disaster in Malaysia is still elusive. Therefore, this study provides a significant input for supporting such efforts, in line with the Target E (2020) by Sendai Framework for Disaster Risk Reduction 2015–2030 (UNISDR 2015).

This paper also presents the implementation of simulation drill to support the Target-G, SFDRR2030 which highlights the importance of early warning systems and dissemination of disaster information to the communities. The simulation drill highlights the involvement of various stakeholders from the community leaders, responders and disaster managers in supporting the ICT utilization in disaster risk management in Malaysia.

Study Area

The State of Selangor has contributed about 24% of the country's GDP in 2018 (Izumi et al. 2019). As the most economically urbanized state in Malaysia, any disruption by disasters may result significant impacts to the nation. Izumi et al. (2019) have reported 86 landslide disaster in Selangor in the period of 2015 until May 2019, with substantial impacts to social, economic and environment.

One of the major recent landslides in Selangor was on November 26, 2016 in Taman Idaman, Serendah. It resulted in an injured victim, 340 residents were evacuated, several vehicles and food stalls were buried under the rubbles, while the main road connecting the area was completely cut off due to the landslide. Figure 1 shows the aerial photo of the Serendah landslide in Taman Idaman, which damaged the main road that connecting many residential houses, recreation areas, government offices, including aborigines settlements.

The result of the investigation indicated that unstable sand pits with the presence of water seepage (JMG 2016). It involved an area of about 45 m wide, 220 m long and 5 m deep. Following the occurrence of the landslide, comprehensive structural measures and stabilization measures have been taken to enhance the stability of the cliffs and road repairs (PWD 2016), that resulted in an non-disrupted use of the road by the Serendah residents. In the study by (Kamarudin and Razak 2018), it stated the urgent request by the local community of Serendah for comprehensive training exposure to disaster preparedness. The development of the bottom-up approach is an important step for enhancing local community's awareness, strengthening local risk governance, and developing a resilient society and sharing knowledge ability on some pertinent issues related to disaster risk reduction (DRR).

This paper presents the implementation of a simulation drill conducted in Taman Idaman, Serendah, Selangor, leading to the formulation of critical factors for the drill simulation. The area is still vulnerable to landslide disasters. In fact, residential and infrastructure at the vicinity were built in the ex-mining area. This is the first recorded simulation drill for landslides disaster at a district level. The landslide disaster simulation drill was based on ICT to support local DRR strategies and establish a community-based disaster resilience program in a complex environment.

Fig. 1 Aerial photo of Taman Idaman, Serendah landslide (SDMU 2016)

Methods

Community-Based Landslide Simulation Drill

This study developed a community-based landslide simulation drill based on the 2016 landslide in Serendah, Selangor. We used a Full Scale Exercise (FSX) simulation drill because it provides a platform to experience in-hands on 'real situation' and provides inputs how to manage disaster by their own self (EMSI 2018; IBRD and IDA 2016). During the drill, the communities were involved, these communities can act as the first responder and become the 'eyes and ears' of the rescue team in the event of a disaster in their place. Reliable early information and correct preliminary actions by local communities are able to reduce the impact of disasters to minimal levels (Simpson 2002). The community-based disaster risk reduction (CBDRR) is crucial to strengthen the preparedness and awareness among local community. The implementation of simulation drill was carried out with comprehensive involvement of all District Disaster Management Committee (DDMC) members and a full support from the State Disaster Management Committee (SDMC). Also, the full involvement of many disaster response players, from public–private-academia-civil society partnership, operating under the NSC Directive No. 20 including local community leaders.

Disaster Management Metamodel

An integrated disaster risk management is increasingly on demand and requires a strong support from various stakeholders to ensure an effective implementation to reduce disaster risk and its impact. Othman et al. (2014) developed the Disaster Management Metamodel (DMM) to support the decision making process for comprehensive disaster management. In this study, we established the metamodel based geospatial multi-disaster prototype system (Razak et al. 2018). It was developed based on the NSC Directive No. 20, disaster-related guidelines and SOPs for the landslide and flood case studies in Selangor. It was later tested and verified during the disaster simulation drill. This simulation drill was designed by testing some scenarios simulation as stated in the NSC Directives No. 20, and was based on the chronology of the Serendah landslide. This DMM represented the unified perspective of unique management process with common methodology and response-oriented applied in various disasters. A series of scientific discussion was carried out with the modellers and the system developers to design the methodological framework and disaster management operational flow. As a domain information delivery platform for the simulation drill, the established metamodel was used to demonstrate the landslide model interface (see Fig. 2). It covers four main phases of disaster management and is translated into the scenarios for the simulation drill, namely (i) early warning, (ii) actions, (iii) response and (iv) investigation-based scenarios.

Disaster Communication

The development of the disaster communication system was tested through the implementation of simulation drill based on the four scenarios that were used in the simulation drill. In each scenario, the roles and responsibilities of communities were identified based on diagnostic landslide features and geomorphological processes, ranging from tell-tale signs of an impending landslide—such as: water puddies at bottom of slopes; leaning trees on retaining wall; tension cracks on face of slope; bulging of slope; cracks on surface of pavement; damaged pipes; crack on building; doors or window cannot open and many more. Later, the community responsible to assist the evacuation process and managing the evacuation center at the early stage before the relief team arrived. Figure 3 shows the example of landslide early sign reported by the community to the local leaders and local authorities via short message services (SMS). The role of community leader is crucial in ensuring the safety of local community before, during and after a disaster. Task distribution was coordinated with all members of the resident committee so that an immediate action could be taken during the actual cases.

The use of proper and accurate communication systems is critical in times of emergency or disaster. Usually, people will use the fastest methods of communication when reporting disasters such as the use of direct calls to rescue agencies or the use of short message services (SMS) or through smartphone applications. It depends on the ability to access the communication network. The response of the rescue team also depends on the speed at which the incident could have been reached based on accurate information from the informers. The operations ended when an ongoing activity resulted in a satisfactory conclusion. The joint technical agency from DDMC and SDMC assessed the damages and destruction due to landslides and fix the error. The final reports of the operation were presented to the stakeholder and reconstruction orders will take place in the future. The After Action Review (AAR) is a qualitative review of actions taken to respond to an emergency as a means of identifying best practices, gaps and lessons learned. Following an emergency response to this simulation drill, an AAR seeks to identify which parts and protocols worked well and which failed to do so and how these practices can be improved, institutionalized and shared with relevant stakeholders.

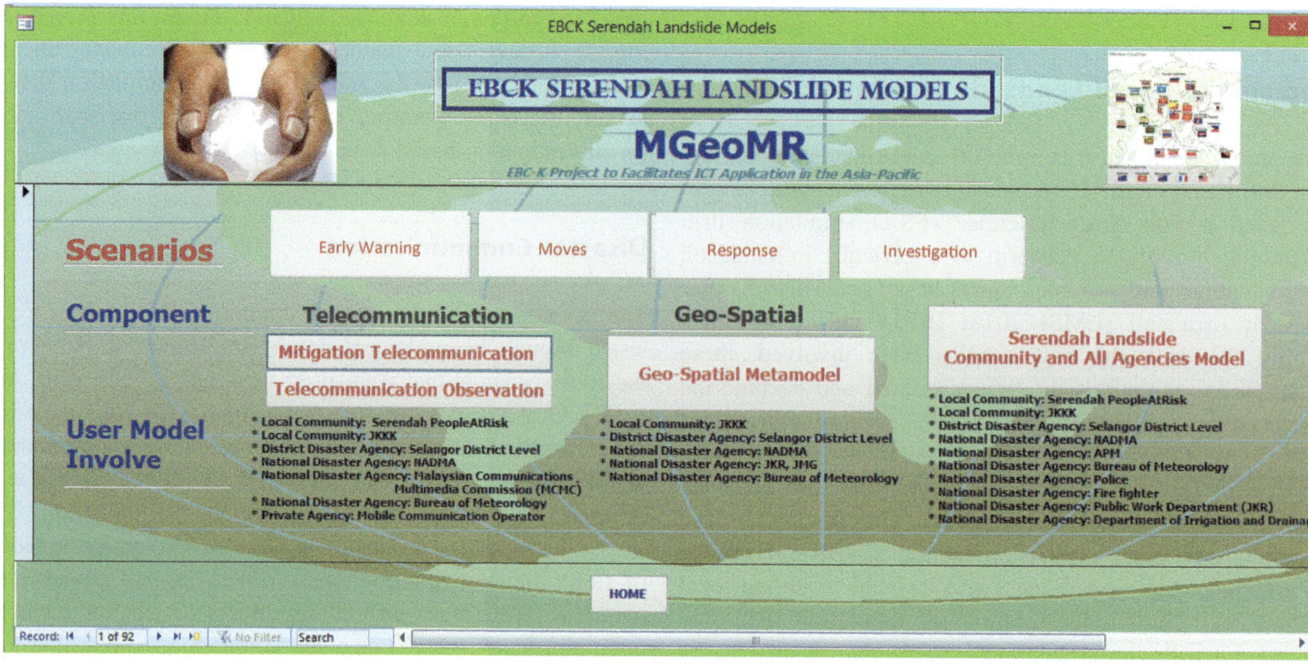

Fig. 2 Disaster management Metamodel (DMM) for Serendah landslide model interface

Fig. 3 Sample format of SMS content of landslide early sign reported by the local community

This process of disaster dissemination information can be simplified and it starts from the collection of information by the community to the information reaction by the rescue team and the decision made by the disaster management council at each level. This process reduces the bureaucracy of disaster information delivery and minimizes the impact of disaster. Therefore a functional framework was developed to synergize a bottom-up approach for community-led DRR activities and to increase coping capacities of vulnerable communities with the help of ICT and geospatial technology in the urban area (Fig. 4).

Discussion

Preparedness and Awareness

Several key findings or gaps were identified during the landslide disaster simulation drill. It is observed that the community leader manually reports the early sign of disaster via calls or short message service to the police station or the district office. Community leaders should make a quick decision for reporting the case to the police or the responder

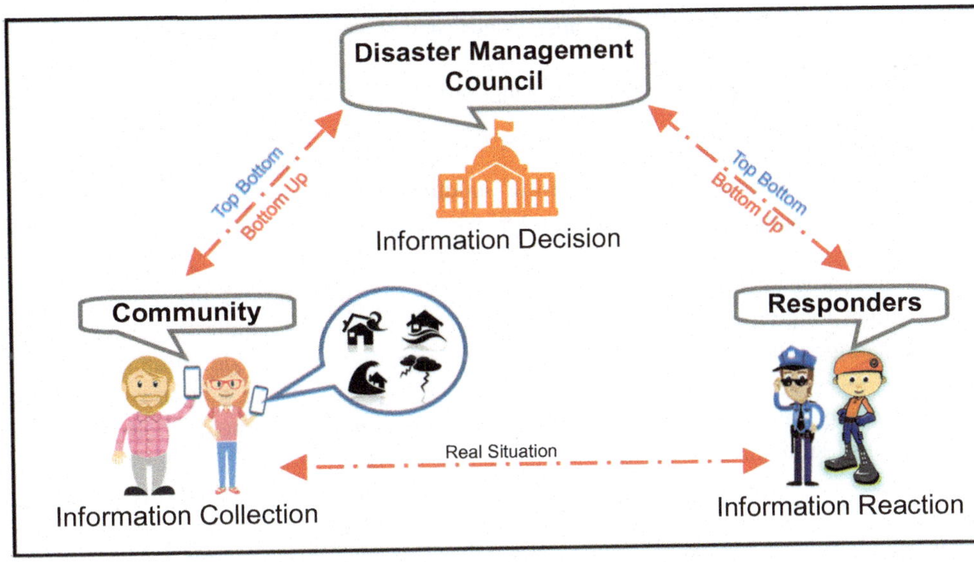

Fig. 4 ICT-driven bottom-up DRR approach

at a district level. However, the community leaders do not have the detailed information on the local families, including the number of family members and is this unable to verify the safety of all family members. Another concern is that there is no communication platform has been established between the community leaders and each family leader prior to the disaster. The need of the emergency warning system by local communities clearly exists should be placed in order to get them prepared for future disaster.

Emergency Response

The community leader logged the emergency report using the Malaysia Emergency Response System 999 (MERS999) were later verified by several responsible officials (police, firefighters, civil defence officers, and medical officers). This causes troublesome since the community leader was busy to disseminate the disaster information to its community. Therefore, the preliminary actions for the evacuation process by the community was delayed. More challenges were faced by the responsible officials due to complexity of the disaster or emergency incidents. Therefore, emergency administrators should seek another method, e.g. to enhance the system performance. A critical factor in the effectiveness of any emergency response agency is the ability to get personnel and equipment to the scene of the disaster or emergency in a timely manner (Sardi and Razak 2019). In the meantime, the responsible officials should understand the "language" with the concurrently rescue operations objectives in order to support the success of the operation. The aforementioned scenario was translated into joint exercises such as this simulation drill to enhance coordination and communication between technical agencies and implementing entities.

Recovery

The use of ICT-based applications as a data inventory for the registration of the evacuees at the evacuation centre was good as practiced by the Welfare Department (Tsai and Yau 2013). However, the improvement can be made to smooth the registration and management process at the evacuation centre, as well as it can help to accelerate the release of disaster relief trust fund. As a proactive act on capacity building, some training sessions for vulnerable and under-priviledged groups are proposed, such as community based disaster or emergency training, first aid training, evacuation drill and many other types of training given to provide awareness to the public and also know the early warning sign due to the event. The template of simulation drill will be in a standard format to be applied in other areas covering 12 districts in Selangor, and other vulnerable states in Malaysia. This simulation drill leads to better support local DRR and resilience strategies, empowering community leaders to provide early warning and multi-stakeholders input.

Critical Factors for Simulation Drill Implementation

In this study, we propose ten (10) critical factors for designing landslide disaster simulation drills and strengthening local DRR resilience strategies as below:-

1. Empowering local disaster experience;
2. Involving 'local actor' participation;
3. Building local capacity and capability of vulnerable community;

4. Promoting participatory-based approaches and local knowledge;
5. Utilizing local resources and preparing basic rescue equipments;
6. Strengthening risk communication;
7. Selecting community-friendly training modules (e.g. disaster imagination games and town watching for DRR);
8. Co-designing local action and timeline planning (short; medium; long-term);
9. Allocating sufficient budget and co-funding;
10. Reviewing outcome-based performance and reflecting impact of simulation drill.

This DRR bottom-up approach utilized the current government structure implementation such as Civil Defence Emergency Response Team—Community (CDERT-Community) (MCDF 2017), as introduced by the Malaysia Civil Defence Force (MCDF). It is significant effort on development of local disaster risk reduction strategies. MCDF is one of the federal agencies that emphasizes the legal functions in line with Malaysia disaster management policy at the state and district level, linking to the national support by the National Disaster Management Agency (NADMA) Malaysia. The local engagement approach in disaster program is essential in training and educating the local communities for establishing community-based emergency response teams in disaster-risk areas. This initiatives strengthened the community resilience by understanding the local risk. The community-led DRR activities contain simple and enjoyable training sessions and easy-to-understand basic knowledge that can be used by the community in the face of disasters. The MCDF aims to ensure that all disaster-prone areas establish the CDERT-Community team to meet the full range of community preparedness needs, based on a bottom up community driven approach.

The community-based approach to disaster preparedness is a key factor to reduce the disaster impacts, as stated on SFDRR2030. In addition to the latest technological developments such as the use of ICT-based in disaster risk management, it is also capable of enhancing the ability of the community to respond rapidly to catastrophic threats. Thus, this study demonstrates that the community-based ICT approach is a viable tool to enhance local community preparedness to future landslide disaster and its cascading geo-disaster. It is crucial to establish the national disaster risk repository as one of the requirements to obtain right information and disaster-related data from various government agencies, and local stakeholders.

Conclusion

This study presents practical approaches in the use of ICT for building disaster resilience, as supported by disaster meta-model characterized by geospatial data. Rapid execution, low cost, and high impact are key factors in promoting ICT for a multi-stakeholder disaster community. A comprehensive interdisciplinary simulation drill program can improve the readiness of the community and local authority to handle the disaster in proper manner. The first recorded landslide simulation drill was held at a district level with strong support of state and federal government agencies, and promoted a bottom-up approach (community-driven). It is worth to mentioning that Selangor Disaster Management Unit (SDMU) is the only dedicated state-level for managing disaster risk and strenghthening risk govenance in Malaysia. This study also demonstrates the most comprehensive and structurally designed simulation drill meant for landslide or slope failure. Beside, activated the current policy able to mobilize the district command centre and later triggered the state- and national disaster command centres to operate. The improvement on disemination risk information at each level can significantly reduce the impact of disaster.

This study demonstrates a multi-stakeholder approach to support locally-led and nationally-supported DRR initiatives and programs. We managed to understand local profiles of landslide disaster risk and build local champions by implementing landslide disaster simulation drill. In order to support the implementation of local DRR and resilience strategy, it is important to develop an integrated framework in the near future for strengthening disaster resilient communities in Malaysia, with significant input and contribution by various stakeholders (public, private, academia, and civil society) at a local level.

Acknowledgements Acknowledgements to Universiti Teknologi Malaysia (EBC-K Grantt, Radis ID: PY/2017/02438), Ministry of Communications and Multimedia Malaysia, Malaysia Civil Defence Force, Selangor State Secretary Office, Hulu Selangor District Office and our international counterpart, National Information Society Agency (NIA) Republic of Korea for the Extra Budgetary Contribution from The Republic of Korea (EBC-K) Project to Facilitate ICT Application in The Asia Pacific, Asia-Pacific Telecomunity (APT).

References

EMSI (2018) Types of exercises. Emergency Management Services International (EMSI)

Gue See-Sew KD, Wong Shiao-Yun (2009) Policy and institutional framework for landslide mitigation and risk reduction. In: Landslides–disaster risk reduction (pp 531–554). Springer, Berlin, Heidelberg (Reprinted from: 978-3-540-69966-8)

IBRD and IDA (2016) Learning from disaster simulation drills in JAPAN. International Bank for Reconstruction and Development (IBRD), International Developement Association (IDA), and the World Bank

Izumi T, Matsuura S, Mohd Yusof A, Razak K, Moriguchi S, Kure S, . . . Supar L (2019) Disaster risk report: understanding and slide and flood for science-based disaster risk reduction in the state of Selangor, p 86

Jaapar A (2006) A framework of a national slope safety system for Malaysia. University of Hong Kong

JMG (2016) Geology investigation report: landslide geohazard incident at Taman Idaman, Serendah, Selangor. Minerals and Geoscience Department Malaysia (JMG)

Kamarudin KH, Razak KA (2018) Comprehensive study on community based disaster risk reduction (CBDRR): the case of Taman Idaman, Serendah, Hulu Selangor, Selangor, p 41 (25 May, 2017)

Kazmi D, Qasim S, Harahap I, Baharom S, Imran M, Moin S (2017) A study on the contributing factors of major landslides in Malaysia. Civil Eng J 2(12):669–678

Komoo I, Aziz S, Sian LC (2011) Incorporating the Hyogo framework for action into landslide disaster risk reduction in Malaysia

Kostis H-N (2019) Global landslide catalogue 2007–2019. National Aeronautics and Space Administration (NASA). ID 4710. https://svs.gsfc.nasa.gov/4710

MCDF (2017) Panduan Pengurusan civil defence emergency response team (CDERT) Angkatan Pertahanan Awam Malaysia. Malaysia Civil Defence Force (MCDF)

Othman SH, Beydoun G, Sugumaran V (2014) Development and validation of a disaster management metamodel (DMM). Inf Proc Manage 50(2):235–271

PWD (27 November, 2016) Laporan Awalan Lawatan Tapak Kejadian Tanah Runtuh di Taman Idaman, Serendah, Sungai Choh, Selangor. Cawangan Kejuruteraan Cerun, Jabatan Kerja Raya Malaysia (JKR)

PWD (2017) Review of the National Slope Master Plan 2016, Cawangan Kejuruteraan Cerun Jabatan Kerja Raya Malaysia. Sec Issue 2016

PWD (2019) List of slopes in Peninsular, Sabah and Sarawak 2019. Slope Engineering Branch, Public Work Department (PWD). Unpublished Books

Razak KA, Othman SH, Dziyauddin RA, Mohamad Kamal NF, Sardi MF (2018) Metamodel-based geopartial multi disaster risk system (MGeoMR) final report. Universiti Teknologi Malaysia (UTM). Unpublished Books

Sardi M, Razak K (2019) Assessment of effectiveness of emergency response time during landslide event in Malaysia. Academy Sci Malaysia Sci J 12. https://doi.org/10.32802/asmscj.2019.360

SDMU (2016) Selangor state secretary office. Taman Idaman Report Serendah Landslide

Sew IDGS, Chin ITY (2006) Landslides: case histories, lessons learned and mitigation measures. Landslide, Sinkhole, Structure Failure: MYTH or SCIENCE

Simpson DM (2002) Earthquake drills and simulations in community-based training and preparedness programmes. Disasters 26(1):55–69

Tsai M-K, Yau N-J (2013) Improving information access for emergency response in disasters. Nat Hazards 66(2):343–354. https://doi.org/10.1007/s11069-012-0485-x

UNISDR (2015) Sendai framework for disaster risk reduction 32 (2015–2030)

A Preliminary Work of Safety Potential Analysis Model for Anchors Used on Freeway Slopes

Sao-Jeng Chao, Chia-Yun Wei, Han-Sheng Liu, Chien-Hua Kao, Hao Yang, and Cheng-Yu Huang

Abstract

In Taiwan, the landforms are mainly dominated by mountains and hills. Thus, countless road sections of the freeways are impossible to avoid the state-of-the-practice problems such as slope cutting. In order to stabilize the freeway slope, ground anchor technique is often employed to improve the stability of cut slope. With the increasing service time of ground anchors, their performance on the freeway slopes is highly required to be assessed. There were many disasters on freeway anchored slopes in recent years, so ground anchor inspection has also received attention to a great extent. This paper introduces the concept of safety potential analysis and then utilizes the proposed model for an anchored slope. Specifically, the proposed safety potential analysis collects plain map, historical slope inspections, monitoring data, ground anchor inspections and maintenance practices, as well as further use the geographic information system to establish the proposed model. Finally, this paper provides the predicted result from safety potential analysis and suggests the dangerous area of the studied anchored slope to the government authorities for immediately response purpose.

Keywords

Anchor • Slope • Freeway • Safety potential analysis • Geographic information system

Introduction

Many freeways in Taiwan are gradually expanding to the hillsides due to increasing economic demands recently. In order to improve the safety of freeway slopes, the ground anchor technique is often employed to increase the slope stability. The use of ground anchors in freeway cut sections could offer substantial benefits in economy and safety. With the increasing of ground anchors service time, it is supposed that the anchors are no longer to hold their original efficiency, thus the anchored slope may be dangerous. At this moment, the ground anchors used on the slope should be identified for any risk as soon as possible. The highway management agency is responsible to regularly execute the slope inspection and ground anchor detection, as well as to entrust professionals to inspect the overall conditions of slopes along the freeways. Besides, the functionality of ground anchors can be judged by appearance inspection, component inspection, and lift-off test.

This study was inspired by the soil liquefaction potential map and thus brought up the concept of safety potential analysis for the anchored slopes. The proposed procedure uses the geographic information system (GIS) software to draw layers, including the ground anchors on slopes, and follows by the procedure of simulating the appearance of slopes. We further consider significance information such as slope inspections, ground anchors detection, and instrument monitoring results within the GIS software. At that stage, the analysis uses interpolation method to derive the low,

S. J. Chao (✉) · H. Yang · C.-Y. Huang
National Ilan University, No.1, Sec. 1, Shennong Rd, Yilan, 260, Taiwan
e-mail: chao@niu.edu.tw

H. Yang
e-mail: b0421024@gmail.com

C.-Y. Huang
e-mail: a8333479@gmail.com

C.-Y. Wei
National Freeway Bureau, Toucheng, Taiwan
e-mail: jubby923@freeway.gov.tw

H.-S. Liu · C.-H. Kao
Taiwan Construction Research Institute, Taipei, Taiwan
e-mail: hansonliu@tcri.org.tw

C.-H. Kao
e-mail: brad@tcri.org.tw

K. Sassa et al. (eds.), *Understanding and Reducing Landslide Disaster Risk*,
ICL Contribution to Landslide Disaster Risk Reduction,
https://doi.org/10.1007/978-3-030-60196-6_52

medium and high potential risk areas, and determines the area where the anchored slope may be harmful. In conclusion, the safety potential analysis can provide the information to the government agency for possible disasters and the timely chance to deal with them as soon as possible.

Geographic Information System

Our life is in contact with GIS all the time, usually closely related to daily life, such as checking maps or searching bus routes that rely on GIS. GIS is a computer system that can build, browse, query, and analyze geospatial data. Its functions range from map display to spatial decision analysis and support. The spatial data processed by GIS is a true surface or underground information, human activities and historical records. It is the result of simplification and digitization, which can visibly express location and information in this world.

This study uses Quantum GIS (QGIS) program and goes with the help of ground anchors detection results, to show the trend around the slope and to observe the high potential risk areas. The program has open source and license under the GNU (General Public License), which is an official project of the Open Source Geospatial Foundation (OSGeo). Under the GNU license, users can review and adjust the code themselves and can also modify the program freely. QGIS is a simple software that provides common features and graphics. With providing basic features and add-ons (Python or C^{++}) to expand, the users are then allowed to view, manage, edit, analyze, and chart.

Principle of Anchor Lift-Off Test

During the execution of the slope stability project, the anchor has an initial design load (T_w). After 4–5 years of service time, the lift-off test can be performed to find the residual load (T_r) to be compared with the design load. During the progress of lift-off test, the minute that anchor has "lift-off" phenomenon, the load–displacement curve will present an obvious turning situation. At this moment, the load–displacement curve presents an apparent turning situation (Fig. 1), which provides the value of the anchor residual load (Chou 2018). The compared result and other monitoring instruments can be used to assess the overall slope safety.

The main tools for the lift-off test include: anchor head fixture, jack chair, hydraulic jack, load cell, displacement gauge, and so on. The anchor lift-off test is operated according to the process of staged pull-out. The results of the lift-off test are accurately significant to show the status of the residual load, and through the load–displacement curve of the lift-off test to understand the situation of the anchor in the slope with the serving time.

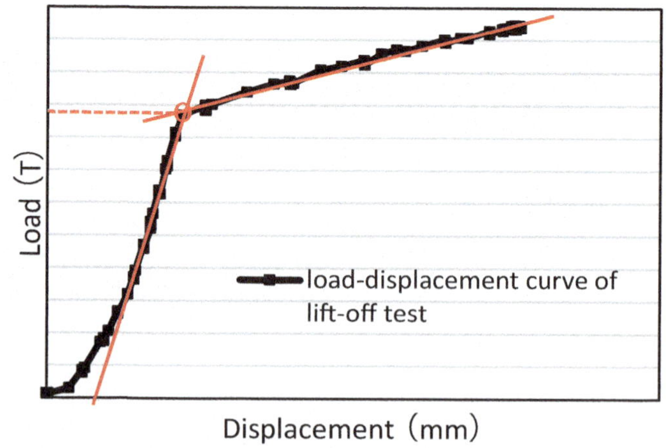

Fig. 1 Load–displacement curve of the lift-off test

It is important to choose the suitable anchor for lift-off testing. In order to confirm the average and randomness of the anchor samples, it is recommended by the freeway bureau to adopt the characteristics of "uniform distribution selection" as the anchor selection principle.

During the test process, the fixtures and clips are used to fix the anchor head first, and then the hydraulic jack is installed so that the tensile strand state is set (Fig. 2). In the case where the strand is stressed (Fig. 3a), P_0 is the residual tensile load of the anchor, which is the strand pull force, so the direction of the force is indicated to the underground anchorage section. P is applied by the hydraulic jack to pull out strand for the lift-off test, thus the force direction of P is opposite. Prior to anchor lift-off test, the strand only bears the residual load. The appearance of the anchor head is totally attached to the bearing plate at present.

However, when the lift-off test has progressed (Fig. 3b), the applied loads on the strand are from two different directions. On the one hand, the residual load of the anchor (P_0), due to the initial pre-stressed loading, the steel strand locked

Fig. 2 Lift-off test apparatus

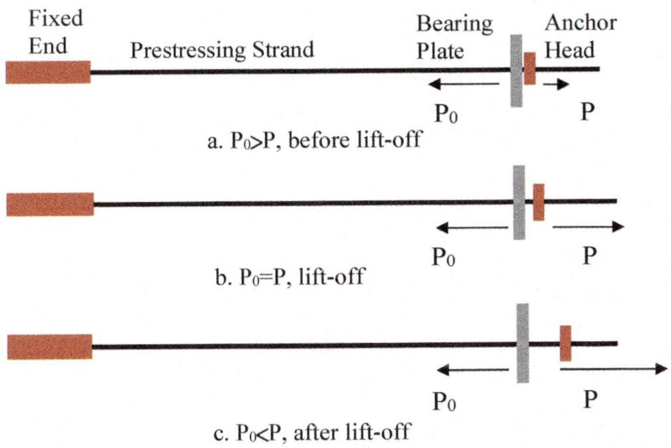

Fig. 3 Loading conditions during anchor lift-off test

to the bearing plate, resulting in the direction of the anchor force for the anchor phase direction. On the other hand, the pull-out force (P) is in the direction of the hydraulic jack to carry out the test in the direction of the anchor head gradually applied to the load. Therefore, the appearance of the anchor head is not change much until $P = P_O$ (Fig. 3b). When the load of the hydraulic jack is equal to the existing residual anchor load, the original position of the anchor head happens to leave the bearing plate by the process of full loading conversion at this moment. This phenomenon is so called the anchor head lift-off.

As a final point, after lift-off (turning point) phenomenon (Fig. 3c), the force–displacement behaviour of the internal strand of the anchor is provided entirely by the hydraulic jack. When the pull-out load is less than the anchor design load (T_w), the test load is stop at the value of 1.1 times the design load (T_w) and then record the existing load (T_r). Alternatively, when the pull-out load is equal to the value of 1.2 times the design load (T_w) but the anchor is still not lift-off, the test is stopped and the existing load (T_r) is recorded as the same value.

Process of Slope Safety Potential Analysis

The concept of safety potential analysis is widely applied, for example, typhoons, landslides, tsunamis, flooded areas and the most recent discussion of soil liquefaction. This section introduces the steps of safety potential analysis procedure for the anchored slope. Drawing of the potential map usually evaluates the stability of the anchored slope on a large scale and the comprehensive inspection, monitoring and ground anchor detection of the slope are scored at the same time. Since ground anchor detection has individual score but each one is closely related to each other on the slope, the detected field should gradually expand to the overall slope. To be precise, it should be extended from the point of interest to the surface of concept, so using the GIS program is a good choice.

The process of safety potential analysis is roughly divided into three steps, which is shown in Table 1 (Sakai and Fujiwara 2017). Step 1 is to select the detected slope and collect relevant information about the slope, such as environmental geological map, geologic sections map, boring logs data, and slope completion drawing. Correspondingly, through the historical inspection data and instrument monitoring data of the slope, the historical disasters and abnormal phenomena of the slope can be known. The slope inspection and ground anchor detection are executed regularly, including appearance inspection, component inspection and lift-off tests. The appearance of ground anchor is overall inspected and checked for obvious phenomenon of abnormality. Based on the results of appearance inspection, the positions were selected for component inspection and lift-off tests. The component inspection and lift-off test evaluate the current status of the anchor and its residual tensile load. The result of each ground anchor test is assessed according to the grading standards X, A, B, C, and D (Freeway Bureau, MOTC 2017).

Step 2 is to check the exact location of the ground anchor on the slope and to establish connection with the ground anchor test results. It follows by creating a GIS shape file of ground anchor points, which enable the selection of categories to be analyzed, including results of appearance inspection, component inspection, lift-off test, and the percentage of residual tensile load $R_{td}(\%)$, which is the ratio of lift-off residual load to the ground anchor design load. Considering the design load as a different value for each ground anchor, the ratio of residual tensile load is used as a comparison standard. After establishing a series of data, the analysis approach uses QGIS built-in function of inverse distance weighting to derive the low, medium and high potential risk areas. Therefore, the grades of the ground anchor is set to D level as 100%, C level as 75%, B level as 50. %, A level as 25%, and X level as 0%.

Step 3 is the most important stage to analyze the safety potential of the anchored slope. The various potential layers drawn in step 2 are nested with various historical inspection data and historical monitoring data to obtain the predicted results of the slope safety potential.

Case Study of Slope Safety Potential Analysis

In this section, an anchored slope is selected here for exploring the functional performance, and the variations in the anchored slope area are discussed. According to the process of drawing the slope safety potential, we first collect basic data of the slope and then use the data to draw the safety potential of the slope, including the appearance safety potential, the component safety

Table 1 Process of safety potential analysis

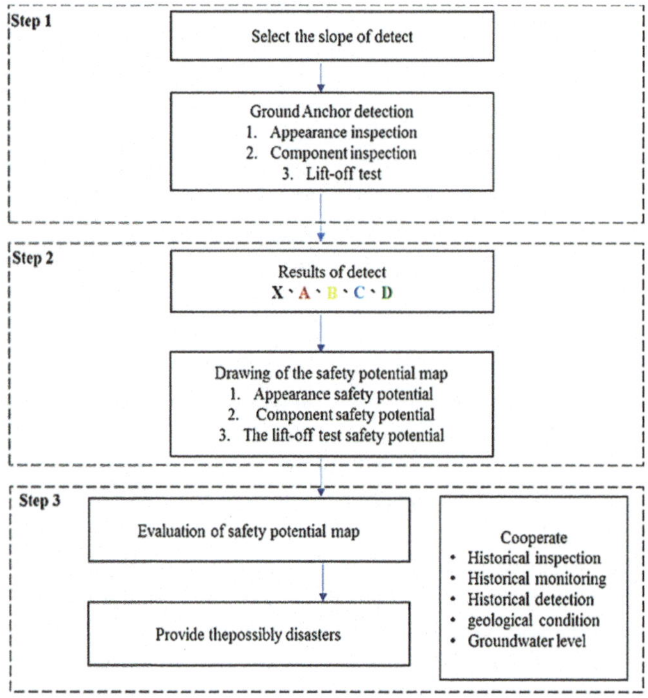

potential, the lift-off test safety potential, and the percentage of residual tensile load R_{td} (%). After combining with historical inspections, monitoring data and ground anchor detection data, the results of the safety potential analysis of the anchored slope can be obtained.

Basic Information on Anchored Slopes

The scene of the studied anchored slope is shown in Fig. 4, while the slope domain is shown in the drawing as shown in Fig. 5. The slope top elevation is 100 m and the slope toe elevation is 60 m. The geological conditions of the slope are divided into layers, which are gravel sand layer, argillaceous sandstone or sandstone mudstone, and gravel interlayer. After the excavation work completed, the slope is 1 V:1.5H and reshaped as 4 tiers. The ground anchors are installed on the first, second and third tiers on the slope. The total number of ground anchors is 474 (Huang 2019).

Fig. 4 The scene of the study anchored slope

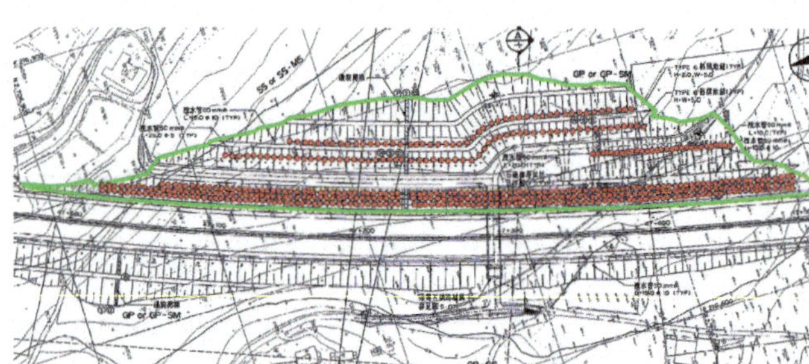

Fig. 5 The drawing of the study anchored slope

Appearance Safety Potential

For the ground anchor detection data, there are 474 anchors in total and the appearances are examined one after the other.

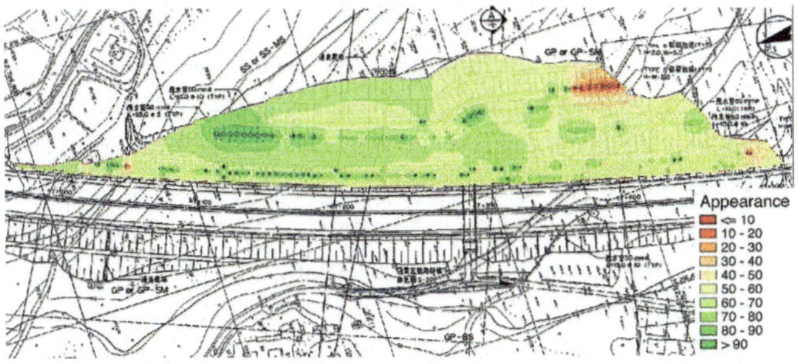

Fig. 6 Appearance safety potential map

The percentage of the inspection results is 13.5% for level D, level C is 15.2%, level B is 70.7%, level A is 0.63%, and level X is 0%. There are many slight damages or cracks in the protective covers, although the damage doesn't affect the operational behaviour, while the appearance mostly belongs to the level B.

The appearance safety potential map is shown in Fig. 6. In QGIS, the customized grade D is 100%, grade C is 75%, grade B is 50%, grade A is 25%, and grade X is 0%. The phase of red indicates serious condition, and the green indicates good condition. It is divided into 10 phases to clearly understand the vary areas. In general, most areas have a light yellow between 40 and 50%. Observing the result of the left side on the study anchored slope, some areas show green between 80 and 90%, and the others show light yellow. However, the right side condition is much poor than the left side, right side areas mostly show orange-red range between 20 and 30%, and some parts of the areas even show red between 0 and 10%, where should be put attention to.

Component Safety Potential

According to the principle of the sampling rule, we selected 28 anchors to carry out the component inspection. The percentage of inspection results is 3.6% for level D, level C is 64.3%, level B is 32.1%, level A and X is 0%. Most of the results fall in the level C, and the component safety potential map is shown in Fig. 7. The phase of red indicates serious condition, and the blue indicates good condition. It is divided into 10 phases to clearly understand the vary areas as well. Most areas show light green between 50 and 60%, and less area shows orange between 30 and 40%. However, in this study anchored slope the orange area indicates that the anchor head is significant rusted and seepage, and the situation may be closely related to the high groundwater level.

Lift-Off Test Safety Potential

The lift-off test can recognize the current residual loads of the ground anchors, and 27 ground anchors were selected to carry out the test in keeping with the sampling rule. The percentage of the lift-off test results is 26% for level D, level C is 44.4%, level B is 18.5%, level A is 11.1% and the level X is 0%. The lift-off test safety potential map is shown in Fig. 8. The colors of residual load are divided into five levels. The level D (■) is between 0.8 and 1.2 of design loads, the level C (■) is between 0.5 and 0.8 of design loads, the level B (■) is between 0.2 and 0.5 of design loads, the level A (■) is more than 1.2 design loads or less than 0.2 design load, and the level X (■) of the strand is broken or the ground anchor is pulled off. On the study anchored slope, the left side residual load is dominated by level C, and the right side residual load is dominated by level A and B.

In addition, R_{td} safety potential map is also shown in Fig. 9. The values R_{td} of the anchor have no one reaching 100% on this slope, which means that each anchor has already loss some load. Therefore, 10 stages are distinguished by 0–100% here, of which dark blue indicates that the load is loss seriously. Overall speaking, the R_{td} of the left side is maintained at 50–80%, and the R_{td} of right side is maintained at 20–50% while the color is getting darker. The existing loads of the anchors are general low on right side, while the loads have a drop trend. As a result, it is necessary to pay attention to the slope safety of the right side.

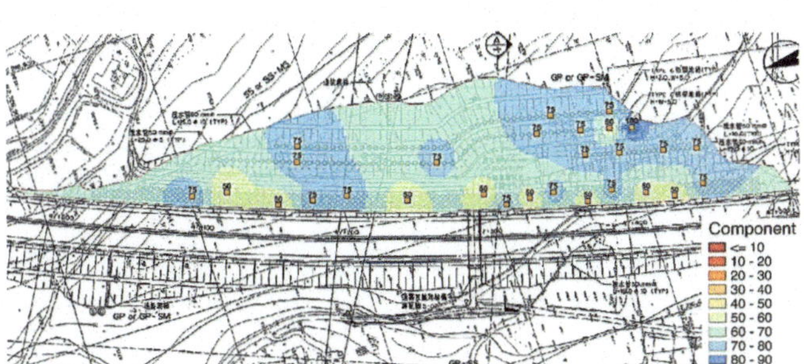

Fig. 7 Component safety potential map

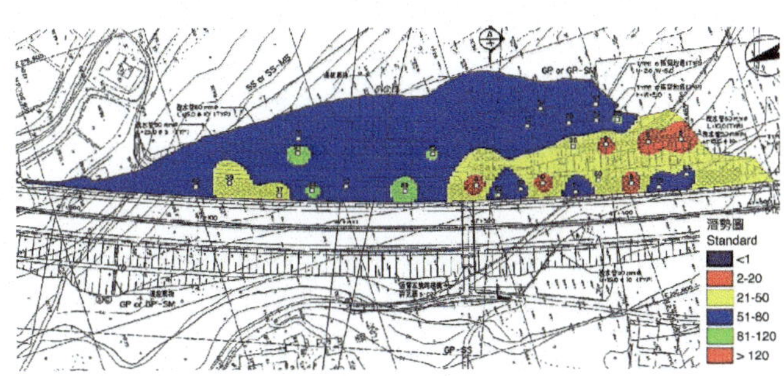

Fig. 8 Lift-off test safety potential map

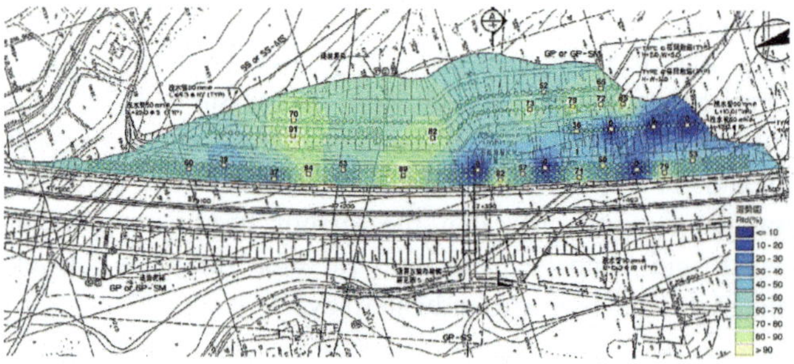

Fig. 9 R_{td} safety potential map

Conclusion and Suggestions

This paper introduces the concept of safety potential approach into the anchored slope and draws the appearance safety potential map, component safety potential map, lift-off test safety potential map, and the percentage of residual tensile load R_{td}(%) safety potential map. By using colors to indicate the low, medium and high potential risk areas, it is possible to predict where the risks may occur, thus provide the government authorities preventive judgment and early warning. Hopefully, the proposed ground anchor slope potential maps can strengthen the attention and maintenance in advance according to the degree of regional hazards, and effectively improve the safety maintenance management of anchored slopes.

There are 2 following suggestions for possible future plans:

1. Since the component inspection and lift-off test are selected according to the principle of the sampling rule, it is not a comprehensive inspection due to insufficient data, the safety potential analysis is thus not delicate enough. If the proposed approach is approved valid in the future, comprehensive inspection is recommended to make the result more confident.
2. At the present time, appearance safety potential map, component safety potential map, lift-off test safety potential map and the percentage of residual tensile load R_{td} (%) safety potential map are four separate ones. Although the various tests are related to each other, there is no appropriate weighting so far. In the future, it is recommended that the weighting should be considered for drawing a complete safety potential map.

Acknowledgements This research was supported by the Project of the state-of-the-art technical consultation on national highway slope safety and maintenance management. We thank our colleagues from Taiwan Construction Research Institute who provided insight and expertise that greatly assisted the research.

References

Chou HT (2018) Study of anchor lift-off test results and assessment of anchored residual loads. Master Thesis, National Ilan University

Freeway Bureau, MOTC (2017) Subgrade and slope, maintenance manual (3rd Chapter)

Huang CY (2019) Exploration of efficiency inspection and maintenance management for the soil anchors on freeway slopes. Master Thesis, National Ilan University

Sakai T, Fujiwara Y (2017) Evaluation of slope stability based on monitoring of residual loads on ground anchors using the SAAM system. GEOMATE, Geotechnique 7(1):10–18

Initial Experiences of Community Involvement in an Early Warning System in Informal Settlements in Medellín, Colombia

Tamara Breuninger, Carolina Garcia-Londoño, Moritz Gamperl, and Kurosch Thuro

Abstract

The project Inform@Risk is developing a landslide early warning system in the informal neighbourhood or barrio Bello Oriente in the city of Medellín, Colombia. Settlements in an already unstable slope increase its instability even more due to several anthropogenic interventions. Since every step in the risk cycle includes human factors, this early warning system must be developed in accordance with and with the full involvement of the residents of the informal settlement. To achieve acceptance and even help from the population they need to be educated about the hazards and about the different aspects of a landslide and a warning system. In the project this education is carried out in several workshops, where the residents are informed, and community walks, in which they apply their new knowledge in the field. This cooperation will continue with the assistance of the residents in the field work and in integrating the early warning system into the barrio.

Keywords

Early warning system • Community participation • Informal settlements • Medellín (Colombia)

T. Breuninger (✉) · M. Gamperl · K. Thuro
Chair of Engineering Geology, Technical University of Munich, Arcisstr. 21, 80333 Munich, Germany
e-mail: tamara.breuninger@tum.de

M. Gamperl
e-mail: moritz.gamperl@tum.de

K. Thuro
e-mail: thuro@tum.de

C. Garcia-Londoño
Geological Society of Colombia, President of Antioquia Chapter, Medellín, Colombia
e-mail: cargalon@gmail.com

K. Sassa et al. (eds.), *Understanding and Reducing Landslide Disaster Risk*, ICL Contribution to Landslide Disaster Risk Reduction,
https://doi.org/10.1007/978-3-030-60196-6_53

The Project Inform@Risk

The city of Medellín, Colombia is subject of the German-Colombian project Inform@Risk, funded by the German government. Its key goal is to develop a landslide early warning system that involves multiple local actors, including the community, in the developing process.

Medellín lies in the Cordillera Central of the northern Andes. Especially the north-eastern slopes of the city are prone to landslides, since these areas are very steep and consist of highly weathered dunite bedrock, that is covered by an up to 50 m thick layer of loose rock made up of grains with sizes ranging from large boulders to clay. In this coverage landslides occur on a regular basis, some of them claiming several lives (Urbam EAFIT, Leibniz University Hannover, Municipality of Medellín 2013).

The project Inform@Risk aims to detect areas of high risk, where monitoring instrumentation and evacuation routes are installed and built with the help of the community living in the study area.

In May 2019 the study site was chosen to be the barrio Bello Oriente in the north-eastern slope of Medellín (Fig. 1). The barrio is an informal settlement at the city border with over 6000 inhabitants and has suffered from at least 10 small to medium landslides of up to 10 m depth in the past 10 years, the last one occurring in early 2017. Luckily, so far none of these landslides have claimed human or farmyard animal life.

Urban Development as a Triggering Factor

Many regions around the world experience a formerly unknown growth in population. Especially larger cities in developing countries are the objective of many people coming in from the countryside to find easier and better paid work than in agriculture (Urbam EAFIT, Leibniz University Hannover, Municipality of Medellín 2013).

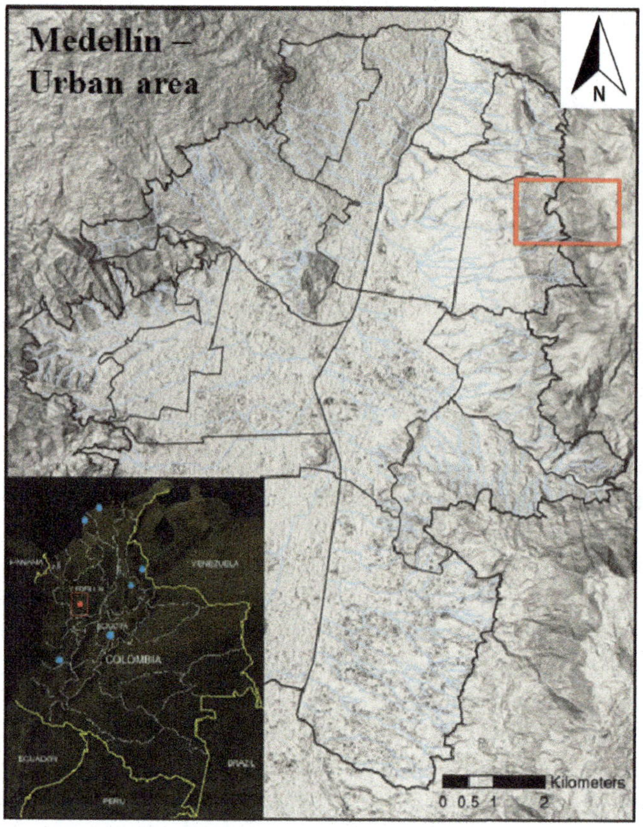

Fig. 1 Position of Medellín in Colombia (bottom left) and of Bello Oriente in Medellín (Gamperl et al. 2020 (in prep.))

Due to a high price level in the city centre, these people settle at the city border. In case the city is situated in a mountain area, like Medellín, the people settle on steep and unstable slopes. Additionally, the constructed houses lack preliminary ground investigation and proper foundation. These constructions weaken the already unstable ground in steep slopes in mountainous areas.

An impressive example of this dynamic is the highland in the south of Sri Lanka (Jayasingha 2016). Here, overpopulation has led to considerable changes in land use, deforestation, embankment cutting for roads and uncontrolled/increased infiltration of water (wastewater and watering for plants). These actions are taken with no or only little understanding of landslide dynamics and their promoting and triggering factors. Especially infiltration of water during or right before monsoon season is a huge problem.

This example shows that it is vital to a functioning disaster management plan to raise awareness and understanding at all stages of early warning (ISRD 2005).

Social Aspects on Landslide Warning

Early warning systems are complex tools for disaster risk reduction and are only effective if they generate an appropriate response in the exposed population (Garcia and Feranley 2012). Every step of the risk cycle includes human factors (Fig. 2). The biggest challenge in the project is to manage those human factors in the best possible way to improve the early warning system and ensure safety for all parties involved.

One of the biggest social problems in early warning is miscommunication between the different stakeholders, not only between the local population and the experts (Nadim and Intrieri 2011). This leads to false responses and in the worst case to devastating losses in case of an event.

Another general problem are false alarms. When issued several times, the people lose trust in the early warning system and in the scientists working with them. This trust is difficult to win back, if it is possible at all (Nadim and Intrieri 2011). Therefore, the selection of threshold levels is important, also on the social aspect. Figure 3 shows a conceptual scheme on how to set thresholds regarding risk and false alarms. It is also necessary to inform the residents, that false alarms can never be erased completely (Anderson-Berry et al. 2018).

A way to reach people in general are social media and TV, already effectively applied in Hong Kong and by the project SafeLand (Nadim and Intrieri 2011). In some countries and societies these measures are not enough. The people need to be addressed directly.

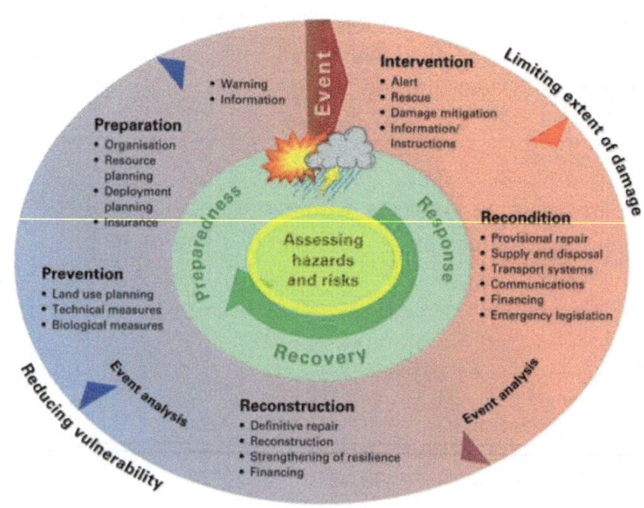

Fig. 2 **Risk Cycle** (Planat 2020)

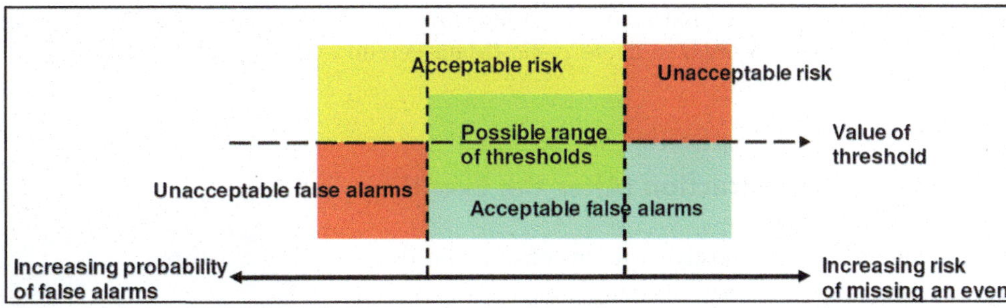

Fig. 3 Conceptual scheme of defining a threshold regarding false alarms and risk (Nadim and Intrieri 2011)

Prevention (Before the Event)

To prevent an event from happening at all, some preventive measures can be applied.

The people on the ground must be made aware of the dangers emerging from their interventions on the slope they are living on (see Section "Urban Development as a Triggering Factor"). This includes education regarding geology, hydrology, morphology and landslide dynamics.

In addition, there should be a program regarding alternative methods of building and planting. This way the risk can be reduced drastically (Jayasingha 2016).

Preparation (Before the Event)

This phase of the process is the most important one. Before an event occurs, the people on the ground needs to be prepared for all the following steps, since the actions to take in case of an event need to be taken fast and without a long phase of preparation or decision making. Everything and everybody needs to be prepared.

Firstly, there must be a clear and precise communication about which area is monitored and which area is not included in the early warning system. The same is necessary for the type of hazard (size, velocity, shape of detachment) and whether they can be detected by the sensors installed in the field or not.

In accordance with preventive measures (see Section "Prevention (Before the Event)") in the phase of preparation there must be programs educating the people on the risk they are living in. These programs should include general information on risk (geology, hydrology, morphology and landslide dynamics) as well as the identification of areas at risk with the help of the people themselves. This interaction strengthens the trust in the scientists, since it gives great meaning to the people's knowledge of their own home.

The education should also include the possible consequences of an event in order to get the residents interested in the program and stress the importance of investigations (Mata-Perello et al. 2005).

Regarding the technology on site, the location of the sensors must be visible to everybody. Also, a chart demonstrating the sensor's functionality is helpful. This way vandalism is reduced because some people understand the importance of the sensor for their wellbeing, feel responsible for it and will protect it against others not interested in the warning system. This dynamic was observed in Sri Lanka (Jayasingha 2016).

To prepare for an actual event, the residents must learn what to do when it occurs and prepare for a possible evacuation. Regarding the reaction, it is of high importance that the people know about the dynamics of landslides (see Section "Prevention (Before the Event)"), especially those in the area they live in in order to react accordingly. The evacuation is planned by the authorities, but the development and building of evacuation routes and safe places must be in cooperation with the community. The implementation of evacuation infrastructures is a significant change in the residents' environment and should not be planned over their heads, otherwise the trust of the community in the authorities continues to weaken (Jayasingha 2016).

There should also be people assigned and trained to special tasks in case of evacuation who help others to get to safety (i.e. kids, elderlies, disabled persons) or give directions to save places in general. This way the community has first responders until help arrives.

In addition, the evacuation could be trained with the whole community to ensure an effective reaction.

Intervention (During the Event)

In case of an event happening, everything must happen very fast. Therefore, most of the work has to be already done in the preparation phase (see Section "Preparation (Before the Event)").

The warning/alarm should be send using multiple channels including app, TV, radio and social media, and include a simple and clear message saying what is happening and what needs to be done. These channels should also inform about evacuation routes, safe places and people in charge of special tasks (first responders) and give reassurance to the people in order to avoid panic (Mata-Perello et al. 2005).

Recondition/Reconstruction (After the Event)

The last phase consists mostly in the reconstruction of the infrastructure and the early warning system. In case of casualties and injuries, psychological support for the victims and their relatives is vital, not only to keep the trust of the community in the authorities and scientists, but for recondition after an event in general.

Most importantly, the community needs the help fast, at best within minutes. Water supply, wastewater treatment, energy, homes, schools and other things need to be reconstructed or fixed within days. Otherwise the community suffers from the event for too long to keep their interest in the early warning system and their trust in the authorities (Jayasingha 2016).

Community Work in the Project

The project Inform@Risk tries to include all the aspects mentioned in Section "Social Aspects on Landslide Warning". This is achieved by several community workshops throughout the timespan of the project by various activities.

Responsible for these activities are the Geological Society of Colombia, Urbam from EAFIT University, the regional early warning agency SIATA, the disaster management agency DAGRD from the government of Medellín and the social organizations Convivamos and Tejearañas.

Workshops

Every month workshops are held in Bello Oriente, sometimes at the schools, sometimes at the community centre (Fig. 4).

In these workshops the Colombian partners educate the residents about preventive measures and preparation in advance of a landslide event (see Sections "Prevention (Before the Event) and Preparation (Before the Event)") using presentations, charts, models and games and by creating flow charts with the people. Most importantly, there are community walks at the end of some workshops, where the residents apply their knowledge of their environment and insert their observations into a map of the barrio (Fig. 5).

Once the geo-sensor network for the landslide detection is completed, the workshops will also include talks about possible evacuation routes, safe houses and how to react to an alarm depending on the area one is in at that time.

Fig. 4 Community workshop in Bello Oriente

Fig. 5 Inserting new observations into the map during a community walk

There are already some residents very eager to work in the program. It is possible, that those will be educated on first responding in case of an emergency.

Assistance in Field Work

During the first field campaign in August 2019, there were several voluntaries from the local community helping with the work, together with voluntaries from the Geological Society of Colombia.

The mapping in the rural part of the study area must be done with at least two of the residents, otherwise it is impossible to find a way through the dense forest and high grass in this area. The people know their own home very well and are eager to show it to the scientists working with them (Fig. 6).

Also, geoelectric measurements were conducted in August 2019 with the help of the residents (Fig. 7).

Fig. 6 Residents of the barrio helping during the field campaign in August 2019

Fig. 7 Geoelectric measurements in August 2019

This cooperation built a strong bond of trust between the residents and the scientists. Every day after the work was done, jam sessions or talks in a local bar took place, which increased the mutual trust even more.

Construction and Installation of Sensors

Starting in May 2020 the sensor network will be installed in the barrio. To increase the residents' risk awareness and their knowledge of the sensors, the people will help building the sensors and the transmission devices. They will also assist installing them in the field in order to stay updated on the locations of the sensors.

Maintenance of Sensor Network

It is planned that the sensor network will be maintained by SIATA and the community. The residents will be trained in basic maintenance activities such as change batteries in the transmission devices and make simple measurements that can be transmitted to SIATA via app.

The aspired outcome of this plan is that these expressions of confidence in the residents increase their sense of responsibility regarding the technology in the barrio and therefore decrease the probability of vandalism (see Section "Preparation (Before the Event)").

Experiences so Far and Lookout

In the project Inform@Risk we try to cover the "last mile" of an early warning system. An early warning system is only as effective as this last mile, no matter how well the technical monitoring system is designed.

The cooperation between the residents and the authorities/scientists is working well. Attendance in the workshops is regular, but more participants would be better, since everybody needs to be informed.

To integrate the sensor system into the environment, the Institute for Landscape Architecture and Urban Development of the Leibniz University Hanover (LUH), leading partner of Inform@Risk, is already working on different designs and is planning workshops with the community to get them involved in the design and building process. The planning for evacuation routes and safe houses and places is carried out until summer 2020 by the Colombian partners, the community and LUH.

Another partner in the project, the Technical University of Deggendorf (THD), is developing an app for the residents, that includes warning in case of increased hazard, a hazard map of the barrio, weather forecast and also an application to report observations directly to the scientists and authorities.

In February/March 2021 the sensor system will be installed in the barrio, in February 2022 it will be fully functional, and the maintenance and supervision will be transferred to SIATA.

References

Anderson-Berry L, Achilles T, Panchuk S, Mackie B, Canterford S, Leck A, Bird DK (2018) Sending a message: how significant events have influenced the warnings landscape in Australia. Int J Disaster Risk Reducti 30:5–17. https://doi.org/10.1016/j.ijdrr.2018.03.005 (29.01.2020)

Gamperl M, Breuninger Menschik BT, Singer J, García-Londoño C, Thuro K (in prep.) (15–19 June, 2020) Development of a landslide early warning system in informal settlements in Medellín, Colombia. In: Proceedings of the 13th international symposium on landslides. Cartagena, Bolívar, Colombia

Garcia C, Fearnley C (2012) Evaluating Critical Links in Early Warning Systems for Natural Hazards. Environ Hazards 11(2):123–137

ISDR (2005) Hyogo Framework for Action 2005–2015: Building the Resilience of Nations and Communities to Disasters. Proceedings of the World Conference on Disaster Reduction, 18–22 January 2005. Kobe, Hyogo, Japan. pp. 1–23.

Jayasingha P (2016) Social Geology and Landslide Disaster Risk Reduction in Sri Lanka. Journal of Tropical Forestry and Environment 6(2):1–13

Mata-Perello JM, Mata-Lleonart R, Vintro-Sanchez C, Restrepo-Martinez C (2005) Social Geology. a New Perspective in Geology. Dyna. 79(171):159–166

Nadim F, Intrieri E (15–17 May, 2011) Early warning system for landslides: challenges and new monitoring technologies. In: Proceedings of the 5th Canadian conference on geotechniques and natural hazard. pp 1–15. Kelowna, BC, Canada

Planat (2020) The Cycle of Integral Risk Management. URL: https://www.planat.ch/en/specialists/risk-management/what-has-to-be-done/ [22.01.2020]

Urbam EAFIT, Leibniz University Hannover, Municipality of Medellín (2013) Rehabitar la Montaña: estrategias y procesos para un hábitat sostenible en las laderas de Medellín.

Capacity Building and Community Preparedness Towards Landslide Disaster in Pagerharjo Village, Kulon Progo Regency of Yogyakarta, Indonesia

Hendy Setiawan, Endah Retnaningrum, Thema Arrisaldi, and Wahyu Wilopo

Abstract

In developing countries, promoting disaster knowledge for disaster mitigation and disaster risk reduction is at best when initiated from a local scale. The bottom-up approach in the landslide disaster management program in Indonesia could catch common awareness of the community for their living neighbourhood. Despite diverse characteristics of the communities, community-based mitigation may effective and often achieving capacity building and community preparedness towards disasters. Landslide disaster mitigation activities have been conducted in Pagerharjo Village, Samigaluh District, Kulon Progo Regency of Yogyakarta Province, Indonesia from April to October 2019. The activities are strongly involved in community participation in risk assessment, disseminating knowledge, establishing a task force team for disaster preparedness and emergency response, monitoring potential landslide disasters, as well as creating evacuation maps and standard operating procedures for evacuation. The main goals of these mitigation activities are well achieved: (1) locations of potential landslides in the village are identified; (2) capacity and preparedness of the community in Pagerharjo Village towards landslide disaster are successfully built through creating a simple risk and evacuation route map; (3) community participation in monitoring of slope mass movement is increased; (4) disaster mitigation knowledge to the community is well promoted, and (5) awareness of the importance of reforestation and slope protection of the area in Pagerharjo Village has become highly concerned.

Keywords

Landslide • Community participation • Preparedness and response team • Evacuation route map • Yogyakarta

Introduction

Disaster management programs on a local scale commonly deal with public participation, capacity building, cooperation among local stakeholders and community preparedness. In particular, delivering disaster knowledge with effective mitigation strategies and guidelines are of challenging tasks when dealing with capacity building and community preparedness in disaster vulnerable areas (Na and Okada 2011; Chen et al. 2006). In addition, good communication and common perception are necessary for all parties involved in disaster management programs (Alcántara-Ayala and Moreno 2016).

Landslide is one of the disasters that frequently occurred in mountainous or hilly areas, resulting in fatalities and affected the community and their local neighbourhood. Therefore, building disaster resilience and community preparedness has become a necessity for local people who live in landslide disaster-prone areas. Such capacity building may effective and sustainable if local people and their community are involved in the assessment and mitigation strategies; they fully understand and aware about the risk of landslide disaster; can identify potential landslides on their living environment; and willing to use and maintain the appropriate early warning instrument, monitoring and tools

H. Setiawan (✉) · T. Arrisaldi · W. Wilopo
Faculty of Engineering, Department of Geological Engineering, Universitas Gadjah Mada, Jl. Grafika No. 2, Kampus UGM, Yogyakarta, 55281, Indonesia
e-mail: hendy.setiawan@ugm.ac.id

T. Arrisaldi
e-mail: thema.a@mail.ugm.ac.id

W. Wilopo
e-mail: wilopo_w@ugm.ac.id

E. Retnaningrum
Faculty of Biology, Universitas Gadjah Mada, Jl. Teknika Selatan, Sekip Utara, Yogyakarta, 55281, Indonesia
e-mail: endahr@ugm.ac.id

K. Sassa et al. (eds.), *Understanding and Reducing Landslide Disaster Risk*, ICL Contribution to Landslide Disaster Risk Reduction, https://doi.org/10.1007/978-3-030-60196-6_54

to reduce landslide disaster risk (Hernández-Moreno and Alcántara-Ayala 2017; Fathani et al. 2014, 2016; Karnawati et al. 2013).

This paper reports a mitigation activity through capacity building and community preparedness programs at the local level towards landslide disaster in Kulon Progo Regency, Yogyakarta Province of Indonesia. The activity refers to the seven sub-systems for landslide early warning proposed by Fathani et al. (2016) that emphasizes active participation from the community, i.e. (1) risk assessment, (2) dissemination and communication with the community, (3) establishment of disaster preparedness and response team, (4) community-based evacuation map, (5) standard operating procedures for evacuation, (6) landslide monitoring with an early warning system and evacuation drill, and (7) building a commitment of local authority and community. Implementation of those sub-systems is complemented with reforestation and protection on the slopes and upstream area as non-structural mitigation support.

Case Study

Pagerharjo village is located in Samigaluh District, Kulon Progo Regency the western border of Yogyakarta Province in Java Island. Pagerharjo village consists of 20 sub-villages (hamlets) with a total area of 1,140.52 ha and inhabited around 4,704 people. Situated on the average altitude of 700 m a.s.l, the economy of Pagerharjo village relies on the sector of agriculture, coffee and tea plantation, and tourism. The annual rainfall rate reaches between 2500 and 3000 mm. According to the Disaster Management Authority (BPBD) of Kulon Progo Regency, about 25 landslide events took place along the year 2016–2017.

Rahardjo and Sukandarrumidi (1995) pointed out that Pagerharjo village regionally located within Menoreh Mountain where tertiary volcanic rocks exist. Specifically, a geological condition in Pagerharjo village consists of sandstone and sandy marl (Nanggulan Formation—the oldest formation in Kulon Progo), andesitic breccia and tuff (Kebobutak Formation), limestone, marl and sandstone (Sentolo Formation), and alluvial sediment along the river (Rahardjo et al. 1995). The topography of Pagerharjo village ranges between steep to very steep with intense deformation, discontinuity and highly weathering in rock-soil layers (Karnawati 2005; Novianto et al. 1997). Moreover, landslide vulnerability map produced by Centre of Vulcanology and Geological Hazard Mitigation of Indonesia (PVMBG 2018) shows that about 92.8% area of Pagerharjo Village is located in the zone of moderate and high vulnerability to have landslides (Fig. 1).

Methodology

Preparation for landslide mitigation activity through capacity building and community preparedness in Pagerharjo Village is conducted by having coordination with villagers and local authorities, in this case, are the government of Pagerharjo Village and the Disaster Management Authority (BPBD) of Kulon Progo Regency. Reconnaissance and surveys are carried out to investigate geological conditions, social, economic and cultural conditions in Pagerharjo Village. Intense communication and build a common perception with people and local government is necessary to schedule a series of landslide mitigation activities.

The seven sub-systems for landslide early warning (Fathani et al. 2016) to build community awareness and preparedness in Pagerharjo Village then performed sequentially with help from the facilitator as follows:

1. Conducting a *risk assessment of landslides* with a representative of the villagers that know in detail the area of Pagerharjo Village.
2. *Disseminating landslide disaster knowledge* and gain common risk perception together with the community in Pagerharjo Village.
3. *Establishment of disaster preparedness and response team* with members from villagers in Pagerharjo. This team then have the training to be capable of handling emergency situation and response when landslide disaster occurs.
4. *The evacuation route map* made by villagers of Pagerharjo. This map is created by combining a community understanding of their living environment with potential landslide disasters that may occur based on the risk assessment. Evacuation route map includes risk zonation of possible landslides, the possible affected area, and information about a simple evacuation route that will be obeyed by all villagers.
5. Create *standard operating procedures for evacuation* if landslides really occurs. This procedure should be made by a disaster response team with all villagers, to reduce panic and to avoid the wrong action when disasters take place.
6. *Monitoring landslide with an early warning system and evacuation drill* should be carried out by all villagers, disaster response team and local authorities. The instrument of landslide early warning system can be prepared with simple, functional with appropriate technology and installed properly. While evacuation drill is necessary to bring a real situation of landslides so that all villagers with response team know how to do and manage when a disaster occurs.

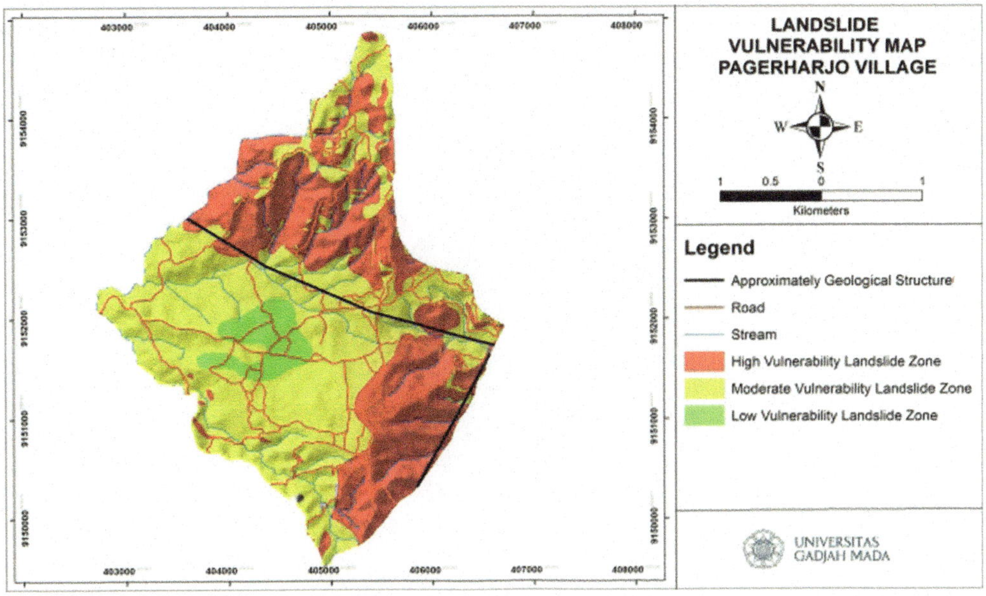

Fig. 1 Landslide vulnerability map in Pagerharjo village (modified after PVMBG 2018)

7. *Build a commitment between local authorities, disaster response team and all villagers* in Pagerharjo Village. At this point, it is crucial to maintain awareness and preparedness for all related parties. Giving repetitive disaster risk knowledge and continuous training to the community is important as well as checking all potentially unstable areas in the village.

Following the above activities, reforestation and slope protection are then conducted at the campground near the top of the hills of Pagerharjo Village. All activities above are conducted from April to October 2019 and implemented through the student community service program in Pagerharjo Village.

Results

Slope Conditions

The slope condition in Pagerharjo Village is investigated during the risk assessment stage. Figure 2 shows that the area of Pagerharjo Village has a gentle slope and valley located at the centre and southwestern side of the hills. The moderate slope to the steep slope area is found on the northern side of the village. Two Sub-Villages, Sarigono and Plono, are located in the steep slopes area (Fig. 3).

Geological Conditions

The geological map is created with a detailed scale of 1/25,000 based on the site investigation and risk assessment at Pagerharjo Village. Mapping results show that there are three lithological units within the area, sequentially from old to young layers which are dominantly andesitic breccia, andesite intrusion on the western side and limestone in the centre (Fig. 4). Steep slopes area consists of highly-weathered andesitic breccia, with soils cover the ground up to 10 m thickness. Such massive and thick weathered soils can be a source of landslides. During the assessment, landslides type of slide, creep and falls are commonly found in local spots.

Social and Economic Condition

According to data from the Central Bureau of Statistics in 2018, population density in Pagerharjo reaches 412 people per km^2. Agriculture area covers about 118.68 Ha, building and settlements approximately 329 Ha, dry land 550.74 Ha and forest about 101.85 Ha. Public facilities such as the village office, schools, mosque, church, and public health centre are located at the gentle slope area, low vulnerability landslide zone. Most of the settlements are located at the moderate to steep slopes, which are highly vulnerable to landslides.

Dissemination of Landslide Disaster Knowledge

Socialization and dissemination about landslide disaster knowledge are conducted in Pagerharjo Village Office. Participants from the community and sub-villages representatives in Pagerharjo are joined in discussion with a facilitator from our team. This socialization and

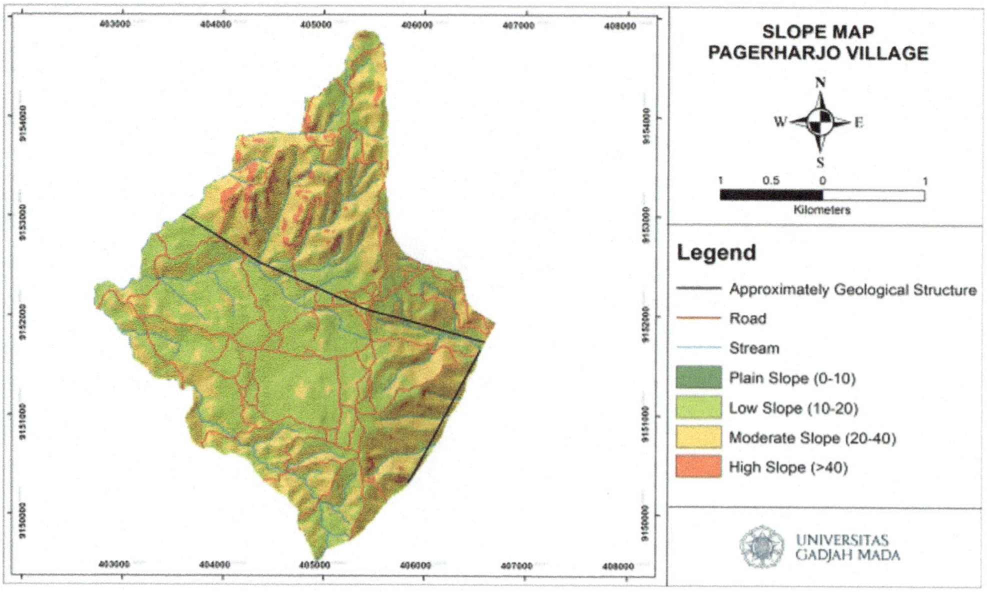

Fig. 2 Slope map of Pagerharjo village

Fig. 3 Slope conditions at Sarigono and Plono sub-villages in Pagerharjo

dissemination session consist of explanation related to the triggering factors and controlling factors of landslides disaster. Results from risk assessment and site investigation of Pagerharjo Village are also reported (Fig. 5).

Response from the community is delivered that most of the landslides and mass movement problems in their village occur during the rainy season, month October to April, with type predominantly creep and slide. These results are then discussed to create evacuation route map and to establish a disaster response team.

Disaster Preparedness and Response Team

Establishment of the disaster preparedness and response team by the community is essential to reduce the disaster risk at the local level. The members of this team are appointed based on the consultation which facilitated by the local government of Pagerharjo Village. All members are trained by a facilitator to understand their tasks related to preparedness, mitigation, handling emergency response and landslide disaster management. In addition, this team should

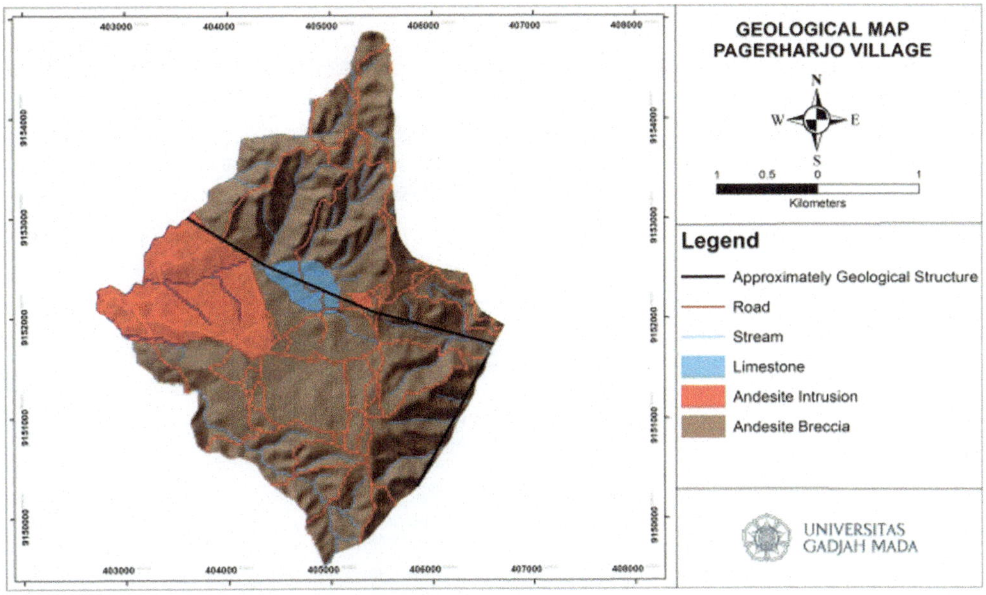

Fig. 4 Geologic map of Pagerharjo village

Fig. 5 Socialization and dissemination of landslide disaster knowledge to the community of Pagerharjo village by facilitator (**a** and **b**) and student (**c** and **d**)

create standard operating procedures for evacuation if landslides really occur. The disaster preparedness and response team of Pagerharjo Village consists of a chairperson, secretary, a logistic division, a communication division, a rapid response team, and shelter division (Fig. 6). This team is supervised by BPBD of Kulon Progo Regency and the local government of Pagerharjo Village.

Evacuation Route Map

A community-based evacuation route map is created by active participation from all villagers. Representatives from all sub-villages in Pagerharjo are trained to create the evacuation route map with a simple approach (Fig. 7). The houses, access roads, public facilities and settlements are

Fig. 6 Disaster preparedness and response team in Pagerharjo village, Samigaluh district, Kulon Progo regency

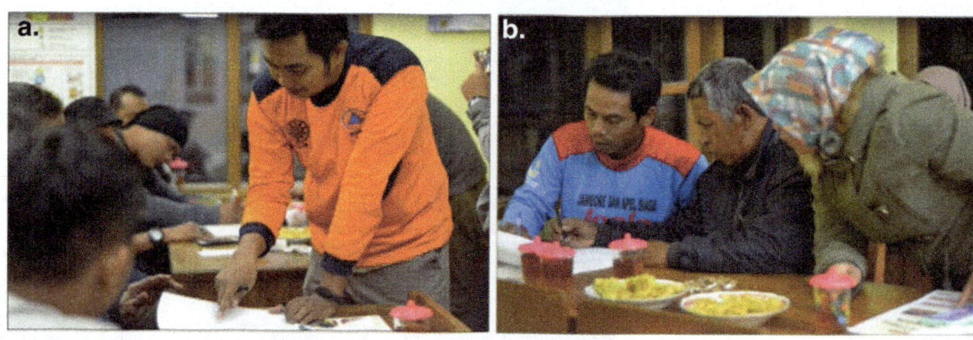

Fig. 7 Community-based evacuation route map facilitating by researcher team with active participation from villagers

sketched and combined with the distribution locations of potential landslides and unstable slopes in the area of Pagerharjo Village from the identification and landslide risk assessment results. A simple community-based evacuation route map is made by connecting the vulnerable areas, roads and access of each settlement in their neighbourhood with appointed evacuation areas.

The process of making an evacuation route map is helped by a facilitator and student. If there are any changes based on the condition on-site, additional public facilities, or new obstacles along the route, then an evacuation route map is allowable to update by the disaster response team and people in Pagerharjo Village. This kind of map may not so technical, but simple, visible, clear and functional so that all communities in Pagerharjo Village are understood easily and can follow the evacuation route that they made by themselves.

Figure 8 shows the example of an evacuation route map of Sarigono Sub-Village. The map in Fig. 8 implies that there are about 55 families in Sarigono Sub-Village which have a high risk of landslide disaster. Sarigono Sub-Village is categorized in a moderate and high vulnerable of landslide disaster. The evacuation route is directed to the meeting point and safer area at neighbouring Sub-Villages of Separang on the south-western side and Jobolawang on the south-eastern side.

Evacuation Drill and Commitment of the Local Government

Evacuation drill is necessary to implement in Pagerharjo Village in order to test the effectiveness of disaster preparedness team, a standard operating procedure for

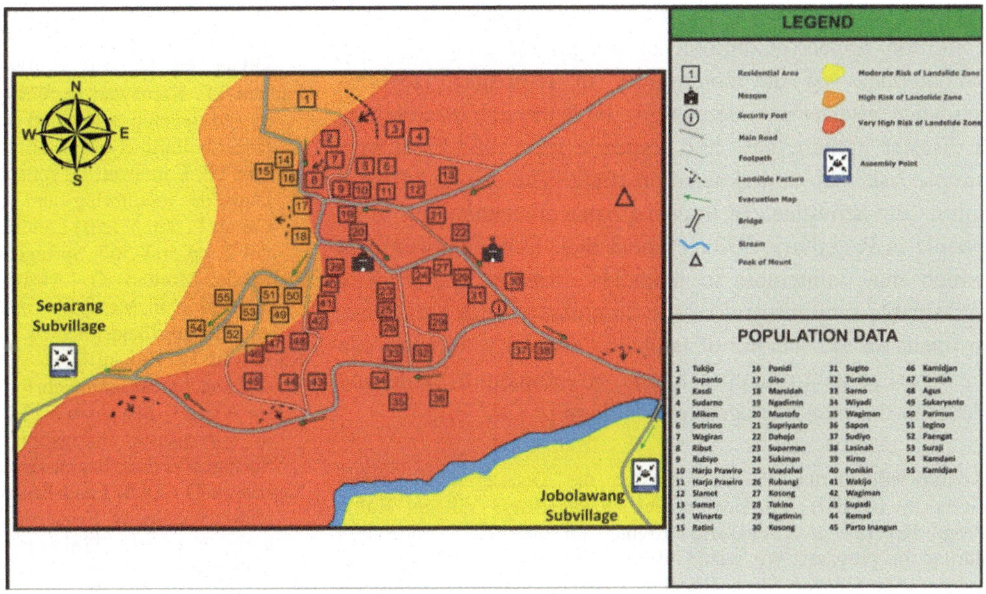

Fig. 8 Evacuation route map of Sarigono sub-village in Pagerharjo village for landslide disaster mitigation

Fig. 9 Trees planting campaigns in Pagerharjo village

evacuation as well as an evacuation route map. The drills also important for all villagers that they fully understand the landslide risk and gain their awareness and how to take action if landslide events really occur. In this case, the government of Pagerharjo Village together with BPBD, Red Cross and Health Centre of Kulon Progo Regency are encouraged to put the strong commitment in maintaining and sustaining this disaster mitigation framework.

At the final stage of this landslide disaster mitigation, reforestation and slope protection are conducted at the campground near the top of the hills through trees planting campaigns, involving students and the community of Pagerharjo Village (Fig. 9).

These non-structural mitigation supports are necessary to remind and encourage all communities that maintaining the function of the upstream area of Pagerharjo Village is important to reduce the possibility of landslide disasters in the area.

Discussion and Conclusion

Disaster management that deals with community participation in planning, decision making and managing the action during preparedness and in an emergency situation when a disaster occurs is a key point to enhance the resilience capacity of people and sustainability of disaster mitigation (Pearce 2003). The implementation of landslide disaster mitigation with such community participation is successfully conducted in Pagerharjo Village, Yogyakarta Province.

Potential landslides in Pagerharjo Village are identified in Sarigono and Plono, two sub-villages that located on the northern side where steep slopes with the weak lithological condition appear. Common awareness, self-capacity and preparedness of the community in Pagerharjo towards landslide disaster are successfully built through establishing a disaster preparedness and response team, and creating a simple risk map and evacuation route. The results of these mitigation activities can improve better understanding to all people in Pagerharjo Village about their living environment which may vulnerable to landslide disaster. Therefore, a sustainable commitment of the community with local government on the scheme of landslide disaster mitigation can be achieved as a reference for the government to build and develop the area of Pagerharjo Village.

Acknowledgements We acknowledge the Disaster Management Authority (BPBD) and Government of Pagerharjo Village, Kulon Progo Regency of Yogyakarta Province for their kind cooperation during the program. We would like to thank our students for their works with villagers in conducting student community service at Pagerharjo Village related to this research. This study was part of the University Program for Community Service based on the Application of Research Output and Appropriate Technologies fiscal year 2019, funded by Universitas Gadjah Mada (BPPTN BH UGM).

References

Alcántara-Ayala I, Moreno AR (2016) Landslide risk perception and communication for disaster risk management in mountain areas of developing countries: a Mexican foretaste. J Mt Sci 13(12):2079–2093

Chen LC, Liu YC, Chan KC (2006) Integrated community-based disaster management program in Taiwan: a case study of Shang-An village. Nat Hazards 37:209–223

Fathani TF, Karnawati D, Wilopo W (2016) An integrated methodology to develop a standard for landslide early warning system. Nat Hazards Earth Syst Sci 16:2123–2135

Fathani TF, Karnawati D, Wilopo W (2014) An adaptive and sustained landslide monitoring and early warning system. In: Sassa K, Canuti P, Yin Y (eds) Landslide science for a safer geoenvironment, vol 2, pp 563–567. Springer International Publishing

Hernández-Moreno G, Alcántara-Ayala I (2017) Landslide risk perception in Mexico: a research gate into public awareness and knowledge. Landslides 14:351–371

Karnawati D, Fathani T F, Wilopo W, Andayani B (2013) Hybrid socio-technical approach for landslide risk reduction in Indonesia. In: Wang F, Miyajima M, Li T, Shan W, Fathani T F (eds) Progress of geo-disaster mitigation technology in Asia, pp 157–169. Springer-Verlag, Berlin Heidelberg (e-ISBN 978-3-642-29107-4)

Karnawati D (2005) Landslide disaster in Indonesia and its mitigation efforts Faculty of Engineering, Universitas Gadjah Mada, Yogyakarta, Indonesia (Department of Geological Engineering) (ISBN 9799581133)

Na J, Okada N (2011) Implementation of the Yonmenkaigi system method for capacity building on disaster risk management in local community of Merapi volcano. Ann Disas Prev Res Inst Kyoto Univ 54B:157–164

Novianto MW, Djaja A, Hermawan W (1997) Engineering geological map of Yogyakarta, scale 1:100,000. Directorate of Geology and Environmental Management, Bandung

Pearce L (2003) Disaster management and community planning, and public participation: how to achieve sustainable hazard mitigation. Nat Hazard 28:211–228

Rahardjo W, Sukandarrumidi RHMD (1995) Geologic map of Yogyakarta, Java. Geology Research Center and Development, Bandung

Protection of a Cultural Heritage Site in Croatia from Rockfall Occurrences

2020 Kyoto Japan

Josip Peranić, Martina Vivoda Prodan, Marin Sečanj, Sanja Bernat Gazibara, Snježana Mihalić Arbanas, and Željko Arbanas

Abstract

The small historical town in the middle part of Dalmatia, City of Omiš, Croatia, was threatened by numerous rockfall occurrences in the past that caused significant damage at residential structures and infrastructure. After several stages and design solutions completed in the past, installation of rockfall protection structures at the slopes above the City of Omiš was conducted from 2016 to 2018. Installed stabilization and protection structures represent the first stage of mitigation measures, while the rockfall hazard from the upper parts of the slope should be mitigated in the future. In this paper, we analyzed the efficiency of installed rockfall barriers in case if rockfall initiations would occur from the upper positions of the slope. The results of conducted parametric rockfall simulations indicate on the probability that some of the rock blocks existing in possible rockfall sources in the upper part of the slope will not be retained by installed rockfall protection barriers and can reach buildings in the old part of the City of Omiš. Conducted analyses suggest that the existing protection measures will not ensure the safety of residents and buildings in the City of Omiš from all possible future rockfall events.

Keywords

Rockfall • Point cloud • Modelling • Rockfall protection • Rockfall barriers

Introduction

Rockfalls are the most frequent and dangerous rock movements affecting mountainous regions, coastal areas, man-made road cuts and excavations, generating high economic and social damage (Emmer 2018). They are extremely dangerous and life-threatening if occurring in populated places, along roads and railways, mainly due to the high-velocity nature of phenomena that make any fast response very difficult (Dorren 2003). There are numerous examples of rockfalls threatening historical cultural heritage sites (e.g. Fanti et al. 2012; Saroglou et al. 2012; Mineo and Pappalardo 2019). Thus, rockfall protection is a major issue in areas exposed to severe rockfall hazard and it includes quantitative rockfall hazard and risk assessment, as well as the design of structural countermeasures (Crosta et al. 2015). Modern approaches imply application of remote sensing techniques and acquiring of point clouds and digital terrain models (DTM); engineering geological mapping of the rock slopes using a combination of remote-sensing techniques and field mapping; rockfall hazard and risk analysis; spatial analysis of rockfall initiating, propagation and run out (Li and Lan 2015), as well as 3D rockfall simulation to identify trajectories, run out and kinetic energy of detached rock block as an input data in rockfall protection barriers designing (Volkwein et al. 2011; Sarro et al. 2018).

In this study, we utilize some of the modern approaches and recent techniques in rockfall hazard analysis and rockfall

J. Peranić · M. Vivoda Prodan · Ž. Arbanas (✉)
Faculty of Civil Engineering, University of Rijeka, Radmile Matejčić 3, Rijeka, 51000, Croatia
e-mail: zeljko.arbanas@gradri.uniri.hr

J. Peranić
e-mail: josip.peranic@uniri.hr

M. Vivoda Prodan
e-mail: martina.vivoda@gradri.uniri.hr

M. Sečanj · S. Bernat Gazibara · S. Mihalić Arbanas
Faculty of Mining, Geology and Petroleum Engineering, University of Zagreb, Pierottijeva 6, Zagreb, 10000, Croatia
e-mail: marin.secanj@rgn.hr

S. Bernat Gazibara
e-mail: sanja.bernat@rgn.hr

S. Mihalić Arbanas
e-mail: snjezana.mihalic@rgn.hr

K. Sassa et al. (eds.), *Understanding and Reducing Landslide Disaster Risk*, ICL Contribution to Landslide Disaster Risk Reduction,
https://doi.org/10.1007/978-3-030-60196-6_55

structural protection design to analyse the performance of a part of the recently installed rockfall mitigation measures at the slopes above the City of Omiš (Arbanas et al. 2019a, b). Firstly, we utilize a high-resolution three-dimensional point cloud (3DPC) of the site surface to analyse a particularly demanding slope section characterized by very steep inclination (over 70°) and possibly unstable rock blocks threatening people and residential buildings present in the toe of the slope. The critical cross-section, which includes overlapping of the 1000 and 2000 kJ capacity rockfall protection barriers installed in 2017 as a part of the protection measures, was determined and analysed.

Study Area

The City of Omiš, Croatia (Fig. 1) is a small historical town situated in the middle part of the Adriatic coast, approximately 25 km SE of Split. Once well-known by the "Corsairs of Almissa", situated at the mouth of the Cetina River and just beneath the steep slope of Omiška Dinara Mountain, this small city, rich with historical monuments, attracts many tourists and visitors every year. However, the city has also experienced numerous rockfalls in the past that have caused significant damage at residential structures and infrastructure, luckily with no human casualties. The rockfall events along the limestone slopes were mostly caused by unfavourable rock mass characteristics, weathering process in combination with heavy rainfalls, and the man-made influences (Arbanas et al. 2012). Although no rockfall inventory or statistical data about the rockfall volumes exist, the documented data indicate that most of the rockfalls had volumes ranging from 1.0 to 5.0 m^3 and could be classified as block falls according to Rochet (1987).

Fig. 1 The City of Omiš and limestone slopes above it

The Omiška Dinara Mountain spreads over 15 km along the Adriatic coast with the highest peak at 865 m a.s.l. It is a part of a large nappe system and it is represented as an overturned anticline striking NW–SE that is the result of compressional tectonics occurred from Cretaceous to Miocene. The core of the anticline is built of Senonian rudist limestone, while the limbs of the anticline are built of Eocene breccia, limestone and flysch (Marinčić et al. 1977).

In the wider area of the City of Omiš, geological contacts between Cretaceous and Paleogene deposits are usually along steep reverse faults, striking E–W, with the tectonic transport top to south. The complexity of the geological-structural setting, caused by faulting and folding led to the formation of numerous discontinuities in the rock mass. Progressive weathering of discontinuities led to the formation of unstable rock blocks of unfavourable orientation that are prone to falling (Sečanj et al. 2017, 2019).

Design of Rockfall Protection Measures

The initial design activities of rockfall protection measures started in 2008, while the main design was finished in 2012. Unfortunately, the main design was based on very poor preliminary data, without preceding of rockfall hazard and risk analysis to identify rockfall potential from the slopes above the City of Omiš as well as the data required for rockfall run out analysis and design of protection measures, including the information about potential rockfall sources, potential rockfall volumes, probable rockfall trajectories and, possible run-out areas (Arbanas et al. 2019b). According to the main design, prevention or protection measures were defined for 22 potentially dangerous locations in total, including rock bolts, rock anchors with steel ropes, steel wire fences and meshes, and the rockfall barriers (Arbanas et al. 2019a).

Despite the limitations provided by the main design, the final design of rockfall prevention and protection measures, which was carried out during the first stage of construction works, followed the modern approaches and recent techniques in rockfall hazard analysis and rockfall structural protection (e.g. Sarro et al. 2018; Volkwein et al. 2011). Collected data have enabled conducting of spatial kinematic analysis (Sečanj et al. 2017) which indicated zones of possible rockfall sources that, along with the determination of rock block volumes, provided quantitative and accurate input data for rockfall simulations (Arbanas et al. 2019a). 2D and 3D rockfall simulations were performed to verify the set-up of rockfall barriers according to the main design, in terms of their location, geometry, and capacity. The obtained results indicated the need for additional rockfall protection barriers at certain positions where the rockfall trajectories tended to leap over barriers (Arbanas et al. 2019a). High-resolution DTM and 3DPC enabled fast positioning of protection measures elements as well as verification of installation conditions once the works were completed, which was found to be another important advantage of adopted methodology in a case of steep and inaccessible terrain. Some details of installed protection measures are shown in Fig. 2.

Methodology

A high-resolution 3DPC of the site surface was provided by combining the terrestrial laser scanning (TLS) in the area close to buildings, and *structure from motion* (SfM) technique from the unmanned aerial vehicle (UAV). A unique georeferenced point cloud (more details in Sečanj et al. 2017) with average point resolution of 2 cm, (i.e., 125.000 points per square meter) was used with CloudCompare software to analyse installation conditions of rockfall protection measures at different locations of the slope. The method was found to be particularly useful to precisely define the exact location and geometrical conditions of installed rockfall barrier (a) elements. A critical section for the analysed part of the slope (Fig. 3) was selected based on the results of kinematic analysis (Sečanj et al. 2017) and due to the presence of the residential buildings in the vicinity of the toe of the slope that is particularly steep in the analysed part (around 70°), with the possible rockfall sources at heights above 120 m a.s.l.

Firstly, the simulation was performed to analyse the efficiency of two rockfall barriers (Fig. 2a), installed in 2017 following the set-up from the main design and modifications from the final design, in terms of the geometrical conditions (precisely determined from the 3DPC and attenuation capacity), the simulations were performed for a total number of 300 rock blocks of different smooth and polygonal shapes, and 0.1, 1, and 2.5-ton masses to consider different block volume sizes (i.e. from 0.04 up to roughly 1 m^3). According to the main design considerations, only rockfall sources bellow 80 m a.s.l. were considered here (RocFall 2019).

The second type of analyses was performed to determine the adequate selection of installed rockfall barriers (both in terms of possible overlapping of installed barriers and their capacity) in the case of rockfalls detached from the higher parts of the slope. Thus, the analyses were re-run but considering additional possible rockfall sources at elevations of 100, 115 and 120 m a.s.l.

Finally, as the calculated trajectories indicated that around 30% of rock blocks could leap over the installed barriers in a case of rockfall detachments occurring at higher parts of the slope, the additional analyses were performed to

Fig. 2 A view on **a** 1000 kJ (upper) and 2000 kJ (lower) rockfall barriers; **b** rock anchors, steel ropes and steel mesh as measures for stabilisation rock block; and **c** rockfall prevention measures at the location of Mirabella Fortress from the thirteenth century

Fig. 3 Point cloud of the analysed part of the slope with selected cross-section for 2D rockfall simulations, including analysis of trajectories, run-out distances and envelopes of kinetic energy for a total of 300 simulated rock blocks of various sizes and detachment heights

determine the appropriate location and set-up for additional barriers that would retain rockfalls initiated from those higher sources.

In all simulations, the initial (horizontal) velocity was set to 0.1 m/s. The coefficients of normal/tangential restitution of 0.35/0.85 for the upper part (elevation higher than 25 m a.s.l.), while 0.30/0.82 was adopted for the lower part of the slope due to the occasional presence of vegetation (adopted from *Rocscience Coefficient of Restitution Table* at www.rocscience.com). Figure 4 shows the selected cross-section that includes two installed barriers transferred from the 3DPC using the CloudCompare software, along with the trajectories calculated for rockfall sources up to 80 m a.s.l.

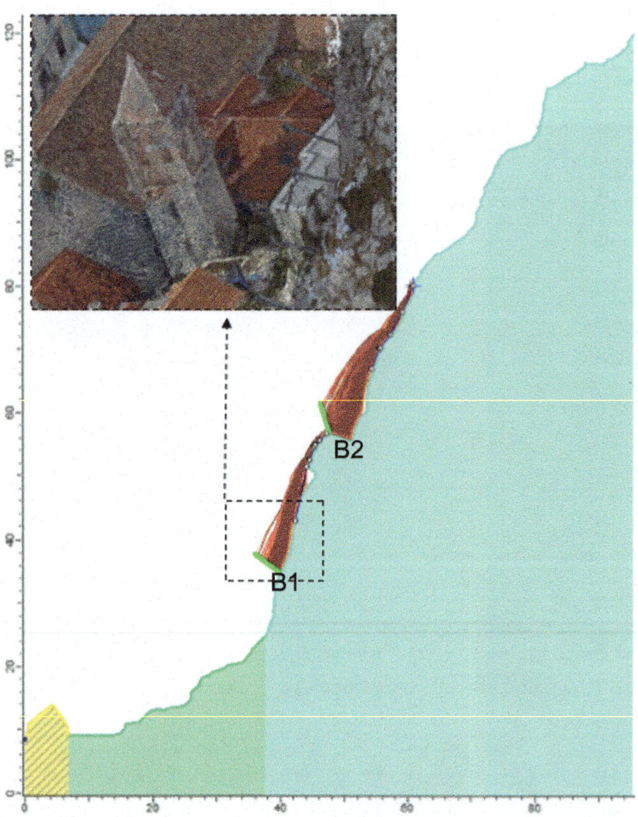

Fig. 4 Model built in Rocfall software with rockfall source at 80 m a.s.l. and 2000 kJ (lower, B1) and 1000 kJ (upper, B2) rockfall barriers transferred from 3DPC

Simulation Results

The run-out distances and total kinetic energy envelopes calculated for different rockfall source elevations (RocFall 2019) are summarized in Figs. 5 and 6, respectively. It seems that, disregarding the size and initial velocity of falling blocks (varied from 0.1 up to 1 m/s), the installed rockfall barriers stop all blocks detached from heights up to 80 m a.s.l. Also, the total kinetic energy envelopes indicate the maximum impact energy for the considered detachment heights is below 500 kJ for both barriers. Thus, the installed rockfall barriers seem to be adequately designed and installed in the case when a rockfall initiation occurs at the lover parts of the slopes, as it is considered in the main design. It is interesting that, if simulations are run without rockfall barriers and with rockfall sources varying from 55 and 80 m a.s.l., around 2/3 of blocks detached from the area above B2 barrier are stopped at the natural berm located at the elevation between 52 and 55 m a.s.l., where the upper barrier is installed.

However, the circumstances are changing if the rockfall sources above 80 m a.s.l. are introduced into the analysis: the rock blocks start to overleap both barriers, reaching the building located at the toe of the slope. For example, for the rockfall source at the elevation of 115 m a.s.l. and for in total 267 calculated rockfall trajectories, 104 rock blocks overleaped both installed barriers. Only 19 out of 291 blocks overleaped both barriers when rockfall detachment source was simulated at 100 m a.s.l. Due to the existing natural berm at the elevation of 115 m a.s.l. and much lower slope angle in the highest part of the analysed section (approximately 30°), calculated trajectories indicate that 243 out of simulated 300

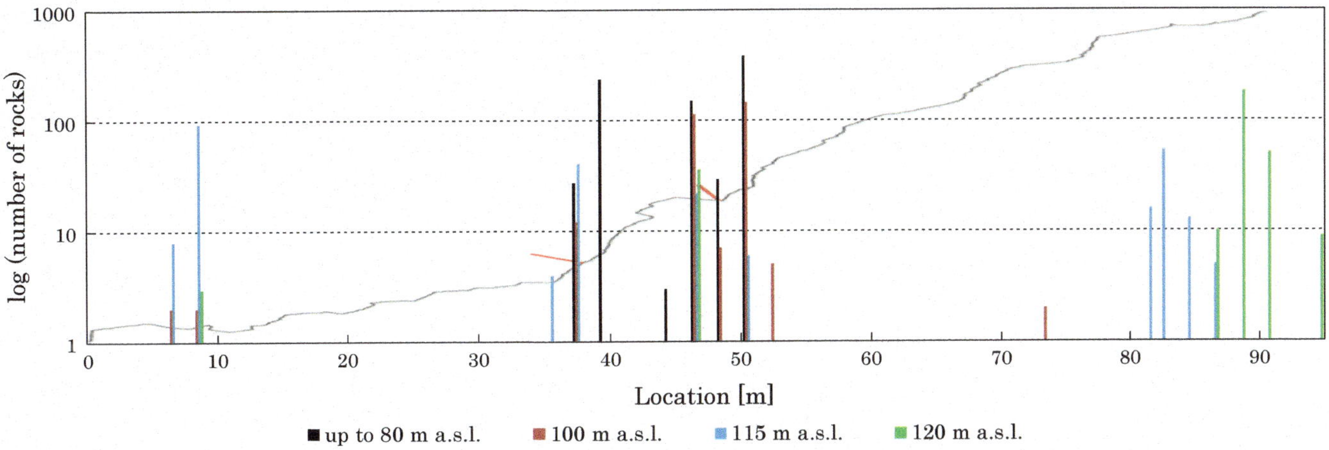

Fig. 5 Run-out distances calculated for rock blocks detached from various parts of the slope

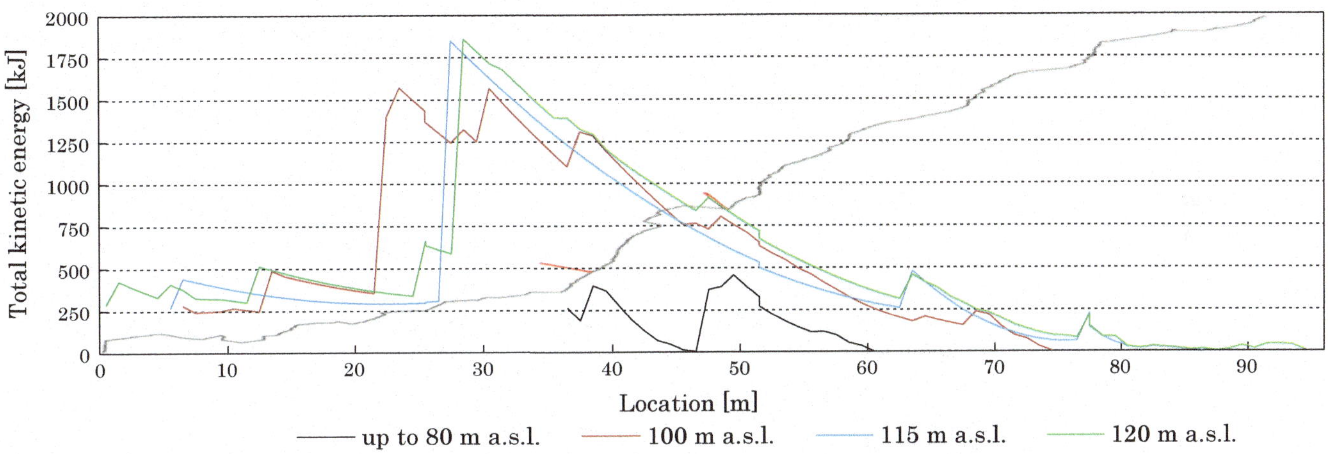

Fig. 6 Envelopes of the total kinetic energy on slope calculated for all block sizes and source areas

rock blocks would not reach altitudes lover than 115 m a.s.l., while only 10 blocks overleaped both of the barriers when rockfall source at 120 m a.s.l. is considered.

Data presented in Fig. 6 indicate that the total kinetic energy distribution along the slope varies only slightly with a change of the rockfall source area. For example, Fig. 7 represents the calculated distribution of the total kinetic energy on the lower (B1) and upper (B2) barriers for the case of rockfall detachment from 115 m a.s.l. The results indicate that the B1 barrier takes a total of 43 hits out of 267 calculated rockfall trajectories, with the highest impact energy of 1323 kJ. There are 2 out of 43 hits with total kinetic energy higher than 1300 kJ, while 79% of impacts have a total kinetic energy of 518 kJ. The B2 barrier takes a total of 23 hits out of 267 calculated rockfall trajectories, with the highest impact energy of 856 kJ. 70% of the hits on the B2 barrier are of total kinetic energy lower than 50 kJ.

Finally, the calculated trajectories point on the location on the analysed profile, located approximately 85 m a.s.l., that could be suitable for installation of an additional rockfall barrier which would stop potential rockfalls detached from the higher part of the slope. As shown in Fig. 8, 5 m high rockfall barrier, inclined at 80° from horizontal, was able to stop all of the 141 out of 265 simulated rock blocks, with detachment height varied between 90 and 120 m a.s.l. Maximum impact energy on this barrier is 484 kJ.

Conclusions

Construction of support systems and installation of rockfall protection barriers above the City of Omiš was completed in November 2018. The installed protection measures are based on design analyses that have analyzed rockfall initiation only from the lower parts of the slope. In this paper, we analyzed the efficiency of installed rockfall barriers in case if rockfall initiations would occur from the upper parts of the slope. Parametric analyses indicate on the probability that some of

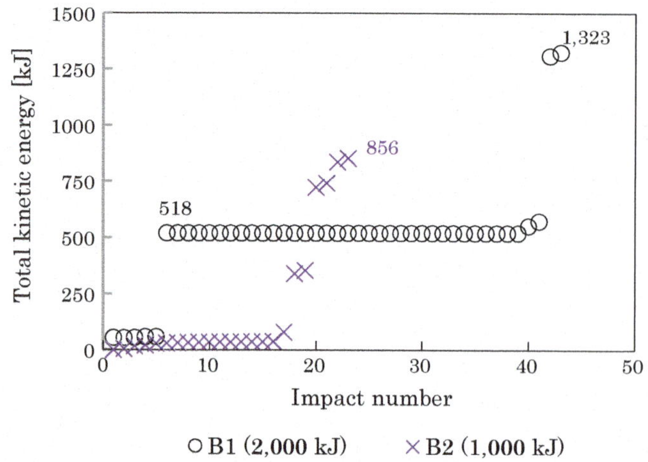

Fig. 7 Total kinetic energy on lower (B1 takes 43 hits) and upper (B2 takes 23 hits) barriers for simulated rockfall source at 115 m a.s.l

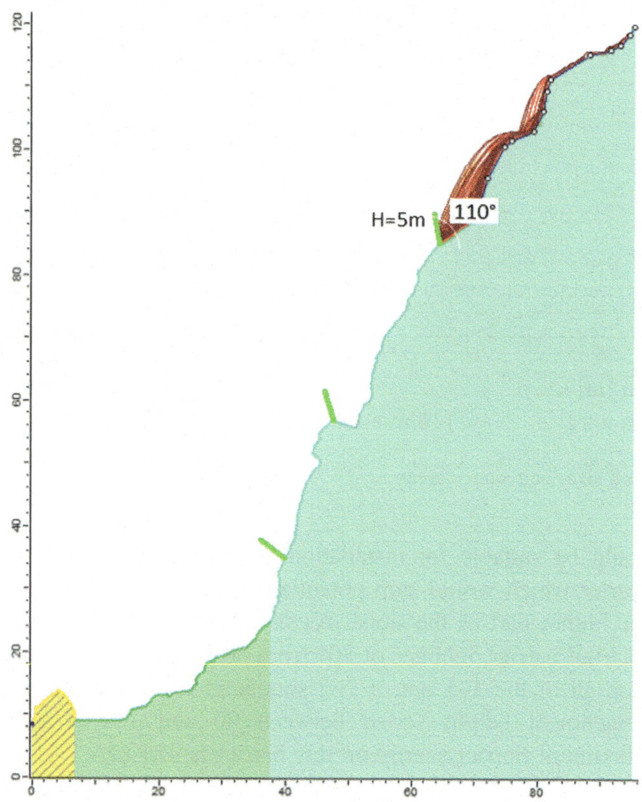

Fig. 8 Analysis of additional barrier installed at 85 m a.s.l. as mitigation measure from possible rockfalls detached from higher parts of the slope

the rock blocks to be released from rockfall sources in the upper part of the slope will not be retained by installed rockfall protection barriers and can reach buildings in the old part of The City of Omiš. Conducted analyses indicated that the existing protection measures will not ensure residents and buildings in the City of Omiš from all possible rockfall events that would occur in the future. To ensure better protection, it would be necessary to conduct a rockfall hazard and risk analysis that would identify rockfall potential from the slopes above the City of Omiš. The results of this study should provide necessary data for the next stage of the rockfall protection design that will reduce the rockfall risk to an acceptable level.

Acknowledgements The research presented in this paper was supported by the University of Rijeka under the Project UniRi-Tehnic-18-276 Research of Rockfall Processes and Rockfall Hazard Assessment. This support is gratefully acknowledged. The authors also want to thank all Projects' members and their help in researches conducting.

References

Arbanas Ž, Grošić M, Udovič D, Mihalić S (2012) Rockfall hazard analyses and rockfall protection along the Adriatic coast of Croatia. J Civil Eng Arch 6(3):344–355

Arbanas Ž, Vivoda Prodan M, Dugonjić Jovančević S, Peranić J, Udovič D, Bernat Gazibara S, Krkač M, Sečanj M, Mihalić Arbanas S (2019a) Rockfall modelling and rockfall protection at the slopes above the City of Omiš, Croatia. In: Proceedings of the ISRM Specialised Conference "Geotechnical challenges in karst", 11–13 April 2019. Omiš, Croatia, pp 121–126

Arbanas Ž, Sečanj M, Vivoda Prodan M, Dugonjić Jovančević S, Peranić J, Bernat Gazibara S, Krkač M, Udovič D, Mihalić Arbanas S (2019b) Protection of the City of Omiš, Croatia, from rockfall threats. In: Proceedings of the 4th Regional Symposium on Landslides in the Adriatic-Balkan Region, 23–25 Oct 2019. Sarajevo, BiH, pp 251–255

CloudCompare (v2.9.1) [GPL software]. 2017. http://www.cloudcompare.org/

Crosta GB, Agliardi F, Frattini P, Lari S (2015) Key issues in rock fall modeling, hazard and risk assessment for rockfall protection. In: Proceedings of the IAEG XII congress: engineering geology for society and territory, vol 2, 15–18 Sept 2014. Torino, Italy, pp 43–58

Dorren LKA (2003) A review of rockfall mechanics and modelling approaches. Progr Phys Geogr Earth Enviro 27(1):69–87

Emmer A (2018) Geographies and scientometrics of research on natural hazards. Geosciences 8(10):382

Fanti R, Gigli G, Lombardi L, Tapete D, Canuti P (2012) Terrestrial laser scanning for rockfall stability analysis in the cultural heritage site of Pitigliano (Italy). Landslides 10:409–420

Li L, Lan H (2015) Probabilistic modeling of rockfall trajectories: a review. Bull Eng Geol Env 74(4):1163–1176

Marinčić S, Korolija B, Mamužić B, Magaš N, Majcen Ž, Brkić M, Benček Đ (1977) Osnovna geološka karta SFRJ 1:100.000. Tumač za list Omiš. Savezni geol. zavod, Beograd, pp 21–35 (in Croatian)

Mineo S, Pappalardo G (2019) Sustainable Fruition of Cultural Heritage in Areas Affected by Rockfalls. Sustainability 12(1):296

Rochet L (1987) Application des modèles numériques de propagationàl'étude deséboulements rocheux. Bull Liaison Pont Chaussée 150(151):84–95 (in French)

RocFall (2019) www.rocscience.com/software/rocfall (accessed on 15 Dec 2019)

Saroglou H, Marinos V, Marinos P, Tsiambaos G (2012) Rockfall hazard and risk assessment: an example from a high promontory at the historical site of Monemvasia, Greece. Nat Hazards Earth Syst Sci 12(6):1823–1836

Sarro R, Riquelme A, García-Davalillo JC, Mateos RM, Tomás R, Pastor JL, Cano M, Herrera G (2018) Rockfall simulation based on UAV photogrammetry data obtained during an emergency declaration: application at a cultural heritage site. Remote Sens 10 (12):1923

Sečanj M, Mihalić Arbanas S, Kordić B, Krkač M, Bernat Gazibara S (2017) Identification of rock prone areas on the steep slopes above the Town of Omiš, Croatia. In: Proceedings of the world landslide forum 4: advancing culture of living with landslides, vol 5, 29 May–2 June 2017. Ljubljana, Slovenia, pp 481–488

Sečanj M, Mihalić Arbanas S, Krkač M, Bernat Gazibara S, Arbanas Ž (2019) Preliminary rockfall susceptibility assessment of the rock slopes above the Town of Omiš (Croatia). In: Proceedings of the ISRM specialised conference "Geotechnical challenges in karst" 11–13 Apr 2019. Omiš, Croatia, pp 347–352

Volkwein A, Schellenberg K, Labiouse V, Agliardi F, Berger F, Bourrier F, Dorren LKA, Gerber W, Jaboyedoff M (2011) Rockfall characterisation and structural protection—a review. Nat Hazards Earth Syst Sci 11(9):2617–2651

Cutting-Edge Technologies Aiming for Better Outcomes of Landslide Disaster Mitigation

Kazuo Konagai

The International Consortium on Landslides (ICL) and The Global Promotion Committee of the International Programme on Landslides (GPC/IPL) have been responsible for organizing the World Landslide Forums (WLFs) every three years since 2008. Ever since the 1st WLF, the forums have long been the arena for landslide researchers and practitioners to exchange up-to-date information of recent devastations caused by landslides, cutting-edge technologies for landslide disaster mitigations and early warnings etc. to establish synergies among all participants worldwide.

Though the upcoming WLF5 has officially been postponed by one year to 2–6 November 2021 due to the global disruption caused by the coronavirus pandemic, the WLF5 will be all the more important with the Kyoto Landslide Commitment 2020 (KLC2020) to be launched as planned in the final online signatory meeting on 5 November 2020; the KLC 2020 is intended to be our action goals as the further advanced successor of the 'Sendai Landslide Partnerships 2015–2025 for Global Promotion of Understanding and Reducing Landslide Disaster Risk' in line with some of 17 Sustainable Development Goals (SDGs), particularly SDG 11, "Make cities and human settlements inclusive, safe, resilient and sustainable," of the United Nations.

For these important goals, the ICL has been inviting sponsorship from industries, businesses, and government agencies; all leading players in landslide science and technologies. They have been supporting a variety of the ICL/IPL activities such as publishing the International full-color journal "Landslides (Journal of the International Consortium on Landslides), full-color books for WLFs, exhibiting their cutting-edge technologies in WLFs, etc. Here follow short introductions of their activities with their names, addresses and contact information:

Marui & Co. Ltd.

1-9-17 Goryo, Daito City, Osaka 574-0064, Japan
URL: http://marui-group.co.jp/en/index.html
Contact: hp-mail@marui-group.co.jp

Marui & Co. Ltd. celebrates its 100th anniversary in 2020. Marui, as one of the leading manufacturers of testing apparatuses in Japan, has been constantly striving to further improve its service since its foundation in 1920, thus contributing to the sustainable development of our nation and society. Our main products cover a wide variety of destructive and non-destructive testing apparatuses in the fields of geotechnical engineering, concrete engineering (mortar, aggregates, etc.), and ceramic engineering. Of special note is that Marui has been helping manufacture ring-shear apparatuses half-century long based on the leading-edge idea of Dr. Kyoji Sassa, Professor Emeritus at the Kyoto University. Marui has delivered total 7 ring-shear apparatuses to the Disaster Prevention Research Institute, Kyoto University, and 2 to the International Consortium on Landslides. Also the apparatuses were exported to the United States of America, China, Croatia and Vietnam.

Marui & Co. Ltd. takes great pleasure in developing, manufacturing, and providing new products of high value sharing the delight of achievement with our customers, and thus contributing to the social development. The whole staff of Marui & Co. Ltd. are determined to devote ceaseless effort to keep its organization optimized for its speedy and high-quality services, by the motto "Creativity and Revolution", and strive hard to take a step further, as a leading manufacturer of testing apparatuses, to answer our customer's expectations for the 22nd century to come.

K. Konagai (✉)
Secretary General, Organizing Committee of the Fifth World Landslide Forum, International Consortium on Landslides, Kyoto, 606-8226, Japan
e-mail: konagai@iclhq.org

K. Sassa et al. (eds.), *Understanding and Reducing Landslide Disaster Risk*, ICL Contribution to Landslide Disaster Risk Reduction,
https://doi.org/10.1007/978-3-030-60196-6_56

Nippon Koei Co., Ltd.

5-4 Kojimachi, Chiyoda-ku, Tokyo 102-8539, Japan
URL: https://www.n-koei.co.jp/english/
Contact: https://www.n-koei.co.jp/english/contact/input

Nippon Koei Co., Ltd. and its group companies conduct many projects to support the growth of developing countries in Asia, Africa, the Middle and Near East, Latin America and other regions. Examples of their efforts include environmental measures to combat global warming, development of regional transportation infrastructure to support the rapid growth of emerging economies, and reconstruction assistance for regions affected by conflict and/or natural disasters.

OSASI Technos, Inc.

65-3 Hongu-cho, Kochi City, Kochi 780-0945, Japan
URL: http://www.osasi.co.jp/en/
Contact: cs@osasi.co.jp

OSASI Technos, Inc. has been making its best efforts to develop its cutting-edge technologies for landslide early warning. Its unique compact and lightweight sensors making up the Landslide Early Warning System enable long-term monitoring of unstable landslide mass movements, precipitations, porewater pressure buildups, etc., in a remote mountainous area where commercial power is often unavailable. OSASI Technos, Inc. is also proud of its advanced technology to transfer observed data even in areas with poor telecom environments as proven in the successful implementations in South Asia.

All stuff members of OSASI Technos work together for mitigation of landslide disasters worldwide.

Godai Corporation

1-35 Kuroda, Kanazawa City, Ishikawa 921-8051, Japan
URL: https://soft.godai.co.jp/En/Soft/Product/Products/LS-RAPID/
Contact: pp-sales@godai.co.jp

Ever since its foundation in 1965, Godai Kaihatsu Co., Ltd., a civil engineering consulting firm, has long been providing a variety of software and measures particularly for natural disaster mitigation. With its rich expertise in both civil engineering and information technology (IT), the company has its primary goal to address real world needs of disaster mitigation. All the staff of Godai Kaihatsu Co., Ltd. feel it more than happy that their cutting-edge technologies help mitigate natural disasters.

Japan Conservation Engineers & Co., Ltd.

3-18-5 Toranomon, Minato-ku, Tokyo 1050001, Japan
URL: https://www.jce.co.jp/en/
Contact: go_info@jce.jp

Japan Conservation Engineers & Co., Ltd. (JCE) is a general consulting firm working on landslide prevention research and consulting. JCE provides various disaster prevention technologies for debris flows, landslides, slope failures, rockfalls, etc. In addition, JCE is proud of its expertise having been conducting surveys and consulting works on coastal erosions and tsunami countermeasures for about 20 years. JCE contributes to the world through its activities in the realm of both structural and non-structural measures to build a resilient society.

OYO Corporation

7 Kanda-Mitoshiro-cho, Chiyoda-ku, Tokyo 101-8486, Japan
URL: https://www.oyo.co.jp/english/
Contact: https://www.oyo.co.jp/english/contacts/

OYO Corporation, the top geological survey company in Japan established in Tokyo in 1957, is well known as one of leading companies providing cutting-edge technologies and measures for natural disasters such as landslides, earthquakes, tsunamis, and floods. Not just developing and selling measuring instruments related to disaster prevention, OYO also delivers a market-leading services in 3D ground/geological modeling and 3D exploration technologies.

Kokusai Kogyo Co., Ltd.

2 Rokubancho, Chiyoda-ku, Tokyo, 102-0085, Japan
URL: https://www.kkc.co.jp/english/index.html
Contact: overseas@kk-grp.jp

Kokusai Kogyo Co., Ltd. as a leading company of geospatial information technologies, has long been providing public services with its comprehensive expertise to address real world needs and cutting-edge measurement technologies. Kokusai Kogyo Co., Ltd. helps rebuild "Green Communities," which has been of our great concern in terms of "environment and energy," "disaster risk reduction" and "asset management". Kokusai Kogyo Co., Ltd. offers the advanced and comprehensive analyses of geospatial information for developing new government policies, maintaining and operating social infrastructures safe and secure, and implementing low-carbon measures in cities.

Influenced by the recent global climate change, extreme rainfall events have become more frequent worldwide and resultant hydro-meteorological hazards are creating more deaths and devastations particularly in many developing countries where effective advanced countermeasures are not readily available. Kokusai Kogyo Co., Ltd. is proud of its achievements in establishing resilient infrastructure systems and implementing effective monitoring/early warning systems in developing countries, which have long been helping reduce the risks from natural hazards.

Geobrugg AG

Aachstrasse 11, 8590 Romanshorn, Switzerland
URL: www.geobrugg.com
Contact: info@geobrugg.com

Swiss company Geobrugg is the global leader in the supply of high-tensile steel wire safety nets and meshes—with production facilities on four continents, as well as branches and partners in over 50 countries. True to the philosophy "Safety is our nature" the company develops and manufactures protection systems made of high-tensile steel wire. These systems protect against natural hazards such as rockfall, landslides, debris flow and avalanches. They ensure safety in mining and tunneling, as well as on motorsport tracks and stop other impacts from falling or flying objects. More than 65 years of experience and close collaboration with research institutes and universities make Geobrugg a pioneer in these fields.

Ellegi srl

Via Petrarca, 55 I-22070 Rovello Porro (CO) Italy
URL: http://www.lisalab.com/engl/?seze=1
Contact: info@lisalab.com

Ellegi srl provides worldwide monitoring services and produces Ground Based synthetic aperture radar (GBInsAR) for remote measurement of displacements and deformations on natural hazards and manmade buildings using its own designed and patented LiSALab system.

Its activities started in 2003 as a spin off project to exploit commercially the Ground Based Linear Synthetic Aperture Radars technology developed by European Commission's Ispra Joint Research Centre and based on the results of more than 10 years of research. Since then Ellegi has industrialized and developed the core technology of the LiSALab system and latest LiSAmobile system represents the 5th generation of development.

In 2003 it was the first commercial company in the world to provide GBInSAR measurements of natural hazards and structure.

Ellegi srl offers:

- Displacement fields measurement, control and monitoring of the deformation caused by natural hazards, like landslides, rockslides, sinkhole, volcanic deformation in every operative condition, including emergencies,
- Structural strain fields measurement, control, monitoring and diagnosis of the deformation affecting buildings, bridges, viaducts, dams.
- GBInSAR monitoring systems, installation, management and maintenance in order to provide information about natural hazards or anthropic activity, that can generate or cause slopes failures or buildings instabilities.

In all the above-mentioned activities Ellegi srl uses the GBInSAR LiSALab technology that represents a real "break-through".

Chuo Kaihatsu Corporation

3-13-5 Nishi-waseda, Shinjuku-ku, Tokyo 169-8612, Japan
URL: https://www.ckcnet.co.jp/global/
Contact: https://www.ckcnet.co.jp/contactus/

Chuo Kaihatsu Corporation (CKC) was founded in 1946, and has been aiming to become the "Only One" consultant for our customers. We engage in the hands-on work that will "Remain with the earth, Remain in people's hearts, and Lead to a prosperous future". We focus on road, river and dam engineering to flesh out industrial infrastructures specifically by means of geophysical/geotechnical/geological investigations, civil engineering surveys and project implementations. In recent years, we make significant efforts on earthquake disaster mitigation, sediment disaster prevention/mitigation and ICT information services. Many achievements of ours have already contributed to mitigation of natural disasters such as landslides, earthquakes and slope failures in Japan, Asia and the Pacific Region.

IDS GeoRadar s.r.l.

Via Augusto Righi, 6, 6A, 8, Loc. Ospedaletto, Pisa, Italy, 56121
URL: https://idsgeoradar.com/
Contact: info@idsgeoradar.com

IDS GeoRadar, part of Hexagon, provides products and solutions, based on radar technology, for monitoring applications including landslides, rockfalls, complex structures, mining and civil engineering. The company is a leading provider of Ground Penetrating Radar (GPR) and Interferometric Radar solutions worldwide. IDS GeoRadar is committed to delivering best-in-class performance solutions and to the pursuit of product excellence, through the creation of application-specific, innovative and cost-efficient systems for a wide range of applications.

METER Group, Inc.

2365 NE Hopkins Court, Pullman, WA 99163, USA
URL: metergroup.com/wlf5
Contact: bryan.wacker@metergroup.com

METER Group provides accurate, rugged, and dependable instrumentation to monitor moisture in all its phases within an unstable slope. METER specializes in instrumentation for near real-time monitoring of incoming moisture in the form of rain and weather. In addition, we provide real-time below-surface monitoring of existing moisture conditions like moisture content and soil suction which show how the soil profile is filling with water to saturation, including the transition to positive pore water pressure.

The ZL6 advanced cloud data logger works together with ZENTRA Cloud data software to simplify and speed up data collection, management, visualization, and alerting. Our well-published instrumentation is used worldwide in universities, research and testing labs, government agencies, and industrial applications.

For almost four decades, scientists and engineers have relied on our instrumentation to understand critical moisture parameters. We've even partnered with NASA to measure soil (regolith) moisture on Mars. Wherever you measure, and whatever you're measuring, rely on METER for accuracy, affordability, and simplicity that will make your job easier.

Asia Air Survey Co., Ltd.

Shinyuri 21 BLDG 3F, 1-2-2 Manpukuji, Asao-Ku, Kawasaki, Kanagawa 215-0004, Japan
URL: https://www.ajiko.co.jp/en/
Contact: service@ajiko.co.jp

Asia Air Survey (AAS), as one of the leading engineering and consulting companies, has long been providing disaster prevention and mitigation services for over 65 years, particularly in the fields of landslide, debris flow, erosion control, etc. AAS is proud of being the inventor of Red Relief Image Map (RRIM), which is a cutting-edge 3D terrain visualization method allowing great geomorphological details to be visualized in one glance, thus has been used in various facets of disaster prevention and mitigation.

Kiso-Jiban Consultants Co., Ltd.

Kinshicho Prime Tower 12 Floor, 1-5-7 Kameido, Koto-ku, Tokyo 36-8577, Japan
URL: https://www.kisojiban.com/
Contact: kisojiban-contactus@kiso.co.jp

Kiso-Jiban Consultants, established in 1953, is an engineering consulting firm especially well known in the field of geotechnical engineering. The areas of its comprehensive services are listed below:

- Geological and Geotechnical Survey
- Geotechnical Analysis and Design
- Disaster Prevention and Management
- GIS (Geographic Information Systems)
- Soil and Rock Laboratory Tests
- Instrumentation and Monitoring
- Geophysical Exploration and Logging
- Distribution of Geosynthetics Products.

Much-talked-about new service is Kiso-SAR System allowing accurate estimation of both extent and rate of landslide movements based upon a comprehensive interpretation of InSAR results from geotechnical and landslide engineering viewpoint (see the one-page introduction of Kiso-Jiban Consultants Co., Ltd.). With Kiso-SAR system, the following pieces of important geotechnical information can be provided:

(1) Extent of a deforming landslide mass (and the rate of its movement
(2) Consolidation buildup in soft clay underlying a fill
(3) Deformation buildups induced by slope cutting.

Okuyama Boring Co., Ltd.

10-39 Shimei-cho, Yokote City, Akita 013-0046, Japan
URL: https://okuyama.co.jp/en/
Contact: info@okuyama.co.jp

Okuyama Boring Co., Ltd. is proud of its achievements in various projects to help solve many landslide problems. The company has been offering services in geological surveys and analyses, developing rational countermeasures against various geotechnical problems as well as safe workflow diagrams, and providing necessary pieces of advice for ensuring safety during landslide countermeasure works. For this purpose, Okuyama Boring Co., Ltd. works on monitoring, observations, field surveys, numerical analyses, countermeasure works, etc. of landslides.

Kawasaki Geological Engineering Co. Ltd.

Mita-Kawasaki Bldg, 2-11-15 Mita, Minato-ku, Tokyo108-8337, Japan
URL: http://www.kge.co.jp/
Contact: post-master@kge.co.jp

Kawasaki Geological Engineering Co., Ltd. as one of the leading members of SAAM Research Group, has proactively been involved in developing "Sustainable Asset Anchor Maintenance (SAAM, hereafter) System," enabling easy maintenance of ground anchors. Its unique jack, weighing about half the weight of a conventional jack, together with a newly developed jig, can be applied to any type of anchor even with a short extra length, thus allowing for in situ lift-off tests on these anchors. The SAAM system also has an optional weight meter that can be installed after performing a lift-off test.

Nissaku Co., Ltd.

4-199-3 Sakuragi-cho, Omiya-ku, Saitama 330-0854, Japan
URL: https://www.nissaku.co.jp/
Contact: survey@nissaku.co.jp

Nissaku Co., Ltd., founded in 1912 as a well drilling company, provides services for far-flung fields of not only groundwater exploitation but also measures for landslides. Having its rich expertise in these fields, Nissaku Co., Ltd. offers general reliable one-stop technical services including designs, investigations, analyses, constructions, and maintenances.

Full-color presentations from the above seventeen exhibitors focusing on their landslide technologies are shown on the following pages. Their cutting-edge technologies have of course been instrumental in the progress that we have made in landslide risk-reduction worldwide, and we want to exert even greater effort to aim high given the KLC 2020 as our new action goals. The International Consortium on Landslides seeks volunteers willing to support our activities introducing their brand-new technologies for landslide disaster mitigation in our international journal "Landslides," full color books for WLFs, exhibitions at WLFs, etc. If you are interested in being engaged in supporting ICL activities, please contact the ICL secretariat secretariat@iclhq.org.

Geohazard Management
Response to natural disasters with various technologies from space to the surface
Integrated technologies and engineers-Application of spaceborne, airborne, and ground-based technologies for disaster rick reduction.
Remote Sensing Technology
Potential hazards around the globe are assessed by optical remote sensing and InSAR which can detect land-resources, topographic features, and ground deformation. Example of InSAR, shown below, is a new effective way to detect deformation of slopes along infrastructures such as roads and railways.
Phase difference between emitted and received waves is analyzed.
LOS Displacement (Unit: cm)
Uplift
Subsidence
Processed by ©RESTEC,Included ©JAXA
Illustration of interferometric SAR (InSAR)
Landslide monitoring using InSAR
A team of 5,497 multidisciplinary experts
Excellent teams, covering advanced and wide range of technologies based on long-standing experiences, are formed to provide optimum solutions customized for each condition and needs.
Field survey by experienced engineers
AI Technology
Our AI technology helps quickly identify morphological features of past and current landslides.
1km
Extracted landslides
Image A
Image B
256pix
Near a volcano, our AI technology can help identify unstable masses of volcanic matters perching on the flanks of the volcano.
Data for machine learning: DEM and landslides identified by an expert
Input
Output
Target
Extraction of landslide topography using AI technology
Numerical simulation
We can predict the extent of damage in the event of a disaster and the effectiveness of countermeasure works by numerical analysis.
Three-dimensional rockfall simulation by R&D center
Numerical simulation for slope excavation by R&D center
R&D center
State-of-the-Art Nippon Koei's R&D Center
Soil & Environmental Science Lab
Hydraulic Model Test Lab
Main Building
Multi-Purpose Lab
River Engineering Yard
NIPPON KOEI
Global Consulting Engineering Firm
Head Office 5-4 Kojimachi, Chiyoda-ku, Tokyo 102-8539, Japan
TEL +81-3-3238-8030
Website www.n-koei.co.jp/english
Japan. Committed to SDGs

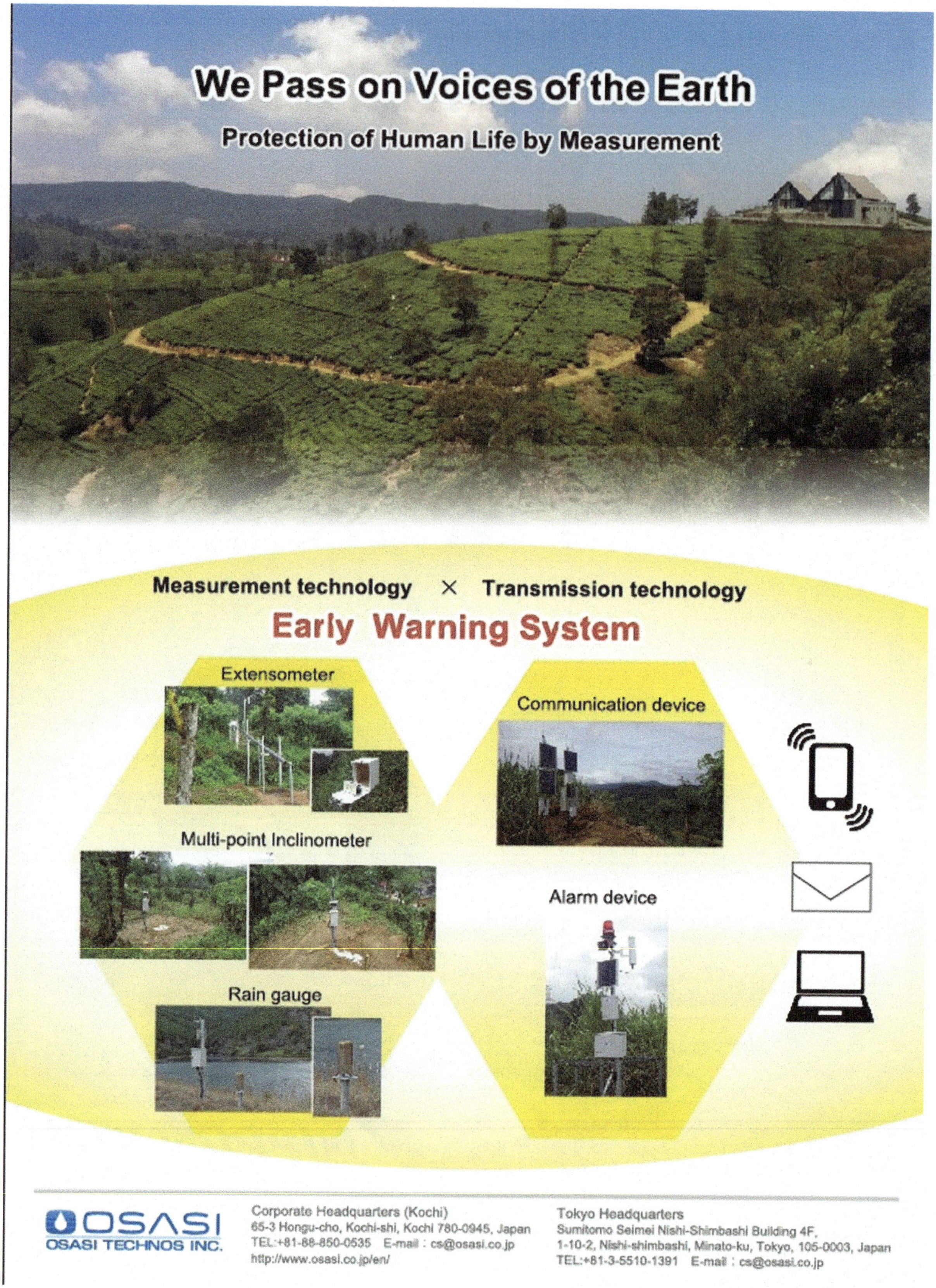
We Pass on Voices of the Earth
Protection of Human Life by Measurement
Measurement technology × Transmission technology
Early Warning System
Extensometer
Communication device
Multi-point Inclinometer
Alarm device
Rain gauge
OSASI
OSASI TECHNOS INC.
Corporate Headquarters (Kochi)
65-3 Hongu-cho, Kochi-shi, Kochi 780-0945, Japan
TEL:+81-88-850-0535 E-mail : cs@osasi.co.jp
http://www.osasi.co.jp/en/
Tokyo Headquarters
Sumitomo Seimei Nishi-Shimbashi Building 4F,
1-10-2, Nishi-shimbashi, Minato-ku, Tokyo, 105-0003, Japan
TEL:+81-3-5510-1391 E-mail : cs@osasi.co.jp

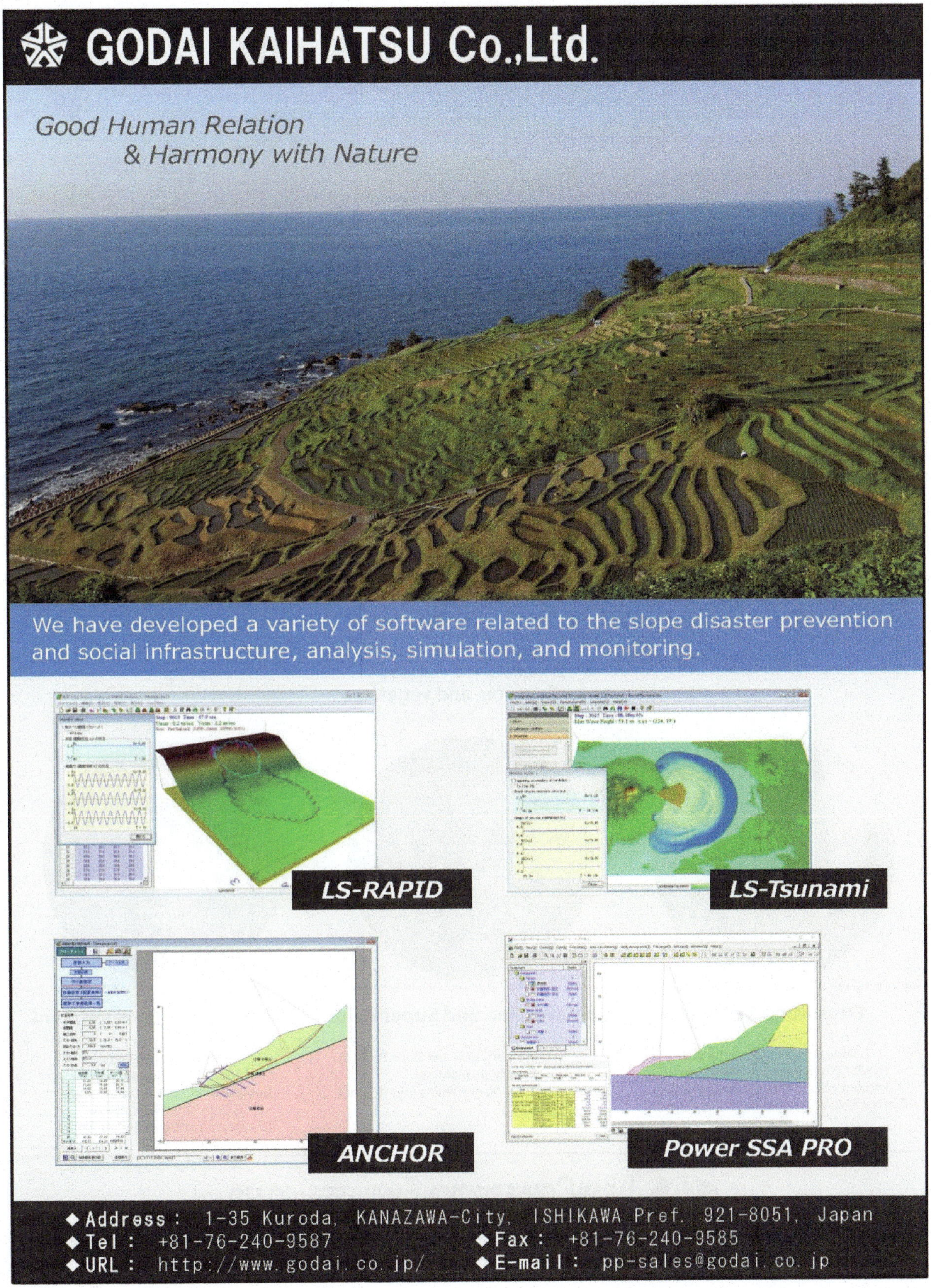
GODAI KAIHATSU Co.,Ltd.
Good Human Relation
& Harmony with Nature
We have developed a variety of software related to the slope disaster prevention and social infrastructure, analysis, simulation, and monitoring.
LS-RAPID
LS-Tsunami
ANCHOR
Power SSA PRO
◆Address : 1-35 Kuroda, KANAZAWA-City, ISHIKAWA Pref. 921-8051, Japan
◆Tel : +81-76-240-9587
◆Fax : +81-76-240-9585
◆URL : http://www.godai.co.jp/
◆E-mail : pp-sales@godai.co.jp

Japan Conservation Engineers is
a consulting firm that specializes
in disaster prevention and mitigation
for the conservation of land and the environment.

OYO
Find the best answer to the future of people and the earth.
To realize a sustainable society.
3D Geological Modeling
GEO-CRE® /GEO-CRE®PRO
OCTAS® Modeler
OCTAS Modeler is a system to support the utilization of subsurface information. It helps to grasp the positional relationship of the borehole logs and create various 3D models including borehole, soil, bearing layer, and terrain.
Inclinometer
Clino-pole
Tilt sensor for Sediment disaster
3D Geophysical Survey
McSEIS-AT
McSEIS-AT is an epoch-making microtremor exploration system that can measure S-wave velocity structures in three dimensions.
oyo corporation
7 kanda Mitoshiro-cho,Chiyoda-ku,Tokyo
101-8486,JAPAN
https://www.oyo.co.jp/english/
Copyright © 2020 OYO Corporation. All rights reserved

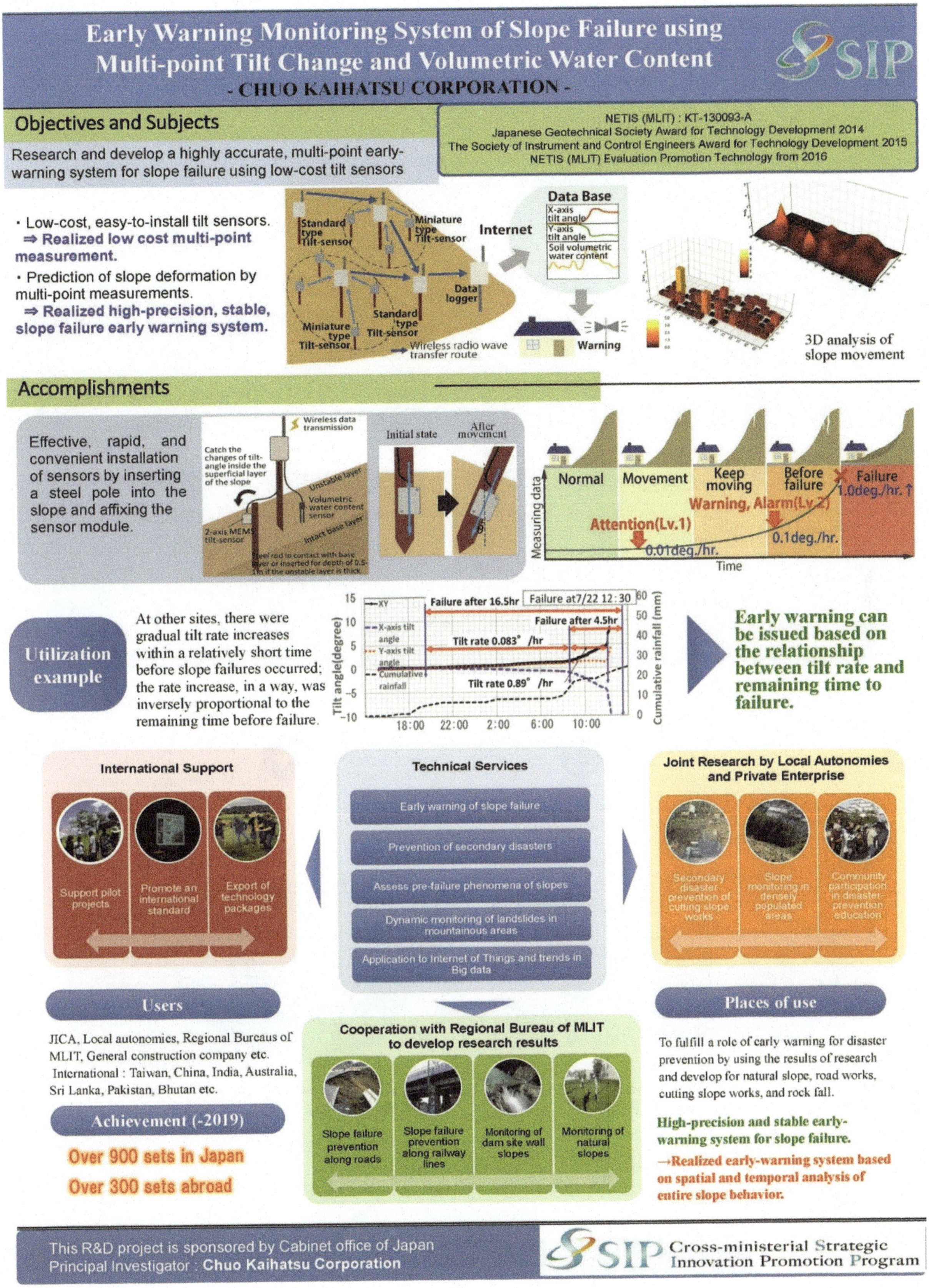

Early Warning Monitoring System of Slope Failure using Multi-point Tilt Change and Volumetric Water Content
- CHUO KAIHATSU CORPORATION -
SIP
Objectives and Subjects
Research and develop a highly accurate, multi-point early-warning system for slope failure using low-cost tilt sensors
NETIS (MLIT) : KT-130093-A
Japanese Geotechnical Society Award for Technology Development 2014
The Society of Instrument and Control Engineers Award for Technology Development 2015
NETIS (MLIT) Evaluation Promotion Technology from 2016
· Low-cost, easy-to-install tilt sensors.
⇒ Realized low cost multi-point measurement.
· Prediction of slope deformation by multi-point measurements.
⇒ Realized high-precision, stable, slope failure early warning system.
Standard type Tilt-sensor
Miniature type Tilt-sensor
Data logger
Wireless radio wave transfer route
Internet
Data Base
X-axis tilt angle
Y-axis tilt angle
Soil volumetric water content
Warning
3D analysis of slope movement
Accomplishments
Effective, rapid, and convenient installation of sensors by inserting a steel pole into the slope and affixing the sensor module.
Wireless data transmission
Catch the changes of tilt-angle inside the superficial layer of the slope
Unstable layer
Volumetric water content sensor
Intact base layer
2-axis MEMS tilt-sensor
Initial state
After movement
Normal
Movement
Keep moving
Before failure
Failure
Measuring data
Attention(Lv.1)
Warning, Alarm(Lv.2)
0.01deg./hr.
0.1deg./hr.
1.0deg./hr.↑
Time
Utilization example
At other sites, there were gradual tilt rate increases within a relatively short time before slope failures occurred; the rate increase, in a way, was inversely proportional to the remaining time before failure.
Failure after 16.5hr
Failure at7/22 12:30
Failure after 4.5hr
Tilt rate 0.083° /hr
Tilt rate 0.89° /hr
XY
X-axis tilt angle
Y-axis tilt angle
Cumulative rainfall
Tilt angle(degree)
Cumulative rainfall (mm)
18:00 22:00 2:00 6:00 10:00
Early warning can be issued based on the relationship between tilt rate and remaining time to failure.
International Support
Support pilot projects
Promote an international standard
Export of technology packages
Technical Services
Early warning of slope failure
Prevention of secondary disasters
Assess pre-failure phenomena of slopes
Dynamic monitoring of landslides in mountainous areas
Application to Internet of Things and trends in Big data
Joint Research by Local Autonomies and Private Enterprise
Secondary disaster prevention of cutting slope works
Slope monitoring in densely populated areas
Community participation in disaster-prevention education
Users
JICA, Local autonomies, Regional Bureaus of MLIT, General construction company etc.
International : Taiwan, China, India, Australia, Sri Lanka, Pakistan, Bhutan etc.
Achievement (-2019)
Over 900 sets in Japan
Over 300 sets abroad
Cooperation with Regional Bureau of MLIT to develop research results
Slope failure prevention along roads
Slope failure prevention along railway lines
Monitoring of dam site wall slopes
Monitoring of natural slopes
Places of use
To fulfill a role of early warning for disaster prevention by using the results of research and develop for natural slope, road works, cutting slope works, and rock fall.
High-precision and stable early-warning system for slope failure.
→Realized early-warning system based on spatial and temporal analysis of entire slope behavior.
This R&D project is sponsored by Cabinet office of Japan
Principal Investigator : Chuo Kaihatsu Corporation
SIP Cross-ministerial Strategic Innovation Promotion Program

Kisojiban

Kiso-Jiban Consultants Co., Ltd.

Monitoring Service for Landslide by Kiso-SAR System

What's SAR ?

SAR (Synthetic Aperture Radar) is a technique that utilizes interference of radio waves for precise determination of distance.
InSAR (interferometric SAR), the phase of the received backscattered signal from two images of the same scene are used to measure path length differences with an accuracy of a few millimeters.

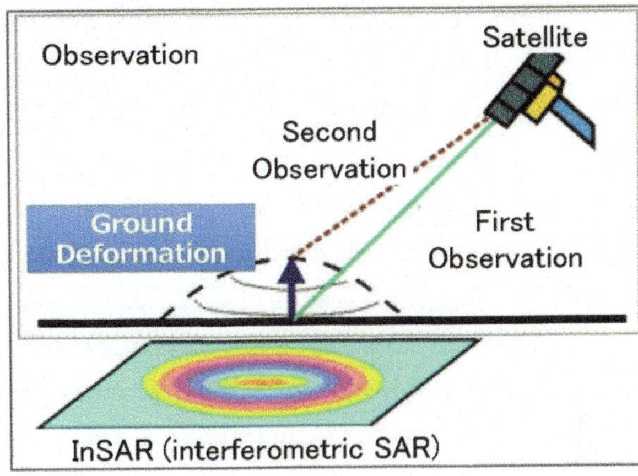

Ground Deformation Estimation in Landslides

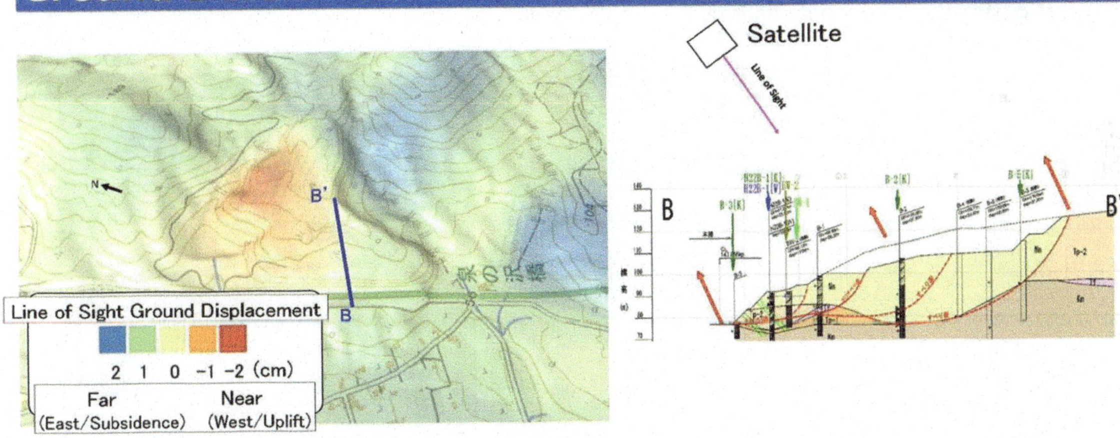

Ground Deformation of Landslide Observed by Kiso-SAR System

Kisojiban
Kiso-Jiban Consultants Co., Ltd.
Kinshicho Prime Tower 12 Floor,1-5-7
Kameido,Koto-ku,Tokyo 136-8577,Japan
Tel.:+81-3-6861-8800

Sustainable Asset Anchor Maintenance System

Earth Doctor

■ What's "SAAM System"?

SAAM stands for "Sustainable Asset Anchor Maintenance".

SAAM Jack, only half the weight of a conventional jack, enables lift-off tests in various onsite conditions.

SAAM, thus, allows for easy maintenance of ground anchors.

SAAM Jack

SAAM System

■ SAAM-A (PAT No. 5971596)

Lift-off tests on anchors with a short extra length can be performed with a newly developed jig.

Installation of Inner Coupler

Removal of Wedge

■ SAAM-L (PAT No. 5440772)

SAAM load cell can be attached to any anchors without loosening them.

Load Meter

■ Planar Distribution of Anchor's Tensioning Force

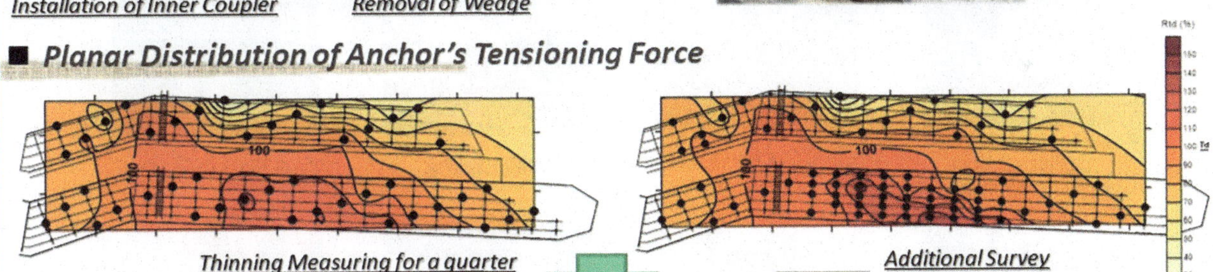

Thinning Measuring for a quarter

Additional Survey

Evaluation based on Soundness of Anchors

	Evaluation	Status	Coping example
0.9 Tys	E+	Risk of Breakage	Emergency measures
1.1 Ta	D+	Risk of Damage	Implementation of countermeasure work
Allowable anchor force (Ta)	C+	Exceeds the allowable value	
Design anchor forcee (Td)	B+		Follow up
Tensile force when fixed (Pt)	A+	Good	
	A−		
0.8 Pt	B−		Follow up
0.5 Pt	C−	Functional deterioration	Implementation of countermeasure work
0.1 Pt	D−	Dysfunction	

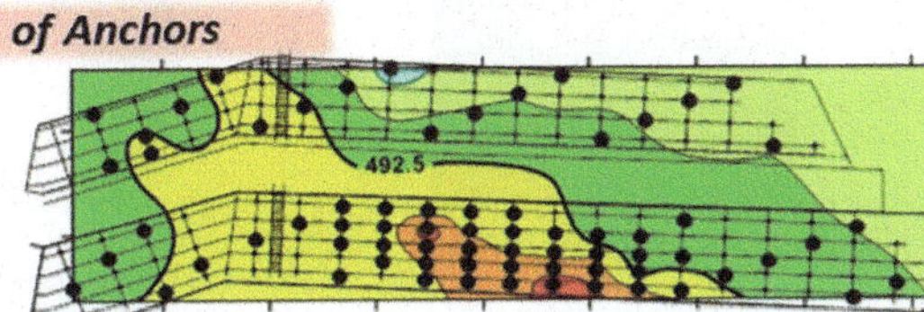

SAAM 合同会社アンカーアセットマネジメント研究会

Society of Anchor Asset Management

川崎地質株式会社

Kawasaki Geological Engineering Co., Ltd.

Marketing & Sales Department

URL : www.kge.co.jp TEL : +81-3-5445-2077

International Consortium on Landslides

International Consortium on Landslides

An international non-government and non-profit scientific organization promoting landslide research and capacity building for the benefit of society and the environment

President: Peter T. Bobrowsky (Geological Survey of Canada)
Vice Presidents: Matjaž Mikoš (University of Ljubljana, Slovenia), Dwikorita Karnawati (Agency for Meteorology, Climatorology, and Geophysics, Indonesia), Nicola Casagli (University of Florence, Italy), Binod Tiwari (California State University, USA), Željko Arbanas (University of Rijeka, Croatia)
Executive Director: Kaoru Takara (Kyoto University, Japan), Treasurer: Kyoji Sassa (Prof. Emeritus, Kyoto University, Japan)

ICL Full Members:

Geotechnical Engineering Office, Hong Kong Special Administrative Region, China
UNESCO Chair for the Prevention and the Sustainable Management of Geo-hydrological Hazards—University of Florence, Italy
Korea Institute of Geoscience and Mineral Resources (KIGAM)
University of Ljubljana, Faculty of Civil and Geodetic Engineering (ULFGG), Slovenia

Albania Geological Survey/The Geotechnical Society of Bosnia and Herzegovina/Center for Scientific Support in Disasters—Federal University of Parana, Brazil/ Geological Survey of Canada/University of Alberta, Canada/Northeast Forestry University, Institute of Cold Regions Science and Engineering, China/China University of Geosciences/Chinese Academy of Sciences, Institute of Mountain Hazards and Environment/Tongji University, College of Surveying and Geo-Informatics, China/The Hong Kong University of Science and Technology, China/Shanghai Jiao Tong University, China/The University of Hong Kong, China/Universidad Nacional de Colombia/Croatian Landslide Group (Faculty of Civil Engineering, University of Rijeka and Faculty of Mining, Geology and Petroleum Engineering, University of Zagreb)/City of Zagreb, Emergency Management Office, Croatia/Charles University, Faculty of Science, Czech Republic/Institute of Rock Structure and Mechanics, Department of Engineering Geology, Czech Republic/Brown Coal Research Institute, Czech Republic/Cairo University, Egypt/Technische Universitat Darmstadt, Institute and Laboratory of Geotechnics, Germany/National Environmental Agency, Department of Geology, Georgia/Universidad Nacional Autonoma de Honduras (UNAH), Honduras/Amrita Vishwa Vidyapeetham, Amrita University/Vellore Institute of Technology, India/National Institute of Disaster Management, India/Agency for Meteorology, Climatology, and Geophysics of the Republic of Indonesia (BMKG Indonesia)/University of Gadjah Mada, Center for Disaster Mitigation and Technological Innovation (GAMA-InaTEK), Indonesia/Parahyangan Catholic University, Indonesia/Building & Housing Research Center, Iran/Italian Institute for Environmental Protection and Research (ISPRA)—Dept. Geological Survey, Italy/University of Calabria, DIMES, CAMILAB, Italy/Istituto de Ricerca per la Protezione Idrogeologica (IRPI), CNR, Italy/DIA–Universita degli Studi di Parma, Italy/University of Torino, Dept of Earth Science , Italy/Centro di Ricerca CERI—Sapienza Università di Roma, Italy/Kyoto University, Disaster Prevention Research Institute, Japan/Japan Landslide Society/Korean Society of Forest Engineering/National Institute of Forest Science, Korea/Korea Infrastructure Safety & Technology Corporation/Korea Institute of Civil Engineering and Building Technology/Slope Engineering Branch, Public Works Department of Malaysia/Institute of Geography, National Autonomous University of Mexico (UNAM)/International Centre for Integrated Mountain Development (ICIMOD), Nepal/University of Nigeria, Department of Geology, Nigeria/Moscow State University, Department of Engineering and Ecological Geology, Russia/JSC "Hydroproject Institute", Russia/University of Belgrade, Faculty of Mining and Geology, Serbia/Comenius University, Faculty of Natural Sciences, Department of Engineering Geology, Slovakia/Geological Survey of Slovenia/University of Ljubljana, Faculty of Natural Sciences and Engineering (ULNTF), Slovenia/Central Engineering Consultancy Bureau (CECB), Sri Lanka/National Building Research Organization, Sri Lanka/Landslide group in National Central University from Graduate Institute of Applied Geology, Department of Civil Engineering, Center for Environmental Studies, Chinese Taipei/National Taiwan University, Department of Civil Engineering, Chinese Taipei/Asian Disaster Preparedness Center, Thailand/Ministry of Agriculture and Cooperative, Land Development Department, Thailand/Institute of Telecommunication and Global Information Space, Ukraine/California State University, Fullerton & Tribhuvan University, Institute of Engineering, USA & Nepal/Institute of Transport Science and Technology, Vietnam/Vietnam Institute of Geosciences and Mineral Resources (VIGMR).

ICL Associates

State Key Laboratory of Geohazard Prevention and Geoenvironment Protection (Chengdu University of Technology), China/Czech Geological Survey, Czech Republic/Department of Earth and Environmental Sciences, University Aldo Moro, Bari, Italy/Department of Sciences and Technologies, University of Sannio, Italy/Department of Earth and Environmental Sciences—University of Pavia, Italy/Geotechnical Engineering Group (GEG), University of Salerno, Italy/Niigata University, Research Institute for Natural Hazards and Disaster Recovery, Japan/Ehime University Center for Disaster Management Informatics Research, Japan/Tian-Shan Geological Society, Kyrgyzstan/Institute of Environmental Geoscience RAS (IEG RAS), Russia/Russian State Geological Prospecting University n.a. Sergo Ordzhonikidze (MGRI-RSGPU)/TEMPOS, environmental civil engineering Ltd., Slovenia/Institute of Earth Sciences—Faculty of Geoscience and Environment, University of Lausanne, Switzerland/Middle East Technical University (METU), Turkey/North Dakota State University, USA

ICL Secretariat:

Secretary General: Kyoji Sassa
International Consortium on Landslides, 138-1 Tanaka Asukai-cho, Sakyo-ku, Kyoto 606-8226, Japan
Web: http://icl.iplhq.org/, E-mail: secretariat@iclhq.org
Tel: +81-75-723-0640, Fax: +81-75-950-0910

K. Sassa et al. (eds.), *Understanding and Reducing Landslide Disaster Risk*,
ICL Contribution to Landslide Disaster Risk Reduction,
https://doi.org/10.1007/978-3-030-60196-6

Printed by Printforce, the Netherlands